S0-BDV-576

BACTERIAL DISEASES (Continued)

Disease	Organism	Type*	Page
verruga peruana (bartonellosis)	*Bartonella bacilliformis*	coccoid, −	661
vibriosis	*Vibrio parahaemolyticus*	R, −	621
whooping cough (pertussis)	*Bordetella pertussis*	CB, −	253, 586–588
yersiniosis	*Yersinia enterocolitica*	R, −	622–623

Disease	Organism	Type*	Page
*Key to types:			
C = coccus	I = irregular	VAR = gram-variable	
CB = coccobacillus	− = gram-negative	A-F = acid-fast	
R = rod	+ = gram-positive	NA = not applicable	
S = spiral			
† Species			

VIRAL DISEASES

Disease	Virus	Reservoir	Page
aplastic crisis in sickle cell anemia	erythrovirus (B19)	humans	276, 666
bronchitis, rhinitis	parainfluenza	humans, some other mammals	580, 582
Burkitt's lymphoma	Epstein-Barr	humans	664
chickenpox	varicella-zoster	humans	276, 277, 560–561
coryza (common cold)	rhinovirus	humans	276, 584–585
	coronavirus	humans	585
cytomegalic inclusion disease	cytomegalovirus	humans	720
Dengue fever	Dengue	humans	325, 326, 662–663
encephalitis	Colorado tick fever	mammals	325, 666
	Eastern equine encephalitis	birds	276, 419, 683
	St. Louis encephalitis	birds	683
	Venezuelan equine encephalitis	rodents	276, 683
	Western equine encephalitis	birds	276, 326, 419, 683
epidemic keratoconjunctivitis	adenovirus	humans	569
fifth disease (erythema infectiosum)	erythrovirus (B19)	humans	280, 667
hantavirus pulmonary syndrome	bunyavirus	rodents	279, 602, 673
hemorrhagic fever, Bolivian	arenavirus	rodents and humans	666
hemorrhagic fever, Korean	bunyavirus (Hantaan)	rodents	279, 666
hemorrhagic fever	Ebola virus (filovirus)	humans (?)	276, 278, 665
	Marburg virus (filovirus)	humans (?)	276, 278, 665
hepatitis A (infectious hepatitis)	hepatitis A	humans	276, 626–628
hepatitis B (serum hepatitis)	hepatitis B	humans	276, 626–628
hepatitis C (non-A, non-B)	hepatitis C	humans	626–628
hepatitis D (delta hepatitis)	hepatitis D	humans	626–628
hepatitis E (enterically transmitted non-A, non-B, non-C)	hepatitis E	humans	626–628
herpes, oral	usually herpes simplex type 1, sometimes type 2	humans	276, 279, 717

Disease	Virus	Reservoir	Page
herpes, genital	usually herpes simplex type 2, sometimes type 1	humans	276, 279, 716–717
HIV disease, AIDS	human immunodeficiency virus (HIV)	humans	276, 539–540
infectious mononucleosis	Epstein-Barr	humans	663–664
influenza	influenza	swine, humans (type A)	276, 278, 598–600
		humans (type B)	276, 278, 598–600
		humans (type C)	598
Lassa fever	arenavirus	rodents	666
measles (rubeola)	measles	humans	276, 558–560
meningoencephalitis	herpes	humans	683, 717–718
molluscum contagiosum	poxvirus group	humans	562
mumps	paramyxovirus	humans	615
pneumonia	adenoviruses, respiratory syncytial virus	humans	253, 254, 258, 588–590
poliomyelitis	poliovirus	humans	276, 689–691
rabies	rabies	all warm-blooded animals	276, 278, 429–430, 681–683
respiratory infections	adenovirus	humans	602
	polyomavirus	none	683–684
Rift Valley fever	bunyavirus (phlebovirus)	humans, sheep, cattle	666
rubella (German measles)	rubella	humans	276, 558
shingles	varicella-zoster	humans	276, 279, 560
smallpox	variola (major and minor)	humans	276, 280, 561
viral enteritis	rotavirus	humans	626
warts, common (papillomas)	human papillomavirus	humans	276, 562, 563
warts, genital (condylomas)	human papillomavirus	humans	276, 562, 719
yellow fever	yellow fever	monkeys, humans, mosquitoes	276, 277, 295, 325, 663

The tables of fungal and parasitic diseases appear on the following page.

FUNGAL DISEASES

Disease	Organism	Page	Disease	Organism	Page
aspergillosis	*Aspergillus* sp.	604	histoplasmosis	*Histoplasma capsulatum*	603–604
blastomycosis	*Blastomyces dermatitidis*	313, 564–565	*Pneumocystis* pneumonia	*Pneumocystis carinii*	604
candidiasis	*Candida albicans*	565	ringworm (tinea)	various species of *Epidermophyton, Trichophyton, Microsporum*	563–564
coccidioidomycosis (San Joaquin valley fever)	*Coccidioides immitis*	602–603			
cryptococcosis	*Filobasidiella neoformans*	604	sporotrichosis	*Sporothrix schenckii*	564
ergot poisoning	*Claviceps purpurea*	631–632, 761	zygomycosis	*Rhizopus* sp., *Mucor* sp.	565

PARASITIC DISEASES

Disease	Organism	Type	Page	Disease	Organism	Type	Page
Acanthamoeba keratitis	*Acanthamoeba culbertsoni*	protozoan	571	malaria	*Plasmodium* sp.	protozoan	74, 303, 306–307, 326, 327, 328, 433, 668–670
African sleeping sickness (trypanosomiasis)	*Trypanosoma brucei gambiense* and *T. brucei rhodesiense*	protozoan	303, 306, 325, 326, 691–692	pediculosis (lice infestation)	*Pediculus humanus*	louse	573
amoebic dysentery	*Entamoeba histolytica*	protozoan	303, 306, 630	pinworm	*Enterobius vermicularis*	roundworm	637
ascariasis	*Ascaris lumbricoides*	roundworm	635–636	river blindness (onchocerciasis)	*Onchocerca volvulus*	roundworm	569
babesiosis	*Babesia microti*	protozoan	670, 672	scabies (sarcoptic mange)	*Sarcoptes scabiei*	mite	573
balantidiasis	*Balantidium coli*	protozoan	630, 631	schistosomiasis	*Schistosoma* sp.	flatworm	647–648
Chagas' disease	*Trypanosoma cruzi*	protozoan	325, 326, 383, 692–693	sheep liver fluke (fascioliasis)	*Fasciola hepatica*	flatworm	316, 317, 632
chigger dermatitis	*Trombicula* sp.	mite	572–573	strongyloidiasis	*Strongyloides stercoralis*	roundworm	637
chigger infestation	*Tunga penetrans*	sandflea	572–573	swimmer's itch	*Schistosoma* sp.	flatworm	566
Chinese liver fluke	*Clonorchis sinensis*	flatworm	633	tapeworm infestation (taeniasis)	*Hymenolepsis nana* (dwarf tapeworm)	flatworm	634
crab lice	*Phthirus pubis*	louse	573				
cryptosporidiosis	*Cryptosporidium* sp.	protozoan	631, 747		*Taenia saginata* (beef tapeworm)	flatworm	316
dracunculiasis (Guinea worm)	*Dracunculus medinensis*	roundworm	566		*Taenia solium* (pork tapeworm)	flatworm	633
elephantiasis (filariasis)	*Wuchereria bancrofti*	roundworm	649		*Diphyllobothrium latum* (fish tapeworm)	flatworm	634
fasciolopsiasis	*Fasciolopsis buski*	flatworm	633		*Echinococcus granulosus* (dog tapeworm)	flatworm	634
giardiasis	*Giardia intestinalis*	protozoan	628–629				
heartworm disease	*Dirofilaria immitis*	roundworm	302, 383, 650	toxoplasmosis	*Toxoplasma gondii*	protozoan	303, 307, 670
hookworm	*Ancylostoma duodenale* (Old World hookworm)	roundworm	635	trichinosis	*Trichinella spiralis*	roundworm	319, 349, 634–635
	Necator americanus (New World hookworm)	roundworm	319, 635	trichomoniasis	*Trichomonas vaginalis*	protozoan	705–706
leishmaniasis	*Leishmania braziliensis*	protozoan	303, 306, 325, 667–668	trichuriasis (whipworm)	*Trichuris trichiura*	roundworm	636–637
kala azar	*L. donovani*			visceral larva migrans	*Toxocara* sp.	roundworm	636
oriental sore	*L. tropica*						
liver/lung fluke (paragonimiasis)	*Paragonimus westermani*	flatworm	317, 604–605				
loaiasis	*Loa loa*	roundworm	570				

MICROBIOLOGY

PRINCIPLES AND APPLICATIONS

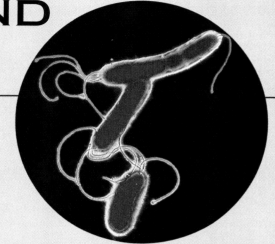

THIRD EDITION

JACQUELYN G. BLACK

MARYMOUNT UNIVERSITY

PRENTICE HALL, UPPER SADDLE RIVER, NEW JERSEY 07458

Library of Congress Cataloging-in-Publication Data

Black, Jacquelyn G.
 Microbiology: principles and applications / Jacquelyn G. Black. –
– 3rd ed.
  p.   cm.
 Includes index.
 ISBN 0–13–190745–X
 1. Medical microbiology. 2. Microbiology. I. Title.
QR46.C877 1996
616′ .01—dc20

95-49966
CIP

EXECUTIVE EDITOR: *David Kendric Brake*
EDITOR-IN-CHIEF: *Paul F. Corey*
EDITORIAL DIRECTOR: *Tim Bozik*
DEVELOPMENT EDITOR: *Karen Karlin*
DIRECTOR OF PRODUCTION
 AND MANUFACTURING: *David W. Riccardi*
EXECUTIVE MANAGING EDITOR: *Kathleen Schiaparelli*
PRODUCTION PROJECT COORDINATOR: *Barbara DeVries*
PRODUCTION PROJECT MANAGERS: *Edward Thomas/*
 Alison Aquino
CREATIVE DIRECTOR: *Paula Maylahn*
ART DIRECTOR: *Heather Scott*
ART MANAGERS: *Patrice Van Acker/Charles Pelletreau*
INTERIOR DESIGN: *Sheree Goodman*
ILLUSTRATORS: *William C. Ober, M.D., & Claire W. Garrison,*
 R.N., Medical and Scientific Illus.; Network Graphics
COVER DESIGN: *Tom Nery*
COVER PHOTO: *A. B. Dowsett/Photo Researchers, Inc.*
EXECUTIVE MARKETING MANAGER: *Kelly McDonald*
PHOTO EDITORS: *Lorinda Morris-Nantz/Melinda Reo*
PHOTO RESEARCHER: *Stuart Kenter Associates*
MANUFACTURING BUYER: *Trudy Pisciotti*
EDITOR-IN-CHIEF OF DEVELOPMENT: *Ray Mullaney*
EDITORIAL ASSISTANT: *Mary Hastings*
COPY EDITOR: *Barbara Liguori*
MARKETING ASSISTANT: *Amy Reed*
MANAGING EDITOR: *Linda Schreiber*

SPECIAL CONTRIBUTORS:
 Jeffrey Pommerville, Glendale Community College
 George A. Wistreich, East Los Angeles College

CONSULTANTS:
 Denise Friedman, Hudson Valley Community College
 William C. Matthai, Tarrant County Junior College
 Donald Stahly, University of Iowa
 Ann Vernon, St. Charles County Community College

OTHER CONTRIBUTORS:
 Stephen Hart
 Bette Weinstein Kaplan
 Kathleen Scogna
 Carol Stone
 Carolyn J. Strange

Photo credits begin on page PC1 and constitute a
continuation of the copyright page.

 ©1996, 1993, 1990 by Prentice-Hall, Inc.
Simon & Schuster / A Viacom Company
Upper Saddle River, New Jersey 07458

All rights reserved. No part of this book may be
reproduced, in any form or by any means,
without permission in writing from the publisher.

Printed in the United States of America

10 9 8 7 6 5 4 3 2 1

ISBN 0-13-190745-X

Prentice-Hall International (UK) Limited, *London*
Prentice-Hall of Australia Pty. Limited, *Sydney*
Prentice-Hall Canada Inc., *Toronto*
Prentice-Hall Hispanoamericana, S.A., *Mexico*
Prentice-Hall of India Private Limited, *New Delhi*
Prentice-Hall of Japan, Inc., *Tokyo*
Simon & Schuster Asia Pte. Ltd., *Singapore*
Editora Prentice-Hall do Brasil, Ltda., *Rio de Janeiro*

ABOUT THE AUTHOR

Jacquelyn G. Black
MARYMOUNT UNIVERSITY, ARLINGTON, VIRGINIA

Jacquelyn Black received her B.A., B.S., and M.S. from the University of Chicago and her Ph.D. from Catholic University of America. She has been teaching microbiology to undergraduates since 1970. She is a member of the American Society for Microbiology, and she has received grants for conducting teacher-training programs.

In addition to her extensive teaching experience, Dr. Black has engaged in fieldwork and studies throughout the globe. Her travels have taken her from the interior of Iceland to Belgium and Portugal to the barrier reef of Belize.

Dr. Black describes herself as an "incorrigible snoop" who is interested in all the various aspects and applications of microbiology. This natural curiosity, coupled with her classroom and laboratory experience, make her uniquely qualified to author an introductory microbiology textbook. This book conveys her sense of excitement for microbiology and offers the most current information on developments and applications within the field.

Dear Reader,

Being an author is a joyous thing. It gives you a sort of "license to snoop." You can call up people and say, "I'm writing a book, and I need to know all about . . . ," or "I'd like to come to visit and see. . . ." Suddenly closed doors swing open, and I'm off on a wonderful adventure, talking to fascinating people, learning things that will stretch my mind out to a dimension from which it will never return. That's my definition of education. I wish I could have taken all of you with me when I went to the U.S. Army Medical Research Institute of Infectious Diseases (USAMRIID)—our former germ warfare center). Once every 2 or 3 years, they "break down" a maximum containment laboratory suite, sterilizing it and getting ready to start a new project in it. It can be visited only on the two days after sterilization, before it will be closed off for the next several years again. I had been, with my nose and camera pressed up against the glass portholes to these rooms,

trying to see everything inside. Then they invited me to come back on one of the two days and go inside. Your heart would have pounded a little, too, I think, if you were beside me walking where only scientists dressed in plastic spacesuits had gone up until the day before. You would have enjoyed my visit to the Carter Foundation in Atlanta. Indeed, had you been able to come with me and see everything yourself, and talk with all the people I did, we might not have needed this book. But because you couldn't, I hope that I have been able to transmit in the pages of this book a sense of the excitement that is microbiology. The longer I am a microbiologist, the more excited I become. It never grows old or boring—it is constantly new and vital. I hope you will love it too.

Sincerely,
Jacquelyn Black

BRIEF CONTENTS

IV

CONTENTS

UNIT *I*

MICROBIOLOGY–THE FUNDAMENTALS

UNIT *II*

MICROBIAL METABOLISM, GROWTH, AND GENETICS

THE ROSTER OF MICROBES AND MULTICELLULAR PARASITES

UNIT *IV*

CONTROL OF MICROORGANISMS

UNIT *V*

HOST–MICROBE INTERACTIONS

UNIT *VI*

INFECTIOUS DISEASES OF HUMAN ORGAN SYSTEMS

UNIT *VII*

ENVIRONMENTAL AND APPLIED MICROBIOLOGY

LIST OF BOXES AND NOTEBOOKS

 CLOSE-UP

 PUBLIC HEALTH

APPLICATIONS

BIOTECHNOLOGY

TRY IT

MICROBIOLOGIST'S NOTEBOOK

PREFACE

"How are you?" How many times have you greeted someone—or been greeted by someone—in this way? This greeting symbolizes our concern about our health and well-being. Such a greeting is no less important today, at the end of the twentieth century, than it was centuries ago, because we still face potentially deadly microorganisms and the diseases they cause. However, to understand the roles microbes play in our lives, including the interplay between microorganisms and humans, we must examine, learn about, and study their world—the world of microbiology.

Microorganisms are everywhere. They exist in a range of environments from mountains and volcanoes to deep-sea vents and hot springs. Microorganisms can be found in the air we breathe, in the food we eat, and even within our own body. In fact, we come in contact with countless numbers of microorganisms every day. Although some microbes can cause disease, most are not disease-producers; rather, they play a critical role in the processes that provide energy and make life possible. Some even prevent disease, and others are used in attempts to cure disease.

Because microorganisms play diverse roles in the world, microbiology continues to be an exciting and critical discipline of study. And because microbes affect our everyday lives, microbiology provides many challenges and offers many rewards. Look at your local newspaper, and you will find items concerning microbiology: to mention a few, reports on diseases such as AIDS, tuberculosis, and cancer; the resurgence of malaria or "new" diseases such as those caused by the Ebola virus and the hantavirus; a bacterium that can cause ulcers; technologies designed to increase food production; genetically engineered tomatoes that resist rotting; microorganisms used to clean up toxic wastes and oil spills; and the Human Genome Project, which will identify the complete set of genetic instructions within the cells of the human body. Yes, microbiology represents a broad field of endless challenges and exciting discoveries.

A theme that permeates this book, then, is that microbiology is a current, relevant, exciting central science that affects all of us. In countless areas—from agriculture to evolution, from ecology to dentistry—microbiology is contributing to scientific knowledge as well as solving human problems. As you read this text, you will get a sense of the history of this science, its methodology, its many contributions to humanity, and the many ways in which it continues to be on the cutting edge of scientific advancement.

This book meets the needs both of students of the health sciences and of students majoring in biology. The book is designed to serve both audiences—in part by using an abundance of clinically important information to illustrate the general principles of microbiology and in part by offering a wide variety of additional applications.

Style and Currency

In countless areas, microbiological studies are contributing both to the knowledge base of biological science and to solving human problems. However, the rapid advances being made in microbiology make teaching about—and learning about—microorganisms challenging. Therefore, every effort has been made in the third edition of *Microbiology: Principles and Applications* to ensure that the writing is simple, straightforward, and functional; that microbiological concepts and methodologies are clearly and thoroughly described; and that the information presented is as accessible as possible to students.

Furthermore, students who enjoy a course are more likely to retain far more of its content for a longer period of time than do those who take the course like a dose of medicine. Students should experience microbiology as an exciting, dynamic, rapidly changing field that is important to human welfare. The development of microbiology—from Leeuwenhoek's astonished observations of "animalcules," to Pasteur's first use of rabies vaccine on a human, to Fleming's discovery of penicillin, to today's race to develop an AIDS vaccine—has been one of the most dramatic stories in the history of science. Whether the topic is the eradication of smallpox, the identification of microorganisms responsible for Lyme disease, or viruses responsible for some forms of cancer, there is no reason for a text to be any less interesting than its subject. Because students find courses interesting when they can relate topics to their everyday life or to career goals, we have emphasized the connection between microbiological knowledge and student experiences.

In a field that changes so quickly—with new research, new drugs, and even new diseases—it is essential that a text be as up to date as possible. This book incorporates the latest information on all aspects of microbiology, including clinical practice. Special attention has been paid to such important, rapidly evolving topics as genetic engineering (Chapter 8); viruses (Chapter 11); drug resistance, especially in tuberculosis (Chapter 14); and immunology (Chapters 18 and 19). Some of the material that is new to the third edition is truly "hot off the presses," such as the Microbiologist's Notebook in Chapter 8, "Puzzling Out an All-New Sequencing Strategy," about the recent success of Craig Venter and Hamilton Smith in sequencing the whole genome of *Haemophilus influenzae,* the first organism to be sequenced completely.

Organization of the Third Edition

The organization of the third edition combines logic with flexibility. The chapter sequence will be useful in most microbiology courses as they are usually taught. Nevertheless, it is not essential that the chapters be assigned in their present order; it should be possible to use this book in courses organized along quite different lines. **Concept link** symbols (∞) are used throughout the chapters to key the student to material that was previously discussed in the text.

The third edition of *Microbiology: Principles and Applications* is organized into seven units:

Unit I: Microbiology—The Fundamentals (Chapters 1–4) provides the basic background information on the scope and history of microbiology, on fundamentals of chemistry, on microscopy, and on cell structure and membrane transport. The information provides the foundation for the rest of the text as well as for the microbiology course. Chapter 4, on cell structure and membrane transport, has been updated to keep pace with the cell biology of prokaryotic and eukaryotic cells and with our current understanding of membrane transport.

Unit II: Microbial Metabolism, Growth, and Genetics (Chapters 5–8) focuses on topics concerned with the metabolism, growth, and genetics of microorganisms. Two full chapters on genetics provide a thorough, clear treatment of such important topics as gene regulation, mutations, genetic engineering, and the polymerase chain reaction (PCR) method. In particular, Chapter 8, on genetic transfer and genetic engineering, has been updated—for example, to include more detail on plasmids.

Unit III: The Roster of Microbes and Multicellular Parasites (Chapters 9–12) examines the major types of microbes and how they are classified. Chapter 11, on viruses, has been extensively revised to reflect the rapidly changing field of virology. This chapter also includes an extensive new section on cancer-causing viruses.

Unit IV: Control of Microorganisms (Chapters 13 and 14) discusses how the growth of microorganisms can be controlled. This unit includes material on chemical and physical control methods (sterilization and disin-

fection) and on antimicrobial chemotherapy, with added coverage of new antimicrobial agents.

Unit V: Host–Microbe Interactions (Chapters 15–19) covers all aspects of the relationship between host and microorganism, including the disease process, epidemiology and nosocomial infections, nonspecific body defense mechanisms, and immunology. Chapters 18 and 19, on immunology, have been extensively rewritten to reflect the current state of knowledge on immunity, autoimmune disorders, immunodeficiency diseases (including AIDS), and immunological tests. The discussion of AIDS is now treated more centrally within Chapter 19.

Unit VI: Infectious Diseases of Human Organ Systems (Chapters 20–25) consists of a survey of infectious diseases, organized by the affected organ or organ system. Chapter 20, on diseases of the skin and eyes as well as wounds and bites, now mentions monkeypox, a new disease involving humans, and necrotizing fasciitis. Chapter 22, on oral and gastrointestinal diseases, now includes updated coverage of the link between *Helicobacter pylori* and gastric ulcers.

Unit VII: Environmental and Applied Microbiology (Chapters 26 and 27) considers environmental microbiology (including water pollution, water purification, and sewage treatment) and applied microbiology (including the microbiology of food). Extra coverage has been added on such "hot topics" as bioremediation (in Chapter 26, on environmental microbiology) and irradiation of food (in Chapter 27, on applied microbiology).

The Art Program

The use of clear, attractive drawings and carefully chosen photographs can significantly contribute to the student's understanding of a scientific subject. In response to the comments of reviewers and users who requested additions and changes to some of the art in the previous edition, new figures have been added to the third edition. Throughout, color has been used not just decoratively but for its pedagogic value. For example, every effort has been made to color similar molecules and structures the same way each time they appear, making them easier to recognize. The availability of full color also makes it possible to present many subjects, such as staining reactions, as they are seen in the laboratory.

We are very pleased to once again have had the services of William C. Ober, M.D., and Claire W. Garrison, R.N., who have expertly executed many of the new and redrawn illustrations for this edition of *Microbiology: Principles and Applications*. Bill and Claire combine technical skill with scientific understanding to a degree rarely found in biological illustration. They have received numerous awards, including the Award of Excellence from the Association of Medical Illustrators, the Certificate of Excellence from the American Institute of Graphic Arts, the art and design award of the Chicago Book Clinic, and the Award of Excellence from the Printing Industries of America. Bill and Claire have also received recognition from Bookbuilders West and are recipients of the Art Directors Award.

Special Pedagogic Features

This book is designed not simply as a vehicle for transmitting information but also as a tool for learning. The features that have made *Microbiology: Principles and Applications* so popular have been retained in the third edition. These features are designed to broaden the scope of topical coverage and to maintain a high level of student interest and involvement with the material.

Boxes

Each chapter contains **boxes** that deal with a wide variety of subjects. In the third edition, all boxes are placed in one of five categories, each of which is distinguished by an icon and a color banner. Nearly 20 percent of the boxes are new (for example, "Ebola Virus Scare" and "The Saga of Typhoid Mary") and replace or supplement those in the previous edition.

Biotechnology boxes report on recent developments in this exciting field, such as triple-stranded DNA, human hemoglobin produced by transgenic pigs, and genetically engineered tomatoes.

Try It boxes encourage students to make the transition from the textbook to their everyday world through hands-on activities that enable them to experience the excitement of discovering things for themselves. Examples include culturing magnetotactic bacteria from local swamps, investigating which organisms grow in the cosmetics and health care products students use daily, and studying parasites found in their backyard.

Applications boxes demonstrate the role of microbiology in the "real world." Some of these applications are clinical; others deal with microbes that grow in your laboratory's emergency eyewash station or with processes such as fermentation or cleaning up toxic materials.

Close-Up boxes contain interesting anecdotes, historical background, quotations from distinguished writers on microbiology, and information about unusual subjects such as a recently discovered macroscopic bacterium,

the origins of viruses, and bacteria that will allow researchers to track the depletion of the ozone layer.

Public Health boxes deal with such issues as the controversy about AIDS testing, food spoilage in poultry, and plague in the United States.

Essays

The text of each chapter ends with an **essay,** a discussion of a supplemental topic that is both interesting in itself and relevant to the content of the chapter. Some instructors may elect to treat these essays as integral parts of the text, whereas other instructors may prefer to assign them as extra reading or make them optional for students. About one-quarter of these essays are new to the third edition, and most of the rest have been updated and revised. Some deal with clinical subjects or public health issues (emerging viruses, hantaviruses); others develop chapter topics in greater detail (using microbes in environmental cleanup) or discuss new techniques (polymerase chain reaction [PCR]); still others focus on current research or unsolved problems in microbiology (the meningitis belt).

Microbiologist's Notebooks

Each of the seven units has a **Microbiologist's Notebook.** This feature invites students into the working world of microbiology. The notebooks are designed to provide insight into the backgrounds and motivations of microbiologists as well as into the nature of their activities, the varied settings in which they work, the types of problems they investigate, and the experiences they have. Our subjects occupy very diverse careers in a range of microbiological fields and disciplines. They include laboratory researchers, zookeepers, physicians, and former president Jimmy Carter. The stories have been selected not only because they are highly interesting in themselves but also because they are timely and relevant to topics discussed in the text. For example, one notebook examines work that has already led to the determination of cancer-causing mutant genes; another examines the control of nosocomial infections. The Carter Foundation's Global 2000 project to eliminate the guinea worm, which causes river blindness, is close to eradicating this scourge.

Endpaper Tables

In addition to the many summary tables used throughout this text to convey large quantities of useful information in a compact and easily accessible form, six pages of endpaper tables appear at the front and back of the book. These endpaper tables list virtually all the medically important microorganisms and diseases discussed in the text. These tables are a valuable learning tool and also enhance the usefulness of the text for students and instructors.

It is not uncommon for a student to remember the name of an infectious organism but not the disease associated with it, or to recall the name of a disease but not the causative organism. Moreover, a given organism or disease may be discussed in several different chapters. The endpaper tables (1) reinforce the association between diseases and their causative organisms and, because they contain page references, (2) make it easy to find the relevant text discussions of a particular organism or disease. The front endpapers list all the major infectious diseases discussed in the text, their special features, the organisms that cause them, and the text pages on which they are discussed. The back endpapers list all the major organisms discussed in the text (classified by type) with their distinguishing characteristics, the diseases they produce, and the relevant text pages.

Other Pedagogy

To enhance learning, this book offers a set of coordinated pedagogic features. Each chapter begins with a list of **focus questions** that define the scope of the chapter and offer the student a preview of the major topics that will be covered. Within each chapter, important terms are highlighted in **boldface type.** Unfamiliar terms that might be difficult to pronounce are accompanied by a **pronunciation guide** based on *Stedman's Medical Dictionary.* (See "A Note on Pronunciation" later in this Preface.) The **derivations** of terms are given whenever they are of special interest or will help students understand and remember the term.

At the end of each chapter is a concise **Chapter Summary** that includes first- and second-level headings to provide a quick review of the chapter's structure. The summary makes it easy to review the essential concepts and terminology of the chapter. Terms that are boldfaced in the chapter are boldfaced in the summary as well. All the boldfaced terms also appear in the **Key Terms** list at the end of the chapter. Page references for the key terms facilitate student review.

The nearly 750 **Questions for Review** are grouped into sections that correspond to the focus questions at the start of each chapter. They test the student's mastery of the factual information presented in the chapter. Many of them have been rewritten and redesigned to test the student's understanding of microbiology concepts, and new objective questions have been added to offer instructors variety. "Visual" questions (diagrams, maps, or tables) now appear in each chapter because today's students are so visually oriented.

More than 150 **Problems for Investigation** typically require critical thinking skills and call for more extended answers, in some cases in the form of an oral report or an essay. They demand more thought than do the questions and generally require the synthesis of information from several parts of the chapter, from several chapters, or from sources outside the text. Problems that describe a series of symptoms and then ask students to identify the disease and causative agent are listed in Appendix F.

The **Some Interesting Reading** sections list books and articles that provide additional information about topics discussed in the chapter. These sections have been thoroughly updated and contain many recent listings in addition to "classic" books and articles in biology and microbiology.

Six **Appendices** are provided to assist students. Appendix A is a review of the **metric system** and **conversions, exponential notation,** and the **pH equation** (new to the third edition). It also now includes a **Most Probable Number Table** that is useful for calculating numbers of microorganisms. Appendix B contains updated versions of the standard **classification of bacteria** (based on *Bergey's Manual of Systematic Bacteriology*) and of a summary of the **classification of viruses.** Appendix C is a listing of **word roots** and their meanings for terms that are commonly encountered in microbiology. Appendix D describes **safety precautions** used in handling clinical specimens in the microbiology laboratory or in health care facilities. The new Appendix E presents more detailed diagrams and explanations of important **metabolic pathways.** Appendix F presents the **answers to selected Problems for Investigation** (of the case study variety).

Instructor's Edition

A special **Instructor's Edition** of this text is available to enhance the usefulness of the text as a teaching tool. For each chapter, the Instructor's Edition provides a wide range of supplemental material designed to assist the instructor in teaching from the text and to allow for deeper and/or more extensive treatment of various topics. Included in the material for each chapter are the following: (1) A **Chapter Overview** explains the organization of the chapter and summarizes its main themes. Its purpose is to place the material of the chapter into a broader context—to focus the attention on the forest rather than on the trees. (2) **Chapter Objectives** are related to the focus questions at the start of each text chapter but are more numerous and detailed. Couched in behavioral terms, they direct the instructor's attention to what students should *be able to do* as well as *know* after successfully completing each chapter. This feature is convenient for instructors who must prepare a course syllabus with chapter-by-chapter goals. (3) A complete **Chapter Outline** includes all levels of headings. For easy reference, it also indicates the locations of all boxes, figures, and tables in the text. (4) Instructional suggestions include **Demonstrations** for class use; **Teaching Tips,** such as suggestions for interesting examples and analogies, points to emphasize, questions and exercises for class use, and effective ways to explain topics that students find especially difficult; **Discussion Topics** to stimulate student interest; **Library Assignments** suitable for student papers or oral reports; and **Audiovisual Aids,** including relevant films, videotapes, and computer software that instructors might find useful in class. (5) A **Review** section contains complete and detailed answers to all end-of-chapter **Questions for Review** and **Problems for Investigation.**

Supplements

To supplement the needs of both students and instructors, the following materials are available to qualified adopters.

For the Instructor

- **Instructor's Edition** with the Instructor's Manual bound in as an integral part of the book, preceding the main text. Written by William Matthai of Tarrant County Junior College, the Instructor's Edition includes overviews, objectives, outlines, demonstration suggestions, teaching tips, discussion topics, activities, audiovisual suggestions, and answers to all text questions and problems.

- **Microbiology Laser Disc,** prepared by Jacquelyn Black, contains more than 2000 images, including full-color micrographs, photographs, illustrations, and animations for use in either a lecture or lab setting. All images are indexed and accessible with or without a bar code scanner.

- **Test Item File,** authored by Denise Friedman of Hudson Valley Community College, contains more than 1000 questions, all referenced to chapter number and section.

- The **Prentice Hall Custom Test** is based on a powerful testing technology developed by Engineering Software Associates, Inc. Available for Windows and Macintosh, Prentice Hall Custom Test allows you to create and tailor the exam to your own needs. With the Online Testing option, exams can also be administered online, and data can then be automatically transferred for evaluation. A comprehensive desk reference guide is included, along with online assistance.

- **Telephone Testing Service** allows you to select questions from the Test Item File, call a toll-free number, and have the test prepared at Prentice Hall for no additional charge. Within 48 hours you will receive a professionally prepared test, an answer key, and answer sheets for your students. Two versions of a test can be furnished.

- **Instructor's Edition to the Laboratory Manual** has answers to all the exercises.

- **Transparency Pack** contains 250 full-color acetates of text art and photos from the book.

- **Microbiology Presentational CD-ROM** provides an ideal opportunity to bring multimedia into the classroom. It features more than 1000 still images, including text art and photomicrographs, as well as video and high-quality three-dimensional animations of important topics in microbiology. With **Prentice Hall's Presentation Manager 2.0 for Microbiology,** included on the same CD, you can navigate easily through the image resources of the text and quickly create an exciting lecture presentation.

- **Director Academic** and **Authorware Academic** are educational adaptations of the leading multimedia authoring tools. Available exclusively through Prentice Hall, these tools put the power of multimedia development into your hands at a fraction of the original cost. Both products also contain templates designed specifically for academic uses. They are available for Macintosh or Windows platforms.

For the Student

- **Study Guide** by George Wistreich of East Los Angeles College is a highly interactive study tool. It features a study outline, key terms and concepts, and concept mastery exercises. At the end of each part are a microbiology trivia quiz, a word-search puzzle, and a mini-review.

- *The New York Times* **Themes of the Times** consists of selected articles from *The New York Times* dealing with topics related to microbiology. This supplement is updated annually and is available free to adopters, who can order as many copies as the number of new texts purchased.

- **Microbiology Laboratory Fundamentals and Applications,** a laboratory manual by George Wistreich, offers a generous selection of microbiology labs. Topics range from basic techniques to real-life applications of microbiology. The labs are supported with thorough student pedagogy, including section overviews, pronunciation guides, learning objectives, procedure diagrams, key terms lists, and self-assessment exercises.

A Note on Pronunciation

The scheme used for pronunciation is simple:

- a single accent mark (') is used for the main accent in a word;
- a double accent mark (") is used for secondary accents, if any;
- any vowel not followed by a consonant is assumed to be long;
- syllables are separated by either a hyphen or an accent mark.

Acknowledgments

I would like to express my gratitude to all the people who shared their experiences with us in the Essay and Microbiologist's Notebook features. For their generous assistance with this edition, special thanks to Jan Curry, MQC, Inc.; Craig Venter, TIGR, and Hamilton Smith, Johns Hopkins University School of Medicine; Arthur Mason, Albert McManus, and Elisabeth Greenfield, Institute of Surgical Research; Paula Fujiwara, NYC Department of Health and CDC, and Wilfrid Desir, New York Medical College and Columbia University College of Physicians & Surgeons; and Douglas Nelson, the University of California, Davis.

This project has consumed large amounts of time and energy over the past 6 years and has required patience and understanding on the part of family members. Special thanks, therefore, to Laura Black for having again shared her mother with "the book" and for working so hard at word-processing it. My student Kanai Hitomi spent many hours poring over manuscripts. Her help was very much needed. Finally, the author gratefully acknowledges the support of Marymount University and especially of Provost Alice S. Mandanis and Dean Robert A. Draghi.

I wish to acknowledge the many Prentice Hall reviewers for their thoughtful suggestions.

FIRST-EDITION REVIEWERS:

Oswald G. Baca, *University of New Mexico*
David L. Balkwill, *Florida State University*
Keith Bancroft, *Southeastern Louisiana University*
Thomas R. Corner, *Michigan State University*
Monica A. Devanas, *Rutgers University*
Lawrence W. Hinck, *Arkansas State University*
Wallace L. Jones, *De Kalb College*
Donald G. Lehman, *Wright State University*
William C. Matthai, *Tarrant County Junior College*
Raymond B. Otero, *Eastern Kentucky University*
Kimberley Pearlstein, *Adelphi University*
Robert A. Pollack, *Nassau Community College*
Dennis J. Russell, *Seattle Pacific University*
Gordon D. Schrank, *St. Cloud State University*
Alan J. Sexstone, *West Virginia University*
Deborah Simon-Eaton, *Santa Fe Community College*
Robert E. Sjogren, *University of Vermont*
Larry Stearns, *Central Piedmont Community College*
Michael R. Yeaman, *University of New Mexico*

SECOND-EDITION REVIEWERS:

D. Andy Anderson, *Utah State University*
Dan C. DeBorde, *University of Montana*
Monica A. Devanas, *Rutgers University*
Von Dunn, *Tarrant County Junior College*
David L. Filmer, *Purdue University*
Eugene Flaumenhaft, *University of Akron*
Denise Y. Friedman, *Hudson Valley Community College*
William R. Gibbons, *South Dakota State University*
Ronald E. Hurlbert, *Washington State University*
Robert J. Janssen, *University of Arizona*
Thomas R. Jewell, *University of Wisconsin–Eau Claire*
Harvey Liftin, *Broward Community College*
William C. Matthai, *Tarrant County Junior College*
Russell A. Normand, *Northeast Louisiana University*

Joseph M. Sobek, *University of Southwestern Louisiana*
Bernice C. Stewart, *Prince George's Community College*
James E. Urban, *Kansas State University*
John Zak, *Texas Tech University*
Thomas E. Zettle, *Illinois Central College*

THIRD-EDITION REVIEWERS:

Ronald W. Alexander, *Tompkins Cortland Community College*
R. L. Bernstein, *San Francisco State University*
David L. Berryhill, *North Dakota State University*
Richard D. Bliss, *Yuba College*
Edward A. Botan, *New Hampshire Technical College*
Sally DeGroot, *St. Petersburg Junior College*
John G. Dziak, *Community College of Allegheny County*
Pamela B. Fouche, *Walters State Community College*
Christine L. Frazier, *Southeast Missouri State University*
Denise Y. Friedman, *Hudson Valley Community College*
William R. Gibbons, *South Dakota State University*
Pamela L. Hanratty, *Indiana University*
Janet Hearing, *State University of New York, Stony Brook*
John J. Iandolo, *Kansas State University*
John W. Kimball, *Harvard University*
Timothy A. Kral, *University of Arkansas*
Caleb Makukutu, *Kingwood College*
Alexsandria Manrov, *Tidewater Community College, Virginia Beach*
Judy D. Marsh, *Emporia State University*
Rosemarie Marshall, *California State University, Los Angeles*
John Martinko, *Southern Illinois University*
William C. Matthai, *Tarrant County Junior College*
Chris H. Miller, *Indiana University*
Rajeev Misra, *Arizona State University*
Christian C. Nwamba, *Wayne State University*
Curtis Pantle, *Community College of Southern Nevada*
Robin K. Pettit, *State University of New York, Potsdam*
Jeff Pommerville, *Glendale Community College*
Russell Robbins, *Drury College*
Richard A. Robison, *Brigham Young University*
Quentin Reuer, *University of Alaska, Anchorage*
Brian R. Shmaefsky, *Kingwood College*
J. Glenn Songer, *University of Arizona*
K. T. Shanmugam, *University of Florida*
Paul M. Steldt, *St. Philips College*
Gerald Stine, *University of Florida*
Teresa Thomas, *Southwestern College*
James E. Urban, *Kansas State University*
Phyllis K. Williams, *Sinclair Community College*
George A. Wistreich, *East Los Angeles College*

Comments and suggestions about the book are most welcome.

Jacquelyn Black
Arlington, Virginia

STUDENT PREFACE

A STUDENT'S GUIDE TO USING THIS TEXT

THIS TEXT WAS DESIGNED WITH YOU, THE STUDENT, IN MIND. EACH FEATURE IS INTENDED TO MAKE YOUR INTRODUCTION INTO THE WORLD OF MICROBIOLOGY A PLEASANT AS WELL AS EDUCATIONAL EXPERIENCE. SHOWN ON THE NEXT FEW PAGES, THESE FEATURES SHOULD HELP YOU LEARN THE BASIC PRINCIPLES OF MICROBIOLOGY AND ENCOURAGE YOU TO APPLY THEM TO THE WORLD AROUND YOU.

■ CHAPTER-OPENING PHOTOGRAPHS

Each chapter opens with a compelling photo that brings to life one topic discussed in the chapter.

■ CHAPTER-OPENING CAPTIONS

The accompanying caption describes the photo and ties it in to the chapter concepts.

■ FOCUS QUESTIONS

Before you begin a chapter, it is important to know what that chapter will cover and what you will be expected to learn from it. Focus questions allow you to preview the major topics.

CHAPTER 8

GENETICS II: TRANSFER OF GENETIC MATERIAL AND GENETIC ENGINEERING

DNA sequencing of recombinant clones in AIDS research. Microbiology has provided the techniques for working with DNA.

THIS CHAPTER FOCUSES ON THE FOLLOWING QUESTIONS:

A What are the nature and significance of gene transfer?
B What are the mechanisms and significance of transformation?
C What are the mechanisms and significance of transduction?
D What are the mechanisms and significance of conjugation?
E What are the characteristics and actions of plasmids?
F How are the following techniques of genetic engineering used: genetic fusion, protoplast fusion, gene amplification, recombinant DNA, and hybridomas?
G Why are scientists concerned about uses of recombinant DNA?

Transfer of genetic material from one organism to another can have far-reaching consequences. In microbes it provides ways for viruses to introduce genetic information into bacteria and mechanisms for bacteria to increase their disease-causing capabilities or to become resistant to antibiotics. Information obtained from studying the transfer of genetic material between microorganisms can be applied to agricultural, industrial, and medical problems and to the unique problems of the prevention and treatment of infectious diseases. In this chapter we will discuss the mechanisms by which genetic transfers occur and the significance of such transfers.

NATURE AND SIGNIFICANCE OF GENE TRANSFER

Gene Transfer

Gene transfer refers to the movement of genetic information between organisms. In most eukaryotes, it is an essential part of the organism's life cycle and usually occurs by sexual reproduction. Male and female parents produce *gametes* (sex cells), which unite to form

the first cell (the *zygote*) of a new individual. Because each parent produces many genetically different gametes, many different combinations of genetic material can be transferred to offspring. In bacteria, gene transfer is not an essential part of the life cycle. When it does take place, generally only a portion of the genetic makeup of the donating cell is transferred to the other participating, or *recipient*, cell. The combining of genes (DNA) from two different cells is called **recombination**, and the resulting cell is referred to as a *recombinant*. Before the 1920s, bacteria were thought to reproduce only by binary fission and to have no means of genetic transfer comparable to that achieved through sexual reproduction in eukaryotes. Since then, three mechanisms of gene transfer in bacteria have been discovered, none of which is associated with reproduction. Each mechanism—*transformation, transduction,* and *conjugation*—is discussed in this chapter.

Gene transfer is significant because it greatly increases the genetic diversity of organisms. As noted in Chapter 7, mutations account for some genetic diversity, but gene transfer between organisms accounts for even more. ∞ (p. 180) When organisms are subjected to changing environmental conditions, genetic diversity increases the likelihood that some organisms will

adapt to any particular condition. Such diversity leads to evolutionary changes. Organisms with genes that allow them to adapt to an environment survive and reproduce, whereas organisms lacking those genes perish. If all organisms were genetically identical, all would survive and reproduce, or all would die.

In *recombinant DNA technology,* genes from one kind of organism are introduced into the genome of another kind of organism (for example, when human genes are inserted into a pig).

TRANSFORMATION

Discovery of Transformation

Bacterial **transformation,** a change in an organism's characteristics because of the transfer of genetic information, was discovered in 1928 by Frederick Griffith, an English physician, while he was studying pneumococcal infections in mice. Pneumococci (bacteria that cause pneumonia) with capsules (Chapter 4) produce smooth, glistening colonies. ∞ (p. 901) Those lacking polysaccharide capsules produce rough colonies with a coarse, nonglistening appearance. Only

the capsule-producing (encapsulated) pneumococci inoculated into mice were *virulent*—that is, they had the power to cause severe disease. One such organism can multiply rapidly and kill a mouse! Capsules help keep molecules produced by the mouse's immune system from reaching the surface of the bacterium. They also make it difficult for white blood cells to engulf the invading bacteria. In other words, the capsule protects the bacteria from the mouse's immune system.

Griffith injected one group of mice with heat-killed smooth pneumococci, a second group with live smooth pneumococci, a third group with live rough pneumococci, and a fourth group with a mixture of live rough and heat-killed smooth pneumococci (Figure 8.1). As expected, mice that received live smooth pneumococci developed pneumonia and died, whereas those that received either heat-killed smooth pneumococci or live rough pneumococci did not develop pneumonia and survived. Surprisingly, those that received the mixture also died of pneumonia. Griffith isolated live smooth organisms from them. The presence of heat-killed, encapsulated organisms in the mixture apparently allowed the live unencapsulated ones to develop capsules and become virulent. Neither Griffith nor his colleagues knew how this transformation occurred.

196

197

■ CONCEPT LINKS

To understand microbiology you must remember many terms and be able to relate concepts introduced in different chapters. Concept links provide a quick visual signal that new material is related to or builds on an earlier discussion. Each "links" icon ∞ is followed by a page reference to help you find the relevant material in the earlier chapter.

■ BOLDFACE TYPE FOR KEY TERMS

Throughout the text, the most important new terms are highlighted in boldface type.

■ TABLES

Numerous tables throughout the text summarize key information in a handy, compact form. Some tables include useful full-color visuals.

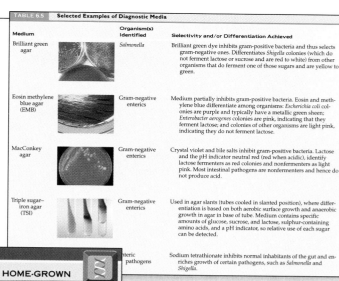

TABLE 6.5	Selected Examples of Diagnostic Media		
Medium		Organism(s) Identified	Selectivity and/or Differentiation Achieved
Brilliant green agar		Salmonella	Brilliant green dye inhibits gram-positive bacteria and thus selects gram-negative ones. Differentiates *Shigella* colonies (which do not ferment lactose or sucrose and are red to white) from other organisms that do ferment one of those sugars and are yellow to green.
Eosin methylene blue agar (EMB)		Gram-negative enterics	Medium partially inhibits gram-positive bacteria. Eosin and methylene blue differentiate among organisms: *Escherichia coli* colonies are purple and typically have a metallic green sheen; *Enterobacter aerogenes* colonies are pink, indicating that they ferment lactose; and colonies of other organisms are light pink, indicating they do not ferment lactose.
MacConkey agar		Gram-negative enterics	Crystal violet and bile salts inhibit gram-positive bacteria. Lactose and the pH indicator neutral red (red when acidic), identify lactose fermenters as red colonies and nonfermenters as light pink. Most intestinal pathogens are nonfermenters and hence do not produce acid.
Triple sugar–iron agar (TSI)		Gram-negative enterics	Used in agar slants (tubes cooled in slanted position), where differentiation is based on both aerobic surface growth and anaerobic growth in agar in base of tube. Medium contains specific amounts of glucose, sucrose, and lactose, sulphur-containing amino acids, and a pH indicator, so relative use of each sugar can be detected.
		enteric pathogens	Sodium tetrathionate inhibits normal inhabitants of the gut and enriches growth of certain pathogens, such as *Salmonella* and *Shigella*.

BIOTECHNOLOGY

SUPERMARKET TOMATOES THAT TASTE LIKE HOME-GROWN

Tired of tomatoes that look beautiful but have all the juice and flavor of cardboard? Ripening on the vine gives tomatoes their wonderful juicy flavor. Home-grown tomatoes are picked at their peak of ripeness and flavor. Tomatoes destined for supermarkets cannot be ripe when picked, or they will arrive at the store as a soggy, spoiled mess. Green tomatoes can withstand the rigors of shipping, but they don't have color, flavor, or juice when they arrive—they didn't stay on the vine long enough. Holding green tomatoes in storage areas in which the gas ethylene has been pumped causes green tomatoes to turn a deceptively beautiful red. However, the juice and flavor never develop.

In May 1994, Calgene, Inc., a plant biotechnology company located in Davis, California, received government approval for their genetically altered "Flavr Savr" tomato. It stays fresh up to 10 days longer than ordinary tomatoes, which allows it to ripen longer on the vine before it is picked and shipped. Five years of development went into the creation of this new tomato. Here is how it is done:

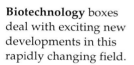

The enzyme polygalacturonase naturally occurs in tomatoes. It breaks down the pectin in cell walls, allowing the fruit to decay and spread seeds.

The polygalacturonase-making gene is isolated.

The gene is cloned and then reversed, to cancel the decay effect of the enzyme.

The reversed gene is inserted into bacteria, which are placed in a dish with pieces of leaf. The pieces absorb the gene with the bacteria.

The leaf pieces sprout roots and are transferred to soil.

The new plants bear seeds that grow into genetically engineered tomatoes.

The Food and Drug Administration (FDA) has given its safety endorsement for Flavr Savr. The tomatoes and their products won't need special labeling—the FDA has determined that no new or unexpected consequences, such as allergic reactions, should result from their use.

Not everyone, however, greeted the government approval with joy. Some people are afraid to eat them. Calgene is placing stickers on each of its Flavr Savr tomatoes and making brochures available in stores to explain its new product to customers.

■ BOXES

The boxes in each chapter show you what a fascinating field microbiology is. Many boxes contain full-color photographs or illustrations. There are five types of boxes, distinguishable by their icon and color banner (displayed at right).

BIOTECHNOLOGY

Biotechnology boxes deal with exciting new developments in this rapidly changing field.

TRY IT

Try It boxes prompt you to discover the wonders of microbiology for yourself— outside the textbook and classroom.

CLOSE-UP

Close-Up boxes cover interesting topics that will make learning microbiology fun.

APPLICATIONS

Applications boxes show how microbiology relates to you and to the world around you.

PUBLIC HEALTH

Public Health boxes give insight into infectious disease topics that are important to anyone involved in health care—as a provider or recipient.

■ ART

You're more likely to remember a concept when you can visualize it. Therefore, attractive, full-color illustrations and photographs are carefully coordinated with the text.

Compound Art—Many of the illustrations in the book offer breakouts to show the relationships among different levels of structure and/or photographs to help you visualize the topic being discussed.

Process Art—Where appropriate, processes are illustrated step by step with accompanying descriptions.

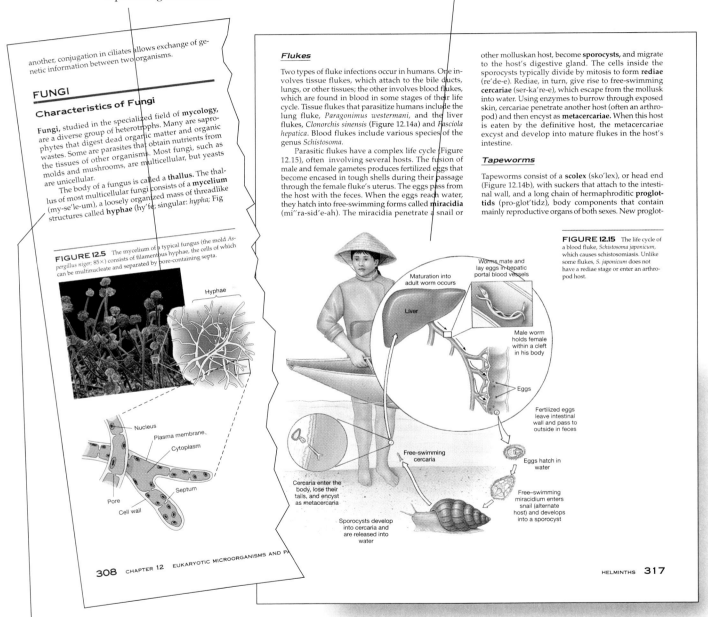

another, conjugation in ciliates allows exchange of genetic information between two organisms.

FUNGI

Characteristics of Fungi

Fungi, studied in the specialized field of **mycology,** are a diverse group of heterotrophs. Many are saprophytes that digest dead organic matter and organic wastes. Some are parasites that obtain nutrients from the tissues of other organisms. Most fungi, such as molds and mushrooms, are multicellular, but yeasts are unicellular.

The body of a fungus is called a **thallus.** The thallus of most multicellular fungi consists of a **mycelium** (my-se'le-um), a loosely organized mass of threadlike structures called **hyphae** (hy'fe; singular: *hypha;* Fig

FIGURE 12.5 The mycelium of a typical fungus (the mold *Aspergillus niger;* 85×) consists of filamentous hyphae, the cells of which can be multinucleate and separated by pore-containing septa.

Hyphae

Nucleus
Plasma membrane
Cytoplasm
Septum
Pore
Cell wall

308 CHAPTER 12 EUKARYOTIC MICROORGANISMS AND P

Flukes

Two types of fluke infections occur in humans. One involves tissue flukes, which attach to the bile ducts, lungs, or other tissues; the other involves blood flukes, which are found in blood in some stages of their life cycle. Tissue flukes that parasitize humans include the lung fluke, *Paragonimus westermani,* and the liver flukes, *Clonorchis sinensis* (Figure 12.14a) and *Fasciola hepatica.* Blood flukes include various species of the genus *Schistosoma.*

Parasitic flukes have a complex life cycle (Figure 12.15), often involving several hosts. The fusion of male and female gametes produces fertilized eggs that become encased in tough shells during their passage through the female fluke's uterus. The eggs pass from the host with the feces. When the eggs reach water, they hatch into free-swimming forms called **miracidia** (mi"ra-sid'e-ah). The miracidia penetrate a snail or

other molluskan host, become **sporocysts,** and migrate to the host's digestive gland. The cells inside the sporocysts typically divide by mitosis to form **rediae** (re'de-e). Rediae, in turn, give rise to free-swimming **cercariae** (ser-ka're-e), which escape from the mollusk into water. Using enzymes to burrow through exposed skin, cercariae penetrate another host (often an arthropod) and then encyst as **metacercariae.** When this host is eaten by the definitive host, the metacercariae excyst and develop into mature flukes in the host's intestine.

Tapeworms

Tapeworms consist of a **scolex** (sko'lex), or head end (Figure 12.14b), with suckers that attach to the intestinal wall, and a long chain of hermaphroditic **proglottids** (pro-glot'tidz), body components that contain mainly reproductive organs of both sexes. New proglot-

FIGURE 12.15 The life cycle of a blood fluke, *Schistosoma japonicum,* which causes schistosomiasis. Unlike some flukes, *S. japonicum* does not have a rediae stage or enter an arthropod host.

Maturation into adult worm occurs

Liver

Worms mate and lay eggs in hepatic portal blood vessels

Male worm holds female within a cleft in his body

Eggs

Fertilized eggs leave intestinal wall and pass to outside in feces

Eggs hatch in water

Free–swimming miracidium enters snail (alternate host) and develops into a sporocyst

Free-swimming cercaria

Cercaria enter the body, lose their tails, and encyst as metacercaria

Sporocysts develop into cercaria and are released into water

HELMINTHS 317

■ PRONUNCIATION GUIDES

Whenever the pronunciation of a new term is not obvious, a pronunciation guide is presented in parentheses. Where appropriate, derivations are also given to help you recognize word roots that are common in the language of science.

MICROBIOLOGIST'S NOTEBOOK

Puzzling Out an All-New Sequencing Strategy

Hamilton O. Smith, M.D. (seated), and J. Craig Venter, Ph.D., have succeeded in deciphering the first complete DNA sequence of an organism's genome—that of Haemophilus influenzae.

Hamilton Smith: We now know the DNA sequence of the entire *Haemophilus influenzae* genome, all 1.9 million base pairs. The collaboration that led to the first complete DNA sequence of an organism's genome began at a Genome Ethics meeting in Spain. A colleague from England first proposed the idea, but I threw cold water on it; a sequencing project of that size would require production-scale work not feasible or affordable in an academic environment. So the whole idea just sat there until several months later, when I ran into Craig Venter for the first time in Spain.

I've worked on *Haemophilus influenzae* for about 25 years—nearly my entire career here at Johns Hopkins University School of Medicine. As far as we know, humans are the only natural host for these bacteria. The strain we've sequenced, serotype *d*, isn't actually very pathogenic. We treat it about like *E. coli* in the lab. But a closely related strain, serotype *b*, causes respiratory tract infections, mostly in children, and it's a significant cause of middle ear infection. Meningitis is the most serious complication, but since a vaccine came into common use in the early 1980s, the incidence of meningitis has dropped dramatically.

Craig Venter: Aside from its clinical significance, this organism has an important historical role—the first restriction endonucleases were isolated from it. Ham discovered them during transformation studies, and for that discovery he and colleagues Daniel Nathans at Hopkins and Werner Arber in Switzerland were awarded the Nobel Prize. This organism has played a fundamental role in molecular biology and biotechnology and has had an outstanding impact on science. Without restriction enzymes, much of what we do would not be feasible.

At The Institute for Genomic Research, or TIGR, where I'm president and director, we're interested in accelerating the sequencing of human, animal, and plant genomes to understand better the role that genes and gene products play in development, evolution, physiology, and disease. It's the world's largest noncommercial facility devoted to large-scale DNA sequencing. We work on characterizing human genes directly through sequencing and indirectly by stimulating the field of bioinformatics [analysis of complex biological data] as the starting point for evolutionary studies. There's been much discussion of model organisms, and we decided to start with human at one end and microorganisms at the other end, so we have both to compare.

We were looking for a prototypical microorganism to test an idea. It didn't much matter which one it was as long as it fit certain criteria. The medium-sized, AT-rich *Haemophilus* genome seemed a good choice. And because Ham had extensive knowledge of the organism, he helped us further develop our ideas on sequencing strategy.

Smith: Craig was doing incredibly exciting work with the human genome. He developed a whole new method of gene discovery, called *expressed sequence tags*, or ESTs, because they're derived from messenger RNA instead of genomic DNA. The EST method has dramatically increased the pace of gene discovery.

I thought I'd have to convince him that we should sequence the *Haemophilus* genome, and I offered to provide a library [a collection of pieces] of the genome for sequencing. It turned out that he already knew exactly how we should do it, because it was a direct extension of the large-scale random-sampling techniques he had so successfully applied to ESTs. We both agreed that shotgun sequencing—looking at totally random sections of the genome—was the way to go. We were going to test a new approach, the whole-genome method.

Here's the project in a nutshell. I made the library by breaking the genome into 25,000–30,000 small overlapping pieces. Then TIGR sequenced those pieces directly, without using any other mapping information. They fed the sequences into their computer, which assembled them in order. It takes about a weekend for a supercomputer to reassemble that many pieces.

Venter: You can't just go and determine the sequence of something nearly 2 million base pairs long. With current sequencing technology, the maximum length of highly accurate sequence you can get in a sequencing run is 500–600 base pairs. You have to sequence these smaller fragments and then rely on finding exact matches in the genetic code from one piece to another.

Large-scale sequencing projects in the past involved sequencing lots and lots of small cloned pieces over and over again to build larger pieces. So, characterizing any genome involved spending months or years first generating a rough map of the genome and a clone set to work from. It's very slow and inefficient. The National Institutes of Health is funding a multiyear project that uses that general method to sequence the genome of *E. coli*. But Ham and I felt that if each genome were to take many years and multimillion dollars, then microbiology would not move forward very quickly.

Smith: While those who were waiting for faster sequencing technology were still sitting in their chairs, Craig waded right in with the current technology. I think TIGR is one of the best places in the world for sequencing. I would liken the approach with *E. coli* as starting at one spot and then working yourself around the whole genome from just one point. It's like an artist starting up in the corner and beginning to draw in the painting with all the details finished in the little corner before moving to the next area of the painting. Instead, TIGR uses a global approach, essentially sketching in the whole picture in outline, then filling it in everywhere at once. The whole picture emerges at one time.

Venter: At 1.9 million base pairs, the *Haemophilus* genome is a little less than half of *E. coli*'s, which is about 4.7 megabases. So they're not comparable-size projects. But we think this strategy would work for either size genome. It's similar to solving a jigsaw puzzle, except without the nice, square borders. Usually with jigsaw puzzles, people build the outside edge and then fill in toward the center. But with whole-genome shotgun sequencing, there are no edges (or maps) to start with. There's nothing to give you any intellectual advantage.

For this global approach to work, the library you sequence from has to be not only a complete representation of the genome but a truly random representation of the genome. If that one [segment] constructing the library was ... our best theories and our best ... cing and the best mathematics ... n't have been able to solve the ... m. That was probably the least ... ive step in the whole process, ... hat had been wrong, the rest of ... d not have worked.

... nith: I think it's fair to say that ... ng to be a brand new era in mi... ogy. Having the complete se... e of virtually any bacterium ... really expedite any investiga... th the microorganism. You'll ... front of you all the gene con... that organism. If you had the se... of, for example, *Neisseria*, you ... ompare that with the known se... e of *Haemophilus* and two or three ... rganisms. You could immedi... ick out those genes that are ... e to *Neisseria*. Or you could select ... nvolved in the virulence [dis... roducing capacity] of the organ... hich would therefore become ... for therapeutics.

Furthermore, you can go in and lift out individual genes by using the powerful PCR techniques. You can do knockouts of individual genes to look at function. You can look at the expression of genes under various growth conditions. There's going to be an enormous change in the way people investigate these organisms.

Just imagine trying to compare two sequences, each over a million base pairs long. It's impossible to deal with this level of information without having sophisticated computers.

Venter: I think we both agree that the computer is going to be the molecular biologist's number one tool for the future. It's going to be the starting point for developing hypotheses, for understanding what is known, and for going forward.

Just to print the *Haemophilus* genome sequence, absolutely packed on a page so that you could read it without a magnifying glass, would take 250 pages. Indicating gene locations and amino acids would require over 1000 pages. And we're just in the infancy of generating this information. This is only the first bacterial genome sequence to be completed. We hope a decade from now there'll be a hundred or more. [In fact, a team led by Dr. Claire Fraser of TIGR has sequenced the genome of a second organism, *Mycoplasma genitalium*. With fewer than 600,000 base pairs, that genome took just 3 months to sequence.]

When we begin comparing entire genomes from one species to another, we'll be able to start rethinking evolution in terms of what genes appeared or disappeared, what mutations occurred that maybe allowed that species to evolve. Evolutionary trees have been drawn up by inferring evolutionary distances based on sequence variations in a few genes. But I suspect that if you look at multiple genes, each will have a slightly different evolutionary tree. And of course, what really evolved were organisms. Having complete gene sets from microorganisms is going to lead to a better understanding of humans.

In fact, it's already happening. In collaboration with Dr. Bert Vogelstein, we recently found three new human cancer genes during a brief search of the TIGR sequence database. Knowing the sequence of a DNA repair enzyme characterized in *E. coli (MutL)*, we had annotated 16 human sequences as DNA repair enzymes. Dr. Vogelstein had speculated that defects in a similar human enzyme, if it existed, might lead to colon cancer. Indeed, mutant forms of the four genes found are responsible for a form of colon cancer.

Our work with *Haemophilus* shows that obtaining the complete genome sequence of an organism is truly feasible. They're not multiyear, multimillion dollar projects. This project took less than 12 months, but much of that time was spent in technology development —this was raw, basic research—so subsequent genomes will take substantially less time. As in any area, you have to have a breakthrough in technology to make new strides. It seemed to both of us that the only way it was going to happen was if we could take this whole-genome approach.

The bacillus Haemophilus influenzae (2800×).

■ MICROBIOLOGIST'S NOTEBOOKS

Each of the seven units of the book contains a Microbiologist's Notebook, a feature that lets you visit the working world of microbiology. You can read about the excitement microbiologists and scientists in related fields have for their work, all in their own words.

IMAGES OF ATOMS

Being able to look inside cells, to see the individual atoms that compose them—what a mind-boggling experience that would have been for Leeuwenhoek! In 1980, Gerd Binnig and Heinrich Rohrer invented the first of a series of rapidly improving scanning tunneling (or scanning probe) microscopes (STM). Five years later they received the Nobel Prize for their discovery.

Instead of using light, Binnig and Rohrer used a thin wire probe made of the metals platinum and iridium to trace over the surface of a substance, much as you might use your finger to feel the bumps while you read Braille. Electron clouds (regions of electron movement) from the surfaces of the probe and the specimen overlap, producing a kind of pathway through which electrons can "tunnel" into each other's clouds. This tunneling sets up an observable electric current. The stronger the current, the closer the top of the atom is to the probe. Running the probe across in a straight line reveals the highs and lows of individual molecules or atoms in a surface. Even movies can be made this way. The first one ever produced showed individual fibrin molecules coming together in forming a clot. Models formed from such studies could serve as starting points for pharmaceutical companies to design clot-inhibiting drugs.

Scanning tunneling microscopy works well under water. Although there is some interference from ion currents, *live* specimens can be probed with an STM. Films have been made of virus-infected cells that explode and release newly formed viruses. Viruses can themselves be examined to determine how the immune system "sees" them. Films of antibodies (defense molecules produced by the immune system) have also been produced.

Biologists will soon be able to see actual molecular events that they previously could only imagine. The physicist Paul Hansma of the University of California at Santa Barbara has hopes that this technique can speed up the Human Genome Project. Current machines are able to identify and sequence 7000 bases in a DNA molecule each day. STMs programmed to recognize labels attached to each of the four different nitrogenous bases of DNA could sequence 10 bases per second, or nearly a million bases per day. At that rate, the Human Genome Project may be finished ahead of its scheduled completion date of 2005, which would save millions of dollars.

Wolfgang Heckl, a professor of experimental physics at the University

Munich, points out that the STM was just the first of a whole family of scanning probe microscopes. The *atomic force microscope* (AFM), which Heckl uses, allows three-dimensional imaging and measurement of structures from atomic size to about 1 μm. In its most basic form, this revolutionary technology measures surfaces by moving a sharp probe across the sample to "feel" its surface details. Heckl claims that the cheapest way to build such a microscope is to alter an inexpensive compact disk player! The AFM uses a most basic and reliable sense of touch. All types of specimens can be examined and imaged in biological fluids or in air, with little or no sample preparation (Figure 3.30). Even cellular and related activities can be observed in action. One problem with AFMs is that the probe may strike and destroy, rather than ride over, a soft molecule. In 1994 Heckl was listed in *The Guiness Book of World Records* for studying the smallest hole in the world—a mere 3.6 × 10⁻⁴ μm across, where a single atom was removed from a crystal.

Heckl also uses the STM routinely to visualize DNA molecules. Unlike Hansma, whose sequencing technique

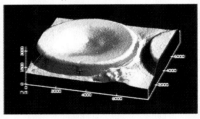
FIGURE 3.30 Surfaces of a human red blood cell, as seen by atomic force microscopy.

ESSAYS

The Essay at the end of each chapter enables you to learn more about a topic discussed in the text or to get information about an interesting supplementary topic that might not otherwise be covered in your microbiology course.

ESSAY CONTINUED

FIGURE 3.31 STM photos of the four nitrogenous bases of DNA: clockwise, from top left—guanine, adenine, thymine, and cytosine.

bonds (electron pairings) rearranging from one position to another.

IBM researchers have found a way to use the STM not just to look at, but to touch and move atoms around. Their first feat was to spell out "IBM" in letters just five atoms tall. The possibilities of literally constructing or engineering molecules has taken on new meaning. In the future, geneticists may use a "molecular syringe" to pick up a specific molecule and replace it with another to perform gene therapy! This interdisciplinary field is now known as nanotechnology (*nanos* is Greek for dwarf).

CHAPTER SUMMARIES

The Chapter Summary offers a concise but detailed listing of the key concepts in the chapter for your review. Important terms that you need to know (and that are boldfaced in the text) appear in boldface here. Each summary is organized by first- and second-level headings to give you a quick overview of chapter structure.

CHAPTER SUMMARY

DEVELOPMENT OF MICROSCOPY

- The existence of microorganisms was unknown until the invention of the microscope. Leeuwenhoek, probably the first to see microorganisms (in the 1600s), set the stage for **microscopy,** the technology of making very small things visible to the human eye.
- Leeuwenhoek's simple microscopes could reveal little detail of specimens. Today, multiple-lens, compound microscopes give us nearly distortion-free images, enabling us to delve further into the study of microbes.

PRINCIPLES OF MICROSCOPY

Metric Units

- The three units most used to describe microbes are the **micrometer** (μm), formerly called a micron, which is equal to 0.000001 m, also written as 10⁻⁶ m; the **nanometer** (nm), formerly called a millimicron (mμ), which is equal to 0.000000001 m, or 10⁻⁹ m; and the **angstrom** (Å), which is equal to 0.0000000001 m, 0.1 nm, or 10⁻¹⁰ m, but is no longer officially recognized.

Properties of Light: Wavelength and Resolution

- The **wavelength,** or the length of light rays, is the limiting factor in resolution.
- **Resolution** is the ability to see two objects as separate, discrete entities. Light wavelengths must be small enough to fit between two objects in order for them to be resolved.
- **Resolving power** can be defined as RP = λ/2NA, where λ = wavelength of light. The smaller the value of λ and

the larger the value of NA, the greater the resolving power of the lens.
- **Numerical aperture** (NA) relates to the extent to which light is concentrated by the condenser and collected by the objective. Its value is engraved on the side of each objective lens.

Properties of Light: Light and Objects

- If light strikes an object and bounces back, **reflection** (which gives an object its color) has occurred.
- **Transmission** is the passage of light through an object. Light must either be reflected from or transmitted through an object for it to be seen with a light microscope.
- **Absorption** of light rays occurs when they neither bounce off nor pass through an object but are taken up by that object. Absorbed light energy is used in performing photosynthesis or in raising the temperature of the irradiated body.
- Reemission of absorbed light as light of longer wavelengths is known as **luminescence.** If reemission occurs only during irradiation, the object is said to **fluoresce.** If reemission continues after irradiation ceases, the object is said to be **phosphorescent.**
- **Refraction** is the bending of light as it passes from one medium to another of different density. **Immersion oil,** which has the same **index of refraction** as glass, is used to replace air and to prevent refraction at a glass-air interface.
- **Diffraction** occurs when light waves are bent as they pass through a small opening, such as a hole, a slit, a space between two adjacent cellular structures, or a small, high-powered, magnifying lens in a microscope. The bent light

KEY TERMS

activation energy (p. 111)
active site (p. 111)
aerobe (p. 120)
aerobic respiration (p. 120)
alcoholic fermentation (p. 119)
allosteric site (p. 114)
amphibolic pathway (p. 129)
anabolic pathway (p. 110)
anabolism (p. 108)
anaerobe (p. 120)
apoenzyme (p. 113)
autotroph (p. 109)
autotrophy (p. 109)
beta oxidation (p. 126)
catabolic pathway (p. 110)
catabolism (p. 108)
chemical equilibrium (p. 116)
chemiosmosis (p. 123)

chemoautotroph (p. 109)
chemoheterotroph (p. 109)
chemolithotroph (p. 129)
coenzyme (p. 113)
cofactor (p. 113)
competitive inhibitor (p. 114)
cyclic photophosphorylation (p. 128)
cytochrome (p. 123)
dark reactions (p. 128)
electron acceptor (p. 109)
electron donor (p. 109)
electron transport (p. 122)
electron transport chain (p. 122)
endoenzyme (p. 113)
enzyme (p. 111)
enzyme–substrate complex (p. 112)

exoenzyme (p. 113)
FAD (p. 113)
feedback inhibition (p. 115)
fermentation (p. 118)
flavoprotein (p. 123)
glycolysis (p. 116)
heterotroph (p. 109)
heterotrophy (p. 109)
holoenzyme (p. 113)
homolactic-acid fermentation (p. 119)
Krebs cycle (p. 120)
light reactions (p. 127)
metabolic pathway (p. 110)
metabolism (p. 108)
NAD (p. 113)
noncompetitive inhibitor (p. 114)

noncyclic photoreduction (p. 128)
oxidation (p. 108)
oxidative phosphorylation (p. 123)
permease (p. 131)
phosphorylation (p. 116)
phosphotransferase system (p. 131)
photoautotroph (p. 109)
photoheterotroph (p. 109)
photolysis (p. 128)
photosynthesis (p. 127)
porin (p. 131)
quinone (p. 123)
reduction (p. 108)
specificity (p. 112)
substrate (p. 111)
surfactant (p. 131)

QUESTIONS FOR REVIEW

A Terms in Metabolism

1. Define metabolism, and distinguish between anabolism and catabolism.

2. Match the following:

_____ Archaeobacteria
_____ fungi
_____ cyanobacteria
_____ humans
_____ pathogenic bacteria

a. chemoheterotrophs
b. photoheterotrophs
c. photoautotrophs
d. chemoautotrophs

B Enzyme Characterisitcs and Functions

3. Label parts (a) through (g) of this enzyme.

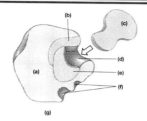

▪ KEY TERMS

The Key Terms list allows you to review essential terminology and includes all boldfaced terms from the chapter. Page references make it easy for you to check the corresponding text discussion for each term.

▪ QUESTIONS FOR REVIEW

A variety of Questions for Review allow you to see how well you learned the factual information and terminology of the chapter. For reinforcement, the questions are organized according to the chapter-opening focus questions. For visual interest, some questions include an illustration, photograph, or table.

PROBLEMS FOR INVESTIGATION

1. Spontaneous combustion caused by bacteria sometimes sets fire to a barn in which damp hay has been stored. How can you explain this?

2. In what sequence might the different kinds of metabolism mentioned in this chapter have evolved? Why didn't just one type of metabolism evolve? Give reasons.

3. Suppose that you had a culture known to contain an *Enterobacter* species and *Escherichia coli*. Devise a way to separate and identify the organisms.

4. Look up the chemical reaction by which certain bacteria change wine into vinegar. Is this an aerobic or anaerobic process?

5. More than 125 human diseases are caused by enzyme deficiencies. Chapters 7 and 8, on bacterial genetics, will explain how bacteria are being used by genetic engineers to remedy these enzyme deficiencies. List some human diseases caused by lack of enzymes, and indicate which enzymes are involved.

SOME INTERESTING READING

Cross, R. L. 1994. "Our primary source of ATP." *Nature* 370 no. 6491 (Aug. 25): 594–95.

Dawes, I. W., and I. W. Sutherland. 1992. *Microbial physiology.* Cambridge, MA.: Blackwell Scientific.

Devlin, T. M. 1992. *Textbook of biochemistry: With clinical correlations.* New York: Wiley.

Dickerson, R. E. 1980. "Cytochrome *c* and the evolution of energy metabolism." *Scientific American* 242 (March):136–153.

Lehninger, A. L., D. L. Nelson, and M. M. Cox. 1993. *Principles of biochemistry,* 2d edition. New York: Worth.

Meighen, E. A. 1991. "Molecular biology of bacterial bioluminescence." *Microbiological Reviews* 55(1):123–42.

Monastersky, R. 1988. "Bacteria alive and thriving at depth." *Science News* 133 (March 5):149.

Neidhardt, F. C., J. L. Ingraham, and M. Schaechter. 1990. *Physiology of the bacterial cell: A molecular approach.* Sunderland, MA.: Sinauer.

Pritchard, P. H. 1991. "Bioremediation as a technology: Experiences with the *Exxon Valdez* oil spill." *J. Hazardous Materials* 28(1–2):115–30.

Slayman, C. L. 1985. "Proton chemistry and the ubiquity of proton pumps." *BioScience* 35(1):16–17.

Trumpower, B. L. 1990. "Cytochrome bc_1 complexes of microorganisms." *Microbiological Reviews* 54(2):101–29.

▪ PROBLEMS FOR INVESTIGATION

The Problems for Investigation check your grasp of concepts, ask you to integrate ideas presented in different parts of the chapter or in several chapters, and suggest research projects to enhance your textbook and classroom work.

▪ SOME INTERESTING READING

A listing of current and classic references encourages you to pursue topics that you find intriguing and gives you a head start on research projects.

OTHER FEATURES:

▪ ENDPAPER TABLES

For easy reference, the six pages of endpaper tables list most of the medically important diseases and organisms discussed in the text. The front endpapers are organized by disease type, and the back endpapers by organism type.

▪ APPENDICES

Six appendices are included for your review or further study:

Appendix A lets you refresh your memory on relevant mathematical topics and offers a most probable number table for calculating numbers of organisms.

Appendix B gives standard classifications for bacteria and viruses.

Appendix C lists word roots of common terms used in microbiology.

Appendix D explains essential safety precautions for specimen handling.

Appendix E illustrates metabolic pathways in detail.

Appendix F allows you to check your answers to case-study-type Problems for Investigation.

CHAPTER 1

SCOPE AND HISTORY OF MICROBIOLOGY

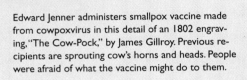

Edward Jenner administers smallpox vaccine made from cowpoxvirus in this detail of an 1802 engraving, "The Cow-Pock," by James Gillroy. Previous recipients are sprouting cow's horns and heads. People were afraid of what the vaccine might do to them.

"It's just some 'bug' going around." You have heard that from others or said it yourself when you have been ill for a day or two. Indeed, the little unidentified illnesses we all have from time to time and attribute to a "bug" are probably caused by viruses, the tiniest of all *microbes*. Other groups of **microorganisms**—bacteria, fungi, protozoa, and some algae—also have disease-causing members. Before studying microbiology, therefore, we are likely to think of microbes as germs that cause disease. Health scientists are concerned with just such microbes and with treating and preventing the diseases they cause. Yet less than 1 percent of known microorganisms cause disease, so focusing our study of microbes exclusively on disease gives us too narrow a view of microbiology.

WHY STUDY MICROBIOLOGY?

If you were to dust your desk and shake your dust cloth over the surface of a medium designed for growing microorganisms, after a day or so you would find a variety of organisms growing on that medium. If you were to cough onto such a medium or make fingerprints on it, you would later find a different assortment of microorganisms growing on the medium. When you have a sore throat and your physician orders a throat culture, a variety of organisms will be present in the culture—perhaps including the one that is causing your sore throat. Thus, microorganisms have a close association with humans. They are in us, on us, and nearly everywhere around us (Figure 1.1). One reason for studying microbiology is that *microorganisms are part of the human environment and are therefore important to human health.*

Microorganisms are essential to the web of life in every environment. Many microorganisms in the ocean and in bodies of fresh water capture energy from sunlight and store it in molecules that other organisms use as food. Microorganisms decompose dead organisms and waste material from living organisms, and they can decompose some kinds of industrial wastes. They make nitrogen available to plants.

These are but a few of many examples of how microorganisms interact with other organisms and help maintain the balance of nature. The vast majority of microorganisms are directly or indirectly beneficial not only to other organisms but also to humans. They form essential links in many food chains that produce plants and animals that humans eat. Aquatic microbes serve as food for small macroscopic animals that, in turn,

THIS CHAPTER FOCUSES ON THE FOLLOWING QUESTIONS:

A Why is the study of microbiology important?

B What is the scope of microbiology?

C What are some major events in the early history of microbiology?

D What is the germ theory of disease, and what historical developments led to its formulation?

E What events mark the emergence of immunology, virology, chemotherapy, genetics, and molecular biology as branches of microbiology?

FIGURE 1.1 A simple experiment shows that microorganisms are almost everywhere in our environment. Soil was added to nutrient agar, a culture medium (dish on top); another dish with agar was exposed to air (bottom left); and a tongue print was made on an agar surface (bottom right). After 3 days of incubation under favorable conditions, abundant microbial growth is easily visible in all three dishes.

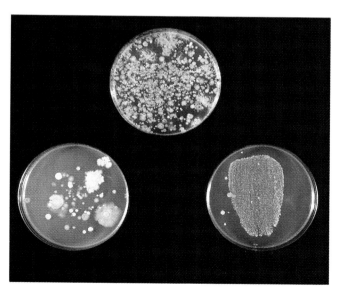

serve as food for fish and shellfish that humans eat. Certain microorganisms live in the digestive tracts of grazing animals such as cattle and sheep and aid in their digestive processes. Without these microbes, cows could not digest grass, and horses would get no nourishment from hay. Humans occasionally eat microbes, such as some algae and fungi, directly. Mushrooms, for instance, are the macroscopic reproductive bodies of masses of microscopic fungi. Biochemical reactions carried out by microbes also are used by the food industry to make pickles, sauerkraut, yogurt and other dairy products, fructose used in soft drinks, and the artificial sweetener aspartame. Fermentation reactions in microorganisms are used in the brewing industry to make beer and wine, and in baking to leaven dough.

One of the most significant benefits that microorganisms provide is their ability to synthesize *antibiotics*, substances derived from one microorganism that kill or restrict the growth of other microorganisms. Therefore, microorganisms can be used to cure diseases as well as cause them. Finally, microorganisms are the major tools of genetic engineering. Several products important to humans, such as interferon and growth hormones, can now be produced economically by microbes because of genetic engineering.

PUBLIC HEALTH

WE ARE NOT ALONE

"We are outnumbered. The average human contains about 10 trillion cells. On that average human are about 10 times as many microorganisms, or 100 trillion microscopic beings. . . . As long as they stay in balance and where they belong, [they] do us no harm. . . . In fact, many of them provide some important services to us. [But] most are opportunists, who if given the opportunity of increasing growth or invading new territory, will cause infection."

—Robert J. Sullivan, 1989

New organisms are being engineered to degrade oil spills, to remove toxic materials from soil, and to digest explosives, which would be too dangerous to handle. They will be major tools in cleaning up our environment. Other organisms will be designed to turn waste products into energy. Still other organisms will receive desirable genes from other types of organisms—for example, crop plants will be given bacterial genes that produce nitrogen-containing compounds needed for plant growth. The citizen of today, and even more so of tomorrow, must be scientifically literate, understanding many microbial products and processes.

Although only certain microbes cause disease, learning how such diseases are transmitted and how to diagnose, treat, and prevent them is of great importance in a health-science career. Such knowledge will help those of you who pursue such a career to care for patients and avoid becoming infected yourself.

Another reason for studying microbiology is that such study *provides insight into life processes in all life forms.* Biologists in many different disciplines use ideas from microbiology and use the organisms themselves. Ecologists draw on principles of microbiology to understand how matter is decomposed and made available for continuous recycling. Biochemists use microbes to study metabolic pathways—sequences of chemical reactions in living organisms. Geneticists use microbes to study how hereditary information is transferred and how such information controls the structure and functions of organisms.

Microorganisms are especially useful in research for at least three reasons:

1. Compared with other organisms, microorganisms have relatively simple structures. It is easier to study most life processes in simple unicellular organisms than in complex multicellular organisms.

2. Large numbers of microorganisms can be used in an experiment to obtain statistically reliable results at reasonable costs. Growing a billion bacteria costs less than maintaining 10 rats. Experiments with large numbers of microorganisms give more reliable results than do those with small numbers of organisms with individual variations.

3. Because microorganisms reproduce quickly, they are especially suitable for studies involving transmission of genetic information. Some bacteria can undergo three cell divisions in an hour, so the effects of genetic transmission can quickly be followed through many generations.

By studying microbes, scientists have already achieved remarkable success in understanding life processes and disease control. For example, within the last few decades, vaccines have nearly eradicated several dreaded childhood diseases—including measles, polio, German measles, and mumps—in developed countries. Smallpox, which once accounted for 10 percent of all deaths in Europe, has not been reported anywhere on earth since 1978. Much also has been learned about genetic changes that lead to antibiotic resistance and about how to manipulate genetic information in bacteria. Much more remains to be learned. For example, how can vaccines be made available on a worldwide basis? How can the development of new antibiotics keep pace with genetic changes in microorganisms? How will increased jet-age world travel affect the spread of infections? Can a vaccine or an effective treatment for AIDS be made available? Therein lie some of the challenges for the next generation of biologists and health scientists.

SCOPE OF MICROBIOLOGY

Microbiology is the study of **microbes,** organisms so small that a microscope is needed to study them. We consider two dimensions of the scope of microbiology: (1) the variety of kinds of microbes and (2) the kinds of work microbiologists do.

The Microbes

The major groups of organisms studied in microbiology are bacteria, algae, fungi, viruses, and protozoa (Figure 1.2a–e). All are widely distributed in nature. For example, a recent study of bee bread (a pollen-derived nutrient eaten by worker bees) showed it to contain 188 kinds of fungi and 29 kinds of bacteria. Most microbes consist of a single cell. (Cells are the basic units of structure and function in living things; they are discussed in Chapter 4.) Viruses, tiny acellular entities on the borderline between the living and the nonliving, behave like living organisms when they gain entry to cells. They, too, are studied in microbiology. Microbes range in size from small viruses 20 nm in diameter to large protozoans 5 mm or more in diameter. In other words, the largest microbes are as much as 250,000 times the size of the smallest ones! (Refer to Appendix A for a review of metric units.)

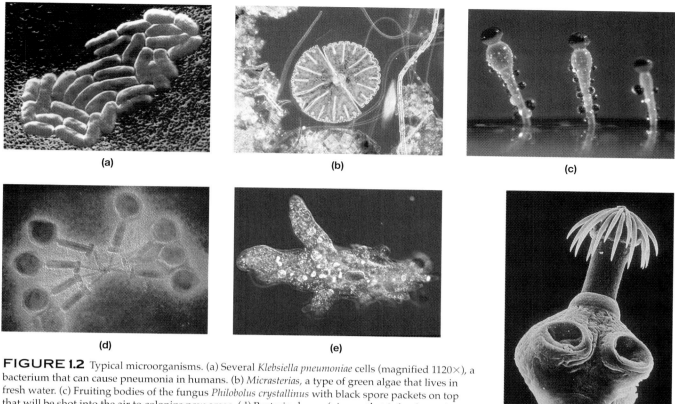

FIGURE 1.2 Typical microorganisms. (a) Several *Klebsiella pneumoniae* cells (magnified 1120×), a bacterium that can cause pneumonia in humans. (b) *Micrasterias,* a type of green algae that lives in fresh water. (c) Fruiting bodies of the fungus *Philobolus crystallinus* with black spore packets on top that will be shot into the air to colonize new areas. (d) Bacteriophages (viruses that infect bacteria; 35,500×). (e) *Amoeba,* a protozoan (175×). (f) Head of the tapeworm *Acanthrocirrus retrirostris* (170×). At the top of the head are hooks and suckers that the worm uses to attach to a host's intestinal tissues.

Among the great variety of microorganisms that have been identified, bacteria probably have been the most thoroughly studied. The majority of **bacteria** (singular: *bacterium*) are single-celled organisms with spherical, rod, or spiral shapes, but a few types form filaments. Most are so small they can be seen with a light microscope only under very high magnification. Although bacteria are cellular, they do not have a cell nucleus, and they lack the membrane-bound intracellular structures found in most other cells. Many bacteria absorb nutrients from their environment, but some make their own nutrients by photosynthesis or other synthetic processes. Some are stationary, and others move about. Bacteria are widely distributed in nature, for example, in aquatic environments and in decaying matter. And some occasionally cause diseases.

In contrast to bacteria, several groups of microorganisms consist of larger, more complex cells that have a cell nucleus. They include algae, fungi, and protozoa, all of which can easily be seen with a light microscope.

Many **algae** (al'je; singular: *alga*) are single-celled microscopic organisms, but some marine algae are large, relatively complex, multicellular organisms. Unlike bacteria, algae have a clearly defined cell nucleus and numerous membrane-bound intracellular structures. All algae photosynthesize their own food as plants do, and many can move about. Algae are widely distributed in both fresh water and oceans. Because they are so numerous and because they capture energy from sunlight in the food they make, algae are an important source of food for other organisms. Algae are of little medical importance; only one species has been found to cause disease in humans.

Like algae, many **fungi** (fun'ji; singular: *fungus*), such as yeasts and some molds, are single-celled microscopic organisms. Some, such as mushrooms, are multicellular, macroscopic organisms. Fungi also have a cell nucleus and intracellular structures. All fungi absorb ready-made nutrients from their environment. Some fungi form extensive networks of branching filaments, but the organisms themselves generally do not move. Fungi are widely distributed in water and soil as decomposers of dead organisms. Some are important in medicine either as agents of disease or as sources of antibiotics.

Viruses are acellular entities too small to be seen with a light microscope. They are composed of specific chemical substances—a nucleic acid and a few proteins (Chapter 2). Indeed, some viruses can be crystallized and stored, but they retain the ability to invade cells. Viruses replicate themselves and display other properties of living organisms only when they have invaded cells. Many viruses can invade human cells and cause disease.

Protozoa (pro-to-zo'ah; singular: *protozoan*) also are single-celled, microscopic organisms with at least one nucleus and numerous intracellular structures. A few species of amoebae are large enough to be seen with the naked eye, but we can study their structure only with a microscope. Many protozoa obtain food by engulfing or ingesting smaller microorganisms. Most protozoa can move, but a few, especially those that cause human disease, cannot. Protozoa are found in a variety of water and soil environments.

In addition to organisms properly in the domain of microbiology, in this text we consider some macroscopic *helminths* (worms) (Figure 1.2f) and *arthropods* (insects and similar organisms). The helminths have microscopic stages in their life cycles that can cause disease, and the arthropods transmit these stages.

We will learn more about the classification of microorganisms in Chapter 9. For now it is important to know only that cellular organisms are referred to by two names: their *genus* and *species* names. For example, a bacterial species commonly found in the human gut is called *Escherichia coli,* and a protozoan species that can cause severe diarrhea is called *Giardia intestinalis.* The naming of viruses is less precise. Some viruses, such as herpesviruses, are named for the group to which they belong. Others, such as polioviruses, are named for the disease they cause.

Disease-causing organisms and the diseases they cause in humans are discussed in detail in Chapters 20–25. Hundreds of infectious diseases are known to medical science. Some of the most important—those diseases that physicians should report to the U.S. Centers for Disease Control and Prevention (CDC)—are listed in Table 1.1 according to the kind of causative organism. The CDC is a federal agency that collects data about diseases and about developing ways to control them.

The Microbiologists

Microbiologists study many kinds of problems that involve microbes. Some study microbes mainly to find out more about a particular type of organism—the life stages of a particular fungus, for example. Other microbiologists are interested in a particular kind of function, such as the metabolism of a certain sugar or the action of a specific gene. Still others focus directly on practical problems, such as how to purify or synthesize a new antibiotic or how to make a vaccine against a particular disease. Quite often the findings from one project are useful in another, as when agricultural scientists use information from microbiologists to control pests and improve crop yields, or when environmentalists attempt to maintain natural food chains and prevent damage to the environment. Some fields of microbiology are described in Table 1.2.

Microbiologists work in a variety of settings (Figure 1.3). Some work in universities, where they are likely to teach, do research, and train students to do research. Microbiologists in both university and commercial laboratories are helping to develop the microorganisms used in genetic engineering. Some law firms are hiring microbiologists to help with the complexities of patenting new genetically engineered organisms. These organisms can be used in such im-

TABLE 1.1	Reportable Diseases Caused by Microorganisms and Parasites[a]			
Bacterial Diseases	**Bacterial Diseases**	**Bacterial Diseases**	**Viral Diseases**	**Algal Diseases**
Anthrax	Legionnaires' disease	Rheumatic fever	AIDS (symptomatic cases)	None
Bacterial meningitis	Leprosy (Hansen's disease)	Rocky Mountain spotted fever	Arbovirus infection	
Botulism	Leptospirosis	Salmonellosis	Aseptic meningitis	**Fungal Diseases**
Brucellosis	Listeriosis	Shigellosis	Chickenpox (varicella)	None
Campylobacteriosis	Lyme disease	Syphilis	Encephalitis	
Chancroid	Lymphogranuloma venereum	Tetanus	Hepatitis A, B, and C	
Chlamydial infections	Meningitis	Toxic shock syndrome	Hepatitis (unspecified)	**Protozoan Diseases**
Cholera	Paratyphoid fever A, B, and C	Trachoma	Influenza	Amebiasis
Cryptosporidiosis	Pertussis (whooping cough)	Tuberculosis	Measles (rubeola)	Giardiasis
Diphtheria	Plague	Tularemia	Mumps	Malaria
Food poisoning	Psittacosis	Typhoid fever	Poliomyelitis	
Gonorrhea	Q fever	Typhus	Rabies (animal and human)	**Helminth Diseases**
Granuloma inguinale	Relapsing fever		Rubella (German measles)	Trichinosis
Haemophilus influenzae infections (invasive)			Yellow fever	

[a]Infectious-disease reporting varies by state. This table lists most of the diseases commonly reported to the U.S. Centers for Disease Control and Prevention.

TABLE 1.2   Fields of Microbiology

Field (pronunciation)	Examples of What Is Studied
Microbial taxonomy	Classification of microorganisms
Fields according to organisms studied	
Bacteriology (bak"ter-e-ol'o-je)	Bacteria
Phycology (fi-kol'o-je)	Algae (*phyco*, seaweed)
Mycology (mi-kol'o-je)	Fungi (*myco*, a fungus)
Protozoology (pro"to-zo-ol'o-je)	Protozoa (*proto*, first; *zoo*, animal)
Parasitology (par"a-si-tol'o-je)	Parasites
Virology (vi-rol'o-je)	Viruses
Fields according to processes or functions studied	
Microbial metabolism	Chemical reactions that occur in microbes
Microbial genetics	Transmission and action of genetic information in microorganisms
Microbial ecology	Relationships of microbes with each other and with the environment
Health-related fields	
Immunology (im"u-nol'o-je)	How host organisms defend themselves against microbial infection
Epidemiology (ep-i-de-me-ol'o-je)	Frequency and distribution of diseases
Etiology (e-te-ol'-o-je)	Causes of disease
Infection control	How to control the spread of nosocomial (nos-o-ko'me-al), or hospital-acquired, infections
Chemotherapy	The development and use of chemical substances to treat diseases
Fields according to applications of knowledge	
Food and beverage technology	How to protect humans from disease organisms in fresh and preserved foods
Environmental microbiology	How to maintain safe drinking water, dispose of wastes, and control environmental pollution
Industrial microbiology	How to apply knowledge of microorganisms to the manufacture of fermented foods and other products of microorganisms
Pharmaceutical microbiology	How to manufacture antibiotics, vaccines, and other health products
Genetic engineering	How to use microorganisms to synthesize products useful to humans

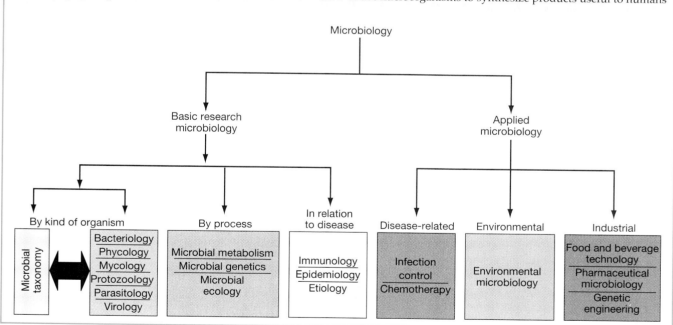

portant ways as cleaning up the environment (*bioremediation*), controlling insect pests, improving foods, and fighting disease. Many microbiologists work in health-related positions. Some work in clinical laboratories, performing tests to diagnose diseases or determining which antibiotics will cure a particular disease.

A few microbiologists develop new clinical tests. Others work in industrial laboratories to develop or manufacture antibiotics, vaccines, or similar biological products. Still others, concerned with controlling the spread of infections and related public health matters, work in hospitals or government labs.

FIGURE 1.3 Microbiology is used in careers as diverse as (a) using genetically engineered bacteria to investigate how diet influences the risk of developing cancer; (b) inspecting plastics made with as much as 40 percent starch (pieces inside baskets) for signs that aquatic microbes are degrading them; (c) using bacteria to decontaminate toxic wastes; (d) using beating nets to survey for ticks that can spread disease to livestock and humans; (e) keeping pets and domestic animals healthy, as well as improving their productivity, by means of advances in veterinary science.

From the point of view of health scientists, today's research is the source of tomorrow's new technologies. Research in *immunology* is greatly increasing our knowledge of how microbes trigger host responses and how the microbes escape these responses. It also is contributing to the development of new vaccines and to the treatment of immunologic disorders. Research in *virology* is improving our understanding of how viruses cause infections and how they are involved in cancer. Research in *chemotherapy* is increasing the number of drugs available to treat infections and is also improving our knowledge of how these drugs work. Finally, research in *genetics* is providing new information about the transfer of genetic information and, especially, about how genetic information acts at the molecular level.

HISTORICAL ROOTS

Many of the ancient Mosaic laws found in the Bible about basic sanitation have been used through the centuries and still contribute to our practices of preventive medicine. In Deuteronomy, Chapter 13, Moses instructed the soldiers to carry spades and bury solid waste matter. The Bible also refers to leprosy and to the isolation of lepers. Although in those days the term *leprosy* probably included other infectious and noninfectious diseases, isolation did limit the spread of the infectious diseases.

The Greeks anticipated microbiology, as they did so many things. The Greek physician Hippocrates, who lived around 400 B.C., set forth ethical standards for the practice of medicine that are still in use today. Hippocrates was wise in human relations and also a shrewd observer. He associated particular signs and symptoms with certain illnesses and realized that diseases could be transmitted from one person to another by clothing or other objects. At about the same time, the Greek historian Thucydides observed that people who had recovered from the plague could take care of plague victims without danger of getting the disease again.

AIDS

Suppose that several close relatives, two of your best friends, and many of your neighbors are suffering from painful illnesses and will soon die. There are no available beds at the hospital, and most of the doctors and nurses have quit work because they fear becoming infected. The local television station has started broadcasting daily the latest figures on deaths and new outbreaks, like the stock market prices or the weather report. Nearly every time you go shopping, you meet a funeral procession.

Over the centuries, many people have found themselves in situations that are similar to this fictional modern epidemic. Again and again, large proportions of the human population have been devastated by infectious diseases such as typhus, smallpox, and bubonic plague. In the mid-fourteenth century (1347–1351), plague alone wiped out 25 million people—one-fourth the population of Europe and neighboring regions—in just 5 years.

We who live in technologically advanced countries tend to think that outbreaks of this sort are a thing of the past. Yet following World War I, within the memory of many people now living, worldwide outbreaks of influenza claimed 20 million lives. Today, acquired immune deficiency syndrome (AIDS) threatens to kill great numbers of people after they have suffered a long and painful illness. Is AIDS in any way comparable to the great killer diseases of the past? More alarmingly, does its appearance mean that all our "triumphs" over infectious disease were an illusion? Will the past few decades, during which vaccines and antibiotics have largely

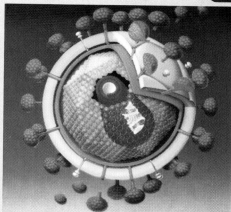

A model of the virus that causes AIDS.

kept contagious disease in check and even have wiped out certain ancient curses such as smallpox, prove to have been just a brief and atypical episode in an endless war that can't be won?

Despite the fears AIDS has aroused, there are many differences between AIDS and the epidemic diseases mentioned above. For one thing, AIDS is not nearly so easily communicable as the epidemic killers of the past. It is not spread by casual contact but largely by certain behaviors, most of which people can learn to avoid (as we will see in Chapter 19).

More important, however, is the state of our knowledge. During past epidemics, people were terrified and demoralized because they had no idea how the disease was caused and spread or how to fight it. Today, we understand far more about the nature of the enemy. The study of AIDS has occupied virologists, chemotherapists, and immunologists from among the world's most talented microbiologists. They have shown that AIDS is caused by a virus that infects cells of the body's immune system. The virus alters genetic information so that instead of fighting the infection, infected cells make viruses and then die. Scientists have determined the precise structure of many of the components of the virus, and they have learned how it attaches itself to its target cells. They have developed some drugs that are being tested to treat AIDS, and they are working on a vaccine. Despite those efforts, however, AIDS remains one of the most threatening infectious diseases and one of the greatest challenges microbiologists have ever faced.

The AIDS Names Project, a memorial to victims of this disease, showing individual quilts on display in Washington, D.C.

The Romans also contributed to microbiology, as early as the first century B.C. The scholar and writer Varro proposed that tiny invisible animals entered the body through the mouth and nose to cause disease. Lucretius, a philosophical poet, cited "seeds" of disease in his *De Rerum Natura (On the Nature of Things).*

Bubonic plague, also called the Black Death, appeared in the Mediterranean region around A.D. 542, where it reached epidemic proportions and killed millions. In 1347 the plague invaded Europe along the caravan routes and sea lanes from central Asia, affecting Italy first, then France, England, and finally northern

Europe. Although no accurate records were kept at that time, it is estimated that tens of millions of people in Europe died during this and successive waves of plague over the next 300 years. The Black Death was a great leveler—it killed rich and poor alike (Figure 1.4). The wealthy fled to isolated summer homes but carried plague-infected fleas with them in unwashed hair and clothing.

One group that escaped the plague's devastation was the Jewish population. Ancient Hebrew laws regarding sanitation offered some protection to those who practiced them. The relatively clean Jewish ghettos harbored fewer rats to spread the disease. When Jews did fall ill, they were carefully nursed and treated with herbal remedies rather than by strenuous purging or excessive bleedings with dirty instruments. As a result, a smaller proportion of Jews than gentiles died of the disease. Ironically, some gentiles regarded the Jews' higher survival rates as proof that Jews were the source of the epidemic.

In his *Diary,* the English writer Samuel Pepys gave a vivid, firsthand account of the plague in London in the 1660s.

> The streets mighty empty all the way now even in London, which is a sad sight. . . . Poor Will, that used to sell us ale, . . . his wife and three children died, all I think in a day. . . . [H]ome to draw over anew my will, which I had bound myself by oath to dispatch by to-morrow night, the town growing so unhealthy that a man cannot depend upon living two days to an end. In the City died this week 7,496, and of them 6,102 of the plague. But it is feared that the true number of the dead is near 10,000; partly from the poor that cannot be taken notice of through the greatness of the number. . . . I saw a dead corps in a coffin lie in the Close unburied; and a watch is constantly kept there night and day to keep the people in, the plague making us cruel as doggs one to another.

Until the seventeenth century, the advance of microbiology was hampered by the lack of appropriate tools to observe microbes. Around 1665, the English scientist Robert Hooke built a compound microscope (one in which light passes through two lenses) and used it to observe thin slices of cork. He coined the term *cell* to describe the orderly arrangement of small boxes that he saw because they reminded him of the cells (small, bare

rooms) of monks. However, it was Anton van Leeuwenhoek (Figure 1.5), a Dutch cloth merchant and amateur lens grinder, who first made and used lenses to observe living microorganisms. The lenses Leeuwenhoek made were of excellent quality; some gave magnifications up to 300× and were remarkably free of distortion. Making these lenses and looking through them were the passions of his life. Everywhere he looked he found what he called "animalcules." He found them in stagnant water, in sick people, and even in his own mouth.

Over the years Leeuwenhoek observed all the major kinds of microorganisms—protozoa, algae, yeast, fungi, and bacteria in spherical, rod, and spiral forms. He once wrote, "For my part I judge, from myself (howbeit I clean my mouth like I've already said), that all the people living in our United Netherlands are not as many as the living animals that I carry in my own mouth this very day." Starting in the 1670s he wrote numerous letters to the Royal Society in London and pursued his studies until his death in 1723 at the age of 91. Leeuwenhoek refused to sell his microscopes.

After Leeuwenhoek's death, microbiology failed to advance for more than a century. Eventually microscopes became more widely available, and progress resumed. Several workers discovered ways to stain microorganisms with dyes to make them more visible. The Swedish botanist Carolus Linnaeus developed a general classification system for all living organisms. The German botanist Matthias Schleiden and the

FIGURE 1.4 A portion of "The Triumph of Death" by Pieter Brueghel the Elder. The picture, painted in the mid-sixteenth century, a time when outbreaks of plague were still common in many parts of Europe, dramatizes the swiftness and inescapability of death for people of all social and economic classes.

FIGURE 1.5 Anton van Leeuwenhoek (1632–1723), shown holding one of his simple microscopes.

German zoologist Theodor Schwann formulated the **cell theory,** which states that cells are the fundamental units of life and carry out all the basic functions of living things. This theory still applies today to all cellular organisms, but not to viruses.

THE GERM THEORY OF DISEASE

The **germ theory of disease** states that microorganisms (germs) can invade other organisms and cause disease. Although this is a simple idea and is generally accepted today, it was not widely accepted when formulated in the mid-nineteenth century. Many people believed that broth, left standing, turned cloudy because of something about the broth itself. Even after it was shown that microorganisms in the broth caused it to turn cloudy, people believed that the microorganisms, like the "worms" (fly larvae, or maggots) in rotting meat, arose from nonliving things, a concept known as **spontaneous generation**. Widespread belief in spontaneous generation, even among scientists, hampered further development of the science of microbiology and the acceptance of the germ theory of disease. As long as they believed that microorganisms could arise from nonliving substances, scientists saw no purpose in considering how diseases were transmitted or how they could

be controlled. Dispelling the belief in spontaneous generation took years of painstaking effort.

Early Studies

For as long as humans have existed, some probably have believed that living things somehow originated spontaneously from nonliving matter. Aristotle's theories about his four "elements"—fire, earth, air, and water—seem to have suggested that nonliving forces somehow contributed to the generation of life. Even some naturalists believed that rodents arose from moist grain, beetles from dust, and worms and frogs from mud. As late as the nineteenth century, it seemed obvious to most people that rotting meat gave rise to "worms."

In the late seventeenth century, the Italian physician Francesco Redi devised a set of experiments to demonstrate that if pieces of meat were covered with gauze so that flies could not reach them, no "worms" appeared in the meat, no matter how rotten it was (Figure 1.6). Maggots did, however, hatch from fly eggs laid on top of the gauze. Despite proof that maggots did not arise spontaneously, some scientists, such as the English clergyman John Needham, still believed in spontaneous generation—at least that of microorganisms. Lazzaro Spallanzani, an Italian cleric and scientist, was more skeptical. He boiled broth infusions containing organic (living or previously living) matter and sealed the flasks to demonstrate that no organisms would develop spontaneously in them. Critics did not accept this as disproof of spontaneous generation. They argued that boiling drove off oxygen (which they thought all organisms required) and that sealing the flasks prevented its return.

FIGURE 1.6 Redi's experiments refuting the spontaneous generation of maggots in meat. When meat is exposed in an open jar, flies lay their eggs on it, and the eggs hatch into maggots (fly larvae). In a sealed jar, however, no maggots appear. If the jar is covered with gauze, maggots hatch from eggs that the flies lay on top of the gauze, but still no maggots appear in the meat.

Several scientists tried different ways of introducing air to counter this criticism. Schwann heated air before introducing it into flasks, and other scientists filtered air through chemicals or cotton plugs. All these methods prevented the growth of microorganisms in the flasks. But the critics still argued that altering the air prevented spontaneous generation.

Even nineteenth-century scientists of some stature continued to argue vociferously in favor of spontaneous generation. They believed that an organic compound previously formed by living organisms contained a "vital force" from which life sprang. The force, of course, required air, and they believed that all the methods of introducing air somehow changed it so that it could not interact with the force.

The proponents of spontaneous generation were finally defeated mainly by the work of the French chemist Louis Pasteur and the English physicist John Tyndall. When the French Academy of Science sponsored a competition in 1859 "to try by well-performed experiment to throw new light on the question of spontaneous generation," Pasteur entered the competition.

During the years Pasteur worked in the wine industry, he had established that alcohol was produced in wine only if yeast was present, and he learned a lot about the growth of microorganisms. Pasteur's experiment for the competition involved his famous "swan-necked" flasks (Figure 1.7). He boiled *infusions* (broths of foodstuffs) in flasks, heated the glass necks, and drew them out into long, curved tubes open at the end. Air could enter the flasks without being subjected to

FIGURE 1.7 A "swan-necked" flask that Pasteur used in refuting the theory of spontaneous generation. Although air could enter the flasks, microbes became trapped in the curved necks and never reached the contents. The contents, therefore, remained sterile—and still are—despite their exposure to the air.

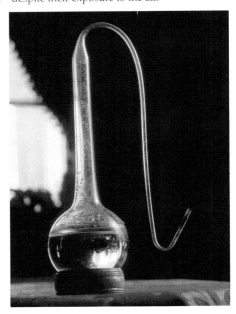

any of the treatments that critics had claimed destroyed its effectiveness. Airborne microorganisms could also enter the necks of the flasks, but they became trapped in the curves of the neck and never reached the infusions. The infusions from Pasteur's experiments remained sterile unless a flask was tipped so that the infusion flowed into the neck and back into the flask. This manipulation allowed microorganisms trapped in the neck to wash into the infusion, where they could grow and cause the infusion to become cloudy. In another experiment Pasteur filtered air through three cotton plugs. He then immersed the plugs in sterile infusions, demonstrating that growth occurred in the infusions from organisms trapped in the plugs.

Tyndall delivered another blow to the idea of spontaneous generation when he arranged sealed flasks of boiled infusion in an airtight box. After allowing time for all dust particles to settle to the bottom of the box, he carefully removed the covers from the flasks. These flasks, too, remained sterile. Tyndall had shown that air could be sterilized by settling, without any treatment that would prevent the "vital force" from acting.

Both Pasteur and Tyndall were fortunate that the organisms present in their infusions at the time of boiling were destroyed by heat. Others who tried the same experiments observed that the infusions became cloudy from growth of microorganisms. We now know that the infusions in which growth occurred contained heat-resistant or spore-forming organisms, but at the time, the growth of such organisms was seen as evidence of spontaneous generation. Still, the works of Pasteur and Tyndall successfully disproved spontaneous generation to most scientists of the time. Recognition that microbes must be introduced into a medium before their growth can be observed paved the way for further development of microbiology—especially for development of the germ theory of disease.

Pasteur's Further Contributions

Louis Pasteur (Figure 1.8) was such a giant among nineteenth-century scientists working in microbiology that we must consider some of his many contributions. Born in 1822, the son of a sergeant in Napoleon's army, Pasteur worked as a portrait painter and a teacher before he began to study chemistry in his spare time. Those studies led to posts in several French universities as professor of chemistry and to significant contributions to the wine and silkworm industries. He discovered that carefully selected yeasts made good wine, but that mixtures of other microorganisms competed with the yeast for sugar and made wine taste oily or sour. To combat this problem, Pasteur developed the technique of pasteurization (heating wine to 56°C in the absence of oxygen for 30 minutes) to kill unwanted organisms. While studying silkworms, he identified three different microorganisms, each of which caused

FIGURE 1.8 Louis Pasteur in his laboratory. The first rabies vaccine, developed by Pasteur, was made from the dried spinal cords of infected rabbits.

a different disease. His association of specific organisms with particular diseases, even though in silkworms rather than in humans, was an important first step in proving the germ theory of disease.

Despite personal tragedy—the deaths of three daughters and a cerebral hemorrhage that left him with permanent paralysis—Pasteur went on to contribute to the development of vaccines. The best known of his vaccines is the rabies vaccine, made of dried spinal cord from rabbits infected with rabies, which was tested in animals. When a 9-year-old boy who had been severely bitten by a rabid dog was brought to him, Pasteur administered the vaccine. The boy, who had been doomed to die, survived and became the first person to be immunized against rabies.

In 1894 Pasteur became director of the Pasteur Institute, which was built for him in Paris. Until his death in 1895, he guided the training and work of other scientists at the Institute. Today the Pasteur Institute is a thriving research center—an appropriate memorial to its founder.

Koch's Contributions

Robert Koch (Figure 1.9), a contemporary of Pasteur's, finished his medical training in 1872 and worked as a physician in Germany throughout most of his career. After he bought a microscope and photographic equipment, he spent most of his time studying bacteria, especially those that cause disease. Koch identified the bacterium that causes anthrax, a highly contagious and lethal disease in cattle and sometimes in humans. He recognized both actively dividing cells and dormant cells (spores) and developed techniques for studying them *in vitro* (outside a living organism).

Koch also found a way to grow bacteria in *pure cultures*—cultures that contained only one kind of organism. He tried streaking bacterial suspensions on

potato slices and then on solidified gelatin. But gelatin melts at incubator (body) temperature; even at room temperature, some microbes liquefy it. Finally, Angelina Hesse, the American wife of one of Pasteur's colleagues, suggested that Koch add agar (a thickener used in cooking) to his bacteriological media. This created a firm surface over which microorganisms could be spread very thinly—so thinly that some individual organisms were separated from all others. Each individual organism then multiplied to make a colony of thousands of descendants. Koch's technique of preparing pure cultures is still used today.

Koch's outstanding achievement was the formulation of four postulates to associate a particular organism with a specific disease. **Koch's postulates,** which provided scientists with a method of establishing the germ theory of disease, are as follows:

1. The specific causative agent must be found in every case of the disease.
2. The disease organism must be isolated in pure culture.
3. Inoculation of a sample of the culture into a healthy, susceptible animal must produce the same disease.
4. The disease organism must be recovered from the inoculated animal.

Implied in Koch's postulates is his one organism–one disease concept. The postulates assume that an infectious disease is caused by a single organism, and they are directed toward establishing that fact. This concept also was an important advance in the development of the germ theory of disease.

After obtaining a laboratory post at Bonn University in 1880, Koch was able to devote his full time to studying microorganisms. He identified the bacterium

FIGURE 1.9 Robert Koch formulated four postulates for linking a given organism to a specific disease.

CLOSE-UP

WHAT'S IN THE LAST DROP?

During the nineteenth century, French and German scientists were fiercely competitive. One area of competition was the preparation of pure cultures. Koch's reliable method of preparing pure cultures from colonies on solid media allowed German microbiologists to forge ahead. The French microbiologists' method of broth dilution, though now often used to count organisms (Chapter 6), hampered their progress. They added a few drops of a culture to fresh broth, mixed it, and added a few drops of the mixture to more fresh broth. After several successive dilutions, they assumed that the last broth that showed growth of microbes had contained a single organism. Unfortunately, the final dilution often contained more than one organism, and sometimes the organisms were of different kinds. This faulty technique led to various fiascos, such as inoculating animals with deadly organisms instead of vaccinating them.

(a)

that causes tuberculosis and developed a complex method of staining this organism. He also guided the research that led to the isolation of *Vibrio cholerae,* the bacterium that causes cholera.

In a few years Koch became professor of hygiene at the University of Berlin, where he taught a microbiology course believed to be the first ever offered. He also developed *tuberculin,* which he hoped would be a vaccine against tuberculosis. Because he underestimated the difficulty of killing the organism that causes tuberculosis, use of tuberculin resulted in several deaths from that disease. Although tuberculin was unacceptable as a vaccine, its development laid the groundwork for a skin test to diagnose tuberculosis. After the vaccine disaster, Koch left Germany. He made several visits to Africa, at least two visits to Asia, and one to the United States.

In the remaining 15 years of his life, his accomplishments were many and varied. He conducted research on malaria, typhoid fever, sleeping sickness, and several other diseases. His studies of tuberculosis won him the Nobel Prize for physiology or medicine in 1905, and his work in Africa and Asia won him great respect on those continents.

Work toward Controlling Infections

Like Koch and Pasteur, two nineteenth-century physicians, Ignaz Philipp Semmelweis of Austria and Joseph Lister of England, were convinced that microorganisms caused infections (Figure 1.10). Semmelweis recognized a connection between autopsies and puerperal

(b)

FIGURE 1.10 Two nineteenth-century pioneers in the control of infections: (a) Ignaz Philipp Semmelweis, who died in an asylum before his innovations were widely accepted, depicted on a 1965 Austrian postage stamp; (b) Joseph Lister, who successfully carried on Semmelweis's work.

(childbed) fever. Many physicians went directly from performing autopsies to examining women in labor without so much as washing their hands. When Semmelweis attempted to encourage more sanitary practices, he was ridiculed and harassed until he had a nervous breakdown and was sent to an asylum.

Ultimately, he suffered the curious irony of succumbing to an infection caused by the same organism that produces puerperal fever. In 1865, Lister, who had read of Pasteur's work on pasteurization and Semmelweis's work on improving sanitation, initiated the use of dilute carbolic acid on bandages and instruments to reduce infection. Lister, too, was ridiculed, but with his imperturbable temperament, resolute will, and tolerance of hostile criticism, he was able to continue his work. His methods, the first *aseptic techniques,* were proven effective by the decrease in surgical wound infections in his surgical wards. At age 75, some 37 years after he introduced the use of carbolic acid, Lister was awarded the Order of Merit for his work in preventing the spread of infection. He is considered the father of antiseptic surgery.

EMERGENCE OF SPECIAL FIELDS OF MICROBIOLOGY

Pasteur, Koch, and most other microbiologists considered to this point were generalists interested in a wide variety of problems. Certain other contributors to microbiology had more specialized interests, but their achievements were no less valuable. In fact, those achievements helped establish the special fields of immunology, virology, chemotherapy, and microbial genetics—fields that are today prolific research areas. Selected fields of microbiology are defined in Table 1.2.

Immunology

Disease depends not only on microorganisms invading a host but also on the host's response to that invasion. Today, we know that the host's response is in part a response of the immune system.

The ancient Chinese knew that a person scarred by smallpox would not again get the disease. They took dried scabs from lesions of people who were recovering from the disease and ground them into a powder that they sniffed. As a result of inhaling weakened organisms, they acquired a mild case of smallpox but were protected against subsequent infection.

Smallpox was unknown in Europe until the Crusaders carried it back from the Near East in the twelfth century. By the seventeenth century it was widespread. In 1717 Lady Ashley Montagu, wife of the British ambassador to Turkey, introduced a kind of immunization to England. A thread was soaked in fluid from a smallpox vesicle (blister) and drawn through a small incision in the arm. This technique, called *variolation,* was used at first by only a few prominent people, but eventually it became widespread.

In the late eighteenth century, Edward Jenner (Figure 1.11) realized that milkmaids who got cowpox did not get smallpox, and he inoculated his own son

FIGURE 1.11 Edward Jenner vaccinating a child against smallpox.

with fluid from a cowpox blister. He later similarly inoculated an 8-year-old and subsequently inoculated the same child with smallpox. The child remained healthy. The word *vaccinia* (*vacca,* the Latin name for cow) gave rise both to the name of the virus that causes cowpox and to the word *vaccine.* In the early 1800s Jenner received grants amounting to a total of 30,000 British pounds to extend his work on vaccination. Today, those grants would be worth more than $1 million. They may have been the first grants for medical research.

Pasteur contributed significantly to the emergence of immunology with his work on vaccines for rabies and cholera. In 1879, when Pasteur was studying chicken cholera, his assistant accidentally used an old chicken cholera culture to inoculate some chickens. The chickens did not develop disease symptoms. When the assistant later inoculated the same chickens with a fresh chicken cholera culture, they remained healthy. Although he had not planned to use the old culture first, Pasteur did realize that the chickens had been immunized against chicken cholera. He reasoned that the organisms must have lost their ability to produce disease but retained their ability to produce immunity. This finding led Pasteur to look for techniques that would have the same effect on other organisms. His development of the rabies vaccine was a successful attempt.

Along with Jenner and Pasteur, the nineteenth-century Russian zoologist Elie Metchnikoff was a pioneer in immunology (Figure 1.12). In the 1880s many scientists believed immunity was due to noncellular substances in the blood. Metchnikoff discovered that certain cells in the body could ingest microbes. He named those cells *phagocytes,* which literally means "cell-eating." The identification of phagocytes as cells that defend the body against invading microorganisms was a first step in understanding immunity. Metchnikoff

FIGURE 1.12 Elie Metchnikoff, one of the first scientists to study the body's defenses against invading microorganisms.

also developed several vaccines. Some were successful, but unfortunately some infected the recipients with the microorganisms against which they were supposedly being immunized. A few of his subjects acquired gonorrhea and syphilis from his vaccines.

CLOSE-UP

A "THORNY" PROBLEM

Metchnikoff's personal life played a role in his discovery of phagocytes. When he married, he took in his wife's young brothers and sisters as part of the marriage agreement. On one occasion, Metchnikoff left under his microscope a starfish he was studying before he went to lunch with his wife. While the starfish was unattended, one of the mischievous children poked thorns into it. Metchnikoff was enraged to discover his mutilated starfish, but he looked through the microscope before he discarded his ruined specimen. He was amazed to discover that cells of the starfish had gathered around the thorns. On further study he identified those cells as *leukocytes* (white blood cells) and found that they could devour foreign particles. Working from what started as a childish prank, Metchnikoff was able to formulate his theory of phagocytosis.

Virology

The science of virology emerged after that of bacteriology because viruses could not be recognized until certain techniques for studying and isolating larger particles such as bacteria had been developed. When Pasteur's collaborator Charles Chamberland developed a porcelain filter to remove bacteria from water in 1884,

he had no idea that any kind of infectious agent could pass through the filter. But researchers soon realized that some filtrates (materials that passed through the filters) remained infectious even after the bacteria were filtered out. The Dutch microbiologist Martinus Beijerinck determined why such filtrates were infectious and was thus the first to characterize viruses. The term *virus* had been used earlier to refer to poisons and to infectious agents in general. Beijerinck used the term to refer to specific *pathogenic* (disease-causing) molecules incorporated into cells. He also believed that these molecules could borrow for their own use existing metabolic and replicative mechanisms of the infected cells, known as *host cells*.

Further progress in virology required development of techniques for isolating, propagating, and analyzing viruses. The American scientist Wendell Stanley

PUBLIC HEALTH

SWAMP AIR OR MOSQUITOES?

During the American effort to dig the Panama Canal in 1905, yellow fever struck the workers as they struggled in the swamps. Yellow fever was a terrible and fatal disease. As Paul de Kruif put it in *Microbe Hunters*, "when folks of a town began to turn yellow and hiccup and vomit black, by scores, by hundreds, every day—the only thing to do was to get up and get out of that town." The entire canal project was in jeopardy because of the disease, and the American physician Walter Reed was assigned the task of controlling the disease. Reed listened to the advice of the Cuban physician Carlos Finlay y Barres, who for years had claimed that yellow fever was carried by mosquitoes. Reed ignored those who called Finlay a theorizing old fool and who insisted that yellow fever was due to swamp air.

Several people, including Reed's longtime associate James Carroll, volunteered to be bitten by mosquitoes known to have bitten yellow-fever patients. Although Carroll survived after his heart nearly stopped, most of the other volunteers died. Jesse Lazear, an American physician working with Reed, was accidentally bitten while working with patients. He began to show symptoms in 5 days and was dead in 12 days. Thus, it became clear that mosquitoes carried the yellow-fever agent. Similar experiments in which volunteers slept on sheets filthy with vomitus of yellow-fever patients demonstrated that bad air, contaminated water, sheets, and dishes were not involved. Later Carroll passed blood from yellow-fever victims through a porcelain filter and used the filtrate to inoculate three people who had not had yellow fever. How he got their cooperation is not known, but it is known that two of them died of yellow-fever. The agent that passed through the porcelain filter was eventually identified as a virus.

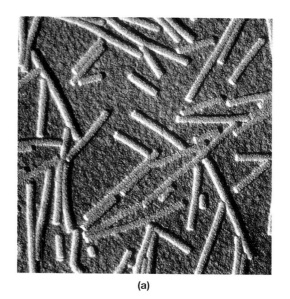

(a)

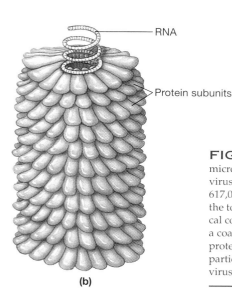

RNA

Protein subunits

(b)

FIGURE 1.13 (a) Electron micrograph of tobacco mosaic virus (magnification approx. 617,000×). (b) The structure of the tobacco mosaic virus. A helical core of RNA is surrounded by a coat that consists of repeating protein units. The structure of the particles is so regular that the viruses can be crystallized.

crystallized tobacco mosaic virus in 1935, showing that an agent with properties of a living organism also behaved as a chemical substance (Figure 1.13). The crystals consisted of protein and ribonucleic acid (RNA). The nucleic acid was soon shown to be important in the infectivity of viruses. Viruses were first observed with an electron microscope in 1939. From that time both chemical and microscopic studies were used to investigate viruses.

By 1952 the American biologists Alfred Hershey and Martha Chase had demonstrated that the genetic material of some viruses is another nucleic acid, deoxyribonucleic acid (DNA). In 1953 the American postdoctoral student James Watson and the English biophysicist Francis Crick determined the structure of DNA. The stage was set for rapid advances in understanding how DNA functions as genetic material both in viruses and in cellular organisms. Since the 1950s hundreds of viruses have been isolated and characterized. Although much remains to be learned about viruses, tremendous progress has been made in understanding their structure and how they function.

Chemotherapy

The Greek physician Dioscorides compiled *Materia Medica* in the first century A.D. This five-volume work listed a number of substances derived from medicinal plants still in use today—digitalis, curare, ephedrine, and morphine—along with a number of herbal medications. Credit for bringing herbal medicine to the United States is given to many groups of settlers, but Native Americans used many medicinal plants before the arrival of white people in the Americas. Many so-called primitive peoples still use herbs extensively, and some pharmaceutical companies finance expeditions into the Amazon Basin and other remote areas to investigate the uses the natives make of the plants around them.

During the Middle Ages virtually no advances were made in the use of chemical substances to treat diseases. Early in the sixteenth century the Swiss physician Aureolus Paracelsus used metallic chemical elements to treat diseases—antimony for general infections and mercury for syphilis. In the mid-seventeenth century Thomas Sydenham, an English physician, introduced cinchona tree bark to treat malaria. This bark, which we now know contains quinine, had been used to treat fevers in Spain and South America. In the nineteenth century morphine was extracted from the opium poppy and used medicinally to alleviate pain.

Paul Ehrlich, the first serious researcher in the field of chemotherapy (Figure 1.14), received his doctoral de-

FIGURE 1.14 Paul Ehrlich, pioneer in the development of chemotherapy for infectious disease.

MICROBIOLOGIST'S NOTEBOOK
Solving Microbial Mysteries in Household Products

After a lifetime of working in applied microbiology, I've seen it all, you might think. I *have* learned a great deal along the way and had some fascinating experiences. Still, my respect for bacteria and other microbes grows every day. Just seeing what they can do continues to amaze me. Like the time a bottle of pancake syrup sat there bubbling right on my desk, yet I couldn't find the bugs—the microbes—that had spoiled it. But I'm getting ahead of myself.

My name is Jan Curry. My career started with 3 years at National Dairy Research (Kraft) and was followed by 30 years with Lever Brothers Research. The past 12 years I've had my own company, MQC, Inc. As a consultant, I help a wide range of companies solve microbial problems. And several times a year I teach professional courses. The microbiologists all come with their own ideas and experiences, so I learn at least as much as they do.

Am I glad I'm a microbiologist? You bet! I've loved every minute of it. Every day is different because there's so much variety, when you consider all the microbiological aspects of a company's products. My experience spans foods, personal, household, and industrial products.

It's such fun to strike out and solve problems involved in manufacturing consumer products and making them safe. Many times I was asked to develop methods to prove a point for a patent or advertising claim. They'd call and say, "We need this certain kind of information. Can you figure out a way to show it?" The test you develop depends on what the product is, how it's used, and the bugs involved. For instance, when you add an antibacterial agent to a deodorant soap, you have to show that the resulting soap works against the microbes on skin when used the way people normally use soap. The situation is entirely different with toothpaste. There's also a lot of testing to see whether certain products

work better than the competition's. These are applied research projects.

In quality control, you're responsible for product integrity. Occasionally you run into trouble with products that you'd never guess would be contaminated, such as household cleaners, fabric softeners—even paint. A contamination problem may take up to a month to solve. It seems easy for me now to plan how best to tackle a problem in a manufacturing plant. It's much like being a detective. You ask, "Who are the most likely culprits, and where are they most likely to grow?" You have to know the whole chain of manufacturing events and the points where the bugs might enter.

Then you walk the line of manufacturing equipment from beginning to end, making sure you know every corner and every time it goes down or up to another floor. In my courses I teach "Put yourself in the bacteria's position. Become the bug and follow yourself through the whole process." I actually pretend I'm a bug running through the pipes, and try to figure out where I'd be most at home. You have to guess at the problem and want all the information you can get. If you guess correctly, you can clean it up quickly. Sometimes the problem goes away by itself and you never know why it happened in the first place. That's frustrating.

Usually I find the contamination in two areas. Complicated equipment such as pumps can be atrocious. Manufacturing machinery is often hard to take apart and clean well, but equipment design is constantly improving. The other common problem is the water used for manufacturing. Even though most products contain over 90 percent water, people often pay the least attention to it as a contamination source. Fortunately, most water-containing products are similar. The kinds of problems that occur are the same for foods, cosmetics, toiletries, paints, or

Jan Curry examining slides—the first step toward identifying contaminants.

household products. Microbes don't care what the product is, as long as they get something to eat. For example, a lot of them stick to the insides of pipes. If a pipe takes a sharp corner instead of a rounded one, there's probably a bunch of them that have been stuck in that corner for years.

Solving problems is a lot quicker and easier if you're flexible and willing to develop new tests or modify traditional ones to get the information you need. One common, industrywide test is the plate count, which tells you how many bugs are in a product. But even with that you have to be flexible, which brings me back to the pancake syrup that bubbled away on my desk. In a plate count test, you take a sample of the product and add various dilutions of it to agar, a gel-like growth medium, in shallow dishes called Petri plates. After a few days, the microbes in the sample grow into visible colonies on the agar that can be counted. But even though that syrup was definitely spoiled—actively fermenting right in front of me—my plates were spotless. Nothing grew on them. It was frustrating to watch that thing bubble and not be able find the bugs responsible.

I wondered if the plate medium was just too different from the syrup, so much so that the bugs causing the problem couldn't grow. So I changed the recipe to create more syruplike growth conditions by adding increasing amounts of a sugar (dextrose) to different batches. I finally found the millions of bugs that were causing the problem. It was a very strange yeast. Lots of microbes can grow with or without highly concentrated growth conditions such as those found in syrup, but one that *requires* those conditions to live is fascinating. The problem went away by itself and was probably caused by a bad lot of corn syrup. But I took very good care of that yeast because it was unique.

This yeast was a problem to the product, not to people. That's usually the case; the major microbial problems are organisms that grow in and spoil the product. Of course you still look for disease producers, or pathogens, because product safety is critical. But product stability is the most common problem.

Speaking of stability, let me tell you another story. One time the company was considering buying a packaged pancake batter from another company. I knew nothing about pancake batter, but the first step in preventing spoilage is to find out how the product spoils. I transferred some of the batter to a glass container so I could hold it in a warm water bath and take a sample every hour for 24 hours to see what grew up. Even though the batter went bad before 24 hours, that was a long night. Except for the night watchman, I was alone in a closed, darkened building. I'll never forget that experience, yet it was well worth it when I got the data back. You could see the different organisms that were in that product grow out, then die, then others move in, grow out, and die. The bugs kept changing. First, ones that grow at neutral pH grew up. Then ones that prefer acidic conditions appeared. Later the yeast moved in. It was really something to see, and it graphed out beautifully. I almost felt as if they did it just for me! The pancake batter was like a little ecosystem with different bugs appearing at different times, depending on what the last bug left behind. When I was done, I knew how that product spoiled—what microbes caused it—so we could develop ways to make sure we had a clean product.

It was an interesting experiment and illustrates an important point. In microbiology we get used to making sure we're working with pure cultures of just one type of organism. We forget that in nature, bacteria live in communities. As soon as you take an organism out of its habitat and isolate it on a growth medium, it's no longer the

same bug. It has the same name, but it behaves differently, partly because it has no hardships. That's very important to remember because we often run tests of antibacterial agents against pure cultures. These agents may not be as active as hoped when actually in use because the flora—the populations of microbes present—are different. As it turned out, the company didn't buy that pancake mix. But it was a great learning experience for me.

Because my staff consisted of only four people, with a lot of quality control and applied research projects to do, we had to develop efficient procedures. I got away from blindly following certain traditional tests because they took too long. We wanted more *information,* not just more plate counts. So I developed some quick and inexpensive methods that I'm still teaching. I try to get people to understand that they don't have to use laborious traditional methods as long as they develop methods that work as well or better. If they've got the data to prove that, they can use any method they want. Also, it's important to focus on the company's line of products and the particular problems you're most likely to encounter.

The most useful method I've developed is the swab streak count test, which saves money, time, and trouble. The traditional pour plate test [in which sequential dilutions are mixed with agar and the final dilution is poured onto a plate] limits the number of samples you can run in a day. Instead, I developed an easy way of swabbing a prepoured agar Petri plate with a cotton swab. It takes a little training, but anybody can learn it. Once you have data to prove your technique is reproducible and equivalent to the old method, you're off and running.

I teach these easy methods because they enable microbiologists to go out and find the bugs. That's much more important than just staying in the lab running routine counts on finished products. Also, you have the flexibility to respond immediately if somebody comes to you with an emergency.

Because you never know what you'll discover, you never get bored. As I said, usually what you find are potential spoilage organisms. We never had a recall or a problem with organisms that were dangerous to people. Sometimes you see a product spoiled on the shelf at the store, but it's rarely the fault of reputable manufacturers. There are many variables in a

manufacturing and transportation chain of events. Companies have little control over getting their products from their plant to your home. We test for shelf life, but a product may go through several different temperatures on its way to your kitchen. We can't control that; nor can we test for every possible variable. We hope and assume refrigeration will be continuous for certain products from the moment they're manufactured until they're used in the home, but it's out of our hands. One time some cases of margarine went bad. Well, it turned out that they'd been left outside a warehouse door for 10 or 12 hours on a hot day. It's no surprise they were ruined.

In all these years, I've never gotten tired of microbiology or wished I'd gone into something else. Microorganisms are all around us. They play an essential role in balancing nature. Bacteria were the first life on earth 3.5 billion years ago, and they're still here. No other forms of life can make that claim. Without them, we'd all be dead. But if we died, they'd carry on just fine. They're absolutely essential parts of our world, and it doesn't make sense to focus only on the bad guys, the pathogens. As far as we're concerned, over 95 percent of the bugs out there are good guys. They're incredibly sturdy and capable of living in any kind of environment they're faced with.

I love microbiology. It has always been fascinating and gets more so every day. I don't think I'll ever stop learning, which is why it remains so exciting.

To obtain a total plate count per gram (the number of organisms present per gram), Jan Curry samples the product being tested and makes dilutions of the sample.

gree from the University of Leipzig, Germany, in 1878. His discovery that certain dyes stained microorganisms but not animal cells suggested that the dyes or other chemicals might selectively kill microbial cells. This led him to search for the "magic bullet," a chemical that would destroy specific bacteria without damaging surrounding tissues. Ehrlich coined the term *chemotherapy* and headed the world's first institute concerned with the development of drugs to treat disease.

Early in the twentieth century the search for the magic bullet continued, especially among scientists at Ehrlich's institute. After testing hundreds of compounds (and numbering each compound), Ehrlich found compound 418 (arsenophenylglycine) to be effective against sleeping sickness and compound 606 (Salvarsan) to be effective against syphilis. For 40 years Salvarsan remained the best available treatment for this disease. In 1922 Alexander Fleming, a Scottish physician, discovered that lysozyme, an enzyme found in tears, saliva, and sweat, could kill bacteria. Lysozyme was the first body secretion shown to have chemotherapeutic properties.

The development of antibiotics began in 1917 with the observation that certain bacteria (actinomycetes) stopped the growth of other bacteria. In 1928 Fleming (Figure 1.15) observed that a colony of *Penicillium* mold contaminating a culture of *Staphylococcus* bacteria had prevented growth of bacteria adjacent to itself. Although he was not the first to observe this phenomenon, Fleming did recognize its potential for countering infections. However, purification of sufficient quantities of the substance he called *penicillin* proved to be very difficult. The great need for such a drug during World War II, money from the Rockefeller Institute, and the

hard work of the German biochemist Ernst Chain, the Australian pathologist Howard Florey, and researchers at Oxford University accomplished the task. Penicillin became available as a safe and versatile chemotherapeutic agent for use in humans.

While this work was going on, sulfa drugs were being developed. In 1935, prontosil rubrum, a reddish dye containing a sulfonamide chemical group, was used in treating streptococcal infections. Further study showed that sulfonamides were converted in the body to sulfanilamides; much subsequent work was devoted to developing drugs containing sulfanilamide. The German chemist Gerhard Domagk played an important role in this work, and one of the drugs, prontosil, saved the life of his daughter. In 1939 he was awarded a Nobel Prize for his work, but Hitler refused to allow him to make the trip to receive it. Extensions of Domagk's work led to the development of isoniazid, an effective agent against tuberculosis. Both sulfa drugs and isoniazid are still used today.

The development of antibiotics resumed with the work of Selman Waksman, who was born in Ukraine and moved to the United States in 1910. Inspired by the 1939 discovery by the French microbiologist Rene Dubos of tyrothricin, an antibiotic produced by soil bacteria, Waksman examined soil samples from all over the world for growth-inhibiting microorganisms or their products. He coined the term *antibiotic* in 1941 to describe actinomycin and other products he isolated. Both tyrothricin and actinomycin proved to be too toxic for general use as antibiotics. After repeated efforts, Waksman isolated the less toxic drug streptomycin in 1943. Streptomycin constituted a major breakthrough in the treatment of tuberculosis. In the same decade Waksman and others isolated neomycin, chloramphenicol, and chlortetracycline.

Examining soil samples proved to be a good way to find antibiotics, and explorers and scientists still collect soil samples for analysis. The more common antibiotic-producing organisms are rediscovered repeatedly, but the possibility of finding a new one always remains. Even the sea has yielded antibiotics, especially from the fungus *Cephalosporium acremonium*. The Italian microbiologist Giuseppe Brotzu noted the absence of disease organisms in sea water where sewage entered, and he determined that an antibiotic must be present. Cephalosporin was subsequently purified, and a variety of cephalosporin derivatives are now available for treating human diseases.

The fact that many antibiotics have been discovered does not stop the search for more. As long as there are untreatable infectious diseases, the search will continue. Even when effective treatment becomes available, it is always possible that a better, less toxic, or cheaper treatment can be found. Of the many chemotherapeutic agents currently available, none can

FIGURE 1.15 Alexander Fleming, who discovered the antibacterial properties of penicillin.

cure viral infections. Consequently, much of today's drug research is focused on developing effective antiviral agents.

Genetics and Molecular Biology

Modern genetics began with the rediscovery in 1900 of Gregor Mendel's principles of genetics. Even after this significant event, for nearly three decades little progress was made in understanding how microbial characteristics are inherited. For this reason, microbial genetics is the youngest branch of microbiology. In 1928 the British scientist Frederick Griffith discovered that previously harmless bacteria could change their own nature so as to become capable of causing disease. The remarkable thing about this discovery was that live bacteria were shown to acquire heritable traits from dead ones. During the early 1940s, Oswald Avery, Maclyn McCarty, and Colin MacLeod of the Rockefeller Institute in New York City demonstrated that the change was produced by DNA. After that finding came the crucial discovery of the structure of DNA by James Watson and Francis Crick. This breakthrough ushered in the modern era of molecular genetics.

About the same time, the American geneticists Edward Tatum and George Beadle used genetic variations in the mold *Neurospora* to demonstrate how genetic information controls metabolism. In the early 1950s the American geneticist Barbara McClintock discovered that some genes (units of inherited information) can move from one location to another on a chromosome. Before McClintock's work, genes were thought to remain stationary. Her revolutionary discovery forced geneticists to revise their thinking about genes.

More recently, scientists have discovered the genetic basis that underlies the human body's ability to make an enormous diversity of *antibodies*, molecules that the immune system produces to combat invading microbes and their toxic products. Within cells of the immune system, genes are shuffled about and spliced together in various combinations, allowing the body to make millions of different antibodies, including some that can protect us from threats that the body has never previously encountered.

TOMORROW'S HISTORY

Today's discovery is tomorrow's history. In an active research field such as microbiology, it is impossible to present a complete history. Some of the microbiologists omitted from this discussion are listed in Table 1.3. The period represented there, 1874–1917, is called the Golden Age of Microbiology. Many terms used to describe these scientists' accomplishments will be unfamiliar, but you will become familiar with them as you pursue the study of microbiology. Since 1900, Nobel Prizes have been awarded annually to outstanding scientists, many of whom were in the fields of physiology or medicine (Table 1.4). In some years the prize has been shared by several scientists, although the scientists may have made independent contributions. Refer to Tables 1.3 and 1.4 as you begin to study each new area of microbiology.

You can see from Table 1.4 that microbiology has been in the forefront of research in medicine and biology for several decades, and probably never more so than today. One reason is the renewed focus on infectious disease brought about by the advent of AIDS. Another is the dramatic progress in genetic engineering that has been made in the past two decades. Microorganisms have been and continue to be an essential part of the genetic engineering revolution. Most of the key discoveries that led to our present understanding of genetics emerged from research with microbes. Today scientists are attempting to redesign microorganisms for a variety of purposes (as we will see in Chapter 8). Bacteria have been converted into factories that produce drugs, hormones, vaccines, and a variety of biologically important compounds. And microbes, viruses in particular, are often the vehicle by which scientists insert new genes into other organisms. Such techniques are beginning to enable us to produce improved varieties of plants and animals such as pest-resistant crops (Figure 1.16) and may even enable us to correct genetic defects in human beings.

In September 1990, a 4-year-old girl became the first gene-therapy patient. She had inherited a defec-

FIGURE 1.16 Plant scientists are using knowledge gained from research with microorganisms to produce superior agricultural crops, such as strains that resist insect pests or that have greater productivity due to soil microorganisms that help the plants grow better. The poinsettia on the right has been inoculated with a combination of fungi and bacteria that help the plant grow better.

Year	Investigator	Achievement
1874	Billroth	Discovery of round bacteria in chains
1876	Koch	Identification of *Bacillus anthracis* as causative agent of anthrax
1878	Koch	Differentiation of staphylococci
1879	Hansen	Discovery of *Mycobacterium leprae* as causative agent of leprosy
1880	Neisser	Discovery of *Neisseria gonorrhoeae* as causative agent of gonorrhea
1880	Laveran and Ross	Identification of life cycle of malarial parasites in red blood cells of infected humans
1880	Eberth	Discovery of *Salmonella typhi* as causative agent of typhoid fever
1880	Pasteur and Sternberg	Isolation and culturing of pneumonia cocci from saliva
1881	Koch	Animal immunization with attenuated anthrax bacilli
1882	Leistikow and Loeffler	Cultivation of *Neisseria gonorrhoeae*
1882	Koch	Discovery of *Mycobacterium tuberculosis* as causative agent of tuberculosis
1882	Loeffler and Schutz	Identification of actinobacillus that causes the animal disease glanders
1883	Koch	Identification of *Vibrio cholerae* as causative agent of cholera
1883	Klebs	Identification of *Corynebacterium diphtheriae* and toxin as causative agent of diphtheria
1884	Loeffler	Culturing of *Corynebacterium diphtheriae*
1884	Rosenbach	Pure culturing of streptococci and staphylococci
1885	Escherich	Identification of *Escherichia coli* as a natural inhabitant of the human gut
1885	Bumm	Pure culturing of *Neisseria gonorrhoeae*
1886	Flugge	Staining to differentiate bacteria
1886	Fraenckel	*Streptococcus pneumoniae* related to pneumonia
1887	Weichselbaum	*Neisseria meningitidis* related to meningitis
1887	Bruce	Identification of *Brucella melitensis* as causative agent of brucellosis in cattle
1887	Petri	Invention of culture dish
1888	Roux and Yersin	Discovery of action of diphtheria toxin
1889	Charrin and Roger	Discovery of agglutination of bacteria in immune serum
1889	Kitasato	Discovery that *Clostridium tetani* produces tetanus toxin
1890	Pfeiffer	Identification of Pfeiffer bacillus, *Haemophilus influenzae*
1890	von Behring and Kitasato	Immunization of animals with diphtheria toxin
1892	Ivanovski	Discovery of filterability of tobacco mosaic virus
1894	Roux and Kitasato	Identification of *Yersinia pestis* as causative agent of bubonic plague
1894	Pfeiffer	Discovery of bacteriolysis in immune serum
1895	Bordet	Discovery of alexin (complement) and hemolysis
1896	Widal and Grunbaum	Development of diagnostic test based on agglutination of typhoid bacilli by immune serum
1897	van Ermengem	Discovery of *Clostridium botulinum* as causative agent of botulism
1897	Kraus	Discovery of precipitins
1897	Ehrlich	Formulation of side-chain theory of antibody formation
1898	Shiga	Discovery of *Shigella dysenteriae* as causative agent of dysentery
1898	Loeffler and Frosch	Discovery of filterability of virus that causes foot-and-mouth disease
1899	Beijerinck	Discovery of intracellular reproduction of tobacco mosaic virus
1901	Bordet and Gengou	Identification of *Bordetella pertussis* as causative agent of whooping cough; development of complement fixation test
1901	Reed and colleagues	Identification of virus that causes yellow fever
1902	Portier and Richet	Work on anaphylaxis
1903	Remlinger and Riffat-Bey	Identification of virus that causes rabies
1905	Schaudinn and Hoffmann	Identification of *Treponema pallidum* as causative agent of syphilis
1906	Wasserman, Neisser, and Bruck	Development of Wasserman reaction for syphilis antibodies
1907	Asburn and Craig	Identification of virus that causes dengue fever
1909	Flexner and Lewis	Identification of virus that causes poliomyelitis
1915	Twort	Discovery of viruses that infect bacteria
1917	d'Herelle	Independent rediscovery of viruses that infect bacteria (bacteriophages)

tive gene that crippled her immune system. Doctors at the National Institutes of Health (NIH) inserted a normal copy of the gene into some of her white blood cells in the laboratory and then injected these gene-treated cells back into her body, where, it is hoped, they will restore her immune system. Critics are worried that a new gene randomly inserted into her white blood cells could damage other genes and cause cancer. The ex-

periment is underway, and thus far the results have been good.

New information is constantly being discovered and sometimes supersedes earlier findings. Occasionally, new discoveries lead almost immediately to the development of medical applications, as occurred with penicillin and as will most certainly occur when a cure or vaccine for AIDS is discovered. However, old ideas

| TABLE 1.4 | Nobel Prize Awards for Research Involving Microbiology |

Year of Prize	Prize Winner	Topic Studied
1901	von Behring	Serum therapy against diphtheria
1902	Ross	Malaria
1905	Koch	Tuberculosis
1907	Laveran	Protozoa and the generation of disease
1908	Ehrlich and Metchnikoff	Immunity
1913	Richet	Anaphylaxis
1919	Bordet	Immunity
1928	Nicolle	Typhus exanthematicus
1939	Domagk	Antibacterial effect of prontosil
1945	Fleming, Chain, and Florey	Penicillin
1951	Theiler	Vaccine for yellow fever
1952	Waksman	Streptomycin
1954	Enders, Weller, and Robbins	Cultivation of polio virus
1958	Lederberg	Genetic mechanisms
	Beadle and Tatum	Transmission of hereditary characteristics
1959	Ochoa and Kornberg	Chemical substances in chromosomes that play a role in heredity
1960	Burnet and Medawar	Acquired immunological tolerance
1962	Watson, Crick, and Wilkins	Structure of deoxyribonucleic acid
1965	Jacob, Lwoff, and Monod	Regulatory mechanisms in microbial genes
1966	Rous	Viruses and cancer
1968	Holley, Khorana, and Nirenberg	Genetic code
1969	Delbruck, Hershey, and Luria	Mechanism of virus infection in living cells
1972	Edelman and Porter	Structure and chemical nature of antibodies
1975	Baltimore, Temin, and Dulbecco	Interactions between tumor viruses and genetic material of the cell
1976	Blumberg and Gajdusek	New mechanisms for the origin and dissemination of infectious diseases
1978	Smith, Nathans, and Arber	Restriction enzymes for cutting DNA
1980	Benacerraf, Snell, and Dausset	Immunological factors in organ transplants
1980	Berg	Recombinant DNA
1984	Milstein, Köhler, and Jerne	Immunology
1987	Tonegawa	Genetics of antibody diversity
1988	Black, Elion, and Hitchings	Principles of drug therapy
1989	Bishop and Varmus	Genetic basis of cancer
1990	Murray, Thomas, and Corey	Transplant techniques and drugs
1993	Mullis	Polymerase chain reaction method to amplify (copy) DNA
1993	Smith	Method to splice foreign components into DNA
1993	Sharp and Roberts	Genes can be discontinuous

such as spontaneous generation and old practices such as unsanitary measures in medicine can take years to replace. Many new bioethics problems will require considerable thought. Decisions regarding AIDS testing and reporting, transplants, environmental cleanup, and related issues will not come easily or quickly. Because of the wealth of prior knowledge, it is likely that you will learn more about microbiology in a single course than many pioneers learned in a lifetime. Yet, those pioneers deserve great credit because they worked with the unknown and had few people to teach them.

Human Genome Project

Microbial genetic techniques have made possible the undertaking of a colossal and controversial scientific plan, the Human Genome Project. At a cost of approximately $3 billion over a period of about 15 years, this project is making progress in identifying the location and chemical sequence of all the genes in the human genome—that is, all the genetic material in the human species—now estimated to consist of about 3 billion base pairs. The project is expected to be completed by the year 2005. When finished, it will be like having the "owner's manual" for humankind. It will make possible an incredible array of manipulations of human genetics and functions. Researchers are now locating and sequencing genes via simple microbes at first, such as *Escherichia coli*, a common organism found in the colon and feces, and have completely sequenced *Haemophilus influenzae*, a cause of ear infections. (See the Unit II Microbiologist's Notebook, "Puzzling Out an All-New Sequencing Strategy," pp. 212–213.) Then they are shifting to human genes as techniques become more efficient. One commercial company uses 50 robotic DNA sequencers, at a cost of about $120,000 each, to find 500 to 1000 new gene sequences per day. Many of these are only partial sequences; they must be fitted together to reveal complete gene sequences. Let us hope that the information gained will be used wisely.

HOW MICROBIOLOGISTS INVESTIGATE PROBLEMS

Like other scientists, microbiologists investigate problems by designing and carrying out experiments. Such experiments have provided the information health scientists apply to solve medical problems. Much of this text is devoted to presenting information obtained from experiments and to showing how that information is used in understanding infectious diseases. However, we believe that all science students will be interested in knowing how scientific problems are investigated.

First, a scientific problem must concern some aspect of the natural world because scientific methods can deal only with natural conditions and events. Microbiological problems deal with natural conditions and events involving microbes. Second, scientific problems must be clearly defined and sufficiently limited in scope so that a hypothesis and a prediction can be formulated. A **hypothesis** is a tentative explanation to account for an observed condition or event. The hypothesis in a particular experiment must be (1) an explanation for the defined problem and (2) testable. A testable hypothesis is one for which evidence can be collected to support or refute the hypothesis. A **prediction** is an outcome or consequence that will result if the hypothesis is true. Before beginning a scientific experiment, one must define the problem and make a hypothesis and a prediction.

A good hypothesis is one that offers the most reasonable explanation and the simplest solution to a problem. The purpose of scientific experiments is to test hypotheses by determining the correctness of predictions derived from the hypotheses. Scientific progress is made by making and testing hypotheses.

For example, suppose that a microbiologist has isolated an organism in pure culture and wants to know the effects of temperature on its growth. On the basis of information from prior research, he or she might (1) hypothesize that the organism's growth rate increases with temperature and (2) predict that the rate of increase in the number of organisms in a culture is proportional to the increase in temperature. After making the hypothesis and a prediction, the investigator designs an experiment to test the hypothesis. The experiment must be designed specifically to test the hypothesis and to collect evidence to determine whether the prediction is true.

To design a good experiment, an investigator must consider all variables that might affect the outcome. A **variable** is anything that can change for the purposes of an experiment. An experiment should have only one **experimental variable,** the factor that is purposely changed for the experiment. For example, in the study of the effects of temperature on the growth of an organism, temperature is the experimental variable. The hypothesis and prediction are related to the experimental variable. All other variables are **control variables,** factors that *can* change but that are prevented from changing for the duration of the experiment. In our example, the control variables include the number and characteristics of the organism, the quantity and properties of the medium, and all environmental factors except temperature.

When all variables have been identified, the investigator establishes the procedures for carrying out the experiment. Once the experiment has been designed, it must be carried out exactly as planned. All observations must be made and recorded accurately and precisely. If problems or unusual situations are encountered, they must be noted carefully. For example, should an incubator fail to maintain certain cultures at the proper temper-ature for the appropriate length of time, this failure should be noted and taken into consideration in interpreting the experiment. When the experiment is completed, the researcher analyzes and interprets the results in light of the hypothesis and prediction. The analysis of the results of an experiment often involves preparation of tables and graphs and usually compares results obtained under experimental and control conditions.

The goal of an experiment is to draw conclusions as to whether the prediction is true. If the experimental results, when analyzed, do not support the hypothesis, they may nevertheless suggest a better alternative hypothesis. The experimenter might wish to design further experiments to test this new hypothesis. Often, it is unexpected experimental results that lead to the most interesting discoveries.

In this textbook, you will find a feature called "Try It" boxes, such as "Does Salt Affect Soil Microbes?" (p. 46) and "What Grows in Your Health and Beauty Aids?" (p. 569). These boxes suggest projects that you might try yourselves. Even if you are not able to try these projects, at least make a mental plan of what hypotheses and steps you could use to investigate such a problem. Perhaps you can discuss your experimental designs during lecture or in lab. What other questions occur to you? How would you go about forming and testing hypotheses for them? In addition to performing experiments, scientists also should report their results so that other scientists can verify and use the information. Scientific knowledge increases by the sharing of information. This allows other scientists to repeat experiments and determine whether the results are reproducible. It also allows them to develop new experiments that build on existing information.

WHY STUDY MICROBIOLOGY?

- Microorganisms are part of the human environment and are therefore important to human health and activities.
- The study of microorganisms provides insight into life processes in all forms of life.

SCOPE OF MICROBIOLOGY

The Microbes

- **Microbiology** is the study of all **microorganisms (microbes)** in the microscopic range. These include **bacteria, algae, fungi, viruses,** and **protozoa.**

The Microbiologists

- Immunology, virology, chemotherapy, and genetics are especially active research fields of microbiology.
- Microbiologists work as researchers or teachers in university, clinical, and industrial settings. They do basic research in the biological sciences; help to perform or devise diagnostic tests; develop and test antibiotics and vaccines; work to control infection, protect public health, and safeguard the environment; and play important roles in the food and beverage industries.

HISTORICAL ROOTS

- The ancient Greeks, Romans, and Jews all contributed to early understandings of the spread of disease.
- Diseases such as bubonic plague and syphilis caused millions of deaths because of the lack of understanding of how to control or treat the infections.
- The development of high-quality lenses by Leeuwenhoek made it possible to observe microorganisms and later to formulate the **cell theory**.

THE GERM THEORY OF DISEASE

- The **germ theory of disease** states that microorganisms (germs) can invade other organisms and cause disease.

Early Studies

- Progress in microbiology and acceptance of the germ theory of disease required that the idea of **spontaneous generation** be refuted. Redi and Spallanzani demonstrated that organisms did not arise from nonliving material. Pasteur, with his swan-necked flasks, and Tyndall, with his dust-free air, finally dispelled the idea of spontaneous generation.

Pasteur's Further Contributions

- Pasteur also studied wine making and disease in silkworms and developed the first rabies vaccine. His association of particular microbes with certain diseases furthered the establishment of the germ theory.

Koch's Contributions

- Koch developed four postulates that aided in the definitive establishment of the germ theory of disease. **Koch's postulates** are as follows:

1. The specific causative agent must be found in every case of the disease.
2. The disease organism must be isolated in pure culture.
3. Inoculation of a sample of the culture into a healthy, susceptible animal must produce the same disease.
4. The disease organism must be recovered from the inoculated animal.

- Koch also developed techniques for isolating organisms, identified the bacillus that causes tuberculosis, developed tuberculin, and studied various diseases in Africa and Asia.

Work toward Controlling Infections

- Lister and Semmelweis contributed to improved sanitation in medicine by applying the germ theory and using aseptic technique.

EMERGENCE OF SPECIAL FIELDS OF MICROBIOLOGY

Immunology

- Immunization was first used against smallpox; Jenner used fluid from cowpox blisters to immunize against it.
- Pasteur developed techniques to weaken organisms so they would produce immunity without producing disease.

Virology

- Beijerinck characterized viruses as pathogenic molecules that could take over a host cell's mechanisms for their own use.
- Reed demonstrated that mosquitoes carry the yellow fever agent, and several other investigators identified viruses in the early twentieth century. The structure of DNA, the genetic material in many viruses and in all cellular organisms, was discovered by Watson and Crick.
- New techniques for isolating, propagating, and analyzing viruses were developed. Viruses could then be observed and in many cases crystallized, and their nucleic acids could be studied.

Chemotherapy

- Substances derived from medicinal plants were virtually the only source of chemotherapeutic agents until Ehrlich began a systematic search for chemically defined substances that would kill bacteria.
- Fleming and his colleagues developed penicillin, and Domagk and others developed sulfa drugs.
- Waksman and others developed streptomycin and other antibiotics derived from soil organisms.

Genetics and Molecular Biology

- Griffith discovered that previously harmless bacteria could change their nature so as to become capable of causing disease. This genetic change was shown by Avery, McCarty, and MacLeod to be due to DNA. Tatum and Beadle studied biochemical mutants of *Neurospora* to show how genetic information controls metabolism.

TOMORROW'S HISTORY

■ Microbiology has been at the forefront of research in medicine and biology, and microorganisms continue to play a critical role in genetic engineering and gene therapy.

Human Genome Project

■ The Human Genome Project will identify the location and sequence of all bases in the human genome. Microbes and microbiological techniques have contributed to this work.

KEY TERMS

algae (p. 3)
bacteria (p. 3)
cell theory (p. 9)
control variable (p. 22)
experimental variable (p. 22)

fungi (p. 3)
germ theory of disease
  (p. 9)
hypothesis (p. 22)
Koch's postulates (p. 11)

microbe (p. 2)
microbiology (p. 2)
microorganism (p. 0)
prediction (p. 22)
protozoa (p. 4)

spontaneous generation
  (p. 9)
variable (p. 22)
viruses (p. 3)

QUESTIONS FOR REVIEW

A **Study of Microbiology**

1. What are the two main reasons for studying microbiology, and why are they important?

B **Scope of Microbiology**

2. Give at least two examples of the importance of microbiology in each of the following:

   a. agriculture   **c.** medicine
   b. pollution control   **d.** basic laboratory research

3. True or False: Animals such as worms and ticks are too large to be included in a microbiology course. Explain.

4. What kinds of organisms are included among microbes?

5. Match the following:

   _____ study of algae   **a.** etiology
   _____ study of fungi   **b.** taxonomy
   _____ causes of disease   **c.** phycology
   ——— frequency and   **d.** mycology
        distribution of disease **e.** epidemiology
   ——— classification of
        microorganisms

6. Distinguish between basic research in microbiology and applied microbiology. Give two examples of each.

C **Early History of Microbiology**

7. What were the major contributions to microbiology of the ancient Greeks, Romans, and Jews?

8. What was Leeuwenhoek's contribution?

9. List three diseases that caused many deaths in the early days of microbiology.

D **Germ Theory of Disease**

10. What is the germ theory of disease?

11. What is the theory of spontaneous generation, and why did it have to be refuted before the germ theory could be accepted?

12. Explain the contributions of Redi, Spallanzani, Pasteur, and Tyndall toward dispelling belief in spontaneous generation.

13. Use the following diagram to explain how Pasteur's swan-necked flasks prevent contamination of sterile broth in the flasks. Describe what happens to the sterile broth in (a) after it has been allowed to cool as in (b). What happens to the broth after the flask has been tipped enough to let the broth come in contact with the dust and microorganisms and is tipped back as in (c)?

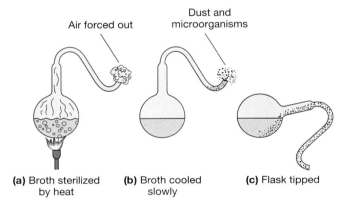

(a) Broth sterilized by heat   **(b)** Broth cooled slowly   **(c)** Flask tipped

14. What were Pasteur's and Koch's contributions to establishing the germ theory?

15. List Koch's postulates, and explain their importance.

16. What did Semmelweis and Lister contribute to microbiology?

E **Branches of Microbiology**

17. How did the use of porcelain filters affect the emergence of virology?

18. What was Beijerinck's contribution to virology?

19. What prevented progress in virology before the 1930s?

20. List the major events that took place in virology between 1930 and 1953.

21. What did chemotherapy consist of before the work of Ehrlich?

22. What were the contributions of Ehrlich, Fleming, Domagk, and Waksman to chemotherapy?

23. How did Griffith, Tatum, and Beadle contribute to microbial genetics?

PROBLEMS FOR INVESTIGATION

1. Suppose that Jenner were alive today and had a vaccine he thought would immunize against a dreaded disease. Would he be allowed to give the vaccine, and if so, under what circumstances? Would he be allowed to inoculate a child he had vaccinated with the disease-causing organism? Should he be allowed to do these things? Why or why not?

2. It is well established that sometimes a disease is not produced after a sample of a culture of microorganisms is inoculated into a healthy, susceptible animal. Does that fact invalidate Koch's postulates? Defend your answer.

3. Select 10 events in the history of microbiology. Explain how each demonstrates the validity of the two reasons for studying microbiology discussed in this chapter.

4. For any one branch of microbiology, prepare a short report on accomplishments not mentioned in your text. (If possible, present an oral report to your class.)

5. Cite at least three examples of delays in the development of microbiology because of the lack of proper tools or procedures.

SOME INTERESTING READING

General References

American Society for Microbiology. Professional journals and other publications. Washington, D.C.: American Society for Microbiology. Most professional microbiologists belong to this society and read its journals to keep up to date in their fields.

Baron, S., et al. 1991. *Medical microbiology*. New York: Churchill Livingstone Inc. An excellent medical-school microbiology textbook.

Brooks, G. F., J. S. Butel, and L. N. Ornston. 1991. *Jawetz, Melnick, and Adelberg's medical microbiology*. East Norwalk, CT: Appleton & Lange. A regularly updated review of microbiology of interest to physicians and health scientists.

Davis, B. D., et al. 1990. *Microbiology*. Philadelphia: Lippincott. A good general reference on most classical topics in microbiology.

Hoeprich, P. D., M. C. Jordan, and A. R. Ronald, eds. 1994. *Infectious diseases: A modern treatise of infectious processes*. Philadelphia: Lippincott. An exhaustive survey of clinical microbiology.

Joklik, W. K., et al., eds. 1992. *Zinsser microbiology*. East Norwalk, CT: Appleton & Lange. A classic textbook of medical microbiology, taxonomically arranged.

U.S. Centers for Disease Control and Prevention. *Morbidity and Mortality Weekly Reports*. Boston: Massachusetts Medical Society. In addition to providing current statistics on death and disease in the United States, these reports include news items on infectious and some other diseases.

Chapter References

Asimov, I. 1982. *Asimov's biographical encyclopedia of science and technology*. Garden City, NY: Doubleday.

Baxby, D. 1981. *Jenner's smallpox vaccine: The riddle of vaccinia virus and its origins*. London: William Heinemann.

Bishop, J. E., and M. Waldholz. 1990. *Genome: The story of the most astonishing scientific adventure of our time — the attempt to map all the genes in the human body*. New York: Simon & Schuster.

Brock, T. D., ed. and trans. 1992. *Milestones in microbiology*. Madison, WI: Science Tech Publishers.

Brock, T. D. 1988. *Robert Koch: A life in medicine and bacteriology*. Madison, WI: Science Tech Publishers.

Brown, W. E., and R. P. Williams. 1990. "Ignaz Semmelweis and the importance of washing your hands." *The American Biology Teacher* 52 (May): 291–94.

Collins, F. and D. Galas. 1993. "A new five-year plan for the U.S. Human Genome Project." *Science* 262 (October 1): 43–46.

Defoe, D. 1722. *A journal of the plague year*. Republished 1992. New York: Norton.

De Kruif, P. 1966. *Microbe hunters*. New York: Harcourt Brace Jovanovich.

Dixon, B. 1994. *Power unseen: How microbes rule the world*. New York: W. H. Freeman.

Dubos, R. 1988. *Pasteur and modern science*. Madison, WI: Science Tech Publishers.

Geison, G. L. 1995. *The private science of Louis Pasteur*. Princeton, NJ: Princeton University Press.

Gest, H. 1987. *The world of microbes*. Madison, WI: Science Tech Publishers.

Harre, R. 1983. *Great scientific experiments: Twenty experiments that changed our view of the world*. Oxford: Oxford University Press.

Karlen, A. 1984. *Napoleon's glands and other ventures in biohistory*. Boston: Little, Brown.

Keller, E. F. 1983. *A feeling for the organism: The life and work of Barbara McClintock*. San Francisco: W. H. Freeman and Co.

Morshead, O. F., and E. H. Shepard, eds. 1926. *Everybody's Pepys: The diary of Samuel Pepys*. New York: Harcourt Brace Jovanovich.

Rosebury, T. 1969. *Life on man*. New York: The Viking Press.

Walker, G. C., ed. 1993. *Frontiers in microbiology*. Washington, D.C.: American Society for Microbiology.

FUNDAMENTALS OF CHEMISTRY

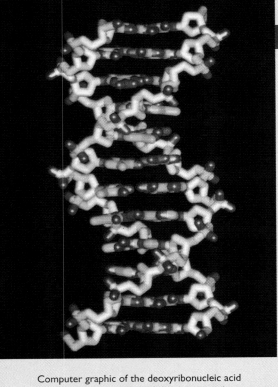

Computer graphic of the deoxyribonucleic acid (DNA) double helix. This molecule stores the genetic material for organisms and for many viruses.

A ll living and nonliving things, including microbes, are composed of matter. Thus, it is not surprising that all properties of microorganisms are determined by the properties of matter.

WHY STUDY CHEMISTRY?

Chemistry is the science that deals with the basic properties of matter. Therefore we need to know some chemistry to begin to understand microorganisms. Chemical substances undergo changes and interact with one another in *chemical reactions*. Metabolism, the use of nutrients for energy or for making the substance of cells, consists of many different chemical reactions. This is true regardless of whether the organism is a human or a microorganism. Thus, understanding the basic principles of chemistry is essential to understanding metabolic processes in living things. A microbiologist uses chemistry to understand the structure and function of microorganisms themselves and to understand how they affect humans in disease processes as well as how they affect all life on earth.

CHEMICAL BUILDING BLOCKS AND CHEMICAL BONDS

Chemical Building Blocks

Matter is composed of very small particles that form the basic chemical building blocks. Over the years, chemists have observed matter and deduced the characteristics of these particles. Just as the alphabet can be used to make thousands of words, the chemical building blocks can be used to make thousands of different substances. The complexity of chemical substances greatly exceeds the complexity of words. Words rarely contain more than 20 letters, whereas some complex chemical substances contain as many as 20,000 building blocks!

The smallest chemical unit of matter is the **atom.** Many different kinds of atoms exist. Matter composed of one kind of atom is called an **element.** Each element has specific properties that distinguish it from other elements. Carbon is an element; a pure sample of carbon consists of a vast number of carbon atoms. Oxygen and nitrogen also are elements; they are found as gases in the earth's atmosphere. Chemists use one-

THIS CHAPTER FOCUSES ON THE FOLLOWING QUESTIONS:

A Why is knowledge of basic chemistry necessary to the understanding of microbiology?

B What terms describe the organization of matter, and which elements are found in living organisms?

C What are the properties of chemical bonds and chemical reactions?

D Which properties of water, solutions, colloidal dispersions, acids, and bases make them important in living things?

E What is organic chemistry, and what are the major functional groups of organic molecules?

F How do the structures and properties of carbohydrates contribute to their roles in living things?

G How do the structures and properties of fats, phospholipids, and steroids contribute to their roles in living things?

H How do the structures and properties of proteins, including enzymes, contribute to their roles in living things?

I How do the structures and properties of nucleotides contribute to their roles in living things?

or two-letter symbols to designate elements—such as C for carbon, O for oxygen, N for nitrogen, and Na for sodium (from its Latin name, *natrium*).

Atoms combine chemically in various ways. Sometimes atoms of a single element combine with each other. For example, carbon atoms form long chains that are important in the structure of living things. Both oxygen and nitrogen form paired atoms, O_2 and N_2. More often, atoms of one element combine with atoms of other elements. Carbon dioxide (CO_2) contains one atom of carbon and two atoms of oxygen; water (H_2O) contains two atoms of hydrogen and one atom of oxygen. (The subscripts in these formulas indicate how many atoms of each element are present.)

When two or more atoms combine chemically, they form a **molecule.** Molecules can consist of atoms of the same element, such as N_2, or atoms of different elements, such as CO_2. Molecules made up of atoms of two or more elements are called **compounds.** Thus, CO_2 is a compound, but N_2 is not. The properties of compounds are different from those of their component elements. For example, in their elemental state, both hydrogen and oxygen are gases at ordinary temperatures. They can combine to form water, however, which is a liquid at ordinary temperatures.

Living things consist of atoms of relatively few elements, principally carbon, hydrogen, oxygen, and nitrogen, but these are combined into highly complex compounds. A simple sugar molecule, $C_6H_{12}O_6$, contains 24 atoms. Many molecules found in living organisms contain thousands of atoms.

Structure of Atoms

Although the atom is the smallest unit of any element that retains the properties of that element, atoms do contain even smaller particles that together account for those properties. Physicists study many such subatomic particles, but we discuss only **protons, neutrons,** and **electrons.** Three important properties of these particles are atomic mass, electrical charge, and location in the atom (Table 2.1). *Atomic mass* is measured in terms of *atomic mass units* (AMU). The mass of a proton or a neutron is almost exactly equal to 1 AMU; electrons have a much smaller mass. With respect to electrical charge, electrons are negatively (−) charged, and protons are

TABLE 2.1	Properties of Atomic Particles			
Particle	Atomic Mass	Electrical Charge	Location	
Proton	1	+	Nucleus	
Neutron	1	None	Nucleus	
Electron	1/1836	−	Orbiting the nucleus	

positively (+) charged. Neutrons are neutral, with no charge. Atoms normally have an equal number of protons and electrons and so are electrically neutral. The heavy protons and neutrons are densely packed into the tiny, central *nucleus* of the atom, whereas the lighter electrons move around the nucleus in what have commonly been described as orbits.

The atoms of a particular element always have the same number of protons; that number of protons is the **atomic number** of the element. Atomic numbers range from 1 to over 100. The numbers of neutrons and electrons in the atoms of many elements can change, but the number of protons—and therefore the atomic number—remains the same for all atoms of a given element.

Protons and electrons are oppositely charged. Consequently, they attract each other. This attraction keeps the electrons near the nucleus of an atom. The electrons are in constant, rapid motion, forming an electron cloud around the nucleus. Because some electrons have more energy than others, chemists use a model with concentric circles, or *electron shells,* to suggest different energy levels. Electrons with the least energy are located nearest the nucleus, and those with more energy are farther from the nucleus. Each energy level corresponds to an electron shell (Figure 2.1).

An atom of hydrogen has only one electron, which is located in the innermost shell. An atom of helium has two electrons in that shell; two is the maximum number of electrons that can be found in the innermost shell. Atoms with more than two electrons always have two electrons in the inner shell and up to eight additional electrons in the second shell. The inner shell is filled before electrons are found in the second shell; the second shell is filled before electrons are found in the third shell, and so on. Very large atoms have several more electron shells of larger capacity, but in elements found in living things, the outer shell is chemically stable if it contains eight electrons. This principle, known as the **rule of octets,** is important for understanding chemical bonding, which we will discuss shortly.

Atoms whose outer electron shells are nearly full (containing six or seven electrons) or nearly empty (containing one or two electrons) tend to form ions. An **ion** is a charged atom produced when an atom gains or loses one or more electrons (Figure 2.2a). When an atom of sodium (atomic number 11) loses the one electron in its outer shell without losing a proton, it becomes a positively charged ion, called a **cation** (kat′i-on). When an atom of chlorine (atomic number 17) gains an electron to fill its outer shell, it becomes a negatively charged ion, called an **anion** (an′i-on). In the ionized state, chlorine is referred to as chloride. Ions of elements such as sodium or chlorine are chemically more *stable* than atoms of these same elements because the ions' outer electron shells are full. Many elements are found in microorganisms or their environments as ions (Table 2.2). Those with one or two electrons in their outer shell tend to lose electrons and form ions with +1 or +2 charges, respectively; those with seven electrons in their outer shell tend to gain an electron and form ions with a charge of −1. Some ions, such as the hydroxyl (hi-drok′sil) ion (OH^-), are compounds—they contain more than one element.

Although all atoms of the same element have the same atomic number, they may not have the same atomic weight. **Atomic weight** is the sum of the number of protons and neutrons in an atom. Many elements consist of atoms with differing atomic weights. For example, carbon usually has six protons and six neutrons; it has an atomic weight of 12. But some naturally occurring carbon atoms have one or two extra neutrons, giving these atoms an atomic weight of 13 or 14. In addition, laboratory techniques are available to create atoms with different numbers of neutrons. Atoms of a particular element that contain differ-

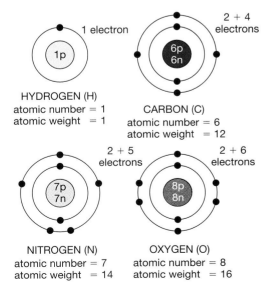

FIGURE 2.1 The structure of four biologically important atoms. Hydrogen, the simplest element, has a nucleus consisting of a single proton and a single electron in the first shell. In carbon, nitrogen, and oxygen the first shell is filled with two electrons and the second shell is partly filled. Carbon, with six protons in its nucleus, has six electrons, four of them in the second shell. Nitrogen has five electrons, and oxygen six electrons in the second shell. It is the electrons in the outermost shell that take part in chemical bonding.

1 electron

1p

HYDROGEN (H)
atomic number = 1
atomic weight = 1

2 + 4 electrons

6p
6n

CARBON (C)
atomic number = 6
atomic weight = 12

2 + 5 electrons

7p
7n

NITROGEN (N)
atomic number = 7
atomic weight = 14

2 + 6 electrons

8p
8n

OXYGEN (O)
atomic number = 8
atomic weight = 16

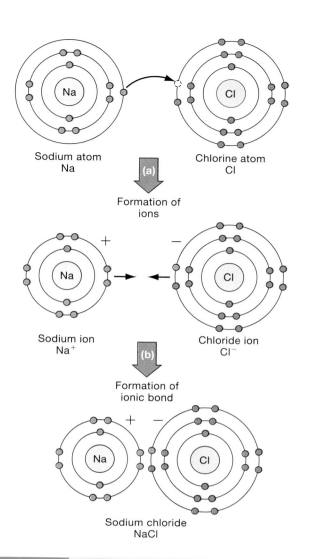

Sodium atom
Na

Chlorine atom
Cl

(a)
Formation of
ions

$+$

Sodium ion
Na$^+$

$-$

Chloride ion
Cl$^-$

(b)
Formation of
ionic bond

$+$ $-$

Sodium chloride
NaCl

FIGURE 2.2 (a) The formation of ions, or electrically charged atoms. When a neutral sodium atom loses the single electron in its outermost shell, the result is a sodium ion, Na$^+$. When a neutral chlorine atom gains an extra electron in its outer shell, the result is a chloride ion, Cl$^-$. (b) Oppositely charged ions attract one another. Such attraction creates an ionic bond and results in the formation of an ionic compound, in this case sodium chloride (NaCl).

ent numbers of neutrons are called **isotopes.** The superscript to the left of the symbol for the element indicates the atomic weight of the particular isotope. For example, carbon with an atomic weight of 14, which is often used to date fossils, is written ^{14}C. The atomic weight of an element that has naturally occurring isotopes is the average atomic weight of the natural mixture of isotopes. Thus, atomic weights are not always whole numbers, even though any particular atom contains a specific number of whole neutrons and protons.

A **gram molecular weight,** or **mole,** is the weight of a substance in grams (g) equal to the sum of the atomic weights of the atoms in a molecule of the substance. For example, a mole of glucose, $C_6H_{12}O_6$, weighs 180 grams: [6 carbon atoms × 12 (atomic weight)] + [12 hydrogen atoms × 1 (atomic weight)] + [6 oxygen atoms × 16 (atomic weight)] = 180 grams. The mole is defined such that one mole of any substance always contains 6.023×10^{23} particles. Table 2.3 summarizes properties of elements found in living things.

Some isotopes are stable, and others are not. The nuclei of unstable isotopes tend to emit subatomic particles and radiation. Such isotopes are said to be *ra-*

TABLE 2.2	Some Common Ions	
Ion	**Name**	**Brief Description**
Na$^+$	Sodium	Contributes to salinity of natural bodies of water and body fluids of multicellular organisms.
K$^+$	Potassium	Important ion that maintains cell turgor.
H$^+$	Hydrogen	Responsible for the acidity of solutions and commonly regulates motility.
Ca^{2+}	Calcium	Often acts as a chemical messenger.
Mg^{2+}	Magnesium	Commonly required for chemical reactions to occur.
Fe^{2+}	Ferrous iron	Carries electrons to oxygen during some chemical reactions that produce energy. Can prevent growth of some microbes that cause human disease.
NH$_4^+$	Ammonium	Found in animal wastes and degraded by some bacteria.
Cl$^-$	Chloride	Often found with a positively charged ion, where it usually neutralizes charge.
OH$^-$	Hydroxyl	Usually present in excess in basic solutions where H$^+$ is depleted.
HCO$_3^-$	Bicarbonate	Often neutralizes acidity of bodies of water and body fluids.
NO$_3^-$	Nitrate	A product of the action of certain bacteria that convert nitrite (NO$_2^-$) into a form plants can use.
SO$_4^{2-}$	Sulfate	Component of sulfuric acid in atmospheric pollutants and acid rain.
PO$_4^{3-}$	Phosphate	Can be combined with certain other molecules to form high-energy bonds where energy is stored in a form living things can use.

TABLE 2.3

Some Properties of Important Elements Found in Living Organisms
(in order of abundance and importance)

Element	Symbol	Atomic Number	Atomic Weight	Electrons in Outer Shell	Biological Occurrence
Oxygen	O	8	16.0	6	Component of biological molecules; required for aerobic metabolism
Carbon	C	6	12.0	4	Essential atom of all organic compounds
Hydrogen	H	1	1.0	1	Component of biological molecules; H^+ released by acids
Nitrogen	N	7	14.0	5	Component of proteins and nucleic acids
Calcium	Ca	20	40.1	2	Found in bones and teeth; regulator of many cellular processes
Phosphorus	P	15	31.0	5	Found in nucleic acids, ATP, and some lipids
Sulfur	S	16	32.0	6	Found in proteins; metabolized by some bacteria
Iron	Fe	26	55.8	2	Carries oxygen; metabolized by some bacteria
Potassium	K	19	39.1	1	Important intracellular ion
Sodium	Na	11	23.0	1	Important extracellular ion
Chlorine	Cl	17	35.4	7	Important extracellular ion
Magnesium	Mg	12	24.3	2	Needed by many enzymes
Copper	Cu	29	63.6	1	Needed by some enzymes; inhibits growth of some microorganisms
Iodine	I	53	126.9	7	Component of thyroid hormones
Fluorine	F	9	19.0	7	Inhibits microbial growth
Manganese	Mn	25	54.9	2	Needed by some enzymes
Zinc	Zn	30	65.4	2	Needed by some enzymes; inhibits microbial growth

dioactive and are called **radioisotopes.** Emissions from radioactive nuclei can be detected by radiation counters. Such emissions can be useful in studying chemical processes, but they also can harm living things.

Chemical Bonds

Chemical bonds form between atoms through interactions of electrons in their outer shells. Energy associated with these bonding electrons holds the atoms together, forming molecules. Three kinds of chemical bonds commonly found in living organisms are ionic, covalent, and hydrogen bonds.

Ionic bonds result from the attraction between ions that have opposite charges. For example, sodium ions, with a positive charge (Na^+) combine with chloride ions, with a negative charge (Cl^-) (Figure 2.2b).

Many compounds, especially those that contain carbon, are held together by **covalent bonds.** Instead of gaining or losing electrons as in ionic bonding, carbon and some other atoms in covalent bonds share pairs of electrons (Figure 2.3). One carbon atom, which has four electrons in its outer shell, can share an electron with each of four hydrogen atoms. At the same time, each of the four hydrogen atoms shares an electron with the carbon atom. Four pairs of electrons are shared, each pair consisting of one electron from carbon and one electron from hydrogen. Such mutual sharing makes a carbon atom stable with eight electrons in its outer shell and a hydrogen atom stable with two electrons in its outer shell. Equal sharing produces *nonpolar compounds*—compounds with no charged regions. Sometimes a carbon atom and an atom such as an oxygen atom share two pairs of electrons to form a double bond. The octet rule still applies, and each atom has eight electrons in its outer shell and is therefore stable. In structural formulas, chemists use a single line to represent a single pair of shared electrons and a double line to represent two pairs of shared electrons (Figure 2.3).

Atoms of four elements, carbon, hydrogen, oxygen, and nitrogen, commonly form covalent bonds that fill their outer electron shells. Carbon shares four electrons, hydrogen one electron, oxygen two electrons, and nitrogen three electrons. Unlike many ionic bonds, covalent bonds are stable and thus are important in molecules that form biological structures.

Hydrogen bonds, though weaker than ionic and covalent bonds, are important in biological structures

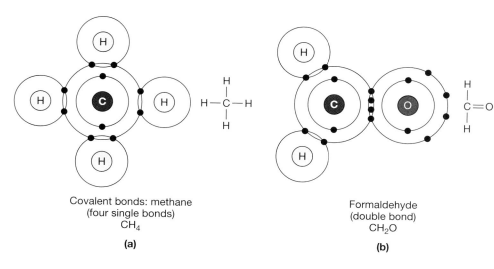

FIGURE 2.3 Covalent bonds are formed by sharing electrons. (a) In methane, a carbon atom, with four electrons in its outermost shell, shares pairs of electrons with four hydrogen atoms. In this way all five atoms acquire stable, filled outer shells. Each shared electron pair constitutes a single covalent bond. (b) In formaldehyde, a carbon atom shares pairs of electrons with two hydrogen atoms and also shares two pairs of electrons with an oxygen atom, forming a double covalent bond.

Covalent bonds: methane
(four single bonds)
CH_4

(a)

Formaldehyde
(double bond)
CH_2O

(b)

and are typically present in large numbers. The atomic nuclei of oxygen and nitrogen attract electrons very strongly. When hydrogen is covalently bonded to oxygen or nitrogen, the electrons of the covalent bond are shared unevenly—they are held closer to the oxygen or nitrogen than to the hydrogen. The hydrogen atom then has a partial positive charge, and the other atom has a partial negative charge. In this case of unequal sharing, the molecule is called a **polar compound** because of its oppositely charged regions. The weak attraction between such partial charges is called a hydrogen bond.

Polar compounds such as water often contain hydrogen bonds. In a water molecule, electrons from the hydrogen atoms stay closer to the oxygen atom, and the hydrogen atoms lie to one side of the oxygen atom (Figure 2.4). Thus, water molecules are polar molecules

that have a positive hydrogen region and a negative oxygen region. Covalent bonds between the hydrogen and oxygen atoms hold the atoms together. Hydrogen bonds between the hydrogen and oxygen regions of different water molecules hold the molecules in clusters.

Hydrogen bonds also contribute to the structure of large molecules such as proteins and nucleic acids, which contain long chains of atoms. The chains are coiled or folded into a three-dimensional configuration that is held together in part by hydrogen bonds.

Chemical Reactions

Chemical reactions in living organisms typically involve the use of energy to form chemical bonds and the release of energy as chemical bonds are broken. For example, the food we eat consists of molecules that have much energy stored in their chemical bonds. During **catabolism** (ka-tab'o-lizm), the breakdown of substances, food is degraded and some of that stored energy is released. Microorganisms use nutrients in the same general way. A catabolic reaction can be symbolized as follows:

$$X—Y \longrightarrow X + Y + \text{energy}$$

where $X—Y$ represents a nutrient molecule and where energy was originally stored in the bond between X and Y.

Catabolic reactions are **exergonic**—that is, they release energy. Conversely, energy is used to form chemical bonds in the synthesis of new compounds. In **anabolism** (a-nab'o-lizm), the buildup, or *synthesis*, of substances, energy is used to create bonds. An anabolic reaction can be symbolized as follows:

$$X + Y + \text{energy} \longrightarrow X—Y$$

where energy is stored in the new substance $X—Y$. Anabolic reactions occur in living cells when small molecules are used to synthesize large molecules. Cells can store small amounts of energy for later use or can ex-

FIGURE 2.4 Water molecules are polar—they have a region with a partial positive charge (the hydrogen atoms) and a region with a partial negative charge (the oxygen atom). Hydrogen bonds, created by the attraction between oppositely charged regions of different molecules, hold the water molecules together in clusters.

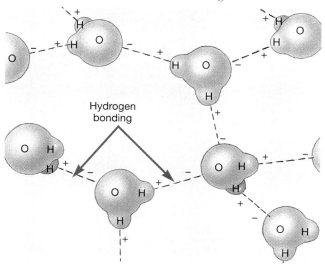

Hydrogen bonding

pend energy to make new molecules. Most anabolic reactions are **endergonic**—that is, they require energy.

WATER AND SOLUTIONS

Water, one of the simplest of chemical compounds, is also one of the most important to living things. It takes part directly in many chemical reactions. Numerous substances dissolve in water or form mixtures called colloidal dispersions. Acids and bases exist and function principally in water mixtures.

Water

Water is so essential to life that humans can live only a few days without it. Many microorganisms die almost immediately if removed from their normal aqueous environments, such as lakes, ponds, oceans, and moist soil. Yet, others can survive for several hours or days without water, and spores formed by a few microorganisms survive for many years away from water. Several bacteria find the moist, nutrient-rich secretions of human skin glands to be an ideal environment.

Water has several properties that make it important to living things. Because water is a polar compound and forms hydrogen bonds, it can form thin layers on surfaces and can act as a *solvent,* or dissolving medium. Water is a good solvent for ions because the polar water molecules surround the ions. The positive region of water molecules is attracted to negative ions, and the negative region of water molecules is attracted to positive ions. Many different kinds of ions can therefore be distributed evenly through a water medium, forming a *solution* (Figure 2.5).

Water forms thin layers because it has a high surface tension. **Surface tension** is a phenomenon in which the surface of water acts as a thin, invisible, elastic membrane (Figure 2.6). The polarity of water molecules gives them a strong attraction for one another

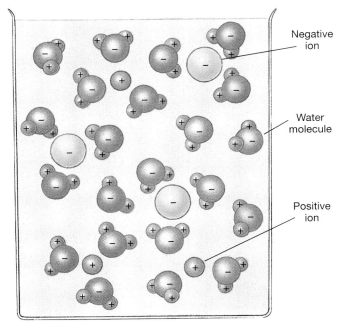

FIGURE 2.5 The polarity of water molecules enables water to dissolve many ionic compounds. The positive regions of the water molecules surround negative ions, and the negative regions of the water molecules surround positive ions, holding the ions in solution.

but no attraction for gas molecules in air at the water's surface. Therefore, surface water molecules cling together, forming hydrogen bonds with other molecules below the surface. In living cells this feature of surface tension allows a thin film of water to cover membranes and to keep them moist.

Water has a high *specific heat,* that is, it can absorb or release large quantities of heat energy with little temperature change. This property of water helps to stabilize the temperature of living organisms, which are composed mostly of water, as well as bodies of water where many microorganisms live.

Finally, water provides the medium for most chemical reactions in cells, and it participates in many

FIGURE 2.6 (a) Hydrogen bonding between water molecules creates surface tension, which causes the surface of water to behave like an elastic membrane. (b) Surface tension is strong enough to support the weight of the insects known as water striders.

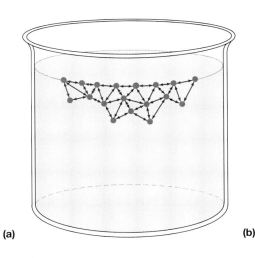

(a)

(b)

of these reactions. Suppose, for example, that substance X can gain or lose H^+ and that substance Y can gain or lose OH^-. The substances that enter a reaction are called **reactants.** In an anabolic reaction, the components of water (H^+ and OH^-) are removed from the reactants to form a larger product molecule:

$$X-H + HO-Y \longrightarrow X-Y + H_2O$$

This kind of reaction, called **dehydration synthesis,** is involved in the synthesis of complex carbohydrates, some lipids (fats), and proteins. Conversely, in many catabolic reactions, water is added to a reactant to form simpler products:

$$X-Y + H_2O \longrightarrow X-H + HO-Y$$

This kind of reaction, called **hydrolysis,** occurs in the breakdown of large nutrient molecules to release simple sugars, fatty acids, and amino acids.

Solutions and Colloids

Solutions and colloidal dispersions are examples of mixtures. Unlike a chemical compound, which consists of molecules whose atoms are present in specific proportions, a **mixture** consists of two or more substances that are combined in any proportion and are not chemically bound. Each substance in a mixture contributes its properties to the mixture. For example, a mixture of sugar and salt could be made using any proportions of the two ingredients. The degree of sweetness or saltiness of the mixture would depend on the relative amounts of sugar and salt present, but both sweetness and saltiness would be detectable.

A **solution** is a mixture of two or more substances in which the molecules of the substances are evenly distributed and ordinarily will not separate out upon standing. In a solution the medium in which substances are dissolved is the **solvent.** The substance dissolved in the solvent is the **solute.** Solutes can consist of atoms, ions, or molecules. In cells and in the medium in which cells live, water is the solvent in nearly all solutions. Typical solutes include the sugar glucose, the gases carbon dioxide and oxygen, and many different kinds of ions. Many smaller proteins also can act as solutes in true solutions.

Few living things can survive in highly concentrated solutions. We make use of this fact in preserving several kinds of foods. Can you think of foods that are often kept unrefrigerated and unsealed for long periods of time? Jellies, jams, and candies do not readily spoil because most microorganisms cannot tolerate the high concentration of sugar. Salt-cured meats are too salty to allow growth of most microorganisms, and pickles are too acidic for most microbes.

Particles too large to form true solutions can sometimes form *colloidal dispersions,* or **colloids.** Gelatin

dessert is a colloid in which the protein gelatin is dispersed in a watery medium. Similarly, colloidal dispersions in cells usually are formed from large protein molecules dispersed in water. The fluid or semifluid substance inside living cells is a complex colloidal system. Large particles are suspended by opposing electrical charges, layers of water molecules around them, and other forces. Media for growing microorganisms sometimes are solidified with agar; these media are colloidal dispersions. Some colloidal systems have the ability to change from a semisolid state, such as gelatin that has "set," to a more fluid state, such as gelatin that has melted. Amoebae seem to move, in part, by the ability of the colloidal material within them to change back and forth between semisolid and fluid states.

Acids, Bases, and pH

In chemical terms, most living things exist in relatively neutral environments, but some microorganisms live in environments that are *acidic* or *basic (alkaline).* Understanding acids and bases is important in studying microorganisms and in studying their effects on human cells. An **acid** is a hydrogen ion (H^+) donor. (A hydrogen ion is a proton.) An acid donates H^+ to a solution. The acids found in living organisms usually are weak acids such as acetic acid (vinegar), although some are strong acids such as hydrochloric acid. Acids release H^+ when carboxyl groups ($-COOH$) ionize to COO^- and H^+. A **base** is a proton acceptor, or a hydroxyl ion donor. It accepts H^+ from, or donates OH^- (hydroxyl ion) to, the solution. The bases found in living organisms usually are weak bases such as the amino (NH_2) group, which accepts H^+ to form NH_3^+.

Chemists have devised the concept of pH to specify the acidity or alkalinity of a solution. The **pH** scale (Figure 2.7), which relates proton concentrations to pH, is a logarithmic scale (see the Close-Up box "Where Have All the Zeros Gone?," p. 34). This means that the concentration of hydrogen ions (protons) changes by a factor of 10 for each unit of the scale. The practical range of the pH scale is from 0 to 14. A solution with a pH of 7 is **neutral**—neither acidic nor **alkaline** (basic). Pure water has a pH of 7 because the concentrations of H^+ and OH^- in it are equal. Figure 2.7 shows the pH of some body fluids, selected foods, and other substances.

COMPLEX ORGANIC MOLECULES

The basic principles of general chemistry also apply to **organic chemistry,** the study of compounds that contain carbon. The study of the chemical reactions that occur in living systems is the branch of organic chemistry known as **biochemistry.** Early in the 1800s it was believed that molecules from living things were filled

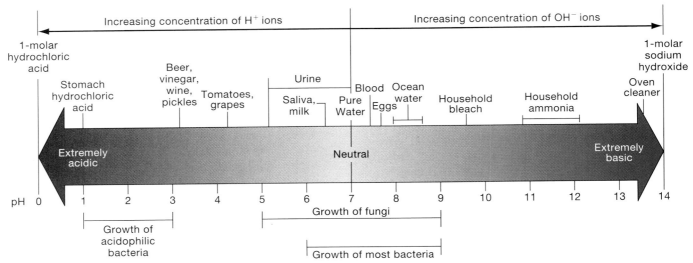

FIGURE 2.7 The pH values of some common substances. Each unit of the pH scale represents a tenfold increase or decrease in the concentration of hydrogen ions. Thus, vinegar, for example, is 10,000 times more acidic than pure water.

CLOSE-UP

WHERE HAVE ALL THE ZEROS GONE?

The term *pH* is defined as the negative logarithm of the hydrogen ion concentration (in moles per liter). To understand pH we need to understand a little about exponents, scientific notation, and logarithms. The *exponent* of a number, a superscript to the right of the number, designates the number of times it is to be multiplied by itself. For example, 2^3 indicates that 2 is to be multiplied by itself 3 times: $2^3 = 2 \times 2 \times 2 = 8$. Similarly, 10^5 indicates that 10 is to be multiplied by itself 5 times. Try it: $10^5 = 100,000$.

Scientists often use *scientific notation* when they work with extremely large or extremely small numbers. Instead of writing all the zeros needed to express one billion (1,000,000,000), they observe that a billion is equal to 10 multiplied by itself 9 times, and they write 1.0×10^9. Likewise, instead of writing one millionth as 1/1,000,000, they note that the denominator of the fraction is equal to 1.0×10^6, and they write 1.0×10^{-6}. A negative exponent means that

the number is less than 1. In scientific notation, the exponent is conventionally chosen so that a single digit appears to the left of the decimal value before the exponential term. For example, the expression 22.6×10^6 is not in scientific notation; it should be expressed as 2.26×10^7.

The logarithms we will use are based on powers of 10. In this system, the logarithm of a number is the power to which 10 must be raised to get the number. For example, the logarithm of 100 is 2, which means that 10 must be multiplied by itself two times to get 100. The logarithm of 1/100 is -2.

Now we can apply the preceding concepts to the definition of pH. The concentration of H^+ in a neutral solution is 1/10,000,000. This number can be expressed in exponential notation as 10^{-7}. The logarithm of this number is -7. To determine its negative logarithm, we multiply -7 by -1 and get 7. Therefore, the pH of a neutral solution is 7.

with a supernatural "vital force" and therefore could not be explained by the laws of chemistry and physics. It was considered impossible to make *organic compounds* outside of living systems. That idea was disproved in 1828 when the German scientist Friedrich Wohler synthesized the organic compound urea, a small molecule excreted as a waste material by many animals. Since that time thousands of organic compounds—plastics, fertilizers, and medicines—have been made in the laboratory. Organic compounds such as carbohydrates, lipids, proteins, and nucleic acids occur naturally in living things and in the products or remains of living things. The ability of carbon atoms

to form covalent bonds and to link up in long chains makes possible the formation of an almost infinite number of organic compounds.

The simplest carbon compounds are the *hydrocarbons*, chains of carbon atoms with their associated hydrogen atoms. The structure of the hydrocarbon propane, C_3H_8, for example, is as follows:

$$
\begin{array}{ccccccc}
 & H & & H & & H & \\
 & | & & | & & | & \\
H - & C & - & C & - & C & - H \\
 & | & & | & & | & \\
 & H & & H & & H &
\end{array}
$$

Carbon chains can have not only hydrogen but other atoms such as oxygen and nitrogen bound to them. Some of these atoms form functional groups. A **functional group** is a part of a molecule that generally participates in chemical reactions as a unit and that gives the molecule some of its chemical properties.

Four significant groups of compounds—alcohols, aldehydes, ketones, and organic acids—have functional groups that contain oxygen (Figure 2.8). An alcohol has one or more hydroxyl groups ($-OH$). An aldehyde has a carbonyl group ($-CO$) at the end of the carbon chain; a ketone has a carbonyl group within the chain. An organic acid has one or more carboxyl groups ($-COOH$). One key functional group that does not contain oxygen is the amino group ($-NH_2$). Found in amino acids, amino groups account for the nitrogen in proteins.

The relative amount of oxygen in different functional groups is significant. Groups with little oxygen, such as alcohol groups, are said to be *reduced;* groups with relatively more oxygen, such as carboxyl groups, are said to be *oxidized* (Figure 2.8). As we shall see in Chapter 5, *oxidation* is the addition of oxygen or the removal of hydrogen or electrons from a substance. Burning is an example of oxidation. *Reduction* is the removal of oxygen or the addition of hydrogen or electrons to a substance. In general, the more reduced a molecule, the more energy it contains. Hydrocarbons, such as gasoline, have no oxygen and thus represent the extreme in energy-rich, reduced molecules. They make good fuels because they contain so much energy. Conversely, the more oxidized a molecule, the less energy it contains. Carbon dioxide (CO_2) represents the extreme in an oxidized molecule because no more than two oxygen atoms can bond to a single carbon atom. As we shall see, oxidation releases energy from molecules.

Let us now consider the major classes of large, complex biochemical molecules of which all living things, including microbes, are composed.

Carbohydrates

Carbohydrates serve as the main source of energy for most living things. Plants make carbohydrates, including structural carbohydrates such as cellulose and energy-storage carbohydrates such as starch. Animals use carbohydrates as food, and many, including humans, store energy in a carbohydrate called *glycogen.* Many microorganisms use carbohydrates from their environment for energy and also make a variety of carbohydrates. Carbohydrates in the membranes of cells can act as markers that make a cell chemically recognizable. Chemical recognition is important in immunological reactions and other processes in living things.

All carbohydrates contain the elements carbon, hydrogen, and oxygen, generally in the proportion of two hydrogen atoms to each carbon and oxygen atom. There are three groups of carbohydrates: monosaccharides, disaccharides, and polysaccharides. **Monosaccharides** consist of a carbon chain or ring with several alcohol groups and one other functional group, either an aldehyde group or a ketone group. Several monosaccharides, such as glucose and fructose, are **isomers**—they have the same molecular formula, $C_6H_{12}O_6$, but different structures and different properties (Figure 2.9). Thus, even at the chemical level we can see that structure and function are related.

Glucose, the most abundant monosaccharide, can be represented schematically as a straight chain, a ring, or a three-dimensional structure. The chain structure, in Figure 2.10a, clearly shows a carbonyl group at carbon 1 (the first carbon in the chain, at the top in this orientation) and alcohol groups on all the other carbons. Figure 2.10b shows how a glucose molecule in solution rearranges and bonds to itself to form a closed ring. The three-dimensional projection in Figure 2.10c more closely approximates the actual shape of the molecule. In studying structural formulas, it is important to imagine each molecule as a three-dimensional object.

FIGURE 2.8 Four classes of organic compounds that incorporate oxygen. Alcohols contain one or more hydroxyl groups ($-OH$); aldehydes and ketones contain carbonyl groups ($-C=O$); and organic acids contain carboxyl groups ($-COOH$).

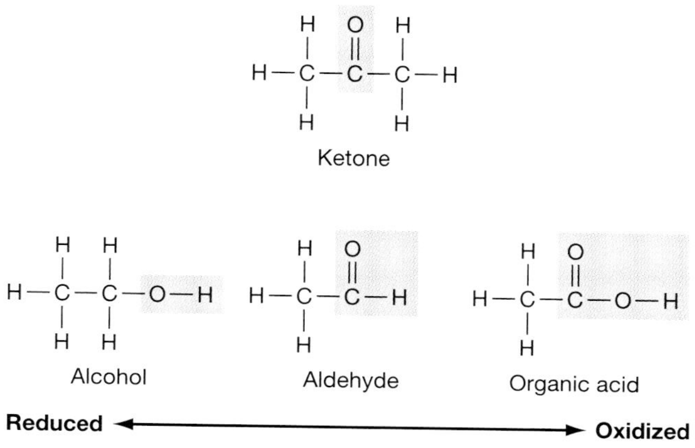

FIGURE 2.9 Glucose and fructose are isomers: They contain the same atoms, but they differ in structure.

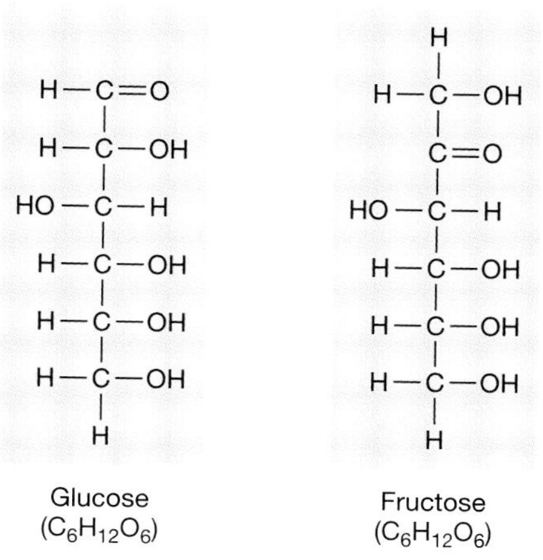

APPLICATIONS

CAN A COW ACTUALLY EXPLODE?

Cows can derive a good deal of nourishment from grass, hay, and other fibrous vegetable matter that are inedible to humans. We can't digest cellulose, the chief component of plants. If you had to live on hay, you would probably starve to death. How, then, do cows and other hooved animals manage on such a diet?

Oddly enough, cows can't digest cellulose either. But they don't need to—it's done for them. Cows and their relatives harbor in their stomachs large populations of microorganisms that do the work of breaking down cellulose into sugars that the animal can use. The same is true of termites: If it weren't for microbes in their guts that help them to digest cellulose, they couldn't dine on the wooden beams in your house.

Cellulose is very similar to starch—both consist of long chains of glucose molecules. The bonds between these molecules, however, are slightly different in their geometry in the two substances. As a result, the enzymes that animals use to break down a starch molecule into its component glucose units have no effect on cellulose. In fact, very few organisms produce enzymes that can attack cellulose. Even the protists (unicellular organisms with a nucleus) that live in the stomachs of cows and termites cannot always do it by themselves. Just as cows and termites depend on the protists in their stomachs, the protists frequently depend on certain bacteria that reside permanently within them. It is these bacteria that actually make the essential digestive enzymes.

The activities of the intestinal microorganisms that perform these digestive services are a mixed blessing, both to the cows and to the humans who keep them. The bacteria also produce methane gas, CH_4—as much as 190 to 380 l per day from a single cow. (Methane production can be so rapid that a cow's stomach may rupture if the cow can't burp. Some ingenious inventors have actually patented cow safety valves to release the gas buildup directly through the animal's side.) When this gas eventually makes its way out of the cow by one route or another, it rises to the upper atmosphere. There it is suspected of contributing to the "greenhouse effect," trapping solar heat and causing an overall warming of the earth's climate (Chapter 26). Scientists have estimated that the world's cows release 50 million metric tons of methane annually; that's not counting the sheep, goats, antelope, water buffalo, and other grass eaters.

FIGURE 2.10 Three ways of representing the glucose molecule. (a) In solution, the straight-chain form is rarely found. (b) Instead, the molecule bonds to itself, forming a six-membered ring. The ring is conventionally depicted as a flat hexagon. (c) The actual three-dimensional structure is more complex. The spheres in this depiction represent carbon atoms.

Monosaccharides can be reduced to form deoxy sugars and sugar alcohols (Figure 2.11). The deoxy sugar *deoxyribose*, which has a hydrogen atom instead of an —OH group on one of its carbons, is a component of DNA. Certain sugar alcohols, which have an additional alcohol group instead of an aldehyde or ketone group, can be metabolized by particular microorganisms. Mannitol and other sugar alcohols are used to identify some microorganisms in diagnostic tests.

Disaccharides are formed when two monosaccharides are connected by the removal of water and the formation of a **glycosidic bond,** a sugar alcohol/sugar linkage (Figure 2.12a). Sucrose, common table sugar, is a disaccharide made of glucose and fructose. **Polysaccharides** are formed when many monosaccharides are linked by glycosidic bonds (Figure 2.12b). Polysaccharides such as starch, glycogen, and cellulose are **polymers**—long chains of repeating units—of glu-

FIGURE 2.11 (a) "Deoxy" indicates one less oxygen atom. One of the carbon atoms of the deoxy sugar deoxyribose lacks a hydroxyl group that (b) ribose has. (c) Glycerol, a three-carbon sugar alcohol that is a component of fats. (d) Mannitol, a sugar alcohol used in diagnostic tests for certain microbes.

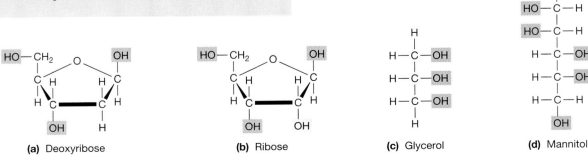

(a) Deoxyribose   (b) Ribose   (c) Glycerol   (d) Mannitol

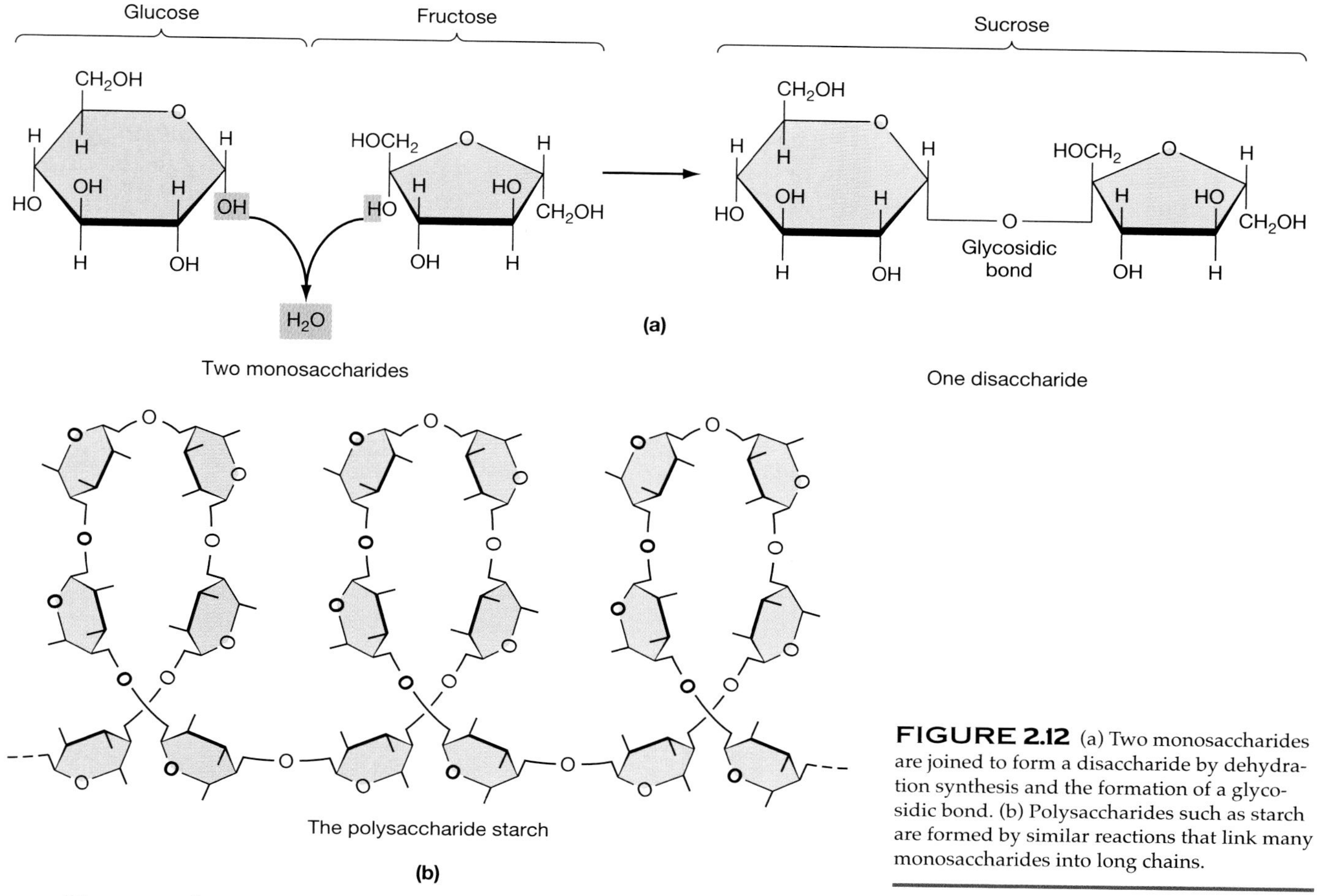

Glucose   Fructose   Sucrose

(a)

Two monosaccharides   One disaccharide

The polysaccharide starch

(b)

FIGURE 2.12 (a) Two monosaccharides are joined to form a disaccharide by dehydration synthesis and the formation of a glycosidic bond. (b) Polysaccharides such as starch are formed by similar reactions that link many monosaccharides into long chains.

cose. However, the glycosidic bonds in each polymer are arranged differently. Plants and most algae make starch and cellulose. Starch serves as a way to store energy, and cellulose is a structural component of cell walls. Animals make and store glycogen, which they can break down to glucose as energy is needed. Microorganisms contain several other important polysaccharides, as we shall see in later chapters.

Table 2.4 summarizes the types of carbohydrates.

Lipids

Lipids constitute a chemically diverse group of substances that includes fats, phospholipids, and steroids. They are relatively insoluble in water but are soluble in nonpolar solvents such as ether and benzene. Lipids form part of the structure of cells, especially cell membranes, and many can be used for energy. Generally, lipids contain relatively more hydrogen and less oxygen than carbohydrates and therefore contain more energy than carbohydrates.

Fats contain the three-carbon alcohol glycerol and one or more fatty acids. A **fatty acid** consists of a long chain of carbon atoms with associated hydrogen atoms and a carboxyl group at one end of the chain. The synthesis of a fat from glycerol and fatty acids involves removing water and forming an ester bond between the

TABLE 2.4	Types of Carbohydrates	
Class of Carbohydrates	**Examples**	**Description and Occurrence**
Monosaccharides	Glucose	Sugar found in most organisms
	Fructose	Sugar found in fruit
	Galactose	Sugar found in milk
	Ribose	Sugar found in RNA
	Deoxyribose	Sugar found in DNA
Disaccharides	Sucrose	Glucose and fructose; table sugar
	Lactose	Glucose and galactose; milk sugar
	Maltose	Two glucose units; product of starch digestion
Polysaccharides	Starch	Polymer of glucose stored in plants, digestible by humans
	Glycogen	Polymer of glucose stored in animal liver and skeletal muscles
	Cellulose	Polymer of glucose found in plants, not digestible by humans; digested by some microbes

carboxyl group of the fatty acid and an alcohol group of glycerol (Figure 2.13a). A **triacylglycerol,** formerly called a *triglyceride,* is a fat formed when three fatty acids are bonded to glycerol. *Monoacylglycerols* (monoglycerides) and *diacylglycerols* (diglycerides) contain one and two fatty acids, respectively, and usually are formed from the digestion of triacylglycerols.

Fatty acids can be saturated or unsaturated. A **saturated fatty acid** contains all the hydrogen it can have; that is, it is saturated with hydrogen (Figure 2.13b). An **unsaturated fatty acid** has lost at least two hydrogen atoms and contains a double bond between the two carbons that have lost hydrogen atoms (Figure 2.13c). "Unsaturated" thus means not completely saturated with hydrogen. Oleic acid is an unsaturated fatty acid. *Polyunsaturated fats,* many of which are vegetable oils

that remain liquid at room temperature, contain many unsaturated fatty acids.

Some lipids contain one or more other molecules in addition to fatty acids and glycerol. For example, **phospholipids,** which are found in all cell membranes, differ from fats by the substitution of phosphoric acid (H_3PO_4) for one of the fatty acids (Figure 2.14a). The charged phosphate group ($-HPO_4^-$) typically attaches to another charged group. Both can mix with water, but the fatty acid end cannot (Figure 2.14b). Such properties of phospholipids are important in determining the characteristics of cell membranes (Chapter 4).

Steroids have a four-ring structure (Figure 2.15a) and are quite different from other lipids. They include cholesterol, steroid hormones, and vitamin D. Cholesterol (Figure 2.15b) is insoluble in water and is found

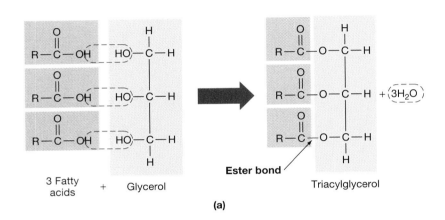

Ester bond

3 Fatty acids + Glycerol → Triacylglycerol + 3H₂O

(a)

FIGURE 2.13 (a) Three fatty acids combine with glycerol to form a molecule of triacylglycerol, a type of fat. The group designated R is a long hydrocarbon chain that varies in length in different fatty acids. It may be saturated or unsaturated. (b) Saturated fatty acids have only single covalent bonds between carbon atoms in their carbon chains and can therefore accommodate the maximum possible number of hydrogens. (c) Unsaturated fatty acids, such as oleic acid, have one or more double bonds between carbons and thus contain fewer hydrogens. The double bond causes a bend in the carbon chain. In (b) and (c), both structural formulas and space-filling models are shown.

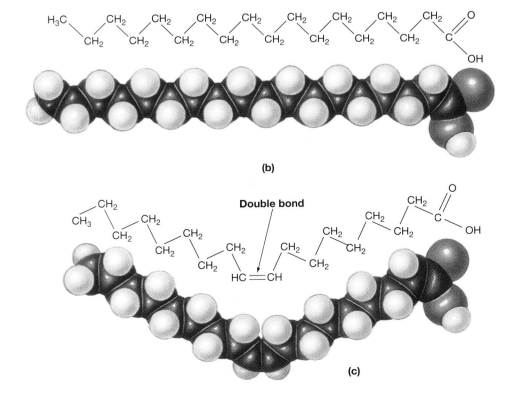

(b)

Double bond

(c)

Hydrogen (H)

Oxygen (O)

Carbon (C)

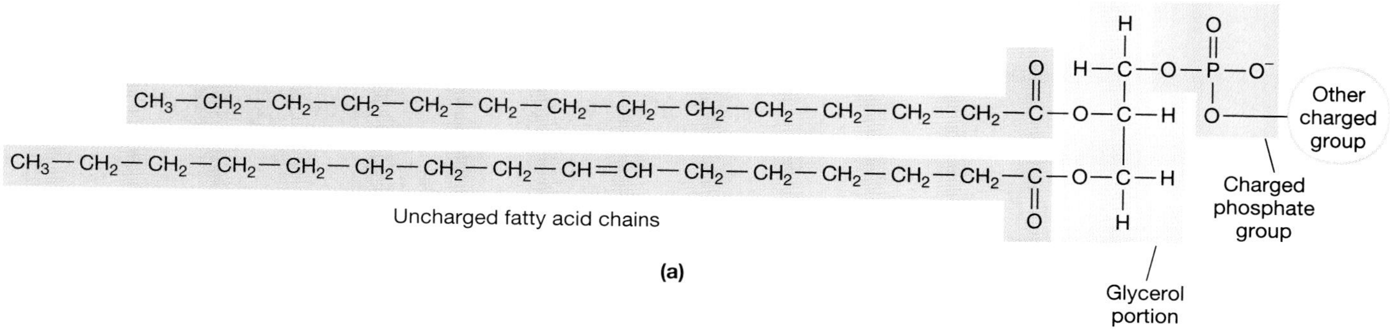

CH₃ — CH₂ — CH₂ — CH₂ — CH₂ — CH₂ — CH₂ — CH₂ — CH₂ — CH₂ — CH₂ — CH₂ — CH₂ — C — O — C — H

CH₃ — CH₂ — CH₂ — CH₂ — CH₂ — CH₂ — CH₂ — CH₂ — CH = CH — CH₂ — CH₂ — CH₂ — CH₂ — CH₂ — C — O — C — H

$$CH_3-CH_2-CH_2-CH_2-CH_2-CH_2-CH_2-CH_2-CH_2-CH_2-CH_2-CH_2-CH_2-\overset{O}{\overset{\|}{C}}-O-\overset{H}{\underset{}{C}}$$

H
 |
H — C — O — P — O⁻
 | ‖
             O

Other charged group

Charged phosphate group

Glycerol portion

Uncharged fatty acid chains

(a)

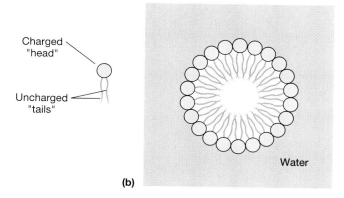

Charged "head"

Uncharged "tails"

Water

(b)

FIGURE 2.14 In phospholipids, (a) one of the fatty acid chains of a fat molecule is replaced by phosphoric acid. The charged phosphate group and another attached group can interact with water molecules, which are polar, but the two long, uncharged fatty acid tails cannot. (b) As a result, phospholipid molecules in water tend to form globular structures with the phosphate groups facing outward and the fatty acids in the interior.

APPLICATIONS

HOW TO GET OIL AND WATER TO MIX

You probably have washed enough greasy dishes to know that soaps and detergents help to get oily substances off the dishes and into the water. Soap and detergent molecules have both a charged and an uncharged region, like a fatty acid with its ionized carboxyl group at one end and its non-ionized, hydrogen-saturated carbons at the other end. In fact, soap can be made by boiling fat with sodium hydroxide (lye) until the fatty acids break away from the glycerol molecules. The charged ends of soap molecules are attracted to the polar water molecules. Their uncharged ends interact with nonpolar fats and oils, thereby allowing water and oil to mix. The fats and oils can then be washed away.

FIGURE 2.15 (a) Steroids are lipids with a characteristic four-ring structure. The specific chemical groups attached to the rings determine the properties of different steroids. (b) One of the most biologically important steroids is cholesterol, a component of the membranes of animal cells and one group of bacteria.

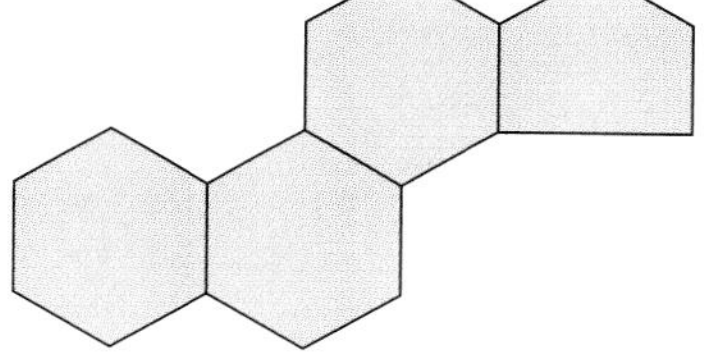

CH₃
 |
HC — CH₃
 |
CH₂
 |
CH₂
 |
CH₂
 |
HC — CH₃

Side chain

(a)

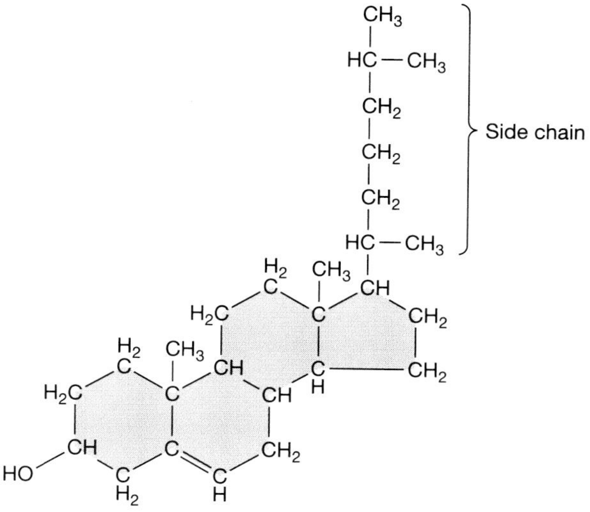

(b)

in the cell membranes of animal cells and the group of bacteria called mycoplasmas. Steroid hormones and vitamin D are important in many animals.

Proteins

Properties of Proteins and Amino Acids

Among the molecules found in living things, proteins have the greatest diversity of structure and function. **Proteins** are composed of building blocks called **amino acids,** which have at least one amino ($-NH_2$) group and one acidic carboxyl ($-COOH$) group. The general structure of an amino acid and some of the 20 amino acids found in proteins are shown in Figure 2.16. Each amino acid is distinguishable by a different chemical group, called an **R group,** attached to the central carbon atom. Because all amino acids contain carbon, hydrogen, oxygen, and nitrogen, and some contain sulfur, proteins also contain these elements.

A protein is a polymer of amino acids joined by **peptide bonds**—that is, covalent bonds that link an amino group of one amino acid and a carboxyl group of another amino acid (Figure 2.17). Two amino acids linked together make a *dipeptide*, three make a *tripeptide*, and many make a **polypeptide.** In addition to the amino and carboxyl groups, some amino acids have an R group called a *sulfhydryl group* ($-SH$). Sulfhydryl groups in adjacent chains of amino acids can lose hydrogen and form *disulfide linkages* ($-S-S-$) from one chain to the other.

Structure of Proteins

Proteins have several levels of structure. The **primary structure** of a protein consists of the specific sequence of amino acids in a polypeptide chain (Figure 2.18a). The **secondary structure** of a protein consists of the folding or coiling of amino acid chains into a particular pattern, such as a helix or pleated sheet (Figure 2.18b). Hydrogen bonds are responsible for such patterns. Further bending and folding of the protein molecule into globular (irregular spherical) shapes or fibrous threadlike strands produces the **tertiary structure** (Figure 2.18c). Some large proteins such as hemoglobin have **quaternary structure,** formed by the association of several tertiary-structured polypeptide chains (Figure 2.19). Tertiary and quaternary structures are maintained by disulfide linkages, hydrogen bonds, and other forces between R groups of amino acids. The three-dimensional shapes of protein molecules and the nature of sites at which other molecules can bind to them are extremely important in determining how proteins function in living organisms.

Several conditions can disrupt hydrogen bonds and other weak forces that maintain protein structure. They include highly acidic or basic conditions and tempera-

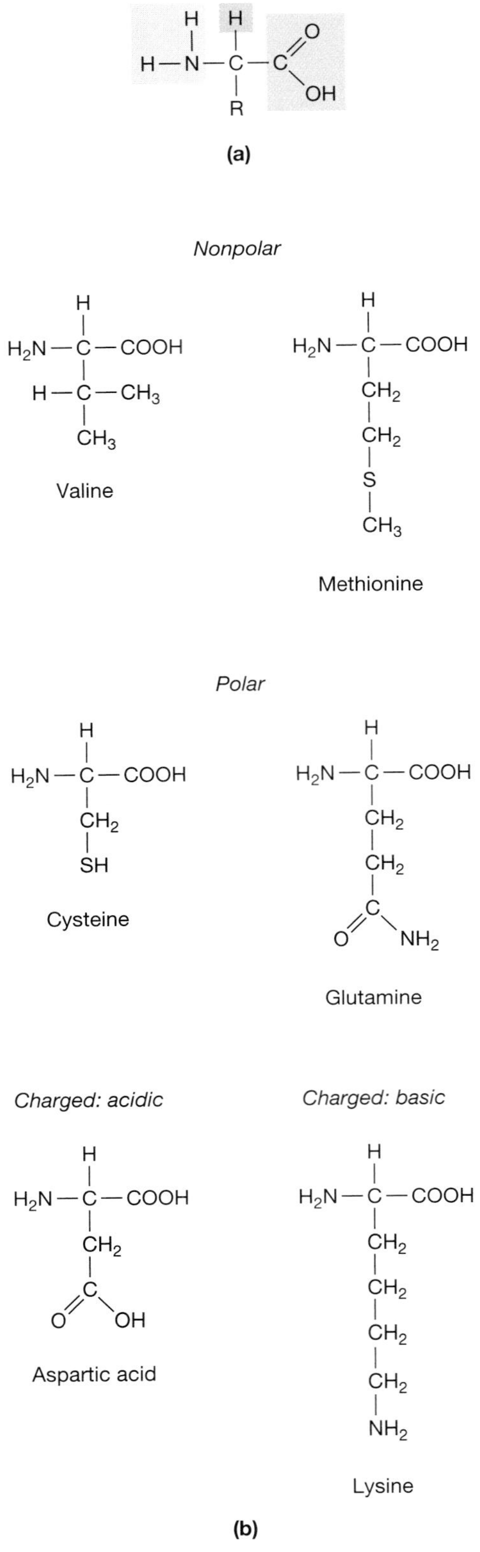

FIGURE 2.16 (a) The general structure of an amino acid, and (b) six representative examples. All amino acids have four groups attached to the central carbon atom: an amino ($-NH_2$) group, a carboxyl ($-COOH$) group, a hydrogen atom, and a group designated R that is different in each amino acid. The R group determines many of the chemical properties of the molecule—for example, whether it is nonpolar, polar, acidic, or basic.

FIGURE 2.17 Two amino acids are joined by the removal of a water molecule (dehydration synthesis) and the formation of a peptide bond between the —COOH group of one and the —NH$_2$ group of the other.

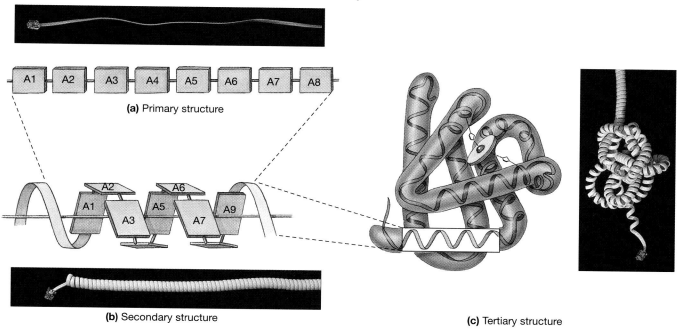

FIGURE 2.18 Three levels of protein structure. (a) Primary structure is the sequence of amino acids (A1, A2, . . .) in a polypeptide chain. Imagine it as a straight telephone cord. (b) Polypeptide chains, especially those of structural proteins, tend to coil or fold into a few simple, regular, three-dimensional patterns called secondary structure. Imagine the telephone cord as a coiled cord. (c) Polypeptide chains of enzymes and other soluble proteins may also exhibit secondary structure. In addition, the chains tend to fold up into complex, globular shapes that constitute the protein's tertiary structure. Imagine the knot formed when a coiled telephone cord tangles.

(a) Primary structure

(b) Secondary structure

(c) Tertiary structure

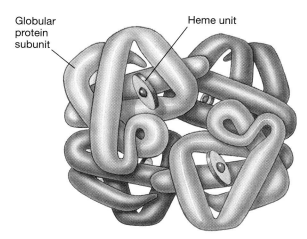

(a) Hemoglobin molecule

FIGURE 2.19 (a) Many large proteins such as hemoglobin, which carries oxygen in human red blood cells, are made up of several polypeptide chains. The arrangement of these chains makes up the protein's quaternary structure. (b) Some structural proteins such as keratin, a component of human skin and hair, also consist of several polypeptide chains and so have quaternary structure.

(b) Keratin fiber

tures above 50°C. Such disruption of secondary, tertiary, and quatenary structures is called **denaturation.** Sterilization and disinfection procedures often make use of heat or chemicals that kill microorganisms by denaturing their proteins. Also, the cooking of meat tenderizes it by denaturing proteins. Therefore, microbes and cells of larger organisms must be maintained within fairly narrow ranges of pH and temperature to prevent disruption of protein structure.

Classification of Proteins

Most proteins can be classified by their major functions as structural proteins or enzymes. **Structural proteins,** as the name implies, contribute to the three-dimensional structure of cells, cell parts, and membranes. Certain proteins, called *motile proteins,* contribute both to structure and to movement. They account for the contraction of animal muscle cells and for some kinds of movement in microbes. **Enzymes** are protein *catalysts*—substances that control the rate of chemical reactions in cells. A few proteins are neither structural proteins nor enzymes. They include proteins that form receptors for certain substances on cell membranes and antibodies that participate in the body's immune responses (Chapter 18).

Enzymes

Enzymes increase the rate at which chemical reactions take place within living organisms in the temperature range compatible with life. We discuss enzymes in more detail in Chapter 5 but summarize their properties here. In general, enzymes speed up reactions by decreasing the energy required to start reactions. They also hold reactant molecules close together in the proper orientation for reactions to occur. Each enzyme has an **active site,** which is the site at which it combines with its **substrate,** the substance on which an enzyme acts. Enzymes have **specificity**—that is, each enzyme acts on a particular substrate or on a certain kind of chemical bond.

Like catalysts in inorganic chemical reactions, enzymes are not permanently affected or used up in the reactions they initiate. Enzyme molecules can be used over and over again to catalyze a reaction, although not indefinitely. Because enzymes are proteins, they are denatured by extremes of temperature and pH.

Nucleotides and Nucleic Acids

The chemical properties of *nucleotides* allow these compounds to perform several essential functions. One key function is storage of energy in **high-energy bonds**—bonds that, when broken, release more energy than do most covalent bonds. Nucleotides joined to form *nucleic acids* are, perhaps, the most remarkable of all biochemical substances. They store information that directs protein synthesis and that can be transferred from parent to progeny.

A **nucleotide** consists of three parts: (1) a nitrogenous base, so named because it contains nitrogen and has alkaline properties; (2) a five-carbon sugar; and (3) one or more phosphate groups, as Figure 2.20a shows for the nucleotide *adenosine triphosphate (ATP).* The

FIGURE 2.20 (a) A nucleotide consists of a nitrogenous base, a five-carbon sugar, and one or more phosphate groups. (b) A nucleoside is comprised of the sugar and base without the phosphates. (c) The nucleotide adenosine triphosphate (ATP), the immediate source of energy for most activities of living cells. In ATP the base is adenine, and the sugar is ribose. Adding a phosphate group to adenosine diphosphate greatly increases the energy of the molecule; removal of the third phosphate group releases energy that can be used by the cell.

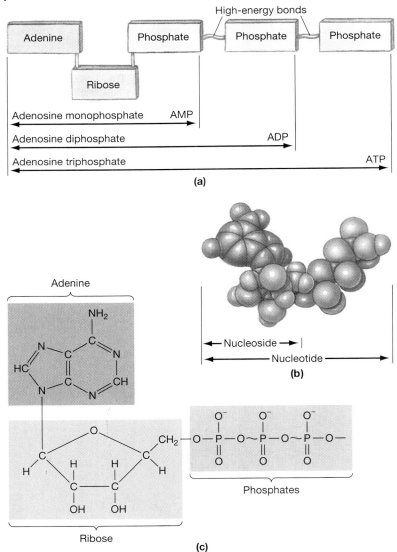

(a)

(b)

(c)

sugar and base alone make up a *nucleoside* (Figure 2.20b).

The nucleotide ATP is the main source of energy in cells because it stores chemical energy in a form cells can use. The bonds between phosphates in ATP that are high-energy bonds are designated by wavy lines (Figure 2.20c). They contain more energy than most covalent bonds, in that more energy is released when they are broken. Enzymes control the forming and breaking of high-energy bonds so that energy is released as needed within cells. The capture, storage, and use of energy is an important component of cellular metabolism (Chapter 5).

Nucleic acids consist of long polymers of nucleotides, called **polynucleotides.** They contain genetic information that determines all the heritable characteristics of a living organism, be it a microbe or a human. Such information is passed from generation to generation and directs protein synthesis in each organism. By directing protein synthesis, nucleic acids determine which structural proteins and enzymes an organism will have. The enzymes determine which other substances the organism can make and which other reactions it can carry out.

The two nucleic acids found in living organisms are **ribonucleic acid (RNA)** and **deoxyribonucleic acid (DNA).** Except in a few viruses, RNA is a single polynucleotide chain, and DNA is a double chain of polynucleotides arranged as a double helix. In both nucleic acids, the phosphate and sugar molecules form a sturdy but inert "backbone" from which nitrogenous bases protrude. In DNA each chain is connected by hydrogen bonds between the bases, so the whole molecule resembles a ladder with many rungs (Figure 2.21a).

DNA and RNA contain somewhat different building blocks (Table 2.5). RNA contains the sugar ribose, whereas DNA contains deoxyribose, which has one less oxygen atom than ribose. Three nitrogenous bases, adenine, cytosine, and guanine, are found in both DNA and

---

FIGURE 2.21 Nucleic acids consist of a backbone of alternating sugar and phosphate groups to which nitrogenous bases are attached. (a) RNA is usually single-stranded. DNA molecules typically consist of two chains held together by hydrogen bonds between bases. (b) The complementary base pairs in DNA, showing how hydrogen bonds are formed.

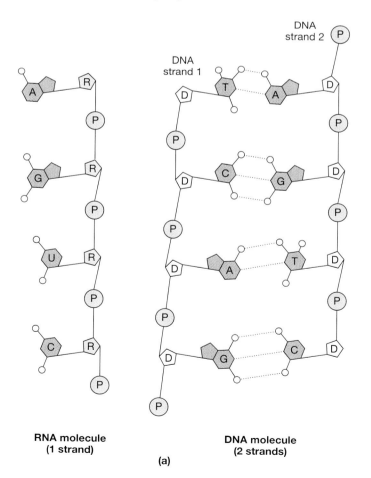

RNA molecule
(1 strand)

DNA molecule
(2 strands)

(a)

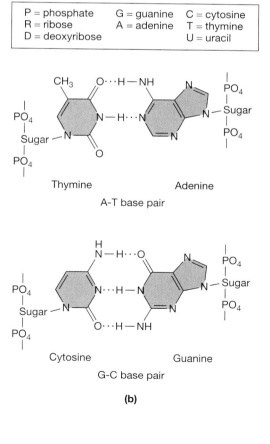

P = phosphate G = guanine C = cytosine
R = ribose A = adenine T = thymine
D = deoxyribose U = uracil

A-T base pair

G-C base pair

(b)

TABLE 2.5   **Components of DNA and RNA**

COMPONENT		DNA	RNA
	Phosphoric acid	X	X
Sugars	Ribose		X
	Deoxyribose	X	
Bases	Adenine	X	X
	Guanine	X	X
	Cytosine	X	X
	Thymine	X	
	Uracil		X

RNA. In addition, DNA contains the base thymine, and RNA contains the base uracil. Of these bases, adenine and guanine are **purines,** nitrogenous base molecules that contain double-ring structures, and thymine, cytosine, and uracil are **pyrimidines,** nitrogenous base molecules that contain a single-ring structure (Figure 2.22). All cellular organisms have both DNA and RNA. Viruses have either DNA or RNA but not both.

The two nucleotide chains of DNA are held together by hydrogen bonds between the bases and by other forces. The hydrogen bonds always connect adenine to thymine and cytosine to guanine, as shown in Figure 2.21b. This linking of specific bases is called **complementary base pairing.** It is determined by the sizes and shapes of the bases. The same kind of complementary base pairing also occurs when information is transmitted from DNA to RNA at the beginning of protein synthesis (Chapter 7). In that situation, adenine in DNA base pairs with uracil in RNA.

DNA and RNA chains contain hundreds or thousands of nucleotides with bases arranged in a particular sequence. This sequence of nucleotides, like the sequence of letters in words and sentences, contains information that determines what proteins an organism will have. As noted earlier, an organism's structural proteins and enzymes, in turn, determine what the organism is and what it can do. Like changing a letter in a word, changing a nucleotide in a sequence can change the information it carries. The number of different possible sequences of bases is almost infinite, so DNA and RNA can contain a great many different pieces of information.

The functions of DNA and RNA are related to their ability to convey information. DNA is transmitted from one generation to the next. It determines the heritable characteristics of the new individual by supplying the information for the proteins its cells will contain. In contrast, RNA carries information from the DNA to the sites where proteins are manufactured in cells. There it directs and participates in the actual assembly of proteins. The functions of nucleic acids are discussed in more detail in Chapters 7 and 8.

FIGURE 2.22 The five bases found in nucleic acids. DNA contains the purines adenine and guanine and the pyrimidines cytosine and thymine. In RNA, thymine is replaced by the pyrimidine uracil.

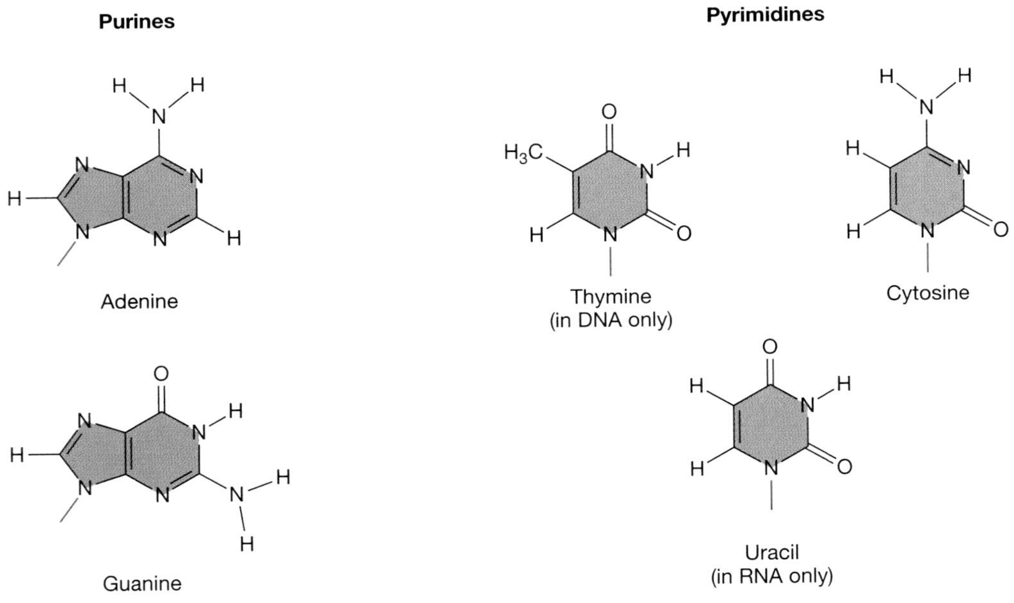

Purines

Adenine

Guanine

Pyrimidines

Thymine
(in DNA only)

Cytosine

Uracil
(in RNA only)

BACTERIAL ACIDS ARE EATING *THE LAST SUPPER*

Past civilizations built their temples, tombs, and monuments out of stone to last forever. Indeed, structures such as the Egyptian pyramids have lasted for thousands of years. But in the last several decades, the ancient stone has begun to crumble into dust. Antiquities experts first believed that the destruction of stone was due to the gaseous pollutants spewed into the air from automobiles and industrial chimneys. Now, however, experts have found that a whole new world—a microsystem of microbes—is living in the stone itself, interacting with the pollutant gases and wreaking chemical havoc.

The prime villain devouring ancient marble is *Thiobacillus thioparis*, a bacterium that converts the pollutant sulfur dioxide (SO_2) gas into sulfuric acid (H_2SO_4). The sulfuric acid acts on the calcium carbonate ($CaCO_3$) in marble, forming carbon dioxide (CO_2) and the salt calcium sulfate ($CaSO_4$), which is a form of plaster. The bacteria use the CO_2 as their source of carbon. The plaster produced by this process is soft and washes away by rain or just crumbles and falls off, taking with it the facial features of statues and the decorative carvings on buildings. On many buildings, such as the Parthenon in Greece, this bacterial "epidemic" has already eroded a 2-in.-thick layer of marble surface into plaster (Figure 2.23). In fact, more marble has been destroyed in the last 35 years than had been destroyed in the previous 300 years.

Not all the damage, however, is due to chemical processes. Some is due to the physical activities of fungi pushing their threadlike growth (hyphae) into the rock, splitting and reducing it to powder. Dozens of kinds of bacteria, yeast, filamentous fungi, and even algae mar the pure surfaces of cathedrals and other buildings both physically and chemically, producing acids that eventually eat into the stone and leave telltale pits and other scars. Even paintings aren't immune to the effects of microbial growth. Frescoes painted on walls, such as Leonardo da Vinci's *Last Supper,* are also being consumed, pieces flaking off and brilliant colors being dulled (Figure 2.24).

Is there a cure for these microbial invasions? At the request of curators at the Museum of Athens, the Italian microbiologist Sergio Curri and his colleagues have devised a treatment plan for the ailing Parthenon treasures. Curri identifies which microorganisms are attacking the stone and then determines which antibiotics will kill them most effectively. Administering antibiotics to a building is a tricky business. A spray gun must be used instead of a hypodermic needle or pills. But some antibiotics cannot be applied by spraying, so it is difficult to find a suitable one. Sometimes disinfectants such as isothiazolinone chloride are used, but such chemicals must be used carefully. In Angkor, Cambodia, for instance, unskilled Army workers armed with hard brushes and fungicidal solutions have scrubbed away ancient temple paintings in their zeal to remove fungal growth.

Unfortunately, antibiotic treatment does not reverse damage already done. Statues cannot grow a new skin, as you might do after recovering from an infection. Therefore, researchers at the Museum of Athens are working on a process to harden the plaster layer formed by microbes by baking the affected pieces at high temperatures.

FIGURE 2.23 Greek authorities are trying to prevent further deterioration of these statues on the Acropolis. Note the loss of facial features and other details due to microbial action and air pollutants.

This method, however, is hardly feasible for an entire statue or a cathedral. And how do we protect these vulnerable structures as long as the pollutants remain in the atmosphere? Stopgap measures have included taking some statues inside and building protective domes over others. The real—but elusive—answer lies in cleaning up our environment.

FIGURE 2.24 A close-up of Leonardo da Vinci's *Last Supper* shows deterioration due to microbial action.

DOES SALT AFFECT SOIL MICROBES?

Consider a situation similar to that described in the Essay. Careless or excessive release of chemicals occurs when winter salt runoff from icy roads affects microbial populations in soil. You might want to investigate this phenomenon in the laboratory by using local soils and various anti-icing products. What effects occur? What concentrations of these products are necessary? Do different products have different effects? Are all soil types equally affected? Are all organisms affected in the same way? How far from the edge of the road, median strip, or sidewalk are the effects found? Do meltwaters in ditches carry these chemicals to distant areas where their effects are also felt? How long do any effects last?

This is a chance for you to practice using the scientific method to design experiments. Even if you do not have time or facilities to carry out the experiments, you can learn much from planning them. Interestingly, when soil is too alkaline, powdered elemental sulfur is sometimes sprinkled into the soil. Sulphur bacteria, such as *Thiobacillus*, will then colonize the sulfur granules and release sulfuric acid, which adjusts the soil pH back toward normal.

CHAPTER SUMMARY

WHY STUDY CHEMISTRY?

- A knowledge of basic chemistry is needed to understand how microorganisms function and how they affect humans and our environment.

CHEMICAL BUILDING BLOCKS AND CHEMICAL BONDS

Chemical Building Blocks

- The smallest chemical unit of matter is an **atom.** An **element** is a fundamental kind of matter, and the smallest unit of an element is an atom. An element is composed of only one type of atom. A **molecule** consists of two or more atoms chemically combined, and a **compound** consists of two or more different kinds of atoms chemically combined.

- The most common elements in all forms of life are carbon (C), hydrogen (H), oxygen (O), and nitrogen (N).

Structure of Atoms

- Atoms consist of positively charged **protons** and neutral **neutrons** in the atomic nucleus and very small, negatively charged **electrons** orbiting the nucleus.

- The number of protons in an atom is equal to its **atomic number.** The total number of protons and neutrons determines the element's **atomic weight.**

- **Ions** are atoms that have gained or lost one or more electrons.

- **Isotopes** are atoms of the same element that contain different numbers of neutrons; some may be **radioisotopes.**

Chemical Bonds

- Atoms of molecules are held together by **chemical bonds.**

- **Ionic bonds** involve attraction of oppositely charged ions. In **covalent bonds,** atoms share pairs of electrons. **Hydrogen bonds** are weak attractions between polar regions of hydrogen atoms and oxygen or nitrogen atoms.

Chemical Reactions

- Chemical reactions involve breaking or forming chemical bonds and associated energy changes.

- **Catabolism,** the breaking down of molecules, releases energy. **Anabolism,** the synthesis of larger molecules, requires energy.

WATER AND SOLUTIONS

Water

- Water is a **polar compound,** acts as a solvent, and forms thin layers because it has high **surface tension.**

- Water also has high specific heat, and it serves as a medium for and participates in many chemical reactions.

Solutions and Colloids

- **Solutions** consist of **mixtures** with one or more **solutes** evenly distributed throughout a **solvent.**

- **Colloids** contain particles too large to form true solutions.

Acids, Bases, and pH

- In most solutions containing acids or bases, **acids** release H^+ ions, and **bases** accept H^+ ions (or release OH^- ions).

- The **pH** of a solution is a measure of its acidity or alkalinity. A pH of 7 is neutral, below 7 is acidic, and above 7 is basic, or **alkaline.**

COMPLEX ORGANIC MOLECULES

- **Organic chemistry** is the study of carbon-containing compounds.
- Organic compounds such as alcohols, aldehydes, ketones, organic acids, and amino acids can be identified by their **functional groups.**

Carbohydrates

- **Carbohydrates** consist of carbon chains in which most of the carbon atoms have an associated alcohol group and one carbon has either an aldehyde or a ketone group.
- The simplest carbohydrates are **monosaccharides,** which can combine to form **disaccharides** and **polysaccharides.** Long chains of repeating units are called **polymers.**
- The body uses carbohydrates primarily for energy.

Lipids

- All **lipids** are insoluble in water but soluble in nonpolar solvents.
- **Fats** consist of glycerol and **fatty acids.**

- **Phospholipids** contain a phosphate group in place of a fatty acid.
- **Steroids** have a complex four-ring structure.

Proteins

- **Proteins** consist of chains of **amino acids** linked by **peptide bonds.**
- Proteins form part of the structure of cells, act as enzymes, and contribute to other functions such as motility, transport, and regulation.
- **Enzymes** are biological catalysts of great **specificity** that increase the rate of chemical reactions in living organisms. Each enzyme has an **active site** to which its **substrate** binds.

Nucleotides and Nucleic Acids

- A **nucleotide** consists of a nitrogenous base, a sugar, and one or more phosphates.
- Some nucleotides contain **high-energy bonds.**
- **Nucleic acids** are important information-containing molecules that consist of chains of nucleotides. The nucleic acids that occur in living organisms are **ribonucleic acid (RNA)** and **deoxyribonucleic acid (DNA).**

KEY TERMS

acid (p. 33)
active site (p. 42)
alkaline (p. 33)
amino acid (p. 40)
anabolism (p. 31)
anion (p. 28)
atom (p. 26)
atomic number (p. 28)
atomic weight (p. 28)
base (p. 33)
biochemistry (p. 33)
carbohydrate (p. 35)
catabolism (p. 31)
cation (p. 28)
chemical bond (p. 30)
colloid (p. 33)
complementary base pairing (p. 44)
compound (p. 27)
covalent bond (p. 30)
dehydration synthesis (p. 33)
denaturation (p. 42)

deoxyribonucleic acid (DNA) (p. 43)
disaccharide (p. 36)
electron (p. 27)
element (p. 26)
endergonic (p. 32)
enzyme (p. 42)
exergonic (p. 31)
fat (p. 37)
fatty acid (p. 37)
functional group (p. 35)
glycosidic bond (p. 36)
gram molecular weight (p. 29)
high-energy bond (p. 42)
hydrogen bond (p. 30)
hydrolysis (p. 33)
ion (p. 28)
ionic bond (p. 30)
isomer (p. 35)
isotope (p. 29)
lipid (p. 37)
mixture (p. 33)

mole (p. 29)
molecule (p. 27)
monosaccharide (p. 35)
neutral (p. 33)
neutron (p. 27)
nucleic acid (p. 43)
nucleotide (p. 42)
organic chemistry (p. 33)
peptide bond (p. 40)
pH (p. 33)
phospholipid (p. 38)
polar compound (p. 31)
polymer (p. 36)
polynucleotide (p. 43)
polypeptide (p. 40)
polysaccharide (p. 36)
primary structure (p. 40)
protein (p. 40)
proton (p. 27)
purine (p. 44)
pyrimidine (p. 44)
quaternary structure (p. 40)
radioisotope (p. 30)

reactant (p. 33)
R group (p. 40)
ribonucleic acid (RNA) (p. 43)
rule of octets (p. 28)
saturated fatty acid (p. 38)
secondary structure (p. 40)
solute (p. 33)
solution (p. 33)
solvent (p. 33)
specificity (p. 42)
steroid (p. 38)
structural protein (p. 42)
substrate (p. 42)
surface tension (p. 32)
tertiary structure (p. 40)
triacylglycerol (p. 38)
unsaturated fatty acid (p. 38)

QUESTIONS FOR REVIEW

A Basic Chemistry

1. How are each of the following topics in chemistry important in the study of microbiology: pH, radioisotopes, surface tension, hydrolysis, denaturation, nucleic acids?

B Organization of Matter; Elements

2. Define *atom* (A), *element* (E), *molecule* (M), and *compound* (C). Place the appropriate letter(s) next to each of the following:

   ____ H_2O ____ sulfur   ____ glucose

   ____ O_2 ____ CH_4 ____ H_2

   ____ salt   ____ sodium   ____ chlorine

3. What are the most common elements in living things, and what are their symbols?

4. Show how protons, electrons, and neutrons are arranged in atoms.

5. The (proton/electron/neutron) is the subatomic particle that (a) weighs the least; (b) has no charge.

6. How do ions differ from atoms?

7. Indicate what charge (1+, 2+, 1−, and so on) the ion formed by each of these elements or compounds will have:

   ____ Ca   ____ H   ____ HPO_4 ____ K

   ____ Cl   ____ Na   ____ OH   ____ NO_3

8. How do isotopes of an element differ from each other?

9. Which of the following "mystery" atoms are isotopes of each other? How do you know this? Why aren't the others also isotopes of these?

	(a)	(b)	(c)	(d)	(e)
Number of protons	6	6	6	8	7
Number of neutrons	6	8	8	8	6
Number of electrons	6	7	6	8	7

C Chemical Bonds and Chemical Reactions

10. Use diagrams to illustrate the differences among ionic, covalent, and hydrogen bonds.

11. Distinguish among metabolism, catabolism, and anabolism and between exergonic and endergonic reactions.

D Water, Solutions, Colloidal Dispersions, Acids, and Bases

12. What properties of water make it important to living things?

13. What are the characteristics of a solution?

14. Distinguish between an acid and a base.

15. (a) How is the pH scale used to measure acidity? Among the pH values 3, 5, 7, 11, and 13, which represents (b) the strongest acid and (c) the strongest base? (d) An organism living in the human mouth would probably be best grown at what pH in the laboratory?

E Organic Chemistry; Functional Groups

16. What is organic chemistry? Which of the following is not an organic compound? Why?

    a. CH_4 b. $C_6H_{12}O_6$ c. NH_3 d. ATP   e. RNA

17. For each of the organic compounds shown, draw a circle around the portion of the structure that constitutes a functional group. Name each group, and list the type(s) of organic compounds that contain that functional group.

F Carbohydrates

18. What is the basic structure of a monosaccharide?

19. How are disaccharides and polysaccharides different from monosaccharides?

20. In what ways are carbohydrates used in living organisms?

G Lipids

21. Describe the structure and uses in the body of fats, phospholipids, and steroids.

22. Which of the following compounds are saturated? Which are unsaturated? How can you tell?

Proteins

23. Define amino acid, peptide bond, and denaturation.

24. What is the chemical nature of this compound? Identify each of the circled parts of the molecule.

a. ____
b. ____
c. ____
d. ____

25. Describe the four levels of protein structure. How is each maintained?

26. In what ways do structural proteins and enzymes differ?

27. List the main properties of enzymes.

I **Nucleotides**

28. Match the following:

____ adenine   **a.** present only in DNA

____ thymine   **b.** present only in RNA

____ phosphate   **c.** present in both DNA and RNA

____ ribose   **d.** present in DNA, RNA, and ATP

____ nucleotide   **e.** present in RNA and ATP

____ deoxyribose

____ uracil

____ guanine

____ nitrogenous base

29. Indicate which base is (or bases are) the complementary paired base(s) for each of the following:

adenine ____

cytosine ____

guanine ____

uracil ____

thymine ____

PROBLEMS FOR INVESTIGATION

1. Virtually all organisms use the nucleotide ATP as a vehicle for chemical energy. Why? What properties of the molecule enable it to store chemical energy? What do you think this universality tells us about the evolutionary history of life?

2. If a protein were able to control its own replication without using DNA or RNA, should we then call it a "genetic material"? Those of you who feel adventurous can look ahead to the discussion of prions in Chapter 11 to consider how these proteinaceous infectious particles may replicate.

3. Lipids play an important part in the cell wall structure of bacteria such as *Mycobacterium tuberculosis* and *Mycobacterium leprae*, the causes of tuberculosis and leprosy, respectively. Read about the waxy materials tuberculin and lepromin. What clinical use is made of them?

4. How can weak interactions such as hydrogen bonds hold together large molecules such as proteins and DNA? Find out more about the various forces involved in creating and maintaining the structure of proteins and nucleic acids.

SOME INTERESTING READING

Alberts, B., et al. 1995. *Molecular biology of the cell*, 3d. ed. New York: Garland Publishing.

Brown, T. L., H. E. LeMay, and B. E. Bursten. 1994. *Chemistry: The central science*. Englewood Cliffs, NJ: Prentice Hall.

Gibbons, A., and M. Hoffman. 1991. "New 3-D protein structures revealed." *Science* 253 (5018) (July 26):382–83. (The shape of cholera toxin and the first protein kinase structure.)

Inoue, M., and M. Koyano. 1991. "Fungal contamination of oil paintings in Japan." *International Biodeterioration & Biodegradation.* 28(1–4):23–35.

McMurry, J., and M. E. Castellion. 1992. *Fundamentals of general, organic, and biological chemistry*. Englewood Cliffs, NJ: Prentice Hall.

"The molecules of life." 1985. *Scientific American* 253, no. 4 (October). (Special issue, various authors.)

Sharon, N, and H. Lis. 1993. "Carbohydrates in cell recognition." *Scientific American* 268 (January):82–89.

Stryer, L. 1995. *Biochemistry,* 4th ed. New York: W. H. Freeman and Co.

Watson, J. D. 1980. *The double helix: A personal account of the discovery of the structure of DNA*. Ed. Gunther S. Stent. New York: W. W. Norton.

Wiggins, P. M. 1990. "Role of water in some biological processes." *Microbiological Reviews* 54(4):432–49.

Zubay, G. L. 1993. *Biochemistry,* 3d. ed. Dubuque, IA: William C. Brown.

CHAPTER 3

MICROSCOPY AND STAINING

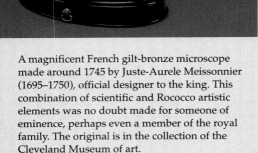

A magnificent French gilt-bronze microscope made around 1745 by Juste-Aurele Meissonnier (1695–1750), official designer to the king. This combination of scientific and Rococco artistic elements was no doubt made for someone of eminence, perhaps even a member of the royal family. The original is in the collection of the Cleveland Museum of art.

For thousands of years, people were aware of the effects of microorganisms—the rotting of food, fermentations by yeasts, infectious diseases, and the like—without being aware of the existence of the microbes themselves. Until microscopes were invented, no one could attribute these effects to microorganisms.

DEVELOPMENT OF MICROSCOPY

Anton van Leeuwenhoek (1632–1723), living in Delft, Holland, was almost certainly the first person to see individual microorganisms. He constructed simple microscopes capable of magnifying objects 100 to 300 times. These instruments were unlike what we commonly think of as microscopes today. Consisting of a single tiny lens, painstakingly ground, they were actually very powerful magnifying glasses (Figure 3.1). It was so difficult to focus one of Leeuwenhoek's micro-scopes that instead of changing specimens, he built a new microscope for each specimen, leaving the previous specimen and microscope together. When foreign investigators came to Leeuwenhoek's laboratory to look through his microscopes, he made them keep their hands behind their backs to prevent them from touching the focusing apparatus!

In a letter to the Royal Society of London in 1676, Leeuwenhoek described his first observations of bacteria and protozoa in water. He kept his techniques secret, however. Even today we are not sure of his methods of illumination, although it is likely that he used indirect lighting, with light bouncing off the sides of specimens rather than passing through them. Leeuwenhoek was also unwilling to part with any of the 419 microscopes he made. It was only near the time of his death that his daughter, at his direction, sent 100 of them to the Royal Society.

Following Leeuwenhoek's death in 1723, no one came forward to continue the work of perfecting the

THIS CHAPTER FOCUSES ON THE FOLLOWING QUESTIONS:

A How is the evolution of microscopy instruments related to progress in microbiology?

B Which metric units are most useful for the measurement of microbes?

C What are the relationships among wavelength, resolution, and numerical aperture?

D How are the following properties of light related to microbiology: transmission, absorption, fluorescence, luminescence, phosphorescence, reflection, refraction, and diffraction?

E What is the function of each part of a compound microscope, and what is total magnification?

F What are the special uses and adaptations of bright-field, dark-field, phase-contrast, differential interference contrast, and fluorescence (UV) microscopes?

G What are the principles of transmission and scanning electron microscopy? How do the advantages and limitations of electron microscopy compare with those of light microscopy?

H Which techniques are used to prepare and heighten contrast in specimens to be viewed with a light microscope?

I What are the uses of the common types of microbial stains?

J What are the functions and results of each of the steps in the Gram staining procedure?

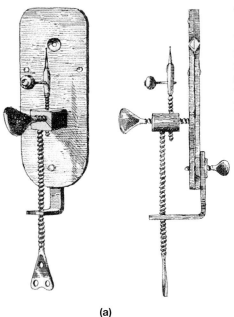

(a)

FIGURE 3.1 (a) Drawing (slightly smaller than life size) of one of Leeuwenhoek's microscopes. This simple microscope—really a very powerful magnifying glass—made use of a single, tiny, almost spherical lens set into the metal plate. The specimen was mounted on the needlelike end of the vertical shaft and examined through the lens from the opposite side. The various screws were used to position the specimen and bring it into focus—a very difficult process. (b) The vinegar eels (nematodes) that so upset Leeuwenhoek's friends.

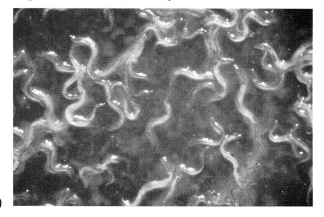

(b)

CLOSE-UP

THE SECRET KINGDOM UNCOVERED

"I have had several gentlewomen in my house, who were keen on seeing the little eels in vinegar; but some of 'em were so disgusted at the spectacle, that they vowed they'd never use vinegar again. But what if one should tell such people in [the] future that there are more animals living in the scum on the teeth in a man's mouth, than there are men in a whole kingdom?"

—Anton van Leeuwenhoek, 1683

design and construction of microscopes, and the progress of microbiology slowed. Still, he had taken the first steps. Through Leeuwenhoek's letters to the Royal Society in the mid-1670s, the existence of microbes was revealed to the scientific community. And in 1683 he described bacteria taken from his own mouth. However, Leeuwenhoek could see very little detail of their structure. Further study required the development of more complex microscopes, as we shall soon see.

PRINCIPLES OF MICROSCOPY

Metric Units

Microscopy is the technology of making very small things visible to the human eye. Because microorganisms are so small, the units used to measure them are likely to be unfamiliar to beginning students used to dealing with a macroscopic world (Table 3.1).

The **micrometer** (μm), formerly called a micron (μ), is equal to 0.000001 m. A micrometer can also be expressed as 10^{-6} m. The second unit, the **nanometer** (nm), formerly called a millimicron (mμ), is equal to 0.000000001 m. It also is expressed as 10^{-9} m. A third unit, the **angstrom** (Å), is found in much of the current

and older literature but no longer has any official recognition. It is equivalent to 0.0000000001 m, 0.1 nm, or 10^{-10} m. Figure 3.2 shows a scale that summarizes the metric system unit equivalents, the ranges of sizes that can be detected by the unaided human eye and by various types of microscopy, and examples of where various organisms fall on this scale.

Properties of Light: Wavelength and Resolution

Light has a number of properties that affect our ability to visualize objects, both with the unaided eye and (more crucially) with the microscope. Understanding these properties will allow you to improve your practice of microscopy.

One of the most important properties of light is its **wavelength,** or the length of a light ray (Figure 3.3). Wavelength is represented by the Greek letter lambda (λ). The sun produces a continuous *spectrum* of electromagnetic radiation with waves of various lengths (Figure 3.4, p. 54). Visible light rays as well as ultraviolet and infrared rays constitute particular parts of this spectrum. The wavelength used for observation is crucially related to the resolution that can be obtained. **Resolution** refers to the ability to see two items as separate and discrete units (Figure 3.5a, p. 54) rather than as a fuzzy, overlapped single image (Figure 3.5b). We can magnify objects, but if the objects cannot be resolved, the magnification is useless. Light must pass between two objects for them to be seen as separate things. If the wavelength of the light by which we see the objects is too long to pass between them, they will appear as one. The key to resolution is to get light of a short-enough wavelength to fit between the objects we want to see separately.

To visualize this phenomenon, imagine a target with a foot-high letter *E* hanging in front of a white background. Suppose that you throw at the target ink-covered objects with diameters corresponding to various wavelengths (Figure 3.6, p. 54). If one object has a diameter smaller than the distance between the "arms" of the letter *E*, the object will pass between the arms, and the arms will be distinguishable as separate

TABLE 3.1	Some Commonly Used Units of Length		
Unit (abbreviation)	**Prefix**	**Metric Equivalent**	**English Equivalent**
meter (m)			3.28 ft
centimeter (cm)	*centi* = one hundredth	0.01 m = 10^{-2} m	0.39 in.
millimeter (mm)	*milli* = one thousandth	0.001 m = 10^{-3} m	0.039 in.
micrometer (μm)	*micro* = one millionth	0.000001 m = 10^{-6} m	0.000039 in.
nanometer (nm)	*nano* = one billionth	0.000000001 m = 10^{-9} m	0.000000039 in.
angstrom (Å)		0.0000000001 m = 10^{-10} m	0.0000000039 in.

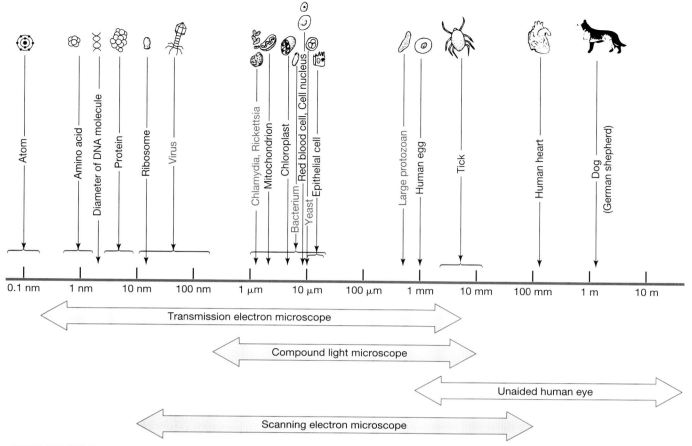

FIGURE 3.2 Relative sizes of microbes. Various organisms and their relation to metric units of measurement are shown; red terms indicate organisms studied in microbiology. Chlamydia and Rickettsia are groups of bacteria that are much smaller in size than the usual bacteria. The range of effective use for various instruments is also depicted.

structures. First, imagine tossing basketballs. Because they cannot fit between the arms, light rays of that size would give poor resolution. Next toss tennis balls at the target. The resolution will improve. Then try jelly beans and, finally, tiny beads. With each decrease in the diameter of the object thrown, the number of such objects that can pass between the arms of the *E* in-

FIGURE 3.3 The distance between two adjacent crests or two adjacent troughs of any wave is defined as one wavelength, designated by the Greek letter lambda (λ).

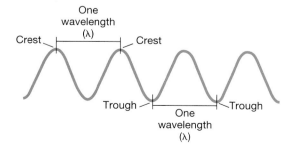

creases. Resolution improves, and the shape of the letter is revealed with greater and greater precision.

Microscopists use shorter and shorter wavelengths of electromagnetic radiation to improve resolution. Visible light, which has an average wavelength of 550 nm, cannot resolve distances less than 220 nm. Ultraviolet light, which has a wavelength of 100 to 400 nm, can resolve distances as small as 110 nm. Thus, microscopes that used ultraviolet light instead of visible light allowed researchers to find out more about the details of cellular structures. But the invention of the electron microscope, which uses electrons rather than light, was the major step in increasing the ability to resolve objects. Electrons behave both as particles and as waves. Their wavelength is about 0.005 nm, which allows resolution of distances as small as 0.2 nm.

The **resolving power** (RP) of a lens is a numerical measure of the resolution of that lens. The smaller the distance between objects that can be distinguished, the greater the resolving power of the lens. We can calculate the RP of a lens if we know its **numerical aperture**

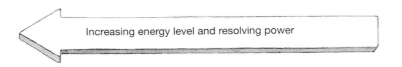

Increasing energy level and resolving power

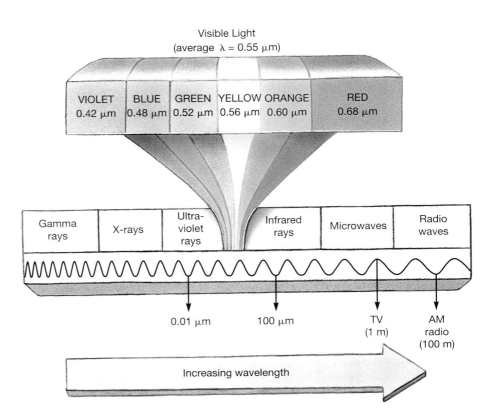

Visible Light
(average λ = 0.55 μm)

VIOLET	BLUE	GREEN	YELLOW	ORANGE	RED
0.42 μm	0.48 μm	0.52 μm	0.56 μm	0.60 μm	0.68 μm

Gamma rays	X-rays	Ultra-violet rays	Infrared rays	Microwaves	Radio waves

0.01 μm   100 μm   TV (1 m)   AM radio (100 m)

Increasing wavelength

FIGURE 3.4 The electromagnetic spectrum. Only a narrow range of wavelengths—those of visible and ultraviolet light—are used in light microscopy. The shorter the wavelength used, the greater the resolution that can be attained.

(NA), a mathematical expression relating to the extent that light is concentrated by the condenser lens and collected by the objective. The formula for calculating resolving power is RP = λ/2NA. As this formula indicates, the smaller the value of λ and the larger the value of NA, the greater the resolving power of the lens.

The NA values of lenses differ in accordance with the power of magnification and other properties. The NA is engraved on the side of each objective lens (the lens nearest the stage) of a light microscope. Look at the NA values on the microscope you use the next time you are in the laboratory. Typical values for the objective lenses commonly found on modern light microscopes are 0.25 for low power, 0.65 for high power, and 1.25 for the oil immersion lens.

FIGURE 3.5 Resolution. (a) The two dots are resolved—that is, they can clearly be seen as separate structures. (b) These two dots are not resolved—they appear to be fused.

(a)   (b)

FIGURE 3.6 An analogy for the effect of wavelength on resolution. Smaller objects (corresponding to shorter wavelengths) can pass more easily between the arms of the letter E, defining it more clearly and producing a sharper image.

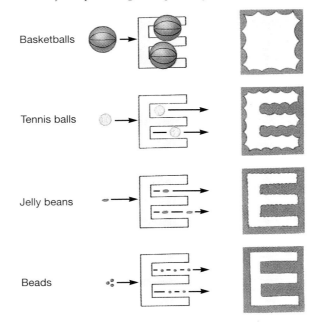

Basketballs

Tennis balls

Jelly beans

Beads

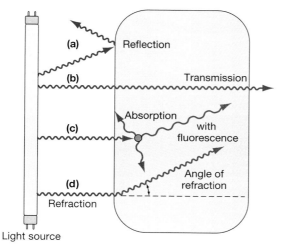

FIGURE 3.7 Various interactions of light with an object that it strikes. (a) Light may be reflected back from the object. The particular wavelengths reflected back to the eye determine the perceived color of the object. (b) Light may be transmitted directly through the object. (c) Light may be absorbed, or taken up, by the object. In some cases, the absorbed light rays are reemitted as longer wavelengths, a phenomenon known as fluorescence. (d) Light passing through the object may be refracted, or bent, by it.

Properties of Light: Light and Objects

Various things can happen to light as it travels through a medium such as air or water and strikes an object (Figure 3.7). We look at some of those things now and consider how they can affect your ability to see through a microscope.

Reflection

If the light strikes an object and bounces back (giving the object color), we say that **reflection** has occurred. For example, light rays in the green range of the spectrum are reflected off the surfaces of the leaves of plants. Those reflected rays are responsible for our seeing the leaves as green.

Transmission

Transmission refers to the passage of light through an object. You cannot see through a rock because light cannot pass through it. In order for you to see objects through a microscope, light must either be reflected from the objects or transmitted through them. Most of your observations of microorganisms will make use of transmitted light.

Absorption

If light rays neither pass through nor bounce off an object but are taken up by the object, **absorption** has occurred. Energy in absorbed light rays can be used in various ways. For example, all wavelengths of the sun's light rays except those in the green range are absorbed by a leaf. Some of the energy in these other light rays is captured in photosynthesis and used by the plant to make food. Energy from absorbed light can also raise the temperature of an object. A black object, which reflects no light, will gain heat much faster than a white object, which reflects all light rays.

In some cases, absorbed light rays, especially ultraviolet light rays, are changed into longer wavelengths and reemitted. This phenomenon is known as **luminescence.** If luminescence occurs only during irradiation (when light rays are striking an object), the object is said to **fluoresce.** Many fluorescent dyes are important in microbiology, especially in the field of immunology, because they help us visualize immune reactions and internal processes in microorganisms. If an object continues to emit light when light rays no longer strike it, the object is **phosphorescent.** Some bacteria that live deep in the ocean are phosphorescent.

Refraction

Refraction is the bending of light as it passes from one medium to another of different density. The bending of the light rays gives rise to an *angle of refraction,* the degree of bending of a transparent material. You have probably seen how the underwater portion of a pole sticking out of water or a drinking straw in a glass of water seems to bend (Figure 3.8). When you remove the object from the water, it is clearly straight. It looks bent because light rays deviate, or bend, when they pass from the water into the air as their speed changes across the water–air interface. The **index of refraction** of a material is a measure of the speed at which light passes through the material. When two substances have different indices of refraction, light will bend as it passes from one material to the other.

Light passing through a glass microscope slide, through air, and then through a glass lens is refracted each time it goes from one medium to another. To avoid this problem, microscopists use **immersion oil,** which has the same index of refraction as glass, to re-

FIGURE 3.8
Refraction of light rays passing from water into air causes the pencil to appear bent.

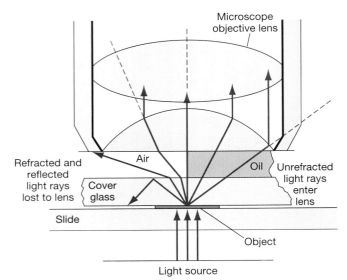

FIGURE 3.9 Use of immersion oil to prevent loss of light due to refraction. The focusing of as much light as possible adds to the clarity of the image. Immersion oil may also be added between the top of the condenser and the bottom of the slide to eliminate another site for refraction.

TRY IT

NOW YOU SEE IT, NOW YOU DON'T

If you want to take some diamonds through customs without declaring them and thus avoid paying duty on them, here's a way you can do it. Obtain an oil with the same refractive index as the diamonds. Pour it into a clear glass bottle labeled "Baby Oil," and drop the diamonds in. The diamonds will be invisible if you make certain that their surfaces are clean. This trick works because when light passes from one medium to another with the same index of refraction, it is not bent. Thus, the boundary between the diamonds and the oil will not be apparent. (However, if a customs agent shakes the bottle, you're out of luck—he or she will hear them rattle, and you will be caught.)

If larceny is not in your heart or the price of diamonds not in your pocket, you can try this entertaining, but legal, activity. Clean a glass rod and dip it in and out of a bottle of immersion oil. Watch it disappear and reappear. Try this little experiment to help you understand what is happening when you use the oil immersion lens.

What has happened to the bottom of the glass dropper on the left? It disappears in immersion oil because glass and immersion oil have the same index of refraction. Therefore, light passing between them is not bent, so we cannot observe the surface of the rod. The bottle on the right contains a liquid with a different index of refraction.

place the air. The slide and the lens are joined by a layer of oil; there is no refraction to cause the image to blur (Figure 3.9). If you forget to use oil with the oil immersion lens of a microscope, it will be impossible to focus clearly on a specimen.

Diffraction

As light passes through a small opening, such as a hole, slit, or space between two adjacent cellular structures, the light waves are bent around the opening. This phenomenon is **diffraction.** Figure 3.10 shows

FIGURE 3.10 Diffraction of light waves passing (a) around the edge of an object and (b) through a small aperture. (c) Water waves being diffracted as they pass through an opening in a breakwater.

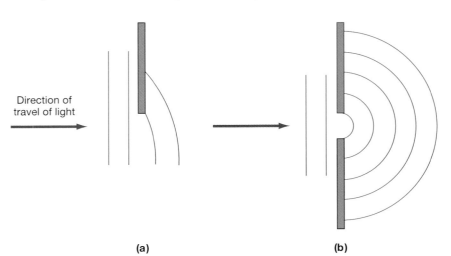

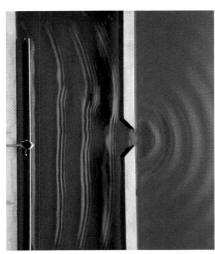

(a)   (b)   (c)

diffraction patterns formed when light passes through a small aperture or around the edge of an object. Similar patterns occur when water passes through an opening in, or around the back of, a breakwater.

Diffraction is a problem for microscopists because the lens acts as a small aperture through which the light must pass. A blurry image results. The higher the magnifying power of a lens, the smaller it must be, and therefore the greater the diffraction and blurring it causes. The oil immersion (100×) lens, with its total magnification capacity of about 1000× (when combined with a 10× ocular lens), represents the limit of useful magnification with the light microscope. The small size of higher-power lenses causes such severe diffraction that resolution is impossible.

LIGHT MICROSCOPY

Light microscopy refers to the use of any kind of microscope that uses visible light to make specimens observable. The modern light microscope is a descendant not of Leeuwenhoek's single lenses but of Hooke's compound microscope—a microscope with more than one lens. ∞(Chapter 1, p. 8) Single lenses produce two problems: They cannot bring the entire field into focus simultaneously, and there are colored rings around objects in the field. Both problems are solved today by the use of multiple correcting lenses placed next to the primary magnifying lens. Used in modern compound microscopes, correcting lenses give us nearly distortion-free images.

Over the years, several kinds of light microscopes have been developed, each adapted for making certain kinds of observations. We look first at the standard light microscope and then at some special kinds of microscopes.

The Compound Light Microscope

The **optical microscope,** or *light microscope,* has undergone various improvements since Leeuwenhoek's time and essentially reached its current form shortly before the turn of the twentieth century. This microscope is a **compound light microscope**—that is, it has more than one lens. The parts of a modern compound microscope and the path light takes through it are shown in Figure 3.11. A compound microscope with a single eyepiece (*ocular*) is said to be **monocular;** one with two eyepieces is said to be **binocular.**

Light enters the microscope from a source in the **base** and often passes through a blue filter, which filters out the long wavelengths of light, leaving the shorter wavelengths and improving resolution. It then goes through a **condenser,** which converges the light beams so that they pass through the specimen. The **iris diaphragm** controls the amount of light that passes through the specimen and into the objective lens. The higher the magnification, the greater the amount of light needed to see the specimen clearly. The **objective lens** magnifies the image before it passes through the **body tube** to the ocular lens in the eyepiece. The **ocular lens** further magnifies the image. A **mechanical**

FIGURE 3.11 Parts of a modern compound light microscope and the path that light takes through it.

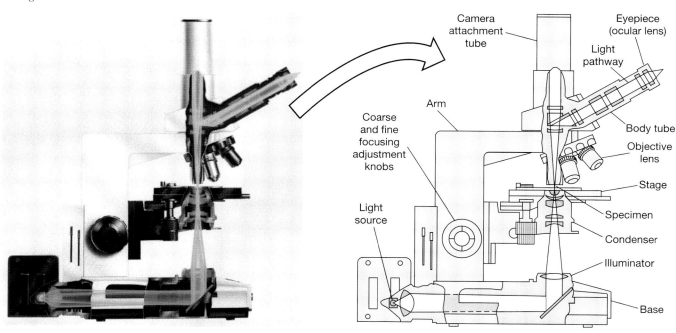

stage allows precise control of moving the slide, which is especially useful in the study of microbes.

The focusing mechanism consists of a **coarse adjustment** knob, which changes the distance between the objective lens and the specimen fairly rapidly, and a **fine adjustment** knob, which changes the distance very slowly. The coarse adjustment knob is used to locate the specimen. The fine adjustment knob is used to bring it into sharp focus.

Compound microscopes have up to six interchangeable objective lenses that have different powers of magnification.

The **total magnification** of a light microscope is calculated by multiplying the magnifying power of the objective lens (the lens used to view your specimen) by the magnifying power of the ocular lens (the lens nearest your eye). Typical values for a $10\times$ ocular are:

- scanning $(3\times) \times (10\times) = 30\times$ magnification
- low power $(10\times) \times (10\times) = 100\times$ magnification
- high "dry" $(40\times) \times (10\times) = 400\times$ magnification
- oil immersion $(100\times) \times (10\times) = 1000\times$ magnification

Most microscopes are designed so that when the microscopist increases or decreases the magnification by changing from one objective lens to another, the specimen will remain very nearly in focus. Such microscopes are said to be **parfocal.** The development of parfocal microscopes greatly improved the efficiency of microscopes and reduced the amount of damage to slides and objective lenses. Most student-grade microscopes are parfocal today.

Some microscopes are equipped with an **ocular micrometer** for measuring objects viewed. This is a glass disk with a scale marked on it that is placed inside the eyepiece between its lenses. This scale must first be calibrated with a stage micrometer, which has metric units engraved on it. When these units are viewed through the microscope at various magnifications, the microscopist can determine the corresponding metric values of the divisions on the ocular micrometer for each objective lens. Thereafter, he or she needs only to count the number of divisions covered by the observed object and multiply by the calibration factor for that lens in order to determine the actual size of the object.

Dark-Field Microscopy

The condenser used in an ordinary light microscope causes light to be concentrated and transmitted directly through the specimen, as shown in Figure 3.12a. This gives **bright-field illumination** (Figure 13.13a). In some cases, however, it is more useful, especially with light-sensitive organisms, to examine specimens that would lack contrast with their background in a bright field under other illumination. Live spirochetes (spi'ro-

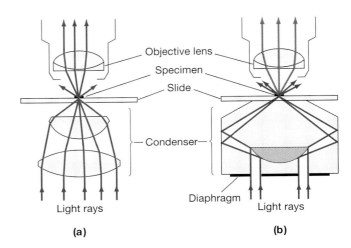

FIGURE 3.12 Comparison of the illumination in (a) bright-field and (b) dark-field microscopy. The condenser of the bright-field microscope concentrates and transmits light directly through the specimen, whereas the dark-field condenser deflects light rays so that they reflect off the specimen at an angle before they are collected and focused into an image.

kets), spiral-shaped bacteria that cause syphilis and other diseases, are just such organisms. In this situation **dark-field illumination** is used. A microscope adapted for dark-field illumination has a condenser that prevents light from being transmitted through the specimen but instead causes the light to reflect off the specimen at an angle (Figure 3.12b). When these rays are gathered and focused into an image, a light object is seen on a dark background (Figure 3.13b).

Phase-Contrast Microscopy

Most living microorganisms are difficult to examine because they cannot be stained by coloring them with dyes—stains usually kill the organisms. To observe them alive and unstained requires the use of **phase-contrast microscopy.** A phase-contrast microscope has a special condenser and objective lenses that accentuate small differences in the refractive index of various structures within the organism. Light passing through objects of different refractive indices is slowed down and diffracted. The changes in the speed of light are seen as different degrees of brightness (Figure 3.14).

Fluorescence Microscopy

In **fluorescence microscopy** ultraviolet light is used to excite molecules so that they release light of a longer wavelength than that originally striking them (see Figure 3.7c). The different wavelengths produced are often seen as brilliant shades of orange, yellow, or yellow green. Some organisms, such as *Pseudomonas,*

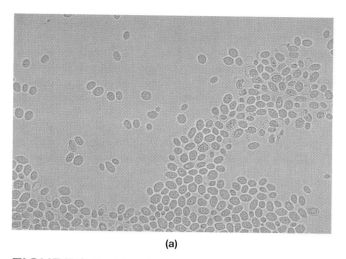

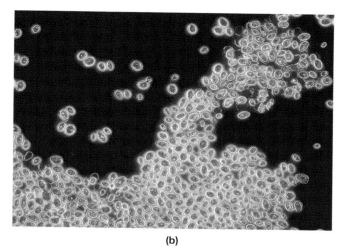

(a)

(b)

FIGURE 3.13 (a) Bright-field and (b) dark-field microscope views of *Saccharomyces cerevisiae* (brewer's yeast; magnified 975×). Dark-field illumination provides an enormous increase in contrast.

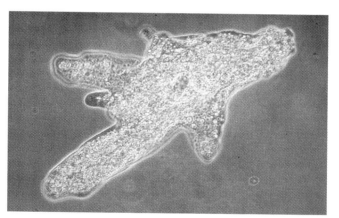

FIGURE 3.14 Phase contrast view of *Amoeba*, a protozoan.

fluoresce naturally when irradiated with ultraviolet light. Other organisms, such as *Mycobacterium tuberculosis* and *Treponema pallidum* (the cause of syphilis), must be treated with a fluorescent dye called a *fluorochrome*. They then stand out sharply against a dark background (Figure 3.15). Acridine orange is a fluorochrome that binds to nucleic acids. In live cells, it colors cell nuclei bright green and cytoplasmic RNA orange. It is sometimes used to screen samples for microbial growth.

Fluorescent antibody staining is now widely used in diagnostic procedures to determine whether an *antigen* (a foreign substance such as a microbe) is present. *Antibodies*—molecules produced by the body as an immune response to an invading antigen—are found in many clinical specimens such as blood and serum. If a patient's specimen contains a particular antigen, that antigen and the antibodies specifically made against it will clump together. However, this reaction is ordinarily not visible. Therefore, fluorescent

dye molecules are attached to the antibody molecules. If the dye molecules are retained by the specimen, the antigen is presumed to be present, and a positive diagnosis can be made. Thus, if fluorescent dye–tagged antibodies against syphilis organisms are added to a specimen containing spirochetes and are seen to bind to the tagged organisms, those organisms can be identified as the cause of syphilis. This technique is especially important in *immunology*, in which the reactions of antigens and antibodies are studied in great detail (see Chapters 18 and 19, especially Figure 19.30 on the technique of fluorescent antibody staining). Diagnoses can often be made in minutes rather than the hours or days it would take to isolate, culture, and identify organisms.

FIGURE 3.15 Bacterial flagella, demonstrated by the fluorescent antibody technique. The fluorescent dye-tagged antibodies clearly show the bacterial cells and their wavy flagella.

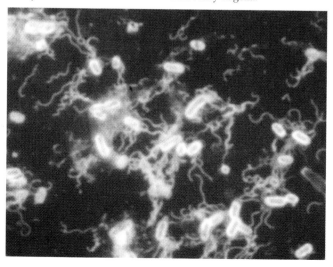

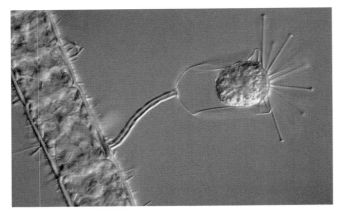

FIGURE 3.16 Nomarski interference microscope image of the protozoan *Paracineta,* attached by a long stalk to the green alga *Spongomorpha* (magnified 400×).

Differential Interference Contrast (Nomarski) Microscopy

Nomarski microscopy, like phase-contrast microscopy, makes use of differences in refractive index to visualize cells and structures. However, the microscope used, a *differential interference contrast microscope,* produces much higher resolution than the standard phase-contrast microscope. It has a very short *depth of field* (the thickness of specimen that is in focus at any one time) and can produce a nearly three-dimensional image (Figure 3.16).

Figure 3.17 illustrates the images produced by four different microscopic techniques.

ELECTRON MICROSCOPY

The light microscope opened doors to the world of microbes. However, the view was limited to observations at the level of whole cells and their arrangements. Few subcellular structures could be seen; neither could viruses. The advent of the **electron microscope (EM)** allowed these small structures to be visualized and studied. The EM was developed in 1932 and was in use in many laboratories by the early 1940s.

The EM uses a beam of electrons instead of a beam of light, and electromagnets rather than glass lenses are used to focus the beam (Figure 3.18). The electrons must travel through a vacuum because collisions with air molecules would scatter the electrons and result in a distorted image. Electron microscopes are much more

FIGURE 3.17 Images of the same organism (*Paramecium;* 600×) produced by means of four different techniques: (a) bright-field microscopy, (b) dark-field microscopy, (c) phase-contrast microscopy, (d) Nomarski microscopy. One microscope may have the optics for all four techniques.

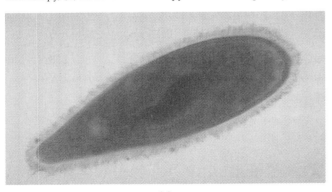

(a)

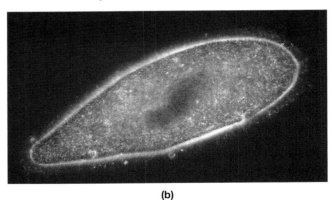

(b)

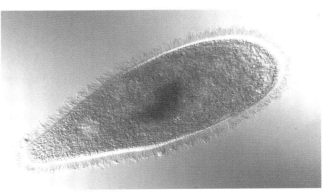

(c)

(d)

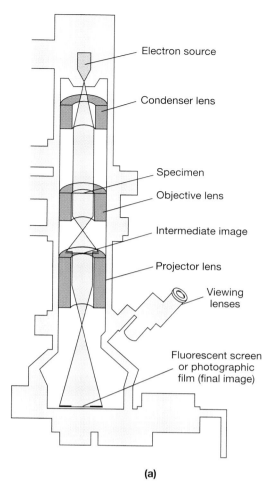

Electron source

Condenser lens

Specimen

Objective lens

Intermediate image

Projector lens

Viewing lenses

Fluorescent screen or photographic film (final image)

(a)

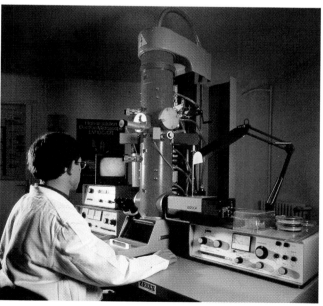

(b)

FIGURE 3.18 (a) Cross-sectional diagram of an electron microscope, showing the pathways of the electron beam as it is focused by electromagnetic lenses. (b) A modern scanning electron microscope in use.

expensive than light microscopes. They also take up much more space and require additional rooms for preparation of specimens and for processing of photographs. Photographs taken on any microscope are called *micrographs;* those taken on an electron microscope are called **electron micrographs.** Nothing else can show the great detail of minute biological structures that EMs can (Figure 3.19).

The two most common types of electron microscopes are the transmission electron microscope and the scanning electron microscope. Both are used to study various life forms, including microbes.

Transmission Electron Microscopy

The **transmission electron microscope (TEM)** gives a better view of the internal structure of microbes than do other types of microscopes. Because of the very short wavelength of illumination (electrons) on which the TEM operates, it can resolve objects as close as 1 nm and magnify microbes (and other objects) up to 500,000×.

FIGURE 3.19 Comparison of (a) light and (b) electron microscope images of a *Didinium* eating a *Paramecium* (425×). Notice how much more detail is revealed by the scanning electron micrograph.

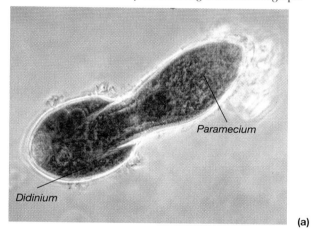

Paramecium

Didinium

(a)

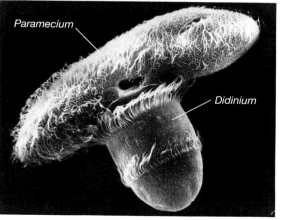

Paramecium

Didinium

(b)

To prepare specimens for transmission electron microscopy, a specimen may be embedded in a block of plastic and cut with a glass or diamond knife to produce very thin slices (*sections*). These sections are placed on thin wire grids for viewing so that a beam of electrons will pass directly through the section. The section must be exceedingly thin (70–90 nm) because electrons cannot penetrate very far into materials. The specimens can also be treated with special preparations that contain heavy metal elements. The heavy metals scatter electrons and contribute to forming an image.

Very small specimens, such as molecules or viruses, can be placed directly on plastic-coated grids. Then a heavy metal such as gold or platinum is sprayed at an angle onto the specimen, a technique known as **shadow casting.** A thin layer of the metal is deposited. Areas behind the specimen that did not receive a coating of metal appear as "shadows," which can give a three-dimensional effect to the image (Figure 3.20). Electron beams are deflected by the densely coated parts of the specimen but, for the most part, pass through the shadows.

It is also possible to view the interior of a cell with a TEM by a technique called **freeze-fracturing.** In this technique the cell is frozen and then fractured with a knife. The cleaving of a specimen reveals the surfaces of structures inside the cell (Figure 3.21a). **Freeze-etching,** which involves the evaporation of water from the frozen and fractured specimen, can then expose additional surfaces for examination (Figure 3.21b). These surfaces must also be coated with a heavy metal layer. This layer, called a *replica,* is viewed by TEM (Figure 3.22).

The image formed by the electron beam is made visible as a light image on a fluorescent screen, or mon-

FIGURE 3.20 Spraying a heavy metal (such as gold or platinum) at an angle over a specimen leaves a "shadow," or darkened area, where metal is not deposited. This technique, known as *shadow casting,* produces images with a three-dimensional appearance, as in this photograph of polio viruses (magnified 324,000×). You can calculate the height of the organisms from the length of their shadows if you know the angle of the metal spray.

itor. (The actual image made by the electron beam is not visible and would burn your eyes if you tried to view it directly.) The electrons are used to excite the phosphors (light-generating compounds) coating the screen. However, the electron beam will eventually burn through the specimen. Therefore, electron micrographs are made, either by photographing the image on the video screen or by replacing the screen itself with a photographic plate (Figure 3.23a). Electron micrographs can be enlarged photographically to obtain an image magnified 20 million times! The micrographs are permanent records of specimens observed and can be

FIGURE 3.21 (a) In freeze-fracture preparation, a specimen is frozen in a block of ice and broken apart with a very sharp knife. The fracture reveals the interiors of cellular structures and typically passes through the center of membrane bilayers, exposing their inner faces. (b) In freeze-etching, water is evaporated directly from the ice and frozen cytoplasm of the specimen, uncovering additional surfaces for observation.

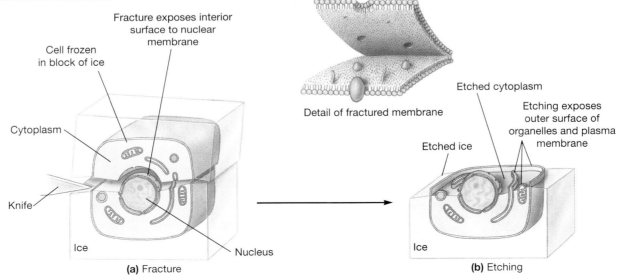

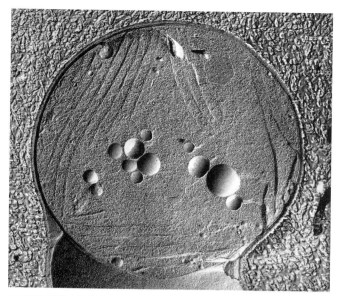

FIGURE 3.22 The toxic cyanobacterium *Microcystis aerugi-nosa* (magnified 18,000×) prepared by the freeze-etch method, showing details of large spherical gas vacuoles.

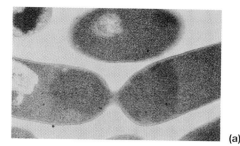

(a)

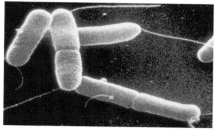

(b)

FIGURE 3.23 Comparison of color-enhanced electron micrographs of *Escherichia coli* produced by (a) transmission electron microscopy and (b) scanning electron microscopy.

studied at leisure. The study of electron micrographs has provided much of our knowledge of the internal structure of microbes.

Scanning Electron Microscopy

The **scanning electron microscope (SEM)** is a more recent invention than the transmission electron microscope and is used to create images of the surfaces of specimens. The SEM can resolve objects as close as 20 nm, giving magnifications up to approximately 50,000×. The SEM gives us wonderful three-dimensional views of the exterior of cells (Figure 3.23b).

Preparing a specimen for the SEM involves coating it with a thin layer of a heavy metal, such as gold or palladium. The SEM is operated by scanning, or sweeping, a very narrow beam of electrons (an electron probe) back and forth across a metal-coated specimen. Secondary, or backscattered, electrons leaving the specimen surface are collected, the current is increased, and the resulting image is displayed on a screen. Photographs of the image can be made and enlarged for further study.

Views of the three-dimensional world of microbes, as shown in Figure 3.24, are breathtakingly beautiful. The various types of microscopy and their uses are summarized in Table 3.2.

FIGURE 3.24 Color-enhanced SEM photos of representative microbes: (a) the fungus *Aspergillus*, a cause of human respiratory disease; (b) *Actinomyces* (40,500×), long thought to be a fungus because of its unusual filamentous form, but now known to be a bacterium; (c) the diatom *Cyclotella meneghiniana* (1800×), one of many that carry on photosynthesis and form the base of many aquatic food chains.

(a)

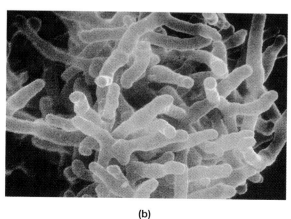

(b)

(c)

TABLE 3.2 Comparison of Types of Microscopy

Type	Special Features	Appearance	Uses
Bright-Field	Uses visible light; simplest to use, least expensive	Colored or clear specimen on light background	Observation of dead stained organisms or live ones with sufficient natural color contrast
Dark-Field	Uses visible light with a special condenser that causes light rays to reflect off specimen at an angle	Bright specimen on dark background	Observation of unstained living or difficult-to-stain organisms. Allows one to see motion
Phase-Contrast	Uses visible light plus phase-shifting plate in objective with a special condenser that causes some light rays to strike specimen out of phase with each other	Specimen has different degrees of brightness and darkness	Detailed observation of internal structure of living unstained organisms
Nomarski	Uses visible light out of phase; has higher resolution than standard phase-contrast microscope	Produces a nearly three-dimensional image	Observation of finer details of internal structure of living unstained organisms
Fluorescence	Uses ultraviolet light to excite molecules to emit light of different wavelengths, often brilliant colors; because UV can burn eyes, special lens materials are used	Bright, fluorescent, colored specimen on dark background	Diagnostic tool for detection of organisms or antibodies in clinical specimens or for immunologic studies
Transmission Electron	Uses electron beam instead of light rays and electromagnetic lenses instead of glass lenses; image is projected on a video screen; very expensive; preparation requires considerable time and practice	Highly magnified, detailed image; not three-dimensional except with shadow casting	Examination of thin sections of cells for details of internal structure, exterior of cells, and viruses, or surfaces when freeze-fracturing is used.
Scanning Electron	Uses electron beam and electromagnetic lenses; expensive; preparation requires considerable time and practice	Three-dimensional view of surfaces	Observation of exterior surfaces of cells or of internal surfaces

TECHNIQUES OF LIGHT MICROSCOPY

Microscopes are of little use unless the specimens for viewing are prepared properly. Here we explain some important techniques used in light microscopy.

Although resolution and magnification are important in microscopy, the degree of contrast between structures to be observed and their backgrounds is equally important. Nothing can be seen without contrast, so special techniques have been developed to enhance contrast.

Preparation of Specimens for the Light Microscope

Wet Mounts

Wet mounts, in which a drop of medium containing the organisms is placed on a microscope slide, can be used to view living microorganisms. The addition of a 2 percent solution of carboxymethyl cellulose, a thick, syrupy solution, helps to slow fast-moving organisms so they can be studied. A special version of the wet mount, called a **hanging drop,** often is used with dark-field illumination (Figure 3.25). A drop of culture is placed on a coverslip that is encircled with petroleum jelly. The coverslip and drop are then inverted over the well of a depression slide. The drop hangs from the coverslip, and the petroleum jelly forms a seal that prevents evaporation. This preparation gives good views of microbial motility.

Smears

Smears, in which microorganisms from a loopful of medium are spread onto the surface of a glass slide, can be used to view killed organisms. Although they are living when placed on the slide, the organisms are killed by the techniques used to fix (attach) them to the slide. Smear preparation often is difficult for beginners. If you make smears too thick, you will have trouble

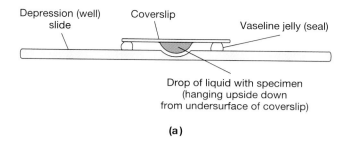

(a)

FIGURE 3.25 (a) The hanging-drop technique. A drop of culture is placed on a coverslip, ringed with petroleum jelly, that is then inverted and placed over the well in a depression slide. The petroleum jelly forms a seal to prevent evaporation. (b) Dark-field micrograph of a hanging-drop preparation (2500×) showing the spiral bacterium *Treponema pallidum*, the cause of syphilis.

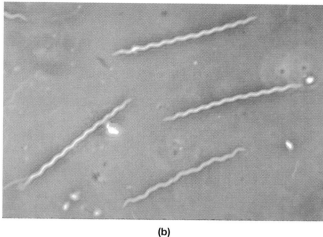

(b)

seeing individual cells; if you make them too thin, you may find no organisms. If you stir the drop of medium too much as you spread it on the slide, you will disrupt cell arrangements. You may see organisms that normally appear in tetrads (groups of four) as single or double organisms. Such variations lead some beginners to imagine that they see more than one kind of organism when, in fact, the organisms are all of the same species.

After a smear is made, it is allowed to air-dry completely. Then it is quickly passed three or four times through an open flame. This process is called **heat fixation.** Heat fixation accomplishes three things: (1) It kills the organisms; (2) it causes the organisms to adhere to the slide; and (3) it alters the organisms so that they more readily accept stains (dyes). If the slide is not completely dry when you pass it through the flame, the organisms will be boiled and destroyed. If you heat-fix too little, the organisms may not stick and will wash off the slide in subsequent steps. Any cells remaining alive will stain poorly. If you heat-fix too much, the organisms may be incinerated, and you will see distorted cells and cellular remains. Certain structures, such as the capsules found on some microbes, are destroyed by heat-fixing, so this step is omitted and these microbes are affixed to the slide just by air-drying.

Principles of Staining

A **stain,** or dye, is a molecule that can bind to a cellular structure and give it color. Staining techniques make the microorganisms stand out against their backgrounds. They are also used to help investigators group major categories of microorganisms, examine the structural and chemical differences in cellular structures, and look at the parts of the cell.

In microbiology the most commonly used dyes are **cationic** (positively charged), or **basic, dyes,** such as methylene blue, crystal violet, safranin, and malachite green. These dyes are attracted to any negatively charged cell components. The cell membranes of most bacteria have negatively charged surfaces and thus attract the positively charged basic dyes. Other stains, such as eosin and picric acid, are **anionic** (negatively charged), or **acidic, dyes.** They are attracted to any positively charged cell materials.

Two main types of stains, simple stains and differential stains, are used in microbiology. They are compared in Table 3.3. A **simple stain** makes use of a single dye and reveals basic cell shapes and cell arrangements. Methylene blue, safranin, carbolfuchsin, and crystal violet are commonly used simple stains. A **differential stain** makes use of two or more dyes and distinguishes between two kinds of organisms or between two different parts of an organism. Common differential stains are the Gram stain, the Ziehl-Neelsen acid-fast stain, and the Schaeffer-Fulton spore stain.

The Gram Stain

The **Gram stain,** probably the most frequently used differential stain, was devised by a Danish physician, Hans Christian Gram, in 1884. Gram was testing new methods of staining biopsy and autopsy materials, and he noticed that with certain methods some bacteria were stained differently than the surrounding tissues. As a result of his experiments with stains, the highly useful Gram stain was developed. In Gram staining, bacterial cells take up crystal violet. Iodine is then added; it acts as a **mordant,** a chemical that helps retain the stain in certain cells. Those structures that cannot retain crystal violet are decolorized with 95 percent

| TABLE 3.3 | Comparison of Staining Techniques |

Type	Examples	Result	Uses
Simple stains Use a single dye; do not distinguish organisms or structures by different staining reactions	Methylene blue Safranin Crystal violet	Uniform blue stain Uniform red stain Uniform purple stain	Shows sizes, shapes, and arrangements of cells
Differential stains Utilize two or more dyes that react differently with various kinds or parts of bacteria, allowing them to be distinguished	Gram stain	Gram+: purple with crystal violet Gram−: red with safranin counterstain Gram-variable: intermediate or mixed colors (some stain + and some − on same slide) Gram-nonreactive: stain poorly or not at all	Distinguish gram+, gram−, gram-variable, and gram-nonreactive organisms
	Ziehl-Neelsen acid-fast stain	Acid-fast bacteria retain carbolfuchsin and appear red. Non-acid-fast bacteria accept the methylene blue counterstain and appear blue.	Distinguishes members of the genera *Mycobacterium* and *Nocardia* from other bacteria
	Negative stain	Capsules appear clear against a dark background.	Allows visualization of organisms with structures that will not accept most stains, such as capsules
Special stains Identify various specialized structures	Flagellar stain	Flagella appear as dark lines with silver, or red with carbolfuchsin.	Indicates presence of flagella by building up layers of stain on their surface
	Schaeffer-Fulton spore stain	Endospores retain malachite green stain. Vegetative cells accept safranin counterstain and appear red.	Allows visualization of hard-to-stain bacterial endospores, as for members of genera *Clostridium* and *Bacillus*

ethanol or an ethanol-acetone solution, rinsed, and subsequently stained (counterstained) with safranin. The steps in the Gram staining procedure are shown in Figure 3.26.

Four groups of organisms can be distinguished with the Gram stain: (1) *gram-positive* organisms, whose cell walls retain crystal violet stain; (2) *gram-negative* organisms, whose cell walls do not retain crystal violet stain; (3) *gram-nonreactive* organisms, which do not stain or which stain poorly; and (4) *gram-variable* organisms, which stain unevenly. The differentiation between gram-positive and gram-negative organisms reveals a fundamental difference in the nature of the cell walls of bacteria, as is explained in Chapter 4. Furthermore, the reactions of bacteria to the Gram stain have helped in distinguishing gram-positive, gram-negative, and gram-nonreactive groups that belong to radically different taxonomic groups (Chapter 10).

Gram-variable organisms have somehow lost their ability to react distinctively to the Gram stain. Organisms from cultures over 48 hours old (and sometimes only 24 hours old) are often gram-variable, probably because of changes in the cell wall with aging. Therefore, to determine the reaction of an organism to the Gram stain, you should use organisms from cultures 18 to 24 hours old.

TRY IT

ARE YOU POSITIVE?– OR, NO STRINGS ATTACHED

The Gram stain is not foolproof. Some anaerobic gram-positive organisms decolorize very easily and may falsely appear to be gram-negative. And gram-negative organisms such as *Streptobacillus moniliformis* (a causative agent of rat-bite fever) can stain gram-positive. When in doubt, is there some way to be sure of the correct Gram reaction of an organism? Actually, there are several ways. Here is one you might want to try: the potassium hydroxide (KOH) test.

Place 2 drops of a 3 percent solution of KOH (3 g KOH dissolved in 100 ml distilled water) on a clean glass slide. Remove an inoculating loopful of the organism in question from a pure colony growing on an agar slant or agar plate. Add it to the KOH on the slide, and mix continuously for 30 seconds. As you stir, occasionally lift the loop up 1 or 2 cm from the surface of the slide to see if "strings" of gooey material hang down from the loop. If the organism is truly gram-negative, the KOH will break down its cell walls, releasing its DNA and forming strings. Gram-positive organisms will not form strings.

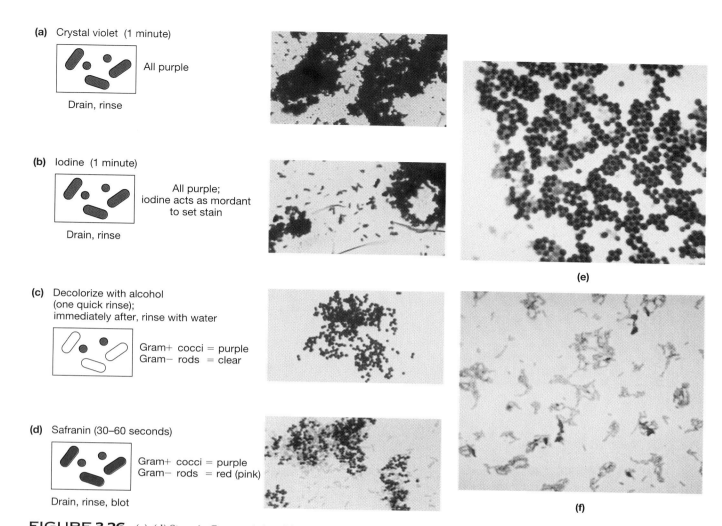

(a) Crystal violet (1 minute)

All purple

Drain, rinse

(b) Iodine (1 minute)

All purple; iodine acts as mordant to set stain

Drain, rinse

(c) Decolorize with alcohol (one quick rinse); immediately after, rinse with water

Gram+ cocci = purple
Gram− rods = clear

(d) Safranin (30–60 seconds)

Gram+ cocci = purple
Gram− rods = red (pink)

Drain, rinse, blot

(e)

(f)

FIGURE 3.26 (a)–(d) Steps in Gram staining. (e) Gram-positive cells retain the purple color of crystal violet, whereas (f) gram-negative cells are decolorized with alcohol and subsequently pick up the red color of the safranin counterstain.

The Ziehl-Neelsen Acid-Fast Stain

The **Ziehl-Neelsen acid-fast stain** is a modification of a staining method developed by Paul Ehrlich in 1882. It can be used to detect tuberculosis- and leprosy-causing organisms of the genus *Mycobacterium* (Figure 3.27). Slides of organisms are covered with carbolfuchsin and heated, rinsed, decolorized with 3 percent hydrochloric acid (HCl) in 95 percent ethanol, rinsed, and then stained with Loeffler's methylene blue. Most genera of bacteria will lose the red carbolfuchsin stain when decolorized. However, those that are *"acid-fast"* retain the bright red color. The lipid components of their walls responsible for this characteristic are discussed in Chapter 4. Bacteria that are not acid-fast lose the red color and can therefore be stained blue with the Loeffler's methylene blue counterstain.

FIGURE 3.27 The Ziehl-Neelsen stain produces vivid red color in acid-fast organisms such as *Mycobacterium leprae* (magnified 1025×), the cause of leprosy.

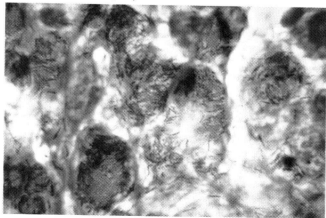

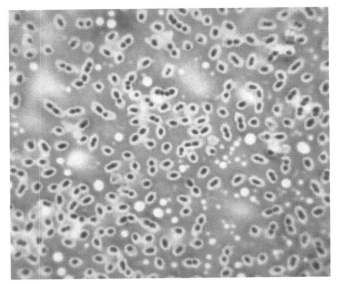

FIGURE 3.28 Negative staining for capsules reveals a clear area (the capsule, which does not accept stain) in a dark pink background of India ink and crystal violet counterstain. The cells themselves are stained deep purple with the counterstain. The bacteria are *Streptococcus pneumoniae* (3300×), which are arranged in pairs.

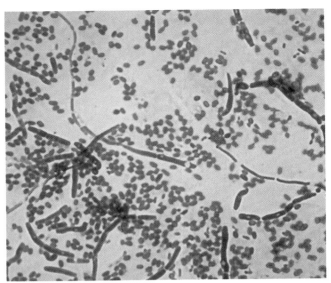

FIGURE 3.29 Schaeffer-Fulton staining of endospores of *Bacillus megaterium* (1000×). The endospores are visible as green, oval structures inside and outside the rod-shaped cells. Vegetative cells, which represent a non-spore-forming stage, and cellular regions without spores stain red.

Special Staining Procedures

NEGATIVE STAINING   **Negative stains** are used when a specimen—or a part of it, such as the capsule—resists taking up a stain. The *capsule* is a layer of polysaccharide material that surrounds many bacterial cells and can act as a barrier to host defense mechanisms. It also repels stains. In negative staining, the background around the organisms is filled with a stain such as India ink or an acidic dye, such as nigrosin. This process leaves the organisms themselves as clear, unstained objects that stand out against the dark background. A second simple or differential stain can be used to demonstrate the presence of the cell inside the capsule. Thus, a typical slide will show a dark background and clear, unstained areas of capsular material, inside of which are purple cells stained with crystal violet (Figure 3.28) or blue cells stained with methylene blue.

FLAGELLAR STAINING   *Flagella,* appendages that some cells have and use for locomotion, are too thin to be seen easily with the light microscope. When it is necessary to determine their presence or arrangement, **flagellar stains** are painstakingly prepared to coat the surfaces of the flagella with dye or a metal such as silver. These techniques are very difficult and time-consuming and so are usually omitted from the beginning course in microbiology. (See Figure 4.11 for some examples of stained flagella.)

ENDOSPORE STAINING   A few types of bacteria produce resistant cells called *endospores.* Endospore walls are very resistant to penetration of ordinary stains. When a simple stain is used, the spores will be seen as clear, glassy, easily recognizable areas within the bacterial cell. Thus, strictly speaking, it is not absolutely necessary to perform an endospore stain to see the spores. However, the differential **Schaeffer-Fulton spore stain** makes spores easier to visualize (Figure 3.29). Heat-fixed smears are covered with malachite green and then gently heated until they steam. Approximately 5 minutes of such steaming causes the endospore walls to become more permeable to the dye. The slide is then rinsed with water for 30 seconds to remove the green dye from all parts of the cell except for the endospores, which retain it. Then a counterstain of safranin is placed on the slide to stain the non-spore-forming, or vegetative, areas of the cells. Cells of cultures without endospores appear red; those with endospores have green spores and red vegetative cells.

Although microscopy and staining techniques can offer valuable information about microorganisms, these methods are usually not enough to permit identification of most microbes. Many species look identical under the microscope—after all, there are only a limited number of basic shapes, arrangements, and staining reactions but thousands of kinds of bacteria. This means that biochemical and genetic characteristics usually must be determined before an identification can be made (Chapter 10).

IMAGES OF ATOMS

Being able to look inside cells, to see the individual atoms that compose them—what a mind-boggling experience that would have been for Leeuwenhoek! In 1980, Gerd Binnig and Heinrich Rohrer invented the first of a series of rapidly improving scanning tunneling (or scanning probe) microscopes (STM). Five years later they received the Nobel Prize for their discovery.

Instead of using light, Binnig and Rohrer used a thin wire probe made of the metals platinum and iridium to trace over the surface of a substance, much as you might use your finger to feel the bumps while you read Braille. Electron clouds (regions of electron movement) from the surfaces of the probe and the specimen overlap, producing a kind of pathway through which electrons can "tunnel" into each other's clouds. This tunneling sets up an observable electric current. The stronger the current, the closer the top of the atom is to the probe. Running the probe across in a straight line reveals the highs and lows of individual molecules or atoms in a surface. Even movies can be made this way. The first one ever produced showed individual fibrin molecules coming together in forming a clot. Models formed from such studies could serve as starting points for pharmaceutical companies to design clot-inhibiting drugs.

Scanning tunneling microscopy works well under water. Although there is some interference from ion currents, *live* specimens can be probed with an STM. Films have been made of virus-infected cells that explode and release newly formed viruses. Viruses can themselves be examined to determine how the immune system "sees" them. Films of antibodies (defense molecules produced by the immune system) have also been produced.

Biologists will soon be able to see actual molecular events that they previously could only imagine. The physicist Paul Hansma of the University of California at Santa Barbara has hopes that this technique can speed up the Human Genome Project. Current machines are able to identify and sequence 7000 bases in a DNA molecule each day. STMs programmed to recognize labels attached to each of the four different nitrogenous bases of DNA could sequence 10 bases per second, or nearly a million bases per day. At that rate, the Human Genome Project may be finished ahead of its scheduled completion date of 2005, which would save millions of dollars.

Wolfgang Heckl, a professor of experimental physics at the University of Munich, points out that the STM was just the first of a whole family of scanning probe microscopes. The *atomic force microscope* (AFM), which Heckl uses, allows three-dimensional imaging and measurement of structures from atomic size to about 1 μm. In its most basic form, this revolutionary technology measures surfaces by moving a sharp probe across the sample to "feel" its surface details. Heckl claims that the cheapest way to build such a microscope is to alter an inexpensive compact disk player! The AFM uses a most basic and reliable sense of touch. All types of specimens can be examined and imaged in biological fluids or in air, with little or no sample preparation (Figure 3.30). Even cellular and related activities can be observed in action. One problem with AFMs is that the probe may strike and destroy, rather than ride over, a soft molecule. In 1994 Heckl was listed in *The Guiness Book of World Records* for studying the smallest hole in the world—a mere 3.6 $\times 10^{-4}$ μm across, where a single atom was removed from a crystal.

Heckl also uses the STM routinely to visualize DNA molecules. Unlike Hansma, whose sequencing technique relies on the identification of tags attached to the bases, Heckl can distinguish bases such as adenine and guanine from each other due to differences in their electron density states (Figure 3.31). As yet he has been unable to make the sugar-phosphate backbone of a single DNA strand lie down flat so that its bases can be imaged for sequencing. Hopefully this preparation problem will soon be overcome, and DNA sequencing by STM will proceed.

Other exciting prospects for STM use include movies of protein synthesis occurring molecule by molecule; of the folding of protein molecules into their proper configurations (tertiary structures) or into alternate shapes; and of electrons being transported along chains of molecules during respiration or being used in photosynthesis. We already have film of ring-shaped molecules that change in successive film frames. You can see the

FIGURE 3.30 Surfaces of a human red blood cell, as seen by atomic force microscopy.

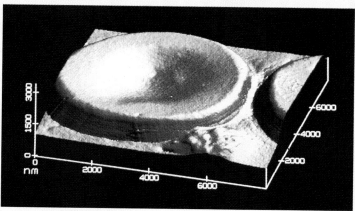

bonds (electron pairings) rearranging from one position to another.

IBM researchers have found a way to use the STM not just to look at, but to touch and move atoms around. Their first feat was to spell out "IBM" in letters just five atoms tall. The possibilities of literally constructing or engineering molecules has taken on new meaning. In the future, geneticists may use a "molecular syringe" to pick up a specific molecule and replace it with another to perform gene therapy! This interdisciplinary field is now known as nanotechnology (*nanos* is Greek for dwarf).

FIGURE 3.31 STM photos of the four nitrogenous bases of DNA: clockwise, from top left—guanine, adenine, thymine, and cytosine.

CHAPTER SUMMARY

DEVELOPMENT OF MICROSCOPY

- The existence of microorganisms was unknown until the invention of the microscope. Leeuwenhoek, probably the first to see microorganisms (in the 1600s), set the stage for **microscopy,** the technology of making very small things visible to the human eye.

- Leeuwenhoek's simple microscopes could reveal little detail of specimens. Today, multiple-lens, compound microscopes give us nearly distortion-free images, enabling us to delve further into the study of microbes.

PRINCIPLES OF MICROSCOPY

Metric Units

- The three units most used to describe microbes are the **micrometer** (μm), formerly called a micron, which is equal to 0.000001 m, also written as 10^{-6} m; the **nanometer** (nm), formerly called a millimicron (mμ), which is equal to 0.000000001 m, or 10^{-9} m; and the **angstrom** (Å), which is equal to 0.0000000001 m, 0.1 nm, or 10^{-10} m, but is no longer officially recognized.

Properties of Light: Wavelength and Resolution

- The **wavelength,** or the length of light rays, is the limiting factor in resolution.

- **Resolution** is the ability to see two objects as separate, discrete entities. Light wavelengths must be small enough to fit between two objects in order for them to be resolved.

- **Resolving power** can be defined as RP = λ/2NA, where λ = wavelength of light. The smaller the value of λ and

the larger the value of NA, the greater the resolving power of the lens.

- **Numerical aperture** (NA) relates to the extent to which light is concentrated by the condenser and collected by the objective. Its value is engraved on the side of each objective lens.

Properties of Light: Light and Objects

- If light strikes an object and bounces back, **reflection** (which gives an object its color) has occurred.

- **Transmission** is the passage of light through an object. Light must either be reflected from or transmitted through an object for it to be seen with a light microscope.

- **Absorption** of light rays occurs when they neither bounce off nor pass through an object but are taken up by that object. Absorbed light energy is used in performing photosynthesis or in raising the temperature of the irradiated body.

- Reemission of absorbed light as light of longer wavelengths is known as **luminescence.** If reemission occurs only during irradiation, the object is said to **fluoresce.** If reemission continues after irradiation ceases, the object is said to be **phosphorescent.**

- **Refraction** is the bending of light as it passes from one medium to another of different density. **Immersion oil,** which has the same **index of refraction** as glass, is used to replace air and to prevent refraction at a glass-air interface.

- **Diffraction** occurs when light waves are bent as they pass through a small opening, such as a hole, a slit, a space between two adjacent cellular structures, or a small, high-powered, magnifying lens in a microscope. The bent light rays distort the image obtained and limit the usefulness of the light microscope.

LIGHT MICROSCOPY

The Compound Light Microscope

The major parts of a compound light microscope and their functions are as follows:

- **Base** Supporting structure that generally contains the light source.
- **Condenser** Converges light beams to pass through the specimen.
- **Iris diaphragm** Controls the amount of light passing through the specimen.
- **Objective lens** Magnifies image.
- **Body tube** Conveys light to the ocular lens.
- **Ocular lens** Magnifies the image from the objective. A microscope with one ocular lens (eyepiece) is **monocular;** a microscope with two oculars is **binocular.**
- **Mechanical stage** Allows precise control in moving the slide.
- **Coarse adjustment** Knob used to locate specimen.
- **Fine adjustment** Knob used to bring specimen into sharp focus.
- The **total magnification** of a light microscope is calculated by multiplying the magnifying power of the objective lens by the magnifying power of the ocular lens. Increased magnification is of no value unless good resolution can also be maintained.

Dark-Field Microscopy

- **Bright-field illumination** is used in the ordinary light microscope, with light passing directly through the specimen.
- **Dark-field illumination** utilizes a special condenser that causes light to reflect off the specimen at an angle rather than pass directly through it.

Phase-Contrast Microscopy

- **Phase-contrast microscopy** utilizes microscopes with special condensers that accentuate small differences in refractive index of structures within the cell, allowing live, unstained organisms to be examined.

Fluorescence Microscopy

- **Fluorescence microscopy** uses ultraviolet light instead of white light to excite molecules within the specimen or dye molecules attached to the specimen. These molecules emit different wavelengths, often of brilliant colors.

Differential Interference Contrast (Nomarski) Microscopy

- Differential interference contrast microscopy, or

Nomarski microscopy, uses microscopes that operate essentially like phase-contrast microscopes but with a much greater resolution and a very short depth of field. They produce a nearly three-dimensional image.

ELECTRON MICROSCOPY

- The **electron microscope** (EM) uses a beam of electrons instead of a beam of light and electromagnets instead of glass lenses for focusing. They are much more expensive and difficult to use but give magnifications of up to $500,000\times$ and a resolving power of less than 1 nm. Viruses can be seen only by using EMs.

Transmission Electron Microscopy

- For the **transmission electron microscope** (TEM), very thin slices (sections) of a specimen are used, revealing the internal structure of microbial and other cells.

Scanning Electron Microscopy

- For the **scanning electron microscope** (SEM), a specimen is coated with a metal. The electron beam is scanned, or swept, over this coating to form a three-dimensional image.

TECHNIQUES OF LIGHT MICROSCOPY

Preparation of Specimens for the Light Microscope

- **Wet mounts** are used to view living organisms. The **hanging-drop** technique is a special type of wet mount, often used to determine whether organisms are motile.
- **Smears** of appropriate thickness are allowed to air-dry completely and are then passed through an open flame. This process, called **heat fixation,** kills the organisms, causing them to adhere to the slide and more readily accept stains.

Principles of Staining

- A **stain,** or dye, is a molecule that can bind to a structure and give it color.
- Most microbial stains are **cationic** (positively charged), or **basic, dyes,** such as methylene blue. Because most bacterial surfaces are negatively charged, these dyes are attracted to them.
- **Simple stains** use one dye and reveal basic cell shapes and arrangements. **Differential stains** use two or more dyes and distinguish various properties of organisms. The **Gram stain,** the **Schaeffer-Fulton spore stain,** and the **Ziehl-Neelsen acid-fast stain** are examples.
- **Negative stains** color the background around cells and their parts, which resist taking up stain.
- **Flagellar stains** add layers of dye or metal to the surfaces of flagella to make those surfaces visible.
- In the Schaeffer-Fulton spore stain, endospores stain green due to the uptake of malachite green, whereas vegetative cells stain red due to safranin uptake.

KEY TERMS

absorption (p. 55)
angstrom (p. 52)
anionic (acidic) dye
    (p. 65)
base (p. 57)
binocular (p. 57)
body tube (p. 57)
bright-field illumination
    (p. 58)
cationic (basic) dye (p. 65)
coarse adjustment (p. 58)
compound light microscope
    (p. 57)
condenser (p. 57)
dark-field illumination
    (p. 58)
differential stain (p. 65)
diffraction (p. 56)
electron micrograph (p. 61)

electron microscope (p. 60)
fine adjustment (p. 58)
flagellar stain (p. 68)
fluoresce (p. 55)
fluorescence microscopy
    (p. 58)
fluorescent antibody stain-
    ing (p. 59)
freeze-etching (p. 62)
freeze-fracturing (p. 62)
Gram stain (p. 65)
hanging drop (p. 64)
heat fixation (p. 65)
immersion oil (p. 55)
index of refraction (p. 55)
iris diaphragm (p. 57)
light microscopy (p. 57)
luminescence (p. 55)
mechanical stage (p. 57)

micrometer (p. 52)
microscopy (p. 52)
monocular (p. 57)
mordant (p. 65)
nanometer (p. 52)
negative stain (p. 68)
Nomarski microscopy
    (p. 60)
numerical aperture (p. 53)
objective lens (p. 57)
ocular lens (p. 57)
ocular micrometer (p. 58)
optical microscope (p. 57)
parfocal (p. 58)
phase-contrast microscopy
    (p. 58)
phosphorescent (p. 55)
reflection (p. 55)
refraction (p. 55)

resolution (p. 52)
resolving power (p. 53)
scanning electron micro-
    scope (p. 63)
Schaeffer-Fulton spore stain
    (p. 68)
shadow casting (p. 62)
simple stain (p. 65)
smear (p. 64)
stain (p. 65)
total magnification (p. 58)
transmission (p. 55)
transmission electron micro-
    scope (p. 61)
wavelength (p. 52)
wet mount (p. 64)
Ziehl-Neelsen acid-fast stain
    (p. 67)

QUESTIONS FOR REVIEW

A Evolution of Microscopy Instruments

1. Leeuwenhoek built simple microscopes. How did these differ from today's modern light microscopes?

B Metric Units

2. If a bacterium is 0.5 μm wide and 15 μm long, how many nanometers wide is it, and how many millimeters long is it?

C Wavelength, Resolution, and Numerical Aperture

3. What is resolution, and why is it important in microscopy?

4. As wavelength decreases, does resolving power increase or decrease? Why?

D Properties of Light

5. Define and contrast absorption, reflection, transmission, and refraction. Identify each on the following diagram:

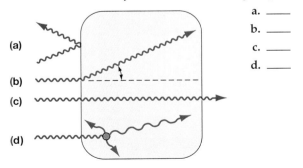

a. _____
b. _____
c. _____
d. _____

6. Define and contrast fluorescence and phosphorescence. In Question 5, which pathway of light represents an event that results in fluorescence?

7. Define and contrast diffraction and refraction.

8. Explain how and why immersion oil is sometimes used in light microscopy.

E Compound Microscope; Total Magnification

9. Label parts (a) through (g) of the diagram.

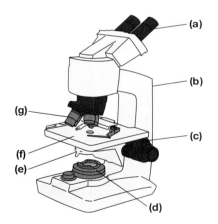

10. What is the total magnification of a microscope that has (a) 20× ocular and 40× objective lenses; (b) 10× ocular and 99× objective lenses?

11. In the days before microscopes were parfocal, what do you think was one of the greatest problems students faced in using the microscope?

12. What is an ocular micrometer used for?

F Other Types of Light Microscopes

13. Explain the differences between bright-field and dark-field illumination, and the advantages of the latter.

14. When is phase-contrast microscopy used? What special advantages does it have?

15. What is fluorescence microscopy? What special advantages does it have?

G **Electron Microscopy**

16. What does the electron microscope use instead of light beams and glass lenses?

17. How can you distinguish TEM from SEM micrographs? Identify each of the following photos as a TEM, an SEM, or a light microscope image.
    - a. [Figure 4.10, p. 86] ____
    - b. [Figure 10.15, p. 260] ____
    - c. [Figure 12.1b, p. 304] ____
    - d. [Figure 4.11a, p. 87] ____
    - e. [Figure 4.20b, p. 94] ____

H **Preparation Methods**

18. What three things does heat fixation accomplish?

19. Explain the hanging-drop technique and why it is useful.

I **Common Microbial Stains**

20. What is the difference between a simple and a differential stain?

21. When would you use the Ziehl-Neelsen acid-fast stain?

22. List some examples of differential stains and the features they can distinguish.

23. Name three special staining techniques, and briefly explain the uses of each.

J **Gram Staining**

24. What is a mordant? Which reagent is a mordant in the Gram staining procedure?

25. Look at Figure 10.15 on page 260. What shape are the gram-positive organisms?

26. What is the difference between gram-nonreactive and gram-variable organisms?

PROBLEMS FOR INVESTIGATION

1. A student failed to complete the Gram stain on her "unknown" culture in her drawer and plans to do a Gram stain on that same slant next week. What has she overlooked that may prevent her from obtaining proper results?

2. What would happen if a student forgot to do the iodine step in the Gram staining procedure of a mixed culture of gram-positive and gram-negative organisms? What would his results look like? Why?

3. Find out something about the comparative costs of the various types of microscopes and any additional facilities needed for their use, such as darkrooms or preparation rooms and equipment. Ask your instructor how much the microscope you use in lab would cost if you had to buy it today. Keep this figure in mind as you use the microscope during the semester—treat it kindly and gently.

4. Why do you suppose a depression slide is used for a hanging-drop preparation? Why would it be difficult to use a regular flat slide?

5. What are the advantages and disadvantages of observing living-specimen slides and heat-fixed-specimen slides of microorganisms?

SOME INTERESTING READING

Binnig, G., and H. Rohrer. 1985. "The scanning tunneling microscope." *Scientific American* 253(2):50–56.

Dobell, C. 1932. *Anthony van Leeuwenhoek and his "little animals."* London: Constable. (Reprinted in paperback 1960 by Dover, New York.)

England, B. M. 1991. "The state of the science: Scanning electron microscopy." *Mineralogical Record* 22, no. 2 (March–April):123–33.

Isenberg, H. D., ed. 1992. *Clinical microbiology procedures handbook,* 2 volumes. Washington, D.C.: American Society for Microbiology.

Lillie, R. D. 1977. *H. J. Conn's biological stains: A handbook on the nature and uses of the dyes employed in the biological laboratory.* 9th ed. Baltimore: Williams & Wilkins.

Molina, T. C., et al. 1990. "Gram staining apparatus for space station applications." *Applied and Environmental Microbiology* 56(March):601–6.

Murray, P. R., et al., eds. 1995. *Manual of clinical microbiology.* 6th edition. Washington, D.C.: American Society for Microbiology.

Powell, C. S. 1990. "Science writ small: A microscope builds an atomic-scale billboard." *Scientific American* 262(6):26.

Slayter, E. M., and H. S. Slayter. 1992. *Light and electron microscopy.* Cambridge, England: Cambridge University Press.

Trifiro, S., A.-M. Bourgault, F. Lebel, and P. Rene. 1990. "Ghost mycobacteria on Gram stain." *Journal of Clinical Microbiology* 28(1):146–47.

Trux, J. 1991. "Through the looking glass (objects viewed through a scanning electron microscope)." *World Magazine* 58, no. 8 (March):58–64.

Wilson, C. 1995. *The invisible world: Early philosophy and the invention of the microscope.* Princeton, NJ: Princeton University Press.

Woeste, S., and P. Demchick. 1991. "New version of the negative stain." *Applied and Environmental Microbiology* 57(6):1858–59.

CHAPTER 4

CHARACTERISTICS OF PROKARYOTIC AND EUKARYOTIC CELLS

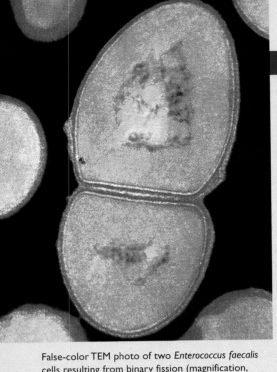

False-color TEM photo of two *Enterococcus faecalis* cells resulting from binary fission (magnification, 120,500×). A gram-positive coccus, it forms short chains and is part of the normal microbial population of the human digestive tract. Under certain conditions, however, it may cause infection of the cardiovascular or urogenital systems.

W e have considered the chemical principles that apply to cells and how to use microscopes and stains to observe cells. We can now look at the structure and function of the cells themselves.

BASIC CELL TYPES

All living cells can be classified as prokaryotic or eukaryotic. **Prokaryotic** (pro-kar"e-ot'ik) **cells** lack a nucleus and other membrane-bound structures, and **eukaryotic** (u-kar"e-ot'ik) **cells** have such structures. Prokaryotic and eukaryotic cells are *similar* in several ways. Both are surrounded by a *cell membrane*, or *plasma membrane*. Although some cells have structures that extend beyond this membrane or surround it, the membrane defines the boundaries of the living cell. Both prokaryotic and eukaryotic cells also encode genetic information in DNA molecules.

These two types of cells are *different* in other, important ways. In eukaryotic cells, DNA is in a nucleus surrounded by a *nuclear envelope*, but in prokaryotic cells, DNA is in a nuclear region not surrounded by a membrane. Eukaryotic cells also have a variety of internal structures called **organelles** (or-ga-nelz'), or "little organs," that are surrounded by one or more membranes. Prokaryotic cells generally lack organelles that are membrane-bound.

In this chapter we examine the similarities and differences of prokaryotic cells and eukaryotic cells, as summarized in Table 4.1. (Refer to this table each time you learn about a new cellular structure.) Viruses do not fit in either category, as they are acellular.

PROKARYOTIC CELLS

All prokaryotic cells are whole unicellular organisms. Prokaryotic organisms include two small groups—the primitive archaeobacteria and the photosynthetic cyanobacteria—and the large group of true bacteria. In short, all bacteria are prokaryotes, and all prokaryotes are bacteria. Eukaryotes include all plants, animals, fungi, and protists (organisms such as *Amoeba, Paramecium,* and the malaria parasite). We take advantage of some of the differences between eukaryotic human cells and prokaryotic bacterial cells when we try to control disease-causing bacteria without harming the human host.

THIS CHAPTER FOCUSES ON THE FOLLOWING QUESTIONS:

A What are the characteristics of prokaryotic and eukaryotic cells?

B How do prokaryotic cells differ in size, shape, and arrangement?

C How are structure and function related in bacterial cell walls and cell membranes?

D How are structure and function related in other bacterial components?

E How are structure and function related in eukaryotic plasma membranes?

F How are structure and function related in other eukaryotic components?

G How do passive transport processes function, and why are they important?
How does active transport function, and why is it important?

H How do exocytosis and endocytosis occur, and why are they important?

Size, Shape, and Arrangement

Size

Prokaryotes are among the smallest of all organisms. Most prokaryotes range from 0.5 to 2.0 μm in diameter. For comparison, a human red blood cell is about 7.5 μm in diameter. Keep in mind, however, that although we often use diameter to specify cell size, many cells are not spherical in shape. Some spiral bacteria have a much larger diameter, and some cyanobacteria are 60 μm long. Because of their small size, bacteria have a large surface-to-volume ratio. For example, spherical bacteria with a diameter of 2 μm have a surface area of about 12 μm^2 and a volume of about 4 μm^3. Their surface-to-volume ratio is 12:4, or 3:1. In contrast, eukaryotic cells with a diameter of 20 μm have a surface area of about 1200 μm^2 and a volume of about 4000 μm^3. Their surface-to-volume ratio is 1200:4000, or 0.3:1—only one-tenth as great. The large surface-to-volume ratio of bacteria means that no internal part of the cell is very far from the surface and that nutrients can easily and quickly reach all parts of the cell.

Shapes

Typically, bacteria display three basic shapes—spherical, rodlike, and spiral (Figure 4.1)—but variations abound. A spherical bacterium is called a **coccus** (kok'us; plural: *cocci* [kok'se]), and a rodlike bacterium is called a **bacillus** (ba-sil'us; plural: *bacilli* [bas-il'e]). Some bacteria, called *coccobacilli,* are short rods intermediate in shape between cocci and bacilli. Spiral bacteria have a variety of curved shapes. A comma-shaped bacterium is called a **vibrio** (vib're-o); a rigid, wavy-shaped one, a **spirillum** (spi-ril'um; plural: *spirilla*); and a corkscrew-shaped one, a **spirochete** (spi'ro-ket). Some bacteria do not fit any of the preceding categories but rather have spindle shapes or irregular, lobed shapes. Square bacteria were discovered on the shores of the Red Sea in 1981. They are 2 to 4 μm on a side and sometimes aggregate in wafflelike sheets.

Even bacteria of the same kind sometimes vary in size and shape. When nutrients are abundant in the environment and cell division is rapid, rods are often twice as large as those in an environment with only a moderate supply of nutrients. Although variations in shape within a single species of bacteria are generally small, there are exceptions to this rule. Some bacteria vary

Characteristic	Prokaryotic Cells	Eukaryotic Cells
Genetic information	Found in single chromosome	Typically found in paired chromosomes
Location of genetic information	Nuclear region (nucleoid)	Membrane-bound nucleus
Nucleolus	Absent	Present
Histones	Absent	Present
Extrachromosomal DNA	In plasmids	In organelles, such as mitochondria and chloroplasts, and in plasmids
Mitotic spindle	Absent	Present during cell division
Plasma membrane	Fluid-mosaic structure lacking sterols	Fluid-mosaic structure containing sterols
Internal membranes	Only in photosynthetic organisms	Numerous membrane-bound organelles
Endoplasmic reticulum	Absent	Present
Respiratory enzymes	Cell membrane	Mitochondria
Chromatophores	Present in photosynthetic bacteria	Absent
Chloroplasts	Absent	Present in some
Golgi apparatus	Absent	Present
Lysosomes	Absent	Present
Peroxisomes	Absent	Present
Ribosomes	70S	80S in cytoplasm and on endoplasmic reticulum, 70S in organelles
Cytoskeleton	Absent	Present
Cell wall	Peptidoglycan found on most cells	Cellulose, chitin, or both found on plant and fungal cells
External layer	Capsule or slime layer	Pellicle, test, or shell in certain protists
Flagella	When present, consist of fibrils of flagellin	When present, consist of complex membrane-bound structure with "9 + 2" microtubule arrangement
Cilia	Absent	Present as structures shorter than, but similar to, flagella in some eukaryotic cells
Pili	Present as attachment or conjugation pili in some prokaryotic cells	Absent

FIGURE 4.1 The most common bacterial shapes.

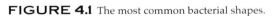

Coccus

Coccobacillus

Bacillus

Vibrio

Spirillum

Spirochete

widely in form even within a single culture, a phenomenon known as **pleomorphism.** Moreover, in aging cultures where organisms have used up most of the nutrients and have deposited wastes, cells not only are generally smaller, but they often display a great diversity of unusual shapes.

Arrangements

In addition to characteristic shapes, many bacteria also are found in distinctive arrangements of groups of cells (Figure 4.2). Such groups occur when cells divide without separating. Cocci can divide in one or two planes, or randomly. Division in one plane produces cells in pairs (indicated by the prefix *diplo*-) or in chains (*strepto*-). Division in two planes produces cells in *tetrads* (four cells arranged in a cube). Division in three planes produces *sarcinae* (singular: *sarcina;* eight cells arranged in a cube). Random division planes produce grapelike clusters (*staphylo*-). Bacilli divide in only one plane, but they can produce cells connected end-to-end (like train cars) or side-by-

FIGURE 4.2 Arrangements of bacteria. (a) Cocci arranged in pairs (diplococci of *Neisseria*) and in chains (*Streptococcus*), formed by division in one plane. (b) Cocci arranged in a tetrad (*Merisopedia,* 100×), formed by division in two planes. (c) Cocci arranged in a sarcina (*Sarcina lutea,* 16,000×), formed by division in three planes. (d) Cocci arranged in a cluster (*Staphylococcus,* 5400×), formed by division in many planes. (e) Bacilli arranged in a rosette (*Caulobacter,* 2400×), attached by stalks to a substrate.

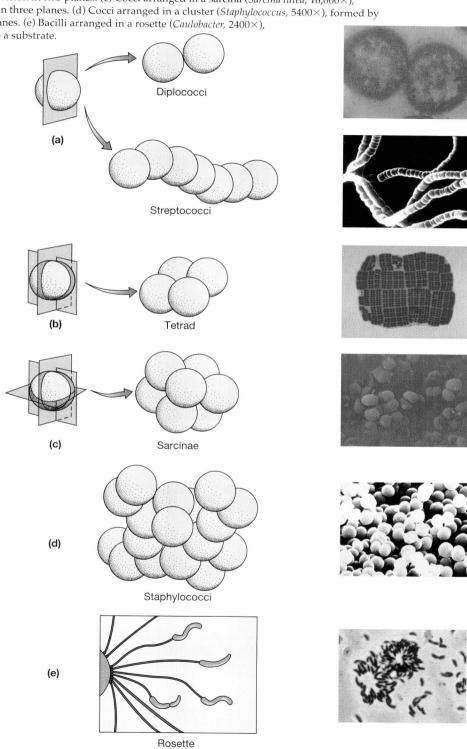

side. Spiral bacteria are not generally grouped together.

Overview of Structure

Structurally, bacterial cells (Figure 4.3) consist of the following:

1. A cell membrane, usually surrounded by a cell wall and sometimes by an additional outer layer.
2. An internal cytoplasm with ribosomes, a nuclear region, and in some cases granules and/or vesicles.
3. A variety of external structures, such as capsules, flagella, and pili.

Let us look at each of these kinds of structures in some detail.

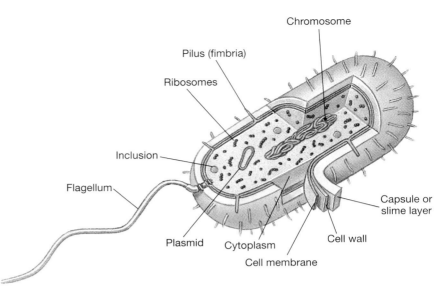

FIGURE 4.3 Diagram of a typical prokaryotic cell. The cell depicted is a bacillus with a polar flagellum (a flagellum at one end).

Cell Wall

The semirigid **cell wall** lies outside the cell membrane in nearly all bacteria. It performs two important functions. First, it maintains the characteristic shape of the cell. If the cell wall is digested away by enzymes, the cell takes on a spherical shape. Second, it prevents the cell from bursting when fluids flow into the cell by *osmosis* (described later in this chapter). Although the cell wall surrounds the cell membrane, in many cases it is extremely porous and does not play a major role in regulating the entry of materials into the cell.

Components of Cell Walls

PEPTIDOGLYCAN Peptidoglycan (pep″ti-do-gly′-kan), also called *murein,* is the single most important component of the bacterial cell wall. It is a polymer so large that it can be thought of as one immense, covalently linked molecule. It forms a supporting net around a bacterium that resembles multiple layers of chain-link fence (Figure 4.4). In the peptidoglycan polymer, molecules of *N*-acetylglucosamine (gluNAc) alternate with molecules of *N*-acetylmuramic acid (murNAc). These molecules are cross-linked by tetrapeptides, chains of four amino acids. In most gram-positive organisms, the third amino acid is lysine; in most gram-negative organisms, it is diaminopimelic acid. Amino acids, like many other organic compounds, have *stereoisomers*—structures that are mirror images of each other, just as a left hand is a mirror image of a right hand. Some of the amino acids in the tetrapeptide chains are mirror images of those amino acids most commonly found in living things. Those chains are not readily broken down because

most organisms lack enzymes that can digest stereoisomeric forms.

Cell walls of gram-positive organisms have an additional molecule, teichoic acid. **Teichoic (tie-ko′ik) acid,** which consists of glycerol, phosphates, and the sugar alcohol ribitol, occurs in polymers up to 30 units long. These polymers extend beyond the rest of the cell wall, even beyond the capsule in encapsulated bacteria. Although its exact function is unclear, teichoic acid furnishes attachment sites for bacteriophages (viruses that infect bacteria).

OUTER MEMBRANE The **outer membrane,** found primarily in gram-negative bacteria, is a bilayer membrane (discussed in the next section). It forms the outermost layer of the cell wall and is attached to the peptidoglycan by an almost continuous layer of small lipoprotein molecules (proteins combined with a lipid). The lipoproteins are embedded in the outer membrane and covalently bonded to the peptidoglycan. The outer membrane acts as a coarse sieve and exerts little control over the movement of substances into and out of the cell. However, it does control the transport of certain proteins from the environment. The outer surface of the outer membrane has surface antigens and receptors. Some receptors bind viruses and thereby help them to infect the bacterium.

Lipopolysaccharide (LPS), also called **endotoxin,** is an important part of the outer membrane and can be used to identify gram-negative bacteria. It is an integral part of the cell wall and is not released until the cell walls of dead bacteria are broken down. LPS consists of polysaccharides and **lipid A** (Figure 4.5, on page 80). The

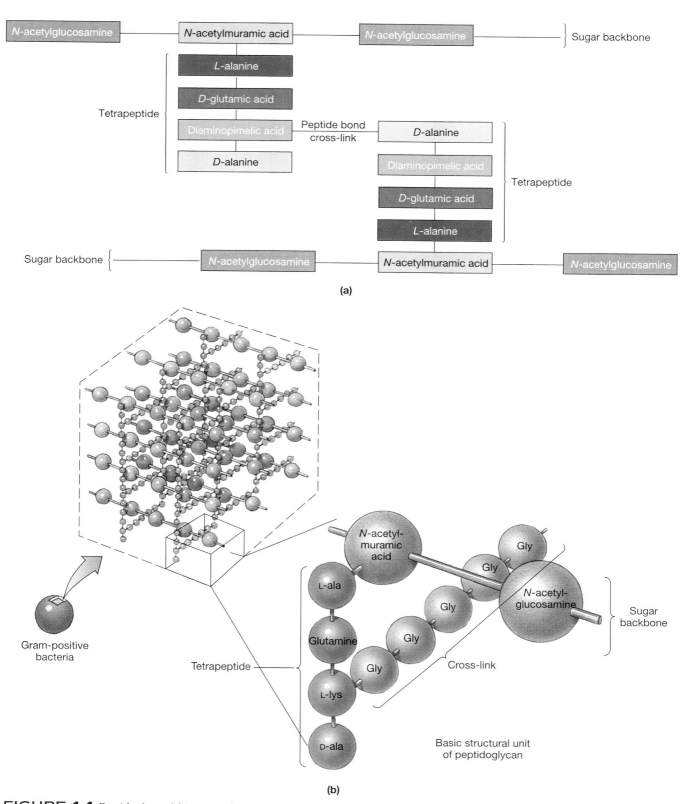

FIGURE 4.4 Peptidoglycan (a) in a two-dimensional view of the gram-negative bacterium *Escherichia coli* is a polymer of two alternating sugar units (purple), *N*-acetylglucosamine and *N*-acetylmuramic acid, both of which are derivatives of glucose. The sugars are joined by short peptide chains (tetrapeptides) that consist of four amino acids (red). The sugars and tetrapeptides are cross-linked by a simple peptide bond. (b) A three-dimensional view of peptidoglycan for the gram-positive bacterium *Staphylococcus aureus*. Compare the components with those in (a). Different organisms can have different amino acids in the tetrapeptide chain as well as different cross-links.

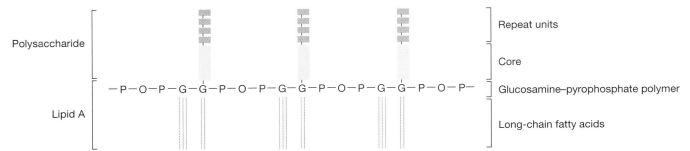

FIGURE 4.5 Lipopolysaccharide, also called endotoxin, an important component of the outer membrane in gram-negative cell walls. The lipid A portion of the molecule consists of a backbone of alternating pyrophosphate units (POP, linked phosphate groups) and glucosamine (G, a glucose derivative), to which long fatty acid side chains are attached. Lipid A is a toxic substance that contributes to the danger of infection by gram-negative bacteria. Polysaccharide side chains extending outward from the glucosamine units make up the remainder of the molecule.

polysaccharides are found in repeating side chains that extend outward from the organism. It is these repeating units that are used to identify different gram-negative bacteria. The lipid A portion is responsible for the toxic properties that make any gram-negative infection a potentially serious medical problem. It causes fever and dilates blood vessels, so the blood pressure drops precipitously. Because bacteria release endotoxin mainly when they are dying, killing them may increase the concentration of this very toxic substance.

PERIPLASMIC SPACE Another distinguishing characteristic of many bacteria is the presence of a gap between the cell membrane and the cell wall. The gap is most easily observed by electron microscopy of gram-negative bacteria. In these organisms the gap is called the **periplasmic** (per'e-plaz"mik) **space.** It represents a very active area of cell metabolism. This space contains not only the cell wall peptidoglycan but also many digestive enzymes and transport proteins that destroy potentially harmful substances and transport metabolites into the bacterial cytoplasm, respectively. The *periplasm*

consists of the peptidoglycan, protein constituents, and metabolites found in the periplasmic space.

Periplasmic spaces are rarely observed in gram-positive bacteria. However, such bacteria must accomplish many of the same metabolic and transport funtions that gram-negative bacteria do. At present most gram-positive bacteria are thought to have only periplasms—not periplasmic spaces—where metabolic digestion occurs and new cell wall peptidoglycan is attached. The periplasm in gram-positive cells is thus part of the cell wall.

Distinguishing Bacteria by Cell Walls

Certain properties of cell walls result in different staining reactions. Gram-positive, gram-negative, and acid-fast bacteria can be distinguished on the basis of these reactions (Table 4.2 and Figure 4.6).

GRAM-POSITIVE BACTERIA The cell wall in gram-positive bacteria has a relatively thick layer of peptidoglycan 20 to 80 nm across. The peptidoglycan

TABLE 4.2	Characteristics of the Cell Walls of Gram-Positive, Gram-Negative, and Acid-Fast Bacteria		
Characteristic	**Gram-Positive Bacteria**	**Gram-Negative Bacteria**	**Acid-Fast Bacteria**
Peptidoglycan	Thick layer	Thin layer	Relatively small amount
Teichoic acid	Often present	Absent	Absent
Lipids	Very little present	Lipopolysaccharide	Mycolic acid and other waxes and glycolipids
Outer membrane	Absent	Present	Absent
Periplasmic space	Absent	Present	Absent
Cell shape	Always rigid	Rigid or flexible	Rigid or flexible
Results of enzyme digestion	Protoplast	Spheroplast	Difficult to digest
Sensitivity to dyes and antibiotics	Most sensitive	Moderately sensitive	Least sensitive

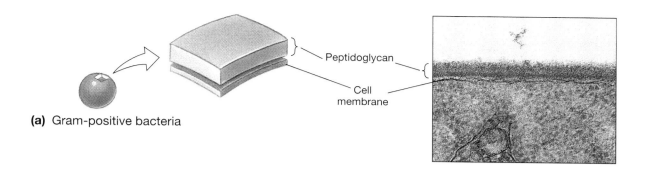

(a) Gram-positive bacteria

Peptidoglycan

Cell membrane

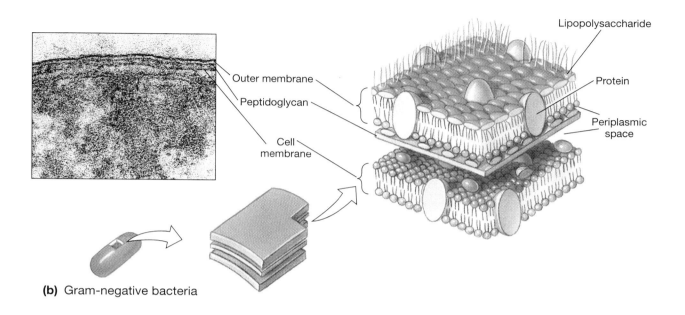

(b) Gram-negative bacteria

Outer membrane

Peptidoglycan

Cell membrane

Lipopolysaccharide

Protein

Periplasmic space

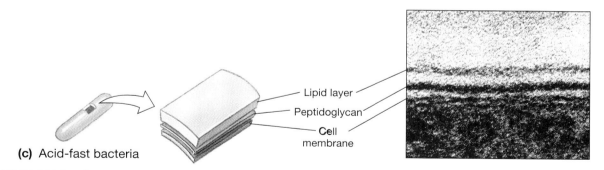

(c) Acid-fast bacteria

Lipid layer

Peptidoglycan

Cell membrane

FIGURE 4.6 Schematic representations of bacterial cell walls, paired with TEM photos of representative bacteria. (a) Gram-positive (*Bacillus fastidosus*). (b) Gram-negative (*Azomonas insignis*). (c) Acid-fast (*Mycobacterium phlei*).

layer is closely attached to the outer surface of the cell membrane. Chemical analysis shows that 60 to 90 percent of the cell wall of a gram-positive bacterium is peptidoglycan. Except for streptococci, most gram-positive cell walls contain very little protein. If peptidoglycan is digested from their cell walls, gram-positive bacteria become **protoplasts,** or cells with a cell membrane but no cell wall. Protoplasts shrivel or burst unless they are kept in an *isotonic* solution—a solution that has the same pressure as that inside the cell.

The thick cell walls of gram-positive bacteria retain such stains as the crystal violet–iodine dye in the cytoplasm, but yeast cells, many of which have thick walls, also retain them. Thus, retention of Gram stain seems to be directly related to wall thickness and not to peptidoglycan. Physiological damage or aging can make a gram-positive cell wall leaky, so the dye complex escapes. Such organisms can become gram-variable or even gram-negative as they age. Therefore, Gram staining must be performed on cultures less than 24 hours old.

Gram-positive bacteria lack both an outer membrane and a periplasmic space. Therefore, digestive enzymes not retained in the periplasm are released into the environment, where they sometimes become so diluted that the organisms derive no benefit from them.

GRAM-NEGATIVE BACTERIA The cell wall of a gram-negative bacterium is thinner but more complex than that of a gram-positive bacterium. Only 10 to 20 percent of the cell wall is peptidoglycan; the remainder consists of various polysaccharides, proteins, and lipids. The cell wall contains an outer membrane, which constitutes the outer surface of the wall, leaving only a very narrow periplasmic space. The inner surface of the wall is separated from the cell membrane by a wider periplasmic space. Toxins and enzymes remain in the periplasmic space in sufficient concentrations to help destroy substances that might harm the bacterium, but they do not harm the organism that produced them. If the cell wall is digested away, gram-negative bacteria become **spheroplasts,** which have both a cell membrane and most of the outer membrane. Gram-negative bacteria fail to retain the crystal violet-iodine dye during the decolorizing procedure partly because of their thin cell walls and partly because of the relatively large quantities of lipoproteins and lipopolysaccharides in the walls.

ACID-FAST BACTERIA The cell wall of *acid-fast bacteria,* the mycobacteria, is thick, like that of gram-positive bacteria. It is approximately 60 percent lipid and contains much less peptidoglycan. In the acid-fast staining process, carbolfuchsin binds to cytoplasm and resists removal by an acid–alcohol mixture. ∞ (Chapter 3, p. 67) The lipids make acid-fast organisms impermeable to most other stains and protect them from acids and alkalis. The organisms grow slowly because the lipids impede entry of nutrients into cells, and the cells must expend large quantities of energy to synthesize lipids.

CONTROL OF BACTERIA BY DAMAGE TO CELL WALLS Some methods of controlling bacteria are based on properties of the cell wall. For example, the antibiotic penicillin blocks the final stages of peptidoglycan synthesis. If penicillin is present when bacterial cells are dividing, the cells cannot form complete

walls, and they die. Similarly, the enzyme lysozyme, found in tears and other human body secretions, digests peptidoglycan. This enzyme helps prevent bacteria from entering the body and is the body's main defense against eye infections.

Cell Membrane

The **cell membrane,** or *plasma membrane,* is a living membrane that forms the boundary between a cell and its environment. Also known as the cytoplasmic membrane, this dynamic, constantly changing membrane is not to be confused with the cell wall. The latter is a static structure external to the cell membrane.

Bacterial cell membranes have the same general structure as the membranes of all other cells. Such membranes, formerly called *unit membranes,* consist mainly of phospholipids and proteins. The **fluid-mosaic model** (Figure 4.7) represents the current understanding of the structure of such a membrane. The model's name is derived from the fact that phospholipids in the membrane are in a fluid state and that proteins are dispersed among the lipid molecules in the membrane, forming a mosaic pattern.

Membrane phospholipids form a *bilayer,* or two adjacent layers. In each layer, the phosphate ends of the lipid molecules extend toward the membrane surface, and the fatty acid ends extend inward. The charged phosphate ends of the molecules are **hydrophilic** (water-loving) and thus can interact with the watery environment (Figure 4.7a). The fatty acid ends, consisting largely of nonpolar hydrocarbon chains, are **hydrophobic** (water-fearing) and form a barrier between the cell and its environment. Some membranes also contain other lipids. The membranes of mycoplasmas, bacteria that lack a cell wall, include lipids called *sterols* that add rigidity.

Interspersed among the lipid molecules are protein molecules (Figure 4.7b). Some extend through the entire membrane and act as carriers or form pores or channels through which materials enter and leave the cell. Proteins on the outer surface include those that make the cell identifiable as a particular organism. Others are embedded in, or loosely attached to, the inner or outer surface of the membrane. Proteins on the inner surface are usually enzymes.

Cell membranes are dynamic, constantly changing entities. Materials constantly move through pores and through the lipids themselves, although selectively. Also, both the lipids and the proteins in membranes are continuously changing positions.

The main function of the cell membrane is to regulate the movement of materials into and out of a cell by transport mechanisms, which are discussed later in this chapter. In bacteria this membrane also performs some functions carried out by other structures in eu-

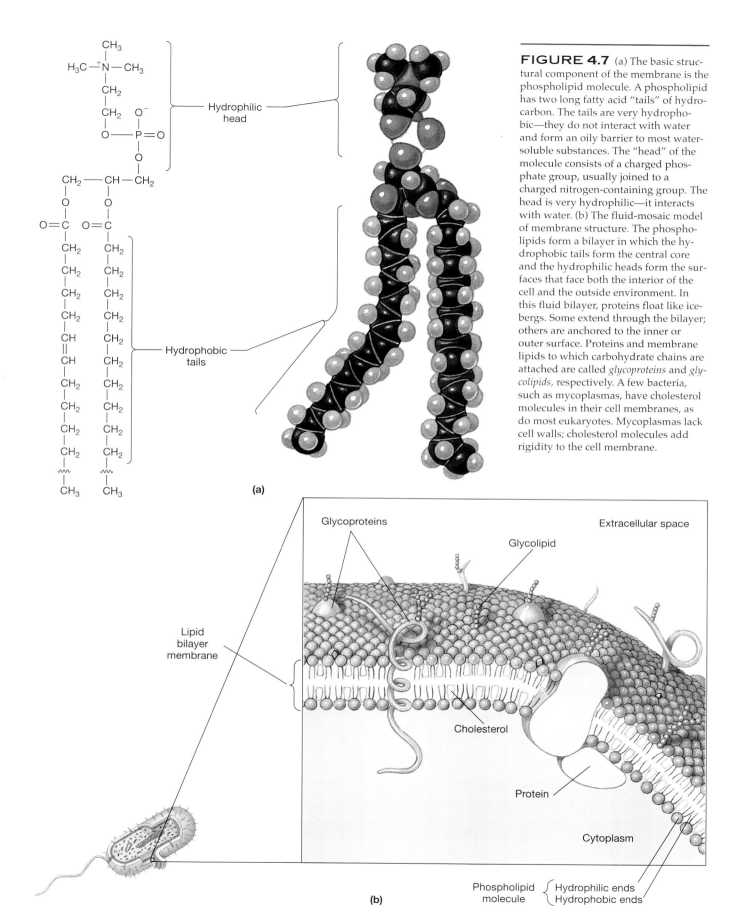

FIGURE 4.7 (a) The basic structural component of the membrane is the phospholipid molecule. A phospholipid has two long fatty acid "tails" of hydrocarbon. The tails are very hydrophobic—they do not interact with water and form an oily barrier to most water-soluble substances. The "head" of the molecule consists of a charged phosphate group, usually joined to a charged nitrogen-containing group. The head is very hydrophilic—it interacts with water. (b) The fluid-mosaic model of membrane structure. The phospholipids form a bilayer in which the hydrophobic tails form the central core and the hydrophilic heads form the surfaces that face both the interior of the cell and the outside environment. In this fluid bilayer, proteins float like icebergs. Some extend through the bilayer; others are anchored to the inner or outer surface. Proteins and membrane lipids to which carbohydrate chains are attached are called *glycoproteins* and *glycolipids,* respectively. A few bacteria, such as mycoplasmas, have cholesterol molecules in their cell membranes, as do most eukaryotes. Mycoplasmas lack cell walls; cholesterol molecules add rigidity to the cell membrane.

Hydrophilic head

Hydrophobic tails

(a)

Glycoproteins

Glycolipid

Extracellular space

Lipid bilayer membrane

Cholesterol

Protein

Cytoplasm

Phospholipid molecule — Hydrophilic ends / Hydrophobic ends

(b)

karyotic cells. It synthesizes cell wall components, assists with DNA replication, secretes proteins, carries on respiration, and captures energy as ATP. It also contains bases of appendages called *flagella;* the actions of the bases cause the flagella to move. Finally, some proteins in the cell membrane respond to chemical substances in the environment.

Internal Structure

Bacterial cells typically contain *ribosomes,* a *nucleoid,* and a variety of *vacuoles* within their *cytoplasm.* Figure 4.3 shows the locations of these structures in a generalized prokaryotic cell. Certain bacteria sometimes contain *endospores* as well.

Cytoplasm

The **cytoplasm** of prokaryotic cells is the semifluid substance inside the cell membrane. Because these cells typically have only a few clearly defined structures, such as a chromosome and some ribosomes, they consist mainly of cytoplasm. Cytoplasm is about four-fifths water and one-fifth substances dissolved or suspended in the water. These substances include enzymes and other proteins, carbohydrates, lipids, and a variety of inorganic ions. Many chemical reactions, both anabolic and catabolic, occur in the cytoplasm.

Ribosomes

Ribosomes consist of ribonucleic acid and protein. They are abundant in the cytoplasm of bacteria, often grouped in long chains called **polyribosomes.** Ribosomes are nearly spherical, stain densely, and contain a large subunit and a small subunit. Ribosomes serve as sites for protein synthesis (Chapter 7).

The relative sizes of ribosomes and their subunits can be determined by measuring their *sedimentation rates*—the rates at which they move toward the bottom of a tube when the tube is rapidly spun in an instrument called a *centrifuge* (Figure 4.8). Sedimentation rates are expressed in terms of *Svedberg* (S) *units,* which generally vary with molecular size. Whole bacterial ribosomes, which are smaller than eukaryotic ribosomes, have a rate of 70S; their subunits have rates of 30S and 50S. Certain antibiotics, such as streptomycin and erythromycin, bind specifically to 70S ribosomes and disrupt protein synthesis. Because those antibiotics do not affect the 80S ribosomes found in eukaryotic cells, they kill bacteria without harming host cells.

Nuclear Region

One of the key features differentiating prokaryotic cells from eukaryotic cells is the absence of a nucleus bounded by a nuclear membrane. Instead of a nucleus, bacteria have a **nuclear region,** or **nucleoid.** The cen-

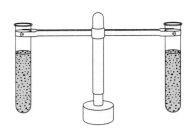

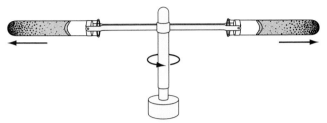

FIGURE 4.8 In a centrifuge, suspended particles in tubes of liquid are whirled around at high speeds, causing them to settle to the bottom of the tubes or to form bands at different levels. The rate of settling or the locations of the bands can be used to determine the size, weight, and shape of the particles.

trally located nuclear region consists mainly of DNA but has some RNA and protein associated with it. The DNA is arranged in one large, circular chromosome. Some bacteria also contain small circular molecules of DNA called *plasmids.* Genetic information in plasmids supplements information in the chromosome (Chapter 8).

Internal Membrane Systems

Photosynthetic bacteria and cyanobacteria contain internal membrane systems, sometimes known as **chromatophores** (Figure 4.9). The membranes of the chromatophores, derived from the cell membrane, contain the pigments used to capture light energy for the synthesis of sugars. Nitrifying bacteria, soil organisms that convert nitrogen compounds into forms usable by green plants, also have internal membranes. They house the enzymes used in deriving energy from the oxidation of nitrogen compounds (Chapter 5).

Electron micrographs of bacterial cells often show large infoldings of the cell membrane, called *mesosomes.* Although these were originally thought to be structures present in living cells, they have now been proven to be artifacts: That is, they were created by the processes used to prepare specimens for electron microscopy.

Inclusions

Bacteria can have within their cytoplasm a variety of small bodies collectively referred to as **inclusions.** Some are called *granules;* others are called *vesicles.*

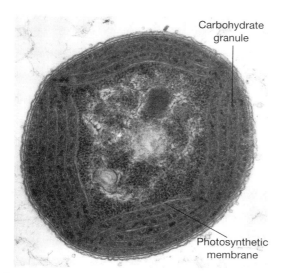

Carbohydrate
granule

Photosynthetic
membrane

FIGURE 4.9 TEM photo of the cyanobacterium *Coccochloris elabens* (47,000×). The cell's outer regions are filled with photosynthetic membranes. The dark spots between the membranes are granules where carbohydrates produced by photosynthesis are stored.

Granules, although not bounded by membrane, contain substances so densely compacted that they do not easily dissolve in cytoplasm. Each granule contains a specific substance, such as glycogen or polyphosphate. *Glycogen,* a glucose polymer, is used for energy. *Polyphosphate,* a phosphate polymer, supplies phosphate for a variety of metabolic processes. Polyphosphate granules are called **volutin** (vo-lu′tin), or **metachromatic granules,** because they display **metachromasia.** That is, although most substances stained with a simple stain such as methylene blue take on a uniform solid color, metachromatic granules exhibit different intensities of color. Although quite numerous in some bacteria, these granules become depleted during starvation.

Certain bacteria have specialized membrane-bound structures called **vesicles** (or *vacuoles*). Some aquatic photosynthetic bacteria and cyanobacteria have rigid gas-filled vacuoles (Figure 3.22). These organisms regulate the amount of gas in vacuoles and, therefore, the depth at which they float to obtain optimum light for photosynthesis. Another type of vesicle,

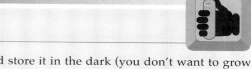

TRY IT

LIVING MAGNETS

Magnetotactic bacteria synthesize magnetite (Fe_3O_4), or lodestone, and store it in membranous vesicles called *magnetosomes.* (Lodestone was the first substance with magnetic properties to be discovered.) The presence of these magnetic inclusions enables these bacteria to respond to magnetic fields. In the Northern Hemisphere, magnetotactic bacteria swim toward the North Pole; in the Southern Hemisphere, they swim toward the South Pole; and near the equator, some swim north and others south. However, they also swim downward in water, because the magnetic force from the earth's poles is deflected through the earth and not over its horizon. This phenomenon is called *magnetotaxis.* It appears to help these anaerobic bacteria move down toward sediments where their food (iron oxide) is abundant and where oxygen, which they cannot tolerate, is deficient.

Magnetotactic bacteria live in mud and brackish waters, and more than a dozen species have been identified. Most have a single flagellum, but *Aquaspirillum magnetotacticum* has two, one at each end, so it can swim forward or backward. When magnetotactic bacteria with one flagellum are placed in the field of an electromagnet, they make U-turns as the poles of the magnet are reversed.

Magnetosomes are nearly constant in size and are oriented in parallel chains like a string of tiny magnets. Experiments are currently under way to use these bacteria in the manufacture of magnets, audiotapes, and videotapes.

TRY IT: You can easily find your own magnetotactic bacteria. They are very common organisms. Bring back a bucket with a few inches of mud from a pond and enough pondwater to fill the rest of the bucket. Try mud from different locations: fresh, salt, and brackish waters. Cover the

bucket and store it in the dark (you don't want to grow algae) for about 1 month. Remove a large beaker of the water, and place a magnet against the outside of the glass. Allow this to stand for a day or two, until you see a whitish spot in the water near the magnet's end. Which attracts more organisms, the north or south end of the magnet? With a pipette, remove a sample of the cloudy liquid and examine it under the microscope. When you lay a magnet on the stage, any magnetotactic bacteria that are present will orient themselves to the field. They can be purified by streaking on agar, a process described in Chapter 6. Microbiology doesn't exist only in books or labs. It's everywhere around you in the real world—just look for it!

TEM photo of the magnetotactic bacterium *Aquaspirillum magnetotacticum.* The dark inclusions, called *magnetosomes,* are composed of iron oxide, Fe_3O_4. The magnetosomes enable these organisms to orient themselves in a magnetic field.

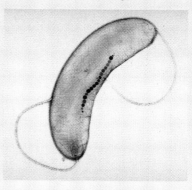

found only in bacteria, contains deposits of poly-β-hydroxybutyrate. These lipid deposits serve as storehouses of energy and as sources of carbon for building new molecules.

Endospores

The properties of bacterial cells just described pertain to **vegetative cells,** or cells that are metabolizing nutrients. However, vegetative cells of some bacteria, such as *Bacillus* and *Clostridium*, produce resting stages called **endospores.** Although bacterial endospores are commonly referred to simply as *spores,* do not confuse them with fungal spores. A bacterium produces a single endospore, which merely helps that organism survive and is not a means of reproduction. A fungus produces numerous spores, which help the organism survive and provide a means of reproduction.

Endospores, which are formed within cells, contain very little water and are highly resistant to heat, drying, acids, bases, certain disinfectants, and even radiation. The depletion of a nutrient will usually induce a large number of cells to produce spores. However, many investigators believe that spores are part of the normal life cycle and that a few are formed even when nutrients are adequate and environmental conditions are favorable. Thus, *sporulation,* or endospore formation, seems to be a means by which some bacteria prepare for the possibility of adverse conditions.

Structurally, an endospore consists of a *core,* surrounded by a *cortex,* a *spore coat,* and in some species a delicately thin layer called the *exosporium* (Figure 4.10). The core has an outer core wall, a cell membrane, nuclear region, and other cell components. Unlike vegetative cells, endospores contain *dipicolinic acid* and a large quantity of calcium ions (Ca^{2+}). These materials, which

are probably stored in the core, appear to contribute to the heat resistance of endospores, as does their very low water content.

Endospores are capable of surviving adverse environmental conditions for long periods of time. Some withstand hours of boiling. When conditions become more favorable, endospores *germinate,* or begin to develop into functional vegetative cells. (The processes of spore formation and germination are discussed in Chapter 6.) Because endospores are so resistant, special methods must be used to kill them during sterilization. Otherwise, they germinate and grow in media thought to be sterile. Methods to ensure that endospores are killed when bacterial media or foods are sterilized are described in Chapter 13.

External Structure

In addition to cell walls, many bacteria have structures that extend beyond or surround the cell wall. *Flagella* and *pili* extend from the cell membrane through the cell wall and beyond it. *Capsules* and *slime layers* surround the cell wall. These structures have a variety of properties and functions.

Flagella

About half of all known bacteria are *motile,* or capable of movement. They often move with speed and apparent purpose, and they usually move by means of long, thin, helical appendages called **flagella** (singular: *flagellum*). A bacterium can have one flagellum or two or many flagella. Bacteria with a single flagellum located at one end, or pole, are said to be **monotrichous** (mon-o-trik'-us; Figure 4.11a); bacteria with two flagella, one at each end, are **amphitrichous** (am-fet-rik'-us; Figure 4.11b); both types are said to be *polar*. Bacteria with two or more flagella at one or both ends are **lophotrichous** (lo-fot-rik'us; Figure 4.11c); and those with flagella all over the surface are **peritrichous** (pe-rit-rik' us; Figure 4.11d). Bacteria without flagella are **atrichous** (a-trik' us).

The diameter of a prokaryote's flagellum is about one-tenth that of a eukaryote's flagellum. It is made of protein subunits called *flagellin*. Each flagellum is attached to the cell membrane by a basal region consisting of a protein other than flagellin (Figure 4.12, on page 88). The basal region has a hooklike structure and a complex *basal body*. The basal body consists of a central rod or shaft surrounded by a set of rings. Gram-negative bacteria have a pair of rings embedded in the cell membrane and another pair of rings associated with the peptidoglycan and lipopolysaccharide layers of the cell wall. Gram-positive bacteria have one ring embedded in the cell membrane and another in the cell wall.

Most flagella rotate like twirling L-shaped hooks, such as a dough hook on a kitchen mixer or the rotat-

FIGURE 4.10 Color-enhanced electron micrograph of an endospore within a *Clostridium perfringens* cell.

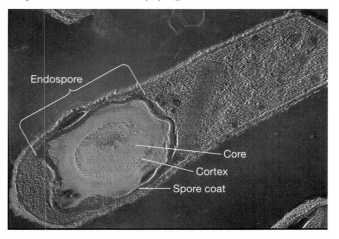

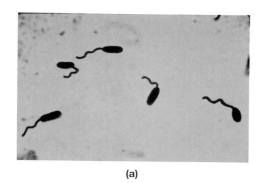
(a)

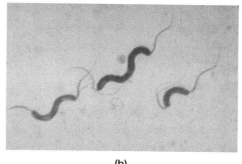
(b)

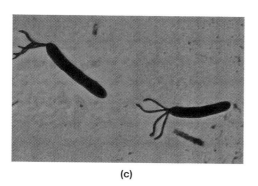
(c)

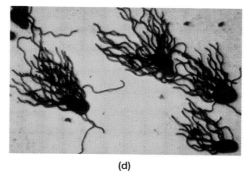
(d)

FIGURE 4.11

Arrangements of bacterial flagella. (a) Polar, monotrichous (single flagellum at one end) *Pseudomonas* (3300×). (b) Polar, amphitrichous (single flagellum at each end) *Spirillum*. (c) Lophotrichous (tuft of flagella at one or both ends) *Spirillum*. (d) Peritrichous (flagella distributed all over) *Salmonella* (1200×).

ing string on a hand-carried grass trimmer. Motion is thought to occur as energy from the expulsion of protons (Chapter 5) is used to make one of the rings in the cell membrane rotate with respect to the other. When flagella rotate counterclockwise, the bacteria *run*, or move in a straight line. When the flagella rotate clockwise, the bacteria *twiddle*, or tumble randomly. Both runs and twiddles are generally random movements; that is, no one direction of movement is more likely than any other direction.

CHEMOTAXIS   Sometimes bacteria move toward or away from substances in their environment by a nonrandom process called **chemotaxis** (Figure 4.13, on page 89). Concentrations of most substances in the environment vary along a gradient—that is, from high to low concentration. When a bacterium is moving in the direction of increasing concentration of an attractant (such as a nutrient), it tends to lengthen its runs and to reduce the frequency of its twiddles. When it is moving away from the attractant, it shortens its runs and increases the frequency of its twiddles. Even though the direction of the individual runs is still random, the net result is movement toward the attractant, or *positive chemotaxis*. Movement away from a repellent, or *negative chemotaxis*, results from the opposite responses: long runs and few twiddles while the bacterium moves in the direction of lower concentration of the harmful substance; short runs and many twiddles while it moves in the direction of higher concen-

tration. The exact mechanism that produces these behaviors is not fully understood, but certain structures on bacterial cell surfaces can detect changes in concentration over time.

PHOTOTAXIS   Some bacteria can move toward or away from light; this response is called **phototaxis.** Bacteria that move toward light exhibit *positive phototaxis*, whereas those that move away from light exhibit *negative phototaxis*. This process may be accomplished by means of flagella. Or, in the case of some photosynthetic aquatic bacteria, oil droplet inclusions in their cytoplasm may give them the buoyancy to rise toward the water surface, where light is more available.

Axial Filaments

Spirochetes have **axial filaments,** or **endoflagella,** instead of flagella that extend beyond the cell wall (Figure 4.14, on page 89). Each filament is attached at one of its ends to an end of the cytoplasmic cylinder that forms the body of the spirochete. Axial filaments cause the rigid spirochete body to rotate like a corkscrew when they twist inside the outer sheath.

Pili

Pili (singular: *pilus*) are tiny, hollow projections. They are used to attach bacteria to surfaces and are not involved in movement. A pilus is composed of subunits of the protein *pilin*. Bacteria can have two kinds of pili

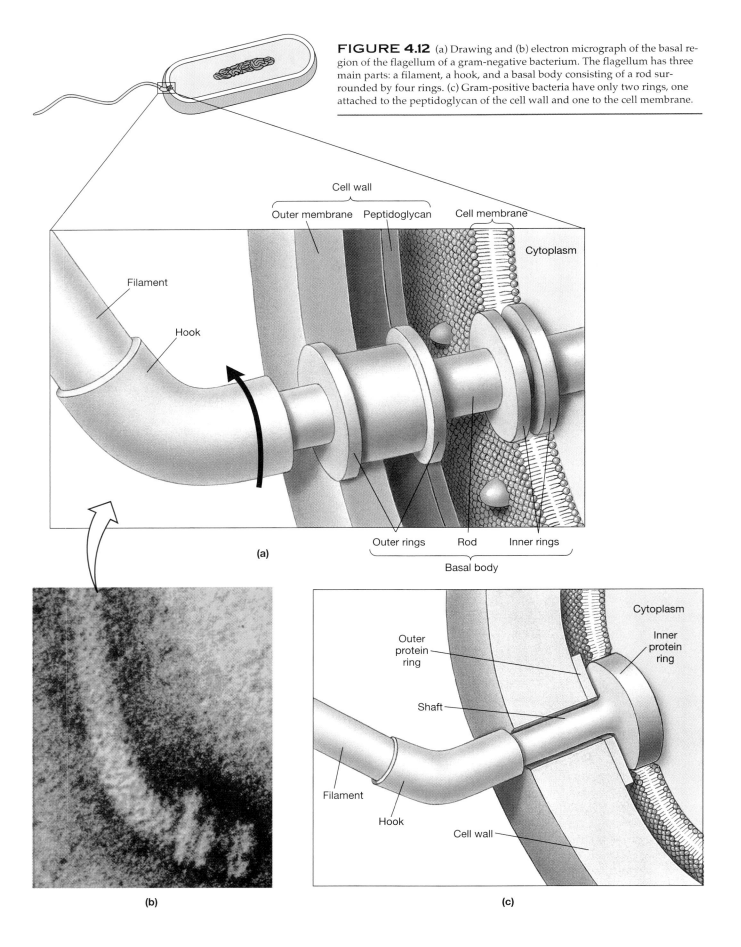

FIGURE 4.12 (a) Drawing and (b) electron micrograph of the basal region of the flagellum of a gram-negative bacterium. The flagellum has three main parts: a filament, a hook, and a basal body consisting of a rod surrounded by four rings. (c) Gram-positive bacteria have only two rings, one attached to the peptidoglycan of the cell wall and one to the cell membrane.

Cell wall

Outer membrane   Peptidoglycan   Cell membrane

Cytoplasm

Filament

Hook

Outer rings   Rod   Inner rings

Basal body

(a)

(b)

Cytoplasm

Outer protein ring

Inner protein ring

Shaft

Filament

Hook

Cell wall

(c)

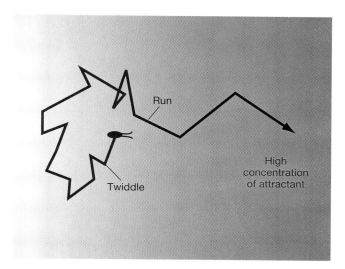

FIGURE 4.13 Movements of a typical bacterium exhibiting chemotaxis. An attractive chemical, such as a nutrient, is concentrated toward the right. When the bacterium is moving toward the attractant, it increases the length of its runs and decreases the number of twiddles. When it is moving away from the attractant, it decreases the length of its runs and increases the frequency of its twiddles.

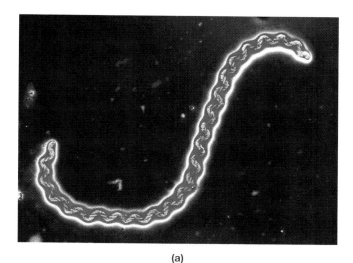

(a)

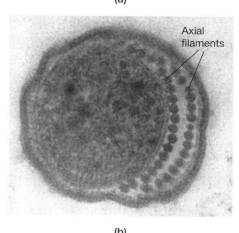

(b)

FIGURE 4.14 (a) Axial filaments made visible by false coloring are clearly seen as yellow ribbons running along the body of the spirochete *Leptospira interrogans* (50,000×). (b) TEM photo (cross section) of a spirochete, showing numerous axial filaments (dark circles). Axial filaments lie between the outer sheath and the cell wall.

(Figure 4.15): (1) long *conjugation pili,* or *F pili* (also called sex pili), and (2) short *attachment pili,* or *fimbriae* (fim'-bre-e; singular: *fimbria*).

CONJUGATION PILI   Conjugation pili (or *sex pili*), found only in certain groups of bacteria, attach two cells and may furnish a pathway for the transfer of the genetic material DNA. This transfer process is called *conjugation* (Figures 4.15 and 8.7). Transfer of DNA furnishes genetic variety for bacteria as sexual reproduction does for many other life forms. Such transfers among bacteria cause problems for humans because antibiotic resistance can be passed on with the DNA transfer. Consequently, more and more bacteria acquire resistance, and humans must look for new ways to control the growth of these bacteria.

ATTACHMENT PILI   Attachment pili, or **fimbriae,** help bacteria adhere to surfaces, such as cell surfaces and the interface of water and air. They contribute to the *pathogenicity* of certain bacteria—their ability to produce disease—by enhancing colonization (the development of colonies) on the surfaces of the cells of other organisms. For example, some bacteria adhere to red blood cells by attachment pili and cause the blood cells to clump, a process called *hemagglutination.* In certain species of bacteria, some individuals have attachment pili and others lack them. In *Neisseria gonorrhoeae,* strains without pili are rarely able to cause gonorrhea, but those with pili are highly infectious because they attach to epithelial cells of the urogenital

FIGURE 4.15 Electron micrograph of an *Escherichia coli* cell (13,000×), showing two kinds of pili. The shorter ones are fimbriae, used for attachment to surfaces. The long tube reaching to another cell is a conjugation pilus, perhaps used to transfer DNA.

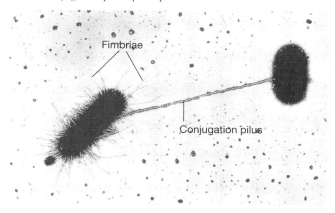

system. Such pili also allow them to attach to sperm cells and thereby spread to the next individual.

Some aerobic bacteria form a shiny or fuzzy, thin layer at the air-water interface of a broth culture. This layer, called a **pellicle,** consists of many bacteria that adhere to the surface by their attachment pili. Thus, attachment pili allow the organisms to remain in the broth, from which they take nutrients, while they congregate near air, where the oxygen concentration is greatest.

Glycocalyx

Glycocalyx is the currently accepted term used to refer to all polysaccharide-containing substances found external to the cell wall, from the thickest *capsules* to the thinnest *slime layers*. All bacteria have at least a thin slime layer.

CAPSULE   A **capsule** is a protective structure outside the cell wall of the organism that secretes it. Only certain bacteria are capable of forming capsules, and not all members of a species have capsules. For example, the bacterium that causes anthrax, a disease found mainly in cattle, does not produce a capsule when it grows outside an organism but does when it infects an animal. Capsules typically consist of complex polysaccharide molecules arranged in a loose gel. However, the chemical composition of each capsule is unique to the strain of bacteria that secreted it. When encapsulated bacteria invade a host, the capsule prevents host defense mechanisms, such as phagocytosis, from destroying the bacteria. If bacteria lose their capsules, they become less likely to cause disease and more vulnerable to destruction.

SLIME LAYER   A **slime layer** is less tightly bound to the cell wall and is usually thinner than a capsule. When present, it protects the cell against drying, helps trap nutrients near the cell, and sometimes binds cells together. Slime layers allow bacteria to adhere to objects in their environments, such as rock surfaces or the root hairs of plants, so that they can remain near sources of nutrients or oxygen (Figure 4.16). Some oral bacteria, for example, adhere by their slime layers and form dental plaque. The slime layer keeps the bacteria in close proximity to the tooth surface, where they can cause dental caries. Plaque is extremely tightly bound to tooth surfaces. If not removed regularly by brushing, it can be removed only by a dental professional in a procedure called *scaling*.

EUKARYOTIC CELLS

Overview of Structure

Eukaryotic cells are larger and more complex than prokaryotic cells. Most eukaryotic cells have a diame-

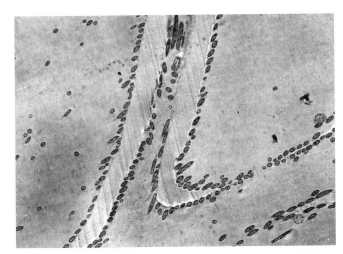

FIGURE 4.16 *Sporocytophaga* bacteria growing on a cellulose fiber, to which they adhere by means of their slime layer.

ter of more than 10 μm, and many are much larger. They also contain a variety of highly differentiated structures. These cells are the basic structural unit of all organisms in the kingdoms Protista, Plantae, Fungi, and Animalia (Chapter 9). Eukaryotic organisms include microscopic protozoa, algae, and fungi and are thus appropriately considered in microbiology. The general structure of the eukaryotic cell is shown diagramatically in Figure 4.17.

Plasma Membrane

The cell membrane, or **plasma membrane,** of a eukaryotic cell has the same fluid-mosaic structure as that of a prokaryotic cell. In addition, eukaryotes also contain several organelles bounded by membranes that have a similar membrane structure.

Eukaryotic membranes differ from prokaryotic membranes in some respects, especially in the greater variety of lipids they contain. Eukaryotic membranes contain sterols, found among prokaryotes only in the mycoplasmas. Sterols add rigidity to a membrane, and this may be important in keeping membranes intact in eukaryotic cells. Because of their larger size, eukaryotic cells have a much lower surface-to-volume ratio than prokaryotic cells. As the volume of cytoplasm enclosed by a membrane increases, the membrane is placed under greater stress. The sterols in the membrane may help it withstand the stress.

Functionally, eukaryotic plasma membranes are less versatile than prokaryotic ones. They do not have respiratory enzymes that capture metabolic energy and store it in ATP; in the course of evolution, that function has been taken over by mitochondria.

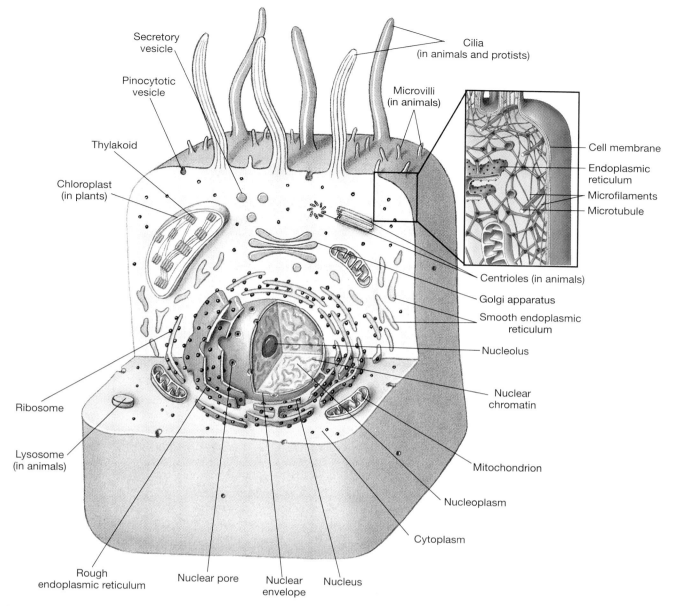

Secretory
vesicle

Pinocytotic
vesicle

Thylakoid

Chloroplast
(in plants)

Cilia
(in animals and protists)

Microvilli
(in animals)

Cell membrane

Endoplasmic
reticulum

Microfilaments

Microtubule

Centrioles (in animals)

Golgi apparatus

Smooth endoplasmic
reticulum

Nucleolus

Nuclear
chromatin

Ribosome

Lysosome
(in animals)

Mitochondrion

Nucleoplasm

Cytoplasm

Rough
endoplasmic reticulum

Nuclear pore

Nuclear
envelope

Nucleus

FIGURE 4.17 Diagram of a generalized eukaryotic cell. Most of the features shown are present in nearly all eukaryotic cells, but some (the centrioles, microvilli, and lysosomes) occur only in animal cells, and others (the chloroplast) are found only in cells capable of carrying out photosynthesis.

Internal Structure

The internal structure of eukaryotic cells is exceedingly more complex than that of prokaryotic cells. It is also much more highly organized and contains numerous organelles.

Cytoplasm

The cytoplasm makes up a relatively smaller portion of eukaryotic cells than of prokaryotic cells because it contains a *nucleus* and many organelles. Like the cyto-

plasm of prokaryotic cells, the cytoplasm of eukaryotic cells is a semifluid substance consisting mainly of water with the same substances dissolved in it. In addition, this cytoplasm contains elements of a *cytoskeleton*, a fibrous network that give these larger cells shape and support.

Cell Nucleus

The most obvious difference between eukaryotic and prokaryotic cells is the presence of a nucleus in the eukaryotic cells. The **cell nucleus** (Figure 4.18) is a

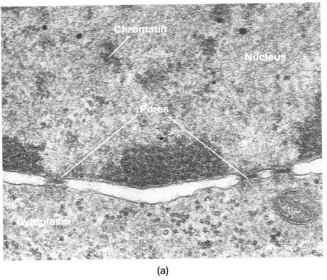
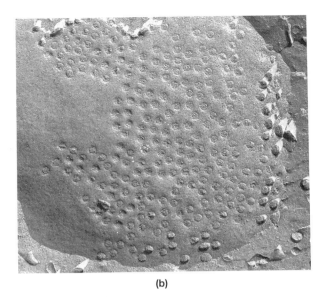

(a)

(b)

FIGURE 4.18 (a) TEM photo of a cell nucleus. The dark, granular material is chromatin. The pores in the nuclear membrane allow for entry and exit of materials. (b) Freeze-fracture TEM photo of a nucleus (compare with Figure 3.22). The many circular structures are nuclear pores.

distinct organelle with a nuclear envelope, nucleoplasm, nucleoli, and (typically paired) chromosomes. The **nuclear envelope** consists of a double membrane, each layer of which is structurally like the plasma membrane. **Nuclear pores** in the envelope allow RNA molecules to leave the semifluid portion of the nucleus, known as **nucleoplasm,** and to participate in protein synthesis. Each nucleus has one or more **nucleoli** (singular: *nucleolus*), which contain a significant amount of RNA and serve as sites for the assembly of ribosomes.

Also present in the nucleus of most eukaryotic organisms are paired **chromosomes,** each of which contains DNA and proteins called **histones.** Histones contribute directly to the structure of chromosomes, and other proteins probably regulate the chromosomes' function. During cell division, the chromosomes are extensively coiled and folded into compact structures. Between divisions, however, the chromosomes are visible only as a tangle of fine threads called **chromatin** that gives the nucleus a granular appearance.

The nuclei of eukaryotic cells divide by the process of **mitosis** (Figure 4.19a). Prior to the actual division of the nucleus, the chromosomes replicate but remain attached, forming **dyads.** In most eukaryotic cells, the nuclear envelope fragments during mitosis, and a system of tiny fibers called the **spindle apparatus** guides the movement of chromosomes. Dyads aggregate in the center of the spindle and separate into single chromosomes as they move along fibers to the poles of the spindle. Each new cell receives one copy of each chromosome that was present in the parent cell. Because the parent cell contained paired chromosomes, the prog-

eny likewise contain paired chromosomes. Cells with paired chromosomes are said to be **diploid** (2*N*) cells.

During sexual reproduction, the nuclei of sex cells divide by a process called **meiosis** (Figure 4.19b). After the chromosomes replicate, forming dyads, pairs of dyads come together. During the course of two cell divisions, the dyads are distributed to four new cells. Thus, each cell receives only one chromosome from each pair. Such cells are said to be **haploid** (1*N*) cells. Haploid cells can become gametes or spores. **Gametes** are haploid cells that participate in sexual reproduction; gametes from each of two parent organisms unite to form a diploid **zygote,** the first cell of a new individual. Some **spores** become dormant, whereas others reproduce by mitosis as haploid vegetative cells. Dormant spores allow for survival during adverse environmental conditions. When conditions improve, the spores germinate and begin to divide. Eventually some of these cells produce gametes, which can unite to form zygotes. Thus, the organism alternates between haploid and diploid generations.

Mitochondria and Chloroplasts

Mitochondria, known as the powerhouses of eukaryotic cells, are exceedingly important organelles. They are quite numerous in some cells and can account for up to 20 percent of the cell volume. Mitochondria are complex structures about 1 μm in diameter with an outer membrane, an inner membrane, and a fluid-filled **matrix** inside the inner membrane (Figure 4.20). The inner membrane is extensively folded to form **cristae,**

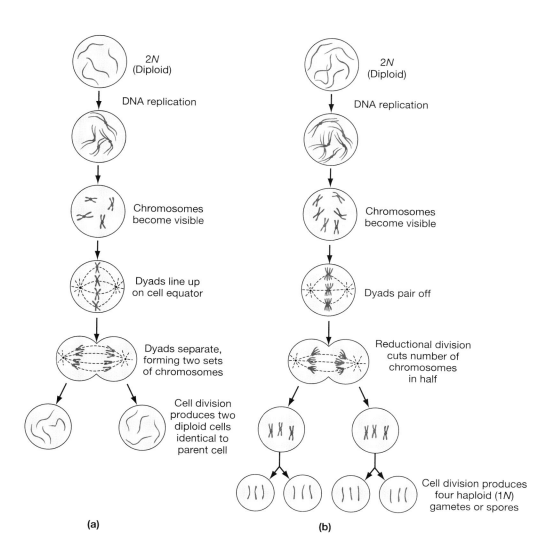

FIGURE 4.19

Comparison of (a) mitosis and (b) meiosis. Both processes are preceded by duplication of DNA; soon after, the chromosomes become visible. In mitosis, two identical daughter cells with the same number and kinds of chromosomes are formed. In meiosis, two divisions give rise to four cells, each with half the number of chromosomes as the original parent cell. For this reason, meiosis is sometimes called *reduction division.*

(a) 2N (Diploid) → DNA replication → Chromosomes become visible → Dyads line up on cell equator → Dyads separate, forming two sets of chromosomes → Cell division produces two diploid cells identical to parent cell

(b) 2N (Diploid) → DNA replication → Chromosomes become visible → Dyads pair off → Reductional division cuts number of chromosomes in half → Cell division produces four haploid (1N) gametes or spores

which extend into the matrix. Mitochondria carry out the oxidative reactions that capture energy in adenosine triphosphate (ATP). Energy in ATP is in a form usable by cells for their activities.

Eukaryotic cells capable of carrying out photosynthesis contain **chloroplasts** (Figure 4.21). They too have an outer and an inner membrane. The inner **stroma** of these organelles corresponds structurally with the matrix of mitochondria. Unlike mitochondria, choloroplasts have separate internal membranes, called **thylakoids,** that contain the pigment *chlorophyll,* which captures energy from light during photosynthesis. Both mitochondria and chloroplasts contain DNA and can replicate independently of the cell in which they function. This and other evidence has led many biologists to speculate that these organelles may have originated as free-living organisms. (See the essay at the end of this chapter.)

Ribosomes

Ribosomes of eukaryotic cells, which are larger than those of prokaryotic cells, are about 60 percent RNA and 40 percent protein. They have a sedimentation rate of 80S, and their subunits have sedimentation rates of 60S and 40S. All ribosomes provide sites for protein synthesis, and some are arranged in chains as polyribosomes. Those that are attached to an organelle called the *endoplasmic reticulum* usually make proteins for secretion from the cell; those that are free in the cytoplasm usually make proteins for use in the cell.

Endoplasmic Reticulum

The **endoplasmic reticulum** (Figure 4.17) is an extensive system of membranes that forms numerous tubes and plates in the cytoplasm. Endoplasmic reticulum can be smooth or rough textured. *Smooth endoplasmic reticulum* contains enzymes that synthesize lipids, especially those to be used in making membranes. *Rough endoplasmic reticulum* has ribosomes bound to its surface, which give it a rough texture. Its function is, along with the ribosomes, to manufacture proteins. Vesicles from this membrane system transport to the Golgi apparatus the lipids and proteins synthesized in or on the endoplasmic reticulum membrane.

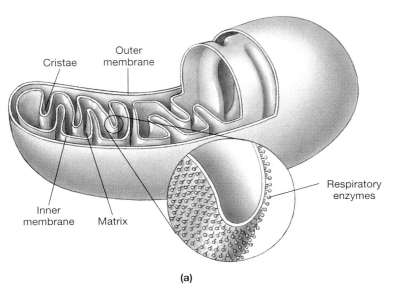

Cristae

Outer
membrane

Inner
membrane

Matrix

Respiratory
enzymes

(a)

FIGURE 4.20 (a) Diagram of a mitochondrion. Respiratory enzymes that make ATP are located on the surfaces of the inner membrane and the cristae, which are infoldings of the inner membrane. (b) TEM photo of two mitochondria, one in longitudinal section and the other in cross section.

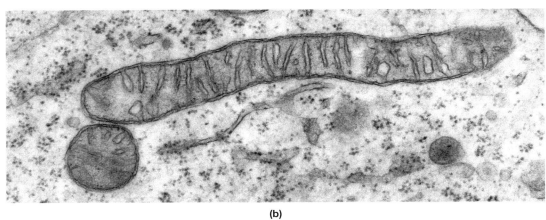

(b)

FIGURE 4.21 (a) Diagram of a eukaryotic chloroplast. The thylakoid membranes contain chlorophyll and other pigments and enzymes needed for photosynthesis. Thylakoids occur in stacks called *grana;* grana are joined by membranous flat sheets called *lamellae.* (b) False-color TEM photo of a chloroplast (magnified 24,000×) from a leaf of corn.

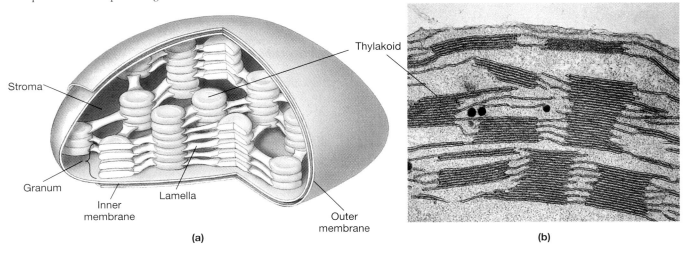

Stroma

Granum

Inner
membrane

Lamella

Thylakoid

Outer
membrane

(a)

(b)

Golgi Apparatus

The **Golgi** (gol'je) **apparatus** (Figure 4.17) consists of a stack of flattened membrane sacs. The Golgi apparatus receives substances transported from the endoplasmic reticulum, stores the substances, and typically alters their chemical structure. It packages these substances in small segments of membrane called **secretory vesicles.** The secretory vesicles fuse with the plasma membrane and release secretions to the exterior of the cell. The Golgi apparatus also helps to form the plasma membrane and membranes of the lysosomes.

Lysosomes

Lysosomes (Figure 4.17) are extremely small, membrane-covered organelles made by the Golgi apparatus in animal cells. They contain digestive enzymes that could destroy a cell if those enzymes were released into the cytoplasm. Lysosomes fuse with *vacuoles* that form as a cell ingests substances, and they release enzymes that digest the substances in the vacuoles.

Peroxisomes

Peroxisomes are small, membrane-bound organelles filled with enzymes. Peroxisomes are found in both plant and animal cells but appear to have different functions in the two kinds of cells. In animal cells their enzymes oxidize amino acids, whereas in plant cells they typically oxidize fats. Peroxisomes are so named because their enzymes convert hydrogen peroxide to water in both plant and animal cells. If hydrogen peroxide were to accumulate in cells, it would kill them just as it kills bacteria when humans use it as an antiseptic.

Vacuoles

In eukaryotic cells **vacuoles** are membrane-bound structures that store materials such as starch, glycogen, or fat to be used for energy. Some vacuoles form when cells engulf food particles. As we have already noted, the contents of these vacuoles are eventually digested by lysosomal enzymes.

Cytoskeleton

The **cytoskeleton** is a network of protein fibers made of **microtubules** (which are hollow tubes) and **microfilaments** (which are filamentous fibers). The cytoskeleton supports and gives rigidity and shape to a cell. It also is involved in cell movements, such as those that occur when cells engulf substances or when they make *amoeboid movements* (to be explained in a later section).

External Structure

Like external structures of prokaryotic cells, external structures of eukaryotic cells either assist with movement or provide a protective covering for the plasma membrane. These structures include flagella, cilia, and cell walls and other coverings. Although *pseudopodia* are not, strictly speaking, external structures, they do achieve movement and so are discussed here. Cells of algae and macroscopic green plants have cell walls, and some protozoa have special cell coverings.

Flagella

Flagella in eukaryotes, which are larger and more complex than those in prokaryotes, consist of two central microtubules and nine pairs of peripheral microtubules (a 9 + 2 arrangement) surrounded by a membrane (Figure 4.22). Each fiber is a microtubule made of the protein *tubulin*. One of these microtubules is about the same size as an entire prokaryotic flagellum. Associated

FIGURE 4.22 Comparison of (a) prokaryotic and (b) eukaryotic flagella. Note the substantial difference in the diameter of these two structures.

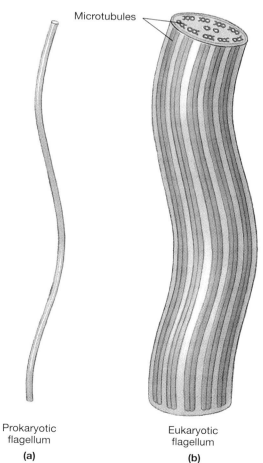

Microtubules

Prokaryotic flagellum
(a)

Eukaryotic flagellum
(b)

with each pair of peripheral microtubules are small molecules of the protein *dynein.* Eukaryotic flagella move like a whip, whereas prokaryotic flagella move like a rotating hook. One mechanism of eukaryotic flagellar movement is a crossbridging among dynein and other flagellar proteins. Through ATP hydrolysis, dynein plays a role in converting chemical energy in ATP to mechanical energy that makes the flagellum move. Microtubules in the flagellum are thought to slide toward or away from the base of the cell and thereby cause the whole flagellum to move.

Flagella are most common among protozoa but are found among algae as well. Most flagellated eukaryotes have one flagellum, but some have two or more.

Cilia

Cilia are shorter and more numerous than flagella, but they have the same chemical composition and basic arrangement of microtubules. Cilia are found mainly among ciliated protozoa, which have 10,000 or more cilia distributed over their cell surface (Figure 4.23). Each cilium passes through a stroke-and-recovery cycle as it beats. Together the cilia of an organism beat in a coordinated pattern, which creates a wave that passes from one end of the organism to the other. The large number of cilia and their coordinated beating allow ciliated organisms, such as paramecia, to move much more rapidly than those with flagella. Cilia on some cells can also propel fluids, dissolved particles, bacteria, mucus, and so on past the cell. This function can be of great importance in host defenses against diseases, particularly in the respiratory tract.

Pseudopodia

Pseudopodia (su"do-po'de-a; singular: *pseudopodium*), or "false feet," are temporary projections of cytoplasm associated with **amoeboid movement.** This kind of movement occurs only in cells without walls, such as amoebas and some white blood cells, and only when the cell is resting on a solid surface. When an amoeba first extends a portion of its body to form a pseudopodium, the cytoplasm is much less dense in the pseudopodium than in other areas of the cell (Figure 4.24). As a result, cytoplasm from elsewhere in the organism flows into the pseudopodium by **cytoplasmic streaming.** Amoeboid movement is a slow, inching-along process.

Cell Walls

A number of unicellular eukaryotic organisms have cell walls, none of which contains the peptidoglycan that is characteristic of bacteria. Algal cell walls consist mainly of cellulose, but some contain other polysaccharides. Cell walls of fungi consist of cellulose or chitin, or both. *Chitin* is a structural polysaccharide that is also common in the exoskeletons of arthropods such as insects and

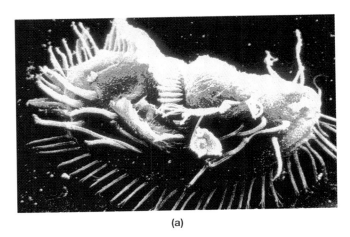

(a)

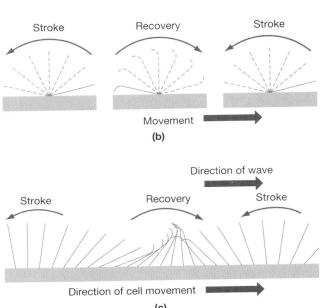

FIGURE 4.23 (a) SEM photo of the ciliated protozoan *Oxytricha.* (b) The stroke-and-recovery motion of a cilium. (c) Cilia on an organism move in a synchronized fashion, creating a wave that propels the organism forward.

crustacea. Regardless of composition, cell walls give cells rigidity and protect them from bursting when water moves into them from the environment.

MOVEMENT OF SUBSTANCES ACROSS MEMBRANES

A living cell, either prokaryotic or eukaryotic, is a dynamic entity. A cell is separated from its environment by a membrane, across which substances constantly move in a carefully controlled manner. Understanding how these movements occur is essential to understanding how a cell functions. Very small polar substances, such as water, small ions, and small water-

FIGURE 4.24 (a) Formation of a pseudopodium, a cytoplasmic extension that allows organisms such as amoebas to move and capture food. (b) Micrographs of an amoeba engulfing food particles (magnified 160×).

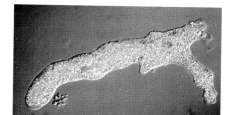

Resting amoeba with cytoplasm distributed evenly.

Newly formed pseudopodium with less dense cytoplasm.

(a)

 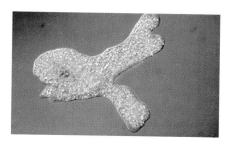

(b)

soluble molecules, probably pass through pores in the membrane. Nonpolar substances, such as lipids and other uncharged particles (molecules or ions), dissolve in and pass through the membrane lipids. Still other substances are moved through the membrane by carrier molecules. Most large molecules are unable to enter cells without the aid of specific carriers.

The mechanisms by which substances move across membranes can be passive or active. In passive transport, the cell expends no energy to move substances down a *concentration gradient,* that is, from higher to lower concentration. Passive processes include *simple diffusion, facilitated diffusion,* and *osmosis.* In active processes, the cell expends energy from ATP, enabling it to transport substances against a concentration gradient. These processes include *active transport.* The processes *endocytosis* and *exocytosis,* which occur only

in eukaryotic cells, are separate mechanisms for moving substances across the plasma membrane.

Simple Diffusion

All molecules have kinetic energy; that is, they are constantly in motion and are continuously redistributed. **Simple diffusion** is the net movement of particles from a region of higher to lower concentration (Figure 4.25). Suppose, for example, that you drop a lump of sugar into a cup of coffee. At first a concentration gradient exists, with the sugar concentration greatest at the lump and least at the rim of the cup. Eventually, though, sugar molecules will become evenly distributed throughout the coffee (they reach *equilibrium*) even without stirring.

Diffusion occurs because of random movement of particles. Although particles move at high velocity, they

FIGURE 4.25 Simple diffusion. The random movements of molecules causes them to spread out (diffuse) from an area of high concentration to areas of lower concentration until eventually they are equally distributed throughout the available space (that is, they reach equilibrium).

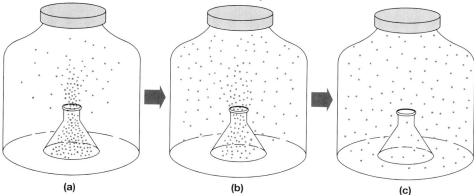

(a)　　　　(b)　　　　(c)

do not travel far in a straight line before they collide with other randomly moving particles. Even so, some particles from a region of high concentration eventually move toward a region of lower concentration. Fewer particles move in the opposite direction for two reasons: (1) There are fewer of them in regions of low concentration to begin with, and (2) they are likely to be repelled by collision with particles from a region of high concentration.

The length of time required for particles to diffuse across a cell increases with cell diameter. Materials can diffuse throughout small prokaryotic cells very quickly and throughout larger eukaryotic cells fast enough to supply nutrients and to remove wastes fairly efficiently. If cells were much larger, diffusion throughout the cell would be too slow to sustain life, so diffusion rates may be responsible in part for limiting the size of cells.

Any membrane severely limits diffusion, but many substances diffuse through the lipids of membranes. Diffusion through the phospholipid bilayer is affected by several factors: (1) the solubility of the diffusing substance in lipid, (2) the temperature, and (3) the difference between the highest and lowest concentration of the diffusing substance. Nonpolar substances such as steroids and gases (CO_2, O_2) cross the membrane rapidly by dissolving in the nonpolar fatty acid tails of the membrane phospholipids.

A few substances also diffuse through pores. Such diffusion is affected by the size and charge of the diffusing particles and the charges on the pore surface. Pores probably have a diameter of less than 0.8 nm, so only water, small water-soluble molecules, and ions such as H^+, K^+, Na^+, and Cl^- pass through them. This is one reason a membrane is said to be **selectively permeable** (*semipermeable*).

Facilitated Diffusion

Facilitated diffusion is diffusion down a concentration gradient and across a membrane with the assistance of special pores or carrier molecules. In fact, membranes contain protein-lined pores for specific ions. These pores have an arrangement of charges that allows rapid passage of a particular ion. The carrier molecules are proteins, embedded in the membrane, that bind to one or to a few specific molecules and assist in their movement. By one possible mechanism for facilitated diffusion, a carrier acts like a revolving door or shuttle that provides a convenient one-way channel for the movement of substances across a membrane (Figure 4.26). Carrier molecules can become saturated, and similar molecules sometimes compete for the same carrier. Saturation occurs when all the carrier molecules are moving the diffusing substance as fast as they can. Under these conditions the rate of diffusion reaches a maximum and cannot increase further. When a carrier molecule can transport more than one substance, the substances compete for the carrier in proportion to their

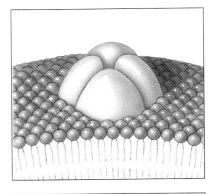

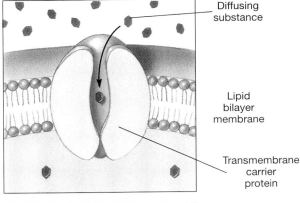

Diffusing substance

Lipid bilayer membrane

Transmembrane carrier protein

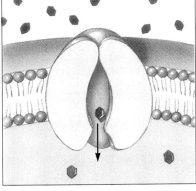

FIGURE 4.26 Facilitated diffusion. Carrier protein molecules aid in the movement of substances through the cell membrane, but only down their concentration gradient (from a region where their concentration is high to one where their concentration is low). This process does not require the expenditure of any energy (ATP) by the cell.

concentrations. For example, if there is twice as much of substance A as substance B, substance A will move across the membrane twice as fast as substance B.

Osmosis

Osmosis is a special case of diffusion—in which water molecules diffuse across a selectively permeable membrane. To demonstrate osmosis, we start with two compartments separated by a membrane permeable only to water. One compartment contains pure water, and the other compartment contains some large, nondiffusible molecules, such as proteins (Figure 4.27a). Water molecules move in both directions, but their net movement is from pure water (concentration 100 percent) toward the water that contains other molecules (concentration less than 100 percent; Figure 4.27b). Thus, osmosis is the net flow of water molecules from a region of higher concentration of those molecules to a region of lower concentration across a semipermeable membrane (Figure 4.27c).

Osmotic pressure is defined as the pressure required to *prevent* the net flow of water by osmosis. The least amount of hydrostatic pressure required to prevent the movement of water from a given solution into pure water is the osmotic pressure of the solution. The osmotic pressure of a solution is proportional to the number of particles dissolved in a given volume of that solution. Thus, NaCl and other salts that form two ions per molecule exert twice as much osmotic pressure as glucose and other substances that do not ionize, provided each compound is present at the same concentration.

FIGURE 4.27 Osmosis: the diffusion of water from (a) an area of higher water concentration (the right side) to an area of lower water concentration (the left side) through a semipermeable membrane. (b, c) Here the net movement of water is into the sugar solution because the concentration of water there is slightly lower than on the other side of the membrane.

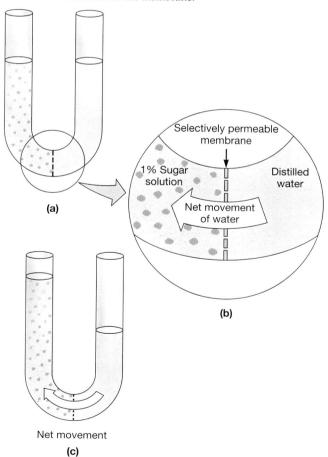

(a)

Selectively permeable membrane

1% Sugar solution

Distilled water

Net movement of water

(b)

Net movement

(c)

The important thing for a microbiologist to know about osmosis and osmotic pressure is how particles dissolved in fluid environments affect microorganisms in those environments (Figure 4.28). For this purpose, tonicity is a useful concept. *Tonicity* describes the behavior of cells in a fluid environment. The cells are the reference point, and the fluid environments are compared to the cells. The fluid surrounding cells is **isotonic** to the cells when no change in cell volume occurs (Figure 4.28a). The fluid is **hypertonic** to the cells if the cells shrivel or shrink as water moves out of them into the fluid environment (Figure 4.28b); it is **hypotonic** to the cells if the cells swell or burst as water moves from the environment into the cells (Figure 4.28c). Although bacteria become dehydrated and their cytoplasm shrinks away from the cell wall in a hypertonic environment, their cell walls usually prevent them from swelling or bursting in the hypotonic environments they typically inhabit.

Active Transport

In contrast to passive processes, **active transport** moves molecules and ions against concentration gradients from regions of lower concentration to those of higher concentration. This process is analogous to rolling something uphill, and it does require the cell to expend energy from ATP. Active transport is important in microorganisms for moving nutrients that are present in low concentrations in the environment of the cells. It requires membrane proteins that act as both carriers and enzymes (Figure 4.29). These proteins display specificity in that each carrier transports a single substance or a few closely related substances. The results of active transport are to concentrate a substance on one side of a membrane and to maintain that concentration against a gradient. As with facilitated diffusion, active transport carriers also are subject to saturation and competition for binding sites by similar molecules.

Group translocation reactions move a substance from the outside of a bacterial cell to the inside while chemically modifying the substance so that it cannot diffuse out. This process allows molecules such as glucose to be accumulated against a concentration gradient. Because the modified molecule inside the cell is different from those outside, no actual concentration exists. Energy for this process is supplied by phosphoenolpyruvate (PEP), a high-energy phosphate compound. Many eukaryotic cells have a similar active transport mechanism for preventing diffusion.

Endocytosis and Exocytosis

In addition to the processes that move substances directly across membranes, eukaryotic cells move substances by forming membrane-bound vesicles. Such vesicles are made from portions of the plasma mem-

Situation	Isotonic	Hypotonic	Hypertonic
A bag, permeable to water but not salt, is placed in a beaker containing one of three different salt solutions.	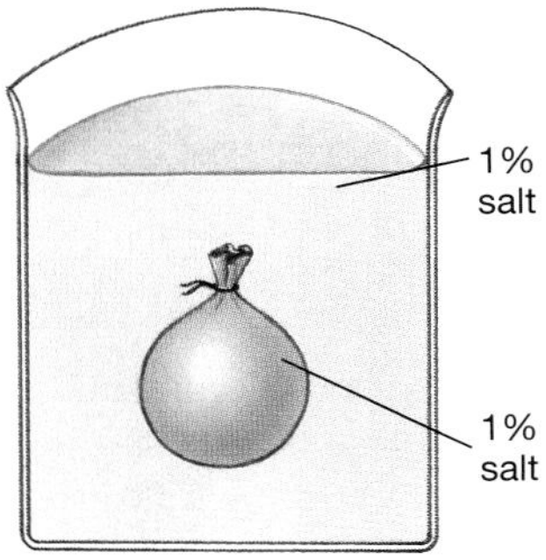 1% salt / 1% salt	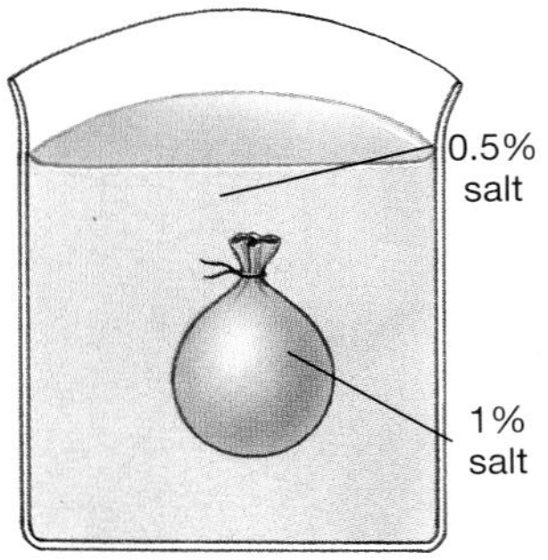 0.5% salt / 1% salt	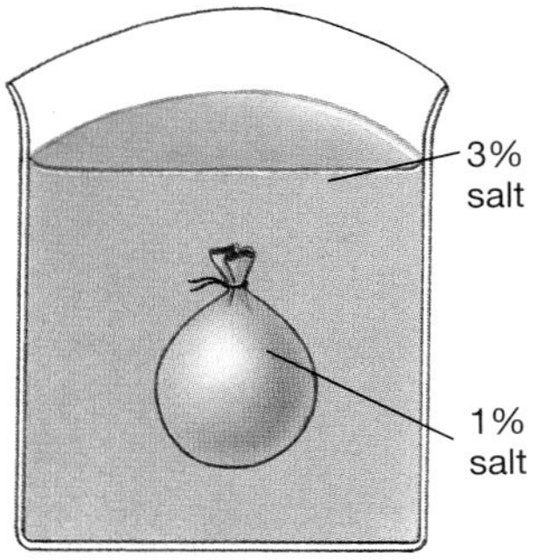 3% salt / 1% salt
Q: How do we know what to name the solution in the beaker (the environment)? **A:** By comparing the concentration of dissolved material in the environment to that dissolved inside the bag.	They have the *same* concentration, so the environment is called *iso*tonic.	0.5% is *less* than 1%, so the solution in the beaker is called *hypo*tonic.	3% is *greater* than 1%, so the solution in the beaker is called *hyper*tonic.
Q: Which way will water flow? **A:** Water flows from an area of greater concentration of water to one of lower concentration (down a concentration gradient).	Environment = 99% H_2O / Inside bag = 99% H_2O / *EQUAL FLOW* into and out of bag / *NO NET CHANGE* There is no concentration gradient.	Environment = 99.5% H_2O / Inside bag = 99.0% H_2O / Water flows *INTO* the bag.	Environment = 97% H_2O / Inside bag = 99% H_2O / Water flows *OUT OF* bag into environment.
Q: If the bag were a cell, what would happen to it? **A:** See diagrams:	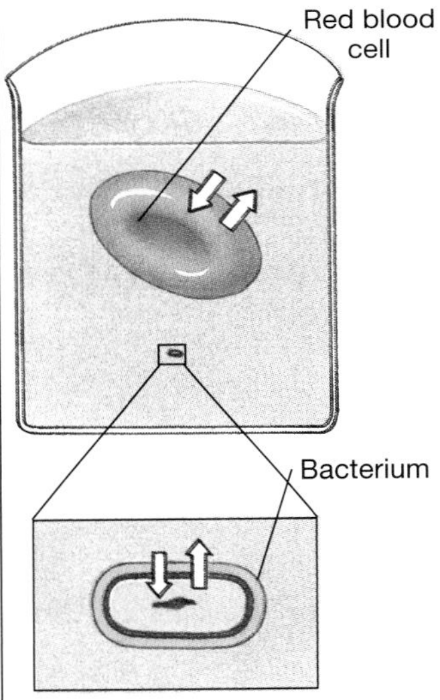 Red blood cell / Bacterium **(a)** Isotonic	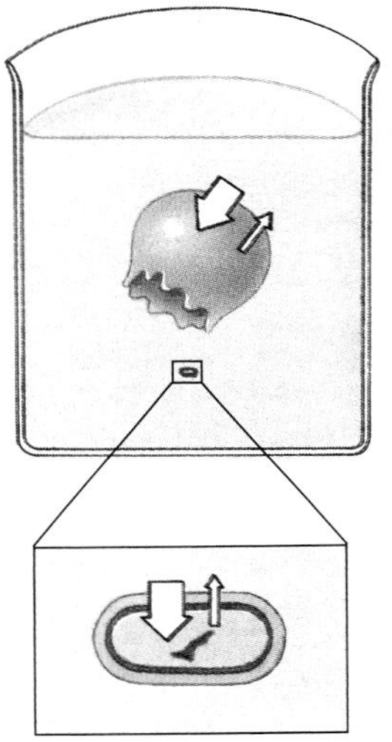 **(b)** Hypotonic	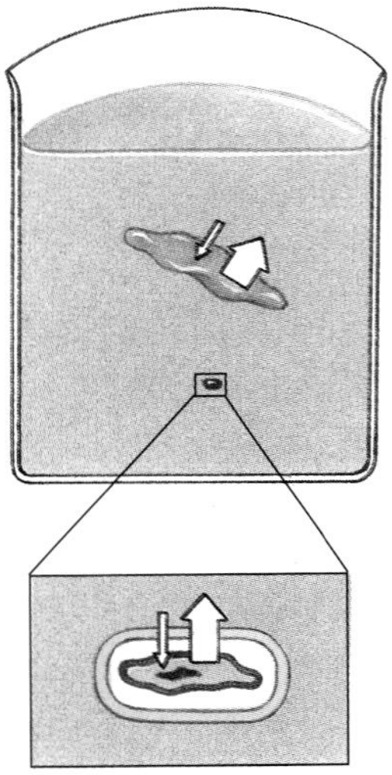 **(c)** Hypertonic

FIGURE 4.28 Experiments that examine the effects of tonicity on osmosis. (a) A cell in an isotonic environment—one that has the same concentration of dissolved material as the interior of the cell—will experience no net gain or loss of water and will retain its original shape. (b) A cell in a hypotonic environment—one with a lower concentration of dissolved material than the interior of the cell—will gain water and swell. Unlike a bacterial cell, a red blood cell will burst because it lacks a cell wall. (c) A cell in a hypertonic environment—one with a higher concentration of dissolved material than the interior of the cell—will lose water and shrink.

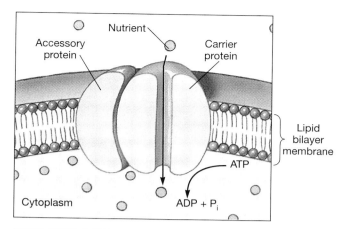

FIGURE 4.29 Active transport. Carrier protein molecules aid in movement of molecules through the membrane. This process can take place against a concentration gradient and so requires the use of energy (in the form of ATP) by the cell. The accessory protein participates in the carrier protein's function. (P_i is inorganic phosphate, HPO_4^{2-}.)

brane. If they form by invagination (poking in) and surround substances outside the cell, the process is called **endocytosis.** These vesicles pinch off from the plasma membrane and enter the cell. If vesicles inside the cell fuse with the plasma membrane and extrude their contents from the cell, the process is called **exocytosis.** Both endocytosis and exocytosis require energy, probably to allow contractile proteins of the cell's cytoskeleton to move vesicles.

Endocytosis

Several types of endocytosis occur. In one type, known as *receptor-mediated endocytosis,* a substance outside the cell binds to the plasma membrane, which invaginates and surrounds the substance. The exact mechanisms that trigger binding and invagination depend on specific receptor sites on the plasma membrane. Once the substance is completely surrounded by plasma membrane to form a vesicle, the vesicle pinches off from the plasma membrane.

Of all the types of endocytosis, only phagocytosis is of special interest to microbiologists. In **phagocytosis,** large vacuoles called *phagosomes* form around microorganisms and debris from tissue injury. These vacuoles enter the cell, taking with them large amounts of the plasma membrane (Figure 4.30). The vacuole membrane fuses with lysosomes, which release their enzymes into the vacuoles. The enzymes digest the contents of the vacuoles (*phagolysosomes*) and release small molecules into the cytoplasm. Often, undigested particles in residual bodies are returned to and fuse with the plasma membrane. The particles are released from the cell by exocytosis. Certain white blood cells are especially adept at phagocytosis and play an important role in defending the body against infection by microorganisms.

Exocytosis

Exocytosis, the mechanism by which cells release secretions, can be thought of as the opposite of endocytosis. Most secretory products are synthesized on ribosomes or smooth endoplasmic reticulum. They are transported through the membrane of the endoplasmic reticulum, packaged in vesicles, and moved to the Golgi apparatus, where their contents are processed to form the final secretory product. Once secretory vesicles form, they move toward the plasma membrane and fuse with it (Figure 4.30). The contents of the vesicles are then released from the cell.

FIGURE 4.30 Endocytosis is the process of taking materials into the cell; exocytosis is the process of releasing materials from the cell. Material taken in by the form of endocytosis called *phagocytosis* is enclosed in vacuoles known as *phagosomes.* The vacuoles fuse with lysosomes, which release powerful enzymes that degrade the vacuolar contents. Reusable components are absorbed into the cell, and debris is released by exocytosis.

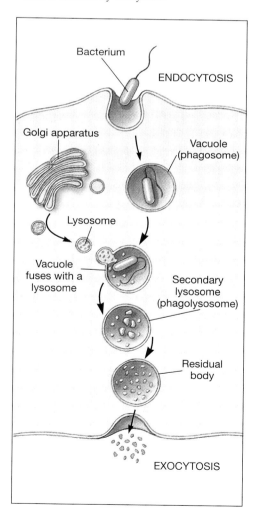

EVOLUTION BY ENDOSYMBIOSIS

Biologists believe that life arose on earth about 4 billion years ago, in the form of simple organisms much like the prokaryotes of today. Not until a billion years later did the first eukaryotic organisms appear. How the development from prokaryote to eukaryote took place is not known precisely—but, given the amount of time required, we can imagine that the transition must have been exceedingly complex, involving numerous small steps.

A plausible explanation for how some of those crucial steps took place is provided by the *endosymbiont theory*. As we have seen, the major difference between prokaryotes and eukaryotes is that eukaryotes possess specialized organelles. The heart of the endosymbiont theory is the idea that the organelles of eukaryotic cells arose from prokaryotic cells that had developed a *symbiotic* relationship with the eukaryote-to-be. **Symbiosis** is a relationship between two organisms that live in close contact. If one lives inside the other, the relationship is known as *endosymbiosis*. The endosymbiont theory proposes that prokaryotic cells—originally free-living organisms—took up residence as endosymbionts within the future eukaryote.

The first eukaryote was probably an amoebalike cell that somehow developed a nucleus. Knowing what we now know about the ease with which bits of membrane pinch off to form vesicles, we can imagine that a primitive chromosome might have become surrounded by membrane, creating a rudimentary nucleus. This primitive eukaryote was probably a phagocytic cell that obtained nutrients by engulfing material from its environment—including, presumably, other cells. Most of these cells would have been digested and used for nourishment. But some, according to the endosymbiont theory, survived their engulfment and became permanent guests within the cytoplasm of the host cell. Eventually, they were incorporated into the cell.

The symbiotic relationship that spawned eukaryotes was one from which both organisms benefited. The engulfed prokaryotes were protected by the eukaryote, which acquired some new capabilities through the presence of its endosymbionts. If the endosymbiont could capture energy from light and synthesize food, that food became available to both organisms. If the endosymbiont could use oxygen to obtain energy from nutrients more efficiently, the energy became available to both. And if the symbiont could move, it might help the eukaryote move.

Evidence for the endosymbiont theory comes from comparing the characteristics of eukaryotic organelles with those of prokaryotic organisms. First, mitochondria and chloroplasts are about the same size as prokaryotic cells. These organelles show a surprising autonomy; for example, they divide independently of their host cells. It is easy to suppose that they once might have been free-living prokaryotes—the mitochondria resembling aerobic bacteria of today, and the chloroplasts resembling photosynthetic organisms like modern cyanobacteria. Furthermore, both the mitochondria and chloroplasts of eukaryotic cells contain their own DNA and ribosomes. Their DNA resembles that of modern bacteria, and the ribosomes are small like bacterial ribosomes. Finally, the process of protein synthesis in these organelles is similar to that found in bacteria and not to the process directed by the nuclear DNA of modern eukaryotes.

Although such circumstantial evidence is suggestive, it relies heavily on historical inference. But the cornerstone of the theory—prokaryotes living in endosymbiotic association with host cells—is not mere speculation; similar relationships abound in nature. Certain eukaryotes living in low-oxygen environments lack mitochondria yet get along quite well, thanks to bacteria that live inside them and serve as "surrogate mitochondria." For example, protists living symbiotically in the intestines of termites are, in turn, colonized by symbi-otic bacteria similar in size, distribution, and function to mitochondria (Figure 4.31). These bacteria oxidize food and supply energy in the form of ATP for their protist partners. Another example is provided by the giant amoeba, *Pelomyxa palustris*. Living in the mud at the bottom of ponds, it too lacks mitochondria but harbors at least two kinds of endosymbionts. If the bacteria are killed with antibiotics, lactic acid accumulates in the host cell. This suggests that the bacteria oxidize the end products of glucose fermentation—a function ordinarily performed by mitochondria.

Lynn Margulis, a pioneer of the endosymbiont theory, has proposed that eukaryotic flagella and cilia originated from symbiotic associations between heterotrophic protists and motile bacteria called spirochetes. The spirochetes would have obtained nutrients that leaked from the eukaryote, while giving the eukaryote motility. Eventually these cells became permanently integrated into their hosts. Again, such associations of present-day species are well known. *Mixotricha paradoxa*, a protist endosymbiont found in the hindgut of one termite, uses the four flagella at its front end to steer, but it depends on the half-million spirochetes covering its surface for driving power. These spirochetes have a natural tendency to coat living or dead surfaces. Once attached, they beat in unison, propelling the host particle along.

FIGURE 4.31 In the cytoplasm of *Pyrsonympha*, a protist that lives symbiotically in the hindgut of termites, bacteria (dark ovals) act as mitochondria for the protist. At lower left, one of the bacteria is dividing.

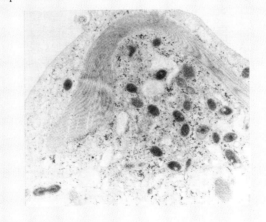

BASIC CELL TYPES

- Both **prokaryotic cells** and **eukaryotic cells** have membranes that define the bounds of the living cell, and both contain genetic information stored in DNA.

- Prokaryotic cells differ from eukaryotic cells in that they lack a defined nucleus and membrane-bound **organelles** (except for a few simple membrane-covered bodies in certain types of prokaryotes).

PROKARYOTIC CELLS

Size, Shape, and Arrangement

- Prokaryotes are the smallest living organisms.

- Bacteria are grouped by shape: **cocci** (spherical), **bacilli** (rod-shaped), **spirilli** (rigid, wavy), **vibrios** (comma-shaped), and **spirochetes** (corkscrew-shaped).

- Arrangements of bacteria include groupings such as pairs, tetrads, grapelike clusters, and long chains.

Overview of Structure

- Bacterial cells include a cell membrane, cytoplasm, ribosomes, a nuclear region, and external structures.

Cell Wall

- The rigid **cell wall** outside the cell membrane is composed mainly of the polymer **peptidoglycan.**

- Cell walls differ in compostition and structure. In gram-positive bacteria, the cell wall consists of a thick, dense layer of peptidoglycan, with **teichoic acid** in it. In gram-negative bacteria, the cell wall has a thin layer of peptidoglycan, separated from the cytoplasmic membrane by the **periplasmic space** and enclosed by an **outer membrane** made of **lipopolysaccharide,** or **endotoxin.** In acid-fast bacteria, the cell wall consists mainly of lipids, some of which are true waxes, and some of which are glyco-lipids.

- Some bacterial cell walls are damaged by penicillin and lysozyme.

Cell Membrane

- The **cell membrane** has a **fluid-mosaic** structure with phospholipids forming a bilayer and proteins interspersed in a mosaic pattern.

- The main function of the cell membrane is to regulate the movement of materials into and out of cells.

- Bacterial cell membranes also perform functions usually carried out by organelles of eukaryotic cells.

Internal Structure

- The **cytoplasm** is the semifluid substance inside the cell membrane.

- **Ribosomes,** which consist of RNA and protein, serve as sites for protein synthesis.

- The **nuclear region** includes a single, large, circular chromosome, which contains the prokaryotic cell's DNA and some RNA and protein.

- Bacteria contain a variety of **inclusions,** including **granules** that store glycogen or other substances and **vesicles** filled with gas.

- Some bacteria form resistant **endospores.** The core of an endospore contains living material and is surrounded by a cortex, spore coat, and exosporium.

External Structure

- Motile bacteria have one or more **flagella,** which propel the cell by the action of rings in their basal body.

- Much bacterial movement is random, but some bacteria exhibit **chemotaxis** (movement toward attractants and away from repellents) and/or **phototaxis** (movement toward or away from light).

- Some bacteria have **pili: Conjugation pili** allow exchange of DNA, whereas **attachment pili (fimbriae)** help bacteria adhere to surfaces.

- The **glycocalyx** includes all polysaccharides external to a bacterial cell wall. **Capsules** prevent host cells from destroying a bacterium; capsules of any species of bacteria have a specific chemical composition. **Slime layers** protect bacterial cells from drying, trap nutrients, and sometimes bind cells together, as in dental plaque.

EUKARYOTIC CELLS

Overview of Structure

- Eukaryotic cells, which are generally larger and more complex than prokaryotic cells, are the basic structural unit of microscopic and macroscopic organisms of the kingdoms Protista, Plantae, Fungi, and Animalia.

Plasma Membrane

- **Plasma membranes** of eukaryotic cells are almost identical to those of prokaryotic cells, except that they contain sterols. The function of eukaryotic plasma membranes, however, is limited primarily to regulating movement of substances into and out of cells.

Internal Structure

- Eukaryotic cells are characterized by the presence of a membrane-bound **cell nucleus,** with a **nuclear envelope, nucleoplasm, nucleoli,** and **chromosomes** (typically paired) that contain DNA and proteins called **histones.**

- In cell division by **mitosis,** each cell receives one of each chromosome found in parent cells. In cell division by **meiosis,** each cell receives one member of each pair of chromosomes, and the progeny can be **gametes** or **spores.**

- **Mitochondria,** the powerhouses of eukaryotic cells, carry out the oxidative reactions that capture energy in ATP.

- Photosynthetic cells contain **chloroplasts,** which capture energy from light.
- Eukaryotic ribosomes are larger than those of prokaryotes and can be free or attached to endoplasmic reticulum. Free ribosomes make protein to be used in the cell; those that are attached to endoplasmic reticulum make proteins to be secreted.
- The **endoplasmic reticulum** is an extensive membrane network. Without ribosomes (smooth ER), the endoplasmic reticulum synthesizes lipids; when combined with ribosomes (rough ER), it produces proteins.
- The **Golgi apparatus** is a set of stacked membranes that receive, modify, and package proteins into **secretory vesicles.**
- **Lysosomes,** in animal cells, are organelles that contain digestive enzymes, which destroy dead cells and digest contents of vacuoles.
- **Peroxisomes** are membrane-bound organelles that convert peroxides to water and oxygen and sometimes oxidize amino acids and fats.
- **Vacuoles** contain various stored substances and materials engulfed by phagocytosis.
- The **cytoskeleton** is a network of **microfilaments** and **microtubules** that support and give rigidity to cells and provide for cell movements.

External Structure

- Most external components of eukaryotic cells are concerned with movement. Eukaryotic flagella are composed of microtubules; sliding of proteins at their bases causes them to move.
- **Cilia** are smaller than flagella and beat in coordinated waves.
- **Pseudopodia** are projections into which cytoplasm flows, causing a creeping movement.
- Eukaryotic cells of the plant and fungi kingdoms have cell walls, as do the algal protists.

MOVEMENT OF SUBSTANCES ACROSS MEMBRANES

- All passive processes involved in movement across membranes involve net movement of substances from a region of higher concentration to a region of lower concentration. These processes do not require expenditure of energy by the cell.

Simple Diffusion

- **Simple diffusion** results from the molecular kinetic energy and random movement of particles. The role of diffusion in living cells depends on the size of particles, nature of membranes, and distances substances must move inside cells.

Facilitated Diffusion

- **Facilitated diffusion** uses protein carrier molecules or protein-lined pores in membranes in moving ions or molecules from high to low concentrations.

Osmosis

- **Osmosis** is the net movement of water molecules through a **selectively permeable** membrane from a region of higher to a region of lower concentration. The **osmotic pressure** of a solution is the pressure required to prevent such a flow.

Active Transport

- Active processes involved in movement of substances across membranes generally result in movement from regions of lower concentration to those of higher concentration and require the cell to expend energy.
- **Active transport** requires a protein carrier molecule in a membrane, a source of ATP, and an enzyme that releases energy from ATP.
- Active transport is important in cell functions because it allows cells to take up substances that are in low concentration in the environment and to concentrate those substances within the cell.

Endocytosis and Exocytosis

- **Endocytosis** and **exocytosis,** which occur only in eukaryotic cells, involve formation of vesicles from fragments of plasma membrane and fusion of vesicles with the plasma membrane, respectively.
- In endocytosis the vesicle enters the cell, as in **phagocytosis**.
- In exocytosis the vesicle leaves the cell, as in secretion.
- Endocytosis and exocytosis are important because they allow the movement of relatively large quantities of materials across plasma membranes.

active transport   (p. 99)
amoeboid movement
   (p. 96)
amphitrichous   (p. 86)
atrichous   (p. 86)
attachment pilus   (p. 89)
axial filament   (p. 87)
bacillus   (p. 75)
capsule   (p. 90)
cell membrane   (p. 82)
cell nucleus   (p. 91)
cell wall   (p. 78)
chemotaxis   (p. 87)
chloroplast   (p. 93)
chromatin   (p. 92)
chromatophore   (p. 84)
chromosome   (p. 92)
cilium   (p. 96)
coccus   (p. 75)
conjugation pilus   (p. 89)
crista   (p. 92)
cytoplasm   (p. 84)
cytoplasmic streaming
   (p. 96)
cytoskeleton   (p. 95)
diploid   (p. 92)
dyad   (p. 92)
endocytosis   (p. 101)

endoflagellum   (p. 87)
endoplasmic reticulum
   (p. 93)
endospore   (p. 86)
endotoxin   (p. 78)
eukaryotic cell   (p. 74)
exocytosis   (p. 101)
facilitated diffusion   (p. 98)
fimbria   (p. 89)
flagellum   (p. 86)
fluid-mosaic model   (p. 82)
gamete   (p. 92)
glycocalyx   (p. 90)
Golgi apparatus   (p. 95)
granule   (p. 85)
group translocation reaction
   (p. 99)
haploid   (p. 92)
histone   (p. 92)
hydrophilic   (p. 82)
hydrophobic   (p. 82)
hypertonic   (p. 99)
hypotonic   (p. 99)
inclusion   (p. 84)
isotonic   (p. 99)
lipid A   (p. 78)
lipopolysaccharide   (p. 78)
lophotrichous   (p. 86)

lysosome   (p. 95)
matrix   (p. 92)
meiosis   (p. 92)
metachromasia   (p. 85)
metachromatic granule
   (p. 85)
microfilament   (p. 95)
microtubule   (p. 95)
mitochondrion   (p. 92)
mitosis   (p. 92)
monotrichous   (p. 86)
nuclear envelope   (p. 92)
nuclear pore   (p. 92)
nuclear region   (p. 84)
nucleoid   (p. 84)
nucleolus   (p. 92)
nucleoplasm   (p. 92)
organelle   (p. 74)
osmosis   (p. 98)
osmotic pressure   (p. 99)
outer membrane   (p. 78)
pellicle   (p. 90)
peptidoglycan   (p. 78)
periplasmic space   (p. 80)
peritrichous   (p. 86)
peroxisome   (p. 95)
phagocytosis   (p. 101)
phototaxis   (p. 87)

pilus   (p. 87)
plasma membrane   (p. 90)
pleomorphism   (p. 76)
polyribosome   (p. 84)
prokaryotic cell   (p. 74)
protoplast   (p. 81)
pseudopodium   (p. 96)
ribosome   (p. 84)
secretory vesicle   (p. 95)
selectively permeable
   (p. 98)
simple diffusion   (p. 97)
slime layer   (p. 90)
spheroplast   (p. 82)
spindle apparatus   (p. 92)
spirillum   (p. 75)
spirochete   (p. 75)
spore   (p. 92)
stroma   (p. 93)
symbiosis   (p. 102)
teichoic acid   (p. 78)
thylakoid   (p. 93)
vacuole   (p. 95)
vegetative cell   (p. 86)
vesicle   (p. 85)
vibrio   (p. 75)
volutin   (p. 85)
zygote   (p. 92)

QUESTIONS FOR REVIEW

A **Characteristics of Prokaryotic and Eukaryotic Cells**

1. How are prokaryotic and eukaryotic cells alike? How are they different?

2. Attribute each of the following to either P, prokaryotes only; E, eukaryotes only; B, both; or N, neither.

   a. ____ single chromo-
      some
   b. ____ membrane-
      bound nucleus
   c. ____ fluid-mosaic
      membrane
   d. ____ viruses
   e. ____ 70S ribosomes
   f. ____ endoplasmic
      reticulum
   g. ____ respiratory en-
      zymes in mito-
      chondria
   h. ____ mitosis

   i. ____ peptidoglycan in
      cell wall
   j. ____ cilia
   k. ____ 80S ribosomes
   l. ____ chloroplasts
   m. ____ "9 + 2" micro-
      tubule arrange-
      ment in flagella
   n. ____ bacteria
   o. ____ can have extra
      chromosomal
      DNA
   p. ____ meiosis

B **Size, Shape, and Arrangement of Prokaryotic Cells**

3. What are the three basic shapes of bacteria?

4. Match the arrangement with the prefix or description below:

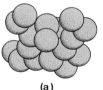

(a)   (b)

(c)   (d)

   I.   strepto-
   II. diplo-
   III. staphylo-
   IV. sarcina
   V.   tetrad

5. How do planes of division affect the arrangement of bacterial cells?

C Bacterial Cell Walls and Cell Membranes

6. Describe the properties of the cell membrane, and relate them to the fluid-mosaic model.

7. What is the major function of the cell membrane, and what additional functions does a prokaryotic membrane perform?

8. What are the distinguishing characteristics of the cell walls of gram-positive, gram-negative, and acid-fast bacteria?

9. Describe the chemical composition of peptidoglycan.

10. Describe the chemical composition of the prokaryote's outer membrane.

11. How do penicillin and lysozyme affect the cell wall?

D Other Bacterial Components

12. Describe the composition and function of ribosomes in prokaryotes, as compared with ribosomes in eukaryotes.

13. What structures and functions are associated with the nuclear region?

14. What materials are found in granules and vesicles of bacteria?

15. (a) Name the components of an endospore. (b) Describe the function of endospores. Under what conditions do they (c) form and (d) germinate?

16. Describe the structure of a bacterial flagellum.

17. How do bacteria move in chemotaxis?

18. How do attachment and conjugation pili differ in structure and function?

19. What is the glycocalyx?

20. What are capsules, and what properties do they impart to bacteria?

21. What are slime layers, and what functions do they perform?

E Eukaryotic Plasma Membranes

22. How do eukaryotic membranes differ from prokaryotic membranes?

23. What is the function of sterols in eukaryotic membranes?

F Other Eukaryotic Components

24. Describe the structure and function of the cell nucleus.

25. Distinguish between mitosis and meiosis and between gametes and spores.

26. Describe the structure of a chloroplast, and briefly describe its function in photosynthesis.

27. Describe the structure, function, and location of ribosomes.

28. List the distinguishing characteristics of endoplasmic reticulum, Golgi apparatus, peroxisomes, lysosomes, vacuoles, and cytoskeleton.

29. How do flagella in eukaryotes differ from those in prokaryotes?

30. Describe the structure and function of cilia and of pseudopodia. What types of organisms have flagella, cilia, and pseudopodia?

31. What different materials are found in cell walls of eukaryotes, and which groups of organisms have cell walls?

G Passive and Active Transport

32. Describe the general characteristics of passive processes by which materials cross membranes.

33. What are the similarities and differences between simple diffusion and facilitated diffusion?

34. What is osmosis?

35. You are lying in a hospital bed, awaiting the arrival of a nurse with intravenous solution that your doctor has ordered for you. The nurse arrives, flustered at having misplaced the doctor's order. However, on the nurse's cart are three bottles, labeled A = distilled water, B = 0.9% saline, and C = 10% saline. Unfortunately the nurse was absent the week that osmosis and tonicity were taught. Confused about these matters, the nurse decides that pure water couldn't possibly harm a patient and so prepares to administer bottle A, the distilled water, to you intravenously. Should you let the nurse do this? Why or why not? If not bottle A, which of the other two should be used, and why?

36. A cell containing 0.8% dissolved materials is placed in a beaker containing 97% water and 3% dissolved materials.
    a. The cell is in a (hypo-, hyper-, iso-)tonic environment.
    b. Water will flow (into, out of, equally into and out of) the cell.
    c. The cell will (shrink, swell, remain the same).

37. In diffusion, molecules move from an area of _____ concentration to an area of _____ concentration.

38. In active transport, molecules move from an area of _____ concentration to an area of _____ concentration.

39. How does active transport differ from passive processes with respect to movement across membranes?

40. How does active transport occur, and why is it important?

H Exocytosis and Endocytosis

41. Distinguish between endocytosis and exocytosis.

42. How are endocytosis and exocytosis important in cell function?

1. Use what you know about the organelles of eukaryotic cells and the general structure of both prokaryotic and eukaryotic cells to explain how prokaryotic cells perform the functions of the organelles they are lacking.

2. Predict what would happen to eukaryotic cells and to prokaryotic cells placed in (a) distilled water, (b) 2 percent saline, and (c) isotonic saline.

3. What might have happened if the flasks Pasteur used to disprove spontaneous generation had contained bacterial endospores?

4. Design an experiment that would demonstrate that flagella are responsible for the movement of bacteria in solutions.

SOME INTERESTING READING

Alberts, B., et al. 1994. *Molecular biology of the cell.* New York: Garland Publishing.

Beveridge, T. J. 1995. "The periplasmic space and the periplasm in gram-positive and gram-negative bacteria." *ASM News* 61(3):125–30.

Blakemore, R. P., and R. B. Frankel. 1981. "Magnetic navigation in bacteria." *Scientific American* 245, no. 6 (December):58–65.

Bretscher, M. S. 1985. "The molecules of the cell membrane." *Scientific American* 253, no. 4 (October):100–8.

Brock, T. D. 1988. "The bacterial nucleus: A history." *Microbiological Reviews* 52:397–411.

Costerton, J. W., R. T. Irvin, and K.-J. Cheng. 1981. "The bacterial glycocalyx in nature and disease." *Annual Review of Microbiology* 35:299–324.

Drlica, K., and M. Riley, eds. 1990. *The bacterial chromosome.* Washington, D.C.: American Society for Microbiology.

Ferris, F. G., and T. J. Beveridge. 1985. "Functions of bacterial cell surface structures." *BioScience* 35(3):172–77.

Hancock, R. E. W. 1991. "Bacterial outer membranes: evolving concepts." *ASM News* 57(4):175–82.

Hill, W. E., et al. 1990. *The ribosome: Structure, function, and evolution.* Washington, D.C.: American Society for Microbiology.

Koch, A. L. 1990. "Growth and form of the bacterial cell wall." *American Scientist* 78, no. 4 (July–August):327–42.

Koppel, T. 1991. "Learning how bacteria swim could set new gears in motion." *Scientific American* 265, no. 3 (September):168–69.

Moir, A., and D. A. Smith. 1990. "The genetics of bacterial spore germination." *Annual Review of Microbiology* 44:531–53.

Nikaido, H., and M. Vaara. 1985. "Molecular basis of bacterial outer membrane permeability." *Microbiological Reviews* 49:1–32.

Shapiro, J. A. 1991. "Multicellular behavior of bacteria." *ASM News* 57(5):247–53.

Unwin, N., and R. Henderson. 1984. "The structure of proteins in biological membranes." *Scientific American* 250 (February):78–94.

Vidal, G. 1984. "The oldest eukaryotic cells." *Scientific American* 250 (February):48–57.

Woese, C. R. 1994. "There must be a prokaryote somewhere: Microbiology's search for itself." *Microbiological Reviews* 58, no. 1 (March):1–9.

CHAPTER **5**

ESSENTIAL CONCEPTS OF METABOLISM

A cow with a "porthole"—a rumen fistula—implanted at the University of Florida to enable researchers to obtain digestive samples easily for nutrition studies.

U ntil the middle of the nineteenth century, people didn't know what caused a fruit juice to become wine, or milk to sour. Then, in 1857, Louis Pasteur proved that alcoholic fermentation was due to microorganisms. A few years later he identified specific organisms from samples of fermenting juices and souring milk. Pasteur was one of the first to study chemical processes in a living organism. Since his time, much has been learned about such processes.

METABOLISM: AN OVERVIEW

Metabolism is the sum of all the chemical processes carried out by living organisms (Figure 5.1). It includes **anabolism,** reactions that require energy to synthesize complex molecules from simpler ones, and **catabolism,** reactions that release energy by breaking complex molecules into simpler ones that can then be reused as building blocks. Anabolism is needed for growth, reproduction, and repair of cellular structures. Catabolism provides the organism with energy for its life processes, including movement, transport, and the synthesis of complex molecules—that is, anabolism.

All catabolic reactions involve *electron transfer,* which allows energy to be captured in high-energy bonds in

ATP and similar molecules. Electron transfer is directly related to oxidation and reduction (Table 5.1). **Oxidation** can be defined as the loss or removal of electrons. Although many substances combine with oxygen and transfer electrons to oxygen, oxygen need not be present if another electron acceptor is available. **Reduction** can be defined as the gain of electrons. When a substance loses electrons, or is oxidized, energy is released, but another substance must gain the electrons, or be reduced, at the same time. For example, during the oxidation of organic molecules, hydrogen

---

FIGURE 5.1 Large, complex molecules are generally richer in energy than are small, simple ones. Catabolic reactions break down large molecules into smaller ones, releasing energy. Organisms capture some of this energy for their life processes. Anabolic reactions use energy to build larger molecules from smaller components. The molecules synthesized in this way are used for growth, reproduction, and repair.

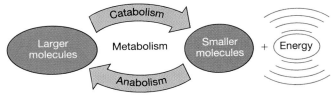

THIS CHAPTER FOCUSES ON THE FOLLOWING QUESTIONS:

A How do the following terms relate to metabolism: autotrophy, heterotrophy, oxidation, reduction, photoautotrophy, photoheterotrophy, chemoautotrophy, chemoheterotrophy, glycolysis, fermentation, aerobic metabolism, and biosynthetic processes?

B What are the characteristics of enzymes, and how do those characteristics contribute to their function?

C What are the main steps and significance of glycolysis and fermentation?

D What are the main steps and significance of the Krebs cycle?

E What are the roles of electron transport and oxidative phosphorylation in energy capture?

F How do microorganisms metabolize fats and proteins for energy?

G What are the main steps and significance of photosynthesis in microbes?

H How do photoheterotrophy and chemoautotrophy differ?

I How do bacteria carry out biosynthetic activities?

J How do bacteria use energy for membrane transport and for movement?

TABLE 5.1	Comparison of Oxidation and Reduction
Oxidation	**Reduction**
Loss of electrons (A)	Gain of electrons (B)
Gain of oxygen	Loss of oxygen
Loss of hydrogen	Gain of hydrogen
Loss of energy (liberates energy)	Gain of energy (stores energy in the reduced compound)
Exothermic; exergonic (gives off heat energy)	Endothermic; endergonic (requires energy, such as heat)

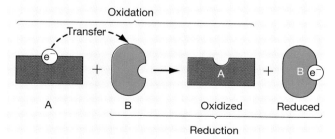

atoms are removed and used to reduce oxygen to form water:

$$2\,H_2 \;+\; O_2 \longrightarrow 2\,H_2O$$

hydrogen   oxygen   water

In this reaction, hydrogen is an **electron donor,** or *reducing agent,* and oxygen is an **electron acceptor,** or *oxidizing agent.* Because oxidation and reduction must occur simultaneously, the reactions in which they occur are sometimes called *redox reactions.*

Among all living things, microorganisms are particularly versatile in the ways in which they obtain energy. The ways different microorganisms capture energy, and obtain carbon, can be classified as **autotrophy** (aw-to-trof'e)—"self-feeding"—or **heterotrophy** (het"er-o-trof'e)—"other-feeding" (Figure 5.2). **Autotrophs** use carbon dioxide to synthesize organic molecules. They include **photoautotrophs,** which obtain energy from light, and **chemoautotrophs,** which obtain energy from oxidizing simple inorganic substances such as sulfides and nitrites. **Heterotrophs** use ready-made organic molecules obtained from other organisms, living or dead. They include **photoheterotrophs,** which obtain chemical energy from light, and **chemoheterotrophs,** which obtain chemical energy from organic compounds.

Autotrophic metabolism (especially photosynthesis) is important as a means of energy capture in many free-living microorganisms. However, such microorganisms do not usually cause disease. We emphasize metabolic processes that occur in chemoheterotrophs because many microorganisms, including nearly all in-

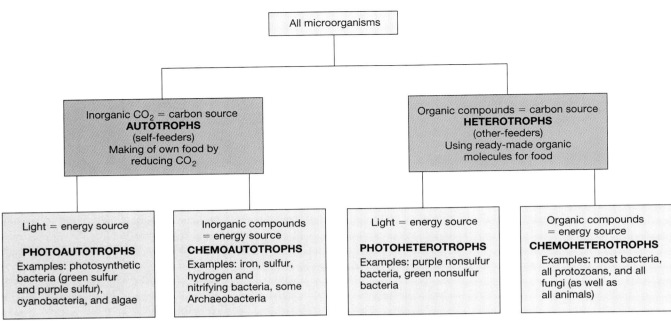

FIGURE 5.2 The main types of energy-capturing metabolism.

fectious ones, are chemoheterotrophs. These processes include *glycolysis* (oxidation of glucose to pyruvic acid), *fermentation* (conversion of pyruvic acid to ethyl alcohol, lactic acid, or other organic compounds), and *aerobic respiration* (oxidation of pyruvic acid to carbon dioxide and water). Glycolysis and fermentation do not require oxygen, and only a small amount of the energy in a glucose molecule is captured in ATP. Aerobic respiration does require oxygen as an electron acceptor and captures a relatively large amount of the energy in a glucose molecule in ATP. Complete oxidation of glucose by glycolysis and aerobic respiration is summarized in the following equation:

$$\underset{\text{glucose}}{C_6H_{12}O_6} + \underset{\text{oxygen}}{6\,O_2} \longrightarrow \underset{\substack{\text{carbon} \\ \text{dioxide}}}{6\,CO_2} + \underset{\text{water}}{6\,H_2O} + \text{energy}$$

A large number of microorganisms obtain energy by *photosynthesis,* the use of light energy and hydrogen from water or other compounds to reduce carbon dioxide to an organic substance containing more energy. The overall synthesis of glucose by photosynthesis in cyanobacteria and algae (and green plants) is summarized in the following equation:

$$\underset{\substack{\text{carbon} \\ \text{dioxide}}}{6\,CO_2} + \underset{\text{water}}{6\,H_2O} \xrightarrow[\text{chlorophyll}]{\text{light energy}} \underset{\text{glucose}}{C_6H_{12}O_6} + \underset{\text{oxygen}}{6\,O_2}$$

(Other photosynthetic bacteria, as we shall see later, use a different version of this process.) Photosynthetic organisms then use the glucose or other carbohydrates made in this way for energy. Figure 5.3 relates respiration and photosynthesis.

Like nearly all other chemical processes in living organisms, glycolysis, fermentation, aerobic respiration, and photosynthesis each consist of a series of chemical reactions in which the *product* of one reaction serves as the *reactant* material for the next: A $\longrightarrow$ B $\longrightarrow$ C $\longrightarrow$ D $\longrightarrow$ E, and so on. Such a chain of reactions is called a **metabolic pathway.** Each reaction in a pathway is controlled by a particular enzyme. In this pathway, A is the initial *substrate*, E is the final product, and B, C, and D are *intermediates*.

Metabolic pathways can be catabolic or anabolic (biosynthetic). **Catabolic pathways** capture energy in a form cells can use. **Anabolic pathways** make the complex molecules that form the structure of cells, enzymes, and other molecules that control cells. These pathways use building blocks such as sugars, glycerol, fatty acids, amino acids, nucleotides, and other molecules to make carbohydrates, lipids, proteins, nucleic acids, or combinations such as glycolipids (made from carbohydrates and lipids), glycoproteins (from carbohydrates and proteins), lipoproteins (from lipids and proteins), and nucleoproteins (from nucleic acids and proteins). ATP molecules are the links that couple catabolic and anabolic pathways. Energy released in catabolic reactions is captured and stored in the form of ATP molecules, which are later broken down to provide the energy needed to build up new molecules in biosynthetic pathways. Bacteria transfer approximately 40 percent of the energy in a glucose molecule to ATP during aerobic metabolism and 5 percent during anaerobic fermentation processes. Yields are higher in aerobic processes because their end products are highly oxidized, whereas end products of anaerobic processes are only partially oxidized.

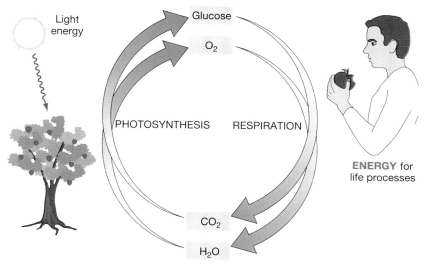

FIGURE 5.3 In photosynthesis, light energy is used to reduce carbon dioxide, forming energy-rich compounds such as glucose and other carbohydrates. In aerobic respiration, energy-rich compounds are oxidized to carbon dioxide and water, and some of the energy released is captured for use in life processes. (The form of photosynthesis depicted here is carried out by cyanobacteria, algae, and green plants. Green and purple bacteria use compounds other than water as a source of hydrogen atoms to reduce CO_2.)

ENZYMES

Enzymes are a special category of proteins found in all living organisms. In fact, most cells contain hundreds of enzymes, and cells are constantly synthesizing proteins, many of which are enzymes. Enzymes act as *catalysts*—substances that remain unchanged while they speed up reactions to as much as a million times the uncatalyzed rate, which is ordinarily not sufficient to sustain life. The only other way to speed up the reaction rate would be to increase the temperature: In general, a 10 degree increase in temperature results in a doubling of the reaction rate. However, most cells would die when exposed to such a rise in temperature. Thus, enzymes are necessary for life at temperatures that cells can withstand. To explain how enzymes do these things, we must consider their properties. ∞ (Chapter 2, p. 42)

Properties of Enzymes

In general, chemical reactions that release energy can occur without input of energy from the surroundings. Nevertheless, such reactions often occur at unmeasurably low rates because the molecules lack the energy to start the reaction. For example, al-

though the oxidation of glucose releases energy, that reaction does not occur unless energy to start it is available. The energy required to start such a reaction is called **activation energy** (Figure 5.4). Activation energy can be thought of as a hurdle over which molecules must be raised to get a reaction started. By analogy, a rock resting in a depression at the top of a hill would easily roll down the hill if pushed out of the depression. Activation energy is like the energy required to lift the rock out of the depression.

A common way to activate a reaction is to raise the temperature, thereby increasing molecular movement, as you do when you strike a match. Matches ordinarily do not burst into a flame spontaneously. If the energy from friction (striking) is added to the reactants on the match head, the temperature increases, and the match bursts into flame. Such a reaction in cells would raise the temperature enough to denature proteins and evaporate liquids. Enzymes lower the activation energy so reactions can occur at mild temperatures in living cells.

Enzymes also provide a surface on which reactions take place. Each enzyme has a certain area on its surface called the **active site,** a binding site. The active site is the region at which the enzyme forms a loose association with its **substrate,** the substance on which the enzyme acts (Figure 5.5a). Like all molecules, a substrate molecule has kinetic energy, and it collides with

FIGURE 5.4 A chemical reaction cannot take place unless a certain amount of activation energy is available to start it. Enzymes lower the amount of activation energy needed to initiate a reaction. They thus make it possible for biologically important reactions to occur at the relatively low temperatures that living organisms can tolerate.

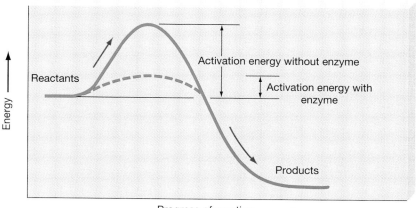

various molecules within a cell. When it collides with the active site of its enzyme, an **enzyme–substrate complex** forms (Figure 5.5b). As a result of binding to the enzyme, some of the chemical bonds in the substrate are weakened. The substrate then undergoes chemical change, and the product or products are formed.

Enzymes generally have a high degree of **specificity;** they catalyze only one type of reaction, and most act on only one particular substrate. An enzyme's shape (its tertiary structure; Chapter 2), especially the shape and electrical charges at its active site, accounts for its specificity. ∞ (p. 40) When an enzyme acts on more than one substrate, it usually acts on substrates with the same functional group or the same kind of chemical bond. For example, *proteolytic,* or protein-splitting, enzymes act on different proteins but always act on the peptide bonds in those proteins.

Enzymes are usually named by adding the suffix *-ase* to the name of the substrate on which they act. For example, phosphatases act on phosphates, sucrase breaks down the sugar sucrose, lipases break down lipids, and peptidases break peptide bonds. Enzymes are commonly placed in one of six categories according to their functions (Table 5.2). They also can be di-

FIGURE 5.5 (a) A computer-generated model of an enzyme (blue) with a substrate molecule (purple) bound to the active site. The active site of the enzyme is a cleft or pocket with a shape and chemical composition that enable it to bind a particular substrate—the molecule on which the enzyme acts. (b) Each substrate binds to an active site, producing an enzyme-substrate complex. The enzyme helps a chemical reaction occur, and one or more products are formed. In this example, the reaction is one that joins two substrate molecules. Other enzyme-catalyzed reactions can involve the splitting of one substrate molecule into two or more parts or the chemical modification of a substrate.

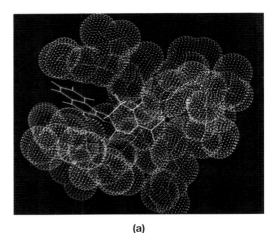

(a)

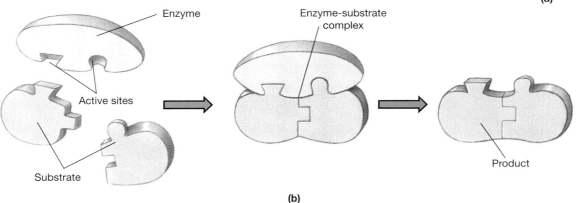

(b)

TABLE 5.2	Classification of Enzymes by Their Function	
Kind of Enzyme	**Functions of Enzyme**	**Examples**
Oxidoreductase	Simultaneously oxidizes one substance and reduces another	Dehydrogenases
Transferase	Transfers a functional group from one molecule to another	Transaminases, kinases
Hydrolase	Adds water and breaks large molecules into two smaller molecules	Peptidases, lipases, maltase, cellulase
Lyase	Removes chemical groups from molecules without adding water	Decarboxylases
Isomerase	Rearranges atoms of a molecule	Aconitase
Ligase	Joins two molecules together and usually requires energy from ATP	Acetyl-CoA synthetase, DNA ligase

vided into two categories on the basis of where they act. **Endoenzymes,** or intracellular enzymes, act within the cell that produced them. **Exoenzymes,** including extracellular enzymes, are synthesized in a cell but cross the cell membrane to act in the periplasmic space or in the cell's immediate environment.

Properties of Coenzymes and Cofactors

Many enzymes can catalyze a reaction only if substances called *coenzymes,* or *cofactors,* are present. Such enzymes consist of a protein portion, called the **apoenzyme,** that must combine with a nonprotein coenzyme or cofactor to form an active **holoenzyme** (Figure 5.6). A **coenzyme** is a nonprotein organic molecule bound to or loosely associated with an enzyme. Many coenzymes are synthesized from *vitamins,* which are essential nutrients precisely because they are required to make coenzymes. For example, *coenzyme A* is made from the vitamin pantothenic acid, and **NAD** (*nicotinamide adenine dinucleotide*) is made from the vitamin niacin. A **cofactor** is usually an inorganic ion, such as magnesium, zinc, or manganese. Cofactors often improve the fit of an enzyme with its substrate, and their presence can be essential in allowing the reaction to proceed.

Carrier molecules such as coenzymes carry hydrogen atoms or electrons in many oxidative reactions (Figure 5.7). When a coenzyme receives hydrogen atoms or electrons, it is reduced; when it releases them, it is oxidized. The coenzyme **FAD** (*flavin adenine dinucleotide*), for example, receives two hydrogen atoms to become $FADH_2$ (reduced FAD). The coenzyme NAD has a positive charge in its oxidized state (NAD^+). In

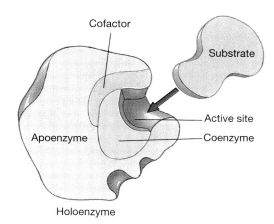

FIGURE 5.6 Many enzymes consist of a protein apoenzyme that must combine with a nonprotein coenzyme (an organic molecule) or cofactor (an inorganic ion) to form the functional holoenzyme.

its reduced state, NADH, it carries a hydrogen atom and an electron from another hydrogen atom, the proton of which remains in the cellular fluids. In all such oxidation–reduction reactions, the electron carries the energy that is transferred from one molecule to another. Thus, for simplicity, we will refer to *electron transfer* regardless of whether "naked" electrons or hydrogen atoms (electrons with protons) are transferred.

Enzyme Inhibition

A molecule similar in structure to a substrate can sometimes bind to an enzyme's active site even though the

FIGURE 5.7 Carrier molecules such as cytochromes and coenzymes carry energy in the form of electrons in many biochemical reactions. Coenzymes such as FAD carry whole hydrogen atoms (electrons together with protons). NAD carries one hydrogen atom and one "naked" electron. When coenzymes are reduced (gain electrons), they increase in energy; when they are oxidized (lose electrons), they decrease in energy.

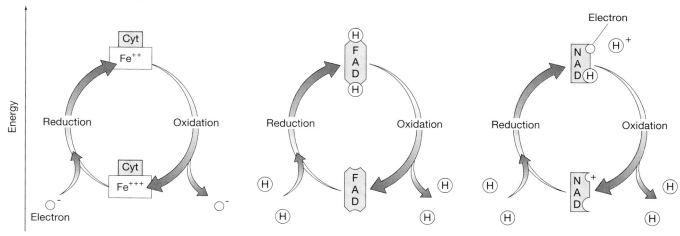

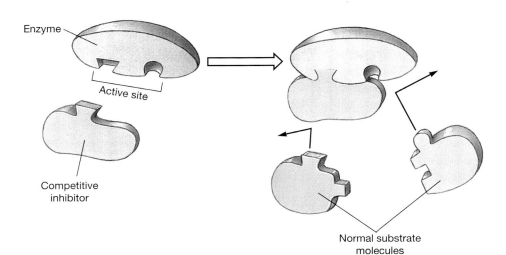

FIGURE 5.8 A competitive inhibitor binds to the active site of an enzyme, preventing the normal substrate from reaching it, but cannot take part in the reaction.

Enzyme

Active site

Competitive inhibitor

Normal substrate molecules

molecule is unable to react. This nonsubstrate molecule is said to act as a **competitive inhibitor** of the reaction because it competes with the substrate for the active site (Figure 5.8). When the inhibitor binds to an active site, it prevents the substrate from binding and thereby inhibits the reaction.

Because the attachment of such a competitive inhibitor is reversible, the degree of inhibition depends on the relative concentrations of substrate and inhibitor. When the concentration of the substrate is high and that of the inhibitor is low, the active sites of only a few enzyme molecules are occupied by the inhibitor, and the rate of the reaction is only slightly reduced. When the concentration of the substrate is low and that of the inhibitor is high, the active sites of many enzyme molecules are occupied by the inhibitor, and the rate of the reaction is greatly reduced.

The *sulfa drugs* (Chapter 14) are competitive inhibitors. Normally, bacterial cells have the enzymes to convert *para-aminobenzoic acid* (PABA) to folic acid, an essential vitamin. If sulfa drugs are present, they compete with PABA for the enzymes' active sites. The greater the sulfa drug concentration, the greater the inhibition of folic acid synthesis.

Enzymes also can be inhibited by substances called **noncompetitive inhibitors.** Some noncompetitive inhibitors attach to the enzyme at an **allosteric site,** which is any site other than the active site (Figure 5.9). Such inhibitors distort the tertiary protein structure and alter the shape of the active site. Any enzyme molecule thus affected no longer can bind substrate, so it cannot catalyze a reaction. Although some noncompetitive inhibitors bind reversibly, others bind irreversibly and permanently inactivate enzyme mole-

FIGURE 5.9 A noncompetitive (allosteric) inhibitor usually binds at a site other than the active site (that is, at an allosteric site). Its presence changes the shape of the enzyme enough to interfere with binding of the normal substrate. Some noncompetitive inhibitors are used in the regulation of metabolic pathways, but others are poisons.

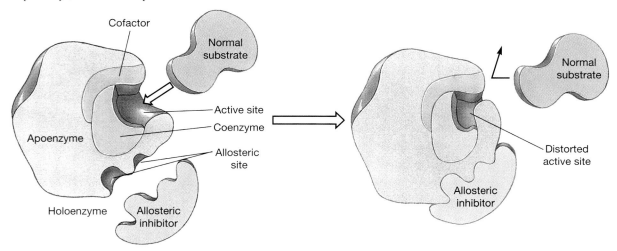

Cofactor

Normal substrate

Active site

Coenzyme

Apoenzyme

Allosteric site

Holoenzyme

Allosteric inhibitor

Normal substrate

Distorted active site

Allosteric inhibitor

cules, thereby greatly decreasing the reaction rate. In noncompetitive inhibition, increasing the substrate concentration does not increase the reaction rate as it does in the presence of a competitive inhibitor. Lead, mercury, and other heavy metals, although not noncompetitive inhibitors, can bind to other sites on the enzyme molecule and permanently change its shape, thus inactivating it.

Feedback inhibition, a kind of reversible noncompetitive inhibition, regulates the rate of many metabolic pathways. For example, when an end product of a pathway accumulates, the product often binds to and inactivates the enzyme that catalyzes the first reaction in the pathway. Feedback inhibition is discussed in more detail in Chapter 7.

Factors That Affect Enzyme Reactions

Factors that affect the rate of enzyme reactions include (1) temperature, (2) pH, and (3) the concentrations of substrate, product, and enzyme.

Temperature and pH

Like other proteins, enzymes are affected by heat and by extremes of pH. Even small pH changes can alter the electrical charge on various chemical groups in enzyme molecules, thereby altering the enzyme's ability to bind its substrate and catalyze a reaction.

Most human enzymes have an *optimum temperature,* near normal body temperature, and an *optimum pH*, near neutral, at which they catalyze a reaction most rapidly. Microbial enzymes likewise function best at optimum temperatures and pHs, which are related to an organism's normal environment. The enzymes of microbes that infect humans have approximately the same optimum temperature and pH requirements as human enzymes.

Changes in *enzyme activity,* the rate at which an enzyme catalyzes a reaction, are shown in Figure 5.10. In this example, enzyme activity increases with temperature up to the enzyme's optimum temperature. Above 40°C, however, the enzyme is rapidly denatured, and its activity decreases accordingly. ∞ (Chapter 2, p. 42) Activity is maximal at an enzyme's optimum pH and decreases as the pH rises or drops from the optimum. Like high temperatures, extremely acidic or alkaline conditions also denature enzymes. Such conditions are used to kill or control the growth of microorganisms.

Concentration

To understand the effects of concentrations of substrates and products on enzyme-catalyzed reactions, we must first note that all chemical reactions are, in theory, reversible. Enzymes can catalyze a reaction to go in either direction: AB $\longrightarrow$ A + B or A + B $\longrightarrow$ AB. The concentrations of substrates and products are among several factors that determine the direction of a reaction. A high concentration of AB drives the reaction toward formation of A and B. Use of A and B in other reactions as fast as these products are formed also drives the reaction toward the formation of more A and B. Conversely, use of AB in another reaction so that its concentration

APPLICATIONS

HOW TO RUIN AN ENZYME

If its target enzyme has a vital metabolic function, an enzyme inhibitor acts as a poison. A competitive inhibitor, competing with the normal substrate for the active site, temporarily poisons enzyme molecules and slows the reaction. If it binds to the active site of all molecules of an enzyme at one time, it can stop the reaction. The enzyme itself is unharmed, however, and resumes function if the poison is removed. If the poison forms a covalent bond to the enzyme, distorting the active site so that it can no longer bind its substrate, it is a permanent poison; the enzyme's ability to function is irreversibly destroyed.

Enzyme inhibition has important medical applications. Some antibiotics, such as penicillin, kill bacteria by damaging their cell wall and causing lysis. The antibiotic binds to and inactivates enzymes that are needed to reseal breaks in the peptidoglycan molecules of the cell wall during cell growth. When the cell becomes sufficiently weakened, lysis occurs. Fluoride, which prevents tooth decay, hardens enamel and poisons enzymes. In low concentrations it kills bacteria in the mouth without damaging human cells, but if the concentration is large enough, it can kill human cells, too. Many pesticides and herbicides exert their effects through competitive inhibition. Certain chemotherapeutic agents used to treat cancer inhibit enzymes that are most active in rapidly dividing cells, including malignant cells, and have lesser effects on normal cells. Heavy metals inactivate enzymes noncompetitively and permanently and so function as active ingredients in many disinfectants. Although they are safe to use on inanimate objects, these metals cause severe toxic effects if ingested or absorbed through the skin.

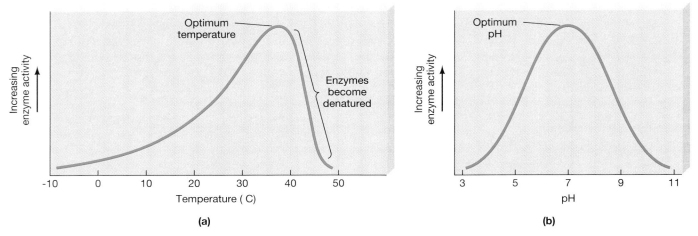

FIGURE 5.10 (a) Enzymes become more active as the temperature rises. Above about 40°C, however, most enzymes become denatured, and their activity falls off sharply. (b) Most enzymes also have an optimal pH at which they function most effectively.

remains low drives the reaction toward formation of AB. When neither AB nor A and B are removed from the system, the reaction will ultimately reach a steady state known as **chemical equilibrium.** At equilibrium, no net change in the concentrations of AB, A, or B occurs.

The quantity of enzyme available usually controls the rate of a metabolic reaction. A single enzyme molecule can catalyze only a specific number of reactions per second, that is, can act on only a specific number of substrate molecules. The reaction rate increases with the number of enzyme molecules and reaches a maximum when all available enzyme molecules are working at full capacity. However, if the substrate concentration is too low to keep all enzyme molecules working at capacity, the substrate concentration will determine the rate of the reaction.

With an overview of metabolic processes and an understanding of enzymes and how they work, we are ready to look at metabolic processes in more detail. We begin with glycolysis, fermentation, and aerobic respiration, the processes used by most microorganisms to capture energy.

ANAEROBIC METABOLISM: GLYCOLYSIS AND FERMENTATION

Glycolysis

Glycolysis (gli-kol'i-sis) is the metabolic pathway used by most autotrophic and heterotrophic organisms, both aerobes and anaerobes, to begin to break down glucose. It does not require oxygen, but it can occur either in the presence or absence of oxygen. Figure 5.11 shows

the 10 steps of the glycolytic pathway, within which four important events occur: (1) substrate-level phosphorylation, (2) the breaking of a six-carbon molecule (glucose) into two three-carbon molecules, (3) the transfer of two electrons to the coenzyme NAD, and (4) the capture of energy in ATP.

Phosphorylation (fos"for-i-la'shun) is the addition of a phosphate group to a molecule, often from ATP. This addition generally increases the molecule's energy. Thus, phosphate groups commonly serve as energy carriers in biochemical reactions. Early in glycolysis, phosphate groups from two molecules of ATP are added to glucose. This expenditure of two ATPs raises the energy level of glucose. It can then participate in subsequent reactions (like the rock pushed out of the depression atop the hill) and is rendered incapable of leaving the cell. The phosphorylated molecule drives the cell's metabolic reactions.

After phosphorylation, glucose is broken into two three-carbon molecules, and each molecule is oxidized as two electrons are transferred from it to NAD. The end products are two molecules of pyruvic acid (called pyruvate in its ionized form), and two molecules of reduced NAD (NADH).

Energy is captured in ATP at the substrate level—that is, in the direct course of glycolysis—in two separate reactions late in the process. With *adenosine diphosphate* (ADP) and *inorganic phosphate* (P_i) available in the cytoplasm, the energy released from substrate molecules is used to form high-energy bonds between ADP and P_i:

$$\text{ADP} + P_i + \text{energy} \longrightarrow \text{ATP}$$

Glycolysis provides cells with a relatively small amount of energy. Energy is captured in two molecules of ATP during the metabolism of each three-carbon

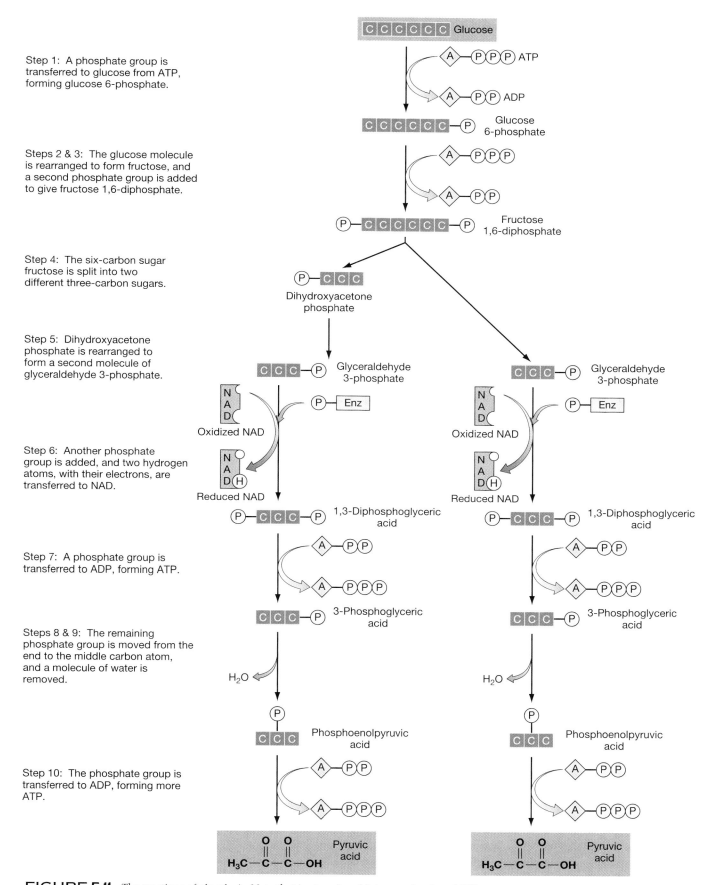

Step 1: A phosphate group is transferred to glucose from ATP, forming glucose 6-phosphate.

Step 2 & 3: The glucose molecule is rearranged to form fructose, and a second phosphate group is added to give fructose 1,6-diphosphate.

Step 4: The six-carbon sugar fructose is split into two different three-carbon sugars.

Step 5: Dihydroxyacetone phosphate is rearranged to form a second molecule of glyceraldehyde 3-phosphate.

Step 6: Another phosphate group is added, and two hydrogen atoms, with their electrons, are transferred to NAD.

Step 7: A phosphate group is transferred to ADP, forming ATP.

Steps 8 & 9: The remaining phosphate group is moved from the end to the middle carbon atom, and a molecule of water is removed.

Step 10: The phosphate group is transferred to ADP, forming more ATP.

FIGURE 5.11 The reactions of glycolysis. Note that in steps 1 and 3, two molecules of ATP are used. (A = adenosine) In steps 7 and 10, two molecules of ATP are formed. Because each glucose molecule yields two of the three-carbon sugars that undergo reactions 7 and 10, four molecules of ATP are actually formed, giving a net yield of two ATP per glucose.

molecule, and a total of four ATPs are formed as one six-carbon glucose molecule is metabolized by glycolysis to two molecules of pyruvic acid. (See Appendix E for a more detailed account of glycolysis.) Because energy from two ATPs was used in the initial phosphorylations, glycolysis results in a net energy capture of only two ATPs per glucose molecule. When atmospheric oxygen is present and the organism has the enzymes to carry out aerobic respiration, electrons from reduced NAD are transferred to oxygen during biological oxidation, as is explained later.

Besides glycolysis, many microorganisms have one or two other metabolic pathways for glucose oxidation. For example, many bacteria, including *Escherichia coli* and *Bacillus subtilis*, have a *pentose phosphate pathway*. This pathway, which can function at the same time as glycolysis, breaks down not only glucose but also five-carbon sugars (pentoses). (Appendix E outlines this pathway.) In a few species of bacteria, including *Pseudomonas*, enzymes carry out the Entner-Doudoroff pathway, which replaces the glycolytic and pentose phosphate pathways. In this pathway, glucose goes through a short series of reactions, one intermediate (glyceraldehyde 3-phosphate) of which goes through the last five steps of a typical glycolysis and produces 2 ATP molecules in the process of forming pyruvic acid.

Two additional features illustrate principles that apply to metabolic pathways in general:

1. Each reaction is catalyzed by a specific enzyme. Although enzyme names have been omitted in our discussion for simplicity, remember that each reaction in a metabolic pathway is catalyzed by an enzyme.

2. When electrons are removed from intermediates in metabolic pathways, they are transferred to one of two coenzymes—NAD or NADP *(nicotinamide adenine dinucleotide phosphate)*. In a reduced form (NADH or NADPH), these coenzymes store a cell's *reducing power*. For example, in

glycolysis, oxidized NAD becomes reduced NAD (NADH). As explained next, electrons are removed from reduced NAD during fermentation, freeing it to remove more electrons from glucose and keep glycolysis operating. Because cells contain limited quantities of both enzymes and coenzymes, the rate at which the reactions of glycolysis and other pathways occur is limited by the availability of these important molecules.

Although glucose is the main nutrient of some microorganisms, other microbes can obtain energy from other sugars. Such organisms usually have specific enzymes to convert a sugar into an intermediate in the glycolytic pathway. Once the sugar has entered glycolysis, it is metabolized to pyruvic acid and then fermented or metabolized aerobically by processes to be described later.

Fermentation

The metabolism of glucose or another sugar by glycolysis is a process carried out by nearly all cells. One process by which pyruvic acid is subsequently metabolized in the absence of oxygen is **fermentation.** Fermentation is the result of the need to recycle the limited amount of NAD by passing off the electrons of reduced NAD to other molecules. It occurs by many different pathways (Figure 5.12). Two of the most important and commonly occurring pathways are homolactic-acid fermentation and alcoholic fermentation. Neither captures energy in ATP from the metabolism of pyruvic acid, but both path-

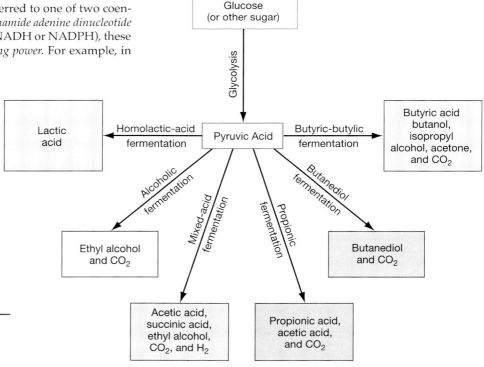

FIGURE 5.12 Some of the many different fermentation pathways found among microorganisms.

ways remove electrons from reduced NAD so that it can continue to act as an electron acceptor. Thus, they indirectly foster energy capture by keeping glycolysis going.

Homolactic-Acid Fermentation

The simplest pathway for pyruvic acid metabolism is **homolactic-acid fermentation,** in which only (homo-) lactic acid is made (Figure 5.13). Pyruvic acid is converted directly to lactic acid, using electrons from reduced NAD. Unlike other fermentations, this type produces no gas. It occurs in some types of the bacteria called lactobacilli, in streptococci, and in mammalian muscle cells. This pathway in lactobacilli is used in making some cheeses.

Alcoholic Fermentation

In **alcoholic fermentation** (Figure 5.14), carbon dioxide is released from pyruvic acid to form the intermediate acetaldehyde, which is quickly reduced to ethyl alcohol by electrons from reduced NAD. Alcoholic fermentation, although rare in bacteria, is common in yeasts and is used in making bread and wine. Chapter 27 deals extensively with these topics.

Other Kinds of Fermentation

The other kinds of fermentation summarized in Figure 5.12 are performed by a great variety of microorganisms. One of the most important things about these processes is that they occur in certain infectious organisms, and their products are used in diagnosis. For example, the Voges-Proskauer test for acetoin, an intermediate in butanediol fermentation, helps detect the bacterium *Klebsiella pneumoniae,* which can cause pneumonia. Anaerobic butyric-butylic fermentation occurs in *Clostridium* species that cause tetanus and

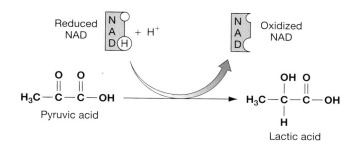

FIGURE 5.13 In homolactic-acid fermentation, pyruvic acid is reduced to lactic acid by the NAD from step 6 of glycolysis (Figure 5.11).

FIGURE 5.14 Alcoholic fermentation is a two-step process. A molecule of carbon dioxide is first removed from pyruvic acid to form acetaldehyde. Acetaldehyde is then reduced to ethyl alcohol by NAD.

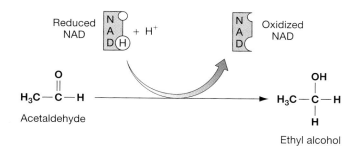

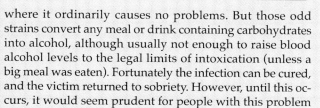

APPLICATIONS

INVOLUNTARY DRUNKENNESS

A man was arrested in Virginia for drunk driving. He offered a most unusual defense: involuntary drunkenness—due to yeast fermenting in food in his stomach, thereby producing alcohol that was absorbed into his bloodstream. The judge didn't think much of his plea and found him guilty. The blood alcohol level present in this particular defendant was considerably higher than that ordinarily found in people with such infections. However, there are documented cases in Japan and in the United States of people who were unable to remain sober due to stomach infections of peculiar strains of the yeast *Candida albicans. Candida* is found in various parts of the digestive tract,

where it ordinarily causes no problems. But those odd strains convert any meal or drink containing carbohydrates into alcohol, although usually not enough to raise blood alcohol levels to the legal limits of intoxication (unless a big meal was eaten). Fortunately the infection can be cured, and the victim returned to sobriety. However, until this occurs, it would seem prudent for people with this problem to refrain from driving.

The yeasts that leaven bread also produce alcohol. Why, then, don't you get drunk from eating your dinner rolls? The reason is that what little alcohol does form evaporates in the oven during baking.

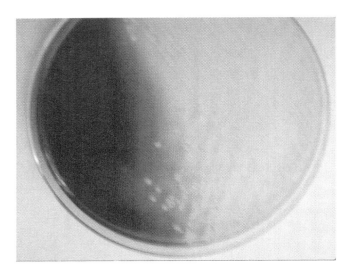

FIGURE 5.15 The mannitol-fermentation test distinguishes the pathogenic *Staphylococcus aureus* (right) from most nonpathogenic *Staphylococcus* species. *S. aureus* ferments mannitol, producing acid that turns the pH indicator (phenol red) in the medium to yellow. The medium before inoculation (left) is light red.

botulism. This fermentation also produces the unpleasant odors of rancid butter and cheese.

The ability to ferment sugars other than glucose forms the basis of other diagnostic tests. One such test (Figure 5.15) uses the sugar mannitol and the pH indicator phenol red. The pathogenic bacterium *Staphylococcus aureus* ferments mannitol and produces acid, which causes the phenol red in the medium to turn yellow. The nonpathogenic bacterium *Staphylococcus epidermidis* fails to ferment mannitol and does not change the color of the medium.

Many of the products formed by these and other fermentations, such as acetic acid, acetone, and glycerol, are of commercial value. Chapter 27 discusses industrial, pharmaceutical, and food products produced by microbial fermentation. Some of these may allow us to reduce our dependence on costly petrochemicals.

AEROBIC METABOLISM: RESPIRATION

As we have noted, most organisms obtain some energy by metabolizing glucose to pyruvate by glycolysis. Among microorganisms, both anaerobes and aerobes carry out these reactions. **Anaerobes** are organisms that do not use oxygen; they include some that are killed by exposure to oxygen. **Aerobes** are organisms that do use oxygen; they include some that must have oxygen. In addition, a significant number of species of microorganisms are *facultative anaerobes* (Chapter 6), which use oxygen if it is available but can function without it. Although aerobes obtain some of their energy from gly-

BIOTECHNOLOGY

PUTTING MICROBES TO WORK

What can you get from *Klebsiella pneumoniae* besides pneumonia? Acrylic plastics, clothing, pharmaceuticals, paints—all products of the parent compound 3-hydroxypropionaldehyde, which *K. pneumoniae* produces by fermentation of glycerol. Glycerol is a common byproduct of the processing of animal fats and vegetable oils, such as those from soybeans. Thus, our nation's surplus farm commodities could eventually help replace costly imported petrochemicals, thanks to microbial fermentations.

Scientists are also exploring the possibility of modifying other microbes such as *Saccharomyces cerevisiae* (baker's yeast) and *S. carlsbergensis* (brewer's yeast) to make them more useful to us. These scientists hope to obtain from yeast a fast-acting enzyme that, when given intravenously, will dissolve blood clots and thus lessen the damage caused by heart attacks and strokes. The biochemical versatility of yeast may also help clear our air of petrochemical pollutants. The yeast *Pachysolen tannaphilus* converts xylose, a sugar found in woody plant parts such as corn stalks, directly into ethyl alcohol. The U.S. Department of Agriculture estimates that such yeasts could produce 4 billion gallons of clean-burning fuel alcohol from agricultural wastes per year.

colysis, they use glycolysis chiefly as a prelude to a much more productive process, one that allows them to obtain far more of the energy potentially available in glucose. This process is **aerobic respiration** via the *Krebs cycle* and *oxidative phosphorylation*.

The Krebs Cycle

The **Krebs cycle,** named for the German biochemist Hans Krebs, who identified its steps in the late 1930s, metabolizes two-carbon units called *acetyl groups* to CO_2 and H_2O. It also is called the *tricarboxylic-acid (TCA) cycle,* because some molecules in the cycle have three carboxyl (COOH) groups, or the *citric-acid cycle,* because citric acid is an important intermediate.

Before pyruvic acid, the product of glycolysis, can enter the Krebs cycle, it must first be converted to *acetyl-CoA*. This complex reaction involves the removal of one molecule of CO_2, transfer of electrons to NAD, and addition of coenzyme A (Figure 5.16). In prokaryotes, these reactions occur in the cytoplasm; in eukaryotes, they occur in the matrix of mitochondria.

The Krebs cycle is a sequence of reactions in which acetyl groups are oxidized to carbon dioxide. Hydrogen atoms are also removed, and their electrons are transferred to coenzymes that serve as electron carriers (Figure 5.17). (The hydrogens, as we will see, are eventually combined with oxygen to form water.) Each reaction in the Krebs cycle is controlled by a specific enzyme, and the molecules are passed from one enzyme to the next as they go through the cycle. The reactions form a cycle because oxaloacetic acid (oxaloacetate), a first reactant, is regenerated at the end of the cycle. As one acetyl group is metabolized, oxaloacetate combines with another and goes through the cycle again. (Appendix E gives more detail on the Krebs cycle.)

Certain events in the Krebs cycle are of special significance: (1) the oxidation of carbon, (2) the removal of electrons to coenzymes, and (3) substrate-level en-

FIGURE 5.16 The doorway into the Krebs cycle. Pyruvic acid loses a molecule of CO_2 and is oxidized by NAD. The resulting two-carbon acetyl group is attached to coenzyme A, forming acetyl-CoA.

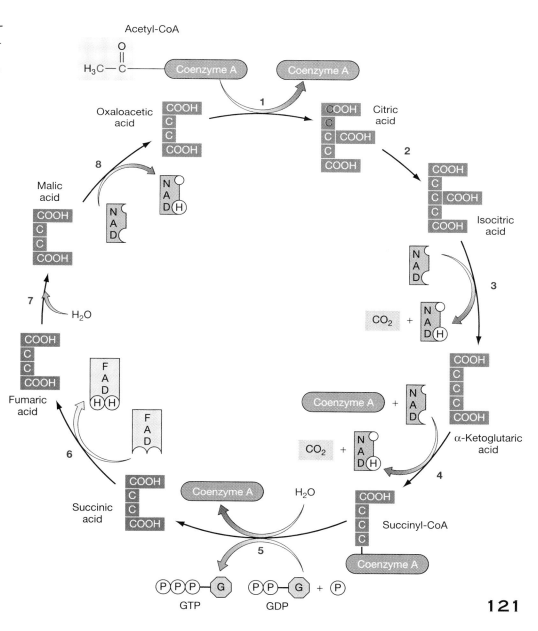

FIGURE 5.17 The reactions of the Krebs cycle. The intermediates are simplified to show only the number of carbon atoms and carboxyl groups for each. A two-carbon acetyl group enters the cycle as acetyl-CoA in step 1, and two carbon atoms leave the cycle as molecules of CO_2 in steps 3 and 4. Energy is captured in guanosine triphosphate (GTP) in step 5 and eventually transferred to adenosine triphosphate (ATP). In addition, electrons are removed by coenzymes in steps 3, 4, 6, and 8. More energy will be extracted from these electrons when they are subsequently fed into the electron transport chain.

ergy capture. As each acetyl group goes through the cycle, two molecules of carbon dioxide arise from the complete oxidation of its two carbons. Four pairs of electrons are transferred to coenzymes: three pairs to NAD, and one pair to FAD. Much energy is derived from these electrons in the next phase of respiration, as we will soon see. Finally, some energy is captured in a high-energy bond in *guanosine triphosphate* (GTP). This reaction takes place at the substrate level. That is, it occurs directly in the course of a reaction of the Krebs cycle. Energy in GTP is easily transferred to ATP. Note that because each glucose molecule produces two molecules of acetyl-CoA, the quantities of the products just given must be doubled to represent the yield from metabolism of a single glucose molecule.

Electron Transport and Oxidative Phosphorylation

Electron transport, the process leading to the transfer of electrons from substrate to O_2, begins during one of the energy-releasing dehydrogenation reactions of catabolism. Two hydrogen atoms (each consisting of one electron and one proton) are transferred to NAD, forming reduced NAD. The resulting compound in turn transfers the pair of atoms to one of a series of other carrier compounds embedded in the cell membrane of bacteria or in the inner membrane of mitochondria. These carrier compounds form an **electron transport chain,** which is often called the *respiratory chain* (Figure 5.18). Through a series of oxidation-reduction reactions, the

FIGURE 5.18 The electron transport chain. The carriers are NAD; FMN (flavin mononucleotide), a derivative of riboflavin (vitamin B_2); iron sulfide (FeS); coenzyme Q (CoQ); and cytochromes a, b, and c. Each carrier is reduced (*red*) as it picks up electrons and is oxidized (*ox*) as it passes them on to the next carrier in the chain. Eventually, the electrons combine with the final electron acceptor, oxygen, to form water. Note that FAD feeds its electrons into the chain at a lower point than does NAD. Because of the arrangement of the various carriers in the cell membrane of prokaryotes or inner mitochondrial membrane of eukaryotes, and because some carriers accept protons (H^+) whereas others carry only electrons, protons are pushed from the bacterial cytoplasm to outside the bacterial cell (or from the matrix into the space between the two mitochondrial membranes) during electron transport. The proton gradient created in this way is used to make ATP (Figure 5.20).

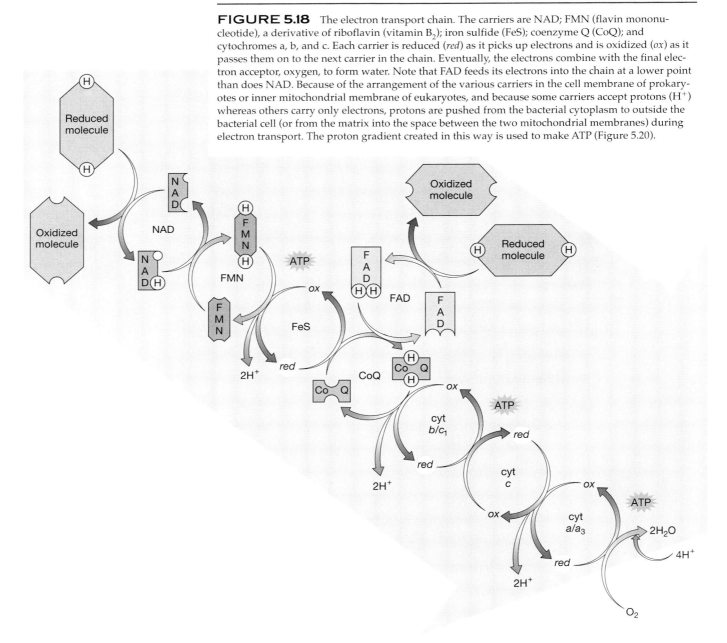

electron transport chain performs two basic functions: (1) accepting electrons from an electron donor and transferring them to an electron acceptor, and (2) conserving for ATP synthesis some of the energy released during the electron transfer. The large amounts of energy obtained from respiration result from the transfer of electrons through the electron transport chain, from a level of high energy to one of low energy, with the formation of ATP (Figure 5.18). Energy is captured in high-energy bonds as P_i combines with ADP to form ATP. This process is known as **oxidative phosphorylation.** Each member of the chain becomes reduced as it picks up electrons; then upon giving up electrons to the next member in line, it is oxidized. In aerobic respiration, oxygen is the final electron acceptor and becomes reduced to water (Figure 5.18)

Several types of enzyme complexes are involved in electron transport. These include NADH dehydrogenase, cytochrome reductase, and cytochrome oxidase. The electron carriers include **flavoproteins** (such as FAD and flavin mononucleotide, FMN), iron-sulfur (FeS) proteins, and **cytochromes,** proteins with an iron-containing ring called *heme.* A group of nonprotein, lipid-soluble electron carriers known as the **quinones,** or *coenzymes Q,* are also found in electron transport systems.

All electron transport chains are not alike; they differ from organism to organism. However, they all have compounds such as flavoproteins and quinones, which accept only hydrogen atoms, and compounds such as cytochromes, which accept only electrons. Unless electrons are continuously transferred from reduced NAD and FAD to oxygen via the electron transport chain, these enzymes cannot accept more electrons from the Krebs cycle, and the entire process will grind to a halt.

From the metabolism of a single glucose molecule, 10 pairs of electrons are transported by NAD (2 pairs from glycolysis, 2 pairs from the pyruvic acid to acetyl-CoA conversion, and 6 pairs from the Krebs cycle). Two additional pairs are transported by FAD (from the Krebs cycle). All these electrons are passed to other electron carriers in the electron transport chain.

Electron transport and oxidative phosphorylation can be likened to a series of waterfalls in which the water makes many small descents and three larger ones (Figure 5.19). In most electron transfers (the small descents), only small amounts of energy are released. At three points (the larger descents), more energy is released, some of which is used to form ATP by the addition of P_i to ADP.

In our waterfall analogy, we can think of water entering the falls at two sites, one higher up the mountain than the other. Water from the higher site falls farther than water entering lower down the mountain. In bacteria, electrons entering the electron transport chain at NAD start at the top, and their descent releases enough energy to make three ATPs. Electrons entering at FAD start partway down the chain and contribute only enough energy to make two ATPs. Thus, during aerobic metabolism of a glucose molecule, the 10 pairs of electrons from NAD produce 30 ATPs, and 2 pairs from FAD produce 4 ATPs, for a total of 34 ATPs. Including the 2 ATP molecules from glycolysis and the 2 GTP molecules (= 2ATPs) from the Krebs cycle gives a total yield of 38 ATPs per glucose molecule.

Oxidative phosphorylation, when compared with fermentation, generates the greatest amount of energy available from glucose. Fermentation, through the substrate-level production of ATP during glycolysis, yields only about 5 percent as much. The net gain of ATP molecules from fermentation is 2.

Chemiosmosis

Electrons for the hydrogen atoms removed from the reactions of the Krebs cycle are transferred through the electron transport system, which generates the high-energy bonds of ATP. The conversion of ADP to ATP occurs via a large ATP-synthesizing complex called *ATP synthase* (or *ATPase*) in a process known as **chemiosmosis** (kem"e-os-mo'sis). This process is the result of a series of chemical reactions that occur in and around a membrane. Although the mechanism for the process, first proposed by the British biochemist Peter Mitchell in

FIGURE 5.19 The electron transport chain modeled as a waterfall. As the electrons are passed from carrier to carrier in the chain, they decrease in energy, and some of the energy they lose is harnessed to make ATP.

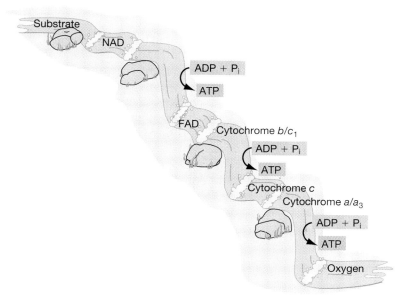

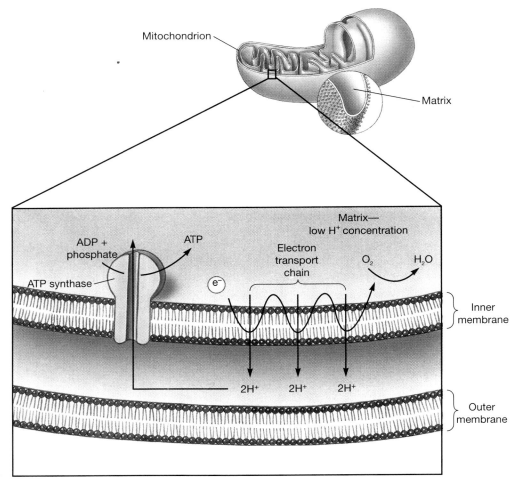

FIGURE 5.20 Energy capture by chemiosmosis in a mitochondrion. Protons "pumped" out of the matrix during electron transport pass back in through channels in the ATP synthases that phosphorylate ADP to make ATP. The flow of protons provides the energy to drive this reaction.

1961, took a number of years to become fully accepted, it is now recognized as a major contribution to the understanding of how ATP is formed during electron transport. Mitchell was awarded the Nobel Prize in 1978 for the development of the chemiosmosis hypothesis.

Chemiosmosis occurs in the cell membrane of prokaryotes such as *Escherichia coli* and in the inner mitochondrial membrane of eukaryotes (Figure 5.20). As electrons are transferred along the electron transport chain, protons are pumped outside the membrane, so the ions' concentration is higher outside the membrane than inside. This process causes a lowering of the proton concentration on the inside and the development of a force that drives the protons back into the cell or mitochondrial matrix to equalize their concentration on both sides of the membrane. Any concentration gradient naturally tends to equalize itself.

In addition to the proton concentration gradient across the membrane, there is also an electrochemical gradient, which makes the membrane a type of biological battery that can power the formation of ATP. The H^+ excess on one side of the membrane gives that side a positive charge compared with the other side.

TRY IT

FUELING THE RACE TO REPRODUCE

Two students, working as a team in the laboratory, observed the growth of yeast in a sugar solution on microscope slides. One student focused her microscope near the edge of the coverslip, where oxygen levels were sufficient for aerobic respiration. The other student focused his microscope on yeast growing under the center of the coverslip. They kept the same fields in focus for the duration of the class period, and they counted the number of yeast cells in the field every 20 to 30 minutes.

Can you predict their results? What would explain these results? With your instructor's consent, you could easily try this experiment while completing your assigned laboratory exercises and then share your data with the rest of the class. Do you think this experiment would give the same results with all types of organisms? Why?

The power generated by the gradient is called the *proton motive force.* The protons flow through special channels within the synthase complex. Energy is thus released and used to form ATP from ADP and inorganic phosphate (P_i).

TABLE 5.3	Energy Captured in ATP Molecules from a Glucose Molecule by Anaerobic and Aerobic Metabolism in Prokaryotes

	NUMBER OF ATP MOLECULES	
Prokaryotic Metabolic Process	Anaerobic Conditions	Aerobic Conditions
Glycolysis		
Substrate level	4	4
Hydrogen to NAD	0	6
Pyruvate to acetyl-CoA		
Hydrogen to NAD	0	6
Krebs cycle		
Substrate level	0	2
Hydrogen to NAD	0	18
Hydrogen to FAD	0	4
Less energy for phosphorylation	−2	−2
Total	2	38

Significance of Energy Capture

In glycolysis and fermentation, as we noted earlier, a net of 2 ATPs is usually produced for every glucose molecule metabolized anaerobically. When glycolysis is followed by aerobic respiration, each glucose molecule produces an additional 2 ATPs at the substrate level in the Krebs cycle and 34 ATPs by oxidative phosphorylation. Thus, a glucose molecule yields 38 ATPs by aerobic metabolism but only 2 ATPs by glycolysis and fermentation (Table 5.3). That is, 19 times as much energy is captured in aerobic metabolism as in fermentation! Hence, aerobic microorganisms in environments with ample oxygen generally grow more rapidly than anaerobes. But aerobes will die if oxygen is depleted, unless they can switch to fermentation. Table 5.4 summarizes the metabolic processes we have studied thus far.

METABOLISM OF FATS AND PROTEINS

For most organisms, including microorganisms, glucose is a major source of energy. However, for almost any organic substance, we can find a type of microor-

TABLE 5.4	Comparison of Metabolic Processes			

	Glycolysis	Fermentation	Krebs Cycle[a]	Electron Transport Chain
Location	In cytoplasm	In cytoplasm	Prokaryotes: in cytoplasm Eukaryotes: in the mitochondrial matrix	Prokaryotes: in cell membrane Eukaryotes: in inner mitochondrial membranes
Oxygen Conditions	Anaerobic, oxygen is not required; does not stop, however, if oxygen is present	Without O_2; presence of oxygen will cause it to stop	Aerobic	Aerobic
Starting Molecule(s)	1 glucose (6C)	Various substrate molecules go through glycolysis, yielding 2 pyruvic acid	2 pyruvic acid	$6 O_2$
Ending Molecules	2 pyruvic acid (3C) 2 NADH	Various, depending on which form of fermentation occurs, e.g., ethanol, lactic acid, CO_2, acetic acid	$6 CO_2$ 8 NADH 2 FADH	$6 H_2O$
Amount of ATP Produced	4 ATP (net 2 ATP)	Various, depending on which form of fermentation occurs, usually 2 or 3 ATP; always far less than is produced in aerobic respiration	2 GTP (= 2 ATP)	34 ATP

[a]Includes the pyruvic acid → acetyl-CoA step.

MICROBIAL CLEANUP

A few species of bacteria, such as some members of the genus *Pseudomonas,* can use crude oil for energy. They can grow in sea water with only oil, potassium phosphate, and urea (a nitrogen source) as nutrients. These organisms act as "bioremediators" by cleaning up oil spills in the ocean. They have also proved useful in degrading oil that remains in the water carried by tankers as ballast after the tankers have unloaded their cargo of oil. Then the water pumped from the tankers into the sea in preparation for a new cargo of oil does not pollute. A detergentlike substance has recently been isolated from these organisms. When the detergent is added to a quantity of oil sludge, it converts 90 percent of the sludge into usable petroleum in about 4 days, thereby reducing waste and providing a convenient means of cleaning oil-fouled tanks.

The effectiveness of microbially produced oil-eating enzymes is being tested on an oil spill in the laboratory.

ganism that can degrade that substance for energy. This attribute of microorganisms, and the fact that microbes are found almost everywhere on our planet, accounts for their ability to degrade dead and decaying remains and wastes of all organisms.

Metabolism of Fats

Most microorganisms, like most animals, can obtain energy from lipids. The following examples give a general idea of how such processes occur. Fats are hydrolyzed to glycerol and three fatty acids. The glycerol is metabolized by glycolysis. The fatty acids, which usually have an even number of carbons (16, 18, or 20), are broken down into two-carbon pieces by a metabolic pathway called **beta oxidation.** In this process a fatty acid first combines with coenyzme A. Oxidation of the beta carbon (second carbon from the carboxyl group) of the fatty acid results in the release of acetyl-CoA and the formation of a fatty acid shorter by two carbon atoms. The process is then repeated, and another acetyl-CoA molecule is released. The newly formed acetyl Co-A is then oxidized via the Krebs cycle to obtain additional energy (Figure 5.21).

Metabolism of Proteins

Proteins also can be metabolized for energy. They are first hydrolyzed into individual

FIGURE 5.21 The catabolism of fats. Triglycerides are hydrolyzed into glycerol and fatty acids. The glycerol is broken down via glycolysis. The fatty acids are broken down into two-carbon units and fed into the Krebs cycle, where they are metabolized to produce additional energy.

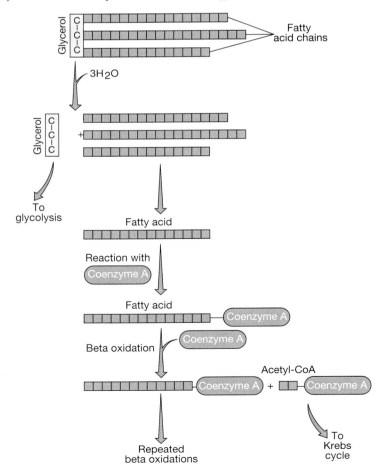

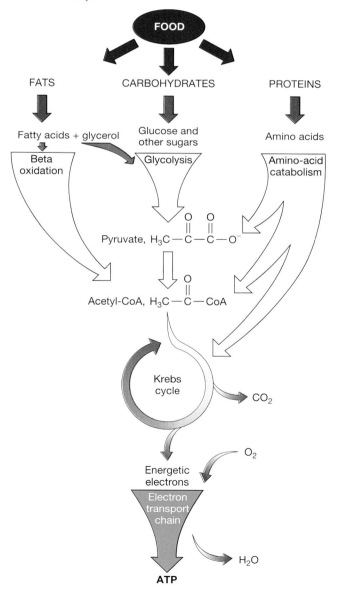

FIGURE 5.22 The catabolism of proteins. Polypeptides are hydrolyzed to amino acids. The amino acids are deaminated, and the resulting molecules enter pathways leading to the Krebs cycle.

amino acids by *proteolytic* (protein-digesting) *enzymes* (Figure 5.22). Then the amino acids are *deaminated*—that is, their amino groups are removed. The resulting deaminated molecules enter glycolysis, fermentation, or the Krebs cycle. The metabolism of all major nutrients (fats, carbohydrates, and proteins) for energy is summarized in Figure 5.23.

OTHER METABOLIC PROCESSES

Having considered energy capture in chemoheterotrophs, we will now briefly consider energy capture in photoautotrophs, photoheterotrophs, and chemoautotrophs.

Photoautotrophy

Organisms called photoautotrophs carry out **photosynthesis,** the capture of energy from light and the use of this energy to manufacture carbohydrates from carbon dioxide. Photosynthesis occurs in green and purple bacteria, in cyanobacteria, in algae, and in higher plants. Photosynthetic bacteria, which probably evolved early in the evolution of living organisms, perform their own version of photosynthesis in the absence of O_2. However, algae and green plants make much more of the world's carbohydrate supply, so we will

consider the process in those organisms first and then see how it is modified in green and purple bacteria.

In green plants, algae, and cyanobacteria, photosynthesis occurs in two parts—the "photo" part, or the *light reactions,* in which light energy is converted to chemical energy, and the "synthesis" part, or the *dark reactions,* in which chemical energy is used to make organic molecules. Each part involves a series of steps.

In the **light reactions,** light strikes the green pigment chlorophyll *a* in thylakoids of chloroplasts. ∞ (Chapter 4, p. 93) Electrons in the chlorophyll become excited—that is, raised to a higher energy level. These electrons participate in generating ATP in cyclic

FIGURE 5.23 Metabolism of the major classes of biomolecules: a summary.

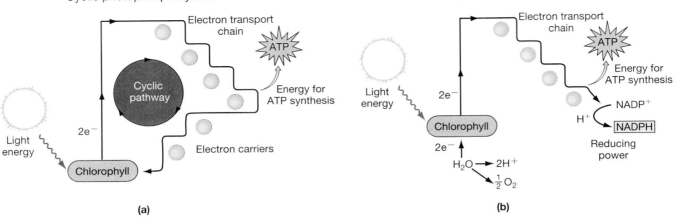

Cyclic photophosphorylation

Noncyclic photoreduction

(a)

(b)

FIGURE 5.24 The light reactions of photosynthesis as performed by cyanobacteria, algae, and green plants. Electrons in chlorophyll receive a boost in energy from light, and their extra energy is used to make ATP. In the pathway of cyclic photophosphorylation, the electrons return to chlorophyll and thus can be used over and over again. In noncyclic photoreduction, the electrons receive a second boost that gives them enough energy to reduce NADP. The electrons are replaced by the splitting of water.

photophosphorylation and in noncyclic photoreduction (Figure 5.24). In **cyclic photophosphorylation,** excited electrons from chlorophyll are passed down an electron transport chain. As they are transferred, energy is captured in ATP by chemiosmosis (as described previously in connection with oxidative phosphorylation). When the electrons return to the chlorophyll, they can be excited over and over again, so the process is said to be cyclic.

In **noncyclic photoreduction,** energy also is captured by chemiosmosis. In addition, membrane proteins and energy from light are used to split water molecules into protons, electrons, and oxygen molecules, a process called **photolysis** (fo-tol'eh-sis). The electrons replace those lost from chlorophyll, which are thus freed to reduce the coenzyme NADP. ATP and reduced NADP (NAPDH)—the products of the light reaction—and atmospheric CO_2 subsequently participate in the dark reactions.

The **dark reactions,** or *carbon fixation,* occur in the stroma of chloroplasts. Carbon dioxide is reduced by electrons from NADPH in a process known as the *Calvin-Benson cycle.* (See Appendix E.) Energy from ATP and electrons from NADPH are required in this synthetic process. Various carbohydrates, chiefly glucose, are the products of the dark reactions (Figure 5.25).

Photosynthesis in green and purple sulfur bacteria differs from that in green plants, algae, and cyanobacteria in ways related to the evolution of organisms. The first photosynthetic organisms probably were purple and green bacteria, which evolved in an atmosphere containing much hydrogen but no oxygen. They differ from green plants, algae, and cyanobacteria as follows:

1. Bacterial chlorophyll absorbs slightly longer wavelengths of light than does chlorophyll *a.*

2. They use hydrogen compounds such as hydrogen sulfide (H_2S), rather than water (H_2O), for reducing carbon dioxide. Electrons from their pigments reach an energy

FIGURE 5.25 The relation between the light and dark reactions. In the dark reactions, ATP and NADPH (the products of the light reactions) are used to reduce carbon dioxide, forming carbohydrates such as glucose. The dark reactions do not require darkness; they are so named because they can take place in the dark, so long as the products of the light reactions are available.

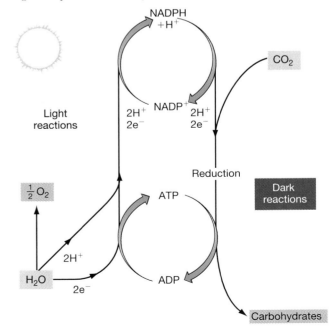

level high enough to split H_2S but not H_2O. (Some purple and green bacteria produce elemental sulfur as a byproduct; a few produce strong sulfuric acid.)

3. They are usually strict anaerobes and can live only in the absence of oxygen. They do not release oxygen as a product of photosynthesis, as green plants do.

TABLE 5.5	Characteristics of Photosynthetic Bacteria	
Group	Family and Representative Genus	Pigments
Green sulfur bacteria	Chlorobiaceae *Chlorobium*	Bacterial chlorophyll
Purple sulfur bacteria	Chromaticeae *Chromatium*	Bacterial chlorophyll and red and purple carotenoid pigments

Characteristics of the groups of bacteria that carry out this primitive form of photosynthesis are summarized in Table 5.5.

The cyanobacteria also are photosynthetic, but they probably evolved after the purple and green bacteria. Although prokaryotic, the cyanobacteria release oxygen during photosynthesis, as do green plants and algae. In fact, cyanobacteria are probably responsible for the addition of oxygen to the primitive atmosphere.

Photoheterotrophy

Photoheterotrophs are a small group of bacteria that can use energy from light but require organic substances such as alcohols, fatty acids, or carbohydrates as carbon sources. These organisms include the non-sulfur, purple or green bacteria.

Chemoautotrophy

Bacteria called *chemoautotrophs,* or **chemolithotrophs,** are unable to carry out photosynthesis but can oxidize inorganic substances for energy. With this energy and carbon dioxide as a carbon source, they are able to synthesize a great variety of substances, including carbohydrates, fats, proteins, nucleic acids, and substances that are required as vitamins by many organisms.

The ability to oxidize, and therefore extract energy from, inorganic substances is probably the most outstanding characteristic of chemolithotrophs, but these bacteria have other noteworthy attributes. The nitrifying bacteria are especially important because they increase the quantity of usable nitrogen compounds available to plants and replace nitrogen that plants remove from the soil. *Thiobacillus* and some other sulfur bacteria produce sulfuric acid by oxidizing elemental sulfur or hydrogen sulfide. Acidity lower than pH 1 has been produced by sulfur bacteria. Sulfur is sometimes added to alkaline soil to acidify it, a practice that works because of the numerous thiobacilli present in most soils. Finally, some chemolithotrophic Archaeobacteria have been found near volcanic vents in the ocean floor, where they grow at extremely high temperatures and sometimes under very acidic conditions. Characteristics of chemolithotrophs are summarized in Table 5.6.

USES OF ENERGY

Microorganisms use energy for such processes as biosynthesis, membrane transport, movement, and growth. Here we will summarize some biosynthetic activities and some mechanisms for membrane transport and movement. We will consider microbial growth in Chapter 6.

Biosynthetic Activities

Microorganisms share many biochemical characteristics with other organisms. All organisms require the same building blocks to make proteins and nucleic acids. Many of these building blocks (amino acids, purines, pyrimidines, and ribose) can be derived from intermediate products of energy-yielding pathways (Figure 5.26). When the energy-yielding pathways were first discovered, they were thought to be purely catabolic. Now that many of their intermediates are known to be involved in biosynthesis, they are more properly called **amphibolic** (am-fe-bol'ik) **pathways** (*amphi-*, either) because they can yield either energy or building blocks for synthetic reactions.

Some biosynthetic pathways are quite complex. For example, synthesis of amino acids in those organisms

TABLE 5.6	Characteristics of Chemolithotrophic Bacteria	
Group and Representative Genus/Genera	Source of Energy	Products after Oxidizing Reaction
Nitrifying bacteria		
Nitrobacter	HNO_2	HNO_3
Nitrosomonas	NH_3	$HNO_2 + H_2O$
Nonphotosynthetic sulfur bacteria		
Thiothrix	H_2S	$H_2O + 2S$
Thiobacillus	S	H_2SO_4
Iron bacteria		
Siderocapsa	Fe^{2+}	$Fe^{3+} + OH^-$
Hydrogen bacteria		
Alcaligenes	H_2	H_2O

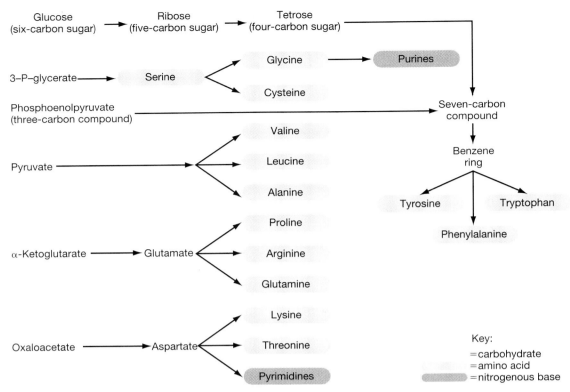

FIGURE 5.26 An outline of some biosynthetic pathways, showing how amino acids, nucleic acid bases, and ribose are made from intermediates in glycolysis and from the Krebs cycle.

that can make them often requires many reactions, with an enzyme for each reaction. Tyrosine synthesis requires no fewer than 10 enzymes, and tryptophan synthesis needs at least 13. The synthetic pathways for making purines and pyrimidines also are complex. The absence of a single enzyme in a synthetic pathway can prevent the synthesis of a substance. Any essential substance that an organism cannot synthesize must be accessible in the environment, or the organism will die. Missing enzymes thus increase the nutritional needs of organisms.

Microorganisms of many different types also synthesize a variety of carbohydrates and lipids. The rate at which they are synthesized also varies and depends on the availability and activity of enzymes. Some organisms, such as the aerobe *Acetobacter*, synthesize cellulose, which is ordinarily found in plants. As strands of cellulose reach the cell surface, they form a mat that traps carbon dioxide bubbles and keeps the cell afloat. Because these organisms must have oxygen, the mat contributes to their survival by keeping them near the surface, where oxygen is plentiful.

Many bacteria synthesize peptidoglycan, lipopolysaccharide, and other polymers associated with cell walls. ∞ (Chapter 4, p. 78) Some bacteria form capsules, especially in media that contain serum or large amounts of sugar. Capsules usually consist of polymers of one or more monosaccharides. However, in *Bacillus anthracis*, the bacterium that causes anthrax,

the capsule is a polypeptide of glutamic acid. The biosynthetic processes in microorganisms are summarized in Figure 5.27.

Membrane Transport and Movement

In addition to using energy for biosynthetic processes, microorganisms also use energy for transporting substances across membranes and for their own movement. These energy uses are as important to the survival of the organisms as are their biosynthetic activities.

Membrane Transport

Microbes use energy to move most ions and metabolites across cell membranes against concentration gradients. For example, bacteria can transport a sugar or an amino acid from a region of low concentration outside the cell to a region of higher concentration inside the cell. This means that they accumulate nutrients within cells in concentrations a hundred to a thousand times the concentration outside the cell. They also concentrate certain inorganic ions by the same means.

Two mechanisms exist in bacteria for concentrating substances inside cells, and both require energy. One active transport mechanism is specific to gram-negative bacteria, such as *E. coli*. Such bacteria have two membranes—the cell membrane, which surrounds the cell's cytoplasm, and the outer membrane, which forms part

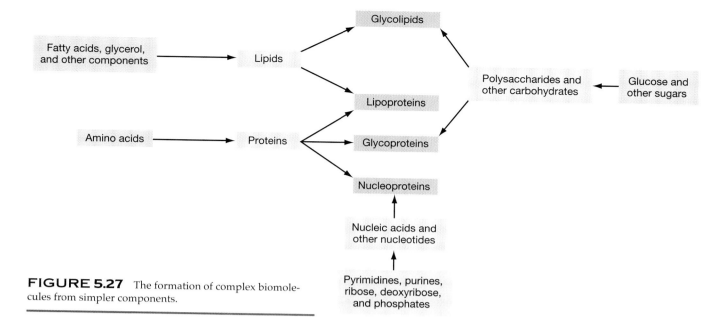

FIGURE 5.27 The formation of complex biomolecules from simpler components.

of the cell wall. ∞ (Chapter 4, p. 78) Transmembrane carrier proteins called **porins** form channels through the outer membrane. Porins allow entry of ions and small hydrophilic metabolites via *facilitated diffusion*. ∞ (Chapter 4, p. 98) After entering the periplasmic space, a specific periplasmic protein combines with one of the diffusing ions or metabolites. The periplasmic protein then facilitates the transport of the substance into the cytoplasm via a specific carrier protein in the cell membrane. Such substances generally gain entry by active transport. Through ATP hydrolysis, the carrier protein changes shape, allowing the metabolite into the cytoplasm (see Figure 4.29).

Another mechanism, present in all bacteria, is called the **phosphotransferase system** (PTS). It consists of sugar-specific enzyme complexes called **permeases** (per'me-a-sez), which form a transport system through the cell membrane. The PTS uses energy from the high-energy phosphate molecule phosphoenolpyruvate (PEP). When PEP is present in the cytoplasm, it can provide energy and a phosphate group to a permease in the membrane. Then the permease transfers the phosphate to a sugar molecule and at the same time moves the sugar across the membrane. A phosphorylated sugar is thus transported inside the cell and is prepared to undergo metabolism. This *group translocation* was discussed in Chapter 4. ∞ (p. 99)

Movement

Most motile bacteria move by means of flagella, but some move by gliding or creeping or in a corkscrew motion. Flagellated bacteria move by rotating their flagella. ∞ (Chapter 4, p. 86) The mechanism for rotation, though not fully understood, appears to involve a proton gradient as in chemiosmosis. As the protons move down the gradient, they drive the rotation. Gliding bacteria move only when in contact with a solid surface, such as decaying organic matter. Rotation of the cell on its own axis often occurs with gliding. A number of mechanisms have been proposed to explain gliding, but the mechanism that propels the gliding bacterium *Myxococcus* is best understood. This organism uses energy to secrete a substance called a **surfactant** (ser-fak'tant), which lowers surface tension at the bacterium's posterior end. The difference in surface tension between the anterior and posterior ends (a passive phenomenon) causes *Myxococcus* to glide.

Spirochetes expend energy for both creeping and writhing motions. On a solid surface they creep along like an inchworm by alternately attaching front and rear ends. Suspended in a liquid medium, they writhe (twist and turn). Both creeping and writhing motions probably occur by waves of contraction within the cell substance that exert force against axial filaments.

CLOSE-UP

RANDOM MOTION VERSUS MOTILITY

Bacteria sometimes appear to engage in random, vibrating movements when viewed under the microscope in a hanging-drop preparation. Such nondirectional movements are due to Brownian movement—the bombardment of the bacteria by molecules of water and other substances in the medium. Brownian movement is thus caused by forces external to the bacteria and is not due to energy expenditure by the cells themselves.

BIOLUMINESCENCE

Bioluminescence, the ability of an organism to emit light, appears to have evolved as a byproduct of aerobic metabolism. Bacteria of the genera *Photobacterium* and *Achromobacter*, fireflies, glowworms, and certain marine organisms living at great depths in the ocean exhibit bioluminescence (Figure 5.28). Many light-emitting organisms have the enzyme *luciferase* (lusif'er-ace), along with other components of the electron transport system. (Luciferase derives its name from Lucifer, which means "morning star.") Luciferase catalyzes a complex reaction in which molecular oxygen is used to oxidize a long-chain aldehyde or ketone to a carboxylic acid. At the same time, $FMNH_2$ from the electron transport chain is oxidized to an excited form of *flavin mononucleotide* (FMN), a carrier molecule derived from riboflavin (vitamin B_2), that emits light as it returns to its unexcited state (Figure 5.29). In this process, phosphorylation reactions are bypassed, and no ATP is generated. Instead, energy is released as light.

Luminescent microorganisms often live on the surface of marine organisms such as some squids and fish. More than 300 years ago, the Irish chemist Robert Boyle observed that the familiar glow of the skin of dead fish lasted only as long as oxygen was available. At that time the electron transport system and the role of oxygen in it were not understood.

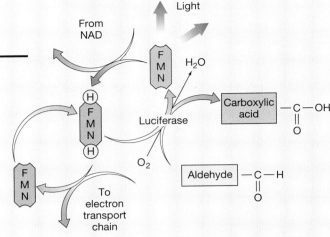

FIGURE 5.29 The reactions that give rise to bioluminescence are catalyzed by the enzyme luciferase. Oxygen is used to oxidize an aldehyde or ketone to a carboxylic acid. In the process, $FMNH_2$ is also oxidized, giving rise to an excited form of FMN that quickly radiates away its extra energy as light. Electrons are thus diverted from the electron transport chain, and light is produced at the expense of generation of ATP.

Bioluminescence exhibited by larger organisms has survival value. It is the sole light source for marine creatures that live at great depths, and it helps land organisms such as fireflies find mates. How bioluminescence came to be established among microorganisms is less clear. One hypothesis is that early in the evolution of living things, bioluminescence served to remove oxygen from the atmosphere as it was produced by some of the first photosynthetic organisms. Although this is not an advantage to aerobes, it is an advantage to strict anaerobes. Because most of the microorganisms in existence at that time were anaerobes susceptible to the toxic effects of oxygen, bioluminescence would have been beneficial to them. Today, many bioluminescent microbes are beneficiaries of symbiotic relationships with their hosts. They provide light in return for a shelter and nutrients.

Scientists have found a way to put bioluminescent bacteria to work. In the Microtox Acute Toxicity Test, bioluminescent bacteria are exposed to a water sample to determine if the sample is toxic. Any change—positive or negative—in the bacteria's growth is observable as a change in their light output. The sample's toxicity is calculated by comparing before-and-after readings of the light levels. The brainchild of Microbics Corp. of Carlsbad, California (Figure 5.30), this toxicity test takes only minutes to perform.

The Microtox Acute Toxicity Test is useful for testing the quality of drinking water and for numerous other industrial applications. For example, waste-water treatment plants use it to determine quickly whether their treated effluent will be able to pass government toxicity compliance tests. Paper mills use the test to determine how much disinfectant is needed to rid their equipment of the microbial growth that slows down the manufacturing process and affects product quality. Makers of household cleansers, shampoos, or cosmetics use the test in place of controversial animal testing in which drops of the products are put into the eyes of rabbits to determine the products' irritancy levels. And unlike cell-culturing techniques, the test requires little skill to perform and to interpret. Bioluminescence could prove to be a very important process to industry in the future.

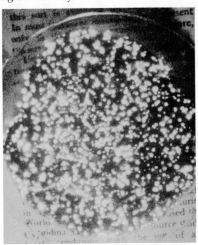

FIGURE 5.28 Bioluminescent bacteria in the Petri dish produce enough light to read by.

FIGURE 5.30 Microbics researchers with bioluminescent bacteria.

METABOLISM: AN OVERVIEW

- **Metabolism** is the sum of all the chemical processes in a living organism. It consists of **anabolism,** reactions that require energy to synthesize complex molecules from simpler ones, and **catabolism,** reactions that release energy by breaking complex molecules into simpler ones.
- **Autotrophs,** which use carbon dioxide to synthesize organic molecules, include **photoautotrophs** (which carry on photosynthesis) and **chemoautotrophs.**
- **Heterotrophs,** which use organic molecules made by other organisms, include **chemoheterotrophs** and **photoheterotrophs.**
- For growth, movement, and other activities, **metabolic pathways** use energy captured in the **catabolic pathways.**

ENZYMES

Properties of Enzymes

- **Enzymes** are proteins that catalyze chemical reactions in living organisms by lowering the **activation energy** needed for a reaction to occur.
- Enzymes have an **active site,** the binding site to which the **substrate** (the substance on which the enzyme acts) attaches to form an **enzyme–substrate complex.** Enzymes typically exhibit a high degree of **specificity** in the reactions they catalyze.

Properties of Coenzymes and Cofactors

- Some enzymes require **coenzymes,** nonprotein organic molecules that can combine with the **apoenzyme,** the protein portion of the enzyme, to form a **holoenzyme.** Some enzymes also require inorganic ions as **cofactors.**

Enzyme Inhibition

- Enzyme activity can be reduced by **competitive inhibitors,** molecules that compete with the substrate for the enzyme's active site, or by **noncompetitive inhibitors,** molecules that bind to an **allosteric site,** a site other than the active site.

Factors That Affect Enzyme Reactions

- Factors that affect the rate of enzyme reactions include temperature, pH, and concentrations of substrate, product, and enzyme.

ANAEROBIC METABOLISM: GLYCOLYSIS AND FERMENTATION

Glycolysis

- **Glycolysis** is a metabolic pathway by which glucose is oxidized to pyruvic acid.
- Under anaerobic conditions, glycolysis yields a net of two ATPs per molecule of glucose.

Fermentation

- **Fermentation** refers to the reactions of metabolic pathways by which NADH is oxidized to NAD.
- Six pathways of fermentation are summarized in Figure 5.12. **Homolactic-acid** and **alcoholic fermentations** are two of the most important and commonly occurring fermentation pathways.

AEROBIC METABOLISM: RESPIRATION

- **Anaerobes** do not use oxygen; **aerobes** use oxygen and obtain energy chiefly via **aerobic respiration.**

The Krebs Cycle

- The **Krebs cycle** involves the metabolism of two-carbon compounds to CO_2 and H_2O, the production of one ATP directly from each acetyl group, and the transfer of hydrogen atoms to the electron transport system.
- In energy production the Krebs cycle processes acetyl-CoA so that (in the electron transport chain) hydrogen atoms can be oxidized for energy.

Electron Transport and Oxidative Phosphorylation

- **Electron transport** is the transfer of electrons to oxygen (the final electron acceptor).
- **Oxidative phosphorylation** involves the **electron transport chain** for ATP synthesis and is a membrane-regulated process not directly related to the metabolism of specific substrates.
- The theory of **chemiosmosis** explains how energy is used to synthesize ATP.

Significance of Energy Capture

- In prokaryotes, aerobic (oxidative) metabolism captures 19 times as much energy as does anaerobic metabolism.

METABOLISM OF FATS AND PROTEINS

- Most organisms get energy mainly from glucose. But for almost any organic substance, there is some microorganism that can metabolize it.
- Fat metabolism involves hydrolysis and the enzymatic formation of glycerol and free fatty acids. Fatty acids are in turn oxidized by **beta oxidation,** which results in the release of acetyl-CoA. Acetyl-CoA then enters the Krebs cycle.
- The metabolism of proteins involves the breakdown of proteins to amino acids, the deamination of the amino acids, and their subsequent metabolism in glycolysis, fermentation, or the Krebs cycle.

OTHER METABOLIC PROCESSES

Photoautotrophy

- **Photosynthesis** is the use of light energy to synthesize carbohydrates: (1) The **light reactions** can involve **cyclic** photophosphorylation or photolysis accompanied by noncyclic photoreduction of NADP; (2) the dark reactions involve the reduction of CO_2 to carbohydrate.

- Photosynthesis in cyanobacteria and algae provides a means of making nutrients, as it does in green plants; however, photosynthetic bacteria generally use some substances besides water to reduce carbon dioxide.

Photoheterotrophy

- **Photoheterotrophy** is the use of light as a source of energy. It requires organic compounds as sources of carbon.

Chemoautotrophy

- Chemoautotrophs, or **chemolithotrophs,** oxidize inorganic substances to obtain energy. Chemolithotrophs require only carbon dioxide as a carbon source.

USES OF ENERGY

Biosynthetic Activities

- An **amphibolic pathway** is a metabolic pathway that can capture energy or synthesize substances needed by the cell.
- Figure 5.26 summarizes the intermediate products of energy-yielding metabolism and some of the building blocks for synthetic reactions that can be made from them.
- Bacteria synthesize a variety of cell wall polymers.

Membrane Transport and Movement

- Membrane transport uses energy derived from the ATP-producing electron transport system in the membrane to concentrate substances against a gradient. It occurs by active transport and by the **phosphotransferase system.**
- Movement in bacteria can be by flagella, by gliding or creeping, or by axial filaments.

KEY TERMS

activation energy (p. 111)
active site (p. 111)
aerobe (p. 120)
aerobic respiration (p. 120)
alcoholic fermentation
  (p. 119)
allosteric site (p. 114)
amphibolic pathway
  (p. 129)
anabolic pathway (p. 110)
anabolism (p. 108)
anaerobe (p. 120)
apoenzyme (p. 113)
autotroph (p. 109)
autotrophy (p. 109)
beta oxidation (p. 126)
catabolic pathway (p. 110)
catabolism (p. 108)
chemical equilibrium (p. 116)
chemiosmosis (p. 123)

chemoautotroph (p. 109)
chemoheterotroph (p. 109)
chemolithotroph (p. 129)
coenzyme (p. 113)
cofactor (p. 113)
competitive inhibitor (p. 114)
cyclic photophosphorylation
  (p. 128)
cytochrome (p. 123)
dark reactions (p. 128)
electron acceptor (p. 109)
electron donor (p. 109)
electron transport (p. 122)
electron transport chain
  (p. 122)
endoenzyme (p. 113)
enzyme (p. 111)
enzyme–substrate complex
  (p. 112)
exoenzyme (p. 113)

FAD (p. 113)
feedback inhibition
  (p. 115)
fermentation (p. 118)
flavoprotein (p. 123)
glycolysis (p. 116)
heterotroph (p. 109)
heterotrophy (p. 109)
holoenzyme (p. 113)
homolactic-acid fermentation (p. 119)
Krebs cycle (p. 120)
light reactions (p. 127)
metabolic pathway (p. 110)
metabolism (p. 108)
NAD (p. 113)
noncompetitive inhibitor
  (p. 114)
noncyclic photoreduction
  (p. 128)

oxidation (p. 108)
oxidative phosphorylation
  (p. 123)
permease (p. 131)
phosphorylation (p. 116)
phosphotransferase system
  (p. 131)
photoautotroph (p. 109)
photoheterotroph (p. 109)
photolysis (p. 128)
photosynthesis (p. 127)
porin (p. 131)
quinone (p. 123)
reduction (p. 108)
specificity (p. 112)
substrate (p. 111)
surfactant (p. 131)

QUESTIONS FOR REVIEW

A Terms in Metabolism

1. Define metabolism, and distinguish between anabolism and catabolism.

2. Match the following:

____ Archaeobacteria
____ fungi
____ cyanobacteria
____ humans
____ pathogenic bacteria

a. chemoheterotrophs
b. photoheterotrophs
c. photoautotrophs
d. chemoautotrophs

B Enzyme Characterisitcs and Functions

3. Label parts (a) through (g) of this enzyme.

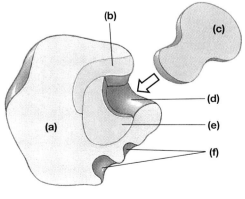

(g)

4. How can enzyme-catalyzed reactions be inhibited?

5. What factors affect the rate of enzyme activity?

C Glycolysis and Fermentation

6. Glycolysis begins with one six-carbon molecule of _____ , which ends up as _____ molecules of _____ , each containing _____ carbon atoms.

7. Glycolysis produces _____ molecules of ATP. However, it requires _____ molecules of ATP to be used up, leaving a net gain of _____ molecules of ATP. Glycolysis takes place (aerobically / anaerobically) and (will / will not) occur in the presence of oxygen.

8. What is fermentation?

9. What are the products of the two main forms of fermentation, and which organisms carry out each type?

10. How are other fermentations useful in diagnosis of infectious organisms?

D The Krebs Cycle

11. The Krebs cycle takes place (aerobically / anaerobically).

12. Which of the following is the correct order of occurrence?
    a. Krebs cycle, glycolysis, electron transport chain
    b. glycolysis, fermentation, Krebs cycle
    c. glycolysis, Krebs cycle, fermentation
    d. Krebs cycle, glycolysis, fermentation
    e. glycolysis, Krebs cycle, electron transport chain

13. The Krebs cycle begins with one molecule of _____ and ends with the production of _____ plus _____ molecules of GTP (ATP).

E Electron Transport and Oxidative Phosphorylation

14. What is electron transport, and what are its products?

15. Explain oxidative phosphorylation and its significance.

16. What is chemiosmosis?

F Metabolism of Fats and Proteins

17. How are fats and proteins used for energy?

G Photosynthesis

18. What is the overall function of photosynthesis?

19. What occurs in the **(a)** light reactions and **(b)** dark reactions?

20. How does photosynthesis in bacteria differ from photosynthesis in plants, algae, and cyanobacteria?

H Photoheterotrophy and Chemoautotrophy

21. What is chemoautotrophy (chemolithotrophy)?

22. What is photoheterotrophy?

I Biosynthetic Activities

23. Define amphibolic, and give an example of an amphibolic process.

24. Which building blocks for biosynthetic activities are derived from intermediates in energy-producing reactions?

J Membrane Transport and Movement

25. What is membrane transport?

26. Describe two ways bacteria transport substances against concentration gradients.

27. By which mechanisms do bacteria move?

PROBLEMS FOR INVESTIGATION

1. Spontaneous combustion caused by bacteria sometimes sets fire to a barn in which damp hay has been stored. How can you explain this?

2. In what sequence might the different kinds of metabolism mentioned in this chapter have evolved? Why didn't just one type of metabolism evolve? Give reasons.

3. Suppose that you had a culture known to contain an *Enterobacter* species and *Escherichia coli*. Devise a way to separate and identify the organisms.

4. Look up the chemical reaction by which certain bacteria change wine into vinegar. Is this an aerobic or anaerobic process?

5. More than 125 human diseases are caused by enzyme deficiencies. Chapters 7 and 8, on bacterial genetics, will explain how bacteria are being used by genetic engineers to remedy these enzyme deficiencies. List some human diseases caused by lack of enzymes, and indicate which enzymes are involved.

SOME INTERESTING READING

Cross, R. L. 1994. "Our primary source of ATP." *Nature* 370, no. 6491 (Aug. 25):594–95.

Dawes, I. W., and I. W. Sutherland. 1992. *Microbial physiology.* Cambridge, MA.: Blackwell Scientific.

Devlin, T. M. 1992. *Textbook of biochemistry: With clinical correlations.* New York: Wiley.

Dickerson, R. E. 1980. "Cytochrome *c* and the evolution of energy metabolism." *Scientific American* 242 (March):136–153.

Lehninger, A. L., D. L. Nelson, and M. M. Cox. 1993. *Principles of biochemistry,* 2d edition. New York: Worth.

Meighen, E. A. 1991. "Molecular biology of bacterial bioluminescence." *Microbiological Reviews* 55(1):123–42.

Monastersky, R. 1988. "Bacteria alive and thriving at depth." *Science News* 133 (March 5):149.

Neidhardt, F. C., J. L. Ingraham, and M. Schaechter. 1990. *Physiology of the bacterial cell: A molecular approach.* Sunderland, MA.: Sinauer.

Pritchard, P. H. 1991. "Bioremediation as a technology: Experiences with the *Exxon Valdez* oil spill." *J. Hazardous Materials* 28(1–2):115–30.

Slayman, C. L. 1985. "Proton chemistry and the ubiquity of proton pumps." *BioScience* 35(1):16–17.

Trumpower, B. L. 1990. "Cytochrome *bc₁* complexes of microorganisms." *Microbiological Reviews* 54(2):101–29.

GROWTH AND CULTURING OF BACTERIA

Bacterial colonies growing on culture media in Petri plates. Each dot is a colony that includes descendants of one original cell.

I n this chapter, we will use what we learned in Chapter 5 about energy in microorganisms to study how to grow them in the laboratory. Growth in bacteria, which has been more thoroughly studied than growth in other microorganisms, is affected by a variety of physical and nutritional factors. Knowing how these factors influence growth is useful in culturing organisms in the laboratory and in preventing their growth in undesirable places. Furthermore, growing the organisms in pure cultures is essential in performing diagnostic tests that are used to identify a number of disease-causing organisms.

GROWTH AND CELL DIVISION

Definition of Growth

In everyday language, growth refers to an increase in size. We are accustomed to seeing children, other animals, and plants grow. Unicellular organisms also grow, but as soon as a cell, called the **mother** (or *parent*) **cell,** has approximately doubled in size and duplicated its contents, it divides into two **daughter cells.** Then the daughter cells grow, and subsequently they also divide. Because individual cells grow larger only

to divide into two new individuals, **microbial growth** is defined not in terms of cell size but as the increase in the number of cells, which occurs by cell division.

Cell Division

Cell division in bacteria, unlike cell division in eukaryotes, usually occurs by *binary fission* or sometimes by *budding*. In **binary fission,** a cell duplicates its components and divides into two cells (Figure 6.1a). The daughter cells become independent when a *septum* (partition) grows between them and they separate (Figure 6.1b). Unlike eukaryotic cells, prokaryotic cells do not have a cell cycle with a specific period of DNA synthesis. Instead, in continuously dividing cells, DNA synthesis also is continuous and replicates the single bacterial chromosome shortly before the cell divides. The chromosome is attached to the cell membrane, which grows and separates the replicated chromosomes. Replication of the chromosome is completed before cell division, when the cell may temporarily contain two or more nucleoids. In some species, incomplete separation of the cells produces linear chains (linked bacilli), **tetrads** (cuboidal groups of four cocci), **sarcinae** (singular: *sarcina*; groups of eight cocci in a cubical packet), or grapelike clusters (staphylococci). Some

THIS CHAPTER FOCUSES ON THE FOLLOWING QUESTIONS:

A How is growth defined in bacteria?

B How does cell division occur in microorganisms?

C What are the phases of growth in a bacterial culture?

D How is bacterial growth measured?

E How do physical factors affect bacterial growth?

F How do biochemical factors affect bacterial growth?

G What occurs in sporulation, and what is its significance?

H What methods are used to obtain a pure culture of an organism for study in the laboratory?

I How are different nutritional requirements supplied by various media?

FIGURE 6.1 (a) Binary fission in a bacterium. (b) Electron micrograph of a thin section of the bacterium *Staphylococcus* undergoing binary fission. (c) Budding in the yeast *Candida albicans,* the major cause of vaginal yeast infections.

bacilli always form chains or filaments; others form them only under unfavorable growth conditions. Streptococci form chains when grown on artificial media but exist as single or paired cells when isolated from a rapidly growing lesion in an infected human host.

Cell division in yeast and a few bacteria occurs through **budding.** In that process, a small, new cell de-

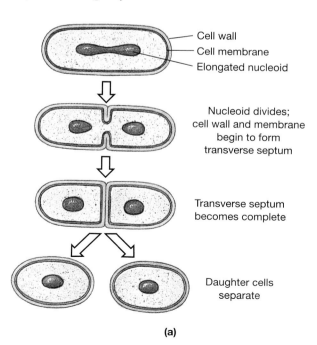

Cell wall
Cell membrane
Elongated nucleoid

Nucleoid divides; cell wall and membrane begin to form transverse septum

Transverse septum becomes complete

Daughter cells separate

(a)

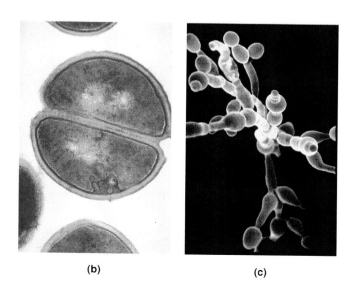

(b)   **(c)**

velops from the surface of an existing cell and subsequently separates from the parent cell (Figure 6.1c).

Phases of Growth

Consider a population of organisms introduced into a fresh, nutrient-rich **medium** (plural: *media*), a mixture of substances on or in which microorganisms grow. Such organisms display four major phases of growth: (1) the lag phase, (2) the log (logarithmic) phase, (3) the stationary phase, and (4) the decline phase, or death phase. These phases form the **standard bacterial growth curve** (Figure 6.2).

Lag Phase

In the **lag phase**, the organisms do not increase significantly in number, but they are metabolically active—growing in size, synthesizing enzymes, and incorporating various molecules from the medium. During this phase the individual organisms increase in size, and they produce large quantities of energy in the form of ATP.

The length of the lag phase is determined in part by characteristics of the bacterial species and in part by conditions in the media—both the medium from which the organisms are taken and the one to which they are transferred. Some species adapt to the new medium in an hour or two; others take several days. Organisms from old cultures, adapted to limited nutrients and large accumulations of wastes, take longer to adjust to a new medium than do those transferred from a relatively fresh, nutrient-rich medium.

Log Phase

Once organisms have adapted to a medium, population growth occurs at an **exponential,** or **logarithmic (log), rate.** When the scale of the vertical axis is logarithmic, growth in this **log phase** appears on a graph as a straight diagonal line, which represents the size of the bacterial population. (On the base-10 logarithmic scale, each successive unit represents a 10-fold increase in the number of organisms.) ∞ (Chapter 2, p. 34) During the log phase, the organisms divide at their most rapid rate—a regular, genetically determined interval called the **generation time.** The population of organisms doubles in each generation time. For example, a culture containing 1000 organisms per milliliter with a generation time of 20 minutes would contain 2000 organisms per milliliter after 20 minutes, 4000 organisms after 40 minutes, 8000 after 1 hour, 64,000 after 2 hours, and 512,000 after 3 hours. Such growth is said to be *exponential,* or *logarithmic* (Figure 6.2).

The generation time for most bacteria is between 20 minutes and 20 hours and is typically less than 1 hour. Some bacteria, such as those that cause tuberculosis and leprosy, have much longer generation times. Some or-

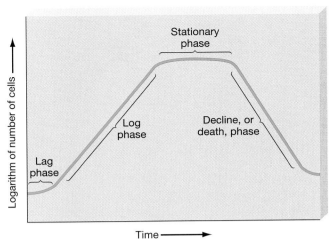

FIGURE 6.2 Phases of growth in a bacterial culture, shown as a standard bacterial growth curve.

ganisms take slightly longer than others to go from the lag phase to the log phase, and they do not all divide precisely together. If they divided together and the generation time was exactly 20 minutes, the number of cells in a culture would increase in a stair-step pattern, exactly doubling every 20 minutes (Figure 6.3)—a hypothetical situation called **synchronous growth.** In an actual culture, each cell divides sometime during the

FIGURE 6.3 Bacterial growth curve for the hypothetical situation in which growth is synchronous. The "stair-step" shape of the curve results when all cells in the population divide at the same time, doubling the population with every division. Compare this with the log phase of the growth curve in Figure 6.2, which is a smooth line for nonsynchronous cultures.

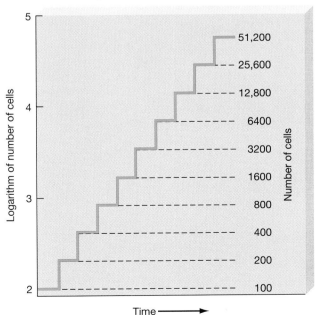

DETERMINING GENERATION TIME

How do scientists determine generation time? Knowing that the generation time is constant for a particular species of bacteria provides a basis for calculating it. The only information needed is the bacterial population at the beginning and at the end of a measured time interval during the phase. The calculations are as follows:

Known Information	Information to be Obtained
B_0 = number of bacteria at zero time	n = number of generations
B_t = number of bacteria at the end of a selected period of time, t	G = generation time
t = time period	

Logarithmic growth can be described by the equation $B_t = B_0 \times 2^n$. To solve for n, take the logarithm of both sides of the equation, and rearrange the terms:

$$\log B_t = \log(B_0 \times 2^n) = \log B_0 + n \log 2$$
$$n = \frac{\log B_t - \log B_0}{\log 2}$$

Use a table of base-10 logarithms to solve the equation. For example,

B_t = 49,000,000	$\log B_t$ = 7.690
B_0 = 12,000	$\log B_0$ = 4.079
	$\log 2$ = 0.301

Thus, $n = (7.690 - 4.079)/0.301 = 12$ generations. This means that it takes 12 generations—that is, 12 population doublings—for the size of a culture to increase from 12,000 to 49,000,000 cells. Note that the number of generations required for this increase in cells is the same regardless of whether the generation time is long or short.

Next, suppose we observe that for a particular species, the time actually elapsed during this population increase is 4 hours (240 minutes). We can then calculate the generation time of that species by dividing the total time by the number of generations. In other words, G (generation time) = t (total time)/n (number of generations). Solving this equation, we see that the generation time is $240/12 = 20$ minutes.

20-minute generation time, with about 1/20 of the cells dividing each minute—a natural situation called **nonsynchronous growth.** Nonsynchronous growth appears as a straight line, not as steps, on a logarithmic graph.

Organisms in a tube of culture medium can maintain logarithmic growth for only a limited period of time. As the number of organisms increases, nutrients are used up, metabolic wastes accumulate, living space may become limited, and aerobes suffer from oxygen depletion. Generally, the limiting factor for logarithmic growth seems to be the rate at which energy can be produced in the form of ATP. As the availability of nutrients decreases, the cells become less able to generate ATP, and their growth rate decreases. The decrease in growth rate is shown in Figure 6.2 by a gradual leveling off of the growth curve (the curved segment to the right of the logarithmic phase).

Leveling off of growth is followed by the stationary phase unless fresh medium is added or organisms are transferred to fresh medium. Logarithmic growth can be maintained by a device called a **chemostat** (Figure 6.4), which has a growth chamber and a reservoir from which fresh medium is continuously added to the growth chamber as old medium is withdrawn. Alternatively, organisms from a culture in the stationary phase can be transferred to a fresh medium. After a brief lag phase, such organisms quickly reenter the log phase of growth.

Stationary Phase

When cell division decreases to the point that new cells are produced at the same rate as old cells die, the number of live cells stays constant. The culture is then in the **stationary phase,** represented by a horizontal straight line in Figure 6.2. The medium contains a limited amount of nutrients and may contain toxic quan-

FIGURE 6.4 A chemostat constantly renews nutrients in a culture, making it possible to grow organisms continuously.

tities of waste materials. Also, the oxygen supply may become inadequate for aerobic organisms, and damaging pH changes may occur.

Decline (Death) Phase

As conditions in the medium become less and less supportive of cell division, many cells lose their ability to divide, and thus the cells die. In this **decline phase,** or **death phase,** the number of live cells decreases at a logarithmic rate, as indicated by the straight, downward-sloping diagonal line in Figure 6.2. During the decline phase, many cells undergo *involution*—that is, they assume a variety of unusual shapes, which makes them difficult to identify. In cultures of spore-forming organisms, more spores than vegetative (metabolically active) cells survive. The duration of this phase is as highly variable as the duration of the logarithmic growth phase. Both depend primarily on the genetic characteristics of the organism. Cultures of some bacteria go through all growth phases and die in a few days; others contain a few live organisms after months or even years.

Growth in Colonies

Growth phases are displayed in different ways in colonies growing on a solid medium. Typically, a cell divides exponentially, forming a small **colony**—all the descendants of the original cell. The colony grows rapidly at its edges; cells nearer the center grow more slowly or begin to die because they have smaller quantities of available nutrients and are exposed to more toxic waste products. All phases of the growth curve occur simultaneously in a colony.

Measuring Bacterial Growth

Bacterial growth is measured by estimating the number of cells that have arisen by binary fission during a growth phase. This measurement is expressed as the number of *viable* (living) organisms per milliliter of culture. Several methods of measuring bacterial growth are available.

Serial Dilution and Standard Plate Counts

One method of measuring bacterial growth is the *standard plate count*. This technique relies on the fact that under proper conditions, a bacterium will divide and form a visible colony on an agar plate. An *agar plate* is a nutrient medium solidified with **agar,** a complex polysaccharide extracted from certain marine algae. Because it is difficult to count more than 300 colonies on one agar plate, it is usually necessary to dilute the original bacterial culture before you plate (transfer) a known volume of the culture onto the solid plate. *Serial dilutions* accomplish this purpose.

To make **serial dilutions** (Figure 6.5), you start with organisms in liquid medium. Adding 1 ml of this medium to 9 ml of sterile water makes a 1:10 dilution; adding 1 ml of the 1:10 dilution to 9 ml of sterile water makes a 1:100 dilution; and so on. The number of bacteria per milliliter of fluid is reduced by 9/10 in each dilution. Subsequent dilutions are made in ratios of 1:1000, 1:10,000, 1:100,000, 1:1,000,000, or even 1:10,000,000 if the original culture contained an extremely large number of organisms.

From each dilution, usually beginning with the 1:100, 1 ml of the culture is transferred to an agar plate. (One milliliter of the 1:10 dilution typically contains too many organisms to yield countable colonies when transferred to a Petri plate.) The transfer can be done by either the pour plate method or the spread plate method (Figure 6.6). A **pour plate** is made by first adding 1.0 ml of a diluted culture from a serial dilution to 9 ml of melted nutrient agar. After the medium is mixed, it is poured into an empty Petri plate. Once the agar medium cools, solidifies, and is incubated, colonies will develop both within the medium and on its surface. Cells suspended in the melted agar during preparation may be heat-damaged, and then they will not form colonies. The **spread plate method** eliminates such problems because all cells remain on the surface of the solid medium. The diluted sample is first placed on the center of a solid, cooled agar medium. The sample is then spread evenly over the medium's surface with a sterile, bent glass rod. After incubation, colonies develop on the agar surface.

FIGURE 6.5 The technique of serial dilution. One milliliter is taken from a broth culture and added to 9 ml of sterile water, thereby diluting the culture by a factor of 10. This procedure is repeated until the desired concentration is reached.

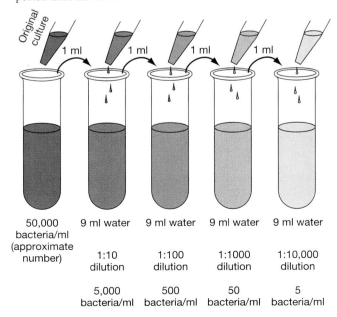

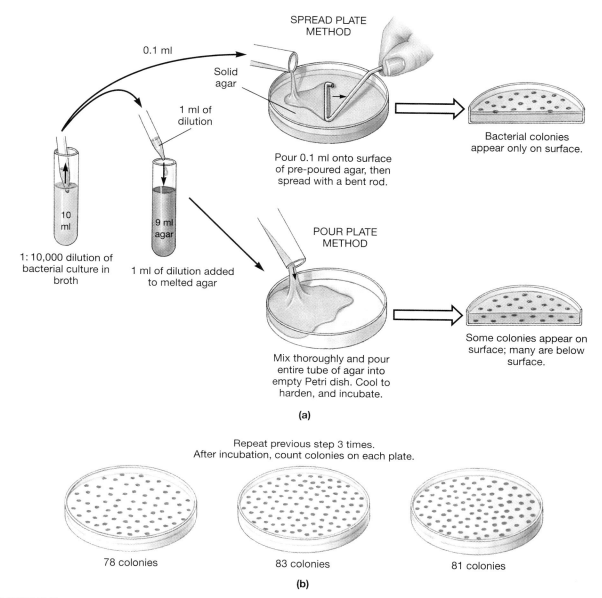

SPREAD PLATE
METHOD

0.1 ml

Solid
agar

Pour 0.1 ml onto surface
of pre-poured agar, then
spread with a bent rod.

Bacterial colonies
appear only on surface.

1 ml of
dilution

10
ml

9 ml
agar

1:10,000 dilution of
bacterial culture in
broth

1 ml of dilution added
to melted agar

POUR PLATE
METHOD

Mix thoroughly and pour
entire tube of agar into
empty Petri dish. Cool to
harden, and incubate.

Some colonies appear on
surface; many are below
surface.

(a)

Repeat previous step 3 times.
After incubation, count colonies on each plate.

78 colonies

83 colonies

81 colonies

(b)

FIGURE 6.6 Calculation of the number of bacteria per milliliter of culture using serial dilution. (a) One milliliter of a 1:10,000 dilution is mixed with 9 ml of melted agar, which is warm enough to stay liquid but not hot enough to kill the organisms being mixed into it. After thorough mixing, the warm agar is quickly poured into an empty, sterile Petri dish (by the pour plate method). When cooled to hardness, it is incubated. Alternatively, 0.1 ml of a 1:10,000 dilution is poured onto a surface of pre-poured agar and then spread with a bent rod (by the spread plate method). Next it is incubated. (b) The colonies that develop are counted. A single measurement is not very reliable, so the procedure is repeated at least three times, and the results are averaged. The average number of colonies is multiplied by the dilution factor to ascertain the total number of organisms per milliliter of the original culture.

Wherever a single living bacterium is deposited on an agar plate, it will divide to form a colony. Each bacterium represents a *colony-forming unit* (CFU). One or more plates should have a small enough number of colonies such that each one is clearly distinguishable and can be counted. If you have made dilutions properly, you should get plates with a **countable number** of colonies (30 to 300 per plate).

To count the actual number of colonies present, you would place the plate under the magnifying lens of a *colony counter* (Figure 6.7), and colonies on the entire plate are counted. To determine the number of colony-forming units in the original culture, multiply the number of colonies found on a plate by the *dilution factor*. A dilution factor of 1000 would be expressed as 1:1000 or 1/1000, and a dilution factor of 10,000 would be expressed as

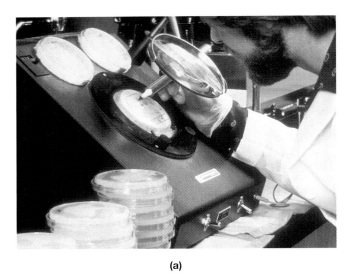

(a)

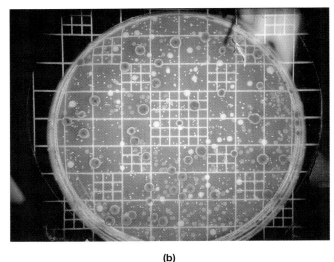

(b)

FIGURE 6.7 (a) Technician using a bacterial colony counter. (b) Bacterial colonies viewed through the magnifying glass against a colony-counting grid. The plate shown was produced by the pour plate method described in Figure 6.6. How many different colony types can you identify on this plate?

1:10,000. A typical calculation for an average colony count of 81 produced by plating a 1/100,000 dilution would be as follows:

$$81 \times 100,000 = 8,100,000, \text{ or } 8.1 \times 10^6 \text{ CFU/ml}$$

The accuracy of the serial dilution and plate count method depends on homogeneous dispersal of organisms in each dilution. Error can be minimized by shaking each culture before sampling and making several plates from each dilution. Accuracy is also affected by the death of cells. Because the number of colonies counted represents the number of living organisms, it does not include organisms that may have died by the time plating was done; nor does it include organisms that cannot grow on the chosen medium. Using young cultures in the log phase of growth minimizes this kind of error.

Direct Microscopic Counts

Bacterial growth can be measured by **direct microscopic counts.** In this method a known volume of medium is introduced into a specially calibrated etched glass slide called a *Petroff-Hausser bacterial counter* (Figure 6.8). A bacterial suspension is introduced onto the chamber with a calibrated pipette. After the bacteria settle and the liquid currents have slowed, the microorganisms are counted in specific calibrated areas. Their number

per unit volume of the original suspension is calculated by using an appropriate formula. The number of bacteria per milliliter of medium can be estimated with a reasonable degree of accuracy. The accuracy of direct microscopic counts depends on the presence of more than 10 million bacteria per milliliter of culture. This is because counting chambers are designed to allow accurate counts only when large numbers of cells are present. An accurate count also requires that the bacteria be homogeneously distributed through-

FIGURE 6.8 The Petroff-Hausser counting chamber. The volume of suspension filling the narrow space between the grid and the cover slide is known, so the number of bacteria per unit of volume can be calculated.

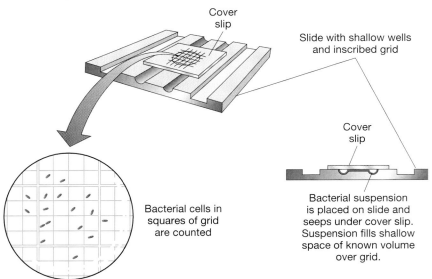

Cover slip

Slide with shallow wells and inscribed grid

Cover slip

Bacterial cells in squares of grid are counted

Bacterial suspension is placed on slide and seeps under cover slip. Suspension fills shallow space of known volume over grid.

out the culture. This technique has the disadvantage of generally not distinguishing between living and dead cells.

Most Probable Number

When samples contain too few organisms to give reliable measures of population size by the standard plate count method, as in food and water sanitation studies, or when organisms will not grow on agar, the **most probable number** (MPN) method is used. With this method, the technician observes the sample, estimates the number of cells in it, and makes a series of progressively greater dilutions. As the dilution factor increases, a point will be reached at which some tubes will contain a single organism and others, none. A typical MPN test consists of five tubes each of three volumes (10, 1, and 0.1 ml) of a dilution (Figure 6.9). Those that contain an organism will display growth by producing gas bubbles when incubated. The number of organisms in the original culture is estimated from a most probable number table. The values in the table, which are based on statistical probabilities, specify that the number of organisms in the original culture has a 95 percent chance of falling within a particular range. A complete MPN table is given in Appendix A. The more tubes that show growth, especially at greater dilutions, the more organisms were present in the sample.

Filtration

Another method of estimating the size of small bacterial populations makes use of **filtration.** A known volume of water or air is drawn through a filter with pores too small to allow passage of bacteria. When the filter is placed on a solid medium, each colony that grows represents originally one organism. Thus, the number of organisms per liter of water or air can be calculated. (Figure 26.18 shows the filtration process and colonies grown on a filter pad.)

Other Methods

Several other methods of monitoring bacterial growth are available. They include simple observation with or without special instruments, measurement of metabolic products by the detection of gas or acid production, and determination of dry weight of cells.

Turbidity (a cloudy appearance) in a culture tube indicates the presence of organisms (Figure 6.10). Fairly accurate estimates of growth can be obtained by measuring turbidity with a photoelectric device, such as a *colorimeter* or a *spectrophotometer* (Figure 6.11). This method is particularly useful in monitoring the rate of growth without disturbing the culture. Samples with very high cell densities, however, must be diluted to ensure accurate readings. Measures of bacterial growth

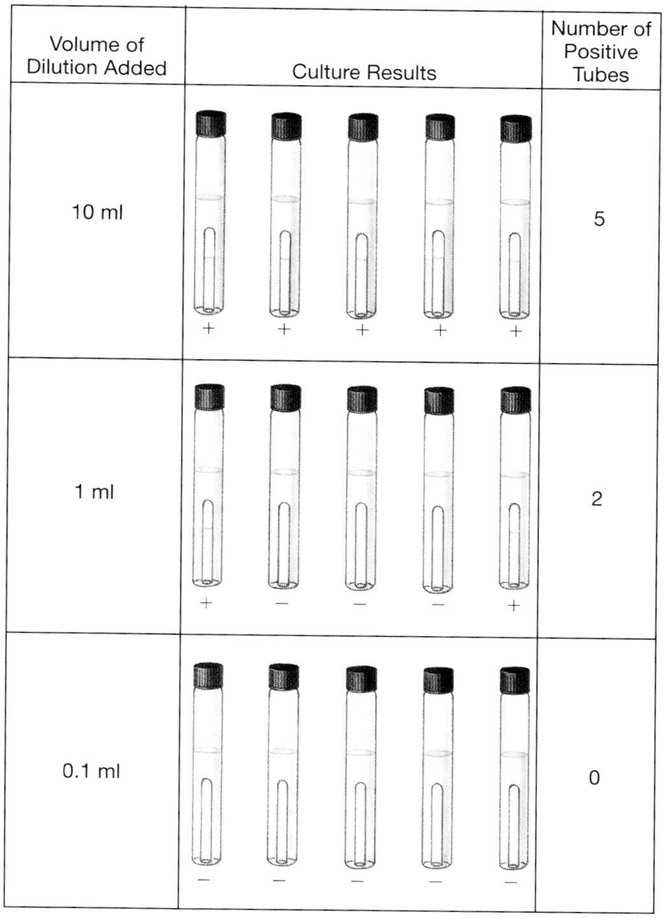

Volume of Dilution Added	Culture Results					Number of Positive Tubes
10 ml	+	+	+	+	+	5
1 ml	+	−	−	−	+	2
0.1 ml	−	−	−	−	−	0

FIGURE 6.9 A typical most probable number test. Those tubes in which gas bubbles are visible (labeled +) contain organisms.

FIGURE 6.10 Turbidity, or a cloudy appearance, is an indicator of bacterial growth in urine in the tube on the left.

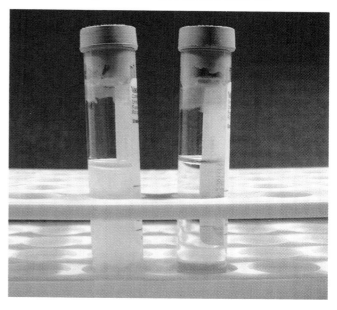

FIGURE 6.11 A spectrophotometer. This instrument can be used to measure bacterial growth by determining the degree of light transmission through the culture. Samples of culture in special optically clear tubes are placed inside the spectrophotometer (inside lid at right of machine) and are measured against standards.

based on turbidity are likewise especially subject to error when cultures contain fewer than 1 million cells per milliliter. Such cultures can display little or no turbidity even when growth is occurring. Conversely, turbidity can be produced by a high concentration of dead cells in a culture.

Measuring the metabolic products of a population can be used to estimate bacterial growth indirectly. The rate at which metabolic products such as gases and/or acids are formed by a culture reflects the mass of bacteria present. Gas production can be detected (rather than measured) by capturing the gas in small inverted tubes placed inside larger tubes of liquid medium containing bacteria. Acid production can be detected by incorporating *pH indicators*—chemical substances that change color with changes in pH—in a liquid medium containing metabolically active bacteria.

The rate at which a substrate such as glucose or oxygen is used up also reflects cell mass. For example, one method for estimating bacterial mass is the *dye reduction test*, which measures the direct or indirect uptake of oxygen. In this test, a dye such as methylene blue is incorporated into a medium containing milk. Bacteria inoculated into the medium use oxygen as they metabolize the milk. Methylene blue is blue in the presence of oxygen and turns colorless in its absence. Thus, the faster the medium loses color, the faster oxygen is being used up, and the more bacteria are presumed to be present. The rate at which the dye is decolorized (dye reduction) is a highly indirect approach; it is not an accurate measure of bacterial mass.

Finally, the number of cells in a culture can be determined by *dry weight measurements*. To calculate the dry weight of cells, they must be separated from the medium by some physical means such as filtration or centrifugation. The cells are then dried, and the resulting mass is weighed.

FACTORS AFFECTING BACTERIAL GROWTH

Microorganisms are found in nearly every environment on earth, including environments in which no other life forms can survive. Microbes can exist in a great many environments because they are small and easily dispersed, occupy little space, need only small quantities of nutrients, and are remarkably diverse in their nutritional requirements. They also have great capacity for adapting to environmental changes. For almost any substance, there is some microbe that can metabolize it as a nutrient; for almost any environmental change, there is some microbe that can survive the change.

As warmblooded, air-breathing, land-dwelling mammals, we tend to forget that 72 percent of our planet's surface is water, that 90 percent of that water is salt water, and that environments containing living organisms have an average temperature of about 5°C. Unlike humans, microorganisms live mostly in water, and many are adapted to temperatures above or below those we consider optimum. The organisms of particular interest in the health sciences account for only a fraction of all microorganisms—those that have adapted to conditions found in or on the human body.

Different species of microorganisms can grow in a wide range of environments—from highly acidic to somewhat alkaline conditions, from Antarctic ice to hot springs, in pure spring water or in salty marshes, in oceans with or without oxygen, and even under great pressure and in boiling steam vents on the ocean floor. Microorganisms use a variety of substances to obtain energy, and some require special nutrients.

The kinds of organisms found in a given environment and the rates at which they grow can be influenced by a variety of factors, both physical and biochemical. **Physical factors** include pH, temperature, oxygen concentration, moisture, hydrostatic pressure, osmotic pressure, and radiation. **Nutritional** (*biochemical*) **factors** include availability of carbon, nitrogen, sulfur, phosphorus, trace elements, and, in some cases, vitamins.

Physical Factors

pH

We saw in Chapter 2 that the acidity or alkalinity of a medium is expressed in terms of pH. ∞ (p. 33) Although the pH scale is now widely used in chemistry, it was invented by the Danish chemist Søren

Sørenson to describe the limits of growth of microorganisms in various media. Microorganisms have an **optimum pH**—the pH at which they grow best. Their optimum pH is usually near neutrality (pH 7). Most microbes do not grow at a pH more than 1 pH unit above or below their optimum pH.

Bacteria are classified as *acidophiles, neutrophiles,* and *alkaliphiles* according to the conditions of acidity or alkalinity they can tolerate. However, no single species can tolerate the full pH range of any of these categories, and many tolerate a pH range that overlaps two categories. **Acidophiles** (a-sid´o-filz), or acid-loving organisms, grow best at a pH below 5.4. *Lactobacillus,* which produces lactic acid, is an acidophile, but it tolerates only mild acidity. Some bacteria that oxidize sulfur to sulfuric acid, however, can create and tolerate conditions as low as pH 1. **Neutrophiles** (nu´tro-filz) exist from pH 5.4 to 8.5. Most of the bacteria that cause disease in humans are neutrophiles. **Alkaliphiles** (al´kah-li-filz), or alkali-loving (base-loving) organisms, exist from pH 7.0 to 11.5. *Vibrio cholerae,* the causative agent of the disease Asiatic cholera, grows best at a pH of about 9. *Alcaligenes faecalis,* which sometimes infects humans already weakened by another disease, can create and tolerate alkaline conditions of pH 9 or higher. The soil bacterium *Agrobacterium* grows in alkaline soil of pH 12.

The effects of pH on organisms can in part be related to the concentration of organic acids in the medium and to the protection that bacterial cell walls sometimes provide. *Lactobacillus* and other organisms that produce organic acids during fermentation inhibit their own growth as acids such as lactic acid and pyruvic acid accumulate in the medium. It appears that the acids themselves rather than the hydrogen ions per se inhibit growth. Changes in pH can lead to denaturing of enzymes and other proteins and can interfere with pumping of ions at the cell membrane. Other organisms have relatively impervious cell walls that prevent the cell membrane from being exposed to an extreme pH in the medium. These organisms appear to tolerate environmental acidity or alkalinity because the cell itself is maintained at a nearly neutral pH.

Many bacteria often produce sufficient quantities of acids as metabolic byproducts that eventually interfere with their own growth. To prevent this situation in the laboratory cultivation of bacteria, *buffers* are incorporated into growth media to maintain the proper pH levels. Phosphate salts are commonly used for this purpose.

Temperature

Most species of bacteria can grow over a 30°C temperature range, but the minimum and maximum temperatures for different species vary considerably. Sea water remains liquid below 0°C, and organisms living there can tolerate below-freezing temperatures. Bacteria can be classified according to growth temperature ranges as *psychrophiles, mesophiles,* and *thermophiles*. Most bacteria, however, do not tolerate the whole temperature range of a category, and some tolerate a range that overlaps categories. Within these groups bacteria are further classified as obligate or facultative. **Obligate** means that the organism *must* have the specified environmental condition. **Facultative** means that the organism is *able* to adjust to and tolerate the environmental condition, but it can also live in other conditions.

Psychrophiles (si´kro-filz), or cold-loving organisms, grow best at temperatures of 15° to 20°C, although some live quite well at 0°C. They can be further divided into **obligate psychrophiles,** such as *Bacillus globisporus,* which cannot grow above 20°C, and **facultative psychrophiles,** such as *Xanthomonas pharmicola,* which grows best below 20°C but also can grow above that temperature. Psychrophiles live mostly in cold water and soil. None can live in the human body, but some are known to cause spoilage of refrigerated foods.

Mesophiles (mes´o-filz), which include most bacteria, grow best at temperatures between 25° and 40°C. Human pathogens are included in this category, and most of them grow best near human body temperature (37°C). *Thermoduric* organisms ordinarily live as mesophiles but can withstand short periods of exposure to high temperatures. Inadequate heating during canning or in pasteurization may leave such organisms alive and therefore able to spoil food.

Thermophiles (therm´o-filz), or heat-loving organisms, grow best at temperatures from 50° to 60°C. Many are found in compost heaps, and a few tolerate temperatures as high as 110°C in boiling hot springs. They can be further divided into **obligate thermophiles,** which can grow only at temperatures above 37°C, and **facultative thermophiles,** which can grow both above and below 37°C. *Bacillus stearothermophilus,* which usually is considered an obligate thermophile, grows at its maximum rate at 65° to 75°C but can display minimal growth and cause food spoilage at temperatures as low as 30°C. Thermophilic sulfur bacteria display zones of optimum growth temperatures in the runoff troughs of geysers (Figure 6.12). Different species collect at various locations along the trough. The most heat-tolerant are near the geyser, and those with lesser heat tolerance are distributed in regions where the water has cooled to their optimum temperature. In deep channels the most heat-tolerant species are found at the greatest depths and the least heat-tolerant near the surface, where the water has cooled. Under laboratory conditions that utilize high pressure to increase water temperature above 100°C,

FIGURE 6.12 Geyser Hot Springs, Black Rock Desert, Nevada. Bacteria can live and grow in the runoff waters from such geysers despite the near-boiling temperatures.

archaeobacteria from deep-sea vents have grown at 115°C (238°F). (Chapter 10 offers more information about these remarkable organisms.)

The temperature range over which an organism grows is determined largely by the temperatures at which its enzymes function. Within this temperature range, three critical temperatures can be identified:

1. The *minimum growth temperature,* the lowest temperature at which cells can divide;

2. The *maximum growth temperature,* the highest temperature at which cells can divide;

3. The *optimum growth temperature,* the temperature at which cells divide most rapidly—that is, have the shortest generation time.

Regardless of the type of bacteria, growth gradually increases from the minimum to the optimum temperature and decreases very sharply from the optimum to the maximum temperature. Furthermore, the optimum temperature is often very near the maximum temperature (Figure 6.13). These growth properties are due to changes in enzyme activity. ∞ (Chapter 5, p. 115) Enzyme activity generally doubles for every 10°C rise in temperature until the high temperature begins to denature all proteins, including enzymes. The sharp decrease in enzyme activity at a temperature only slightly higher than the optimum temperature occurs as enzyme molecules become so distorted by denaturation that they cannot catalyze reactions.

Temperature is important not only in providing conditions for microbial growth, but also in preventing such growth. The refrigeration of food, usually at 4°C, reduces the growth of psychrophiles and prevents the growth of most other bacteria. However, food and other materials, such as blood, can support growth of some bacteria even when refrigerated. For this reason, materials that can withstand freezing are stored at temperatures of −30°C if they are to be kept for long periods of time. High temperatures also can be used to prevent bacterial growth (Chapter 13). Laboratory equipment and media are generally sterilized with heat, and food is frequently preserved by heating and storing in closed containers. Bacteria are more apt to survive extremes of cold than extremes of heat; enzymes are not denatured by chilling but can be permanently denatured by heat.

Oxygen

Bacteria, especially heterotrophs, can be divided into aerobes, which require oxygen to grow, and anaerobes, which do not require it. ∞ (Chapter 5, p. 120) Among the aerobes, cultures of rapidly dividing cells require more oxygen than do cultures of slowly dividing cells.

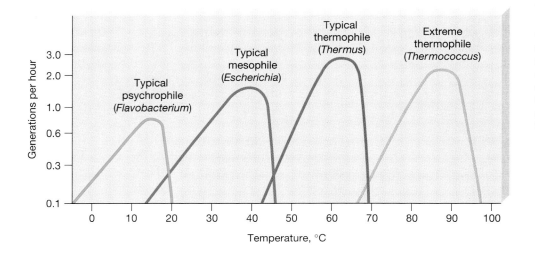

FIGURE 6.13
Comparison of the growth rates of typical psychrophilic, mesophilic, and thermophilic organisms. Note that there is some overlap of ranges at which these organisms can survive, but rates of growth are much lower at the extreme ends of the ranges.

Obligate aerobes, such as *Pseudomonas,* which is a common cause of hospital-acquired infections, must have free oxygen for aerobic respiration, whereas obligate anaerobes, such as *Bacteroides* are killed by free oxygen. In a culture tube containing nutrient broth, obligate aerobes grow near the surface, where atmospheric oxygen diffuses into the medium; obligate anaerobes grow near the bottom of the tube, where little or no free oxygen reaches them (Figure 6.14).

For aerobes, oxygen is often the environmental factor that limits growth rate. Oxygen is poorly soluble in water, and a variety of methods are sometimes employed to maintain a high O_2 concentration in cultures, including vigorous mixing or forced aeration by bubbling air through a culture, as is done in a fish tank. This is especially important in such commercial processes as the production of antibiotics and in sewage treatment.

Between the extremes of obligate aerobes and obligate anaerobes are the *microaerophiles,* the *facultative anaerobes,* and the *aerotolerant anaerobes.* Microaerophiles (mi"kro-aer'o-filz) appear to grow best in the presence of a small amount of free oxygen. They grow below the surface of the medium in a culture tube at the level where oxygen availability matches their needs. Microaerophiles such as *Campylobacter,* which can cause intestinal disorders, also are capnophiles, or carbon dioxide–loving organisms. They thrive under conditions of low oxygen and high carbon dioxide concentration. Facultative anaerobes ordinarily carry on aerobic metabolism when oxygen is present, but they shift to anaerobic metabolism when oxygen is absent. *Staphylococcus* and *Escherichia coli* are facultative anaerobes; they often are found in the intestinal and urinary tracts, where only a small amount of oxygen is available. Aerotolerant anaerobes can survive in the presence of oxygen but do not use it in their metabolism. *Lactobacillus,* for example, always captures energy by fermentation, regardless of whether the environment contains oxygen.

Compared with other groups of organisms defined according to oxygen requirements, facultative anaerobes have the most complex enzyme systems. They have one set of enzymes that enables them to use oxygen as an electron acceptor and another set that enables them to use another electron acceptor when oxygen is not available. In contrast, the enzymes of the other groups defined here are limited to either aerobic or anaerobic respiration.

Obligate anaerobes are killed not by gaseous oxygen but by a highly reactive form of oxygen called superoxide (O_2^-). Superoxide is formed by certain oxidative enzymes and is converted to molecular oxygen (O_2) and hydrogen peroxide (H_2O_2) by an enzyme called superoxide dismutase. Hydrogen peroxide is converted to water and molecular oxygen by the enzyme catalase. Obligate aerobes and most facultative anaerobes have both enzymes. Some facultative and aerotolerant anaerobes have superoxide dismutase but lack catalase. Most obligate anaerobes lack both enzymes and succumb to the toxic effects of superoxide and hydrogen peroxide.

Moisture

All actively metabolizing cells generally require a water environment. Unlike larger organisms that have protective coverings and internal fluid environments, single-celled organisms are exposed directly to their environment. Most vegetative cells can live only a few hours without moisture; only the spores of spore-forming organisms can exist in a dormant state in a dry environment.

Hydrostatic Pressure

Water in oceans and lakes exerts hydrostatic pressure, pressure exerted by standing water, in proportion to its depth. Such pressure doubles with every 10 m increase in depth. For example, in a lake 50 m deep, the pressure is 32 times the atmospheric pressure. Some ocean valleys have depths in excess of 7000 m, and certain bacteria are the only organisms known to survive the extreme pressure at such depths. Bacteria that live at high pressures, but die if left in the laboratory for only a few hours at standard atmospheric pressure, are called barophiles. It appears that their membranes and enzymes do not simply tolerate pressure but require pressure to function properly. The high pressure is necessary to keep their enzyme molecules in the proper three-dimensional configuration. Without it, the enzymes lose their shape and denature, and the organisms die.

FIGURE 6.14 Different organisms incubated for 24 hours in tubes of a nutrient broth accumulate in different regions depending on their need for, or sensitivity to, oxygen.

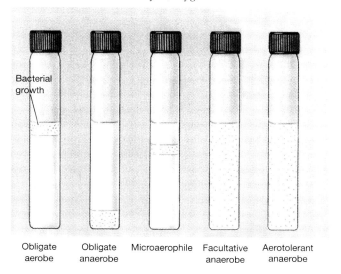

Bacterial growth

| Obligate aerobe | Obligate anaerobe | Microaerophile | Facultative anaerobe | Aerotolerant anaerobe |

WHEN THE GOING GETS TOUGH, HIDE INSIDE A ROCK

A few bacteria live in bitterly cold, dry valleys of Antarctica where very few other organisms can survive. The relative humidity is so low that water passes directly from the frozen to the vapor state and is rarely found as a liquid. Organisms that live there nevertheless manage to carry out their metabolic activities, either by using water vapor or by melting tiny amounts of ice with their metabolic heat. But they cannot survive the harsh conditions of the Antarctic atmosphere. The bacteria must hide inside translucent rocks (such as quartz, feldspar, and certain marbles), which allow the sun's rays to penetrate so these bacteria can carry out photosynthesis. Because they do not produce mineral-dissolving chemicals, these endolithic organisms must colonize only porous rocks. They are usually able to invade several millimeters into the rock, where they find safe refuge until the rock is eroded by wind.

Mars was originally a warm planet, but it cooled down when it lost its atmosphere. If life had evolved on Mars during its warm phase, would that life have sought shelter inside surface rocks? Perhaps examination of Martian rocks will reveal evidence of early life forms entombed in their last refuge.

Osmotic Pressure

We saw in Chapter 4 that the membranes of all microorganisms are selectively permeable. The cell membrane allows water to move by osmosis between the cytoplasm and the environment. ∞ (p. 98) Environments that contain dissolved substances exert osmotic pressure, and the pressure can exceed that exerted by dissolved substances in cells. Cells in such *hyperosmotic* environments lose water and undergo **plasmolysis** (plas-mol'e-sis), or shrinking of the cell. In microorganisms with a cell wall, the cell or plasma membrane separates from the cell wall. Conversely, cells in distilled water have a higher osmotic pressure than their environment and, therefore, gain water. In bacteria, the rigid cell wall prevents cells from swelling and bursting, but the cells fill with water and become *turgid* (distended).

Most bacterial cells can tolerate a fairly wide range of concentrations of dissolved substances. Their cell membranes contain transport systems that regulate the movement of dissolved substances across the membrane. ∞ (Chapter 5, p. 130) Yet, if concentrations outside the cells become too high, water loss can inhibit growth or even kill the cells.

The use of salt as a preservative in curing hams and bacon and in making pickles is based on the fact that high concentrations of dissolved substances exert sufficient osmotic pressure to kill or inhibit microbial growth. The use of sugar as a preservative in making jellies and jams is based on the same principle.

Bacteria called **halophiles** (hal'o-filz), or salt-loving organisms, require moderate to large quantities of salt (sodium chloride). Their membrane transport systems actively transport sodium ions out of the cells and concentrate potassium ions inside them. Two possible explanations for why halophiles require sodium have been proposed. One is that the cells need sodium to maintain a high intracellular potassium concentration so that their enzymes will function. The other is that they need sodium to maintain the integrity of their cell walls.

Halophiles are typically found in the ocean, where the salt concentration (3.5 percent) is optimum for their growth. Extreme halophiles require salt concentrations of 20 to 30 percent (Figure 6.15). They are found in exceptionally salty bodies of water, such as the Dead Sea, and sometimes even in brine vats, where they cause spoilage of pickles being made there.

Radiation

Radiant energy, such as gamma rays and ultraviolet light, can cause mutations (changes in DNA) and even kill organisms. However, some microorganisms have pigments that screen radiation and help to prevent DNA damage. Others have enzyme systems that can repair certain kinds of DNA damage.

Nutritional Factors

The growth of microorganisms is affected by nutritional factors as well as by physical factors. Nutrients needed by microorganisms include carbon, nitrogen, sulfur, phosphorus, certain trace elements, and vitamins. Although we are concerned with ways microorganisms satisfy their own nutritional needs, we can note that in satisfying such needs, they also help to recycle elements in the environment. Activities of microbes in the carbon, nitrogen, sulfur, and phosphorus cycles are described in Chapter 26. A few microbes are **fastidious**—that is, they have special nutritional needs that can be difficult to meet in the laboratory. Some fastidious organisms, including those that cause gonorrhea, grow quite well in the human body but still cannot be easily grown in the laboratory on nutrient media.

Carbon Sources

Most bacteria use some carbon-containing compound as an energy source, and many use carbon-containing compounds as building blocks to synthesize cell com-

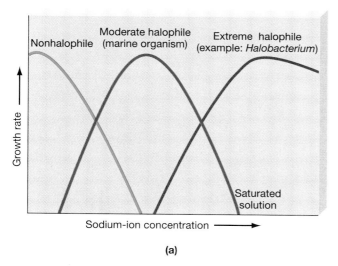

(b)

FIGURE 6.15 (a) Growth rates of halophilic ("salt-loving") and nonhalophilic organisms are related to sodium ion concentration. (b) The Great Salt Lake in Utah, an example of an environment in which halophilic organisms thrive. Note the white areas of dried salt around the edges of the lake.

ponents. Photoautotrophic organisms reduce carbon dioxide to glucose and other organic molecules. Both autotrophic and heterotrophic organisms can obtain energy from glucose by glycolysis, fermentation, and the Krebs cycle. They also synthesize some cell components from intermediates in these pathways.

Nitrogen Sources

All organisms, including microorganisms, need nitrogen to synthesize enzymes, other proteins, and nucleic acids. Some microorganisms obtain nitrogen from inorganic sources, and a few even obtain energy by metabolizing inorganic nitrogen-containing substances. Many microorganisms reduce nitrate ions (NO_3^-) to amino groups (NH_2) and use the amino groups to

make amino acids. Some can synthesize all 20 amino acids found in proteins, whereas others must have one or a few amino acids provided in their medium. Certain fastidious organisms require all 20 amino acids and other building blocks in their medium. Many disease-causing organisms obtain amino acids for making proteins and other nitrogenous molecules from the cells of humans and other organisms they invade.

Once amino acids are synthesized or obtained from the medium, they can be used in protein synthesis. Similarly, purines and pyrimidines can be used to make DNA and RNA. The processes by which proteins and nucleic acids are synthesized are directly related to the genetic information contained in a cell. Thus, the synthesis of proteins and of nucleic acids will be discussed in Chapters 7 and 8.

Sulfur and Phosphorus

In addition to carbon and nitrogen, microorganisms need a supply of certain minerals, especially sulfur and phosphorus, which are important cell components. Microorganisms obtain sulfur from inorganic sulfate salts and from sulfur-containing amino acids. They use sulfur and sulfur-containing amino acids to make proteins, coenzymes, and other cell components. Some organisms can synthesize sulfur-containing amino acids from inorganic sulfur and other amino acids. Microorganisms obtain phosphorus mainly from inorganic phosphate ions (PO_4^{3-}). They use phosphorus (as phosphate) to synthesize ATP, phospholipids, and nucleic acids.

Trace Elements

Many microorganisms require a variety of **trace elements,** tiny amounts of minerals such as copper, iron, zinc, and cobalt, usually in the form of ions. Trace elements often serve as cofactors in enzymatic reactions. All organisms require some sodium and chloride, and halophiles require large amounts of these ions. Potassium, zinc, magnesium, and manganese are used to activate certain enzymes. Cobalt is required by organisms that can synthesize vitamin B_{12}. Iron is required for the synthesis of heme-containing compounds (such as the cytochromes of the electron transport system) and for certain enzymes. Although little iron is required, a shortage severely retards growth. Calcium is required by gram-positive bacteria for synthesis of cell walls and by spore-forming organisms for synthesis of spores.

Vitamins

A **vitamin** is an organic substance that an organism requires in small amounts and that is typically used as a coenzyme. Many microorganisms make their own vitamins from simpler substances. Other microorgan-

PICKY EATERS

Spiroplasma species, tiny spiral bacteria that lack cell walls, are among the most nutritionally fastidious organisms known. Recently, Kevin Hackett, a U.S. Department of Agriculture (USDA) scientist, devised an exact formula of 80 ingredients, including lipids, carbohydrates, amino acids, salts, vitamins, organic acids, and penicillin (to suppress potential competitors) to meet their needs. In his laboratory, he uses this medium to keep more than 30 species of spiroplasmas alive and well, making it possible for researchers to study them outside the more than 100 insect, tick, and plant species that they normally inhabit. Until now, most *Spiroplasma* species have been impossible to keep alive outside their hosts.

The spiroplasmas are responsible for hundreds of crop and animal diseases. Medical researchers are particularly interested in one species that experimentally causes tumors in animals. Another species kills honey bees, and a third lives harmlessly in the Colorado potato beetle, an insect that damages potato, eggplant, and tomato plants. Scientists hope to alter this last species genetically so that it will kill its potato beetle host.

USDA scientists are now trying to formulate complex media to grow mycoplasmalike organisms, a related group of bacteria that also lack cell walls. These bacteria cause hundreds of crop diseases and millions of dollars in economic losses each year. They are spread from plant to plant by infected insects. Another medium being designed would grow the bacterium *Mycoplasma pneumoniae,* which is the cause of a form of "walking pneumonia" in humans.

USDA scientist Kevin Hackett working on his microbial "witch's brew"—a mix of some 80 ingredients that will support the growth of nutritionally fastidious spiroplasmas outside their hosts.

Spiroplasma (magnified 29,000×).

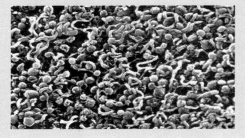

isms require several vitamins in their media because they lack the enzymes to synthesize them. Vitamins required by some microorganisms include folic acid, vitamin B_{12}, and vitamin K. Human pathogens often require a variety of vitamins and thus are able to grow well only when they can obtain these substances from the host organism. Growing such organisms in the laboratory requires a complex medium that contains all the nutrients they normally obtain from their hosts. Microbes living in the human intestine manufacture vitamin K, which is necessary for blood clotting, and some of the B vitamins, thus benefiting their host.

Nutritional Complexity

An organism's **nutritional complexity,** the number of nutrients it must obtain to grow, is determined by the kind and number of its enzymes. The absence of a single enzyme can render an organism incapable of synthesizing a specific substance. The organism therefore must obtain the substance as a nutrient from its environment. Microorganisms vary in the number of enzymes they possess. Those with many enzymes have simple nutritional needs because they can synthesize nearly all the substances they need. Those with fewer enzymes have complex nutritional requirements because they lack the ability to synthesize many of the substances they need for growth. Thus, nutritional complexity reflects a deficiency in biosynthetic enzymes.

Locations of Enzymes

Most microorganisms move a variety of small molecules across their cell or plasma membranes and metabolize them. These substances include glucose, amino acids, small peptides, nucleosides, and phosphates as well as various inorganic ions. In addition to the endoenzymes that are produced for use within the cell (Chapter 5), many bacteria (and fungi) produce *exoenzymes* and release them through the cell or plasma membrane. ∞ (p. 113) These enzymes include **extracellular enzymes,** usually produced by gram-positive

APPLICATIONS

BACTERIA THAT MEASURE

The discovery that certain organisms require vitamins or other special nutrients led to the development of *bioassay techniques*, the use of living organisms to measure the amount of a particular substance in a food or other material. To conduct a bioassay, a medium is prepared that contains all the nutrients an organism needs except the one to be assayed. Then a known quantity of the substance to be assayed and the organism that requires it are added to the medium. For example, suppose that you wanted to know how much folic acid a food contained. You could add a known amount of the food to a medium that lacks folic acid (but is rich in all other nutritional requirements) and inoculate it with an organism whose folic acid requirements are known. Folic acid would be the limiting factor for growth, and the growth would therefore be proportional to the amount of folic acid in the medium.

TABLE 6.1	Examples of Exoenzymes
Enzymes	**Action**
Enzymes That Act on Complex Carbohydrates	
Carbohydrases	Break down large carbohydrate molecules into smaller ones
Amylase	Breaks down starch to maltose
Cellulase	Breaks down cellulose to cellobiose
Enzymes That Act on Sugars	
Sucrase	Breaks down sucrose to glucose and fructose
Lactase	Breaks down lactose to glucose and galactose
Maltase	Breaks down maltose to two glucose molecules
Enzymes That Act on Lipids	
Lipases	Break down fats to glycerol and fatty acids
Enzymes That Act on Proteins	
Proteases	Break down proteins to peptides and amino acids
Caseinase	Breaks down milk protein to amino acids and peptides
Gelatinase	Breaks down gelatin to amino acids and peptides

rods, which act in the medium around the organism, and **periplasmic enzymes,** usually produced by gram-negative organisms, which act in the periplasmic space. Most exoenzymes are hydrolases; they add water as they split large molecules of carbohydrate, lipid, or protein into smaller ones that can be absorbed (Table 6.1). Although microbes cannot move large molecules across membranes, in nature they use large molecules from other organisms by digesting those molecules with exoenzymes before absorbing them.

Adaptation to Limited Nutrients

Microorganisms adapt to limited nutrients in several ways:

1. Some synthesize increased amounts of enzymes for uptake and metabolism of limited nutrients. This allows the organisms to obtain and use a larger proportion of the few nutrient molecules that are available.

2. Others have the ability to synthesize enzymes needed to use a different nutrient. For example, if glucose is in short supply, some microorganisms can make enzymes to take up and use a more plentiful nutrient such as lactose.

3. Many organisms adjust the rate at which they metabolize nutrients and the rate at which they synthesize molecules required for growth to fit the availability of the least plentiful nutrient. Both metabolism and growth are slowed, but no energy is wasted on synthesizing products that cannot be used. Growth is as rapid as conditions will allow.

SPORULATION

Sporulation, the formation of endospores, occurs in *Bacillus, Clostridium,* and a few other gram-positive genera but has been studied most carefully in *B. subtilis* and *B. megaterium*. Bacteria that form endospores generally do so during the stationary phase in response to environmental, metabolic, and cell cycle signals.

When nutrients such as carbon or nitrogen become limiting, highly resistant endospores form inside mother cells. (With very low frequency, some bacteria form endospores even when nutrients are available.) Although endospores are not metabolically active, they can survive long periods of drought and are resistant to killing by extreme temperatures, radiation, and some toxic chemicals. Some endospores can withstand much higher temperatures than vegetative cells can. The endospore itself cannot divide, and the parent cell can produce only one endospore, so sporulation is a protective or survival mechanism, not a means of reproduction.

As endospore formation begins, DNA is replicated and forms a long, compact, *axial nucleoid* (Figure 6.16). The two chromosomes formed by replication separate and move to different locations in the cell. In some bacteria the endospore forms near the middle of the cell, and in others it forms at one end (Figure 6.17). The DNA where the endospore will form directs endospore formation. Most of the cell's RNA and some cytoplasmic protein molecules gather around the DNA to make

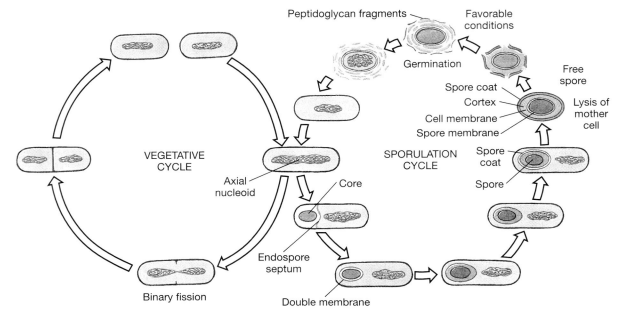

FIGURE 6.16 The vegetative and sporulation cycles in bacteria capable of sporulation.

FIGURE 6.17 Bacterial endospores in two *Clostridium* species. (a) Cells with centrally located endospores. (b) False-color TEM photo of cells with terminally located endospores, which give the organisms a club-shaped appearance.

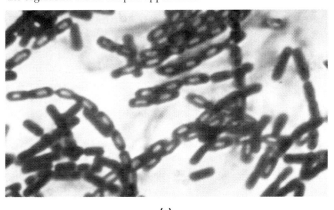

(a)

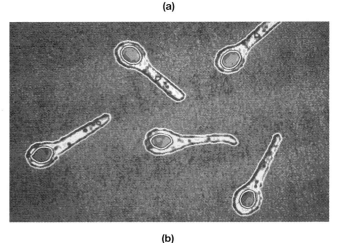

(b)

the **core,** or living part, of the endospore. The core contains **dipicolinic** (di-pik-o-lin'ik) **acid** and calcium ions, which probably contribute to an endospore's heat resistance by stabilizing protein structure. An **endospore septum,** consisting of a cell membrane but lacking a cell wall, grows around the core, enclosing it in a double thickness of cell membrane. Both layers of this membrane synthesize peptidoglycan and release it into the space between the membranes. Thus, a laminated layer called the **cortex** is formed. The cortex protects the core against changes in osmotic pressure, such as those that result from drying. A **spore coat** of keratin-like protein, which is impervious to many chemicals, is laid down around the cortex by the mother cell. Finally, in some endospores an **exosporium,** a lipid–protein membrane, is formed outside the coat by the mother cell. The function of the exosporium is unknown. Under laboratory conditions, sporulation takes about 7 hours.

Once favorable conditions return, an endospore develops into a vegetative cell, which lacks the endospore's resistant properties. **Germination,** in which a spore returns to its vegetative state, occurs in three stages. The first stage, *activation,* usually requires some traumatic agent such as low pH or heat, which damages the coat. Without such damage, some endospores germinate slowly, if at all. The second stage, *germination proper,* requires water and a germination agent (such as the amino acid alanine or certain inorganic ions) that penetrates the damaged coat. During this process, much of the cortical peptidoglycan is broken down, and its fragments are released into the medium. The living cell (which occupied the core) now takes in large quantities of water and loses its resistance to heat and staining as well as its *refractility* (ability to bend light rays). Finally, *outgrowth* occurs in a medium with adequate nutrients. Proteins and RNA are synthesized, and in about an hour, DNA synthesis begins. The cell is now a vegetative cell and undergoes binary fission.

Thus, bacterial cells capable of sporulation display two cycles—the *vegetative cycle* and the *sporulation cycle* (Figure 6.16). The vegetative cycle is repeated at intervals of 20 minutes or more, and the sporulation cycle is initiated periodically. Endospores known to be 300 or more years old have been observed to undergo germination when placed in a favorable medium.

Certain bacteria, such as *Azotobacter,* form resistant **cysts,** or spherical, thick-walled cells, that resemble endospores. Like endospores, cysts are metabolically inactive and resist drying. Unlike endospores, they lack dipicolinic acid and have only limited resistance to high temperatures. Cysts germinate into single cells and therefore are not a means of reproduction.

Some filamentous bacteria, such as *Micromonospora* and *Streptomyces,* form asexually reproduced **conidia** (ko-nid'e-ah), or chains of aerial spores with thick outer walls. These spores are temporarily dormant but are not especially resistant to heat or drying. When the spores, which are produced in large numbers, are dispersed to a suitable environment, they form new filaments. Unlike endospores, these spores do contribute to reproduction of the species.

CULTURING BACTERIA

Culturing of bacteria in the laboratory presents two problems. First, a pure culture of a single species is needed to study an organism's characteristics. Second, a medium must be found that will support the growth of the desired organism. Let us look at some of the ways these problems are solved.

Methods of Obtaining Pure Cultures

To study bacteria in the laboratory, it is important to obtain a **pure culture,** a culture that contains only a single species of organism. Prior to the development of pure culture techniques, scientists studied *mixed cultures,* or cultures containing several different kinds of organisms. Researchers could make observations of different shapes and sizes of organisms, but they could find out little about the nutritional needs or growth characteristics of individual species. Today, pure cultures are obtained by isolating the progeny of a single cell.

Simple as it seems now, the technique of isolating pure cultures was difficult to develop. Attempts to isolate single cells by serial dilution were often unsuccessful because two or more organisms of different species were often present in the highest dilutions. Koch's technique of spreading bacteria thinly over a solid surface was more effective because it deposited a single bacterium at some sites. However, he tried several different solid substances. Using the discovery of Angelina Hesse, the wife of an associate, he settled on agar as the ideal solidifying agent. Only a very few organisms digest it, and in 1.5 percent solution it does not melt below 95°C. Furthermore, after being melted, agar remains in the liquid state until it has cooled to about 40°C, a temperature cool enough to allow the addition of nutrients and living organisms that might be destroyed by heat.

Streak Plate Method

Today, the accepted way to prepare pure cultures is the **streak plate method,** which uses agar plates. Bacteria are picked up on a sterile wire loop, and the wire is moved lightly along the agar surface, depositing streaks of bacteria on the surface. The inoculating loop is flamed, and a few bacteria are picked up from the region already deposited and streaked onto a new region (Figure 6.18). Fewer and fewer bacteria are de-

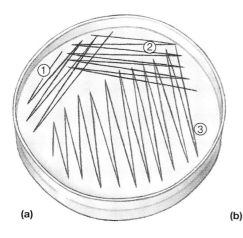

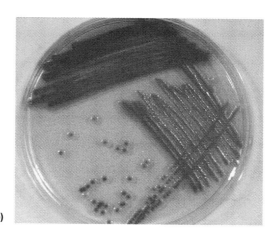

FIGURE 6.18 The streak plate method of obtaining pure cultures. (a) A drop or bit of culture on a wire inoculating loop is lightly streaked across the top of the agar in region 1. The loop is flamed, the plate is rotated, and a few organisms are picked up from region 1 and streaked out into region 2. The loop is flamed again, and the process is repeated in region 3. The plate is then incubated. (b) A streak plate of *Serratia marcescens* after incubation. Note the greatly reduced numbers of colonies in each successive region.

posited as the streaking continues, and the loop is flamed after each streaking. Individual organisms are deposited in the region streaked last. After the plate is incubated at a suitable growth temperature for the organism, small colonies—each derived from a single bacterial cell—appear. The wire loop is used to pick up a portion of an isolated colony and transfer it to any appropriate sterile medium for further study. The use of sterile (aseptic) technique assures that the new medium will contain organisms of a single species.

Pour Plate Method

Another way to obtain pure cultures, the **pour plate method,** makes use of serial dilutions (see Figure 6.6a). A series of dilutions are made such that the final dilution contains about a thousand organisms. Then 1 ml of liquid medium from the final dilution is placed in 9 ml of melted agar medium (45°C), and the medium is quickly poured into a sterile plate. The resulting pour plate will contain a small number of bacteria, some of which will form isolated colonies on the agar. This method allows some organisms to be embedded in the medium. It is particularly useful for growing microaerophiles that cannot tolerate exposure to oxygen in the air at the surface of the medium.

Culture Media

In nature many species of bacteria and other microorganisms are found growing together in oceans, lakes, and soil and on living or dead organic matter. These materials might be thought of as *natural media.* Although soil and water samples are often brought into the laboratory, organisms from them are typically isolated, and pure cultures are prepared for study.

Growing bacteria in the laboratory requires knowledge of their nutritional needs and the ability to provide the needed substances in a medium. Through years of experience in culturing bacteria in the laboratory, microbiologists have learned what nutrients must be supplied to each of many different organisms. Certain organisms, such as those that cause syphilis and leprosy, still cannot be cultured in laboratory media. They must be grown in cultures that contain living human or other animal cells. Many other organisms whose nutritional needs are reasonably well known can be grown in one or more types of media.

Types of Media

Laboratory media are generally synthetic media, as opposed to the natural media mentioned previously. A

TABLE 6.2	A Defined Synthetic Medium for Growing *Proteus vulgaris*		
Ingredient	Amount	Ingredient	Amount
Water	1 liter	K_2HPO_4	1 g
$MgSO_4 \cdot 7H_2O$	200 mg	$FeSO_4 \cdot 7H_2O$	10 mg
$CaCl_2$	10 mg	Glucose	5 g
NH_4Cl	1 g	Nicotinic acid	0.1 mg

Trace elements (Mn, Mo, Cu, Co, Zn as inorganic salts, known quantities of 0.02–0.5 mg each)

SOURCE: Adapted from R. Y. Stanier et al. 1986. *The microbial world.* 5th ed. Englewood Cliffs, NJ: Prentice Hall.

synthetic medium is a medium prepared in the laboratory from materials of precise or reasonably well defined composition. A **defined synthetic medium** is one that contains known specific kinds and amounts of chemical substances. Examples of defined synthetic media are given in Tables 6.2 and 6.3. A **complex medium,** or **chemically nondefined medium,** is one that contains certain reasonably familiar materials but that varies slightly in chemical composition from batch to batch. Such media contain blood or extracts from beef, yeasts, soybeans, and other organisms. A common ingredient is **peptone,** a product of enzyme digestion of proteins. It provides small peptides that microorganisms can use. Although the exact concentrations are not known, trace elements and vitamins are present in sufficient quantities in complex media to support the growth of many organisms. Both liquid nutrient broth and solidified agar medium used to culture many organisms are complex media. An example of a complex medium is given in Table 6.4.

Commonly Used Media

Most routine laboratory cultures use media containing peptone from meat or fish in nutrient broth or solid agar medium. Such media are sometimes enriched with **yeast extract,** which contains a number of vitamins, coenzymes, and nucleosides. **Casein hydrolysate** made from milk protein contains many amino acids and is used to enrich certain media. Because blood contains many nutrients needed by fastidious pathogens, **serum** (the liquid part of the blood after clotting factors have been removed), whole blood, and heated whole blood can be useful in enriching media. **Blood agar** is useful in identifying organisms that can cause hemolysis, or breakdown of red blood cells. Sheep's blood is used because its hemolysis is more clearly defined than when human blood is used in the agar medium. **Chocolate agar,** made with heated blood, is so named because it turns

TABLE 6.3	A Defined Synthetic Medium for Growing the Fastidious Bacterium *Leuconostoc mesenteroides*

Ingredient	Amount	Ingredient	Amount
Water	1 liter		
Energy Source			
Glucose	25 g		
Nitrogen Source			
NH$_4$Cl	3 g		
Minerals			
KH$_2$PO$_4$	600 mg	FeSO$_4$ · 7H$_2$O	10 mg
K$_2$HPO$_4$	600 mg	MnSO$_4$ · 4H$_2$O	20 mg
MgSO$_4$ · 7H$_2$O	200 mg	NaCl	10 mg
Organic Acid			
Sodium acetate	20 g		
Amino Acids			
DL-α-Alanine	200 mg	L-Lysine · HCl	250 mg
L-Arginine	242 mg	DL-Methionine	100 mg
L-Asparagine	400 mg	DL-Phenylalanine	100 mg
L-Aspartic acid	100 mg	L-Proline	100 mg
L-Cysteine	50 mg	DL-Serine	50 mg
L-Glutamic acid	300 mg	DL-Threonine	200 mg
Glycine	100 mg	DL-Tryptophan	40 mg
L-Histidine · HCl	62 mg	L-Tyrosine	100 mg
DL-Isoleucine	250 mg	DL-Valine	250 mg
DL-Leucine	250 mg		
Purines and Pyrimidines			
Adenine sulfate · H$_2$O	10 mg	Uracil	10 mg
Guanine · HCl · 2H$_2$O	10 mg	Xanthine · HCl	10 mg
Vitamins			
Thiamine · HCl	0.5 mg	Riboflavin	0.5 mg
Pyridoxine · HCl	1.0 mg	Nicotinic acid	1.0 mg
Pyridoxamine · HCl	0.3 mg	*p*-Aminobenzoic acid	0.1 mg
Pyridoxal · HCl	0.3 mg	Biotin	0.001 mg
Calcium pantothenate	0.5 mg	Folic acid	0.01 mg

SOURCE: H. E. Sauberlich and C. A. Baumann. "A factor required for the growth of *Leuconostoc citrovorum*." *J. Biol. Chem.* 176(1948):166.

TABLE 6.4	A Complex Medium Suitable for Many Heterotrophic Organisms

Nutrient Broth Ingredient	Amount
Water	1 liter
Peptone	5 g
Beef extract	3 g
NaCl	8 g
Solidified Medium	
Agar	15 g

Above ingredients in amounts specified.

chocolate brown. It is used to culture fastidious pathogens.

Selective, Differential, and Enrichment Media

To isolate and identify particular microorganisms, especially those from patients with infectious diseases, *selective, differential,* or *enrichment media* are often used. Such special media are an essential part of modern diagnostic microbiology. Table 6.5 shows some examples of special diagnostic media.

A **selective medium** is one that encourages the growth of some organisms but suppresses the growth of others. For example, to identify *Clostridium botulinum* in food samples suspected of being agents of food poisoning, the antibiotics sulfadiazine and polymyxin sulfate (SPS) are added to anaerobic cultures of *Clostridium* species. This culture medium is called *SPS agar*. It allows growth of *Clostridium botulinum* while inhibiting growth of most other *Clostridium* species.

A **differential medium** has a constituent that causes an observable change (a color change or a change in pH) in the medium when a particular biochemical reaction occurs. This change allows microbiologists to distinguish a certain type of colony from others growing on the same plate. SPS agar also serves as a differential medium. Colonies of *Clostridium botulinum* formed on this medium are black because of hydrogen sulfide made by the organisms from the sulfur-containing additives.

Many media, such as SPS agar and MacConkey agar, are both selective and differential. *MacConkey agar* contains crystal violet and bile salts, which inhibit growth of gram-positive bacteria while allowing growth of gram-negative bacteria. MacConkey agar also contains the sugar lactose plus a pH indicator that turns colonies of lactose fermenters red and leaves colonies of nonfermenters colorless and translucent. Although there are some exceptions, most organisms that are normally found in the human intestines ferment lactose, whereas most pathogens (disease-causing microorganisms) do not.

An **enrichment medium** contains special nutrients that allow growth of a particular organism that might not otherwise be present in sufficient numbers to allow it to be isolated and identified. Unlike a selective medium, an enrichment medium does not suppress others. For example, because *Salmonella typhi* organisms may not be sufficiently numerous in a fecal sample to allow positive identification, they are cultured on a medium containing the trace element selenium, which supports growth of the organism. After incubation in the enrichment medium, the greater numbers of the organisms increase the likelihood of a positive identification.

TABLE 6.5   **Selected Examples of Diagnostic Media**

Medium		Organism(s) Identified	Selectivity and/or Differentiation Achieved
Brilliant green agar	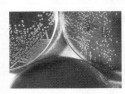	*Salmonella*	Brilliant green dye inhibits gram-positive bacteria and thus selects gram-negative ones. Differentiates *Shigella* colonies (which do not ferment lactose or sucrose and are red to white) from other organisms that do ferment one of those sugars and are yellow to green.
Eosin methylene blue agar (EMB)	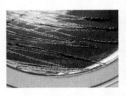	Gram-negative enterics	Medium partially inhibits gram-positive bacteria. Eosin and methylene blue differentiate among organisms: *Escherichia coli* colonies are purple and typically have a metallic green sheen; *Enterobacter aerogenes* colonies are pink, indicating that they ferment lactose; and colonies of other organisms are light pink, indicating they do not ferment lactose.
MacConkey agar	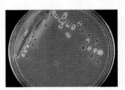	Gram-negative enterics	Crystal violet and bile salts inhibit gram-positive bacteria. Lactose and the pH indicator neutral red (red when acidic), identify lactose fermenters as red colonies and nonfermenters as light pink. Most intestinal pathogens are nonfermenters and hence do not produce acid.
Triple sugar–iron agar (TSI)		Gram-negative enterics	Used in agar slants (tubes cooled in slanted position), where differentiation is based on both aerobic surface growth and anaerobic growth in agar in base of tube. Medium contains specific amounts of glucose, sucrose, and lactose, sulphur-containing amino acids, and a pH indicator, so relative use of each sugar can be detected.
Sodium tetrathionate broth		Enteric pathogens	Sodium tetrathionate inhibits normal inhabitants of the gut and enriches growth of certain pathogens, such as *Salmonella* and *Shigella*.

Controlling Oxygen Content of Media

Obligate aerobes, microaerophiles, and obligate anaerobes require special attention to maintain oxygen concentrations suitable for growth. Most obligate aerobes obtain sufficient oxygen from nutrient broth or on the surface of solidified agar medium, but some need more. Oxygen gas is bubbled through the medium or into the incubation environment with filters between the gas source and the medium to prevent contamination of the culture. Microaerophiles can be incubated in tubes of a nutrient medium or agar plates in a jar in which a candle is lit before the jar is sealed (Figure 6.19). (Scented candles should not be used because oils from them inhibit bacterial growth.) The burning candle uses the oxygen in the jar and adds carbon dioxide to it. When the carbon dioxide extinguishes the flame, conditions are optimum for the growth of microorganisms that require small amounts of carbon dioxide,

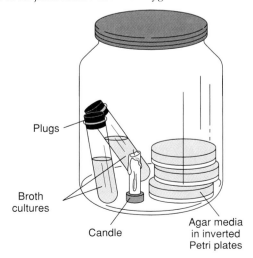

FIGURE 6.19 Microaerophiles are growing in culture tubes and on Petri plates in a sealed jar in which a candle burned until it was extinguished by carbon dioxide accumulation in the atmosphere of the jar. A small amount of oxygen remains.

Plugs

Broth cultures

Candle

Agar media in inverted Petri plates

APPLICATIONS

DON'T LEAVE HOME WITHOUT YOUR CO_2!

The transport of specimens from patients to the laboratory in a viable condition sometimes presents special problems. The organisms must not be subjected to drying conditions or to too much or too little oxygen. And, of course, specimen handlers must be protected from infection. Cultures that may contain *Neisseria gonorrhoeae* from patients with gonorrhea pose one such problem—that of providing an atmosphere relatively high in carbon dioxide. Various commercial systems, such as the widely used JEMBEC (John E. Martin Biological Environmental Chamber), are available for this purpose. This system consists of a small plastic plate of selective medium and a tablet of sodium bicarbonate and citric acid. The plate is inoculated, and the tablet is placed in it. The plate is then placed in a plastic bag and sealed. Moisture from the medium causes the tablet to release carbon dioxide; an appropriate concentration of 5 to 10 percent is obtained. The culture is incubated for 18 to 24 hours to allow growth to begin before shipment to the laboratory.

The JEMBEC system is used to culture gonorrhea specimens.

FIGURE 6.20 CO_2 incubator. When activated, these chemicals remove oxygen and are enclosed with cultures in a sealed jar to create an anaerobic chamber. These are useful for the small laboratory that has only a few plates needing anaerobic incubation.

that colonial growth can be studied, special jars are used that can hold both plates and tubes. Agar plates are incubated in sealed jars containing chemical substances that remove oxygen from the air and generate carbon dioxide (Figure 6.20). *Stab cultures* can be made by stabbing a straight inoculating wire coated with organisms into a tube of agar-solidified medium. In laboratories where anaerobes are regularly handled, an *anaerobic transfer chamber* is often used (Figure 6.21). Equipment and cultures are introduced through an air lock, and the technician uses glove ports to manipulate the cultures.

Maintaining Cultures

Once an organism has been isolated, it can be maintained indefinitely in a pure culture called a **stock culture.** When needed for study, a sample from a stock culture is inoculated into fresh medium. The stock culture itself is never used for laboratory studies. However, organisms in stock cultures go through growth phases, deplete nutrients, and accumulate wastes just as those in any culture do. As the culture ages, the organisms may acquire odd shapes or other altered characteristics. Stock cultures are maintained by making subcultures in fresh medium at frequent intervals to keep the organisms growing.

such as the bacterium *Neisseria gonorrhoeae,* which causes gonorrhea.

To culture obligate anaerobes, all molecular oxygen must be removed and kept out of the medium. Addition of oxygen-binding agents such as thioglycollate, the amino acid cysteine, or sodium sulfide to the medium prevents oxygen from exerting toxic effects on anaerobes. Media can be dispensed in sealed screw-cap tubes, completely filled to exclude air, or in Petri plates. When cultures must be grown in plates so

FIGURE 6.21 A large anaerobic transfer chamber with an air lock for introducing equipment and cultures and with ports to allow manipulation of the cultures.

The use of careful aseptic techniques is important in all manipulations of cultures. **Aseptic techniques** minimize the chances that cultures will be contaminated by organisms from the environment or that organisms, especially pathogens, will escape into the environment. Such techniques are especially important in making subcultures from stock cultures. Otherwise an undesirable organism might be introduced, and the stock organism would have to be reisolated. Even with regular transfers of organisms from stock cultures to fresh media, organisms can undergo mutations (changes in DNA) and develop altered characteristics.

Preserved Cultures

To avoid the risk of contamination and to reduce the mutation rate, stock culture organisms also should be kept in a **preserved culture,** a culture in which organisms are maintained in a dormant state. The most commonly used technique for preserving cultures is *lyophilization* (freeze-drying), in which cells are quickly frozen, dehydrated while frozen, and sealed in vials under vacuum (Chapter 13). Such cultures can be kept indefinitely at room temperature.

Because microorganisms frequently undergo genetic changes, reference cultures are maintained. A **reference culture** is a preserved culture that maintains the organisms with the characteristics as originally defined. Reference cultures of all known species and strains of bacteria and many other microorganisms are maintained in the American Type Culture Collection, and many also are maintained in universities and research centers. Then if stock cultures in a particular laboratory undergo change or if other laboratories wish to obtain certain organisms for study, reference cultures are always available.

METHODS OF PERFORMING MULTIPLE DIAGNOSTIC TESTS

Many diagnostic laboratories use culture systems that contain a large number of differential and selective media, such as the Enterotube Multitest System or the Analytical Profile Index (API). These systems allow simultaneous determination of an organism's reaction to a variety of carefully chosen diagnostic media from a single inoculation. The advantages of these systems are that they use small quantities of media, occupy little space in an incubator, and provide an efficient and reliable means of making positive identification of infectious organisms.

The Enterotube System® is used to identify enteric pathogens, or organisms that cause intestinal diseases such as typhoid and paratyphoid fevers, shigellosis, gastroenteritis, and some kinds of food poisoning. The causative organisms are all gram-negative rods indistinguishable from one another without biochemical tests. The Enterotube System consists of a tube with compartments, each of which contains one or more different media, and a sterile inoculating rod (Figure 6.22a). Each compartment is inoculated when the tip of the rod is touched to a colony and the rod is drawn through the tube. After the tube has been incubated for 24 hours at 37°C, the results of 15 biochemical tests can be obtained by observing (1) whether gas was produced

and (2) the color of the medium in each compartment. Tests are grouped in sets of three; within each group, tests are assigned a number 1, 2, or 4 (Figure 6.22b). The sum of the numbers of positive tests in each group indicates which tests are positive. The sum 3 means that tests 1 and 2 were positive; the sum 5 means that tests 1 and 4 were positive; the sum 6, that tests 2 and 4 were positive; and the sum 7, that all tests were positive. The single-digit sums for each of the five sets of tests are combined to form a five-digit identification number for a particular organism. For example, 36601 is *Escherichia coli*, and 34363 is *Klebsiella pneumoniae*. A list of identification numbers and the corresponding organisms is provided with the system.

API consists of a plastic tray with 20 microtubes called *cupules*, each containing a different kind of dehydrated medium (Figure 6.23). Each cupule medium is rehydrated and inoculated with a suspension of bacteria from an isolated colony. As with Enterotubes, the tray is incubated, test results are determined, and the values 1, 2, and 4 are summed for sets of three tests. The seven-digit profile number identifies the organism.

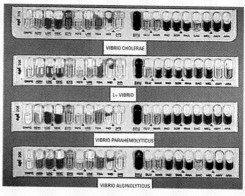

FIGURE 6.23 The Analytical Profile Index (API) 20E System. Various species of the genus *Vibrio* are shown here, with the differences in reactions that enable them to be identified. This system allows identification to species level of 125 gram-negative intestinal bacilli.

In this brief discussion of diagnostic systems, we have considered only the tip of the iceberg. Of the many other available tests, a large number are based on immunological properties of organisms. We will consider some of them when we discuss immunology or particular infectious agents. Also, much is known about which organisms are likely to infect certain human organs and tissues, and many diagnostic tests are designed to distinguish among organisms found in respiratory secretions, fecal samples, blood, other tissues, and body fluids.

FIGURE 6.22 (a) The Enterotube Multitest System. (b) After inoculation and incubation, compartments with positive test results are assigned a number. The numbers are summed within zones to get a definitive index number that identifies an organism on the list in the coding manual. Any necessary confirmatory tests are also noted there. By numbering each test in a zone with a digit equal to a power of 2 (1, 2, 4, 8, and so on), the sum of any set of positive reactions results in a unique number. A given species may, however, be coded for many different numbers, as individual strains of that species will vary somewhat in their characteristics.

(a)

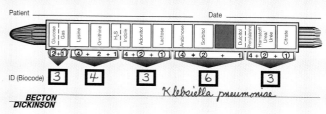

(b)

GROWTH AND CELL DIVISION

Definition of Growth

- Microbial growth can be defined as the orderly increase in quantity of all cell components and in the number of cells of an organism.

- Because of limited increase in cell size and the frequency of cell division, growth in microorganisms is measured by increase in cell number.

Cell Division

- Most cell divisions in bacteria occur by **binary fission,** in which the nuclear body divides and the cell forms a transverse septum that separates the original cell into two cells.

- Yeast cells and some bacteria divide by **budding,** in which a small, new cell develops from the surface of an existing cell.

Phases of Growth

- In a nutrient-rich **medium** (a mixture of substances on or in which microorganisms grow), bacteria divide rapidly. The length of time required for one division is called the **generation time.** Such growth is said to occur at an **exponential,** or **logarithmic, rate.**

- Bacteria introduced into a fresh, nutrient-rich medium display four major phases of growth: (1) In the **lag phase,** the organisms are metabolically active—growing and synthesizing various substances but not increasing in number. (2) In the **log phase,** organisms divide at an exponential, or logarithmic, rate and with a constant generation time. These properties of growth in the log phase can be used to calculate both the number of generations and the generation time. Cultures can be maintained by the use of a **chemostat,** which allows continuous addition of fresh medium. (3) In the **stationary phase,** the number of new cells produced equals the number of cells dying. The medium contains limited amounts of nutrients and may contain toxic quantities of waste materials. (4) In the **decline phase,** or **death phase,** many cells lose their ability to divide and eventually die. A logarithmic decrease in the number of cells results.

- Growth in colonies parallels that in liquid medium, except that most growth occurs at the edge of the colony, and all phases of growth occur simultaneously somewhere in the colony.

Measuring Bacterial Growth

- Growth can be measured by **serial dilution,** in which successive 1:10 dilutions of a liquid culture of bacteria are made and transferred onto an **agar plate;** the colonies that arise are counted. Each colony represents one live cell from the original sample.

- Growth also can be measured by **direct microscopic counts,** the **most probable number** technique, **filtration,** observing or measuring **turbidity,** measuring products of metabolism, and obtaining the dry weight of cells.

FACTORS AFFECTING BACTERIAL GROWTH

Physical Factors

- Acidity and alkalinity of the medium affect growth, and most organisms have an **optimum pH** range of no more than one pH unit.

- Temperature affects bacterial growth. (1) Most bacteria can grow over a 30°C temperature range. (2) Bacteria can be classified according to growth temperature into three categories: **psychrophiles,** which grow at low temperatures (below 25°C); **mesophiles,** which grow best at temperatures between 25° and 40°C; and **thermophiles,** which grow at high temperatures (above 40°C). (3) The temperature range of an organism is closely related to the temperature at which its enzymes function best.

- The quantity of oxygen in the environment affects the growth of bacteria. (1) **Obligate aerobes** require relatively large amounts of free molecular oxygen to grow. (2) **Obligate anaerobes** are killed by free oxygen and must be grown in the absence of free oxygen. (3) **Facultative anaerobes** can metabolize substances aerobically if oxygen is available or anaerobically if it is absent. (4) **Aerotolerant anaerobes** metabolize substances anaerobically but are not harmed by free oxygen. (5) **Microaerophiles** must have only small amounts of oxygen to grow.

- Actively metabolizing bacteria require some water in their environment.

- Some bacteria, but no other living things, can withstand extreme **hydrostatic pressures** in deep valleys in the ocean.

- Osmotic pressure affects bacterial growth, and water can be drawn into or out of cells according to the relative osmotic pressure created by dissolved substances in the cell and the environment. (1) Active transport minimizes the effects of high osmotic pressure in the environment. (2) Bacteria called **halophiles** require moderate to large amounts of salt and are found in the ocean and in exceptionally salty bodies of water.

Nutritional Factors

- All organisms require a carbon source: (1) Autotrophs use CO_2 as their carbon source and synthesize other substances they need. (2) Heterotrophs require glucose or another organic carbon source from which they obtain energy and intermediates for synthetic processes.

- Microorganisms require an organic or inorganic nitrogen source from which to synthesize proteins and nucleic acids. They also require a source of other elements found within them, including sulfur, phosphorus, potassium, iron, and many **trace elements.**

- Microorganisms that lack the enzymes to synthesize particular **vitamins** must obtain those vitamins from their environment.

- The nutritional requirements of an organism are determined by the kind and number of its enzymes. **Nutritional complexity** reflects a deficiency in biosynthetic enzymes.

- Bioassay techniques use metabolic properties of organisms to determine quantities of vitamins and other compounds in foods and other materials.

- Most microorganisms move substances of low molecular weight across their cell membranes and metabolize them internally. Some bacteria (and fungi) also produce exoenzymes that digest large molecules outside the cell membrane of the organism.

- Microorganisms adjust to limited nutrient supplies by increasing the quantities of enzymes they produce, by making enzymes to metabolize another available nutrient, or by adjusting their metabolic activities to grow at a rate consistent with availability of nutrients.

SPORULATION

- **Sporulation,** which occurs in *Bacillus, Clostridium,* and a few other gram-positive genera, involves the steps summarized in Figure 6.16.

- Sporulation lets the bacterium withstand long periods of dry conditions and extreme temperatures.

- When more favorable conditions are restored, **germination** occurs—endospores begin to develop into vegetative cells.

CULTURING BACTERIA

Methods of Obtaining Pure Cultures

- The **streak plate method** of obtaining a **pure culture** involves spreading bacteria across a sterile, solid surface such as an agar plate so that the progeny of a single cell

can be picked up from the surface and transferred to a sterile medium.

- The **pour plate method** of obtaining a pure culture involves serial dilution, transferring to melted agar a specific volume of the dilution containing a few organisms and picking up cells from a colony on the agar.

Culture Media

- In nature, microorganisms grow on natural media, or the nutrients available in water, soil, and living or dead organic material.

- In the laboratory, microorganisms are grown in **synthetic media:** (1) **Defined synthetic media** consist of known quantities of specific nutrients. (2) **Complex media** consist of nutrients of reasonably well known composition that vary in composition from batch to batch.

- Most routine laboratory cultures make use of **peptones,** or digested meat or fish proteins. Other substances such as **yeast extract, casein hydrolysate, serum,** whole blood, or heated whole blood are sometimes added.

- Diagnostic media are (1) **selective media** if they encourage growth of some organisms and inhibit growth of others, (2) **differential media** if they allow different kinds of colonies on the same plate to be distinguished from one another, or (3) **enrichment media** if they provide a nutrient that fosters growth of a particular organism.

- Cultures are maintained as **stock cultures** for routine work, as **preserved cultures** to prevent risk of contamination or change in characteristics, and as **reference cultures** to preserve specific characteristics of species and strains.

KEY TERMS

acidophile (p. 145)
aerotolerant anaerobe (p. 147)
agar (p. 140)
alkaliphile (p. 145)
aseptic technique (p. 158)
barophile (p. 147)
binary fission (p. 136)
blood agar (p. 154)
budding (p. 137)
capnophile (p. 147)
casein hydrolysate (p. 154)
catalase (p. 147)
chemically nondefined medium (p. 154)
chemostat (p. 139)
chocolate agar (p. 154)
colony (p. 140)
complex medium (p. 154)
conidium (p. 153)
core (p. 152)
cortex (p. 152)
countable number (p. 141)
cyst (p. 153)
daughter cell (p. 136)
death phase (p. 140)
decline phase (p. 140)

defined synthetic medium (p. 154)
differential medium (p. 155)
dipicolinic acid (p. 152)
direct microscopic count (p. 142)
endospore septum (p. 152)
enrichment medium (p. 155)
exosporium (p. 152)
exponential rate (p. 138)
extracellular enzyme (p. 150)
facultative (p. 145)
facultative anaerobe (p. 147)
facultative psychrophile (p. 145)
facultative thermophile (p. 145)
fastidious (p. 148)
filtration (p. 143)
generation time (p. 138)
germination (p. 152)
halophile (p. 148)
hydrostatic pressure (p. 147)
lag phase (p. 138)
logarithmic rate (p. 138)
log phase (p. 138)
medium (p. 138)

mesophile (p. 145)
microaerophile (p. 147)
microbial growth (p. 136)
most probable number (p. 143)
mother cell (p. 136)
neutrophile (p. 145)
nonsynchronous growth (p. 139)
nutritional complexity (p. 150)
nutritional factor (p. 144)
obligate (p. 145)
obligate aerobe (p. 147)
obligate anaerobe (p. 147)
obligate psychrophile (p. 145)
obligate thermophile (p. 145)
optimum pH (p. 145)
peptone (p. 154)
periplasmic enzyme (p. 151)
physical factor (p. 144)
plasmolysis (p. 148)
pour plate (p. 140)
pour plate method (p. 154)
preserved culture (p. 158)
psychrophile (p. 145)

pure culture (p. 153)
reference culture (p. 158)
sarcina (p. 136)
selective medium (p. 155)
serial dilution (p. 140)
serum (p. 154)
spore coat (p. 152)
sporulation (p. 151)
spread plate method (p. 140)
standard bacterial growth curve (p. 138)
stationary phase (p. 139)
stock culture (p. 157)
streak plate method (p. 153)
superoxide (p. 147)
superoxide dismutase (p. 147)
synchronous growth (p. 138)
synthetic medium (p. 154)
tetrad (p. 136)
thermophile (p. 145)
trace element (p. 149)
turbidity (p. 143)
vitamin (p. 149)
yeast extract (p. 154)

A **Bacterial Growth**

1. What is growth, and how is it defined in microorganisms?

B **Microbial Cell Division**

2. How does cell division occur in microorganisms?

C **Bacterial Growth Phases**

3. What events occur in each of the four phases of bacterial growth?

4. Identify the position of each of the following on the accompanying graph:

____ organisms divide at their most rapid rate

____ new cells are produced at same rate as old cells die

____ lag phase

____ log phase

____ many cells undergo involution

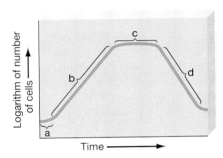

5. How does growth in a colony differ from growth in a liquid medium?

D **Measuring Bacterial Growth**

6. Name three ways of measuring bacterial growth.

7. What kinds of errors are likely to occur in measuring bacterial growth?

8. a. How many different types of colonies can you distinguish in Figure 6.7b?

   b. Does this plate have a countable number of colonies?

   c. What could cause the total number of colonies on the plate to be different from the total number of live cells originally deposited in the plate?

   d. If you had taken a sample of the same culture equal in volume to that which was placed in the pour plate but had counted it by using a Petroff-Hausser counting chamber, would you get the same number as you did by counting colonies? Why, or why not?

E **Effects of Physical Factors**

9. How does each of the following factors affect growth in bacteria: pH, temperature, moisture, hydrostatic pressure, osmotic pressure, and radiation?

10. Why would you expect facultative organisms to have more different enzymes than other groups of organisms?

11. How do microaerophiles differ from aerotolerant anaerobes?

12. What is superoxide, and how is it destroyed?

F **Effects of Biochemical Factors**

13. What uses do bacteria make of carbon sources?

14. What other nutrients might bacteria require, and why?

15. Under what conditions would bacteria require vitamins?

16. How are nutritional requirements related to an organism's enzymes?

G **Sporulation**

17. What are the steps in the production of endospores?

18. Under what conditions do endospores form?

19. How and under what conditions do endospores germinate?

H **Pure Cultures**

20. How would you isolate a pure culture from a sample of mixed organisms, such as a soil sample or a throat culture?

21. Where does your school obtain the cultures you work with in the laboratory? How much does a single culture cost? Can anyone buy any culture? How would you check to be sure that a stock culture has not become contaminated? If microbiology is not taught during summer school, will the stock cultures from spring still be good in fall?

I **Various Media**

22. What is the difference between a defined synthetic medium and a complex medium? Which of the following would be found *only* in a complex medium?

   ____ glucose   ____ peptone

   ____ yeast extract   ____ NaCl

   ____ vitamins   ____ serum

   ____ casein hydrolysate ____ whole blood

   ____ agar   ____ water

23. How do selective, differential, and enrichment media differ?

24. True or false:

   a. An enrichment medium suppresses growth of organisms other than the desired ones.

   b. A single medium can be both selective and differential.

   c. When grown on nutrient agar plates, colonies of both species A and B appear white. A medium on which colonies of species A appear green and colonies of species B appear pink is a selective medium.

PROBLEMS FOR INVESTIGATION

1. An attempt to transfer bacteria to new media during the death phase of a culture resulted in actual growth of the organisms. What is the most likely explanation for this phenomenon?

2. Devise an experiment to determine whether an unknown organism is a psychrophile, a mesophile, or a thermophile. If it is a psychrophile or a thermophile, show how to determine whether it is facultative or obligate.

3. If 100 bacteria with a generation time of 30 minutes are transferred to new media at 10 A.M., how many organisms will be present by 3 P.M.? How many generations will have been produced by 5 P.M.?

4. Tuberculosis organisms (*Mycobacterium tuberculosis*) are notoriously slow growers and can be overlooked if media are discarded prematurely. They also need special nutrients. Read about the special culture and diagnostic methods for these organisms and other members of this genus.

5. Prepare a library research paper on modern diagnostic media.

SOME INTERESTING READING

American Type Culture Collection. 1984. *Media handbook.* Rockville, MD: American Type Culture Collection.

Atlas, R. M. 1993. *Handbook of microbiological media* (L. C. Parks, ed.). Boca Raton, FL: CRC Press.

Baron, E. J., L. R. Peterson, and S. M. Finegold. 1994. *Bailey and Scott's diagnostic microbiology,* 9th ed. St. Louis: Mosby.

Benathen, I. A. 1990. "Isolation of pure cultures from mixed cultures: A modern approach." *American Biology Teacher* 52(1):46–47.

Brock, T. D. 1985. "Life at high temperatures." *Science* 230, no. 4722 (October 11):132–38.

Difco manual: Dehydrated culture media and reagents for microbiology. 1984, 10th ed. Detroit: Difco Laboratories.

Doyle, A. E., and A. Doyle, eds. 1991. *Maintenance of microorganisms and cultured cells.* San Diego: Academic Press.

Gerhardt, P., ed. 1993. *Methods for general and molecular bacteriology.* Washington, D.C.: American Society for Microbiology.

Hoch, J. A. 1993. "Regulation of the phosphorelay and the initiation of sporulation in *Bacillus subtilis.*" *Annual Review of Microbiology* 47:441–65.

Meyer, H-P., O. Käppeli, and A. Fiechter. 1985. "Growth control in microbial cultures." *Annual Review of Microbiology* 39:299–319.

Murray, P. R., ed., et al. 1995. *Manual of clinical microbiology,* 6th ed. Washington, D.C.: American Society for Microbiology.

Postgate, J. R. 1994. *The outer reaches of life.* New York: Cambridge University Press.

Power, D. A., and P. J. McCuen. 1988. *Manual of BBL products and laboratory procedures.* Cockeysville, MD: Becton-Dickinson Microbiology Systems.

Robert, F. M. 1990. "Impact of environmental factors on populations of soil microorganisms." *American Biology Teacher* 52, no. 6 (September):364–69.

"Salty life on Mars." 1991. *Discover* 12 (6):12.

Tunnicliffe, V. 1992. "Hydrothermal-vent communities of the deep sea." *American Scientist* 80(July–August):336–49.

Yarmolinsky, M. B. 1995. "Programmed cell death in bacterial populations." *Science* 267, no. 5199 (February 10):836–37.

GENETICS I: GENE ACTION, GENE REGULATION, AND MUTATION

DNA spills from a ruptured *Escherichia coli* bacterium in this TEM photo (magnification, 20,800×), revealing the great length of the single circular chromosome as it folds back and forth over itself.

We have considered many aspects of metabolism and growth, but we have yet to consider the synthesis of nucleic acids and proteins. Synthesis of these complex molecules is the basis of **genetics,** the study of heredity. The genetics of microorganisms is an exciting and active research area; it also has been a rewarding area for microbiologists. Since the inception of the Nobel Prize in physiology or medicine in 1900, more than 30 annual prizes have been awarded in microbiology-related fields, especially molecular biology and microbial genetics. Because of this intensive investigation, much is now known about microbial genetics. We will begin our study of genetics by seeing how bacteria synthesize nucleic acids—DNA and RNA—and how the nucleic acids are involved in the synthesis of proteins. We will also see how *genes* (specific segments of DNA) act, how they are regulated, and how they are altered in mutations. In the next chapter we will discuss the mechanisms by which genetic information is transferred among microorganisms.

OVERVIEW OF GENETIC PROCESSES

The Basis of Heredity

All information necessary for life is stored in an organism's genetic material, DNA, or, for many viruses, RNA. To explain **heredity**—the transmission of information from an organism to its progeny (offspring)—we must consider the nature of chromosomes and genes.

A *chromosome* is a circular or linear, threadlike molecule of DNA. Recall that DNA consists of a double chain of nucleotides arranged in a helix with the nucleotide base pairs held together by hydrogen bonds (Figure 7.1, page 166). ∞ (Chapter 2, p. 42) The particular nucleotide sequence in DNA provides information for the synthesis of new DNA and for the synthesis of proteins.

The typical prokaryotic cell contains a single circular chromosome. When a prokaryotic cell reproduces

THIS CHAPTER FOCUSES ON THE FOLLOWING QUESTIONS:

A How are genes, chromosomes, and mutations involved in heredity in prokaryotic organisms?

B How do nucleic acids store and transfer information?

C How is DNA replicated in prokaryotic cells?

D What are the major steps in protein synthesis?

E How do mechanisms that regulate enzyme activity differ from those that regulate gene expression?

F What happens in feedback inhibition, enzyme induction, and enzyme repression?

G What changes in DNA occur as a result of mutations, and how do mutations affect organisms?

H How do spontaneous and induced mutations differ?

I How do the fluctuation test, replica plating, and the Ames test make use of bacteria in studying mutations?

by binary fission, the chromosome reproduces, or *replicates*, itself, and each daughter cell receives one of the chromosomes. This mechanism provides for the orderly transmission of genetic information from parent cell to daughter cells.

A **gene,** the basic unit of heredity, is a linear sequence of nucleotides of DNA that form a functional unit of a chromosome. All information for the structure and function of an organism is coded in its genes. The genetic makeup of an organism is called its **genotype.** The genotype represents *potential* characteristics but not the characteristics themselves. The observable physical traits of an organism are collectively known as its **phenotype.** For example, the ability of an organism to carry out a particular metabolic pathway is part of its phenotye—a physical expression of the organism's genotype. In many cases, one gene determines a single trait. Thus, each enzyme in a metabolic pathway may be the result of information contained in a single gene. However, eukaryotic cells have pairs of identical chromosomes. Although both members of

a pair have a gene for the same trait at a particular **locus,** or location on the chromosome, they may have different forms of the trait. Genes with different information at the same locus are called **alleles** (al-elz'). For example, in human blood types any one of three genes, A, B, and O, can occupy a certain locus. Allele A causes red blood cells to have a certain glycoprotein, which we will designate as molecule A, on their surfaces. Allele B causes them to have molecule B, and allele O does not cause them to have any glycoprotein molecule on the cell surfaces. People with type AB blood produce both molecules A and B because they have both alleles A and B.

Heritable variations in the characteristics of progeny can arise from mutations. A **mutation** is a permanent alteration in DNA. Mutations usually change the sequence of nucleotides in DNA and thereby change the information in the DNA. When the mutated DNA is transmitted to a daughter cell, the daughter cell can differ from the parent cell in one or more characteristics. We will see in Chapter 8 that heritable variations

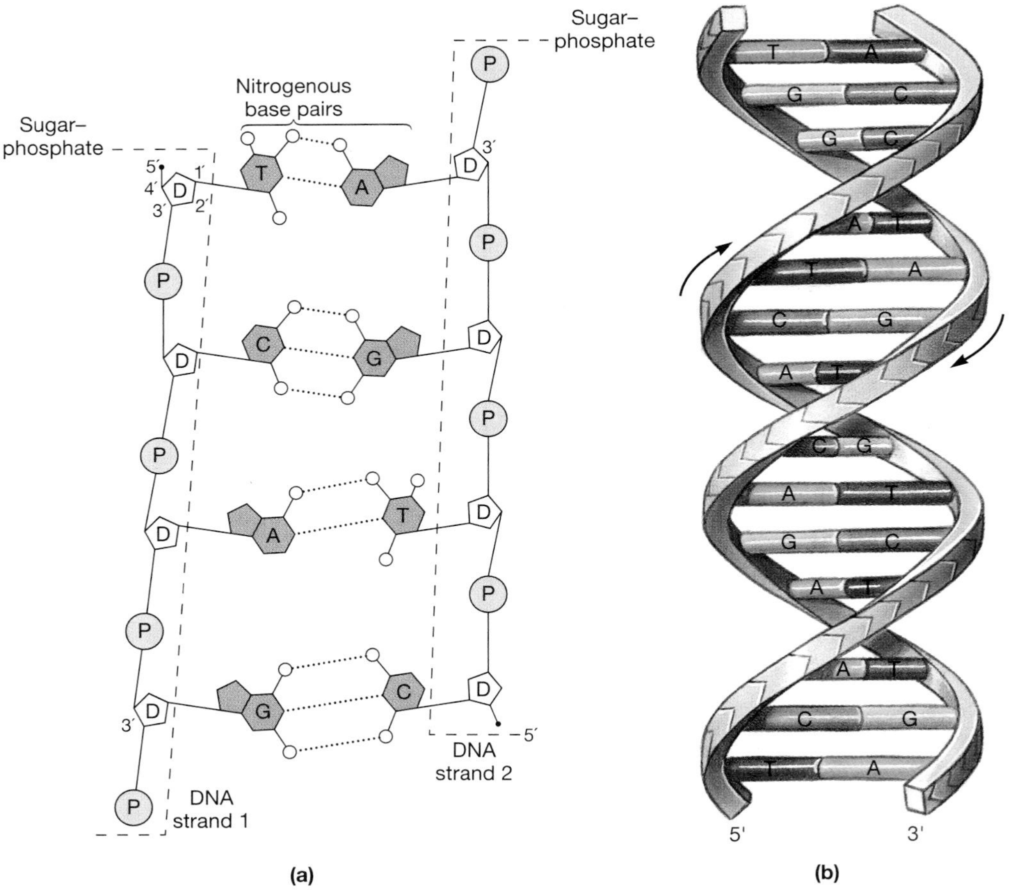

FIGURE 7.1 The structure of DNA. (a) The two upright strands, composed of the sugar deoxyribose (D) and phosphate groups (P), are held together by hydrogen bonding between complementary bases. Adenine (A) always pairs with thymine (T), and guanine (G) always pairs with cytosine (C). Each strand can thus provide the information needed for the formation of a new DNA molecule. (b) The DNA molecule is twisted into a double helix. The two sugar-phosphate strands run in opposite (antiparallel) directions. Each new strand grows from the 5' end toward the 3' end.

in the characteristics of prokaryotic organisms can occur by a variety of other mechanisms as well.

Nucleic Acids in Information Storage and Transfer

Information Storage

All the information for the structure and functioning of a cell is stored in DNA. For example, in the chromosome of the bacterium *Escherichia coli*, each of the paired strands of DNA contains about 5 million bases arranged in a particular linear sequence. The information in those bases is divided into units of several hundred bases each. Each of these units is a gene. Some of the genes and their locations on the chromosome of *E. coli* are shown in Figure 7.2.

We might think of a gene as a sentence in the language of a cell. Each sentence in this language is constructed from a four-letter alphabet corresponding to the four nitrogenous bases in DNA: adenine (A), thymine (T), cytosine (C), and guanine (G). When these four letters combine to make "sentences" several hundred letters long, the number of possible sentences becomes almost infinite. Likewise, an almost infinite number of possible genes exists. If each gene contained 500 bases, a chromosome containing 5 million bases could contain 10,000 different genes. Thus, the information storage capacity of DNA is exceedingly large!

Information Transfer

Information stored in DNA is used both to guide the replication of DNA in preparation for cell division and to direct protein synthesis. Both in DNA **replication** and in the first step of protein synthesis, DNA serves as a **template,** or pattern, for the synthesis of a new nucleotide polymer. The sequence of bases in each new polymer is complementary to that in the original DNA. Such an arrangement is accomplished by base pairing. Recall from Chapter 2 that in complementary base pairing in DNA, adenine always pairs with thymine (A–T), and guanine always pairs with cytosine (G–C). Recall also that when DNA serves as a template for synthesis of RNA, the pairing is different: In RNA, thymine is replaced by uracil (U), which pairs with adenine.

In the replication of DNA, the new polymer is also DNA. In protein synthesis, the new polymer is a particular type of RNA called *messenger RNA (mRNA),*

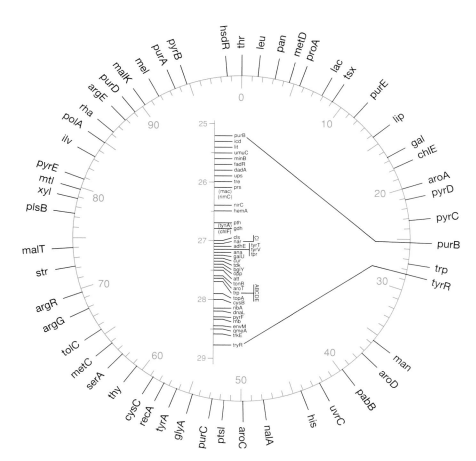

FIGURE 7.2 Part of the chromosome map of *Escherichia coli*. The entire genome of *E. coli* consists of approximately 3000 genes. The outer circle is a simplified representation of the chromosome, with a number of the most commonly studied genes marked on it. It takes about 100 minutes to transfer the entire chromosome from a donor to a recipient cell in conjugation (Chapter 8), a mechanism by which genes are transferred between bacteria. The numbers marked inside the circle represent the number of minutes of transfer required to reach that point on the chromosome. The insert is a small segment of the *E. coli* map, enlarged to show some of the additional genes that have been located within that region (after Bachman).

which then serves as a second template that dictates the arrangement of amino acids in a protein. For example, some proteins form the structure of a cell, others (enzymes) regulate its metabolism, and still others transport substances across a membrane.

In the overall process of protein synthesis, the synthesis of mRNA from a DNA template is called **transcription,** and the synthesis of protein from information in mRNA is called **translation.** By analogy, transcription transfers information from one nucleic

APPLICATIONS

ENEMY DNA

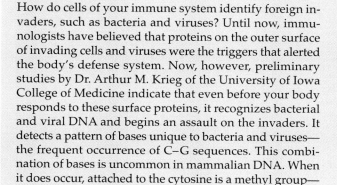

How do cells of your immune system identify foreign invaders, such as bacteria and viruses? Until now, immunologists have believed that proteins on the outer surface of invading cells and viruses were the triggers that alerted the body's defense system. Now, however, preliminary studies by Dr. Arthur M. Krieg of the University of Iowa College of Medicine indicate that even before your body responds to these surface proteins, it recognizes bacterial and viral DNA and begins an assault on the invaders. It detects a pattern of bases unique to bacteria and viruses—the frequent occurrence of C–G sequences. This combination of bases is uncommon in mammalian DNA. When it does occur, attached to the cytosine is a methyl group—a group that bacterial and viral C–G sequences do not have.

There is some evidence that people with systemic lupus erythematosus, an autoimmune disease in which the immune system attacks the body's own DNA, may not have normal ability to add methyl groups to their DNA. Thus their DNA may look like foreign DNA to the immune system. Perhaps the cure for this disease lies in increasing the patient's ability to add methyl groups. Another clinical application may lie in administering artificially produced C–G sequences to patients whose immune systems need stimulation. In the laboratory the addition of such C–G sequences to flasks of B cells (immune cells that make antibodies, or proteins that respond to foreign invaders) causes 95 percent of the cells to begin multiplying within a half hour. Further research is needed to see if the same effect occurs in whole organisms.

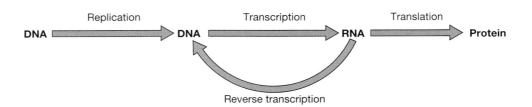

FIGURE 7.3 Information transmission from DNA to protein. As we shall see later, certain viruses, such as the one that causes AIDS, can direct synthesis of DNA from their RNA (reverse transcription).

DNA → Replication → DNA → Transcription → RNA → Translation → Protein

Reverse transcription

acid to another as you might transcribe handwritten sentences to typewritten sentences in the same language. Translation transfers information from the language of nucleic acids to the language of amino acids as you might translate English sentences into another language.

DNA replication, transcription, and translation all transfer information from one molecule to another (Figure 7.3). These processes allow information in DNA to be transferred to each new generation of cells and to be used to control the functioning of cells through protein synthesis.

REPLICATION OF DNA

To understand DNA replication, we need to recall from Chapter 2 that pairs of helical DNA strands are held together by base pairing of adenine with thymine and cytosine with guanine. We also need to know that the ends of each strand are different. At one end, called the 3' (3 prime) end, carbon 3 of deoxyribose is free to bind to other molecules. At the other end, the 5' (5 prime) end, carbon 5 of deoxyribose is attached to a phosphate (Figure 7.1). This structure is somewhat analogous to that of a freight train, with the 3' end the engine and

CLOSE-UP

IF DNA MAKES ONLY PROTEINS, WHO MAKES CARBOHYDRATES AND LIPIDS?

If genetic information in DNA is used specifically to determine the structure of proteins, how are the structures of carbohydrates and lipids determined? Stop and think of the kinds of proteins a cell has. Many are enzymes, and, of course, some of those enzymes direct the synthesis of carbohydrates and lipids. The entire cell is controlled by DNA—either directly, in DNA replication and synthesis of structural proteins, or indirectly, by the synthesis of enzymes that in turn control the synthesis of carbohydrates and lipids.

the 5' end the caboose. When the two strands of a double helix combine by base pairing, they do so in a head-to-tail, or **antiparallel,** fashion. The arrangement of the strands is somewhat like two trains pointed in opposite directions, and base pairing is like passengers in the two trains shaking hands.

DNA replication begins at a specific location in the circular chromosome of a prokaryotic cell and usually proceeds simultaneously in both directions, creating two moving **replication forks,** the points at which the two strands of DNA separate to allow replication of DNA (Figure 7.4). Various enzymes break the hydrogen bonds between the bases in the two DNA strands, unwind the strands from each other, and stabilize the exposed single strands. Molecules of the enzyme **DNA polymerase** then move along behind each replication fork, synthesizing new DNA strands complementary to the original ones at a speed of approximately 1000 nucleotides per second.

The enzyme DNA polymerase can add nucleotides only to the 3' end of a growing DNA strand. Consequently, only one strand, the **leading strand** of original DNA, can serve as template for synthesis of a continuous new strand. Along the other strand, the **lagging strand,** synthesis of new DNA must be *discontinuous;* that is, the polymerase must continually jump ahead and work backward, making a series of short DNA segments called **Okazaki fragments.** The fragments are then joined together by another enzyme called a **ligase.** Ultimately, two separate chromosomes are formed (Figure 7.4).

In newly synthesized chromosomes, each double helix consists of one strand of old, or parent, DNA and one strand of new DNA. Such replication is called **semiconservative replication** because one strand is always conserved (Figure 7.5; page 170). Replication was shown to be semiconservative by the radioisotope experiments of Matthew Meselson and Franklin Stahl in 1958 and was later confirmed by autoradiography experiments (see the box titled "Using Autoradiography"). When the cells were incubated with a radioactive building block, they incorporated radioactive nucleotides into their DNA. All chromosomes were found to contain approximately the same amount of radioactivity, and none lacked radioactivity. Thus, each chromosome was proved to consist of part old and part

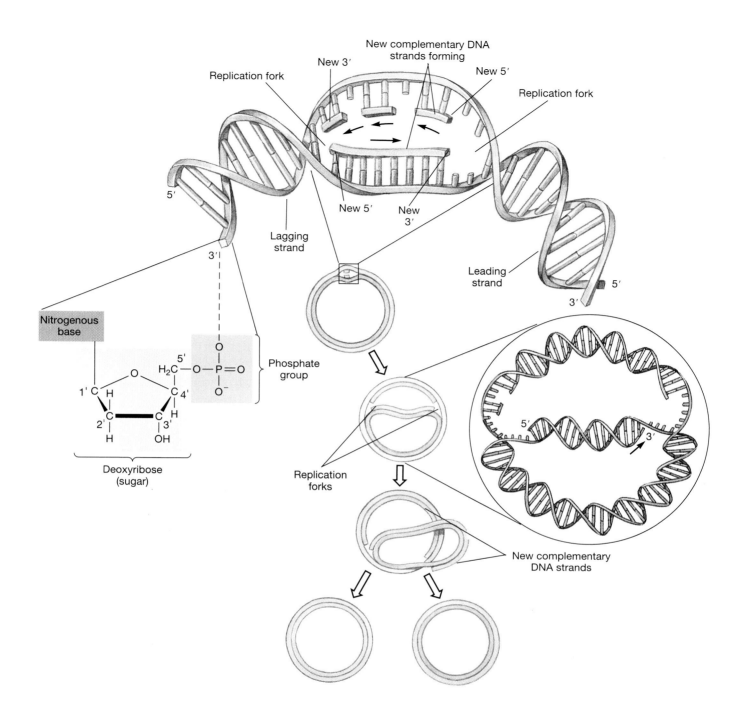

FIGURE 7.4 In the replication of DNA in a prokaryote, DNA strands separate, and replication begins at a replication fork on each strand. As synthesis proceeds, each strand of DNA serves as a template for the replication of its partner. Note the antiparallel arrangement of the complementary strands of the DNA double helix. Because synthesis of new DNA can take place in only one direction, the process must be discontinuous along one strand. Short segments are formed and then spliced together, as the arrows indicate. Each new cell can undergo subsequent replications. But no matter how many daughter cells form in a population, only two of the cells in each generation will have one of the original strands from the mother cell.

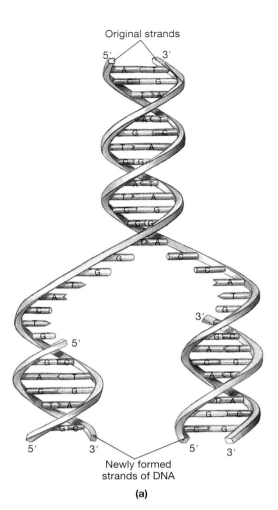

Original strands

5′   3′

3′
5′

5′   3′   5′   3′

Newly formed
strands of DNA

(a)

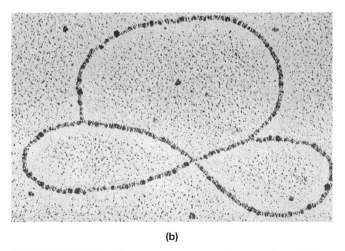

(b)

FIGURE 7.5 (a) Semiconservative replication of DNA. The old strands of DNA are not destroyed but become part of the next generation of DNA molecules. (b) By the technique of autoradiography, an EM photo shows that DNA replication is semiconservative. Both chromosomes contain approximately the same amount of radioactive material from which the new DNA was synthesized. Blackened silver grains in the film emulsion overlie and thereby indicate the radioactive structures.

USING AUTORADIOGRAPHY

Autoradiography can be used to determine where synthetic reactions take place in cells. For example, when dividing cells are incubated with radioactive thymidine, one of the building blocks of DNA, DNA molecules become radioactive wherever they incorporate the radioactive base. After incubation, some cells are placed on microscope slides, dried, and taken into a darkroom, where they are dipped in melted photographic emulsion like that found on camera film. When the emulsion coating is dry, the slides are sealed in a lightproof container that is kept in refrigerated storage for a period of days or weeks. During this time, radioactive emissions from the DNA strike silver grains nearest them in the emulsion. The exposed grains turn black when the slides are subsequently developed with photographic solutions. Finally, the slides are stained and examined under a microscope. Cells are visible under the clear emulsion, with black dots appearing in the emulsion exactly over radioactive cell parts. Thus, we can see where the radioactive thymidine was incorporated.

new DNA. If half the chromosomes had been made completely of old DNA and half of new DNA, the new ones would have contained more radioactivity, and the old ones would have contained no radioactivity.

PROTEIN SYNTHESIS

Transcription

All cells must constantly synthesize proteins to carry out their life processes: reproduction, growth, repair, and regulation of metabolism. This synthesis involves the accurate transfer of linear information of the DNA strands into a linear sequence of amino acids in proteins. To set the stage for protein synthesis, hydrogen bonds between bases in DNA strands are broken enzymatically in certain regions so that the strands separate. Short sequences of unpaired DNA bases are thus exposed to serve as templates in transcription. Only one strand directs the synthesis of mRNA for any one gene; the complementary strand is used as a template during DNA replication or during the transcription of some other gene. Recall that RNA contains the base uracil instead of thymine. ∞ (Chapter 2, p. 44) Thus, when mRNA is transcribed from DNA, uracil pairs with adenine; otherwise, base pairing occurs just as it does in DNA replication. Messenger RNA is formed in the 5′ to 3′ direction.

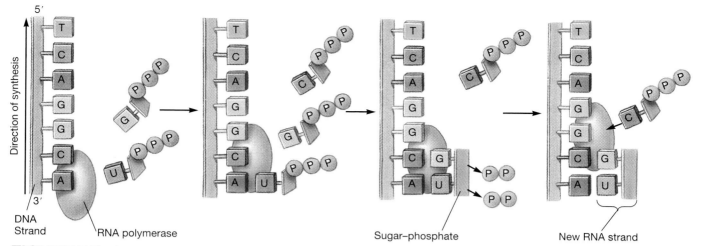

FIGURE 7.6 The transcription of RNA from template DNA. The —PPP represents a triphosphate, and PP represents pyrophosphate. In RNA, U (rather than T) pairs with A.

For transcription to occur, a cell must have sufficient quantities of nucleotides containing high-energy phosphate bonds, which provide energy for the nucleotides to participate in subsequent reactions. The enzyme **RNA polymerase** binds to one strand of exposed DNA. As shown in Figure 7.6, after an enzyme binds to the first base in DNA (adenine, in this case), the appropriate nucleotide joins the DNA base–enzyme complex. The new base then attaches by base pairing to the template base of DNA. The enzyme moves to the next DNA base, and the appropriate phosphorylated nucleotide joins the complex. The phosphate of the second nucleotide is linked to the ribose of the first nucleotide, and *pyrophosphate* (two attached molecules of phosphate) is released. This forms the first link in a new polymer of RNA. Energy to form this link comes from the hydrolysis of ATP and the release of two more phosphate groups. This process is repeated until the RNA molecule is completed.

In eukaryotes transcription takes place in the cell nucleus. The mRNA of transcription must be completely formed and transported through the nuclear envelope to the cytoplasm before translation can begin. Moreover, the mRNA molecule undergoes additional processing before it is ready to leave the nucleus. In eukaryotic cells the regions of genes that code for proteins are called **exons.** Exons are typically separated within a gene by DNA segments that do not code for proteins. Such noncoding *intervening regions* are called **introns.** In the nucleus, RNA polymerase first forms mRNA from the entire gene, including all exons and introns. The newly formed long mRNA molecule is streamlined by other enzymes, which remove the introns and splice together the exons. The resulting mRNA is ready to direct protein synthesis and to leave the nucleus.

Kinds of RNA

Three kinds of RNA—*ribosomal RNA, messenger RNA,* and *transfer RNA*—are involved in protein synthesis. Each RNA consists of a single strand of nucleotides and is synthesized by transcription, using DNA as a template. To complete the story of protein synthesis, we will need more information about these types of RNA.

Ribosomal RNA (rRNA) binds closely to certain proteins to form two kinds of ribosome subunits. A subunit of each kind combines to form a ribosome. Recall that ribosomes are sites of protein synthesis in a cell. ∞ (Chapter 4, p. 84) They serve as binding sites for transfer RNA, and some of their proteins act as enzymes that control protein synthesis. Prokaryotic ribosomes are made of a small (30S) and a large (50S) subunit. (Eukaryotic ribosomes are formed from a 40S and a 60S subunit.) After the two subunits join together around the strand of mRNA (Figure 7.7), peptide synthesis occurs. The newly formed polypeptide chain grows out via a tunnel in the 50S subunit.

Messenger RNA (mRNA) is synthesized in units that contain sufficient information to direct the synthesis of one or more polypeptide chains. One mRNA molecule corresponds to one or more genes, the functional units of DNA. Each mRNA molecule becomes associated with one or more ribosomes. At the ribosome, the information coded in mRNA acts during translation to dictate the sequence of amino acids in the protein.

In translation, each triplet (sequence of three bases) in mRNA constitutes a **codon** (ko'don). Codons are the "words" in the language of nucleic acids. Each codon specifies a particular amino acid or acts as a terminator codon. The first codon in a molecule of mRNA acts as a "start" codon. It always codes for the amino acid methionine, even though the methionine may be re-

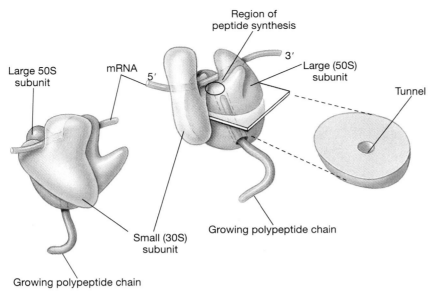

Region of peptide synthesis

mRNA

Large 50S subunit

5'

3'

Large (50S) subunit

Tunnel

Small (30S) subunit

Growing polypeptide chain

Growing polypeptide chain

FIGURE 7.7 Prokaryotic ribosomal structure. The small (30S) and large (50S) subunits are shown from two different angles. The subunits enfold the mRNA strand. The region of peptide synthesis is the junction of these three components. The growing polypeptide chain passes through a tunnel in the 50S subunit, which can be seen in cross section.

moved from the protein later. The last codon in a molecule of mRNA is a *terminator*, or "stop," *codon*. It acts as a kind of punctuation mark to indicate the end of a protein molecule. Using a sentence as an analogy, the methionine codon is the capital letter at the beginning of the sentence, and the terminator codon is the punctuation mark at the end.

At least one codon exists for each of the 20 amino acids found in proteins; several codons exist for some amino acids. The relationship between each codon and a specific amino acid constitutes the **genetic code** (Figure 7.8). Those codons that code for an amino acid are called **sense codons.** Early in the study of the genetic code, investigators found a few codons that did not code for any amino acid. Those codons were therefore named **nonsense codons.** It was later found that they were stop codons. Although genetic information is stored in DNA, the genetic code is written in codons of mRNA. Of course, the information in the codons is derived *directly* from DNA by complementary base pairing during transcription.

Comparisons of the codons among different organisms have shown them to be nearly the same in all organisms, from bacteria to humans. This universality of the genetic code allows research on other organisms to be applied to the understanding of information transmission in human cells. Much of what is known about how the genetic code operates has been learned from research on bacteria.

The function of **transfer RNA (tRNA)** is to transfer amino acids from the cytoplasm to the ribosomes

FIGURE 7.8 The genetic code, with standard three-letter abbreviations for amino acids. To find the amino acid for which the mRNA codon AGU codes, go down the left column to the block labeled A, move across to the fourth square labeled G at the top of the figure, and find the first line in the square labeled U on the right side of the figure. There you will find Ser, the abbreviation for serine. Stop designates a terminator codon. The Start codon is AUG, which also codes for methionine. Therefore, protein synthesis always begins with methionine. The methionine is usually removed later, however, so not all proteins actually start with methionine. When found in the middle of an mRNA strand, AUG codes for methionine.

First position	Second position				Third position
	U	**C**	**A**	**G**	
U	**UUU** Phe	**UCU** Ser	**UAU** Tyr	**UGU** Cys	U
	UUC Phe	**UCC** Ser	**UAC** Tyr	**UGC** Cys	C
	UUA Leu	**UCA** Ser	**UAA** *Stop*	**UGA** *Stop*	A
	UUG Leu	**UCG** Ser	**UAG** *Stop*	**UGG** Trp	G
C	**CUU** Leu	**CCU** Pro	**CAU** His	**CGU** Arg	U
	CUC Leu	**CCC** Pro	**CAC** His	**CGC** Arg	C
	CUA Leu	**CCA** Pro	**CAA** Gln	**CGA** Arg	A
	CUG Leu	**CCG** Pro	**CAG** Gln	**CGG** Arg	G
A	**AUU** Ile	**ACU** Thr	**AAU** Asn	**AGU** Ser	U
	AUC Ile	**ACC** Thr	**AAC** Asn	**AGC** Ser	C
	AUA Ile	**ACA** Thr	**AAA** Lys	**AGA** Arg	A
	AUG Met	**ACG** Thr	**AAG** Lys	**AGG** Arg	G
G	**GUU** Val	**GCU** Ala	**GAU** Asp	**GGU** Gly	U
	GUC Val	**GCC** Ala	**GAC** Asp	**GGC** Gly	C
	GUA Val	**GCA** Ala	**GAA** Glu	**GGA** Gly	A
	GUG Val	**GCG** Ala	**GAG** Glu	**GGG** Gly	G

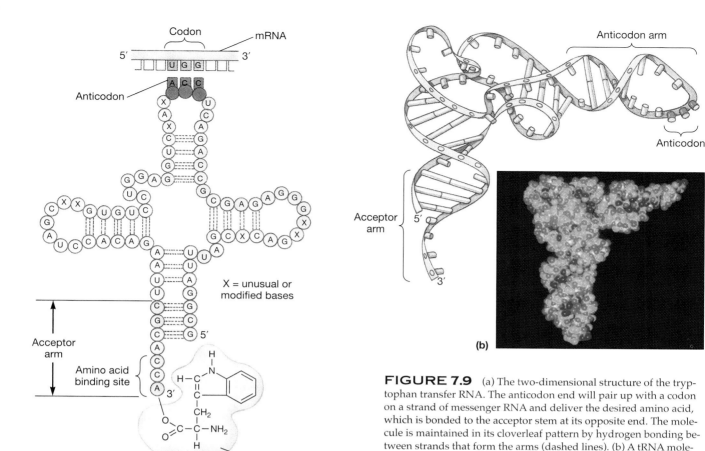

(a)

X = unusual or modified bases

Amino acid (Trp)

FIGURE 7.9 (a) The two-dimensional structure of the tryptophan transfer RNA. The anticodon end will pair up with a codon on a strand of messenger RNA and deliver the desired amino acid, which is bonded to the acceptor stem at its opposite end. The molecule is maintained in its cloverleaf pattern by hydrogen bonding between strands that form the arms (dashed lines). (b) A tRNA molecule folded into its complex three-dimensional shape, in diagram form and as a computer-generated model.

for placement in a protein molecule. Many different kinds of tRNAs have been isolated from the cytoplasm of cells. A tRNA molecule consists of 75 to 80 nucleotides and is folded back on itself to form several loops that are stabilized by complementary base pairing (Figure 7.9). Each tRNA has a three-base **anticodon** (an'ti-ko"don) that is complementary to a particular mRNA codon. It also has a binding site for an amino acid—the particular amino acid specified by the mRNA codon. (The mRNA codon, of course, got its information directly from DNA.) Thus, the tRNAs are the link between the codons and the corresponding amino acids. Amino acid attachment to specific tRNA molecules is achieved by the action of amino acid-activating enzymes and energy derived from ATP.

The anticodon attaches by complementary base pairing to the appropriate mRNA codon such that its amino acid is aligned for incorporation into a protein. The accuracy of amino acid placement in protein synthesis depends on this precise pairing of codons and anticodons. The properties of the three types of RNA are summarized in Table 7.1.

TABLE 7.1	Properties of Kinds of RNA
Kind of RNA	**Properties**
Ribosomal	Combines with specific proteins to form ribosomes.
	Serves as a site for protein synthesis.
	Associated enzymes function in controlling protein synthesis.
Messenger	Carries information from DNA for synthesis of a protein.
	Molecules correspond in length to one or more genes in DNA.
	Has base triplets called codons that constitute the genetic code.
	Attaches to one or more ribosomes.
Transfer	Found in the cytoplasm, where they pick up amino acids and transfer them to mRNA.
	Molecules have an attachment site for a specific amino acid.
	Each has a single triplet of bases called an anticodon, which pairs complementarily with the corresponding codon in mRNA.

Translation

Protein synthesis, an important process in bacterial growth, uses 80 to 90 percent of a bacterial cell's energy. Generally, during protein synthesis, the various RNAs and amino acids are available in sufficient quantities. The RNAs can be reused many times before they lose their ability to function. Of the types of RNA, mRNA is produced in the most precise quantity in accordance with the cell's need for a particular protein. Figure 7.10 shows the three types of RNA and how they function in protein synthesis.

Once an mRNA molecule has been transcribed and has combined with a ribosome, the ribosome initiates protein synthesis and provides the site for protein assembly. Each ribosome attaches first to the end of the mRNA that corresponds to the beginning of a protein.

BIOTECHNOLOGY

RIBOZYMES

The American biochemist Thomas Cech, 1989 corecipient of the Nobel Prize for chemistry, recently patented pieces of RNA that he called *ribozymes*. Ribozymes act like enzymes to cut selected pieces out of strands of viral RNA, thereby inactivating the virus. Like enzymes, ribozymes are not used up in the reaction. Therefore, they can go on to snip up one virus after the next until all the viruses are neutralized. Cech's work has been very successful in laboratory settings. However, the human body is far more complicated than a culture flask of viruses, and it may be some time before ribozymes can be used therapeutically.

FIGURE 7.10   (a) Transcription from DNA to RNA. (b) Translation from RNA to protein. Many ribosomes that are connected to and read the same piece of mRNA are called a polyribosome.

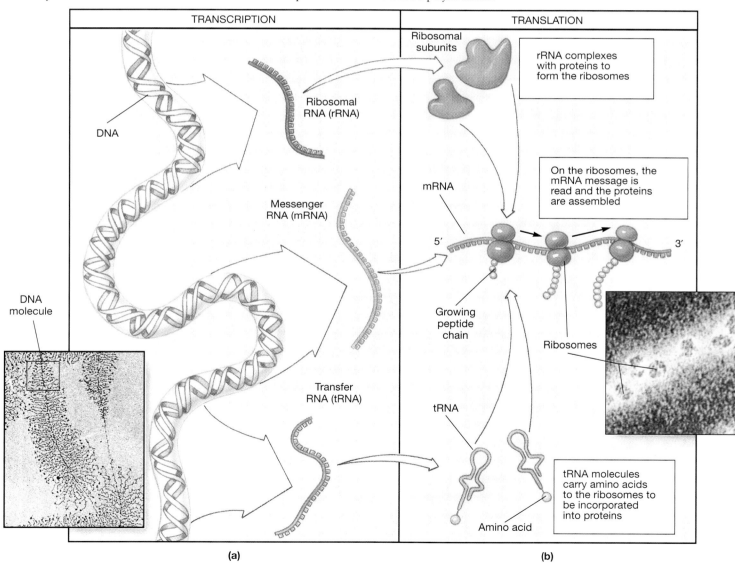

(a)   (b)

The length of each polypeptide chain extending from a ribosome corresponds to the amount of mRNA the ribosome has "read." Several ribosomes can be attached at different points along an mRNA molecule to form a **polyribosome** (or *polysome*).

The main steps in protein synthesis (Figure 7.11) can be summarized as follows: The process begins when a molecule of mRNA becomes properly oriented on a ribosome. As each codon of the mRNA is read, the appropriate tRNA combines with it and thereby delivers a particular amino acid to the protein assembly site. The location on the ribosome where the first tRNA pairs is called the *P site.* The second codon of the mRNA then pairs with a tRNA that transports the second amino acid to the *A site,* which is next to the P site. Matching of codon and anticodon by base pairing allows coded information in mRNA to specify the sequence of amino acids in a protein. Any tRNAs with nonmatching anticodons simply do not bind to the ribosome. As amino acids are delivered one after another and peptide bonds form between them, the length of the polypeptide chain increases. This process continues until the ribosome recognizes a stop codon. The three codons (designated in Figure 7.8 by the word *stop*) that signal the end of the information for a protein are called **terminator codons** (or *nonsense codons*). When the ribosome reads a terminator codon at the A site, it releases the finished protein from the P site.

Any mRNA molecule can direct simultaneous synthesis of many identical protein molecules—one for each ribosome passing along it. Ribosomes, mRNAs, and tRNAs are reusable. The tRNAs shuttle back and forth between the cytoplasm, where they pick up amino acids, and the ribosome, where the amino acids are incorporated into protein.

REGULATION OF METABOLISM

Significance of Regulatory Mechanisms

Bacteria use most of their energy to synthesize substances needed for growth. These substances include structural proteins, which form cell parts, and enzymes, which control both energy production and synthetic reactions. The survival of bacteria depends on their ability to grow even when conditions are less than ideal—for example, when nutrients are in short supply. In their evolution, cells of bacteria (and all other organisms) have developed mechanisms to turn reactions on and off in accordance with their needs. Energy and materials are too valuable to waste. Also, the cell has a limited amount of space for storing excesses of materials it synthesizes. Thus, cells use energy to synthesize substances in the amounts needed and shut off these processes before wasteful excesses are produced.

All living organisms are presumed to have control mechanisms that regulate their metabolic activities. However, more research on control mechanisms has been done on bacteria than on all other organisms. Bacteria are ideal for such studies for several reasons:

1. They can be grown in large numbers relatively inexpensively under a variety of controlled environmental conditions.
2. They produce many new generations quickly.
3. Because they reproduce so rapidly, a variety of mutations can be observed in a relatively short time.

Mutant organisms that have an alteration in their control mechanisms can be isolated and studied along with nonmutated organisms to understand better the operation of control mechanisms.

Categories of Regulatory Mechanisms

The mechanisms that control metabolism either regulate enzyme activity directly or regulate enzyme synthesis by turning on or off genes that code for particular enzymes. Of the various mechanisms that regulate metabolism, three have been extensively investigated in bacteria. Where enzyme activity is regulated directly, the control mechanism determines how rapidly enzymes already present will catalyze reactions. *Feedback inhibition* is an example of such regulation of enzyme activity. Where regulation occurs indirectly by enzyme synthesis, the control mechanism determines which enzymes will be synthesized and in what amounts. *Enzyme induction* and *enzyme repression* are examples of such regulation of gene expression.

Feedback Inhibition

In **feedback inhibition,** also called **end-product inhibition,** the end product of a biosynthetic pathway directly inhibits the first enzyme in the pathway. This mechanism was discovered when it was observed that the addition of one of several amino acids to a growth medium could cause a bacterium suddenly to stop synthesizing that particular amino acid. Synthesis of the amino acid threonine, for example, is regulated by feedback inhibition. Threonine is made from aspartate, and the allosteric enzyme that acts on aspartate is inhibited by threonine (Figure 7.12; page 177). (Aspartate is derived from oxaloacetate formed in the Krebs cycle.) When an inhibitor (threonine) attaches to the allosteric site, it alters the enzyme's shape so the substrate (aspartate) cannot attach to the active site. ∞ (Chapter 5, p. 114) Thus, feedback inhibition occurs when the end product of a reaction sequence binds to the allosteric site of the enzyme for the first step in the sequence.

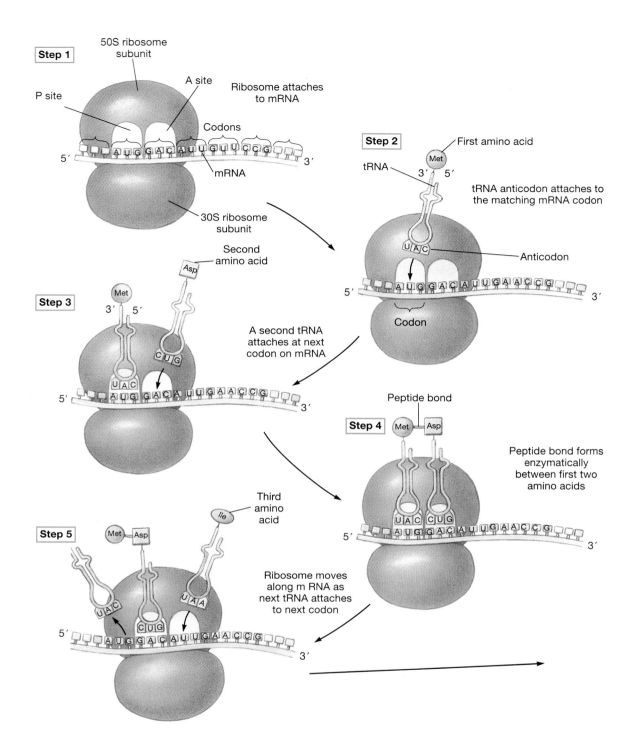

FIGURE 7.11 Steps 1–5: The main steps in protein synthesis. Steps 6 and 7 (page 177): Many ribosomes can "read" the same strand of mRNA simultaneously. The ribosomes are shown moving from left to right.

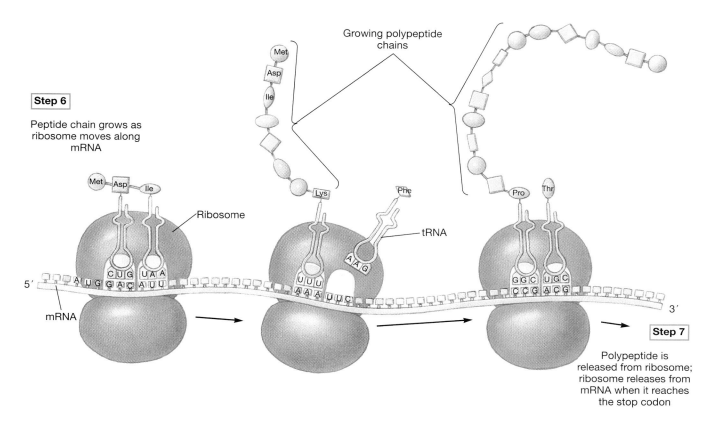

Step 6

Peptide chain grows as ribosome moves along mRNA

Growing polypeptide chains

Step 7

Polypeptide is released from ribosome; ribosome releases from mRNA when it reaches the stop codon

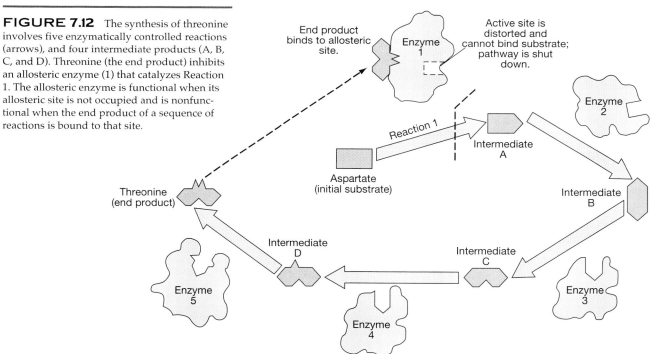

FIGURE 7.12 The synthesis of threonine involves five enzymatically controlled reactions (arrows), and four intermediate products (A, B, C, and D). Threonine (the end product) inhibits an allosteric enzyme (1) that catalyzes Reaction 1. The allosteric enzyme is functional when its allosteric site is not occupied and is nonfunctional when the end product of a sequence of reactions is bound to that site.

Feedback inhibition regulates the synthesis of various substances other than amino acids (pyrimidines, for example). This regulatory mechanism also occurs in many organisms other than bacteria. Because feedback inhibition acts quickly and directly on a metabolic process, it allows the cell to conserve energy in two ways:

1. When it is plentiful, the inhibitor (end product) attaches to the enzyme; when it is in short supply, it is released from the enzyme. Thus, the cell expends energy to synthesize the end product only when it is needed.

2. Regulation of enzyme activity requires less energy than the more complex processes involved in the regulation of gene expression.

Enzyme Induction

At one point in the investigation of metabolic regulation, it was discovered that certain organisms always contain active enzymes for glucose metabolism even when glucose is not present in the medium. Such enzymes are called **constitutive enzymes;** they are synthesized continuously regardless of the nutrients available to the organism. The genes that make these enzymes are always active. In contrast, enzymes that are synthesized by genes that are sometimes active and sometimes inactive depending on the presence or absence of substrate are called **inducible enzymes.**

When bacteria such as *E. coli* are grown on a nutrient medium that contains no lactose, the cells do not make any of the enzymes that they would need to utilize lactose as an energy source. When lactose is present, however, the cells synthesize the enzymes needed for its metabolism. This phenomenon is an example of **enzyme induction.** Enzyme induction controls the breakdown of nutrients as they become available in the growth medium. Such a system is turned on when a nutrient is available and turned off when it is depleted. The nutrient itself acts as an **inducer** of enzyme production.

The *operon* (op'er-on) *theory,* a model that explains the regulation of protein synthesis in bacteria, was proposed in 1961 by French scientists Francois Jacob and Jacques Monod, who received a Nobel Prize in 1965 for their work. Although the theory applies to several operons, we will illustrate it with the *lac* operon, which regulates lactose metabolism. An **operon** is a sequence of closely associated genes that regulate enzyme production. An operon includes one or more **structural genes,** which carry information for the synthesis of specific proteins such as enzyme molecules, and **regulatory sites,** which control the expression of the structural genes. A **regulator** (*i*) **gene** works in conjunction with the operon but may be located some distance from it. In prokaryotes several structural genes are controlled by one operon—a more efficient method than that of eukaryotes, in which each gene is controlled by its own regulatory site.

The *lac* operon (Figure 7.13) consists of regulatory sites called a *promoter* and an *operator,* and three structural genes *Z, Y,* and *A,* which direct synthesis of specific enzymes. An RNA polymerase molecule must bind to the promoter before transcription can begin. When lactose is not present in the medium, the separate *i* gene directs synthesis of a substance called the *lac repressor.* The **repressor** is a protein that binds to the operator and prevents transcription of the adjacent *Z, Y,* and *A* genes. Consequently, the enzymes that metabolize lactose are not synthesized. The *i* gene represents a *constitutive gene*—it is always undergoing protein synthesis to produce more repressor protein and is not controlled by the promoter.

APPLICATIONS

INDUCIBLE ENZYMES

Inducible enzyme systems play an important role in humans as well as in microbes. Patients suspected of having diabetes mellitus are usually sent for a glucose tolerance test, in which they are fed a measured load of glucose. The subsequent rise and fall of blood glucose level is measured. Failure to remove glucose from the bloodstream at the normal rate *could* indicate diabetes—but could it be something else? Patients fearing a diagnosis of diabetes sometimes plan to "beat" the test: They won't eat anything sugary for days before the test, for they don't want any excess sugar around on that day. But they also aren't going to have very good supplies of enzymes to utilize the test load of sugar. These enzymes are inducible, and the patients haven't been eating very much inducer (sugar, in this case). They can't instantly produce the enzymes, and so their blood glucose level will drop more slowly than normal. A patient should eat at least the amount of sugar in a candy bar each day for the three days preceding the test to induce as much enzyme as possible the day of the test.

When present in the medium, lactose acts as the inducer by binding to and inactivating the *lac* repressor. The repressor then no longer blocks the operator. The RNA polymerase then binds to the promoter and transcribes the *Z, Y,* and *A* genes as a single long strand of mRNA. This mRNA becomes associated with ribosomes and directs synthesis of three enzymes: β-galactosidase (*Z* gene), permease (*Y* gene), and transacetylase (*A* gene). Permease transports lactose into cells, and β-galactosidase breaks down lactose into glucose and galactose. The role of the transacetylase is under debate. When the available lactose has been broken down, it is no longer available to bind to the repressor. The active repressor again binds to the operator, and the operon is turned off.

Enzyme Repression

In contrast to enzyme induction, which typically regulates catabolism, **enzyme repression** typically regulates anabolism. It controls processes in which substances needed for growth are synthesized. Synthesis of the amino acid tryptophan, for example, is regulated by enzyme repression through actions of the *trp* operon, which consists of five structural genes.

When tryptophan is available to a bacterial cell, the amino acid binds to an inactive repressor. Binding activates the repressor protein, which can now bind to the promoter and repress synthesis of the enzymes needed to make tryptophan. When tryptophan is not available, the repressor protein remains inactive, and

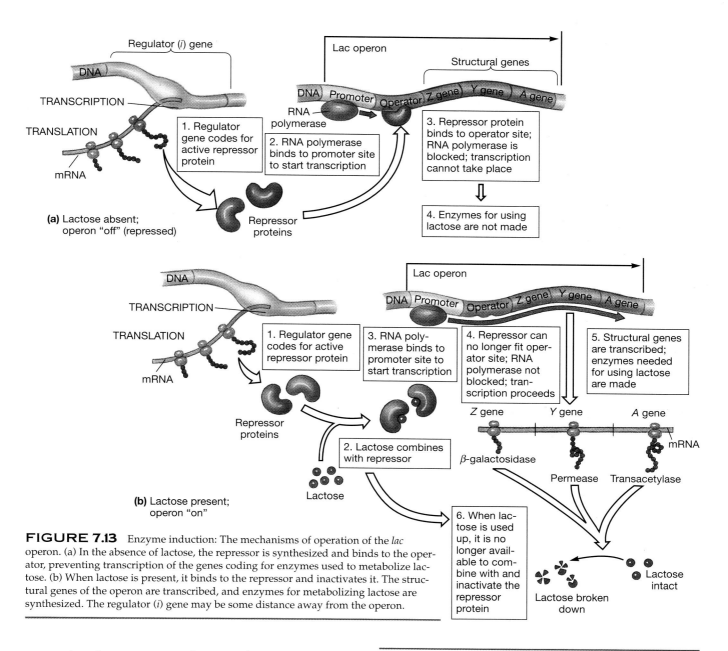

FIGURE 7.13 Enzyme induction: The mechanisms of operation of the *lac* operon. (a) In the absence of lactose, the repressor is synthesized and binds to the operator, preventing transcription of the genes coding for enzymes used to metabolize lactose. (b) When lactose is present, it binds to the repressor and inactivates it. The structural genes of the operon are transcribed, and enzymes for metabolizing lactose are synthesized. The regulator (*i*) gene may be some distance away from the operon.

repression does not occur. Structural genes are transcribed, and tryptophan is synthesized. When tryptophan becomes plentiful, it again represses the operon. An even finer control mechanism, called **attenuation,** allows transcription of the *trp* operon to begin but terminates it prematurely by a complex process when sufficient amounts of tryptophan are already present in the cell. Several operons, especially those for amino acid synthesis, have attenuation mechanisms.

A slightly different kind of repression operates in connection with some catabolic pathways. When certain bacteria (*E. coli,* for example) are grown on a nutrient medium containing both glucose and lactose, they grow at a logarithmic rate as long as glucose is available. When the glucose is depleted, they enter a stationary phase but soon begin to grow again at a logarithmic rate, though not quite as rapidly (Figure 7.14).

FIGURE 7.14 The growth curve for bacteria in a medium initially containing both glucose and lactose. When glucose is used up, growth stops temporarily but begins again at a slower rate, using lactose as an energy source.

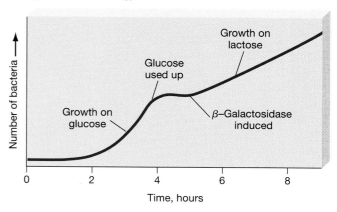

Regulatory Mechanism (example)	Type of Pathway Regulated	Regulating Substance	Condition That Leads to Gene Expression
Enzyme induction (*lac* operon)	Catabolic (degradational) and releases energy	Nutrient (lactose)	Presence of nutrient (lactose)
Enzyme repression (*trp* operon)	Anabolic (biosynthetic) and uses energy	End product (tryptophan)	Absence of end product (tryptophan)

This time the logarithmic growth rate results from the metabolism of lactose. The stationary phase is the period during which the enzymes needed to utilize lactose are being synthesized.

Why was the synthesis of these enzymes not induced before the glucose was depleted, since lactose was present in the medium from the start? Because bacteria use glucose as a nutrient with high efficiency. The enzymes for metabolizing glucose, being constitutive, are always present in the cell. Thus, when glucose is abundant, there is no advantage in making enzymes for metabolizing lactose even if lactose is also available. Consequently, the *lac* operon that we described previously is repressed when glucose is present in adequate quantities, an effect known as **catabolite repression.** In this way the cell saves energy by not making enzymes it doesn't need. When glucose supplies fall, the repression is lifted, the *lac* operon genes are transcribed, and the cell is ready to switch over to using lactose. In short, the transcription of the *lac* operon requires both that lactose be present and that glucose *not* be present.

Both enzyme induction and enzyme repression are regulatory mechanisms that control enzyme production by altering gene expression. Although these two mechanisms have different effects, they actually represent two examples of the operation of a single mechanism for turning genes on and off (Table 7.2).

MUTATIONS

We can now define mutations, or changes in DNA, more precisely as heritable changes in the sequence of nucleotides in DNA. Mutations account for evolutionary changes in microorganisms (and larger organisms) and for alterations that produce different strains within species. Here we will consider what kinds of changes occur in DNA during mutations and how these changes affect the organisms.

Types of Mutations and Their Effects

Before we can consider mutations and their effects, we need to review the differences between an organism's genotype and its phenotype. Genotype refers to the genetic information contained in the DNA of the organism. Phenotype refers to the specific characteristics displayed by the organism. Mutations always change the genotype. Such a change may or may not be expressed in the phenotype, depending on the nature of the mutation.

Two important kinds of mutations are *point mutations,* which affect a single base, and *frameshift mutations,* which can affect more than one base in DNA. Mutations can make an organism unable to synthesize one or more proteins. The absence of a protein may lead to changes in the organism's structure or in its ability to metabolize a particular substance.

A **point mutation** involves base substitution, or nucleotide replacement, in which one base is substituted for another at a specific location in a gene (Figure 7.15). The mutation changes a single codon in mRNA, and it may or may not change the amino acid sequence in a protein. Let's look at some examples.

FIGURE 7.15 The effects of base substitution (a point mutation). The resulting protein may or may not be significantly affected, depending on whether the new codon specifies the same amino acid, one with similar properties, or an entirely different one.

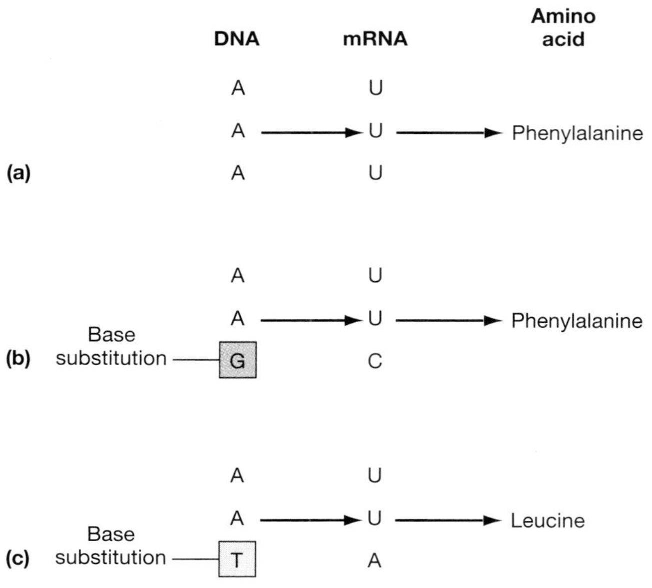

Suppose that a three-base sequence of DNA is changed from AAA to AAG. During transcription the mRNA code becomes UUC instead of UUU. (Recall that uracil in RNA pairs with adenine in DNA.) ∞ (Chapter 2, p. 44) Because both the UUC and UUU codons code for phenylalanine, the mutation has no effect on the protein being synthesized. In this case, though the genotype has changed, the phenotype is unaffected.

If the sequence in DNA is changed from AAA to AAT, the mRNA codon will change from UUU to UUA. When the information in the mRNA is used to synthesize protein, the amino acid leucine will be substituted for phenylalanine in the protein. (To verify this for yourself, refer to the genetic code in Figure 7.8.) Because of the single amino acid substitution, the new protein will be different from the normal protein. The effects on the phenotype of the organism will be negligible if the new protein functions as well as the original one. The effects will be significant if the new protein functions poorly or not at all. In rare instances the new protein may function better and produce a phenotype that is better adapted to its environment than the original phenotype.

Sometimes the substitution of a single base in DNA produces a terminator codon in mRNA. If the terminator codon is introduced in the middle of a molecule of mRNA destined to produce a single protein, synthesis will be terminated part way through the molecule. A polypeptide that will most likely be unable to function in the cell will be released, and the appropriate protein will not be synthesized. If the missing protein is essential to cell structure or function, the effect can be lethal.

A **frameshift mutation** is a mutation in which there is a **deletion** or an **insertion** of one or more bases (Figure 7.16). Such mutations alter all the three-base sequences beyond the deletion or insertion. When mRNA transcribed from such altered DNA is used to synthesize a protein, many amino acids in the sequence may be altered. (Remember, a ribosome reads an mRNA in codons, sets of three bases.) Such mutations also commonly introduce terminator codons and cause protein synthesis to stop when only a short polypeptide has been made. Frameshift mutations usually prevent synthesis of a particular protein, and they change both the genotype and the phenotype. Their effect on the organism depends on the role of the missing protein in the organism's function. These types of mutations and their effects are summarized in Table 7.3. If three bases, or a multiple of three bases, were inserted or lost, what would happen? One or more amino acids would be gained or lost.

Phenotypic Variation

Phenotypic variations frequently seen in mutated bacteria include alterations in colony morphology, colony color, or nutritional requirements. For example, instead of being a normal smooth, glossy, raised colony, a colony with mutated DNA may have a flat, rough appearance. The mutation has impaired synthesis of certain cell surface

FIGURE 7.16 The effects of frameshift mutations. The addition or deletion of one or more nucleotides changes the amino acid sequence coded for by the entire gene from that point on. (The addition or deletion of three nucleotides might not affect the resulting protein very much. Can you see why?)

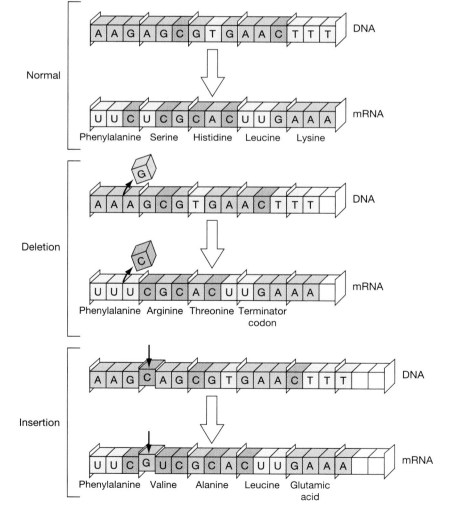

TABLE 7.3	Types of Mutations and Their Effects on Organisms	

Types of Mutations	Effects on Organisms
Point mutation	
Single base change in DNA with no change in the amino acid specified by the mRNA codon.	No effect on protein; a "silent" mutation.
Change in DNA with change in the amino acid sequence specified by the mRNA codon.	Change in protein by substitution of one amino acid for another; can significantly alter function of protein.
Change in DNA that creates a terminator codon in mRNA.	Produces polypeptide of no use to organism and prevents synthesis of normal protein.
Frameshift mutation	
Deletion or insertion of one or more bases in DNA.	Changes entire sequence of codons and greatly alters amino acid sequence; can introduce terminator codon and produce useless polypeptides instead of normal proteins.

substances. In organisms that typically form capsules, the mutation can prevent synthesis of capsular polysaccharides. Mutations that alter nutritional requirements generally increase the nutritional needs of an organism, usually by impairing the organism's ability to synthesize one or more enzymes. As a result, the organism may require certain amino acids or vitamins in its medium because it can no longer make them itself.

Studies of bacteria that have lost the ability to synthesize a particular enzyme have played an important role in our understanding of metabolic pathways. Such nutritionally deficient mutants are called **auxotrophs** (awks'o-trofs"; *auxo,* increase, and *trophos,* food); they require special substances in their medium to maintain growth. In contrast to auxotrophs, normal nonmutant forms are called **prototrophs** (pro'to-trofs), or *wild types.* Comparisons of characteristics of auxotrophs and prototrophs show the effects of a mutation on metabolism. By observing which metabolites accumulate and which nutrients must be added to the medium of auxotrophs, microbiologists have determined the specific steps in the metabolism of certain substances.

Still another type of phenotypic variation of genetic origin is temperature sensitivity. For example, suppose that an organism at one time could grow over

a wide range of environmental temperatures. As a result of a mutation, it loses the ability to grow at the higher temperatures of its former range. It can still grow at 25° C, but it can no longer grow at 40° C. This phenomenon may be due to a point mutation that changed a single amino acid in an enzyme. The slightly altered enzyme may function at moderate temperatures but may be easily denatured and inactivated at higher temperatures.

Some phenotypic variations are caused by environmental factors and occur without any change in the genotype (alteration in DNA). For example, large amounts of sugar or irritants in the medium can cause some organisms to form a larger-than-normal capsule. Some organisms, such as the anthrax bacterium, form spores in open air, in spilled blood, or on tissue surfaces but not inside tissues. Variations in environmental temperature can affect pigment synthesis. *Serratia marcescens* usually produces pigment at room temperature but may not do so at higher temperatures. It has the gene for pigment production, but the gene is expressed only at certain temperatures.

Spontaneous versus Induced Mutations

Mutations appear to be random or chance events; it is usually impossible to predict when a mutation will occur or which genes will be altered. Although all mutations result from permanent changes in DNA, they can be spontaneous or induced. **Spontaneous mutations** occur in the absence of any agent known to cause changes in DNA. They arise during the replication of DNA and appear to be due to errors in the base pair-

CLOSE-UP

TURNING GENES ON AND OFF

In organisms, phenotypic variations that are not due to genotypic variations—changes in DNA—are not heritable. For example, humans whose hair loses pigment and turns gray with aging still have the genes for hair color they received from their parents and can pass on those genes on to their offspring. The genes for hair color that are not expressed in pigment formation in the aging adults will be expressed in their offspring—the children will not be born with gray hair. Similarly, genes for making capsules, spores, or pigments in bacteria fail to be expressed under certain conditions. During an organism's life, regulatory mechanisms turn on and off the expression of certain genes, but the genes are still transmitted to that organism's offspring.

| TABLE 7.4 | Some Mutagens and Their Effects |

Mutagen	Effects
Chemical Agents	
Base analog *Examples:* caffeine, 5-bromouracil	Substitutes "look-alike" molecule for the normal nitrogenous base during DNA replication $\longrightarrow$ point mutation.
Alkylating agent *Example:* nitrosoguanidine	Adds an alkyl group, such as methyl group ($-CH_3$), to nitrogenous base, resulting in incorrect pairing $\longrightarrow$ point mutation.
Deaminating agent *Example:* nitrous acid	Removes an amino group ($-NH_2$) from a nitrogenous base $\longrightarrow$ point mutation.
Acridine derivative *Examples:* acridine dyes, quinacrine	Inserts into DNA ladder between backbones to form a new rung, distorting the helix $\longrightarrow$ frameshift mutation.
Radiation	
Ultraviolet	Links adjacent pyrimidines to each other, as in thymine dimer formation, and thereby impairs replication.
X-ray and gamma ray	Ionize and break molecules in cells to form free radicals, which in turn break DNA.

ing of nucleotides in the old and new strands of DNA. Various genes in the DNA of bacteria have different spontaneous *mutation rates,* ranging from 10^{-3} to 10^{-9} per cell division. In other words, one gene might undergo a spontaneous mutation once in every thousand ($1/10^3$) cell divisions, whereas another gene might undergo a spontaneous mutation only once in every billion ($1/10^9$) cell divisions. **Induced mutations** are produced by agents called **mutagens,** which increase the mutation rate above the spontaneous mutation rate. Mutagens include chemical agents and radiation (Table 7.4).

Chemical Mutagens

Chemical mutagens act at the molecular level to alter the sequence of bases in DNA. They include base analogs, alkylating agents, deaminating agents, and acridine derivatives.

A **base analog** is a molecule quite similar in structure to one of the nitrogenous bases normally found in DNA. A cell may incorporate a base analog into its DNA in place of the normal base. For example, 5-bromouracil can be inserted in DNA instead of thymine (Figure 7.17). When DNA containing 5-bromouracil is replicated, the analog can cause an error in base pairing. The 5-bromouracil that replaced thymine may pair with guanine instead of with adenine, which normally pairs with thymine. When DNA is replicated in the presence of a significant quantity of 5-bromouracil, the analog can be incorporated at many sites in the DNA molecule. A mutation occurs wherever the analog causes the insertion of guanine instead of adenine in the subsequent replication. Another purine base analog, caffeine, can cause mutations in an unborn child. For this reason pregnant women are advised to avoid or limit their caffeine intake.

Alkylating agents are substances that add alkyl groups (such as a methyl group, $-CH_3$) to other molecules. Adding an alkyl group to a nitrogenous base alters the shape of the base and can cause an error in base pairing. For example, the addition of a methyl group to guanine can cause it to pair with thymine instead of cytosine. Such a change can give rise to a point mutation.

Deaminating agents such as nitrous acid (HNO_2) remove an amino group ($-NH_2$) from a nitrogenous base. Removing an amino group from adenine causes it to resemble guanine, and the deaminated base pairs with cytosine instead of thymine. Nitrates (NO_3^-) and nitrites (NO_2^-) are sometimes added to foods such as hot dogs and cold cuts for coloring, flavoring, or antibacterial action. The hazard of such additives is that, in the body, they form nitrosamines—deaminating agents known to cause birth defects, cancer, and other mutations in laboratory animals.

In contrast to these alterations, which cause point mutations, **acridine derivatives** cause frameshift mutations. The acridine molecule contains one pyrimidine

FIGURE 7.17 Similarity of the structure of the base analog 5-bromouracil to the structure of the normal base thymine allows it in some cases to be taken up in place of thymine. The bromine (Br) group occupies an area about the same size as the methyl (CH_3) group.

Thymine   5-Bromouracil

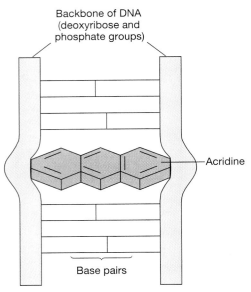

FIGURE 7.18 Insertion of acridine into a DNA helix can produce a frameshift mutation.

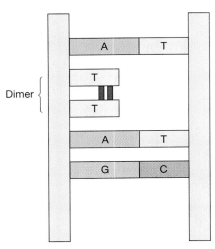

FIGURE 7.19 The formation of a dimer prevents the affected bases from pairing with bases in the complementary chain of DNA, impairing replication and preventing transcription.

ring and two benzene rings (Figure 7.18). This molecule or one of its derivatives can become inserted in the DNA double helix, displacing both members of a base pair. Such a modification distorts the helix and causes partial unwinding of the DNA strands. The distortion allows one or more bases to be added or deleted, and a frameshift mutation results. The drug quinacrine (Atabrine) is an acridine derivative that was used to treat malaria until other drugs with less unpleasant side effects were developed. It causes mutations in the malarial parasite and possibly in the human host that receives the drug.

Radiation as a Mutagen

Radiation such as X-rays and ultraviolet rays can act as mutagens. Ultraviolet rays affect only the skin of humans because the rays lack energy for deeper penetration, but they have significant effects on microorganisms, which they penetrate easily. Ultraviolet lights are sometimes mounted in hospitals and laboratories to kill airborne bacteria. When ultraviolet rays strike DNA, they can cause pyrimidine dimers to form. A **dimer** (di′mer) consists of two adjacent pyrimidines (two thymines, two cytosines, or thymine and cytosine) bonded together in a DNA strand (Figure 7.19). Binding of pyrimidines to each other prevents base pairing during replication of the adjacent DNA strand, so a gap is produced in the replicated DNA. Transcription of mRNA stops at the gap, and the affected gene fails to transmit information.

X-rays and gamma rays, which are more energetic than ultraviolet radiation, easily break chemical bonds in molecules. ∞ (Chapter 3, p. 54) The product is often a *free radical,* a highly reactive atom, molecule, or ion that in turn attacks other cell molecules, including DNA.

Until recently, microbiologists had no control over which genes underwent mutation when organisms were treated with mutagens. Now certain enzymes are available that greatly facilitate such studies. **Restriction endonucleases** cut DNA at precise base sequences, and **exonucleases** remove segments of DNA. These enzymes allow individual genes to be isolated and mutated at predetermined sites. The mutated gene can be inserted into a host's chromosome, and the effect of the specific mutation studied.

Repair of DNA Damage

Many bacteria, and other organisms as well, have enzymes that can repair certain kinds of damage to DNA. Two mechanisms, *light repair* and *dark repair,* are known to repair damage caused by dimers.

Light repair, or **photoreactivation,** occurs in the presence of visible light in bacteria previously exposed to ultraviolet light. When organisms containing dimers are kept in visible light, the light activates an enzyme that breaks the bonds between the pyrimidines of a dimer (Figure 7.20a). Thus, mutations that might have been passed along to daughter cells are corrected, and the DNA is returned to its normal state. This mechanism contributes to the survival of the bacteria but creates a problem for microbiologists. Cultures that are irradiated with ultraviolet light to induce mutations must be kept in the dark for the mutations to be retained.

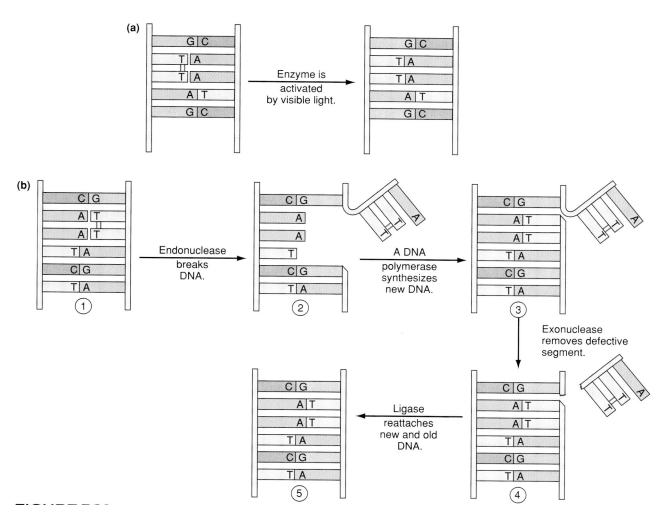

(a)

Enzyme is activated by visible light.

(b)

Endonuclease breaks DNA.

A DNA polymerase synthesizes new DNA.

Exonuclease removes defective segment.

Ligase reattaches new and old DNA.

① ② ③ ④ ⑤

FIGURE 7.20 (a) Light repair of DNA (photoreactivation) removes dimers. (b) In dark repair, a defective segment of DNA is cut out and replaced.

Dark repair, which occurs in some bacteria, requires several enzyme-controlled reactions (Figure 7.20b). First, an endonuclease breaks the defective DNA strand near the dimer. Second, a DNA polymerase synthesizes new DNA to replace the defective segment, using the normal complementary strand as a template. Third, an exonuclease removes the defective DNA segment. Finally, a ligase connects the repaired segment to the remainder of the DNA strand. These reactions were identified in *E. coli* but are now known to occur in many other bacteria. Human cells have similar mechanisms; some human skin cancers, such as xeroderma pigmentosum (Figure 7.21), are due to a defect in the cellular DNA repair mechanism.

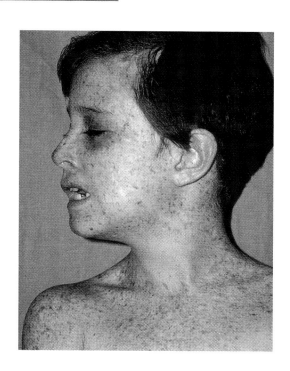

FIGURE 7.21 Xeroderma pigmentosum is a human genetic disease in which the enzymes to repair UV damage to DNA are defective, and exposure to sunlight results in multiple skin cancers.

OZONE BIOSENSORS

Because ozone filters out harmful ultraviolet radiation, the discovery of holes in the ozone layer of the earth's atmosphere has raised concern about how much ultraviolet light reaches the earth's surface. Of particular concern are the questions of how deeply into sea water ultraviolet radiation penetrates and how it affects marine organisms, especially plankton (floating microorganisms) and viruses that attack plankton. Plankton form the base of the marine food chains and are believed to affect our planet's temperature and weather by their uptake of CO_2 for photosynthesis.

Deneb Karentz, a researcher at the Laboratory of Radiobiology and Environmental Health (University of California in San Francisco) has devised a simple method for measuring ultraviolet penetration and intensity. Working in the Antarctic Ocean, she submerged to various depths thin plastic bags containing special strains of *E. coli* that are almost totally unable to repair ultraviolet radiation damage to their DNA. Bacterial death rates in these bags were compared with rates in unexposed control bags of the same organism. The bacterial "biosensors" revealed constant significant ultraviolet damage at depths of 10 m and frequently at 20 and 30 m. Karentz plans additional studies of how ultraviolet may affect seasonal plankton blooms (growth spurts) in the oceans.

Collecting *E. coli* samples in the Antarctic Ocean to measure ultraviolet penetration and intensity.

The Study of Mutations

Microorganisms are especially useful in studying mutations because of their short generation time and the relatively small expense of maintaining large populations of mutant organisms for study. Comparisons of normal and mutant organisms have led to important advances in the understanding of both genetic mechanisms and metabolic pathways. Microorganisms continue to be important to researchers attempting to fur-

ther our knowledge of these processes. However, the study of mutations is not without its problems. Two common problems are (1) distinguishing between spontaneous and induced mutations and (2) isolating particular mutants from a culture containing both mutated and normal organisms. The *fluctuation test* and the technique of *replica plating* are used to distinguish between spontaneous and induced mutations; replica plating also is used to isolate mutants.

Why is it important to differentiate between spontaneous and induced mutations? Making this distinction helps us understand mechanisms in the evolution of microorganisms and presumably other organisms as well. For example, some organisms are penicillin resistant—they grow in the presence of penicillin despite its antibiotic properties. Theoretically, there are two ways organisms can acquire such resistance: Either the penicillin *induces* a change in the organism that enables it to grow in the presence of penicillin, or a mutation occurs *spontaneously* that will allow the organism to grow if it is later exposed to penicillin. In the latter case, penicillin will kill nonresistant organisms, thereby *selecting* for the resistant mutant. Various experiments, two of which are described next, have shown that the second mechanism, the selection of spontaneous mutants, is the primary means of evolution in microorganisms.

The **fluctuation test,** designed by Salvador Luria and Max Delbruck in 1943, is based on the following hypothesis: If mutations that confer resistance occur spontaneously and at random, we would expect great fluctuation in the number of resistant organisms per culture among a large number of cultures. This fluctuation would occur regardless of whether the substance to which resistance develops is present. A mutation might occur early in the incubation period, late in that period, or not at all. Cultures with early mutations would contain many mutated progeny. Those with late mutations would have few mutated progeny, and those without a mutation would have none. An alternative hypothesis is that mutations conferring resistance to a substance occur only in the presence of the substance. Then cultures containing the substance would be expected to have approximately equal numbers of resistant organisms, whereas cultures lacking the substance would have no resistant organisms.

To test these hypotheses, Luria and Delbruck inoculated a large flask of liquid medium with a bacterial species that was sensitive to the antibiotic streptomycin. At the same time they inoculated 100 small tubes of liquid medium with the same bacteria. No streptomycin was present in either the flask or the tubes. Both the flask and the tubes were allowed to reach maximum growth (10^9 organisms per milliliter). One-milliliter samples were then used to inoculate agar plates containing streptomycin; a plate was made from each tube, and many plates were made from the flask.

After 24 hours the colonies on each plate were counted. Each colony represented a resistant mutant that could grow in the presence of streptomycin. There was far greater fluctuation in the number of colonies among the plates inoculated from the tubes than among the plates inoculated from the flask (Figure 7.22). Therefore, mutations must have occurred at different times or not at all in the various tubes. Mutations also must have occurred at different times in the flask, but progeny of mutated organisms became distributed through the medium, so that the number of mutants in each sample did not vary greatly. Luria and Delbruck concluded that resistance was conferred from random mutations occurring at different times among the organisms in the tubes and not from exposure to streptomycin. (Can you predict what results would have been obtained if resistance arose only from exposure to streptomycin?)

The technique of **replica plating,** devised by Joshua and Esther Lederberg in 1952, is also used to study mutations. Based on the same reasoning as the fluctuation test, it hypothesizes that resistance to a substance arises spontaneously and at random without the need for exposure to the substance. In the original replica plating studies (Figure 7.23), bacteria from a liquid culture were evenly spread on a master agar plate and allowed to grow for 4 to 5 hours. A sterile velveteen pad was then gently pressed against the surface of the master plate to pick up organisms from each colony. The tiny fibers of velveteen acted like hundreds of tiny inoculating needles. The pad was carefully kept in the same orientation and used to inoculate an agar plate containing a substance such as penicillin to which

bacteria might be resistant. After incubation, the exact positions of corresponding colonies on the two plates were noted. The bacteria in colonies found on the penicillin plate had resistance to penicillin without ever having been exposed to it.

Replica plating not only demonstrates spontaneity of mutations that confer resistance, it also provides a means of isolating resistant organisms without exposing them to a substance. By keeping the velveteen pad in perfect alignment during the transfer process, colonies on the original plate that contain resistant organisms can be identified by their location relative to colonies on the master penicillin plate.

Replica plating is now widely used to study changes in the characteristics of many bacteria. The velveteen pads have been replaced by other materials that are easier to sterilize and manipulate. The technique is especially useful for identifying mutants whose nutritional needs have changed. Replicas can be transferred to a variety of different media, each deficient in a particular nutrient. Failure of particular colonies to grow on the deficient medium indicates that a mutation prevented the organism from synthesizing that nutrient.

The Ames Test

Human cancers can be induced by environmental substances that act by altering DNA. Much research effort is now being devoted to determining which substances are **carcinogens** (cancer-producing compounds). Carcinogens tend to be mutagenic, so determining whether a substance is mutagenic is often a first step

---

FIGURE 7.22 The fluctuation test of Luria and Delbruck proves that mutations conferring antibiotic resistance are random—they are not induced by exposure to the antibiotic.

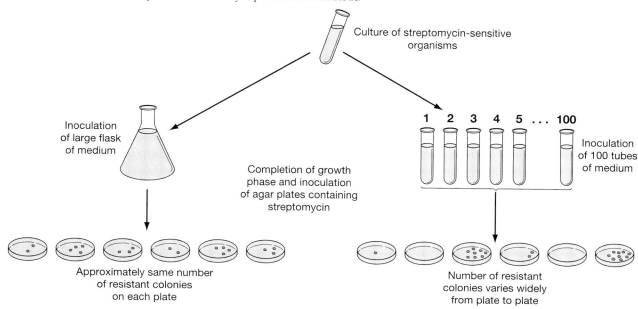

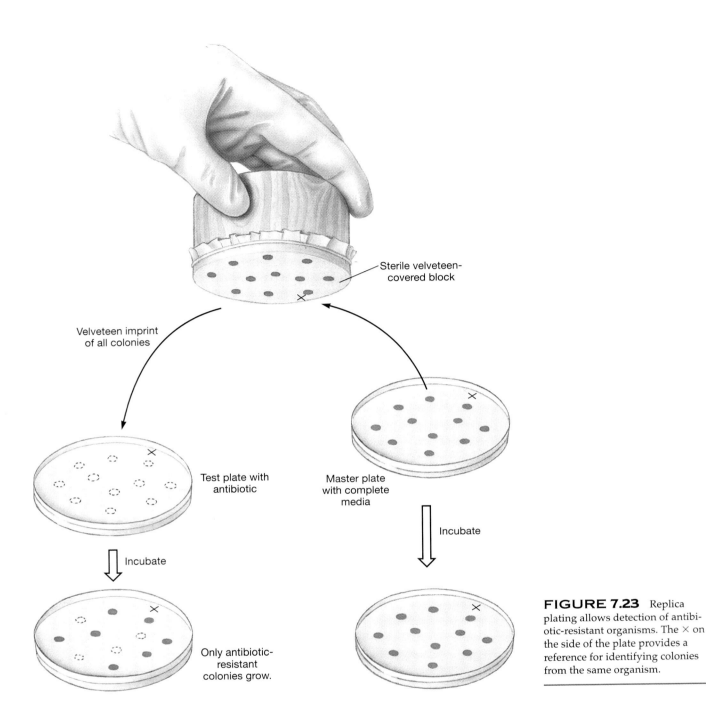

Velveteen imprint
of all colonies

Sterile velveteen-
covered block

Test plate with
antibiotic

Master plate
with complete
media

Incubate

Incubate

Only antibiotic-
resistant
colonies grow.

FIGURE 7.23 Replica plating allows detection of antibiotic-resistant organisms. The × on the side of the plate provides a reference for identifying colonies from the same organism.

in identifying it as a carcinogen. Bacteria, being subject to mutation and being easier and cheaper to study than larger organisms, are ideal organisms to use in screening substances for mutagenic properties. Proving that a substance causes mutations in bacteria does not prove that it does so in human cells. Even proving that a substance causes mutations in human cells does not prove that the mutations will lead to cancer. Additional tests, including tests in animals, are necessary to identify carcinogens, but initial screening with bacteria can eliminate some substances from further study. If a substance induces no mutations in a large population of bacteria, most researchers believe that it is not likely to be a carcinogen.

The **Ames test** (Figure 7.24a), devised by the American microbiologist Bruce Ames, is used to test whether substances induce mutations in certain strains of *Salmonella* (auxotrophs) that have lost their ability to synthesize the amino acid histidine. These strains easily undergo another mutation that restores their

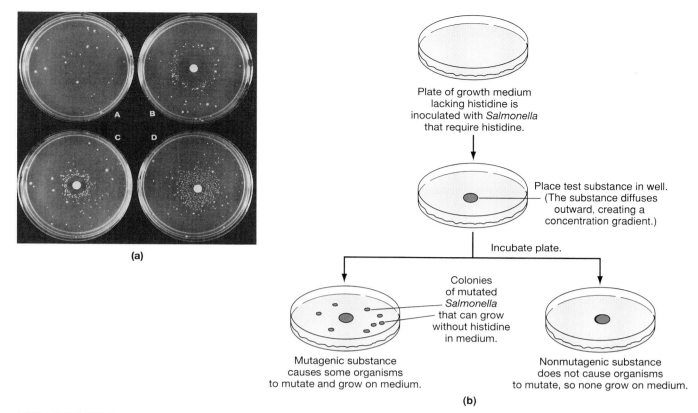

FIGURE 7.24 (a) Plates used in the Ames test. (b) The test is used to determine whether a substance is a mutagen and therefore a potential carcinogen.

ability to synthesize histidine. The Ames test is based on the hypothesis that if a substance is a mutagen, it will increase the rate at which these organisms revert to histidine synthesizers (Figure 7.24b). Furthermore, the more powerful a substance's mutagenic capacity, the greater the number of reverted organisms it causes to appear. In practice, the organism is grown in the presence of a test substance. If any organisms regain the ability to synthesize histidine, the substance is suspected of being a mutagen. The larger the number of organisms that regain the synthetic ability, the stronger the substance's mutagenic capacity is likely to be.

POLYMERASE CHAIN REACTION (PCR)– KEY TO PAST AND FUTURE WORLDS OF DNA

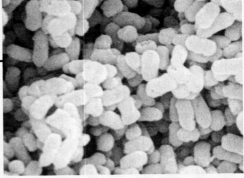

(a)

Is there privacy in the grave? Not anymore. Ancient DNA, sometimes as old as 17 million years, is being recovered, and its bases sequenced. Which genes did past organisms have? How many were handed down to us, and how many mutant versions do we have? DNA has been extracted from the brains of 91 prehistoric Native Americans buried and mummified in Florida peat bogs 7500 years ago. That DNA is now undergoing analysis, thanks to **polymerase chain reaction (PCR)**, a technique that first became available in 1985. PCR allows us to produce rapidly (amplify) a billion copies of DNA without needing a living cell. These large quantities are then easily analyzed.

Scientists have applied this tool to many questions concerning the past. For example, they have extracted DNA from fossil flies embedded in amber— and, ironically, did so shortly after the book *Jurassic Park* was written. They have also studied DNA from fossil leaves embedded in Idaho shale 17 million years ago (the DNA is very similar to that of modern magnolias) and from the bloodstains, hair, and bone chips preserved by doctors attending Abraham Lincoln at the time of his assassination. Lincoln is suspected of having had the hereditary disease Marfan's syndrome, which causes weakened arteries that can rupture and cause death. Most people with Marfan's syndrome die before they reach Lincoln's age. Would Lincoln have died soon had he not been assassinated? We can now create a library of Lincoln's DNA. As the Human Genome Project identifies the sequence of various genes (including those for Marfan's), we can match them to Lincoln's DNA and know with certainty which genes he had.

A modern forensic problem has brought the term PCR to the lips of the average American. Attorneys in the O. J. Simpson trial argued about DNA analysis and about the reliability of the PCR techniques in front of millions worldwide. Seldom has a scientific advance entered public awareness so rapidly. Prisoners already incarcerated for many years began asking for DNA

analyses of evidence from their trials. PCR made it possible to bring forth evidence that was not available earlier.

DNA analysis can be used to protect and free the innocent as well as to convict the guilty. After PCR and DNA analysis were conducted on semen samples, one man convicted of rape was shown not to have been the rapist. His family had never lost faith in his innocence throughout the 10 years he had been locked up. (Analy-sis of semen in sexual assault cases has led to the freeing of 30 percent of initial suspects.)

These techniques are powerful forensic tools. Enough DNA can be recovered, and amplified by PCR, from the sweatband of a baseball cap to identify its wearer with extremely high certainty. DNA analysis of saliva on the back of a stamp on an envelope can identify the person who licked the stamp.

Rising from an 11,000-year-old grave in Ohio are cultures of two strains of *Enterobacter cloacae*, recovered alive but frozen in a state of suspended animation, from the intestine of a 4-ton mastodon that had been killed by prehistoric hunters, butchered, and then sunk into a peat bog (a primitive form of food preservation) (Figure 7.25). And preserved they are! No mutations have occurred over the past 11,000 years. PCR analysis will reveal how these ancient organisms differ from today's strains of *E. cloacae*. Botanists await the tantalizing analysis of DNA from the mastodon's last meal: pollen grains stuck in his teeth, swamp grass, mosses, leaves, and a water lily. Chapters of evolutionary history may need to be rewritten.

For the living, PCR-amplified DNA analysis can reveal the presence of organisms that are difficult, dangerous, slow-growing, or require extra skill to culture in standard clinical laboratories. Tuberculosis cultures require 8 weeks to grow; PCR techniques will confirm the presence of the

(b)

FIGURE 7.25 (a) *Enterobacter cloacae*. These bacteria were isolated and cultured from the remains of an 11,000-year-old mastodon fossil found in Newark, Ohio, in 1989. (b) The mastodon's remains. Its digestive tract could be identified as a darker-colored, discrete cylindrical mass bent into the shape of intestinal loops. No organisms were found in areas sampled adjacent to the intestine.

DNA of tuberculosis organisms in just hours. Medical-technology programs will need to train students in these techniques of the future, and current personnel will need to be retrained.

How does PCR amplification of DNA work? A large piece, or mixture of pieces, of DNA is cut up into smaller pieces by restriction endonuclease enzymes. You need to know the base composition of the ends of the exact piece of DNA you wish to replicate. In less than 24 hours, automated synthesizing equipment can make oligonucleotide (*oligos*, few) primer molecules that will pair complementarily with the ends of the desired section of DNA. A *primer* is a molecule that serves as a starting point for DNA synthesis.

The sequence of events in PCR amplification is shown in Figure 7.26. The heating process (thermal cycling), which converts newly formed DNA

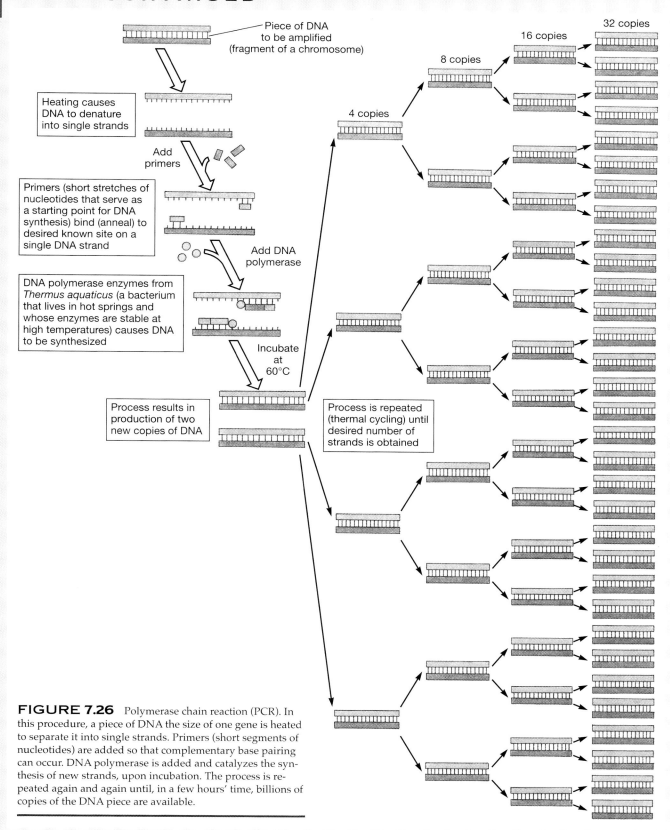

Piece of DNA
to be amplified
(fragment of a chromosome)

Heating causes
DNA to denature
into single strands

Add
primers

Primers (short stretches of
nucleotides that serve as
a starting point for DNA
synthesis) bind (anneal) to
desired known site on a
single DNA strand

Add DNA
polymerase

DNA polymerase enzymes from
Thermus aquaticus (a bacterium
that lives in hot springs and
whose enzymes are stable at
high temperatures) causes DNA
to be synthesized

Incubate
at
60°C

Process results in
production of two
new copies of DNA

Process is repeated
(thermal cycling) until
desired number of
strands is obtained

4 copies

8 copies

16 copies

32 copies

FIGURE 7.26 Polymerase chain reaction (PCR). In
this procedure, a piece of DNA the size of one gene is heated
to separate it into single strands. Primers (short segments of
nucleotides) are added so that complementary base pairing
can occur. DNA polymerase is added and catalyzes the synthesis of new strands, upon incubation. The process is repeated again and again until, in a few hours' time, billions of
copies of the DNA piece are available.

into single strands, is repeated until it
has produced billions of copies of the
desired piece of DNA. The DNA is
then easily detected (as in a clinical diagnostic test) or analyzed for total base
sequence. Cutting a large piece of
DNA into smaller pieces, sequencing
the PCR-amplified quantities of the
pieces, and then looking for overlaps
at the ends allows us finally to determine the sequence of the entire original DNA piece.

OVERVIEW OF GENETIC PROCESSES

The Basis of Heredity

- **Heredity** involves the transmission of information from an organism to its progeny.

- **Genes** are linear sequences of DNA that carry coded information for the structure and function of an organism.

- Prokaryotic chromosomes are threadlike circular structures made of DNA. Transmission of information in prokaryotes typically occurs during asexual reproduction in which the chromosome is reproduced (replicates), and each daughter cell receives a chromosome like the one in the parent cell.

- The genetic makeup of an organism is its **genotype;** the physical expression of the genotype is the **phenotype.**

- **Mutations** (alterations in DNA) transmitted to progeny account for much of the variation in organisms.

Nucleic Acids in Information Storage and Transfer

- All information for the functioning of a cell is stored in DNA in a specific sequence of the nitrogenous bases: adenine, thymine, cytosine, and guanine.

- Information stored in DNA is used for two purposes: (1) to replicate DNA in preparation for cell division and (2) to provide information for protein synthesis. In both of these processes, information is transferred by base pairing.

REPLICATION OF DNA

- **Replication** of bacterial DNA begins at a specific point in the circular chromosome and usually proceeds in both directions simultaneously. The main steps in DNA replication are summarized in Figure 7.4.

- DNA replication is **semiconservative**—each chromosome consists of one strand of old (parent) DNA and one of newly synthesized DNA.

PROTEIN SYNTHESIS

Transcription

- In **transcription, messenger RNA (mRNA)** is transcribed from DNA, as summarized in Figure 7.6, and serves as a **template** for protein synthesis.

Kinds of RNA

- Besides mRNA, two other kinds of RNA are similarly produced: (1) **ribosomal RNA (rRNA),** which combines with specific proteins to form ribosomes, the sites for protein assembly; and (2) **transfer RNA (tRNA),** which carries amino acids to the assembly site.

Translation

- In the process of **translation,** three-base sequences in mRNA act as **codons** and are matched by base pairing with **anticodons** of tRNA. The mRNA codons constitute the **genetic code**—a code that is essentially the same for all living organisms.

- Once the mRNA and ribosomes are aligned, the process of protein synthesis proceeds as summarized in Figure 7.11 until a **terminator codon** is reached.

REGULATION OF METABOLISM

Significance of Regulatory Mechanisms

- Mechanisms that regulate metabolism turn reactions on and off in accordance with the needs of cells, allowing the cells to use various energy sources and to limit synthesis of substances to the amounts needed.

Categories of Regulatory Mechanisms

- The two basic categories of regulatory mechanisms are: (1) mechanisms that regulate the activity of enzymes already available in the cell and (2) mechanisms that regulate the action of genes, which determine what enzymes and other proteins will be available.

Feedback Inhibition

- In **feedback inhibition,** the end product of a biochemical pathway directly inhibits the first enzyme in the pathway (Figure 7.12).

- Enzymes subject to such regulation are generally allosteric.

- Feedback inhibition regulates the activity of existing enzymes and is a quick-acting control mechanism.

Enzyme Induction

- In **enzyme induction** (Figure 7.13), the presence of a substrate activates an **operon,** a sequence of closely associated genes that includes **structural genes** and **regulatory sites:** (1) In the absence of lactose, a **repressor,** a product of the **regulator** (*i*) gene, attaches to the operator and prevents transcription of the genes of the *lac* operon. (2) When lactose is present, it inactivates the repressor and allows transcription of the genes of the *lac* operon.

Enzyme Repression

- In **enzyme repression,** the presence of a synthetic product inhibits its further synthesis by inactivating an operon: (1) When tryptophan is present, it attaches to the repressor protein and represses genes of the *trp* operon. (2) In the absence of tryptophan, the repressor is not activated, and genes of the *trp* operon are transcribed.

- In **catabolite repression,** the presence of a preferred nutrient (often glucose) represses the synthesis of enzymes that would be used to metabolize some alternative substance.

- Both enzyme induction and enzyme repression regulate by altering gene expression. The effect on enzyme synthesis in both cases depends on the presence or absence of the regulatory substance—lactose, tryptophan, or glucose in the preceding examples.

MUTATIONS

Types of Mutations and Their Effects

- Mutations cause a change in the genotype, of an organism; the change may or may not be expressed in the phenotype.

- Two major classes of mutations are (Table 7.3): (1) **point mutations,** which consist of changes in a single nucleotide, and (2) **frameshift mutations,** which consist of the **insertion** or **deletion** of one or more nucleotides.

Phenotypic Variation

- Phenotypic variations produced by mutations can involve alterations in colony morphology, nutritional requirements, and temperature sensitivity.

Spontaneous versus Induced Mutations

- **Spontaneous mutations** occur in the absence of any known mutagen and appear to be due to errors in base pairing during DNA replication. Various genes have different rates of mutation.

- **Induced mutations** are mutations produced by agents called **mutagens;** mutagens increase the mutation rate.

Chemical Mutagens

- Chemical mutagens include **base analogs, alkylating agents, deaminating agents,** and **acridine derivatives.**

Radiation as a Mutagen

- **Radiation** often causes the formation of **dimers**—two adjacent pyrimidine bases that are bound to each other and interfere with DNA replication.

Repair of DNA Damage

- Many bacteria have enzymes that can repair certain damages to DNA (Figure 7.20). (1) **Light repair** involves an enzyme that is activated by visible light and that breaks bonds between pyrimidines of a dimer. (2) **Dark repair** involves several enzymes that do not require light for activation; they excise defective DNA and replace it with DNA complementary to the normal DNA strand.

The Study of Mutations

- Microorganisms are useful in studying mutations because many generations can be produced quickly and inexpensively.

- The **fluctuation test** demonstrated that resistance to chemical substances occurs spontaneously rather than being induced.

- **Replica plating** likewise demonstrated the spontaneous nature of mutations; it also can be used for isolating mutants without exposing them to the substance to which they are resistant.

The Ames Test

- The **Ames test** is based on the ability of **auxotrophic** bacteria to mutate by reverting to their original synthetic ability. It is used for screening chemicals for mutagenic properties, which indicate potential **carcinogens.**

KEY TERMS

acridine derivative (p. 183)
alkylating agent (p. 183)
allele (p. 165)
Ames test (p. 188)
anticodon (p. 173)
antiparallel (p. 168)
attenuation (p. 179)
auxotroph (p. 182)
base analog (p. 183)
carcinogen (p. 187)
catabolite repression (p. 180)
codon (p. 171)
constitutive enzyme (p. 178)
dark repair (p. 185)
deaminating agent (p. 183)
deletion (p. 181)
dimer (p. 184)
DNA polymerase (p. 168)
end-product inhibition (p. 175)

enzyme induction (p. 178)
enzyme repression (p. 178)
exon (p. 171)
exonuclease (p. 184)
feedback inhibition (p. 175)
fluctuation test (p. 186)
frameshift mutation (p. 181)
gene (p. 165)
genetic code (p. 172)
genetics (p. 164)
genotype (p. 165)
heredity (p. 164)
induced mutation (p. 183)
inducer (p. 178)
inducible enzyme (p. 178)
insertion (p. 181)
intron (p. 171)
lagging strand (p. 168)
leading strand (p. 168)
ligase (p. 168)
light repair (p. 184)

locus (p. 165)
messenger RNA (mRNA) (p. 171)
mutagen (p. 183)
mutation (p. 165)
nonsense codon (p. 172)
Okazaki fragment (p. 168)
operon (p. 178)
phenotype (p. 165)
photoreactivation (p. 184)
point mutation (p. 180)
polymerase chain reaction (p. 190)
polyribosome (p. 175)
prototroph (p. 182)
radiation (p. 184)
regulator gene (p. 178)
regulatory site (p. 178)
replica plating (p. 187)
replication (p. 166)
replication fork (p. 168)
repressor (p. 178)

restriction endonuclease (p. 184)
ribosomal RNA (rRNA) (p. 171)
RNA polymerase (p. 171)
semiconservative replication (p. 168)
sense codon (p. 172)
spontaneous mutation (p. 182)
structural gene (p. 178)
template (p. 166)
terminator codon (p. 175)
transcription (p. 167)
transfer RNA (tRNA) (p. 172)
translation (p. 167)

A **Genes, Chromosomes, and Mutations**

1. Define chromosome, gene, allele, and mutation.

B **Nucleic Acids and Information Storage**

2. How do nucleic acids function in information storage? In information transfer?

3. Which nucleotide is present in DNA but not in RNA? In RNA but not in DNA?

4. Which nucleotide will pair with cytosine? With thymine?

C **DNA Replication in Prokaryotes**

5. Match the following processes and the names for those processes:

____ DNA makes DNA   **a.** translation

____ RNA makes DNA   **b.** transcription

____ DNA makes RNA   **c.** reverse transcription

____ RNA makes protein   **d.** replication

6. Describe the steps in the replication of DNA.

D **Protein Synthesis**

7. Describe the steps in protein synthesis.

8. What are the three types of RNA, and how do they differ in structure and function?

9. Use the genetic code in Figure 7.8 to answer the following:

   a. What is the sequence of amino acids for which the following piece of mRNA codes: AUGCUUAGGUAU-UAG? Is there a terminator codon in this sequence? If so, which is it, and what is its effect?

   b. What is the sequence of DNA nucleotides that coded for the piece of mRNA in part (a)?

10. Write out two different mRNA nucleotide sequences that would code for the following polypeptide chain:

Val-Trp-Asn-Leu-Met-Ser-Stop

11. Match the following:

____ anticodon   **a.** DNA

____ codons that contain thymine   **b.** mRNA

____ contains no codons   **c.** tRNA

____ contains codon that matches to anticodon   **d.** rRNA

   e. protein

E **Regulatory Mechanisms**

12. What is the significance of the presence of mechanisms to regulate metabolism?

13. What factors distinguish the two basic regulatory mechanisms?

F **Feedback Inhibition, Enzyme Induction, and Enzyme Repression**

14. How does feedback inhibition operate, and what does it accomplish?

15. How does enzyme induction operate, and what does it accomplish?

16. For the *lac* operon, match the following:

____ inducer   **a.** regulator gene

____ place where repressor binds to shut off operon   **b.** promoter

____ substance that binds to promoter site to start transcription   **c.** structural genes

   d. lactose

   e. operator

____ combines with repressor to keep operon "on"   **f.** RNA polymerase

____ Z, Y, A   **g.** repressor

____ may be located some distance from the operon and is not under control of the promoter

17. How does enzyme repression operate, and what does it accomplish?

18. In what ways are enzyme induction and repression similar, and in what ways are they different?

G **Mutations**

19. Name and describe the effects of the following mutations (read from left to right):

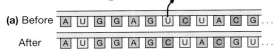

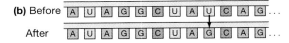

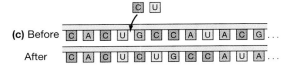

20. How might each kind of mutation affect an organism?

H **Spontaneous and Induced Mutations**

21. How do spontaneous and induced mutations differ?

22. How do chemical mutagens alter DNA?

23. How might radiation alter DNA?

24. Briefly describe two mechanisms by which bacteria can sometimes repair damage to DNA.

I **Fluctuation Test, Replica Plating, and Ames Test**

25. Why are bacteria useful in the study of mutations?

26. How is the fluctuation test performed, and what does it show?

27. How is replica plating done, and what can it accomplish?

28. What is the Ames test, and what is it used for?

1. How does the role of the repressor differ in induction and repression?

2. Devise an experiment to produce and isolate a mutant organism. If possible, carry out your experiment and perform tests that will confirm your hypothesis that the organism is, in fact, a mutant.

3. During the early stages of development of the earth's atmosphere, the planet was exposed to greater amounts of ultraviolet radiation than it is today. What do you suppose were the effects of this radiation on the longevity of individual organisms and on the rate of evolution of life forms?

4. Read about and prepare a report on xeroderma pigmentosum, the genetic disease in which people lack the ability to repair ultraviolet damage to their DNA.

5. What is the amino acid sequence of the polypeptide coded by the DNA sequence ATAGCAAAACCGATG?

SOME INTERESTING READING

Alberts, B., et al. 1995. *Molecular biology of the cell,* 3d edition. New York: Garland.

Bachman, B. J. 1990 "Linkage map of *Escherichia coli* K–12, Edition 8." *Microbiological Reviews* 54, no. 2 (June):130–97.

Beardsley, T. 1991. "Smart genes." *Scientific American* 265(August):86–95.

Brock, T. D., et al. 1994. *Biology of microorganisms,* 7th edition. Englewood Cliffs, NJ: Prentice Hall.

Cathcart, R. 1990. "Advances in automated DNA sequencing." *Nature* 347, no. 6290 (September 20):310.

Cherfas, J. 1991. "Ancient DNA: Still busy after death." *Science* 253, no. 5026 (September 20):1354–56.

Devoret, R. 1979. "Bacterial tests for potential carcinogens." *Scientific American* 241, no. 2 (August):40.

Folger, T. 1992. "Oldest living bacteria tell all." *Discover* 13(January):30–31.

Krawiec, S., and M. Riley. 1990. "Organization of the bacterial chromosome." *Microbiological Reviews* 54, no. 4 (December):502–39.

Levine, A. J. 1988. "Oncogenes of DNA tumor viruses." *Cancer Research* 48 (February 1):493–96.

Maloy, S. R., J. Cronan, and D. Freifelder. 1994. *Microbial genetics,* 2d edition. Boston: Jones and Bartlett.

Moss, R. 1991. "Genetic transformation of bacteria." *The American Biology Teacher* 53(March):179–80.

Mullis, K. B. 1990. "The unusual origin of the polymerase chain reaction." *Scientific American* 262(April):56–61.

Mullis, K. B., F. Ferre, and R. A. Gibbs, eds. 1994. *The polymerase chain reaction.* Boston: Birkhauser.

Osawa, S., T. H. Jukes, K. Watanabe, and A. Muto. 1992. "Recent evidence for evolution of the genetic code." *Microbiological Reviews* 56, no. 1 (March):229–64.

"PCR diagnostics: Ownership, technology are changing." 1991. *ASM News* 57(10):503–4.

Persing, D. H. 1991. "Polymerase chain reaction: Trenches to benches." *Journal of Clinical Microbiology* 29(7):1281–85.

Sykes, B. 1991. "Ancient DNA: The past comes alive." *Nature* 352(August 1):381–82.

Waldrop, M. M. 1989. "Did life really start out in an RNA world?" *Science* 246, no. 4935 (December 8):1248–49.

Watson, J. D., et al. 1988. *Molecular biology of the gene,* 4th edition. Redwood City, CA: Benjamin Cummings.

Weinberg, R. A. 1983. "A molecular basis of cancer." *Scientific American* 249(November):126–42.

Weintraub, H. M. 1990. "Antisense RNA and DNA." *Scientific American* 262(January):40–46.

Weiss, J. B. 1995. "DNA probes and PCR for diagnosis of parasitic infections." *Clinical Microbiology Reviews* 8, no. 1 (January):113–30.

Witkin, E. M. 1976. "Ultraviolet mutagenesis and inducible DNA repair in *Escherchia coli.*" *Bacteriological Reviews* 40:869.

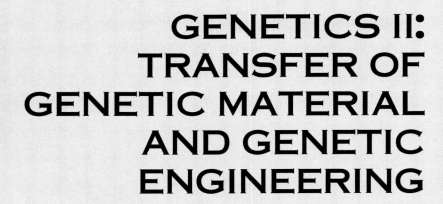

CHAPTER 8

GENETICS II: TRANSFER OF GENETIC MATERIAL AND GENETIC ENGINEERING

DNA sequencing of recombinant clones in AIDS research. Microbiology has provided the techniques for working with DNA.

Transfer of genetic material from one organism to another can have far-reaching consequences. In microbes it provides ways for viruses to introduce genetic information into bacteria and mechanisms for bacteria to increase their disease-causing capabilities or to become resistant to antibiotics. Information obtained from studying the transfer of genetic material between microorganisms can be applied to agricultural, industrial, and medical problems and to the unique problems of the prevention and treatment of infectious diseases. In this chapter we will discuss the mechanisms by which genetic transfers occur and the significance of such transfers.

NATURE AND SIGNIFICANCE OF GENE TRANSFER

Gene Transfer

Gene transfer refers to the movement of genetic information between organisms. In most eukaryotes, it is an essential part of the organism's life cycle and usually occurs by sexual reproduction. Male and female parents produce *gametes* (sex cells), which unite to form

the first cell (the *zygote*) of a new individual. Because each parent produces many genetically different gametes, many different combinations of genetic material can be transferred to offspring. In bacteria, gene transfer is not an essential part of the life cycle. When it does take place, generally only a portion of the genetic makeup of the donating cell is transferred to the other participating, or *recipient,* cell. The combining of genes (DNA) from two different cells is called **recombination,** and the resulting cell is referred to as a *recombinant*. Before the 1920s, bacteria were thought to reproduce only by binary fission and to have no means of genetic transfer comparable to that achieved through sexual reproduction in eukaryotes. Since then, three mechanisms of gene transfer in bacteria have been discovered, none of which is associated with reproduction. Each mechanism—*transformation, transduction,* and *conjugation*—is discussed in this chapter.

Gene transfer is significant because it greatly increases the genetic diversity of organisms. As noted in Chapter 7, mutations account for some genetic diversity, but gene transfer between organisms accounts for even more. ∞ (p. 180) When organisms are subjected to changing environmental conditions, genetic diversity increases the likelihood that some organisms will

THIS CHAPTER FOCUSES ON THE FOLLOWING QUESTIONS:

A What are the nature and significance of gene transfer?

B What are the mechanisms and significance of transformation?

C What are the mechanisms and significance of transduction?

D What are the mechanisms and significance of conjugation?

E What are the characteristics and actions of plasmids?

F How are the following techniques of genetic engineering used: genetic fusion, protoplast fusion, gene amplification, recombinant DNA, and hybridomas?

G Why are scientists concerned about uses of recombinant DNA?

adapt to any particular condition. Such diversity leads to evolutionary changes. Organisms with genes that allow them to adapt to an environment survive and reproduce, whereas organisms lacking those genes perish. If all organisms were genetically identical, all would survive and reproduce, or all would die.

In *recombinant DNA technology,* genes from one kind of organism are introduced into the genome of another kind of organism (for example, when human genes are inserted into a pig).

TRANSFORMATION

Discovery of Transformation

Bacterial transformation, a change in an organism's characteristics because of the transfer of genetic information, was discovered in 1928 by Frederick Griffith, an English physician, while he was studying pneumococcal infections in mice. Pneumococci (bacteria that cause pneumonia) with capsules (Chapter 4) produce smooth, glistening colonies. ∞ (p. 90) Those lacking capsules produce rough colonies with a coarse, nonglistening appearance. Only the capsule-produc-

ing (encapsulated) pneumococci inoculated into mice were *virulent*—that is, they had the power to cause severe disease. One such organism can multiply rapidly and kill a mouse! Capsules help keep molecules produced by the mouse's immune system from reaching the surface of the bacterium. They also make it difficult for white blood cells to engulf the invading bacteria. In other words, the capsule protects the bacteria from the mouse's immune system.

Griffith injected one group of mice with heat-killed smooth pneumococci, a second group with live smooth pneumococci, a third group with live rough pneumococci, and a fourth group with a mixture of live rough and heat-killed smooth pneumococci (Figure 8.1). As expected, mice that received live smooth pneumococci developed pneumonia and died, whereas those that received either heat-killed smooth pneumococci or live rough pneumococci did not develop pneumonia and survived. Surprisingly, those that received the mixture also died of pneumonia. Griffith isolated live smooth organisms from them. The presence of heat-killed, encapsulated organisms in the mixture apparently allowed the live unencapsulated ones to develop capsules and become virulent. Neither Griffith nor his colleagues knew how this transformation occurred.

197

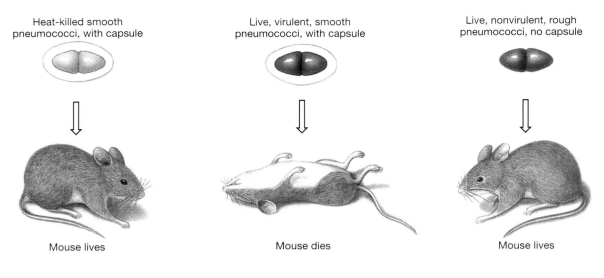

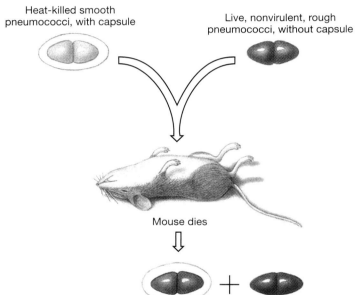

Live smooth pneumococci (with capsule) plus live rough pneumococci
(without capsule) isolated from dead mouse

FIGURE 8.1 Griffith's experiment with pneumococcal infections in mice, which led to the discovery of transformation. When S-type pneumococci (which produce smooth-appearing colonies, due to the presence of capsules) are injected into mice, the mice die of pneumonia. The mice survive when R-type pneumococci (which produce rough-appearing colonies, due to the lack of capsules) or heat-killed S-type pneumococci are injected. But when a mixture of live R-type and heat-killed S-type pneumococci—neither of which is lethal by itself—are injected, the mice die, and live S-type organisms as well as R-types are recovered from the dead animals. Griffith had no way of knowing exactly what had happened, but he realized that some cells had been "transformed" from type R to type S. Moreover, the change was heritable. We now know that the R-type bacteria picked up naked DNA, liberated from disintegrated dead S-type bacteria, and incorporated it into their own DNA. R-type bacteria that picked up DNA that had the genes for capsule production were genetically transformed into S-type organisms.

In subsequent studies of transformation, Oswald Avery discovered that a capsular polysaccharide was responsible for the virulence of pneumococci. In 1944, Avery, Colin MacLeod, and Maclyn McCarty isolated the substance responsible for the transformation of pneumococci and determined that it was DNA. In retrospect, this discovery marked the "birth" of molecular genetics, but at the time DNA was not known to carry genetic information. Researchers working with plant and animal chromosomes had isolated both DNA and protein from them, but they thought the genetic information was in the protein. Only when James

Watson and Francis Crick determined the structure of DNA did it become clear that DNA encodes genetic information. After this original work with pneumococci (now called *Streptococcus pneumoniae*), transformation was observed in a wide variety of gram-positive and gram-negative bacteria, including *Acinetobacter, Bacillus, Haemophilus, Neisseria,* and *Staphylococcus,* and in the yeast *Saccharomyces cerevisiae.*

Mechanism of Transformation

To study the mechanism of transformation, scientists extract DNA from donor organisms by a biochemical process that yields hundreds of naked DNA fragments from each bacterial chromosome. (*Naked DNA* is DNA that is not incorporated into chromosomes or other structures.) When the extracted DNA is placed in a medium with organisms capable of incorporating it, most organisms can take up a maximum of about 10 fragments, which is less than 5 percent of the amount of DNA normally present in the organism.

Uptake of DNA occurs only at a certain stage in a cell's growth cycle, probably prior to the completion of cell wall synthesis. In this stage, a protein called **competence factor** is released into the medium and apparently facilitates the entry of DNA. When competence factor from one culture is used to treat a culture that lacks it, cells in the treated culture become *competent* to receive DNA—they can now take up DNA fragments. However, not all bacteria can become competent; thus not all bacteria can be transformed. DNA entry depends on such factors as modifications of the cell wall and the formation of specific receptor sites on the plasma membrane that can bind DNA. Receptor sites seem to recognize different sources of DNA. They bind DNA from the same genus and species, or from closely related genera and species, but generally reject DNA from distantly related organisms (Figure 8.2).

Once DNA reaches the entry sites, endonucleases cut double-stranded DNA into units of 7000 to 10,000 nucleotides. The strands separate, and only one strand enters the cell. Single-stranded DNA is vulnerable to attack by various nucleases and can enter a cell only if the nucleases on the cell surface somehow have been inactivated. Inside the cell, the donor single-stranded DNA must combine by base pairing with a portion of the recipient chromosome immediately or else be destroyed. In transformation, as well as in other mechanisms of gene transfer, the donor single-stranded DNA is positioned alongside the recipient DNA such that identical loci are next to one another. Splicing of a DNA strand involves breaking the strand, removing a segment, inserting a new segment, and attaching the ends. Enzymes in the recipient cell excise (cut out) a portion of the recipient's DNA and replace it with the donor DNA, which becomes a permanent part of the recipient's chromosome. The leftover recipient DNA is subsequently broken down, so the number of nucleotides in the cell's DNA remains constant.

Significance of Transformation

Although transformation has been observed mainly in the laboratory, it occurs in nature. It probably fol-

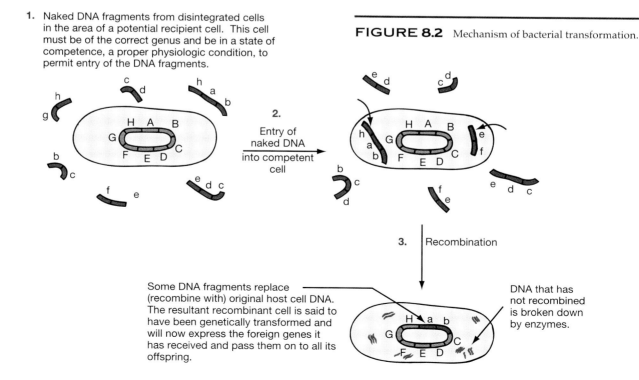

1. Naked DNA fragments from disintegrated cells in the area of a potential recipient cell. This cell must be of the correct genus and be in a state of competence, a proper physiologic condition, to permit entry of the DNA fragments.

2. Entry of naked DNA into competent cell

FIGURE 8.2 Mechanism of bacterial transformation.

3. Recombination

Some DNA fragments replace (recombine with) original host cell DNA. The resultant recombinant cell is said to have been genetically transformed and will now express the foreign genes it has received and pass them on to all its offspring.

DNA that has not recombined is broken down by enzymes.

lows the breakdown of dead organisms in an environment where live ones of the same or a closely related species are present. However, the degree to which transformation contributes to the genetic diversity of organisms in nature is not known. In the laboratory, researchers induce transformation in order to study the effects of DNA that differs from the DNA that the organism already has. Transformation also can be used to study the locations of genes on a chromosome and to insert DNA from one species into that of another species, thereby producing *recombinant* DNA.

TRANSDUCTION

Discovery of Transduction

Transduction, like transformation, is a method of transferring genetic material from one bacterium to another. Unlike transformation, in which naked DNA is transferred, in transduction DNA is carried by a **bacteriophage** (bak-ter'-e-o-faj)—a virus that can infect bacteria. The phenomenon of transduction was discovered in *Salmonella* in 1952 by Joshua Lederberg and Norton Zinder and has since been observed in many different genera of bacteria.

Mechanisms of Transduction

To understand the mechanisms of transduction, we need to describe briefly the properties of bacteriophages, also called **phages** (faj'ez). Phages, which are described in more detail in Chapter 11, are composed of a core of nucleic acid covered by a protein coat. They infect bacterial cells (*hosts*) and reproduce within them, as shown in Figure 8.3. A phage capable of infecting a bacterium attaches to a receptor site on the cell wall. The phage nucleic acid enters the bacterial cell after a phage enzyme weakens the cell wall. The protein coat remains outside, attached to the cell wall. Once the nucleic acid is in the cell, further events follow one of two pathways, depending on whether the phage is *virulent* or *temperate*.

A **virulent phage** is capable of causing infection and, eventually, the destruction and death of a bacterial cell. Once the phage nucleic acid enters the cell, phage genes direct the cell to synthesize phage-specific nucleic acids and proteins. Some of the proteins de-

FIGURE 8.3 When a bacteriophage injects its viral DNA into a host bacterial cell, at least two different outcomes are possible. In the lytic cycle, characteristic of virulent phages, the phage DNA takes control of the cell and ① causes it to synthesize new viral components, which are assembled into whole viral particles. The cell lyses, releasing the infective viruses, which can then enter new host cells. In the lysogenic cycle, the DNA of a temperate phage enters the host cell, ② becomes incorporated into the bacterial chromosome as a prophage, and replicates along with the chromosome through many cell divisions. However, a lysogenic phage can suddenly revert to the lytic life cycle. A prophage is thus a sort of "time bomb" sitting inside the infected cell.

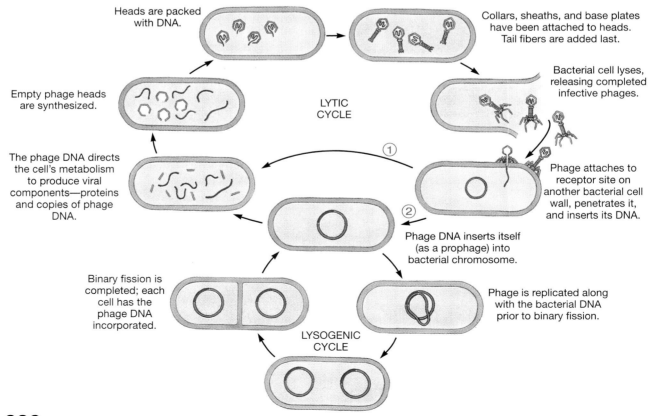

Heads are packed with DNA.

Collars, sheaths, and base plates have been attached to heads. Tail fibers are added last.

Bacterial cell lyses, releasing completed infective phages.

Empty phage heads are synthesized.

LYTIC CYCLE

The phage DNA directs the cell's metabolism to produce viral components—proteins and copies of phage DNA.

Phage attaches to receptor site on another bacterial cell wall, penetrates it, and inserts its DNA.

Phage DNA inserts itself (as a prophage) into bacterial chromosome.

Binary fission is completed; each cell has the phage DNA incorporated.

Phage is replicated along with the bacterial DNA prior to binary fission.

LYSOGENIC CYCLE

stroy the host cell's DNA, whereas other proteins and the nucleic acids assemble into complete phages. When the cell becomes filled with a hundred or more phages, phage enzymes rupture the cell, releasing newly formed phages, which can then infect other cells. Because this cycle results in **lysis** (li'sis), or rupture, of the infected (host) cell, it is called a **lytic** (lit'ik) **cycle.**

A **temperate phage** ordinarily does not cause a disruptive infection. Instead the phage DNA is incorporated into a bacterium's DNA and is replicated with it. This phage also produces a repressor substance that prevents the destruction of bacterial DNA, and the phage's DNA does not direct the synthesis of phage particles. Phage DNA that is incorporated into the host bacterium's DNA is called a **prophage** (pro'faj). Persistence of a prophage without prophage replication and destruction of the bacterial cell is called **lysogeny** (li-soj'e-ne), and cells containing a prophage are said to be **lysogenic** (li"so-jen'ik). Several ways to induce such cells to enter the lytic cycle are known, and most involve inactivation of the repressor substance.

Temperate phages can replicate themselves either as a prophage in a bacterial chromosome or independently by assembling into new phages. Temperate phages can carry out both generalized and specialized forms of transduction. In *generalized transduction,* any bacterial gene can be transferred by the phage; in *specialized transduction,* only specific genes are transferred.

Specialized Transduction

Several lysogenic phages are known to carry out specialized transduction, but lambda (λ) phage in *Escherichia coli* has been extensively studied. Phages usually insert at a specific location when they integrate with a chromosome. Lambda phage inserts into the *E. coli* chromosome between the *gal* gene, which controls galactose use, and the *bio* gene, which controls biotin synthesis. The *gal* gene and *bio* gene are operons. ∞ (Chapter 7, p. 178) When cells containing lambda phage are induced to enter the lytic cycle, genes of the phage form a loop and are excised from the bacterial chromosome (Figure 8.4). Lambda phage

FIGURE 8.4 Specialized transduction by lambda phage in *E. coli.* In this process, phage DNA always inserts itself into the bacterial host chromosome at a particular site. When the phage DNA replicates, it takes bacterial genes from either side of the site and packages them with its own DNA into new phages. Only genes adjacent to the insertion site, not genes from other parts of the host chromosome, are transduced. These genes can then be introduced into the phage's next host cell, where they will confer new genetic traits.

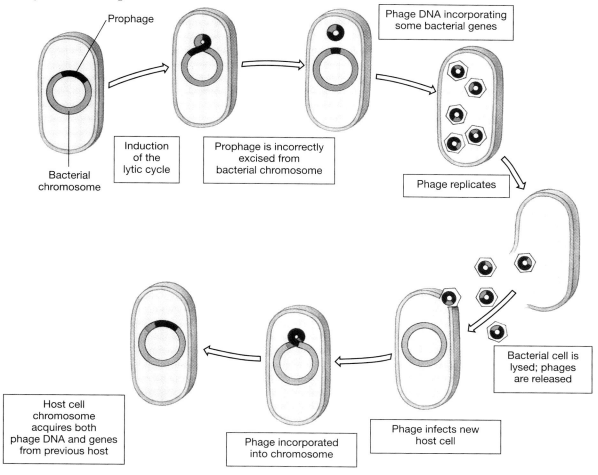

then directs the synthesis and assembly of new phage particles, and the cell lyses.

In most cases, the new phage particles released contain only phage genes. Occasionally (about one excision in a million) the phage contains one or more bacterial genes that were adjacent to the phage DNA when it was part of the bacterial chromosome. For example, the *gal* gene might be incorporated into a phage particle. When it infects another bacterial cell, the particle transfers not only the phage genes but also the *gal* gene. This process, in which a phage particle transduces (transfers) specific genes from one bacterial cell to another, is called **specialized transduction.** In specialized transduction, the bacterial DNA transduced is limited to one or a few genes lying adjacent to the prophage.

Generalized Transduction

When bacterial cells containing phage DNA enter the lytic cycle, phage enzymes break host cell DNA into many small segments (Figure 8.5). As the phage directs synthesis and assembly of new phage particles, it packages DNA by the "headful" (enough DNA to fill the head of a virus). This allows a bacterial DNA fragment occasionally to be incorporated into a phage particle. Once this phage particle, with its newly acquired bacterial DNA, leaves the infected host, it may infect another susceptible bacterium, thereby transferring the bacterial genes through the process of **generalized transduction.** Each bacterial fragment from the host cell chromosome has an equal chance of accidentally becoming a part of phage particles during the phage's replication cycle.

Significance of Transduction

Transduction is significant for several reasons. First, it transfers genetic material from one bacterial cell to another and alters the genetic characteristics of the recipient cell. As demonstrated by the specialized transduction of the *gal* genes, a cell lacking the ability to metabolize galactose could acquire that ability. Other characteristics also can be transferred either by specialized or generalized transduction.

Second, the incorporation of phage DNA into a bacterial chromosome demonstrates a close evolutionary relationship between the prophage and the host bacterial cell. The DNA of the prophage and that of the host chromosome must have regions of quite similar base sequences. Otherwise, the prophage would not bind to the bacterial chromosome.

Third, the discovery that a prophage can exist in a cell for a long period of time suggests a similar possible mechanism for the viral origin of cancer. If a prophage can exist in a bacterial cell and at some point alter the expression of the cell's DNA, this could explain how animal viruses cause malignant changes.

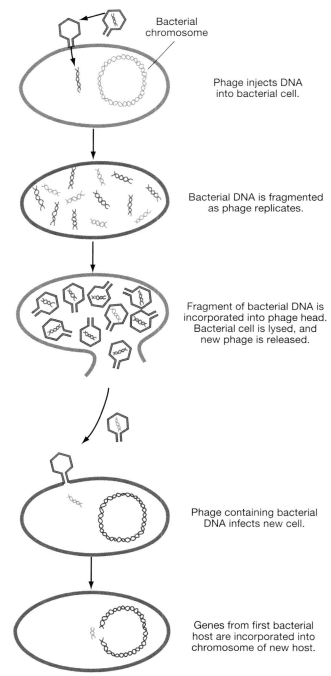

FIGURE 8.5 Generalized transduction. Bacteriophage infection of a host bacterium initiates the lytic cycle. The bacterial chromosome is broken into many fragments, any of which can be picked up and packaged along with phage DNA into new phage particles. When those particles are released and infect another bacterial cell, the new host acquires the genes that were brought along (transduced) from the previous bacterial host cell.

For example, viral genes inserted into a human gene could disrupt regulation of that gene, allowing it to be active at the wrong times, continuously, or maybe even not at all. Fetal genes cause rapid proliferation of cells during early development, but growth soon slows down and eventually stops in adulthood. If these fetal genes were turned on again later in life in a group of cells, they could rapidly grow into a tumor. (Viruses and cancer are discussed in Chapter 11.)

Finally, and most importantly to molecular geneticists, transduction provides a way to study gene linkage. Genes are said to be *linked* when they are so close together on a DNA segment that they are likely to be transferred together. Different phages can be incorporated into a bacterial chromosome, each kind usually entering at a specific site. By studying many different phage transductions, scientists can determine where they were inserted on the chromosome and which adjacent genes they are capable of transferring. The combined findings of many such studies eventually allow identification of the sequence of genes in a chromosome. This procedure is called **chromosome mapping.**

CONJUGATION

Discovery of Conjugation

In **conjugation,** like transformation and transduction, genetic information is transferred from one bacterial cell to another. Conjugation differs from those other mechanisms in two ways: (1) It requires contact between donor and recipient cells, and (2) it transfers much larger quantities of DNA (occasionally whole chromosomes). It is found mainly among gram-negative bacteria.

Conjugation was discovered in 1946 by Joshua Lederberg, who at that time was still a medical student. In his experiments, Lederberg used mutated strains of *E. coli* that were unable to synthesize certain substances. He selected two strains, each defective in a different synthetic pathway, and grew them in a nutrient-rich medium (Figure 8.6). He removed cells from each culture and washed them to remove the residue of the nutrient medium. He then attempted to culture cells of each strain on agar plates that lacked the special nutrients needed by the strain. He also mixed cells from the two strains and plated them on the same medium. Whereas cells from the original cultures failed to grow, some from the mixed cultures did grow. The latter must have acquired the ability to synthesize all the substances they needed. Lederberg and others continued to study this phenomenon and eventually discovered many of the details of the mechanism of conjugation.

Lederberg was indeed fortunate in his choice of organisms, because similar studies of other strains of *E. coli* failed to demonstrate conjugation. In addition to the mutations that led to synthetic deficiencies in Lederberg's organisms, other changes had modified their cell surfaces so that conjugation could occur.

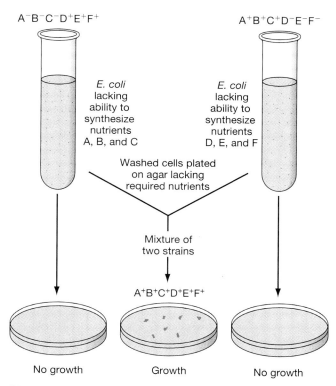

FIGURE 8.6 A schematic diagram of Lederberg's initial experiment that led to the discovery of conjugation.

Mechanisms of Conjugation

The mechanisms involved in conjugation were clarified through several important experiments, each of which built on the findings of the preceding one. Of those experiments, we will consider three: transfer of F plasmids, high-frequency recombinations, and transfer of F′ plasmids. Recall from Chapter 4 that **plasmids** are small *extrachromosomal DNA* molecules. ∞ (p. 84) Bacterial cells often contain several different plasmids that carry genetic information for various nonessential cell functions.

Transfer of F Plasmids

After Lederberg's initial experiment, an important discovery about the mechanism of conjugation was made. Two types of cells, called F$^+$ and F$^-$, were found to exist in any population of *E. coli* capable of conjugating. F$^+$, or *donor*, cells contain extrachromosomal DNA called **F** (fertility) **plasmids;** F$^-$, or *recipient*, cells lack F plasmids. (Lederberg coined the term *plasmid* in the 1950s to describe these fragments of DNA.)

Among the genetic information carried on the F plasmid is information for the synthesis of proteins that form F pili. The F$^+$ cell makes an **F pilus** (or *sex pilus* or *conjugation pilus*), a bridge by which it attaches to the F$^-$ cell when F$^+$ and F$^-$ cells conjugate (Figure 8.7). ∞ (Chapter 4, p. 89) A copy of the F plasmid is then trans-

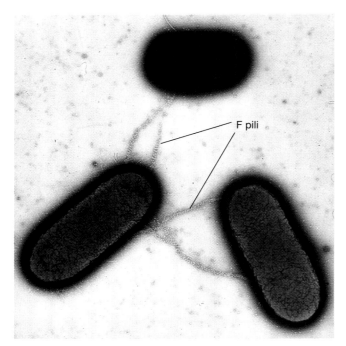

FIGURE 8.7 TEM photo of F pili of *E. coli* (1 F$^+$ and 2 F$^-$; 18,000×). Phages along the pili make them visible. Unlike the shorter attachment pili (fimbriae), this long type of pilus is used for transfer of genes in conjugation and is often called a sex pilus.

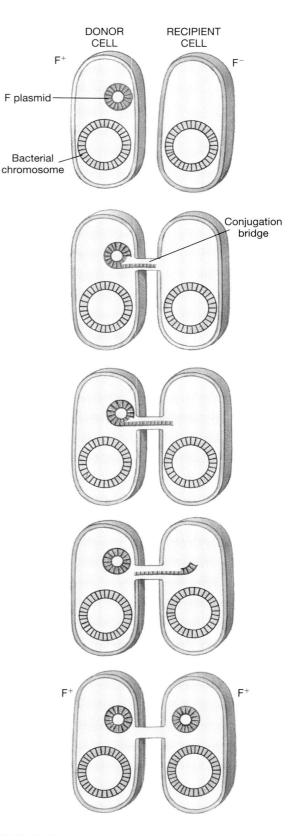

FIGURE 8.8 An F$^+$ × F$^-$ mating. The F$^+$ cell transfers one strand of DNA from its F plasmid to the F$^-$ cell via the conjugation bridge. As this occurs, the complementary strands of F plasmid DNA are synthesized. Thus, the recipient cell gets a complete copy of the F plasmid, and the donor cell retains a complete copy.

ferred from the F$^+$ cell to the F$^-$ cell (Figure 8.8). F$^+$ cells also are called *male* cells, and F$^-$ cells also are called *female* cells. Because pili are formed only on gram-negative bacteria, the transfer of F plasmids is specific to those organisms.

Although the exact transfer process remains unknown, the DNA is transferred as a single strand via a *conjugation bridge.* This bridge might represent a channel through the F pilus or, as some evidence suggests, a temporary fusion between the F$^+$ and F$^-$ cells, during which time the DNA is transferred. Each cell then synthesizes the complementary strand of DNA, so both have a complete F plasmid. Because all F$^-$ cells in a mixed culture of F$^+$ and F$^-$ cells receive the F plasmid, the entire population quickly becomes F$^+$.

High-Frequency Recombinations

The mechanisms of conjugation were further clarified when the Italian scientist L. L. Cavalli-Sforza isolated a **clone,** a group of identical cells descended from a single parent cell, from an F$^+$ strain that could induce more than a thousand times the number of genetic recombinations seen in the F$^+$ and F$^-$ conjugations. Such a donor strain is called a **high frequency of recombination (Hfr) strain.**

Hfr strains arise from F$^+$ strains when the F plasmid is incorporated into the bacterial chromosome at

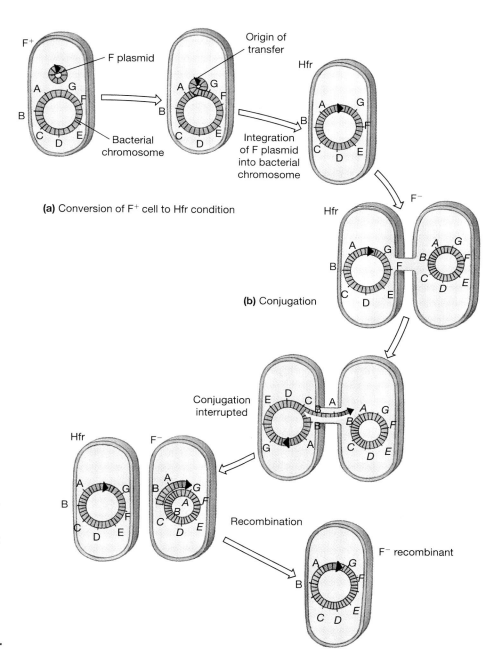

(a) Conversion of F⁺ cell to Hfr condition

(b) Conjugation

FIGURE 8.9 (a) Conversion of F⁺ cells to the Hfr condition. Hfr cells arise from F⁺ cells when their F plasmid is incorporated into a bacterial chromosome at one of several possible sites. (b) During conjugation, the (pink) initiating site of the F plasmid and adjacent genes are transferred to a recipient cell. Genes are transferred in linear sequence, and the number of genes transferred depends on the duration of conjugation and whether the DNA strand breaks or remains intact.

one of several possible sites (Figure 8.9a). When an Hfr cell serves as a donor in conjugation, the F plasmid initiates transfer of chromosomal DNA. Usually, only part of the F plasmid, called the **initiating segment,** is transferred, along with some adjacent chromosomal genes (Figure 8.9b).

In the 1950s the French scientists Elie Wollman and Francois Jacob studied this process in a series of interrupted mating experiments. They combined cells of an Hfr strain with cells of an F⁻ strain and removed samples of cells at short intervals. Each cell sample was subjected to mechanical agitation through vibration or whirling in a blender to disrupt the conjugation

process. Cells from each sample were plated on a variety of media, each of which lacked a particular nutrient, to determine their nutrient requirements. By careful observation of the genetic characteristics of cells from many experiments, the investigators determined that transfer of DNA in conjugation occurred in a linear fashion and according to a precise time schedule. When conjugation was disrupted after 8 minutes, most recipient cells had received one gene. When it was disrupted after 120 minutes, recipient cells had received much more DNA, sometimes an entire chromosome. At intermediate intervals, the number of donor genes transferred was proportional to the length of time con-

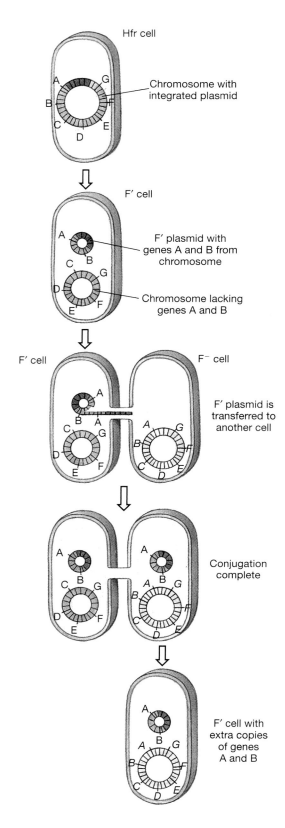

Hfr cell

Chromosome with integrated plasmid

F' cell

F' plasmid with genes A and B from chromosome

Chromosome lacking genes A and B

F' cell   F⁻ cell

F' plasmid is transferred to another cell

Conjugation complete

F' cell with extra copies of genes A and B

FIGURE 8.10 When the F plasmid in an Hfr cell separates from the bacterial chromosome, it may carry some chromosomal DNA with it. Such an F' plasmid may then be transferred by conjugation to an F⁻ cell. The recipient cell will then have two copies of some genes—one on its chromosome and one on the plasmid.

jugation was allowed to proceed. However, because of a tendency of chromosomes to break during transfer, some cells received fewer genes than would have been predicted by the time allowed. Whatever the number of genes transferred, they were always transferred in linear sequence from the initiation site created by the incorporation of the F plasmid.

Transfer of F' Plasmids

The process of incorporating an F plasmid into a bacterial chromosome is reversible. In other words, DNA incorporated into a chromosome can separate from it and again become an F plasmid. In some cases this separation occurs imprecisely, and a fragment of the chromosome is carried with the F plasmid, creating what is called an **F'** (F prime) **plasmid** (Figure 8.10). Cells containing such plasmids are called *F' strains*. When F' cells conjugate with F⁻ cells, the whole plasmid (including the genes from the chromosome) is transferred. Hence, recipient cells have two of some chromosomal genes—one on the chromosome and one associated with the plasmid. F' plasmids generally do not become part of the recipient cell's chromosome.

In the transfer of F⁺ and F' plasmids, as in all other transfers during conjugation, the donor cell retains all the genes it had prior to the transfer, including copies of the F plasmid. Single-stranded DNA is transferred, and both donor and recipient cells synthesize a complementary strand for any single-stranded DNA they contain.

The results of conjugation with F⁺, Hfr, and F' transfers are summarized in Table 8.1.

Significance of Conjugation

Like other mechanisms for gene transfer, conjugation is significant because it contributes to genetic variation. Larger amounts of DNA are transferred in conjugation than in other transfers, so conjugation is especially important in increasing genetic diversity. In fact, conju-

TABLE 8.1		Results of Selected Conjugations	
Donor	**Recipient**	**Molecule(s) Transferred**	**Product**
F⁺	F⁻	F plasmid	F⁺ cells
Hfr	F⁻	Initiating segment of F plasmid and variable quantity of chromosomal DNA	F⁻ with variable quantity of chromosomal DNA
F'	F⁻	F' plasmid and some chromosomal genes it carries with it	F' plasmid with some duplicate chromosomal genes

gation may represent an evolutionary stage between the asexual processes of transduction and transformation and the actual fusion of whole cells (the gametes) that occur during sexual reproduction in eukaryotes. For the microbial geneticist, conjugation is of special significance because precise linear transfer of genes is useful in chromosome mapping.

COMPARISON OF GENE TRANSFER MECHANISMS

The most fundamental differences among the major types of transfers of genetic information concern the quantity of DNA transferred and the mechanism by which the transfer takes place. In transformation, less than 1 percent of the DNA in one bacterial cell is transferred to another, and the transfer involves only chromosomal DNA.

In transduction, the quantity of DNA transferred varies from a few genes to large fragments of the chromosome, and a bacteriophage is always involved in the transfer. In specialized transduction, the phage inserts into a bacterial chromosome and carries a few host genes with it when it separates. In generalized transduction, the phage causes fragmentation of the bacterial chromosome; some of the fragments are accidentally packed into viruses as they are assembled.

In conjugation, the quantity of DNA transferred is highly variable, depending on the mechanism. A plasmid is always involved in the transfer. An F plasmid itself can be transferred, as occurs in F^+ to F^- conjugation. An initiating segment of a plasmid and any quantity of chromosomal DNA—from a few genes to the whole chromosome—is transferred in Hfr conjugation. A plasmid and whatever chromosomal genes it has carried with it from the chromosome are transferred in F′ conjugation. These characteristics are summarized in Table 8.2.

PLASMIDS

Characteristics of Plasmids

The F plasmid just described was the first plasmid to be discovered. Since its discovery, many other plasmids have been identified. Most are circular, double-stranded extrachromosomal DNA. They are self-replicating by the same mechanism that any other DNA uses to replicate itself. Most plasmids have been identified by virtue of some recognizable function that they serve in a bacterium. These functions include the following:

TABLE 8.2	Summary of the Effects of Various Transfers of Genetic Information
Kind of Transfer	**Effects**
Transformation	Transfers less than 1% of cell's DNA. Requires competence factor. Changes certain characteristics of an organism depending on which genes are transferred.
Transduction	Transfer is effected by a bacteriophage.
Specialized	Only genes near the prophage are transferred to another bacterium.
Generalized	Fragments of host bacterial DNA of variable length and number are packed into the head of a virus.
Conjugation	Transfer is effected by a plasmid.
F⁺	A single plasmid is transferred.
Hfr	An initiating segment of a plasmid and a linear sequence of bacterial DNA that follows the initiating segment are transferred.
F′	A plasmid and whatever bacterial genes associated with it when it leaves a bacterium are transferred.

1. F plasmids direct the synthesis of proteins that self-assemble into conjugation pili.
2. *Resistance (R) plasmids* carry genes that provide resistance to various antibiotics such as chloramphenicol and tetracycline and to heavy metals such as arsenic and mercury.
3. Other plasmids direct the synthesis of bacteriocidal (bacteria-killing) proteins called *bacteriocins*.
4. Virulence plasmids, such as those in *Salmonella*, cause disease signs and symptoms.
5. Tumor-inducing (Ti) plasmids can cause tumor formation in plants.
6. Some plasmids contain genes for catabolic enzymes.

Generally, plasmids carry genes that code for functions not essential for cell growth; the chromosome carries the genes that code for essential functions.

Resistance Plasmids

Resistance plasmids, also known as *R plasmids* or *R factors,* were discovered when it was noted that some enteric bacteria—bacteria found in the digestive tract—had acquired resistance to several commonly used antibiotics. We don't know how resistance plasmids arise, but we know that they are not induced by antibiotics. This has been demonstrated by the observation that cultures kept in storage from a time prior to the use of antibiotics exhibited antibiotic resistance on first exposure to the drugs. However, antibiotics contribute to the survival of strains that contain resistance plasmids. That is, when a population of organisms containing both resistant and nonresistant organisms is exposed to an antibiotic, the resistance organisms will survive and multiply, whereas the nonresistant ones will be killed. The resistant organisms are thus said to be *selected* to survive. Such selection is a major force in evolutionary change, as Charles Darwin realized.

According to Darwin, all living organisms are subject to *natural selection,* the survival of organisms on the basis of their ability to adapt to their environment. After studying many different kinds of plants and animals, Darwin drew two important conclusions. First, living organisms have certain heritable—that is, genetic—characteristics that help them adapt to their environment. Second, when environmental conditions change, those organisms with characteristics that allow them to adapt to the new environment will survive and reproduce. Organisms lacking such characteristics will perish and leave no offspring. A change in environmental conditions does not directly cause organisms to change. It merely provides a test of their ability to adapt. Only the organisms that can carry out their life processes under the new conditions will survive.

Resistance plasmids (Figure 8.11) generally contain two components: a **resistance transfer factor (RTF)** and one or more **resistance (R) genes.** The DNA in an RTF is similar to that in F plasmids. The RTF implements transfer by conjugation of the whole resistance plasmid and is essential for the transfer of resistance from one organism to another. Each R gene carries information that confers resistance to a specific antibiotic or to a toxic metal. For antibiotic resistance, such genes usually direct synthesis of an enzyme that inactivates the antibiotic. Some resistance plasmids carry R genes for resistance to four widely used antibiotics: sulfanilamide, chloramphenicol, tetracycline,

and streptomycin. Transfer of such a plasmid to any recipient confers resistance to all four antibiotics. Other resistance plasmids carry genes for resistance to one or more of these antibiotics. A few plasmids carry genes for resistance to more than four antibiotics.

The transfer of resistance plasmids from resistant to nonresistant organisms is rapid, so large numbers of previously nonresistant organisms can acquire resistance quickly. Furthermore, transfer of resistance plasmids occurs not only within a species, but also between closely related genera such as *Escherichia, Klebsiella, Salmonella, Serratia, Shigella,* and *Yersinia.* Transfer has even been observed between less closely related genera. Transfer of resistance plasmids is of great medical significance because it accounts for increasingly large populations of resistant organisms and reduces the effective use of antibiotics.

As scientists accumulate information on plasmids and how they confer antibiotic resistance, they become more concerned about the development of resistant strains and their potential danger to public health. As we shall see in Chapter 14, penicillin-resistant strains of *Neisseria gonorrhoeae, Haemophilus influenzae,* and some species of *Staphylococcus* already exist. Other antibiotics must now be used to treat the diseases caused by those strains, and the day may come when no antibiotic will effectively treat them. The more frequently antibiotics are used, the greater the selection is for resistant strains. Therefore, it is extremely important to

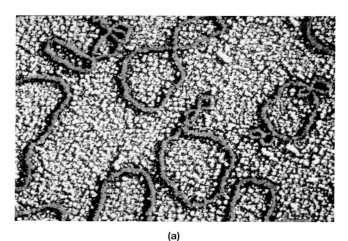

(a)

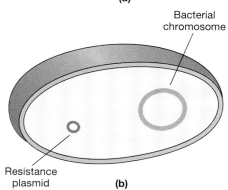

Bacterial chromosome

Resistance plasmid **(b)**

FIGURE 8.11   (a) Resistance plasmids (magnified 268,000×). These circular pieces of DNA are much smaller than a bacterial chromosome (b). A typical resistance plasmid (c) can carry genes for resistance to various antibiotics and to inorganic toxic substances, sometimes used in disinfectants. The resistance transfer factor includes genes needed for the plasmid to undergo conjugation.

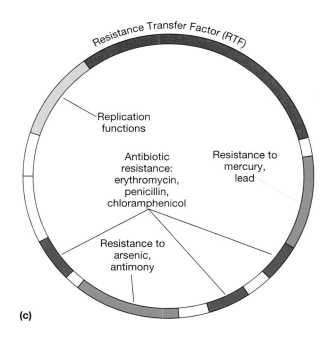

Resistance Transfer Factor (RTF)

Replication functions

Antibiotic resistance: erythromycin, penicillin, chloramphenicol

Resistance to mercury, lead

Resistance to arsenic, antimony

(c)

identify the antibiotic to which an organism is most sensitive before using any antibiotic to treat a disease.

Transposons

In addition to being transferred on resistance plasmids by conjugation, R genes also can move from one plasmid to another in a cell or even become inserted in the chromosome. The ability of a genetic sequence to move from one location to another is called **transposition.** Such a mobile genetic sequence is called a **transposable element.** The simplest type of transposable element, an *insertion sequence,* contains a gene that codes for an enzyme (transposase) needed to transpose the insertion sequence; that gene is flanked on either side by a sequence of 9 to 41 nucleotides called *inverted repeats.* Transposable elements replicate only when in a plasmid or in a chromosome. During transposition, the insertion sequence is copied by the transposase and cellular enzymes. The copy randomly inserts itself into the bacterial chromosome or into another plasmid; the original insertion element remains in its original position. However, the ability to move among plasmids or to a chromosome greatly increases the ways a transposable element can affect the genetic makeup of a cell. The coding sequence or regulatory regions of any gene into which a transposable element inserts can be disrupted. Transposable elements are known to cause mutations and may be responsible for some spontaneous mutations.

A **transposon** (tranz-pos′on) is a transposable element that contains the genes for transposition and one or more other genes as well (Figure 8.12). Typically, these other genes represent R genes, conferring resis-

tance to antibiotics such as tetracycline, chloramphenicol, or ampicillin. Thus, a transposon can move R genes from one plasmid to another or to the bacterial chromosome. The transposition of a transposon may disrupt gene function.

In 1983 Barbara McClintock won the Nobel Prize for her work on transposons, using corn. Transposons were next found in microorganisms and are now considered a universal phenomenon. Transposition is a relatively rare event and is not easily detected in eukaryotes. It is easier to detect in bacteria because researchers can work with large populations that can be tested more easily for particular characteristics.

Bacteriocinogens

In 1925 the Belgian scientist André Gratia observed that some strains of *E. coli* release a protein that inhibits growth of other strains of the same organism. This allows them to compete more successfully for food and space against the other strains. About 20 such proteins, called **colicins** (ko′leh-sinz), have been identified in *E. coli,* and similar proteins have been identified in many other bacteria. All these growth-inhibiting proteins are now called **bacteriocins** (bak-ter″e-o′sinz). Typically, bacteriocins inhibit growth only in other strains of the same species or in closely related species.

Bacteriocin production is directed by a plasmid called a **bacteriocinogen** (bak-ter″e-o-sin′o-jen). Although in most situations bacteriocinogens are repressed, in some cases the plasmid escapes repression and causes synthesis of its bacteriocin. Ultraviolet radiation can induce the formation and release of bacteriocin. When a bacteriocin is released, it can have a very

FIGURE 8.12 A typical transposon is bounded by inverted repeat terminals—DNA segments with base sequences that are identical when read in opposite directions on different strands. A gene coding for a transposase enzyme that cuts DNA at these insertion sequences allows the transposon to cut itself into and out of plasmids and chromosomes. There is also a gene for a repressor protein that can keep the transposase gene from being transcribed.

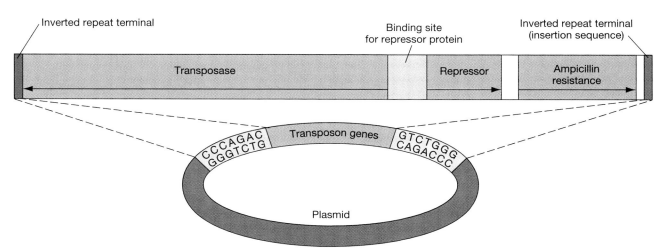

potent effect on susceptible cells; one molecule of bacteriocin can kill a bacterium.

The mechanisms of action of bacteriocins are quite variable. Some enter a bacterial cell and destroy DNA. Others arrest protein synthesis by disrupting the molecular structure of ribosomes required for protein synthesis. Still others act on cell membranes by inhibiting active transport or by increasing membrane permeability to ions.

GENETIC ENGINEERING

Genetic engineering refers to the purposeful manipulation of genetic material to alter the characteristics of an organism in a desired way (Table 8.3). Various methods of genetic manipulation allow microbial geneticists to create new combinations of genetic material in microbes. Transfer of genes between different members of the same species occurs in nature and has been done in the laboratory for several decades. Lederberg's experiment (Figure 8.6) provides one example of such a technique. Transfer of genes between different species also is now possible. We have selected five techniques of genetic engineering for discussion here. They are *genetic fusion, protoplast fusion, gene amplification, recombinant DNA technology,* and creation of *hybridomas.*

Genetic Fusion

Genetic fusion allows transposition of genes from one location on a chromosome to another. It can also involve deletion of a DNA segment, resulting in the coupling of portions of two operons. For example, sup-

TABLE 8.3	Some Products and Applications of Genetic Engineering	
Pharmaceutical Products	**Use**
Human insulin	Treat diabetes
Human growth hormone	Prevent pituitary dwarfism
Blood clotting factor VIII	Treat hemophilia
Erythropoietin	Treat anemia; stimulate formation of new red blood cells
Alpha-, beta-, and gamma-interferon	Treat cancer and viral disease
Tumor necrosis factor	Disintegrate cancer cells
Interleukin-2	Treat cancer and immunodeficiencies
Tissue plasminogen activator	Treat heart attacks, dissolve clots
Taxol	Treat ovarian and breast cancers
Bone growth factor	Heal bone fractures, treat osteoporosis, stimulate bone growth
Epidermal growth factor	Heal wounds
Monoclonal antibodies	Diagnose and treat diseases
Hepatitis A and B vaccines	Prevent hepatitis
AIDS subunit vaccine (in clinical trials)	Incomplete virus vaccine
Human hemoglobin	Blood substitute in emergencies (produced in gene-altered pigs)
Antibiotics	Inhibit or kill microbial growth (increase yields by gene amplification)
Genetic Studies	
DNA and RNA probes	Identify organisms, diseases, genetic defects in fetuses and adults
Gene therapy	Insert missing gene, or replace defective gene, in adults or in egg and sperm; treat cystic fibrosis
Gene libraries	Understand gene structure and function, relatedness of organisms, Human Genome Project
Industrial Applications	
Oil-eating recombinant bacteria	Clean up oil spills, remove oil residue from empty tankers
Pollutant/toxic materials–degrading recombinant bacteria	Clean up contaminated sites
Enzymes, vitamins, amino acids, industrial chemicals	Various uses (yield increased by gene amplification in producing microbes)
Agricultural Applications	
Frostban bacteria	Prevent frost damage to strawberry crops
Breeding new types of plants and animals	Provide food, decoration, other uses
Herbicide-resistant crop plants	Allow crop plants to survive weeding done by spraying with herbicides; only crop plants survive
Viruses used as insecticides	Infect and kill insect pests

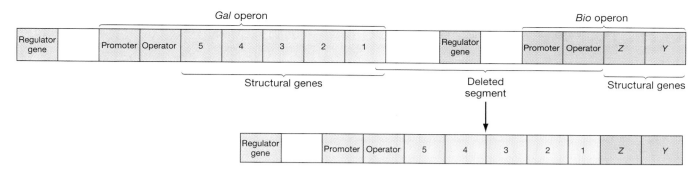

| Regulator gene | | Promoter | Operator | 5 | 4 | 3 | 2 | 1 | | Regulator gene | | Promoter | Operator | Z | Y |

Gal operon — Structural genes

Bio operon — Deleted segment — Structural genes

| Regulator gene | | Promoter | Operator | 5 | 4 | 3 | 2 | 1 | Z | Y |

FIGURE 8.13 A possible example of genetic fusion, in which the deletion of a part of a chromosome causes two different adjacent operons to be joined together. The control mechanisms of the first operon will now govern the expression of the genes that were originally part of the second operon.

pose that the *gal* operon, which regulates galactose use, and the *bio* operon, which regulates biotin synthesis, lie adjacent to each other on a chromosome (Figure 8.13). Deletion of the control genes of the *bio* operon and subsequent coupling of the operons would constitute genetic fusion. Such fusion would allow the genes that control the use of galactose to control the entire operon, including the making of the enzymes involved in biotin synthesis.

The major applications of genetic fusion within a species, as just described, are in research studies on the properties of microbes. However, the techniques developed for genetic fusion experiments have been extended and modified in the development of other kinds of genetic engineering.

One application of genetic fusion involves *Pseudomonas syringae,* a bacterium that grows on plants. Genetically altered strains have been developed that increase the resistance of plants, such as potatoes and strawberries, to frost damage. Strains of this bacterium that occur naturally on the leaves of plants produce a protein that forms a nucleus for the formation of ice crystals. The ice crystals damage the plants by causing cracks in the cells and leaves. By removing part of the gene that produces the "ice crystal" protein, scientists have engineered strains of *P. syringae* that cannot make the protein. (See the Essay at the end of this chapter.) When organisms of this strain are sprayed on the leaves of plants, they crowd out the naturally occurring strain. The treated plants then become resistant to frost damage at temperatures as low as −5°C.

Protoplast Fusion

A **protoplast** is an organism with its cell wall removed. **Protoplast fusion** (Figure 8.14) is accomplished by removing the cell walls of organisms of two strains and mixing the resulting protoplasts. This allows fusion of the cells and their genetic material; that is, material

from one strain recombines with that from the other strain before new cell walls are produced. Although genetic recombination occurs in nature in about one in a million cells, it occurs in protoplast fusion in as many as one in five cells. Thus, protoplast fusion simply speeds up a process that occurs in a very limited way in nature.

By mixing two strains, each of which has a desirable characteristic, new strains that have both characteristics can be produced. For example, a slow-growing strain that produces large quantities of a desired substance can be mixed with a fast-growing, poor producer. After protoplast fusion, some organisms will probably be fast-growing, good producers of the substance. Other organisms that turn out to be slow-growing, poor producers are discarded. Alternatively, two

FIGURE 8.14 False-color TEM photo of two tobacco plant leaf cells (magnified 775×) undergoing protoplast fusion. Protoplast fusion involves the use of enzymes to digest away the cell walls of cells of organisms from two different strains. When placed together, the cells fuse and develop a new cell wall around the hybrid cell containing the genes of both organisms.

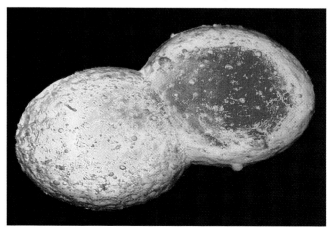

MICROBIOLOGIST'S NOTEBOOK

Puzzling Out an All-New Sequencing Strategy

Hamilton O. Smith, M.D. (seated), and J. Craig Venter, Ph.D., have succeeded in deciphering the first complete DNA sequence of an organism's genome—that of Haemophilus influenzae.

Hamilton Smith: We now know the DNA sequence of the entire *Haemophilus influenzae* genome, all 1.9 million base pairs. The collaboration that led to the first complete DNA sequence of an organism's genome began at a Genome Ethics meeting in Spain. A colleague from England first proposed the idea, but I threw cold water on it; a sequencing project of that size would require production-scale work not feasible or affordable in an academic environment. So the whole idea just sat there until several months later, when I ran into Craig Venter for the first time in Spain.

I've worked on *Haemophilus influenzae* for about 25 years—nearly my entire career here at Johns Hopkins University School of Medicine. As far as we know, humans are the only natural host for these bacteria. The strain we've sequenced, serotype *d*, isn't actually very pathogenic. We treat it about like *E. coli* in the lab. But a closely related strain, serotype *b*, causes respiratory tract infections, mostly in children, and it's a significant cause of middle ear infection. Meningitis is the most serious complication, but since a vaccine came into common use in the early 1980s, the incidence of meningitis has dropped dramatically.

Craig Venter: Aside from its clinical significance, this organism has an important historical role—the first restriction endonucleases were isolated from it. Ham discovered them during transformation studies, and for that discovery he and colleagues Daniel Nathans at Hopkins and Werner Arber in Switzerland were awarded the Nobel Prize. This organism has played a fundamental role in molecular biology and biotechnology and has had an outstanding impact on science. Without restriction enzymes, much of what we do would not be feasible.

At The Institute for Genomic Research, or TIGR, where I'm president and director, we're interested in accelerating the sequencing of human, animal, and plant genomes to understand better the role that genes and gene products play in development, evolution, physiology, and disease. It's the world's largest noncommercial facility devoted to large-scale DNA sequencing. We work on characterizing human genes directly through sequencing and indirectly by stimulating the field of bioinformatics [analysis of complex biological data] as the starting point for evolutionary studies. There's been much discussion of model organisms, and we decided to start with human at one end and microorganisms at the other end, so we have both to compare.

We were looking for a prototypical microorganism to test an idea. It didn't much matter which one it was as long as it fit certain criteria. The medium-sized, AT-rich *Haemophilus* genome seemed a good choice. And because Ham had extensive knowledge of the organism, he helped us further develop our ideas on sequencing strategy.

Smith: Craig was doing incredibly exciting work with the human genome. He developed a whole new method of gene discovery, called *expressed sequence tags,* or ESTs, because they're derived from messenger RNA instead of genomic DNA. The EST method has dramatically increased the pace of gene discovery.

I thought I'd have to convince him that we should sequence the *Haemophilus* genome, and I offered to provide a library [a collection of pieces] of the genome for sequencing. It turned out that he already knew exactly how we should do it, because it was a direct extension of the large-scale random-sampling techniques he had so successfully applied to ESTs. We both agreed that shotgun sequencing—looking at totally random sections of the genome—was the way to go. We were going to test a new approach, the whole-genome method.

Here's the project in a nutshell. I made the library by breaking the genome into 25,000–30,000 small overlapping pieces. Then TIGR sequenced those pieces directly, without using any other mapping information. They fed the sequences into their computer, which assembled them in order. It takes about a weekend for a supercomputer to reassemble that many pieces.

Venter: You can't just go and determine the sequence of something nearly 2 million base pairs long. With current sequencing technology, the maximum length of highly accurate sequence you can get in a sequencing run is 500–600 base pairs. You have to sequence these smaller fragments and then rely on finding exact matches in the genetic code from one piece to another.

Large-scale sequencing projects in the past involved sequencing lots and lots of small cloned pieces over and over again to build larger pieces. So, characterizing any genome involved spending months or years first generating a rough map of the genome and a clone set to work from. It's very slow and inefficient. The National Institutes of Health is funding a multiyear project that uses that general method to sequence the genome of *E. coli*. But Ham and I felt that if each genome were to take many years and multimillion dollars, then microbiology would not move forward very quickly.

We recognized that the real problem with larger-scale sequencing has nothing to do with sequencing technology. In fact, we rely on off-the-shelf instruments. It has to do with the mathematical limitations of assembling bits of sequence into longer stretches. We had considered a whole-genome shotgun approach when we sequenced the smallpox genome but rejected the idea because of the assembly problem. TIGR has the combination of a high-production sequencing facility and, more importantly, the software to enable us to compare and align sequences.

It becomes quite a surmountable computational problem to find out how 25,000–30,000 individual sequences, each 300–500 nucleotides long, all go together. But it was out of our work sequencing human DNA that we had to develop that ability. We're always looking for ways to approach genome sequencing at new, highly efficient levels.

Smith: While those who were waiting for faster sequencing technology were still sitting in their chairs, Craig waded right in with the current technology. I think TIGR is one of the best places in the world for sequencing. I would liken the approach with *E. coli* as starting at one spot and then working yourself around the whole genome from just one point. It's like an artist starting up in the corner and beginning to draw in the painting with all the details finished in the little corner before moving to the next area of the painting. Instead, TIGR uses a global approach, essentially sketching in the whole picture in outline, then filling it in everywhere at once. The whole picture emerges at one time.

Venter: At 1.9 million base pairs, the *Haemophilus* genome is a little less than half of *E. coli*'s, which is about 4.7 megabases. So they're not comparable-size projects. But we think this strategy would work for either size genome. It's similar to solving a jigsaw puzzle, except without the nice, square borders. Usually with jigsaw puzzles, people build the outside edge and then fill in toward the center. But with whole-genome shotgun sequencing, there are no edges (or maps) to start with. There's nothing to give you any intellectual advantage.

For this global approach to work, the library you sequence from has to be not only a complete representation of the genome but a truly random representation of the genome. If that one step of constructing the library was wrong, our best theories and our best sequencing and the best mathematics wouldn't have been able to solve the problem. That was probably the least expensive step in the whole process, but if that had been wrong, the rest of it would not have worked.

Smith: I think it's fair to say that it's going to be a brand new era in microbiology. Having the complete sequence of virtually any bacterium should really expedite any investigation with the microorganism. You'll have in front of you all the gene content of that organism. If you had the sequence of, for example, *Neisseria,* you could compare that with the known sequence of *Haemophilus* and two or three other organisms. You could immediately pick out those genes that are unique to *Neisseria.* Or you could select genes involved in the virulence [disease-producing capacity] of the organism, which would therefore become targets for therapeutics.

Furthermore, you can go in and lift out individual genes by using the powerful PCR techniques. You can do knockouts of individual genes to look at function. You can look at the expression of genes under various growth conditions. There's going to be an enormous change in the way people investigate these organisms.

Just imagine trying to compare two sequences, each over a million base pairs long. It's impossible to deal with this level of information without having sophisticated computers.

Venter: I think we both agree that the computer is going to be the molecular biologist's number one tool for the future. It's going to be the starting point for developing hypotheses, for understanding what is known, and for going forward.

Just to print the *Haemophilus* genome sequence, absolutely packed on a page so that you could read it without a magnifying glass, would take 250 pages. Indicating gene locations and amino acids would require over 1000 pages. And we're just in the infancy of generating this information. This is only the first bacterial genome sequence to be completed. We hope a decade from now there'll be a hundred or more. [In fact, a team led by Dr. Claire Fraser of TIGR has sequenced the genome of a second organism, *Mycoplasma genitalium*. With fewer than 600,000 base pairs, that genome took just 3 months to sequence.]

When we begin comparing entire genomes from one species to another, we'll be able to start rethinking evolution in terms of what genes appeared or disappeared, what mutations occurred that maybe allowed that species to evolve. Evolutionary trees have been drawn up by inferring evolutionary distances based on sequence variations in a few genes. But I suspect that if you look at multiple genes, each will have a slightly different evolutionary tree. And of course, what really evolved were organisms. Having complete gene sets from microorganisms is going to lead to a better understanding of humans.

In fact, it's already happening. In collaboration with Dr. Bert Vogelstein, we recently found three new human cancer genes during a brief search of the TIGR sequence database. Knowing the sequence of a DNA repair enzyme characterized in *E. coli (MutL)*, we had annotated 16 human sequences as DNA repair enzymes. Dr. Vogelstein had speculated that defects in a similar human enzyme, if it existed, might lead to colon cancer. Indeed, mutant forms of the four genes found are responsible for a form of colon cancer.

Our work with *Haemophilus* shows that obtaining the complete genome sequence of an organism is truly feasible. They're not multiyear, multimillion dollar projects. This project took less than 12 months, but much of that time was spent in technology development—this was raw, basic research—so subsequent genomes will take substantially less time. As in any area, you have to have a breakthrough in technology to make new strides. It seemed to both of us that the only way it was going to happen was if we could take this whole-genome approach.

The bacillus *Haemophilus influenzae* (2800×).

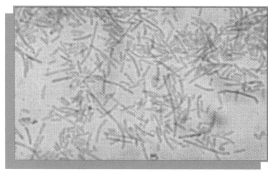

SUPERMARKET TOMATOES THAT TASTE LIKE HOME-GROWN

Tired of tomatoes that look beautiful but have all the juice and flavor of cardboard? Ripening on the vine gives tomatoes their wonderful juicy flavor. Home-grown tomatoes are picked at their peak of ripeness and flavor. Tomatoes destined for supermarkets cannot be ripe when picked, or they will arrive at the store as a soggy, spoiled mess. Green tomatoes can withstand the rigors of shipping, but they don't have color, flavor, or juice when they arrive—they didn't stay on the vine long enough. Holding green tomatoes in storage areas in which the gas ethylene has been pumped causes green tomatoes to turn a deceptively beautiful red. However, the juice and flavor never develop.

In May 1994, Calgene, Inc., a plant biotechnology company located in Davis, California, received government approval for their genetically altered "Flavr Savr" tomato. It stays fresh up to 10 days longer than ordinary tomatoes, which allows it to ripen longer on the vine before it is picked and shipped. Five years of development went into the creation of this new tomato. Here is how it is done:

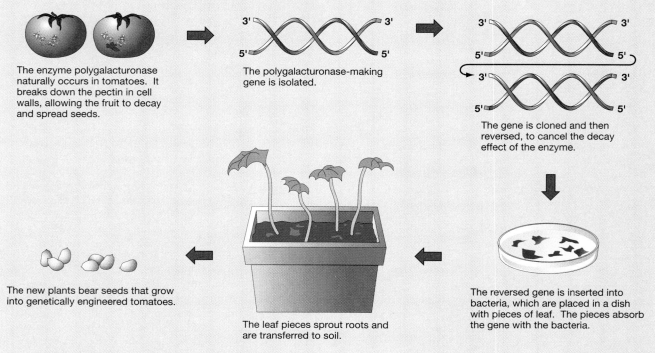

The enzyme polygalacturonase naturally occurs in tomatoes. It breaks down the pectin in cell walls, allowing the fruit to decay and spread seeds.

The polygalacturonase-making gene is isolated.

The gene is cloned and then reversed, to cancel the decay effect of the enzyme.

The reversed gene is inserted into bacteria, which are placed in a dish with pieces of leaf. The pieces absorb the gene with the bacteria.

The leaf pieces sprout roots and are transferred to soil.

The new plants bear seeds that grow into genetically engineered tomatoes.

The Food and Drug Administration (FDA) has given its safety endorsement for Flavr Savr. The tomatoes and their products won't need special labeling—the FDA has determined that no new or unexpected consequences, such as allergic reactions, should result from their use.

Not everyone, however, greeted the government approval with joy. Some people are afraid to eat them. Calgene is placing stickers on each of its Flavr Savr tomatoes and making brochures available in stores to explain its new product to customers.

good producers can be mixed to obtain a super producer. This has been done with two strains of the bacterium *Nocardia lactamdurans,* which produce the antibiotic cephalomycin. The new strains produce 10 to 15 percent more antibiotic than the best of the parent strains.

Protoplast fusion works best between strains of the same species. It has been accomplished in molds, however, between two species of the same genus (*Aspergilus nidulans* and *A. rugulosus*) and even between two genera of yeasts (*Candida* and *Endomycopsis*).

Microbiologists are exploring possible applications of protoplast fusion. It offers great promise for the future as procedures are refined and useful strains are developed.

Gene Amplification

Gene amplification is a process by which plasmids, or in some cases bacteriophages, are induced to reproduce within cells at a rapid rate. If the genes required for the production of a substance are in the plasmids or can be moved to them, increasing the number of plasmids will increase production of the substance by the host cells.

Most bacteria and many fungi, including those that produce antibiotics, contain plasmids. Such plasmids, which often carry genes for antibiotic synthesis, provide many opportunities for using gene amplification to increase antibiotic yields. Even when genes concerned with antibiotic production are in the chromosome, scientists can transfer them to plasmids. Increased reproduction of plasmids would then greatly increase the number of copies of genes that act in antibiotic synthesis. This, in turn, would significantly increase the amount of antibiotic such cells could produce.

The possible applications of gene amplification are not limited to increasing antibiotic production. In fact, gene amplification may turn out to be even more effective in increasing production of substances that are synthesized by somewhat simpler pathways. These substances include enzymes and other products such as amino acids, vitamins, and nucleotides.

Rapid reproduction of bacteriophages already can be used to make the amino acid tryptophan. Bacteriophages carrying the *trp* operon (genes that control synthesis of enzymes to make tryptophan) of *E. coli* are induced to reproduce rapidly. Thus, cells containing large numbers of copies of the *trp* operon synthesize large quantities of the enzymes. Subsequent analysis of such cells has shown that half the intracellular proteins are enzymes for tryptophan synthesis.

Recombinant DNA Technology

One of the most useful of all techniques of genetic engineering is the production of **recombinant DNA**—DNA that contains information from two different species of organisms. If these genes integrate permanently into the egg or sperm cells such that the genes can be transferred to offspring, the resulting organism is said to be **transgenic** or a *recombinant organism*. Making recombinant DNA involves three processes:

1. The manipulation of DNA *in vitro*—that is, outside cells.
2. The recombination of another organism's DNA with bacterial DNA in a phage or a plasmid.
3. The *cloning,* or production of many genetically identical progeny, of phages or plasmids that carry foreign DNA.

These processes were first carried out in 1972 by Paul Berg and A. D. Kaiser, who inserted other prokaryotic DNA into bacteria, and then by S. N. Cohen and Herbert Boyer, who inserted eukaryotic DNA into bacteria.

DNA from either prokaryotic or eukaryotic cells is removed from the cells and cut into small segments. The donor DNA segments are then incorporated into a **vector,** or self-replicating carrier such as a phage or a plasmid (Figure 8.15). First, a restriction endonuclease is used to cut double-stranded DNA in the vector and donor DNA. The cuts leave overlapping ends. A particular restriction endonuclease always produces the same complementary ends. Then, donor DNA is incorporated into the vector by an enzyme called a *ligase,* which reunites the ends of nucleotide chains. Thus, the vector contains all the original DNA plus a new segment of donor DNA.

Once this new segment of DNA is inserted into the vector, it can be introduced into cells such as *E. coli* that have been rendered competent by heating (annealing) in a solution of calcium chloride or by **electroporation.** This technique uses a brief electrical pulse to produce temporary pores in the cell membrane through which the vector can pass. As the *E. coli* cells divide, the vectors in them also are reproduced by cloning. Such vector-containing cells can be identified, grown, and lysed, and the vectors containing a specific cloned segment of donor DNA can be retrieved. In other cases, large quantities of the protein product expressed by the donor gene are required.

Medical Applications of Recombinant DNA

One of the most medically significant applications of recombinant DNA technology is the modification of bacterial cells to make substances useful to humans. To make bacterial cells produce human proteins, a human DNA gene with the information for synthesizing the protein is inserted into the vector. Interferon, a substance used to treat certain viral infections and cancers (Chapter 17), and the hormone insulin were among the first products made with recombinant DNA. Human growth hormone now can be made that way, and new products—vaccines, blood coagulation proteins for people with hemophilia, and enzymes such as cholesterol oxidase to diagnose disorders in cholesterol metabolism—have been developed. Many of the products in Table 8.3 are the result of recombinant DNA technology.

The use of recombinant DNA technology to make substances useful to humans makes certain treatments potentially safer, cheaper, and available to more patients. For example, prior to the use of recombinant DNA to manufacture human insulin, the insulin for diabetic patients came exclusively from slaughtered cattle and pigs. Some patients develop allergies to such insulin, and the number of patients requiring insulin is increasing. Making nonallergenic human insulin and increasing the insulin supply are two important benefits of insulin production by recombinant DNA technology. Most patients in the United States now use a preparation containing genetically engineered insulin.

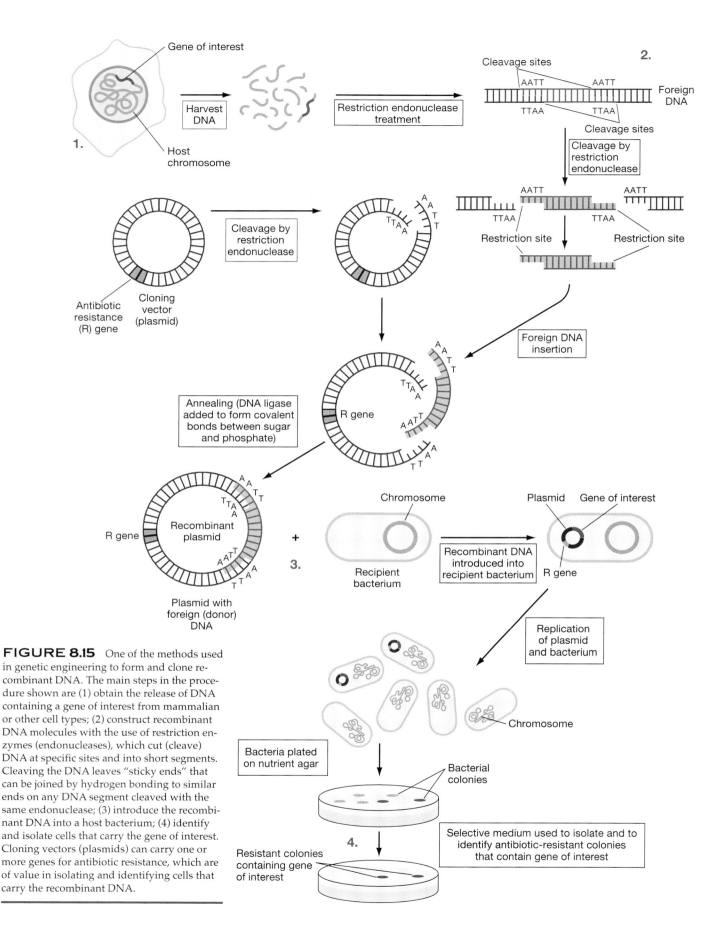

FIGURE 8.15 One of the methods used in genetic engineering to form and clone recombinant DNA. The main steps in the procedure shown are (1) obtain the release of DNA containing a gene of interest from mammalian or other cell types; (2) construct recombinant DNA molecules with the use of restriction enzymes (endonucleases), which cut (cleave) DNA at specific sites and into short segments. Cleaving the DNA leaves "sticky ends" that can be joined by hydrogen bonding to similar ends on any DNA segment cleaved with the same endonuclease; (3) introduce the recombinant DNA into a host bacterium; (4) identify and isolate cells that carry the gene of interest. Cloning vectors (plasmids) can carry one or more genes for antibiotic resistance, which are of value in isolating and identifying cells that carry the recombinant DNA.

NEED A BLOOD TRANSFUSION? CALL ON A GENETICALLY ALTERED PIG

By the year 2000, researchers expect to get government approval of a blood-substitute product composed primarily of human hemoglobin that is produced by transgenic (genetically altered) pigs. The biotechnology firm DNX, located in Princeton, New Jersey, injected thousands of copies of two human hemoglobin genes into 1-day-old pig embryos that had been removed from their mothers' uteri. The embryos were then implanted into a second pig's uterus to grow until delivery. Two days after birth, the piglets were tested to see if they produced human hemoglobin along with pig hemoglobin—that is, if they were transgenic. Only about 0.5 percent of such transfers succeed.

It costs $50,000 to $75,000 to make one transgenic animal. DNX was initially successful in making three such pigs. Then the company bred the one transgenic male with more than 1000 regular females and interbred the offspring of these crosses. DNX now has hundreds of transgenic pigs. Such crossing has continued for four generations. The human hemoglobin genes still function perfectly in the altered pigs, allaying fears that the genes would mutate or disintegrate in their new environment. The transgenic pigs currently produce blood containing more than 30 percent human hemoglobin, about 20 percent hybrid human/pig hemoglobin, and almost 50 percent pig hemoglobin. DNX hopes eventually to reach a blood level of 50 percent human hemoglobin in the pigs.

This pig has genes for production of human hemoglobin. Such pigs will be bred and bled to collect human hemoglobin to save people's lives by transfusion. This is an example of biotechnologic "pharming" of molecules.

To obtain the blood substitute, the pigs are bled and the red blood cells are ruptured. Pure human hemoglobin is separated from hybrid human/pig and pig hemoglobin by a multistep purification process that uses all available forms of chromatography—to safeguard against reliance on a single method that might fail and retain impurities.

The substitute product has several advantages over actual human blood:

1. It has a storage life of months instead of weeks.
2. Because naked hemoglobin does not stimulate the immune system to act against it, as do intact red blood cells containing hemoglobin, it can be transfused into anyone without the need for blood typing and matching.
3. It can ensure safety from human pathogens (including the AIDS virus) that might now contaminate human blood.
4. Because of the oxygen-carrying capacity of hemoglobin, it can serve as an immediate source of oxygen in earthquakes, in accidents, or in times of war, thereby enabling the injured to survive the trip to the hospital.

Human blood costs have recently risen to $300 to $500 per unit. By the year 2000, the blood substitute may cost the same as, or even less than, a unit of blood. Storage will be cheaper, and blood-typing costs would be eliminated.

One drawback to this procedure is that, once transfused, naked hemoglobin lasts only hours or days instead of 6 months, but this might be long enough to treat emergency cases. Another problem with this product is possible contamination with pig molecules or pig pathogens if purification processes fail. One issue that does *not* appear to be a problem is the use of this product by Jewish people who, for religious reasons, do not eat pork. The director of the Rabbinical Council of America has said that there would probably be no religious objection. By Judaic law, pigs may be used for purposes other than eating (such as sources of heart-valve replacements and insulin), and kosher laws are suspended in cases of life or death.

DNX had hoped to have government approval by now, but it has been unable to raise the $50 million needed to do so. In the meantime, it has been working on increasing its herd and techniques for large-scale production. Future use of the hybrid hemoglobin has not been ruled out, but further research is needed to determine how well it would function. DNX is also looking into the production of transgenic pig hearts for transplant into humans.

Likewise, prior to the manufacture of human growth hormone by recombinant DNA technology, the hormone was obtained from the pituitary glands of cadavers at autopsy, and several cadavers were needed to obtain a single dose with which to treat children with the congenital condition of pituitary dwarfism. This treatment was suspended in April 1985 in both the United States and the United Kingdom due to reports of several cases of the fatal, degenerative neurological Creutzfeldt-Jakob disease. But later in 1985, the FDA approved a genetically engineered form of the hormone that produces the same effects as the human cadaver hormone. In addition to correcting congenital disorders, this product of recombinant DNA technology might prove useful in the treatment of delayed wound or fracture healing and in the metabolic problems associated with aging.

The manufacture of certain blood coagulation proteins by recombinant DNA technology makes these substances more readily available to individuals with hemophilia or other blood disorders. It also assures that the recipient will not acquire AIDS or hepatitis B from a contaminated blood product.

Recombinant DNA is being used to make vaccines more economically and in larger quantities than before. In this application, some microorganisms are used to combat the disease-causing capacity of others. Genes that direct the synthesis of specific substances, called *antigens,* from a disease-causing bacterium, virus, or protozoan parasite are inserted into another organism. The organism then makes a pure antigen. When the antigen is introduced into a human, the human immune system makes another specific substance, called an *antibody,* which takes part in the body's defense against the disease-causing organism (Chapter 18).

Recombinant DNA procedures for making vaccines for hepatitis A and B and influenza are already available. The vaccines are not only cheaper than conventional ones but also purer and more specific, and they cause fewer undesirable side effects. Vaccines have proved to be highly specific and extremely effective against hepatitis A and B.

Many other applications of recombinant DNA techniques are being developed. An especially important one is the diagnosis of genetic defects in a fetus, which can be done by studying enzymes in fetal cells from amniotic fluid. Such defects are detected by using recombinant DNA with a known nucleotide sequence to find errors in the nucleotide sequence in fetal DNA segments. Such errors in fetal DNA denote genetic defects that can be responsible for absent or defective enzymes. Application of these techniques could greatly improve prenatal diagnosis of many genetic defects. Ultimately, as techniques for preparing recombinant DNA in animal cells improve, it may become possible to insert a missing gene or to replace a defective one in human cells (*gene therapy*). In fact, insertion of functional genes in appropriate cells may have cured the genetic disease severe combined immunodeficiency disease (Chapter 19). Insertion of such a gene in a defective gamete (egg or sperm) might prevent offspring from inheriting a genetic disease.

Forensic applications of DNA technology are rapidly coming into use in the courtroom. In paternity cases, for example, experts can now determine with about 99 percent certainty that a given man is the father of a particular child, on the basis of DNA comparison (refer to the Polymerase Chain Reaction Essay in Chapter 7). ∞ (p. 190) Likewise, rapists and murderers can be identified by the "DNA fingerprints" they leave behind at the scene of the crime in the form of semen, blood, hair, or tissue under their victim's fingernails. Sufficient DNA can even be collected for analysis from the rim of a used drinking glass or from the sweatband of a hat.

Industrial Applications of Recombinant DNA

Fermentation processes used in making wine, antibiotics, and other substances might be greatly improved by the use of recombinant DNA. For example, addition of genes for the synthesis of amylase to the yeast *Saccharomyces* could allow these organisms to produce alcohol from starch. Malting of grain to make beer would be unnecessary, and wines could be made from juices containing starches instead of sugars. Still other applications might include degradation of cellulose and lignin (plant materials often wasted), manufacture of fuels, clean-up of environmental pollutants, and leaching of metals from low-grade ores. Strains of *Pseudomonas putida,* already known to degrade different components of oil, might be engineered so that one strain degrades all components. Industrial leaching, or extraction, of metals from copper and uranium ores is already carried out by certain bacteria of the genus *Thiobacillus.* If these organisms could be made more resistant to heat and to the toxicity of the metals that they leach, the extraction process could be greatly accelerated.

Agricultural Applications of Recombinant DNA

Some bacteria are being engineered to control insects that destroy crops. The Monsanto Company has recently modified the genetic makeup of a strain of *Pseudomonas fluorescens,* which colonizes the roots of corn. This bacterium has been induced to carry genetic information inserted into it from *Bacillus thuringiensis,* allowing *P. fluorescens* to synthesize a protein that kills insects (Figure 8.16). Toxin made by *B. thuringiensis* has

FIGURE 8.16 SEM photo (118,750×) of crystals of a substance toxic to many insects. The genes for production of this toxin are being taken from *Bacillus thuringiensis*, which produces it naturally, and are being incorporated into other organisms through genetic engineering techniques. Imagine the benefits of crop plants that have built-in pesticide: no expense for chemical pesticides, no danger during application, no buildup in soil or water, and no entry or magnification of pesticides in the food chain.

been extracted and used for many years as an insecticide. Now the pseudomonads, applied to the surface of corn seeds, can make the toxin as they grow around the roots of the corn.

If the pseudomonads survive in corn fields as well as they have in greenhouses, they could replace the use of chemical insecticides to control black cutworm and probably other insect larvae that damage crops. With further research, other bacteria normally present on crops might be modified to control additional pests.

BIOTECHNOLOGY

VIRUS WITH A SCORPION'S STING

Caterpillars that devour valuable crop plants are sometimes infected by a naturally occurring virus that is harmless to all other organisms. The caterpillars develop a long illness that culminates in their climbing to the top of a plant and bursting—and releasing a shower of infectious viral particles. But during their long illness, they still eat quite a bit of the crops. Now genetic engineers have inserted the gene for scorpion venom into the virus. Caterpillars infected with the recombinant virus manufacture scorpion venom and quickly develop convulsions, paralysis, and death—all without stopping for a bite to eat. Any altered virus that remains uneaten is destroyed in sunlight, and no chemical pesticide residues enter the food chain.

Some optimistic scientists believe chemical pesticides may be phased out in favor of these safer and cheaper methods of pest control.

Also under development are genetically engineered seeds for crop plants that have high yield and other desirable characteristics and that will resist herbicides that kill weeds. The Environmental Protection Agency has begun steps to approve growth of genetically altered crops to build up supplies of seeds that will produce their own insecticides. Potatoes, corn, and cotton have received genes for the production of insect-killing toxin from *B. thuringiensis*. Their seeds will kill the Colorado potato beetle, the European corn borer and other mothlike insects, and the cotton bollworm, pink bollworm, and tobacco budworm, respectively. Farmers will be able to buy seeds that will solve many cultivation problems. Attempts have also been made to introduce nitrogen-fixing genes into nonleguminous plants. (We will discuss nitrogen fixation in Chapter 26.) This work has been successful in some plants but not yet in any important crop plants. If it can be extended to crop plants, many agricultural crops could be made to satisfy their nitrogen needs and thrive without commercial fertilizers, which are expensive and tend to pollute groundwater. This would be especially beneficial in some developing nations, where famine is an ever-present threat, and money for expensive fertilizers is not available.

Hybridomas

Along with the study of genetic recombinations in microorganisms came studies of such combinations in higher organisms. The first combination useful in industrial microbiology was the fusion of a myeloma (bone marrow cancer) cell with an antibody-producing white blood cell. Such a fusion of two cells is called a **hybridoma** (hi-brid-o'ma), or hybrid cell (Figure 8.17). This particular hybridoma can be grown in the laboratory, and it produces pure specific antibodies, called **monoclonal** (mon-o-klo'nal) **antibodies,** against any antigen to which the white blood cell was previously sensitized. Prior to the production of hybridomas and their monoclonal antibodies, no source of pure antibodies existed. Now many different kinds of monoclonal antibodies are produced commercially, and they represent a major advance in immunology. The production and uses of monoclonal antibodies are discussed in more detail in Chapter 18.

The ability to produce hybridomas may in the future lead to other important advances. Recently, agricultural scientists used this technique to fuse cells from a commercial potato plant with cells from an extremely rare wild strain. The wild strain was selected because it contains the gene for production of a natural insect repellent. The researchers were able to grow the fused cells

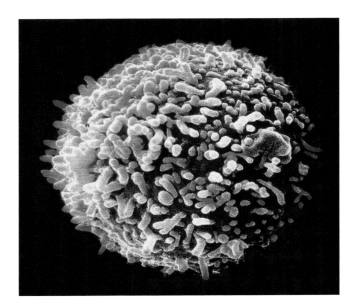

FIGURE 8.17 False-color SEM photo (9500×) of a hybridoma—a single cell fused from two other cells. Hybridomas are often made by fusing an antibody-producing plasma cell and a cancer cell. The latter keeps the culture dividing and growing indefinitely; the former causes the hybridoma to produce pure antibody against the antigen to which it was sensitized.

into a plant that has the necessary commercial properties but also synthesizes the insect repellent in its leaves.

Weighing the Risks and Benefits of Recombinant DNA

Despite the many potential benefits of recombinant DNA research, at first some scientists working with it were concerned about its hazards. They feared that some recombinants might prove to be new and especially virulent pathogens for which humans would have no natural defenses and no effective treatments. In 1974 they called for a moratorium on certain experiments until the hazards could be assessed. From this assessment emerged the idea of biological containment—the practice of making recombinant DNA only

in organisms with mutations that prevent them from surviving outside the laboratory.

In 1981 the constraints on recombinant DNA research were relaxed because of the following observations:

1. No illnesses in laboratory workers could be traced to recombinants.
2. The strain of *E. coli* used in the experiments failed to infect humans who voluntarily received large doses.
3. Incorporation of mammalian genes into *E. coli* was observed in nature, and these genes invariably impaired the organism's ability to adapt to the environment. This suggested that if laboratory organisms did escape, they probably would not survive in the natural environment.
4. Mutants of *E. coli* containing recombinant DNA were subject to control by accepted sanitary practices.

Most scientists now agree that, as currently practiced, recombinant DNA techniques offer significant benefits and exceedingly small risks to humans.

BIOTECHNOLOGY

VIRAL INSECTICIDES

The baculoviruses, a group of fairly well known insect viruses, have been used as natural insecticides since the 1970s. Recently they have become powerful tools in genetic engineering. One of these viruses was used to make a protein from the AIDS virus, which in turn was used to make the first AIDS vaccine to be approved for human trials in the United States. Baculoviruses also are used to make an insect neurotoxin, which can kill crop pests more rapidly than simple virus infections. Such a toxin may soon become the best available weapon against the gypsy moth. Another application of genetic engineering in baculoviruses is rapid protein synthesis that uses regulatory signals from the virus and cellular enzymes. The proteins produced can then be used in diagnosis and therapy.

REDESIGNING BACTERIA

Anyone who's ever tried to grow tomatoes knows how damaging the first frost can be. Beautiful red tomatoes turn into blackened, spongy masses overnight. If that were your main source of food, you'd be in trouble. So the concept of making plants—especially food plants—frost-resistant becomes very attractive.

Many people are surprised to learn that frost is partly a microbiological phenomenon—certain bacteria are largely responsible for the formation of ice crystals. These "ice-plus bacteria" produce a protein that has the same structure as that of water molecules in an ice crystal, and water molecules tend to align themselves on it. As the ice crystals grow, they enter the spaces inside the leaf or flower, causing frost injury. This damage makes the plant susceptible to disease caused by *Pseudomonas syringae*, the predominant frost-causing strain.

Chemicals can control the bacteria that cause frost to form. But such chemicals are expensive, harm the environment, are of limited effectiveness, and may lead to diseases. These chemicals kill beneficial bacteria as well as harmful ones. Hence they destroy the delicate balance of the plant's ecology.

One of the microbiologists who is working on this problem is Trevor Suslow, director of microbial pesticides at the DNA Plant Technology Corporation in Oakland, California. Suslow conducts studies to improve the production of food for developing countries. His research focuses on alternatives to chemical pesticides—on how to use the biological controls that already exist in nature and on evolving new ones. The goal of Suslow's team in controlling frost-causing bacteria was to leave the beneficial bacteria untouched and remove just the ice-plus strains. The strategy was to locate or develop "ice-minus bacteria" to compete successfully with ice-plus strains. Then if the ice-minus bacteria were applied to a frost-susceptible crop, they might protect the plant by keeping the ice-plus bacteria from becoming established—the effect of competitive exclusion.

Suslow's team focused on altering ice-plus bacteria so that they could no longer produce the ice-building protein. Using genetic engineering techniques, the team located the gene that controls this production and transferred it to *Escherichia coli* cells. With the aid of restriction enzymes, about one-third of the gene was moved. This was enough not only to inactivate the gene but also to ensure that no mutation could ever restore it to functional form. Next, the team cloned the defective gene, put it on a plasmid, and reinserted the plasmid into *P. syringae*. A plasmid and a chromosome in the same cell can exchange similar genes that are present on both. So Suslow's team looked for bacteria in which this natural process (called homologous recombination) had taken place, leaving the bacterial cells with a nonfunctional copy of the ice-forming gene in its chromosome. These were ice-minus bacteria. Deletion mutations of this kind occur constantly in nature. But it's more accurate, predictable, and cost effective to perform the genetic surgery in the laboratory. The altered ice-minus strain—whether naturally or genetically engineered—is identical in every other respect to the unaltered ice-plus variety.

Suslow and his team conducted several hundred tests of the ice-minus strain, which they called Frostban. Then it was time to do field trials (Figure 8.18). Suslow obtained the necessary permits and approvals for field testing but was not prepared for the commotion such tests would cause. Protests came from several quarters. Many organizations opposed the release of the bacteria into the environment. It didn't matter that Frostban was safe. They objected to the genetic engineering of a living organism—any organism. Frostban became the symbol in their fight against a future full of recombinant DNA.

Environmentalists were concerned about the destruction of the ecological balance of an increasingly fragile earth. Local farmers had more immediate concerns. Some simply weren't sure what these bacteria were and what they might do. Others worried about potential risks to their own crops from a product that offered them no benefit. In fact, some California farmers like frost on their crops! A good frost sweetens up certain vegetables, such as carrots and Brussels sprouts. These farmers were concerned that Frostban bacteria

FIGURE 8.18 A Frostban field test in Brentwood, California.

would drift over their fields and keep their crops from getting the frost they needed. Research showing that there was virtually no likelihood that Frostban would spread in this way didn't convince the farmers.

Suslow and his team were plagued by an increasing number of lawsuits and even vandalism. The world didn't seem ready for strawberries served with a large dollop of recombinant DNA technology. Discontinuing the field trials, they went back to the laboratory to isolate natural ice-minus bacteria—those in which the ice-building gene is already missing. Although such organisms have the same effect as the bioengineered ones, they're more acceptable to many people because they occur naturally.

Why did the field trials of Frostban cause such controversy? Probably because Frostban was a first—the first commercially available, genetically engineered microbial agent. As such, it attracted much publicity. More fundamentally, people are unaware of the roles that microorganisms play in our lives. If people aren't comfortable with naturally occurring microbes, how can they be comfortable with bioengineered ones? Still, there is tremendous potential for genetically engineered biological controls. According to Suslow, some type of genetic manipulation must take place in order to give growers what they need. He plans to continue the educational process that began with Frostban. Suslow feels that the public, if involved in what he and his team are doing early on, will be more supportive—and that we'll be able to begin controlling some of the blights that reduce the world's food supply. And scientists, together with governmental officials, are developing protocols to ensure that safety is the main concern.

NATURE AND SIGNIFICANCE OF GENE TRANSFER

- **Gene transfer** refers to the movement of genetic information between organisms. It occurs in bacteria by transformation, transduction, and conjugation.

- Gene transfer is significant because it increases genetic diversity within a population, thereby increasing the likelihood that some members of the population will survive environmental changes.

TRANSFORMATION

Discovery of Transformation

- Bacterial **transformation** was discovered in 1928 by Griffith, who showed that a mixed culture of live rough and heat-killed smooth pneumococci could produce live smooth pneumococci capable of killing mice.

- Avery later showed that a capsular polysaccharide was responsible for virulence and that DNA was the substance responsible for transformation. Watson and Crick determined the structure of DNA, which led to studies showing that a cell's genetic information is encoded in its nucleic acids.

Mechanism of Transformation

- Transformation involves the release of naked DNA fragments and their uptake by other cells at a certain stage in their growth cycle: (1) Uptake of DNA requires a protein called **competence factor** to make recipient cells ready to bind DNA. (2) Endonucleases cut double-stranded DNA into units; the strands separate, and only one strand is transferred. (3) Ultimately, donor DNA is spliced into recipient DNA. Leftover recipient DNA is broken down, so a cell's total DNA remains constant.

Significance of Transformation

- Transformation is significant because (1) it contributes to genetic diversity; (2) it can be used to introduce DNA into an organism, observe its effects, and study gene locations; (3) it can be used to create recombinant DNA.

TRANSDUCTION

Discovery of Transduction

- In **transduction,** genetic material is carried by a **bacteriophage** (phage).

Mechanisms of Transduction

- **Phages** can be virulent or temperate. (1) **Virulent phages** destroy a host cell's DNA, direct synthesis of phage particles, and cause lysis of the host cell in the lytic cycle. (2) **Temperate phages** can replicate themselves as a **prophage**—part of a bacterial chromosome—or eventually produce new phage particles and lyse the host cell. Persistence of the phage in the cell without the destruction of the host cell is called **lysogeny.**

- Prophage can be incorporated into the bacterial chromosome. Cells containing a prophage are called **lysogenic** cells because they have the potential to enter the **lytic cycle.**

- Transduction can be specialized or generalized. (1) In **specialized transduction,** the phage is incorporated into the chromosome and can transfer only genes adjacent to the phage. (2) In **generalized transduction,** the phage exists as a plasmid and can transfer any DNA fragment attached to it.

Significance of Transduction

- Transduction is significant because it transfers genetic material and demonstrates a close evolutionary relationship between prophage and host cell DNA. Also, its persistence in a cell suggests a mechanism for the viral origins of cancer, and it provides a possible mechanism for studying gene linkage.

CONJUGATION

Discovery of Conjugation

- In **conjugation** large quantities of DNA are transferred from one organism to another during contact between donor and recipient cells.

- Conjugation was discovered by Lederberg in 1946 when he observed that mixing strains of *E. coli* with different metabolic deficiencies allowed the cells to overcome deficiencies.

- **Plasmids** are extrachromosomal DNA molecules.

Mechanisms of Conjugation

- Three mechanisms of conjugation have been observed: (1) In the transfer of **F plasmids,** a piece of extrachromosomal DNA (a **plasmid**) is transferred. (2) In high-frequency recombinations, parts of F plasmids that have been incorporated into the chromosome (the **initiating segment**) are transferred along with adjacent bacterial genes. (3) An F plasmid incorporated into the chromosome and subsequently separated becomes an **F' plasmid** and transfers chromosomal genes attached to it.

Significance of Conjugation

- The significance of conjugation is that it increases genetic diversity, it may represent an evolutionary stage between asexual and sexual reproduction, and it provides a means of mapping genes in bacterial chromosomes.

COMPARISON OF GENE TRANSFER MECHANISMS

- Genetic transfer mechanisms differ in the quantity of DNA transferred.

PLASMIDS

Characteristics of Plasmids

- Plasmids are circular, self-replicating, double-stranded extrachromosomal DNA that carry information that is generally not essential for cell growth.

Resistance Plasmids

- **Resistance (R) plasmids** carry genetic information that confers resistance to various antibiotics and to certain heavy metals. They generally consist of a **resistance transfer factor (RTF)** and one or more **resistance (R) genes.**

Transposons

- R genes that move from one plasmid to another in a cell or become inserted in the chromosome are part of a **transposon** because they transpose, or change, their locations.

Bacteriocinogens

- **Bacteriocinogens** are plasmids that produce **bacteriocins,** which are proteins that inhibit growth of other strains of the same species or closely related species.

GENETIC ENGINEERING

- **Genetic engineering** is the manipulation of genetic material to alter the characteristics of an organism.

Genetic Fusion

- **Genetic fusion** allows transposition of genes from one location on a chromosome to another, or the joining of genes from two different operons.

Protoplast Fusion

- **Protoplast fusion** combines **protoplasts** (organisms without cell walls) and allows mixing of genetic information.

Gene Amplification

- **Gene amplification** involves the addition of plasmids to microorganisms to increase yield of useful substances.

Recombinant DNA Technology

- **Recombinant DNA** is DNA produced when genes from one kind of organism are introduced into the genome of a different kind of organism. The resulting organism is a **transgenic,** or recombinant, organism.
- Recombinant DNA has proven especially useful in medicine, industry, and agriculture.

Hybridomas

- **Hybridomas** are genetic recombinations involving cells of higher organisms.

Weighing the Risks and Benefits of Recombinant DNA

- When recombinant DNA techniques were first developed, scientists were concerned that virulent pathogens might be created, and they developed containment procedures. As research proceeded and no illnesses caused by recombinants were observed, most scientists came to believe that the benefits of recombinant DNA techniques outweigh the risks.

KEY TERMS

bacteriocin (p. 209)
bacteriocinogen (p. 209)
bacteriophage (p. 200)
chromosome mapping
 (p. 203)
clone (p. 204)
colicin (p. 209)
competence factor (p. 199)
conjugation (p. 203)
electroporation (p. 215)
F pilus (p. 203)
F plasmid (p. 203)
F′ plasmid (p. 206)
gene amplification (p. 214)
gene transfer (p. 196)

generalized transduction
 (p. 202)
genetic engineering (p. 210)
genetic fusion (p. 210)
high frequency of recombination (Hfr) strain (p. 204)
hybridoma (p. 219)
initiating segment (p. 205)
lysis (p. 201)
lysogenic (p. 201)
lysogeny (p. 201)
lytic cycle (p. 201)
monoclonal antibody
 (p. 219)
phage (p. 200)

plasmid (p. 203)
prophage (p. 201)
protoplast (p. 211)
protoplast fusion (p. 211)
recombinant DNA (p. 215)
recombination (p. 196)
resistance (R) gene (p. 208)
resistance plasmid (p. 207)
resistance transfer factor
 (RTF) (p. 208)
specialized transduction
 (p. 202)
temperate phage (p. 201)
transduction (p. 200)
transformation (p. 197)

transgenic (p. 215)
transposable element (p. 209)
transposition (p. 209)
transposon (p. 209)
vector (p. 215)
virulent phage (p. 200)

QUESTIONS FOR REVIEW

A Gene Transfer

1. List the main characteristics of gene transfer.
2. Match the following:

   ____ uptake of naked DNA   a. conjugation

   ____ virus involved   b. transformation

   ____ F⁺, F⁻, Hfr   c. transduction

   ____ competence factor

   ____ F pilus

3. What is the significance of gene transfer?

B Transformation

4. How was transformation discovered?
5. What events led to the determination that genetic information was being transferred in transformation?
6. Describe the steps in transformation.
7. What is the significance of transformation?

8. Identify the type of gene transfer represented in the accompanying diagram, and label parts (a) and (b).

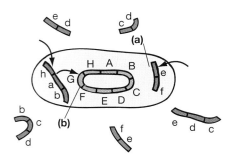

C Transduction
9. What is transduction?
10. What is the role of viruses in transduction?
11. How do generalized and specialized transduction differ?
12. Identify processes (a) through (e) in the following diagram:

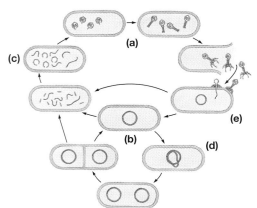

13. What is the significance of transduction?

D Conjugation
14. What is conjugation?
15. Describe the steps in the process of conjugation.
16. How do the following kinds of conjugation differ: F plasmid, high-frequency recombination, and F′ plasmid?
17. What is the significance of conjugation?

E Plasmids
18. List the characteristics of plasmids.
19. How are plasmids classified?
20. What are resistance plasmids, and what do they do?
21. What are bacteriocins, and what do they do?
22. What are bacteriocinogens, and what do they do?

F Genetic Engineering
23. What is genetic engineering?
24. What is genetic fusion, and how has it been used?
25. What is protoplast fusion, and what benefits might be derived from this technique?
26. Distinguish between genetic fusion and protoplast fusion.
27. How is gene amplification accomplished, and what are its applications?
28. How is recombinant DNA produced?
29. What applications have been found for recombinant DNA?
30. What are hybridomas, and how have they been used?

G Recombinant DNA
31. Compare the risks and benefits of recombinant DNA techniques.

PROBLEMS FOR INVESTIGATION

1. Hold a class debate concerning the benefits and hazards of gene manipulation. If you were addressing a meeting of chefs and restaurant owners, how would you explain genetically engineered tomatoes to them? Should they identify such tomatoes on menus, for customers' information? How should they respond to customer concerns? Why or why wouldn't you want to eat such tomatoes?

2. Once the Human Genome Project (described in Chapter 1) has identified and sequenced genes that are important to human health, which types of genetic engineering should be permitted on the human genome? If we allow people to "fix" genes in their own body, should they also be allowed to "fix" their eggs or sperm so as to affect all future generations of their offspring? Should we be allowed to transfer nonhuman genes into humans?

3. You have applied for a job and have had a company-requested physical exam. Should the potential employer have the right to know which genes you have? What constitutes the right to privacy? If the company knows that you carry a gene for cancer, will they discriminate against you and give the job to someone without a cancer gene? Have a class discussion of these problems of DNA testing in the workplace.

4. Research how human insulin is now being made in bacteria via genetic engineering techniques. Why is this so important to human diabetics? What kind(s) of insulin had previously been used?

5. Research the connection between plasmids and bacteria that produce toxins harmful to humans.

6. Bring in some news articles to share in class about current developments in genetic engineering.

Beardsley, T. 1994. "La Ronde, what goes around comes around for life's master molecule." *Scientific American* 270(June):28–29.

"Biotechnology federal budget initiative shows growth." 1992. *ASM News* 58(5):52–54.

Brill, W. J. 1985. "Safety concerns and genetic engineering in agriculture." *Science* 227(January 25):381–84.

Chauthaiwale, V. M., A. Therwath, and V. V. Deshpande. 1992. "Bacteriophage lambda as a cloning vector." *Microbiology Reviews* 56, no. 4 (December):577–91.

Cohrssen, J. J. 1988. "United States biotechnology policy." *American Biotechnology Laboratory* 6(January):22.

Crawford, M. 1987. "California field test goes forward." *Science* 236(May 1):511.

"EPA considers commercial release of engineered microbe." 1995. *ASM News* 61(3):111–12.

Erickson, D. 1991. "Gene rush: Companies seek profits in the genome project." *Scientific American* 264(January):112–13.

Erickson, D. 1992. "Hacking the genome." *Scientific American* 266(April):128–137.

Fleischmann, R. D., et al. 1995. "Whole-genome random sequencing and assembly of *Haemophilus influenzae* Rd." *Science* 269(July 28):496–512.

Frost, L. S. 1992. "Bacterial conjugation: Everybody's doin' it." *Canadian Journal of Microbiology* 38 (November):1091–96.

Gasser, C. S., and R. T. Fraley. 1992. "Transgenic crops." *Scientific American* 266(June):62–69.

Koncz, C., et al. 1990. "Bacterial and firefly luciferase genes in transgenic plants: Advantages and disadvantages of a reporter gene." *Developments in Genetics* 11(3):224–32.

Maloy, S. R., J. Cronan, and D. Freifelder. 1994. *Microbial genetics.* Boston: Jones and Bartlett.

Marx, J. L. 1987. "Assessing the risks of microbial release." *Science* 237 (September 18):1413–17.

Miller, H. I. 1988. "FDA regulation of products of the new biotechnology." *American Biotechnology Laboratory* 6(1) (January):38.

Neufeld, P. J., and N. Colman. 1990. "When science takes the witness stand." *Scientific American* 262(May):46–53.

Pendick, D. 1992. "Better than the real thing: Industry serves up the fruits of tomato biotechnology." *Science News* 142, no. 22 (November 28):376–77.

Pennisi, E. 1993. "Mouse of a different YAC: Yeast artificial chromosomes make possible bigger gene transfers." *Science News* 143, no. 23 (June 5):360–63.

Peters, P. M. 1993. *Biotechnology: A guide to genetic engineering.* Dubuque, IA: Wm. C. Brown.

Stahl, F. W. 1987. "Genetic recombination." *Scientific American* 256(February):90–101.

Stewart, G. J., and C. A. Carlson. 1986. "The biology of natural transformation." *Annual Review of Microbiology* 40:211–35.

Verma, I. M. 1990. "Gene therapy." *Scientific American* 263(November):68–84.

Watson, J., et al. 1992. *Recombinant DNA,* 2d ed. New York: Scientific American Books.

Weintraub, H. M. 1990. "Antisense RNA and DNA." *Scientific American* 262(January):40–46.

MICROBES IN THE SCHEME OF LIFE: AN INTRODUCTION TO TAXONOMY

Color-enhanced SEM photo (140×) of various types of radiolaria—protozoa with shells of silica. This assortment shows the unity and diversity of organisms.

Humans appear to have an innate need to name things. In many primitive societies, a person who knows the true name of an object or of another person is believed to have power over that object or person. Naming helps us to understand our world and to communicate with others about it.

TAXONOMY–THE SCIENCE OF CLASSIFICATION

In science accurate and standardized names are essential. All chemists must mean the same thing when they talk about an element or a compound; physicists must agree on terms when they discuss matter or energy; and biologists must agree on the names of organisms, be they tigers or bacteria.

Faced with the great number and diversity of organisms, biologists use the characteristics of different organisms to describe specific forms of life and to identify new ones. The grouping of related organisms together is the basis of *classification*. The most obvious reasons for classification are (1) to establish the criteria for identifying organisms; (2) to arrange related organisms into groups; and (3) to provide important information

on how organisms evolved. **Taxonomy** is the science of classification. It provides an orderly basis for the naming of organisms and for placing organisms into a category, or **taxon** (plural: *taxa*).

Another important aspect of taxonomy is that it makes use of and makes sense of the fundamental concepts of unity and diversity among living things. Organisms classified in any particular group have certain common characteristics—that is, they have unity with respect to these characteristics. For example, humans walk upright and have a well-developed brain; *Escherichia coli* cells are rod-shaped and have a gram-negative cell wall. The organisms within taxonomic groups exhibit diversity as well. Even members of the same species display variations in size, shape, and other characteristics. Humans vary in height, weight, hair and eye color, and facial features. Certain kinds of bacteria vary somewhat in shape and in their ability to form specific structures, such as endospores. A basic principle of taxonomy is that members of higher-level groups share fewer characteristics than those in lower-level groups. Like all other vertebrates, humans have backbones, but humans share fewer characteristics with fish and birds than with other mammals. Likewise, nearly all bacteria have a cell wall, but in

THIS CHAPTER FOCUSES ON THE FOLLOWING QUESTIONS:

A How are microorganisms named?

B What did Linnaeus contribute to taxonomy?

C How is a dichotomous taxonomic key used to identify organisms?

D What types of problems and major developments in taxonomy have occurred since Linnaeus founded that science?

E What are the main characteristics of the kingdoms in the five-kingdom system of taxonomy?

F How are viruses classified?

G What special methods are needed for determining evolutionary relationships among prokaryotes?

some the wall is gram-positive and in others it is gram-negative.

Linnaeus—The Father of Taxonomy

The eighteenth-century Swedish botanist Carolus Linnaeus is credited with founding the science of taxonomy (Figure 9.1). He originated **binomial nomenclature,** the system that is still used today to name all living things. In the binomial, or "two-name," system, the first name designates the **genus** (plural: *genera*) of an organism, and its first letter is capitalized. The second name is the **specific epithet,** and it is not capitalized even when derived from the name of the person who discovered it. Together the genus and specific epithet identify the **species** to which the organism belongs. Both words are italicized in print but underlined when handwritten. When there is no danger of confusion, the genus name may be abbreviated to a single letter. Thus, *Escherichia coli* is often written *E. coli,* and humans (*Homo sapiens*) may be identified as *H. sapiens.*

The name of an organism often tells something about it, such as its shape, where it is found, what nutrients it uses, who discovered it, or what disease it

FIGURE 9.1 Carolus Linnaeus (1707–1778), the father of taxonomy.

causes. Some examples of names and their meanings are shown in Table 9.1.

The members of a species generally have several common characteristics that distinguish that species from all other species. As a rule, members of the species cannot be divided into significantly different groups on the basis of a particular characteristic, but there are exceptions to this rule. Sometimes members of a species are divided on the basis of a small but permanent genetic difference, such as a need for a particular nutrient, resistance to a certain antibiotic, or the presence of a particular antigen. When organisms in one pure culture of a species differ from the organisms in another pure culture of the same species, the organisms in each culture are designated as strains. A **strain** is a subgroup of a species with one or more characteristics that distinguish it from other subgroups of the same species. Each strain is identified by a name, number, or letter that follows the specific epithet. For example, *E. coli* strain K12 has been extensively studied because of its plasmids and other genetic characteristics, and *E. coli* strain 0157:H7 causes hemorrhagic inflammation of the colon in humans.

In addition to introducing the binomial system of nomenclature, Linnaeus also established a hierarchy of taxonomic ranks: species, genus, family, order, class, phylum or division, and kingdom. At the highest level, Linnaeus divided all living things into two *kingdoms*— plant and animal. In his taxonomic hierarchy, which is still used today, each organism is assigned a species name, and species of very similar organisms are grouped into a genus. As we proceed up the hierarchy, several similar genera are grouped to form a *family*, several families to form an *order*, and so on to the top of the hierarchy. Some hierarchies today have additional levels, such as *subphyla*. Also, it has become accepted practice to refer to the first categories within the animal kingdom as *phyla* and to those within other kingdoms (we now have five) as *divisions*. The classifications of a human and a bacterium are shown in Figure 9.2.

APPLICATIONS

WHERE DO STOCK CULTURES COME FROM?

Particular strains of organisms are often sufficiently valuable to be preserved because their characteristics are important in research or in industrial applications such as winemaking. Preservation of these organisms in a dormant (inactive) state prevents them from undergoing genetic changes that might alter their characteristics. One method of preserving organisms is *lyophilization* (freeze-drying), as is explained in Chapter 13. Preserved cultures can be deposited in a central type culture collection, in which each culture is directly descended from the original organism to which a particular strain or species designation was first assigned. The American Type Culture Collection (ATCC) in Rockville, Maryland, keeps some dormant cultures in a secured vault, which also protects them against theft. Such organisms are of value to manufacturers of wine or cheese because they create distinctive flavors or other characteristics of products. If the organisms were lost, the ability to make particular products would also be lost. Stock cultures of many organisms are available to qualified scientists. The existence of such a collection allows different researchers studying a particular strain to be confident that they truly are dealing with the same organism and that their results can be meaningfully compared.

The American Type Culture Collection uses lyophilization (freeze-drying) to preserve and store organisms.

TABLE 9.1	Meanings of Names of Some Microorganisms
Name of Microorganism	**Meaning of Name**
Entamoeba histolytica	*Ent* = intestinal, *amoebae* = shape and means of movement, *histo* = tissue, *lytic* = lysing, or digesting, tissue
Escherichia coli	Named after Theodor Escherich in 1888; found in the colon
Haemophilus ducreyi	*Hemo* = blood, *phil* = love; named after Augusto Ducrey in 1889
Neisseria gonorrhoeae	Named after Albert L. Neisser in 1879; causes gonorrhea
Saccharomyces cerevisiae	*Saccharo* = sugar, *myco* = mold, *cerevisia* = beer or ale
Staphylococcus aureus	*Staphylo* = cluster, *kokkus* = berry, *aureus* = golden
Lactococcus lactis	*Lacto* = milk, *kokkus* = berry

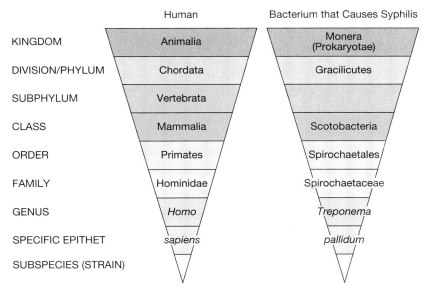

	Human	Bacterium that Causes Syphilis
KINGDOM	Animalia	Monera (Prokaryotae)
DIVISION/PHYLUM	Chordata	Gracilicutes
SUBPHYLUM	Vertebrata	
CLASS	Mammalia	Scotobacteria
ORDER	Primates	Spirochaetales
FAMILY	Hominidae	Spirochaetaceae
GENUS	*Homo*	*Treponema*
SPECIFIC EPITHET	*sapiens*	*pallidum*
SUBSPECIES (STRAIN)		

FIGURE 9.2 Classification of a human and a bacterium.

Using a Taxonomic Key

Biologists often use a taxonomic *key* to identify organisms according to their characteristics. The most common kind of key is a **dichotomous key,** which has paired statements describing characteristics of organisms. Paired statements present an "either-or" choice, such that only one statement is true. Each statement is followed by directions to go to another pair of statements until the name of the organism finally appears. Figure 9.3 is a dichotomous key that will identify each of the four most common U.S. coins: quarters, dimes, nickels, and pennies. Read statements 1a and 1b, and decide which statement applies to a given coin. Look at the number to the right of the statement; it tells you which pair of statements to look at next. Continue in this manner until you reach a group designation. If you have followed the key carefully, that designation will name the coin.

Of course, you don't need a taxonomic key to identify something as simple and as familiar as coins. But identifying all the many kinds of bacteria in the world

FIGURE 9.3 A dichotomous key for classifying typical U.S. coins.

1a	Smooth-edged	Go to 2
1b	Rough-edged	Go to 3

2a	Silver-colored	Nickel
2b	Copper-colored	Penny

3a	Large (about 1-in. diameter)	Quarter
3b	Small (about 3/4-in. diameter)	Dime

is a more difficult task. Major groups of bacteria can be identified with the key in Figure 9.4. More detailed keys use staining reactions, metabolic reactions (fermentation of particular sugars or release of different gases), growth at different temperatures, properties of colonies on solid media, and similar characteristics of cultures. By proceeding step by step through the key, one should be able to identify an unknown organism, or even a strain, if the key is sufficiently detailed.

Problems in Taxonomy

Among the aims of a taxonomic system are organizing knowledge about living things and establishing standard names for organisms so that we can communicate about them. Ideally, we would like to classify organisms according to their **phylogenetic,** or evolutionary, relationships, but this is not always easy. Evolution occurs continuously and at a relatively rapid rate in microorganisms, and our knowledge of the evolutionary history of

FIGURE 9.4 A dichotomous key for classifying major groups of bacteria.

1a	Gram-positive	Go to 2
1b	Not gram-positive	Go to 3

2a	Cells spherical in shape	Gram-positive cocci
2b	Cells not spherical in shape	Go to 4

3a	Gram-negative	Go to 5
3b	Not gram-negative (lack cell wall)	Mycoplasma

4a	Cells rod-shaped	Gram-positive bacilli
4b	Cells not rod-shaped	Go to 6

5a	Cells spherical in shape	Gram-negative cocci
5b	Cells not spherical in shape	Go to 7

6a	Cells club-shaped	Corynebacteria
6b	Cells variable in shape	Propionibacteria

7a	Cells rod-shaped	Gram-negative bacilli
7b	Cells not rod-shaped	Go to 8

8a	Cells helical with several turns	Spirochetes
8b	Cells comma-shaped	Vibrioids

organisms is incomplete. Taxonomy must change with evolutionary changes and new knowledge. *It is far more important to have a taxonomic system that reflects our current knowledge than to have a system that never changes.*

Creating a taxonomic system that provides an organized overview of all living things and how they are related to each other poses certain problems. Two such problems arise at opposite ends of the taxonomic hierarchy: (1) deciding what constitutes a species and (2) deciding what constitutes a kingdom. In the first case, taxonomists try to decide how much diversity can be tolerated within the unity of a species. In the second, taxonomists try to decide how to sort the diverse characteristics of living things into categories that reflect fundamental differences of evolutionary significance. In most advanced organisms, such as plants and animals, species that reproduce sexually are distinguished primarily by their reproductive capabilities. A male and a female of the same species are capable of DNA transfer through mating and producing fertile offspring, whereas members of different species ordinarily either cannot mate successfully or will have sterile offspring. *Morphology* (structural characteristics) and geographic distribution also are considered in defining species.

In bacteria such criteria can normally not be used in defining a species, primarily because genetic transfer (genetic recombination) among bacteria is relatively rare and morphological differences are minor. A bacterial species is defined by the similiarities found among its members. Properties such as biochemical reactions, chemical composition, cellular structures, genetic characteristics, and immunological features are used in defining a bacterial species. Identifying a species and determining its limits present the most challenging aspects of biological classification—for any type of organism.

Before taxonomists turned their attention to microorganisms, the two-kingdom system of plants and animals worked reasonably well. Anyone can tell plants from animals—for example, trees from dogs. Plants make their own food but cannot move, and animals move but cannot make their own food. Simple enough, or is it? In this scheme, how do you classify *Euglena,* a mobile microorganism that makes its own food? How would you classify jellyfishes and sponges, which are motile or immotile depending on their stage of life? And how do you classify colorless fungi that neither move nor make their own food? Finally, how do you classify slime molds, organisms that can be unicellular or multicellular and mobile or immobile? Obviously, microorganisms pose a number of problems when one tries to use a two-kingdom system.

Developments Since Linnaeus's Time

The problem of classifying microorganisms was first addressed by the German biologist Ernst H. Haeckel in 1866 when he created a third kingdom, the Protista.

He included among the protists all "simple" forms of life such as bacteria, many algae, protozoa, and multicellular fungi and sponges. Haeckel's original term, Protista, is still used in taxonomic schemes today, but it is now limited mainly to unicellular eukaryotic organisms.

Classification of bacteria has posed taxonomic problems over the centuries and still does (see the Essay at the end of this chapter). Until recently, many taxonomists regarded bacteria as small plants that lacked chlorophyll. As late as 1957, the seventh edition of *Bergey's Manual of Determinative Bacteriology,* a work devoted to the identification of bacteria, considered bacteria to be unicellular plants. Changes in this viewpoint came as the tools to study bacteria were developed. First, light microscopy and staining techniques were used to describe the basic structure of cells. Second, electron microscopy was used to study the ultrastructure of cells. And, third, biochemical techniques were used to study chemical composition and chemical reactions in cells. One of the most important discoveries from these various studies was that DNA looked and behaved differently during cell division in bacteria than in cells whose DNA is organized into chromosomes within a nucleus.

Studies of the structure and function of cells also led to the recognition of two general patterns of cellular organization, prokaryotic and eukaryotic. Basing taxonomy on these two different patterns of cellular organization was proposed as early as 1937. Various taxonomists such as H. F. Copeland, R. Y. Stanier, C. B. van Niel, and R. H. Whittaker, working in the late 1950s, placed bacteria in a separate kingdom of anucleate (lacking a cell nucleus) organisms rather than with organisms that have true nuclei. In 1962 Stanier and van Niel stated, "The distinctive property of bacteria is the prokaryotic nature of their cells."

In 1956, Lynn Margulis and H. F. Copeland proposed a scheme of classifying prokaryotes and eukaryotes by the following four-kingdom system of classification:

1. Monera: all prokaryotes, including true bacteria and blue-green algae
2. Protoctista: all eukaryotic algae, protozoa, and fungi
3. Plantae: all green plants
4. Animalia: all animals derived from a zygote, a cell formed by the union of an egg and a sperm

These taxonomists also proposed that evolution from prokaryotic to eukaryotic life forms had taken place by endosymbiosis. ∞ (Chapter 4, p. 102)

R. H. Whittaker felt that endosymbiosis could not account for all the differences between prokaryotes and eukaryotes. He also felt that a taxonomic system should give more consideration to the methods organisms use to obtain nourishment. Autotrophic nutrition by photosynthesis and heterotrophic nutrition by ingestion of substances from other organisms had been considered in earlier taxonomies. Absorption as a sole means of acquiring nutrients had been overlooked. To Whittaker, fungi, which acquire nutrients solely by absorption, were sufficiently different from plants to justify placing them in a different kingdom. Also, fungi have certain reproductive processes not shared with any other organisms. Consequently, Whittaker proposed a taxonomic system in 1969 that separated the Protoctista into two kingdoms—Protista (pro-tis'tah) and Fungi—but retained the Monera, Plantae, and Animalia. Finally, through refinements of Whittaker's system by several taxonomists over the past few decades, the five-kingdom system was created.

THE FIVE-KINGDOM CLASSIFICATION SYSTEM

Before we discuss the five-kingdom classification system and how it applies to microorganisms, we must emphasize that all living organisms, regardless of the kingdom to which they are assigned, display certain characteristics that define the unity of life. All organisms are composed of cells, and all carry out certain functions, such as obtaining nutrients and getting rid of wastes. The cell is the basic structural and functional

unit of all living things. The fact that viruses are not cells is one reason they are not considered to be living organisms. All cells are bounded by a cell or plasma membrane, carry genetic information in DNA, and have ribosomes where proteins are made. All cells also contain the same kinds of organic compounds—proteins, lipids, nucleic acids, and carbohydrates. They also selectively transport material between their cytoplasm and their environment. Thus, although organisms may be classified in very diverse taxonomic groups, their cells have many similarities in structure and function.

No single classification system is completely accepted by all biologists. For this text we have elected to use the **five-kingdom system** (Figure 9.5). A major advantage of this system is the clarity with which it deals with microorganisms. It places all **prokaryotes,** microorganisms that lack a cell nucleus, in the kingdom Monera (Prokaryotae). ∞ (Chapter 4, p. 74) It places most unicellular **eukaryotes,** organisms whose cells contain a distinct nucleus, in the kingdom Protista. (Margulis proposed a very similar five-kingdom system in 1982, but she refers to the kingdom of simple eukaryotes as Protoctista instead of Protista.) The five-kingdom system also places fungi in the separate kingdom Fungi.

Some taxonomists have recommended the creation of a sixth kingdom for **archaeobacteria** (ar'ke-o-bak-ter''e-ah). These microbes differ in important ways from the **eubacteria** (u'bak-ter''e-ah), or true bacteria, and may be of very ancient origin. (See the Essay at the end of this chapter.)

The properties and members of each of the five kingdoms are described below and summarized in Table 9.2. A more detailed classification of bacteria is provided in Appendix B.

Kingdom Monera

The kingdom **Monera** (mo-ner'ah) is also called the kingdom **Prokaryotae,** as suggested by the French marine biologist Edouard Chatton in 1937. It consists of all prokaryotic organisms, including the true bacteria and the cyanobacteria—and for the present, the archaeobacteria, too (Figure 9.6).

All monerans are unicellular; they lack true nuclei and generally lack membrane-bound organelles. Their DNA has little or no protein associated with it. Reproduction in the kingdom Monera occurs mainly by binary fission. Of all monerans, the true bacteria are of greatest concern in the health sciences and will be considered in detail in several chapters of this book.

FIGURE 9.5   The five-kingdom system of classification.

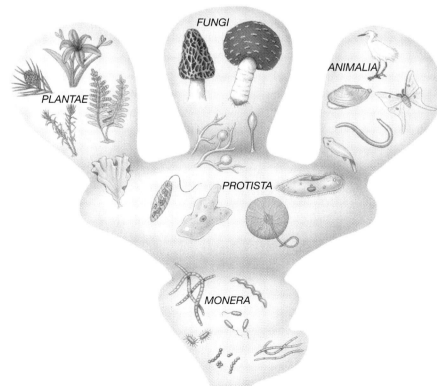

Kingdom	Characteristics
Monera (Prokaryotae)	Prokaryotic; unicellular, but sometimes cells are grouped; nutrition by absorption, but in some forms by photosynthesis or chemosynthesis; reproduction asexual, usually by fission.
Protista	Eukaryotic; unicellular, but in some cases cells are grouped; nutrition varies among phyla and can be by ingestion, photosynthesis, or absorption; reproduction asexual and in some forms both sexual and asexual.
Fungi	Eukaryotic; unicellular or multicellular; nutrition by absorption; reproduction usually both sexual and asexual and often involves a complex life cycle.
Plantae	Eukaryotic; multicellular; nutrition by photosynthesis; reproduction both sexual and asexual.
Animalia	Eukaryotic; multicellular; nutrition by ingestion but in some parasites by absorption; reproduction primarily sexual.

TABLE 9.2 The Five-Kingdom System of Classification

The **cyanobacteria** (si″an-o-bak-ter′e-ah), formerly known as blue-green algae, are of special importance in the balance of nature. They are photosynthetic, typically unicellular, organisms, although cells may sometimes be connected to form threadlike filaments. Being autotrophs, cyanobacteria do not invade other organisms, so they pose no health threat to humans, except for toxins (poisons) some release into water.

Cyanobacteria grow in a great variety of habitats, including anaerobic ones, where they often serve as food sources for more complex heterotrophic organisms. Some "fix" atmospheric nitrogen, converting it to nitrogenous compounds that algae and other organisms can use. Certain cyanobacteria also thrive in nutrient-rich water and are responsible for algal blooms—a thick layer of algae on the surface of water that prevents light from penetrating to the water below. Such blooms release toxic substances that can give the water an objectionable odor and even harm fish and livestock that drink the water.

Archaeobacteria surviving today are primitive prokaryotes adapted to extreme environments. The methanogens reduce carbon-containing compounds to the gas methane. The extreme halophiles live in excessively salty environments, and the thermoacidophiles live in hot acidic environ-

ments, such as volcanic vents in the ocean floor (Figure 9.7).

Kingdom Protista

Although the modern protist group is very diverse, it contains fewer kinds of organisms than when first defined by Haeckel. All organisms now classified in the kingdom **Protista** (Figure 9.8) are eukaryotic. Most are unicellular, but some are organized into colonies. Protists have a true membrane-bound nucleus and organelles within their cytoplasm, as do other eukaryotes. Many protists live in fresh water, some live in sea water, and a few live in soil. They are distinguished more by what they don't have or don't do than by what they have or do. Protists do not develop from an embryo, as plants and animals do, and they do not develop from distinctive spores, as the fungi do. Yet, among the protists are the algae, which resemble plants; the protozoa, which resemble animals; and the euglenoids, which have both plant and animal characteristics. The protists of greatest interest to health scientists are the protozoa that can cause disease.

Kingdom Fungi

The kingdom **Fungi** (Figure 9.9 on page 234) includes mostly multicellular and some unicellular organisms. Fungi obtain nutrients solely by absorption of organic matter from dead organisms. Even when they invade

FIGURE 9.6 Some typical monerans, prokaryotic organisms without a cell nucleus and other internal, membrane-bound structures.

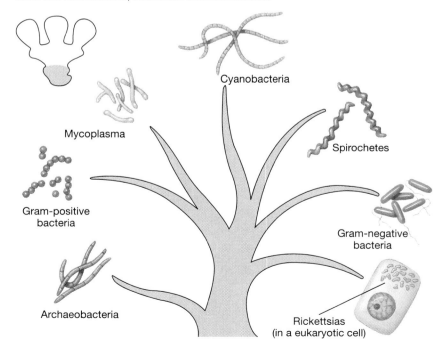

Cyanobacteria

Mycoplasma

Spirochetes

Gram-positive bacteria

Gram-negative bacteria

Archaeobacteria

Rickettsias (in a eukaryotic cell)

FIGURE 9.7 Organisms living at deep-ocean vents, where hot volcanic gases are released from the earth's interior, survive in one of the most extreme environments known. Temperature and pressure are extremely high, yet these vents are the most productive ecosystem on our planet.

living tissues, fungi typically kill cells and then absorb nutrients from them. Although the fungi have some characteristics in common with plants, their structures are much simpler in organization than true leaves or stems. Fungi form spores but do not form seeds. Many fungi pose no threat to other living things, but some attack plants and animals, even humans.

Kingdom Plantae

The placement of most microscopic eukaryotes with the protists leaves only macroscopic green plants in the kingdom **Plantae.** Most plants live on land and contain chlorophyll in organelles called chloroplasts. Plants are of interest to microbiologists only because some contain medicinal substances such as quinine, which has been used to treat microbial infections.

Kingdom Animalia

The kingdom **Animalia** includes all animals derived from zygotes. Although nearly all members of this kingdom are macroscopic and therefore of no concern to microbiologists, several groups of animals live in or on other organisms, and some serve as carriers of microorganisms (Figure 9.10).

Certain *helminths* (worms) are parasitic in humans and other animals. Helminths include flukes, tapeworms, and roundworms, which live inside the body of their host. They also include leeches, which live on the surface of their hosts. Microbiologists often need to identify both microscopic and macroscopic forms of helminths.

Certain *arthropods* live on the surface of their hosts, and some spread disease. Ticks, mites, lice, and fleas are arthropods that live on their hosts for at least part of their lives. Ticks, lice, fleas, and mosquitoes can spread infectious microorganisms from their bodies to those of humans or other animals.

CLASSIFICATION OF VIRUSES

Viruses are acellular infectious agents that are smaller than cells. They contain nucleic acid (DNA or RNA) and are coated with protein. They have not been assigned to a kingdom. In fact, they display only a few characteristics associated with living organisms.

Initially viruses were classified according to the hosts they invaded and

FIGURE 9.8 Some typical protists, unicellular eukaryotic organisms.

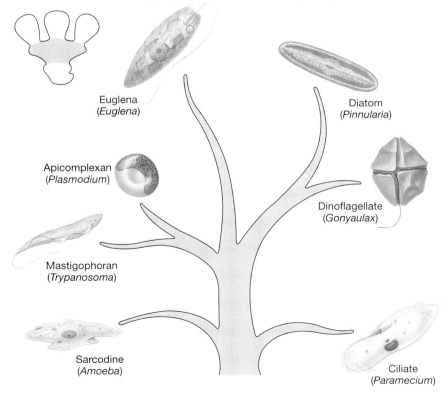

Euglena
(*Euglena*)

Diatom
(*Pinnularia*)

Apicomplexan
(*Plasmodium*)

Dinoflagellate
(*Gonyaulax*)

Mastigophoran
(*Trypanosoma*)

Sarcodine
(*Amoeba*)

Ciliate
(*Paramecium*)

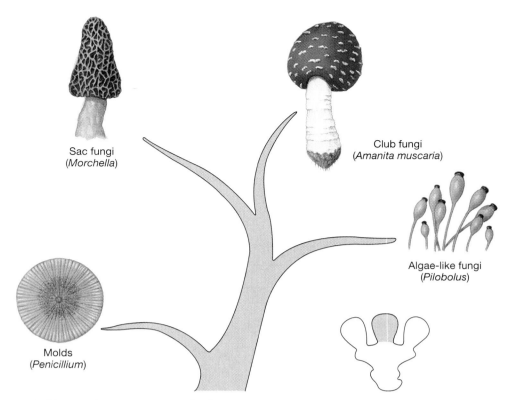

FIGURE 9.9 Some typical fungi, eukaryotic organisms that have cell walls and do not carry out photosynthesis. Fungi take their food from other organic sources (that is, they are chemoheterotrophs).

FIGURE 9.10 Groups from the kingdom Animalia that are relevant to microbiology.

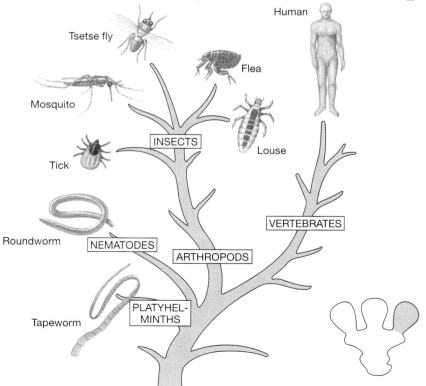

by the diseases they caused. As more was learned about viruses, the early concept of "one virus, one disease" used in classification was found to be invalid for many viruses. Today viruses are classified by chemical and physical characteristics such as the type and arrangement of their nucleic acids, shape (cubical or tubular), the symmetry of the protein coat that surrounds the nucleic acid, and the presence or absence of such things as a membrane covering (called an envelope), enzymes, tail structures, or lipids (Figure 9.11). These groupings reflect only common characteristics and are not intended to represent evolutionary relationships. A classification of viruses is presented in Appendix B.

The study of viruses, or *virology*, is extremely important in any microbiology course for two reasons: (1) Virology is a recognized branch of microbiology, and techniques to study viruses are derived from microbiological techniques; and (2) viruses are of concern to health scientists because many cause diseases in humans, other animals, plants, and even microorganisms.

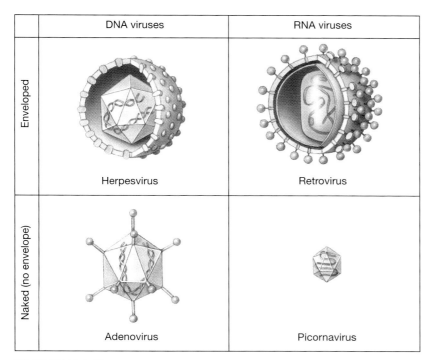

	DNA viruses	RNA viruses
Enveloped	Herpesvirus	Retrovirus
Naked (no envelope)	Adenovirus	Picornavirus

FIGURE 9.11 Some categories of viruses.

THE SEARCH FOR EVOLUTIONARY RELATIONSHIPS

Many biologists are interested in how living things evolved and how they are related to one another. In fact, most people have some curiosity about how life originated and gave rise to the diverse assortment of living things we see today. Although the details of the search for evolutionary relationships are of interest mainly to taxonomists, they are of some significance to health scientists. For example, many of the biochemical properties used to establish evolutionary relationships also can be used in identifying microorganisms. In addition, infectious agents, their hosts, and the relationships between them generally evolve together. Some knowledge of such evolution is useful in understanding the circumstances under which one organism becomes able to infect another and how the disease process occurs.

Special Methods Needed for Prokaryotes

The taxonomy of most eukaryotes is based on morphology (structural characteristics) of living organisms, genetic features, and on knowledge of their evolutionary relationships from fossil records. However, morphology and fossil records provide little information about prokaryotes. For one thing, prokaryotes have left few fossil records. Fossilized mats of prokaryotes, or *stromatolites* (stro-mat'o-litz), have been found mainly at sites where the environment millions of years ago allowed the deposition of dense layers of bacteria (Figures 9.12a and b). Unfortunately, most bacteria do not form such mats, so most ancestral prokaryotes have disappeared without a trace.

Some rocks containing fossils of individual cells of cyanobacteria have been discovered (Figure 9.12c), but they have failed to reveal much information about the organisms. Moreover, prokaryotes have few structural characteristics, and these characteristics are subject to rapid change when the environment changes. Large organisms tend to require a fairly long period of time to reproduce, but prokaryotes reproduce rapidly. Rapid reproduction allows for mutations in each generation and much change in a relatively short time.

Because morphology and evolution are of little use in classifying prokaryotes, metabolic reactions, genetic relatedness, and other specialized properties have been used instead. Health scientists use these properties to identify infectious prokaryotes in the laboratory, but such identification does not necessarily reflect evolutionary relationships among the organisms. The methods described next are of use in exploring evolutionary relationships. Although the methods are particularly appropriate for eukaryotes, they can be used for prokaryotes as well.

CLOSE-UP

VIROIDS AND PRIONS

Particles even smaller than viruses have recently been discovered, and some appear to serve as infectious agents. They include *viroids* (vir'oids) and *prions* (pre'onz). A viroid consists of a fragment of RNA. One particular viroid (the viroid that causes potato spindle tuber disease) contains only 359 bases—enough information to specify the location of only 119 amino acids if all the bases function as codons. This amount of nucleic acid is only 1/10 that found in the smallest viruses. Prions, or *proteinaceous infectious particles,* are only 1/10 the size of a virus and consist of a protein molecule that folds incorrectly as a result of mutation. These particles are thought to be self-replicating. (How might a protein replicate itself?) Prions may be responsible for some mysterious brain infections in humans.

FIGURE 9.12 (a) Mats of bacteria, growing as stromatolites, in western Australia. (b) A cross section through fossil stromatolites from Montana, showing horizontal layers of bacterial growth. (c) Filamentous cyanobacteria (*Paleolyngbya*) from the Lakhanda Formation in eastern Siberia. These microfossils date from the late Precambrian and are approximately 950 million years old.

(a)

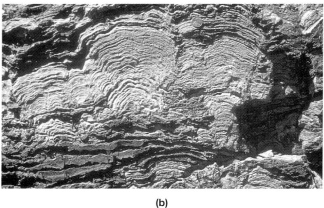

(b)

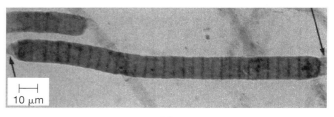

(c)

Numerical Taxonomy

Numerical taxonomy is based on the idea that increasing the number of characteristics of organisms that we observe increases the accuracy with which we can detect similarities among them. Although the idea of numerical taxonomy was developed before computers were available, computers allow us to compare large numbers of organisms rapidly and according to many different characteristics. In a simple example of numerical taxonomy, each characteristic is assigned a value of 1 if present and 0 if not present. Characteristics such as reaction to Gram staining, oxygen requirements, presence or absence of a capsule, properties of nucleic acids and proteins, and the presence or absence of particular enzymes and chemical reactions can be evalu-

ated. Organisms are then compared, and patterns of similarities and differences are detected (Figure 9.13). If two organisms match on 90 percent or more of the characteristics studied, they are presumed to belong to the same species. Provided that the characteristics are genetically determined, the more characteristics two organisms share, the closer the evolutionary relationship the two have. Computerized numerical taxonomy offers great promise for improving our understanding of relationships among all organisms.

Genetic Homology

The discovery of the structure of DNA by James Watson and Francis Crick in 1953 provided new knowledge that was quickly applied by taxonomists, especially those studying taxonomic relationships and the evolution of eukaryotes. These scientists began to study the **genetic homology,** or the similarity of DNA, among organisms. Several techniques for determining genetic homology are now available. Similarities in DNA can be studied directly by determining the base composition of the DNA, by sequencing the bases in DNA or RNA, and by DNA hybridization. Because an organism's proteins are determined by its DNA, similarities in DNA can be studied indirectly by preparing *protein profiles* and by analyzing amino acid sequences in proteins.

Base Composition

Organisms can be grouped by comparing the relative percentages of bases present in the DNA of their cells. DNA contains four bases, abbreviated as A (adenine), T (thymine), G (guanine), and C (cytosine). ∞ (Chapter 2, p. 43) Base pairing occurs only between A and T and between G and C. In making base comparisons, the total amount of G and C in a sample of DNA is determined and expressed as a percentage of total DNA. By subtracting this percentage from 100, we get the percent of total A and T in the sample. For example, if the DNA is 60 percent G–C, then it is 40 percent A–T. The base composition of an organism is generally stated in terms of the percent of guanine plus cytosine and is referred to as the G–C content.

Studies of base composition have shown that the G–C content varies from 23 to 75 percent in bacteria. These studies also have shown that certain species of bacteria, such as *Clostridium tetani* and *Staphylococcus aureus,* have very similar DNA compositions, but that *Pseudomonas aeruginosa* has a very different DNA composition. Thus, *C. tetani* and *S. aureus* are probably more closely related to each other than either is to *P. aeruginosa.* Similar percentages of bases do not in themselves prove that the organisms are closely related, because the *sequence* of bases may be quite different. (Human beings and *Bacillus subtilis,* for example, have nearly identical G–C percentages.) We can say, however, that

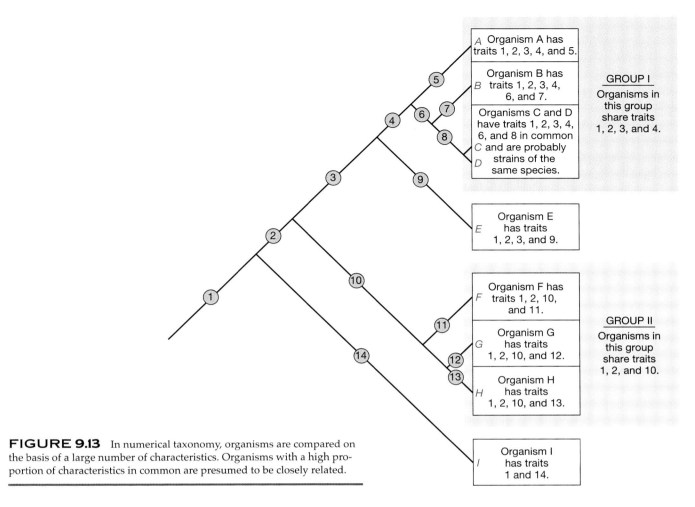

FIGURE 9.13 In numerical taxonomy, organisms are compared on the basis of a large number of characteristics. Organisms with a high proportion of characteristics in common are presumed to be closely related.

if the percentages in two organisms are quite different, they are not likely to be closely related.

DNA and RNA Sequencing

Automated equipment for identifying the base sequences in DNA or RNA is now available at reasonable cost (Figure 9.14). It is therefore easier than before to search a culture for base sequences known to be unique to certain species. Using PCR techniques and a DNA synthesizer, one can produce a large number of **probes,** single-stranded DNA fragments that have sequences complementary to those being sought. ∞ (Chapter 7, p. 190) A fluorescent dye or a radioactive tag (reporter molecules) can be attached to the probe. When the probe finds its target DNA, it complementarily binds to it and does not wash off when rinsed. The specimen is then examined for fluorescing dye or for radioactivity. The presence or absence of the unique DNA sequence helps in identification of the specimen.

DNA Hybridization

In **DNA hybridization** the double strands of DNA of each of two organisms are split apart, and the split strands from the two organisms are allowed to combine (Figure 9.15). The strands from different organisms will anneal (bond to each other) by base pairing—A with T and G with C. The amount of annealing is directly proportional to the quantity of identical base sequences in the two DNAs. A high degree of homology (similarity) exists when both organisms have long identical sequences of bases. Close DNA homology indicates that the two organisms are closely related and that they probably evolved from a common ancestor. A small degree of homology indicates that the organisms are not very closely related. Ancestors of such organisms probably diverged from each other thousands of centuries ago and have since evolved along separate lines.

Protein Profiles and Amino Acid Sequences

Every protein molecule consists of a specific sequence of amino acids and has a particular shape with an assortment of surface charges. Modern laboratory methods allow cells or organisms to be compared according to these properties of their proteins. Although variations in proteins among cells make these techniques dif-

FIGURE 9.14 A DNA sequencer. Automated systems can identify the sequence of nucleotide bases in a piece of DNA.

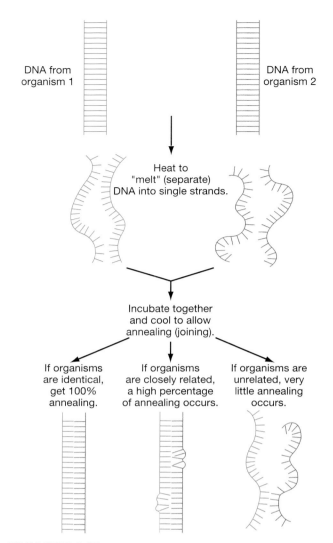

DNA from organism 1

DNA from organism 2

Heat to "melt" (separate) DNA into single strands.

Incubate together and cool to allow annealing (joining).

If organisms are identical, get 100% annealing.

If organisms are closely related, a high percentage of annealing occurs.

If organisms are unrelated, very little annealing occurs.

FIGURE 9.15 In DNA hybridization, strands of DNA are separated, and individual strands from two different organisms are allowed to anneal (join by hydrogen bonding at sites where there are many complementary base pairs). The degree of annealing reflects the degree of relatedness between the organisms, on the basis of the assumption that annealing takes place only where genes, or parts of genes, are identical.

ficult to apply to multicellular organisms, they are quite helpful in studying unicellular organisms.

A **protein profile** is a laboratory-prepared pattern of the proteins found in a cell (Figure 9.16a). Because a cell's proteins are the products of its genes, the cells of each species synthesize a unique array of proteins—as distinctive as a fingerprint is for humans. Analysis of the profiles of one or more proteins of different bacterial species provides a reasonable basis for comparisons.

Protein profiles are produced by the **polyacrylamide gel electrophoresis** (PAGE) method, which separates proteins on the basis of molecular size (Figure 9.16b). In this method, samples of protein obtained from lysed cells are dissolved in a detergent and poured into wells (depressions) of a thin slab of polyacrylamide gel. The slab is inserted into a buffer-filled chamber. An electric current is then passed through the gel for a period of time. The current causes the protein molecules to migrate to the opposite end of the gel. Large protein molecules migrate more slowly than do smaller ones. After the smallest proteins have migrated, the current is turned off, and the gel slab is then removed. Next it is stained so that the various proteins show up as separate stained bands in the gel slab.

Each band in the profile from one kind of cell represents a different protein in that cell. Bands at the same location in profiles from different kinds of cells indicate that the same protein is present in the different cells.

Determination of amino acid sequences in proteins also identifies similarities and differences among organisms. Certain proteins, such as cytochromes, which contribute to oxidative metabolism in many organisms, are commonly used to study amino acid sequences. The amino acid sequences in the same kind of protein from several organisms are determined. As with DNA hybridization, the extent of matching sequences of amino acids in the proteins indicates the relatedness of the organisms.

The proteins an organism contains are determined directly by the information in that organism's DNA. Thus, both protein profiles and determinations of amino acid sequences are as significant a measure of the relatedness of organisms as are DNA homologies. All are also related to the evolutionary history of the organisms.

Other Techniques

Other techniques for studying evolutionary relatedness include determining properties of ribosomes, immunological reactions, and phage typing.

FIGURE 9.16 (a)

Protein profiles, which provide a "fingerprint" of the proteins present in particular cells, can be used to compare different organisms to determine their degree of relatedness. (b) The PAGE process.

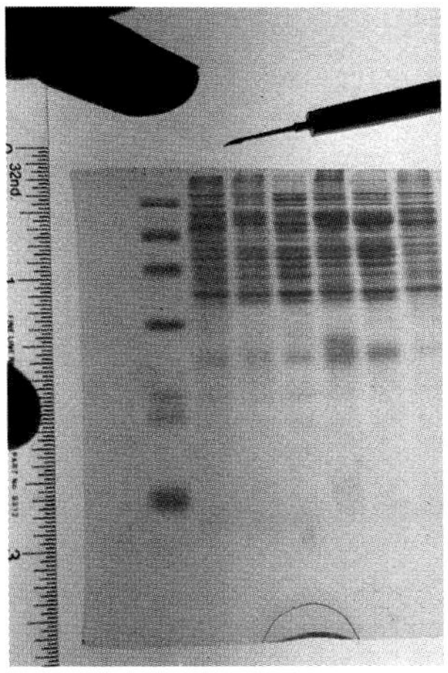

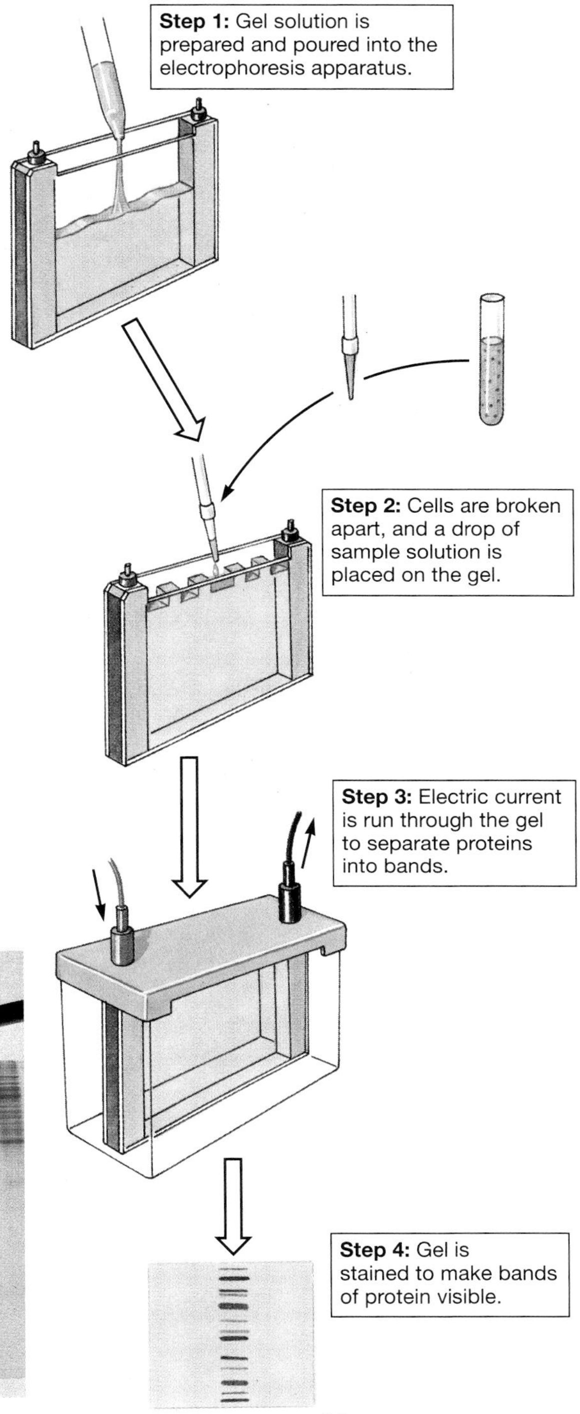

Step 1: Gel solution is prepared and poured into the electrophoresis apparatus.

Step 2: Cells are broken apart, and a drop of sample solution is placed on the gel.

Step 3: Electric current is run through the gel to separate proteins into bands.

Step 4: Gel is stained to make bands of protein visible.

(a)

(b)

rRNA changes very slowly through evolution because the molecule is easily rendered nonfunctional—and are universally distributed among organisms. It is the degree of similarity in 16S rRNA sequences between two organisms that indicates their evolutionary relatedness. If the nucleotide sequences of 16S rRNA molecules from two types of organisms are very similar, those organisms are likely to be quite close evolutionarily. Although direct sequencing of 16S rRNA is used to show evolutionary relationships between species, newer methods, such as PCR, are beginning to replace it. The PCR technique, which is being used to amplify rRNA genes, requires less cell material and is more rapid and convenient for large studies than is direct rRNA sequencing.

Immunological Reactions

Immunological reactions also are used to identify and study surface structures and the composition of microorganisms, as explained in Chapter 18. As we shall see, one highly specific and sensitive technique involves proteins called *monoclonal antibodies*. Monoclonal antibodies can be created so that they will bind to a specific protein, usually a protein found on a cell surface. If the antibodies bind to the surfaces of more than one kind of organism, the organisms have that protein in common. This technique promises to be particularly useful in identifying specific biochemical properties of microorganisms. In turn, identification of such properties will be extremely useful in determining taxonomic relationships.

Phage Typing

Phage typing involves the use of bacteriophages, viruses that attack bacteria, to determine similarities among different bacteria. A separate agar plate is inoculated for each bacterium being studied. A sterile cotton swab or bent glass rod is used to spread the inoculum over the agar surface. After incubation, a *lawn*, or continuous sheet, of *confluent* bacterial growth is produced. The underside of the plate is marked with numbered squares so that drops of known phages can be spotted onto specific zones of the lawn and later identi-

Properties of Ribosomes

Ribosomes serve as sites of protein synthesis in both prokaryotic and eukaryotic cells. RNA in ribosomes can be separated into several types according to the size of the RNA units. A particular RNA unit, the 16S rRNA component, has proven especially useful in studying evolutionary relationships for several reasons. Ribosomal RNAs are functionally constant in bacteria—that is,

IDENTIFICATION BY "BREATHPRINTS"

Rapid identification of organisms is not always easy. A new system produced by Biolog provides 96 different kinds of substrates, each in its own well on a plastic plate. Suspensions of an unknown bacterial culture are added to each well, and the plate is incubated for 4 to 24 hours. Each well also contains an indicator dye, tetrazolium violet. If the organism is able to use the substrate in a given well for its carbon source, respiratory gases released turn that well violet. The plate makes a quick trip through a plate reader, a computer compares the 96 test results with a library of results (called "breathprints," in analogy to fingerprints) for known species and identifies the organism in less than 10 seconds. Think how long it would take you to run 96 tests on your unknown specimen in the lab—especially if you had to make up all 96 media!

This machine allows rapid clinical diagnosis but is also very valuable in environmental studies. Many organisms in soil, animals, and other materials are unidentified. Such a system makes it possible to identify the characteristics of new organisms rapidly and to ascertain their possible relationship to known species.

Biolog system of identifying organisms. Organisms grown in different kinds of media, each of which is located in a separate well in the large plastic plate at the right, are placed in a reading device to assess color changes that result from biochemical reactions. Computerized software analyzes the patterns of these reactions and identifies the organism as one of 300 clinically or environmentally important gram-negative species.

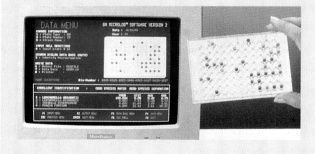

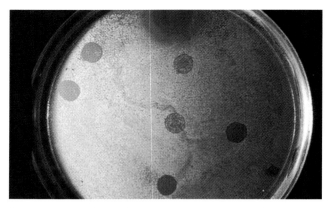

FIGURE 9.17 Phage typing. Receptor sites for bacteriophages are highly specific; certain strains of a species of bacterium are attacked only by particular types of phages. Clear sites (plaques) are left when phages have killed bacterial cells. On the basis of which phages have attacked a bacterial culture, one can determine which strain of that bacterial species is present.

to group closely related organisms and to separate them from less closely related ones. When groups of closely related organisms are identified, it is presumed that they probably had a common ancestor and that small differences among them have arisen by *divergent evolution*. **Divergent evolution** occurs as certain subgroups of a species with common ancestors undergo sufficient mutation to be identified as separate species.

Within the eubacteria, an early divergence gave rise to two important subgroups, the gram-positive bacteria and the gram-negative ones. Subsequent divergence within each group has given rise to many modern species of bacteria. Among the gram-negative bacteria, the purple nonsulfur bacteria gave rise to modern bacteria that inhabit animal digestive tracts. One proposed scheme of divergent evolution in this group is shown in Figure 9.18.

fied. After a suitable incubation period, zones of lysis (*plaques*) appear in the bacterial lawn (Figure 9.17). By observing which phages cause holes in the lawn, researchers can identify the strain. Strains lysed by the same phages are presumed to be more closely related than strains that show different patterns of lysis by phages.

Significance of Findings

The main significance of methods of determining evolutionary relationships is that these methods can be used

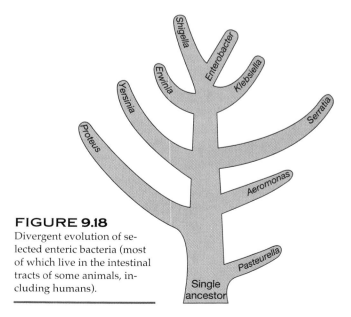

FIGURE 9.18 Divergent evolution of selected enteric bacteria (most of which live in the intestinal tracts of some animals, including humans).

EVOLUTION FROM A UNIVERSAL COMMON ANCESTOR

For centuries humans have pondered the questions of how life arose and how living things are related. Before microorganisms were discovered, living things were classified as either plants or animals. When microscopes first allowed scientists to observe microbes, some scientists tried to classify microorganisms as either plants or animals, and others put them in one or more separate categories. As microscopy became more powerful, fundamental differences in cells were recognized. Living organisms were then classified as prokaryotes and eukaryotes on the basis of the properties of their cells. The prokaryotes included the then-known bacteria and the cyanobacteria; the eukaryotes included all organisms with nucleated cells. It was eventually presumed that prokaryotes had a universal common ancestor from which all living things evolved and that an ancestral eukaryote arose by endosymbiosis from among various groups of prokaryotes (Figure 9.19).

TABLE 9.3	Comparison of Archaeobacteria, Eubacteria, and Eukaryotes		
Characteristic	Archaeobacteria	Eubacteria	Eukaryotes
Cell wall	Lack peptidoglycan	Contain peptidoglycan	Absent or made of other materials
Lipids in membranes	Branched-chain fatty acids	Straight-chain fatty acids	Straight-chain fatty acids and sterols
Protein synthesis	Not impaired by antibiotics such as chloramphenicol	Impaired by antibiotics such as chloramphenicol	Most not impaired by antibiotics such as chloramphenicol
First amino acid in a protein	Methionine	Formylmethionine	Methionine
Habitat	Usually found only in extreme environments	Found in a wide range of environments	Found in a wide range of environments

Then came the discovery of archaeobacteria. Studies of these organisms in the late 1970s by Carl Woese, G. E. Fox, and others suggested that the archaeobacteria represented a third cell type (Table 9.3). These investigators proposed another scheme for the evolution of living things from a universal common ancestor (Figure 9.20). They hypothesized a group of *urkaryotes*, the earliest or original cells, that gave rise to the eukaryotes directly rather than

FIGURE 9.19 A model of major evolutionary lines of descent proposed before the discovery of archaeobacteria.

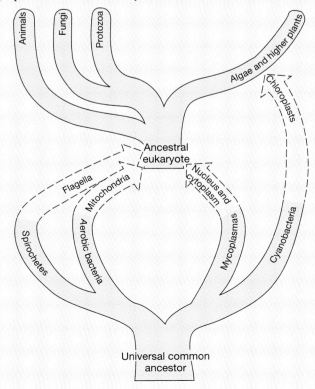

FIGURE 9.20 A model of major evolutionary lines of descent proposed after the discovery of archaeobacteria.

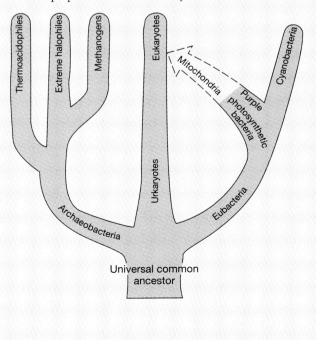

by way of the prokaryotes. These investigators did not, however, dispense with the idea of endosymbiosis. Rather, they proposed that nucleated urkaryotes became true eukaryotes by acquiring organelles by endosymbiosis from certain eubacteria.

At about the same time that archaeobacteria were first being investigated, studies of stromatolites indicated that life had arisen nearly 4 billion years ago. These studies indicated that an "age of microorganisms," in which there were no multicellular living organisms, lasted about 3 billion years. Combined evidence from studies of archaeobacteria and the most ancient stromatolites convinced many scientists that three branches of the tree of life formed during the age of microorganisms and that each branch gave rise to distinctly different groups of organisms.

The three-branch tree was not accepted by all scientists. In 1977 the English taxonomist T. Cavalier-Smith proposed instead that the archaeobacteria arose later than the eubacteria by divergent evolution from a group of gram-positive bacteria similar to present-day actinomycetes, which were once thought to be fungi (Figure 9.21). The American taxonomist J. A. Lake proposed in 1988 still another model with two main branches (Figure 9.22).

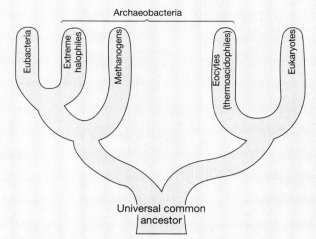

FIGURE 9.22   A model of evolution proposed by Lake.

In Lake's model, one branch gave rise to the eubacteria and to two groups of archaeobacteria—those that live in extremely salty environments and those that release methane. The other branch gave rise to the eukaryotes and to a group of archaeobacteria Lake calls the *eocytes*, which grow in hot, acidic environments.

Woese recently suggested that a new taxonomic category, the **domain**, be erected above the level of kingdom. He bases this suggestion on comparative studies, at the molecular level, of prokaryotes and eukaryotes and their probable evolutionary relationships. Woese concludes that the archaeobacteria may be more closely related to eukaryotes than they are to eubacteria. The three domains he proposes are shown in Figure 9.23.

Which, if any, of these models is correct remains to be seen. Further analysis of nucleotide and amino acid sequences promises to provide much better information about relationships among organisms than has ever been available before. It also may help clarify the nature and time of origin of the universal common ancestor.

FIGURE 9.21   A model of evolution proposed by Cavalier-Smith.

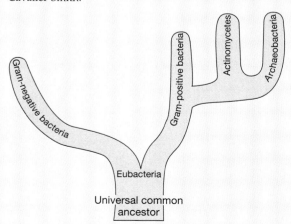

FIGURE 9.23   Three domains, a new taxonomic level above kingdom, have been proposed by Carl Woese. He believes that archaeobacteria may be more closely related to eukaryotes than they are to eubacteria.

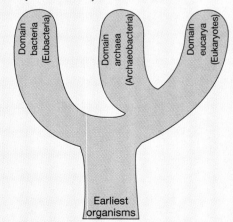

TAXONOMY—THE SCIENCE OF CLASSIFICATION

- Organisms are named according to their characteristics, where they are found, who discovered them, or what disease they cause. **Taxonomy** is the science of classification, and each category is a **taxon.**

Linnaeus—the Father of Taxonomy

- Linnaeus developed the system of **binomial nomenclature,** a two-name identification system for each living organism.
- The **genus** and **specific epithet** of each organism identify the **species** to which it belongs.
- Linnaeus also established the hierarchy of taxonomy and classified organisms into two kingdoms, Plantae and Animalia.

Using a Taxonomic Key

- A **dichotomous** taxonomic **key** consists of a series of paired statements presented as either-or choices that describe characteristics of organisms. By selecting appropriate statements to progress through the key, one can classify organisms and, if the key is sufficiently detailed, identify them by genus and species.

Problems in Taxonomy

- Ideally, organisms should be classified by their **phylogenetic,** or evolutionary, relationships.
- Problems in taxonomy include the rapid pace of evolutionary change in microorganisms and the difficulty in deciding what constitutes a kingdom and what constitutes a species.

Developments since Linnaeus's Time

- Since Linnaeus's time, several taxonomists have proposed three- and four-kingdom systems on the basis of various fundamental characteristics of living things. Whittaker proposed a five-kingdom system in 1969.
- Since 1925 *Bergey's Manual of Determinative Biology* has served as an important tool in identifying bacteria.

THE FIVE-KINGDOM CLASSIFICATION SYSTEM

- The kingdoms of the **five-kingdom system** are **Monera** (**Prokaryotae**), **Protista, Fungi, Plantae,** and **Animalia.** The characteristics of members of each kingdom are summarized in Table 9.2.
- Some taxonomists think that a sixth kingdom should be proposed for the **archaeobacteria,** to distinguish them from the **eubacteria,** or true bacteria.

Kingdom Monera

- All monerans are unicellular **prokaryotes:** They generally lack organelles, have no true nuclei, and their DNA has little or no protein associated with it.
- The **cyanobacteria** are photosynthetic monerans of great ecological importance.

Kingdom Protista

- The protists are a diverse group of mostly unicellular **eukaryotes.**

Kingdom Fungi

- The fungi include some unicellular and many multicellular organisms that obtain nutrients solely by absorption.

Kingdom Plantae

- Most plants live on land and contain chlorophyll in organelles called chloroplasts.

Kingdom Animalia

- All animals are derived from zygotes; most are macroscopic.

CLASSIFICATION OF VIRUSES

- **Viruses** are acellular infectious agents that share only a few characteristics with living organsims, are not included in any of the five kingdoms. Viruses are classified by their nucleic acids, chemical composition, and their morphology.

THE SEARCH FOR EVOLUTIONARY RELATIONSHIPS

Special Methods Needed for Prokaryotes

- Special methods are needed for determining evolutionary relationships among prokaryotes because they have few morphological characteristics and have left only a sparse fossil record.
- Several methods, including numerical taxonomy and genetic homology, are currently used to determine evolutionary relationships among organisms.

Numerical Taxonomy

- In **numerical taxonomy,** organisms are compared on the basis of a large number of characteristics and grouped according to the percentage of shared characteristics.

Genetic Homology

- **Genetic homology** is the similarity of DNA among different organisms, which provides a measure of their relatedness. Several techniques that determine genetic homology are available.
- The relative percentages of bases in DNA are a measure of relatedness. The base composition of DNA is determined, and G–C percentages are compared among organisms.
- Base sequences can be identified by DNA **probes.**
- In **DNA hybridization,** the degree of matching between strands of DNA is compared among organisms.
- **Protein profiles,** made by **polyacrylamide gel electrophoresis,** are used to indicate whether the same proteins are present in different organisms.

- The amino acid sequences of related organisms are similar, so the determination of amino acid sequences is another measure of relatedness.

Other Techniques

- Other methods make use of properties of ribosomes, immunological reactions, and **phage typing.**

Significance of Findings

- Evolutionary relationships can be used to group closely related organisms. Small differences among organisms descended from a common ancestor arise by **divergent evolution.** An early divergence gave rise to the two major subgroups of eubacteria, the gram-positive bacteria and the gram-negative ones.

KEY TERMS

Animalia (p. 233)
archaeobacteria (p. 231)
binomial nomenclature
   (p. 227)
cyanobacteria (p. 232)
dichotomous key (p. 229)
divergent evolution (p. 240)
DNA hybridization (p. 237)
domain (p. 242)

eubacteria (p. 231)
eukaryote (p. 231)
five-kingdom system
   (p. 231)
Fungi (p. 232)
genetic homology   (p. 236)
genus (p. 227)
Monera (p. 231)
numerical taxonomy   (p. 236)

phage typing (p. 239)
phylogenetic (p. 229)
Plantae (p. 233)
polyacrylamide gel elec-
   trophoresis (PAGE)
   (p. 238)
probe (p. 237)
Prokaryotae (p. 231)
prokaryote (p. 231)

protein profile (p. 238)
Protista (p. 232)
species (p. 227)
specific epithet (p. 227)
strain (p. 228)
taxon (p. 226)
taxonomy (p. 226)
virus (p. 233)

QUESTIONS FOR REVIEW

A **Naming Microorganisms**

1. How are organisms named?

B **Linnaeus and Taxonomy**

2. Define taxonomy, and describe Linnaeus's contributions to it.

C **Dichotomous Taxonomic Key**

3. What is a dichotomous key, and how is it used?

D **Problems and Developments in Taxonomy**

4. True or False: A good taxonomic system should never change, no matter what new evolutionary data are discovered.

5. Describe some of the problems associated with developing a good taxonomic system.

6. How have taxonomists since Linnaeus's time attempted to solve the problems of developing a good taxonomic system?

E **Five-Kingdom System**

7. List the main characteristics of organisms in each of the kingdoms in the five-kingdom system.

F **Virus Classification**

8. How are viruses classified?

G **Determining Evolutionary Relationships**

9. Why are special methods needed to determine evolutionary relationships among prokaryotes?

10. How do the following methods help determine evolutionary relationships: numerical taxonomy, genetic homology, protein profiles, and amino acid sequencing?

11. In a numerical taxonomic study, seven cultures of bacteria (A through G) were analyzed for the presence or absence of 11 characteristics. The results were plotted as follows:

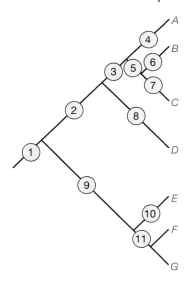

a. Which organism has characteristics 1, 2, 3, 5, and 7?

b. Which organisms have characteristics 1 and 2 in common?

c. Which organisms have characteristic 5 in common?

d. Which organisms are probably just different strains of the same species?

e. Is organism E more closely related to D or to F?

PROBLEMS FOR INVESTIGATION

1. What is a species? How is "species" defined in organisms that do not generally reproduce sexually?

2. DNA hybridization experiments were performed in which the double strands of DNA of two organisms were separated into single strands. The single strands of both organisms were then incubated together, and the percentages that hybridized (combined with that of the other species) were determined. From the following data, determine which two species are probably most closely related?

Species	Percentage Hybridization
A and B	46
A and C	58
B and C	75

3. When DNA strands are "melted" (separated into single strands by heating), more heat is required to break a G–C base pairing, which is held together by three hydrogen bonds, than is required to break an A–T base pairing, which is held together by only two hydrogen bonds. The DNA from four species of organisms was melted, and the melting temperature used for each was recorded.

Species	Melting Temperature (°C)
A	90.2
B	86.3
C	87.1
D	94.7

Which species has the highest G–C content? Which has the lowest? Which two species are most likely to be closely related? Why?

4. Cite examples of how the concepts of unity and diversity among living organisms are handled in taxonomic schemes.

5. Select five members of your class to go to the front of the room. Make a dichotomous key to identify them. Which characteristics are good or bad to use in your key? Would height, weight, or hair color be good if you tried to use this key again at your 25th class reunion?

SOME INTERESTING READING

Bergey's manual of systematic bacteriology. Vol. 1, 1984, N. R. Krieg and J. G. Holt, eds.; Vol. 2, 1986, J. G. Holt et al., eds.; Vol. 3, 1989, J. G. Holt et al., eds.; Vol. 4, 1989, J. G. Holt et al., eds. Baltimore: Williams & Wilkins.

Cavalier-Smith, T. 1987. "The origin of eukaryote and archaebacterial cells." *Annals of the New York Academy of Sciences* 503:17–54.

Doolittle, W. F. 1987. "The evolutionary significance of the archaebacteria." *Annals of the New York Academy of Sciences* 503:72–77.

"Germs in space?" 1991. *Sky and Telescope* 81(April): 357.

Knoll, A. M. 1991. "End of the proterozoic eon." *Scientific American* 265(October):64–73.

Margulis, L. and R. Guerrero. 1991. "Kingdoms in turmoil." *New Scientist* 129, no. 1761 (March 23):46–50.

Margulis, L., and K. V. Schwartz. 1988. *Five kingdoms: An illustrated guide to the phyla of life on earth*. New York: W. H. Freeman and Co.

Palleroni, N. J. 1994. "Some reflections on bacterial diversity." *ASM News* 60(October):537–40.

Persing, D. H., et al. 1993. *Diagnostic molecular microbiology: Principles and applications*. Washington, D.C.: American Society for Microbiology.

Schopf, J. W. 1993. "Microfossils of the early Archean Apex chert: New evidence of the antiquity of life." *Science* 260, no. 5108 (April 30):640–46.

Whittaker, R. H. 1969. "New concepts of kingdoms of organisms." *Science* 163, no. 3863 (January 10):150–160.

Woese, C. R. 1981. "Archaebacteria." *Scientific American* 244 (June):98–122.

———. 1987. "Bacterial evolution." *Microbiological Reviews* 51, no. 2 (June):221–71.

Zillig, W. 1987. "Eukaryotic traits in archaebacteria: Could the eukaryotic cytoplasm have arisen from archaebacterial origin?" *Annals of the New York Academy of Sciences* 503: 78–82.

10

THE BACTERIA

SEM photo of *Pseudomonas putida* (magnified 8100×) joined by F pili. This bacterium, which is used to decompose oil, is just one among the thousands of bacterial species that have been recognized.

I n Chapter 9 we considered general aspects of classification and the characteristics of members of each kingdom according to the five-kingdom system. Here we will see how bacteria are classified and why there are problems in classifying bacteria.

BACTERIAL TAXONOMY AND NOMENCLATURE

Criteria for Classifying Bacteria

Most macroscopic organisms can preliminarily be classified according to observable structural characteristics. It is more difficult to classify microscopic organisms, especially bacteria, because many of them have similar structures. Separating them according to cell shape, size, and arrangement does not produce a very useful classification system. Nor does the presence of specific structures such as flagella, endospores, or capsules allow identification of particular species. Therefore, other criteria must be used. Staining reactions, especially the Gram stain, were among the first properties

other than morphology to be used in classification of bacteria. Other properties now in use include features related to growth, nutritional requirements, physiology, biochemistry, genetics, and molecular analysis. These features include properties of DNA and proteins, as explained in Chapter 9. Important criteria used in classifying bacteria are summarized in Table 10.1, and biochemical tests used in classifying and identifying them are described in Table 10.2 (page 248).

By using various classification criteria, we can identify an organism as belonging to a particular genus and species. For bacteria, a species is regarded as a collection of strains that share many common features and differ significantly from other strains. ∞ (Chapter 9, p. 228) A bacterial *strain* consists of descendants of a single isolation in pure culture. Bacteriologists designate one strain of a species as the **type strain.** Usually this is the first strain described. It is the name-bearer of the species and is preserved in one or more type culture collections. The American Type Culture Collection (ATCC), a nonprofit scientific organization established in 1925, collects, preserves, and distributes authenticated type cultures of microorganisms. Many impor-

THIS CHAPTER FOCUSES ON THE FOLLOWING QUESTIONS:

A What criteria are used for classifying bacteria?

B What problems are associated with bacterial taxonomy?

C What are the history and significance of *Bergey's Manual of Systematic Bacteriology*?

D What are the characteristics of those bacterial genera that have medical significance, and what infectious diseases do they cause?

TABLE 10.1	Criteria for Classifying Bacteria
Morphology	Size and shape of cells, arrangement in pairs, clusters or filaments, presence of flagella, pili, endospores, capsules
Staining	Gram-positive, gram-negative, acid-fast
Growth	Characteristics in liquid and solid cultures, colony morphology, development of pigment
Nutrition	Autotrophic, heterotrophic, fermentative with different products, energy sources, carbon sources, nitrogen sources, needs for special nutrients
Physiology	Temperature (optimum and range), pH (optimum and range), oxygen requirements, salt requirements, osmotic tolerance, antibiotic sensitivities and resistances
Biochemistry	Nature of cellular components such as cell wall, RNA molecules, ribosomes, storage inclusions, pigments, antigens; biochemical tests
Genetics	Percentages of DNA bases, DNA hybridization

tant research studies dealing with classification, identification, and industrial uses of microorganisms would be seriously hampered without the services of the ATCC.

For many strains of bacteria, scientists are able to determine that they are members of a particular species. For other strains, however, difficult judgments must be made to decide whether the strain belongs to an existing species or differs sufficiently to be defined as a separate species. In recent years similarities of DNA and proteins among organisms have proved a reliable means of assigning a strain to an existing species or establishing the basis for a new species.

Curiously, assigning bacterial genera to higher taxonomic levels—families, orders, classes, and divisions (or phyla)—can be even more difficult than organizing species and strains *within* genera. Many macroscopic organisms are classified by establishing their evolutionary relationships to other organisms from fossil records. Efforts are being made to classify bacteria by evolutionary relationships, too, but these efforts are hampered by the incompleteness of the fossil record and by the limited information gleaned from

Biochemical Test	Nature of Test
Sugar fermentation	Organism is inoculated into a medium containing a specific sugar; growth and end products of fermentation, including gases, are noted. Anaerobic fermentations can be detected by inoculating organisms via a "stab" culture into solid medium.
Gelatin liquefaction	Organism is inoculated (stabbed) into a solid medium containing gelatin; liquefaction at room temperature or inability to resolidify at refrigerator temperature indicates the presence of proteolytic (protein-digesting) enzymes.
Starch hydrolysis	Organism is inoculated onto an agar medium containing starch; after the plate is flooded with Gram's iodine, clear areas around colonies indicate the presence of starch-digesting enzymes.
Litmus milk	Organism is inoculated into litmus milk medium (10 percent powdered skim milk plus litmus indicator); characteristic changes such as alteration of pH to acid or alkaline, denaturation of the protein casein (curdling), and gas production can be used to help identify specific organisms.
Catalase	Hydrogen peroxide (H_2O_2) is poured over heavy growth of an organism on an agar slant; release of O_2 gas bubbles indicates the presence of catalase, which oxidizes H_2O_2 to H_2O and O_2.
Oxidase	Two or three drops (or a disk) of an oxidase test reagent is added to an organism growing on an agar plate; a color change of the test reagent to blue, purple, or black indicates the presence of cytochrome oxidase.
Citrate utilization	Organism is inoculated into citrate agar medium in which citrate is the sole carbon source; an indicator in the medium changes color if citrate is metabolized; use of citrate indicates the presence of the permease complex that transports citrate into the cell.
Hydrogen sulfide	Organism is inoculated into peptone iron medium; formation of black iron sulfide indicates the organism produces hydrogen sulfide (H_2S).
Indole production	Organism is inoculated into a medium containing the amino acid tryptophan; production of indole, a nitrogenous breakdown product of tryptophan, indicates the presence of a set of enzymes that convert tryptophan to indole.
Nitrate reduction	Organism is inoculated into a medium containing nitrate (NO_3^-); presence of nitrite (NO_2^-) indicates that the organism has the enzyme nitrate reductase; absence of nitrite indicates either absence of nitrate reductase or presence of nitrite reductase (which reduces nitrite to N_2 or NH_3).
Methyl red	Organism is cultured in MR–VP broth; a methyl red indicator is added; presence of acid causes an indicator color change (red).
Voges-Proskauer	Organism is cultured in MR–VP broth; alpha naphthol and KOH–creatine are added; presence of the enzyme cytochrome oxidase causes color change in an indicator (rose color).
Phenylalanine deaminase	Organism is inoculated into a medium containing phenylalanine and ferric ions; formation of phenylpyruvate and its reaction with ferric ions produces a color change that demonstrates the presence of the enzyme phenylalanine deaminase.
Urease	Organism is inoculated into a medium containing urea; production of ammonia, usually detected by an indicator for alkaline pH, indicates the presence of the enzyme urease.
Specific nutrient	Organism is inoculated into a medium containing a specific nutrient, such as a particular amino acid (for example, cysteine) or vitamin (for example, niacin); growth of an organism that fails to grow in media lacking the specific nutrient can be used to identify some auxotrophs.

what fossils have been found. Even a complete fossil record might supply only morphological information and would thus be inadequate for determining evolutionary relationships.

Problems of Bacterial Taxonomy

Despite the tremendous effort spent in classifying bacteria, no complete classification of bacteria from kingdom to species has been established. The plight of bacteriologists looking at the taxonomy of bacteria might be described as follows: Those looking from the top level down can propose at least plausible divisions of the kingdom Monera, or kingdom Prokaryotae, as it is commonly designated by bacteriologists. The Prokaryotae include eubacteria, cyanobacteria, and archaeobacteria. Those bacteriologists looking from the bottom up can establish strains, species, and genera and can sometimes assign bacteria to higher-level groups. But too little is

known about evolutionary relationships to establish clearly defined taxonomic classes and orders for many bacteria.

Taxonomy from the Top Down

Taxonomists generally agree that true bacteria belong in the kingdom Prokaryotae. Like other prokaryotes, they lack nuclear membranes. Taxonomists do not agree, however, on how to separate the true bacteria into divisions or phyla. In 1968 the bacteriologist R. G. E. Murray proposed four divisions: gram-negative, gram-positive, gram-variable, and organisms lacking a cell wall. In 1982 Lynn Margulis grouped bacteria as fermenting heterotrophs, respiring heterotrophs, or autotrophs.

Bergey's Manual of Systematic Bacteriology, the accepted reference on the identification of bacteria, currently divides the Kingdom Procaryotae into the following four divisions on the basis of cell wall properties:

Division I.	Gracilicutes (gras"ih-lik-yoo'teez)	Prokaryotes with thin cell walls, which are typical of gram-negative organisms
Division II.	Firmicutes (fer-mik"yoo-teez)	Prokaryotes with thick cell walls, indicating a gram-positive type of cell wall
Division III.	Tenericutes (ten"er-ik'yoo-teez)	Prokaryotes of a pliable, soft nature, and lacking cell walls
Division IV.	Mendosicutes (men-doh-sik'yoo-teez)	Prokaryotes having cell walls that lack peptidoglycan

Taxonomy from the Bottom Up

Within each division, bacteria have been classified into 33 groups, called *sections*, in *Bergey's Manual*. The sections are based on relatively easily observable characteristics such as shape, staining reactions, presence or absence of a cell wall, motility, reproduction by budding, and mode of metabolism. Most sections were established many years ago as a practical means of classifying bacteria. The criteria were established before modern techniques for molecular analysis (especially of DNA and RNA) and biochemical studies were available and before any fossil bacteria had been found.

As more and more bacteria have been subjected to modern analyses, certain discrepancies in assignments to sections have become apparent. For example, the cell walls of some genera have unusual properties that prevent their being classified as either gram-positive or gram-negative. The organisms assigned to a particular section of *Bergey's Manual* are not necessarily closely related.

The Muddle in the Middle

The difficulties of classifying bacteria are greatly magnified as one proceeds up the taxonomic hierarchy. Although families, orders, and classes can be clearly defined for a few bacteria, too little information is available to do so for many. Until discrepancies in section assignments can be resolved and taxonomic levels between genera and divisions can be more precisely determined, it is practical to continue to use section designations. A complete listing of the sections is provided in Appendix B.

Bacterial Nomenclature

Despite all these taxonomic problems, there is an established nomenclature for bacteria. *Bacterial nomenclature* refers to the naming of species according to internationally agreed upon rules. Both taxonomy and

CLOSE-UP

WHO CARES ABOUT BACTERIAL TAXONOMY?

"It also seems worthwhile to emphasize . . . [that] bacterial classifications are devised for microbiologists, not for the entities being classified. Bacteria show little interest in the matter of their classification. For the systematist [taxonomist], this is sometimes a very sobering thought!"

—J. T. Staley and N. R. Krieg (1985)

nomenclature are subject to change as new information is obtained. Organisms are sometimes moved from one category to another, and their official names are sometimes changed. For example, the bacterium that causes tularemia, a fever acquired by handling infected rabbits, was for many years called *Pasteurella tularensis*. Its genus name was changed to *Francisella* after DNA hybridization studies revealed that hybridization between its DNA and that of *Pasteurella* species essentially did not occur. It does, however, have a 78 percent match with the DNA of *Francisella novicida*.

CLASSIFICATION ACCORDING TO *BERGEY'S MANUAL OF SYSTEMATIC BACTERIOLOGY*

History and Significance of *Bergey's Manual*

The first edition of *Bergey's Manual of Determinative Bacteriology* was published in 1923 by the American Society for Microbiology; David H. Bergey (Figure 10.1) was chairperson of its editorial board. Since then eight editions, an abridged version, and several supplements have been published. *Bergey's Manual* has become an internationally recognized reference for bacterial taxonomy. It has also served as a reliable standby for medical workers interested in identifying causative agents of infections.

Bergey's Manual contains names and descriptions of organisms and diagnostic keys and tables for identifying organisms. It is organized into sections, with each section devoted to a group of similar organisms. The term "section" is used rather than established taxonomic terms such as "family" and "order" because evolutionary relationships between sections (and even within sections) are not yet clearly understood.

A new manual, *Bergey's Manual of Systematic Bacteriology,* has a much broader scope than previous

FIGURE 10.1 David H. Bergey, originator of *Bergey's Manuals*, the first of which was published in 1923. In 1936 he set up an educational trust to which all rights and royalties from the *Manuals* would be transferred for preparing, editing, and publishing future editions, as well as providing funds for research to clarify problems arising in the process. This nonprofit trust ensures that *Bergey's Manual* will be a self-perpetuating publication.

manuals. All the kinds of information found in earlier determinative manuals are included, and information on taxonomy, ecology, cultivation, maintenance, and preservation of organisms has been expanded. This manual is a four-volume work. The first volume was published in 1984, the second in 1986, and the final two in 1989. Then *determinative information* (information used to identify bacteria) was collected into a single volume, the ninth edition of *Bergey's Manual of Determinative Bacteriology*, which was published in 1994. And in 1995 *Bergey's Manual* became available in CD-ROM format as well.

Bergey's Manual is widely accepted as a reference for identifying bacteria. The four-volume *Bergey's Manual of Systematic Bacteriology* retains that significance; it also is of great significance to those interested in bacterial taxonomy. An important motivation for publishing the larger work is to stimulate research to improve the understanding of taxonomic relationships among bacteria.

As a beginning student, you will doubtless find it difficult to remember many characteristics of specific microorganisms that we cover in this course. A four-volume set of *Bergey's Manual* weighs about 21 pounds and costs about $400—not something you could carry to class and back. However, you can use the endpapers, inside the front and back covers of this textbook. If you wish to find out whether a given organism is gram-positive or gram-negative, its shape, the disease(s) it causes, and so on, look it up by name in the back endpapers. Microorganisms are grouped as bacteria, viruses, fungi, and parasites (protozoa and helminths). Or if you are discussing a disease but can't remember which organism(s) cause it, look in the front endpapers under the name of the disease (again grouped as bacterial, viral, fungal, and parasitic diseases), and you will find the organism's name and some of its characteristics. Page numbers are given to direct you to further information. Thumbing through the book or index can be frustrating when you need some little piece of information. But flipping to the cover of your book is easy, and we encourage you to do so often, until the information gradually becomes more and more familiar.

Bergey's Manual of Systematic Bacteriology provides many kinds of information, such as descriptions and photos of species, tests to distinguish among genera and species, DNA relatedness among organisms, and various numerical taxonomy studies (Figure 10.2). ∞ (Chapter 9, pp. 235–240)

Bacteria by Section of *Bergey's Manual*

Of the 33 sections of prokaryotes described in *Bergey's Manual of Systematic Bacteriology,* some contain only a few organisms, whereas others contain hundreds. Certain sections contain organisms of medical significance, whereas others are of interest mainly to ecologists, taxonomists, or researchers. In the remainder of this chapter, we consider characteristics of organisms in the major sections of *Bergey's Manual,* emphasizing those with medical significance. When relevant, order and family names are noted. Such names have consistent endings: Orders always end in *ales* and families in *aceae.*

Spirochetes (Section 1)

The **spirochetes** are helical, motile bacteria (Figure 10.3, p. 252). They have a multilayered, membranous *outer sheath* that surrounds the cell wall, inside of which is the coiled *protoplasmic cylinder*. Axial filaments are structurally similar to other bacterial flagella, but they lie between the outer sheath and the cell wall. ∞ (Chapter 4, p. 87) Each axial filament is attached at one pole and permanently wound around the cylinder. Filaments vary in number from 2 in some species to as many as 100 in others. In fluid environments, spirochetes display three kinds of movement: locomotion (straight-line motion), rotation on their longitudinal axis, and flexing (bending). Axial filaments enable spirochetes to move in relatively viscous (syrupy) fluids that prevent movement of bacteria with exposed flagella.

All spirochetes divide by binary fission, and division separates an elongated cell into two shorter ones. None form endospores. Among the spirochetes are aerobic, facultatively anaerobic, and anaerobic organisms. Free-living spirochetes inhabit a variety of aqueous environments, including sewage and muds containing hydrogen sulfide. Pathogenic spirochetes live mainly in body fluids.

The spirochetes include three genera that contain pathogens: *Treponema, Borrelia,* and *Leptospira,* all mem-

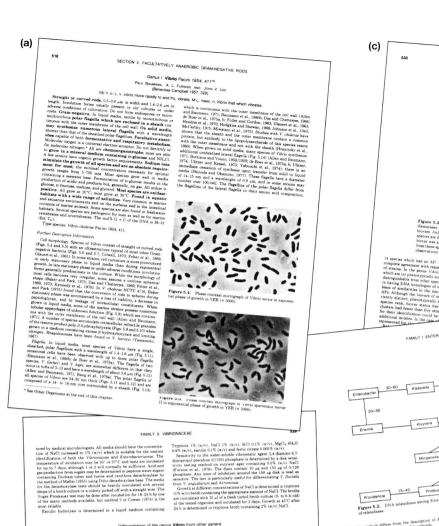

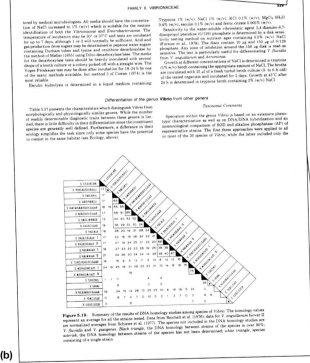

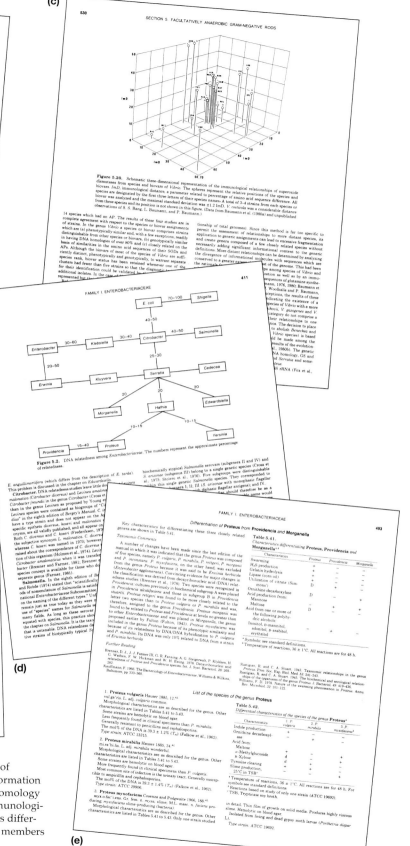

FIGURE 10.2 Pages reproduced from the four-volume set of *Bergey's Manual of Systematic Bacteriology,* showing the types of information found there. (a) Description of a genus. (b) Comparison of DNA homology among different species of the same genus. (c) Comparison of immunological relationships of enzymes within a genus. (d) Key characteristics differentiating three closely related genera. (e) DNA relatedness among members of the family Enterobacteriaceae.

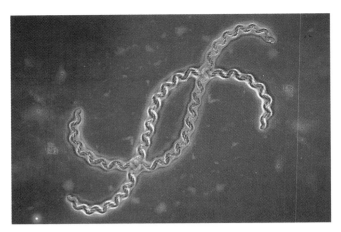

FIGURE 10.3 False-color TEM photo of *Leptospira interrogans*. Both the tightly coiled shape and axial filaments are characteristic of spirochetes. This organism causes leptospirosis in humans.

bers of the order Spirochaetales. The pathogenic **treponemes** have not yet been grown on laboratory media. The type species of the treponemes is *Treponema pallidum,* the causative agent of the sexually transmitted disease syphilis. Other strains of this same species cause nonvenereal infections such as bejel (a mild rashlike infection of the skin) and yaws (deep, oozing skin lesions that can erode underlying bone). Treponemes such as *T. denticola* and *T. vincentii* are often found in the mouth. Although they usually are not pathogenic, under certain conditions they contribute to periodontal (gum) disease. Several other spirochetes can cause human disease. *Borrelia recurrentis,* which is carried by ticks, causes relapsing fever, in which the patient suffers repeated bouts of fever and chills. *B. burgdorferi,* also tick-borne, is responsible for Lyme disease. Several species of *Leptospira* can cause leptospirosis, a disease characterized by fever and liver and kidney damage in humans who ingest contaminated food or water. In laboratory cultures, leptospiras oxidize long-chain fatty acids, but borrelias do not.

Aerobic/Microaerophilic, Motile, Helical/Vibrioid, Gram-Negative Bacteria (Section 2)

As the title indicates, bacteria of this group are helical, or vibrioid (comma-shaped); as such they are similar to the spirochetes. The helix can have half a turn (comma) or many turns. All these bacteria have flagella and swim in a straight line with a corkscrew motion. Many are small spiral organisms found in soil and in fresh and stagnant water where oxygen to satisfy aerobes or microaerophiles is available (Chapter 6); some grow in plant roots. ∞ (p. 147) All are gram-negative. Genera representative of this group are *Spirillum, Aquaspirillum,* and *Azospirillum.*

Bacteria with a variety of spiral morphologies are common inhabitants of the gastrointestinal tracts of animals. The helical bacterium *Helicobacter pylori* has been found to lead to peptic ulcers in the human stomach and upper portion of the small intestine and has been linked to some cancers (see the book cover). Various species of *Campylobacter,* which are slender spiral rods, inhabit the reproductive and intestinal tracts and the mouth of humans. *C. jejuni* (Figure 10.4) causes enteritis (an inflammation of the digestive tract), resulting in diarrhea and sometimes vomiting. Bacteremia—bacteria in the blood—can lead to infections elsewhere in the body. Other *Campylobacter* species have been found to cause spontaneous abortion, stillbirth, and premature delivery in animals. In 1994 a species present in healthy and diarrheic cats and dogs, *C. upsaliensis,* was found to infect and cause miscarriage in humans.

Gram-Negative Aerobic Rods and Cocci (Section 4)

Among the gram-negative aerobic rods and cocci are a large number of bacteria—eight families and some genera that have not been assigned to families. Many of the genera include significant human pathogens.

The **pseudomonads** (su-do-mo'-nadz), members of the genus *Pseudomonas,* are generally aerobic motile rods with polar flagella (Figure 10.5). Many species synthesize a soluble yellow-green pigment that fluoresces under ultraviolet light. Most contain an oxidase enzyme and give positive oxidase test results. Free-living forms are found in soils and in freshwater and marine environments, where they are important decomposers of organic material. *Pseudomonas aeruginosa* is the main human pathogen; it is often seen in urinary tract infections and in wounds and burns.

FIGURE 10.4 *Campylobacter jejuni.* False-color TEM photo (14,000×) shows spiral, gram-negative organisms with flagella at one or both ends. *C. jejuni* is part of the normal intestinal microbiota of birds and other animals. It is a common cause of enteritis in humans, causing fever, abdominal pain, diarrhea, and vomiting.

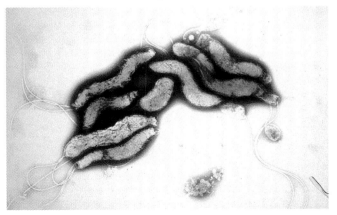

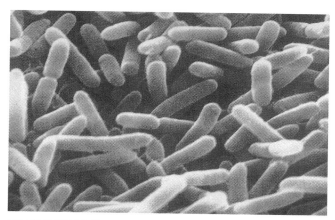

FIGURE 10.5 SEM photo of *Pseudomonas aeruginosa*. Pseudomonads, aerobic motile rods, are widely distributed in nature. They are very hardy organisms, and some can even grow in surgical scrub solutions, where they can cause hospital-acquired infections.

Among the pseudomonads are a variety of animal and plant pathogens. One such animal pathogen causes glanders, a debilitating disease with ulceration of lymph nodes in horses. Another causes melioidosis, a disease similar to pneumonia, in both animals and humans in Southeast Asia. The plant pathogens cause rots, scabs, and wilts on a wide variety of plants, including some common house plants. Pseudomonads that infect plants require quite different growth conditions than those that infect humans, so you needn't worry about getting such an infection while caring for a sick plant.

Identifying and naming *Legionella* (le"jun-el'lah) *pneumophila* as the causative agent of Legionnaires' disease, which affected 200 veterans staying in the same hotel while attending an American Legion convention in Philadelphia in 1976, required that a new family (Legionellaceae) be established. *L. micdadei*, the causative agent of Pittsburgh pneumonia, and several other *Legionella* species have since been added to that family. These organisms had not been previously identified because fastidious nutritional requirements make them difficult to culture and isolate. They are pleomorphic (that is, they can have many different shapes), varying in shape from coccoid to filamentous. ∞ (Chapter 4, p. 76) After **legionellas** were identified in the laboratory, they were found in water and soil, usually associated with other microorganisms, such as cyanobacteria, algae, protozoa, and other bacteria. They are often found living inside various species of amoebas.

Members of the family Neisseriaceae constitute another important group of medically important aerobic gram-negative rods and cocci. Most are found on mucous membranes of humans and animals. *Neisseria gonorrhoeae* (ni-se're-a gon-o-re'-e) thrives in the mucous membranes of the human urogenital tract and causes gonorrhea. *N. meningitidis* first colonizes nasopharyn-

geal mucous membranes, but it can invade blood and cerebrospinal fluid, where it causes meningitis, an inflammation of the membranes that cover the brain. Several species of the genus *Moraxella* can cause conjunctivitis, an inflammation of membranes of the eyes.

Several genera among the gram-negative aerobic rods and cocci have been identified but not assigned to a family. Most members of the three genera *Brucella*, *Bordetella*, and *Francisella* are pathogens. All *Brucella* are *obligate intracellular parasites* (they can replicate only inside a living host) that usually multiply in phagocytic white blood cells. In humans they cause brucellosis, or undulant fever, in which the patient has daily episodes of fever and chills. They cause similar diseases in domestic animals. *Bordetella pertussis* causes whooping cough. One species, *Francisella tularensis*, causes tularemia in rabbits, some other wild animals, and humans. This pathogen's special need for the amino acid cysteine helps identify it in laboratory cultures. Because *F. tularensis* is highly infectious, specimens suspected of containing this bacterium should be sent to specially equipped laboratories.

Several other gram-negative aerobic organisms are of interest in agriculture and environmental sciences. Some are important in nitrogen metabolism in plants (Figure 10.6); others cause some kinds of tumorlike growths called *galls* and other plant diseases. A few are specialized to metabolize methane, and many are important decomposers in mineral cycles. Another small group, the *halophiles*, tolerate high salt concentrations for growth. ∞ (Chapter 6, p. 148) Members of the Acetobacteraceae family produce acetic acid (vinegar) and are important in the food industry. They are found on fruits and vegetables and in alcoholic beverages. Some are a nuisance in the brewing industry because they change the alcohol produced by yeasts into acetic acid.

FIGURE 10.6 Nitrogen-fixing bacteria (*Rhizobium*; 1675×) inside nodules on alfalfa plant roots. These organisms capture N_2 gas from air and make it available to plants in more useful forms.

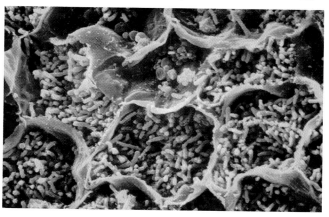

Facultatively Anaerobic Gram-Negative Rods (Section 5)

The facultatively anaerobic gram-negative rods constitute another large and medically important group of bacteria. The group is divided into three families and includes several genera not assigned to families. Among them are many important human pathogens, including the enterics, the vibrios, and the *Pasteurella–Haemophilus* group. The distinguishing characteristics of these organisms, summarized in Figure 10.7, are based on biochemical properties. These properties include kinds of fermentation reactions, percentages of bases in DNA, and the presence of certain enzymes (Table 10.2).

THE ENTERICS Members of the family Enterobacteriaceae are sometimes referred to as **enteric bacteria** because many of them inhabit the intestine of humans or other animals. ("Enteric" means pertaining to the intestine.) In addition, some are free-living organisms found in soil and water. Others live in cooperation with, or at the expense of, their hosts; a few decompose dead organic matter. Enteric bacteria are small, morphologically similar gram-negative rods. Some move with peritrichous flagella. They are facultative anaerobes that ferment glucose and other sugars and sometimes produce carbon dioxide and other gases. (Various fermentation pathways are described in Chapter 5.) ∞ (p. 118)

FIGURE 10.7 Distinguishing characteristics of facultatively anaerobic gram-negative rods.

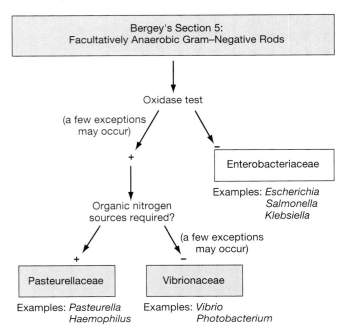

Escherichia coli is an important enteric bacterium for three reasons:

1. *E. coli* is always present among the normal microbiota of the human intestine. Finding it in water indicates that the water is contaminated with fecal matter and possibly with pathogenic enteric bacteria.

2. *E. coli* can be a pathogen in its own right; different strains produce various toxins and can cause opportunistic infections in a variety of body sites.

3. *E. coli* is an important organism in genetic engineering. Many important biological products can now be made by recombinant DNA technology, and *E. coli* frequently is the bacterium into which new genetic material is introduced. Consequently, this bacterium might be considered the workhorse of today's biotechnology.

Most members of the family Enterobacteriaceae, including *E. coli*, are found in the colon or large intestine and are sometimes referred to as "coliforms."

Members of the genera *Shigella* and *Salmonella* are closely related to *E. coli* in that all carry out mixed-acid fermentation (an anaerobic metabolic pathway that produces a combination of acids). *Shigella* can be distinguished from other enterics by its inability to produce gas during glucose fermentation. *Shigella* species cause a severe diarrhea called bacillary dysentery. *Salmonella* can be distinguished from *Escherichia* by its inability to produce acid and gas during lactose fermentation. Various species of *Salmonella* cause typhoid fever, enteritis, and food poisoning.

Another closely related set of genera are *Klebsiella, Enterobacter,* and *Erwinia,* all of which carry out butanediol fermentation. In laboratory cultures, the ability of *Klebsiella* to break down urea distinguishes that genus from *Enterobacter.* Not all members of the family Enterobacteriaceae cause enteric diseases. *Klebsiella pneumoniae* causes a bacterial pneumonia, and other *Klebsiella* species cause a variety of chronic respiratory diseases. *Enterobacter* species are opportunistic pathogens—that is, they can infect humans with lowered resistance but ordinarily do not cause disease. *Erwinia* species are of no medical significance, but they are plant pathogens and are thus important agriculturally.

Proteus, Providencia, and *Yersinia* constitute yet another related set of genera; all are mixed-acid fermenters, and they have a much lower percentage of G and C bases in their DNA than do other enterics. They can be distinguished in the laboratory by two enzyme reactions: *Proteus* has both enzymes phenylalanine deaminase and urease, but *Yersinia* lacks phenylalanine deaminase, and *Providencia* lacks urease.

Most species of *Proteus* and *Providencia* are associated with urinary tract infections and with burns and wounds. They are especially prominent as causative agents of nosocomial (hospital-acquired) infections. *Proteus mirabilis* is frequently found in urine, blood, and other specimens. Some species of *Proteus* and a few other genera can lose their cell wall and swell into irregularly shaped cells called **L forms** (Figure 10.8). L forms can arise spontaneously or can be induced by lysozyme or penicillin, which remove the cell wall or prevent its formation. They can persist and divide repeatedly or spontaneously revert to normal-walled cells.

Yersinia pestis has a long history as an agent of human misery. When introduced into the body through the bite of an infected flea, *Y. pestis* causes bubonic plague, which is characterized by *buboes,* or abscesses of lymph nodes. When it reaches the lungs, it causes pneumonic plague, a pneumonia-like disease. Other species of *Yersinia* cause mesenteric lymphadenitis, an inflammation of lymph nodes in membranes that support abdominal organs, and a variety of other diseases.

Among enterics, *Serratia marcescens* seems to be in a class by itself. It is very different from common enteric pathogens and is also unusual in its ability to produce a characteristic pigment. Strains that grow at room temperature are generally harmless, free-living rods that usually produce the red pigment prodigiosin (Figure 10.9). Strains that grow at human body temperature fail to produce pigment and are less innocuous. In fact, they are avidly opportunistic! They seem to infect almost any system except the digestive tract; they often infect heart valves and have been reported to cause pneumonia, urinary tract infections, meningitis, and wound infections. Many strains are resistant to several antibiotics. Because of their opportunism and antibiotic resistance, *S. marcescens* strains cause formidable problems among surgical patients, especially the immunodeficient, debilitated, and the elderly.

THE VIBRIOS The vibrios are curved or comma-shaped, gram-negative facultative anaerobes with polar flagella. Many are found in fresh- or saltwater

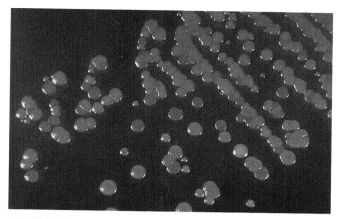

FIGURE 10.9 Colonies of *Serratia marcescens*. Note the distinctive red pigment produced by this organism.

environments, but some inhabit the intestinal tract of humans and other animals. Among the vibrios are two important pathogens: *Vibrio cholerae* causes cholera, and *V. parahaemolyticus* causes a foodborne enteritis. Both are waterborne pathogens; the latter is common in coastal marine environments, and humans who eat contaminated shellfish often become infected. Also included with the vibrios are the luminescent marine bacteria of the genus *Photobacterium.*

THE *PASTEURELLA–HAEMOPHILUS* GROUP Members of the *Pasteurella–Haemophilus* **group** are very small gram-negative bacilli and coccobacilli. They lack flagella and are nutritionally fastidious. ∞ (Chapter 6, p. 148) Most are parasites of animals, and a few are human pathogens, usually attacking mucous membranes of the respiratory tract. The genus *Pasteurella* is named after Louis Pasteur, who identified *P. multocida* as the causative agent of fowl cholera. Most members of the *Pasteurella* genus infect animals. Humans become infected from cat and dog bites and can develop abscesses around the wound and septicemia (blood poisoning). Members of the genus *Haemophilus* ("blood-loving") received their name from the fact that they grow in the laboratory only in media enriched with blood (Figure 10.10). *Haemophilus* species sometimes inhabit the human respiratory membranes as normal microbiota and can be responsible for opportunistic infections. *H. influenzae* does not itself cause influenza but at the time of its discovery was thought to do so, because the causative viruses could not be seen then. In patients with viral influenza, *H. influenzae* frequently causes a *secondary infection* (an infection that follows one already established). Some species of *Haemophilus* attack other membranes; *H. ducreyi* causes chancroid, a sexually transmitted disease, and *H. aegyptius* causes acute conjunctivitis ("pinkeye").

FIGURE 10.8 L forms of *Proteus vulgaris.*

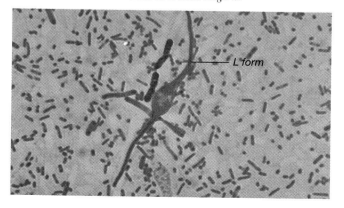

L form

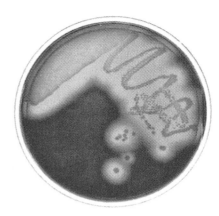

FIGURE 10.10 *Haemophilus* species growing on blood agar. The clear areas surrounding the colonies are due to complete destruction of red blood cells (beta hemolysis).

Anaerobic Gram-Negative Straight, Curved, and Helical Rods (Section 6)

The anaerobic gram-negative rods are members of a single family, Bacteroidaceae. They grow only in strictly anaerobic conditions and were cultured in the laboratory only after methods to maintain anaerobic conditions were developed. Although referred to as rods, many of these organisms are pleomorphic. Some have peritrichous flagella. Members of this group are found mainly in the intestinal and respiratory tracts of humans and other animals. Rod-shaped *Bacteroides* produce propionic acid or succinic acid. Spindle-shaped *Fusobacterium* and comma-shaped *Butyrivibrio* produce butyric acid. *Leptotrichia* produces lactic acid. Anaerobic gram-negative rods cause a variety of opportunistic human infections not only in the intestine but also in the mouth, upper respiratory tract, and urogenital tract. They are especially likely to be found in deep puncture wounds. *Butyrivibrio* is important in digestive processes in cows and other animals that chew their cud. Among members of this section, species of *Bacteroides* are the most frequently encountered in human infections.

Anaerobic Gram-Negative Cocci (Section 8)

Except for shape, the anaerobic gram-negative cocci are quite similar to their rod-shaped cousins described earlier. Although several members of this group are found in animal intestines, only those of the genus *Veillonella* are of medical significance. Certain capnophilic (carbon dioxide–loving) species of *Veillonella* are found in the crevices between teeth and gums, where they often cause tooth abscesses and gum disease. Cocci occur in pairs, chains, or large masses, and they ferment lactic acid.

Rickettsias and Chlamydias (Section 9)

Rickettsias and **chlamydias** (Figure 10.11) are obligate intracellular parasites. Many are human pathogens. Although once thought to be viruses because they were first found in eukaryotic host cells, they are now known to have typical bacterial cell walls and to contain both DNA and RNA. Most lack certain enzymes necessary for life outside host cells. These organisms probably arose from more typical bacteria by losing certain enzymes and other cell components. As they lost enzymes, they became more dependent on their hosts. These organisms also lack flagella and are nonmotile. Rickettsias and chlamydias constitute separate taxonomic orders. Rickettsias and chlamydias are compared with several bacterial groups and with viruses in Table 10.3.

The rickettsias are small rods (1–2 µm), coccobacilli, or cocci that parasitize cells of mammals and arthropods. Many live in both human cells and cells of ticks,

FIGURE 10.11 (a) TEM photo of a *Rickettsia* species (95,000×), an obligate intracellular parasite. (b) TEM photo of *Chlamydia trachomatis* within an infected cell (30,000×). Damage to the reproductive tract caused by this organism can eventually lead to sterility.

(a)

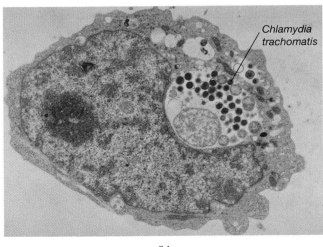

Chlamydia trachomatis

(b)

Characteristic	Typical Bacteria	Rickettsias	Chlamydias	Mycoplasmas	Ureaplasmas	Viruses
Cell wall	Yes	Yes	Yes	No	Sometimes	No
Grow only in cells	No	Yes	Yes	No	No	Yes
Require sterols	No	No	No	Sometimes	Yes	No
Contain DNA and RNA	Yes	Yes	Yes	Yes	Yes	No
Have metabolic systems	Yes	Yes	Yes	Yes	Yes	No

lice, fleas, and mites, which harbor and spread rickettsial diseases. Most human rickettsial pathogens belong to the genus *Rickettsia*. Variations in their characteristics are related to the diseases they cause. For example, members of the typhus fever group, which cause severe headache, chills, and fever, grow mainly in the nuclei of cells and have an optimum growth temperature of 35°C. In patients they cause **hemolysis,** the destruction of red blood cells. Members of the spotted fever group, which cause a more severe but similar disease, grow mainly in cytoplasm and have an optimum growth temperature of 32° to 34° C. They do not cause hemolysis in patients. Other rickettsias cause a variety of fevers. *Rickettsia tsutsugamushi* causes scrub typhus, a fever with rash and lymph node inflammation; it grows mostly in nuclei and has an optimum temperature of 35° C. *Rochalimaea quintana* causes trench fever, which spreads especially among soldiers in trenches. *Rochalimaea* can be grown on artificial media and is an

exception to the rule that all rickettsias are obligate intracellular parasites. The rickettsia *Bartonella henselae* causes bacillary angiomatosis, a complication of AIDS and other conditions that lessen the effectiveness of the immune system. Individuals with this infection experience fever, diarrhea, weight loss, and the formation of swellings (nodules) under the skin. *Bartonella bacilliformis* causes Oroya fever, which is accompanied by a wartlike rash. And *Coxiella burnetii* causes Q fever, a pneumonia-like disease.

The chlamydias are tiny (0.2–1.0 μm) rod-shaped or coccoid bacteria that invade host cells, where they are protected from host immune defense mechanisms. Like rickettsias, chlamydias also were once thought to be viruses because they are so small and because they multiply only in host cells. They have an unusual life cycle, in which the organisms appear in two different nonmotile forms (Figure 10.12). When they are multiplying inside the host cell in membrane-bound cytoplasmic

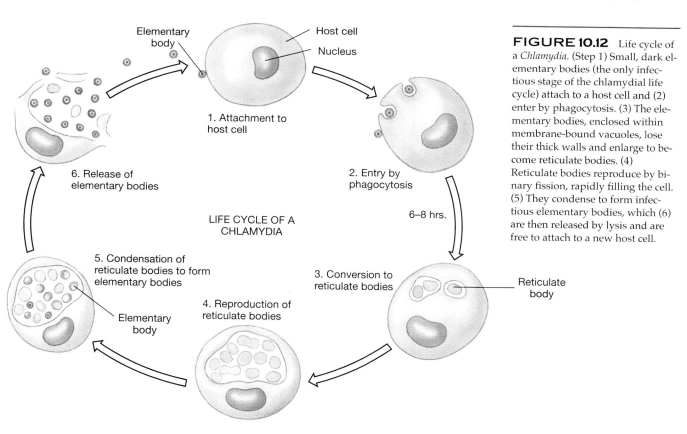

1. Attachment to host cell

2. Entry by phagocytosis

3. Conversion to reticulate bodies

4. Reproduction of reticulate bodies

5. Condensation of reticulate bodies to form elementary bodies

6. Release of elementary bodies

LIFE CYCLE OF A CHLAMYDIA

6–8 hrs.

Elementary body

Host cell

Nucleus

Reticulate body

FIGURE 10.12 Life cycle of a *Chlamydia*. (Step 1) Small, dark elementary bodies (the only infectious stage of the chlamydial life cycle) attach to a host cell and (2) enter by phagocytosis. (3) The elementary bodies, enclosed within membrane-bound vacuoles, lose their thick walls and enlarge to become reticulate bodies. (4) Reticulate bodies reproduce by binary fission, rapidly filling the cell. (5) They condense to form infectious elementary bodies, which (6) are then released by lysis and are free to attach to a new host cell.

TABLE 10.4	Human Disease Caused by *Chlamydia trachomatis*
Disease	**Organs and Systems Affected**
Trachoma	Eyes
Inclusion conjunctivitis	Eyes
Otitis media	Middle ear
Urethral syndrome	Urinary system of females
Nongonococcal urethritis	Urinary system of both males and females
Mucopurulent cervicitis	Reproductive system of females
Salpingitis	Reproductive system of females
Bartholinitis	Reproductive system of females
Epididymitis	Reproductive system of males
Proctitis	Lower intestinal tract
Pneumonia	Respiratory system, especially in the young, the elderly, and the debilitated or immunosuppressed
Peritonitis	Inflammation of the body cavity
Hepatitis	Inflammation of the liver
Endocarditis	Inflammation of the valves of the heart

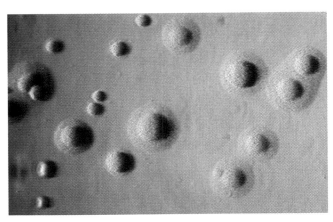

FIGURE 10.13 The "fried egg" appearance of mycoplasma colonies (magnified 270×) is unique to this group.

vacuoles, chlamydias consist of relatively large, metabolically active **reticulate bodies,** a noninfectious form with flexible cell walls. Aggregations of reticulate bodies are sometimes referred to as **inclusion bodies.** Reticulate bodies divide many times by binary fission, rupture, and release large numbers of small, dense **elementary bodies,** an infectious form with rigid cell walls. Elementary bodies are adapted for extracellular existence and are infectious; that is, they can attack other cells of the same host or be transmitted to new hosts.

Three species of chlamydia can cause human infections. *Chlamydia psittaci* causes severe pneumonia in humans and parrot fever in birds. A relatively new species, *C. pneumoniae,* is a major cause of human respiratory infections. *C. trachomatis,* long recognized as the cause of the eye infection trachoma, is also known to mimic *Neisseria gonorrhoeae* infections. One strain of *C. trachomatis* causes a severe form of sexually transmitted disease known as lymphogranuloma venereum. *C. trachomatis* can infect many different systems and organs, as summarized in Table 10.4. The ability of this pathogen to attack such a large number of different kinds of tissues and organs serves to illustrate an important concept: Although the "one organism, one disease" concept applies to many pathogens, it has many exceptions, as we shall see in chapters devoted to diseases of particular body systems.

Mycoplasmas (Section 10)

The **mycoplasmas** are one of the few groups of bacteria for which an entire sequence of taxonomic categories from division to species has been defined (Appendix B).

These organisms are so very small (0.1–0.8 μm) that they were once classified as viruses, but they are now known to have cell membranes, DNA, and RNA, all of which identify them as bacteria. They are sometimes said to be gram-negative, not because they have the typical gram-negative cell wall layers, but because they entirely lack a cell wall. Mycoplasmas are pleomorphic because of their flexible cell membrane, and they often form slender branched filaments. Most have sterols, such as cholesterol, in their cell membrane; in this respect they resemble fungi and protists. Mycoplasmas lack flagella. However, a few display a gliding movement on a wet surface. Most are facultatively anaerobic; a few are obligately anaerobic.

Although mycoplasmas can be grown in a special agar medium, where they form colonies with a "fried egg" appearance (Figure 10.13), in nature they have several modes of existence. Some must live at the expense of host organisms; some live on a host without damaging it; and others live on decaying organic matter. Human pathogens among the mycoplasmas are found in the genera *Mycoplasma* and *Ureaplasma*. *M. pneumoniae* causes primary atypical pneumonia, whereas other species are usually opportunists of the gums, oropharynx, and urogenital tract. *M. fermentans* may play a role in rheumatoid arthritis. *U. urealyticum* is unusual in that it metabolizes urea; it is found in the urogenital tract, where it sometimes causes sexually transmitted disease. Other mycoplasmas are plant pathogens. Some infect orange trees, which are treated by pumping solutions of the antibiotic tetracycline into the trees.

Gram-Positive Cocci (Section 12)

The gram-positive cocci represent one end of a spectrum of gram-positive organisms distinguishable by shape and by whether they form spores (Figure 10.14). As we shall see when we consider the next few sections of bacteria, these organisms have other distinguishing properties.

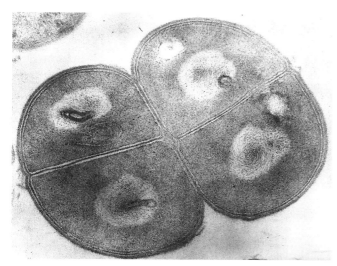

FIGURE 10.14 *Sporosarcina ureae,* gram-positive cocci that produce spores. The cocci (57,600×) are arranged in cubical packets.

The gram-positive cocci constitute a relatively large and medically important group of bacteria. These cocci are divided into families on the basis of metabolic properties and cellular arrangements. The **micrococci** are aerobes or facultative anaerobes that form irregular clusters by dividing in two or more planes. The **streptococci** are aerotolerant anaerobes that obtain energy from fermenting sugars to lactic acid. They form pairs or chains by dividing in one or two planes. Most lack the enzyme catalase. The **peptococci** are anaerobes that lack both catalase and the enzymes to ferment lactic acid. They also form pairs or irregular clusters.

MICROCOCCI Many micrococci are free-living saprophytes (organisms that feed on dead matter) found in soil, fresh water, and marine environments. They are easily transmitted on the surfaces of plants and animals to meat, dairy products, and other foods. The micrococci include an aerobic genus, *Micrococcus,* and a facultatively anaerobic genus, *Staphylococcus.* Various species of *Micrococcus* are found on the skin and in the mouth and upper respiratory tract. *Staphylococcus aureus* is a common human pathogen. It is often responsible for skin abscesses and boils. If it invades the blood, it can travel to other tissues and cause pneumonia, meningitis, and osteomyelitis (infection of the marrow cavity of bones). *Staphylococcus epidermidis* normally inhabits skin and mucous membranes; it often causes opportunistic infections.

STREPTOCOCCI The streptococci include a large number of species, which are differentiated by whether they have certain antigens on their cell surfaces and whether they carry out hemolysis. For example, some strains of *Streptococcus pyogenes* with so-called group-A antigens completely hemolyze red blood cells in lab-oratory cultures. Such complete hemolysis is called **beta hemolysis** (Figure 10.10). These strains are the causative agent of strep throat, scarlet fever, rheumatic fever, and a variety of other infections. In contrast, streptococci of the **viridans group,** which often infect the valves and lining of the heart, cause incomplete hemolysis, or **alpha hemolysis,** in laboratory cultures. *S. agalactiae,* long known to cause mastitis (inflammation of the udders) in cattle, is now known to be sexually transmitted and responsible for female urogenital infections in humans. If transmitted to infants during the birth process, it can cause fatal illness.

PEPTOCOCCI The peptococci are obligate anaerobes. Although the genus *Sarcina* consists of saprophytes found in soil and on plants, members of the genera *Peptococcus* and *Peptostreptococcus* are frequently found among the microbes that normally inhabit the digestive, respiratory, and urogenital tracts. These organisms are versatile opportunists; they can cause peritonitis (inflammation of the body cavity and its membranes), postpartum sepsis (blood poisoning after the birth of a child), accumulation of pus in deep wounds, osteomyelitis, vaginitis, and sinus and dental infections.

Endospore-Forming Gram-Positive Rods and Cocci (Section 13)

The endospore-forming gram-positive rods and cocci consist mainly of rod-shaped bacteria of the genera *Bacillus* and *Clostridium.* Cocci of the genus *Sporosarcina* are also included in this group, but they are soil bacteria of no medical significance. The ability to form endospores is the main distinguishing characteristic of these organisms.

All members of the genus *Bacillus* obtain nutrients from dead organic matter; some are obligate aerobes, and others are facultative anaerobes. They can be distinguished by the location of spores and by growth temperatures. For example, *B. cereus* and *B. subtilis* have centrally located endospores and grow at moderate temperatures. In contrast, *B. stearothermophilus* has terminal spores and grows at temperatures of 65° C or higher.

Several species of *Bacillus*—*B. subtilis, B. licheniformis, B. polymyxa,* and *B. brevis*—produce antibiotics. *B. anthracis* causes anthrax, a severe blood infection of cattle, sheep, and horses that can infect humans. *B. cereus* is sometimes implicated in food poisoning.

Members of the genus *Clostridium* are strictly anaerobic motile rods found in soil, water, and the intestinal tract of humans and other animals. Several species are important human pathogens that exert their effects mainly through the production of potent toxins. *C. tetani* causes tetanus (lockjaw), and *C. botulinum* causes botulism (a kind of food poisoning). *C. perfringens* also can cause food poisoning. Several species of *Clostridium*

CLOSE-UP

GIANT BACTERIA

In 1994, a giant bacterium believed to be related to members of the genus *Clostridium* was reported by Norman Pace, Esther Angert, and Kendall Clements of Indiana University. The bacterium, which was placed in the genus *Epulopiscium*, was initially mistaken for a protozoan. An inhabitant of the intestines of marine surgeonfish, *Epulopiscium* is about one million times larger than the average bacterium. Some of these giant bacteria grow to be longer than 0.5 mm, making them visible to the naked eye. Thus, this "micro"organism is macroscopic—an oxymoron of science!

Bright-field micrograph of *Epulopiscium*, one of the largest bacterial species known (magnified 70×). Three protozoa (*Paramecium* species) are also present. One million *Escherichia coli* cells could fit into one *Epulopiscium*, which can be seen with the naked eye.

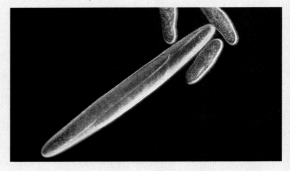

produce gas, predominantly hydrogen, under the anaerobic conditions of deep tissue wounds and cause gangrene, or tissue necrosis (death). Gangrene due to an infection is called *gas gangrene* to distinguish it from dry gangrene, which occurs when a tissue lacks an adequate blood supply. Other clostridia also infect wounds, and some can cause low-grade bacteremias, especially in debilitated patients.

Regular Nonsporing Gram-Positive Rods (Section 14)

The regular nonsporing (non-spore-forming) gram-positive rods are called "regular" because they stain evenly and are of uniform shape rather than being pleomorphic. They are obligate or facultative anaerobes found in fermenting plant and animal products. They include three genera, *Lactobacillus, Listeria,* and *Erysipelothrix.* **Lactobacilli** (Figure 10.15) are found in a wide variety of foods. They are used in the production of cheeses, yogurt, sourdough, and many other

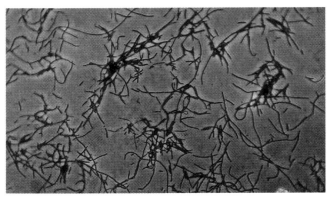

FIGURE 10.15 *Lactobacillus acidophilus* (dark purple rods) forms the predominant microbiota in the vagina of sexually mature women. These gram-positive bacteria produce lactic acid by fermenting the glycogen in the epithelial cells lining the vagina. The acidic environment that they produce and live in inhibits the growth of many other types of bacteria and thus helps protect the vagina from infection.

fermented foods. *L. acidophilus* and *L. casei* are found among the natural microbiota of the digestive and urogenital tracts of humans and are generally nonpathogenic. A few lactobacilli have been linked to tooth decay. *Listeria monocytogenes* causes listeriosis, an inflammation of the brain and the membranes that cover it, in humans and animals. *Erysipelothrix rhusiopathiae* causes erysipeloid lesions (red, painful skin lesions) in humans and more severe erysipelas in swine.

Irregular Nonsporing Gram-Positive Rods (Section 15)

The irregular nonsporing gram-positive rods are irregularly shaped and unevenly staining; they include club-shaped **corynebacteria,** pleomorphic **propionibacteria,** and filamentous **actinomycetes.** Nearly all are facultative anaerobes, but a few propionibacteria are strict anaerobes.

The genus *Corynebacterium* includes a large and diverse assortment of saprophytes, many of which are found in air, soil, and water or as pathogens in plants and a few animals. *C. xerosis* is a normal inhabitant of human conjunctiva; *C. pseudodiphtheriticum* is a normal inhabitant of the human pharynx. Neither produces damaging toxin. The most medically significant species is *C. diphtheriae.* It produces a potent toxin that causes diphtheria.

The family Propionibacteriaceae includes the genera *Propionibacterium* and *Eubacterium. P. freudenreichii* is found in dairy products. *P. acnes,* which is found in wounds and abscesses, commonly infects but does not cause acne lesions. Most species of *Eubacterium* are obligate anaerobes; they are found in soil and plant products and sometimes infect body cavities of humans and

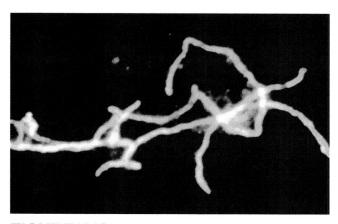

FIGURE 10.16 *Actinomyces israelii* from a brain abscess. Note the branching, filamentous type of growth, which formerly led the actinomycetes to be classified as fungi.

other animals. *E. foedans* is found in dental tartar and in a variety of infections.

Once classified as fungi, the order Actinomycetales, which means "ray fungus," includes mainly soil bacteria that tend to form branching filaments (Figure 10.16). Some of these bacteria provide nitrogen to plants; a few are normal inhabitants of the human mouth. The most important human pathogens in this group are members of the genus *Actinomyces*. *A. israelii* causes actinomycosis (jaw abscesses and sometimes lung disease), and several other species are found in dental caries (cavities) and periodontal infections. *A. bovis* causes a disease called lumpy jaw in cattle.

Mycobacteria (Section 16)

The **mycobacteria** are unusual in that their cell walls contain large amounts of lipids. The lipids resist basic dyes, and so mycobacteria can be stained with carbolfuchsin in hot phenol and washed with acid alcohol. The red stain remains; thus, these organisms are said to be acid-fast (refer to Figure 3.27). Mycobacteria are slender rods, usually without clubbed ends, and frequently form filaments. Most are soil saprophytes, but some are human pathogens. *Mycobacterium tuberculosis* causes tuberculosis, and *M. leprae* causes Hansen's disease, formerly known as leprosy. Another organism, *M. bovis*, causes tuberculosis in cattle and can be transmitted to humans. *M. avium-intracellulare*, or the *M. avium complex*, is recognized as an extremely important opportunist. It commonly causes lung and systemic infections in persons with AIDS and in others with poorly functioning immune systems. *M. avium-intracellulare* is widespread in soil, dust, and water.

Nocardioforms (Section 17)

The **nocardioforms** comprise nine genera of aerobes, including *Nocardia*. They are closely related to corynebacteria, actinomycetes, and mycobacteria. Nocardioforms are gram-positive, nonmotile, pleomorphic, aerobic, and generally acid-fast and filamentous. Although many species of *Nocardia* are found in soil, some are known pathogens. For example, *N. asteroides* causes skin abscesses and lung infections, and *N. brasiliensis* also causes skin abscesses.

Actinomycetes That Divide in More Than One Plane (Section 27)

The actinomycetes that divide in more than one plane are similar to but distinct from the actinomycetes described earlier. These bacteria nearly always form masses of filaments. Some are free-living in soil; others live in plant root nodules, where they oxidize molecular nitrogen. The medically important species belong to the genus *Dermatophilus*, which causes skin lesions.

Streptomycetes and Related Genera (Section 29)

Like the actinomycetes, the **streptomycetes** and their allies are soil-dwelling organisms that resemble fungi. They develop extensive branching filaments with spores of many different shapes. The characteristics of spores are used to separate the streptomycetes into genera. None are pathogenic, but many are medically significant because of the antibiotics they produce. More than 500 different antibiotic substances have been isolated from various species of the genus *Streptomyces*. Among these antibiotics are substances effective against bacteria, viruses, protozoa, and fungi.

Characteristics of bacteria in 16 medically important sections of *Bergey's Manual of Systematic Bacteriology* are summarized in Table 10.5.

Bacteria in Ecology and Evolution: The "Nonmedical" Groups

Seventeen of the 33 sections of bacteria defined in *Bergey's Manual* are not known to have medical significance either as pathogens or as antibiotic producers. Yet these organisms are of great importance in ecosystems and suggest some evolutionary trends. Let us look at examples of the unusual and important characteristics of these groups.

The *nonmotile, gram-negative curved bacteria* (Section 3) are aerobic, free-living organisms found in soil, fresh water, and ocean water. The genus *Spirosoma* is an example. These bacteria are a source of food for larger soil organisms.

The *dissimilatory sulfate- or sulfur-reducing bacteria* (Section 7) give off sulfur-containing compounds as wastes. (Assimilation refers to taking in; dissimilation refers to giving off.) These bacteria, which include the genus *Desulfovibrio* (Figure 10.17), are anaerobes found

Section (number)	Medically Important Members	Diseases or Products
Spirochetes (1): gram-negative, helical, move by axial filaments	*Treponema*	Syphilis
	Borrelia	Relapsing fever, Lyme disease
	Leptospira	Leptospirosis
Aerobic, Motile, Helical Gram-Negative Bacteria (2): move by flagella, helical or comma-shaped	*Campylobacter*	Urogenital and digestive tract infections
	Helicobacter	Peptic ulcers
Gram-Negative Aerobic Rods and Cocci (4): Some contain pigments or oxidase, some have fastidious nutritional requirements, some are obligate parasites.	*Pseudomonas*	Urinary tract infections, burns, and wounds
	Legionella	Pneumonia and other respiratory infections
	Neisseria	Gonorrhea, meningitis, and nasopharyngeal infections
	Moraxella	Conjunctivitis
	Brucella	Brucellosis
	Bordetella	Whooping cough
	Francisella	Tularemia
Facultatively Anaerobic Gram-Negative Rods (5): Some have peritrichous flagella, many can be distinguished by their characteristic fermentation reactions.	*Escherichia*	Opportunistic infections of colon and other sites
	Shigella	Bacillary dysentery
	Salmonella	Typhoid fever, enteritis, and food poisoning
	Klebsiella	Respiratory and urinary tract infections
	Enterobacter	Opportunistic infections
	Serratia	Opportunistic infections
	Proteus	Urinary tract infections (especially nosocomial)
	Providencia	Wound and burn infections, urinary tract infections
	Morganella	Summer diarrhea, opportunistic infections
	Yersinia	Plague, mesenteric lymphadenitis, septicemia
	Vibrio	Cholera, acute gastroenteritis
	Pasteurella	Cat- and dog-bite wounds
	Haemophilus	Respiratory infections, meningitis, conjunctivitis, chancroid
	Calymmatobacterium	Granuloma inguinale
	Gardnerella	Vaginitis
	Eikenella	Wound infections
	Streptobacillus	Rat-bite fever
Anaerobic Gram-Negative Rods (6): straight, curved, or helical, motile	*Bacteroides,* *Fusobacterium*	Oral, digestive, respiratory, urogenital infections, wounds, and abscesses

mainly in water sediments, sewage, and polluted water. They are very important in the environmental sulfur cycle, and some members of the group can survive at extremes of temperature and high salinity. Although they can be found in the human intestine, they are not known to cause disease. They produce hydrogen sulfide gas, which is toxic and corrosive and smells like rotten eggs. In sufficiently high concentrations, it kills fish and plants and corrodes pipes.

The *endosymbionts* (Section 11) are a diverse assortment of gram-negative bacteria that live in the bod-

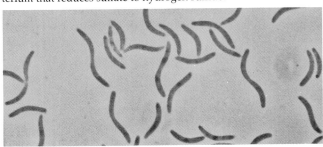

FIGURE 10.17 *Desulfovibrio gigas* (5000×), an anaerobic bacterium that reduces sulfate to hydrogen sulfide.

Section (number)	Medically Important Members	Diseases or Products
Anaerobic Gram-Negative Cocci (8): nonmotile	*Veillonella*	Oral microbiota and abscesses
Rickettsias and Chlamydias (9): intracellular parasites; chlamydias form reticulate and elementary bodies.	*Rickettsia*	Typhus, Rocky Mountain spotted fever, rickettsialpox
	Rochalimaea	Trench fever
	Coxiella	Q fever
	Bartonella	Oroya fever
	Chlamydia	Trachoma, inclusion conjunctivitis, non-gonococcal urethritis, lymphogranuloma venereum, parrot fever
Mycoplasmas (10): lack cell walls, extremely small	*Mycoplasma*	Atypical pneumonia, urogenital infections
	Ureaplasma	Opportunistic urogenital infections
Gram-Positive Cocci (12): aerobic to strictly anaerobic; non-spore-forming; typically pyogenic (pus-forming)	*Staphylococcus*	Skin abscesses, opportunistic infections
	Streptococcus	Strep throat and other respiratory infections, skin and other abscesses, puerperal fever, opportunistic infections
	Peptococcus	Postpartum septicemia, visceral lesions
	Peptostreptococcus	Puerperal fever and various pyogenic infections
Endospore-Forming Gram-Positive Rods and Cocci (13): aerobic to strictly anaerobic; some motile and some nonmotile	*Bacillus*	Anthrax, food poisoning, source of the antibiotic bacitracin
	Clostridium	Tetanus, botulism, gas gangrene, bacteremia
Regular Nonsporing Gram-Positive Rods (14): facultatively or strictly anaerobic; nonmotile	*Lactobacillus*	Microflora of the digestive tract and vagina
	Listeria	Listeriosis
	Erysipelothrix	Erysipeloid
Irregular Nonsporing Gram-Positive Rods (15): club-shaped, pleomorphic, filamentous; aerobic to anaerobic	*Corynebacterium*	Diphtheria and skin opportunists
	Propionibacterium	Wound infections and abscesses
	Eubacterium	Oral and other infections
	Actinomyces	Actinomycoses
Mycobacteria (16): gram-positive, acid-fast	*Mycobacterium*	Tuberculosis, leprosy, and chronic infections
Nocardioforms (17): gram-positive, filamentous, some acid-fast	*Nocardia*	Nocardiosis, mycetoma, abscesses
Actinomycetes That Divide in More Than One Plane (27): gram-positive, filamentous	*Dermatophilus*	Skin lesions
Streptomycetes and Their Allies (29): gram-positive, filamentous	*Streptomyces*	Produce over 500 different antibiotics

ies of protozoa, fungi, algae, other bacteria, and insects. Little is known of how these organisms relate to their hosts. Even less is known about their possible roles in the evolution of microorganisms. Certain endosymbionts of protozoa produce toxins. Some toxins pass from one protozoan to another during conjugation, killing the recipient. Other endosymbionts give toxin resistance to their hosts.

The *gliding nonfruiting bacteria* (Section 23) glide along a solid surface, leaving a slime trail behind them. They are said to be nonfruiting because they do not form cell aggregates called *fruiting bodies,* as do another group called the gliding fruiting bacteria. Gliding non-fruiting bacteria are important in ecological cycles. Members of the genus *Cytophaga* digest cellulose, chitin, agar, and other complex organic materials. *Beggiatoa* species oxidize hydrogen sulfide to sulfur, thereby contributing to the sulfur cycle.

As the name suggests, *anoxygenic phototrophic bacteria* (Section 18) capture energy from light but do not release oxygen as photosynthetic plants do (Figure 10.18). These bacteria contain a special bacterial chlorophyll

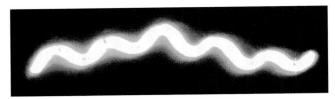

FIGURE 10.18 *Rhodospirillum rubrum,* an anoxygenic phototrophic bacterium.

(called bacteriochlorophyll) and a variety of other pigments: purple, red, orange, and brown. They include the purple sulfur bacteria, the purple nonsulfur bacteria, the green sulfur bacteria, and the green nonsulfur bacteria. The sulfur varieties use sulfur rather than water in photosynthesis; the nonsulfur varieties use organic compounds. The anoxygenic phototrophic bacteria contribute to ecosystems by providing nutrients for other organisms.

The *budding and/or appendaged bacteria* (Section 21) reproduce by *budding,* or dividing unequally. Small buds that contain DNA and a small amount of cytoplasm pinch off from the parent cell and grow to become parent cells themselves. Some of these bacteria, such as members of the genus *Caulobacter,* have an appendage, or stalk, that generally has a sticky holdfast at its tip (Figure 10.19). The holdfast is a device that allows a bacterium to anchor itself on a solid surface. These bacteria are found in soil and water. Both budding reproduction and development of stalks are examples of interesting evolutionary developments in this group.

The *archaeobacteria* (Section 25) are a diverse group, mostly of anaerobes, that live in extremely harsh environments. These organisms share properties—such as

FIGURE 10.19 *Caulobacter* is a nonmotile stalked bacterium anchored to surfaces by a thin filament during part of its life cycle. However, it begins life as a free-swimming "swarmer" cell. Eventually the swarmer cell loses its flagellum and grows an attachment stalk at the site of its former flagellum. After settling down, it grows, elongates, and forms a new flagellum at the end opposite the attachment stalk. It then divides unequally by binary fission to produce a new swarmer cell plus the old attached cell.

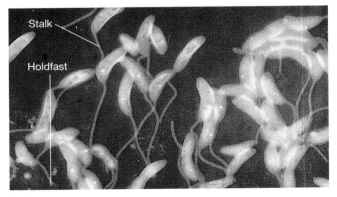

unusual transfer RNA molecules, unusual lipids in their plasma membranes, and distinctive RNA polymerases—that are significantly different from those of the eubacteria. The archaeobacteria can be gram-positive or gram-negative, motile or nonmotile, and rods or cocci. Three distinct groups are currently recognized: methanogens, halobacteria, and thermoacidophiles. The **methanogens** produce methane and, in fact, are the sole natural source of natural methane, or marsh gas. They are found in soil, water, sewage, and even the digestive tract of humans and other animals. The **halobacteria** live in very concentrated salt solutions. In some places they can be seen from an airplane as pink patches on salt flats.

Thermoacidophiles metabolize sulfur and live in extremely hot, acidic environments. These bacteria and other organisms found near deep-sea "black smoker" vents (so named because of the black sulfur compounds that spew up from the sea bottom like plumes of black smoke) have been studied from the submersible vessel *Alvin.* The organisms are found in surprisingly dense populations. The bacteria provide food for many organisms, including mussels, giant clams big enough to provide 20 lb of meat per clam, and tube worms 3 m long and as thick as a human arm but lacking mouth, digestive tract, and anus (Figure 10.20). Some of the bacteria are free-living and form white, fluffy, tennis ball–sized masses in a bacterial mat. Others live symbiotically in the worm or clam tissues. There they contribute food that they have made with the use of energy from the sulfur compounds of the vent. These bacteria make these volcanic vents the most productive environments of all in terms of energy capture.

The *sheathed bacteria* (Section 22) form filaments and surround themselves with a secreted sheath made of lipoproteins and polysaccharides. If these gram-negative organisms are grown in the presence of minerals such as iron or manganese, they incorporate the metals into their sheath. Sheathed bacteria reproduce by releasing flagellated swarmer cells, free-swimming cells that leave the filament and start new colonies. These water dwellers are responsible for the slimy coating on rocks in and along streams and for blooms (dense floating masses of organisms) when the nutrient supply suddenly increases. They provide evidence of the evolutionary trend toward multicellular structures.

The *gliding fruiting bacteria* (Section 24), also referred to as the slime bacteria, are gram-negative aerobes. Found in well-oxygenated soil, these organisms are important decomposers. Their fruiting bodies consist of large, often highly colored, cell aggregates. Cells released from the fruiting bodies are released and migrate to form new colonies—more evidence for the evolutionary trend toward multicellular structures.

The *chemolithotrophs* (Section 20) are another diverse group of bacteria that obtain energy from inorganic substances. They include nitrogen bacteria found in soil, such as *Nitrosomonas,* which oxidizes

FIGURE 10.20 Thermoacidophilic archaeobacteria thrive in hot, acidic environments, including regions of the deep-ocean floor near volcanic vents. There they serve as the base of food chains, living symbiotically inside the tissues of animals such as these giant tube worms. The worms lack a mouth and an anus and thus rely on the bacterial capture of chemical energy to feed their tissues.

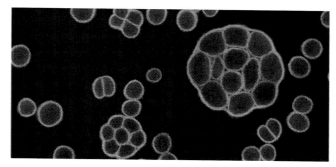

FIGURE 10.21 *Prochloron,* a photosynthetic prokaryote with some similarities to plants. Each cell in a cluster contains chlorophyll arranged on membranous layers.

ammonia to nitrite, and *Nitrobacter,* which oxidizes nitrite to nitrate. Also in this group are *Thiobacillus,* which oxidizes sulfur, and certain other bacteria that reduce sulfur. Sulfur bacteria are found in soil, water, sewage, sulfur springs, and acid mine wastes. Still other chemolithotrophs oxidize iron and/or manganese and are responsible for large deposits of minerals on the ocean floor. A few of these organisms, known as *magnetotactic bacteria,* absorb iron into intracellular crystals. When placed in a magnetic field, they orient in a north-south direction. ∞ (Chapter 4, p. 85) Chemolithotrophs are important in the cycling of nitrogen and sulfur. Exactly where the chemolithotrophs fit into the evolution of life is not yet clear; they present some interesting taxonomic problems.

The *cyanobacteria* (Section 19), formerly called blue-green algae and classified with algae, are now known to be prokaryotes. Yet they are remarkably similar to plants: Their chlorophyll is much like plant chlorophyll, and they release oxygen during photosynthesis, as do plants. Cyanobacteria may represent a step in the evolution of photosynthetic mechanisms.

The *prochlorophytes* (Section 19) are similar to the cyanobacteria, but they are placed in a different group because they have two distinctly plantlike chlorophylls found in structures much like chloroplasts. The only two currently known genera of this group are *Prochloron* (Figure 10.21), which lives with marine invertebrates called ascidians, and *Prochlorothrix.* Some scientists believe that *Prochloron* represents a "missing link" between photosynthetic prokaryotes and eukaryotic plants.

Some *sporangiate actinomycetes* (Sections 28 and 30) are soil bacteria that form relatively extensive filaments, complete with *sporangia* (spore sacs) filled with spores. Although these organisms do not infect humans, they are similar to the genera *Actinomyces* and *Dermatophilus,* which do.

Other *conidiate* genera (Section 28) have even more complex mycelial structures, including chains of spores borne on ends of hyphae, and more closely resemble fungi than other funguslike bacteria. ∞ (Chapter 6, p. 153) The species *Micromonospora purpurea* produces the antibiotic gentamicin.

If you have found the funguslike bacteria somewhat confusing, you are not alone. For centuries, they confused taxonomists, who classified them with the fungi until they were found to be prokaryotic. Indeed, the editors of *Bergey's Manual* changed their minds between Volumes 2 and 4 as to how to classify some funguslike bacteria—the Actinomycetes and Nocardioforms (Sections 17 and 26).

Some bacteria not yet classified have recently been isolated from deep aquifers—porous rock layers that hold underground water. These bacteria produce carbon dioxide, which makes some mineral waters effervescent. The carbon dioxide creates carbonic acid, which dissolves limestone and increases the size of the aquifers and thereby the quantity of water they can hold. Some such bacteria appear to feed selectively on molecules containing carbon-13, a heavy isotope of carbon, if it is available. Why they do this is not known.

We can characterize the "nonmedical" bacteria as bacteria that are important in ecosystems and challenging to biologists interested in evolutionary relationships. In ecosystems they capture energy by photosynthesis and chemolithotrophy; they serve as food for larger organisms, and they decompose the dead bodies and wastes of many organisms. They also provide important links in the recycling of nitrogen, sulfur, and other minerals. With respect to evolution, they display trends toward multicellularity and a diverse assortment of metabolic processes that may represent some of nature's experiments with the origin of life itself.

ARE WE STILL DISCOVERING NEW ORGANISMS?

Are there any new worlds to discover, or new creatures in them? Yes! In recent years scientists have discovered living organisms in such diverse environments as submarine hot vents, inside volcanoes, and in deep oil wells. In 1990 a joint U.S. and Soviet team discovered hot vents in fresh water for the first time, complete with an associated community of archaeobacteria, worms, sponges, and other organisms. The vents lie more than 400 m deep in a most unusual Russian lake, Lake Baikal, which is the deepest lake in the world and holds the greatest quantity of fresh water in the world. Located in central Asia, in Siberia, it lies in a pocket between two continental plates. Asia was formed as a solid mass when several plates collided one after another and remained together. The area of Lake Baikal is being pulled apart, forming a rift valley and eventually a new ocean. This region is comparable to the sea-floor spreading centers (ridges) of the Pacific Ocean, where other vent communities have been found. In both locations, hot materials from deep inside the earth are emerging. Lake Baikal is a unique treasure for studying the evolution of life and microbial forms. Most lakes are only thousands of years old, but Lake Baikal may be 25 million years old. Microbes similar to the early evolving stages of life may still exist in its depths.

What lives inside a volcano? Studies following the 1980 volcanic eruptions of Mount St. Helens (Figure 10.22) have raised some interesting questions. Archaeobacteria, which had previously been known from the deep-sea volcanic vents (black smokers) located 2200 m below the sea surface, have been found living on and in Mount St. Helens at temperatures of 100°C. Where did they come from? Some scientists think that they may have been present deep inside the volcano. For that matter, where do the archaeobacteria in the submarine vents come from? Do they point out a linkage between terrestrial and submarine vol-

FIGURE 10.22 Mount St. Helens, shown here erupting in July 1980, is home to archaeobacteria.

canic activity? We tend to think of life as being present only on the *surface* of the earth, but perhaps there is a whole different range of life that we know nothing about, deep *within* the crust of the earth. Daily, more and more evidence accumulates in favor of the idea of "continuous crustal culture."

Cores taken from the deepest oil wells being bored within the earth reveal archaeobacteria at sites not connected with volcanic activities. Ancient bacteria from the early stages of our planet's development may still be colonizing the hot anaerobic interior of the earth—places with conditions that resemble those formerly on the earth's surface.

Various ecological problems have sent out scientists from universities, government, and industry to hunt for new microbes with properties that make the organisms useful in cleaning up the environment. Scientists from the Woods Hole Oceanographic Institution, in Massachusetts, took their search to a depth of more than 1800 m in the Gulf of California, where they have discovered anaerobic bacteria that can degrade naphthalene and possibly other hydrocarbons that might be found in oil spills. Sites in need of bioremediation often lack oxy-

gen, making it impossible to utilize aerobic organisms for cleanup; hence the hunt in deep anaerobic environments. General Electric has also found an anaerobic bacterium that it plans to use to destroy polychlorinated biphenyls (PCBs), industrial byproduct chemicals that accumulate in animal tissues and cause damage, including cancer and birth defects.

A new bacterium, so far referred to as GS-15, discovered in the Potomac River by U.S. Geological Survey (USGS) scientists, ordinarily changes iron from one form to another. It seems, however, that these bacteria can just as easily feed on uranium, getting twice as much energy in the process and transforming the uranium into an insoluble precipitate. The USGS team plans to use GS-15 to remove uranium from contaminated well and irrigation water found in much of the U.S. West and at uranium mining, processing, and nuclear waste sites.

Yes, there are many new microbes yet to be discovered. In addition to the naturally occurring species, new ones will be designed by scientists using genetic engineering techniques. All these species will need to be classified and named. Clearly, *Bergey's Manual* will never be "finished."

BACTERIAL TAXONOMY AND NOMENCLATURE

Criteria for Classifying Bacteria

- The criteria used for classifying bacteria are summarized in Table 10.1. These criteria can be used to classify bacteria into species and even into strains within species.

- For many species a particular strain is designated as the **type strain**, which is preserved in a type culture collection.

Problems of Bacterial Taxonomy

- Taxonomists do not agree on how members of the kingdom Prokaryotae (Monera) should be divided. Many species of bacteria have been grouped into genera and some, into families. Four *divisions* (the equivalent of phyla) have been established. Much information is needed to determine evolutionary relationships and establish classes and orders.

CLASSIFICATION ACCORDING TO BERGEY'S MANUAL OF SYSTEMATIC BACTERIOLOGY

History and Significance of *Bergey's Manual*

- *Bergey's Manual* was first published in 1923 and has been revised several times; a ninth edition was published in 1994.

- *Bergey's Manual* provides definitive information on the identification and classification of bacteria.

Bacteria by Section of *Bergey's Manual*

- The genera of medical significance, their characteristics, and the diseases they cause are summarized in Table 10.5.

- Key groups of medically significant bacteria include **spirochetes** (helical, motile bacteria), **pseudomonads** (gram-negative aerobic rods with polar flagella), **legionellas** (fastidious gram-negative aerobic rods), **enteric bacteria** (small, facultatively anaerobic gram-negative rods), *Pasteurella–Haemophilus* **group** members (very small gram-negative bacilli and coccobacilli), **mycoplasmas** (very small bacteria with no cell walls), the **rickettsias** and **chlamydias** (obligate intracellular parasites), **mycobacteria** (slender, acid-fast rods), **nocardioforms** (gram-positive, nonmotile, pleomorphic, aerobic bacteria), and **streptomycetes** (gram-positive, filamentous, spore-forming soil bacteria).

- "Nonmedical" bacteria are important in ecosystems and may represent some of nature's experiments with the origin of life itself. They include the sulfate- or sulfur-reducing bacteria, endosymbionts, budding and/or appendaged bacteria, archaeobacteria, chemolithotrophs, and cyanobacteria.

KEY TERMS

actinomycetes (*p. 260*)
alpha hemolysis (*p. 259*)
beta hemolysis (*p. 259*)
chlamydias (*p. 256*)
corynebacteria (*p. 260*)
elementary body (*p. 258*)
enteric bacteria (*p. 254*)
halobacteria (*p. 264*)

hemolysis (*p. 257*)
inclusion body (*p. 258*)
lactobacilli (*p. 260*)
legionellas (*p. 253*)
L form (*p. 255*)
methanogen (*p. 264*)
micrococci (*p. 259*)
mycobacteria (*p. 261*)

mycoplasmas (*p. 258*)
nocardioforms (*p. 261*)
Pasteurella–Haemophilus group (*p. 255*)
peptococci (*p. 259*)
propionibacteria (*p. 260*)
pseudomonads (*p. 252*)
reticulate body (*p. 258*)

rickettsias (*p. 256*)
spirochetes (*p. 250*)
streptococci (*p. 259*)
streptomycetes (*p. 261*)
thermoacidophile (*p. 264*)
treponemes (*p. 252*)
type strain (*p. 246*)
viridans group (*p. 259*)

QUESTIONS FOR REVIEW

A **Criteria for Classifying Bacteria**

1. Describe at least three different kinds of bacteria, using the characteristics in Table 10.1.

2. How are the biochemical tests in Table 10.2 used?

3. True or False: Morphology is the most important criterion for classifying bacteria.

B **Problems of Bacterial Taxonomy**

4. True or False: Assigning bacteria to higher taxonomic levels such as families, orders, and classes is easier than assigning them to the correct genus or species.

5. Why is it difficult to classify bacteria by their evolutionary relationships?

C *Bergey's Manual*

6. When was *Bergey's Manual of Determinative Bacteriology* first published, and which is the current edition?

7. Of what use is *Bergey's Manual of Systematic Bacteriology* to medical personnel? To ecologists? To undergraduate microbiology students?

D **Bacterial Genera of Medical Significance**

8. For any six diseases, identify the organism and the section of *Bergey's Manual of Systematic Bacteriology* in which each belongs.

9. Name some organisms that cause disease in more than one body system.

10. Name some sections of *Bergey's Manual* that include organisms responsible for opportunistic infections.

11. Name some sections of *Bergey's Manual* that include organisms normally found on or in the human body.

12. Match the following (more than one may apply to each):

____ treponemes	**a.** Enterobacteriaceae
____ lack cell wall	**b.** mycoplasmas
____ elementary and	**c.** spirochetes
reticulate bodies	**d.** chlamydias
____ once classified as viruses	**e.** rickettsias
____ *Escherichia coli*	
____ colonies with a "fried	
egg" appearance	
____ *Klebsiella pneumoniae*	
____ vibrios	
____ obligate intracellular parasites	

13. In the following diagram of the life cycle of *Chlamydia trachomatis,* identify numbered stages 1–6 and parts (a)–(d).

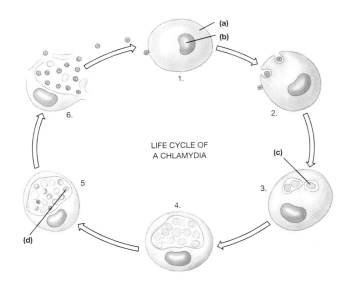

LIFE CYCLE OF A CHLAMYDIA

14. True or False: Viruses are not classified in *Bergey's Manual*.

PROBLEMS FOR INVESTIGATION

1. Identify and describe at least three genera of bacteria in which some species infect humans and others infect other animals. Suggest how this situation might have arisen. Assess the likelihood that the animal diseases might someday appear in humans.

2. Use information about bacteria to present an argument to support each of the following hypotheses: (a) Autotrophs evolved before heterotrophs, (b) heterotrophs evolved before autotrophs, (c) bacteria and fungi have common ancestors, (d) bacteria and protists have common ancestors, and (e) bacteria and algae have common ancestors.

3. What benefits might bacteria derive from the antibiotics they produce?

4. Use *Bergey's Manual of Systematic Bacteriology* to look up the following information:

a. Are members of the genus *Serratia* motile?

b. What lab test could you use to distinguish *Proteus vulgaris* from *P. mirabilis*?

c. What percentage of DNA to DNA hybridization (homology) exists between *Treponema pallidum* and *T. pertenue*? What conclusion is drawn on the basis of this percentage?

d. What shape(s) are members of the genus *Mycoplasma*?

e. Is *Staphylococcus aureus* able to ferment mannitol with production of acid?

5. Explain some of the ways bacteria contribute to recycling in ecosystems.

6. Explain how some particular physical or biochemical characteristic of a given bacterium has helped it adapt to a given niche in the environment.

SOME INTERESTING READING

Balows, A., et al., eds. 1992. *The prokaryotes: A handbook on the biology of bacteria*. 2d ed. New York: Springer-Verlag.

Folger, T. 1994. "Primordial landlubbers." *Discover* 15 (April):28–29.

Fox, J. L. 1994. "Microbial diversity: Low profile, immense breadth." *ASM News* 60(October):533–35.

Gober, J. W., and M. V. Marques. 1995. "Regulation of cellular differentiation in *Caulobacter crescentus*." *Microbiological Reviews* 59, no. 1 (March):31–47.

Goodfellow, M., and D. E. Minnikin. 1985. *Chemical methods in bacterial systematics*. Orlando: Academic Press.

Goodfellow, M., and A. G. O'Donnell. 1993. *Handbook of new bacterial systematics*. San Diego: Academic Press.

Holt, J. G., P. H. A. Sneath, N. S. Mair, and M. E. Sharpe, eds. 1986. *Bergey's manual of systematic bacteriology*. Vol. 2. Baltimore: Williams & Wilkins. (Contains sections 12–17.)

Holt, J. G., P. H. A. Sneath, N. S. Mair, M. E. Sharpe, and J. T. Staley, eds. 1989. *Bergey's manual of systematic bacteriology*. Vol. 3. Baltimore: Williams & Wilkins. (Contains sections 18–23.)

Holt, J. G., P. H. A. Sneath, N. S. Mair, M. E. Sharpe, S. T. Williams, and J. T. Staley, eds. 1989. *Bergey's manual of systematic bacteriology*. Vol. 4. Baltimore: Williams & Wilkins. (Contains sections 27–30.)

Holt, J. G., et al., eds. 1994. *Bergey's manual of determinative bacteriology*. 9th ed. Baltimore: Williams & Wilkins.

Holt, J. G., et al., eds. 1992. *Stedman's Bergey's bacteria words*. Baltimore: Williams & Wilkins.

Krieg, N. R., and J. G. Holt, eds. 1984. *Bergey's manual of systematic bacteriology*. Vol. 1. Baltimore: Williams & Wilkins. (Contains sections 1–11, along with a useful essay on classification of bacteria by J. T. Staley and N. R. Krieg.)

Lederberg, J. 1992. "Bacterial variation since Pasteur." *ASM News* 58(5):261–65.

Lipkin, R. 1995. "Bacterial chatter: How patterns reveal clues about bacteria's chemical communication." *Science News* 147 (March 4):136–37, 142.

Logan, N. A. 1994. *Bacterial systematics*. Oxford and Boston: Blackwell Scientific Publications.

Monastersky, R. 1988. "Bacteria alive and thriving at depth." *Science News* 133 (March 5):149.

Pennisi, E. 1994. "Static evolution: Is pond scum the same now as billions of years ago?" *Science News* 145 (March 12):168–69.

Priest, F. G., A. Ramos-Cormenzana, and B. J. Tindall, eds. 1994. *Bacterial diversity and systematics*. New York: Plenum Press.

Smith, S. 1990. "Afterlife of a whale." *Discover* 11 (February):46–49.

Stanier, R. Y., et al. 1986. *The microbial world*. 5th ed. Englewood Cliffs, NJ: Prentice Hall.

Weisburd, S. 1986. "First fossils of slime bacteria studied." *Science News* 130 (November 29):347.

CHAPTER 11

VIRUSES

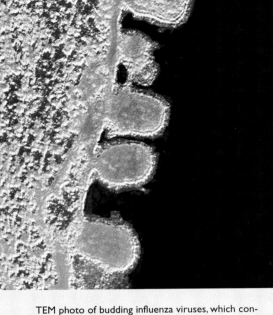

TEM photo of budding influenza viruses, which contain RNA genomes.

The sudden emergence of AIDS and the identification of the viruses that cause the syndrome have made us very aware of the devastating effect such microbes can have on humans—medically, socially, politically, and financially. The appearance of other "new" viruses, such as the Ebola viruses in Africa and the hantavirus in the United States, clearly indicate that viruses are emerging that we are not aware of and that we may not be able to fight effectively. As antibiotics have reduced the incidence of bacterial infections, viral infections, which usually do not respond to antibiotics, have become increasingly apparent. In fact, we have much to learn about viruses and how to prevent viral infections in humans. Research since the 1950s on the molecular biology of viruses continues to provide us with a better understanding of viral structure, viral replication, and viral diseases—an important point for health scientists. Such research is also extending the understanding of the fundamental nature of life—an important theme for all biologists.

This chapter examines the structure and behavior of viruses. By the end of this chapter, you will have a better understanding of and appreciation for one of nature's smallest but most dangerous groups of microbes.

GENERAL CHARACTERISTICS OF VIRUSES

What Are Viruses?

Viruses are infectious agents that are too small to be seen with a light microscope and that are not cells. They have no cell nucleus, organelles, or cytoplasm. When they invade susceptible cells, viruses display some properties of living organisms and so appear to be on the borderline between living and nonliving. Viruses can *replicate*, or multiply, only inside a living host cell. As such they are called **obligate intracellular parasites,** a distinction they share with chlamydias and rickettsias. ∞ (Chapter 10, p. 256) (We may have to reconsider the traditional definition of viruses following the announcement in December 1991 by E. Wimmer, A. Molla, and A. Paul that they successfully grew polioviruses in test tubes containing ground up human cells, but no live cells.)

Viruses differ from cells in important ways. Whereas prokaryotic and eukaryotic cells contain both DNA and RNA, individual virus particles contain only one kind of nucleic acid—either DNA or RNA, never both. Cells grow and divide, but viruses

270

THIS CHAPTER FOCUSES ON THE FOLLOWING QUESTIONS:

A What are the general characteristics of viruses?

B How are viruses classified?

C How do bacterial and animal viruses replicate?

D How are animal viruses cultured for experimental studies?

E What is a teratogen, and how do viruses act as teratogens?

F What are the properties of viruslike agents?

G How do viruses cause cancer?

do neither. Viral multiplication requires that a virus particle infect a cell and program the host cell's machinery to synthesize the components required for the assembly of new virus particles. The infected cell may produce hundreds to thousands of new viruses and usually dies. Tissue damage as a result of cell death accounts for the destructive effects seen in many viral diseases. The term "virus," which was first used by Pasteur, comes from the Latin word for "poison."

Components of Viruses

Typical viral components are shown in Figure 11.1. These components are a nucleic acid core and a surrounding protein coat called a **capsid.** In addition, some viruses have a surrounding lipid bilayer membrane called an **envelope.** A complete virus particle, including its envelope, if it has one, is

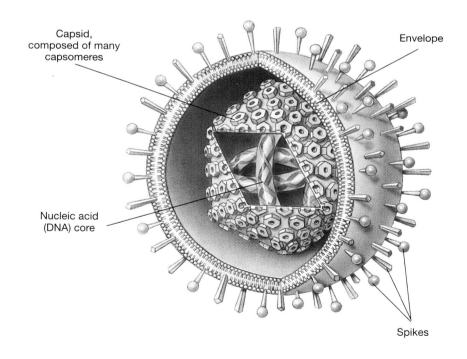

FIGURE 11.1 The components of an animal virus (a herpesvirus).

called a **virion.** Certain viruses may also contain a few enzymes.

Nucleic Acids

The British immunologist Peter Medawar, who shared the 1960 Nobel Prize in physiology or medicine, once described viruses as "a piece of bad news wrapped up in protein." The nucleic acid is the "bad news" because viruses use their **genome,** their genetic information, to replicate themselves in host cells. The result is often a disruption of host cellular activities or death of the host. Viral genomes are either DNA or RNA. Viral replication depends on the expression of the viral genome for the formation of viral proteins and the replication of new viral genomes within the infected host cell. Viral nucleic acid can be single-stranded or double-stranded, and linear, circular, or segmented.

Capsids

The nucleic acid of an individual virion is in most cases enclosed within a capsid that protects it and determines its shape. Capsid structure is determined in large part by the viral nucleic acid enclosed and accounts for most of the virus' mass. Each capsid is composed of protein subunits called **capsomeres** (Figure 11.1). In some viruses, the proteins found in the capsomere are of a single type. In other viruses a number of different proteins may be present. The number of proteins and the arrangement of viral capsomeres are characteristic of specific viruses and thus can be useful in virus identification and classification.

Envelopes

Enveloped viruses, in which individual capsids are enclosed in an envelope, acquire their envelope after they are assembled in a host cell as they *bud*, or move through, one or several membrane systems. The viruses use these systems, including the endoplasmic reticulum, Golgi apparatus, and nuclear or plasma membrane, for viral replication. The composition of an envelope generally is determined by the viral nucleic acid and by the substances derived from host membranes. Combinations of lipids, proteins, and carbohydrates make up most envelopes. Depending on the virus, projections referred to as **spikes** (Figure 11.1) may or may not extend from the viral envelope. These surface projections are **glycoproteins** that serve to attach virions to specific receptor sites on susceptible host cell surfaces. In certain viruses the possession of spikes causes various types of red blood cells to clump, or *hemagglutinate*—a property that is useful in viral identification.

A virion's **nucleocapsid** comprises the viral genome together with the capsid. Viruses with only a nucleocapsid and no envelope are known as **naked,** or nonenveloped, **viruses.**

What advantages might envelopes have for viruses? Because envelopes are similar to host cell membranes, viruses may be "hidden" from attack by the host's immune system. Also, envelopes help viruses infect new cells by fusion of the envelope with the host's cell or plasma membrane. Conversely, enveloped viruses are damaged easily. Environmental conditions that destroy membranes—increased temperature, freezing and thawing, pH below 6 or above 8, lipid solvents, and some chemical disinfectants such as chlorine, hydrogen peroxide, and phenol—will also destroy the envelope. Naked viruses generally are more resistant to such environmental conditions.

Enzymes

As described earlier, most virions contain only nucleic acid within the core of the virus. However, some viruses package an inactive enzyme within the core. After infection, these enzymes become active and help copy the genetic information. For example, the virus responsible for AIDS is the **human immunodeficiency virus (HIV).** HIV is an enveloped RNA virus. Once the virus has infected a cell, the viral enzymes become active and form temporary DNA to carry on the replication cycle. When new viruses are made in the infected host cell, inactive enzymes are packaged with the RNA into the virion.

Sizes and Shapes

All viruses are too small to be seen with a light microscope, but Figure 11.2 shows that they have a range of sizes. The largest are the orthopoxviruses, which are about 240 nm by 300 nm—the size of the smallest bacteria. The complex bacteriophages are about 65 nm by 200 nm. Among the smallest viruses known are the enteroviruses, which are less than 30 nm in diameter. But as Figure 11.2 shows, most viruses are quite small when compared with bacteria or eukaryotic cells. To put things in perspective, consider that typical ribosomes are about 25–30 nm in diameter.

Although some viruses are variable in shape, most viruses have a specific shape that is determined by the capsomeres or the envelope. Figure 11.2 shows several examples of viral symmetry. A *helical* capsid consists of a ribbonlike protein that forms a spiral around the nucleic acid. The tobacco mosaic virus is a helical virus. ∞ (Chapter 1, p. 15) *Polyhedral* viruses are many-sided. The picornaviruses (pi-kor'na-vi"rus-ez) and the human adenoviruses (ad'e-no-vi"rus-ez) are polyhedral viruses. One of the most common polyhedral capsid shapes is the icosahedron; *icosahedral* viruses have 20 triangular faces. A *complex* capsid is a combination of helical and icosahedral shapes, and some viruses have a *bullet-shaped* capsid.

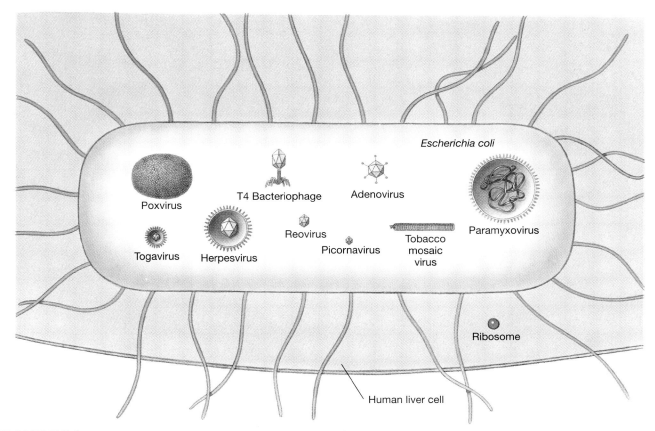

FIGURE 11.2 Variations in shapes and sizes of viruses compared with a bacterial cell, an animal cell, and a eukaryotic ribosome.

Viruses with envelopes have a somewhat spherical shape. For example, the herpesvirus shown in Figure 11.1 has a polyhedral capsid and an envelope.

The poxviruses and many bacterial viruses are called **complex viruses** because they have a more elaborate coat or capsid (Figure 11.2). Many **bacteriophages,** or viruses that infect bacteria, have a complex shape that incorporates specialized structures such as heads, tails, and tail fibers (Figure 11.2). Like spikes, the tail fibers are used by virions to attach to host bacteria. Other specialized bacteriophage structures are used to infect the bacterial cells.

Host Range and Specificity of Viruses

Although viruses are quite small and differ from one another structurally and in their replication strategies, they are capable of infecting all forms of life (hosts). The **host range** of a virus refers to the spectrum of hosts that a virus can infect. Different viruses can infect bacteria, fungi, algae, protozoa, plants, vertebrates, or even invertebrates. However, most viruses are limited to only one host and to only specific cells and/or tissues of that

host. Polioviruses, for example, can be grown in the laboratory in monkey kidney cells but have never been observed to cause a natural infection in any animals other than humans. In contrast, the rabies virus can attack the central nervous system of many warmblooded animals. The host range of the rabies virus is much more extensive than that of polioviruses.

Viral specificity, another important property of viruses, refers to the specific kinds of cells a virus can infect. For example, certain papillomaviruses, which cause warts, are so specific in their replication strategy that they infect only skin cells. In contrast, cytomegaloviruses, known for their lethal effects, attack cells of the salivary glands, gastrointestinal tract, liver, lungs, and other organs. They can also cross the placenta and attack fetal tissues, especially those of the central nervous system.

Viral specificity takes into account a number of factors, including a virus's ability to attach to, penetrate, and establish an infection in a host cell that will result in the release of new virions. Attachment depends on the presence of specific receptor sites on the surfaces of host cells and on specific attachment structures on viral capsids or envelopes. Specificity is also affected

PLANT VIRUSES

Besides the specificity shown by some viruses for bacteria and humans, other viruses are specific to and infect plants. Most viruses enter plant cells through damaged areas of the cell wall and spread through cytoplasmic connections called *plasmodesmata*.

Because plant viruses cause serious crop losses, much research has been done on them. The tobacco mosaic virus infects tobacco plants. Other plant viruses, which have either DNA or RNA genomes, infect various ornamental plants, including carnations and tulips. Food crops are not immune to viral infections. Lettuce, potatoes, beets, cucumbers, tomatoes, beans, corn, cauliflower, and turnips are all subject to infection by specific plant viruses.

Insects are known to cause serious crop losses because of their voracious eating habits. But many insects carry and transmit plant viruses as well. By damaging plants as they eat, insects provide an excellent infection mechanism for the plant viruses they harbor. Researchers now hope to control some crop-destroying insects by infecting them with specific insect viruses.

The infectious yellows virus, carried by whiteflies, causes discoloration and stunted growth in cherry trees.

The beautiful streaks in these tulips are caused by a viral infection. Unfortunately, the infection (which can spread from plant to plant) also weakens the tulips somewhat.

by the availability of appropriate host enzymes, other proteins, and organelles for viral replication. If all goes well, the new progeny of virions are released from the infected host cell to spread to other similar cells.

CLASSIFICATION OF VIRUSES

Before they knew much about the structure or chemical properties of viruses, virologists classified viruses by the type of host infected or by the type of host structures infected. Thus, viruses have been classified as bacterial viruses, plant viruses, or animal viruses. And animal viruses are grouped by the tissues they attack as *dermotropic* if they infected the skin, *neurotropic* if they infected nerve tissue, *viscerotropic* if they infected organs of the digestive tract, or *pneumotropic* if they infected the respiratory system.

As more was learned about the structure of viruses at the biochemical and molecular levels, classification of viruses came to be based on the type and structure of their nucleic acids, method of replication, host range, and other chemical and physical characteristics. Thus, as more viruses were discovered, conflicting classification systems came into use, resulting in much confusion. A single, universal taxonomic scheme for viruses was clearly needed. The problem of viral taxonomy was first addressed in 1966 at a meeting of the International Congress of Microbiology in Moscow. Then in 1973 virologists and other biologists formally established the International Committee on Taxonomy of Viruses (ICTV). This committee, which meets every 4 years, establishes the rules for classifying viruses. Virus classification is summarized in Appendix B.

Because viruses are so different from cellular organisms, it is difficult to classify them according to typical taxonomic categories—kingdom, phylum, and the like. The family has been the highest taxonomic category used by the ICTV. Viral genera also have been established, but most are new and slow to gain acceptance. Despite advances in classification, the problems of defining and naming *viral species*—a group of viruses that share the same genome and the same relationships with organisms— have not yet been resolved completely. Currently, the ICTV requires that the English common name, rather than a Latinized binomial term, be used to designate a viral species. For example, the formal taxonomic designations for the rabies virus and HIV would be as follows: family: Rhabdoviridae; genus: *Lyssavirus;* species: rabies virus; and family: Retroviridae; genus: *Lentivirus;* species: human immunodeficiency virus (HIV).

The names of specific viruses often consist of a group name and a number, such as HIV-1 or HIV-2. Virus families are often distinguished first on the basis

CLOSE-UP

NAMING VIRUSES

Although the ICTV approves all virus names, virologists often are creative in naming newly discovered viruses. As the following family names show: The Picornaviridae received their name from the fact that they are extremely small viruses (*piccolo,* Italian for "very small") and contain <u>RNA</u> as their genetic information. The Retroviridae have RNA as their genome and use it to direct the synthesis of DNA, reversing (<u>*retro,*</u> Latin for "backward") the usual direction of transcription. The Parvoviridae are very small viruses (<u>*parvus,*</u> Latin for "small"), whereas the Togaviridae received their name from someone who must have thought that the envelope of a togavirus resembled a <u>toga</u>, or cloak. And the arboviruses are a collection of <u>ar</u>thropod<u>bo</u>rne viruses (including togaviruses, flaviviruses, bunyaviruses, and arenaviruses). Can you see how the Coronaviridae got their name (<u>*corona,*</u> Latin for "crown")?

When it became possible to culture viruses, some were isolated that could not be linked to a known disease and that did not cause disease in laboratory animals. These viruses were dubbed "orphans." Thus, the Reoviridae are <u>r</u>espiratory <u>e</u>nteric <u>o</u>rphan viruses. Although the ICTV does not allow a person's name to be used in a virus name, geographical locations can be used. The Bunyaviridae received their name from <u>*Bunya*</u>mwere, Uganda, where they were first discovered.

Can you see why this virus is called a coronavirus?

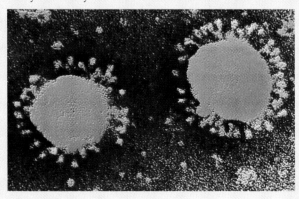

of their nucleic acid type, capsid symmetry (shape), envelope, and size (Tables 11.1 and 11.2). As of 1995 the ICTV had assigned more than 4000 member viruses to 71 families, 11 subfamilies, and 175 genera. Many of the virus families contain viruses that cause important infections of humans and some other animals. Additional families contain viruses that infect only other animals, plants, fungi, algae, or bacteria.

Nucleic Acid Classification

Viruses are distinguished first by their nucleic acid content as either RNA or DNA viruses. Subsequent subdivisions are based largely on other properties of nucleic acids. The RNA viruses can be single-stranded (*ssRNA*) or double-stranded (*dsRNA*), although most are single-stranded (Table 11.1). Because most eukaryotic cells do not have the enzymes to copy viral RNA molecules, the RNA viruses must either carry the enzymes or have the genes for those enzymes as part of their genome. Table 11.1 identifies two types of single-stranded RNA viruses—positive sense and negative sense RNA viruses. Many ssRNA viruses contain **positive (+) sense RNA,** meaning that during an infection the RNA acts like mRNA and can be translated by the host's ribosomes. Other ssRNA viruses have **negative (−) sense RNA.** In such viruses the RNA acts as a template during transcription to make a complementary (+) sense mRNA after a host cell has been entered. ∞ (Chapter 7, p. 170) This strand is translated by host ribosomes. In order to perform the transcription step, (−) sense RNA viruses must carry an RNA polymerase within the virion.

Like RNA viruses, DNA viruses can also occur in single-stranded or double-stranded form (Table 11.2). For example, the human adenoviruses, responsible for some common colds, and the herpesviruses are double-stranded DNA (*dsDNA*) viruses. Only one single-stranded DNA (*ssDNA*) virus is currently known to produce human disease.

With this background, let us briefly examine several families of animal RNA and DNA viruses.

RNA VIRUSES

General Properties of RNA Viruses

The different families of RNA viruses are distinguished from one another by their nucleic acid content, their capsid shape, and the presence or absence of an envelope (Table 11.1; Figure 11.3). Most families of RNA viruses contain either one (+) sense RNA or one (−) sense RNA molecule. However, some RNA viruses are placed in separate families if the RNA exists as two complete copies of (+) sense RNA or contains small segments of (−) sense RNA. Finally, one family has segmented dsRNA.

Important Groups of RNA Viruses

PICORNAVIRIDAE The **picornaviruses** are very small (30 nm in diameter), naked, polyhedral, (+) sense RNA viruses. They include more than 150 species that cause disease in humans. After infection these viruses quickly interrupt all functions of DNA

Family	Envelope and Capsid Shape	Typical Size (nm)	Example (Genus or Species)	Infection or Disease
(+) Sense RNA Viruses				
Picornaviridae (1 copy)	Naked, polyhedral	18–30	Enterovirus Rhinovirus Hepatovirus	Polio Common cold Hepatitis A
Togaviridae (1 copy)	Enveloped, polyhedral	40–90	Rubella virus Equine encephalitis virus	Rubella (German measles) Equine encephalitis
Flaviviridae (1 copy)	Enveloped, polyhedral	40–90	Flavivirus	Yellow fever
Retroviridae (2 copies)	Enveloped, spherical	100	HTLV-I HIV	Adult leukemia, tumors, AIDS
(−) Sense RNA Viruses				
Paramyxoviridae (1 copy)	Enveloped, helical	150–200	Morbillivirus	Measles
Rhabdoviridae (1 copy)	Enveloped, helical	70–180	Lyssavirus	Rabies
Orthomyxoviridae (1 copy in 8 segments)	Enveloped, helical	100–200	Influenzavirus	Influenza A and B
Filoviridae (1 copy)	Enveloped, filamentous	80	Filovirus	Marburg, Ebola
Bunyaviridae (1 copy in 3 segments)	Enveloped, spherical	90–120	Hantavirus	Respiratory distress, hemorrhagic fevers
Double-Stranded RNA Viruses				
Reoviridae (1 copy in 10–12 segments)	Naked, polyhedral	70	Rotavirus	Respiratory and gastrointestinal infections

Family	Envelope and Capsid Shape	Typical Size (nm)	Example (Genus or Species)	Infection or Disease
Double-Stranded DNA Viruses				
Adenoviridae (linear DNA)	Naked, polyhedral	75	Human adenoviruses	Respiratory infections
Herpesviridae (linear DNA)	Enveloped, polyhedral	120–200	Simplexvirus Varicellovirus	Oral and genital herpes Chickenpox, shingles
Poxviridae (linear DNA)	Enveloped, complex	230 × 270 shape	Orthopoxvirus	Smallpox, cowpox
Papovaviridae (circular DNA)	Naked, polyhedral	45–55	Human papilloma-viruses	Warts, cervical and penile cancers
Hepadnaviridae (circular DNA)	Enveloped, polyhedral	40–45	Hepatitis B virus	Hepatitis B
Single-Stranded DNA Viruses				
Parvoviridae (linear DNA)	Naked, polyhedral	22	B19	Fifth disease (erythema infectiosum) in children

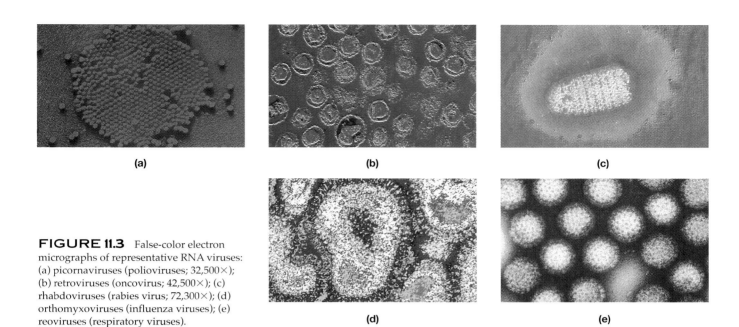

FIGURE 11.3 False-color electron micrographs of representative RNA viruses: (a) picornaviruses (polioviruses; 32,500×); (b) retroviruses (oncovirus; 42,500×); (c) rhabdoviruses (rabies virus; 72,300×); (d) orthomyxoviruses (influenza viruses); (e) reoviruses (respiratory viruses).

and RNA in the host cell. The Picornaviridae are divided into several groups, including the genera *Enterovirus, Hepatovirus,* and *Rhinovirus.*

Enteroviruses (*entero,* Greek for "intestine") include the polioviruses (Figure 11.3a). These viruses are resistant to many chemical substances and can replicate in and pass through the digestive tract unharmed. Unless inactivated by host defense mechanisms, the viruses invade the blood and lymph, spreading throughout the body but especially into the nervous system. Poor sanitation increases the numbers of enteroviruses, and human overcrowding helps their spread.

Poor sanitation is also responsible for the spread of certain **hepatoviruses** (*hepato,* Greek for "liver"). The hepatitis A virus, for instance, is transmitted via the fecal–oral route, with disease arising from the ingestion of contaminated food or water. The major organ infected is the liver.

The genus *Rhinovirus* (*rhino,* Greek for "nose"), which includes more than 100 types of human **rhinoviruses,** is one of the genera of viruses responsible for the common cold. Human rhinoviruses do not cause digestive tract diseases because they cannot survive the acidic conditions in the stomach. Instead, they enter the body through the mucous membranes of the nasal passages and replicate in the epithelial cells of the upper respiratory tract. Much has recently been learned about the capsids of rhinoviruses (Figure 11.4). Virologists have discovered that these capsids attach to only a few receptors in the nasal mucous membranes. Thus, it may be possible in the future to prevent colds by designing chemicals that cover these receptors so that rhinoviruses cannot attach.

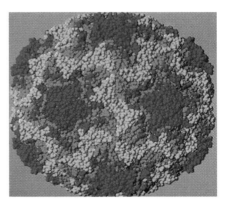

FIGURE 11.4 Computer-generated model of a human rhinovirus, cause of the common cold. The colors represent different capsomeres of the capsid.

TOGAVIRIDAE The **togaviruses** are small, enveloped, polyhedral, (+) sense RNA viruses that multiply in the cytoplasm of many mammalian and arthropod host cells. Togaviruses known as arthropodborne viruses are transmitted by mosquitoes and cause several kinds of encephalitis (plural: *encephalitides*) in humans and in horses. The rubella virus, which causes German measles (rubella), is in this family but is not transmitted by arthropods; rather, it is spread person to person.

FLAVIVIRIDAE The **flaviviruses** are enveloped, polyhedral, (+) sense RNA viruses that are transmitted by mosquitoes and ticks. The viruses produce a variety of encephalitides or fevers in humans. The yellow fever virus is a flavivirus that causes a hemorrhagic fever—

in which blood vessels in the skin, mucous membranes, and internal organs bleed uncontrollably. Hepatitis C infection is also caused by a flavivirus.

RETROVIRIDAE The **retroviruses** are enveloped viruses that have two complete copies of (+) sense RNA (Figure 11.3b). They also contain the enzyme **reverse transcriptase,** which uses the viral RNA to form a complementary strand of DNA that is then replicated to form a dsDNA. This reaction is exactly the reverse of the typical transcription step (DNA ⟶ RNA) in protein synthesis. For virus replication to continue, the newly formed DNA must be transcribed into viral RNA that will function as mRNA for viral protein synthesis and be incorporated into new virions. To do so, the DNA must first migrate to the host cell nucleus and become incorporated into chromosomes of host cells. Such integrated viral DNA is known as a **provirus.** Retroviruses cause tumors and leukemia in rodents and birds as well as in humans. The human retroviruses invade immune defense cells called *T lymphocytes* and are referred to as <u>h</u>uman <u>T</u> cell <u>l</u>eukemia <u>v</u>iruses (HTLV). HTLV-1 and HTLV-2 are associated with malignancies (leukemia and other tumors), whereas the <u>h</u>uman <u>i</u>mmunodeficiency <u>v</u>irus (HIV-1 and HIV-2 strains) causes <u>a</u>cquired <u>i</u>mmune <u>d</u>eficiency <u>s</u>yndrome (AIDS). AIDS is discussed in Chapter 19.

PARAMYXOVIRIDAE The **paramyxoviruses** (*para,* Latin for "near"; *myxo,* Greek for "mucus") are medium-sized, enveloped, (−) sense RNA viruses, with a helical nucleocapsid. Different genera of paramyxoviruses are responsible for mumps, measles, viral pneumonia, and bronchitis in children and mild upper respiratory infections in young adults.

RHABDOVIRIDAE Another (−) sense RNA virus group, the **rhabdoviruses** (*rhabdo,* Greek for "rod") consists of medium-sized, enveloped viruses. Although these viruses have an envelope, the capsid is helical and makes the viruses nearly rod- or bullet-shaped (Figure 11.3c). Rhabdoviridae virions contain an RNA-dependent RNA polymerase that uses the (−) sense strand to form a (+) sense strand. The newly produced strand serves as mRNA and as a template for the synthesis of new viral RNA. Human rabies almost always results from a bite by a rabid animal carrying the rabies virus. Rhabdoviruses also infect other vertebrates, invertebrates, or plants. The Lago virus, which produces disease in bats, and the Mokolo virus, which infects shrews in Africa, are closely related to rabies viruses.

ORTHOMYXOVIRIDAE The **orthomyxoviruses** (*ortho,* Greek for "straight") are medium-sized, enveloped, (−) sense RNA viruses that vary in shape

PUBLIC HEALTH

RECENT VIRUS DISEASES IN ANIMALS

Several viruses have been identified as the cause of disease in animals other than humans. A retrovirus found in cats, called *feline immunodeficiency virus* (FIV), is very similar to the human immunodeficiency virus (HIV). Approximately 1 to 3 percent of randomly tested U.S. cats are infected. FIV infects the cat's lymph nodes but causes no immediate symptoms. However, like HIV, FIV gradually attacks the immune system over 3 to 6 years, leading to frequent infections of the mouth, skin, and respiratory systems. Loss of immune defenses also leads to diarrhea, weight loss, pneumonia, fever, and neurologic disease. Unlike HIV, FIV is not transmitted sexually; nor can it be transmitted to humans or other animals. Transmission between cats is usually through a bite. Veterinarians believe that the virus, like HIV, has been around for decades, and there is no cure.

Several other viruses also are responsible for recent disease outbreaks in animals. A virus closely related to FIV has been found in zoo lions and tigers and in wild panthers. *Caliciviruses,* members of a family similar to the picornaviruses, have been identified as the cause of disease outbreaks in swine, sea lions, and cats. These viruses cause a skin eruption similar to foot-and-mouth disease in cattle. In swine, viral infection causes weight loss, fever, and the formation of skin blisters on the feet, snout, and tongue.

Between December 1993 and January 1995, more than 100 African lions in Tanzania died from canine distemper. The canine distemper virus, which is very unusual in large cats but common in dogs and wolves, probably was transmitted to the lions by domestic dogs that live near the Tanzanian game parks. African researchers believe the virus might spread to other animals, including leopards and wild dogs.

from spherical to helical (Figure 11.3d). Their genome is segmented into eight pieces. Like the paramyxoviruses, orthomyxoviruses have an affinity for mucus. Influenza virus A, with which we are all too familiar, is an orthomyxovirus that also infects birds, swine, and horses. Influenza virus B appears to be specific to humans.

FILOVIRIDAE The **filoviruses** are enveloped, filamentous, single (−) sense RNA viruses. These viruses can be transmitted from person to person by close contact with blood, semen, or other secretions and by using contaminated needles. The filoviruses include the viruses responsible for Marburg and Ebola diseases, which are hemorrhagic fevers.

BUNYAVIRIDAE The **bunyaviruses** also are enveloped, (−) sense RNA viruses whose genome has three segments. Bunyaviruses can be transmitted by arthropods, but rodents typically are the principal host. The most recently recognized member of the Bunyaviridae is the hantavirus responsible for hantavirus pulmonary syndrome (HPS). Other forms of the genus *Hantavirus* cause hemorrhagic fevers.

ARENAVIRIDAE Like the bunyaviruses, the **arenaviruses** are enveloped, (−) sense RNA viruses, but their genome has only two segments. Arenaviruses are carried by rodents. Human infections occur via aerosols, exposure to infectous urine or feces, or rat bites. Argentinean and Bolivian hemorrhagic fevers and Lassa fever are arenavirus infections.

REOVIRIDAE The **reoviruses** have a naked, polyhedral capsid (Figure 11.3e). They are medium-sized dsRNA viruses. They replicate in the cytoplasm and form distinctive inclusions that stain with eosin. Reoviruses include the orthoreoviruses, orbiviruses, and rotaviruses. The rotaviruses are the most common cause of severe diarrhea in infants and in young children under age 2. They also are responsible for minor upper respiratory and gastrointestinal infections in adults. The other reoviruses infect other animals.

DNA Viruses

General Properties of DNA Viruses

Like the RNA viruses, the animal DNA viruses are grouped into families according to their DNA organization (Table 11.2; Figure 11.5). The dsDNA viruses are further separated into families on the basis of the shape of their DNA (linear or circular), their capsid shape, and the presence or absence of an envelope. Only one family of viruses has ssDNA.

Important Groups of DNA Viruses

ADENOVIRIDAE The **adenoviruses** (*adeno*, Greek for "gland") are medium-sized, naked viruses with linear dsDNA. First identified in adenoid tissue, they are highly resistant to chemical agents and are stable from pH 5 to 9 and from 36° to 47°C. Freezing causes little loss of infectivity. More than 80 different types of adenoviruses have been identified, many being responsible for human respiratory disease. Adenovirus types 40 and 41 cause 10 to 30 percent of all cases of severe diarrhea in babies and young children. Only half the children carrying the virus in their throat actually become ill.

Diseases caused by adenoviruses are generally acute (that is, have sudden onset and short duration). Soon after entering the body, the virus appears in the blood and a measleslike rash may develop. Sources of adenoviruses are respiratory secretions and feces from infected persons.

HERPESVIRIDAE The **herpesviruses** (*herpes*, Greek for "creeping") are relatively large, enveloped viruses with linear dsDNA (Figure 11.5a). Herpesviruses are widely distributed in nature, and most animals are infected with one or more of the 100 types discovered. These viruses cause a broad spectrum of diseases, which are summarized in Table 11.3.

The core of the virion contains proteins around which the DNA is coiled. In cells infected with herpesviruses, the viral dsDNA can exist as a provirus. Therefore, a universal property of herpesviruses is **latency,** the ability to remain in host cells, usually in neurons, for long periods and to retain the ability to replicate. For example, a child who has recovered from chickenpox will still have the virus in a latent form. Years or decades later the virus may be reactivated as a result of psychological or physical factors. This adult disease, which can be quite painful and debilitating, is called *shingles.*

FIGURE 11.5 False-color electron micrographs of representative DNA viruses (arrows point to one virion): (a) herpesviruses (pink spheres within the cell); (b) papovaviruses (human papillomaviruses; 13,000×); (c) parvoviruses (70,000×).

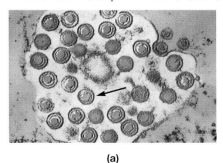
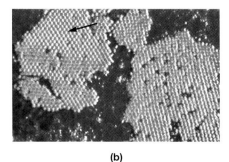
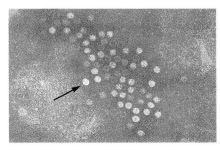

(a)   (b)   (c)

TABLE 11.3   Herpesviruses That Cause Human Disease

Genus	Virus Type	Infection or Disease
Simplexvirus	Herpes simplex type 1	Oral herpes (sometimes genital and neonatal herpes), encephalitis
	Herpes simplex type 2	Genital and neonatal herpes (sometimes oral herpes), meningo-encephalitis
Varicellovirus	Varicella-zoster	Chickenpox (varicella) and shingles (zoster)
Cytomegalovirus	Cytomegaloviruses (salivary gland virus)	Acute febrile illness; infections in AIDS patients, transplant recipients, and others with reduced immune system function; a leading cause of birth defects
Roseolovirus	Roseola infantum (formerly called human herpesvirus 6)	Exanthema subitum (roseola infantum), a common disease of infancy, featuring rash and fever
Lymphocryptovirus	Epstein–Barr virus	Infectious mononucleosis and Burkitt's lymphoma (cancer of the jaw seen mainly in African children); also linked to Hodgkin's disease (cancer of lymphocytes) and B cell lymphomas, and to nasopharyngeal cancer in Asians

POXVIRIDAE   The **poxviruses,** another group of enveloped, linear dsDNA viruses, are the largest and most complex of all viruses. They are widely distributed in nature; nearly every animal species can be infected by a form of poxvirus. The human poxviruses (orthopoxviruses) are large, enveloped, brick-shaped viruses 250 to 450 nm long and 160 to 260 nm wide. These viruses multiply in specialized portions of the host cell cytoplasm called *viroplasm,* where they can cause skin lesions typical of smallpox, molluscum contagiosum, and cowpox. Other poxviruses, such as monkeypox, can infect humans who have close contact with infected animals.

The smallpox virus represents the first human pathogen purposely eradicated from the face of the earth (as we shall see in the Chapter 20 Essay, "Smallpox—The Eradication of a Human Scourge").

PAPOVAVIRIDAE   The **papovaviruses** (pa-po'va-vi"rus-ez) are named for three related viruses, the *pa*pilloma, *po*lyoma, and *va*cuolating viruses. These are small, naked, polyhedral dsDNA viruses that replicate in the nuclei of their host cells. Papovaviruses are widely distributed in nature; more than 25 human papillomaviruses and 2 human polyomaviruses have been found. Papillomaviruses are frequently found in host cell nuclei without being integrated into host DNA (Figure 11.5b); polyomaviruses are nearly always integrated as a provirus. The papillomaviruses cause both benign and malignant warts in humans, and some papillomaviruses are associated with cervical cancer. The most thoroughly studied vacuolating virus is simian virus 40 (SV-40). This virus has been used by virologists to study the mechanisms of viral replication, integration, and oncogenesis (the development of cancerous cells).

HEPADNAVIRIDAE   The **hepadnaviruses** (he'pa-dee-en-ay-vi"rus-ez) are small, enveloped, mostly dsDNA (partially ssDNA) viruses. Their name comes from the infection of the liver—*hepa*titis—by a *DNA* virus. The hepadnaviruses can cause chronic (that is, of long and continued duration) liver infections in humans and other animals, including ducks. In humans, the hepatitis B virus causes hepatitis B, which can progress to liver cancer. We will discuss in Chapter 22 other forms of hepatitis that are caused by viruses.

PARVOVIRIDAE   The **parvoviruses** are small, naked, linear ssDNA viruses (Figure 11.5c). Their genetic information is so limited that they must enlist the aid of an unrelated helper virus or a dividing host cell to replicate. Three genera, *Dependovirus, Parvovirus,* and *Erythrovirus,* have been identified in vertebrates. The dependoviruses are often called adeno-associated viruses because, to replicate more virions they require coinfection with adenoviruses (or herpesviruses). No known human disease is associated with this genus. Members of the genus *Parvovirus* can cause disease in rats, mice, swine, cats, and dogs. Rat parvovirus causes congenital defects in the unborn. Canine parvovirus is responsible for severe and sometimes fatal gastroenteritis in dogs and puppies. The only known parvovirus to infect humans (predominantly children) is *Erythrovirus,* which is also called B19. This virus, identified in 1974, is responsible for "fifth disease" (erythema infectiosum), which causes a deep red rash on children's cheeks and ears and both a rash and arthritis in adults. B19 can cross the placenta and damage blood-forming cells in the fetus, leading to anemia, heart failure, and even fetal death.

VIRAL REPLICATION

General Characteristics of Replication

In general, viruses go through the following five steps in their **replication cycles** to produce more virions:

1. **Adsorption,** the attachment of viruses to host cells.
2. **Penetration,** the entry of virions (or their genome) into host cells.
3. **Biosynthesis,** the synthesis of new nucleic acid molecules, capsid proteins, and other viral components within host cells while using the metabolic machinery of those cells.
4. **Maturation,** the assembly of newly synthesized viral components into complete virions.
5. **Release,** the departure of new virions from host cells. Release generally, but not always, kills (lyses) host cells.

Replication of Bacteriophages

Bacteriophages, or simply *phages,* are viruses that infect bacterial cells. Phages were first observed in 1915 by Frederic Twort in England and in 1917 by Felix d'Herelle in France. D'Herelle named them bacteriophages, which means "eaters of bacteria." Over the years, attempts have been made to use phages to fight bacterial infections. These attempts, though not successful until recently (see the box "Viruses Instead of Drugs," page 283), did provide useful information about phages. Phages have been studied in great detail because it is much easier to manipulate bacterial cells and their viruses in the laboratory than to work with viruses that have multicellular hosts.

Properties of Bacteriophages

Like other viruses, bacteriophages can have their genetic information in the form of either double-stranded or single-stranded RNA or DNA. They can be relatively simple or complex in structure. To understand phage replication, we will examine the *T-even phages*.

These phages, designated T2, T4, and T6 (T stands for "type") are complex but well-studied naked phages that have dsDNA as their genetic material. The most widely studied is the T4 phage, an obligate parasite of the common enteric bacterium *Escherichia coli.* T4 has a distinctly shaped capsid made of a head, collar, and tail (Figure 11.6; Table 11.4). The DNA is packaged in the polyhedral head, which is attached to a helical tail.

Replication of T-Even Phages

Infection with and replication of new T4 phages occurs in the series of steps illustrated in Figure 11.7.

ADSORPTION If T4 phages collide in the correct orientation with host cells, the phages will attach to, or adsorb onto, the host cell surface. Adsorption is a chemical attraction; it requires specific protein recognition factors found in the phage tail fibers that bind to specific receptor sites on the host cells. The fibers

TABLE 11.4	The Functions of Bacteriophage Structural Components
Component	**Function**
Genome	Carries the genetic information necessary for replication of new phage particles
Tail sheath	Retracts so that the genome can move from the head into the host cell's cytoplasm
Plate and tail fibers	Attach phage to specific receptor sites on the cell wall of a susceptible host bacterium

FIGURE 11.6 (a) Structure and electron micrograph of a T-even (T4) bacteriophage (magnified 191,500×). (b) DNA normally is packaged into the phage head. Osmotic lysis has released the DNA from this phage, showing the large amount of DNA that must be packaged into a phage (or into an animal or plant virus).

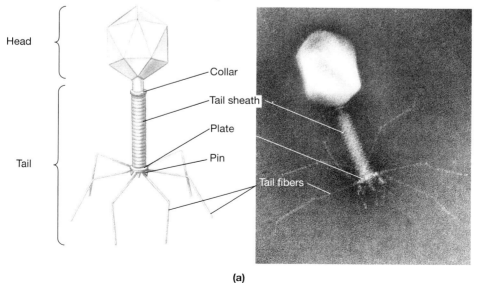
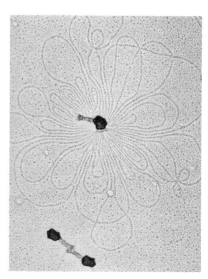

Head

Tail

Collar

Tail sheath

Plate

Pin

Tail fibers

(a)

(b)

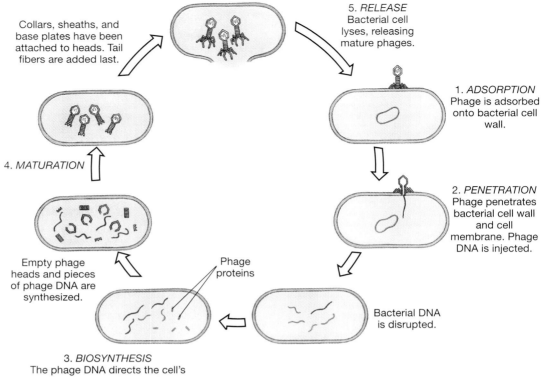

FIGURE 11.7
Replication of a virulent bacteriophage. A virulent phage undergoes a lytic cycle to produce new phage particles within a bacterial cell. Cell lysis releases new phage particles that can infect more bacteria.

Collars, sheaths, and base plates have been attached to heads. Tail fibers are added last.

4. *MATURATION*

Empty phage heads and pieces of phage DNA are synthesized.

3. *BIOSYNTHESIS*
The phage DNA directs the cell's metabolism to produce viral components—proteins and copies of phage DNA.

Phage proteins

Bacterial DNA is disrupted.

5. *RELEASE*
Bacterial cell lyses, releasing mature phages.

1. *ADSORPTION*
Phage is adsorbed onto bacterial cell wall.

2. *PENETRATION*
Phage penetrates bacterial cell wall and cell membrane. Phage DNA is injected.

bend and allow the pins to touch the cell surface. Although many phages, including T4, attach to the cell wall, other phages can adsorb to flagella or to pili.

PENETRATION The enzyme *lysozyme,* which is present within phage tails, weakens the bacterial cell wall. When the tail sheath contracts, the hollow tube (core) in the tail is forced to penetrate the weakened cell wall and come into contact with the bacterial cell membrane. The viral DNA then moves from the head through the tube into the bacterial cell. It is not clear whether the DNA is introduced directly into the cytoplasm; according to recent evidence, T4 phages introduce their DNA into the periplasmic space, between the cell membrane and the cell wall. Either way, the phage capsid remains outside the bacterium.

BIOSYNTHESIS Viral genomes are too small to contain all the genetic information to replicate themselves. Therefore, they must use the biosynthetic machinery present in host cells. Once the phage DNA enters the host cell, phage genes take control of the host cell's metabolic machinery. Usually, the bacterial DNA is disrupted so that the nucleotides of hydrolyzed nucleic acids can be used as building blocks for new phage. Phage DNA is transcribed to mRNA, using the host cell's machinery. The mRNA, translated on host ribosomes, then directs the synthesis of capsid proteins and viral enzymes. Some of these enzymes are DNA poly-

merases that replicate the phage DNA. Thus, phage infection directs the host cell to make only viral products—that is, viral DNA and viral proteins.

MATURATION The head of a T4 phage is assembled in the host cell cytoplasm from newly synthesized capsid proteins. Then, a viral dsDNA molecule is packed into each head. At the same time, phage tails are assembled from newly formed base plates, sheaths, and collars. When the head is properly packed with DNA, each head is attached to a tail. Only after heads and tails are attached are the tail fibers added to form mature, infective phages.

RELEASE The enzyme lysozyme, which is coded for by a phage gene, breaks down the cell wall, allowing viruses to escape. In the process the bacterial host cell is lysed. Thus, phages such as T4 are called **virulent (lytic) phages** because they lyse and destroy the bacteria they infect. ∞ (Chapter 8, p. 201) The released phages can now infect more susceptible bacteria, starting the infection process all over again. Such infections by virulent phages represent a **lytic cycle** of infection.

The time from adsorption to release is called the **burst time;** it varies from 20 to 40 minutes for different phages. The number of new virions released from each bacterial host represents the **viral yield,** or **burst size.** In phages such as T4, anywhere from 50 to 200 new phages may be released from one infected bacterium.

BIOTECHNOLOGY

VIRUSES INSTEAD OF DRUGS

Recently, virologists successfully used bacteriophages to treat 138 patients with long-term antibiotic-resistant bacterial infections. Each patient received a bacteriophage that specifically attacks the bacterium causing the infection. All patients benefited from the treatment, and 88 percent were completely cured.

The use of phages instead of antibiotics to treat infection has several advantages. The phages are highly specific: They attack the infectious agent without harming potentially beneficial bacteria that normally inhabit the human digestive tract. They are also cheap and easy to produce, are effective in small doses, and rarely cause side effects. Furthermore, the bacteriophages are effective against bacteria that have developed antibiotic resistance. One potential disadvantage is that bacteria might also develop resistance to bacteriophage infection.

Viruses are also being used in attempts to cure or combat noninfectious diseases. For example, in patients suffering from the congenital disease cystic fibrosis, one malfunctioning gene is responsible for the respiratory disease symptoms (accumulation of mucus in the lungs and pancreas). Molecular biologists have added a good copy of the cystic fibrosis gene to human adenoviruses and have modified them so that they cannot replicate in host cells. Trials are underway in which volunteers breathe the modified virus into their lungs. Ideally, once inhaled, the adenoviruses will infect lung cells, delivering the therapeutic gene—and curing the disease.

Phage Growth and the Estimation of Phage Numbers

Like bacterial growth, viral growth (biosynthesis and maturation) can be described by a **growth curve,** which generally is based on observations of phage-infected bacteria in laboratory cultures (Figure 11.8). The growth curve of a phage includes an **eclipse period,** which spans from penetration through biosynthesis. During the eclipse period, mature virions cannot be detected in host cells. The **latent period** spans from penetration up to the point of phage release. As Figure 11.8 shows, the latent period is longer than—and includes—the eclipse period. The number of viruses per infected host cell rises after the eclipse period and eventually levels off.

If you had a phage suspension in a test tube, how could you determine the number of viruses in the tube? Phages cannot be seen with the light microscope, and it is not feasible to count such viruses from electron micrographs. Therefore, virologists and microbiologists use a different approach to estimate phage number. The viral assay method used is called a **plaque assay.** To perform a plaque assay, virologists start with a suspension of phages. Serial dilutions, like those described for bacteria, are prepared. ∞ (Chapter 6, p. 140) A sample of each dilution is inoculated onto a plate containing a susceptible **bacterial lawn**—a layer of bacteria. Ideally, virologists want a dilution that will permit only one phage to infect one bacterial cell. As a result of infection, new phages are produced from each infected bacterial cell, lysing the cell. These phages then infect surrounding susceptible cells and lyse them. After incubation and several rounds of lysis, the bacterial lawn shows clear areas called **plaques** (Figure 11.9). Plaques represent areas where viruses have lysed host cells. In other parts of the bacterial lawn, uninfected bacteria multiply rapidly and produce a turbid growth layer.

Each plaque should represent the progeny from one infectious phage. Therefore, by counting the number of plaques and multiplying that number by the dilution factor, virologists can estimate the number of phages in a milliliter of suspension. Sometimes, however, two phages are deposited so close together that they produce a single plaque. And not all phages are infective. Thus, counting the number of plaques on a plate will approximate, but may not exactly equal, the number of infectious phages in the suspension. Therefore, such counts

FIGURE 11.8 Growth curve for a bacteriophage. The eclipse period represents the time after penetration through the biosynthesis of mature phages. The latent period represents the time after penetration through the release of mature phages. The number of viruses per infected cell is the viral yield, or burst size.

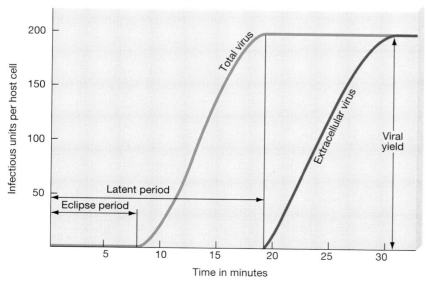

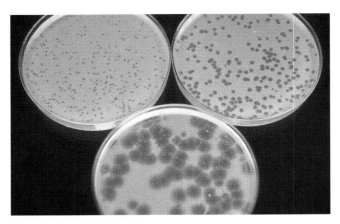

FIGURE 11.9 Plaque assay of the number of bacteriophages in a sample is done by spreading the sample out over a "lawn" of solid bacterial growth. When the phages replicate and destroy the bacterial cells, they leave a clear spot, called a plaque, in the lawn. The number of plaques corresponds roughly to the number of phages that were initially present in the sample. Different kinds of phages produce plaques of different size or shape when replicating in the same bacterial species—in this case, *Escherichia coli*. The upper left-hand plate was inoculated with T2 phage; the upper right-hand plate, with T4 phage; and the lower plate, with lambda phage.

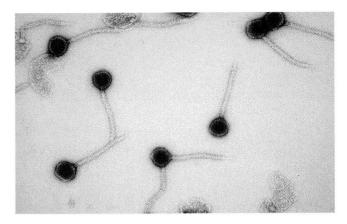

FIGURE 11.10 False-color TEM photo of the temperate phage lambda (84,000×), which infects the bacterium *Escherichia coli*.

usually are reported as **plaque-forming units** (pfu) rather than as the number of phages.

Lysogeny

General Properties of Lysogeny

The bacteriophages we have been discussing, virulent phages, destroy their host cells. **Temperate phages,** however, also exhibit **lysogeny,** a stable long-term relationship between the phage and its host. ∞ (Chapter 8, p. 201) Such participating bacteria are called *lysogenic cells.* One of the most widely studied lysogenic phages is the lambda (λ) phage of *Escherichia coli* (Figure 11.10). Lambda phages attach to bacterial cells and insert their linear DNA into the bacterial cytoplasm (Figure 11.11). However, once in the cytoplasm, the phage DNA circularizes and then integrates into the circular bacterial chromosome at a specific location. This viral DNA within the bacterial chromosome is called a **prophage.** The combination of a bacterium and a temperate phage is called a **lysogen.**

Insertion of a lambda phage into a bacterium alters the genetic characteristics of the bacterium. Two genes present in the prophage produce proteins that repress virus replication. The prophage also contains another gene that provides "immunity" to infection by another phage of the same type. This process, called **lysogenic conversion,** prevents the adsorption or biosynthesis of phages of the type whose DNA is already carried by the lysogen. The gene responsible for

such immunity does not protect the lysogen against infection by a different type of temperate phage or by a virulent phage.

Lysogenic conversion can be of medical significance because the toxic effects of some bacterial infections are caused by the prophages they contain. For example, the bacteria *Corynebacterium diphtheriae* and *Clostridium botulinum* contain prophages that have a gene that codes for the production of a toxin. The conversion from non-toxin production to toxin production is largely responsible for the tissue damage that occurs in diphtheria and botulism, respectively. Without the prophages, the bacteria do not cause disease.

Once established as a prophage, the virus can remain dormant for a long time. Each time a bacterium divides, the prophage is copied and is part of the bacterial chromosome in the progeny bacteria. Thus, this period of bacterial growth with a prophage represents a **lysogenic cycle** (Figure 11.11). However, either spontaneously or in response to some outside stimulation, the prophage can become active and initiate a typical lytic cycle. This process, called **induction,** may be due to a lack of nutrients for bacterial growth or the presence of chemicals toxic to the lysogen. The provirus seems to sense that "living" conditions are deteriorating and that it is time to find a new home. Through induction, the provirus removes itself from the bacterial chromosome. The phage DNA then codes for viral proteins to assemble new temperate phages in a manner similar to that used by lytic phages. As a result, new temperate phages mature and are released through cell lysis.

The French microbiologist André Lwoff first described lysogeny in 1950. He also discovered that only a small proportion of lysogens produce phages at any one time. Those that do are lysed as a result of phage release. The remaining lysogens do not undergo induction and, due to lysogenic conversion, remain protected from infection by phages of the same type. In

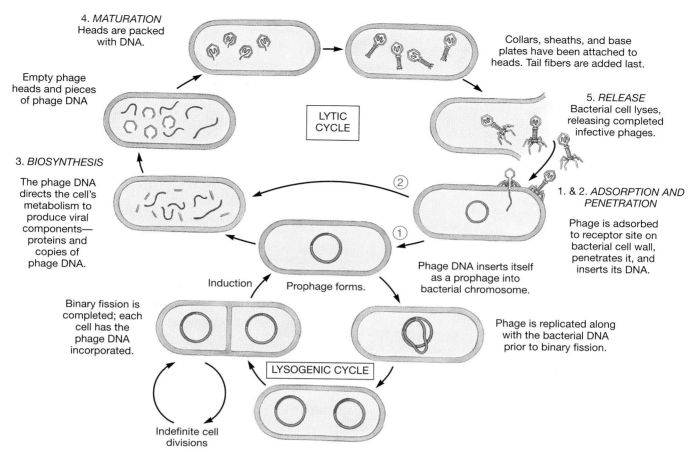

4. MATURATION
Heads are packed with DNA.

Collars, sheaths, and base plates have been attached to heads. Tail fibers are added last.

Empty phage heads and pieces of phage DNA

LYTIC CYCLE

5. RELEASE
Bacterial cell lyses, releasing completed infective phages.

3. BIOSYNTHESIS
The phage DNA directs the cell's metabolism to produce viral components—proteins and copies of phage DNA.

②

①

1. & 2. ADSORPTION AND PENETRATION
Phage is adsorbed to receptor site on bacterial cell wall, penetrates it, and inserts its DNA.

Phage DNA inserts itself as a prophage into bacterial chromosome.

Induction   Prophage forms.

Binary fission is completed; each cell has the phage DNA incorporated.

Phage is replicated along with the bacterial DNA prior to binary fission.

LYSOGENIC CYCLE

Indefinite cell divisions

FIGURE 11.11 Replication of a temperate bacteriophage. Following adsorption and penetration, the virus undergoes prophage formation. ① In the lysogenic cycle, temperate phages can exist harmlessly as a prophage within the host cell for long periods of time. Each time the bacterial chromosome is replicated, the prophage also is replicated; all daughter bacterial cells are "infected" with the prophage. Induction involves either a spontaneous or environmentally induced excision of the prophage from the bacterial chromosome. ② A typical lytic cycle, involving biosynthesis and maturation, occurs, and new temperate phages are released.

1965 Lwoff shared the Nobel Prize in physiology or medicine with François Jacob and Jacques Monod.

The majority of bacteriophages undergo lysogeny. The reason may have to do with replication. Remember, virulent phages can move from host to host only by forming new phages that are released from one cell and that infect another. Rather, a lysogenic cycle allows temperate phages to "infect" more bacteria without forming new bacteriophages. As a result of binary fission, a copy of the phage DNA is distributed to each new bacterial cell.

Replication of Animal Viruses

Like other viruses, animal viruses invade and replicate in animal cells by the processes of adsorption, penetration, biosynthesis, maturation, and release. However, animal viruses perform these processes in ways that differ from those employed by bacteriophages—and in a number of different ways among themselves. A complete replication cycle for an animal DNA virus is summarized in Figure 11.12, and two mechanisms for the replication of (+) sense RNA animal viruses are summarized in Figure 11.13.

Adsorption

As we have seen, bacteriophages have specialized structures for attaching to bacterial cell walls. Although animal cells lack cell walls, animal viruses have ways of attaching to host cells. Specificity involves a combination of virus and host cell recognition.

Naked viruses have attachment sites (proteins) on the surfaces of their capsids that bind to corresponding sites on appropriate host cells. For example, virologists have shown that rhinoviruses have "canyons,"

FIGURE 11.12 Replication of an enveloped dsDNA animal virus, such as a herpesvirus.

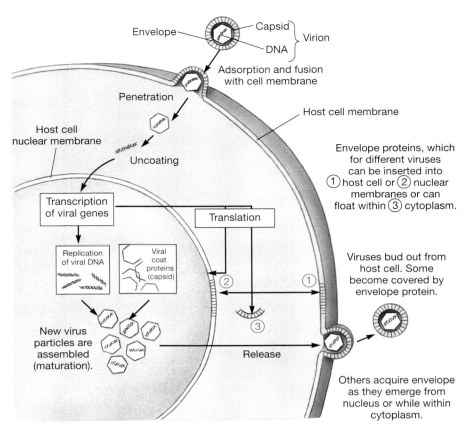

Envelope — Capsid ⟩ Virion
DNA

Adsorption and fusion with cell membrane

Penetration

Host cell membrane

Host cell nuclear membrane

Uncoating

Transcription of viral genes

Translation

Envelope proteins, which for different viruses can be inserted into ① host cell or ② nuclear membranes or can float within ③ cytoplasm.

Replication of viral DNA

Viral coat proteins (capsid)

②

①

Viruses bud out from host cell. Some become covered by envelope protein.

New virus particles are assembled (maturation).

③

Release

Others acquire envelope as they emerge from nucleus or while within cytoplasm.

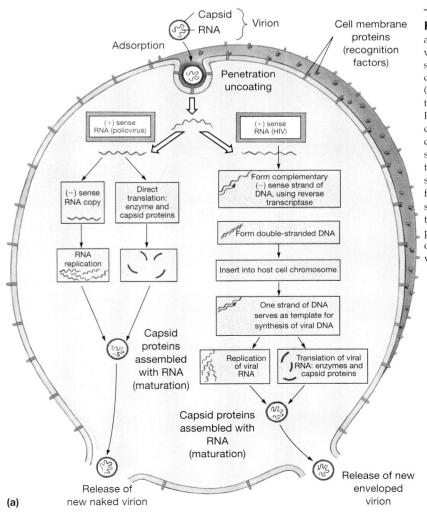

Capsid — RNA ⟩ Virion

Adsorption

Penetration uncoating

Cell membrane proteins (recognition factors)

(+) sense RNA (poliovirus)

(+) sense RNA (HIV)

(−) sense RNA copy

Direct translation: enzyme and capsid proteins

Form complementary (−) sense strand of DNA, using reverse transcriptase

RNA replication

Form double-stranded DNA

Insert into host cell chromosome

One strand of DNA serves as template for synthesis of viral DNA

Capsid proteins assembled with RNA (maturation)

Replication of viral RNA

Translation of viral RNA: enzymes and capsid proteins

Capsid proteins assembled with RNA (maturation)

Release of new naked virion

Release of new enveloped virion

(a)

FIGURE 11.13 (a) Two of the replication mechanisms used by different (+) sense RNA animal viruses. *Left:* In the poliovirus the viral (+) sense RNA serves as mRNA—it is translated immediately to produce proteins needed for reproduction of the virus. A (−) sense RNA copy is then made, which serves as a template for the production of more viral (+) sense RNA molecules. Mature polioviruses lyse the cell during release. *Right:* In HIV each (+) sense RNA, copied with the help of reverse transcriptase, forms a ssDNA, which serves as template for the synthesis of the complementary strand. The dsDNA is then inserted into the host chromosome, where it can remain for some time. When virus replication occurs, one strand of the DNA becomes the template for the synthesis of viral (+) sense RNA molecules. Mature HIV particles usually do not lyse the cell but rather bud off the cell surrounded by an envelope. (b) HIV viruses budding from a T-4 lymphocyte.

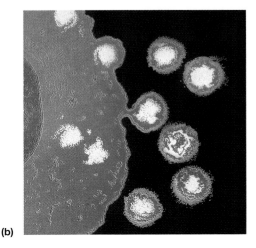

(b)

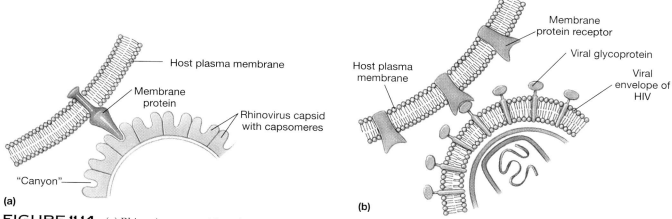

FIGURE 11.14 (a) Rhinovirus recognition of an animal host cell. Rhinoviruses have "canyons," or depressions, in the capsid that attach to specific membrane proteins on the host cell membrane. (b) HIV has specific envelope spikes (viral glycoproteins) that attach to a membrane protein receptor on the surface of specific host immune defense cells.

or depressions, in their capsids that bind to a specific membrane protein normally involved with cell adhesion (Figure 11.14a). Conversely, enveloped viruses, such as HIV, have spikes that recognize, in part, a membrane protein receptor on the surface of certain specific immune defense cells (Figure 11.14b).

Penetration

Penetration follows quite quickly after adsorption of the virion to the host's plasma membrane. Unlike bacteriophages, animal viruses do not have a mechanism for injecting their nucleic acid into host cells. Thus, both the nucleic acid and the capsid usually penetrate animal host cells. Most naked viruses enter the cell by endocytosis in which virions are captured by pitlike regions on the surface of the cell and enter the cytoplasm within a membranous vesicle (Figure 11.15). Enveloped viruses may fuse their envelope with the host's plasma membrane or enter by endocytosis. In the latter case the envelope fuses with the vesicle membrane.

Once the animal virus enters the host cell's cytoplasm, the viral genome must be separated from its protein coat (released) through a process called **uncoating.** Naked viruses are uncoated by proteolytic enzymes from host cells or from the viruses themselves. The uncoating of viruses such as the poxviruses is completed by a specific enzyme that is encoded by viral DNA and formed soon after infection. Polioviruses begin uncoating even before penetration is complete.

Biosynthesis

The synthesis of new genetic material and proteins depends on the nature of the infecting virus.

BIOSYNTHESIS IN DNA ANIMAL VIRUSES
Generally, DNA animal viruses replicate their DNA in

FIGURE 11.15 Animal virus penetration of host cells. Many naked virions adhere to the cell surface and become trapped in pits of the cell membrane. These pits invaginate to form separate cytoplasmic vesicles. In the electron micrograph, coronaviruses are being taken into the cytoplasm of a host cell.

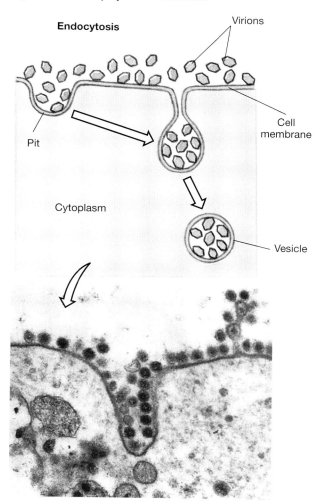

the host cell nucleus with the aid of viral enzymes and synthesize their capsid and other proteins in the cytoplasm by using host cell enzymes. The new viral proteins move to the nucleus, where they combine with the new viral DNA to form virions (Figure 11.12). This pattern is typical of adenoviruses, hepadnaviruses, herpesviruses, and papovaviruses. Poxviruses are the only exception; their parts are synthesized in the host cell's cytoplasm (viroplasm).

In DNA viruses, replication proceeds in a complex series of steps designated as *early* and *late* transcription and translation. The early events take place before the synthesis of viral DNA and result in the production of the enzymes and other proteins necessary for viral DNA replication. The late events occur after the synthesis of viral DNA and result in the production of structural proteins needed for building new capsids. Compared with bacteriophage replication, biosynthesis in animal virus replication can take much longer. The capsids of the herpesviruses, for example, contain so many proteins that their synthesis requires 8 to 16 hours.

BIOSYNTHESIS IN RNA ANIMAL VIRUSES

Biosynthesis among RNA animal viruses takes place in a greater variety of ways than is found among DNA animal viruses. In RNA viruses such as the picornaviruses, the $(+)$ sense RNA acts as mRNA, and viral proteins are made immediately after penetration and uncoating (Figure 11.13). The nucleus of the host cell is not involved. Viral proteins also play key roles in the synthesis of these viruses. One protein inhibits synthetic activities of the host cell. For biosynthesis an enzyme uses the $(+)$ sense RNA as a template to make a $(-)$ sense RNA. This $(-)$ sense RNA in turn acts as a template RNA to replicate many $(+)$ sense RNA molecules for virion formation.

In the retroviruses, such as HIV, the two copies of $(+)$ sense RNA do not act as mRNA. Rather, they are transcribed into ssDNA with the help of reverse transcriptase (Figure 11.13). The ssDNA then is replicated through complementary base pairing to make dsDNA molecules. Once in the cell nucleus, this molecule inserts itself as a provirus into a host cell chromosome. The provirus can remain there for an indefinite period of time. When infected cells divide, the provirus is replicated along with the rest of the host chromosome. Thus, the viral genetic information is passed to progeny host cells.

Unlike prophages, however, the provirus cannot be excised. If an event occurs that activates the provirus, its genes are expressed; that is, the genes are used to make viral mRNA, which directs synthesis of viral proteins. Full-length $(+)$ sense RNA molecules also are transcribed from the prophage. Two copies of the $(+)$ sense RNA are packaged into each virion.

In $(-)$ sense RNA animal viruses, such as the viruses causing measles and influenza A, a packaged transcriptase uses the $(-)$ sense RNA to make $(+)$ sense RNA molecules (mRNA). Prior to assembly, new $(-)$ sense RNA is made from $(+)$ sense RNA templates. The process is essentially the same regardless of whether the viral RNA is in one segment (measles) or in many segments (influenza A).

In the reoviruses the dsRNA codes for several viral proteins. Each strand of the dsRNA acts as a template for its partner. Like DNA replication, RNA replication is semiconservative, so the molecules produced have one strand of old RNA and one strand of new RNA. These viruses have a double-walled capsid that is never completely removed, and replication takes place within the capsid.

Maturation

Once an abundance of viral nucleic acid, enzymes, and other proteins have been synthesized, assembly of components into complete virions starts. This step constitutes maturation or assembly of progeny viruses. The cellular site of maturation varies depending on the virus type. For example, human adenovirus nucleocapsids are assembled in the cell nucleus (Figure 11.12), whereas viruses such as HIV are assembled at the inner surface of the host cell's plasma membrane. The poxviruses and picornaviruses are assembled in the cytoplasm.

Maturation of enveloped viruses is a longer and more complex process than that of most bacteriophages. As we have seen, both the infecting virus and nucleic acids and enzymes made in the host cell participate in synthesizing components. Among the components destined for the progeny viruses, the proteins and glycoproteins are coded by the viral genome; envelope lipids and glycoproteins are synthesized by host cell enzymes and are present in the host plasma membrane. If the virus is to have an envelope, the virion is not complete until it buds through a host membrane—either the nuclear, endoplasmic reticulum, Golgi, or plasma membrane—depending on the specific virus (Figure 11.12).

Release

The budding of new virions through a membrane may or may not kill the host cell. Human adenoviruses, for example, bud from the host cell in a controlled manner. This *shedding* of new virions does not lyse the host cells. Other types of animal viruses kill the host cell. When an infected animal cell is filled with progeny virions, the plasma membrane lyses and the progeny are released. Lysis of cells often produces the clinical symptoms of the infection or disease. The herpesviruses that cause cold sores and the poxviruses destroy skin cells as a result of virion release. And the polioviruses destroy nerve cells during the release process.

CULTURING OF ANIMAL VIRUSES

Development of Culturing Methods

Initially, if a virologist wanted to study viruses, the viruses had to be grown in whole animals. This made it difficult to observe specific effects of the viruses at the cellular level. In the 1930s virologists discovered that embryonated (intact, fertilized) chicken eggs could be used to grow herpesviruses, poxviruses, and influenza viruses. Although the chick embryo is simpler in organization than a whole mouse or rabbit, it is still a complex organism. The use of embryos did not completely solve the problem of studying cellular effects caused by viruses. Another problem was that bacteria also grow well in embryos, and the effects of viruses often could not be determined accurately in bacterially contaminated embryos. Virology progressed slowly during these years until techniques for growing viruses in cultures improved.

Two discoveries greatly enhanced the usefulness of cell cultures for virologists and other scientists. First, the discovery and use of antibiotics made it possible to prevent bacterial contamination. Second, biologists found that proteolytic enzymes, particularly trypsin, can free animal cells from the surrounding tissues without injuring the freed cells. After the cells are washed, they are counted and then dispensed into plastic flasks, tubes, or Petri dishes. Cells in such suspensions will attach to the plastic surface, multiply, and spread to form sheets one cell thick, called **monolayers.** These monolayers can be subcultured. **Subculturing** is the process by which cells from an existing culture are transferred to new containers with fresh nutrient media. A large number of separate subcultures can be made from a single tissue sample, thereby assuring a reasonably homogeneous set of cultures with which to study viral effects.

The term **tissue culture** remains in widespread usage to describe the preceding technique, although the term **cell culture** is perhaps more accurate. Today, the majority of cultured cells are in the form of monolayers grown from enzymatically dispersed cells. With a wide variety of cell cultures available and with antibiotics to control contamination, virology entered its "Golden Age." In the 1950s and 1960s, more than 400 viruses were isolated and characterized. Although new viruses are still being discovered, emphasis is now on characterizing the viruses in more detail and on determining the precise steps in viral infection and viral replication.

Plaque assays similar to those used to study phages can be used for animal viruses. For example, cultures of susceptible human cells are grown in cell monolayers and then inoculated with viruses. If the viruses lyse cells, several rounds of infection will produce plaques.

Types of Cell Cultures

Three basic types of cell cultures are widely used in clinical and research virology: (1) primary cell cultures, (2) diploid fibroblast strains, and (3) continuous cell lines. **Primary cell cultures** come directly from the animal and are not subcultured. The younger the source animal, the longer the cells will survive in culture. They typically consist of a mixture of cell types, such as muscle and epithelial cells. Although such cells usually do not divide more than a few times, they support growth of a wide variety of viruses.

If primary cell cultures are repeatedly subcultured, one cell type will become dominant, and the culture is called a **cell strain.** In cell strains all the cells are genetically identical to one another. They can be subcultured for several generations with only a very small likelihood that changes in the cells themselves will interfere with the determination of viral effects.

Among the most widely used cell strains are **diploid fibroblast strains** (Figure 11.16). *Fibroblasts* are immature cells that produce collagen and other fibers as well as the substance of connective tissues, such as the dermis of the skin. Derived from fetal tissues, these strains retain the fetal capacity for rapid, repeated cell division. Such strains support growth of a wide range of viruses and are usually free of contaminating viruses often found in cell strains from mature animals. For this reason they are used in making viral vaccines.

The third type of cell culture in extensive use is the continuous cell line. A **continuous cell line** consists of cells that will reproduce for an extended number of generations. The most famous of such cultures is the HeLa cell line, which has been maintained and grown

---

FIGURE 11.16 A culture of diploid fibroblasts. These immature cells, derived from fetal tissue, multiply rapidly and provide an ideal cell culture for many viruses.

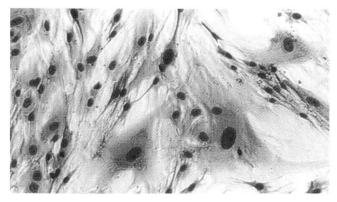

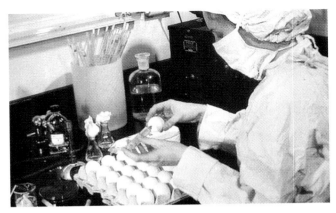

FIGURE 11.17 Some viruses, such as influenza viruses, are grown in embryonated chicken eggs.

in culture since 1951 and has been used by many researchers worldwide. The original cells of the HeLa cell line came from <u>He</u>len <u>La</u>mbert, a woman with cervical cancer. In fact, many of the early continuous cell lines used malignant cells because of their capacity for rapid growth. Such immortal cell lines grow in the laboratory without aging, divide rapidly and repeatedly, and have simpler nutritional needs than normal cells. The HeLa cell line, for example, contains two viral genes necessary for its own immortality. Immortal cell lines are heteroploid (have different numbers of chromosomes) and are therefore genetically diverse.

Cell cultures have largely replaced animals and embryonated eggs for studies in animal virology. Yet, the embryonated chicken egg remains one of the best host systems for influenza A viruses (Figure 11.17). In

addition, young albino Swiss mice are still used to culture *arboviruses* (<u>ar</u>thropod<u>bo</u>rne viruses), and other mammalian cell lines—as well as mosquito cell lines—have been used for some time.

The Cytopathic Effect

The visible effect viruses have on cells is called the **cytopathic effect** (CPE). Cells in culture show several common effects, including changes in cell shape and detachment from adjacent cells or the culture container (Figure 11.18). However, CPE can be so distinctive that an experienced virologist often can use it to make a preliminary identification of the infecting virus. For example, human adenoviruses and herpesviruses cause infected cells to swell because of fluid accumulation, whereas picornaviruses arrest cell functions when they enter and lyse cells when they leave. The paramyxoviruses cause adjacent cells in culture to fuse, forming giant, multinucleate cells called **syncytia** (sinsish'e-a; singular: *syncytium*). Syncytia can contain 4 to 100 nuclei in a common cytoplasm. Another type of CPE produced by some viruses is *transformation:* the conversion of normal cells into malignant ones, which we discuss later in this chapter.

VIRUSES AND TERATOGENESIS

Teratogenesis is the induction of defects during embryonic development. A **teratogen** (ter'a-to-jen) is a drug or other agent that induces such defects. Certain viruses are known to act as teratogens and can be transmitted across the placenta and infect the fetus. The ear-

FIGURE 11.18 (a) Normal and (b) transformed (malignant) cells in culture. Such transformation is an example of a cytopathic effect (CPE) caused by infection with the Rous sarcoma virus (RSV). In the transformed state, the cells become rounded and do not adhere to the culture container.

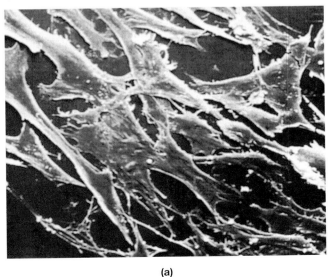

(a)

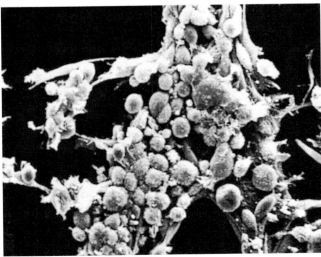

(b)

lier in pregnancy the embryo is infected, the more extensive the damage is likely to be. During the early stages of embryologic development, when an organ or body system may be represented by only a few cells, viral damage to those cells can interfere with the development of that organ or body system. Viral infections occurring later in development may damage fewer cells and thus have a proportionately smaller effect. That is because, by then, the total cell population in the fetus has greatly increased, and each organ or body system consists of thousands of cells.

Three human viruses—cytomegalovirus (CMV), herpes simplex virus (HSV) types 1 and 2, and rubella—account for a large number of teratogenic effects. Cytomegalovirus (CMV) infections are found in about 1 percent of live births; of those, about 1 in 10 eventually die from the CMV infection. Most of the defects are neurological, and the children have varying degrees of mental retardation. Some also have enlarged spleens, liver damage, and jaundice. HSV infections usually are acquired at or shortly after birth. Infections acquired before birth are rare. In cases of disseminated infections (those that spread through the body), some infants die, and survivors have permanent damage to the eyes and central nervous system.

Rubella virus infections in the mother during the first 4 months of pregnancy are most likely to result in fetal defects referred to as the "rubella syndrome." These defects include deafness, damage to other sense organs, heart and other circulatory defects, and mental retardation. The degree of impairment is highly variable. Some children adapt to their disabilities and live productive lives; in other cases the fetus is so impaired that death and natural abortion occur. Congenital rubella is discussed in Chapter 20.

A series of blood tests often referred to as the **TORCH series** is sometimes used to identify teratogenic diseases in pregnant women and newborn infants. These tests detect antibodies made against _T_oxoplasma (a protozoan), _o_ther disease-causing viruses (including the hepatitis B virus and the varicella, or chickenpox, virus), _r_ubella virus, _C_MV, and _H_SV.

VIRUSLIKE AGENTS: VIROIDS AND PRIONS

Viruses represent the smallest microbes that, in most cases, have the genetic information to produce new virions in a host cell. The exceptions are those viruses that do not have all their own genetic information to produce new virions. We mentioned earlier that viruses such as the dependoviruses (family Parvoviridae) must use a helper virus to supply the necessary components to produce more virions. However, there are even smaller infectious agents that can cause disease.

Viroids

In 1971 the plant pathologist T. O. Diener described a new type of infectious agent. He was studying potato tuber spindle disease, which was thought to be caused by a virus. However, no virions could be detected. Rather, Diener discovered molecules of RNA in the nuclei of diseased plant cells (Figure 11.19a). He proposed the concept of a **viroid,** an infectious RNA particle smaller than a virus. Since then viroids have been found to differ from viruses in six ways:

1. Each viroid consists of a single circular RNA molecule of low molecular weight.

2. Viroids exist inside cells as particles of RNA without capsids or envelopes.

3. Unlike viruses such as the parvoviruses, viroids do not require a helper virus.

FIGURE 11.19 (a) Viroid particles that cause potato spindle tuber disease (shown as yellow rods in this artist's rendition of an electron micrograph) are very short pieces of RNA containing only 300 to 400 nucleotides. The much larger (blue and purple) strand is DNA from a T7 bacteriophage. Such comparisons make it easy to see how viroids were overlooked for many years. (b) The tomato plant on the left is normal; the one on the right is infected with a viroid causing tomato apical stunt disease.

(a)

(b)

ORIGINS OF VIRUSES

Viruses are clearly quite different from cellular microbes. Free viruses are incapable of reproduction—they must infect host cells, uncoat their genetic material, and then use the host's machinery to copy or transcribe the viral genetic material. Thus, some debate remains as to whether viruses are living or are nonliving chemical aggregates. Because viruses cannot reproduce or metabolize or perform metabolic functions on their own, some scientists say that they are not living. Other scientists claim that because viruses have the genetic information for replication, and this information is active after infection, they are living. Much of the genetic regulation of viral genes is similar to the regulation of host genes. In addition, viruses use the host's ribosomes for their replication metabolism.

At present, we cannot definitively say whether viruses are living or nonliving. But we can ask, What are the origins of viruses? We do not know that either. However, because viruses cannot replicate without a host cell, it is likely that viruses were not present before primitive cells evolved. One hypothesis proposes that viruses and cellular organisms evolved together, with both viruses and cells originating from self-replicating molecules present in the precellular world. Another idea is that viruses were once cells that lost all cell functions, retaining only that information to replicate themselves using another cell's metabolic machinery. A third hypothesis proposes that viruses evolved from plasmids, the independently replicating DNA molecules found in many bacterial cells. ⌒⌒ (Chapter 8, p. 203) Plasmids are self-replicating and occur in both DNA and RNA forms. They do not, however, have genes to make capsids. In fact, it has been proposed that plasmids evolved from viroids. As some viroids moved from cell to cell, the viroid RNA may have picked up several pieces of genetic information, including the information for making a protein coat.

In trying to understand the origins of viruses, virologists have uncovered some nucleotide sequence relationships common to certain viruses. On the basis of this information, these viruses have been placed into families with similar nucleotide sequences and genetic organization. Thus, it may be possible to predict the potential disease effects of newly discovered viruses by analyzing the nucleotide sequences of their genomes and comparing them with sequences found in other, known viruses.

---

4. Viroid RNA does not produce proteins.

5. Unlike virus RNA, which may be copied in the host cell's cytoplasm or nucleus, viroid RNA is always copied in the host cell nucleus.

6. Viroid particles are not apparent in infected tissues without the use of special techniques to identify nucleotide sequences in the RNA.

Viroids must disrupt host cell metabolism in some way, but because no protein products are produced, it is not clear how viroids and their RNA cause disease. They may interfere with the cell's ability to process mRNA molecules. Without mature mRNA molecules, proteins cannot be synthesized. If so, cell metabolism would be so disturbed that cell death could result. Although some viroids cause no apparent effect or only mild pathogenic effects in the host, other viroids are known to cause several lethal plant diseases, such as potato spindle tuber disease, Chrysanthemum stunt disease, and tomato apical stunt disease (Figure 11.19b).

At least two hypotheses have been proposed to account for the origin of viroids. One suggests that they originated early in precellular evolution when the primary genetic material probably consisted of RNA. A second suggests that they are relatively new infectious agents that represent the most extreme example of parasitism.

Prions

In the 1920s several cases of a slow but progressive dementing illness in humans were observed independently by Hans Gerhard Creutzfeldt and Alfons Maria Jakob. The disease, now called *Creutzfeldt–Jakob disease* (CJD), is characterized by mental degeneration, loss of motor function, and eventual death. Since that time several similar neurological degenerative diseases have been described (Chapter 24). One is *kuru*, which caused loss of voluntary motor control and eventual death of natives of New Guinea. These deaths were attributed to an infective agent transmitted as a result of cannibalism. In other animals, *scrapie* in sheep and *bovine spongiform encephalopathy* (BSE)—commonly called mad cow disease—in dairy cattle have been observed to cause slow loss of neuronal function that leads to death.

Although some recent research into these similar diseases indicates that they may be caused by viruses, other evidence points to a different type of infective agent. The infective agent may be an exceedingly small proteinaceous infectious particle. In 1982 Stanley Prusiner proposed that such an infectious particle be called a **prion** (pre'on). Prions have the following characteristics:

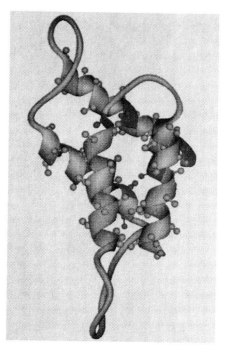

FIGURE 11.20 Protein structure model of the harmless form of the prion protein (PrP). The protein helices are represented as spiral ribbons.

1. Prions are resistant to inactivation by heating to 90°C, which will inactivate viruses.
2. Prion infection is not sensitive to radiation treatment that damages virus genomes.
3. Prions are not destroyed by enzymes that digest DNA or RNA.
4. Prions are sensitive to protein denaturing agents, such as phenol and urea.

Prusiner's research and that of others suggest that prions are normal proteins that become folded incorrectly, possibly as a result of a mutation (Figure 11.20). The harmless, normal proteins are found on the plasma membrane of many mammalian cells, especially brain cells. The prion proteins (*PrP*) are thought to stick together inside cells, forming small fibers, or fibrils. Because the fibrils cannot be organized in the plasma membrane correctly, such aggregations eventually kill the cell.

The most urgent question is to determine how a prion-caused disease spreads. Researchers believe that prion diseases are spread by the prions' causing other copies of the normal protein to fold improperly. In an outbreak of mad cow disease early in the 1990s, the infectious prion originally came from a protein supplement in the feed. This supplement included byproducts from scrapie-infected sheep! In fact, experiments show that when mice are injected with prion extracts, disease results.

VIRUSES AND CANCER

Cancer is known as a set of diseases that perturb the normal behavior and functioning of cells. We can define **cancer** as an uncontrolled, invasive growth of abnormal cells—in other words, cancer cells divide repeatedly. In many cases they cannot stop dividing; the result is a **neoplasm,** or localized accumulation of cells known as a **tumor.** A neoplasm can be **benign**— a noncancerous growth. But if the cells invade and interfere with the functioning of surrounding normal tissue, the tumor is **malignant.** Malignant tumors and their cells can **metastasize,** or spread, to other tissues in the body.

That viruses could cause some cancers in animals was discovered in 1911 by F. Peyton Rous. He showed that certain *sarcomas* (neoplasms of connective tissue) in chickens were caused by a virus, named the *Rous sarcoma virus* (RSV). Therefore, it was not surprising to discover that viruses can be associated with cancer in humans as well. Although most human cancers arise from genetic mutations, cellular damages from environmental chemicals, or both, epidemiologists estimate that about 15 percent of human cancers arise from viral infections.

Human Cancer Viruses

After many years of research and testing, we now know of at least six viruses that are associated with human cancers. There probably are many more yet to be identified.

The Epstein–Barr virus (EBV) perhaps is the best understood of the human cancer viruses. This DNA virus is a herpesvirus that was first discovered in African children suffering from Burkitt's lymphoma, a malignant tumor that causes swelling and eventual destruction of the jaw. In fact, evidence points to three other tumors also associated with EBV.

Several of the human papillomaviruses (HPV) have shown a strong correlation with some human cancers. Although some of these DNA viruses cause only benign warts, other types (HPV-8 and HPV-16) lead to a *carcinoma* (neoplasm of epithelial tissues) of the uterine cervix. Another potential cancer-causing DNA virus is hepatitis B virus (HBV). It causes inflammation of the liver and can lead to liver cancer. *Kaposi's sarcoma,* a cancer of the endothelial cells of the blood vessels or lymphatic system, is thought to be associated with a herpesvirus.

The major human cancer viruses discovered so far are dsDNA viruses. However, some (+) sense RNA viruses, specifically the retroviruses, are also associated with cancers. HTLV-I causes *adult T cell leukemia/lymphoma.*

How Cancer Viruses Cause Cancer

Like bacteriophages, some animal viruses that infect animal cells often cause cell death through cell lysis. Other animal viruses can infect cells and form proviruses. In some cases these infections result in physical and genetic changes to the host cells—the CPE discussed earlier. For example, RSV causes cells in culture to detach themselves from the culture flask and round up (Figure 11.18). In the case of **DNA tumor viruses,** which can exist as proviruses, the major CPE is the uncontrollable division of the infected cells. This process, called **neoplastic transformation,** is typical of DNA tumor viruses. Many insert all or part of their DNA at random sites into the host DNA. However, only a few of these viral genes are necessary for transformation.

The papillomaviruses (family Papovaviridae) that cause human cancers infect cells, but their viral DNA remains free in the cytoplasm of the host (Figure 11.21). A few genes of the papillomavirus are active so that the virus can replicate with each cell division. Should the viral DNA accidentally integrate into the host cell DNA, unregulated replication of viral proteins can occur. These proteins cause host cells to divide uncontrollably. Some of these viral proteins block the effects of tumor-suppressor genes, which prevent uncontrolled cell divisions. Without the products of these genes, the host experiences uncontrolled cell divisions—and a tumor develops.

Many of the retroviruses are **RNA tumor viruses.** In fact, they are the only RNA viruses known to cause cancer. Recall that retroviruses use their own reverse transcriptase to transcribe (+) sense RNA into DNA that then integrates as a provirus into the host chromosome. The provirus of HTLV-I codes for proteins that transform the host cells into neoplastic ones. Infection also leads to the production of new virions by budding, which does not kill the infected cell. Thus, RNA tumor viruses can continue to infect other uninfected cells or sex cells. In the latter case, the presence of virus particles ensures transmission of virions to offspring.

Oncogenes

The proteins produced by tumor viruses that cause uncontrolled host cell division come from segments of DNA called **oncogenes** (*onco,* Greek for "mass"). In DNA tumor-causing viruses, not only do oncogenes cause a neoplasm, they also contain the information for synthesizing viral proteins needed for viral replication. The oncogenes in RNA tumor viruses are quite different. Virologists and cell biologists have shown that some RNA tumor viruses pick up "extra" genes from normal host cells during viral replication. These genes, which are similar to oncogenes, are called protooncogenes. A **proto-oncogene** is a normal gene that,

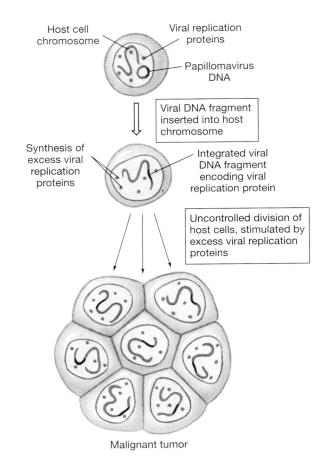

FIGURE 11.21 Malignant tumor formation caused by a papillomavirus (DNA tumor virus). Integration of the provirus causes synthesis of viral replication proteins that promote host cell divisions, leading to cancer.

when under the control of a virus, can cause uncontrolled cell division; that is, it can act as an oncogene. Such oncogenes carried by these viruses are not needed for virus replication.

Many oncogenes have been discovered in oncogenic viruses, and most code for information leading to unlimited cell divisions. Such oncogenes are mutant genes containing deletions or substitutions. ∞ (Chapter 7, p. 181) These mutations cause structural changes in the proteins for which the genes code. Such oncogenes work in one of two ways: (1) The product of the oncogene can disrupt normal cell function, leading to cell divisions. (2) The oncogene is controlled by viral regulators near the site of their integration into the host cell's chromosome. These regulators "turn on" the gene so normal protein is made—but in excessive amounts or at the wrong time in the host cell's life. Again, excessive cell divisions occur. The discovery of oncogenes in viruses has had a major impact on our understanding of cancer. Although there is still much to be learned about cancer in humans, perhaps in the future effective antiviral drugs will prevent virus-induced cancers.

EMERGING VIRUSES

Viruses have been infecting humans for thousands of years, and the diseases they cause have been responsible for millions of deaths. Microbiologists believe that many recent, unexpected viral diseases have been caused by *emerging viruses*—viruses that were previously *endemic* (low levels of infection in localized areas) or had "crossed species barriers"—that is, expanded their host range to other species. For example, although the poliovirus has been endemic since ancient times, only since 1900 have *pandemics* (high levels of infection worldwide) resulted from this virus, with numerous annual outbreaks. Why the increase in disease?

The poliovirus has not mutated over the centuries into a more pathogenic form. Rather, virologists and epidemiologists suggest that the urban populations that developed after the Industrial Revolution provided an ideal environment for viral spread. The large numbers of nonimmune people who had emigrated from nonendemic areas were exposed to immune people who carried the poliovirus. Thus, polio spread rapidly—and with deadly results. Only through the development of polio vaccines in the 1950s was the epidemic halted. Although there still are areas where polio is endemic, the World Health Organization hopes to eradicate the poliovirus through vaccination programs by the year 2000.

Other endemic viral diseases that are transmitted among humans, such as measles, also are recurring due to changes in human population densities and travel. But several viral diseases involve other animals that act as *reservoirs* (a "healthy" organism harboring an infectious agent that is available to infect another host) or *vectors* (carriers) for a virus. In these cases the virus could cross species barriers if the vector transmitted the virus from a reservoir species to humans. For example, prior to 1930, yellow fever was thought to be carried solely by one species of mosquito, *Aedes aegypti*. By controlling that mosquito in urban areas (through vaccination and spraying with DDT), the disease could be controlled. However, in the late 1950s

an outbreak of yellow fever occurred that was not carried by *A. aegypti*. Rather, jungle (sylvan) yellow fever was carried by another mosquito, of the genus *Haemagogus*. In the jungle these mosquitoes transmit the virus among monkeys that live high in the forest canopy. Forest clearing through tree cutting had brought the *Haemagogus* mosquito from the treetops to the forest floor, where it passed the virus to people involved in timber cutting and agricultural activities.

This yellow fever situation represents an excellent example of how viral diseases can remain endemic to those parts of the world where an insect vector lives. However, in tropical areas where once-uninhabited lands are being converted for use in agriculture and farming, contact with the insects (and the viruses they carry) is inevitable.

Of the more than 500 known arboviruses, about 80 cause disease in humans; of these, 20 are considered emerging viruses. The most dangerous are the yellow fever virus, which has been endemic for more than a century—and is reemerging—and the Dengue fever viruses. Both are carried by mosquitoes.

Many virologists believe that a similar event may have occurred in the case of the human immunodeficiency virus (HIV). Retroviruses similar to HIV exist in domesticated cats (feline immunodeficiency virus, FIV) and in monkeys (simian immunodeficiency virus, SIV). It is possible that a mutated form of SIV crossed over to humans from contact with an infected monkey. However, antibodies to SIV have been found in humans. Therefore, SIV itself may initially have infected humans and later mutated. Natural selection could have favored these mutations because they were better adapted than SIV to the new human host.

In the United States one recent emerging virus is the hantavirus, which is transmitted from rodent feces and urine to humans. The virus, which causes hantavirus pulmonary syndrome (HPS), struck first in New Mexico in May 1993. Although we do not know how the virus got into the rodent population, genetic analysis sug-

gests that the virus has been endemic in rodents for years. We do know that conditions were right for an explosion in the rodent population, guaranteeing more contact between rodent feces and humans. Other hantaviruses are known to cause hemorrhagic fever in millions of people worldwide. The first, called the Hantaan virus, was isolated in Korea in 1978 but is believed to have been causing disease since the 1930s.

One of the disease transmission mechanisms of great concern today is air travel. Since the 1950s the annual number of international air travel passengers has risen from 2 million to almost 300 million, and almost 600 million are expected by the year 2000. In the confined air system of a jet airplane (an ideal atmosphere for the rapid spread of disease), a person or mosquito harboring an emerging virus could carry the virus around the world overnight.

So, how do we protect ourselves from such potential viral threats? Many virologists suggest that "virus outposts" be set up to try to detect emerging viruses before they spread. Viruses such as HIV that spread slowly might be hard to detect because they take years to emerge on a global scale. Conversely, the yellow fever virus shows clinical symptoms in nonimmunized individuals within days, if not hours. Perhaps quarantine will be required for people visiting or working in areas suspected of harboring emerging viruses.

Several factors have been proposed that can contribute to the emergence of viral (and other infectious) diseases. These include ecological changes and development (human contact with natural hosts or reservoirs), changes in human demographics (typically due to famine or war), international travel and commerce (allowing rapid introduction of viruses to new habitats and hosts), technology and industry (for example, during processing and rendering of infected animals), microbial adaptations and change (high mutation rates or shifts in genetic composition), and environmental changes (such as extension of vector ranges by global warming).

GENERAL CHARACTERISTICS OF VIRUSES

What Are Viruses?

- **Viruses** are submicroscopic **obligate intracellular parasites**—they replicate only inside a living host cell.

Components of Viruses

- Viruses consist of a nucleic acid core and a protein **capsid.** Some viruses also have a membranous **envelope.**
- Viral genetic information is contained in either DNA or RNA—not both.
- Capsids are made of subunits called **capsomeres.**
- A viral capsid and genome form a **nucleocapsid.** Such viruses are called **naked viruses;** those with a nucleocapsid surrounded by an envelope are **enveloped viruses.**

Sizes and Shapes

- Viruses have polyhedral, helical, binal, bullet, or complex shapes and vary in size from 20 to 300 nm in diameter.

Host Range and Specificity of Viruses

- Viruses vary in **host range** and **viral specificity.** Many viruses infect a specific kind of cell in a single host species; others infect several kinds of cells, several hosts, or both.

CLASSIFICATION OF VIRUSES

- Viruses are classified by the nucleic acid (DNA or RNA) they contain, other chemical and physical properties, their mode of replication, shape, and host range. Some of these characteristics are summarized in Tables 11.1 and 11.2.
- Similar viruses are grouped into genera, and genera are grouped into families. Viruses that share the same genome and relationships with organisms generally constitute a viral species.

RNA Viruses

- Among the (+) sense RNA virus families are the Picornaviridae, which include the poliovirus, the hepatitis A virus and the rhinoviruses; the Togaviridae, which include the virus that causes rubella; the Flaviviridae, which include the yellow fever virus; and the Retroviridae, which cause some cancers and AIDS. The (−) sense RNA viruses include the Paramyxoviridae, which cause measles, mumps, and several respiratory disorders; the Rhabdoviridae, one of which causes rabies; the Orthomyxoviridae, which include the influenza viruses; the Filoviridae, which cause Marburg and Ebola diseases; the Arenaviridae, which cause Lassa fever; and the Bunyaviridae, one of which causes hantavirus pulmonary syndrome. The double-stranded (ds) RNA virus families include the Reoviridae, which cause a variety of upper respiratory and gastrointestinal infections.

DNA Viruses

- The dsDNA virus families include the Adenoviridae, some of which cause respiratory infections; the Herpesviridae, which cause oral and genital herpes, chicken pox, shingles, and infectious mononucleosis; the Poxviridae, which cause smallpox and similar, milder infections. The Papovaviridae cause warts; some papovaviruses are associated with certain cancers. The Hepadnaviridae cause human hepatitis B infection, and the Parvoviridae cause relatively rare human infections.

VIRAL REPLICATION

General Characteristics of Replication

- Viruses generally go through five steps in the replication process: **adsorption, penetration, biosynthesis, maturation,** and **release.** These steps are somewhat different in bacteriophages and animal viruses.

Replication of Bacteriophages

- **Bacteriophage** replication has been thoroughly studied in T-even phages, which are **virulent phages.**
- T-even phages have recognition factors that attach to specific receptors on bacterial cell walls during adsorption. Enzymes weaken the bacterial wall so viral nucleic acid can penetrate it.
- During biosynthesis viral DNA directs the making of viral components.
- In the maturation stage, the viral components are assembled into complete virions.
- Release, the final stage, is facilitated by the enzyme lysozyme. **Burst time** is the time from adsorption to release of progeny virions; **burst size** is the number of phage progeny released from one host cell.
- The growth curve of a phage includes an **eclipse period** (the time following penetration through biosynthesis) and a **latent period** (the time after penetration up to release).
- The number of phages produced in an infection can be determined by counting the number of **plaques** produced on a plate of virus-infected bacteria (**plaque assay**). Each plaque represents a **plaque-forming unit.**
- Phages carrying out these stages of replication, leading to host cell destruction, represent a **lytic cycle** of infection.

Lysogeny

- **Lysogeny,** a stable long-term relationship between certain phages and host bacteria, occurs in **temperate phages.** Temperate phage DNA can exist as a **prophage** or revert through **induction** to a lytic cycle.
- Prophages such as the lambda phage insert into a bacterial chromosome at a specific location.

Replication of Animal Viruses

- On the surface of some viruses, proteins are used for attachment to host plasma membranes during adsorption; animal viruses thus gain entry into the cell. **Uncoating** (loss of the capsid) occurs at the plasma membrane or in the cytoplasm.

- Synthesis and maturation differ in DNA and RNA viruses. In most DNA viruses, DNA is synthesized in an orderly sequence in the nucleus, and proteins are synthesized in the cytoplasm of the host cell. In RNA viruses, RNA can act as a template for protein synthesis, for making mRNA, or for making DNA by reverse transcription. Virions are assembled in the cell; sometimes viral DNA is incorporated as a **provirus** into the host cell's chromosome.
- Release can occur through direct lysis of the host cell or by budding through the host membrane.

CULTURING OF ANIMAL VIRUSES

Development of Culturing Methods

- The discovery of antibiotics to prevent bacterial contamination of chicken embryos and **cell cultures** and the use of trypsin to separate cells in culture systems into **monolayers** provided an important impetus to the study of virology.

Types of Cell Cultures

- **Primary cell cultures** come directly from animals and are not subcultured.
- All the cells in a **cell strain,** which are derived from subcultured primary cell cultures, are very similar. **Diploid fibroblast strains** from primary cultures of fetal tissues produce stable cultures that can be maintained for years; they are used to produce vaccines.
- **Continuous cell lines,** usually derived from cancer cells, grow in the laboratory without aging, can divide repeatedly, have greatly reduced nutritional needs, and display heteroploidy.
- The visible effects that viruses produce in infected host cells are collectively called the **cytopathic effect.**

VIRUSES AND TERATOGENESIS

- A **teratogen** is an agent that induces defects during embryonic development.
- Viruses can act as teratogens by crossing the placenta and infecting embryonic cells. The earlier in pregnancy an infection occurs, the more extensive damage is likely to be.
- The rubella virus can be responsible for the death of fetuses and severe birth defects in others; cytomegaloviruses and occasionally herpesviruses also act as teratogens.

VIRUSLIKE AGENTS: VIROIDS AND PRIONS

Viroids

- **Viroids** are very different from viruses; each viroid is solely a small RNA molecule.
- Viroids may cause plant diseases by interfering with mRNA processing.

Prions

- **Prions** are infectious particles made of protein. Research suggests that prions are normal proteins that become folded incorrectly.
- Prions cause neurological degenerative diseases, including Creutzfeldt-Jakob disease and kuru.

VIRUSES AND CANCER

- **Cancer** is generally an uncontrolled and/or invasive growth of abnormal cells.

Human Cancer Viruses

- **Tumors,** or **neoplasms,** are **benign** (noncancerous) or **malignant** (cancerous). Malignant tumors spread by **metastasis.**
- Several animal viruses are thought to cause some forms of cancer, including the Epstein–Barr virus, certain human papillomaviruses, the hepatitis B virus, and some retroviruses, such as HTLV-I.

How Cancer Viruses Cause Cancer

- **DNA tumor viruses** contain viral genes whose protein products disrupt the activities of normal host cell proteins that control cell division.
- **RNA tumor viruses** contain viral genes used for **neoplastic transformation** and viral replication.

Oncogenes

- **Oncogenes** are viral genes that cause host cells to divide uncontrollably.
- **Proto-oncogenes** are normal genes that, when under the control of a virus, act as oncogenes causing uncontrolled cell division.
- Oncogenes in RNA tumor viruses produce proteins in excessive amounts or produce proteins at the wrong times. In either case, infected host cells start uncontrolled cell division.

KEY TERMS

adenovirus   (p. 279)
adsorption   (p. 281)
arenavirus   (p. 279)
bacterial lawn   (p. 283)
bacteriophage   (p. 273)
benign   (p. 293)
biosynthesis   (p. 281)

bunyavirus   (p. 279)
burst size   (p. 282)
burst time   (p. 282)
cancer   (p. 293)
capsid   (p. 271)
capsomere   (p. 272)
cell culture   (p. 289)

cell strain   (p. 289)
complex virus   (p. 273)
continuous cell line   (p. 289)
cytopathic effect   (p. 290)
diploid fibroblast strain
   (p. 289)
DNA tumor virus   (p. 294)

eclipse period   (p. 283)
enterovirus   (p. 277)
envelope   (p. 271)
enveloped virus   (p. 272)
filovirus   (p. 278)
flavivirus   (p. 277)
genome   (p. 272)

glycoprotein   *(p. 272)*
growth curve   *(p. 283)*
hepadnavirus   *(p. 280)*
hepatovirus   *(p. 277)*
herpesvirus   *(p. 279)*
host range   *(p. 273)*
human immunodeficiency
   virus (HIV)   *(p. 272)*
induction   *(p. 284)*
latency   *(p. 279)*
latent period   *(p. 283)*
lysogen   *(p. 284)*
lysogenic conversion
   (p. 284)
lysogenic cycle   *(p. 284)*
lysogeny   *(p. 284)*
lytic cycle   *(p. 282)*
lytic phage   *(p. 282)*
malignant   *(p. 293)*

maturation   *(p. 281)*
metastasize   *(p. 293)*
monolayer   *(p. 289)*
naked virus   *(p. 272)*
negative (−) sense RNA
   (p. 275)
neoplasm   *(p. 293)*
neoplastic transformation
   (p. 294)
nucleocapsid   *(p. 272)*
obligate intracellular parasite   *(p. 270)*
oncogene   *(p. 294)*
orthomyxovirus   *(p. 278)*
papovavirus   *(p. 280)*
paramyxovirus   *(p. 278)*
parvovirus   *(p. 280)*
penetration   *(p. 281)*
picornavirus   *(p. 275)*

plaque   *(p. 283)*
plaque assay   *(p. 283)*
plaque-forming unit
   (p. 284)
positive (+) sense RNA
   (p. 275)
poxvirus   *(p. 280)*
primary cell culture
   (p. 289)
prion   *(p. 292)*
prophage   *(p. 284)*
proto-oncogene   *(p. 294)*
provirus   *(p. 278)*
release   *(p. 281)*
reovirus   *(p. 279)*
replication cycle   *(p. 280)*
retrovirus   *(p. 278)*
reverse transcriptase   *(p. 278)*
rhabdovirus   *(p. 278)*

rhinovirus   *(p. 277)*
RNA tumor virus   *(p. 294)*
spike   *(p. 272)*
subculturing   *(p. 289)*
syncytia   *(p. 290)*
temperate phage   *(p. 284)*
teratogen   *(p. 290)*
teratogenesis   *(p. 290)*
tissue culture   *(p. 289)*
togavirus   *(p. 277)*
TORCH series   *(p. 291)*
tumor   *(p. 293)*
uncoating   *(p. 287)*
viral specificity   *(p. 273)*
viral yield   *(p. 282)*
virion   *(p. 272)*
viroid   *(p. 291)*
virulent phage   *(p. 282)*
virus   *(p. 270)*

QUESTIONS FOR REVIEW

A Virus Characteristics

1. Describe the shapes of viruses.

2. What is a virion?

3. List the main characteristics of viruses.

B Virus Classification

4. How are viruses classified?

5. For each of the following characteristics, identify the virus family, and give an example of a typical virus:

   a. enveloped, double-stranded DNA with a complex shape

   b. two (+) sense RNA molecules and reverse transcriptase

   c. cause influenza

   d. among the smallest RNA viruses with a naked, polyhedral capsid

   e. one (−) sense RNA, enveloped and filamentous

C Viral Replication

6. List the general steps in viral replication.

7. How do T-even phages attach to and enter cells?

8. Match each of the following descriptions with the correct step of viral replication:

   _____ Viral parts are assembled into whole virions.

   _____ Spikes or fibers are important in this step.

   _____ Viral genetic information is duplicated.

   _____ Burst size can be determined.

   _____ Viral DNA is transcribed into mRNA, which is used to produce new viral proteins and viral DNA.

   a. adsorption
   b. penetration
   c. biosynthesis
   d. maturation
   e. release

9. Identify parts (a) through (c) of the following phage growth curve, and describe what happens in each phase.

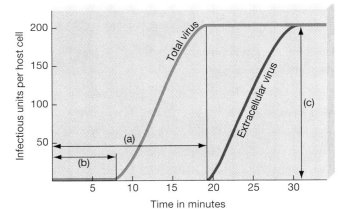

10. What is lysogeny?

11. Determine whether each of the following characteristics describes events in the (lytic cycle/lysogenic cycle/both/neither):

   a. Induction stimulates new viral synthesis and maturation.

   b. Viral components assemble into complete virions.

   c. Prophages form during this cycle.

   d. This is the reproductive cycle of a virulent phage.

   e. The virus may remain dormant in the host bacterium for an extended period of time.

12. What are some differences in the replication process among RNA viruses?

D Culturing Animal Viruses

13. What problems had to be solved before cell culturing methods could contribute significantly to virology?

14. Match the replication characteristic for RNA viruses with the appropriate virus type:

—— (+) ssRNA acts as mRNA.

—— They can be copied into ssDNA before replication occurs.

—— An RNA polymerase uses a (−) ssRNA to make mRNA.

—— (+) ssRNA forms a provirus prior to replication.

a picornaviruses

b. reoviruses

c. orthomyxoviruses

d. retroviruses

15. What characteristics of diploid fibroblast strains make them useful?

16. List the characteristics of continuous cell lines.

E Teratogens

17. What is a teratogen?

18. What kinds of damage do viruses cause in embryos?

F Viruslike Agents

19. How do viroids differ from viruses?

20. What are prions, and what diseases do they cause?

G Viruses and Cancer

21. What makes cancer cells different from normal cells?

22. How are DNA tumor viruses different from RNA tumor viruses?

23. How do oncogenes work?

PROBLEMS FOR INVESTIGATION

1. Prepare two lists—one describing properties that identify viruses as living, and the other describing properties that identify viruses as nonliving. Are viruses living or nonliving?

2. In a lytic cycle caused by a virulent phage, biosynthesis and maturation involve the synthesis of enzymes to aid phage release. What would happen if these enzymes were synthesized very early (in the first 10 minutes) in the infection process?

3. Lysogenic phages, such as lambda, are specific to certain bacteria; DNA viruses, such as polyomaviruses, are specific to certain animal cells. How do these very different viruses use very similar infection and replication processes?

4. Some animal viruses replicate in the host cell's cytoplasm, whereas others replicate in the cell nucleus. Specifically, why must these viruses replicate in these parts of the host cell?

5. Could you use Koch's postulates to identify a latent herpesvirus infection? Explain.

6. It was once stated that "the death of the host is a result as harmful to the virus's future as to that of the host itself." Explain the significance of this statement.

7. Make a "classification summary" similar to Figure 10.7 for (a) the RNA viruses and (b) the DNA viruses. Use the characteristics in Tables 11.1 and 11.2 to generate your summary.

SOME INTERESTING READING

Bass, T. 1990. "Virus hunting on the Niger." In *Camping with the prince*. New York: Houghton Mifflin.

Beardsley, T. 1990. "Oravske kuru: A human dementia raises the stakes in mad cow disease." *Scientific American* 263 (July):24–26.

Cann, A. J. 1993. *Principles of molecular virology*. San Diego: Academic Press.

"The cervical cancer virus." 1995. *Discover* 16 (September): 26–28.

Chang, Y., et al. 1994. "Identification of herpesvirus-like DNA sequences in AIDS-associated Kaposi's sarcoma." *Science* 266, no. 5192 (December 16):1865–69.

Fields, B. N., and Knipe, D. M. 1990. *Fields' virology*, 2d ed. New York: Raven Press.

Gallo, R. C. 1987. "The AIDS virus." *Scientific American* 256 (January):47–56.

Henig, R. M. 1993. *A dancing matrix: Voyages along the viral frontier*. New York: Alfred A. Knopf.

Hogle, J. M., M. Chow, and D. J. Filman. 1987. "The structure of poliovirus." *Scientific American* 256 (March):42–49.

Jaret, P. 1994. "Viruses." *National Geographic* 186(1):59–91.

Le Guenno, B. 1995. "Emerging viruses." *Scientific American* 273 (October):56–64.

Levine, A. J. 1992. *Viruses*. New York: Scientific American Library.

Levy, J. A., H. Fraenkel-Conrat, and R. A. Owens. 1994. *Virology*, 3d ed. Englewood Cliffs, NJ: Prentice Hall.

Maurice, J. 1993. "Fever in the urban jungle." *New Scientist* 140(1895):25–29.

Morse, S., ed. 1993. *Emerging viruses*. New York: Oxford University Press.

Prusiner, S. B. 1995. "The prion diseases." *Scientific American* 272 (January):48–57.

Simons, K., H. Garoff, and A. Helenuis. 1982. "How an animal virus gets into and out of its host cell." *Scientific American* 246 (February):58–66.

Varmus, H. 1987. "Reverse transcription." *Scientific American* 257 (September):56–64.

EUKARYOTIC MICROORGANISMS AND PARASITES

False-color SEM photo of a potato plant leaf (840×).
Emerging from the leaf are potato blight sporangia
of *Phytophthora infestans*, a parasitic fungus.

I n our survey of microbes, we have devoted significant attention to bacteria of the kingdom Monera and to viruses. However, some members of eukaryotic kingdoms are also of interest to microbiologists and health scientists. The kingdoms Protista and Fungi contain large numbers of microscopic species, some of which supply food and antibiotics, and some of which cause disease. The kingdom Animalia contains helminths that cause disease and arthropods that cause or transmit diseases. Study of the microscopic eukaryotes, as well as the helminths and arthropods, constitutes a significant part of a health scientist's training. Unless health scientists take a course in parasitology, their only opportunity to learn about helminths and arthropods is in conjunction with the study of microscopic infectious agents.

PRINCIPLES OF PARASITOLOGY

A **parasite** is an organism that lives at the expense of another organism, called the **host.** Parasites vary in the degree of damage they inflict on their hosts. Although some cause little harm, others cause moderate to se-

vere damage. Parasites that cause disease are called **pathogens. Parasitology** is the study of parasites.

Although few people realize it, among all living forms, there are probably more parasitic than nonparasitic organisms. Many of these parasites are microscopic throughout their life cycle or at some stage of it. Historically, in the development of the science of biology, parasitology came to refer to the study of protozoa, helminths, and arthropods that live at the expense of other organisms. We will use the term *parasite* to refer to these organisms. Strictly speaking, bacteria and viruses that live at the expense of their hosts also are parasites.

The manner in which parasites affect their hosts differs in some respects from that described in earlier chapters for bacteria and viruses. Special terms also are used to describe parasites and their effects. This introduction to parasitology will make discussions of parasites here and in later chapters more meaningful.

Significance of Parasitism

Parasites have been a scourge throughout human history. In fact, even with modern technology to treat and control parasitic diseases, there are more parasitic infections than there are living humans. It has been estimated that among the 60 million people dying each year, fully one-fourth die of parasitic infections or their complications.

Parasites play an important, though negative, role in the worldwide economy. For example, less than half the world's cultivable land is under cultivation, primarily because parasites endemic to (always present in) those lands prevent humans and domesticated animals from inhabiting some of them. As the world population increases, and the need for food with it, cultivation of such lands will become more important. In some inhabited regions, many people are near starvation and severely debilitated by parasites. Furthermore, parasitic infections in wild and domestic animals provide sources of human infection and cause debilitation and death among the animals, thus preventing the raising of cattle and other animals for food. Given the many human problems created by parasites, all citizens—and especially health scientists—need to understand the problems associated with the control and treatment of parasitic diseases.

Parasites in Relation to Their Hosts

Parasites can be divided into **ectoparasites,** such as ticks and lice, which live on the surface of other organisms, and **endoparasites,** such as some protozoa and worms, which live within the bodies of other organisms. Most parasites are **obligate parasites:** They must spend at least some of their life cycle in or on a host. For example, the protozoan that causes malaria invades red blood cells. A few parasites are **facultative parasites:** They normally are free-living, such as some soil fungi, but they can obtain nutrients from a host, as many fungi do when they cause skin infections. Hosts that are invaded by parasites usually lack effective defenses against them, so such diseases can be serious and sometimes fatal.

Parasites are also categorized according to the duration of their association with their hosts. **Permanent parasites,** such as tapeworms, remain in or on a host once they have invaded it. **Temporary parasites,** such as many biting insects, feed on and then leave their hosts. **Accidental parasites** invade an organism other than their normal host. Ticks that ordinarily attach to dogs or to wild animals sometimes attach to humans; the ticks are then accidental parasites. **Hyperparasitism** refers to a parasite itself having parasites. Some mosquitoes, which are temporary parasites, harbor the malaria parasite or other parasites. Such insects serve as **vectors,** or agents of transmission, of many human parasitic diseases.

An organism that transfers a parasite to a new host is a vector. A vector in which the parasite goes through part of its life cycle is a **biological vector**. The malaria mosquito is both a host and a biological vector. A **mechanical vector** is a vector in which the parasite does not go through any part of its life cycle during transit. Flies that carry parasite eggs, bacteria, or viruses from feces to human food are mechanical vectors.

Hosts are classified as **definitive hosts** if they harbor a parasite while it reproduces sexually; they are said to be **intermediate hosts** if they harbor the parasite during some other developmental stages. The mosquito is the definitive host for the malaria parasite because that parasite reproduces sexually in the mosquito; the human is an intermediate host, even though humans suffer greater damage from the parasite. **Reservoir hosts** are infected organisms that make parasites available for transmission to other hosts. Reservoir hosts for human parasitic diseases typically are wild or domestic animals. **Host specificity** refers to the range of different hosts in which a parasite can mature. Some parasites are quite host specific—they mature in only one host. The malaria parasite matures primarily in *Anopheles* mosquitoes. Other parasites can mature in many different hosts. The worm that causes trichinosis can mature in almost any warmblooded animal, but the parasite is most often acquired by humans from pigs through the consumption of inadequately cooked, contaminated pork.

Over thousands of years of evolution, parasites tend to become less injurious to their hosts. Such an arrangement preserves the host so that the parasites are guaranteed a continuous supply of nutrients. A parasite that destroys its host also destroys its own means of support. The adjustment of parasites and hosts to each other is closely related to the host's defense mechanisms. Many parasites have one or more of the following mechanisms for evading host defense mechanisms:

1. *Encystment,* the formation of an outer covering that protects against unfavorable environmental conditions. These resistant cyst stages also sometimes provide a site for internal reorganization of the organism and cell division, help attach a parasite to a host, or serve to transmit a parasite from one host to another.

2. Changing the parasite's surface antigens (molecules that elicit immunity) faster than the host can make new antibodies (molecules that recognize and attack antigens).

3. Causing the host's immune system to make antibodies that cannot react with the parasite's antigens.

4. Invading host cells, where the parasites are out of reach of host defense mechanisms.

When parasites successfully evade host defenses, they can cause several kinds of damage. All parasites rob their hosts of nutrients. Some take such a large share of nutrients or damage so much surface area of the host's intestines that the host receives too little nourishment. Many parasites cause significant trauma to host tissues. They cause open sores on the skin, destroy cells in tissues and organs, clog and damage blood vessels, and may even cause internal hemorrhages. Parasites that do not evade defense mechanisms sometimes trigger severe inflammatory and immunological reactions. For example, treatment to rid human hosts of some worm infections effectively kills the worms, but toxins from the dead worms cause more tissue damage than do the living parasites. The dog heartworm, *Dirofilaria immitis*, perforates the heart wall and leaves holes in the heart when the worms die and decay. It is therefore important for a veterinarian to test all dogs for the presence of heartworms before administering preventive heartworm medication.

A hallmark of many parasites is their reproductive capability. Parasitism, although an easy life once the parasite is established, is a hazardous existence during transfers from one host to another. For example, many parasites that leave the human body through feces die from desiccation (drying out) before they reach another host. If several hosts are required to complete the life cycle, the hazards are greatly multiplied. Consequently, many parasites have exceptional reproductive capacities. Some parasites, such as certain protozoa, undergo **schizogony** (skiz-og'o-ne), or multiple fission, in which one cell gives rise to many cells, all of which are infective. Others, such as various worms, produce large numbers of eggs. Some worms are **hermaphroditic**—that is, one organism has both male and female reproductive systems. In fact, certain worms, such as tapeworms, lack a digestive tract and consist almost exclusively of reproductive systems.

PROTISTS

Characteristics of Protists

The **protists,** members of the kingdom Protista, are a diverse assortment of organisms that share certain common characteristics. Protists are unicellular (though sometimes colonial), eukaryotic organisms with cells that have true nuclei and membrane-bound organelles. Although most protists are microscopic, they vary in diameter from 5 μm to 5 mm.

Importance of Protists

Protists have captured the fancy of biologists since Leeuwenhoek made his first microscopes. In fact, most of the "animalcules" he observed were protists. Like Leeuwenhoek, many people find protists inherently interesting, and biologists have learned much about life processes from protists.

Protists also are important to humans for other reasons. For instance, they are a key part of food chains. Autotrophic protists capture energy from sunlight. Some heterotrophic protists ingest autotrophs and other heterotrophs. Others decompose, or digest, dead organic matter, which then can be recycled to living organisms. Protists also serve as food for higher-level consumers. Ultimately, some energy originally captured by protists reaches humans. For example, energy from the sun is transferred to protists, protists are eaten by oysters, and the oysters are eaten by humans.

Protists can be economically beneficial or detrimental. Certain protists have **tests,** or shells, of calcium carbonate. Carbonate shells deposited in great numbers by such protists that lived in ancient oceans formed the white cliffs of Dover and the limestone used in building the pyramids of Egypt. Because different test-forming protists gained prominence during different geological eras, the identification of the protists in rock layers helps determine the age of the rocks. Certain test-forming protists tend to occur in rock layers near petroleum deposits, so geologists looking for oil are pleased to find them. Some autotrophic protists produce toxins that do not harm oysters that eat the protists, but the accumulated toxins can cause disease or even death in people who subsequently eat the oysters. Oyster beds infected with such protists can cause great economic losses to oyster harvesters. Other autotrophic protists multiply very rapidly in abundant inorganic nutrients and form a "bloom," a thick layer of organisms over a body of water. This process, called **eutrophication** (u''tro-fi-ka'shun), blocks sunlight, killing plants beneath the bloom and causing fish to starve. Microbes that decompose dead plants and animals use large quantities of oxygen, and the lack of oxygen leads to more deaths. Together these events result in great economic losses in the fishing industry.

Finally, some protists are parasitic. They cause debilitation in large numbers of people, and sometimes death, especially in poor countries that lack the resources to eradicate those protists. Parasitic diseases caused by protozoa include amoebic dysentery, malaria, sleeping sickness, leishmaniasis, and toxoplasmosis. Together, these diseases account for severe losses in human productivity, incalculable human misery, and many deaths.

Classification of Protists

Like all groups of living things, the protists display great variation, which provides a basis for dividing the kingdom Protista into sections and phyla. However, taxonomists do not agree about how these classifications should be made. We can accomplish our main purpose of illustrating diversity and avoid taxonomic problems by grouping protists according to the king-

TABLE 12.1	Properties of Protists	
Group	Characteristics	Examples
Plantlike protists	Have chloroplasts; live in moist, sunny environments	Euglenoids, diatoms, dinoflagellates
Funguslike protists	Most are saprophytes; may be unicellular or multicellular	Water molds, plasmodial and cellular slime molds
Animal-like protists	Heterotrophs; most are unicellular, most are free-living, but some are commensals or parasites	Mastigophorans, sarcodines, apicomplexans, and ciliates

dom of macroscopic organisms they most resemble (Table 12.1). Thus, we speak of protists that resemble plants (Figure 12.1), protists that resemble fungi (Figure 12.2), and protists that resemble animals (Figure 12.3).

The Plantlike Protists

The plantlike protists, or algae, have chloroplasts and carry on photosynthesis. They are found in moist, sunny environments. Most have cell walls and one or two flagella, which allow them to move. The **euglenoids** (u-gle'noidz) usually have a single flagellum and a pigmented eyespot called a *stigma*. The stigma may orient flagellar movement so that the organism moves toward light. A typical euglenoid, *Euglena gracilis* (Figure 12.1a), has an elongated, cigar-shaped, flexible body. Instead of a cell wall, it has a **pellicle,** or outer membranous cover. Euglenoids usually reproduce by binary fission. Most live in fresh water, but a few are found in the soil.

Another group of plantlike protists have other pigments in addition to chlorophyll. These protists usually have cell walls surrounded by a loosely attached secreted test that contains silicon or calcium carbonate. Most reproduce by binary fission. They include the **diatoms** (di'ah-tomz), which lack flagella (Figure 12.1b), and several other groups, which have flagella and are distinguished by their yellow and brown pigments. Diatoms are an especially numerous group and are important as producers in both freshwater and marine environments. Fossil deposits of diatoms, known as diatomaceous earth, are used as filtering agents and abrasives in various industries.

The **dinoflagellates** (di'no-flaj''el-atz) are plantlike protists that usually have two flagella—one extending behind the organism like a tail, and the other lying in a transverse groove (Figure 12.1c). They are small organisms that may or may not have a cell wall. Some have a *theca*, a tightly affixed secreted layer that typically contains cellulose. Cellulose is an uncommon substance in protists, although it is abundant in plants.

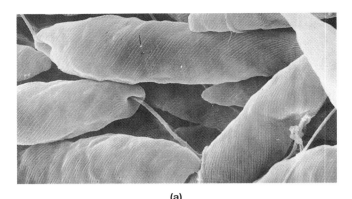

(a)

(b)

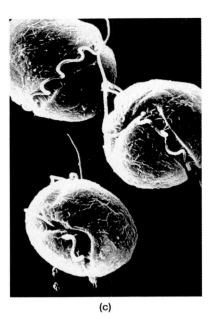

(c)

FIGURE 12.1 SEM photos of representative algae, or plant-like protists. (a) *Euglena*, a euglenoid. (b) The diatom *Campylodiscus hibernicus* (magnified 250×). (c) *Gonyaulax*, a dinoflagellate that causes red tides.

PUBLIC HEALTH

RED TIDES

Some species of *Gonyaulax* and some other dinoflagellates produce toxins. When these marine organisms appear seasonally in large numbers, they cause a bloom known as a *red tide*. The toxins accumulate in the bodies of shellfish such as oysters and clams that feed on the protists. Although the toxin does not harm the shellfish, it causes paralytic shellfish poisoning in some fish and humans who eat the infected shellfish. Even animals as large as dolphins have been killed in large numbers by this toxin. Inhaling air that contains small quantities of the toxin can irritate respiratory membranes, so sensitive individuals should avoid the sea and its products during red tides.

A red tide, caused by proliferation of toxin-producing dinoflagellates.

Whereas most dinoflagellates have chlorophyll and are capable of carrying on photosynthesis, others are colorless and feed on organic matter. Several dinoflagellates exhibit bioluminescence. The photosynthetic dinoflagellates are second only to the diatoms as producers in marine environments.

The Funguslike Protists

The funguslike protists, or water molds and slime molds, have some characteristics of fungi and some of animals.

WATER MOLDS   The **water molds** and related protists that cause mildew—the **Oomycota**—are sometimes classified as fungi. These molds, mildews, and plant blights produce flagellated spores, called *zoospores*, during asexual reproduction and large motile gametes during sexual reproduction. The most prominent phase of their life cycle consists of diploid cells from the union of gametes. These protists live freely in fresh water or as plant parasites; they cause such diseases as downy mildew on grapes and sugar beets and late blight in potatoes. A member of the Oomycota was responsible for the Irish potato famine in the 1840s. With a few exceptions, water molds are not medically significant to humans. They do, however, cause disease in fish and other aquatic organisms.

SLIME MOLDS   **Slime molds** are commonly found as glistening, viscous masses of slime on rotting logs; they also live in other decaying organic matter or in

EUKARYOTIC ALGAE

Taxonomists disagree on how to classify eukaryotic algae. These algae are sometimes classified as protists, sometimes classified as plants, and sometimes divided between the protist and plant kingdoms. Eukaryotic algae should not be confused with blue-green algae (now called cyanobacteria), which are prokaryotes.

Although both eukaryotic and prokaryotic algae are important as producers in many environments, they are generally not of medical significance. (However, cyanobacteria of the genus *Prototheca*, which have lost their chlorophyll, have been reported to cause skin lesions.) Agar, which is of great importance in the microbiology laboratory, is a product extracted from smaller seaweeds (red algae). Some eukaryotic algae, such as kelps (brown algae), are used as food and in the manufacture of products such as cheese spreads, toothpaste, and mayonnaise, to which they add smoothness and spreadability. They also enable beer to retain a foamy "head."

The eukaryotic alga *Macrocystis*. This giant kelp can grow several centimeters per day. Air bladders along the stalk keep the photosynthetic leaflike blades afloat where they can obtain sunlight. This organism is a source of algin, which is used in many foods and other products.

soil. Most slime molds are **saprophytes** (sap'ro-fitz), or organisms that feed on dead or decaying matter. A few are parasites of algae, fungi, or flowering plants, but not of humans. Slime molds occur as plasmodial slime molds and as cellular slime molds.

Plasmodial slime molds (Figure 12.2a) form a multinucleate, amoeboid mass called a **plasmodium,** which moves about slowly and phagocytizes dead matter. Sometimes a plasmodium stops moving and forms *fruiting bodies*. Each fruiting body develops *sporangia*, sacs that produce spores. When spores are released, they germinate into flagellated gametes. Two gametes fuse, lose their flagella, and form a new plasmodium. As a plasmodium feeds and grows, it can also divide and produce new plasmodia directly.

The **cellular slime molds** (Figure 12.2b) produce pseudoplasmodia, fruiting bodies, and spores with characteristics that are quite different from those of plasmodial slime molds. A **pseudoplasmodium** is a slightly motile aggregation of cells. It produces fruiting bodies, which in turn produce spores. The spores germinate into amoeboid phagocytic cells that divide repeatedly, producing more independent amoeboid cells. Depletion of the food supply causes the cells to aggregate into loosely organized new pseudoplasmodia.

The Animal-like Protists

The animal-like protists, or **protozoa,** are heterotrophic, mostly unicellular organisms, but a few form colonies. Most are free-living. Some are **commensals,** which live

---

FIGURE 12.2 Representative funguslike protists. (a) A plasmodial slime mold of the genus *Hemitrichia,* on a decaying log. (b) Pseudoplasmodia of a cellular slime mold, *Dictyostelium discoideum.*

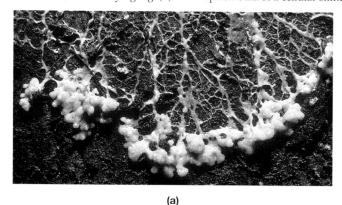

(a)

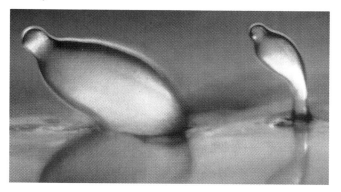

(b)

FIGURE 12.3

Representative protozoa, or animal-like protists. (a) *Trichonympha*, a mastigophoran, an endosymbiont from a termite gut. Particles seen inside the body are ingested wood particles. (b) *Amoeba proteus* (350×), a sarcodine, free-living inhabitant of ponds. (c) *Plasmodium vivax* (inside red blood cells), an apicomplexan, one of the parasites that causes malaria. (d) *Paramecium caudatum* (135×), a ciliate.

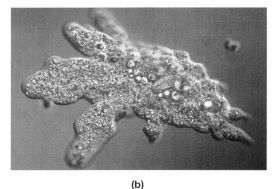

(a)

(b)

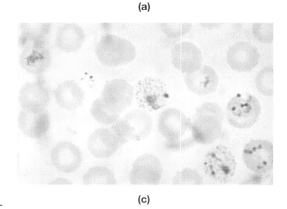

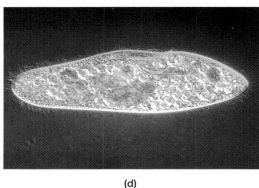

(c)

(d)

in or on other organisms without harming them, and a few are parasites. The parasitic protozoa are of particular interest in the health sciences. Many protozoa live in watery environments, and encyst when conditions are not favorable. Some protozoa are protected by a tough outer pellicle. Many are motile and are further classified on the basis of their means of locomotion (Figure 12.3). The protozoa that you will encounter in this book belong to the groups Mastigophora, Sarcodina, Apicomplexa (also known as Sporozoa), or Ciliata (also known as Ciliophora).

MASTIGOPHORANS The **mastigophorans** (mas″ti-gof′or-anz) have flagella. A few species are free-living in either fresh or salt water, but most live in symbiotic relationships with plants or animals. The symbiont *Trichonympha* (Figure 12.3a) lives in the termite gut and contributes enzymes that digest cellulose. Mastigophorans that parasitize humans include members of the genera *Trypanosoma, Leishmania, Giardia,* and *Trichomonas.* Trypanosomes cause African sleeping sickness, leishmanias cause skin lesions or systemic disease with fever, giardias cause diarrhea, and trichomonads cause vaginal inflammation.

SARCODINES The **sarcodines** are usually amoeboid—they move by means of pseudopodia (Figure 12.3b). ∞ (Chapter 4, p. 96) A few sarcodines have flagella at some stage in their life cycle. They feed mainly on other microorganisms, including other protozoa

and small algae. The sarcodines include foraminiferans and radiolarians, which have shells and are found mainly in marine environments, and amoebas, which have no shells and are typically parasites.

Numerous species of amoebas are capable of inhabiting the human intestinal tract. Most form cysts that help them withstand adverse conditions. The more commonly observed genera—*Entamoeba, Dientamoeba, Endolimax,* and *Iodamoeba*—cause amoebic dysenteries of varying degrees of severity. *Entamoeba gingivalis* is found in the mouth. *D. fragilis,* which is unusual in that it has two nuclei and does not form cysts, is found in the large intestine of about 4 percent of the human population. Its means of transmission is unknown. Although usually considered a commensal, it can cause chronic, mild diarrhea.

APICOMPLEXANS The **apicomplexans** (or sporozoans) are parasitic and immobile (Figure 12.3c). Enzymes present in groups (complexes) of organelles at the tips (apices) of their cells digest their way into host cells, giving the group the name Apicomplexa. These parasites usually have complex life cycles. An important example is the life cycle of the malaria parasite, *Plasmodium,* which requires both a human and a mosquito host (Figure 12.4). (Do not confuse this apicomplexan with the plasmodium form of slime molds.) The parasites, which are present as **sporozoites** (sporo-zo′itz) in the salivary glands of an infected mosquito, enter human blood through the mosquito's bite.

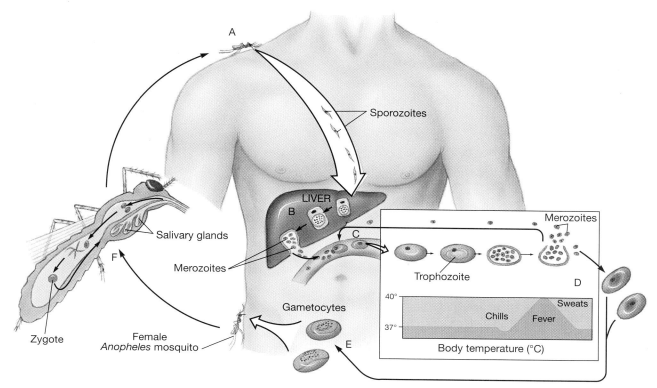

FIGURE 12.4 Life cycle of the malaria parasite, *Plasmodium*. A: Female *Anopheles* mosquito transmits sporozoites from its salivary glands when it bites a human. The sporozoites travel in human blood to the liver. B: In the liver, the sporozoites multiply and become merozoites, which are shed into the bloodstream when liver cells rupture. C: The merozoites enter red blood cells and become trophozoites, which feed and eventually form many more merozoites. D: Merozoites are released by the rupture of the red blood cells, accompanied by chills, high fever (40°C), and sweating. They can then infect other red blood cells. E: After several such asexual cycles, gametocytes (sexual stages) are produced. F: Upon ingestion by a mosquito, the gametocytes form a zygote, which gives rise to more infective sporozoites in the salivary glands. These can then infect other people.

The sporozoites migrate to the liver and become **merozoites** (meh-ro-zo'itz). After about 10 days, they emerge into the blood, invade red blood cells, and become **trophozoites** (tro-fo-zo'itz). Trophozoites reproduce asexually, producing many more merozoites, which are released into the blood by the rupture of red blood cells. Multiplication and release of merozoites is repeated several times during a bout of malaria. Some merozoites enter the sexual reproductive phase and become **gametocytes,** or sex cells. When a mosquito takes a blood meal from an infected human, it also takes in gametocytes, most of which mature and unite to form zygotes in the lining of the mosquito's stomach. Zygotes pass through the stomach wall and produce sporozoites, which eventually make their way to the salivary glands.

Several species of *Plasmodium* cause malaria, and each displays variations in the life cycle just described and in the particular species of mosquito that serves as a suitable host. Another apicomplexan, *Toxoplasma gondii,* causes lymphatic infections and blindness in adults and severe neurological damage to the fetuses

of infected pregnant women. Contact with infected domestic cats and their feces, consumption of contaminated raw meat, and failure to wash hands after handling such meat are means of transmitting the parasite.

CILIATES The largest group of protozoans, the **ciliates,** have cilia over most of their surfaces. Cilia have a basal body near their origin that anchors them in the cytoplasm and enables them to extend from the surface of the cell. Cilia allow the organisms to move, and in some genera, such as *Paramecium* (Figure 12.3d), cilia assist in food gathering. *Balantidium coli,* the only ciliate that parasitizes humans, causes dysentery.

Ciliates have several highly specialized structures. Most ciliates have a well-developed contractile vacuole, which regulates cell fluids. Some have a strengthened pellicle. Others have **trichocysts,** tentacles that can be used to capture prey, or long stalks by which they attach themselves to surfaces. Ciliates also undergo **conjugation.** Unlike bacterial conjugation, in which one organism receives genetic information from another, conjugation in ciliates allows exchange of genetic information between two organisms.

FUNGI

Characteristics of Fungi

Fungi, studied in the specialized field of **mycology,** are a diverse group of heterotrophs. Many are saprophytes that digest dead organic matter and organic wastes. Some are parasites that obtain nutrients from the tissues of other organisms. Most fungi, such as molds and mushrooms, are multicellular, but yeasts are unicellular.

The body of a fungus is called a **thallus.** The thallus of most multicellular fungi consists of a **mycelium** (my-se′le-um), a loosely organized mass of threadlike structures called **hyphae** (hy′fe; singular: *hypha;* Figure 12.5). The mycelium is embedded in decaying organic

matter, soil, or the tissues of a living organism. Mycelial cells release enzymes that digest the *substratum* (the surface on which the fungus grows) and absorb small nutrient molecules. The cell walls of a few fungi contain cellulose, but those of most fungi contain **chitin** (ki′tin), a polysaccharide also found in the exoskeletons (outer coverings) of arthropods such as ticks and spiders. All fungi have lysosomal enzymes that digest damaged cells and help parasitic fungi to invade hosts. Many fungi synthesize and store granules of the nutrient polysaccharide glycogen. Some fungi, such as yeasts, are known to have plasmids. These plasmids can be used to clone foreign genes into the yeast cells, a technique of great use in genetic engineering. ∞ (Chapter 8, p. 218)

The hyphal cells of most fungi have one or two nuclei, and many hyphal cells are separated by crosswalls called **septa** (singular: *septum*). Pores in septa allow both cytoplasm and nuclei to pass between cells. Some fungi have septa with so many pores that they are sievelike, and a few lack septa entirely. Certain fungi with a single septal pore have an organelle called a *Woronin* (weh-ro′nin) *body*. When a hyphal cell ages or is damaged, the Woronin body moves to and blocks the pore so that materials from the damaged cell cannot enter a healthy cell.

Many fungi reproduce both sexually and asexually, but a few have only asexual reproduction. Asexual reproduction always involves mitotic cell division, which in yeast occurs by budding (Figure 12.6). Sexual reproduction occurs in several ways. In one way, haploid gametes unite, and their cytoplasm mingles in a process called **plasmogamy** (plaz-mog′am-e). However, if the nuclei fail to unite, a **dikaryotic** ("two-nucleus") cell forms; it can persist for several cell divisions. Eventually, the nuclei fuse in a process called **karyogamy** (kar″-e-og′am-e) to produce a diploid cell. Such cells or their progeny later produce new haploid cells. Some fungi also can reproduce sexually during dikaryotic (diploid) phases of their life cycle. Fungi

---

FIGURE 12.5 The mycelium of a typical fungus (the mold *Aspergillus niger;* 85×) consists of filamentous hyphae, the cells of which can be multinucleate and separated by pore-containing septae.

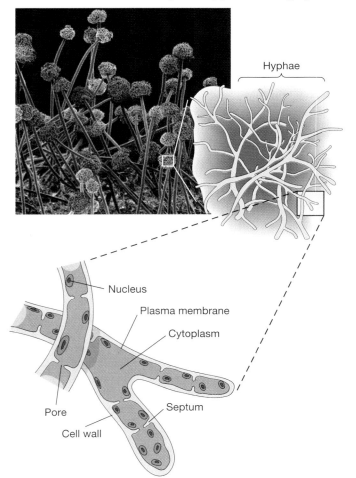

Hyphae

Nucleus

Plasma membrane

Cytoplasm

Pore

Septum

Cell wall

FIGURE 12.6 Budding yeast (7000×). Circular scars seen on the surface of the cell on the right represent sites of previous budding.

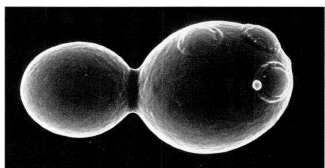

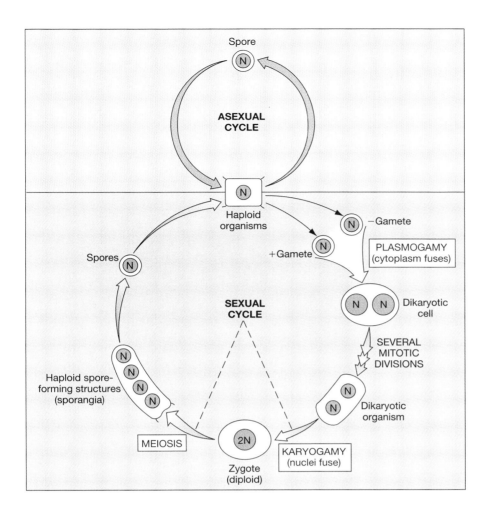

FIGURE 12.7 One method of sexual reproduction in fungi. Haploid organisms may maintain themselves by asexual spore formation (beige background) or budding. Alternatively (blue background), they may produce gametes that initially undergo plasmogamy (fuse their cytoplasmic portions). After several mitotic divisions of the still-separate nuclei, the two nuclei undergo karyogamy (fuse their nuclei) to form a diploid zygote. The zygote then undergoes meiosis to return to the haploid state and produces reproductive spores.

usually go through haploid, dikaryotic, and diploid phases in their life cycle (Figure 12.7).

Fungi can produce spores both sexually and asexually, and spores can have one or several nuclei (Figure 12.8). Typically, aquatic fungi produce motile spores with flagella, and terrestrial fungi produce spores with thick protective walls. Germinating spores produce either single cells or germ tubes. *Germ tubes* are filamentous structures that break through weakened spore walls and develop into hyphae.

FIGURE 12.8 (a) Formation of asexual spores (conidiospores) in the fungus *Penicillium* (1400×). (b) Spores of the rose rust fungus *Phragmidium* (1000×).

(a)

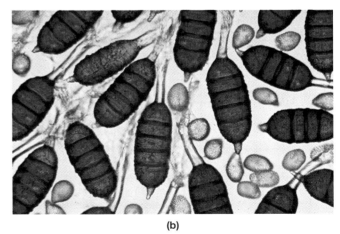

(b)

LICHENS: DUAL MICROORGANISMS

A *lichen* is not a single organism but a fungus living in symbiosis with either a cyanobacterium or a green alga. The members of the pair can be separated, and each will live a perfectly normal life by itself. Presumably, though, there are advantages to living together. Although not all scientists agree, most feel that lichens represent a mutualistic relationship, in which each member of the partnership benefits. The fungus obtains food from the photosynthetic organism while providing it with structure and protection from the elements (especially from dehydration).

Lichen-forming organisms take on different and very specific shapes when they grow in association. *Crustose lichens* grow on surfaces and resemble a crust. *Foliose lichens* are leaflike, growing in crinkled layers that project up from a substratum, usually a rock or tree trunk. *Fruticose lichens* are the tallest, sometimes looking like miniature forests. Reindeer moss is not a true moss but rather an example of a fruticose lichen.

Lichens are pioneer organisms, being among the first to colonize bare rock. Gradually they erode the rock with acids they produce. They also accumulate tiny bits of dust and humus around their bases and begin the process of soil formation, making it possible for seeds or spores of plants to get a start. Lichens are highly sensitive to air pollution, so they may be hard to find in the city.

Several species of lichens growing on a rock along the shore of Acadia National Park in Maine. At the lower right corner is a gray foliose type. The pink, purple, and green species are crustose types.

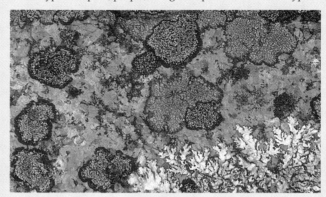

Importance of Fungi

In ecosystems, fungi are important decomposers. In the health sciences, they are important as facultative parasites—they can obtain nutrients from nonliving organic matter or from living organisms. Fungi are never obligate parasites because all fungi can obtain nutrients from dead organisms. Even when fungi parasitize living organisms, they kill cells and obtain nutrients as saprophytes. Nearly every form of life is parasitized by some type of fungus. Some fungi produce antibiotics that inhibit the growth of or kill bacteria. Parasitic fungi vary in the damage they inflict. Fungi such as those that cause athlete's foot are nearly always present on the skin and rarely cause severe damage. However, the fungus that causes histoplasmosis can spread through the lymphatic system to cause fever, anemia, and death.

Saprophytic fungi are beneficial as decomposers and as producers of antibiotics. The digestive activities of such fungi provide nutrients not only for the fungi themselves but for other organisms, too. The carbon and nitrogen compounds they release from dead organisms contribute significantly to the recycling of substances in ecosystems. Fungi are essential for decomposing lignins and other woody substances. Some fungi excrete metabolic wastes that are toxic to other organisms, especially soil microorganisms. In the soil, the production of such toxins, which are antibiotics, is called **antibiosis.** These toxins presumably help the species that produce them compete and survive. The antibiotics, when extracted and purified, are used to treat human infections (Chapter 14).

Parasitic fungi can be destructive when they invade other organisms. These fungi have three require-

FUNGI AND ORCHIDS

When explorers first brought orchids from South America back to England during the nineteenth century, the English were delighted to have such handsome specimens in their conservatories. However, they suffered great disappointment when the plants failed to thrive, no matter how carefully they were potted in fresh soil and new pots. It took the English several years of experimentation, and perhaps the fortuitous importation of a few orchids in their native medium, to learn to grow orchids out of their natural environment. Eventually, it was discovered that orchids require certain fungi to thrive. These fungi form symbiotic associations with the orchid roots; such associations are called *mycorrhizae* (my''ko-ri'ze). When medium in which orchids had grown was used to pot fresh specimens, the orchids and the fungi formed mycorrhizae, and both thrived.

ments for invasion: (1) proximity to the host, (2) the ability to penetrate the host, and (3) the ability to digest and absorb nutrients from host cells. Many fungi reach their hosts by producing spores that are carried by wind or water. Other fungi arrive on the bodies of insects or other animals. For example, wood-boring insects spread spores of the fungal Dutch elm disease (Figure 12.9) throughout North America in the decades following World War I, killing almost all elm trees in some parts of the United States. Fungi penetrate plant cells by forming hyphal pegs that press on and push through cell walls. How fungi penetrate animal cells, which lack cell walls, is not fully understood, but lysosomes apparently play an important role. Once fungi have entered cells, they digest cell components and absorb nutrients. As cells die, the fungus invades adjacent cells and continues to digest and absorb nutrients.

Fungal parasites in plants cause diseases such as wilts, mildews, blights, rusts, and smuts and thereby produce extensive crop damage and economic losses. Fungal infections of domestic birds and mammals are also responsible for extensive economic losses. Those fungi that invade humans cause human suffering, decreased productivity, and sometimes long-term medical expenses. Human fungal diseases, or **mycoses,** often are caused by more than one organism. Mycoses can be classified as superficial, subcutaneous, or systemic. *Superficial* diseases affect only keratinized tissue in the skin, hair, and nails. *Subcutaneous* diseases affect skin layers beneath keratinized tissue and can spread to lymph vessels. *Systemic* diseases invade internal or-

BIOTECHNOLOGY

ROACH TRAP USES FUNGUS

A company specializing in pest control by means of naturally occurring microbes has patented a baited roach trap lined with the soil fungus *Metarhizium anisopliae*. As roaches crawl through the trap, the fungus sticks to their back. Within 12 hours, the fungus has penetrated the body wall, grown throughout the interior, and is digesting the tissue. Because roaches groom each other by licking, a single roach can spread the fungus throughout a colony. The fungus attacks only roaches and is safer than chemical pesticides.

gans and cause significant destruction. Some fungi are opportunistic; they do not ordinarily cause disease but can do so in individuals whose defenses are impaired, such as AIDS patients and transplant recipients who are receiving immunosuppressive drugs. More individuals with fungal infections are seeking hospital treatment than ever before.

Culturing and identification of the causative agents of mycoses require special laboratory techniques. Acidic, high-sugar media with antibiotics added help prevent bacterial growth and allow fungal growth. The medium Sabouraud agar, which was developed nearly a century ago by a French mycologist, is still used in many laboratories. Under the best conditions, most pathogenic fungi in cultures grow slowly; some may take 2 to 4 weeks to grow as much as bacteria do in 24 hours.

Classification of Fungi

Fungi are classified according to the nature of the sexual stage in their life cycles. Such classification is complicated by two problems: (1) No sexual cycle has been observed for some fungi, and (2) it is often difficult to match the sexual and asexual stages of some fungi. For instance, one researcher may work out an asexual phase and give the fungus a name; another researcher may work out a sexual phase and give the same fungus a different name. Because the relationship between the sexual and asexual phases is not always apparent, a particular species of fungi may have two names until someone discovers that the two phases occur in the same organism. Another problem is that many fungi look quite different when growing in tissues (yeastlike) and when growing in their natural habitats (filamentous). The ability of an organism to alter its structure when it changes habitats is called **dimorphism** (di-mor'fizm) (Figure 12.10). Dimorphism in fungi has

---

FIGURE 12.9 American elm (*Ulmus americana*) killed by Dutch elm disease.

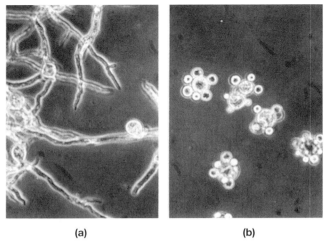

(a)   (b)

FIGURE 12.10 Dimorphism in fungi. (a) Hyphae of *Mucor*. (b) Yeast form of *Mucor*.

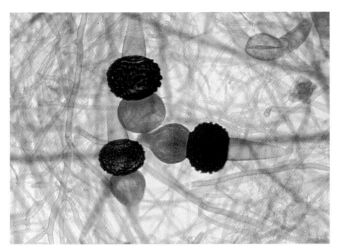

FIGURE 12.11 The black bread mold, *Rhizopus nigricans*, magnification 370×. Sexual zygospores (black, spiny structures) are the result of the joining and fusion of genetic materials at the tips of special hyphal side branches. The zygospores germinate to produce a sporangium that, in turn, produces many asexual spores.

complicated the problem of identifying causative agents in fungal diseases. We will consider bread molds, sac fungi, club fungi, and the so-called Fungi Imperfecti, which are believed to have lost their sexual cycle (Table 12.2).

Bread Molds

The **bread molds, Zygomycota** or conjugation fungi, have complex mycelia composed of hyphae (lacking septa) with chitinous walls. The black bread mold, *Rhizopus* (Figure 12.11), has hyphae that grow rapidly along a surface and into the substratum. Some bread-mold hyphae produce spores that are easily carried by air currents. When the spores reach an appropriate substratum, they germinate to produce new hyphae. Sometimes short branches of the hyphae of two different strains, called plus and minus strains, grow together. This joining of hyphae gave rise to the name conjugation fungi. Chemical attractants are involved in attracting hyphae to each other. Multinucleate cells form where the hyphae join, and many pairs of plus and minus nuclei fuse to form zygotes. Each zygote is enclosed in a **zygospore,** a thick-walled, resistant structure that also produces spores. Genetic information in

zygospores comes from two strains, whereas that in hyphal spores comes from a single strain.

Although bread molds interest mycologists and frustrate bacteriologists whose cultures they contaminate, they usually do not cause human disease. *Rhizopus,* however, is an opportunistic human pathogen; it is especially dangerous to people with diabetes mellitus that is not well controlled.

Sac Fungi

The **sac fungi** are a diverse group containing over 30,000 species, including yeasts. Sac fungi have chitin in their cell walls and produce no flagellated spores. With the exception of some yeasts, which do not form hyphae, the hyphae of sac fungi have septa with a central pore. These fungi are properly called **Ascomycota** (as'ko-mi-ko"ta); unlike other fungi, they produce a saclike **ascus** (plural: *asci*) during sexual reproduction (Figure 12.12). Yeasts are included among the ascomycetes, even though most yeasts have no known sexual stage. In species that reproduce both sexually and asexually, the

TABLE 12.2	Properties of Fungi		
Phylum	**Common Name**	**Characteristics**	**Examples**
Zygomycota	Bread molds	Display conjugation	*Rhizopus* and other bread molds
Ascomycota	Sac fungi	Produce asci and ascospores during sexual reproduction	*Neurospora, Penicillium, Saccharomyces,* and other yeasts; *Candida, Trichophyton,* and several other human pathogens
Basidiomycota	Club fungi	Produce basidia and basidiospores	*Amanita* and other mushrooms; *Claviceps* (which produces ergot); *Cryptococcus*
Deuteromycota	Fungi Imperfecti	Sexual stage nonexistent or unknown	Soil organisms, various human pathogens

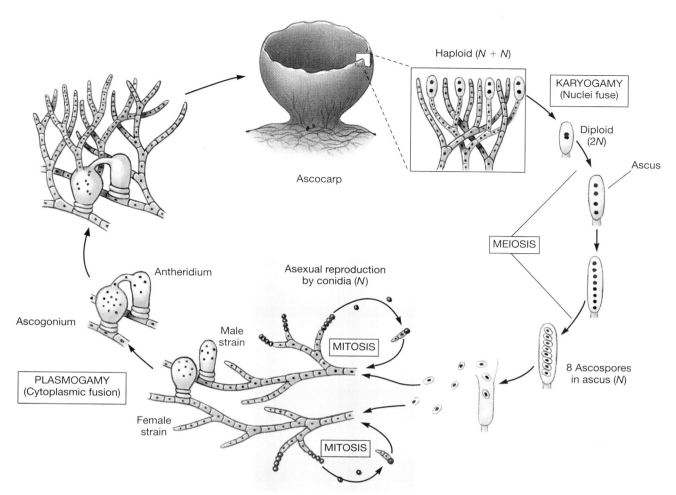

FIGURE 12.12 The life cycle of an ascomycete. In the asexual phase, spores called conidia are formed at the tips of modified hyphae. In the sexual phase, the mycelium that produces conidia also forms the gamete-producing structures, antheridia (male) and ascogonia (female). After cytoplasmic fusion of those structures occurs, dikaryotic hyphal cells develop and interweave into an ascocarp. There saclike asci grow. In each ascus, the dikaryotic nuclei fuse to form a zygote, and the zygote nucleus divides into eight nuclei. From them, eight ascospores form and are forcefully released.

asexual phase forms spores called **conidia** at the ends of modified hyphae. In the sexual phase, one strain has a large *ascogonium,* and an adjacent strain has a smaller *antheridium.* These structures fuse; their nuclei mingle, and hyphal cells with dikaryotic nuclei grow from the fused mass. Eventually, dikaryotic nuclei fuse to form a zygote, and the zygote nucleus divides to form eight nuclei in each ascus. Each ascus forms eight **ascospores,** sometimes releasing them forcefully.

Several sac fungi are of interest in microbiology. *Neurospora* is significant because studies of its ascospores have provided important genetic information. *Penicillium notatum* produces the antibiotic penicillin. *P. roquefortii* and *P. camemberti* are responsible for the color, texture, and flavor of Roquefort and Camembert cheeses. Yeasts, especially those of the genus *Saccharomyces,* release carbon dioxide and alcohol as metabolic products of fermentation and are

used to leaven bread and to make alcohol in beer and wine (Chapter 27). A number of sac fungi are human pathogens. *Candida albicans* causes vaginal yeast infections. *Trichophyton* is associated with athlete's foot, and *Aspergillus* with opportunistic respiratory infections. Species of *Blastomyces* and *Histoplasma* cause respiratory infections and can spread throughout the body.

Club Fungi

The **club fungi** include mushrooms, toadstools, rusts, and smuts. The rusts and smuts parasitize plants and cause significant crop damage. In addition to having hyphae aggregated to form mycelia, the club fungi have club-shaped sexual structures called **basidia,** from which the name **Basidiomycota** is derived (Figure 12.13). In a typical basidiomycete life cycle,

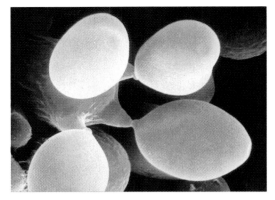

FIGURE 12.13 (a) The gills on the bottom of a mushroom (*Leucoagaricus naucinus*) cap have microscopic, club-shaped structures called basidia. (b) Each basidium (of *Psilocybe mexicana*, 225×) produces four balloonlike structures called basidiospores.

(a)

(b)

CLOSE-UP

ARE THE BIGGEST AND OLDEST ORGANISMS ON EARTH FUNGI?

Weighing about 100 tons (more than a blue whale) and extending through nearly 40 acres of soil in a forest near Crystal Falls in upper Michigan is a gigantic individual of the fungus *Armillaria bulbosa*. Between 1500 and 10,000 years ago, most likely at the end of the last Ice Age, a single pair of compatible spores blew in from parent mushrooms, germinated, and mated. They began growth that continues today. The fungus grows primarily under the soil, so it is usually not visible to the casual observer. The hyphae of its mycelium probe through the soil, seeking woody debris to decompose and recycle. Experimental measurements of its growth rate through soil enabled scientists to estimate the time required to reach its present size.

DNA analysis of 12 genes from the organism's fruiting structures—commonly called *button* or *honey mushrooms*—and its stringlike underground colonizing structures—called *rhizomorphs*—revealed the huge fungus to be a giant clone. All parts of the clone are identical in genetic composition. Although there are minor breaks in its continuity, it is still regarded as a single individual.

Despite the massive size of this fungus, its discoverers, Myron Smith and James Anderson of the University of Toronto, and Johann Bruhn of Michigan Technological University, predicted that it might not be the largest organism of its kind. Writing in the journal *Nature* in April 1992, they explained that they found the fungus in a mixed forest containing many kinds of trees. In a single-type forest such as a large stand of birch or aspen, a fungus with a preference for that type of tree could reach even greater size. This one, however, has probably reached its maximum size, as it collides with competing fungi along its borders.

The scientists' prediction quickly proved to be prophetic. About a month after the *Nature* article was published, two forest pathologists—Ken Russell of the State

These mushrooms are only a small part of the immense 100-ton fungal organism *Armillaria bulbosa* extending throughout nearly 40 acres of a Michigan forest.

Department of Natural Resources and Terry Shaw of the U.S. Forest Service—announced that they had been studying an even larger fungus near Mount Adams in southwestern Washington. This organism, an individual of *Armillaria ostoyae*, covers 1500 acres (about 2.5 square miles), making it almost 40 times as large as the Michigan fungus. The Washington fungus grows in a region populated largely by a single type of tree—in this case, pine—and therefore enjoys a vast source of nourishment. Although the Washington fungus dwarfs its Michigan counterpart in size, it is actually younger, having an estimated age of 400 to 1000 years. Thus, the Michigan fungus retains the title "oldest" (at least for now) but not "largest."

Will scientists eventually discover fungi that are even bigger than the Washington fungus? Most likely, yes. In fact, in an interview, Shaw referred to an *A. ostoyae* in Oregon that might be larger than the one he discovered in Washington. And still bigger ones may remain to be found. The search for the "biggest and oldest" promises to be an exciting episode in the field of microbiology.

SPORE PRINTS

Mushroom identification often requires you to know something about the spores of your unknown specimen. Some keys for mushroom identification are arranged according to color of spores. How do you obtain such information? Try this simple method of obtaining spores for study. Generations of children, as well as professional scientists, have enjoyed making spore prints.

First, collect fresh mushrooms whose caps are just opening—that is, whose undersides are pulling away from the stem—or that are fully open. Cut off the stem flush with the bottom of the cap. Place the cap, gill-side down, on a piece of paper in a place where it can remain undisturbed overnight or for a few days to dry. When you return, gently lift the cap and see the sunburstlike pattern of spores that have been shed from the surfaces of the gills. If you have two mushrooms of the same variety, use a dark piece of paper under one and white paper under the other before the drying process, because you won't know what color spores to expect. Spores may range in color from black to white, tan, or even pink. The spore prints may be made permanent by gently spraying with clear lacquer. Before you spray the spores, scrape some off onto a wet mount to examine them under a microscope.

A spore print made from the mushroom *Psathyrella foenisecii.*

sexual spores called **basidiospores** germinate to form septate mycelia, and cells of mycelia unite into dikaryotic forms. The dikaryotic mycelium grows and produces basidia, which in turn produce basidiospores. Some mushrooms, such as *Amanita,* produce toxins that can be lethal to humans. *Claviceps purpurea,* a parasite of rye, produces the toxic substance ergot. This substance can be used in small quantities to treat migraine headaches and induce uterine contractions, but in larger quantities it can kill (Chapter 22). The yeast *Cryptococcus* causes opportunistic respiratory infections, which can be fatal if they spread to the central nervous system, causing meningitis and brain infection. This organism is increasingly being seen in AIDS patients.

Fungi Imperfecti

The **Fungi Imperfecti,** or **Deuteromycota,** are called "imperfect" because no sexual stage has been observed in their life cycles. Without information on the sexual cycle, taxonomists cannot assign them to a taxonomic group. However, by their vegetative characteristics and the production of asexual spores, most of these fungi seem to belong with the sac fungi. Many of the Fungi Imperfecti have recently been placed in other phyla and given new genus names. We have kept the older designations, however, because the new ones are not yet familiar or widely used in clinical work.

HELMINTHS

Characteristics of Helminths

Helminths, or worms, are bilaterally symmetrical—that is, they have left and right halves that are mirror images. A helminth also has a head and tail end, and its tissues are differentiated into three distinct tissue layers: ectoderm, mesoderm, and endoderm. Helminths that parasitize humans include flatworms and roundworms (Table 12.3).

Flatworms

Flatworms (Platyhelminthes) are primitive worms usually no more than 1 mm thick, but some, such as large tapeworms, can be as long as 10 m. Flatworms lack a **coelom** (se'lom), a cavity that lies between the digestive tract and the body wall in higher animals. Most flatworms have a simple digestive tract with a single opening, but some parasitic flatworms, the tapeworms, have lost their digestive tracts. Most flatworms are hermaphroditic, each individual having both male and female reproductive systems. They have an aggregation of neurons in the head end, representing an early stage in the evolution of a brain. Flatworms lack

TABLE 12.3	Properties of Helminths	
Group	**Characteristics**	**Examples**
Flatworms	Worms live in or on hosts	*Taenia* and other tape worms are internal parasites; flukes can be internal or external parasites.
Roundworms (nematodes)	Most worms live in the intestine or circulatory system of hosts	Hookworms, pinworms, and several other roundworms live in intestines or lymph system.

circulatory systems, and most absorb nutrients and oxygen through their body walls.

More than 15,000 species of flatworms have been identified. They include free-living, mostly aquatic organisms such as *planarians* and two groups of parasitic organisms, the **flukes** (*trematodes*) and the **tapeworms** (*cestodes*). Both parasitic groups have highly specialized reproductive systems and suckers or hooks by which they attach to their host. The flukes can be internal or external parasites. *Fasciola hepatica* and several other flukes parasitize humans. Tapeworms parasitize the small intestine of animals almost exclusively but occasionally occur in the eye or brain. The beef tapeworm, *Taenia saginata*, and several other worms parasitize humans.

Roundworms

Roundworms, or **nematodes,** share many characteristics with the flatworms, but they have a **pseudocoelom,** a primitive, fluid-filled body cavity that lacks the complete lining found in higher animals. The roundworms have cylindrical bodies with tapered ends and are covered with a thick, protective cuticle. They vary in length from less than 1 mm to more than 1 m. Contractions of strong muscles in the body wall exert pressure on the fluid in the pseudocoelom and stiffen the body. Pointed ends and stiff bodies allow roundworms to move through soil and tissues easily. Roundworm females are larger than males. Breeding is enhanced by chemical at-

tractants released by females that attract males. Females can lay as many as 200,000 eggs per day. The large number of eggs, well protected by hard shells, assures that some will survive and reproduce.

Over 80,000 species of roundworms have been described. They occur free-living in soil, fresh water, and salt water and as parasites in every plant and animal species ever studied. A single acre of soil can contain billions of roundworms. Many parasitize insects and plants; only a relatively small number of species infect humans, but they cause significant debilitation, suffering, and death. Most roundworms that parasitize humans, such as hookworms and pinworms, live mainly in the intestinal tract, but a few, such as *Wuchereria*, have larval forms that live in blood or lymph. The effects of roundworms on humans were first recorded in ancient Chinese writings and have been recorded by nearly every civilization since then. (For an account of modern American experiences with sushi and other forms of raw fish, which may harbor roundworms, see the box "Sushi" in Chapter 22.)

Parasitic Helminths

We will concern ourselves only with parasitic helminths and consider four groups: flukes, tapeworms, adult roundworms of the intestine, and roundworm larvae (Figure 12.14). Because helminths have complex life cycles related to their ability to cause diseases, we consider a typical life cycle for each group.

FIGURE 12.14 Representative helminths. (a) *Clonorchis sinensis,* the Chinese liver fluke, stained to show internal organs. It infests the gallbladder, bile ducts, and pancreatic ducts, where it causes biliary cirrhosis and jaundice. (b) Head (scolex) of a tapeworm (220×). The hooked spines and suckers are used for attachment to intestinal surfaces. (c) Mouth of the Old World hookworm *Ancylostoma duodenale*. The muscular pharynx of this roundworm pumps blood from the intestinal lining of its host. (d) The microfilarial (miniature larval) stage of the heartworm *Dirofilaria immitis,* in a sample of dog blood (370×), is transmitted by mosquito bites. The larger stages live inside the heart and perforate its walls.

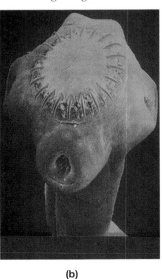
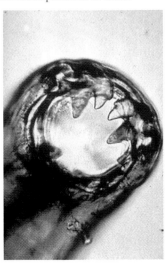
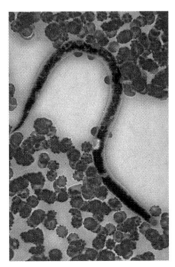

(a)　　　　　(b)　　　　　(c)　　　　　(d)

Flukes

Two types of fluke infections occur in humans. One involves tissue flukes, which attach to the bile ducts, lungs, or other tissues; the other involves blood flukes, which are found in blood in some stages of their life cycle. Tissue flukes that parasitize humans include the lung fluke, *Paragonimus westermani,* and the liver flukes, *Clonorchis sinensis* (Figure 12.14a) and *Fasciola hepatica.* Blood flukes include various species of the genus *Schistosoma.*

Parasitic flukes have a complex life cycle (Figure 12.15), often involving several hosts. The fusion of male and female gametes produces fertilized eggs that become encased in tough shells during their passage through the female fluke's uterus. The eggs pass from the host with the feces. When the eggs reach water, they hatch into free-swimming forms called **miracidia** (mi''ra-sid'e-ah). The miracidia penetrate a snail or other molluskan host, become **sporocysts,** and migrate to the host's digestive gland. The cells inside the sporocysts typically divide by mitosis to form **rediae** (re'de-e). Rediae, in turn, give rise to free-swimming **cercariae** (ser-ka're-e), which escape from the mollusk into water. Using enzymes to burrow through exposed skin, cercariae penetrate another host (often an arthropod) and then encyst as **metacercariae.** When this host is eaten by the definitive host, the metacercariae excyst and develop into mature flukes in the host's intestine.

Tapeworms

Tapeworms consist of a **scolex** (sko'lex), or head end (Figure 12.14b), with suckers that attach to the intestinal wall, and a long chain of hermaphroditic **proglottids** (pro-glot'tidz), body components that contain mainly reproductive organs of both sexes. New proglot-

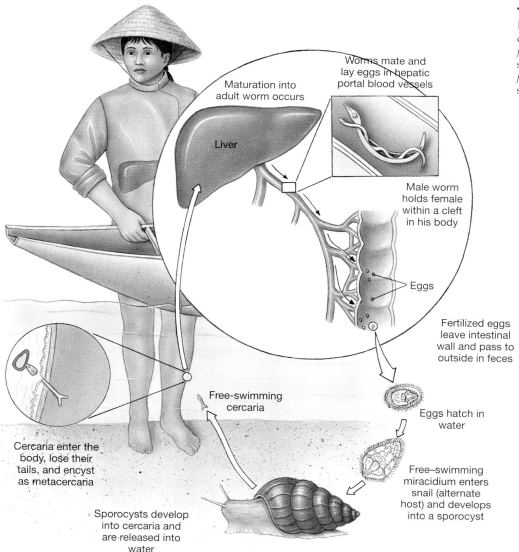

Maturation into adult worm occurs

Liver

Worms mate and lay eggs in hepatic portal blood vessels

Male worm holds female within a cleft in his body

Eggs

Fertilized eggs leave intestinal wall and pass to outside in feces

Free-swimming cercaria

Cercaria enter the body, lose their tails, and encyst as metacercaria

Sporocysts develop into cercaria and are released into water

Eggs hatch in water

Free–swimming miracidium enters snail (alternate host) and develops into a sporocyst

FIGURE 12.15 The life cycle of a blood fluke, *Schistosoma japonicum,* which causes schistosomiasis. Unlike some flukes, *S. japonicum* does not have a rediae stage or enter an arthropod host.

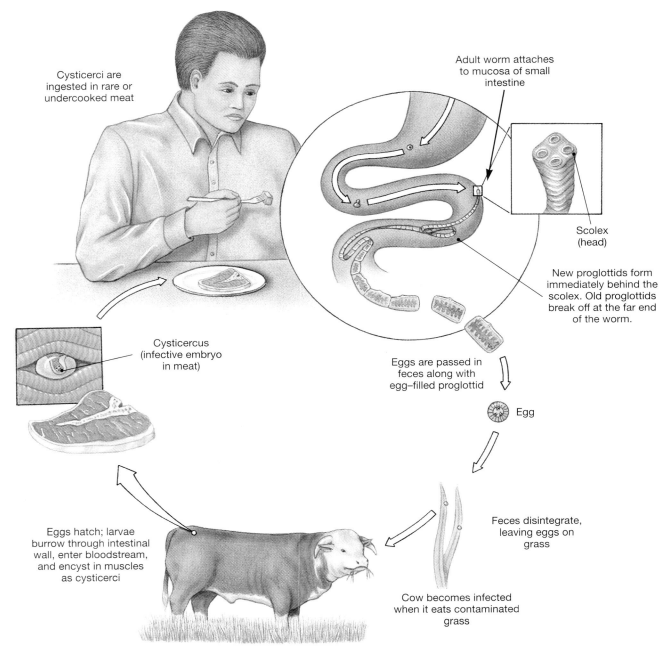

Cysticerci are ingested in rare or undercooked meat

Adult worm attaches to mucosa of small intestine

Scolex (head)

New proglottids form immediately behind the scolex. Old proglottids break off at the far end of the worm.

Cysticercus (infective embryo in meat)

Eggs are passed in feces along with egg–filled proglottid

Egg

Eggs hatch; larvae burrow through intestinal wall, enter bloodstream, and encyst in muscles as cysticerci

Feces disintegrate, leaving eggs on grass

Cow becomes infected when it eats contaminated grass

FIGURE 12.16 The life cycle of the beef tapeworm, *Taenia saginata.*

tids develop behind the scolex, mature, and fertilize themselves. Old ones disintegrate and release eggs at the rear end. Among the tapeworms that can infect humans are beef and pork tapeworms that are species of *Taenia,* dwarf and rat tapeworms that are species of *Hymenolepis,* the hydatid worm *Echinococcus,* the dog tapeworm *Dipylidium,* and the broad fish tapeworm *Diphyllobothrium.*

Although different species display minor variations, the life cycle of tapeworms (Figure 12.16) usually includes the following stages: Embryos develop inside eggs and are released from proglottids; the proglottids and eggs leave the host's body with the feces. When another animal ingests vegetation or water contaminated with eggs, the eggs hatch into larvae, which invade the intestinal wall and can migrate to other tissues. A larva can develop into a **cysticercus** (sis-ti-ser′kus), or bladder worm, or it can form a cyst. A cysticercus can remain in the intestinal wall or migrate through blood vessels to other organs. A cyst can enlarge and develop many

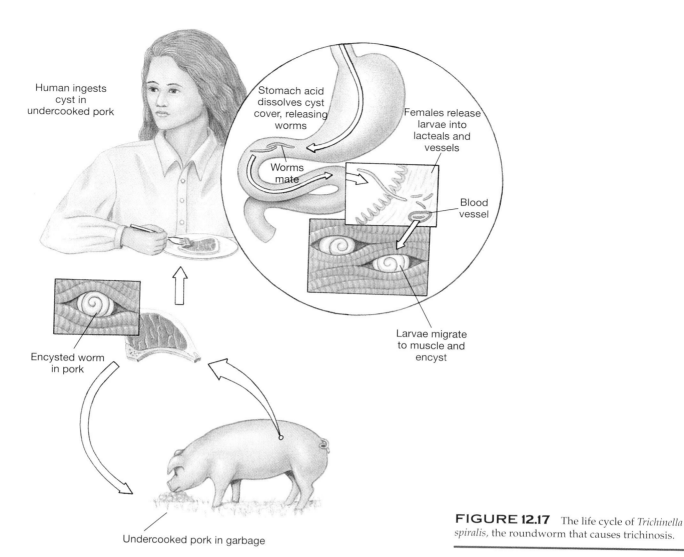

Human ingests cyst in undercooked pork

Stomach acid dissolves cyst cover, releasing worms

Females release larvae into lacteals and vessels

Worms mate

Blood vessel

Larvae migrate to muscle and encyst

Encysted worm in pork

Undercooked pork in garbage

FIGURE 12.17 The life cycle of *Trichinella spiralis,* the roundworm that causes trichinosis.

tapeworm heads within it, becoming a **hydatid** (hi-da′tid) **cyst** (Chapter 22). If an animal eats flesh containing such a cyst, each scolex can develop into a new tapeworm.

Adult Roundworms

Most roundworms that parasitize humans live much of their life cycle in the digestive tract. They usually enter the body by ingestion with food or water, but some, such as the hookworm, penetrate the skin. These helminths include the pork roundworm *Trichinella spiralis,* the common roundworm *Ascaris lumbricoides,* the guinea worm *Dracunculus medinensis,* the pinworm *Enterobius vermicularis,* and the hookworms *Ancylostoma duodenale* (Figure 12.14c) and *Necator americanus.*

The life cycles of intestinal roundworms show considerable variation. We use the life cycle of *Trichinella*

spiralis as an example (Figure 12.17). These worms enter humans as encysted larvae in the muscle of infected pigs when undercooked pork is eaten. The cyst walls are digested with the meat, and the larvae are released into the intestine. They mature sexually in about 2 days and then mate. Females burrow into the intestinal wall and produce eggs that hatch inside the adult worm and emerge as larvae. The larvae migrate to lymph vessels and are carried to the blood. From the blood the larvae burrow into muscles and encyst. These cysts can remain in muscles for years. The same processes occur in the pigs themselves, so cysts are present in their tissues.

Roundworm Larvae

Whereas most roundworms cause much of their tissue damage as adults in the intestine, some cause their damage mainly as larvae in other tissues. These round-

MICROBIOLOGIST'S NOTEBOOK

Carter Center Expects Eradication of Guinea Worm from World by 1996

Former President Jimmy Carter: Once you've seen a small child with a 2- or 3-foot-long live Guinea worm protruding from her body, right through her skin, you never forget it. I first saw the devastating effects of Guinea worm disease [dracunculiasis] in two villages near Accra, Ghana, in March 1988. In just a few minutes, Rosalynn and I saw nearly 200 victims, including people with worms coming out of their ankles, knees, groins, legs, arms, and other parts of their bodies. One woman, in great agony, was cradling her breast as if it were an infant. On it was an abscess the size of a fist where a Guinea worm was about to emerge. I have seen people with as many as a dozen or more of these worms emerging at the same time. I was shocked to find that this debilitating disease, which strikes nearly 10 million people each year, could be easily prevented. The disease is caused by contaminated drinking water; prevention is simply a matter of showing people how to make their water supply safe.

We are a compassionate nation, and that is one reason why those people still suffering needlessly from this disease should matter to us. The world is full of difficult problems that we cannot yet solve. This is one we can solve, and we can do it quickly if we put our minds to it. Here in Atlanta, The Carter Center's Global 2000 program is targeting December 31, 1995, as the date for global eradication of Guinea worm. Dr. Donald Hopkins, Senior Consultant for Global 2000's health programs, directs the Center's Guinea worm eradication efforts. He has helped enlist the support of governments and nonprofit organizations in this fight, and he has helped implement eradication programs in several African countries.

Dr. Donald Hopkins: The international community didn't realize until about 1980 that it was possible to eradicate Guinea worm completely. In fact, they didn't even realize the extent of the problem in their own countries, especially in rural areas. Before 1988, Ghana reported an average of 4500 cases of Guinea worm to the World Health Organization each year. Then in 1989, Ghana conducted its first nationwide search for cases and as a result reported more than 170,000 cases. A UNICEF study in southeastern Nigeria estimated that the rice farmers in that area of 1.6 million people are losing $20 million each year in potential profits because so many of them are crippled by Guinea worm at harvest time. In one heavily affected district of Ghana's northern region, the production of yams reportedly increased by 33 percent during the first 9 months of 1991 because so many farmers were restored to full productivity by the sharp reduction of Guinea worm.

Mr. Carter: The value is not just in terms of money. Healthy mothers can take better care of their children. Children can get a better education. Guinea worm disease causes children to miss an average of 12 weeks of school in a year when they are infected, compared to a total loss of 1 to 2 weeks lost due to all other causes combined. Often children are so badly behind when they return to school, especially if this happens several years in a row, that they become permanent dropouts. Additional income means that people can improve their homes and do all kinds of things they couldn't do before. They become more self-reliant. Tackling the next problem becomes easier when they have pride in the success of this effort.

Guinea worm disease is contracted by drinking from ponds, step wells, cis-

Former President Jimmy Carter, on a trip to Africa, examining a water hole contaminated with Guinea worms.

terns and other sources of stagnant water that have been contaminated by the worm larvae. Guinea worm, *Dracunculus medinensis,* affects only humans, and it actually uses its human host to further its life cycle. Contaminated water contains water fleas that have eaten immature Guinea worm larvae. The larvae escape when the digestive juices in the person's stomach kill the flea. The larvae penetrate the stomach wall, wander around the abdomen, mature in a few months, and mate, after which the male worms die. It is only the female worm that grows to 2 or 3 feet in length and, about a year later, secretes a toxin that causes a blister on the skin. When the blister ruptures, usually when the infected part of the body is immersed in cool water, the worm starts to emerge. This process can take 30 to 100 days before the worm finally finishes making its way out of the body. When an infected person enters the village pond or watering hole, the worm discharges hundreds of thousands of tiny larvae into the water, beginning the cycle again.

Dr. Hopkins: A small incision made before the emergence blister is raised allows the worm to be wound out gradually, wrapped around a stick. This may have been the origin of the symbol of the medical profession, the caduceus, a serpent coiled around a stick. Many scholars think the Guinea worm is the "fiery serpent" of the Bible. It takes several weeks of daily gentle winding to complete the removal of a worm. If the worm breaks and dies, it will decompose inside the host, causing festering and infection. If a portion of it retracts into the tissues, it can carry tetanus spores back with it, leading to fatal tetanus disease. The local practice in some countries of putting cow dung on the wound makes tetanus especially common. In Upper Volta and Nigeria, Guinea worm is the third leading cause of acquiring tetanus. Other types of microorganisms can also enter the wound, and secondary infections are frequent, even if tetanus is avoided. If the worm emerges near a major joint, permanent scarring leads to stiff and crippled joints. One man died of starvation when a worm came out under his tongue and he couldn't eat. Although most worms emerge from the lower limbs, they can be found anywhere: scrotum, scalp, chest, face.

Imagine the suspense of living in an area where Guinea worms abound. Will you have worms again this year? How many will you have? Where will they emerge? Will you become crippled, infected, or die of tetanus? Will you be able to go to work or school? Transmission season is only during wet periods. In sub-Saharan Africa these are the months of June, July, and August. The worms emerge about 12 months later during the next wet season. Years when there are droughts help reduce the number of infections.

Mr. Carter: There is no cure, only prevention. No matter how many times you have been infected with Guinea worms, no immunity to them develops. Prevention involves providing safe sources of drinking water such as borehole wells, which people with emerging worms cannot enter to contaminate. It can also be prevented by teaching villages to boil their water (although many cannot afford enough fuel to do so) or to filter their drinking water through a clean, finely woven cloth. The E. I. DuPont Company, in association with Precision Fabrics Group, donated 1.4 million monofilament nylon cloth filters to The Carter Center for use in the eradication campaign. The American Cyanamid Company has donated over $2 million of the larvicide Abate, which is used to treat water safely, killing the water fleas and Guinea worms but harming nothing else. The company has pledged to continue donating Abate until the goal of eradication is reached. Using the nylon filters and Abate, the two villages I visited in Ghana in March 1988 reduced the number of cases of Guinea worm there by more than 90 percent in one year.

Dr. Hopkins: Money is the limiting factor in our fight against the Guinea worm. We have the training and the manpower. Mr. and Mrs. Carter have been very active, making several trips to Africa to talk with leaders and to go into the villages and explain how the worm is transmitted and to teach people how to break that chain of transmission. This is a problem that affects 19 nations and that will probably require $65 million to eradicate. The countries themselves cannot afford the entire cost themselves, so help has to come from wealthier nations. The Japanese have put in 191 wells. The Swedes, Dutch, and Danes are the only European countries helping.

I was part of the international team that eradicated smallpox. It is exciting to be working on another eradication project that is so close to success. We do expect to make our target date of the end of 1995. We believe we succeeded in totally eliminating Guinea worm from Pakistan by the end of 1991. So far we haven't heard of any new cases there.

Mr. Carter: A while back, I was asked what I felt my top presidential achievements were. Of course there were things such as the Panama Canal treaties, the Camp David accord between Israel and Egypt, and normalization of relations with China. But, if in 1996 we find that where there were 10 million cases of Guinea worm there are none, I would say that eradication of the Guinea worm is more important.

Update

At press, December 31, 1995, is no longer the target date for eradication of the Guinea worm. The eradication program is behind schedule—in large part due to political unrest in participating nations. Ethnic fighting in northern Ghana early in 1994, for example, temporarily slowed down the eradication efforts there. And a 12-year-old civil war threatens the eradication program in Sudan, which now has the highest incidence of dracunculiasis in the world. To allow the Sudanese program to continue, Former President Carter negotiated a "Guinea worm cease-fire" between the government of Sudan and opposing forces. Both sides have agreed to stop fighting long enough to permit accelerated eradication efforts before onset of the rainy season in southern Sudan. The cease fire went into effect on March 29, 1995, and has been extended indefinitely.

Ninety-five percent of all cases of Guinea worm infestation have been eliminated. Most of the last 5 percent of cases are identified and under medical surveillance. Infested persons who step into water sources are fined. Rewards are being offered for reporting new cases. It is believed that the final eradication of the Guinea worm will occur during 1996.

Female Guinea worm emerging from a blister on the foot of a victim.

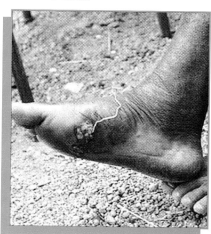

321

worms include *Wuchereria bancrofti,* which lives in lymphatic tissue and causes elephantiasis; *Loa loa,* which infects the eyes and eye membranes; and *Onchocerca volvulus,* which infects both the skin and eyes.

The life cycles of roundworms that parasitize humans as larvae also require a mosquito host (Figure 12.18). These worms enter the human body as immature larvae called **microfilariae** (mi''kro-fi-lar'e-e) with the bite of an infected mosquito. The microfilariae migrate through the tissues to lymph glands and ducts and mature and mate as they migrate. Females produce large numbers of new microfilariae, which enter the blood (Figure 12.14d), usually at night. The microfilariae are ingested by mosquitoes as they bite infected humans. Any one of several species of mosquitoes can serve as host. When the microfilariae reach the midgut of the mosquito, they penetrate its wall and migrate first to the thoracic muscles and

APPLICATIONS

PARASITES ON PARASITES

Parasites can themselves have parasites. A large pinworm, *Heterakis gallinarum,* infects chickens, turkeys, and other birds. Earthworms eat the parasites from bird feces. The parasites cause no disease in the earthworm, but the parasite eggs become infected with a flagellated protozoan, *Histomonas meleagridis.* When a bird eats the earthworm, it becomes infected with both pinworms and the protozoan. The protozoan causes the serious disease histomoniasis in turkeys. Phenothiazine is usually added to chicken and turkey feed to prevent these and other infections. How much of the drug goes into eggs and meat eaten by humans, and what its effects might be, are unknown.

FIGURE 12.18 The life cycle of *Wuchereria bancrofti,* a roundworm that produces microfilariae and causes elephantiasis (a chronic edema; see Figure 23.4), especially of the legs and scrotum.

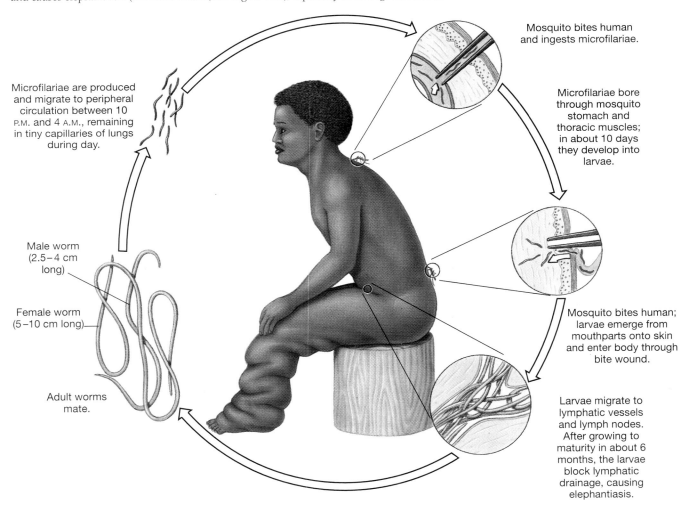

Microfilariae are produced and migrate to peripheral circulation between 10 P.M. and 4 A.M., remaining in tiny capillaries of lungs during day.

Male worm (2.5–4 cm long)

Female worm (5–10 cm long)

Adult worms mate.

Mosquito bites human and ingests microfilariae.

Microfilariae bore through mosquito stomach and thoracic muscles; in about 10 days they develop into larvae.

Mosquito bites human; larvae emerge from mouthparts onto skin and enter body through bite wound.

Larvae migrate to lymphatic vessels and lymph nodes. After growing to maturity in about 6 months, the larvae block lymphatic drainage, causing elephantiasis.

PARASITES IN YOUR YARD?

Does your family pet have parasites? Are the neighbors' dogs and cats leaving parasite-laden feces on your lawn? Parasites are not just figures in a textbook. They may be a very real (but not obvious) part of your daily surroundings. Try the following procedure to discover who may be living in your neighborhood.

Use aseptic technique (so that you won't become an alternate host!): Wear protective gloves and safety glasses. Using disposable glassware, collect a *fresh* fecal sample from an outdoor pet such as a dog, cat, or horse. Carefully mix about the amount that would fit on a thumbnail into 15 ml of a saturated solution of sodium chloride. Strain through several layers of folded cheesecloth or gauze to remove solids. Carefully pour or transfer by bulb pipette some of the strained fluid to a very small test tube; fill it to the top without letting it overflow. Next, place a clean coverslip over the top of the test tube so that it is in contact with the surface of the liquid. Allow it to sit undisturbed for 15 minutes. During this time, parasite eggs, if present, will float to the top of the tube and adhere to the underside of the coverslip. Using forceps, transfer the coverslip, wet side down, to a slide. Examine this wet mount under low-power (10×) and high-power (40×) objective lenses. How many kinds of helminth eggs can you find? Try to identify them with the help of diagrams in veterinary medicine texts. Carefully dispose of all contaminated materials.

then to the mosquito's mouthparts. There they can be transferred to a new human host, where the cycle is repeated.

ARTHROPODS

Characteristics of Arthropods

Arthropods constitute the largest group of living organisms; as many as 80 percent of all animal species belong to the phylum Arthropoda. Arthropods are characterized by jointed chitinous exoskeletons, segmented bodies, and jointed appendages associated with some or all of the segments. The name arthropod is derived from *arthros,* joint, and *podos,* foot. The exoskeleton both protects the organism and provides sites for the attachment of muscles. These organisms have a true coelom, which is filled with fluid that supplies nutrients, as blood does in higher organisms. Arthropods have a small brain and an extensive network of nerves. Various groups have different structures that extract oxygen from air or from aquatic environments. The sexes are distinct in arthropods, and females lay many eggs. Arthropods are found in nearly all environments—free-living in soil, on vegetation, in fresh and salt water, and as parasites on many plants and animals.

Classification of Arthropods

Certain members of three subgroups (classes) of arthropods, the arachnids, insects, and crustaceans (Table

TABLE 12.4	Properties of Three Classes of Arthropods	
Group	Identifying Characteristic	Examples
Arachnids	Have eight legs	Spiders, scorpions, ticks, mites
Insects	Have six legs	Lice, fleas, flies, mosquitoes, true bugs
Crustaceans	A pair of appendages on each body segment	Crabs, crayfish, copepods

12.4), are important either as parasites or as disease vectors (Figure 12.19). The diseases transmitted by arthropods are summarized in Table 12.5.

Arachnids

Arachnids have two body regions—a cephalothorax and an abdomen, four pairs of legs, and mouthparts that are used in capturing and tearing apart prey. They include spiders, scorpions, ticks, and mites. Spider bites and scorpion stings can produce localized inflammation and tissue death, and their toxins can produce severe systemic effects. Ticks and mites are external parasites on many animals; some also serve as vectors of infectious agents.

Infected ticks transmit several human diseases. Certain species of *Ixodes* carry viruses that cause en-

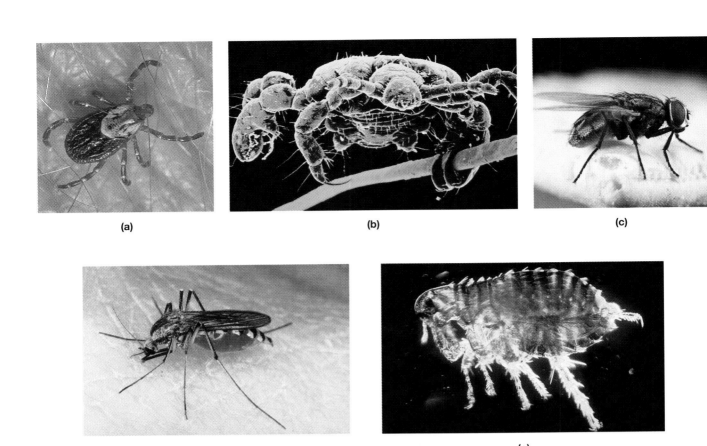

FIGURE 12.19 Representative arthropods that are parasitic or can serve as disease vectors. (a) A wood tick, *Dermacentor andersoni*. (b) False-color SEM photo of the pubic louse, *Phthirus pubis*, also known as a crab louse, clinging to a human pubic hair (55×). The lice suck blood, feeding about five times a day. (c) The housefly, *Musca domestica* (4×), can carry microbes on its body. (d) The *Aedes* mosquito. (e) A flea, *Ctentocephalidis canis* (33×).

TABLE 12.5	**Diseases Transmitted by Arthropods**		
Disease	**Causative Agents**	**Principal Vectors**	**Endemic Areas**
Plague	*Yersinia pestis*	Fleas	Only sporadic in modern times; reservoir of infection maintained in rodents
Tularemia	*Francisella tularensis*	Fleas and ticks	Western United States
Salmonellosis	*Salmonella* species	Flies	Worldwide
Lyme disease	*Borrelia burgdorferi*	Ticks	Parts of United States, Australia, and Europe
Relapsing fever	*Borrelia* species	Ticks and lice	Rocky Mountains and Pacific Coast of United States; many tropical and subtropical regions
Typhus fever	*Rickettsia prowazekii*	Lice	Asia, North Africa, and Central and South America
Tick-borne typhus fever	*Rickettsia conorii*	Ticks	Mediterranean area; parts of Africa, Asia, and Australia
Scrub typhus fever	*Rickettsia tsutsugamushi*	Mites	Asia and Australia
Murine typhus fever	*Rickettsia typhi*	Fleas	Tropical and subtropical regions
Rocky Mountain spotted fever	*Rickettsia rickettsii*	Ticks	United States, Canada, Mexico, and parts of South America

cephalitis and the spirochete *Borrelia burgdorferi,* which causes Lyme disease. The common tick *Dermacentor andersoni,* which can cause tick paralysis, can also carry the viruses that cause encephalitis and Colorado tick fever, the rickettsias that cause Rocky Mountain spotted fever, and the bacterium that causes tularemia. Several species of *Amblyoma* ticks also carry the Rocky Mountain spotted fever rickettsia, and *Ornithodorus* ticks transmit the spirochete responsible for relapsing fever. Mites serve as vectors for the rickettsial diseases scrub typhus fever and Q fever.

Insects

Insects have three body regions—head, thorax, and abdomen; three pairs of legs; and highly specialized mouthparts. Some insects have specialized mouthparts for piercing skin and for sucking blood and can inflict painful bites. Insects that can serve as vectors of disease include all lice and fleas and certain flies, mosquitoes, and true bugs, such as bedbugs and reduviid bugs. Although we often refer to all insects as "bugs," entomologists—scientists who study insects—use the term *true bug* to refer to certain insects that typically have thick, waxy wings and sucking, rather than biting, mouthparts.

The body louse is the main vector for the rickettsias that cause typhus and trench fevers and a spirochete that causes relapsing fever. (This spirochete is a different species of *Borrelia* from the one carried by ticks.) Epidemics of all louseborne diseases usually occur under crowded, unsanitary conditions. All louseborne disease agents enter the body when louse feces are scratched into bite wounds.

The human flea, *Pulex irritans,* lives on other hosts and can transmit plague. However, fleas that normally parasitize rats and other rodents are more likely to transmit plague to humans. This bacterial disease still occurs in the United States in individuals who have had contact with wild rodents and their fleas.

Several kinds of flies feed on humans and serve as vectors for various diseases. The common housefly, *Musca domestica,* is not part of the life cycle of any pathogens, yet it is an important carrier of any pathogens found in feces. This fly is attracted to both human food and human excreta, and it leaves a trail of bacteria, vomit, and feces wherever it goes. Other insects, such as blackflies, serve as vectors for *Onchocerca volvulus.* Sandflies serve as vectors for leishmanias, for bacteria that cause bartonellosis, and for viruses that cause sandfly fever and several other diseases. Tsetse flies are vectors for trypanosomes that cause African sleeping sickness, and deer flies are vectors for the worm that causes loaiasis. Eye gnats, which look like tiny houseflies, may be responsible for transmission of bacterial conjunctivitis and the spirochete that causes yaws.

Many species of mosquitoes serve as vectors for diseases. *Culex pipiens,* a common mosquito, breeds in any water and feeds at night. It is a vector for *Wuchereria.* Another mosquito, *C. tarsalis,* breeds in water in sunny locations and also feeds at night. It is

Disease	Causative Agents	Principal Vectors	Endemic Areas
Q fever	*Coxiella burnetii*	Ticks and mites	Worldwide
Trench fever	*Rochalimaea quintana*	Lice	Known only in fighting armies
Viral encephalitis	Togaviruses	Mosquitoes	Worldwide but varies by virus and vector
Yellow fever	Togavirus	Mosquitoes	Tropics and subtropics
Dengue fever	Togavirus	Mosquitoes	India, Far East, Hawaii, Caribbean Islands, and Africa
Sandfly fever	A virus, probably of bunyavirus family	Female sandfly	Mediterranean region, India, and parts of South America
Colorado tick fever	An orbivirus	Ticks	Western United States
Tick-borne encephalitis	Various viruses	Ticks	Europe and Asia
African sleeping sickness	Trypanosomes	Tsetse fly	Africa
Chagas' disease	*Trypanosoma cruzi*	True bug	South America
Kala azar and other leishmaniases	*Leishmania* species	Sandfly	Tropical and subtropical regions

a vector for viruses that cause western equine encephalitis (WEE) and St. Louis encephalitis. Although WEE most often causes severe illness in horses, it also can cause severe encephalitis in children and a milder disease with fever and central nervous system infections in adults. (The latter form is sometimes called sleeping sickness, but it should not be confused with African sleeping sickness.) Many species of *Aedes* play a role in human discomfort and disease. *Aedes aegypti* is a vector of a variety of viral diseases, including dengue fever (breakbone fever), yellow fever, and epidemic hemorrhagic fever. Several species of *Anopheles* serve as vectors for malaria. They have a variety of breeding habits, and thus control of them requires the application of several different eradication methods.

Several species of reduviid bugs transmit the parasite that causes Chagas' disease, which is a leading cause of cardiovascular disorders in Central and South America. Bedbugs cause dermatitis and may be responsible for spreading one kind of hepatitis, a liver infection.

Crustaceans

Crustaceans are generally aquatic arthropods that typically have a pair of appendages associated with each segment. Appendages include mouthparts, claws, walking legs, and appendages that aid in swimming or in copulation. Crustaceans that are hosts for disease agents that infect humans include some crayfish, crabs, and smaller crustaceans called copepods.

THE WAR AGAINST MALARIA: MISSTEPS AND MILESTONES

Malaria, caused by protozoa of the genus *Plasmodium*, is one of the most serious parasitic infections that afflicts humans. Despite massive efforts to control the spread of this disease, in any given year it strikes up to 500 million people and claims the lives of 1.5 to 3 million a year, many of them children. Malaria is an ancient human scourge. Written records on Egyptian papyrus from 1550 B.C. describe a disease with high intermittent fever that must have been malaria. They refer to the use of tree oils as mosquito repellents—although not until some 30 centuries later was the mosquito proven to be the vector (or carrier) of the disease. In some low-lying ancient cities, nearly entire populations succumbed to the disease. Only the wealthy could afford to "head for the hills" in summer to escape the heat, mosquitoes, and fevers. Many of the Crusaders of medieval times also died of malaria, and slave trading in later years contributed significantly to its spread. The one bright spot in this dismal history was the discovery in the sixteenth century that malaria could be treated with quinine, a drug from the *Chinchona* tree.

A connection between swamps and fevers had long been recognized but was misinterpreted. Most early investigations of malaria's cause focused on air—indeed, the disease was named for the "bad air" (*mal,* bad) thought to be responsible—and water. The association between the disease and mosquitoes suggested in Egyptian records was ignored. By the 1870s, with the advent of the germ theory, some scientists came to believe that malaria was caused by a bacterium, which they named *Bacillus malariae*.

Alphonse Laveran, a French army physician in North Africa, was not convinced that the malaria organism had been found. With unstained preparations and a poor-quality, low-power microscope, he continued to search for it. He eventually found what we now know to be male sex cells of the malarial parasite in human blood. Most scientists rejected Laveran's findings in favor of the *Bacillus* theory, until he demonstrated the male sex cells to Pasteur in 1884. Only then did the scientific community accept that protozoa of the genus *Plasmodium* cause malaria.

The Italian physiologist Camillo Golgi added to Laveran's work in 1885 by identifying several species of *Plasmodium*. And in 1891 Russian researchers developed the Romanovsky staining procedure (methylene blue and eosin) for malarial blood smears. With slight modifications, it is still used.

Although the causative agent of malaria had been found, its transmission was not yet understood. Ronald Ross, a medical officer in India, spent years searching in his spare time for proof that malaria parasites are carried by mosquitoes. Eventually, he found the parasites in *Anopheles* mosquitoes, though he never fully explained the mode of transmission. His efforts were neither appreciated nor supported by his superiors, but in 1902 he was awarded the Nobel Prize in physiology or medicine.

Once the vector for malaria had been identified, researchers turned to controlling transmission of the disease by controlling vector populations. Credit for developing the first effective mosquito control measures is given to the American physician William Crawford Gorgas, chief medical officer for sanitation during the building of the Panama Canal early in the twentieth century. By draining wet areas and instituting the use of mosquito netting, Gorgas significantly reduced the incidence of both malaria and yellow fever among canal workers.

In the 1930s and 1940s, advances in malarial treatment and control led many people to believe that malaria was no longer a threat. However, when World War I soldiers thought to have been cured with quinine treatment suffered relapses after they had left malaria-infested areas, researchers began a new round of investigations. This time, they searched for sites in which the parasites become sequestered in human tissues. In 1938, S. P. James and P. Tate discovered the parasite outside red blood cells in certain birds; in 1948 H. C. Shortt and P. C. C. Garnham made a similar discovery of *P. vivax* in humans. The parasite is now known to disappear from blood circulation, out of the reach of drugs, by invading cells of the liver and other tissues.

The war against malaria continues, but with limited success. Massive insecticide (DDT) spraying in the 1960s appeared to have eradicated the disease from many regions of the world, but DDT-resistant mosquitoes soon emerged. Incidence of the disease increased, sometimes to epidemic proportions. Some strains of the parasite also became resistant to chloroquine, one of the best drugs for treating malaria.

Another factor contributing to malaria's recent comeback may be environmental changes in developing countries. In the late 1950s, for instance, the inhabitants of Kenya's Karo Plain turned from subsistence farming of maize and cattle grazing to a cash-producing rice culture. Rice must grow in wet conditions, so the dry plains were flooded and the cattle were banished from the area. But along with rice, the Karo Plain soon supported *Anopheles* mosquitoes. Attracted by the water and the increased humidity, malaria-carrying mosquitoes came to outnumber non-malaria-carrying mosquitoes 2 to 1. And because the favorite host of the *Anopheles* mosquitoes—cattle—had been removed from the fields, the mosquitoes turned to humans for their meals. The result of all these environmental changes was a frightening increase in malaria incidence. In fact, global warming is predicted to bring about the spread of *P. falciparum,* the species that causes the most life-threatening form of malaria. Computer models predict it will spread to the eastern United States and Canada, to most of Europe, and to Australia.

Several new methods for controlling malaria are currently being studied. One of the most hopeful is a malaria vaccine. Manuel Patarroyo and his colleagues at the Immunology Institute in Bogota, Colombia, have developed a vaccine against *P. falciparum.* Called SPf66, the vaccine is a genetically engineered replica of several of the proteins that make up *P. falciparum.* The "acid test" for SPf66 came in 1993, when Patarroyo took his vaccine to Tanzania, where people are bitten by infected mosquitoes almost daily. Patarroyo reported that his vaccine achieved a protectiveness of 31 percent. (Most successful vaccines, such as those against polio and diphtheria, have an 80 percent efficacy.) Many researchers find Patarroyo's results encouraging. Given malaria's high death toll, even a 31 percent reduction in malaria transmission is noteworthy. A subsequent study in 1995 reported, however, that the new vaccine had no significant effect on episodes of malaria.

Health education and affordable replacements for chloroquine are paramount in controlling malaria. Modern mosquito controls that improve on the DDT "bombings" of the 1960s, such as mosquito nets impregnated with mosquito-killing chemicals, could also be effective. In Gambia, childhood mortality from malaria fell by an astonishing 63 percent after the introduction of these nets. But because a magic bullet has proved elusive, a combination of approaches may be necessary to reduce the transmission of *Plasmodium* to more acceptable levels.

CHAPTER SUMMARY

PRINCIPLES OF PARASITOLOGY

- A **parasite** is an organism that lives at the expense of another organism, the **host. Pathogens** are parasites that cause disease.
- **Parasitology** is the study of parasites, which typically includes protozoa, helminths, and arthropods.

Significance of Parasitism

- Parasites are responsible for much disease and death of humans, plants, and animals and for extensive economic losses.

Parasites in Relation to Their Hosts

- Parasites can live on or in hosts. Parasites can be **obligate** or **facultative** and **permanent, temporary,** or **accidental. Vectors** are agents of parasite transmission.
- Parasites reproduce sexually in **definitive hosts** and spend other life stages in **intermediate hosts. Reservoir hosts** can transmit parasites to humans.
- **Host specificity** refers to the number of different hosts in which a parasite can mature.
- Over time parasites become more adapted to and less destructive of their hosts. Most parasites have mechanisms to evade host defenses and exceptionally adept reproductive capacities.

PROTISTS

Characteristics of Protists

- **Protists** are eukaryotic, and most are unicellular. They can be autotrophic or heterotrophic, and some are parasitic.

Importance of Protists

- Protists are important in food chains as producers and decomposers; they can be economically beneficial or detrimental.

Classification of Protists

- Protists include plantlike organisms (such as **euglenoids, diatoms,** and **dinoflagellates**), funguslike organisms (the **water molds** and **slime molds**), and animal-like organisms (the **protozoa,** such as **mastigophorans, sarcodines, apicomplexans,** and **ciliates**). The groups of protists are summarized in Table 12.1.
- **Saprophytes** are organisms that feed on dead matter.

FUNGI

Characteristics of Fungi

- **Fungi** are saprophytes or parasites that generally have a **mycelium,** a loosely organized mass consisting of threadlike **hyphae.** Most fungi reproduce both sexually and asexually, and their sexual stages are used to classify them.

Importance of Fungi

- Fungi are important as decomposers in ecosystems and as parasites in the health sciences.

Classification of Fungi

- Fungi include **bread molds, sac fungi, club fungi,** and the **Fungi Imperfecti,** which cannot be classified in another group because either they lack a sexual stage or none has yet been identified. The groups of fungi are summarized in Table 12.2.

HELMINTHS

Characteristics of Helminths

- **Helminths,** or worms, are bilaterally symmetrical and have head and tail ends and differentiated tissue layers.

Parasitic Helminths

- Only two groups of helminths, the flatworms and the roundworms (nematodes), contain parasitic species.
- **Flatworms** lack a **coelom,** have a simple digestive tract with one opening, and are **hermaphroditic.** They include **tapeworms** and **flukes.**

- **Roundworms** have a **pseudocoelom,** separate sexes, and a cylindrical body. They include hookworms, pinworms, and other parasites of the intestinal tract and lymphatics.

ARTHROPODS

Characteristics of Arthropods

- **Arthropods** have jointed, chitinous exoskeletons, segmented bodies, and jointed appendages.

Classification of Arthropods

- Parasitic and vector arthropods include some arachnids and insects; a few crustacea also serve as intermediate hosts for human parasites. Arthropod vectors of disease are summarized in Table 12.5.
- **Arachnids** have eight legs; they include scorpions, spiders, ticks, and mites.
- **Insects** have six legs; they include lice, fleas, flies, mosquitoes, and true bugs.
- **Crustaceans** are generally aquatic arthropods, typically with a pair of appendages on each segment; they include crayfish, crabs, and copepods.

KEY TERMS

accidental parasite *(p. 301)*
antibiosis *(p. 310)*
apicomplexan *(p. 306)*
arachnid *(p. 323)*
arthropod *(p. 323)*
Ascomycota *(p. 312)*
ascospore *(p. 313)*
ascus *(p. 312)*
Basidiomycota *(p. 313)*
basidiospore *(p. 315)*
basidium *(p. 313)*
biological vector *(p. 302)*
bread mold *(p. 312)*
cellular slime mold *(p. 305)*
cercaria *(p. 317)*
chitin *(p. 308)*
ciliate *(p. 307)*
club fungus *(p. 313)*
coelom *(p. 315)*
commensal *(p. 305)*
conidium *(p. 313)*
conjugation *(p. 307)*
crustacean *(p. 326)*
cysticercus *(p. 318)*
definitive host *(p. 302)*

Deuteromycota *(p. 315)*
diatom *(p. 303)*
dikaryotic *(p. 308)*
dimorphism *(p. 311)*
dinoflagellate *(p. 303)*
ectoparasite *(p. 301)*
endoparasite *(p. 301)*
euglenoid *(p. 303)*
eutrophication *(p. 303)*
facultative parasite *(p. 301)*
flatworm *(p. 315)*
fluke *(p. 316)*
Fungi Imperfecti *(p. 315)*
fungus *(p. 308)*
gametocyte *(p. 307)*
helminth *(p. 315)*
hermaphroditic *(p. 302)*
host *(p. 300)*
host specificity *(p. 302)*
hydatid cyst *(p. 319)*
hyperparasitism *(p. 301)*
hypha *(p. 308)*
insect *(p. 325)*
intermediate host *(p. 302)*
karyogamy *(p. 308)*

mastigophoran *(p. 306)*
mechanical vector *(p. 302)*
merozoite *(p. 307)*
metacercaria *(p. 317)*
microfilaria *(p. 322)*
miracidium *(p. 317)*
mycelium *(p. 308)*
mycology *(p. 308)*
mycosis *(p. 311)*
nematode *(p. 316)*
obligate parasite *(p. 301)*
Oomycota *(p. 304)*
parasite *(p. 300)*
parasitology *(p. 300)*
pathogen *(p. 300)*
pellicle *(p. 303)*
permanent parasite *(p. 301)*
plasmodial slime mold *(p. 305)*
plasmodium *(p. 305)*
plasmogamy *(p. 308)*
proglottid *(p. 317)*
protist *(p. 302)*
protozoan *(p. 305)*
pseudocoelom *(p. 316)*

pseudoplasmodium *(p. 305)*
redia *(p. 317)*
reservoir host *(p. 302)*
roundworm *(p. 316)*
sac fungus *(p. 312)*
saprophyte *(p. 305)*
sarcodine *(p. 306)*
schizogony *(p. 302)*
scolex *(p. 317)*
septum *(p. 308)*
slime mold *(p. 304)*
sporocyst *(p. 317)*
sporozoite *(p. 306)*
tapeworm *(p. 316)*
temporary parasite *(p. 301)*
test *(p. 303)*
thallus *(p. 308)*
trichocyst *(p. 307)*
trophozoite *(p. 307)*
vector *(p. 301)*
water mold *(p. 304)*
Zygomycota *(p. 312)*
zygospore *(p. 312)*

A Parasites; Parasitology

1. Define parasite and parasitology.
2. Match the following (more than one may apply):
   - ____ lice
   - ____ tapeworm
   - ____ biting mosquito
   - ____ housefly walking on manure
   - ____ ringworm fungus
     - a. ectoparasite
     - b. endoparasite
     - c. facultative parasite
     - d. permanent parasite
     - e. temporary parasite
     - f. biological vector
     - g. mechanical vector
3. How do definitive, intermediate, and reservoir hosts differ?
4. What is host specificity?
5. How do host–parasite relationships change over time?
6. How do parasites evade host defenses, and what kinds of damage do they cause?
7. True or False: The most successful parasite is the one that kills its host. Explain your answer.

B Characteristics of Protists

8. List the characteristics that distinguish protists from other organisms.
9. In what ways are protists beneficial, and in what ways are they harmful?

C Groups of Protists

10. List the main characteristics of plantlike protists.
11. Name (a) the two types of funguslike protists and (b) the two types of slime molds. (c) Describe the characteristics of the funguslike protists.
12. List the main characteristics of animal-like protists.
13. Match the following:
    - ____ Diatoms
    - ____ Amoebas
    - ____ Cellular slime molds
    - ____ Most saprophytes
    - ____ Dinoflagellates
    - ____ Malaria parasites
      - a. Plantlike protists
      - b. Funguslike protists
      - c. Animal-like protists
14. What protists parasitize humans?

D Characteristics of Fungi

15. What characteristics distinguish fungi from other organisms?
16. In what ways are fungi beneficial, and in what ways are they harmful?
17. True or False: Most fungi reproduce both sexually and asexually.
18. True or False: Fungi are classified on the basis of their asexual stages.

E Groups of Fungi

19. How do bread molds differ from sac fungi?
20. How do sac fungi differ from club fungi?
21. True or False: The Fungi Imperfecti are so named because they lack septa in their hyphae.
22. Which fungi parasitize humans?

F Characteristics of Parasitic Helminths

23. What characteristics distinguish helminths from other organisms?
24. How do the two groups of helminths that parasitize humans differ?

G Groups of Parasitic Helminths

25. How do tapeworms and flukes differ?
26. In the following diagram of a tapeworm, identify parts (a) and (b), the oldest proglottids, and the newest proglottids.

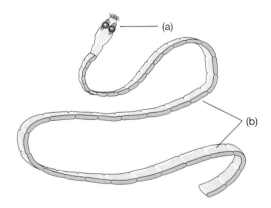

27. What is the only way that a pig could acquire trichinosis from a pig farmer?
28. How do adult nematodes differ from larval nematodes in their behavior as parasites?
29. Give some examples of helminths that parasitize humans.

H Characteristics of Parasitic and Vector Arthropods

30. What characteristics distinguish arthropods from other organisms?
31. How would you expect the behavior of a parasitic arthropod to differ from that of a vector arthropod? Can an arthropod be both a parasite and a vector? If so, how?

I Groups of Parasitic and Vector Arthropods

32. Which groups of arthropods are human parasites?
33. Which groups of arthropods are vectors of disease?

PROBLEMS FOR INVESTIGATION

1. Locate information on the incidence of parasitic diseases in a developing country and in the United States. Explain which factors contributed to the differences between the two countries, and suggest ways the incidence of parasitic diseases could be decreased in both places.

2. Obtain specimens or photographs of 10 protists, and classify them as plantlike, funguslike, or animal-like.

3. Obtain slides or photographs of the eggs of 10 helminths, and devise a classification scheme based on characteristics of the eggs.

4. List several examples of economically important fungi.

5. If you were making champagne, could you use the same strain of yeast you would use to make a Chianti? Why or why not?

6. Find out the proper method to remove a tick that is biting you. Which precautions should you observe? Why?

7. A previously healthy girl who had never traveled outside South Carolina developed generalized seizures. A CAT scan revealed a single lesion in her brain. Surgery showed it to be a cysticercus (larval stage) of the pork tapeworm *Taenia solium*. The patient remains asymptomatic on anticonvulsant medication. All family members and contacts of the patient were negative when tested for *T. solium*. However, a neighbor who had immigrated from Mexico was often visited by other Mexican immigrant friends when the girl was at his house. Three of his friends tested positive for *T. solium* cysticercosis and had proglottids in their stools. Given that *T. solium* cysticercosis is virtually unknown in swine in the United States, transmission through the pig–human cycle was unlikely. In addition, the girl denied ever having eaten raw or undercooked pork. How then did she acquire the infection? (The answer to this problem appears in Appendix F.)

SOME INTERESTING READING

Anderson, D. M. 1994. "Red tides." *Scientific American* 271 (August):62–68.

Barnett, J. A., R. W. Payne, and D. Yarrow. 1990. *Yeasts: Characteristics and identification*, 2d ed. New York: Cambridge University Press.

Beaver, P. C., and R. C. Jung, eds. 1985. *Animal agents and vectors of human disease*, 5th ed. Philadelphia: Lea and Febiger.

Bold, H. C., and M. J. Wynne. 1985. *Introduction to the algae: Structure and reproduction*. Englewood Cliffs, NJ: Prentice Hall.

Donelson, J. E., and M. J. Turner. 1985. "How the trypanosome changes its coat." *Scientific American* 252 (February):44–51.

Farmer, J. N. 1980. *The protozoa: Introduction to protozoology*. St. Louis: C. V. Mosby.

"Fungus routs gypsy moth outbreak." 1990. *Science News* 138, no. 5 (August 4):77.

Garcia, L. S., et al. 1993. "Diagnostic parasitology: Parasitic infections and the compromised host." *Laboratory Medicine* 24 (April):205–15.

Janerette, C. A. 1991. "An introduction to mycorrhizae." *The American Biology Teacher* 53 (January):13–19.

Larone, D. H. 1993. *Medically important fungi: A guide to identification*, 2d ed. Washington, D.C.: American Society for Microbiology.

Newhouse, J. R. 1990. "Chestnut blight." *Scientific American* 263 (July):106–11.

Rennie, J. 1991. "Proteins 2, malaria 0: Malaria-free mice offer clues for developing a human vaccine." *Scientific American* 265 (July):24–25.

Ruiz-Tiben, E., et al. 1995. "Progress towards the eradication of dracunculiasis (Guinea worm disease)." *Emerging Infectious Diseases* 1(2):58–60.

Schmidt, G. D., and L. S. Roberts. 1989. *Foundations of parasitology*, 4th ed. St. Louis: Times Mirror/Mosby.

Sieburth, J. M. 1975. *Microbial seascapes: A pictorial essay on marine microorganisms and their environments*. Baltimore: University Park Press.

Sze, P. 1986. *A biology of the algae*. Dubuque, Iowa: Wm. C. Brown.

STERILIZATION AND DISINFECTION

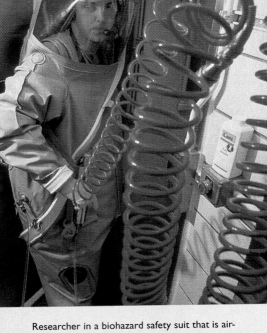

Researcher in a biohazard safety suit that is air-proof. Such a suit must be worn in any high-security laboratory, including CDC Level 4 labs, where dangerous microbes are housed.

D o you like spicy foods? Perhaps you won't like the original reasons for their popularity. Before modern methods of food preservation, such as canning and refrigeration, were available, control of microbial growth in foods was a difficult problem. Inevitably after a short while, food began to take on the "off" flavors of spoilage. Spices were used to mask these unpleasant tastes. Some spices were also effective as preservatives. The antimicrobial effects of garlic have long been known. Fortunately we need not eat spoiled food today, and we can use spices solely to enhance our enjoyment of safely preserved foods.

Medical care, especially in the operating room, is safer, too, today. As we have seen from the work of Ignaz Semmelweis and Joseph Lister, careful washing and the use of chemical agents are effective in controlling many infectious microorganisms. ∞ (Chapter 1, p. 12) In this chapter we will consider the properties of various chemical and physical agents used to control microorganisms in laboratories, in medical facilities, and in homes.

PRINCIPLES OF STERILIZATION AND DISINFECTION

Agents called **disinfectants** are typically applied to inanimate objects, and agents called **antiseptics** are applied to living tissue. A few agents are suitable as both disinfectants and antiseptics, although most disinfectants are too harsh for use on delicate skin tissue. *Antibiotics*, though often applied to skin, are considered separately in Chapter 14.

Sterilization is the killing or removal of all microorganisms in a material or on an object. There are no degrees of sterility—**sterility** means that there are *no* living organisms in or on a material. When properly carried out, sterilization procedures ensure that even highly resistant bacterial endospores and fungal spores are killed. Much of the controversy regarding spontaneous generation in the nineteenth century resulted from the failure to kill resistant cells in materials that were thought to be sterile. In contrast with sterilization, **disinfection** means reducing the number of path-

CHAPTER **13**

332

THIS CHAPTER FOCUSES ON THE FOLLOWING QUESTIONS:

A How do sterilization and disinfection differ, and what terms are used to describe these processes?

B What important principles apply to the processes of sterilization and disinfection?

C What factors affect the potency of antimicrobial chemical agents?

D How is the effectiveness of an antimicrobial chemical agent assessed?

E By what mechanisms do antimicrobial chemical agents act?

F What are the properties of commonly used antimicrobial chemical agents?

G How are dry heat, moist heat, and pasteurization used to control microorganisms?

H How are refrigeration, freezing, drying, and freeze-drying used to control and to preserve microorganisms?

I How are radiation, sonic and ultrasonic waves, filtration, and osmotic pressure used to control microorganisms?

ogenic organisms on objects or in materials so that they pose no threat of disease. Terms related to sterilization and disinfection are defined in Table 13.1.

Control of Microbial Growth

As explained in the discussion of the growth curves in Chapter 6, both the growth and death of microorganisms occur at logarithmic rates. ∞ (p. 138) Here we are concerned with the death rate and the effects of antimicrobial agents—substances that kill microbes or inhibit their growth—on it.

Organisms treated with antimicrobial agents obey the same laws regarding death rates as those declining in numbers from natural causes. We will illustrate this principle with heat as the agent because its effects have been the most thoroughly studied. When heat is applied to a material, the death rate of the organisms in or on it remains logarithmic but is greatly accelerated. Heat acts as an antimicrobial agent. If 20 percent of the organisms die in the first minute, 20 percent of those remaining alive will die in the second minute, and so

on (Figure 13.1a). If, at a different temperature, 30 percent die in the first minute, 30 percent of the remaining ones will die in the second minute, and so on (Figure 13.1b). From these observations we can derive the principle that *a definite proportion of the organisms die in a given time interval.*

Consider now what happens when the number of live organisms that remain becomes small—100, for example. At a death rate of 30 percent per minute, 70 will remain after 1 minute, 49 after 2 minutes, 34 after 3 minutes, and only 1 after 12 minutes (Figure 13.1c). Soon the probability of finding even a single live organism becomes very small. Most laboratories say a sample is sterile if the probability is no greater than one chance in a million of finding a live organism.

The total number of organisms present when disinfection is begun affects the length of time required to eliminate them. We can state a second principle: *The fewer organisms present, the shorter the time needed to achieve sterility.* Thoroughly cleaning objects before attempting to sterilize them is a practical application of this principle. Clearing objects of tissue debris and

TABLE 13.1	Terms Related to Sterilization and Disinfection
TERM	DEFINITION
Sterilization	The killing or removal of all microorganisms in a material or on an object.
Disinfection	The reduction of the number of pathogenic microorganisms to the point where they pose no danger of disease.
Antiseptic	A chemical agent that can safely be used externally on living tissue to destroy microorganisms or to inhibit their growth.
Disinfectant	A chemical agent used on inanimate objects to destroy microorganisms. Most disinfectants do not kill spores.
Sanitizer	A chemical agent typically used on food-handling equipment and eating utensils to reduce bacterial numbers so as to meet public health standards. Sanitization may simply refer to thorough washing with only soap or detergent.
Bacteriostatic agent	An agent that inhibits the growth of bacteria.
Germicide	An agent capable of killing microbes rapidly; some such agents effectively kill certain microorganisms but only inhibit the growth of others.
Bactericide	An agent that kills bacteria. Most such agents do not kill spores.
Viricide	An agent that inactivates viruses.
Fungicide	An agent that kills fungi.
Sporocide	An agent that kills bacterial endospores or fungal spores.

FIGURE 13.1 Microbial death rates on a logarithmic scale. (a) At a death rate of 20%/min, 20% die in the first minute; 20% of the survivors die in the second minute, and so on. A death rate of 30%/min (b) from the same initial number of organisms and (c) from a much smaller initial number.

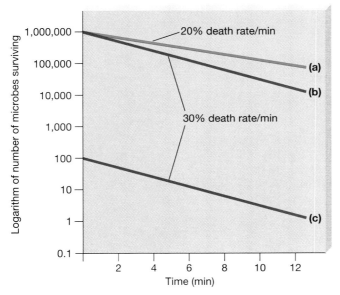

blood is also important because such organic matter impairs the effectiveness of many chemical agents.

Different antimicrobial agents affect various species of bacteria and their endospores differently. Furthermore, any given species may be more susceptible to an antimicrobial agent at one phase of growth than at another. The most susceptible phase for most organisms is the logarithmic growth phase, because during that phase many enzymes are actively carrying out synthetic reactions, and interfering with even a single enzyme might kill the organism. From these observations, we can state a third principle: *Microorganisms differ in their susceptibility to antimicrobial agents.*

CHEMICAL ANTIMICROBIAL AGENTS

Potency of Chemical Agents

The potency, or effectiveness, of a chemical antimicrobial agent is affected by time, temperature, pH, and concentration. The death rate of organisms is affected by the length of time the organisms are exposed to the antimicrobial agent, as was explained earlier for heat. Thus, adequate time should always be allowed for an agent to kill the maximum number of organisms. The death rate of organisms subjected to a chemical agent is accelerated by increasing the temperature. Increasing temperature by 10°C roughly doubles the rate of chemical reactions and thereby increases the potency of the chemical agent. Acidic or alkaline pH can increase or decrease the agent's potency. A pH that increases the degree of ionization of a chemical agent often increases its ability to penetrate a cell. Such a pH also can alter the contents of the cell itself. Finally, increasing concentration may increase the effects of most antimicrobial chemical agents. High concentrations may be **bactericidal** (killing), whereas lower concentrations may be **bacteriostatic** (growth-inhibiting).

Both ethyl and isopropyl alcohol are exceptions to the rule regarding increasing concentrations. They have long been believed to be more potent at 70 percent than at higher concentrations, although they are also effective up to 99 percent concentration. Some water must be present for alcohols to disinfect because they act by coagulating (permanently denaturing) proteins, and water is needed for the coagulation reactions. Also, a 70 percent alcohol–water mixture penetrates more deeply than pure alcohol into most materials to be disinfected.

Evaluation of Effectiveness of Chemical Agents

Many factors affect the potency of chemical antimicrobial agents, so evaluation of effectiveness is difficult. No entirely satisfactory method is available.

However, we need some way to compare the effectiveness of disinfecting agents, especially as new ones come on the market.

Phenol Coefficient

Since Lister introduced *phenol* (carbolic acid) as a disinfectant in 1867, it has been the standard disinfectant to which other disinfectants are compared under the same conditions. The result of this comparison is called the **phenol coefficient**. Two organisms, *Salmonella typhi*, a pathogen of the digestive system, and *Staphylococcus aureus*, a common wound pathogen, are typically used to determine phenol coefficients. A disinfectant with a phenol coefficient of 1.0 has the same effectiveness as phenol. A coefficient less than 1.0 means that the disinfectant is less effective than phenol; a coefficient of greater than 1.0 means that it is more effective. Phenol coefficients are reported separately for the different test organisms (Table 13.2). Lysol, for instance, has a coefficient of 5.0 against *Staphylococcus aureus* but only 3.2 when used on *Salmonella typhi*, whereas ethyl alcohol has a value of 6.3 against both.

The phenol coefficient can be determined by the following steps. Prepare several dilutions of a chemical agent, and place the same volume of each in different test tubes. Prepare an identical set of test tubes, using phenol dilutions. Put both sets of tubes in a 20°C water bath for at least 5 minutes to ensure that the contents of all tubes are at the same temperature. Transfer 0.5 ml of a culture of a standard test organism to each tube. After 5, 10, and 15 minutes, use a sterile loop to transfer a specific volume of liquid from each tube into a separate tube of nutrient broth, and incubate the tubes. After 48 hours, check cultures for cloudiness, and find the smallest concentration (highest dilution) of the agent that killed all organisms in 10 minutes but not in 5 minutes. Find the ratio of this dilution to the dilution of phenol that has the same effect. For example, if a 1:1000 dilution of a chemical agent has the same effect as a 1:100 dilution of

TABLE 13.2	Phenol Coefficients of Various Chemical Agents	
Chemical Agent	Staphylococcus aureus	Salmonella typhi
Phenol	1.0	1.0
Chloramine	133.0	100.0
Cresols	2.3	2.3
Ethyl alcohol	6.3	6.3
Formalin	0.3	0.7
Hydrogen peroxide	——	0.01
Lysol	5.0	3.2
Mercury chloride	100.0	143.0
Tincture of iodine	6.3	5.8

phenol, the phenol coefficient of that agent is 10 (1000/100). If you performed this test on a new disinfectant and obtained these results, you would have found a very good disinfectant! The phenol coefficient provides an acceptable means of evaluating the effectiveness of chemical agents derived from phenol, but it is less acceptable for other agents. Another problem is that the materials on or in which organisms are found may affect the usefulness of a chemical agent by complexing with it or inactivating it. These effects are not reflected in the phenol coefficient number.

Filter Paper Method

The **filter paper method** of evaluating a chemical agent is simpler than determining a phenol coefficient. It uses small filter paper disks, each soaked with a different chemical agent. The disks are placed on the surface of an agar plate that has been inoculated with a test organism. A different plate is used for each test organism. After incubation, a chemical agent that inhibits growth of a test organism is identified by a clear area around the disk where the bacteria have been killed (Figure 13.2). Note: What is effective against one organism may have little or no effect on the others.

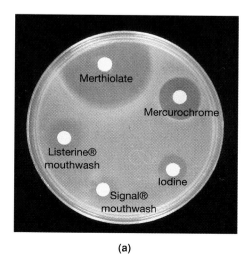

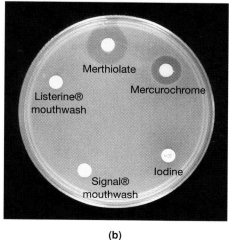

(a)   (b)

FIGURE 13.2 The filter paper method of evaluating disinfectants and antiseptics shows the difference in response of (a) *Staphylococcus aureus* (gram positive) and (b) *Escherichia coli* (gram negative) to several common chemical agents. In both cases, the greatest inhibition of growth is seen near the top of the Petri dish surrounding the merthiolate-soaked paper disk. The various disks were soaked in mercurochrome, iodine, Signal® mouthwash, or Listerine® mouthwash before being placed on the surface of the nutrient medium, which had first been confluently inoculated with one of the test organisms.

Use-Dilution Test

A third way of evaluating chemical agents, the **use-dilution test,** uses standard preparations of certain test bacteria. A broth culture of one of these bacteria is coated onto small stainless steel cylinders and allowed to dry. Each cylinder is then dipped into one of several dilutions of the chemical agent for 10 minutes, removed, rinsed with water, and placed into a tube of broth. The tubes are incubated and then observed for the presence or absence of growth. Agents that prevent growth at the greatest dilutions are considered the most effective. Many microbiologists feel that this measurement is more meaningful than the phenol coefficient.

Disinfectant Selection

Several qualities should be considered in deciding which disinfectant to use. An ideal disinfectant should

1. Be fast acting even in the presence of organic substances, such as those in body fluids;
2. Be effective against all types of infectious agents without destroying tissues or acting as a poison if ingested;
3. Easily penetrate material to be disinfected without damaging or discoloring the material;
4. Be easy to prepare and stable even when exposed to light, heat, or other environmental factors;
5. Be inexpensive and easy to obtain and use;
6. Not have an unpleasant odor.

No disinfectant is likely to satisfy all these criteria, so the agent that meets the greatest number of criteria for the task at hand is chosen.

In practice, many agents are tested in a wide range of situations and are recommended for use where they are most effective. Thus, some agents are selected for sanitizing kitchen equipment and eating utensils, whereas other agents are chosen for rendering pathogenic cultures harmless. Furthermore, certain agents can be used in dilute concentration on the skin and in stronger concentration on inanimate objects.

Mechanisms of Action of Chemical Agents

Chemical antimicrobial agents kill microorganisms by participating in one or more chemical reactions that damage cell components. Although the kinds of reactions are almost as numerous as the agents, agents can be grouped by whether they affect proteins, membranes, or other cell components.

Reactions That Affect Proteins

Much of a cell is made of protein, and all its enzymes are proteins. Alteration of protein structure is called *de-*

naturation. ∞ (Chapter 2, p. 42) In denaturation, hydrogen and disulfide bonds are disrupted, and the functional shape of the protein molecule is destroyed. Any agent that denatures proteins prevents them from carrying out their normal functions. When treated with mild heat or with some dilute acids, alkalis, or other agents for a short time, proteins are temporarily denatured. After the agent is removed, some proteins can regain their normal structure. However, most antimicrobial agents are used in a strong enough concentration over a sufficient length of time to denature proteins permanently. Permanent denaturation of a microorganism's proteins kills the organism. Denaturation is bactericidal if it permanently alters the protein such that the protein's normal state cannot be restored. Denaturation is bacteriostatic if it temporarily alters the protein, and the normal structure can be recovered (Figure 13.3).

Reactions that denature proteins include hydrolysis, oxidation, and the attachment of atoms or chemical groups. (Recall that hydrolysis is the breaking down of a molecule by the addition of water and that oxidation is the addition of oxygen to, or the removal of hydrogen from, a molecule.) ∞ (Chapter 2, p. 33) Acids, such as boric acid, and strong alkalis destroy protein by hydrolyzing it. Oxidizing agents (electron acceptors), such as hydrogen peroxide and potassium permanganate, oxidize disulfide linkages (—S—S—) or sulfhydryl groups (—SH). Agents that contain halogens—the elements chlorine, fluorine, bromine, and iodine—also sometimes act as oxidizing agents. Heavy metals, such as mercury and silver, attach to sulfhydryl groups. Alkylating agents, which contain methyl (—CH$_3$) or similar groups, donate these groups to proteins. Formaldehyde and some dyes are alkylating agents. Halogens can be substituted for hydrogen in carboxyl (—COOH), sulfhydryl, amino (—NH$_2$), and alcohol (—OH) groups. All these reactions can kill microorganisms.

Reactions That Affect Membranes

Membranes contain proteins and so can be altered by all the preceding reactions. Membranes also contain lipids and thus can be disrupted by substances that dissolve lipids. **Surfactants** (sur-fak'tantz) are soluble compounds that reduce surface tension, just as soaps and detergents break up grease particles in dishwater (Figure 13.4). Surfactants include alcohols, detergents, and *quaternary ammonium compounds,* such as benzalkonium chloride, which dissolve lipids. Phenols, which are alcohols, dissolve lipids and also denature proteins. Detergent solutions, also called **wetting agents,** are often used with other chemical agents to help the agent penetrate fatty substances. Although detergent solutions themselves usually do not kill microorganisms, they do help get rid of lipids and other organic materials so that antimicrobial agents can reach the organisms.

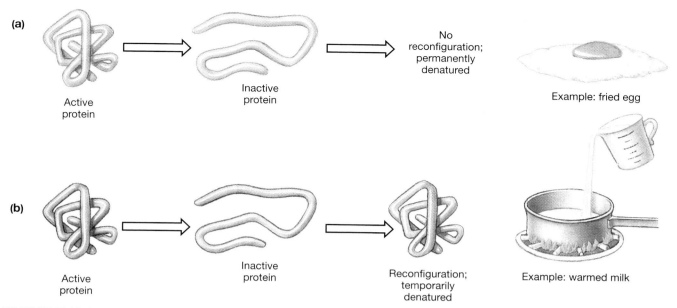

FIGURE 13.3 (a) A permanently denatured protein, like that of a fried egg, cannot return to its original configuration. (b) A temporarily denatured protein, like that in warmed milk, can refold into its original configuration. The protein structure of milk that has been warmed is recovered when the milk is cooled.

FIGURE 13.4 The action of a surfactant. Here the surfactant molecule has ionized into sodium ions and long hydrocarbon chains, whose zigzagged, covalently bonded tails are able to enter an insoluble substance such as grease. The other end of these molecules has a carboxyl group with a negatively charged oxygen. These negative charges attract the positively charged sides of water molecules, thereby making the attached insoluble substance soluble in the water so the substance can be washed away.

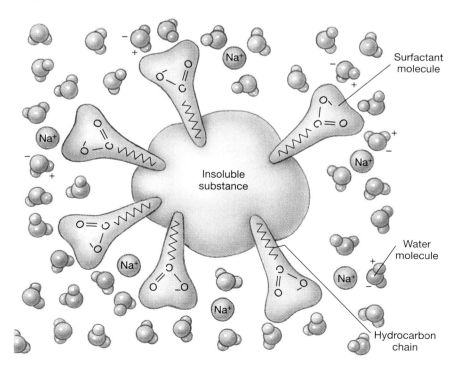

Reactions That Affect Other Cell Components

Other cell components affected by chemical agents include nucleic acids and energy-producing systems. Alkylating agents can replace hydrogen on amino or alcohol groups in nucleic acids. Certain dyes, such as crystal violet, interfere with cell wall formation. Some substances, such as lactic acid and propionic acid (end products of fermentation), inhibit fermentation and thus prevent energy production in certain bacteria, molds, and some other organisms.

Reactions That Affect Viruses

Like many cellular microorganisms, viruses can cause infections and must be controlled. Control of viruses requires that they be inactivated—that is, rendered permanently incapable of infecting or replicating in cells. Inactivation can be effected by destroying either the viruses' nucleic acid or their proteins.

Alkylating agents, such as ethylene oxide, nitrous acid, and hydroxylamine act as chemical mutagens—

APPLICATIONS

NO MORE FROGS IN FORMALIN

Formalin, a 37 percent aqueous solution of formaldehyde, was for many years the standard material used to preserve laboratory specimens for dissection. But formaldehyde is toxic to tissues and may cause cancer, so it is rarely used today as a preservative. A variety of other preservatives are now used. Although less toxic to students performing dissections, they are also less effective for long-term preservation. Molds growing on the surface of specimens are now a common problem.

they alter DNA or RNA. If the alteration prevents the DNA or RNA from directing the synthesis of new viral particles, the alkylating agents are effective inactivators. Detergents, alcohols, and other agents that denature proteins act on bacteria and viruses in the same way. Certain dyes, such as acridine orange and methylene blue, render viruses susceptible to inactivation when exposed to visible light. This process disrupts the structure of the viral nucleic acid.

Viruses sometimes remain infective even after their proteins are denatured, so methods used to rid materials of bacteria may not be as successful with infectious viruses. Also, use of an agent that does not inactivate viruses can lead to laboratory-acquired infections.

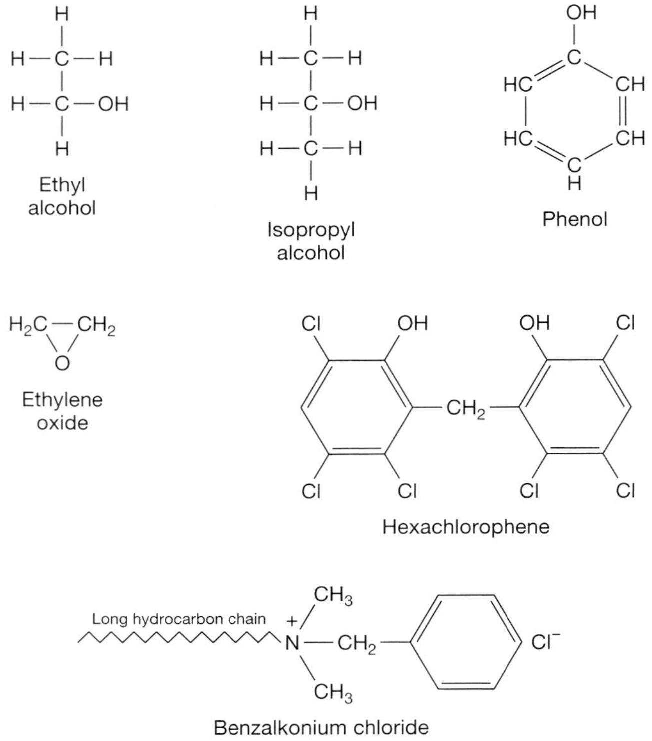

Specific Chemical Antimicrobial Agents

Now that we have considered general principles of sterilization and disinfection and the kinds of reactions caused by such agents, we can look at some specific agents and their applications. The structural formulas of some of the most important compounds discussed are shown in Figure 13.5.

Soaps and Detergents

Soaps and detergents remove microbes, oily substances, and dirt. Mechanical scrubbing greatly enhances their action. In fact, vigorous hand washing is one of the easiest and cheapest means of preventing the spread of disease among patients in hospitals, in medical and dental offices, among employees and patrons in food establishments, and among family members. Unlike surgical scrubs, germicidal soaps usually are not significantly better disinfectants than ordinary soaps.

Soaps contain alkali and sodium and will kill many species of *Streptococcus, Micrococcus,* and *Neisseria* and will destroy influenza viruses. Many pathogens that

FIGURE 13.5 Structural formulas of some important disinfectants.

survive washing with soap can be killed by a disinfectant applied after washing. A common practice after washing and rinsing hands and inanimate objects is to apply a 70 percent alcohol solution. Even these measures do not necessarily rid hands of all pathogens. Consequently, disposable gloves are used where there is a risk that health care workers may become infected or may transmit pathogens to other patients.

Detergents, when used in weak concentrations in wash water, allow the water to penetrate into all crevices and cause dirt and microorganisms to be lifted out and washed away. Detergents are said to be *cationic* if they are positively charged and *anionic* if they are negatively charged. Cationic detergents are used to sanitize food utensils. Although not effective in killing endospores, they do inactivate some viruses. Anionic detergents are used for laundering clothes and as household cleaning agents. They are less effective sanitizing agents than cationic detergents, probably because the negative charges on bacterial cell walls repel them.

Many cationic detergents are **quaternary ammonium compounds,** or **quats,** which have four organic groups attached to a nitrogen atom. A variety of quats are available as disinfecting agents; their chemical structures vary according to their organic groups. One problem with quats is that their effectiveness is decreased in the presence of soap, calcium or magnesium

PUBLIC HEALTH

SOAP AND SANITATION

Washing and drying clothing in modern public laundry facilities is generally a safe practice because the clothing is almost disinfected if the water temperature is high enough. Soaps, detergents, and bleaches kill many bacteria and inactivate many viruses. Agitation of the clothes in the washer provides good mechanical scrubbing. Many microbes that survive this action are killed by heat in the dryer. The use of bar soap in a public washroom is not such a safe practice: The soap may be a source of infectious agents. In a study of 84 samples of bar soap taken from public washrooms, every sample contained microorganisms. More than 100 strains of bacteria and fungi were isolated from the soap samples, and some of the organisms were potential pathogens. Many restaurants and other establishments have installed soap dispensers for this very reason. In fact, many jurisdictions have made the use of bar soap in such facilities illegal.

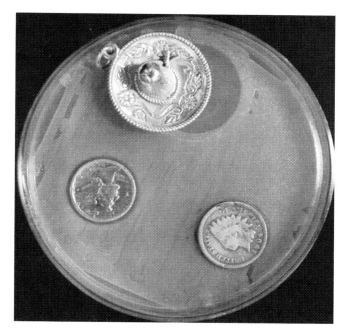

FIGURE 13.6 The inhibitory effects of silver ions (a heavy metal) can be seen as clear zones in which no growth has occurred around the silver charm (which has been pushed aside) and silver dime. The nonsilver coin (a copper penny) has not inhibited growth of the organisms as effectively as the silver objects have.

ions, or porous substances such as gauze. An even more serious problem with these agents is that they support the growth of some bacteria of the genus *Pseudomonas* rather than killing them. Zephiran (benzalkonium chloride) was once widely used as a skin antiseptic. It is no longer recommended because it is less effective than originally thought and is subject to the same problems as other quats. Quats are now often mixed with another agent to overcome some of these problems and to increase their effectiveness.

Acids and Alkalis

Soap is a mild alkali, and its alkaline properties help destroy microbes. A number of organic acids lower the pH of materials sufficiently to inhibit fermentation. Several are used as food preservatives. Lactic and propionic acids retard mold growth in breads and other products. Benzoic acid and several of its derivatives are used to prevent fungal growth in soft drinks, catsup, and margarine. Sorbic acid and sorbates are used to prevent fungal growth in cheeses and a variety of other foods. Boric acid, formerly used as an eyewash, is no longer recommended because of its toxicity.

Heavy Metals

Heavy metals used in chemical agents include selenium, mercury, copper, and silver. Even tiny quantities of such metals can be very effective in inhibiting bacterial growth (Figure 13.6). Silver nitrate was once widely used to prevent gonococcal infection in newborn infants. A few drops of silver nitrate solution were placed in the baby's eyes at the time of delivery to protect against infection by gonococci entering the eyes during passage through the birth canal. For a time, many hospitals replaced silver nitrate with antibiotics such as erythromycin. However, the development of antibiotic-resistant strains of gonococci has led some localities to require the use of silver nitrate, to which gonococci do not develop resistance.

Organic mercury compounds, such as merthiolate and mercurochrome, are used to disinfect surface skin wounds. Such agents kill most bacteria in the vegetative state but do not kill spores. They are not effective against *Mycobacterium*. Merthiolate is generally prepared as a **tincture** (tingk'chur), that is, dissolved in alcohol. The alcohol in a tincture may have a greater germicidal action than the heavy metal compound. Thimerosal, another organic mercury compound, can be used to disinfect skin and instruments and as a preservative for vaccines. Phenylmercuric nitrate and mercuric naphthenate inhibit both bacteria and fungi and are used as laboratory disinfectants.

Selenium sulfide kills fungi, including spores. Preparations containing selenium are commonly used to treat fungal skin infections. Shampoos that contain selenium are effective in controlling dandruff. Dandruff,

APPLICATIONS

FOR A CLEAR AQUARIUM

If you have an aquarium, you probably have contended with water that looks like pea soup because of the large numbers of algae growing in it. This problem can be corrected by placing a few pennies in the tank. Enough copper to inhibit algal growth dissolves from the pennies into the water. For this small investment, you can greatly increase visibility and enjoyment of your fish.

a crusting and flaking of the scalp, is often, though not always, caused by fungi.

Copper sulfate is used to control algal growth. Although algal growth usually is not a direct medical problem, it is a problem in maintaining water quality in heating and air-conditioning systems and outdoor swimming pools. (The Environmental Protection Agency, however, is evaluating copper sulfate as an environmental hazard.)

Halogens

Hypochlorous acid, formed by the addition of chlorine to water, effectively controls microorganisms in drinking water and swimming pools. It is the active ingredient in household bleach and is used to disinfect food utensils and dairy equipment. It is effective in killing bacteria and inactivating many viruses. However, chlorine itself is easily inactivated by the presence of organic materials. That is why a substance such as copper sulfate is used to control algal growth in water to be purified with chlorine.

Iodine also is an effective antimicrobial agent. Tincture of iodine was one of the first skin antiseptics to come into use. Now *iodophors,* slow-release compounds in which the iodine is combined with organic molecules, are more commonly used. In such preparations, the organic molecules act as surfactants. Betadine and Isodine are used for surgical scrubs and on skin where an incision will be made. These compounds take several minutes to act and do not sterilize the skin. Betadine in concentrations of 3 to 5 percent destroys fungi, amoebas, and viruses as well as most bacteria, but it does not destroy bacterial endospores. Contamination of Betadine with *Pseudomonas cepacia* has been reported.

Bromine is sometimes used in the form of gaseous methyl bromide to fumigate soil that will be used in the propagation of bedding plants. It is also used in some pools and indoor hot tubs because it does not give off the strong odor that chlorine does.

Alcohols

When mixed with water, alcohols denature protein. They are also lipid solvents and dissolve membranes. Ethyl and isopropyl alcohols can be used as skin antiseptics. Isopropyl alcohol is more often used due to legal regulation of ethyl alcohol. It disinfects skin where injections will be made or blood drawn. Alcohol disinfects but does not sterilize skin because it evaporates quickly and stays in contact with microbes for only a few seconds. It also does not penetrate deeply enough into pores in the skin. It kills vegetative microorganisms on the skin surface but does not kill endospores, resistant cells, or cells deep in skin pores.

Phenols

Phenol and phenol derivatives called *phenolics* disrupt cell membranes, denature proteins, and inactivate enzymes. They are used to disinfect surfaces and to destroy discarded cultures because their action is not impaired by organic materials. Amphyl, which contains amylphenol, destroys vegetative forms of bacteria and fungi and inactivates viruses. It can be used on skin, instruments, dishes, and furniture. When used on surfaces, it retains its antimicrobial action for several days. The orthophenylphenol in Lysol gives it similar properties. A mixture of phenol derivatives called *cresols* is found in creosote, a substance used to prevent the rotting of wooden posts, fences, railroad ties, and such. However, because creosote is irritating to skin and a carcinogen, its use is limited. The addition of halogens to phenolic molecules usually increases their effectiveness. Hexachlorophene and dichlorophene, which are halogenated phenols, inhibit staphylococci and fungi, respectively, on the skin and elsewhere. Chlorhexidine gluconate (Hibiclens), which is chlorinated and similar in structure to hexachlorophene, is effective against a wide variety of microbes even in the presence of organic material. It is a good agent for surgical scrubs.

Oxidizing Agents

Oxidizing agents disrupt disulfide bonds in proteins and thus disrupt the structure of membranes and proteins. Hydrogen peroxide (H_2O_2), which forms highly reactive superoxide (O_2^-), is used to clean puncture wounds. When hydrogen peroxide breaks down into oxygen and water, the oxygen kills obligate anaerobes present in the wounds. Hydrogen peroxide is quickly inactivated by enzymes from injured tissues. A recently developed method of sterilization that uses vaporized hydrogen peroxide can now be used for small rooms or areas, such as glove boxes and transfer hoods (Figure 13.7). Another oxidizing agent, potassium permanganate, is used to disinfect instruments and, in low concentrations, to clean skin.

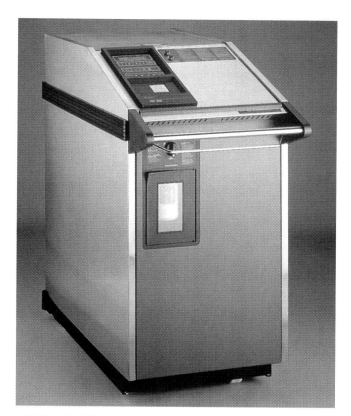

FIGURE 13.7 Recently developed biodecontamination equipment uses vaporized hydrogen peroxide to sterilize small, sealable enclosures such as glove boxes, hoods, or transfer rooms. It is not sufficient, however, to sterilize a larger space such as an operating room.

Alkylating Agents

Alkylating agents disrupt the structure of both proteins and nucleic acids. Because they can disrupt nucleic acids, these agents may cause cancer and should not be used in situations where they might affect human cells. Formaldehyde, glutaraldehyde, and β-propiolactone are used in aqueous solutions. Ethylene oxide is used in gaseous form.

Formaldehyde inactivates viruses and toxins without destroying their antigenic properties. Glutaraldehyde kills all kinds of microorganisms, including spores, and sterilizes equipment exposed to it for 10 hours. Betapropiolactone destroys hepatitis viruses, as well as most other microbes, but penetrates materials poorly. It is, however, used to inactivate viruses in vaccines.

Gaseous ethylene oxide has extraordinary penetrating power. Used at a concentration of 500 mg/l at 50°C for 4 hours, it sterilizes rubber goods, mattresses, plastics, and other materials destroyed by higher temperatures. Special equipment used during ethylene oxide sterilization is shown in Figure 13.8. As will be explained when we discuss autoclaving, an *ampule* (a sealed glass container) of endospores should be processed with ethylene oxide sterilization to check the effectiveness of sterilization.

FIGURE 13.8 Equipment used for ethylene oxide sterilization. This equipment must be used very carefully, as ethylene oxide is both explosive and carcinogenic. An aerator is used to remove all traces of the gas from sterilized material.

PUBLIC HEALTH

HEXACHLOROPHENE

Hexachlorophene is an excellent skin disinfectant. In a 3 percent solution, it kills staphylococci and most other gram-positive organisms, and its residue on skin is strongly bacteriostatic. Because staphylococcal skin infections can spread easily among newborn babies in hospitals, this antiseptic was used extensively in the 1960s for daily bathing of infants. The unforeseen price paid for controlling infections was permanent brain damage in infants bathed in it over a period of time. Hexachlorophene is absorbed through the skin and travels in the blood to the brain. Baby powder containing hexachlorophene killed 40 babies in France in 1972. Available only by prescription in the United States today, hexachlorophene is used routinely, though very cautiously, in hospital neonatal units because it is still the most effective agent for preventing the spread of staphylococcal infections.

| TABLE 13.3 | Properties of Chemical Antimicrobial Agents |

Agent	Actions	Uses
Soaps and detergents	Lower surface tension, make microbes accessible to other agents	Hand washing, laundering, sanitizing kitchen and dairy equipment
Surfactants	Dissolve lipids, disrupt membranes, denature proteins, and inactivate enzymes in high concentrations; act as wetting agents in low concentrations	Cationic detergents are used to sanitize utensils; anionic detergents to launder clothes and clean household objects; quaternary ammonium compounds are sometimes used as antiseptics on skin.
Acids	Lower pH and denature proteins	Food preservation
Alkalis	Raise pH and denature proteins	Found in soaps
Heavy metals	Denature proteins	Silver nitrate is used to prevent gonococcal infections, mercury compounds to disinfect skin and inanimate objects, copper to inhibit algal growth, and selenium to inhibit fungal growth.
Halogens	Oxidize cell components in absence of organic matter	Chlorine is used to kill pathogens in water and to disinfect utensils; iodine compounds are used as skin antiseptics.
Alcohols	Denature proteins when mixed with water	Isopropyl alcohol is used to disinfect skin; ethylene glycol and propylene glycol can be used in aerosols.
Phenols	Disrupt membranes, denature proteins, and inactivate enzymes; not impaired by organic matter	Phenol is used to disinfect surfaces and destroy discarded cultures; amylphenol destroys vegetative organisms and inactivates viruses on skin and inanimate objects; chlorhexidine gluconate is especially effective as a surgical scrub.
Oxidizing agents	Disrupt disulfide bonds	Hydrogen peroxide is used to clean puncture wounds, potassium permanganate to disinfect instruments.
Alkylating agents	Disrupt structure of proteins and nucleic acids	Formaldehyde is used to inactivate viruses without destroying antigenic properties, glutaraldehyde to sterilize equipment, betapropiolactone to destroy hepatitis viruses, and ethylene oxide to sterilize inanimate objects that would be harmed by high temperatures.
Dyes	May interfere with replication or block cell wall synthesis	Acridine is used to clean wounds, crystal violet to treat some protozoan and fungal infections.

All articles sterilized with ethylene oxide must be well ventilated with sterile air to remove all traces of this toxic gas, which can cause burns if it reaches living tissues and is also highly explosive. After exposure to ethylene oxide, articles such as catheters, intravenous lines, in-line valves, and rubber tubing must be thoroughly flushed with sterile air. Both the toxicity and flammability of ethylene oxide can be reduced by using it in gas containing 90 percent carbon dioxide. *It is exceedingly important that workers be protected from ethylene oxide vapors, which are toxic to skin, eyes, and mucous membranes and may also cause cancer.*

Dyes

The dye acridine, which interferes with cell replication by causing mutations in DNA (Chapter 7), can be used to clean wounds. ∞ (p. 183) Methylene blue inhibits growth of some bacteria in cultures. Crystal violet (gentian violet) blocks cell wall synthesis, possibly by the same reaction that causes this dye to bind to cell wall material in Gram staining. It effectively inhibits growth of gram-positive bacteria in cultures and in skin infections. It can be used to treat protozoan (*Trichomonas*) and yeast (*Candida albicans*) infections.

Other Agents

Certain plant oils have special antimicrobial uses. Thymol, derived from the herb thyme, is used as a preservative, and eugenol, derived from oil of cloves, is used in dentistry to disinfect cavities. A variety of other agents are used primarily as food preservatives. They include sulfites and sulfur dioxide, used to preserve dried fruits and molasses; sodium diacetate, used to retard mold in bread; and sodium nitrite, used to preserve cured meats and some cold cuts. Foods containing nitrites should be eaten in moderation because the nitrites are converted during digestion to substances that may cause cancer.

The properties of chemical antimicrobial agents are summarized in Table 13.3.

PHYSICAL ANTIMICROBIAL AGENTS

For centuries, physical antimicrobial agents have been used to preserve food. Ancient Egyptians dried perishable foods to preserve them. Scandinavians made holes in the centers of pieces of dry, flat, crisp bread in order to hang them in the air of their homes during the winter; likewise they kept seed grains in a dry place. Otherwise, both flour and grains would have molded during the long and very moist winters. Europeans used heat in the food-canning process 50 years before Pasteur's work explained why heating prevented food from spoiling. Today, physical agents that destroy microorganisms are still used in food preservation and preparation. Such agents remain a crucial weapon in the prevention of infectious disease. Physical antimicrobial agents include various forms of heat, refrigeration, desiccation (drying), irradiation, and filtration.

Principles and Applications of Heat Killing

Heat is a preferred agent of sterilization for all materials not damaged by it. It rapidly penetrates thick materials not easily penetrated by chemical agents. Several measurements have been defined to quantify the killing power of heat. The **thermal death point** is the temperature that kills all the bacteria in a 24-hour-old broth culture at neutral pH in 10 minutes. The **thermal death time** is the time required to kill all the bacteria in a particular culture at a specified temperature. The **decimal reduction time,** also known as the **DRT** or **D value,** is the length of time needed to kill 90 percent of the organisms in a given population at a specified temperature. (The temperature is indicated by a subscript: $D_{80^\circ C}$, for example.)

These measurements have practical significance in industry as well as in the laboratory. For example, a food-processing technician wanting to sterilize a food as quickly as possible would determine the thermal death point of the most resistant organism that might be present in the food and would employ that temperature. In another situation it might be preferable to make the food safe for human consumption by processing it at the lowest possible temperature. This could be important in processing foods containing proteins that would be denatured, thereby altering their flavor or consistency. The processor would then need to know the thermal death time at the desired temperature for the most resistant organism likely to be in the food.

Dry Heat, Moist Heat, and Pasteurization

Dry heat probably does most of its damage by oxidizing molecules. *Moist heat* destroys microorganisms mainly by denaturing proteins; the presence of water molecules helps disrupt the hydrogen bonds and other weak interactions that hold proteins in their three-dimensional shapes. ∞ (Chapter 2, p. 40) Moist heat may disrupt membrane lipids as well. Heat also inactivates many

viruses, but those that can infect even after their protein coats are denatured require extreme heat treatment, such as steam under pressure, that will disrupt nucleic acids.

Dry Heat

Dry (oven) heat penetrates substances more slowly than moist (steam) heat. It is usually used to sterilize metal objects and glassware and is the only suitable means of sterilizing oils and powders. Objects are sterilized by dry heat when subjected to 171°C for 1 hour, 160°C for 2 hours or longer, or 121°C for 16 hours or longer, depending on the volume.

An open flame is a form of dry heat used to sterilize inoculating loops and the mouths of culture tubes by incineration and to dry the inside of pipettes. When flaming objects in the laboratory, you must avoid the formation of floating ashes and **aerosols** (droplets released into the air). These substances can be a means of spreading infectious agents if the organisms in them are not killed by incineration as intended. For this reason, specially designed loop incinerators with deep throats are often used for sterilizing inoculating loops.

Moist Heat

Moist heat, because of its penetrating properties, is a widely used physical agent. Boiling water destroys vegetative cells of most bacteria and fungi and inactivates some viruses, but it is not effective in killing all kinds of spores. The effectiveness of boiling can be increased by adding 2 percent sodium bicarbonate to the water. However, if water is heated under pressure, its boiling point is elevated, so temperatures above 100°C can be reached. This is normally accomplished by using an **autoclave** (aw'to-klav), as shown in Figure 13.9, in which a pressure of 15 lb/in.² above atmospheric pressure is maintained for 15 to 20 minutes,

FIGURE 13.9 A small counter-top autoclave.

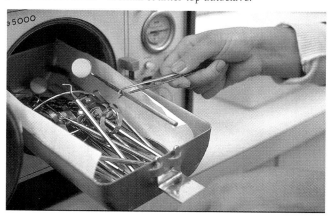

depending on the volume of the load. At this pressure, the temperature reaches 121°C, which is high enough to kill spores as well as vegetative organisms and to disrupt the structure of nucleic acids in viruses. In this procedure it is the increased temperature, and not the increased pressure, that kills microorganisms.

Sterilization by autoclaving is invariably successful if properly done and if two common-sense rules are followed: First, articles should be placed in the autoclave so that steam can easily penetrate them; second, air should be evacuated so that the chamber fills with steam. Wrapping objects in aluminum foil is not recommended because it may interfere with steam penetration. Steam circulates through an autoclave from a steam outlet to an air evacuation port (Figure 13.10). In preparing items for autoclaving, containers should be unsealed and articles should be wrapped in materials that allow steam penetration. Large packages of dressings and large flasks of media require extra time for heat to penetrate them. Likewise, packing many articles close together in an autoclave lengthens the processing time to as much as 60 minutes to ensure sterility. It is more efficient and safer to run two separate, uncrowded loads than one crowded one.

Several methods are available to ensure that autoclaving achieves sterility. Modern autoclaves have devices to maintain proper pressure and record internal temperature during operation. Regardless of the presence of such a device, the operator should check pressure periodically and maintain the appropriate pressure. Tapes impregnated with a substance that causes the word "sterile" to appear when they have been exposed to an effective sterilization temperature can be placed on packages. These tapes are not fully reliable because they do not indicate how long appropriate conditions were maintained. Tapes or other sterilization indicators should be placed inside and near the center of large packages to determine whether heat penetrated them. This precaution is necessary because when an object is exposed to heat, its surface becomes hot much more quickly than its center. (When a large piece of meat is roasted, for example, the surface can be well done while the center remains rare.)

The Centers for Disease Control and Prevention recommends weekly autoclaving of a culture containing heat-resistant endospores, such as those of *Bacillus stearothermophilus*, to check autoclave performance. Endospore strips are commercially available to make this task easy (Figure 13.11). The spore strip and an ampule of medium are enclosed in a soft plastic vial. The vial is placed in the center of the material to be sterilized and is autoclaved. Then the inner ampule is broken, releasing the medium, and the whole container is incubated. If no growth appears in the autoclaved culture, sterilization is deemed effective.

In large laboratories and hospitals, where great quantities of materials must be sterilized, special autoclaves, called *prevacuum autoclaves*, are often used

FIGURE 13.10 Steam is heated in the jacket of an autoclave, enters the sterilization chamber through an opening at the upper rear, and is exhausted through a vent at the bottom front.

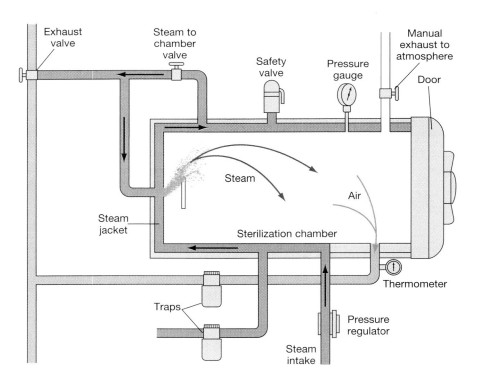

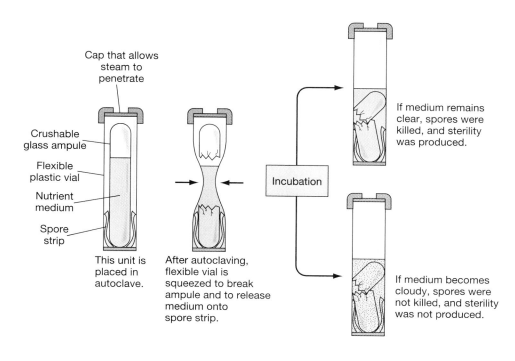

Cap that allows steam to penetrate

Crushable glass ampule

Flexible plastic vial

Nutrient medium

Spore strip

This unit is placed in autoclave.

After autoclaving, flexible vial is squeezed to break ampule and to release medium onto spore strip.

Incubation

If medium remains clear, spores were killed, and sterility was produced.

If medium becomes cloudy, spores were not killed, and sterility was not produced.

FIGURE 13.11 To check if an autoclave is operating properly, a commercially prepared spore test ampule is placed in the autoclave and run with the rest of the load. Afterward, the vial is crushed to release medium onto a strip containing spores. If the load was truly sterilized, the spores will have been killed, and growth will not occur in the medium. Sometimes an indicator dye is added to the medium, which will turn color if microbial growth occurs, due to the accumulation of acid byproducts. This is faster than waiting for sufficient growth to turn the medium cloudy.

APPLICATIONS

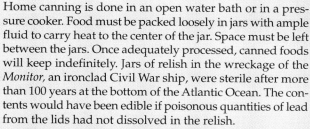

HOME CANNING

Home canning is done in an open water bath or in a pressure cooker. Food must be packed loosely in jars with ample fluid to carry heat to the center of the jar. Space must be left between the jars. Once adequately processed, canned foods will keep indefinitely. Jars of relish in the wreckage of the *Monitor,* an ironclad Civil War ship, were sterile after more than 100 years at the bottom of the Atlantic Ocean. The contents would have been edible if poisonous quantities of lead from the lids had not dissolved in the relish.

The water bath reaches a temperature of 100°C and is adequate for preventing spoilage of acidic foods such as fruits and tomatoes. Acid in these foods inhibits the germination of most spores, should some survive the boiling-water treatment. However, meats and alkaline vegetables, such as corn and beans, must be processed in a pressure cooker. Adding onions or green peppers to a jar of tomatoes increases the pH, so such mixtures also must be cooked under pressure. Because acid-tolerant spores do exist, home canning is safest when it is done by pressure cooking. All commercially canned foods are processed in pressurized equipment. A pressure cooker functions like an autoclave. Foods in jars are processed at least 15 minutes at 15 lb/in.² pressure. Any spores that might be present in these foods are killed, so the food is sterile.

Failure to process alkaline foods at the high temperature reached in a pressure cooker can lead to the accumulation of toxin produced by still-living *Clostridium botulinum* bacteria while the food is stored. Even a tiny amount of this toxin can be lethal. Any home-canned or commercially canned foods should be discarded if they have an unpleasant odor or if the lids of the containers bulge, because this indicates that gas is being produced by living organisms inside the container. Unfortunately, toxins can be present even when the lids don't bulge and there is no noticeable odor, so great care must be used in home canning to be sure that adequate pressure is maintained for a long enough time. As a safety precaution, after a jar of home-canned food is opened, 15 to 20 minutes of vigorous boiling should destroy any botulism toxin. However, the best rule to follow is "When in doubt, throw it out".

Canning peaches at home by the hot water bath method.

FIGURE 13.12 A large automatic hospital autoclave. Recording charts keep records of the actual temperatures and pressures reached during the time of operation. These would alert the operator to some malfunctions.

(Figure 13.12). The chamber is emptied of air as steam flows in, creating a partial vacuum. The steam enters and heats the chamber much more rapidly than it would without the vacuum, so the proper temperature is reached quickly. The total sterilization time is cut in half, and the costs of sterilization are greatly decreased.

Pasteurization

Pasteurization, a process invented by Pasteur to destroy organisms that caused wine to sour, does not achieve sterility. It does kill pathogens, especially *Salmonella* and *Mycobacterium,* that might be present in milk, other dairy products, and beer. *Mycobacterium* used to cause many cases of tuberculosis among children who drank raw milk. Milk is pasteurized by heating it to 71.6°C for at least 15 seconds in the *flash method* or by heating it to 62.9°C for 30 minutes in the *holding method.* Some years ago certain strains of bacteria of the genus *Listeria* were found in pasteurized milk and cheeses. This pathogen causes diarrhea and encephalitis and can lead to death in pregnant women. A few such infections have prompted questions about the need to revise standard procedures for pasteurization. However, finding these pathogens in pasteurized milk has not become a persistent problem, and no action has been taken.

Although most milk for sale in the United States is pasteurized fresh milk, sterile milk also is available. All evaporated or condensed canned milk is sterile, and some milk packaged in cardboard containers also is sterile. The canned milk is subjected to steam under pressure and has a "cooked" flavor. Sterilized milk in cardboard containers is widely available in Europe and can be found in some stores in the United States. It is subjected to a process that is similar to pasteurization but uses higher temperatures. It too has a "cooked" flavor but can be kept unrefrigerated as long as the container remains sealed. Such milk is often flavored with vanilla, strawberry, or chocolate. **Ultrahigh temperature** (UHT) **processing** raises the temperature from 74° to 140°C and then drops it back to 74°C in less than 5 seconds. A complex cooling process that prevents the milk from ever touching a surface hotter than itself prevents development of a "cooked" flavor. Some, but not all, small containers of coffee creamer are treated by this method.

Refrigeration, Freezing, Drying, and Freeze-drying

Cold temperature retards the growth of microorganisms by slowing the rate of enzyme-controlled reactions but does not kill many microbes. Heat is much more effective than cold at killing microorganisms. *Refrigeration* is used to prevent food spoilage. *Freezing, drying,* and *freeze-drying* are used to preserve both foods and microorganisms, but these methods do not achieve sterilization.

Refrigeration

Many fresh foods can be prevented from spoiling by keeping them at 5°C (ordinary refrigerator temperature). However, storage should be limited to a few days because some bacteria and molds continue to grow at this temperature. To convince yourself of this, recall some of the strange things you have found growing on leftovers in the back of your refrigerator. In rare instances strains of *Clostridium botulinum* have been found growing and producing lethal toxins in a refrigerator when the organisms were deep within a container of food, where anaerobic conditions exist.

APPLICATIONS

YOGURT

Certain foods such as yogurt are made by introducing into milk organisms such as lactobacilli that ferment the milk. The fermented foods are often heat-treated after initial pasteurization to kill the fermenting organisms and to increase the shelf life of the products. The labels of such products indicate whether they contain live fermenting organisms. If you make yogurt at home, be sure to purchase a live-culture brand of yogurt to use as your starter.

Freezing

Freezing at −20°C is used to preserve foods in homes and in the food industry. Although freezing does not sterilize foods, it does significantly slow the rate of chemical reactions so that microorganisms do not cause food to spoil. Frozen foods should not be thawed and refrozen. Repeated freezing and thawing of foods causes large ice crystals to form in the foods during slow freezing. Cell membranes in the foods are ruptured, and nutrients leak out. The texture of foods is thus altered, and they become less palatable. It also allows bacteria to multiply while food is thawed, making the food more susceptible to bacterial degradation.

Freezing can be used to preserve microorganisms, but this requires a much lower temperature than that used for food preservation. Microorganisms are usually suspended in glycerol or protein to prevent the formation of large ice crystals (which could puncture cells), cooled with solid carbon dioxide (dry ice) to a temperature of −78°C, and then held there. Alternatively, they can be placed in liquid nitrogen and cooled to −180°C.

Drying

Drying can be used to preserve foods because the absence of water inhibits the action of enzymes. Many foods, including peas, beans, raisins, and other fruits, are often preserved by drying (Figure 13.13). Yeast used in baking also can be preserved by drying. Endospores present on such foods can survive drying, but they do not produce toxins. Dried pepperoni sausage and

---

FIGURE 13.13 Sun drying is an ancient means of preventing growth of microorganisms. These grapes will remain edible as raisins because microbes need more water than remains inside the dried fruit.

APPLICATIONS

HOME FREEZING

Home freezing of foods is probably a more common practice today than home canning. Before freezing fresh fruits and vegetables, they should be blanched, or immersed in boiling water for about a minute. Blanching helps kill microorganisms on the foods, but its main purpose is to denature enzymes in the foods that can cause discoloration or changes in texture even at freezer temperatures. The foods should then be cooled quickly in cold water and placed in clean containers. Finally, they should be placed in the coldest part of the home freezer with space around the containers so that they freeze as quickly as possible.

---

smoked fish retain enough moisture for microorganisms to grow. Because smoked fish is not cooked, eating it poses a risk of infection. Sealing such fish in plastic bags creates conditions that allow anaerobes such as *Clostridium botulinum* to grow.

Drying also naturally minimizes the spread of infectious agents. Some bacteria, such as *Treponema pallidum,* which causes syphilis, are extremely sensitive to drying and die almost immediately on a dry surface; thus they can be prevented from spreading by keeping toilet seats and other bathroom fixtures dry. Drying of laundry in dryers or in the sunshine also destroys pathogens.

Freeze-drying

Freeze-drying, or **lyophilization** (li-of″i-li-za′shun), is the drying of a material from the frozen state (Figure 13.14). This process is used in the manufacture of some brands of instant coffee; freeze-dried instant coffee has a more natural flavor than other kinds. Microbiologists use lyophilization for long-term preservation rather than for destruction of cultures of microorganisms. Organisms are rapidly frozen in alcohol and dry ice or in liquid nitrogen and are then subjected to a high vacuum to remove all the water while in the frozen state. Rapid freezing allows only very tiny ice crystals to form in cells, so the organisms survive this process. Organisms so treated can be kept alive for years, stored under vacuum in the freeze-dried state.

Radiation

Four general types of radiation—ultraviolet light, ionizing radiation, microwave radiation, and strong visible light (under certain circumstances)—can be used to control microorganisms and to preserve foods. Refer

(a)

FIGURE 13.14 Freeze-drying (lyophilization) equipment. (a) A stoppering tray dryer, in which the trays supply the heat needed to remove moisture. After completion of lyophilization (about 24 hours), the device automatically stoppers the vials. (b) A manifold dryer, in which prefrozen samples in vials of different sizes can be attached via ports to the dryer, which supplies a vacuum to remove water. The sample will dry in 4 to 20 hours, depending on its initial thickness. Unlike the tray dryer, this device allows samples to be added and removed.

(b)

to the electromagnetic spectrum (Figure 3.4) to review their wavelengths and positions relative to one another along the spectrum.

Ultraviolet Light

Ultraviolet light consists of light of wavelengths between 40 and 390 nm, but wavelengths in the 200-nm range are most effective in killing microorganisms. Ultraviolet light is absorbed by the purine and pyrimidine bases of nucleic acids. Such absorption can permanently destroy these important molecules. Ultraviolet light is especially effective in inactivating viruses. However, it kills far fewer bacteria than one might expect because of DNA repair mechanisms. Once DNA is repaired, new molecules of RNA and protein can be synthesized to replace the damaged molecules. ∞ (Chapter 7, p. 184)

Ultraviolet light is of limited use because it does not penetrate glass, cloth, paper, or most other mate-

rials, and it does not go around corners or under lab benches. It does penetrate air, effectively reducing the number of airborne microorganisms and killing them on surfaces in operating rooms and rooms that will contain caged animals (Figure 13.15). Ultraviolet lights lose effectiveness over time and should be monitored often. To help sanitize the air without irradiating humans, these lights can be turned on when the rooms are not in use. Exposure to ultraviolet light can cause burns, as anyone who has had a sunburn knows, and can also damage the eyes; years of exposure of skin can lead to skin cancer. Hanging laundry outdoors on bright, sunny days takes advantage of the ultraviolet light present in sunlight. Although the quantity of ultraviolet rays in sunlight is small, these rays may help kill bacteria on clothing, especially diapers.

In some communities, ultraviolet light is replacing chlorine in sewage treatment. When chlorine-treated sewage effluent is discharged into streams or other bodies of water, carcinogenic compounds form and may enter the food chain. The cost of removing chlorine before discharging treated effluent could add as much as $100 per year to the sewage bills of the average American family, and very few sewage plants do this. Running the sewage effluent under ultraviolet light before discharging it can destroy microorganisms without altering the odor, pH, or chemical composition of the water and without forming carcinogenic compounds.

Ionizing Radiation

X-rays, which have wavelengths of 0.1 to 40 nm, and gamma rays, which have even shorter wavelengths, are forms of *ionizing radiation*—so called because it can

FIGURE 13.15 Effects of ultraviolet radiation can be seen in this Petri plate of *Serratia marcescens*; the left side was exposed to ultraviolet rays while the right side was shielded. Most of the organisms on the left side have been killed.

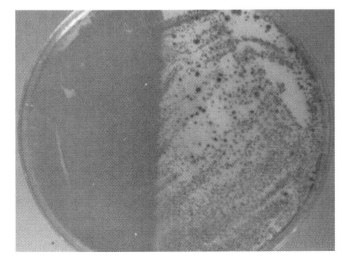

dislodge electrons from atoms, creating ions. (Longer wavelengths comprise *nonionizing radiation*.) These forms of radiation also kill microorganisms and viruses. Many bacteria are killed by absorbing 0.3 to 0.4 millirads of radiation; polioviruses are inactivated by absorbing 3.8 millirads. (A **rad** is a unit of radiation energy absorbed per gram of tissue; a millirad is one-thousandth of a rad. Humans usually do not become ill from radiation unless they are subjected to doses greater than 50 rad.) Ionizing radiation damages DNA and produces peroxides, which act as powerful oxidizing agents in cells. This radiation can also kill or cause mutations in human cells if it reaches them. It is used to sterilize plastic laboratory and medical equipment and pharmaceutical products. It can be used to prevent spoilage in seafoods by doses of 100 to 250 kilorads, in meats and poultry by doses of 50 to 100 kilorads, and in fruits by doses of 200 to 300 kilorads. (One kilorad equals 1000 rads.) Many consumers in the United States reject irradiated foods for fear of receiving radiation, but such foods are quite safe—free of both pathogens and radiation. In Europe, milk and other foods are often irradiated to achieve sterility.

Microwave Radiation

Microwave radiation, in contrast with gamma, X-ray, and ultraviolet radiation, falls at the long-wavelength end of the electromagnetic spectrum. It has wavelengths of approximately 1 mm to 1 m, a range that includes television and police radar wavelengths. Microwave oven frequencies are tuned to match energy levels in water molecules. In the liquid state, water molecules quickly absorb the microwave energy and then release it to surrounding materials as heat. Thus, materials that do not contain water, such as plates made of paper, china, or plastic, remain cool while the moist food on them becomes heated. For this reason the home microwave cannot be used to sterilize items such as bandages and glassware. Conduction of energy in metals leads to problems such as sparking, which makes most metallic items also unsuitable for microwave sterilization. Moreover, bacterial endospores, which contain almost no water, are not destroyed by microwaves. However, a specialized microwave oven has recently become available that can be used to sterilize media in just 10 minutes (Figure 13.16). It has 12 pressure vessels, each of which holds 100 ml of medium. Microwave energy increases the pressure of the medium inside the vessels until sterilizing temperatures are reached.

Caution should be observed in cooking foods in the home microwave oven. Geometry and differences in density of the food being cooked can cause certain regions to become hotter than others, sometimes leaving very cold spots. Consequently, to cook foods thoroughly

FIGURE 13.16 The MikroClave™ system is specifically designed for rapid sterilization of microbiological media and solutions. Using microwave energy, it can sterilize 1.2 liters of media in 6.5 minutes, or 100 ml in 45 seconds. Agar need not be boiled prior to sterilization.

in a microwave oven, it is necessary to rotate the items either mechanically or by hand. For example, pork roasts must be turned frequently and cooked thoroughly to kill any cysts of the pork roundworm *Trichinella* (Chapter 22). Failure to kill such cysts could lead to the disease trichinosis, in which cysts of the worm become embedded in human muscles and other tissues. All experimentally infected pork roasts, when microwaved without rotation, showed live worms remaining in some portion at the end of standard cooking time.

Strong Visible Light

Sunlight has been known for years to have a bactericidal effect, but the effect is due primarily to ultraviolet rays in the sunlight. Strong visible light, which contains light of wavelengths from 400 to 700 nm (violet to red light), can have direct bactericidal effects by oxidizing light-sensitive molecules such as riboflavin and porphyrins (components of oxidative enzymes) in bacteria. For that reason, bacterial cultures should not be exposed to strong light during laboratory manipulations. The fluorescent dyes eosin and methylene blue can denature proteins in the presence of strong light because they absorb energy and cause oxidation of proteins and nucleic acids. The combination of a dye and strong light can be used to rid materials of both bacteria and viruses.

Sonic and Ultrasonic Waves

Sonic, or sound, waves in the audible range can destroy bacteria if the waves are of sufficient intensity. Ultrasonic waves, or waves with frequencies above 15,000 cycles per second, can cause bacteria to cavitate. **Cavitation** (kav″i-ta′shun) is the formation of a partial vacuum in a liquid—in this case, the fluid cytoplasm in the bacterial cell. Bacteria so treated disintegrate, and their proteins are denatured. Enzymes used in detergents are obtained by cavitating the bacterium *Bacillus subtilis*. The disruption of cells by sound waves is called **sonication** (son"i-ka'shun). Neither sonic nor ultrasonic waves are a practical means of sterilization. We mention them here because they are useful in fragmenting cells to study membranes, ribosomes, enzymes, and other components.

Filtration

Filtration is the passage of a material through a filter, or straining device. Sterilization by filtration requires filters with exceedingly small pores. Filtration has been used since Pasteur's time to separate bacteria from media and to sterilize materials that would be destroyed by heat. Over the years, filters have been made of porcelain, asbestos, diatomaceous earth, and sintered glass (glass that has been heated without melting). *Membrane filters* (Figure 13.17), thin disks with pores that prevent the passage of anything larger than the pore size, are widely used today. They are usually made of nitrocellulose and have the great advantage that they can be manufactured with specific pore sizes from 25 μm to less than 0.025 μm. Particles filtered by various pore sizes are summarized in Table 13.4.

Membrane filters have certain advantages and disadvantages. Except for those with the smallest pore sizes, membrane filters are relatively inexpensive, do not clog easily, and can filter large volumes of fluid rea-

TABLE 13.4	Pore Sizes of Membrane Filters and Particles That Pass Through Them
Pore Size (in μm)	**Particles That Pass through Them**
10	Erythrocytes, yeast cells, bacteria, viruses, molecules
5	Yeast cells, bacteria, viruses, molecules
3	Some yeast cells, bacteria, viruses, molecules
1.2	Most bacteria, viruses, molecules
0.45	A few bacteria, viruses, molecules
0.22	Viruses, molecules
0.10	Medium-sized to small viruses, molecules
0.05	Small viruses, molecules
0.025	Only the very smallest viruses, molecules
Ultrafilter	Small molecules

sonably rapidly. They can be autoclaved or purchased already sterilized. A disadvantage of membrane filters is that many of them allow viruses and some mycoplasmas to pass through. Other disadvantages are that they may absorb relatively large amounts of the filtrate and may introduce metallic ions into the filtrate.

Membrane filters are used to sterilize materials likely to be damaged by heat sterilization. These materials include media, special nutrients that might be added to media, and pharmaceutical products such as drugs, sera, and vitamins. Some filters can be attached to syringes so that materials can be forced through them relatively quickly. Filtration can also be used instead of pasteurization in the manufacture of beer. When using filters to sterilize materials, it is important

---

FIGURE 13.17 (a) Various types of membrane filters are available to sterilize large or small quantities of liquids. Some can be vacuum-filtered, ensuring that what is forced into the bottle or flask will be sterile. (b) Scanning electron micrograph of *Staphylococcus epidermidis* cells trapped on the surface of a 0.22-μm Millipore membrane filter. Membrane pore size can be selected to allow viruses, but not bacteria, to pass through or to prevent both from passing.

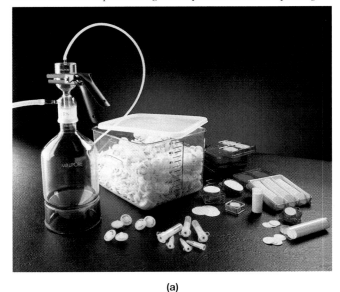

(a)

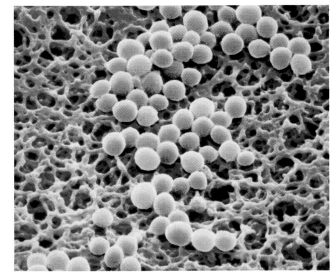

(b)

to select a filter pore size that will prevent any infectious agent from passing into the product.

In the manufacture of vaccines that require the presence of live viruses, it is important to select a filter pore size that will allow viruses to pass but prevent bacteria from doing so. By selecting a filter with a proper pore size, scientists can separate polioviruses from the fluid and debris in tissue cultures in which they were grown. This procedure simplifies the manufacture of polio vaccine. Cellulose acetate filters with extremely tiny pores are now available and are capable of removing many viruses (although not the very smallest) from liquids. However, these filters are expensive and clog easily.

Membrane filters used to trap bacteria from air and water samples can be transferred directly to agar plates, and the quantity of bacteria in the sample can be determined. Alternatively, the filters can be transferred from one medium to another, so organisms with different nutrient requirements can be detected. Filtration is also used to remove microorganisms and other small particles from public water supplies and in sewage treatment facilities. This technique, however, cannot sterilize; it merely reduces contamination.

High-efficiency particulate air (HEPA) filters are used in the ventilation systems of areas where microbial control is especially important, such as in operating rooms, burn units, and laminar flow transfer hoods in laboratories. HEPA filters also capture organisms released in rooms occupied by patients with tuberculosis or in laboratories where especially dangerous microbes are studied, such as the maximum containment units shown in Figure 16.14. These filters remove almost all organisms larger than 0.3 μm in diameter. Used filters are soaked in formalin before they are disposed of.

Osmotic Pressure

High concentrations of salt, sugar, or other substances create a hyperosmotic medium, which draws water from microorganisms by osmosis. ∞ (Chapter 4, p. 98) **Plasmolysis** (plaz-mol'i-sis), or loss of water, severely interferes with cell function and eventually leads to cell death. The use of sugar in jellies, jams, and syrups or salt solutions in curing meat and making pickles plasmolyzes most organisms present and prevents growth of new organisms. A few halophilic organisms, however, thrive in these conditions and cause spoilage, especially of pickles, and some fungi can live on the surface of jams.

Properties of physical antimicrobial agents are summarized in Table 13.5.

TABLE 13.5	Properties of Physical Antimicrobial Agents	
Agent	**Action**	**Use**
Dry heat	Denatures proteins	Oven heat used to sterilize glassware and metal objects; open flame used to incinerate microorganisms.
Moist heat	Denatures proteins	Autoclaving sterilizes media, bandages, and many kinds of hospital and laboratory equipment not damaged by heat and moisture; pressure cooking sterilizes canned foods.
Pasteurization	Denatures proteins	Kills pathogens in milk, dairy products, and beer.
Refrigeration	Slows the rate of enzyme-controlled reactions	Used to keep fresh foods for a few days; does not kill most microorganisms.
Freezing	Greatly slows the rate of enzyme-controlled reactions	Used to keep fresh foods for several months; does not kill most microorganisms; used with glycerol to preserve microorganisms.
Drying	Inhibits enzymes	Used to preserve some fruits and vegetables; sometimes used with smoke to preserve sausages and fish.
Freeze-drying	Dehydration inhibits enzymes	Used to manufacture some instant coffees; used to preserve microorganisms for years.
Ultraviolet light	Denatures proteins and nucleic acids	Used to reduce the number of microorganisms in air in operating rooms, animal rooms, and where cultures are transferred.
Ionizing radiation	Denatures proteins and nucleic acids	Used to sterilize plastics and pharmaceutical products and to preserve foods.
Microwave radiation	Absorbs water molecules, then releases microwave energy to surroundings as heat	Cannot be used reliably to destroy microbes except in special media-sterilizing equipment.
Strong visible light	Oxidation of light-sensitive materials	Can be used with dyes to destroy bacteria and viruses; may help sanitize clothing.
Sonic and ultrasonic waves	Cause cavitation	Not a practical means of killing microorganisms but useful in fractionating and studying cell components.
Filtration	Mechanically removes microbes	Used to sterilize media, pharmaceutical products, and vitamins, in manufacturing vaccines, and in sampling microbes in air and water.
Osmotic pressure	Removes water from microbes	Used to prevent spoilage of foods such as pickles and jellies.

HOSPITAL SANITATION AND THE INFECTION CONTROL PRACTITIONER

Maintaining sanitation in a modern hospital is becoming an increasingly complex task for several reasons. First, far more patients with infectious diseases are treated in hospitals today than a decade ago. Second, more techniques are available to assist with maintaining sanitation—disposable equipment and supplies, more complex isolation procedures, and advances in sanitizing agents and equipment. Using these techniques improves sanitation, but it also makes more work for the hospital staff. Third, and most important, increasing numbers of pathogens are becoming resistant to antibiotics, and these organisms are especially common in hospitals, where antibiotics are in continuous use. Hospital-acquired, or *nosocomial*, infections are therefore a constant threat to the lives of already seriously ill patients. In Chapter 16 we discuss the nature of nosocomial infections; here we focus on sterilization and disinfection procedures used to limit the spread of nosocomial and other infections in hospitals.

The hospital environment is a particularly likely place to acquire an infection. Because people with infections come to hospitals for treatment, the density of pathogens is greater in hospitals than in most other environments. The movements of hospital personnel and visitors and even air currents from elevator shafts tend to spread microbes throughout the hospital. The most likely mode of transmission of infections in a hospital is direct contact, as when a health-care worker touches an infected patient and fails to wash his or her hands before touching another patient. Indirect contact, as with contaminated equipment, and airborne transmission of pathogens are less frequent modes of transmission.

Several procedures are used to minimize the spread of infection in a hospital. Each room is disinfected after a patient is discharged and before another patient occupies it. Floors are regularly mopped with disinfectant solutions, and carpets are kept dry and vacuumed often. Linens, especially those from patients with infections, are placed in plastic bags before being dropped into a laundry chute. The linens are then laundered in strong detergents and very hot water (71°C for 25 minutes). Food is heated to an internal temperature of 74°C and kept covered and above 60°C until it is served. Dishes are washed at 60°C for 20 seconds and rinsed at 82°C for 10 seconds. Electronic air filters that ionize airborne microbes are installed in ventilating systems, especially those that serve critical-care units, burn units, and nurseries.

Hospital personnel can minimize the risk of infecting themselves and others in several ways. The single most important way is by thoroughly washing their hands between patient contacts. Another way is by receiving appropriate immunizations. Personnel should be immunized against diphtheria and tetanus and against hepatitis B if they will have contact with blood or other potentially infectious fluids. They should also be immunized against measles and mumps if they have not had those diseases and against influenza if they are susceptible to frequent pulmonary infections. Personnel also can learn and practice good aseptic techniques and carry out recommendations of the hospital's employee health programs.

Hospitals are ethically and legally responsible for patients who acquire nosocomial infections. In fact, to maintain accreditation by the American Hospital Association, hospitals must have a program that includes surveillance of nosocomial infections in both patients and staff, a microbiology laboratory, isolation procedures, accepted procedures for the use of catheters and other instruments, general sanitation procedures, and a nosocomial disease education program for staff members. The program's goal is to engage all hospital personnel in active measures to prevent infections. Most hospitals have an infection-control practitioner (ICP) to manage such a program.

ICP candidates must pass a registry examination offered by the Association for Practitioners in Infection Control (APIC). This specialty requires knowledge of both microbiology and patient-care techniques. Among currently registered specialists, some were first trained as microbiologists and many as nurses. The duties of an ICP include surveillance and identification of infections; supervision of, or collaboration with, the hospital's employee health program; keeping up-to-date on newly available immunizations to determine which ones hospital personnel should receive; assisting with studies of antibiotic use in infection control and detection of resistant organisms; and providing instruction to new staff on aseptic techniques and the hospital's infection-control program, including isolation procedures (Chapter 16).

Some specific and effective means of infection control are as follows: (1) thorough washing of hands with soap and water, which can greatly reduce the risk of physicians, nurses, and other staff members spreading diseases among patients; (2) scrupulous care in sterilizing equipment and maintaining its sterility while inserting and using catheters and other invasive instruments; (3) the use of gloves when drawing blood or handling infectious materials, such as dressings and bedpans.

Other techniques are needed to minimize the development of antibiotic-resistant pathogens. The routine use of antimicrobial agents to prevent infections has turned out to be a misguided effort becaues it contributes to the development of resistant organisms (Chapter 14). Therefore, some hospitals maintain surveillance of antibiotic use. Antibiotics are given for known infections but are given prophylactically (as a preventive measure) only in special situations. Prophylactic antibiotics are justified in surgical procedures such as hysterectomies, colorectal surgeries, and repair of traumatic injuries, in which the surgical field is invariably contaminated with potential pathogens, and in immunosuppressed patients and excessively debilitated patients, whose natural defense mechanisms may fail.

If all the known techniques for preventing nosocomial infections were rigorously practiced, the incidence of such infections could probably be reduced to half the present level.

PRINCIPLES OF STERILIZATION AND DISINFECTION

- **Sterilization** refers to the killing or removal of all organisms in any material or on any object.
- **Disinfection** refers to the reduction in numbers of pathogenic organisms on objects or in materials so that the organisms no longer pose a disease threat.
- Important terms related to sterilization and disinfection are defined in Table 13.1.

Control of Microbial Growth

- Because of the logarithmic death rate of microorganisms, a definite proportion of organisms die in a given time interval.
- The fewer organisms present, the less time will be needed to achieve sterility.
- Microorganisms differ in their susceptibility to antimicrobial agents.

CHEMICAL ANTIMICROBIAL AGENTS

Potency of Chemical Agents

- The potency, or effectiveness, of a chemical antimicrobial agent is affected by time, temperature, pH, and concentration of the agent.
- Potency increases with the length of time organisms are exposed to the agent, increased temperature, acidic or alkaline pH, and usually increased concentration of the agent.

Evaluation of Effectiveness of Chemical Agents

- Evaluation of effectiveness is difficult, and no entirely satisfactory method is available.
- For agents similar to phenol, the **phenol coefficient** is determined; it is the ratio of the dilution of the agent to the dilution of phenol that will kill all organisms in 10 minutes but not in 5 minutes.

Disinfectant Selection

- A variety of criteria are considered in selecting a **disinfectant.** In practice, most chemical agents are tested in various situations and used in situations where they produce satisfactory results.

Mechanisms of Action of Chemical Agents

- Actions of chemical antimicrobial agents can be grouped according to their effects on proteins, cell membranes, and other cell components.
- Reactions that alter proteins include hydrolysis, oxidation, and attachment of atoms or chemical groups to protein molecules. Such reactions denature proteins, rendering them nonfunctional.
- Membranes can be disrupted by agents that denature proteins and by **surfactants,** which reduce surface tension and dissolve lipids.
- Reactions of other chemical agents damage nucleic acids and energy-producing systems. Damage to nucleic acids is an important means of inactivating viruses.

Specific Chemical Antimicrobial Agents

- Soaps and detergents aid in the removal of microbes, oils, and dirt but do not sterilize.
- Acids are commonly used as food preservatives; alkali in soap helps destroy microorganisms.
- Among the agents containing heavy metals, silver nitrate is used to kill gonococci, and mercury-containing compounds are used to disinfect instruments and skin.
- Among the agents containing halogens, chlorine is used to kill pathogens in water, and iodine is a major ingredient in several skin disinfectants.
- Alcohols are used to disinfect skin.
- Phenol derivatives can be used on skin, instruments, dishes, and furniture, and to destroy discarded cultures; they work well in the presence of organic materials.
- Oxidizing agents are particularly useful in disinfecting puncture wounds.
- Alkylating agents can be used to disinfect or to sterilize a variety of materials, but all are carcinogens.
- Some dyes, plant oils, sulfur-containing substances, and nitrates can be used as disinfectants or food preservatives.

PHYSICAL ANTIMICROBIAL AGENTS

Principles and Applications of Heat Killing

- Heat destroys microorganisms by denaturing protein, melting lipids, and, when open flame is used, by incineration.

Dry Heat, Moist Heat, and Pasteurization

- Dry heat is used to sterilize metal objects and glassware.
- Flame is used to sterilize inoculating loops and the mouths of culture tubes.
- The **autoclave,** which uses moist heat under pressure, is a common instrument for sterilization and is very effective when proper procedures are followed.
- **Pasteurization** kills most pathogens in milk, other dairy products, and beer but does not sterilize.

Refrigeration, Freezing, Drying, and Freeze-drying

- Refrigeration, freezing, drying, and freeze-drying can be used to retard the growth of microorganisms.
- **Lyophilization,** drying in the frozen state, can be used for long-term preservation of live microorganisms.

Radiation

- Radiation used to control microorganisms includes ultraviolet light, ionizing radiation, and sometimes microwaves and strong sunlight.

Sonic and Ultrasonic Waves

■ Sonic and ultrasonic waves can kill microorganisms but are used mostly for **sonication,** the disruption of cells by sound waves.

Filtration

■ **Filtration** can be used to sterilize substances that are de-stroyed by heat, to separate viruses, and to collect microorganisms from air and water samples.

Osmotic Pressure

■ High concentrations of sugar or salt create osmotic pressure that results in the **plasmolysis** of cells (causes them to lose water) and prevents growth of microorganisms in highly sweetened or salted foods.

KEY TERMS

aerosol *(p. 343)*
antiseptic *(p. 332)*
autoclave *(p. 343)*
bactericidal *(p. 334)*
bacteriostatic *(p. 334)*
cavitation *(p. 349)*
decimal reduction time
    (DRT) *(p. 343)*

disinfectant *(p. 332)*
disinfection *(p. 332)*
D value *(p. 343)*
filter paper method *(p. 335)*
filtration *(p. 350)*
lyophilization *(p. 347)*
pasteurization *(p. 346)*
phenol coefficient *(p. 335)*

plasmolysis *(p. 351)*
quaternary ammonium com-
    pound (quat) *(p. 338)*
rad *(p. 349)*
sonication *(p. 350)*
sterility *(p. 332)*
sterilization *(p. 332)*
surfactant *(p. 336)*

thermal death point
    (p. 343)
thermal death time *(p. 343)*
tincture *(p. 339)*
ultrahigh temperature proc-
    essing *(p. 346)*
use-dilution test *(p. 336)*
wetting agent *(p. 336)*

QUESTIONS FOR REVIEW

A **Sterilization and Disinfection**

1. How do sterilization and disinfection differ?

2. Define sterilization, disinfection, antiseptic, disinfectant, sanitizer, bacteriostatic agent, germicide, bactericide, viricide, fungicide, and sporocide.

3. True or False: There are degrees of sterility. Explain.

B **Principles of Sterilization and Disinfection**

4. True or False: A given antimicrobial agent may not affect all species of bacteria equally. Explain.

5. What three principles apply to the processes of sterilization and disinfection?

C **Potency of Antimicrobial Chemical Agents**

6. How do each of the following affect the potency of chemical antimicrobial agents: time, temperature, pH, and concentration of the chemical agent?

D **Effectiveness of Antimicrobial Chemical Agents**

7. What is the phenol coefficient, and how is it determined?

8. Explain how antimicrobial agents A through D in the following diagram compare in effectiveness against gram-positive and gram-negative bacteria. The control filter paper was soaked in sterile water.

9. True or False: A single antimicrobial chemical can be both bacteriostatic and bactericidal. Explain.

10. Why is a 70 percent alcohol–water mixture more effective at killing microbes than is a 99 percent alcohol–water mixture?

E **Mechanisms of Action of Antimicrobial Chemical Agents**

11. How do chemical agents affect (a) proteins and (b) cell membranes?

12. What other effects do chemical agents have on cells?

F **Properties of Antimicrobial Chemical Agents**

13. How do each of the following agents affect cells: acids and alkalis, heavy metals, halogens, alcohols, phenols, oxidizing agents, alkylating agents, surfactants, and dyes?

14. Match the following:

    ——— Hibiclens   **a.** heavy metal

    ——— mercurochrome   **b.** halogen

    ——— household bleach **c.** phenols

    ——— hydrogen peroxide **d.** oxidizing agent

    ——— selenium shampoo **e.** dye

    ——— Betadine   **f.** alkylating agent

    ——— cresol

    ——— acridine

    ——— ethylene oxide

    ——— tincture of iodine

    ——— formaldehyde

G **Dry Heat; Moist Heat; Pasteurization**

15. Compare and contrast the effects of different ways to control microorganisms with heat.

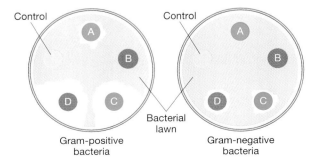

Gram-positive bacteria   Gram-negative bacteria

Control   Control   A   B   D   C   Bacterial lawn

H Refrigeration; Freezing; Drying; Freeze-Drying

16. What benefits and hazards are associated with refrigerating foods?

17. Explain how freezing, drying, and freeze-drying can be used to control or to preserve organisms.

I Radiation; Sonic and Ultrasonic Waves; Filtration; Osmotic Pressure

18. How are cells affected by various types of radiation?

19. What is sonication, and how is it used in microbiology?

20. List (a) some advantages and disadvantages and (b) some applications of membrane filters.

21. What is plasmolysis, and how can it be used to prevent growth of microorganisms?

PROBLEMS FOR INVESTIGATION

1. Suppose you were a member of a committee in charge of maintaining the best possible control of microorganisms in a hospital, a laboratory, or a food-handling industry. Show what kinds of problems of controlling microorganisms might exist and how you would effectively solve these problems.

2. Devise a method of evaluating the effectiveness of one of the following groups of chemical agents: skin antiseptics, food–utensil disinfectants, or spore-killing agents.

3. Evaluate the autoclaving procedures used in your laboratory.

4. A technician was testing a new disinfectant to determine its phenol coefficient and obtained the following results. What is the phenol coefficient? Is this likely to be a good disinfectant?

EXPOSURE TIME	PHENOL DILUTION				NEW DISINFECTANT DILUTION			
	1:100	1:110	1:120	1:130	1:50	1:60	1:70	1:80
5 min.	+	+	+	−	+	+	−	−
10 min.	+	+	−	−	+	−	−	−

5. Look up the recommended procedures for a surgical scrub. Prepare a report and demonstration for your class.

6. Find out which procedures are used on a patient's skin prior to surgery.

7. Report on the industrial use of lasers for sterilization.

8. The central sterilization facilities in a large hospital were located in a basement where air exchange was not very good. One day all employees working in this unit began to complain of headache, burning eyes, nausea, and dizziness. Then two of them fainted. Everyone quickly evacuated the area. (a) What could have caused this problem? (b) What further damage may employees have suffered? (The answer to this question appears in Appendix F.)

SOME INTERESTING READING

Block, S. S., ed. 1991. *Disinfection, sterilization, and preservation.* 4th ed. Philadelphia: Lea and Febiger.

Bryan, R. M., and L. A. Bland. 1981. "Occupational exposure to ethylene oxide: Its effect and control." *Journal of Environmental Health* 43:254–59.

Centers for Disease Control and Prevention. 1993. "Recommended infection control practices for dentistry." *MMWR* 41 (no. RR-8):1–12.

Favero, M. S., and W. W. Bond. 1991. "Sterilization, disinfection, and antisepsis in the hospital." In *Manual of clinical microbiology,* 5th ed., by A. Balows, et al. Washington, D.C.: American Society for Microbiology.

Fleming, D. O., et al., eds. 1995. *Laboratory safety: Principles and practices.* 2d ed. Washington, D.C.: ASM Press.

Jensen, M. M., and D. N. Wright. 1993. *Introduction to medical microbiology.* Englewood Cliffs, NJ: Prentice Hall.

Kelly, M. C. 1993. "Sterilization and disinfection choices and methods." *Practical Hygiene* (July/August):37–40.

Litsky, W. 1990. "Wanted: Plastics with antimicrobial properties." *American Journal of Public Health* 80 (January): 13–15.

Reinhardt, P. A., and J. G. Gordon. 1990. *Infectious and medical waste management.* Chelsea, MI: Lewis Publishers.

Rutala, W. A. 1993. "Disinfection, sterilization, and waste disposal." In *Prevention and control of nosocomial infections,* 2d ed., edited by R. P. Wenzel. Baltimore: Williams and Wilkins.

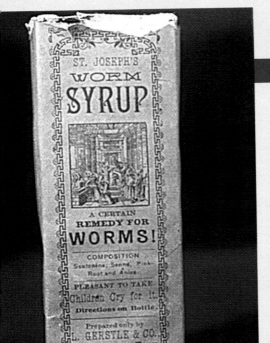

CHAPTER

14

ANTIMICROBIAL THERAPY

St. Joseph's Worm Syrup—proclaimed to be "a certain remedy for worms!" More modern drugs have decreased the death rate due to microbial infections, but overuse of such medications can promote growth of resistant organisms.

As you lie in your sickbed, suffering from some infectious disease, how reassuring it is to be able to reach over, swallow some capsules, and look forward to getting well soon. But people have not always been able to do this. Over the ages anxious parents have watched their children die of fevers, diarrhea, infected wounds, and other maladies. Whole villages have been wiped out by plagues. During the middle of the fourteenth century, more than a quarter of the population of Europe died of the Black Death (bubonic plague) in just a few years. More people have died of infection in wartime than from swords or bullets. Until relatively recently, the only defenses against infectious diseases were such things as herbal teas and poultices (soaks), or fleeing the disease area. Modern and effective weapons against microbes—antibiotics and sulfa drugs—did not become available until the twentieth century.

Think back through your life. Have there been times when you might have died had it not been for antimicrobial drugs? At what age would you have died? Which of your family members might not be alive now? Happily, we live in a better time with re-

spect to illnesses and deaths caused by infectious organisms. In the United States the life expectancy of a baby born in 1850 was less than 40 years; in 1900, about 50 years; and in 1990 over 70 years. Infectious diseases claimed the lives of about 1 of every 100 U.S. residents per year as late as 1900 but only about 1 of every 300 in 1990. Although antimicrobial agents still don't save all patients, they have drastically lowered the death rate from infectious disease. A period of increased infectious diseases could return, however, if patients and the medical community fail to protect the effectiveness of antimicrobial agents. Our ability to fight infectious diseases is dwindling as many pathogens develop resistance to available antimicrobial drugs.

ANTIMICROBIAL CHEMOTHERAPY

The term **chemotherapy** was coined by the German medical researcher Paul Ehrlich to describe the use of chemical substances to kill pathogenic organisms without injuring the host. Today, chemotherapy refers to

356

THIS CHAPTER FOCUSES ON THE FOLLOWING QUESTIONS:

A What terms are used to discuss chemotherapy and antibiotics, and what do they mean?

B How have chemotherapeutic agents been developed?

C How do the terms selective toxicity, spectrum of activity, and modes of action apply to antimicrobial agents?

D What kinds of side effects are associated with antimicrobial agents?

E What is resistance to antibiotics, and how do microorganisms acquire it?

F How are sensitivities of microbes to chemotherapeutic agents determined?

G What are the attributes of an ideal antimicrobial agent?

H What are the properties, uses, and side effects of antibacterial agents?

I What are the properties, uses, and side effects of antifungal agents, antiviral agents, antiprotozoan agents, and antihelminthic agents?

J How do drug-resistant hospital infections arise, and how can they be treated and prevented?

---

the use of chemical substances to treat various aspects of disease—aspirin for headache and inflammation, drugs to regulate heart function, and agents to rid the body of malignant cells. With this broad modern definition of chemotherapy, we describe a **chemotherapeutic agent** as any chemical substance used in medical practice. Such agents also are referred to as **drugs.**

In microbiology we are concerned with **antimicrobial agents,** a special group of chemotherapeutic agents used to treat diseases caused by microbes. Thus, in modern terms, an antimicrobial agent is synonymous with a chemotherapeutic agent as Erhlich originally defined it. In this chapter we consider a variety of antimicrobial agents and a few agents used to treat helminth infections.

Antibiosis literally means "against life." In the 1940s Selman Waksman, the discoverer of streptomycin, defined an **antibiotic** as "a chemical substance produced by microorganisms which has the capacity to inhibit the growth of bacteria and even destroy bacteria and other microorganisms in dilute solution." In contrast, agents synthesized in the laboratory are called **synthetic drugs.** Some antimicrobial agents are syn-

thesized by chemically modifying a substance from a microorganism. More often a synthetic precursor different from the natural one is supplied to a microorganism, which then completes synthesis of the antibiotic. Antimicrobial agents made partly by laboratory synthesis and partly by microorganisms are called **semisynthetic drugs.**

HISTORY OF CHEMOTHERAPY

Throughout history humans have attempted to alleviate suffering by treating disease—often by taking concoctions of plant substances. Although ancient Egyptians used moldy bread to treat wounds, they had no knowledge of the antibiotics it contained. Extracts of willow bark, now known to contain a compound closely related to aspirin, were used to alleviate pain. Parts of the foxglove plant were used to treat heart disease in the sixteenth century, although the active ingredient, digitalis, had not been identified. Likewise, quinine-containing extracts from the cinchona tree were used to treat malaria.

Despite their reputation for using rituals irrelevant to the cure of disease, traditional healers of primitive societies, especially in the tropics, are quite knowledgeable about medicinal properties of plants. Their knowledge has been passed down from generation to generation. Because these healers are disappearing, pharmaceutical companies are attempting to learn from them and make written records of their treatments as well as test the plants they use.

In Western civilization, the first systematic attempt to find specific chemical substances to treat infectious disease was made by Paul Ehrlich. ∞ (Chapter 1, p. 15) Although his discovery in 1910 of Salvarsan to treat syphilis was of great therapeutic benefit, even more important were the concepts he developed in the new science of chemotherapy. He was interested in the mechanisms by which chemical substances bind to microorganisms and to animal tissues. His studies of chemicals that bind to tissues led to histological (tissue) stains that are still used today.

The next advances in chemotherapy were the nearly concurrent development of sulfa drugs and antibiotics. In 1935 Gerhard Domagk discovered that prontosil, a red dye, inhibits growth of many gram-positive bacteria. The following year Ernest Fourneau found that the antimicrobial activity was due to the sulfanilamide portion of the prontosil molecule. These discoveries stimulated the development of a group of substances called *sulfonamides,* or *sulfa drugs.* As the number of sulfa drugs grew, it became possible to use them to attack directly a variety of pathogens. However, the usefulness of sulfa drugs is limited. They do not attack all pathogens, and they sometimes cause kidney damage and allergies. But they have saved many lives and continue to do so today.

Alexander Fleming reasoned that the ability of the mold *Penicillium* to inhibit growth of microorganisms might be exploited. This idea led him to identify the inhibitory agent and name it *penicillin.* In 1928 Fleming had observed the contamination of his bacterial cultures with this fungus many times, as had many other microbiologists. However, instead of grumbling about another contaminated culture and tossing it out, Fleming saw the tremendous potential in this accidental finding. If only the substance (penicillin) could be extracted and collected in large quantities, it could be used to combat infection.

Fleming's idea did not come to fruition until the early 1940s, when Ernst Chain and Howard Florey finally isolated penicillin and worked with other researchers to develop methods of mass production. Such mass production occurred during World War II and saved the lives of many people whose wounds became infected. Supplies of the drug were limited, however, and it was not readily available to civilians until after the war. Following the war, research proceeded rapidly, and new antibiotics were discovered one after another.

CLOSE-UP

WHEN DOCTORS LEARNED TO CURE

"We were provided with a thin, pocketsize book called *Useful Drugs,* one hundred pages or so, and we carried this around in our white coats when we entered the teaching wards and clinics . . . , but I cannot recall any of our instructors ever referring to this volume. Nor do I remember much talk about treating disease at any time in the four years of medical school except by the surgeons. . . . Our task for the future was to be diagnosis and explanation. Explanation was the real business of medicine. What the ill patient and his family wanted most was to know the name of the illness, and then, if possible, what had caused it, and finally, most important of all, how it was likely to turn out. . . . It gradually dawned on us that we didn't know much that was really useful, that we could do nothing to change the course of the great majority of the diseases we were so busy analyzing. . . . Then came the explosive news of sulfanilamide, and the start of the real revolution in medicine. I remember with astonishment when the first cases of pneumococcal and streptococcal septicemia were treated in Boston in 1937. The phenomenon was almost beyond belief. Here were moribund patients, who would surely have died without treatment, improving in their appearance within a matter of hours of being given the medicine and feeling entirely well within the next day or so."

—Lewis Thomas, 1983

The introduction of penicillin and sulfonamides in the 1930s can be said to mark the beginning of modern medicine. As the medical writer Lewis Thomas said, "Doctors could now *cure* disease, and this was astonishing, most of all to the doctors themselves."

GENERAL PROPERTIES OF ANTIMICROBIAL AGENTS

Antimicrobial agents share certain common properties. We can learn much about how these agents work and why they sometimes do not work by considering such properties as selective toxicity, spectrum of activity, mode of action, side effects, and resistance of microorganisms to them.

Selective Toxicity

Some chemical substances with antimicrobial properties are too toxic to be taken internally and are used only for topical application—application to the skin's surface. For internal use, an antimicrobial drug must have **selective toxicity**—that is, it must harm the microbes without causing significant damage to the host. Some drugs, such as penicillin, have a wide range between the **toxic dosage level,** which causes host damage, and the **therapeutic dosage level,** which successfully eliminates the pathogenic organism if the level is maintained over a period of time. The relationship between an agent's toxicity to the body and its toxicity to an infectious agent is expressed in terms of its chemotherapeutic index. For any particular agent, the **chemotherapeutic index** is defined as the maximum tolerable dose per kilogram of body weight, divided by the minimum dose per kilogram of body weight, that will cure the disease. Thus an agent with a chemotherapeutic index of 8 would be more effective and less toxic to the patient than an agent with a chemotherapeutic index of 1.

For drugs such as those containing arsenic, mercury, and antimony, the dosage must be calculated very precisely because these substances are highly toxic to human and other animal hosts as well as to pathogens. Treatment of worm infections is especially difficult because what damages the parasite will also damage the host. In contrast, bacterial pathogens often can be treated by interfering with metabolic pathways not shared by the host. For example, penicillin interferes with cell wall synthesis; it is not toxic to human cells, which lack walls, though some patients are allergic to it.

Spectrum of Activity

The range of different microbes against which an antimicrobial agent acts is called its **spectrum of activity.** Those agents that are effective against a great number of microorganisms from a wide range of taxonomic groups, including both gram-positive and gram-negative bacteria, are said to have a **broad spectrum** of activity. Those that are effective against only a small number of microorganisms or a single taxonomic group have a **narrow spectrum** of activity (Figure 14.1). Some common antibiotics are classified according to their spectrum of activity in Table 14.1.

A broad-spectrum drug is especially useful when a patient is seriously ill with an infection caused by an unidentified organism. Using such a drug increases the chance that the organism will be susceptible to it. However, if the identity of the organism is known, a narrow-spectrum drug should be used. Using such a drug minimizes the destruction of the host's *microbiota,* or

TABLE 14.1	The Spectrum of Activity of Selected Antimicrobial Agents	
Organisms Affected	Broad-Spectrum Agents*	Narrow-Spectrum Agents
Bacteroides and other anaerobes	Cephalosporins	Lincomycin
Yeasts	Chloramphenicol	Nystatin
Gram-positive bacteria	Gentamicin	Penicillin G
	Ampicillin	Erythromycin
Gram-negative bacteria	Kanamycin	Polymyxins
Streptococci and some gram-negative bacteria	Tetracyclines	Streptomycin
Staphylococci, enterococci, and some clostridia	Tetracyclines	Vancomycin

*Broad-spectrum agents affect most bacteria.

FIGURE 14.1 Broad-spectrum drugs, such as tetracycline, affect a variety of different organisms. Narrow-spectrum drugs, such as isoniazid, affect only a few specific types of organisms.

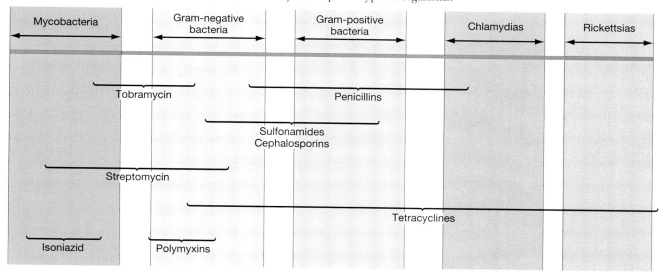

normal flora—the indigenous microbes that naturally occur in or on the host—that sometimes compete with and help destroy infectious organisms. The use of narrow-spectrum drugs also decreases the likelihood that organisms will develop drug resistance.

Modes of Action

Like other medicines, antimicrobial agents are sometimes used simply because they work, without our always knowing how they work. Many people's lives have been saved by medicines whose actions at the cellular level have never been understood. However, it is always desirable to know the mode of action of an agent. With that knowledge, effects of actions on patients can be better monitored and controlled, and ways of improving them may be found.

Antimicrobial drugs generally act on an important microbial structure or function that usually differs from its counterpart in animals. This difference is exploited in exerting a *bactericidal,* or killing, effect or a *bacteriostatic,* or growth-inhibiting, effect on bacteria while having minimal effects on host cells. ∞ (Chapter 13, p. 334) However, the host's immune systems or phagocytic defenses must still complete the elimination of the invading microbes.

Five different modes of action of antimicrobials are discussed here: (1) inhibition of cell wall synthesis, (2) disruption of cell membrane function, (3) inhibition of protein synthesis, (4) inhibition of nucleic acid synthesis, and (5) action as antimetabolites (Figure 14.2).

Inhibition of Cell Wall Synthesis

Many bacterial and fungal cells have rigid external cell walls, whereas animal cells lack cell walls. Consequently, inhibiting cell wall synthesis selectively damages bacterial and fungal cells. Bacterial cells, especially gram-positive ones, have a high internal osmotic pressure. Without a normal, sturdy cell wall, these cells burst when subjected to the low osmotic pressure of body fluids. ∞ (Chapter 4, p. 99) Antibiotics such as penicillin and cephalosporin contain a chemical structure called a *β-lactam ring*, which attaches to the enzymes that cross-link peptidoglycans. ∞ (Chapter 4, p. 78) By interfering with the cross-linking of tetrapeptides, these antibiotics prevent cell wall synthesis (Figure 14.3). Fungi, whose cell walls lack peptidoglycan, are unaffected by these antibiotics.

Disruption of Cell Membrane Function

All cells are bounded by a membrane. Although the membranes of all cells are quite similar, those of bacteria and fungi differ sufficiently from those of animal cells to allow selective action of antimicrobial agents. Certain polypeptide antibiotics, such as polymyxins, act as detergents and distort bacterial cell membranes, probably by binding to phospholipids in the membrane. (With this distortion, the membrane is no longer regulated by membrane proteins, and the cytoplasm and cell substances are lost.) These antibiotics are especially effective against gram-negative bacteria,

---

FIGURE 14.2 The five major modes of action by which drugs exert their antimicrobial effects on bacterial cells.

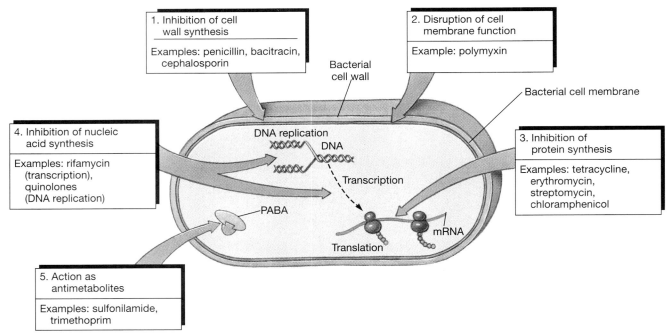

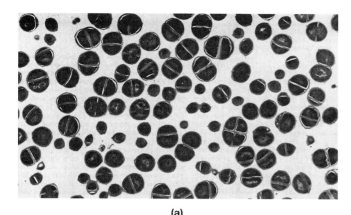

(a)

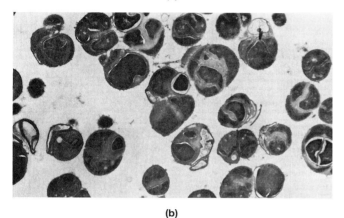

(b)

FIGURE 14.3 SEM photos of bacteria (a) before and (b) after exposure to penicillin (magnified 785×). Note the distortion of the cell shape due to the disruption by penicillin of the cross-linking tetrapeptides in the peptidoglycan layer of the cell wall.

which have an outer membrane rich in phospholipids. ∞ (Chapter 4, p. 78) Polyene antibiotics, such as amphotericin B, bind to particular sterols present in the membranes of fungal (and animal) cells. Thus, polymyxins do not act on fungi, and polyenes do not act on bacteria.

Inhibition of Protein Synthesis

In all cells, protein synthesis requires not only the information stored in DNA, plus several kinds of RNA, but also ribosomes. Differences between bacterial (70S) and animal (80S) ribosomes allow antimicrobial agents to attack bacterial cells without significantly damaging animal cells—that is, with selective toxicity. Aminoglycoside antibiotics, such as streptomycin, derive their name from the amino acids and glycosidic bonds they contain. They act on the 30S portion of bacterial ribosomes by interfering with the accurate reading (translation) of the mRNA message—that is, the incorporation of the correct amino acids. ∞ (Chapter 7, p. 167)

CLOSE-UP

SHARKS: GERM WARFARE EXPERTS

Two years is a long time to be pregnant—especially when your reproductive tract allows seawater to leak into it. Infection by the many microbes that enter the uteruslike oviducts of the female dogfish shark (*Squalus acanthias*) would seem a likely end of the mother and/or most of her seven baby "pups." However, the sharks do not develop infections. The secret of their resistance lies in a broad-spectrum antibiotic called squalamine, which is present in all their tissues. It attacks cell membranes of microbes, forming holes in them. Squalamine has been synthesized in the laboratory and is under investigation for possible human use.

Chloramphenicol and erythromycin act on the 50S portion of bacterial ribosomes, inhibiting the formation of the growing polypeptide. Because animal cell ribosomes consist of 60S and 40S subunits, these antibiotics have little effect on host cells. (Mitochondria, however, which have 70S ribosomes, can be affected by such drugs.)

Inhibition of Nucleic Acid Synthesis

Differences between the enzymes used by bacterial and animal cells to synthesize nucleic acids provide a means for selective action of antimicrobial agents. Antibiotics of the rifamycin family bind to a bacterial RNA polymerase and inhibit RNA synthesis. ∞ (Chapter 7, p. 171)

Action as Antimetabolites

The normal metabolic processes of microbial cells involve a series of intermediate compounds called *metabolites* that are essential for cellular growth and survival. **Antimetabolites** are substances that affect the utilization of metabolites and therefore prevent a cell from carrying out necessary metabolic reactions. Antimetabolites function in two ways: (1) by competitively inhibiting enzymes and (2) by being erroneously incorporated into important molecules such as nucleic acids. Antimetabolites are structurally similar to normal metabolites. The actions of antimetabolites are sometimes called **molecular mimicry** because they mimic, or imitate, the normal molecule, preventing a reaction from occurring or causing it to go awry.

In competitive inhibition an enzymatic reaction is inhibited by a substrate that binds to the enzyme's

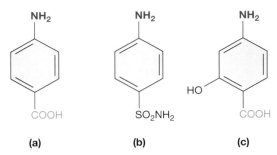

FIGURE 14.4 (a) *Para*-aminobenzoic acid (PABA), a metabolite required by many bacteria. (b) Sulfanilamide, a sulfa drug. (c) *Para*-aminosalicylic acid (PAS). Sulfanilamide and PAS act as competitive inhibitors to PABA. Note the similarity in the structures of the three compounds.

active site but cannot react. ∞ (Chapter 5, p. 114) While this competing substrate occupies the active site, the enzyme is unable to function, and metabolism will slow or even cease if enough enzyme molecules are inhibited. Consider sulfanilamide and *para*-aminosalicylic acid (PAS), which are chemically very similar to *para*-aminobenzoic acid (PABA) (Figure 14.4). They competitively inhibit an enzyme that acts on PABA. Many bacteria require PABA in order to make folic acid, which they use in synthesizing nucleic acids and other metabolic products. When sulfanilamide or PAS instead of PABA is bound to the enzyme, the bacterium cannot make folic acid. Animal cells lack the enzymes to make folic acid and must obtain it from their diets; thus their metabolism is not disturbed by these competitive inhibitors.

Antimetabolites such as the purine analogue vidarabine and the pyrimidine analogue idoxuridine are erroneously incorporated into nucleic acids. These molecules are very similar to the normal purines and pyrimidines of nucleic acids (Figure 14.5). When incorporated into a nucleic acid, they garble the information that it encodes because they cannot form the correct base pairs during replication and transcription. Purine and pyrimidine analogues are generally as toxic to animal cells as to microbes because all cells use the same purines and pyrimidines to make nucleotides. These agents are most useful in treating viral infections, because viruses incorporate analogues more rapidly than do cells and are more severely damaged.

Kinds of Side Effects

The side effects of antimicrobial agents on infected persons (hosts) fall into three general categories: (1) toxicity, (2) allergy, and (3) disruption of normal microbiota. The development of resistance to antibiotics can also be thought of as a side effect on the microorganisms. As is explained later, resistance produces infections that can be difficult to treat.

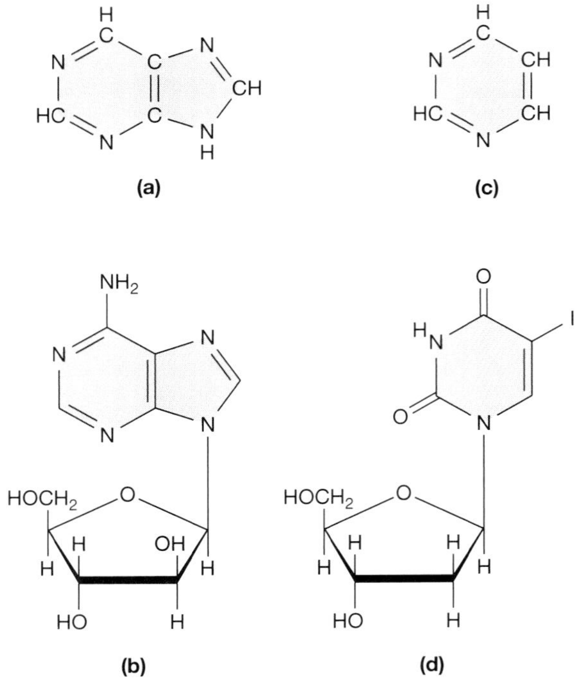

FIGURE 14.5 Nucleic acid bases and their analogues: molecules so similar in structure that they can be incorporated in place of the correct molecule, thus acting as antimetabolites. (a) Basic structure of a purine. (b) The purine analogue vidarabine. (c) Basic structure of a pyrimidine. (d) The pyrimidine analogue idoxuridine.

Toxicity

By their selective toxicity and modes of action, antimicrobial agents kill microbes without seriously harming host cells. However, some antimicrobials do exert toxic effects on the patients receiving them. The toxic effects of antimicrobial agents are discussed later in connection with specific agents.

Allergy

An *allergy* is a condition in which the body's immune system responds to a foreign substance, usually a protein. For example, breakdown products of penicillins combine with proteins in body fluids to form a molecule that the body treats as a foreign substance. Allergic reactions can be limited to mild skin rashes and itching, or they can be life-threatening. One kind of life-threatening allergic reaction, called *anaphylactic shock* (Chapter 19), occurs when an individual is subjected to a foreign substance to which his or her body has already become sensitized—that is, a substance to which the individual has been exposed and developed antibodies against.

Disruption of Normal Microbiota

Antimicrobial agents, especially broad-spectrum antibiotics, may exert their adverse effects not only on

BIOTECHNOLOGY

PHARMACY OF THE FUTURE: TRIPLE-STRANDED DNA

Why struggle to destroy or counteract a harmful microbial product, such as a toxin or a pyrogen (a substance that produces fever), when you could shut it off at its source? Such pathogenic substances either are proteins or are produced by reactions that require specific proteins (enzymes). Either way, a pathogen cannot produce these substances without transcribing and translating specific genes.

To short-circuit these processes, pharmacists may someday soon dispense preparations containing a third strand of nucleotides. When taken, this strand will nestle between the pair of DNA strands at a specific location on a chromosome to form triple-stranded (triplex) DNA. Bonding to both strands, it will prevent them from separating to manufacture messenger RNA, in essence "turning off" the gene needed to produce the harmful substance. Moreover, if vital genes are inactivated in this way, the pathogen will die. This process should work against viruses as well.

The same strategy might also make it possible to turn off defective human genes. Then, by means of genetic engineering techniques, the patient could be given good genes to replace them. Researchers are currently working with drugs that form triplex DNA. They are also experimenting with another genetic medicine called an *antisense drug*, made of short pieces of nucleic acids called oligonucleotides that bind to mRNA and prevent its translation. Several of these new "magic bullets" have already reached clinical trials.

pathogens but also on indigenous microbiota—the microorganisms that normally inhabit the skin and the digestive, respiratory, and urogenital tracts. When these microbiota are disturbed, other organisms not susceptible to the antimicrobial agent, such as *Candida* yeast, invade the unoccupied areas and multiply rapidly. Invasion by replacement microbiota is called **superinfection.** Superinfections are difficult to treat because they are susceptible to few antibiotics.

Although short-term use of penicillins generally does not severely disrupt normal microbiota, oral ampicillin sometimes allows overgrowth of toxin-producing *Clostridia*. Long-term use of penicillin or aminoglycosides can abolish natural microbiota and allow colonization of the gut with resistant gram-negative bacteria and fungi such as *Candida*. (A preparation called Lactinex, which contains normal microbiota, can be given to counteract the effects of antibiotics.) Oral and vaginal superinfections with species of *Candida*

yeasts are common after prolonged use of antimicrobial agents such as cephalosporins, tetracyclines, and chloramphenicol. The risk of serious superinfections is greatest in hospitalized patients receiving broad-spectrum antibiotics, for two reasons. First, patients often are debilitated and less able to resist infection. Second, they are in an environment in which drug-resistant pathogens are prevalent.

Resistance of Microorganisms

Resistance of a microorganism to an antibiotic means that a microorganism formerly susceptible to the action of the antibiotic is no longer affected by it. An important factor in the development of drug-resistant strains of microorganisms is that many antibiotics are bacteriostatic rather than bactericidal. Unfortunately, the most resilient microbes evade defenses (Chapter 15) and are likely to develop resistance to the antibiotic.

How Resistance Is Acquired

Microorganisms generally acquire antibiotic resistance by genetic changes, but sometimes they do so by nongenetic mechanisms. Nongenetic resistance occurs when microorganisms such as those that cause tuberculosis persist in the tissues out of reach of antimicrobial agents. If the sequestered microorganisms start to multiply and release their progeny, the progeny are still susceptible to the antibiotic. This type of resistance might more properly be called *evasion*. Another type of nongenetic resistance occurs when certain strains of bacteria temporarily change to L forms that lack most of their cell walls. ∞ (Chapter 10, p. 255) For several generations, while the cell wall is lacking, these organisms are resistant to antibiotics that act on cell walls. However, when they revert to producing cell walls, they again become susceptible to the antibiotics.

Genetic resistance to antimicrobial agents develops from genetic changes followed by natural selection (Figure 14.6). ∞ (Chapter 8, p. 207) For example, in

FIGURE 14.6 Method for detecting genetic resistance. (a) A mixed population of bacteria of varying resistance to a new antibiotic is present. (b) Antibiotic is added to the Petri plate. Only those organisms with sufficient resistance will survive. Introduction of the antibiotic represents a change in the environment, but it does not create the resistant organisms—they were already there.

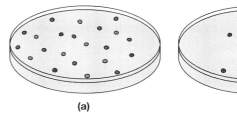

(a) (b)

ANTIBIOTIC RESISTANCE I: SLUDGE

Some cities dump sludge from waste treatment plants on farmland as fertilizer. Although the practice is an economically attractive method of sludge disposal and helps agricultural productivity, it has risks. Large numbers of antibiotic-resistant bacteria in the sludge, such as *Escherichia coli*, are released into the environment and may later be found on vegetables harvested from the land. Although these organisms are not themselves pathogenic, they are likely to become incorporated into the natural microbial populations of the human gut. There they can serve as a reservoir of antibiotic resistance genes, which could be transferred to pathogens. Other risks include accumulation of heavy metals and toxic compounds in the soil. For these reasons, the EPA requires food plants whose edible parts contact the ground not be grown in sludge-treated soil.

Dumping sludge from waste treatment on farm fields in Switzerland.

most bacterial populations, mutations occur spontaneously at a rate of about 1 per 10 million to 10 billion organisms. Bacteria reproduce so rapidly that billions of organisms can be produced in a short period of time, and among them there will always be a few mutants. If a mutant happens to be resistant to an antimicrobial agent in the environment, that mutant and its progeny will be most likely to survive, whereas the nonresistant organisms will die. After a few generations, most survivors will be resistant to the antimicrobial agent. Antibiotics do *not* induce mutations, but they can create environments that favor the survival of mutant resistant organisms.

Genetic resistance in bacteria, where it is best understood, can be due to changes in the bacterial chromosome or to the acquisition of extrachromosomal

DNA, usually in plasmids. (The mechanisms by which genetic changes occur were described in Chapters 7 and 8.) **Chromosomal resistance** is due to a mutation in chromosomal DNA and will usually be effective only against a single type of antibiotic. Such mutations often alter the DNA that directs the synthesis of ribosomal proteins. **Extrachromosomal resistance** is usually due to the presence of particular kinds of **resistance (R) plasmids,** or **R factors.** ∞ (Chapter 8, p. 207) How R plasmids originated is unknown, but they were first discovered in *Shigella* in Japan in 1959. Since that time many different R plasmids have been identified. Some R plasmids carry as many as six or seven genes, each of which confers resistance to a different antibiotic. R plasmids can also be transferred from one strain or species of bacteria to another. Most transfers occur by transduction (the transfer of plasmid DNA in a bacteriophage), and some occur by conjugation. ∞ (Chapter 8, pp. 200 and 203)

Mechanisms of Resistance

Five mechanisms of resistance have been identified, each of which involves the alteration of a different microbial structure. One involves the alteration of the target to which antimicrobial agents bind, a process that generally is caused by a mutation in the bacterial chromosome. The other mechanisms involve alterations in membrane permeability, enzymes, or metabolic pathways, which usually are caused by the acquisition of R plasmids. The five mechanisms are explained next.

(1) The alteration of targets usually affects bacterial ribosomes. The mutation alters the DNA such that the protein produced or target is modified. Antimicrobial agents can no longer bind to the target. Resistance to erythromycin, rifamycin, and antimetabolites has developed by this mechanism.

(2) The alteration of membrane permeability occurs when new genetic information changes the nature of proteins in the membrane. Such alterations change a membrane transport system or pores in the membrane, so an antimicrobial agent can no longer cross the membrane. In bacteria, resistance to tetracyclines, quinolones, and some aminoglycosides has occurred by this mechanism. The presence of penicillin or cephalosporin can partially overcome such resistance because these agents interfere with cell wall synthesis.

(3) The development of enzymes that can destroy or inactivate antimicrobial agents is a common cause of resistance. One enzyme of this type is β-lactamase. Several β-lactamases exist in various bacteria; they are capable of breaking the β-lactam ring in penicillins and some cephalosporins. Similar enzymes that can destroy various aminoglycosides and chloramphenicol have been found in certain gram-negative bacteria.

(4) The alteration of an enzyme such that a formerly inhibited reaction can occur is exemplified by a mechanism found among certain sulfonamide-resistant bacteria. These organisms have developed an enzyme that has a very high affinity for PABA and a very low affinity for sulfonamide. Consequently, even in the presence of sulfonamide, the enzyme works well enough to allow the bacterium to function.

(5) The alteration of a metabolic pathway to bypass a reaction inhibited by an antimicrobial agent occurs in other sulfonamide-resistant bacteria. These organisms have acquired the ability to use ready-made folic acid from their environment and no longer need to make it from PABA.

First-Line, Second-Line, and Third-Line Drugs

As a strain of microorganism acquires resistance to a drug, another drug must be found to treat resistant infections effectively. If resistance to a second drug develops, a third drug is needed, and so on. Drugs used to treat gonorrhea illustrate this point. Before the 1930s no effective treatment was available for gonorrhea. Then sulfonamides were found to cure the disease. After a few years, sulfonamide-resistant strains developed, but penicillin was soon available as a "second-line" drug. Over several decades penicillin-resistant strains developed but were combatted with very large doses of penicillin. By the 1970s some strains of gonococci developed the ability to produce a β-lactamase enzyme, which completely counteracted the effects of penicillin (Figure 14.7). "Third-line" spectinomycin is now being used. As spectinomycin-resistant strains start to appear, forcing physicians to resort to "fourth-line" drugs, we have to wonder whether the development of new drugs can go on indefinitely.

Drug-resistant organisms have most frequently been encountered within hospitals, where seriously ill patients with lowered resistance to infections serve as convenient hosts. However, more and more resistant organisms are being isolated from infections among the general population, and the risk of acquiring a drug-resistant infection is increasing for everyone. Moreover, many organisms are resistant to multiple antibiotics. Infections with such organisms are particularly difficult to treat. A new use for genetic probes will be to look for resistance genes in organisms, in order to avoid delays in effective treatment.

Cross-Resistance

Cross-resistance is resistance to two or more similar antimicrobial agents via a common mechanism. The action of β-lactamases provides a good example of cross-resistance. In many instances, an enzyme that will break down one β-lactam antibiotic also will break down several other β-lactam antibiotics. The presence of such an enzyme would give a microorganism resistance to all the antibiotics it can break down.

Limiting Drug Resistance

Although, as we have seen, drug resistance is not induced by antibiotics, it is fostered by environments containing antibiotics. The progress of microbes in acquiring resistance can be thwarted in three ways. First, high levels of an antibiotic can be maintained in the bodies of patients long enough to kill all pathogens, including resistant mutants, or to inhibit them so that body defenses can kill them. This is the reason your doctor admonishes you to be sure to take all of an antibiotic prescription and not to stop taking it once you begin to feel better. The development of resistance when medication is discontinued before all pathogens are killed is illustrated in Figure 14.8.

Second, two antibiotics can be administered simultaneously so that they can exert an additive effect called **synergism.** For example, when streptomycin and penicillin are combined in therapy, the damage to the cell wall caused by the penicillin allows better penetration by streptomycin. A variation on this principle is the use of one agent to destroy the resistance of microbes to another agent. When clavulanic acid and a penicillin called amoxicillin are given together (Augmentin), the clavulanic acid binds tightly to β-lactamases and prevents them from inactivating the amoxicillin. However, some drugs are less effective when used in combination than when used alone. This decreased effect, called **antagonism,** can be observed when bacteriostatic drugs such as tetracyclines, which inhibit

FIGURE 14.7 The effect of β-lactamase on penicillin. Numerous bacteria (staphylococci, streptococci, and gonococci) produce this enzyme, which inactivates penicillin. The enzyme can be transmitted by plasmids. Cephalosporins, although similar in action to penicillin, have a different cyclic ring structure and are more resistant to the effects of the enzyme.

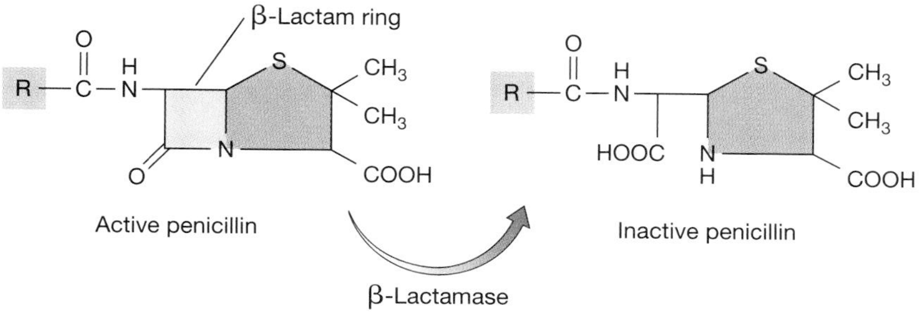

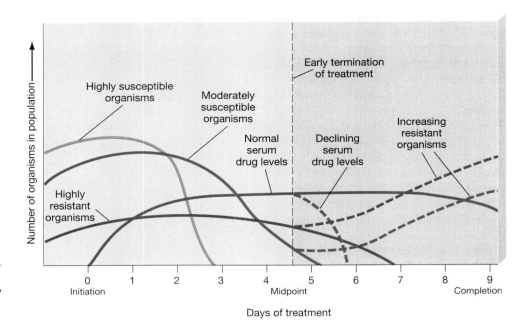

FIGURE 14.8 The development of drug resistance when a patient stops taking medication too early in the course of treating an infection.

growth, are combined with bactericidal penicillins, which require growth to be effective.

Third, antibiotics can be restricted to essential uses only. For example, most physicians do not prescribe antibiotics for colds and other viral diseases, except in the case of patients at high risk of secondary bacterial infections, because such diseases do not respond to antibiotics. Restrictions on antibiotic use would be particularly valuable in hospitals, where microbes "just waiting to acquire resistance" lurk in antibiotic-filled environments. In addition, the use of antibiotics in animal feeds could be banned; see the box "Drugs in Animal Feeds."

DETERMINATION OF MICROBIAL SENSITIVITIES TO ANTIMICROBIAL AGENTS

Microorganisms vary in their susceptibility to different chemotherapeutic agents, and susceptibilities can change over time. Ideally, the appropriate antibiotic to treat any particular infection should be determined before any antibiotics are given. Sometimes an appropriate agent can be prescribed as soon as the causative organism is identified from a laboratory culture. Often

PUBLIC HEALTH

ANTIBIOTIC RESISTANCE II: DRUGS IN ANIMAL FEEDS

Soon after the discovery of chlortetracycline in 1948, scientists learned that wastes from its manufacture were useful as animal food supplements. Such wastes contained small quantities of the antibiotic. The antibiotic was found to serve as a powerful growth stimulant and therefore was added to animal feeds. By the early 1950s, antibiotics were being added to feeds all over the world. Today, penicillin, tetracycline, and many other antibiotics are added to animal feeds. Nearly half the antibiotics manufactured in the United States are used for purposes other than human medicines; most go into animal feeds.

Unfortunately, the widespread use of antibiotics in animal feeds encourages the growth of resistant organisms. In response, some European governments have taken steps to limit antibiotic use. For example, after a rise in *Salmonella* in-

fections in Britain, antibiotics used to treat humans were banned from use in animal feeds. The proportion of *E. coli* strains resistant to antibiotics subsequently dropped from 31 percent to 18 percent. When the use of tetracycline in animal feeds was banned in the Netherlands, the incidence of tetracycline-resistant *Salmonella* strains dropped from 35 percent to 8 percent.

Widespread use of small doses of an antibiotic provides an ideal environment for the natural selection of mutant organisms resistant to the antibiotic. Once evolved, such organisms tend to thrive. They are a health threat, microbiologists believe, to humans and farm animals, and the Food and Drug Administration agrees. So far in the United States, however, little action to control nonhuman consumption of antibiotics has been taken.

tests are needed to show which antibiotic kills the organism. Several methods—disk diffusion, dilution, and automated methods—are available to do this.

Disk Diffusion Method

In the **disk diffusion method,** or **Kirby-Bauer method,** a standard quantity of the causative organism is uniformly spread over an agar plate. Then several filter paper disks impregnated with specific concentrations of selected chemotherapeutic agents are placed on the agar surface (Figure 14.9a). Finally, the culture with the antibiotic disks is incubated.

During incubation, each chemotherapeutic agent diffuses out from the disk in all directions. Agents with lower molecular weights diffuse faster than those with higher molecular weights. Clear areas called **zones of inhibition** appear on the agar around disks where the agents inhibit the organism. The size of a zone of inhibition is not necessarily a measure of the degree of inhibition because of differences in the diffusion rates of chemotherapeutic agents. An agent of large molecular size might be a powerful inhibitor even though it might diffuse only a small distance and produce a small zone of inhibition. Standard measurements of zone diameters for particular media, quantities of organisms, and drug concentrations have been established and correlated to zone diameters in order to determine whether the organisms are *sensitive, moderately sensitive,* or *resistant* to the drug.

Even when inhibition has been properly interpreted in a disk diffusion test, the most inhibitory chemotherapeutic agent may not cure an infection. The agent will probably inhibit the causative organism, but it may not kill sufficient numbers of the organism to control the infection. A bactericidal agent is often needed to eliminate an infectious organism, and the disk diffusion method does not assure that a bactericidal agent will be identified. Moreover, results obtained *in vivo* (in a living organism) often differ from those obtained *in vitro* (in a laboratory vessel). Metabolic processes in the body of a living organism may inactivate or inhibit an antimicrobial compound.

Dilution Method

The **dilution method** of testing antibiotic sensitivity was first performed in tubes of culture broth; it is now performed in shallow wells on standardized plates (Figure 14.9b). In this method a constant quantity of microbial inoculum (specimen) is introduced into a series of broth cultures containing decreasing concentrations of a chemotherapeutic agent. After incubation (for 16 to 20 hours) the tubes or wells are examined, and the lowest concentration of the agent that prevents visible growth (indicated by turbidity or dots of grow-

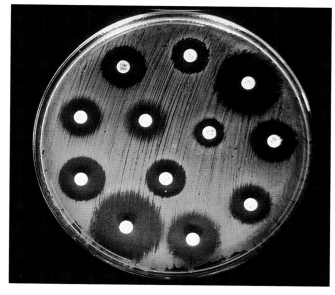

(a)

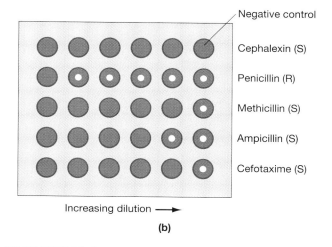

Increasing dilution ⟶

(b)

FIGURE 14.9 (a) Disk diffusion (Kirby-Bauer) method of determining microbial sensitivities to various antibiotics. A Petri plate is prestreaked on an agar medium with the organism to be tested. Paper disks, each containing a measured amount of a particular antibiotic, are placed firmly in contact with the medium and diffuse outward. After the plate is incubated, a clear area of no growth around a disk (a zone of inhibition) represents inhibition of the test organism by the antibiotic. Nonclear areas indicate resistance to that antibiotic. The largest zone of inhibition does not always indicate the most effective antibiotic, as different molecules do not diffuse at the same speed into the medium. Also, some drugs do not behave the same way in living organisms as they do on agar. However, the diameters of zones of inhibition, when compared with measured standards, help indicate whether an organism is sensitive or resistant to a drug. (b) Minimal inhibitory concentration (MIC) microbial susceptibility testing. A standardized microdilution plate with shallow wells that contain increasing dilutions (decreasing concentrations) of selected antibiotics in a broth is inoculated with a test bacterium. The plate is incubated; the lowest concentration that prevents growth (dots in well) is the MIC. The test bacterium on this plate is sensitive (S) to all the antibiotics except penicillin (R). The negative control well contains only broth.

ing organisms) is noted. This concentration is the **minimum inhibitory concentration** (MIC) for a particular agent acting on a specific microorganism. This test can be done for several agents simultaneously by using several sets of tubes or wells, but it is time-consuming and therefore expensive.

Finding an inhibitory agent by the dilution method does no more to prove that it will kill the infectious organism in the patient than finding one by the disk diffusion method. However, the dilution method allows a second test to distinguish between bactericidal agents, which kill microorganisms, and bacteriostatic agents, which merely inhibit their growth. Samples from tubes that show no growth but that might contain inhibited organisms can be used to inoculate broth that contains no chemotherapeutic agent. In this test the lowest concentration of the chemotherapeutic agent that yields no growth following this second inoculation, or *subculturing*, is the **minimum bactericidal concentration** (MBC). Thus, both an effective chemotherapeutic agent and an appropriate concentration to control an infection can be determined. That concentration should be maintained at the sites of infection because it is the minimum concentration that will cure the disease.

Serum Killing Power

Still another method of determining the effectiveness of a chemotherapeutic agent is to measure its **serum killing power.** This test is performed by obtaining a sample of a patient's blood while the patient is receiving an antibiotic. A bacterial suspension is added to a known quantity of the patient's **serum** (blood plasma minus the clotting factors). Growth (turbidity) in the serum after incubation means that the antibiotic is ineffective. Inhibition of growth suggests that the drug is working, and more quantitative determinations can be made to identify the lowest concentration that still provides serum killing power.

Automated Methods

Automated methods (Figure 14.10) are now available to identify pathogenic organisms and to determine which antimicrobial agents will effectively combat them. One such method makes use of prepared trays with small wells into which a measured quantity of inoculum is automatically dispensed. As many as 36 organisms from different patients can be inoculated onto the same tray. Trays containing several kinds of media suitable for identifying members of different groups of organisms—such as gram-positive bacteria, gram-negative bacteria, anaerobic bacteria, and yeasts—are available. Trays are also available to determine the sensitivity of organisms to a variety of antimicrobial agents.

The trays are inserted into a machine that measures microbial growth. Some machines do this by using a beam of light to measure turbidity. Others use media containing radioactive carbon. Organisms growing on such media release radioactive carbon dioxide into the air, and a sampling device automatically detects it. Machines vary in their degree of automation and the speed with which results become available. Some require technicians to perform some steps; others pro-

FIGURE 14.10 (a) An automated system for identifying microorganisms and determining their sensitivity to various antimicrobial agents. A sample containing the organism(s) is automatically inoculated (b) into wells on a thin plastic tray, each containing a specific chemical reagent. Tests are carried out in an incubation chamber, and the results are read and recorded by computer. Trays for a wide variety of different identification and antimicrobial sensitivity tests are available.

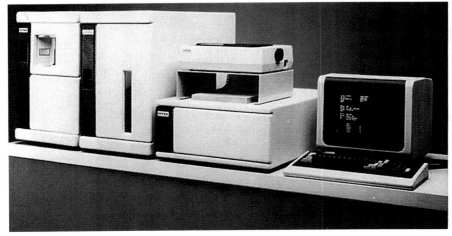

(a)

(b)

vide a computerized printout of results that is relayed to the patient's chart. Some machines provide results in 3 to 6 hours, and most provide them overnight, although slow-growing organisms may require 48 hours.

Automated methods make laboratory identification of organisms and their sensitivities to antimicrobials more efficient and less expensive. Once the results of laboratory tests are available, the physician can then choose an appropriate drug on the basis of the nature of the pathogen, the location of the infection, and other factors such as the patient's allergies. Automated methods allow physicians to prescribe an appropriate antibiotic early in an infection rather than prescribing a broad-spectrum antibiotic while awaiting laboratory results.

ATTRIBUTES OF AN IDEAL ANTIMICROBIAL AGENT

Having considered various characteristics of antimicrobial agents and methods of determining microbial sensitivities to them, we can now list the characteristics of an ideal antimicrobial agent:

1. *Solubility in body fluids.* Agents must dissolve in body fluids to be transported in the body and reach the infectious organisms. Even agents used topically must dissolve in the fluids of injured tissue to be effective; however, they must not bind too tightly to serum proteins.

2. *Selective toxicity.* Agents must be more toxic to microorganisms than to host cells. Ideally, a great difference should exist between the low concentration toxic to microorganisms and the concentration that damages host cells.

3. *Toxicity not easily altered.* The agent should maintain a standard toxicity and not be made more or less toxic by interactions with foods, other drugs, or abnormal conditions such as diabetes and kidney disease in the host.

4. *Nonallergenic.* The agent should not elicit an allergic reaction in the host.

5. *Stability—maintenance of a constant, therapeutic concentration in blood and tissue fluids.* The agent should be sufficiently stable in body fluids to have therapeutic activity over many hours; it should be degraded and excreted slowly.

6. *Resistance by microorganisms not easily acquired.* There should be few, if any, microorganisms with resistance to the agent.

7. *Long shelf life.* The agent should retain its therapeutic properties over a long period of time with a minimum of special procedures such as refrigeration or shielding from light.

8. *Reasonable cost.* The agent should be affordable to patients who need it.

Many antimicrobial agents meet these criteria reasonably well. But few, if any, meet all the criteria for the ideal antimicrobial agents. As long as better drugs might be found, the search for them will continue.

ANTIBACTERIAL AGENTS

Most antimicrobial agents are *antibacterial agents*, so we will start our "catalog" of antimicrobial agents with them, keeping in mind that some are effective against other microbes as well. Antibacterial agents can be categorized in several ways; we have chosen to use their modes of action. Another way of grouping antibiotics is by the microorganism that produces them (Table 14.2).

Inhibitors of Cell Wall Synthesis

Penicillins

Natural **penicillins,** such as *penicillin G* and *penicillin V,* are extracted from cultures of the mold *Penicillium notatum.* The discovery in the 1950s that certain strains of *Staphylococcus aureus* are resistant to penicillin provided the impetus to develop semisynthetic penicillins. The first of these was *methicillin,* which is effective against penicillin-resistant organisms because it is not broken down by β-lactamase enzymes. Other semisynthetic penicillins, including *nafcillin, oxacillin, ampicillin, amoxicillin, carbenicillin,* and *ticarcillin,* emerged in rapid succession. Each is synthesized by adding a particular side chain to a penicillin nucleus

TABLE 14.2	Selected Microbes That Serve as Sources of Antibiotics
Microbe	**Antibiotic**
Fungi	
Aspergillus fumigatus	Fumagillin
Cephalosporium species	Cephalosporins
Penicillium griseofulvum	Griseofulvin
Penicillium notatum and *P. chrysogenum*	Penicillin
Streptomycetes	
Streptomyces nodosus	Amphotericin B
Streptomyces venezuelae	Chloramphenicol
Streptomyces erythreus	Erythromycin
Streptomyces griseus	Streptomycin
Streptomyces kanamyceticus	Kanamycin
Streptomyces fradiae	Neomycin
Streptomyces noursei	Nystatin
Streptomyces antibioticus	Vidarabine
Actinomycetes	
Micromonospora species	Gentamicin
Other Bacteria	
Bacillus licheniformis	Bacitracin
Bacillus polymyxa	Polymyxins
Bacillus brevis	Tyrocidin

Microbiologist's Notebook
Fighting Disease at the Zoo

Rob Schumaker: As soon as I walked into the monkey house that morning I knew that one of them was really sick. The first thing I noticed was a large amount of liquid stool in the cage. Colobus monkeys have stools like little pellets, so one of them was definitely in trouble.

I looked at the animals to see who it was. We keepers look at each animal every day, you know, and we really get to know them. I realized that it was Emily. Her face was very swollen—so much so that her eyes were nearly swollen shut. Her tail was filthy. I was really concerned. Emily is the oldest female in the group, which makes her the core of the females. Also, we suspected that she was pregnant.

I'm Rob Schumaker. I work as a keeper of monkeys and great apes at the Smithsonian's National Zoo in Washington, D.C. I'm studying for a graduate degree in zoology, and I have a particular interest in microbiology. The monkeys I'm talking about are the black and white colobus: *Colobus abyssinicus*. Their range is Africa, from Ethiopia to Tanzania and Zaire. Their natural diet consists of leaves and fruits. Colobus monkeys live in groups of two to 15 or more. The males are somewhat solitary, but the females have a highly structured society with many rules.

I watched Emily some more. The veterinarians here go on rounds every day, but I was concerned about Emily and didn't want to wait for them to come to us. Even though it could be very stressful for her, we decided to put her in her transport cage and take her to the hospital. We had to find out what this illness was. Our groups had always been okay; up until then we had escaped the difficulties that other monkey colonies have experienced.

Dr. Richard Montali: Those other difficulties were outbreaks of dysentery caused by *Shigella* bacteria. It's a problem in many primate colonies, and even in good zoos, monkeys and apes are

One of the zoo's black and white colobus monkeys—ruminants that depend on gut microbes to help them digest their food.

often carriers. I'm Dr. Richard Montali, the veterinary pathologist at the National Zoo. Our zoo never had a *Shigella* problem until 1984. Now Dr. Mitchell Bush, the zoo's clinical veterinarian, and I are doing an intensive study of *Shigella* infection. Twice a year, we screen cultures of each animal's feces on three consecutive days. We've found *S. flexneri* and *S. sonnei*. We've found carrier animals. So far none of the keepers has caught it—although one of the veterinarians did.

The *Shigella* could have been introduced by humans, or it could have been brought in by a new animal. We really don't know. However, we are considering the possibility of using human vaccine for the animals. We're talking to Johns Hopkins about a *Shigella* vaccine they're developing that might be applicable to our primates. Walter Reed Army Medical Center has also been helpful. We identified the dysentery bacterium in our own labs, but now they do special cultures for us, to characterize organisms by species and even by strain. It's fortunate that these medical centers are close by.

We've also identified other pathogens from the zoo animals in our labs:

Salmonella, Campylobacter, parasites. Timing is very important. For example, to isolate *Yersinia pseudotuberculosis* and *Y. enterocolitica,* carried by rats, we have to take cultures 3 to 5 days in a row. Even then, we may find the pathogen on only one of those days.

It's terrible when a zoo population falls victim to disease like this. We lost two gibbons to *Shigella*. One was old, the other had other underlying problems. But even if the mortality is low, it's upsetting to the staff *and* to the other animals—their whole social structure can be thrown off.

Once Emily was diagnosed as having a *Shigella* infection, the treatment was clear. Colobus monkeys have ruminant-type stomachs, similar to those of cows. That is, the stomach has several compartments and is populated with symbiotic bacteria that help break down the cellulose in the plant material the monkeys eat. [See Chapter 2.] These very special gastrointestinal microorganisms are acquired shortly after birth from handling the feces of other animals, and they are essential for life. A colobus monkey that lacks the proper microbes cannot get enough nutrients from its food to survive.

When these animals are given antibiotics, the bacterial populations that help them digest their food are destroyed. Saving them from possible death by infection involves exposing them to possible illness from malnutrition. So after the *Shigella* infection is cured, we try to repopulate the gut with stomach juices aspirated from other, healthy animals. Unfortunately, this technique doesn't always work. In Emily's case, it proved effective. But Emily was a very weak animal for a long time afterward. She lost the baby she was carrying, and also miscarried in her next two pregnancies. This year, though, she has a healthy baby.

Schumaker: Even though we knew what organism was making Emily so sick, we were still very worried about her. Her separation from her family group could be very stressful for both her and the group, and that stress could be the difference in her recovering or not. Being in a hospital is also stressful for animal patients—they don't understand where they are or why. The hospital keepers are wonderful, but the animals don't know them. And they may not always know how to talk to individual animals or know their specific needs. So the primate keepers go up to the hospital and visit their sick charges. They hand-feed them, if necessary.

Montali: The keepers in the zoo and the veterinary staff work closely together. It's important that the staff be educated, and aware of potential problems. For example, we have learned to avoid mixed exhibits of different species of animals, because some may be endangered by the food of others. We saw some losses in the small mammal house because of such a situation. Some tamarins died from *Streptococcus zooepidemicus,* Group C. This is an opportunistic pathogen in humans and is carried by horses. It is usually present in the respiratory and genital tracts of the horses and can sometimes cause serious infection even in them. In a small mammal, such as a tamarin, infection begins in the lymph nodes and then develops into septicemia.

How did the tamarins get an equine infection? Probably from the food that was given to the armadillos they lived with. We fed the armadillos a commer-

Rob Schumaker at the National Zoo in Washington, D.C.

cially sold raw horsemeat product that we received frozen. In a mixed exhibit such as the one we're talking about, the tamarin may have gone over to the armadillo's food dish and tasted the food. Or it may just have handled the meat and then put its fingers in its mouth. Although the pathogen didn't affect the armadillos, it killed several of the tamarins.

It's easy to understand how the horsemeat could be contaminated in this way. Do many *healthy* horses go to slaughter? Probably not. And the quality controls on food intended for animal consumption are much less stringent than those on food for human consumption. Our solution was to switch to beef and to reorganize the mixed exhibits.

Schumaker: The Zoo has developed a cooperative program to reintroduce golden lion tamarins to their native Brazil. They are an endangered species in the wild but breed well in zoos, where a surplus number of animals have accumulated. We run a staging program here in which we keep the tamarins in an outdoor, uncaged setting. This helps them adjust to the conditions they will find in the wild. We want them to survive when they get there.

Montali: And we want to be sure that they don't bring any diseases into the wild population that might kill what remains of the endangered populaton. In the zoo, animals can pick up parasites and infections not found in their native countries. A cockroach might scurry from the cage of an Asian animal into an adjacent cage, where an animal from South America will eat it, thus acquiring a parasite from a different continent. The German cockroach has been found to carry an intestinal nematode parasite from its indigenous host, the slow lorus, to tamarins. Thus, we examine these tamarins *very* carefully before we send them to Brazil. We also eliminate animals with genetic defects from the program.

Schumaker: We keepers have to be extraordinarily careful to avoid contamination of our animals and of ourselves. Whenever there is illness, we take extra precautions. For example, we always wear waterproof boots around the enclosures—for washing them down and so forth. When Emily got sick, we set up a footbath, containing a diluted bleach solution, at the door to her cage. The keepers stepped in it going into and coming out of the cage. (We didn't want to cont-

A pair of squirrel-sized golden lion tamarin monkeys becoming acclimated to uncaged life at the National Zoo before being released in Brazil.

aminate anything outside that enclosure.) We kept special tools only for that particular cage. We tended to that cage last and washed all tools in a bleach solution. We took extra care in scrubbing the cage, using a special antiseptic solution.

Those were the precautions we took over and above our regular procedures. We ordinarily clean the inside of the cages daily. They're scrubbed, hosed, and disinfected. We rotate among three products: an antiseptic detergent; bleach, which is a good stain remover; and a deodorizing agent.

As for the keepers, we always wear filter masks so that we don't inhale any aerosol material. We wear gloves, boots, and coveralls on the job, and we don't wear any work clothes out of the building—there are a washer and dryer in the basement of the building that we use. We wear our street clothes in and out.

Just as we check the monkeys to make sure they're healthy on a routine basis, we check ourselves too. All keepers' stools are cultured for three consecutive days twice a year. There are a number of pathogens that can be transmitted back and forth between humans and animals. Early in this century, for example, tuberculosis was a problem among both animals and staff. The keepers gave it to the animals; the animals gave it to the keepers. Even now, mammalian tuberculosis can be a danger to keepers, so we undergo regular TB testing. We have all come to the realization that disease transmission in a zoo can be a two-way problem. And since the animals are totally dependent on us, it is our problem to solve.

TRY IT

INTERMICROBIAL WARFARE

You, too, can discover an antibiotic. Chances are that it will be missing several of the attributes of an ideal antimicrobial agent and thus would not be of commercial value. Still, it's fun to find something yourself instead of just reading about it in books.

Soil microbes are good sources of antibiotics. Pharmaceutical companies pay people to bring back soil samples from all over the world in hopes of finding new and useful antibiotic-producing microbes. Old cemeteries, wildlife refuges, swamps, and airfield landing strips have all been sampled. Try collecting soil samples from some places that interest you. Using aseptic technique, swab the surface of a Petri plate of nutrient agar with a broth culture of an organism such as *E. coli*. Make confluent strokes so that the entire surface is inoculated evenly, like a lawn. You could try different organisms on different plates. Then sprinkle tiny bits of your soil sample over each inoculated plate. Incubate the plates or leave them at room temperature, and examine them daily. Do clear areas develop around some of the colonies that have grown from soil particles? If so, these are areas where natural antibiotics have diffused out into the agar and prevented the background "lawn" bacteria from growing. Congratulations, you've found some antibiotics!

Next, make stained slides of some of the antibiotic-producing colonies. Are they bacteria or fungi? Fungi may be especially abundant in acid soils. If circumstances permit, subculture one or more of your antibiotic-producing colonies and see if it inhibits all types of microbes equally. Does the antibiotic have a broad spectrum or a narrow spectrum? If you were to give your culture to a pharmaceutical company, which further tests would they need to perform to determine whether it has potential commercial value?

(Figure 14.11). Both the natural and semisynthetic penicillins are bactericidal.

Penicillin G, the most frequently used natural penicillin, is administered *parenterally*—that is, by some means other than via the gut, such as intramuscularly or intravenously. When administered orally, most of it is broken down by stomach acids. Penicillin is rapidly absorbed into the blood, reaches its maximum concentration, and is excreted unless it is combined with an agent such as procaine, which slows excretion and prolongs activity.

Penicillin G is the drug of choice in treating infections caused by streptococci, meningococci, pneumococci, spirochetes, clostridia, and aerobic gram-positive rods. It is also suitable for treating infections caused by

a few strains of staphylococci and gonococci that are not resistant to it. Because it retains activity in urine, it is suitable for treating some urinary tract infections. Infections caused by organisms resistant to penicillin G can be treated with semisynthetics such as nafcillin, oxacillin, ampicillin, or amoxicillin. Carbenicillin and ticarcillin are especially useful in treating *Pseudomonas* infections. Allergy to penicillin is rare among children but occurs in 1 to 5 percent of adults. Penicillins are generally nontoxic, but large doses can have toxic effects on the kidneys, liver, and central nervous system.

In addition to their use as treatment for infections, penicillins also are used prophylactically—that is, to *prevent* infection. For example, patients with heart defects (in particular, malformed or artificial valves) or heart disease are especially susceptible to endocarditis, an inflammation of the lining of the heart, caused by a bacterial infection. Organisms tend to attack surfaces of damaged valves. To prevent such infections, susceptible patients often receive penicillin before surgery or dental procedures (even cleanings) that could release bacteria into the bloodstream.

Cephalosporins

Natural **cephalosporins** (sef"a-lo-spor'-inz), derived from several species of the fungus *Cephalosporium*, have limited antimicrobial action. Their discovery led to the development of a large number of bactericidal, semisynthetic derivatives of natural cephalosporin C. The nucleus of a cephalosporin is quite similar to that of penicillin; both contain β-lactam rings (Figure 14.11). Semisynthetic cephalosporins, like semisynthetic penicillins, differ in the nature of their side chains. Frequently used cephalosporins include *cephalexin* (Keflex), *cephradine*, and *cefadroxil*, all of which are fairly well absorbed from the gut and so can be administered orally. Other cephalosporins, such as *cephalothin* (Keflin), *cephapirin*, and *cefazolin*, must be administered parenterally, usually into muscles or veins.

Although cephalosporins usually are not the first drug considered in the treatment of an infection, they are frequently used when allergy or toxicity prevents the use of other drugs. But because cephalosporins are structurally similar to penicillin, some patients who are allergic to penicillin may also be sensitive to the cephalosporins. Nevertheless, cephalosporins account for one-fourth to one-third of pharmacy expenditures in American hospitals, mainly because they have a fairly wide spectrum of activity, rarely cause serious side effects, and can be used prophylactically in surgical patients. Unfortunately, they are often used when a less expensive and narrower-spectrum agent would be just as effective.

The development of new varieties of cephalosporins seems to be a race against the ability of bacteria to ac-

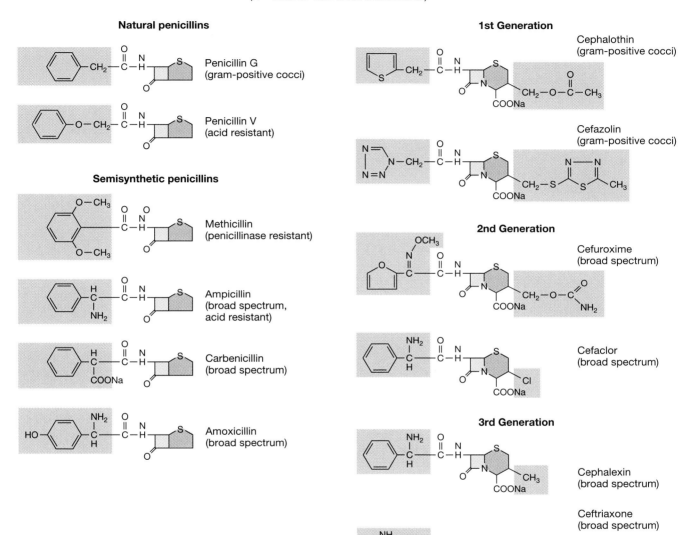

FIGURE 14.11 A comparison of the penicillin and cephalosporin molecules with β-lactam rings (blue). Cephalosporin differs slightly in the attached ring (red) and has two sites for side-chain attachments (purple) rather than one, as on the penicillin molecule.

quire resistance to older varieties. When organisms became resistant to early "first-generation" cephalosporins, new "second-generation" cephalosporins, including *cefuroxime* and *cefaclor,* were produced (Figure 14.11). Now "third-generation" cephalosporins, such as *ceftriaxone* and *cephalexin,* are used against organisms resistant to older drugs. These drugs are especially effective (for now) in dealing with hospital-acquired infections resistant to many antibiotics. They are being tried in patients with AIDS and other immunodeficiencies. (Do not confuse these with second- and third-line drugs, described earlier, which are not derivatives of one another.)

Adverse effects from cephalosporins tend to be local reactions, such as irritation at the injection site or nausea, vomiting, and diarrhea when the drug is administered orally. Four to 15 percent of patients allergic to penicillin also are allergic to cephalosporins. Moreover, newer cephalosporins have little effect on gram-positive organisms, which can cause superinfections during the treatment of gram-negative infections.

Other Antibacterial Agents That Act on Cell Walls

Carbapenems (kar′ba-pen-emz) represent a new group of bactericidal antibiotics with two-part structures. *Primaxin,* a typical carbapenem, consists of a β-lactam antibiotic (*imipenem*) that interferes with cell wall synthesis and *cilastatin sodium,* a compound that prevents degradation of the drug in the kidneys. As a group, the carbapenems have an extremely broad spectrum of activity.

Bacitracin, a small bactericidal polypeptide derived from the bacterium *Bacillus licheniformis,* is used only on lesions and wounds of the skin or mucous membranes because it is poorly absorbed and toxic to the kidneys. *Vancomycin* is a large, complex molecule produced by the soil actinomycete *Streptomyces orientalis.* It can be used to treat infections caused by methicillin-resistant staphylococci and enterococci. It is also the drug of choice against antibiotic-induced pseudomembranous colitis (enteritis with the formation of false membranes in stool). Because it is poorly absorbed through the gastrointestinal tract, it must be administered intravenously. Vancomycin is fairly toxic, causing hearing loss and kidney damage, especially in older patients, if the drug is not monitored carefully.

Disrupters of Cell Membranes

Polymyxins

Five **polymyxins,** designated A, B, C, D, and E, have been obtained from the soil bacterium *Bacillus polymyxa.* Polymyxins B and E are the most common clinically. They are usually applied topically, often with bacitracin, to treat skin infections caused by gram-negative bacteria such as *Pseudomonas.* Used internally, polymyxins can cause numbness in the extremities, serious kidney damage, and respiratory arrest. They are administered by injection when the patient is hospitalized and kidney function can be monitored.

Tyrocidins

Tyrocidins are cyclic polypeptides obtained from *Bacillus brevis.* The first to be discovered was *tyrothricin,* from which *tyrocidine* and *gramicidin* (named for Hans Gram, the originator of the Gram stain) have been derived. All are highly toxic, but they can be used topically to prevent bacterial growth, especially that of gram-positive cocci.

Inhibitors of Protein Synthesis

Aminoglycosides

Aminoglycosides are obtained from various species of the genera *Streptomyces* and *Micromonospora.* The first, **streptomycin,** was discovered in the 1940s and was effective against a variety of bacteria. Since then, many bacteria have become resistant to it. Moreover, streptomycin can damage kidneys and the inner ear, sometimes causing permanent ringing in the ears and dizziness. Consequently, this compound is now used only in special situations and generally in combination with other drugs. For example, it can be used with tetracyclines to treat plague and tularemia and with isoniazid and rifampin to treat tuberculosis.

Other aminoglycosides, such as *neomycin, kanamycin, amikacin, gentamicin, tobramycin,* and *netilmicin,* also have special uses and display varying degrees of toxicity to the kidneys and inner ear. At lower, less toxic, doses aminoglycosides tend to be bacteriostatic. They are usually administered intramuscularly or intravenously because they are poorly absorbed when given orally.

An important property of aminoglycosides is their ability to act synergistically with other drugs—an aminoglycoside and another drug together often control an infection better than either could alone. For example, gentamicin and penicillin or ampicillin are effective against penicillin-resistant streptococci. In other synergistic actions, gentamicin or tobramycin work with carbenicillin or ticarcillin to control *Pseudomonas* infections, especially in burn patients, and aminoglycosides work with cephalosporins to control *Klebsiella* infections.

Other applications of aminoglycosides include the treatment of bone and joint infections, peritonitis (inflammation of the lining of the abdominal cavity), pelvic abscesses, and many hospital-acquired infections. In bone and joint infections, gentamicin and tobramycin are especially useful because they can penetrate joint cavities. Because peritonitis and pelvic abscesses are severe and are often caused by a mixture of enterococci and anaerobic bacteria, aminoglycoside treatment is usually started before the organisms are identified. Amikacin is especially effective in treating hospital-acquired infections resistant to other drugs. It should not be used in less demanding situations lest organisms become resistant to it, too.

Aminoglycosides often damage kidney cells, causing protein to be excreted in the urine, and prolonged use can kill kidney cells. These effects are most pronounced in older patients and those with preexisting kidney disease. Some aminoglycosides damage the eighth cranial nerve: Streptomycin causes dizziness and disturbances in balance, and neomycin causes hearing loss.

Tetracyclines

Several **tetracyclines** are obtained from species of *Streptomyces*, in which they were originally discovered. Commonly used tetracyclines include tetracycline itself, *chlortetracycline* (Aureomycin), and *oxytetracycline* (Terramycin). Newer semisynthetic tetracyclines include *minocycline* (Minocin) and *doxycycline* (Vibramycin). All are bacteriostatic at normal doses, are readily absorbed from the digestive tract, and become widely distributed in tissues and body fluids (with the exception of cerebrospinal fluid).

The fact that tetracyclines have the widest spectrum of activity of any antibiotics is a two-edged sword. They are effective against many gram-positive and gram-negative bacterial infections and are suitable for treating rickettsial, chlamydial, mycoplasmal, and some fungal infections. But because they have such a wide spectrum of activity, they destroy the normal intestinal microbiota and often produce severe gastrointestinal disorders. Recalcitrant superinfections of tetracycline-resistant *Proteus, Pseudomonas,* and *Staphylococcus,* as well as yeast infections, also can result.

Tetracyclines can cause a variety of mild to severe toxic effects. Nausea and diarrhea are common, and extreme sensitivity to light is sometimes seen. The drug can also cause pustules to form on the skin. Effects on the liver and kidneys are more serious. Liver damage can be fatal, especially in patients with severe infections or during pregnancy. Kidney damage can lead to acidosis (low blood pH) and to excretion of protein and glucose. Anemia can occur, but it is rare.

Staining of the teeth (Figure 14.12) occurs when children under 5 years of age receive tetracycline or when their mothers received it during the last half of their pregnancy. Both deciduous teeth (baby teeth) and permanent teeth will be mottled because the buds of

both types of teeth form before birth. Tetracycline taken during pregnancy also can lead to abnormal bone formation in the fetal skull and a permanent abnormal skull shape. The ability of calcium ions to form a complex with tetracycline is responsible for its effects on bones and teeth. Because this reaction destroys the antibiotic effect of the drug, patients should not consume milk or other dairy products with the drug or for a few hours after taking it.

Chloramphenicol

Chloramphenicol, originally obtained from cultures of *Streptomyces venezuelae,* is now fully synthesized in the laboratory. Like tetracyclines it is bacteriostatic, is rapidly absorbed from the digestive tract, is widely distributed in tissues, and has a broad spectrum of activity. It is used to treat typhoid fever, infections due to penicillin-resistant strains of meningococci and *Haemophilus influenzae,* brain abscesses, and severe rickettsial infections.

Chloramphenicol damages bone marrow in two ways. It causes a dose-related, reversible aplastic anemia, in which bone marrow cells produce too few erythrocytes and sometimes too few leukocytes and platelets as well. Terminating use of the drug usually allows the bone marrow to recover normal function. It also causes a non-dose-related, permanent aplastic anemia due to destruction of bone marrow. Aplastic anemia appears days to months after treatment is discontinued and is most common in newborns. Unless a successful bone marrow transplant can be performed, aplastic anemia is usually fatal. It is seen in only one in 25,000 to 40,000 patients treated with chloramphenicol. Long-term use of chloramphenicol can cause inflammation of the optic and other nerves, confusion, delirium, and mild to severe gastrointestinal symptoms.

FIGURE 14.12 Staining of teeth caused by tetracycline. If the condition results from ingestion of the antibiotic during pregnancy, both the deciduous (baby) and permanent teeth will be affected, as both sets of tooth buds are forming in the fetus at that time.

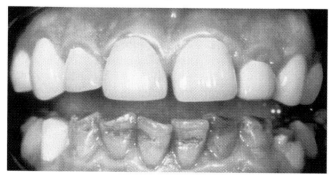

Other Antibacterial Agents That Affect Protein Synthesis

MACROLIDES **Erythromycin** (e-rith"ro-mi'sin), a commonly used **macrolide** (large-ring compound), is produced by several strains of *Streptomyces erythreus*. Erythromycin exerts a bacteriostatic effect, is readily absorbed, and reaches most tissues and body fluids (with the exception of cerebrospinal fluid). It is recommended for infections caused by streptococci, pneumococci, and corynebacteria but is also effective against *Mycoplasma* and some *Chlamydia* and *Campylobacter* infections. Erythromycin is most valuable in treating infections caused by penicillin-resistant organisms or in patients allergic to penicillin. Unfortunately, resistance to erythromycin often emerges during treatment. Dual antibiotic treatment—erythromycin and some other drug—is often used on patients with a pneumonia-like disease that might be Legionnaires' disease. Several antibiotics combat other pneumonias, but erythromycin is the only common antibiotic that will combat Legionnaires' disease. Erythromycin is one of the least toxic of commonly used antibiotics. Mild gastrointestinal disturbances are seen in 2 to 3 percent of patients receiving it.

LINCOSAMIDES *Lincomycin* is produced by *Streptomyces lincolnensis*, and *clindamycin* is a semisynthetic derivative that is more completely absorbed and less toxic than lincomycin. Both drugs, which are collectively called lincosamides, exert a bacteriostatic effect. Lincomycin can be used to treat a variety of infections but is not significantly better than other widely used antibiotics, and organisms quickly become resistant to it. Clindamycin is effective against *Bacteroides* and other anaerobes, except *Clostridium difficile*, which often becomes established as a superinfection during clindamycin therapy. Toxins from *C. difficile* can cause a severe, and sometimes fatal, colitis (inflammation of the large intestine) unless diagnosed early and treated with oral vancomycin. (See the Essay at the end of this chapter.)

Inhibitors of Nucleic Acid Synthesis

Rifampin

From among the **rifamycins** produced by *Streptomyces mediterranei*, only the semisynthetic *rifampin* is currently used. Easily absorbed from the digestive tract except when taken directly after a meal, it reaches all tissues and body fluids. Rifampin blocks RNA transcription. Although it is bactericidal and has a wide spectrum of activity, it is approved in the United States only for treating tuberculosis and eliminating meningococci from the nasopharynx of carriers.

Rifampin can cause liver damage but usually does so only when excessive doses are given to patients with preexisting liver disease. It is unusual among antibi-

PUBLIC HEALTH

ANTIBIOTIC RESISTANCE III: ANTIBIOTICS AND ACNE

Low doses of tetracyclines and erythromycin suppress skin bacteria, mostly *Propionibacterium acnes*, and reduce the release of microbial lipases, which contribute to skin inflammation. This therapy is used to treat acne, but its effectiveness has not been proved. Studies intended to assess antibiotic effectiveness in acne therapy have been useless because they lack suitable controls, fail to characterize adequately the type and severity of cases, and employ other concurrent therapies. Some studies have shown that low doses of many antibiotics can lead to the appearance of antibiotic-resistant strains. Are the benefits of antibiotics to acne patients worth the risk of promoting development of resistant organisms?

APPLICATIONS

RED MAN SYNDROME

Rifampin has been shown to cause the so-called red man syndrome. In this disorder, which occurs with high doses of the antibiotic, colored metabolic products of the drug accumulate in the body and are eliminated through sweat glands. It is characterized by bright orange or red urine, saliva, and tears, as well as skin that looks like a boiled lobster's. The red skin secretions can be washed away, but liver damage caused by the drug is only slowly repaired.

Red man syndrome due to rifampin. The red secretions can be washed away, but they would certainly frighten an unwarned patient.

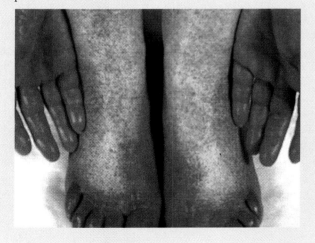

otics in its ability to interact with other drugs, and possibilities of such interactions should be considered before the drug is given. Taking rifampin concurrently with oral contraceptives has been implicated in an increased risk of pregnancy and menstrual disorders. Dosages of anticoagulants must be increased while a patient is taking rifampin to achieve the same degree of reduction in blood clotting. Finally, drug addicts receiving methadone sometimes suffer withdrawal symptoms if they are given rifampin without an increase in methadone dosage. One explanation for these diverse effects is that rifampin stimulates the liver to produce greater quantities of enzymes that are involved in the metabolism of a variety of drugs.

Quinolones

Quinolones, a new group of synthetic bactericidal analogues of *nalidixic acid,* are effective against many gram-positive and gram-negative bacteria. Quinolones' mode of action is to inhibit bacterial DNA synthesis by blocking DNA gyrase, the enzyme that unwinds the DNA double helix preparatory to its replication. *Norfloxacin, ciprofloxacin,* and *enoxacin* are examples of this group of antibiotics. They are especially effective in the treatment of traveler's diarrhea and in urinary tract infections caused by multiply resistant organisms.

A recent advance has produced a hybrid class of antibiotics. One of these, a quinolone–cephalosporin combination, is currently being tested. When the β-lactamase enzymes act on the cephalosporin component, the quinolone is released from the hybrid molecule and is available to kill the cephalosporin-resistant organisms. The use of such a dual-acting synergistic antibiotic may also prevent or delay development of antibiotic resistance in organisms.

Antimetabolites and Other Antibacterial Agents

Sulfonamides

The **sulfonamides,** or *sulfa drugs,* are a large group of entirely synthetic, bacteriostatic agents. Many are derived from *sulfanilamide* (sul-fa-nil'a-mid), one of the first sulfonamides (Figure 14.4b). In general, orally administered sulfonamides are readily absorbed and become widely distributed in tissues and body fluids. They act by blocking the synthesis of folic acid, which is needed to make the nitrogenous bases of DNA. Sulfonamides have now been largely replaced by antibiotics because antibiotics are more specific in their actions and less toxic than sulfonamides.

When sulfonamides first came into use in the 1930s, they frequently led to kidney damage. Newer forms of these drugs usually do not damage kidneys, but they do occasionally produce nausea and skin rashes. Certain sulfonamides are still used to suppress intestinal microbiota prior to colon surgery. They also are used to treat some kinds of meningitis because they enter cerebrospinal fluid more easily than do antibiotics. *Cotrimoxazole* (Septra), a combination of *sulfamethoxazole* and *trimethoprim,* is used to treat urinary tract infections and a few other infections. Cotrimoxazole is the primary drug of choice to control *Pneumocystis* pneumonia, a common fungal complication of AIDS patients. Unfortunately, both drugs are toxic to bone marrow and may cause nausea and skin rashes.

Isoniazid

Isoniazid (i-so-ni'a-zid) is an antimetabolite for two vitamins—nicotinamide and pyridoxal. It binds to and inactivates the enzyme that converts the vitamins to useful molecules. This bacteriostatic synthetic agent, which has little effect on most bacteria, is effective against the mycobacterium that causes tuberculosis. Isoniazid is completely absorbed from the digestive tract and reaches all tissues and body fluids. Because the mycobacteria present in any such infection usually include some isoniazid-resistant organisms, isoniazid usually is given with another agent such as rifampin or ethambutol (discussed in the following section). Dietary supplements of nicotinamide and pyridoxal also should be given with isoniazid.

Ethambutol

The synthetic agent **ethambutol** is effective against certain strains of mycobacteria that do not respond to isoniazid. Ethambutol is well absorbed and reaches all tissues and body fluids. However, mycobacteria acquire resistance to it fairly rapidly, so it is used with other drugs such as isoniazid and rifampin.

Nitrofurans

Nitrofurans (ni"tro-fyu'ranz) are antibacterial drugs that enter susceptible cells and apparently damage sensitive microbial respiratory systems. Several hundred nitrofurans have been synthesized since the first one was made in 1930. Only a few of these are currently used. Oral doses of *nitrofurantoin* (Furadantin) are bacteriostatic in low doses, easily absorbed, and quickly metabolized. This drug is especially useful in treating acute and chronic urinary infections. The low incidence of resistance to it makes it an ideal prophylactic agent to prevent recurrences. Unfortunately, 10 percent of patients experience nausea and vomiting as a side effect and must then be treated with an antibiotic instead. Another nitrofuran, *nifuratel,* has been shown to have the same clinical advantages as nitrofurantoin with a much lower incidence of gastrointestinal side effects. Unfortunately, this cheap, effective drug is not currently available.

The chemical structures, uses, and side effects of antibacterial agents are summarized in Figure 14.13.

Agent	Used to Treat	Common Method of Administration*	Side Effects
Agents that inhibit cell wall synthesis			
Penicillin (natural)	Wide variety of infections, mostly of gram-positive bacteria	IM, O	Relatively few side effects, but allergies do occur
Penicillin (semisynthetic)	Infections resistant to natural penicillin	O, IV	Same as natural penicillin
Cephalosporins	Wide variety of infections when allergy or toxicity make other agents unsuitable	IV, IM, O	Relatively nontoxic but can lead to superinfections
Carbapenems	Mixed infections, nosocomial infections, infections of unknown etiology	IV	Allergic reactions, superinfections, seizures, gastrointestinal disturbances
Bacitracin	Skin infections (topical application)	T	Internal use toxic to kidneys

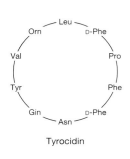

Imipenem
(a carbapenem)

Bacitracin

Agents that interfere with cell membrane function			
Polymyxins	Skin infections (topical application, with bacitracin)	T, IV	Internal use highly toxic
Tyrocidins	Skin infections caused by gram-positive cocci (topical application)	T, IV	Internal use highly toxic

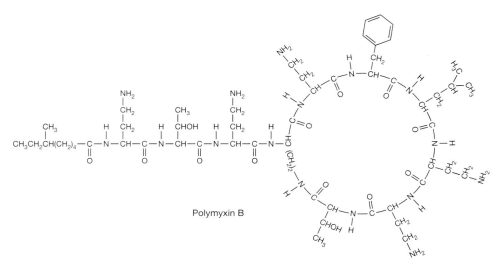

Polymyxin B

Tyrocidin

Antimetabolites and other agents			
Sulfonamides	Some kinds of meningitis and to supress intestinal flora before colon surgery	O, IV	Early forms caused kidney damage, but ones now in use do not
Isoniazid	Tuberculosis (used with ethambutol)	O	May cause pyridoxine deficiency
Ethambutol	Tuberculosis (used with isoniazid)	O	
Nitrofurantoin	Urinary tract infections	O	Nausea and vomiting

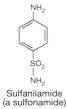

Sulfanilamide
(a sulfonamide)

Isoniazid

Ethambutol

Nitrofurantoin

* IM = intramuscular   O = oral
IV = intravenous   T = topical

FIGURE 14.13 Selected antibacterial drugs.

Agent	Used to Treat	Common Method of Administration*	Side Effects
Agents that inhibit protein synthesis			
Streptomycin	Tuberculosis (used with isoniazid and rifampin)	IM, O	Damages kidneys and inner ear
Gentamicin and other aminoglycosides	Antibiotic-resistant and hospital-acquired infections (used synergistically with other drugs)	IM, T (burns)	Varying degrees of kidney and inner ear damage
Tetracyclines	A broad spectrum of bacterial infections and some fungal infections	O	Stain teeth; cause gastrointestinal symptoms; can lead to super-infections
Chloramphenicol	A broad spectrum of bacterial infections, brain abscesses and penicillin-resistant infections	O	Can damage bone marrow and cause aplastic anemia
Erythromycin	Gram-positive bacterial infections, some penicillin-resistant infections, and Legionnaires' disease	O	One of the least toxic of commonly used antibiotics

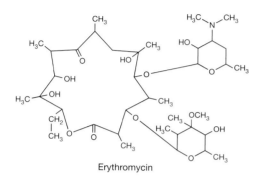

Erythromycin

Gentamicin

Tetracycline

Chloramphenicol

Streptomycin

Agents that inhibit nucleic acid synthesis			
Rifampin	Tuberculosis and to eliminate meningococci from the nasopharynx	O	Bright orange or red urine, saliva, tears, and skin; liver damage; many disorders when used with other agents
Quinolones	Urinary tract infections, traveller's diarrhea; effective against many drug-resistant organisms	O	Nausea; headaches and other nervous system disturbances

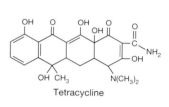

Rifampin

Ciprofloxacin
(a quinolone)

ANTIFUNGAL AGENTS

Antifungal agents are being used with greater frequency because of the emergence of resistant strains and an increase in the number of immunosupressed patients, especially those with AIDS. Because fungi are eukaryotes and thus similar to human cells, antifungal treatment often causes toxic side effects. At less toxic levels, many systemic fungal infections are slow to respond. Furthermore, laboratory tests are not available to determine appropriate susceptibility and therapeutic levels. Despite these difficulties, numerous effective drugs are now becoming available, many without prescription.

Imidazoles and Triazoles

The **imidazoles** (im"id-az'olz) and *triazoles* comprise a large group of related synthetic fungicides. Several agents, including *clotrimazole, ketoconazole, miconazole,* and *fluconazole,* are currently in use; many are available without prescription. The imidazoles and triazoles appear to affect fungal plasma membranes by disrupting the synthesis of membrane sterols. All these agents are used topically in creams and solutions to control fungal skin infections (dermatomycoses) and *Candida* yeast infections of the skin, nails, mouth, and vagina. Ketoconazole has also been given orally to treat systemic fungal infections, especially when other antifungal agents have not been effective. Some patients, however, have experienced mild to severe skin irritations with the topical agents. Furthermore, potentially severe drug interactions may occur, especially with certain antihistamines and immunosuppresents.

Polyenes

The **polyene** family of antibiotics consists of antifungal agents that contain at least two double bonds. Amphotericin B and nystatin are two of the most common polyene antibiotics.

AMPHOTERICIN B   The fungicidal antibiotic *amphotericin* (am"fo-ter'i-sin) *B* [Fungizone] is derived from *Streptomyces nodosus.* This drug binds to plasma membrane ergosterol (a crystallizing sterol) found in fungi and some algae and protozoa but not in human cells. Amphotericin B increases membrane permeability such that glucose, potassium, and other essential substances leak from the cell. The drug is poorly absorbed from the digestive tract and so is given intravenously. Even then, only 10 percent of the dose given is found in the blood. Excretion persists for up to 3 weeks after treatment is discontinued, but it is not known where the drug is sequestered in the meantime.

   Amphotericin B is the drug of choice in treating most systemic fungal infections, especially cryptococ-

PUBLIC HEALTH

NONPRESCRIPTION VAGINAL YEAST INFECTION REMEDY

Three out of four women suffer at least one vaginal yeast (*Candida albicans*) infection during their lifetime. Some have several recurrences per year. This naturally occurring fungus overgrows when the normal balance with bacteria is upset by diabetes, antibiotics, birth control pills, or any of a number of other factors. Symptoms include a white to yellow vaginal discharge along with vaginal itch and discomfort. Fortunately, women can now get faster, cheaper treatment for this very common condition. Clotrimazole is now available over the counter, without prescription, under the trade name Gyne-Lotrimin. A similar drug, miconazole, available as Monistat, is also effective.

   Some people worry about making antibiotics available on a nonprescription basis. One reason is that symptoms of some sexually transmitted diseases resemble those of a vaginal yeast infection. Women unfamiliar with yeast infection symptoms may mistakenly try to treat other diseases and thus delay proper treatment. Others may overtreat themselves when it is not necessary. But for the majority of women, being able to skip the wait for and cost of an office visit is most welcome.

cosis, coccidioidomycosis, and aspergillosis. Although fungi are not known to develop resistance to this agent, side effects are numerous and sometimes severe. They include abnormal skin sensations, fever and chills, nausea and vomiting, headache, depression, kidney damage, anemia, abnormal heart rhythms, and even blindness. Because some of the fungal infections are fatal without treatment, patients, especially those who are immunocompromised or have AIDS, have little choice but to risk these unfortunate side effects.

NYSTATIN   The polyene antibiotic *nystatin* (Mycostatin) is produced by *Streptomyces noursei.* This drug has the same mode of action as amphotericin B but is also effective topically in the treatment of *Candida* yeast infections. Because it is not absorbed through the intestinal wall, it can be given orally to treat fungal superinfections in the intestine, which often occur after long-term treatment with antibiotics. Nystatin was named for the New York State Health Department, where it was discovered.

Griseofulvin

Griseofulvin (gris"e-o-ful'vin), originally derived from *Penicillium griseofulvum,* is used primarily for superficial fungal infections. This fungistatic drug is incorporated into new cells that replace infected cells; it interferes with fungal growth, probably by impairing the mitotic spindle apparatus used in cell division. Although griseofulvin (Fulvicin) is poorly absorbed from the intestinal tract, it is given orally and appears to reach the target tissues through perspiration. It is ineffective against bacteria and most systemic fungal agents but is very useful topically in treating fungal infections of the skin, hair, and nails. Most infections are cured within 4 weeks, but recalcitrant infections associated with fingernails and toenails may persist even after a year of treatment. Reactions to griseofulvin are usually limited to mild headaches but can include gastrointestinal disturbances, especially when prolonged treatment is required.

Other Antifungal Agents

Flucytosine is a synthetic drug used in treating infections caused by *Candida* and several other fungi. This fluorinated pyrimidine is transformed in the body to fluorouracil, an analogue of uracil, and thereby interferes with nucleic acid and protein syntheses. The drug can be given orally and is easily absorbed, but 90 percent of the amount given is found unchanged in the urine within 24 hours. Because it is less toxic and causes fewer side effects than amphotericin B, flucytosine should be given instead of amphotericin B whenever possible.

Tolnaftate (Tinactin) is a common topical fungicide that is readily available without prescription. Although its mode of action is still not clear, it is effective in the treatment of various skin infections, including athlete's foot and jock itch.

Terbinafine (Lamisil), a relatively new fungicide, has been approved for topical use in skin infections and cutaneous candidiasis. Because it is absorbed directly through the skin, it reaches therapeutic levels in much less time than do orally administered agents such as griseofulvin.

ANTIVIRAL AGENTS

Until recent years no chemotherapeutic agents effective against viruses were available. One reason for the difficulty in finding such agents is that the agent must act on viruses within cells without severely affecting the host cells. Currently available *antiviral agents* inhibit some phase of viral replication, but they do not kill the viruses.

Purine and Pyrimidine Analogues

Several purine and pyrimidine analogues are effective antiviral agents. All cause the virus to incorporate erroneous information (the analogue) into a nucleic acid and thereby interfere with the replication of viruses. ∞ (Chapter 7, p. 183) The drugs include idoxuridine, vidarabine, ribavirin, acyclovir, and ganciclovir.

Idoxuridine and *trifluridine,* both analogues of thymine, are administered in eye drops to treat inflammation of the cornea caused by a herpesvirus. They should not be used internally because they suppress bone marrow.

Vidarabine (ARA-A), an analogue of adenine, has been used effectively to treat viral encephalitis, an inflammation of the brain caused by herpesviruses and by cytomegaloviruses. It is not effective against cytomegalovirus infections acquired before birth. Vidarabine is less toxic than either idoxuridine or cytarabine, but it sometimes causes gastrointestinal disturbances.

Ribavirin (Virazole), a synthetic nucleotide analogue of guanine, blocks replication of certain viruses. In an aerosol spray it can combat influenza viruses; in an ointment it can help to heal herpes lesions. Although it has low toxicity, it can induce birth defects and should not be given to pregnant women.

Acyclovir (Zovirax), an analogue of guanine, is much more rapidly incorporated into virus-infected cells than into normal cells. Thus, it is less toxic than other analogues. It can be applied topically or given orally or intravenously. It is especially effective in reducing pain and promoting healing of primary lesions in a new case of genital herpes. It is given prophylactically to reduce the frequency and severity of recurrent lesions, which appear periodically after a first attack. It does not, however, prevent the establishment of latent viruses in nerve cells. Acyclovir is more effective than vidarabine against herpes encephalitis and neonatal herpes, an infection acquired at birth, but is not effective against other herpesviruses.

Ganciclovir is an analogue of guanine similar to acyclovir. The drug is active against several kinds of herpesvirus infections and in particular cytomegalovirus eye infections in AIDS cases.

Amantadine

The tricyclic amine **amantadine** prevents influenza A viruses from penetrating cells. Given orally, it is readily absorbed and can be used from a few days before to a week after exposure to influenza A viruses to reduce the incidence and severity of symptoms. Unfortunately, it causes insomnia and ataxia (inability to coordinate voluntary movements), especially in elderly patients, who also are often severely affected by influenza infections. *Rimantadine,* a drug similar to amantadine, may be effective against a wider variety of viruses and be less toxic as well.

APPLICATIONS

DRUG-RESISTANT VIRUSES

Evidence is accumulating that viruses, like bacteria, can develop resistance to chemotherapeutic agents. Herpesviruses and cytomegaloviruses with resistance to acyclovir have been observed in AIDS patients. Some laboratory strains of the virus that causes AIDS have become resistant to azidothymidine (AZT), the most effective drug currently available to treat the disease. Resistance to chemotherapeutic agents is a greater problem in viruses than in bacteria because so few antiviral agents are available. When a bacterium becomes resistant to one antibiotic, another usually can be found to which the bacterium is susceptible. Unfortunately, this is not the case with viruses, and we must hope that biotechnology can help us battle drug-resistant viruses.

Treatment of AIDS

Several agents are being tested for the treatment of AIDS. New information about AIDS, its complications, and its treatment is becoming available with great rapidity. We consider AIDS, agents used to treat it, and ramifications for health care workers in Chapter 19.

Interferons and Immunoenhancers

Cells infected with viruses produce one or more proteins collectively referred to as *interferons* (Chapter 17). When released, these proteins induce neighboring cells to produce antiviral proteins, which prevent these cells from becoming infected. Thus, interferons represent a natural defense against viral infection. Some interferons are currently being genetically engineered and tested as antiviral agents. Some positive results have been obtained in controlling chronic viral hepatitis and warts and arresting virus-related cancers, such as Kaposi's sarcoma.

Because cells produce interferons naturally, a possible way to combat viruses is to induce cells to produce interferons. Synthetic double-stranded RNA has been shown to increase the quantity of interferon in the blood. Experiments with one such substance in virus-infected monkeys have shown sufficient increase in interferon to prevent viral replication.

Two other agents, *levamisole* and *inosiplex,* appear to stimulate the immune system to resist viral and other infections. Both seem to stimulate activity of leukocytes called T lymphocytes rather than to stimulate interferon release. Levamisole appears to be effective prophylactically in reducing the incidence and severity of chronic upper respiratory infections, which are probably viral in nature. It also reduces symptoms of autoimmune disorders such as rheumatoid arthritis, in which the body reacts against its own tissues. Inosiplex has a more specific action: It stimulates the immune system to resist infection with certain viruses that cause colds and influenza.

Although efforts to improve antiviral therapies by enhancing natural defenses have been somewhat successful, none is yet in widespread use. More research is needed to identify or synthesize effective agents, to determine how they act, and to discover how they can be most effectively used.

ANTIPROTOZOAN AGENTS

Although many protozoa are free-living organisms, a few are parasitic in humans. The parasite that causes malaria invades red blood cells and causes the patient to suffer alternating fever and chills. Other protozoan parasites cause intestinal or urinary tract infections. Several *antiprotozoan agents* have been found that are successful in controlling or even curing most protozoan infections, but some have rather unpleasant side effects.

Quinine

Quinine, from the bark of the chinchona tree (native to Peru and Bolivia, but now cultivated exclusively in Indonesia), was used for centuries to treat malaria. One of the first chemotherapeutic agents to come into widespread use, it is now used only to treat malaria caused by strains of the parasite resistant to other drugs.

Chloroquine and Primaquine

Currently the most widely used antimalarial agents are the synthetic agents **chloroquine** (Aralen) and **primaquine.** Chloroquine appears to interfere with protein synthesis, especially in red blood cells, which it enters more readily than it does other cells. The drug may concentrate in vacuoles within the parasite and prevent it from metabolizing hemoglobin. Chloroquine is used to combat active infections. The malarial parasite persists in red blood cells and can cause relapses when it multiplies and is released into blood plasma. A combination of chloroquine and primaquine can be used prophylactically to protect people who visit or work in regions of the world where malaria occurs and are thus at risk of becoming infected. However, the drugs must be taken both before and after a malarial zone is entered. A new prophylactic agent, *mefloquine* (Lariam), has proved effective against resistant strains.

Metronidazole

The synthetic imidazole **metronidazole** (met-ro-ni'da-zol) is effective in treating *Trichomonas* infections, which typically cause a vaginal discharge and itching. It also

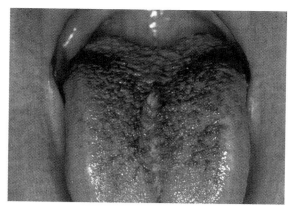

FIGURE 14.14 Black hairy tongue is a reaction to the drug metronidazole (Flagyl). The papilli on the tongue surface become elongated and filled with breakdown products of hemoglobin, which darken the tongue.

is effective against intestinal infections caused by parasitic amoebas and *Giardia*. Although metronidazole (Flagyl) controls these infections, it does not prevent overgrowth of *Candida* yeast infections. It also can cause birth defects and cancer and can be passed to infants in breast milk. Metronidazole sometimes causes an unusual side effect called "black hairy tongue," or "brown furry tongue," because it breaks down hemoglobin and leaves deposits in papillae (small projections) on the surface of the tongue (Figure 14.14).

Other Antiprotozoan Agents

A variety of other organic compounds have been found effective in treating certain infections caused by protozoa. *Pyrimethamine* (pir-i-meth′a-men) interferes with the synthesis of folic acid, which pathogenic protozoa need in greater quantities than do host cells. It is used with sulfanilamide to treat some protozoan infections such as toxoplasmosis. Pyrimethamine (Daraprim) can also be used prophylactically to prevent malaria.

Suramin sodium, a sulfur-containing compound, can be given intravenously to treat African sleeping sickness (trypanosomiasis) and other trypanosome infections. *Nifurtimox*, a nitrofuran, is used against the trypanosomes that cause Chagas' disease. Arsenic and antimony compounds, although very toxic, have been used with some success against stubborn amoebic infections and leishmanias. *Pentamidine isethionate* is used to treat African trypanosomiasis and as a drug of second choice for *Pneumocystis* pneumonia, a fungal complication of AIDS patients.

ANTIHELMINTHIC AGENTS

Various helminths can infect humans. A variety of *antihelminthic agents* are available to help rid the body of these unwelcome parasites.

Niclosamide

Niclosamide interferes with carbohydrate metabolism, thereby causing a parasite to release large quantities of lactic acid. This drug may also inactivate products made by the worm to resist digestion by host proteolytic enzymes. It is effective mainly in the treatment of tapeworm infections.

Mebendazole

The imidazole **mebendazole** (Vermox) blocks the uptake of glucose by parasitic roundworms. It is useful in treating whipworm, pinworm, and hookworm infections. However, it can damage a fetus and thus should not be given to pregnant women.

Other Antihelminthic Agents

Piperazine (Antepar), a simple organic compound, is a powerful neurotoxin that paralyzes body wall muscles of roundworms and is useful in treating *Ascaris* and pinworm infections. Although piperazine exerts its effect on worms in the intestine, if absorbed it can reach the human nervous system and cause convulsions, especially in children.

The compound *ivermectin*, originally developed for the treatment of parasitic nematodes in horses (and widely used to prevent heartworm infections in dogs), has been found to be extremely effective against *Onchocerca volvulus* in humans. Infection with this roundworm, widespread in many parts of Africa, causes a progressive loss of sight known as onchocerciasis, or river blindness.

Figure 14.15 provides the chemical structures, uses, and side effects of antifungal, antiviral, antiprotozoan, and antihelminthic agents.

SPECIAL PROBLEMS WITH DRUG-RESISTANT HOSPITAL INFECTIONS

As soon as antibacterial agents became available, resistant organisms began to appear. One of the first successes in treating bacterial infections was the use of sulfanilamide to treat infections caused by hemolytic streptococci. It was then discovered that sulfadiazine is useful in preventing recurrent streptococcal infections of rheumatic fever. Strains of streptococci resistant to sulfonamides soon emerged. Epidemics (mostly in military installations during World War II) caused by resistant strains led to many deaths. These epidemics were brought under control when penicillin became available, but soon penicillin-resistant streptococci were seen.

This chain of events has been repeated again and again. As new antibiotics were developed, strains of

Agent	Used to Treat	Common Method of Administration*	Side Effects
Antifungal Agents			
Clotrimazole	Skin and nail infections	O	Skin irritation
Miconazole	Skin infections and systemic infections resistant to other agents	T, IV	Severe itching, nausea, fever, thrombophlebitis
Amphotericin B	Systemic infections	IV	Fever, chills, nausea, vomiting, anemia, kidney damage, blindness
Nystatin	*Candida* yeast infections, intestinal superinfections	T	
Griseofulvin	Infections of skin, hair, and nails	T, O	Mild headaches, nerve inflammations, gastrointestinal disturbances
Flucytosine	*Candida* and some systemic infections	O	Less toxic than many fungal agents

Clotrimazole   Miconazole   Amphotericin B

Nystatin   Griseofulvin   Flucytosine

Agent	Used to Treat	Common Method of Administration*	Side Effects
Antihelminthic agents			
Niclosamide	Tapeworm infections	O	Irritation of gut
Piperazine	Pinworm and *Ascaris* infections	O	Can cause convulsions in children
Mebendazole	Whipworm, pinworm, and hookworm infections	O	Can damage fetus if given to pregnant women
Ivermectin	*Onchocerca volvulus* infections (cause of river blindness), heartworm infections in animals	O	Minimal

Niclosamide   Piperazine   Mebendazole

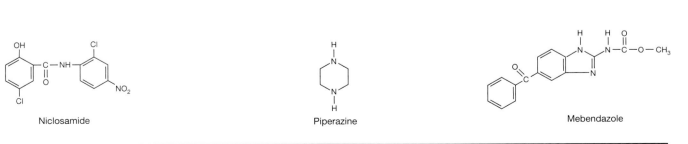

* IM = intramuscular   O = oral
IV = intravenous   T = topical

FIGURE 14.15 Selected antifungal, antihelminthic, antiviral, and antiprotozoan drugs.

Agent	Used to Treat	Common Method of Administration*	Side Effects
Antiviral Agents			
Idoxuridine	Corneal infections	T	Suppresses bone marrow
Ganciclovir	CMV eye infections in AIDS	IV	Suppresses bone marrow
Vidarabine	Viral encephalitis	T, IV	Less toxic than other antiviral agents
Ribavirin	Herpes lesions (topical application), influenza (in aerosol)	T	Can cause birth defects if given to pregnant women
Acyclovir	Herpesvirus infections; lessens severity of symptoms	IV, O, T	Less toxic than other analogues
Amantadine	Infections of influenza A viruses from entering cells (preventive)	O	Insomnia and ataxia

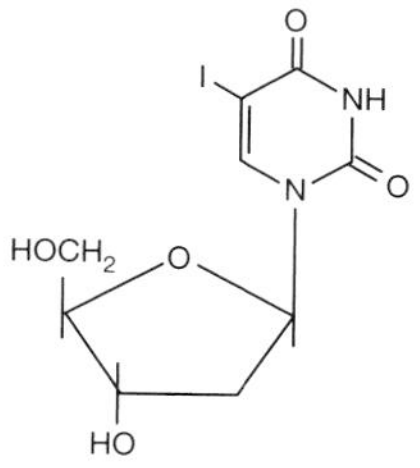

Idoxuridine

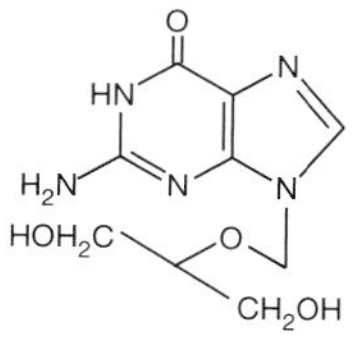

Ganciclovir

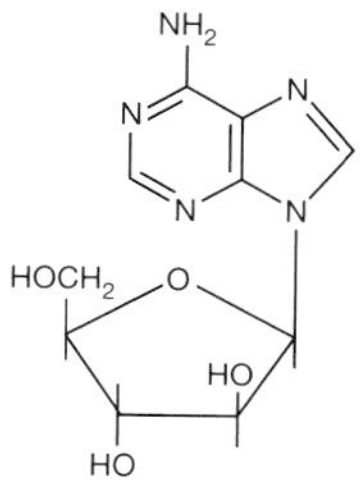

Vidarabine

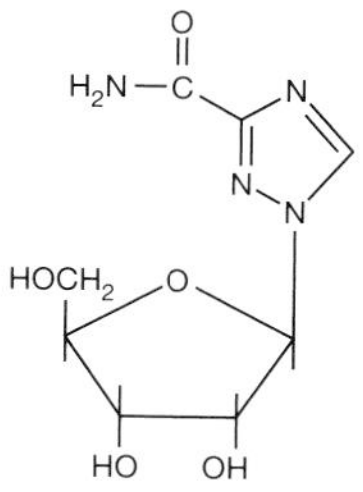

Ribavirin

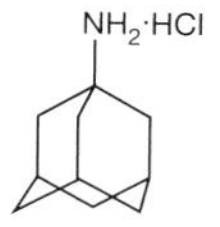

Amantadine

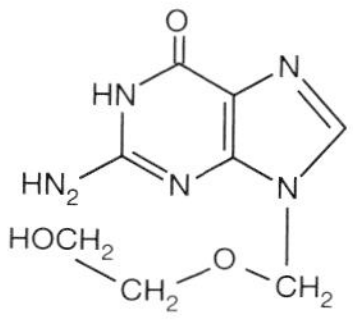

Acyclovir

Agent	Used to Treat	Common Method of Administration*	Side Effects
Antiprotozoan Agents			
Quinine	Malaria resistant to other agents	O	
Chloroquine	Malaria	O	Headache, itching
Primaquine	With chloroquine to prevent relapse of malaria	O	Slight nausea and abdominal pain
Pyrimethamine	Various protozoan infections	O	Large doses damage bone marrow
Metronidazole	*Trichomonas, Giardia,* and amoebic infections	O, IV, T	Black hairy tongue

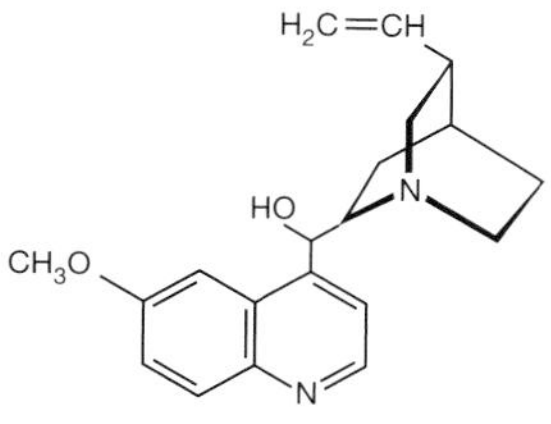

Quinine

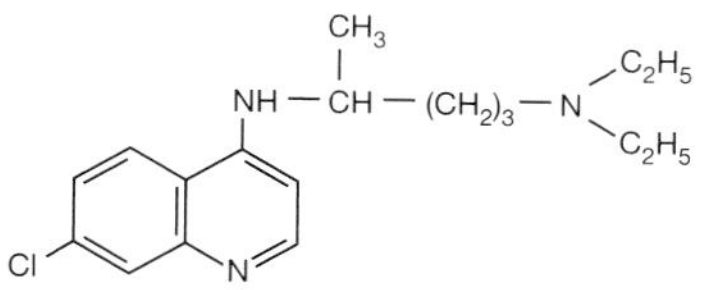

Chloroquine

Primaquine

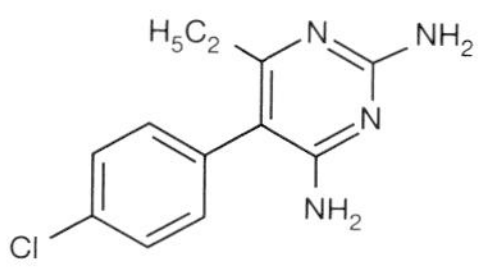

Pyrimethamine

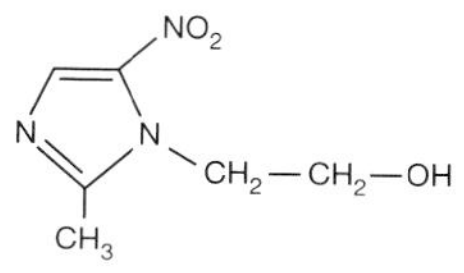

Metronidazole

streptococci resistant to many of them evolved. Similar events led to the emergence of antibiotic-resistant strains of many other organisms, including staphylococci, gonococci, *Salmonella, Neisseria,* and especially *Pseudomonas. Pseudomonas* infections are now a major problem in hospitals. Many of these organisms are now resistant to several different antibiotics, and new resistant strains are constantly being encountered.

Why are resistant organisms found more often in hospitalized patients than among outpatients? This question can be answered by looking at the hospital environment and the patients likely to be hospitalized. First, despite efforts to maintain sanitary conditions, a hospital provides an environment where sick people live in close proximity and where many different kinds of infectious agents are constantly present and are easily spread. Second, hospitalized patients tend to be more severely ill than outpatients; many have lowered resistance to infection because of their illnesses or because they have received immunosuppressant drugs. Finally, and most importantly, hospitals typically make intensive use of a variety of antibiotics. Because many infections are being treated and different antibiotics are used, organisms resistant to one or more of the antibiotics are likely to emerge. The resistant strains can readily spread among patients.

Treatment of resistant infections creates a vicious cycle. If an antibiotic can be found to which an organism is susceptible, that drug can be used to treat the infection. However, some strains of the organism that are resistant to the new antibiotic may then proliferate and require treatment with another new drug. A recurrent cycle in which new antibiotics are used and the organisms subsequently develop resistance to them is established.

Preventing infections caused by antibiotic-resistant strains of microorganisms is a difficult task, but several guidelines should be followed. First, the use of antibiotics should be limited to situations in which the patient is unlikely to recover without antibiotic treatment. Second, sensitivity tests should be done, and patients should receive only an antibiotic to which the organism is known to be sensitive. Third, when antibiotics are used, they should be continued until the organism is completely eradicated from the patient's body. Double antibiotic use, as described earlier under quinolones, is especially helpful (Figure 14.16). Finally, any patient with an infectious disease should be isolated from other patients.

FIGURE 14.16 Use of double antibiotic therapy to eradicate resistant-strain infections.

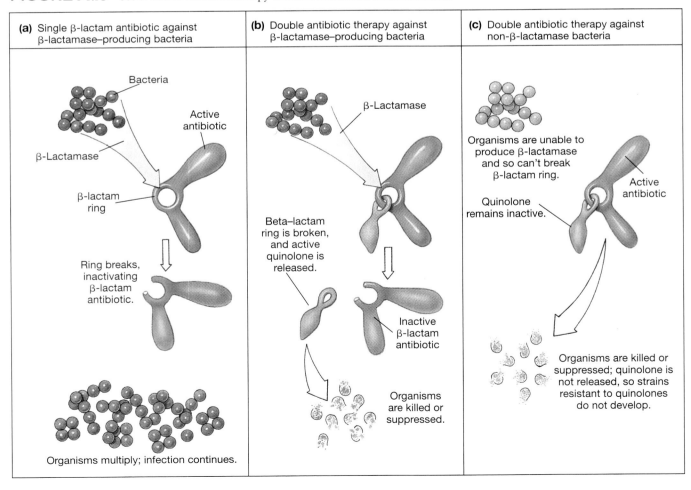

(a) Single β-lactam antibiotic against β-lactamase–producing bacteria

Bacteria
Active antibiotic
β-Lactamase
β-lactam ring
Ring breaks, inactivating β-lactam antibiotic.
Organisms multiply; infection continues.

(b) Double antibiotic therapy against β-lactamase–producing bacteria

β-Lactamase
Beta–lactam ring is broken, and active quinolone is released.
Inactive β-lactam antibiotic
Organisms are killed or suppressed.

(c) Double antibiotic therapy against non-β-lactamase bacteria

Organisms are unable to produce β-lactamase and so can't break β-lactam ring.
Quinolone remains inactive.
Active antibiotic
Organisms are killed or suppressed; quinolone is not released, so strains resistant to quinolones do not develop.

THE ROLE OF *CLOSTRIDIUM DIFFICILE* IN ANTIBIOTIC-ASSOCIATED INTESTINAL DISEASE

Antibiotic therapy sometimes has its dark side. Today, *Clostridium difficile* is one of the most important intestinal bacterial pathogens in the developed world, in terms of prevalence and severity of disease. Yet *C. difficile* causes intestinal disease only in patients to whom antibiotics have been administered. (A few exceptions to this rule existed in the preantibiotic era, and no completely convincing explanation is available to explain these cases.)

This pathogen is a gram-positive anaerobic rod (Figure 14.17). *C. difficile* was first described in 1935, but it was not definitively associated with disease until 1978. During the 1950s, especially, *Staphylococcus aureus* was blamed for what are now known to have been *C. difficile* infections. *C. difficile* is now understood to be a major nosocomial pathogen and is currently the only organism recognized as a common cause of antibiotic-associated colitis. *C. difficile* is found in 15 to 25 percent of patients with antibiotic-associated diarrhea; 50 to 75 percent of patients with colitis; and more than 90 percent of patients with pseudomembranous colitis (PMC). In individuals with PMC, a fibrous pseudomembrane covers the mucosa of the colon due to fibrin-containing fluid that collects there.

The organism is rarely found in healthy people except for newborns, among whom it occurs frequently but without harm. These newborns can act as reservoirs for *C. difficile* and can easily spread it to others in hospitals or at home. It disappears as babies approach 6 to 12 months of age. Older children rarely develop *C. difficile* problems, despite frequent use of antibiotics. Older adults are most likely to develop disease, but this age factor has not yet been explained. PMC can develop in patients who have undergone abdominal or intestinal surgery and in those whose gastrointestinal microbiota has been changed by the use of antibiotics.

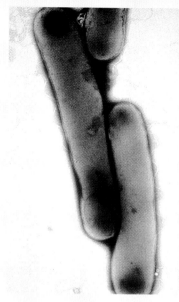

FIGURE 14.17   False-color TEM photo of the bacterium *Clostridium difficile,* which is an anaerobic, spore-forming, gram-positive bacillus (magnified 11,800×).

C. difficile is a moderately strict anaerobe. It is widely distributed in nature, being found in soil and in feces of animals such as cows and horses. Once it has established itself in an environment, *C. difficile* is difficult to remove. It may persist in hospital wards or nursing homes for months or years despite vigorous efforts at eradication. Organisms have been recovered from the floor, bedpans, linens, and even walls of rooms of infected patients.

Clindamycin is the antibiotic most frequently associated with *C. difficile* infections, followed by ampicillin and cephalosporins. These drugs are often administered to patients who will undergo abdominal surgery, in order to decrease normal microbiota. The *C. difficile* organisms are usually very susceptible to these antibiotics but survive by forming endospores. After the drug is discontinued, they regenerate and can overgrow the intestinal tract, but they do not invade its tissues.

Virtually all strains of *C. difficile* produce two toxins, but the strains vary widely in the amount of toxin produced. Toxin A, often called the enterotoxin, has a cytotoxic (cell-killing) effect and is responsible for most of the symptoms observed. Toxin B, called the cytotoxin, is considerably more cytotoxic but is not active in the gastrointestinal tract. The mechanism of action of these toxins is still not completely known.

Symptoms of the colitis include abdominal cramps, diarrhea, fever (up to 106°F, or 41.1°C), electrolyte imbalance, toxic megacolon (an acute dilation of the colon), and even perforation of the colon. The antibiotic vancomycin is the single most effective treatment and is almost 100 percent effective when perforation has not occurred. Fever drops within 24 to 48 hours, and normal bowel action returns in 5 to 7 days. However, relapse is frequent due to the sporulating forms. Vancomycin costs four times as much as an equivalent weight of gold. Cost of the treatment ranges from $200 to $600. Metronidazole is also used due to its anti-anaerobic properties.

Endoscopic observation of pseudomembranes in the colon is considered diagnostic for a severe *C. difficile* disease. Confirmation usually requires isolating the organism from stool specimens and identifying the toxins by ELISA and tissue culture assay. Toxin A is so potent that one molecule is enough to cause the changes seen in cells cultured.

ANTIMICROBIAL CHEMOTHERAPY

- **Chemotherapy** is the use of any chemical agent in the treatment of disease.
- A **chemotherapeutic agent,** or **drug,** is any chemical agent used in medical practice.
- An **antimicrobial agent** is a chemical agent used to treat a disease caused by a microbe.
- An **antibiotic** is a chemical substance produced by microorganisms that inhibits the growth of or destroys other microorganisms.
- A **synthetic drug** is one made in the laboratory.
- A **semisynthetic drug** is one made partly by microorganisms and partly by laboratory synthesis.

HISTORY OF CHEMOTHERAPY

- The first chemotherapeutic agents were concoctions from plant materials used by primitive societies.
- Paul Ehrlich's search for the "magic bullet" was the first systematic attempt to find chemotherapeutic agents. Subsequent events included the development of sulfa drugs, penicillin, and many other antibiotics.

GENERAL PROPERTIES OF ANTIMICROBIAL AGENTS

Selective Toxicity

- **Selective toxicity** is the property of antimicrobial agents that allows them to exert greater toxic effects on microbes than on the host.
- The **therapeutic dosage level** of an antimicrobial agent is the concentration over a period of time required to eliminate a pathogen.
- The **chemotherapeutic index** is a measure of the toxicity of an agent to the body relative to its toxicity for an infectious organism.

Spectrum of Activity

- The **spectrum of activity** of an antimicrobial agent refers to the variety of microorganisms sensitive to the agent. A **broad-spectrum** agent attacks many different organisms. A **narrow-spectrum** agent attacks only a few different organisms.

Modes of Action

- Agents that kill bacteria are bactericidal; those that inhibit bacterial growth are bacteriostatic.
- Agents that inhibit cell wall synthesis allow the membrane of the affected microbe to rupture and release the cell contents.
- Agents that disrupt membrane function dissolve the membrane or interfere with the movement of substances into or out of cells.
- Agents that inhibit protein synthesis prevent the growth of microbes by disrupting ribosomes or otherwise interfering with the process of translation.

- Agents that inhibit nucleic acid synthesis interfere with synthesis of RNA (transcription) or DNA (replication) or disrupt the information these molecules contain.
- Agents that act as **antimetabolites** affect normal metabolites by competitively inhibiting microbial enzymes or by being erroneously incorporated into important molecules such as nucleic acids.

Kinds of Side Effects

- Side effects of antimicrobial agents on the host include toxicity, allergy, and disruption of normal microbiota.
- Allergic reactions to antimicrobial agents occur when the body reacts to the agent as a foreign substance.
- Many antimicrobial agents attack not only the infectious organism but also normal microbiota. **Superinfections** with new pathogens can occur when the defensive capacity of normal microbiota is destroyed.

Resistance of Microorganisms

- **Resistance** to an antibiotic means that a microorganism formerly susceptible to the action of an antibiotic is no longer affected by it.
- Nongenetic resistance occurs when microorganisms are sequestered from antibiotics or undergo a temporary change, such as the loss of their cell walls, that renders them nonsusceptible to antibiotic action.
- Genetic resistance occurs when organisms survive exposure to an antibiotic because of their genetic capacity to avoid damage by the antibiotic. As susceptible organisms die, the resistant survivors multiply unchecked and increase in numbers.
- **Chromosomal resistance** is due to a mutation in microbial DNA; **extrachromosomal resistance** is due to **resistance (R) plasmids,** or **R factors.**
- Mechanisms of resistance include alterations of receptors, cell membranes, enzymes, or metabolic pathways.
- **Cross-resistance** is resistance against two or more similar antimicrobial agents.
- Drug resistance can be minimized by (1) continuing treatment with an appropriate antibiotic at therapeutic dosage level until all the disease-causing organisms are destroyed; (2) using two antibiotics that exert **synergism,** an additive effect; and (3) using antibiotics only when absolutely necessary.

DETERMINATION OF MICROBIAL SENSITIVITIES TO ANTIMICROBIAL AGENTS

- Sensitivity of microbes to chemotherapeutic agents is determined by exposing them to the agents in laboratory cultures.

Disk Diffusion Method

- In the **disk diffusion (Kirby-Bauer) method,** antibiotic-impregnated filter paper disks are placed on agar plates inoculated with a lawn of the test organism. Sensitivities to the drugs are determined by comparing the size of clear zones around the disks to a table of standard measurements.

Dilution Method

- In the **dilution method,** a constant inoculum is placed into broth cultures or wells with differing known quantities of chemotherapeutic agents. The minimum inhibitory concentration (MIC) of the agent is the lowest concentration in which no growth of the organism is observed. The **minimum bactericidal concentration** (MBC) of the agent is the lowest concentration in which subculturing of broth yields no growth.

Serum Killing Power

- In the **serum killing power** method, a bacterial suspension is added to a patient's **serum** drawn while the patient is receiving an antibiotic, and it is noted whether the organisms are killed.

Automated Methods

- Automated methods allow rapid identification of microorganisms and determination of their sensitivities to antimicrobial agents.

ATTRIBUTES OF AN IDEAL ANTIMICROBIAL AGENT

- An ideal antimicrobial agent is soluble in body fluids, selectively toxic, and nonallergenic; can be maintained at a constant therapeutic concentration in blood and body fluids; is unlikely to elicit resistance; has a long shelf life; and is reasonable in cost.

ANTIBACTERIAL AGENTS

- Antibacterial agents inhibit cell wall synthesis, disrupt cell membrane functions, inhibit protein synthesis, inhibit nucleic acid synthesis, or act by some other means to kill bacteria.

ANTIFUNGAL AGENTS

- Antifungal agents increase plasma membrane permeability, interfere with nucleic acid synthesis, or otherwise impair cell functions.

ANTIVIRAL AGENTS

- Antiviral agents have been difficult to find because they must damage intracellular viruses without severely damaging host cells.
- Most antiviral agents are analogues of purines or pyrimidines.
- Interferon is released by virus-infected cells and stimulates neighboring cells to produce antiviral proteins. Interferons are being made by genetic engineering and being tested in treatment of viral infections and cancer.

ANTIPROTOZOAN AGENTS

- Some antiprotozoan agents interfere with protein synthesis or folic acid synthesis. The mechanism of action of others is not well understood.

ANTIHELMINTHIC AGENTS

- Antihelminthic agents interfere with carbohydrate metabolism or act as neurotoxins.

SPECIAL PROBLEMS WITH DRUG-RESISTANT HOSPITAL INFECTIONS

- Resistant hospital infections are due largely to intensive use of a variety of antibiotics, which fosters the growth of resistant strains. Treatment and prevention of such infections are extremely difficult.

KEY TERMS

A Terms for Chemotherapy and Antibiotics

1. Define and give an example of each of the following: antibiotic, synthetic drug, semisynthetic drug.

B Development of Chemotherapeutic Agents

2. How were the first chemotherapeutic agents developed?
3. How has chemotherapy developed since Ehrlich's time?

C Selective Toxicity, Spectrum of Activity, and Modes of Action

4. Define selective toxicity, and relate it to antimicrobial agents.
5. Explain therapeutic dosage levels and chemotherapeutic indexes.
6. What is meant by the spectrum of activity of an antimicrobial agent?
7. A broad-spectrum agent is indicated when ——————, whereas a narrow-spectrum agent should be used when ——————.
8. Identify five modes of action of antimicrobial drugs, and briefly explain how each mode affects microbes.

D Side Effects of Antimicrobial Agents

9. Describe at least three common side effects of antimicrobial drugs.

E Resistance to Antibiotics

10. What is drug resistance?
11. Compare nongenetic with genetic drug resistance.
12. What is cross-resistance?
13. Drug resistance can be minimized by (1) ——————, (2) ——————, and (3) ——————.

F Microbial Sensitivities to Chemotherapeutic Agents

14. How does the disk diffusion method determine an organism's sensitivity to antimicrobial agents?
15. List the advantages and disadvantages of the tube dilution method of determining sensitivities over the disk diffusion method.
16. How can sensitivity be determined by automated methods?
17. A gram-negative coliform was isolated from a patient with a urinary tract infection and tested for susceptibility. The following MIC was determined:

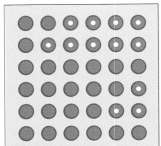

Cephalothin
Penicillin
Primaxin
Bacitracin
Clindamycin
Cotrimoxazole

Increasing dilution ⟶

(a) To which drug(s) is the organism most sensitive? (b) Least sensitive? (c) Which drug(s) would be the best choice for treating the infection? (d) Which drug(s) would not be good choices for therapy, and why not?

G Ideal Antimicrobial Agents

18. List the attributes of an ideal drug, and evaluate any three drugs according to these criteria.

H Antibacterial Agents

19. Summarize the properties, uses, and side effects of antibacterial agents that inhibit cell wall synthesis.
20. True or False: Cephalosporins are usually not the first antibiotic tried in treating an infection. Explain why or why not.
21. Summarize the properties, uses, and side effects of antibacterial agents that interfere with protein synthesis.
22. Give examples of synergistic drug action.
23. What are some precautions in the use of streptomycin, tetracyclines, chloramphenicol, and sulfonamides?
24. Given that rifampin has a broad spectrum of activity, why is it used in only a few specific situations?

I Antifungal, Antiviral, Antiprotozoan, and Antihelminthic Agents

25. Summarize the properties, uses, and side effects of antifungal agents.
26. Summarize the properties, uses, and side effects of antiviral agents.
27. Why has it been more difficult to develop antiviral agents than to develop antibacterial agents?
28. Summarize the properties, uses, and side effects of antiprotozoan agents.
29. Summarize the properties, uses, and side effects of antihelminthic agents.
30. Match the following items:
    —— aminoglycoside
    —— antimetabolite
    —— causes "black hairy" tongue
    —— possesses β-lactam ring
    —— purine analogue
    —— antimalarial drug
    —— controls tapeworm infections
    —— affects cell membranes
    —— prevents influenza A
    —— topical antifungal drug

    a. acyclovir
    b. metronidazole
    c. amphotericin B
    d. chloroquine
    e. amantadine
    f. streptomycin
    g. sulfonilamide
    h. penicillin
    i. griseofulvin
    j. niclosamide

J Drug-Resistant Hospital Infections

31. Why do antibiotic-resistant infections occur so often in hospitals?
32. Why are treatment and prevention of such infections so difficult?

1. Explain how you would design a program to help prevent the emergence of resistance to antimicrobial agents, especially multiply resistant *Staphylococcus aureus*.

2. Design a procedure for the isolation of a microorganism that will produce a new antibiotic useful to humans.

3. Discuss the advantages and disadvantages of using more than one drug simultaneously in treatment.

4. Study the structural formulas of antimicrobial agents given in this chapter, and cite at least three examples of similar structures that have similar modes of action. Also cite an example of agents with similar structures that have different actions.

5. Read about the combined efforts of the United States and Britain to develop penicillin production during World War II. What were some of the major problems? Why was the research begun in Britain but later moved to the United States?

6. An outbreak of Legionnaires' disease kills six members of a small rural community. Proper diagnosis is made only at autopsy. What could have prevented this tragedy? (The answer to this question appears in Appendix F.)

SOME INTERESTING READING

"Attacking HIV with antisense and catalytic RNA." 1990. *ASM News* 56(2):73–74.

Bean, B. 1992. "Antiviral therapy: Current concepts and practices." *Clinical Microbiology Reviews* 5(2):146–82.

Cohen, J. S., and M. E. Hogan. 1994. "The new genetic medicines." *Scientific American* 271, no. 6 (December):76–82.

Johnson, H. M., et al. 1994. "How interferons fight disease." *Scientific American* 270 (May):68–75.

Kessler, D. A., and K. L. Feiden. 1995. "Faster evaluation of vital drugs." *Scientific American* 272 (March):48–54.

Knoop, F. C., and M. Owens. 1993. "*Clostridium difficile*: Clinical disease and diagnosis." *Clinical Microbiology Reviews* 6, no. 3 (July):251–65.

Levy, S. B. 1992. "The antibiotic paradox: How miracle drugs are destroying the miracle." New York: Plenum Press.

Livermore, D. M. 1993. "Carbapenemases: The next generation of β-lactamases?" *ASM News* 59, no. 3 (March):129–35.

Parry, M. F. 1987. "The penicillins." *Medical Clinics of North America* 71, no. 6 (November):1093.

Peterson, P. K., and J. Verhoef. 1986. *The antimicrobial agents annual*, Vol. 1. New York: Elsevier.

Physician's desk reference. Annual. Oradell, NJ: Medical Economics Co.

Pratt, W. B., and R. Fekety. 1986. *The antimicrobial drugs*. New York: Oxford University Press.

Rex, J. H., et al. 1993. "Antifungal susceptibility testing." *Clinical Microbiology Reviews* 6, no. 4 (October):367–81.

Rezabek, G. H., and A. D. Friedman. 1992. "Superficial fungal infections of the skin: Diagnosis and current treatment recommendations." *Drugs* 43:674–82.

Russell, A. D., et al. 1986. "Bacterial resistance to antiseptics and disinfectants." *Journal of Hospital Infections* 7, no. 3 (May):213.

Silberner, J. 1987. "Drug resistance: Malaria–cancer similarity?" *Science News* 131, no. 10 (March 7):148.

Skolnick, A. 1991. "New insights into how bacteria develop antibiotic resistance." *Journal of the American Medical Association* 265, no. 1 (January 2):14–16.

Temple, R. 1993. "Trends in pharmaceutical development." *Drug Information Journal* 27, no. 2 (April):355–66.

Weiss, R. 1988. "Delivering the goods." *Science News* 133, no. 23 (June 4):360–62. (Problems with getting drugs to disease-causing agents.)

Wexler, H. M. 1991. "Susceptibility testing of anaerobic bacteria: Myth, magic, or method?" *Clinical Microbiology Reviews* 4(4):470–84.

In this woodcut from Willem van den Bosche's *Historia medica* (1638), a woman is applying leeches to her arm for bloodletting. Such misguided efforts could not halt the disease process of an infectious disease.

CHAPTER 15

HOST-MICROBE RELATIONSHIPS AND DISEASE PROCESSES

How is it that every now and then, no matter how careful you are, you "catch" an infectious disease? You become ill. With or without antimicrobial agents, you generally recover from the disease. In the process, you *may* develop *immunity*—that is, if you are exposed to the disease agent at another time, you may be protected from contracting the disease again.

Recall from Chapter 12 that a **pathogen** is a parasite capable of causing disease in a host. ∞ (p. 300) The ability of a pathogen to cause a disease in you depends on whether the pathogen or you—the host—wins the battle. Pathogens have certain invasive capabilities, and you have a variety of defenses. For example, in many countries the measles virus is present in a portion of the population at all times. Those who are infected release the virus, and it makes its way into the tissues of susceptible individuals. There, the virus can overcome defenses, invade tissues, and cause disease. But some individuals do not become infected.

If a virus does make its way into your tissues, your immune-system defenses may destroy it before it can cause disease. You could become immune to further exposures without actually becoming sick. Even when your first defenses fail and the disease occurs, you may develop immunity and will not be susceptible to the disease on subsequent exposures.

To begin the study of host–microbe interactions, we will look at a variety of relationships between host and microbe and see how some of these relationships result in disease. We will then characterize diseases and look at the disease process brought on by pathogens.

HOST–MICROBE RELATIONSHIPS

Microorganisms display a variety of complex relationships with other microorganisms and with larger forms of life that serve as hosts for them. A **host** is any organism that harbors another organism.

Symbiosis

Symbiosis is an association between two (or more) species. Meaning "living together" the term *symbiosis* encompasses a spectrum of relationships. These include *mutualism, commensalism,* and *parasitism.*

At one end of the spectrum is **mutualism,** in which both members of the association living together benefit from the relationship (Figure 15.1). For example, the ability of termites to digest wood, or cellulose, depends on protozoa they harbor in their intestines. These protozoa secrete into the intestine enzymes that digest the chewed wood. The protozoa themselves gain a safe, stable environment in which to live. For the termites, the relationship is obligatory; they would starve without their protozoan partners. Similarly, large numbers of *Escherichia coli* live in the large intestine of humans. These bacteria release useful products such as vitamin K, which we use to make certain blood-clotting factors. Although the relationship is not obligatory, *E. coli* does make a modest contribution to satisfying our need for vitamin K. The bacteria, in turn, get a favorable environment in which to live and obtain nutrients.

At the other end of the spectrum is **parasitism,** in which one organism, the parasite, benefits from the relationship, whereas the other organism, the host, is harmed by it (Figure 15.2). By this broad definition of the term **parasite,** bacteria, viruses, protozoa, fungi, and helminths are parasites. (Some biologists use "parasite" to refer only to protozoa, helminths, and arthropods that live on or in their host.) Parasitism encompasses a wide range of relationships, from those in which the host sustains only slight harm to those in which the host is killed.

FIGURE 15.1 Lichens, a mutualistic relationship between algae and fungi. Algae provide nutrients through the process of photosynthesis, and the fungi provide moisture and a framework for the association. The different colors of the lichens show the involvement of different algal species.

Some parasites obtain comfortable living arrangements by causing only modest harm to their host. Other parasites kill their hosts, thereby rendering themselves homeless. ∞ (Chapter 12, p. 300) The most successful

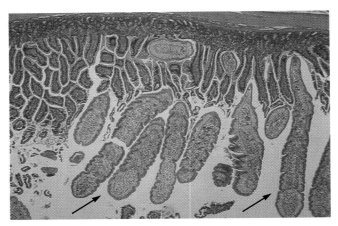

FIGURE 15.2 Micrograph of the intestine of a porcupine infected with helminths (see arrows)—a parasitic relationship.

parasites are those that maintain their own life processes without severely damaging their hosts.

Somewhere in the middle lies **commensalism,** in which two species live together in a relationship such that one benefits and the other one neither benefits nor is harmed (Figure 15.3). For example, many microorganisms live on our skin surfaces and utilize metabolic products secreted from pores in the skin. Because those products are released whether or not they are used by microorganisms, the microorganisms benefit, and ordinarily we are neither benefited nor harmed.

The line between commensalism and mutualism is not always a clear one. By taking up space and utilizing nutrients, microbes that show mutualistic or commensalistic behavior may prevent colonization of the skin by other, potentially harmful, disease-causing microbes—a phenomenon known as *microbial competition.* Hence these symbiotic relationships confer an indirect benefit on the host.

There is also a fine line between parasitism and commensalism. In healthy hosts, many microbes of the large intestine form harmless associations, simply feeding off digested food materials. But a "harmless" microbe could act as a parasite if it gains access to a part of the body where it would not normally exist.

Contamination, Infection, and Disease

Contamination, infection, and disease can be viewed as a sequence of conditions in which the severity of the effects microorganisms have on their hosts increases. **Contamination** means that the microorganisms are present. Inanimate objects and the surfaces of skin and mucous membranes can be contaminated with a wide variety of microorganisms. Commensals do no harm, but parasites have the capacity to invade tissues. **Infection** refers to the multiplication of any parasitic organism within or upon the host's body. (Sometimes the term **infestation** is used to refer to the presence of larger parasites, such as worms or arthropods, in or on the body.) If an infection disrupts the normal functioning of the host, disease occurs. **Disease** is a disturbance in the state of health wherein the body cannot carry out all its normal functions.

Both infection and disease result from interactions between parasites and their hosts. Sometimes an infection produces no observable effect on the host even though organisms have invaded tissues. More often an infection produces observable disturbances in the host's state of health; that is, disease occurs. When an infection causes disease, the effects of the disease range from mild to severe.

Let us look at some examples to understand the differences among contamination, infection, and disease. A health care worker who fails to follow aseptic

FIGURE 15.3 Bacteria on human skin (40,800×). Most of these organisms are commensals, which indirectly benefit us by competing with harmful organisms for nutrients and preventing those organisms from finding a site to attach to and invade tissues.

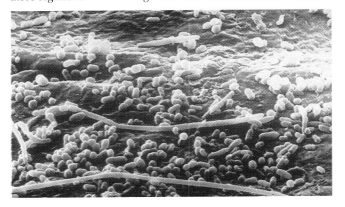

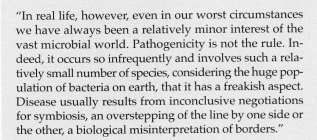

CLOSE-UP

PATHOGENS: UNSUCCESSFUL ATTEMPTS AT SYMBIOSIS

"In real life, however, even in our worst circumstances we have always been a relatively minor interest of the vast microbial world. Pathogenicity is not the rule. Indeed, it occurs so infrequently and involves such a relatively small number of species, considering the huge population of bacteria on earth, that it has a freakish aspect. Disease usually results from inconclusive negotiations for symbiosis, an overstepping of the line by one side or the other, a biological misinterpretation of borders."

—Lewis Thomas, 1974

procedures while dressing a skin wound contaminates her hands with staphylococci. However, after she finishes her task, she washes her hands properly and suffers no ill effects. Although her hands were contaminated, she did not develop an infection. Another worker performing the same task on another patient fails to wash his hands properly after treating the patient, and the organisms gain entrance to the body and infect a small cut. Soon the skin around the cut becomes reddened for a day or so. This worker was contaminated and infected. In a similar situation, a third worker develops a reddened area on her skin; she ignores it and in a few days has a large boil. This worker has experienced contamination, infection, and disease.

Disease, or illness, is characterized by changes in the host that interfere with normal function. These changes can be mild, severe but reversible, or irreversible. For example, if you become infected with one of the viruses that cause the common cold, you may have just a runny nose for a few days. Or you may have a severe cold with a sore throat, cough, fever, and headache, but the disease runs its course in a week or so without any permanent effects. The changes in your state of health are reversible. But if you develop trachoma, a bacterial infection of the eye, without treatment scarring of the cornea can occur, leading to permanent vision impairment and sometimes to blindness. Likewise, if you fail to get proper treatment for streptococcal infections, you might suffer irreversible damage to your heart or kidneys.

Pathogens, Pathogenicity, and Virulence

Pathogens vary in their abilities to disrupt the state of an individual's health—that is, they display different degrees of pathogenicity. **Pathogenicity** is the capacity to produce disease. An organism's pathogenicity depends on its ability to invade a host, multiply in the host, and avoid being damaged by the host's defenses. Some disease agents, such as *Mycobacterium tuberculosis*, frequently cause disease upon entering a susceptible host. Other agents, such as *Staphylococcus epidermidis*, cause disease only in rare instances and usually only in hosts with poor defenses. Most infectious agents exhibit a degree of pathogenicity between these extremes.

An important factor in pathogenicity is the number of infectious organisms that enter the body. If only a small number enter, the host's defenses may be able to eliminate the organisms before they can cause disease. If a large number enter, they may overwhelm the host's defenses and cause disease.

Virulence refers to the intensity of the disease produced by pathogens, and it varies among different microbial species. For example, *Bacillus cereus* causes mild gastroenteritis, whereas the rabies virus causes neuro-

logical damage that is nearly always fatal. Virulence also varies among members of the same species of pathogen. For example, organisms freshly discharged from an infected individual tend to be more virulent than those from a carrier, who characteristically shows no signs of disease. The virulence of a pathogen can increase by **animal passage,** the rapid transfer of the pathogen through animals of a species susceptible to infection by the pathogen. As one animal becomes diseased, organisms released from that animal are passed to a healthy animal, which then also gets sick. If this sequence is repeated two or three times, each newly infected animal suffers a more serious case of the disease than the one before it. Presumably the microbe becomes better able to damage the host with each animal passage. Sometimes an infectious disease spreads through human populations in this fashion, and an epidemic of the disease results. Influenza epidemics often proceed in this manner; the first people to become infected have a mild illness, but those infected later have a much more severe form of the disease. This process does not continue forever; the microbe reaches the height of its virulence, and the exposed population acquires immunity.

The virulence of a pathogen can be decreased by **attenuation,** the weakening of the disease-producing ability of the pathogen. Attenuation can be achieved by repeated subculturing on laboratory media or by transposal of virulence. **Transposal of virulence** is a laboratory technique in which a pathogen is passed from its normal host to a new host species and then passed sequentially through many individuals of the new host species. Eventually, the pathogen adapts so completely to the new host that it is no longer virulent for the original host. In other words, virulence has been transposed to another organism. Pasteur made use of transposal of virulence in preparing rabies vaccines. By repeated passage through rabbits, the virus eventually became harmless to humans and was safe to use in a human vaccine. We will see in Chapter 18 that attenuation is an important step in the production of some vaccines in use today.

Normal (Indigenous) Microbiota

As we have described, microorganisms found in various symbiotic associations with humans do not necessarily cause disease. An adult human body consists of approximately 10^{13} (10 trillion) eukaryotic cells. It harbors an additional 10^{14} (100 trillion) prokaryotic and eukaryotic microorganisms on the skin surface, on mucous membranes, and in the passageways of the digestive, respiratory, and reproductive systems. Thus, there are 10 times more microbial cells on or in the human body than there are cells making up the body!

Before birth, a fetus exists in a sterile environment. During passage through the birth canal, the fetus acquires certain microorganisms that may become permanently or temporarily associated with it. Organisms that live on or in the body but do not cause disease are referred to collectively as **normal microbiota,** or *normal flora* (Table 15.1). Many such organisms have well-established associations with humans. Most organisms among the normal microbiota are commensals—they obtain nutrients from host secretions—waste substances found on the surfaces of skin and mucous membranes. Two categories of organisms can be distinguished: *resident microbiota* and *transient microbiota*.

The **resident microbiota** (Figure 15.4) comprise microbes that are always present on or in the human body. They are found on the skin and conjunctiva, in the mouth, nose, and throat, in the large intestine, and in passageways of the urinary and reproductive systems, especially near their openings. In each of these body regions, resident microbiota are adapted to prevailing conditions. The mouth and the lower part of the large intestine provide warm, moist conditions and ample nutrients. Mucous membranes of the nose, throat, urethra, and vagina also provide warm, moist conditions, although nutrients are in shorter supply. The skin provides ample nutrients but is cooler and less moist.

Other regions of the body lack resident microbiota either because these regions provide conditions unsuitable for microorganisms, are protected by host de-

TABLE 15.1	Major Normal Microbiota (Unless Otherwise Noted, Bacteria) of the Human Body

Skin
*Staphylococcus epidermidis**
Staphylococcus aureus
Lactobacillus species
*Propionibacterium acnes**
Pityrosporon ovale (fungus)*

Mouth
*Streptococcus salivarius**
Streptococcus pneumoniae
*Streptococcus mitis**
*Staphylococcus epidermidis**
Staphylococcus aureus
Moraxella catarrhalis
*Veillonella alcalescens**
Lactobacillus species*
Klebsiella species
*Haemophilus influenzae**
*Fusobacterium nucleatum**
*Treponema denticola**
Candida albicans (fungus)*
Entamoeba gingivalis (protozoan)*
Trichomonas tenax (protozoan)*

Upper respiratory tract
*Staphylococcus epidermidis**
Staphylococcus aureus
*Streptococcus mitis**
Streptococcus pneumoniae
Moraxella catarrhalis
Lactobacillus species
Haemophilus influenzae

Intestine
*Staphylococcus epidermidis**
Staphylococcus aureus
*Streptococcus mitis**
Enterococcus species*
Lactobacillus species*
Clostridium species*
*Eubacterium limosum**
*Bifidobacterium bifidum**
Actinomyces bifidus
*Escherichia coli**
Enterobacter species*
Klebsiella species
Proteus species
Pseudomonas aeruginosa
Bacteroides species*
Fusobacterium species
Treponema denticola
Endolimax nana (protozoan)
Giardia intestinalis (protozoan)

Urogenital tract
*Streptococcus mitis**
Streptococcus species*
*Staphylococcus epidermidis**
Lactobacillus species*
Clostridium species
Actinomyces bifidus
Candida albicans (fungus)*
Trichomonas vaginalis (protozoan)

*Well-established associations

fenses, or are inaccessible to microorganisms (Table 15.2). For example, conditions in the stomach are too acidic to permit survival of microbiota. Under normal conditions the nervous system is inaccessible to microbes. Blood has no resident microbiota because it is relatively inaccessible, and host defense mechanisms normally destroy microorganisms before they become established.

Transient microbiota are microorganisms that can be present under certain conditions in any of the locations where resident microbiota are found. They persist for hours to months, but only as long as the necessary conditions are met. Transient microbiota appear on mucous membranes when greater than normal quantities of nutrients are available or on the skin when it is warmer and more moist than usual. Even pathogens can be transient microbiota. For example, suppose that you come in contact with a child infected with measles, and some of the viruses enter your nose

APPLICATIONS

CAN A BLOODHOUND FIND THE CORRECT IDENTICAL TWIN?

As we walk around, we shed a "dandruff cloud" of skin flakes that lands on the ground or on nearby objects. A bloodhound sniffs these flakes to follow a trail. Normal microbiota on humans metabolize oils and other secretions into byproducts with particular odors. Identical twins are not colonized by identical normal microbiota, and so their skin flakes develop slightly different odors. A bloodhound can distinguish among these odors, leading the dog to the right twin.

Following the unique mixture of odors produced by normal microbiota present on skin flakes that have been shed, a bloodhound trails his quarry.

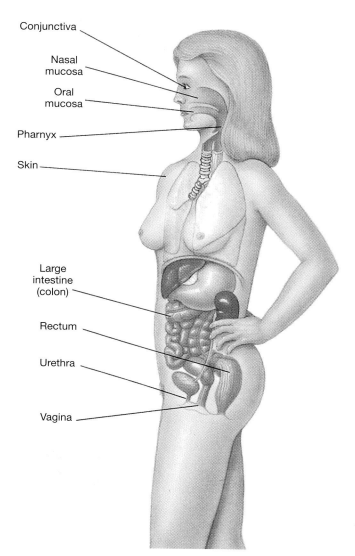

Conjunctiva

Nasal
mucosa

Oral
mucosa

Pharnyx

Skin

Large
intestine
(colon)

Rectum

Urethra

Vagina

FIGURE 15.4 Locations of resident microbiota of the human body.

TABLE 15.2	Body Tissues, Organs, and Fluids That Are Normally Microbe-Free
Internal tissues and organs	*Body fluids*
Middle and inner ear	Blood
Sinuses	Cerebrospinal fluid
Internal eye	Saliva prior to secretion
Bone marrow	Urine in kidneys and in
Muscles	bladder
Glands	Semen prior to entry into
Organs	the urethra
Circulatory system	
Brain and spinal cord	
Ovaries and testes	

and throat. You had measles years ago and are immune to the disease, so your body's defenses prevent the viruses from invading cells. But you harbor the viruses as transients for a short time.

Among the resident and transient microbiota are some species of organisms that do not usually cause disease but can do so under certain conditions. These organisms are called **opportunists** because they take advantage of particular opportunities to cause disease. Conditions that create opportunities for such organisms include:

1. *Failure of the host's normal defenses.* Individuals with weakened immune defenses are said to be **immunocompromised.** Factors such as advanced malnutrition, the presence of another disease, advanced age, treatment with radiation or immunosuppressive drugs, and physical or mental stress can lead to this state. The failure of host defenses in AIDS patients, for example, allows several different opportunistic infections to develop.

2. *Introduction of the organisms into unusual body sites.* The bacterium *Escherichia coli* is a normal resident of the human large intestine, but it can cause disease if it gains entrance to unusual sites such as the urinary tract, surgical wounds, or burns.

3. *Disturbances in the normal microbiota.* Thriving populations of normal microbiota compete with pathogenic organisms and in some instances actively combat their growth, an effect known as **microbial antagonism.** The normal microbiota interfere with the growth of pathogens by competing for and depleting nutrients the pathogens need or by producing substances that create environments in which the pathogens cannot grow. As we saw in Chapter 14, antibiotics sometimes destroy or disturb the normal microbiota as they bring a pathogen under control. ∞ (p. 362) This disturbance allows other potential pathogens, such as yeasts that are not harmed by the antibiotic, to thrive in the absence of their antagonists, the normal microbiota.

Although in later chapters we will focus on microorganisms that cause human disease, we must not lose sight of the importance of the many nonpathogenic microorganisms associated with the human body. In addition, we must remember that disease can result from disturbances in the normal ecological balance between resident populations and the host.

KOCH'S POSTULATES

The work of Robert Koch and the role of his postulates in relating causative agents to specific diseases was described briefly in Chapter 1. ∞ (p. 11) Now we can use our understanding of infection and disease to look at those postulates more carefully. For example, we now know that infection with an organism does not necessarily indicate that disease is present. With that knowledge, we can better appreciate the need for all four of

Koch's postulates to be met to prove that a specific organism is the causative agent of a particular disease:

1. The specific causative agent must be observed in every case of a disease.
2. The agent must be isolated from a diseased host and must be grown in pure culture.
3. When the agent from the pure culture is inoculated into healthy, but susceptible, experimental hosts, the agent must cause the same disease.
4. The agent must be reisolated from the inoculated, diseased experimental host and identified as identical to the original specific causative agent.

It is relatively easy today to demonstrate that each postulate is met for a variety of diseases caused by bacteria (Figure 15.5). Some bacteria, however, are difficult to culture because they have fastidious nutritional requirements or other special needs for growth. For ex-

ample, although the causative agent of syphilis, *Treponema pallidum,* has been known for many years, it has not been successfully grown on artificial media. Moreover, parasites such as viruses and rickettsias cannot be grown in artificial media and must instead be grown in living cells. For some agents that cause disease in humans, no other host has been found. Consequently, in such cases inoculation into a susceptible host is impossible unless human volunteers can be found. There are obvious ethical problems associated with inoculating humans with infectious agents, even if volunteers might be available.

KINDS OF DISEASES

Human diseases are caused by infectious agents, structural or functional genetic defects, environmental factors, or any combination of these causes.

FIGURE 15.5 Demonstration that a bacterial disease satisfies Koch's postulates.

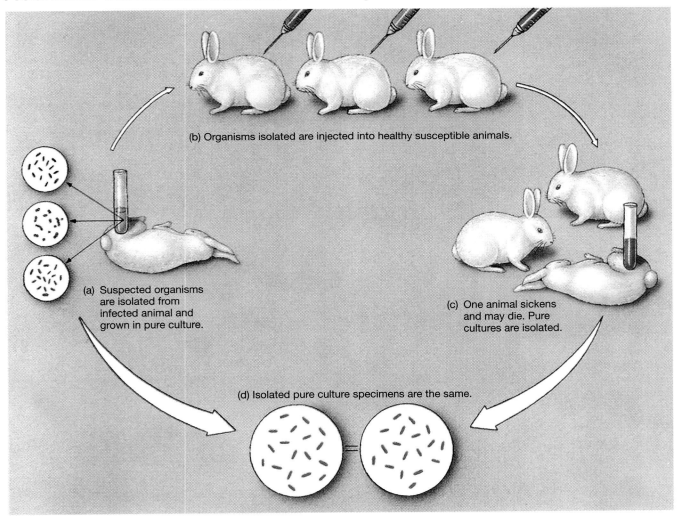

(b) Organisms isolated are injected into healthy susceptible animals.

(a) Suspected organisms are isolated from infected animal and grown in pure culture.

(c) One animal sickens and may die. Pure cultures are isolated.

(d) Isolated pure culture specimens are the same.

ARMADILLO: CULTURE VESSEL FOR LEPROSY

The organism that causes Hansen's disease (leprosy) is difficult to culture. Many different methods had been tested and found unsatisfactory until someone tried inoculating the organism into the footpads of the nine-banded armadillo. There it grows very well; in fact, the organism multiplies faster there than in human tissues. When the organism does infect humans, it can have an incubation period of up to 30 years before disease symptoms appear. Before the armadillo was used to culture the organism, Koch's third postulate could not be fulfilled. No one wanted to hold out an arm and say, "Here, try to give me leprosy." Also, 30 years was a long time to wait to determine the results of such an experiment. The use of the armadillo has made it possible to confirm Koch's postulates for *Mycobacterium leprae* as the causative agent of leprosy. This bacterium, seen by Armauer Hansen in 1878, was one of the first infectious agents to be identified and associated with a disease but one of the last to satisfy Koch's postulates. The armadillo was chosen as an experimental host after naturally occurring leprosy infections were found in the armadillo populations of Texas and Louisiana.

DNA studies have shown the strains of *M. leprae* in the armadillo to be identical with those infecting humans. Cases of humans' having acquired infection from armadillos have been confirmed, and we now designate armadillos as a reservoir for the disease in the U.S. Southwest. Cases of leprosy in African chimpanzees and mangabey monkeys have also been found. Injection of *M. leprae* collected from armadillos into mangabeys has caused the monkeys to develop leprosy.

Armadillos are used to culture the organism that causes Hansen's disease (leprosy) in humans.

Infectious and Noninfectious Diseases

Infectious diseases are diseases caused by infectious agents such as bacteria, viruses, fungi, protozoa, and helminths. Chapters 20 through 25 of this text are devoted to discussions of particular infectious agents and the diseases they cause. **Noninfectious diseases** are caused by any factor other than infectious organisms.

Classification of Diseases

Classification of diseases as infectious or noninfectious gives a very limited view of human disease. The following scheme for classifying diseases provides a more comprehensive view. More importantly, it shows that infectious agents can interact with other factors in causing disease.

1. *Inherited diseases* are caused by errors in genetic information. The resulting developmental disorders may be caused by abnormalities in the number and distribution of chromosomes or by the interaction of genetic and environmental factors. Although inherited diseases have a noninfectious cause, some are associated with microbial activities. Sickle-cell anemia weakens patients and makes them more susceptible to infectious diseases. However, sickle-cell patients or carriers of the defect tend to be resistant to malaria. The abnormal hemoglobin S of sickle-cell patients gives up stored oxygen. With less oxygen, red blood cells change into a sickle shape and are removed by the spleen. Malaria-causing parasites that enter red blood cells make the cells sickle and are thereby killed before they complete their life cycle.

2. *Congenital diseases* are structural and functional defects present at birth, caused by drugs, excessive X-ray exposure, or certain infections. When a mother has a rubella (German measles) or a syphilis infection, the infectious agent may cross the placenta and cause congenital defects. Some medicines, such as the antiwrinkle drug retinoid-A and the antibiotic tetracycline, may cause congenital defects when taken by pregnant women.

3. *Degenerative diseases* are disorders that develop in one or more body systems as aging occurs. Patients with degenerative diseases such as emphysema or impaired kidney function are susceptible to infections. Conversely, infectious agents can cause tissue damage that leads to degenerative disease, as occurs in bacterial endocarditis, rheumatic heart disease, and some kidney diseases.

4. *Nutritional deficiency diseases* lower resistance to infectious diseases and contribute to the severity of infections. For example, the bacterium that causes diphtheria (*Corynebacterium diphtheriae*) produces more toxin in people with iron deficiencies than in those with normal amounts of iron. Poor nutrition also increases the severity of measles and contributes to deaths from the disease. Nutritional

deficiencies can themselves develop from the action of, for example, helminths that severely damage the intestinal lining.

5. *Endocrine diseases* are due to excesses or deficiencies of hormones. Viral infection has been linked to pancreatic damage that leads to insulin-dependent diabetes.

6. *Mental disease* can be caused by a variety of factors, including those of an emotional, or psychogenic (si-ko-jen'-ik), nature as well as certain infections. For example, psychological stress may give rise to several gastrointestinal disorders, skin irritations, and even breathing difficulties. Mental disease can also result from brain infections such as in cases of neurosyphilis and the prion-caused Creutzfeld-Jakob disease.

7. *Immunological diseases* such as allergies, autoimmune diseases, and immunodeficiencies are caused by malfunction of the immune system. AIDS is a consequence of a viral infection and destruction of certain cells of the immune system.

8. *Neoplastic diseases* involve abnormal cell growth that leads to the formation of various types of generally harmless or cancerous growths or tumors. Causes of such diseases include chemicals, physical agents such as various forms of radiation, and microorganisms, especially viruses. Papillomaviruses, which are known to cause warts, have been associated with the development of cervical cancer, and other viruses are known to cause tumorous growths in plants. ∞ (Chapter 11, p. 293)

9. *Iatrogenic* (i-at"ro-jen'ik) *diseases* (*iatros*, Greek for "physician") are caused by medical procedures and/or treatments. Examples include surgical errors, drug reactions, and infections acquired from hospital treatment. The latter are called *nosocomial infections*. For example, *Staphylococcus aureus* is a common bacterium associated with surgical wound infections. Nosocomial infections are discussed in Chapter 16.

10. *Idiopathic* (id"e-o-path'ik) *diseases* are diseases whose cause is unknown. Some researchers believe that Alzheimer's disease, which causes mental deterioration, has an infectious basis.

Communicable and Noncommunicable Diseases

Some infectious diseases can be spread from one host to another and are said to be **communicable infectious diseases.** Some communicable diseases are more easily spread than others. Rubeola (red measles) and rubella are highly communicable, or **contagious diseases,** especially among young children. Vaccines protect children in developed countries, but nearly all unimmunized children in developing nations still get these diseases. Influenza is highly communicable among adults, especially the elderly. Gonorrhea and genital herpes infections are easily spread among unprotected sexual partners. Although also communicable, certain other diseases such as *Klebsiella* pneumonia are less contagious. Some diseases that normally affect other animals are transmissible to humans (Chapter 16), whereas diseases such as Hansen's disease (leprosy) can also be transmitted from humans to other animals.

Noncommunicable infectious diseases are not spread from one host to another. You cannot "catch" a noncommunicable disease from another person. Such diseases may result from (1) infections caused by an individual's normal microbiota, such as an inflammation of the abdominal cavity lining following rupture of the appendix; (2) poisoning following the ingestion of preformed toxins, such as staphylococcal enterotoxin, a common cause of food poisoning; and (3) infections caused by certain organisms found in the environment, such as tetanus, a bacterial infection resulting from spores in the soil gaining access to a wound. Other noncommunicable infectious diseases, such as legionellosis, a form of pneumonia, can spread through contaminated air-conditioning systems.

THE DISEASE PROCESS

How Microbes Cause Disease

Microorganisms act in certain ways that allow them to cause disease. These actions include gaining access to the host, adhering to and colonizing cell surfaces, invading tissues, and producing toxins and other harmful metabolic products. However, host defense mechanisms tend to thwart the actions of microorganisms. The occurrence of a disease depends on whether the pathogen or the host wins the battle; if it is a draw, a chronic, long-lasting disease may result.

Most of the pathogens considered in this text are prokaryotic microorganisms and viruses, which together account for the majority of human disease agents. However, several eukaryotes such as fungi, protozoa, and multicellular parasites (mostly worms) display pathogenicity ∞ (Chapter 12, p. 315) Eukaryotic pathogens can be present in a host without causing disease signs or symptoms, or they can cause severe disease. The extent of damage caused by these pathogens, like that caused by prokaryotic infectious agents, is determined by the properties of the pathogens and by the host's response to them.

How Bacteria Cause Disease

Bacterial pathogens often have special structures or physiological characteristics that improve the chances of successful host invasion and infection. **Virulence factors** are structural or physiological characteristics that help organisms cause infection and disease. These factors include structures such as pili for adhesion to cells and tissues, enzymes that either help in evading host defenses or protect the organism from host defenses, and toxins that can directly cause disease.

DIRECT ACTIONS OF BACTERIA Bacteria can enter the body by penetrating the skin or mucous membranes, by sexual transmission, by being ingested with food, by being inhaled in aerosols, or by transmission on a *fomite* (any inanimate object contaminated with an infectious agent). If the bacteria are immediately swept out of the body in urine or feces or by coughing or sneezing, they cannot initiate an infection.

A critical point in the production of bacterial disease is the organism's **adherence,** or attachment, to a host cell's surface. The occurrence of certain infections depends in part on the interaction between host plasma membranes and bacterial adherence factors. **Adhesins** are proteins or glycoproteins found on attachment pili (fimbriae) and capsules. ∞ (Chapter 4, p. 89) Most adhesins that have been identified permit the pathogen to adhere only to receptors on membranes of certain cells or tissues (Table 15.3). For example, an adhesin on attachment pili of certain strains of *Escherichia coli* attaches to receptors on certain host epithelial cells. (Host leukocytes also have receptors for this adhesin, so the same adhesin that helps the bacterium attach may also help the host destroy it.) However, very often the capsules and attachment pili are also antiphagocytic structures. It is difficult for phagocytic cells to engulf bacteria that have capsules or attachment pili, so these structures make excellent virulence factors.

Attachment to a host cell surface is not enough to cause an infection. The microbes must also be able to colonize the cell's surface or to penetrate it. **Colonization** refers to the growth of microorganisms on epithelial surfaces, such as skin or mucous membranes or other host tissues. For colonization to occur after adherence, the pathogens must survive and reproduce despite host defense mechanisms. For example, pathogenic bacteria on the skin's surface must withstand environmental conditions and bacteriostatic skin secretions. Those on respiratory membranes must escape the action of mucus and cilia. Those on the lining of portions of the digestive tract must withstand peristaltic movements, mucus, digestive enzymes, and acid.

Only a few pathogens cause disease by colonizing surfaces; most have additional virulence factors that enable the pathogen to invade tissues. The degree of **invasiveness** of a pathogen—its ability to invade and grow in host tissues—is related to the virulence factors the pathogen possesses and determines the severity of disease it produces. Some bacteria, such as pneumococci and other streptococci, release digestive enzymes that allow them to invade tissues rapidly and cause severe illnesses. Streptococci produce the enzyme **hyaluronidase,** or *spreading factor*. This enzyme digests hyaluronic acid, a gluelike substance that helps hold the cells of certain tissues together (Figure 15.6a). Digestion of hyaluronic acid allows streptococci to pass between epithelial cells and invade deeper tissues.

In some cases the same pathogen can display varying degrees of invasiveness and pathogenicity in different tissues. Both bubonic plague and pneumonic plague are caused by the bacterium *Yersinia pestis*. In bubonic plague, the organisms enter the body by means of a flea bite, migrate through the blood, and infect many organs and tissues. Untreated, this disease has a mortality rate of about 55 percent. As a victim of pneumonic plague coughs or sneezes, the bacteria are spread by aerosols to other individuals. *Yersinia pestis* can cause a severe infection of the lungs with a mortality rate as high as 98 percent.

Most bacteria that invade tissues damage cells and are found around cells. Thus, enzymes that contribute to tissue damage are another important virulence factor. **Coagulase** is a bacterial enzyme that accelerates the coagulation (clotting) of blood. When blood plasma, the fluid portion of blood, leaks out of vessels

TABLE 15.3	Examples of Adhesive Virulence Factors	
Bacterium	**Disease**	**Adhesion Mechanism**
Upper respiratory tract		
Mycoplasma pneumoniae	Atypical pneumonia	Adhesin on cell surface adheres to receptor in respiratory lining
Neisseria meningitidis	Meningitis	Adhesin on pili
Streptococcus pneumoniae	Pneumonia	Surface adhesins attach to carbohydrate on respiratory lining
Mouth		
Streptococcus mutans	Dental caries	Capsule attaches to tooth enamel
Intestinal tract		
Shigella species	Dysentery	Unknown mechanism for attachment to intestinal lining
Escherichia coli	Diarrhea	Adhesins on pili attach to receptor on intestinal lining
Campylobacter jejuni	Diarrhea	Adhesins on flagella attach to intestinal lining
Vibrio cholerae	Cholera	Adhesins on flagella bind to receptors on intestinal lining
Urogenital tract		
Treponema pallidum	Syphilis	Bacterial protein attaches to cells
Neisseria gonorrhoeae	Gonorrhea	Adhesins on pili attach to lining of genital tract

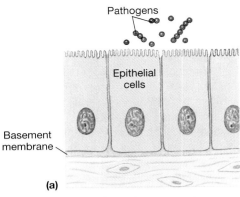
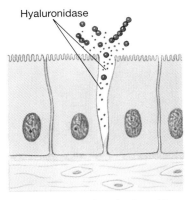
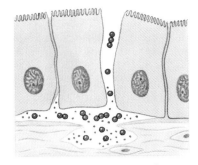

1. Invasive pathogens reach epithelial surface

2. Pathogens produce hyaluronidase

3. Pathogens invade deeper tissues

Pathogens

Epithelial cells

Basement membrane

Hyaluronidase

(a)

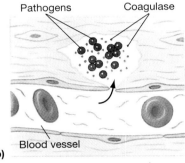
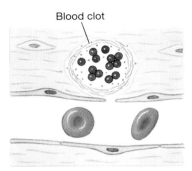
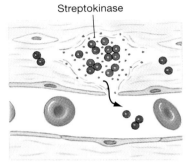

1. Pathogens produce coagulase

2. Blood clot forms around pathogens

3. Pathogens produce streptokinase, dissolving clot and releasing bacteria

Pathogens   Coagulase

Blood clot

Streptokinase

Blood vessel

(b)

FIGURE 15.6 Enzymatic virulence factors help bacteria invade tissues and evade host defenses. (a) Hyaluronidase dissolves the "cement" that holds together the cells that line the intestinal tract. Bacteria that produce hyaluronidase can then invade deeper cells within the intestinal tissues. (b) Coagulase triggers blood plasma clotting, allowing bacteria protection from immune defenses. Streptokinase dissolves blood clots. Bacteria trapped within a clot can free themselves and spread the infection by producing streptokinase.

into tissues, coagulase causes the plasma to clot. *Staphylococcus aureus* produces coagulase to aid in infection (Figure 15.6b). Coagulase is a two-edged sword: It keeps organisms from spreading but also helps wall them off from immune defenses that might otherwise destroy them. Conversely, the bacterial enzyme **streptokinase** dissolves blood clots. Pathogens trapped in blood clots free themselves to spread to other tissues by secreting these virulence factors.

Some bacterial pathogens actually enter cells. The rickettsias, chlamydias, and a few other pathogens must invade cells to grow, reproduce, and produce disease. In other situations, organisms that can survive within host phagocytic cells not only escape destruction by the phagocytes but also obtain free transportation to deeper body tissues. Such organisms include *Mycobacterium tuberculosis* and *Neisseria gonorrhoeae*.

BACTERIAL TOXINS A **toxin** is any substance that is poisonous to other organisms. Some bacteria produce toxins, which are synthesized inside bacterial cells and are classified according to how they are released. **Exotoxins** are soluble substances secreted into host tissues. **Endotoxins** are part of the cell wall and are released into host tissues—sometimes in large quantities—from gram-negative bacteria, often when the bacteria die or divide. ∞ (Chapter 4, p. 78) Giving antibiotics that kill such bacteria can release sufficient toxin to cause the patient to die of severely reduced blood pressure (*endotoxic shock*). Let us look at some of the properties and effects of endotoxins and exotoxins (Table 15.4).

Relatively weak (except in large doses), endotoxins are produced by certain gram-negative bacteria. All endotoxins consist of lipopolysaccharide (LPS) complexes, the components of which vary among genera. They are relatively stable molecules that do not display affinities for particular tissues. Bacterial endotoxins have nonspecific effects such as fever or a sudden drop in blood pressure. They also cause tissue damage in diseases such as typhoid fever and epidemic menin-

TABLE 15.4   **Properties of Toxins**

Property	Exotoxins	Endotoxins
Organisms producing	Almost all gram-positive, some gram-negative	Almost all gram-negative
Location in cell	Extracellular, excreted into medium	Bound within bacterial cell wall; released upon death of bacterium
Chemical nature	Mostly polypeptides	Lipopolysaccharide complex
Stability	Unstable; denatured above 60°C and by ultraviolet light	Relatively stable; can withstand several hours above 60°C
Toxicity	Among the most powerful toxins known (some are 100 to 1 million times as strong as strychnine)	Weak, but can be fatal in relatively large doses
Effect on tissues	Highly specific; some act as neurotoxins or cardiac muscle toxins	Nonspecific; ache-all-over systemic effects or local site reactions
Fever production	Little or no fever	Rapid rise in temperature to high fever
Antigenicity	Strong; stimulates antibody production and immunity	Weak; recovery from disease often does not produce immunity
Toxoid conversion and use	By treatment with heat or chemicals; toxoid used to immunize against toxin	Cannot be converted to toxoid; cannot be used to immunize
Examples	Botulism, gas gangrene, tetanus, diphtheria, staphylococcal food poisoning, cholera, enterotoxins, plague	Salmonellosis, tularemia, endotoxic shock

gitis (an inflammation of membranes that cover the brain and spinal cord).

Exotoxins are more powerful toxins produced by several gram-positive and a few gram-negative bacteria. Most are polypeptides, which are denatured by heat, ultraviolet light, and chemicals such as formaldehyde. Species of *Clostridium, Bacillus, Staphylococcus, Streptococcus,* and several other bacteria produce exotoxins.

Some exotoxins are enzymes. **Hemolysins** were first discovered in cultures of bacteria grown on blood–agar plates. The action of these exotoxins is to lyse (rupture) red blood cells. Two kinds of hemolysins were identified from bacteria grown on blood–agar plates. **Alpha-hemolysins** hemolyze blood cells, partially break down hemoglobin, and produce a greenish ring around colonies; **β-hemolysins** also hemolyze blood cells but completely break down hemoglobin and leave a clear ring around colonies (Figure 15.7). Streptococci and staphylococci produce different hemolysins that are helpful in identifying them in laboratory cultures. There is no evidence that red blood cell lysis plays a role in the disease syndrome. Rather, the hemolysins release iron from the hemoglobin molecules in the red blood cells. Iron is a critical element for growth of all cells, both host and microbe. But there is very little free iron within the human body. Most of it is bound in a form such as hemoglobin, and the microbe must enzymatically release it. Bacteria that can produce hemolysins can grow better than those that do not produce these enzymes. Especially in the staphylococci, the hemolysins can damage other types of cells as well. Alpha-hemolysin damages smooth muscle and kills skin cells.

Virulence factors called **leukocidins** are exotoxins produced by many bacteria, including the strep-

FIGURE 15.7   (a) Alpha, or partial, hemolysis of red blood cells results in a greenish zone around colonies of *Streptococcus pneumoniae* grown on blood agar. (b) *S. pyogenes* colonies release β-hemolysins, which produce complete breakdown of hemoglobin, causing clear zones to form around colonies grown on blood agar.

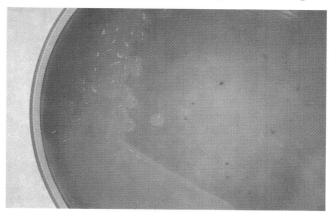

(a)

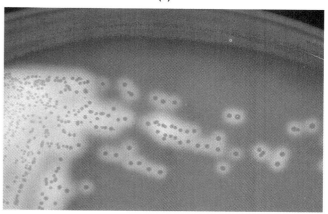

(b)

tococci and staphylococci. These toxins damage or destroy certain kinds of white blood cells called *neutrophils* and *macrophages*. Leukocidins are most effective when released by microbes that have been engulfed by a neutrophil. Because of the action of leukocidins, the number of white blood cells decreases in certain diseases, although most infections are characterized by an elevated white cell count. A similar substance, called **leukostatin,** interferes with the ability of leukocytes to engulf microorganisms that secrete the exotoxin.

In the preceding examples, the spreading of exotoxins by blood from the site of infection is called **toxemia.** But some diseases caused by microbes are due not to infection and invasion of tissues by pathogens but instead to the ingestion of preformed toxins made by pathogens. For example, botulism food poisoning strikes within hours of ingesting food that contains a significant amount of a toxin produced by *Clostridium botulinum*—too short a time for the microbe to invade tissues and cause disease. The toxins accumulate during the storage of an improperly sterilized jar or can of food and have an immediate and often ultimately

lethal effect on the consumer. Diseases that result from the ingestion of a toxin are termed **intoxications** rather than infections.

Many exotoxins have a special attraction for particular tissues. **Neurotoxins,** such as the botulism and tetanus toxins, are exotoxins that act on tissues of the nervous system to prevent muscle contraction (botulism) or muscle relaxation (tetanus). **Enterotoxins,** such as the toxin that causes cholera, are exotoxins that act on tissues of the gut. Many exotoxins can act as *antigens,* foreign substances against which the immune system reacts. Antigenic exotoxins inactivated by treatment with chemical substances such as formaldehyde are called *toxoids*. A **toxoid** (*-oid* is Latin for "like") is an altered toxin that has lost its ability to cause harm but that retains antigenicity. Toxoids can be used to stimulate the development of immunity without causing disease. For example, when you get a tetanus booster shot, you are receiving the tetanus toxoid. It stimulates your body to produce immunity so that if you are exposed to active tetanus toxin through a cut or puncture of the skin, you will not get tetanus. The effects of bacterial exotoxins in human disease are sum-

APPLICATIONS

CLINICAL USE OF BOTULINUM TOXIN

The powerful effects of the neurotoxins produced by *Clostridium botulinum*, best known as a cause of lethal food poisoning, have been harnessed to help victims of dystonia. *Dystonia* refers to a group of neurologic disorders characterized by abnormal, sustained involuntary movements, often twisting, which are of unknown origin. In one form, blepharospasm, the patient's eyes remain tightly closed at all times. Physicians inject small quantities of botulinum toxin (trade name, Oculinum A) at several sites around each eye. The toxin blocks nerve impulses to muscles, thereby relieving the spasms of the eyelids. Injections are needed every 2 to 3 months. Some people receive such treatment for 4 to 5 years without problems; others develop antibodies (immune defenses) against the toxin after many injections. Because there are several different types of the toxin, it is hoped that patients can switch to a different form once they develop antibodies against one form. A round of injections can cost from $400 to $1800.

Other dystonias being helped by botulinum toxin include oromandibular dystonia, in which the patient's jaws are clenched so tightly that the jaw bones may break, causing midfacial collapse. Eating and speaking are difficult, and some patients starve to death. Vocal cord spasms, causing a cracked tremulous voice, and "stenographers' cramp," causing the middle finger to extend rigidly, are also being treated experimentally with the toxin.

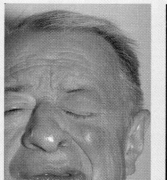

A patient with hemifacial spasm before (left) and 2 weeks after (right) treatment with botulinum toxin.

Oculinum A is licensed for treatment of adults with strabismus (cross-eye, lazy eye). Small amounts are injected into the overcontracted eye muscle, which then relaxes and lengthens. The antagonistic muscles on the other side of the eye contracts to take up the slack, and the eye can then look straight ahead.

marized in Table 15.5. The specific effects of such diseases are discussed in later chapters.

How Viruses Cause Disease

Viruses can replicate only after they have attached to cells and then penetrated specific host cells. In tissue culture systems, once inside a cell, viruses cause observable changes collectively called the **cytopathic effect** (CPE). ∞ (Chapter 11, p. 290) CPE can be cytocidal when the viruses kill the cell and noncytocidal when they do not. Cytocidal viruses can kill cells by causing enzymes from cellular lysosomes to be released or by diverting the host cell's synthetic processes, thereby stopping the synthesis of host proteins and other essential macromolecules. CPE can be observed in laboratory tissue cultures with a compound microscope (Figure 15.8). CPE can be so distinctive that an experienced clinical virologist can make a tentative identification by looking at infected cells through the microscope, even though further tests are needed to confirm the identification (Table 15.6).

Many viruses produce pathogenic effects in host cells. These include *inclusion bodies,* which consist of nucleic acids and proteins not yet assembled into viruses, masses of viruses, or remnants of viruses. Rabiesviruses make inclusion bodies that are so distinctive they can be used to diagnose rabies. Retroviruses and oncoviruses integrate into host chromosomes and can remain in cells indefinitely, sometimes leading to the expression of their antigens on host-cell surfaces. Influenza and parainfluenza viruses produce hemagglutinins, which cause agglutination, or clumping together, of erythrocytes. This feature is of value in laboratory testing.

Viral infections can be productive or abortive. A **productive infection** occurs when viruses enter a cell and

TABLE 15.5	Effects of Exotoxins		
Bacterium	**Name of Toxin or Disease**	**Action of Toxin**	**Host Symptoms**
Bacillus anthracis	Anthrax (cytotoxin)	Increases vascular permeability	Hemorrhage and pulmonary edema
Bacillus cereus	Enterotoxin	Causes excessive loss of water and electrolytes	Diarrhea
Clostridium botulinum	Botulism (eight serological types; neurotoxins)	Blocks release of acetylcholine at nerve endings	Respiratory paralysis, double vision
Clostridium perfringens	Gas gangrene (alpha toxin, a hemolysin)	Breaks down lecithin in cell membranes	Cell and tissue destruction
	Food poisoning (enterotoxin)	Causes excessive loss of water and electrolytes	Diarrhea
Clostridium tetani	Tetanus (lockjaw) (neurotoxin)	Inhibits antagonists of motor neurons of brain; 1 ng can kill 2 tons of cells	Violent skeletal muscle spasms, respiratory failure
Corynebacterium diphtheriae	Diphtheria; produced by virus-infected (cytotoxin) bacteria	Inhibits protein synthesis	Heart damage can cause death weeks after apparent recovery
Escherichia coli	Traveler's diarrhea (enterotoxin)	Causes excessive loss of water and electrolytes	Diarrhea
Escherichia coli	O157:H7 (enterotoxin)	Hemolytic uremic syndrome	Destroys intestinal lining and causes hemorrhages in kidney Bleeding and kidney hemorrhage and failure
Pseudomonas aeruginosa	Various infections (exotoxin A)	Inhibits protein synthesis	Lethal, necrotizing lesions
Shigella dysenteriae	Bacillary dysentery (enterotoxin)	Cytotoxic effects; as potent as botulinum toxin	Diarrhea, causes paralysis in rabbits from spinal cord hemorrhage and edema
Staphylococcus aureus	Food poisoning (enterotoxin)	Stimulates brain center that causes vomiting	Vomiting
	Scalded skin syndrome (exfoliatin)	Causes intradermal separation of cells	Redness and sloughing of skin
Streptococcus pyogenes	Scarlet fever (erythrogenic, or red-producing toxin)	Causes vasodilation	Maculopapular (slightly raised, discolored) lesions
Vibrio cholerae	Cholera (enterotoxin)	Causes excessive loss of water (up to 30 l/day) and electrolytes	Diarrhea; can kill within hours

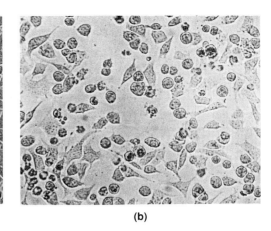

FIGURE 15.8 An example of the cytopathic effect (CPE). (a) Uninfected mouse cells; (b) the same cells 24 hours after infection with vesicular stomatitis virus. A large number have died, and many others have rounded up into abnormal shapes.

(a)

(b)

produce infectious offspring. An **abortive infection** occurs when viruses enter a cell but are unable to express all their genes to make infectious offspring. Productive infections vary in the degree of damage they cause, depending on the kind and number of cells the virus invades. An enterovirus, such as a human rotavirus or human adenovirus, that infects the gut can destroy millions of intestinal epithelial cells. Because these cells are rapidly replaced, the infection causes temporary, though sometimes severe, symptoms such as diarrhea, but no permanent damage. But a poliovirus that infects motor neurons of the central nervous system can destroy these cells. Destroyed neurons cannot be replaced, so permanent paralysis may result. Human papillomaviruses that cause warts are limited to cells in localized areas. In contrast, the measles virus replicates and spreads throughout the body and may damage many tissues.

Latent viral infections are characteristic of herpesviruses. For example, chickenpox infections occur during childhood and usually are brought under control by host immune defenses. However, the virus may retreat into the nervous system and remain inactive, or latent. Later in life, factors such as stress, other infections, or fever can reactivate the virus. A weakened immune system allows the virus to multiply. Whatever the cause, the disease appears as shingles.

TABLE 15.6	Examples of Changes (Cytopathic Effects) to Virus-Infected Cells
Virus Family	**Cytopathic Effect**
Adenoviridae	Cells swell
Herpesviridae	Cells swell
Picornaviridae	Cells swell and lyse
Paramyxoviridae	Cell membranes fuse, and up to 100 nuclei accumulate in a newly formed giant cell
Rhabdoviridae (rabies)	Inclusion bodies called Negri bodies form (site of viral replication or the accumulation of viral antigens)
Orthomyxoviridae	Produce hemagglutinins that cause erythrocytes to agglutinate, or clump together

Persistent viral infections involve a continued production of viruses over many months or years. The hepatitis B virus (HBV) infects the liver in such a chronic fashion that there may be no outward signs of an infection. However, such persistent infections can lead to cirrhosis of the liver or even liver cancer.

How Fungi, Protozoa, and Helminths Cause Disease

In addition to bacteria and viruses, infectious diseases can also be caused by eukaryotes—specifically fungi, protozoa, and helminths. Even a few algae produce neurotoxins, and one alga (*Prototheca*) directly invades skin cells.

Most fungal diseases result from fungal spores that are inhaled or from fungal cells and/or spores that enter cells through a cut or wound. Fungi damage host tissues by releasing enzymes that attack cells. As the first cells are killed, the fungi progressively digest and invade adjacent cells. Some fungi also release toxins or cause allergic reactions in the host. Certain fungi that parasitize plants produce *mycotoxins,* which cause disease if ingested by humans. Ergot, from a fungus that grows on rye, and aflatoxins, highly carcinogenic compounds that can be found in grains, cereals, even peanut butter made from moldy peanuts, are mycotoxins (see Chapter 22).

Pathogenic protozoans and helminths cause human disease in several ways. Some protozoans, including those that cause malaria, invade and reproduce in red blood cells. ∞ (Chapter 12, p. 306) The protozoan *Giardia intestinalis* attaches to tissues and ingests cells and tissue fluids of the host. *Giardia's* virulence factor is an *adhesive disc* by which it attaches to cells that line the small intestine (see Figure 22.12). While burrowing into the tissue, the parasite uses its flagella to expel tissue fluids. This process creates so strong a suction that the parasite is not disturbed by peristaltic contractions.

Most helminths are extracellular parasites, inhabiting the intestines or other body tissues. Many release toxic waste products and antigens in their excretions that often cause allergic reactions in the host. The outer surface of many helminths is quite tough and resistant to immune attacks.

Signs, Symptoms, and Syndromes

Most diseases are recognized by their signs and symptoms. A **sign** is a characteristic of a disease that can be observed by examining the patient. Signs of disease include such things as swelling, redness, rashes, coughing, pus formation, runny nose, fever, vomiting, and diarrhea. A **symptom** is a characteristic of a disease that can be observed or felt only by the patient. Symptoms include such things as pain, shortness of breath, nausea, sore throat, headache, and malaise (discomfort).

A **syndrome** is a combination of signs and symptoms that occur together and are indicative of a particular disease or abnormal condition. For example, most infectious diseases cause the body to mount an acute inflammatory response. This response, which is discussed in Chapter 17, is characterized by a syndrome of fever, malaise, swollen lymph nodes, and **leukocytosis** (an increase in the numbers of white blood cells circulating in the blood).

In addition to the inflammatory response, many infectious diseases cause other signs and symptoms. Infections of the gut called enteric infections often cause nausea, vomiting, and diarrhea. Upper respiratory infections are usually characterized by coughing, sneezing, sore throat, and runny nose. Unfortunately, the signs and symptoms of diseases caused by different pathogens may be too similar to allow a specific diagnosis to be made. Thus, laboratory tests to identify infectious agents are an important component of modern medicine.

Even after recovery, some diseases leave aftereffects, called **sequelae** (se-kwel′e; singular: *sequela*). Bacterial infections of heart valves often cause permanent valve damage, and poliovirus infections leave permanent paralysis.

Types of Infectious Disease

Infectious diseases vary in duration, location in the body, and other attributes. Several important terms, summarized in Table 15.7, are used to describe these attributes.

An **acute disease** develops rapidly and runs its course quickly. Measles and colds are examples of acute diseases. A **chronic disease** develops more slowly than an acute disease, is usually less severe, and persists for a long, indeterminate period. Tuberculosis and Hansen's disease (leprosy) are chronic diseases. A **subacute disease** is intermediate between an acute and a chronic disease. Gingivitis, or gum disease, can exist as a subacute disease. A **latent disease** is characterized by periods of inactivity either before signs and symptoms appear or

TABLE 15.7	Terms Used to Describe Infections
Term	**Characteristic of Infection**
Acute disease	Disease in which symptoms develop rapidly and that runs its course quickly
Chronic disease	Disease in which symptoms develop slowly and disease is slow to disappear
Subacute disease	Disease with symptoms intermediate between acute and chronic
Latent disease	Disease in which symptoms appear and/or reappear long after infection
Local infection	Infection confined to a small region of the body, such as a boil or bladder infection
Focal infection	Infection in a confined region from which pathogens travel to other regions of the body, such as an abscessed tooth or infected sinuses
Systemic infection	Infection in which the pathogen is spread throughout the body, often by traveling through blood or lymph
Septicemia	Presence and multiplication of pathogens in blood
Bacteremia	Presence but not multiplication of bacteria in blood
Viremia	Presence but not multiplication of viruses in blood
Toxemia	Presence of toxins in blood
Sapremia	Presence of metabolic products of saprophytes in blood
Primary infection	Infection in a previously healthy person
Secondary infection	Infection that immediately follows a primary infection
Superinfection	Secondary infection that is usually caused by an agent resistant to the treatment for the primary infection
Mixed infection	Infection caused by two or more pathogens
Inapparent infection	Infection that fails to produce full set of signs and symptoms

between attacks. The herpes simplex virus and several other viral infections produce latent disease.

A **local infection** is confined to a specific area of the body. Boils and bladder infections are local infections. A **focal infection** is confined to a specific area, but pathogens from it, or their toxins, can spread to other areas. Abscessed teeth and sinus infections are focal infections. A **systemic infection,** or *generalized infection*, affects most of the body, and the pathogens are widely distributed in many tissues. Typhoid fever is a systemic infection. When focal infections spread, they become systemic infections. For example, organisms from an abscessed tooth can enter the bloodstream and be carried to other tissues, including the kidneys. The organisms can then infect the kidneys and other parts of the urinary tract.

Pathogens can be present in the blood with or without multiplying there. In **septicemia,** once known as blood poisoning, pathogens are present in and multiply

APPLICATIONS

TO SQUEEZE OR NOT TO SQUEEZE

There is good cause for the common warning not to squeeze pimples and boils. Left alone, the body's defense mechanisms ordinarily confine these lesions to the skin. However, squeezing them can disperse microorganisms into the blood and cause septicemia—a far worse condition. In septicemia the organisms are spread throughout the body and can cause severe infections.

in the blood. In **bacteremia** and **viremia,** bacteria and viruses, respectively, are transported in the blood but do not multiply in transit. Such spread of organisms often occurs in cases of injury, such as a cut, abrasion, or even teeth cleaning. As we have seen, some pathogens release toxins into the blood; the presence of toxins in blood is called *toxemia.* Saprophytes feed on dead tissues. Fungi behave as parasites when they destroy cells and as saprophytes when they feed on them or on other dead or decaying matter. They release metabolic products into the blood, thereby causing a condition called **sapremia.**

A **primary infection** is an initial infection in a previously healthy person. Most primary infections are acute. A **secondary infection** follows a primary infection, especially in individuals weakened by the primary infection. A person who catches the common cold as a primary infection, for instance, might come down with a middle-ear infection as a secondary infection. A **superinfection** (Chapter 14) is a secondary infection that results from the destruction of normal microbiota and often follows the use of broad-spectrum antibiotics. Although many infections are caused by a single pathogen, **mixed infections** are caused by several species of organisms present at the same time. Dental caries and periodontal disease are due to mixed bacterial infections. An **inapparent,** or **subclinical, infection** is one that fails to produce the full range of signs and symptoms either because too few organisms are present or because host defenses effectively combat the pathogens. People with inapparent infections, such as carriers of the hepatitis B virus, can spread the disease to others.

Stages of an Infectious Disease

At one time or another, all of us have suffered from infectious diseases such as the common cold, for which there is no cure. We simply must let the disease "run its course." Most diseases caused by infectious agents have a fairly standard course, or series of stages. These stages include the *incubation period*, the *prodromal phase*, the *illness phase* (which includes the *acme*), the *decline*

phase, and the *convalescence period* (Figure 15.9). Even when treatment is available to eliminate the pathogen, the disease process usually passes through most of the stages. Treatment commonly lessens the severity of symptoms because pathogens can no longer multiply. It shortens the duration of the disease and the time required for recovery. The signs and symptoms associated with the stages of an infectious disease and the resulting tissue damage caused by the pathogen are summarized in Table 15.8.

Incubation Period

The **incubation period** for an infectious disease is the time between infection and the appearance of signs and symptoms. Although the infected person is not aware of the presence of an infectious agent, he or she can spread the disease to others. Each infectious disease has a typical incubation period (Figure 15.10 on page 410). The length of the incubation period is determined by the properties of the pathogen and the response of the host to the organism.

Properties that affect the incubation period include the nature of the organism, its virulence, how many organisms enter the body, and where they enter in relation to the tissues they affect. For example, if large numbers of an extremely virulent strain of *Shigella* quickly reach the intestine, profuse diarrhea can appear in a day. In contrast, if only small numbers of a less virulent strain enter the digestive tract with a large quantity of food, the disease will develop more slowly. In fact, host defenses might be able to destroy the small number of organisms such that the disease will not occur at all. In the course of a lifetime, we undoubtedly have many more exposures than infections and more infections than we have overt diseases. As we will see in Chapter 17, host defenses frequently attack

---

FIGURE 15.9 Stages in the course of an infectious disease.

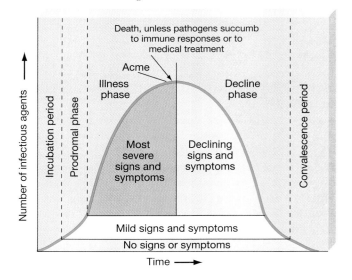

| TABLE 15.8 | Correlation of Signs and Symptoms with Tissue Damage |

Signs and Symptoms	Probable Nature of Tissue Damage
Incubation period	
None	None
Prodromal phase	
Local redness and swelling	Pathogen has damaged tissue at site of invasion and caused release of chemicals that dilate blood vessels (redness) and allow fluid from blood to enter tissues (swelling).
Headache	Chemicals from tissue injury dilate blood vessels in the brain.
General aches and pains	Chemicals from tissue injury stimulate pain receptors in joints and muscles.
Illness phase	
Cough	Mucosal cells of respiratory tract have been damaged by pathogens; excess mucus is released, and neural centers in the brain elicit coughing to remove mucus.
Sore throat	Lymphatic tissue of the pharynx is swollen and inflamed by substances released by pathogens and by leukocytes.
Fever	Leukocytes release pyrogens that reset the body's thermostat and cause temperature to rise.
Swollen lymph nodes	Leukocytes release other substances that stimulate cell division and fluid accumulation in lymph nodes; lymph nodes themselves release substances that flow to and affect other lymph nodes; some pathogens multiply in lymph nodes.
Skin rashes	Leukocytes release substances that damage capillaries and allow small hemorrhages; some pathogens invade skin cells and cause pox, vesicles, and other skin lesions.
Nasal congestion	Nasal mucosal cells have been damaged by pathogens (usually viruses) that release fluids and increase mucous secretions.
Pain at specific sites (earache, local pain at a wound site)	Substances from pathogens or leukocytes have stimulated pain receptors; messages are relayed to the brain, where they are interpreted as pain.
Nausea	Toxins from pathogens have stimulated neural centers; you interpret the stimuli as nausea.
Vomiting	Toxins in food have stimulated the brain's vomiting center; vomiting helps rid the body of toxins.
Diarrhea	Toxins in food cause fluids to enter the digestive tract; some pathogens directly injure the intestinal epithelium; both toxins and pathogens stimulate peristalsis; frequent watery stools result.
Acme	
All signs and symptoms are at peak intensity	Full development of all signs and symptoms
Decline Phase	
Signs and symptoms subside.	Host defense mechanisms (and treatment, if applicable) have contributed to overcoming the pathogen.
Convalescence Period	
Patient regains strength	Tissue repair occurs; substances that caused signs and symptoms no longer released.

pathogens as they start to invade tissues, thus averting potential diseases.

Prodromal Phase

The **prodromal phase** of disease is a short period during which nonspecific, often mild, symptoms such as malaise and headache sometimes appear. A **prodrome** (*prodromos*, Greek for "forerunner") is a symptom indicating the onset of a disease. You wake up one morning feeling bad, and you know you're coming down with something, but you don't know yet whether you will break out in spots, start to cough, develop a sore throat, or experience other signs or symptoms. Many diseases lack a prodromal phase and begin with a sudden onset of symptoms such as fever and chills. During the prodromal phase, infected individuals are contagious and can spread the disease to others.

Illness Phase

The **illness phase** is the period during which the individual experiences the typical signs and symptoms of the disease. These may include fever, nausea, headache, rash, and swollen lymph nodes. During this phase, the time when the signs and symptoms reach their greatest intensity is known as the **acme**. During the acme, pathogens invade and damage tissues. In some diseases, such as some kinds of meningitis, this phase is referred to as being **fulminating** (*fulmen*, Latin for "lightning"), or sudden and severe. In other diseases, such as hepatitis B, it can be persistent or chronic or appear gradually with inapparent symptoms. A period of chills followed by fever marks the acme of many diseases. As signs and symptoms appear, the form the infection will take becomes clear. Individuals at this critical stage are still contagious. The battle be-

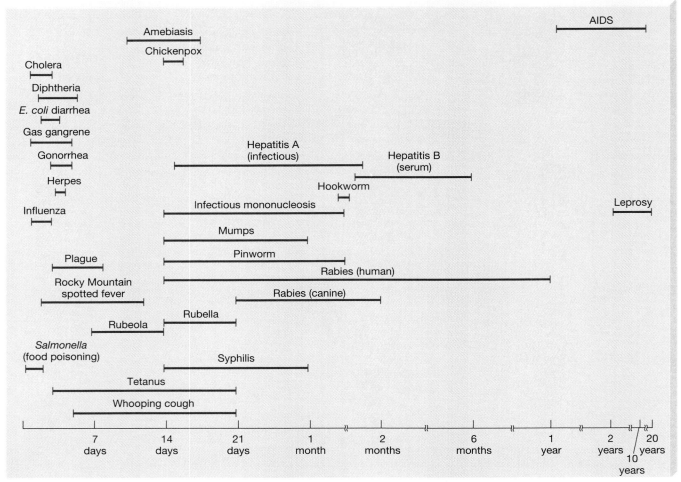

FIGURE 15.10 Incubation periods of selected infectious diseases.

tween pathogens and host defenses is at its height. A pathogen victory could lead to severe impairment of body function; if treatment is not available or provided in time, death can result.

Fever is an important component of the acme of many diseases. Certain pathogens produce substances called **pyrogens** that act on a center in the hypothalamus sometimes referred to as the body's "thermostat." Pyrogens set the thermostat at a higher-than-normal temperature. The body responds with involuntary muscle contraction that generates heat and constriction (narrowing) of blood vessels in the skin to prevent heat loss. Because our bodies function at lower temperatures than the newly set temperature, we feel cold and have chills at this stage. We shiver and get "goose bumps" as muscles contract involuntarily.

As the effects of the pyrogens diminish, the thermostat is reset to the normal, lower temperature, and the body responds to reach and maintain that temperature. This response includes sweating and dilation (widening) of blood vessels in the skin to increase heat loss. Because our bodies are warmer than the newly set temperature, we feel hot and say that we

have fever. Our skin gets moist from sweat as more blood circulates near the skin surface. In many infectious diseases, repeated episodes of pyrogen release occur, thereby accounting for bouts of fever and chills. Fever and its function are described in more detail in Chapter 17.

Decline Phase

As symptoms begin to subside, the disease enters the **decline phase**—the period of illness during which host defenses and the effects of treatment finally overcome the pathogen. The body's thermostat and other body activities gradually return to normal. Secondary infections may occur during this phase.

Convalescence Period

During the **convalescence period,** tissues are repaired, healing takes place, and the body regains strength and recovers. Individuals no longer have disease symptoms. In some diseases, however, especially those in which scabs form over lesions, persons recovering from the disease can still transmit pathogens to others.

INFECTIOUS DISEASES–PAST, PRESENT, AND FUTURE

In all the centuries of human history up to the twentieth, recovery or death from infectious diseases was determined largely by whether the human host or the pathogen won the war they waged against each other. Assorted potions and palliative (pain-reducing) treatments were available, but none could cure infectious diseases. Sometimes treatment was based on the notion that imbalances in body fluids caused disease. Depending on which fluid was judged to be in excess, efforts were made to remove some of it. Blood was removed by opening a vein or by applying blood-sucking leeches to the patient's skin. In eighteenth-century Europe, patients were bled until they lost consciousness. In 1774 when King Louis XV of France came down with smallpox, his desperate physicians bled him for 3 days in a row, each time removing "four large basinfuls of blood." In other cases, harsh laxatives were given to rid the body of excess bile. In most instances, these treatments failed to rid the patient of infectious agents, which at the time were unknown. At best, such treatments probably reduced suffering by hastening death.

Even after microorganisms came to be recognized as agents of disease, many years of painstaking research were required to relate specific diseases with the agents that caused them. More tedious research was needed to find antimicrobial agents that could cure diseases and to develop vaccines that could prevent them. The effects of these medical advances are clearly reflected in changes in the death rate in the United States. That death rate has decreased from 1560 per 100,000 people in 1900 to 505 per 100,000 people in the 1990s (Figure 15.11). The greatest single factor in the decrease was the control of infectious diseases by better treatment or by immunization. Better sanitation has also helped. Figure 15.11 shows that in 1900 the proportion of the population dying from infectious diseases was 27 times the current rate. Deaths from typhoid fever, syphilis, and childhood diseases (measles, whooping cough, diphtheria) have nearly been elimi-

nated, and deaths from pneumonia, influenza, and tuberculosis have been greatly reduced.

On a global scale, however, we have failed to eliminate diseases via technology that is already available. Measles kills nearly 1.5 million children each year, mostly in developing nations. Single doses of vaccine that

cost less than 12 cents could save most of these lives. In fact, as of 1995 each year 14 million children under the age of 5 die from infectious diseases such as measles, whooping cough, tetanus, diarrhea, and pneumonia—all of which are vaccine-preventable diseases. That is, one child dies every 2 seconds from these diseases.

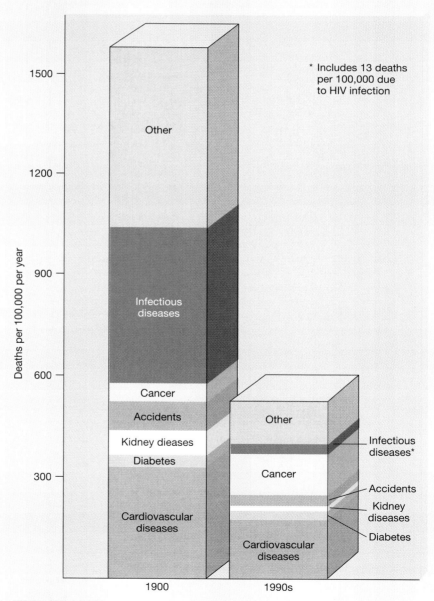

FIGURE 15.11 Changes in the causes of death in the United States from 1900 to the 1990s.

Look again at Figure 15.11. Because deaths from infectious diseases have decreased, average lifespans have increased. More people live long enough to develop degenerative diseases, and as they age their likelihood of developing a malignant disease increases. Thus, the death rate for cancer has increased by more than 240 percent—from 55 per 100,000 in 1900 to about 133 per 100,000 in the 1990s.

Past successes in treating infectious diseases suggest that disease eradication should be possible. But at least four factors make eradication difficult:

1. *Available medical expertise is not always applied.* Preventable diseases such as measles and mumps still occur in the United States because parents fail to have their children immunized. Also, some people, both young and old, fail to obtain treatment for curable diseases, a problem that could be solved by improved access to health care.

2. *Infectious agents are often highly adaptable.* Many strains of microorganisms have developed resistance to several of the available antibiotics. The use of antibiotics has prevented so many deaths. However, the misuse and/or the overuse of antibiotics through the years has contributed to the development of mutant drug-resistant bacterial strains. Treating diseases caused by such microbes presents a challenge that will not disappear or be solved quickly.

3. *Previously unknown or rare diseases become significant as a result of changes in human activities and/or social conditions.* The epidemic of legionellosis that marred the festivities of a Bicentennial celebration in 1976 was eventually found to be caused by a microorganism that was not commonly known and had existed and occasionally caused disease in the past. However, this time it was spread through a hotel air-conditioning system—something that could not have happened before air conditioning was invented. In the early 1980s many cases of toxic shock syndrome (TSS) suddenly appeared. This disease was shown to be caused by a staphylococcal toxin that usually reached the blood from organisms growing in certain rough, high-absorbency tampons used by women during menstruation. TSS was very rare before such tampons were invented; since then they have been modified by the manufacturer so that they no longer pose a health threat.

4. *Immigration and international travel and commerce introduce new or recurrent strains of the pathogen.* The high influx of legal and illegal immigrants often brings specific disease, such as tuberculosis, into areas where the disease had been eradicated. And the ease of international airline travel helps reintroduce previously eradicated diseases.

Earlier in the twentieth century, microbiologists and public health officials thought that the use of antibiotics and vaccines would eliminate infectious disease. Although this dream has been realized for smallpox and probably will be true for polio shortly after the turn of the century, many infectious diseases remain a serious health threat. Tuberculosis and cholera represent diseases again on the rise, mostly due to antibiotic resistance and the absence of sanitary control measures. Other diseases have arisen from new infectious agents; AIDS, caused by HIV, is perhaps the most prominent. Some scientists believe that HIV arose as a mutant strain in another animal species—an African monkey—and has "jumped species" to humans. Such jumps have probably occurred many times before, and with limited mobility of the human population, the virus would have disappeared without escaping into the rest of the world. However, the infection is slow to develop, and so unsuspecting carriers of the disease now take it with them from one part of the world to another, thanks to the ease of international air travel.

In 1992 the Institute of Medicine stated that emerging infectious diseases in the United States should be taken seriously and that public health agencies were not prepared to deal with possible epidemics. In fact, AIDS might be the best example of our unpreparedness—and an indication of other diseases yet to come.

CHAPTER SUMMARY

- A **pathogen** is a parasite capable of causing disease.

HOST–MICROBE RELATIONSHIPS

- A **host** is an organism that harbors another organism.

Symbiosis

- **Symbiosis** means "living together" and includes **commensalism,** in which one organism benefits and the other neither benefits nor is harmed; **mutualism,** in which both organisms benefit; and **parasitism,** in which one organism (the **parasite**) benefits and the other (the host) is harmed.

Contamination, Infection, and Disease

- **Contamination** refers to the presence of microorganisms. In **infection,** pathogens invade the body; in **disease,** pathogens or other factors disturb the state of health such that the body cannot perform its normal functions. **Infestation** refers to the presence of worms or arthropods in or on the body.

Pathogens, Pathogenicity, and Virulence

- **Pathogenicity** is the capacity of a pathogen to produce disease. **Virulence** is the intensity of a disease caused by a pathogen. **Attenuation** is weakening of a pathogen's disease-producing capacity.

Normal (Indigenous) Microbiota

- **Normal microbiota** (normal flora) are microorganisms found in or on the body that do not normally cause disease.

- **Resident microbiota** are those organisms that are always present on or in the body; **transient microbiota** are those present temporarily and under certain conditions. **Opportunists** are resident or transient microbiota that can cause disease under certain conditions or in certain locations in the body.

KOCH'S POSTULATES

- **Koch's postulates** provide a way to link a pathogen with a disease:

  1. A specific causative agent must be observed in every case of a disease.
  2. The agent must be isolated from a host displaying the disease and grown in pure culture.
  3. When the agent from the pure culture is inoculated into an experimental healthy, susceptible host, the agent must cause the disease.
  4. The agent must be reisolated from the inoculated, diseased experimental host and identified as the original specific causative agent.

- When Koch's postulates are met, an organism has been proved to be the causative agent of an infectious disease.

KINDS OF DISEASES

Infectious and Noninfectious Diseases

- **Infectious diseases** are caused by infectious agents; **noninfectious diseases** are caused by other factors.

Classification of Diseases

- Although many diseases are caused by noninfectious agents, some of these diseases may be associated with an infectious agent.

Communicable and Noncommunicable Diseases

- A **communicable,** or **contagious, infectious disease** can be spread from one host to another. A **noncommunicable infectious disease** cannot be spread from host to host and may be acquired from soil, water, or contaminated foods.

THE DISEASE PROCESS

How Microbes Cause Disease

- Many microbes have **virulence factors** that enable the establishment of infections. These factors include adhesion molecules, enzymes, and toxins.

- Bacteria cause disease by **adhering** to a host, by **colonizing** and/or invading host tissues, and sometimes by in-

vading cells. The ability of a pathogen to invade and grow in host tissues, called **invasiveness,** is related to particular virulence factors. **Hyaluronidase** helps bacteria invade tissues.

- Bacteria release other substances, most of which damage host tissues. **Hemolysins** lyse red blood cells in cultures and may or may not directly cause tissue damage in the host. **Leukocidins** destroy neutrophils. **Coagulase** accelerates blood clotting. **Streptokinase** digests blood clots and helps pathogens spread to body tissues.

- Many bacteria also produce **toxins. Endotoxins** are part of the cell wall of gram-negative bacteria and are released when cells divide or are killed. **Exotoxins** are produced by and released from bacteria; they are called **neurotoxins** if they affect the nervous system and **enterotoxins** if they affect the digestive system. **Toxoids** are inactivated exotoxins that retain antigenic properties and are used for immunization.

- Viruses damage cells and produce a variety of observable changes called the **cytopathic effect** (CPE). A **productive infection** leads to the release of virus progeny, whereas an **abortive infection** does not produce infectious progeny.

- Pathogenic fungi can invade and progressively digest cells, and some produce toxins.

- Protozoa and helminths damage tissues by ingesting cells and tissue fluids, releasing toxic wastes, and causing allergic reactions.

Signs, Symptoms, and Syndromes

- A **sign** is an observable effect of a disease. A **symptom** is an effect of a disease felt by the infected person. A **syndrome** is a group of signs and symptoms that occur together.

Types of Infectious Disease

- Terms used to describe types of diseases are defined in Table 15.7.

Stages of an Infectious Disease

- The **incubation period** is the time between infection and the appearance of signs and symptoms of a disease.

- The **prodromal phase** is the stage during which pathogens begin to invade tissues; it is marked by early nonspecific symptoms.

- The **illness phase** is the period during which the individual experiences the typical signs and symptoms of the disease. During this phase the signs and symptoms reach their greatest intensity at the **acme**.

- The **decline phase** is the stage during which host defenses overcome pathogens; signs and symptoms subside during this phase, and secondary infections may occur.

- The **convalescence period** is the stage during which tissue damage is repaired and the patient regains strength. Recovering individuals may still transmit pathogens to others.

abortive infection (p. 406)
acme (p. 409)
acute disease (p. 407)
adherence (p. 401)
adhesin (p. 401)
α-hemolysin (p. 403)
animal passage (p. 395)
attenuation (p. 395)
bacteremia (p. 408)
β-hemolysin (p. 403)
chronic disease (p. 407)
coagulase (p. 401)
colonization (p. 401)
commensalism (p. 394)
communicable infectious
    disease (p. 400)
contagious disease (p. 400)
contamination (p. 394)
convalescence period
    (p. 410)
cytopathic effect (p. 405)
decline phase (p. 410)
disease (p. 394)
endotoxin (p. 402)
enterotoxin (p. 404)

exotoxin (p. 402)
focal infection (p. 407)
fulminating (p. 409)
hemolysin (p. 403)
host (p. 392)
hyaluronidase (p. 401)
illness phase (p. 409)
immunocompromised
    (p. 397)
inapparent infection
    (p. 408)
incubation period (p. 408)
infection (p. 394)
infectious disease (p. 399)
infestation (p. 394)
intoxication (p. 404)
invasiveness (p. 401)
Koch's postulates (p. 398)
latent disease (p. 407)
latent viral infection
    (p. 406)
leukocidin (p. 403)
leukocytosis (p. 407)
leukostatin (p. 404)
local infection (p. 407)

microbial antagonism
    (p. 397)
mixed infection (p. 408)
mutualism (p. 392)
neurotoxin (p. 404)
noncommunicable infectious
    disease (p. 400)
noninfectious disease
    (p. 399)
normal microbiota (p. 396)
opportunist (p. 397)
parasite (p. 393)
parasitism (p. 393)
pathogen (p. 392)
pathogenicity (p. 395)
persistent viral infection
    (p. 406)
primary infection (p. 408)
prodromal phase (p. 409)
prodrome (p. 409)
productive infection
    (p. 405)
pyrogen (p. 410)
resident microbiota (p. 396)
sapremia (p. 408)

secondary infection (p. 408)
septicemia (p. 407)
sequela (p. 407)
sign (p. 407)
streptokinase (p. 402)
subacute disease (p. 407)
subclinical infection
    (p. 408)
superinfection (p. 408)
symbiosis (p. 392)
symptom (p. 407)
syndrome (p. 407)
systemic infection (p. 407)
toxemia (p. 404)
toxin (p. 402)
toxoid (p. 404)
transient microbiota
    (p. 396)
transposal of virulence
    (p. 395)
viremia (p. 408)
virulence (p. 395)
virulence factor (p. 400)

QUESTIONS FOR REVIEW

A Host–Microbe Relationships

1. What types of relationships distinguish the various kinds of symbiosis?

2. How does infection differ from contamination?

3. Relate the terms pathogen and disease.

4. How do pathogenicity and virulence differ, and how does attenuation affect virulence?

5. Describe the three classes of normal microbiota.

B Koch's Postulates

6. What are Koch's postulates?

7. In what ways are Koch's postulates sometimes difficult to demonstrate?

C Infectious vs. Noninfectious and Communicable vs. Noncommunicable Diseases

8. How are infectious and noninfectious diseases distinguished?

9. What kinds of diseases have both infectious and noninfectious components?

10. How are communicable and noncommunicable diseases distinguished?

D How Microbes Cause Disease

11. Describe the mechanisms used by bacteria to adhere to specific cells and tissues.

12. How do bacterial enzymes act as virulence factors?

13. Match each of the following characteristics with the type of toxins associated with it:

_____ may belong to gram-negative organisms

_____ cause high fever

_____ easily inactivated by heat

_____ consist of lipopolysaccharide (LPS) complexes

_____ are examples of entero-toxins

_____ toxoid can be used for immunization

a. exotoxins only

b. endotoxins only

c. both exotoxins and endotoxins

14. How do viruses cause disease?

15. Describe the mechanisms by which fungi, protozoa, and helminths cause infections and disease.

E Terms for Describing Diseases

16. How do signs, symptoms, and syndromes differ from one another?

17. Write a paragraph that correctly uses all the terms in Table 15.7.

F Stages of Infectious Diseases

18. Trace the course of a disease in the accompanying graph. Identify stages (a) through (f), and relate each to signs and symptoms and to activities of a pathogen.

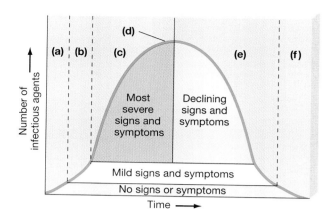

PROBLEMS FOR INVESTIGATION

1. In Table 15.1, the diversity of the normal microbiota increases from the mouth to the intestine. Explain why there is a greater diversity of the microbiota in the intestine than in the mouth.

2. What factors lead to the wide variation in the incubation period for rabies (14 days to 1 year)? Would it be better to be bitten on the foot or the shoulder?

3. Identify the type of symbiotic relationship represented in each of the following cases:

   a. bacteria in the rumen of a cow's digestive system that break down the cellulose in grass and hay

   b. tapeworms in the human intestine

   c. AIDS virus in a human

4. A microbiology student clad in shorts overturns his motorcycle on a patch of gravel. He sustains considerable abrasions and deep lacerations of the right leg. Emergency-room personnel have great difficulty trying to remove embedded gravel and dirt. Many areas of infection develop, and dead tissue can be seen at the sites of abrasion. Are his infections most likely due to pathogens or opportunists? Why is it difficult to determine?

5. Recall the last time you had an infectious disease. Relate the signs and symptoms you experienced to the steps in the course of an infection and the ways in which the pathogen was damaging tissues.

6. Read about opportunistic infections. Under what conditions do they cause infections? Why are such infections often difficult to treat?

7. A young mother tried to save money by buying cracked eggs at a local farm. The eggs were labeled "not for human consumption—for pets only." She cooked some of the eggs and fed them to her toddler son, who became intensely ill within hours. He was hospitalized with a diagnosis of salmonellosis. He nearly died and is now permanently retarded.

   a. Did the boy suffer from an infection or from an intoxication? What piece of clinical data supports your answer?

   b. Why didn't cooking the eggs protect him?

   c. Can he be immunized against further occurrences?

   d. Where did the *Salmonella* organisms come from?

   (The answer to this question appears in Appendix F.)

SOME INTERESTING READING

Eisenstein, B. I. 1990. "New opportunistic infections—more opportunities." *The New England Journal of Medicine* 323(December 6):1625–27.

Ewald, P. W. 1993. "The evolution of virulence." *Scientific American* 268(April):86–93.

Finlay, B. B. 1995. "Bacterial virulence factors." *Scientific American: Science and Medicine* 2, no. 3 (May/June):16–25.

Lederberg, J., R. E. Shope, and S. C. Oaks, Jr., eds. 1992. *Emerging infections: Microbial threats to health in the United States.* Washington, D.C.: National Academy Press.

Lewis, R. 1992. "The bugs within us." *FDA Consumer* 26(7):37–42.

McKeown, T. 1988. *The origins of human disease.* Oxford: Basil Blackwell.

Miller, V. L. 1992. "*Yersinia* invasion genes and their products." *ASM News* 58(1):26–33.

Morrison, D. C., et al. 1994. "Current status of bacterial endotoxins." *ASM News* 60 (9):479–84.

Ranger, T., and P. Slack, eds. 1992. *Epidemics and ideas: Essays on the historical perception of pestilence.* Cambridge: Cambridge University Press.

Schantz, E. J., and E. A. Johnson. 1992. "Properties and use of botulinum toxin and other microbial neurotoxins in medicine." *Microbiological Reviews* 56, no. 1 (March):80–99.

Segal, M. 1993. "Parasitic invaders and the reluctant human host." *FDA Consumer* 27(6):7–13.

Vines, G. 1994. "The enemy within." *New Scientist* 144, no. 1947 (October 15):12–14.

CHAPTER 16

EPIDEMIOLOGY AND NOSOCOMIAL INFECTIONS

Quarantine, one method of controlling disease transmission, is intended to prevent the spread of a disease during its incubation period.

U p to this point in our study of host–microbe interactions, we have examined the characteristics of pathogens that lead to infectious diseases, and we have seen how the disease process occurs in individuals. But individuals with infectious diseases are members of a population—they acquire infectious diseases and transmit them within a population. Therefore, to further our understanding of such diseases, we must consider their effects on populations, including hospital populations.

EPIDEMIOLOGY

What Is Epidemiology?

Epidemiology (ep"i-de-me-ol'o-je) is the study of factors and mechanisms involved in the frequency and spread of diseases and other health-related problems within populations of humans, other animals, or plants. The term is derived from the Greek words *epidemios*, meaning "among the people," and *logos*, meaning "study." Although **epidemiologists,** scientists who study epidemiology, may consider such health-related

problems as automobile accidents, lead poisoning, or cigarette smoking, we will limit our discussion to factors and mechanisms that concern the transmission of infectious diseases as well as their **etiology** (e-te-ol'o-je), or cause, in a population. We will also see how epidemiologists use this information to design ways to control and prevent the spread of infectious agents.

Epidemiology is a branch of microbiology because many diseases that concern epidemiologists, such as AIDS, tuberculosis, and malaria, are caused by microbes. Parasitic worm diseases, such as hookworm disease and ascariasis, also are studied by epidemiologists. Epidemiology includes relationships among pathogens, their hosts, and the environment. It is related to public health because it provides information and methods used to understand and control the spread of disease within the human population. Agricultural and environmental scientists often are concerned with the epidemiology of animal and plant diseases. When such diseases can be transmitted to humans, they, too, become a public health problem.

In following diseases and their spread, epidemiologists are especially interested in the frequencies of diseases within populations.

416

THIS CHAPTER FOCUSES ON THE FOLLOWING QUESTIONS:

A What is epidemiology, and what special terms are used by epidemiologists?

B How are diseases classified according to their spread in populations?

C What are the purposes and methods of epidemiologic studies?

D How do various kinds of reservoirs of infection contribute to human disease?

E What are the roles of portals of entry and exit and modes of transmission in the spread of human disease?

F What is an infectious disease cycle, and how is herd immunity related to disease cycles?

G What methods are used to control disease transmission?

H How do the functions of public health organizations and the reporting of notifiable diseases contribute to public health?

I What are nosocomial infections, and how are they studied epidemiologically?

J How can nosocomial infections be prevented and controlled?

Incidence and Prevalence Rates

The **incidence rate** of a disease is the number of *new* cases contracted within a set population during a specific period of time (usually expressed as new cases per 100,000 people per year). The **prevalence rate** is the *total number* of people infected within the population at any time. Prevalence includes both old and newly diagnosed cases. For example, if epidemiologists conduct weekly surveys regarding a disease that lasts four weeks, an individual infected in week one could be counted as many as four times in a prevalence study but only once in an incidence survey. Thus, incidence data are reliable indicators of the spread of a disease—a drop in incidence suggests a reduction in the spread of the disease. In contrast, prevalence data measure how seriously and how long the disease is affecting a population (Figure 16.1).

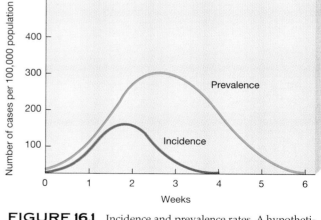

FIGURE 16.1 Incidence and prevalence rates. A hypothetical disease lasting 4 weeks illustrates the difference between the incidence and prevalence numbers.

Morbidity and Mortality Frequencies

Frequencies also are expressed as proportions of the total population. The **morbidity rate** is the number of individuals affected by a disease during a set period in relation to the total number in the population. It usually is expressed as the number of cases per 100,000 people per year. The **mortality rate** is the number of deaths due to a disease in a population during a specific period in relation to the total population. It is expressed as deaths per 100,000 people per year.

Diseases in Populations

When studying the frequency of diseases in populations, epidemiologists must consider the geographic areas affected and the degree of harm the diseases cause in the population. On the basis of their findings, they classify diseases as *endemic, epidemic, pandemic,* or *sporadic*.

An infectious disease agent is **endemic** if it is present continually in the population of a particular geographic area but both the number of reported cases and the severity of the disease remain too low to constitute a public health problem. For example, mumps is endemic in the entire United States, and valley fever is endemic in the southwestern United States. Chickenpox is an endemic disease with seasonal variation; that is, many more cases are seen from late winter through spring than at other times (Figure 16.2). Endemic diseases also can vary in incidence in different parts of the endemic region.

An **epidemic** arises when a disease suddenly has a higher-than-normal incidence in a population. Then the morbidity rate or the mortality rate or both become high enough to pose a public health problem. Endemic diseases can give rise to epidemics, especially when a particularly virulent strain of a pathogen appears or when most of a population lacks immunity. For example, St. Louis encephalitis, a viral inflammation of the brain, reached epidemic proportions in the United States in 1975 (Figure 16.3). It arose due to the presence of both a large nonimmune bird population that carried the virus and a large mosquito population that transferred the virus from birds to humans.

With the breakup of the former Soviet Union into the New Independent States (NIS), epidemic diphtheria has emerged. Since 1990, overcrowding due to population migrations has resulted in a dramatic increase in the diphtheria incidence rate (Figure 16.4). In 1994, 47,802 cases were reported, with 1746 deaths. World health officials consider the expanding diphtheria epidemic an in-

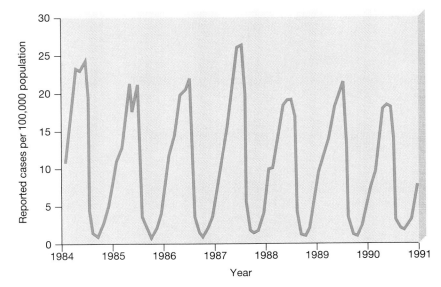

FIGURE 16.2 Incidence rate of chickenpox in the United States. Chickenpox, an endemic disease, shows obvious seasonal variation, with most cases occurring during the spring. (Chickenpox is no longer reportable to the CDC, so current data are not available.)

ternational health emergency. Diphtheria cases arising in the NIS have been reported in eastern and northern Europe. Health officials outlined a strategy in early 1995 to ensure that all NIS children are vaccinated against diphtheria.

A **pandemic** occurs when an epidemic spreads worldwide. In 1918 the swine flu reached pandemic proportions (as we shall see in Chapter 21). Cholera has been responsible for seven pandemics over the centuries. Its spread through the Americas during the epidemic of 1991–1992 is shown in Figure 16.5.

A **sporadic disease** occurs in a random and unpredictable manner, involving several isolated cases that pose no great threat to the population as a whole.

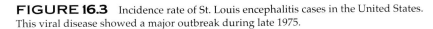

FIGURE 16.3 Incidence rate of St. Louis encephalitis cases in the United States. This viral disease showed a major outbreak during late 1975.

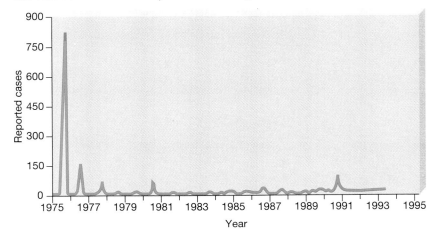

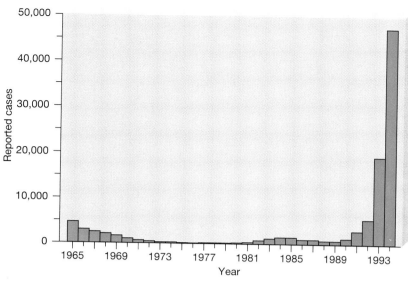

FIGURE 16.4 Diphtheria cases in the New Independent States, 1965 to 1994. The number of reported cases increased from 1991 to 1994 because of insufficient health precautions as a result of the breakup of the Soviet Union. (Source: World Health Organization)

Eastern equine encephalitis (EEE) is a sporadic disease in the Americas. Figure 16.6 contrasts the sporadic nature of EEE with endemic California encephalitis (CE) in certain parts of the United States and epidemic western equine encephalitis (WEE) in the Americas.

FIGURE 16.5 Cholera spread through South and Central America during the period 1991–1992, beginning in Peru in January 1991. It then moved into Colombia and Ecuador. By late 1992 the epidemic had spread to Venezuela, Bolivia, Chile, and Brazil in South America and to Guatemala, Honduras, Panama, Nicaragua, and El Salvador in Central America. (Map from *Morbidity and Mortality Weekly Record*.)

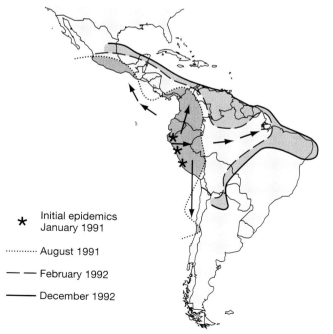

* Initial epidemics
January 1991

.... August 1991

— — February 1992

——— December 1992

The nature and spread of epidemics can vary according to the source of the pathogen and how it reaches susceptible hosts. A **common-source outbreak** is an epidemic that arises from contact with contaminated substances. The term *outbreak* does not evoke the fear that the word *epidemic* can. The spread of a common-source outbreak can typically be traced to a water supply contaminated with fecal material or to improperly handled food. Many individuals become ill quite suddenly. For example, on a cruise in 1994 from San Pedro, California, to Ensenada, Mexico, 586 of 1589 passengers acquired gastrointestinal illness due to *Shigella flexneri*. Such outbreaks subside quickly once the source of infection is eradicated.

A **propagated epidemic** arises from direct person-to-person contacts

CLOSE-UP

CITY TALES

"Hamburg persisted in postponing costly improvements to its water supply. . . . [It] drew its water from the Elbe without special treatment. Adjacent lay the town of Altona . . . where a solicitous government installed a water filtration plant. In 1892, when cholera broke out in Hamburg, it ran down one side of the street dividing the two cities and spared the other completely. . . . A more clearcut demonstration of the importance of the water supply in defining where the disease struck could not have been devised. Doubters were silenced; and cholera has, in fact, never returned to European cities, thanks to systematic purification of urban water supplies from bacteriological contamination."

—William H. McNeill, 1976

"Cities are essential to civilization, and until well into the nineteenth century, all cities were the spawning grounds of infectious disease. How could it be otherwise? For centuries all the precautions we now know to be necessary to prevent the spread of bacteria and animal parasites were unthought of. City streets were littered with human and animal filth, water came from contaminated wells; rats, fleas, and lice were universal. Crowded together in such filthy environments, every city dweller was inevitably exposed to infection every day of his life. It is no wonder that the population of cities through all history has had to be recruited periodically from the country."

—F. Macfarlane Burnet and David White, 1972

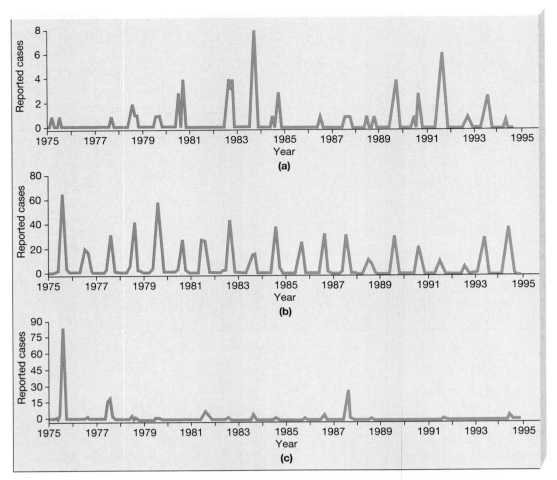

FIGURE 16.6 Incidence rate of three different types of encephalitis. (a) The sporadic pattern of eastern equine encephalitis is compared with the (b) endemic pattern of California encephalitis and (c) epidemic outbreaks of western equine encephalitis.

(horizontal transmission). The pathogen moves from infected people to uninfected but susceptible individuals. In a propagated epidemic, the number of cases rises and falls more slowly, and the pathogen is more difficult to eliminate, than in a common-source outbreak. Differences between common-source outbreaks and propagated epidemics are illustrated in Figure 16.7.

Epidemiologic Studies

Collecting frequency data and drawing conclusions is the foundation of any **epidemiologic study.** The British physician John Snow made what may have been the first epidemiologic study. In 1854 Snow investigated the cause of a cholera epidemic then sweeping through London. He eventually traced the source of the epidemic to the Broad Street pump in Golden Square (Figure 16.8). He proved that people became infected by drinking water contaminated with human feces.

FIGURE 16.7 Differences in incidence patterns of common-source outbreaks and propagated epidemics. In common-source outbreaks, all cases occur within a fairly short time period after exposure to the single source and then stop, whereas in propagated epidemics, new cases are continually seen.

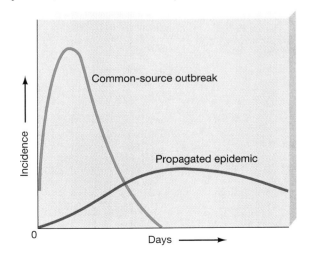

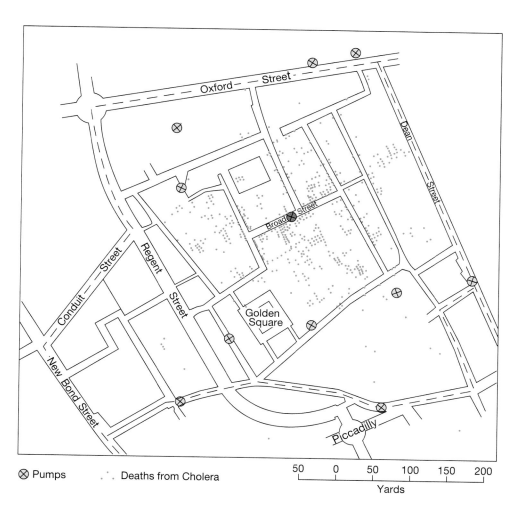

⊗ Pumps   ∴ Deaths from Cholera

50   0   50   100   150   200
Yards

FIGURE 16.8 In 1854 John Snow, a London physician, recorded the locations of cholera cases in the city. He found that they were clustered around the Broad Street pump, which supplied water to the nearby area. He tracked the epidemic's source to contamination of the pump by sewage. When the handle of the pump was removed, the outbreak subsided. Snow's study was the first modern, systematic, scientific epidemiologic study.

Since Snow's landmark study, many other investigators have conducted epidemiologic studies to learn more about the spread of disease in populations. Such studies can be *descriptive, analytical,* or *experimental.*

DESCRIPTIVE STUDIES   A **descriptive study** is concerned with the physical aspects of an existing disease and disease spread. Such a study records (1) the number of cases of a disease, (2) those segments of the population that were affected, and (3) the locations and time period of the cases. The age, gender, race, marital status, socioeconomic status, and occupation of each patient also are recorded. From careful analysis of data accumulated from several studies, epidemiologists can determine whether people of a certain age group, males or females, or members of a certain race are particularly susceptible to the disease. Data on marital status or sexual behavior can help show whether the disease is transmitted sexually. Data on socioeconomic status might show that a disease is most easily transmitted among the undernourished or among those living in substandard conditions. Noting the occupations of infected individuals can help epidemiologists trace

diseases to certain factories, slaughterhouses, or hide-processing plants. If most cases are among veterinarians, for example, the disease probably is transmitted by the animals they handle.

The geographic distribution of cases is also important, as Snow's study showed. Some of Snow's modern counterparts have traced disease outbreaks to contaminated water supplies, to restaurants where workers are infected with viral hepatitis, and to areas where particular infectious agents thrive.

Multiple sclerosis (MS) is a disease in which one's own immune system attacks the myelin sheath surrounding nerve cells in the spinal cord and brain. It eventually leads to a loss of muscle control and paralysis. Epidemiologically, most cases of MS are found in clusters, in temperate climates, and among Caucasians. In the United States, the incidence of MS increases significantly northward from Mississippi to Minnesota. However, an important determining factor seems to be where a person spent the first 14 years of life. For example, for an individual who grew up in a northern region with a high incidence of MS, moving south later would not reduce the likelihood of acquiring the disease. Various studies have led some investigators to propose that an infectious virus may contribute to the development of MS. Thus, genetic factors coupled with an infection by one or more specific microbes increases the likelihood that a person will acquire MS later in life. (Chapter 19 discusses MS and other autoimmune disorders.)

Finally, the time period over which cases appear and the season of the year are important considerations in descriptive studies. To study the role of time in an epidemic, epidemiologists define an **index case** as the first case of the disease to be identified. As we have seen, common-source outbreaks can be distinguished from propagated epidemics by the rate at which the number of cases increases and the time required for the epidemic to subside. The season of the year in which epidemics occur may help identify the causative agent. Arthropodborne infections usually occur in relatively warm weather, and certain respiratory infections usually occur in cold weather, when people often are crowded together indoors. The seasonal nature of encephalitis, which is most prevalent in the fall, was shown in Figure 16.6.

ANALYTICAL STUDIES An **analytical study** focuses on establishing cause-and-effect relationships in the occurrence of diseases in populations. Such studies can be retrospective or prospective. A *retrospective* study takes into account factors that preceded an epidemic. For example, the investigator might ask patients where they had been and what they had done in the month or so before they became ill. The patients are then compared with a *control group*—individuals in the same population who are unaffected by the disease. Thus, if most patients had hiked in a certain wooded area, had contact with horses, or shared another common activity in which the control group did not participate, that activity might provide a clue to the source of the infection. Several investigations of this sort are described in Berton Roueché's fascinating book *The Medical Detectives* (see Some Interesting Reading at the end of this chapter).

A *prospective* study considers factors that occur as an epidemic spreads. Which children in a population get chickenpox, at what age, and under what living conditions, for instance, are factors used to determine susceptibility and resistance to infection. As the 1993 hantavirus outbreak spread in the U.S. Southwest, epidemiologists had to determine what was causing the disease and how to change living conditions so that the infectious agent would stop spreading.

EXPERIMENTAL STUDIES An **experimental study** designs experiments to test a hypothesis, often about the value of a particular treatment. Such studies are limited to animals or to humans in which participants

are not subjected to harm. For example, an investigator might test the hypothesis that a particular treatment will be effective in controlling a disease for which no accepted cure is available. One group (the experimental group) from a population receives the treatment, and another group (the control group) receives a placebo. A **placebo** (pla-se'bo) is a nonmedicinal substance that has no effect on the recipient but that the recipient *believes* is a treatment. From the results of the study, the investigator could learn whether the new treatment was effective.

However, in some circumstances control groups might not be used. In the early days of AIDS research, drug treatments were carried out without a control group because the U.S. government considered it unethical to allow control subjects to be deprived of potentially lifesaving treatment.

Infectious diseases threaten human populations only when the diseases can be spread or transmitted. Factors important to the spread of disease or of infectious agents include (1) reservoirs of infection, (2) portals by which organisms enter and leave the body, and (3) mechanisms of transmission. Let's look at each of these factors in more detail.

Reservoirs of Infection

Most pathogens infecting humans cannot survive outside the body of a host long enough to serve as a source of infection. Therefore, sites in which organisms can persist and maintain their ability to infect are essential for new human infections to occur. Such sites are called **reservoirs of infection.** Examples are humans, other animals (including insects), plants, and certain nonliving materials, such as water and soil.

Human Reservoirs

Humans with active infections are important reservoirs because they can easily transmit organisms to other humans. **Carriers,** individuals who harbor an infectious agent without having any observable clinical signs or symptoms, also are important reservoirs. Thus, a disease may be transmitted by a person (or animal) with a **subclinical,** or **inapparent, infection**—an infection with signs and symptoms too mild to be recognized, except by special tests. Many cases of whooping cough (pertussis) in adults, for example, are never diagnosed; yet such *"healthy"* carriers harbor and transmit infectious agents. Infectious diseases are said to be *communicable* if they can be transmitted during the incubation period (before symptoms are apparent) and during recovery from the disease. A *chronic carrier* is a reservoir of infection for a long time after he or she has recovered from a disease. An *intermittent carrier* periodically releases organisms.

Depending on the disease, carriers can discharge organisms from the mouth or nose, in urine, or in feces. Diseases commonly spread by carriers include diphtheria, typhoid fever, amoebic and bacillary dysenteries, hepatitis, streptococcal infections, and pneumonia. Our cholera example shows how international jet travel can increase the risk of introducing infectious agents from reservoirs in one region to populations visiting from another region. (See also the Essay at the end of Chapter 15.)

Animal Reservoirs

About 150 pathogenic microorganisms can infect both humans and some other animals. In such instances, the animals can serve as reservoirs of infection for humans. Animals that are physiologically similar to humans are most likely to serve as reservoirs for human infections. Therefore, monkeys are important reservoirs for malaria, yellow fever, and many other human infections. Once humans are infected, they, too, can serve as reservoirs for the infections.

Diseases that can be transmitted under natural conditions to humans from other vertebrate animals are called **zoonoses** (zo-o-no'ses; singular: *zoonosis*). Selected zoonoses are summarized in Table 16.1. Of these diseases, rabies is perhaps the greatest threat in the United States because of the severity of the disease and because both domestic pets and wild animals can serve as reservoirs for the rabies virus. Where vaccination of dogs and cats is widespread, humans are more likely to acquire rabies from wild animals in which the disease is endemic, such as skunks, raccoons, bats, and foxes.

The larger the animal reservoir, both in the number of species and in the total number of susceptible animals, the more unlikely it is that a disease can be eradicated. This is especially true if the reservoir contains wild animals in which a disease is epidemic. It is impossible to find all the infected animals and to control the disease among them. Even today *Yersinia pestis*, the bacterium responsible for the plague, persists among gophers, ground squirrels, and other wild rodents in the American West and occasionally causes human cases.

TABLE 16.1	Selected Zoonoses (with emphasis on those that occur in pets)	
Disease	**Animals Infected**	**Modes of Transmission**
Bacterial diseases		
Avian tuberculosis	Birds	Respiratory aerosols
Anthrax	Domestic animals, including dogs and cats	Direct contact with animals, contaminated soil, and hides; ingestion of contaminated milk or meat; inhalation of spores
Brucellosis (undulant fever)	Domestic animals	Direct contact with infected tissues, ingestion of milk from infected animals
Bubonic plague	Rodents	Fleas
Leptospirosis	Primarily dogs; also pigs, cows, sheep, rodents, and other wild animals	Direct contact with urine, infected tissues, and contaminated water
Psittacosis	Parrots, parakeets, and other birds	Respiratory aerosols
Relapsing fever	Rodents	Ticks and lice
Rocky Mountain spotted fever	Dogs, rodents, and other wild animals	Ticks
Salmonellosis	Dogs, cats, poultry, turtles, and rats	Ingestion of contaminated food or water
Viral diseases		
Equine encephalitis (several varieties)	Horses, birds, and other domestic animals	Mosquitoes
Rabies	Dogs, cats, bats, skunks, and wolves	Bites, infectious saliva in wounds, aerosols
Lassa fever, hantavirus pulmonary syndrome, hemorrhagic fevers	Rodents	Urine
Fungal diseases		
Histoplasmosis	Birds	Aerosols of dried infected feces
Ringworm (several varieties)	Cats, dogs, and other domestic animals	Direct contact
Parasitic diseases		
African sleeping sickness	Wild game animals	Tsetse flies
Tapeworms	Cattle, swine, rodents	Ingestion of cysts in meat or via proglottids in feces
Toxoplasmosis	Cats, birds, rodents, and domestic animals	Aerosols, contaminated food and water, and placental transfer

Humans, their pets, and other domestic animals similarly serve as reservoirs of infection for wild animals. Distemper, an infectious viral disease in dogs, has spread to and killed many black-footed ferrets. This animal, already an endangered species, is now in even greater jeopardy. Thus, pets and domestic animals should not be allowed to enter wildlife refuges.

Nonliving Reservoirs

Soil and water can serve as reservoirs for pathogens. Soil, for example, is the natural environment of several bacterial species. *Clostridium tetani* (the cause of tetanus) and *C. botulinum* (the cause of botulism) are found everywhere, but especially where animal fecal matter is used as fertilizer. They are part of the normal intestinal microbiota of cattle, horses, and some humans. Many fungi, including the organism that causes valley fever, also are common soil inhabitants. Often soil fungi can invade human tissues and cause ringworm, other skin diseases, or systemic infections. Water contaminated by human or animal feces can contain a variety of pathogens. Most of these cause gastrointestinal diseases. Improperly prepared or stored food also can serve as a temporary nonliving reservoir of disease. Poorly cooked contaminated meats can be a source of infection with *Salmonella* species and a variety of helminths. Failure to refrigerate foods can lead to growth of microorganisms and to the production of toxins that cause food poisoning. Even with proper refrigeration, helminth larvae remain infectious unless the foods that contain them are cooked thoroughly.

Portals of Entry

To cause an infection, a microorganism must enter body tissues. The sites at which microorganisms can enter the body are called **portals of entry.** Common portals of entry include the skin and the mucous membranes of the digestive, respiratory, and urogenital systems (Figure 16.9). Although intact skin usually prevents the entry of microorganisms, some enter through the ducts of sweat glands, mammary glands, or through hair follicles. Some fungi invade cells on the skin's surface, and a few can pass on to other tissues. Larvae of some parasitic worms such as the hookworm can bore through the skin to enter other tissues.

Openings to the outside of the body, such as the ears, nose, mouth, eyes, anus, urethra, and vagina, allow microbes to enter. Organisms that infect the respiratory system typically enter in inhaled air, on dust particles, or in airborne droplets. Those that infect the digestive system typically enter in food or water but can also enter from contaminated fingers. As a result of fluid discharge from mucous membranes, sexual intercourse provides a portal of entry for urogenital system infections. Some infectious microbes also travel through the urethra and vagina from skin surfaces.

CLOSE-UP

WHAT'S IN DUST?

Household dust typically contains an amazing assortment of microbes, spores, a few larger organisms, dandruff and other debris from the human body, and other nonliving materials. Spores of *Clostridium perfringens*, which causes gas gangrene in deep wounds, have been found in air conditioner filters. Many different genera of fungi, including *Penicillium, Rhizopus* (bread mold), and *Aspergillus*, which can cause swimmer's ear, have been found in dust. Bacterial endospores, fungal spores, plant pollens, insect parts, and hundreds of mites (small organisms related to spiders) also abound in dust. Mites feed on skin particles sloughed from our body surfaces; fortunately they do not feed on living skin. Dust also contains large quantities of human and animal hair and sometimes nail clippings as well, all of which have microorganisms on their surfaces. Finally, dust contains nonliving particles from sources as diverse as flaking paint and meteorites.

SEM photo of the house dust mite (genus *Dermatophagoides*). Household dust provides a home for dust mites.

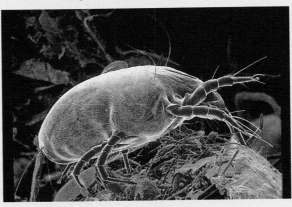

Sometimes microorganisms are introduced directly into the damaged tissues created by bites, burns, injections, and accidental or surgical wounds. *Pseudomonas aeruginosa* infections are especially common in hospitals among burn and surgical patients. Insect bites are a common portal of entry for a variety of parasitic protozoan and helminthic diseases carried by insects.

Finally, a few infectious organisms, mostly viruses, can cross the placenta from an infected mother and cause an infection in the fetus. Congenital infectious diseases such as cytomegalovirus infection, toxoplasmosis, syphilis, AIDS, and rubella (German measles) are acquired in this way.

Some organisms can enter the body through only a single portal. Others can enter through any of sev-

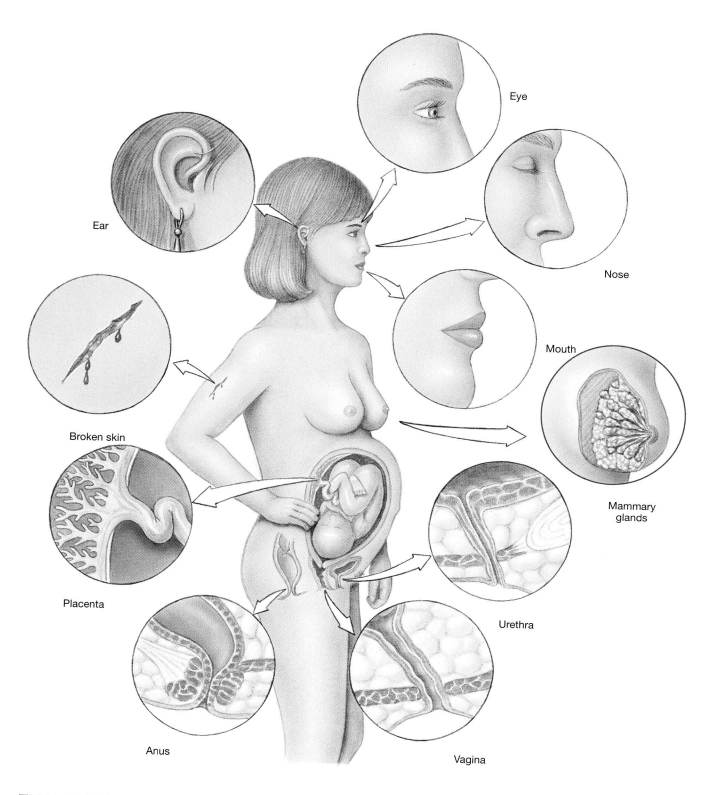

FIGURE 16.9 Portals of entry for human pathogens. Besides those portals common to females and males, additional portals in females include the placenta and the milk ducts of the mammary glands.

Eye

Nose

Ear

Mouth

Broken skin

Mammary glands

Placenta

Urethra

Anus

Vagina

eral portals, and their pathogenicity may depend on the portal of entry. For example, many pathogens that cause diseases of the digestive tract cause no disease at all if they happen to enter the respiratory tract. Similarly, most pathogens that cause respiratory diseases do not infect the skin or the tissues of the digestive tract. However, a few organisms cause illnesses no matter where they enter the body—but quite different illnesses, depending on which portal they enter. For example, the plague bacterium (*Yersinia pestis*) causes *bubonic plague,* which has a mortality of about 50 percent if untreated, and is acquired through a flea bite. But when inhaled into the lungs, the same organism causes *pneumonic plague,* which has a mortality of close to 100 percent.

Even when a pathogen enters the body, it may not reach an appropriate site to cause an infection. Almost everyone unknowingly carries *Klebsiella pneumoniae* in the pharynx at some time during winter. Why, then, do so few of us come down with pneumonia? The reason is that the organism must enter the lungs before it can cause disease. There is no such thing as pneumonia of the pharynx.

Portals of Exit

How infectious agents get out of their hosts is important to the spread of diseases. The sites where organisms leave the body are called **portals of exit** (Figure 16.10).

Generally, pathogens exit with body fluids or feces. Respiratory pathogens exit through the nose or mouth in fluids expelled during coughing, sneezing, or speaking. Saliva from dogs, cats, insects, and other animals can transmit infectious organisms. Pathogens of the gastrointestinal tract exit with fecal material. Some of these pathogens are helminth eggs, which are exceedingly resistant to drying and other environmental conditions. Urine and, in males, semen from the urethra carry urogenital pathogens. Semen is an important, though sometimes ignored, means by which pathogens, especially viruses, exit the body. For example, the AIDS virus (HIV) can be carried from the body by white blood cells and sperm present in semen, as can the hepatitis C virus.

Blood from patients also sometimes contains infectious organisms, such as HIV or hepatitis viruses. As such, blood can be a source of infection for health care workers or others rendering aid to an injured person. Another person's blood should always be considered potentially infectious.

Modes of Transmission of Diseases

For new cases of infectious diseases to occur, pathogens must be transmitted from a reservoir or portal of exit to a portal of entry. Transmission can occur by several modes, which we have grouped into three categories:

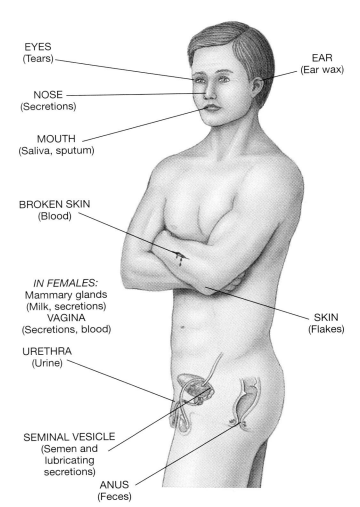

FIGURE 16.10   Portals of exit for human pathogens. Besides those portals common to males and females, additional portals in males include the seminal vesicles and in females, the mammary glands and vagina.

(1) contact transmission, (2) transmission by vehicles, and (3) transmission by vectors. Figure 16.11 presents an overview of these transmission modes.

Contact Transmission

Contact transmission can be direct, indirect, or by droplets. **Direct contact transmission** requires body contact between individuals. Such transmission can be horizontal or vertical. In **horizontal transmission,** individuals pass on pathogens by shaking hands, kissing, touching sores, or having sexual contact. Pathogens can also be spread from one part of the body to another through unhygienic practices. For example, touching genital herpes lesions and then touching other parts of the body, such as the eyes, can spread the infection. Pathogens from fecal matter also can be spread by unwashed hands to the mouth; this is **direct fecal–oral transmission.** In **vertical transmission,**

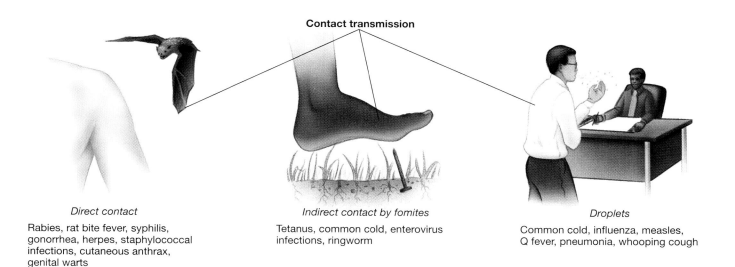

Contact transmission

Direct contact

Rabies, rat bite fever, syphilis, gonorrhea, herpes, staphylococcal infections, cutaneous anthrax, genital warts

Indirect contact by fomites

Tetanus, common cold, enterovirus infections, ringworm

Droplets

Common cold, influenza, measles, Q fever, pneumonia, whooping cough

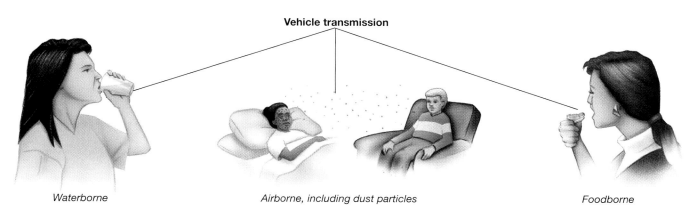

Vehicle transmission

Waterborne

Cholera, shigellosis, leptospirosis, *Campylobacter* infections

Airborne, including dust particles

Chickenpox, tuberculosis, coccidioidomycosis, histoplasmosis, influenza, measles

Foodborne

Intoxication with aflatoxins and botulinum toxin, paralytic shellfish poisoning, staphylococcal food poisoning, typhoid fever, salmonellosis, listeriosis, toxoplasmosis, tapeworms, hepatitis A

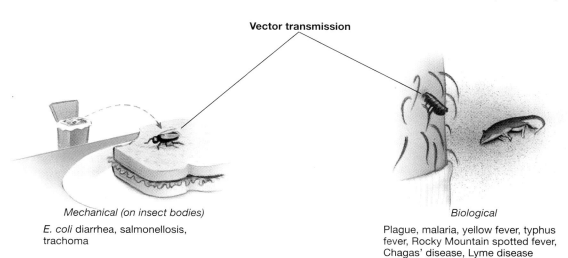

Vector transmission

Mechanical (on insect bodies)

E. coli diarrhea, salmonellosis, trachoma

Biological

Plague, malaria, yellow fever, typhus fever, Rocky Mountain spotted fever, Chagas' disease, Lyme disease

FIGURE 16.11 Overview of mechanisms by which diseases can be transmitted.

pathogens are passed from parent to offspring in an egg or sperm, across the placenta, in breast milk, or in the birth canal (as can happen with syphilis and gonorrhea).

Indirect contact transmission occurs through fomites (fo'mi-tez), nonliving objects that can harbor and transmit an infectious agent. Examples of fomites include soiled handkerchiefs, dishes, eating utensils, doorknobs, toys, bar soap, and money. (U.S. paper currency is treated with an antimicrobial agent that reduces transmission of microorganisms.)

Droplet transmission, a third kind of contact transmission, occurs when a person coughs, sneezes, or speaks near others (Figure 16.12). Droplet nuclei consist of dried mucus, which protects microorganisms embedded in it. These particles can be inhaled directly, can collect on the floor with dust particles, or can become airborne. Droplet transmission over a distance of less than 1 m is not considered airborne.

Transmission by Vehicles

A vehicle is a nonliving carrier of an infectious agent from its reservoir to a susceptible host. Common vehicles include water, air, and food. Blood, other body fluids, and intravenous fluids also can serve as vehicles of disease transmission.

WATERBORNE TRANSMISSION   Although waterborne pathogens do not grow in pure water, some survive transit in water with small quantities of nutrients or in water polluted with fertilizer. Waterborne pathogens usually thrive in and are transmitted in water contaminated with untreated or inadequately

---

FIGURE 16.12   A cough is photographed with the Schlieren technique, which detects air-speed differences as different colored areas. Such dispersal is most important within a radius of about 1 m; however, the smallest particles can be dispersed much farther and kept aloft by air currents. Even a surgical mask will not prevent spread of all droplets.

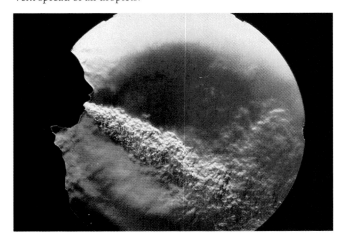

treated sewage. Such indirect fecal–oral transmission occurs when pathogens from feces of one organism infect another organism. Pathogens have been isolated from public water supplies, semiprivate water supplies (camps, parks, and hotels that have their own water systems), and private water supplies (springs and wells). Polioviruses and enteroviruses as well as several bacteria are waterborne microbes that infect the digestive system and cause gastrointestinal symptoms. Waterborne infections can be prevented by proper treatment of water and sewage, although enteroviruses are especially difficult to eradicate from water.

AIRBORNE TRANSMISSION   Airborne microorganisms are mainly transients from soil, water, plants, or animals. They do not grow in air, but some reach new hosts through air despite dryness, temperature extremes, and ultraviolet radiation. In fact, dry air actually enhances transmission of many viruses. Pathogens are said to be airborne if they travel more than 1 m through the air. Both airborne pathogens and those suspended in droplets have the best chance of reaching new hosts when people are crowded together indoors. Increased incidence of airborne infections is associated with nearly sealed modern buildings in which temperatures are controlled with heating and air-conditioning systems, and little fresh air enters.

Airborne pathogens fall to the floor and combine with dust particles or become suspended in aerosols. An aerosol is a cloud of tiny water droplets or fine solid particles suspended in air. Microorganisms in aerosols need not come directly from humans; they can also come from dust particles stirred by dry mopping, changing bedding, or even changing clothing. In the microbiology laboratory, flaming a transfer loop full of bacteria can disperse microorganisms into aerosols.

Dust particles can harbor many pathogens. Bacteria with sturdy cell walls, such as staphylococci and streptococci, can survive for several months in dust particles. Naked viruses as well as bacterial and fungal spores can survive for even longer periods.

Hospitalized patients are at great risk of getting airborne diseases because they often have lowered resistance and because former patients may have left pathogens deposited in dust particles. Cleaning floors with a wet mop, wiping surfaces with a damp cloth, and carefully unfolding bed linens and towels help reduce aerosols. Masks and special clothing are used in operating rooms, burn wards, and other areas where patients are at greatest risk of infection. Some hospitals also use ultraviolet lights and special air flow devices to prevent exposure of patients to airborne pathogens.

FOODBORNE TRANSMISSION   Pathogens are most likely to be transmitted in foods that are inspected improperly, processed unsanitarily, cooked incom-

ARE THERE AMOEBAS IN YOUR EMERGENCY EYEWASH STATION?

Emergency eyewash stations are found in most laboratories, where they are intended to supply large volumes of clean water in case of a chemical accident. The American National Standards Institute recommends that eyewash stations be flushed weekly to keep them clean. Eyewash stations are often neglected, however, and water stands in them for long periods of time at room temperature—ideal conditions for growth of biofilms of bacteria and fungi. Some of these microbes can serve as food for water-dwelling amoebas, such as *Acanthamoeba*, which can cause serious eye infections, leading to blindness. Many species of amoebas supply growth factors and an intracellular home for *Legionella* bacteria. *Legionella pneumophila*, the cause of Le-

gionnaires' disease, is spread by airborne water droplets. As few as 100 cells can cause infection. Other bacteria commonly found in emergency eyewash stations include *Pseudomonas*, a dangerous pathogen. It can cause considerable tissue destruction and is difficult to treat.

In a 1995 study of 30 dual-spray eyewash units connected to building plumbing, 60 percent of the units contained more than one genus of amoebas. Even regular flushing could not eliminate them. Some laboratories use eyewash bottles that are not connected to plumbing. Would you expect more or less contamination of water in such bottles? How would it compare with tap water?

---

pletely, or refrigerated poorly. As with waterborne pathogens, foodborne pathogens are most likely to produce gastrointestinal symptoms.

Transmission by Vectors

As you learned in Chapter 12, **vectors** are living objects that transmit disease to humans. ∞ (p. 301) Most vectors are arthropods such as ticks, flies, fleas, lice, and mosquitoes. However, the mechanism of vector transmission can be mechanical or biological.

MECHANICAL VECTORS Insects act as *mechanical vectors* when they transmit pathogens passively on their feet and body parts. Houseflies and other insects, for example, frequently feed on animal and, if available, human fecal matter. If they then move on to feed on human fare, they can deposit pathogens in the process. Disease transmission by mechanical vectors does not require that the pathogen multiply on or in the vector.

This method of disease transmission can be prevented simply by keeping these vectors out of areas where food is prepared and eaten. The fly that walked across dog feces in the park should not be allowed to walk across your picnic potato salad. The use of screened areas to keep insects out also reduces disease transmission by mechanical vectors. Unfortunately, in some poverty-ridden areas of the world, screens are lacking on windows—even those that open into hospital operating rooms!

BIOLOGICAL VECTORS Insects act as *biological vectors* when they transmit pathogens actively; that is, the infectious agent must complete part of its life cycle in the vector before the insect can transmit the infective

form of the microbe. Compared with direct transmission through animal bites, the transmission of zoonoses through vectors is much more common. In most vector-transmitted diseases, such as malaria and schistosomiasis, a biological vector is the host for some phase of the life cycle of the pathogen. Control of zoonoses transmitted by biological vectors can often be achieved by controlling or eradicating the vectors. The spraying of standing water with oil kills many insect larvae. Spraying breeding grounds with pesticides also can be an effective control, at least until the vectors become resistant to the pesticides.

Special Problems in Disease Transmission

Transmission of disease by carriers poses special epidemiologic problems because carriers are often difficult to identify. The carriers themselves usually do not know they are carriers and sometimes cause sudden outbreaks of disease. Depending on the pathogen they carry, carriers can transmit disease by direct or indirect contact or through vehicles such as water, air, or food; they can even be a source of pathogens for vectors.

Another special transmission problem arises with people who have *sexually transmitted diseases* (STDs). Such diseases are most often transmitted by direct sexual contact, including kissing, but some can be transmitted by oral or anal sex. STDs present epidemiologic problems because infected individuals sometimes have contact with multiple sexual partners. In fact, the incidences of AIDS, genital herpes, genital warts, syphilis, and *Chlamydia* infections are rapidly increasing.

Zoonoses are another epidemiologic problem. They can be transmitted by direct contact, as when humans get rabies from the bite of an infected domestic

"DOG GERMS!"

When Snoopy kisses Lucy, she always screams, "Dog germs!" However, Snoopy is the one who should be worried. Some infections can be transmitted in either direction between humans and pets, but dogs may be in greater danger of becoming infected from humans than the other way around. A human mouth is more likely to contain pathogens than is a dog's mouth because dog saliva is much more acidic and is thus a less hospitable environment for microorganisms.

Although kissing probably does little harm, swimming with man's best friend may be more dangerous for humans. For example, *Leptospira* spirochetes ordinarily infect animal kidneys and are passed out in their urine. However, as one man discovered when he went swimming in a river with his dog, water contaminated with such urine can lead to human leptospirosis. This disease is characterized by fever, headache, and kidney damage. An infected dog would be even more hazardous in a crowded swimming pool. Although vigilant chlorination of pools and vaccination of pets against leptospirosis might prevent them from transmitting the disease, it is best not to allow pets in pools.

A variation on this chain of transmission occurred when some teenagers drove their swamp buggy through

Is it the child or the dog who is more likely to become infected?

water contaminated with urine from leptospirosis-infected deer. Droplets splashed up by the tires apparently infected the passengers.

or wild animal. An oral vaccine for rabies has been developed, to be administered to wildlife.

Disease Cycles

Many diseases occur in cycles. For years or even decades, only a few cases are seen, but then many cases suddenly appear in epidemic or pandemic proportions. Let's look at one example. Bubonic plague—or the Black Death, as it was called—has occurred in pandemic outbreaks followed by recurrent cycles for centuries. Between A.D. 543 and 548, the disease spread from India or Africa, through Egypt, to Constantinople (now Istanbul, Turkey), where it killed 200,000 people in only 4 months. The disease quickly spread via fleas on ship rats to Europe and the Mediterranean basin. About a half century later it appeared in China, with equally devastating results. After the initial outbreaks, it occurred in cycles of 10 to 24 years over the next two centuries.

All of Europe breathed a sigh of relief for about the next 500 years, which were plague-free. But in 1346, a second pandemic, worse than the first, afflicted North Africa, the Middle East, and most of Europe. Nearly one-third of the population of Europe died, and in many cities three-fourths of the population lost their lives to the dreaded disease. Then cyclic recurrences claimed

more lives in epidemics of the seventeenth century in England and the eighteenth century in France. Near the beginning of the twentieth century, a pandemic killed more than a million people in India and spread to many parts of the world, including San Francisco.

Thus, cyclic diseases pose special epidemiologic problems. Epidemiologists still cannot predict when one will break out and reach epidemic proportions. It is difficult to be prepared to treat sudden, large increases in the incidence of a disease and nearly impossible to persuade people to be immunized against a disease they have never seen.

Herd Immunity

An important factor in cyclic disease is **herd immunity** (or *group immunity*), which is the proportion of individuals in a community or population who are immune to a particular disease. If herd immunity is high—that is, if most of the individuals in a population are immune to a disease—then the disease can spread only among the small number of susceptible individuals in the population (Figure 16.13). Even when a member of the population becomes infected, the likelihood that that person will transmit the disease to others is small. Thus, a sufficiently high herd immunity protects the entire population, including its susceptible members.

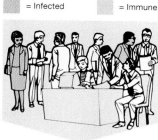 = Infected  = Immune = Susceptible

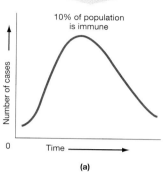

10% of population
is immune

Number of cases

0   Time ⟶

(a)

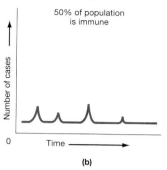

50% of population
is immune

Number of cases

0   Time ⟶

(b)

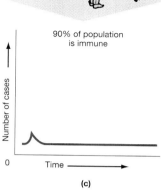

90% of population
is immune

Number of cases

0   Time ⟶

(c)

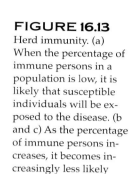

FIGURE 16.13
Herd immunity. (a) When the percentage of immune persons in a population is low, it is likely that susceptible individuals will be exposed to the disease. (b and c) As the percentage of immune persons increases, it becomes increasingly less likely that a susceptible individual will be exposed to the disease.

PUBLIC HEALTH

PLAGUE IN THE 1990S

The many epidemics and pandemics of the plague are not relics of the past. Plague has become endemic in the western United States, where *Yersinia pestis* survives in animal reservoirs such as wild rodents. Outbreaks of the disease still occur; for example, in August 1994, outbreaks of bubonic and pneumonic plague occurred in parts of India where the bacterium was epidemic in rats. Widespread panic erupted, and health care workers worldwide became deeply concerned that the disease could become pandemic. By October 1994 the number of suspected cases of bubonic and pneumonic plague reported in India had reached 693, with 56 deaths. Due to rapid identification and control measures, the epidemic subsided.

In the age of jet travel, it is easy for a disease to become pandemic almost overnight. Thus, the reports from India prompted the New York City Department of Health, along with other state and national agencies, to develop an emergency plan for detecting and managing any suspected cases arriving by plane. In fact, 11 air travelers arriving in New York were suspected of having the plague, but they were later found to be suffering from other diseases. Although this time we were lucky, plague would be a monumental problem in the rat-infested cities of the world, including those in the United States. The available vaccine is not completely effective, and many rats have become resistant to the best available rat poisons.

It is easy to see, then, why public health officials want to maintain the highest possible herd immunity, especially against common cyclic diseases. They encourage parents to have children immunized against measles and other communicable diseases. In many cities in the United States, children are required to be immunized against measles before they can start school. As a result, about 95 percent of elementary-age children are immune to measles. Although high school and college students who have neither had measles nor received vaccine are protected by herd immunity, many school systems and some colleges and universities are requiring immunization against measles for all students, whatever their age.

The loss of herd immunity could lead to a reemergence of a disease. The rise in diphtheria incidence in the New Independent States of the former Soviet Union is due in part to a lack of childhood vaccinations. Herd immunity in these populations has dwindled. Some people fear that herd immunity to smallpox is bound to drop now that vaccinations have ceased as a result of the eradication of the smallpox virus. If the smallpox virus were to appear again, it could be devastating—few people would be immune to the disease.

Control of Disease Transmission

Several methods are currently available for full or partial control of communicable diseases. They include *isolation, quarantine, immunization,* and *vector control.*

In **isolation,** a patient with a communicable disease is prevented from having contact with the general population. Isolation generally is accomplished in a hospital. There appropriate procedures can be carried out to reduce the spread of disease among susceptible individuals and to prevent the spread of disease in the general population. In all, there are seven categories of isolation (Table 16.2). Strict isolation makes use of all available procedures to prevent transmission of organisms or virulent infections to medical personnel and visitors. Even medical researchers working with highly pathogenic strains of microbes must use high levels of isolation and special laboratories equipped to contain such agents (Figure 16.14).

Quarantine is the separation of "healthy" human or animal carriers from the general population when they have been exposed to a communicable disease. Quarantine prevents spread of the disease during its incubation period. Although it is one of the oldest methods of controlling communicable diseases, it is now used mainly for serious diseases such as cholera and yellow fever. Quarantine differs from isolation in two ways: (1) It is applied to healthy people who were exposed to a disease during the incubation period, and (2) it pertains to limiting the movements of such people and not necessarily to precautions during treatment. Quarantine is rarely used today because it is very difficult to carry out. To ensure that no infected persons spread a disease, everyone who had been exposed to it would have to be quarantined for the disease's incubation period. This would mean, for example, that all travelers returning to the United States from a region of the world where cholera is endemic would have to be quarantined for 3 days in accommodations provided at airports and seaports—not likely to be a popular idea.

Large-scale *immunization* programs are an extremely effective means of controlling communicable diseases for which safe vaccines are available. Such programs

| TABLE 16.2 | A Summary of Important Isolation Procedures |

Isolation Categories

Strict	Contact	Respiratory	Tuberculosis	Enteric Precautions	Drainage/ Secretion Precautions	Blood and Body Fluid Precautions
Visitors must check in at nursing station before entering patient's room						
Yes	Yes	Yes	Yes	Yes	Yes	Yes
Hands must be washed on entering and on leaving patient's room						
Yes	Yes	Yes	Yes	Yes	Yes	Yes
Gowns must be worn by personnel and visitors						
Yes	Yes	No	Yes	Only for direct patient contact	Only if soiling is likely	Only if soiling is likely
Masks must be worn by personnel and visitors						
Yes	Yes	Unless not susceptible to disease	Only if coughing	No	No	No
Gloves must be worn by personnel and visitors						
Yes	Only for direct patient contact	No	No	Only for direct contact with patient or feces	Only for direct contact with lesion site	Only for direct contact with lesion site
Private room required with door closed						
Yes	Yes	Yes	Yes	Only for children	No	If hygiene is poor
Examples of diseases						
Pneumonic plague, rabies, diphtheria, disseminated herpes zoster, Lassa fever, chickenpox, draining *Staphylococcus aureus* wounds	Severe noninfected dermatitis, noninfected burns	Measles, mumps, rubella, pertussis	Pulmonary tuberculosis	Typhoid fever, cholera, salmonellosis, shigellosis, hepatitis	Bubonic plague, gas gangrene, localized herpes, puerperal sepsis	AIDS, hepatitis B

FIGURE 16.14 Scientists who work with very dangerous, and often easily disseminated, microbes do so in an isolation lab. This laboratory worker is in the highest level of isolation, a biosafety level 4 lab. Extreme precautions, such as the use of special ventilation systems and the wearing of "space suits," are required of personnel to avoid contact with microbes and to prevent the escape of the microbes from the facility.

greatly increase herd immunity and thus greatly decrease human suffering and deaths from infectious diseases. In the United States, immunizations have nearly eradicated polio, measles, mumps, diphtheria, and whooping cough. Unfortunately, as the incidence of these diseases becomes very small, people become complacent about getting immunized. Such complacency can lead to a sufficient decrease in herd immunity, resulting in outbreaks of vaccine-preventable diseases.

Vector control is an effective means of controlling infectious diseases if the vector, such as an insect or rodent, can be identified and its habitat, breeding habits, and feeding behavior determined. Places where a vec-

tor lives and breeds can be treated with insecticides or rodenticides. Window screens, mosquito netting, insect repellents, and other barriers can be used to protect humans from becoming victims of the bites of feeding vectors. Unfortunately, vectors have their own defenses. Some escape or become resistant to pesticides or make their way through barriers.

Malaria control in the United States was accomplished mainly by the control of mosquito vectors. Consequently, Americans have little herd immunity to malaria because most have never had the disease. Since the 1940s new cases of malaria in the United States have occurred mainly among people infected in other countries (Figure 16.15). As long as infected individuals do not reintroduce the parasite into the mosquito population, a malaria epidemic in the United States is most unlikely despite low herd immunity. In recent years, however, laborers from areas where malaria still is common have brought the parasite northward, and the disease has become endemic in some parts of California.

Although communicable diseases are theoretically preventable, some still have a high incidence in every human population. In countries with relatively high living standards, the common cold and many sexually transmitted diseases occur with great frequency. In countries with lower standards of living—especially those in the tropics—the prevalence of malaria and a variety of other diseases, including some nearly eradicated in other countries, is extremely high. And certainly AIDS has become a worldwide threat that now extends beyond the special high-risk groups with which it was once associated.

Public Health Organizations

In the United States and many other countries, the importance of controlling infectious diseases and reduc-

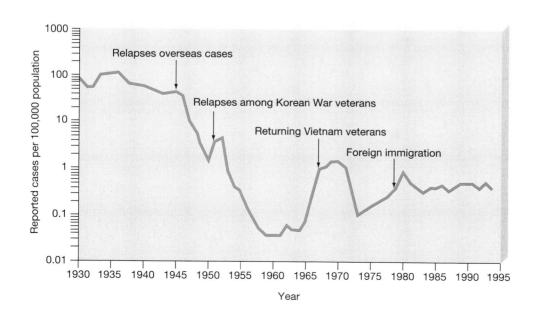

FIGURE 16.15 Incidence rate of malaria cases, as related to sources of infection into the United States.

ing other health hazards has led to the creation of public health agencies. City and county health departments provide immunizations, inspect restaurants and food stores, and work with other local agencies to ensure that water and sewage are treated properly. State health departments deal with problems that extend beyond cities and counties. They often do laboratory tests, such as identifying rabies in animals, and hepatitis and toxins in water.

The Centers for Disease Control and Prevention

In the United States, the federal government operates the U.S. Public Health Service (USPHS), which has several branches. Of these branches, the *Centers for Disease Control and Prevention* (CDC) in Atlanta, Georgia (Figure 16.16), has major responsibilities for the control and prevention of infectious diseases and other preventable conditions. Among the microbiology-related activities of CDC are (1) providing guidelines for occupational health and safety, quarantines, tropical medicine, co-operative activities with national agencies in other countries and with international agencies, and public health education; (2) making recommendations to the medical community regarding the use of antibiotics, especially for the treatment of diseases caused by antibiotic-resistant organisms; (3) storing infrequently used drugs and providing them to physicians who encounter patients with tropical parasitic diseases and other diseases rarely seen in the United States; and (4) making recommendations regarding the administration of vaccines—which should be used, who should receive them, and at what ages.

The CDC carries out epidemiologic studies, which are published as the *Morbidity and Mortality Weekly Report* (MMWR). This publication provides statistics for specific diseases in various parts of the United States and the world (Figure 16.17). Other CDC periodical reports are *Recommendations and Reports* and *Surveillance Summaries,* which provide in-depth coverage of specific issues, many related to infectious diseases.

The World Health Organization

The *World Health Organization* (WHO) is an international agency based in Geneva, Switzerland, that co-ordinates and sets up programs to improve health in more than 100 member countries. Its basic objective is that all peoples attain the highest possible level of health. This aim is embodied in WHO's current theme: "Health for All by the Year 2000." Specific activities are carried out by six regional organizations in Africa, the Eastern Mediterranean, Europe, Southeast Asia, the Western Pacific, and the Americas. WHO works closely

FIGURE 16.16   CDC headquarters in Atlanta, Georgia.

with the United Nations on population control, management of food supplies, and various other scientific and educational activities.

WHO sets health standards for international disease control; helps developing nations establish effec-

FIGURE 16.17   The *Morbidity and Mortality Weekly Report* is issued weekly by the CDC. It reports new and unusual cases and trends. It also lists, by state, the number of cases of notifiable diseases recorded that week, the number recorded for the same week in the previous year, and the cumulative totals. It is available from the Massachusetts Medical Society; C.S.P.O. Box 9120; Waltham, MA 02254-9120.

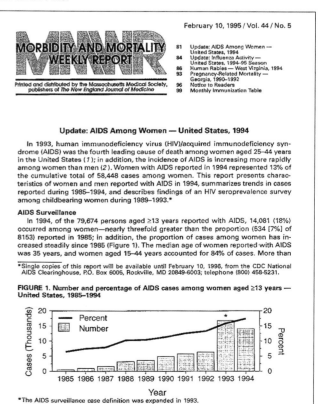

(a)

(b)

FIGURE 16.18 Some typical activities of the World Health Organization. (a) Eye tests for the prevention of blindness are given high in the Andes Mountains of Peru. (b) A medical assistant travels by horseback to treat patients in remote villages of China.

tive control and immunization programs; collects, analyzes, and distributes health data; and maintains surveillance of potential epidemics (published in WHO's *Weekly Epidemiological Record*). It also provides training and research programs for health personnel and information for individuals (Figure 16.18). The agency has helped more than 100 countries in immunizing against diphtheria, measles, whooping cough, poliomyelitis, tetanus, and tuberculosis and hopes eventually to eradicate measles worldwide. It conducts research and training to combat widespread tropical diseases such as leprosy, malaria, and several diseases caused by helminths. WHO has been instrumental in coordinating the eradication of smallpox worldwide.

Notifiable Diseases

Cooperation among state and national health organizations in the United States has led to the establishment of a list of **notifiable diseases,** which are infectious diseases that are potentially harmful to the public's health and must be reported by physicians. As of July 1994, 49 infectious diseases were listed as notifiable at the national level (Table 16.3). On the basis of CDC suggestions, each year the Council of State and Territorial Epidemiologists (CSTE) adds diseases to, or deletes diseases from, the list. If a specific disease shows a decline in incidence, it may be removed from the list.

Although reporting infectious diseases is mandatory only at the state level, the reporting of notifiable diseases at the national level is intended to accomplish two things: (1) to ensure that public health officials learn of diseases that jeopardize the health of populations, and (2) to provide consistency and uniformity in the reporting of those diseases. Various kinds of information about notifiable diseases in the United States are available from the CDC; a sample of this information is provided in Table 16.4.

TABLE 16.3	Infectious Diseases Designated as Notifiable at the National Level—United States, 1994		
AIDS	Hepatitis, non-A, non-B	Rabies, human	
Amebiasis	Hepatitis, unspecified	Rheumatic fever	
Anthrax	Legionellosis	Rocky Mountain spotted fever (Typhus fever, tickborne)	
Aseptic meningitis	Leprosy (Hansen disease)		
Botulism			
Brucellosis	Leptospirosis	Rubella	
Chancroid	Lyme disease	Salmonellosis	
Cholera	Lymphogranuloma venereum	Shigellosis	
Congenital rubella syndrome	Malaria	Syphilis	
Diphtheria	Measles	Syphilis, congenital	
Encephalitis	Meningococcal infection	Tetanus	
Escherichia coli O157:H7	Mumps	Toxic shock syndrome	
Gonorrhea	Pertussis	Trichinosis	
Granuloma inguinale	Plague	Tuberculosis	
Haemophilus influenzae	Poliomyelitis	Tularemia	
Hepatitis A	Psittacosis	Typhoid fever	
Hepatitis B	Rabies, animal	Varicella (chickenpox)	
		Yellow fever	

SOURCE: *Morbidity and Mortality Weekly Report* 43(43):801.

TABLE 16.4A
Summary of Cases of Notifiable Diseases, United States, Weeks Ending December 31, 1994, and January 1, 1994 (52nd Week)

AIDS	Aseptic Meningitis	Encephalitis		Gonorrhea		Hepatitis (Viral), by type				Legionellosis	Lyme disease
		Primary	Post-infectious			A	B	NA, NB	Unspecified		
Cum.* 1994	Cum. 1994	Cum. 1994	Cum. 1994	Cum. 1994	Cum. 1993	Cum. 1994	Cum. 1994	Cum. 1994	Cum. 1994	Cum. 1994	Cum. 1994
78,126	8050	674	107	400,592	398,684	23,507	11,402	4233	409	1535	11,424

Malaria	Measles (Rubeola)					Meningococcal Infections	Mumps		Pertussis			Rubella		
	Indigenous		Imported		Total									
Cum.* 1994	This Week 1994	Cum. 1994	This Week 1994	Cum. 1994	Cum. 1993	Cum. 1994	This Week 1994	Cum. 1994	This Week 1994	Cum. 1994	Cum. 1993	This Week 1994	Cum. 1994	Cum. 1993
1065	10	707	–	188	312	2638	9	1322	105	3590	6586	1	209	192

Syphilis (Primary & Secondary)		Toxic-Shock Syndrome	Tuberculosis		Tularemia	Typhoid Fever	Typhus Fever (Tickborne) (RMSF)	Rabies, Animal
Cum. 1994	Cum. 1993	Cum. 1994	Cum. 1994	Cum. 1993	Cum. 1994	Cum. 1994	Cum. 1994	Cum. 1994
20,183	26,470	183	22,152	24,324	85	410	441	7347

*Cumulative

TABLE 16.4B
Notifiable Diseases of Low Frequency, United States, 1994

	Cum. 1994		Cum. 1994
Anthrax	—	Hansen's disease	111
Botulism: Foodborne	59	Leptospirosis	35
Infant	76	Plague	14
Other	7	Poliomyelitis, paralytic	1
Brucellosis	95	Psittacosis	41
Cholera	39	Rabies, human	5
Congenital rubella syndrome	6	Syphilis, congenital, age < 1 year	1123
Diphtheria	1	Tetanus	29
Haemophilus influenzae (invasive disease)	1126	Trichinosis	35

NOSOCOMIAL INFECTIONS

A **nosocomial** (nos-o-ko'me-al) **infection** is an infection acquired in a hospital or other medical facility. The term nosocomial is derived from the Greek words *nosos*, meaning "disease," and *komeo*, meaning "to take care of." Nosocomial diseases are acquired during medical treatment. Although many such infections occur in patients, infections acquired at work by staff members also are considered nosocomial infections.

Among patients admitted to American hospitals each year, about 2 million (10 percent) acquire an infection that increases the risk of death, the duration of the hospital stay, and the cost of treatment. Of these, over 20,000 people per year die of their nosocomial infections. Curiously, nosocomial infections are largely a product of advances in medical treatment. Intravenous, urinary, and other catheters, invasive diagnostic tests, and complex surgical procedures increase the likelihood that pathogens will enter the body. Intensive use of antibiotics contributes to the development of resistant strains of pathogens. And, therapies to reduce the chances of rejection of transplanted organs impair the immune response to pathogens. Despite the risk of nosocomial infections, the medical treatments now available save far more patients than are lost to such infections.

Epidemiology of Nosocomial Infections

The epidemiology of nosocomial infections, like the epidemiology of diseases acquired in the community, considers sources of infection, modes of transmission, susceptibility to infection, and prevention and control. In addition, it focuses on medical procedures that increase the risk of infection, the sites at which infections often occur, and the correlation between procedures and sites of infection.

Sources of Infection

Nosocomial infections can be exogenous or endogenous. **Exogenous infections** are caused by organisms that enter the patient from the environment. The organisms can come from other patients, staff members, or visitors. They can also be passed on by insects (ants, roaches, flies) from fomites (toilet, trash can) to patient. Other inanimate objects, such as equipment used in respiratory or intravenous therapy, catheters, bathroom fixtures and soap, and water systems, also can be a source of exogenous infections. Some nosocomial infections have even been traced to disinfectants such as quaternary ammonium compounds, to which certain organisms are resistant. **Endogenous infections** are caused by opportunists among the patient's own normal microbiota. Opportunists are most likely to cause infection if the patient has lowered resistance or if normal microbiota that compete with pathogens have been eliminated by antibiotics.

A small group of organisms, including *Escherichia coli, Enterococcus* species, *Staphylococcus aureus,* and *Pseudomonas* species, are responsible for about half of all nosocomial infections (Figure 16.19). These organisms are particularly likely to cause such infections because they are ubiquitous (present everywhere) and can survive outside the body for long periods. In addition, some strains of these organisms are resistant to many antibiotics; methicillin- and vancomycin-resistant

---

PUBLIC HEALTH

AIDS TESTING: MANDATORY FOR HEALTH CARE WORKERS?

If your physician, dentist, or nurse had AIDS, would you still want to be treated by him or her? If you tested positive for HIV, the virus that causes AIDS, should you be forced to reveal this? Does it matter whether you are the provider or the recipient of health care?

By the end of 1994, the CDC had received reports of 40 American health care workers with documented cases of HIV infections acquired "on the job." Of these, 12 have developed AIDS, and 1 has died of the disease. Of the approximately 460,000 cases of AIDS among adults, adolescents, and children, about 16,000 are health care workers. Of these, about 1900 nurses and 820 physicians have died.

Many people are increasingly uneasy that the dental drill in their mouth, or another invasive instrument being used on them, may have previously been used on an HIV-positive individual. Estimates suggest that there is about 1 chance in 250 that a health care worker will acquire an HIV infection from a patient and a remote chance that a health care worker will pass the virus to a patient. Even so, the questions are, Who should be tested, when, how often, and with what consequences?

The current AIDS test does not test for the AIDS virus but rather for the presence of antibodies to the virus in the blood. Thus, there is about a 0.5% chance of getting a false-positive test result; that is, in 1 out of 200 cases, the test detects antibodies to HIV when, in fact, there are none. Such results can come from the presence of antibodies to other viruses, such as the flu virus, that cross-react in the AIDS test. Also, when a person becomes infected with HIV, it takes at least 6 weeks for antibodies to become detectable and to produce a positive blood test.

In July 1991 the CDC issued new guidelines recommending that health care workers doing certain invasive surgical procedures should undergo voluntary (not mandatory) tests for the viruses that cause AIDS and hepatitis B. Those found to be infected should stop doing such procedures until they obtain permission from a panel of experts and inform their patients. The procedures include tooth extractions and other dental procedures, vaginal and Caesarean deliveries, hysterectomies, and operations on bones, joints, the colon, and the rectum. Cardiac catheterization and angiography (X-rays and tests of the heart and blood vessels) also require permission.

Reactions to these recommendations have been mixed. So far, professional societies have refused to cooperate with the CDC in agreeing to enforce the new regulations. Some people fear that a strict interpretation would prevent an infected health care worker from ever practicing again. Others fear that a hospital's concerns regarding liability would cause individual hospitals to require testing. Insurers might require testing to obtain malpractice insurance. Critics point out that 4 days after the CDC made its recommendation, a New York State physician revealed his HIV-positive status, and his hospital demanded—and received—his resignation.

Although most public health agencies and health professional societies, such as the American Medical Association, oppose mandatory testing, most health care workers do not. A July 1991 survey found that 57 percent of the doctors and 63 percent of the nurses questioned favored mandatory testing of health care workers. Three-quarters of those surveyed also favored mandatory AIDS testing of surgical patients and pregnant women as well as mandatory reporting of the names of patients who test positive to local health departments.

Deciding who should be tested is a complex matter. How do you feel about this issue? What does a poll of your classmates reveal?

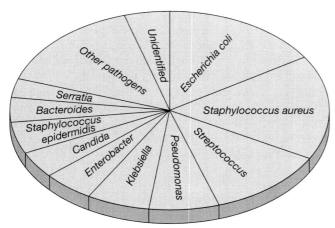

FIGURE 16.19 Common causative agents of nosocomial infections.

staphylococci and carbapenem-resistant *Pseudomonas aeruginosa* are especially problematic.

Susceptibility and Transmission

Compared with the general population, patients in hospitals are much more susceptible to infection; that is, they are **compromised hosts.** Many patients have breaks in the skin from lesions, wounds (surgical and accidental), or bed sores. Some also have breaks in mu-

cous membranes that line the digestive, respiratory, urinary, or reproductive systems. The lack of intact skin and mucous membranes provides easy access for infectious organisms. Also, most patients are debilitated; their resistance to infectious organisms is lower than normal. Patients undergoing organ transplants receive immunosuppressant drugs, and patients with AIDS and other disorders of the immune system also have reduced resistance. Factors that contribute to host resistance are discussed in Chapters 17 and 18.

Theoretically, nosocomial infections can be transmitted by all modes of transmission that occur in the community. However, direct person-to-person transmission between an infected patient, staff member, or visitor and noninfected patients; indirect transmission through equipment, supplies, and hospital procedures; and transmission through air are most common in hospitals (Figure 16.20). Some organisms can be transmitted by more than one route.

Universal Precautions

In 1988, the CDC, concerned about the possibilities that the AIDS virus would be transmitted in the health care setting, issued guidelines to reduce the risks. These guidelines, called **Universal Precautions,** are summarized in Appendix D. Some hospitals and other medical facilities choose to exercise even more caution than

FIGURE 16.20 Some common modes of transmission of nosocomial infections in a hospital setting.

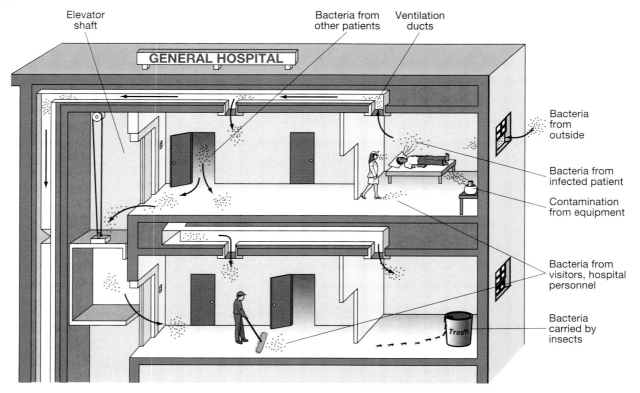

1. Wear gloves and gowns if soiling of hands, exposed skin, or clothing with blood or body fluids is *likely*.
2. Wear masks *and* protective eyewear or chin-length plastic faceshields whenever splashing or splattering of blood or body fluids is *likely*. A mask alone is not sufficient.
3. Wash hands before and after patient contact, and after removal of gloves. Change gloves between *each* patient.
4. Use disposable mouthpiece/airway for cardiopulmonary resuscitation.
5. Discard contaminated needles and other sharp items *immediately* into a *nearby*, special puncture-proof container. Needles must *not* be bent, clipped, or recapped.
6. Clean spills of blood or contaminated fluids by: (1) Putting on gloves and any other barriers needed, (2) wiping up with disposable towels, (3) washing with soap and water, and (4) disinfecting with a 1:10 solution of household bleach and water. Allow it to stand on surface for at least 10 minutes. Bleach solution should not have been prepared more than 24 hours beforehand

the CDC recommends. A shortened list of these precautions is given in Table 16.5. Universal Precautions apply to *all* patients, not just those infected with the viruses that cause AIDS or hepatitis B—hence the term *universal*. Universal Precautions apply to the following body fluids: blood, semen, and vaginal, tissue, cerebrospinal, synovial (joint cavity), pleural, peritoneal, pericardial, and amniotic fluids. The CDC has stated that Universal Precautions do not apply to feces, nasal secretions, sputum, sweat, tears, urine, and vomitus, as long as these do not contain visible blood. This is not to imply that no viruses are present in these fluids but rather that the risk of transmission is either very low or unproved. For example, studies have shown that HIV is present in all tested samples of saliva from AIDS patients. However, the levels are so low that it takes PCR techniques to detect them. ∞ (Chapter 7, p. 190) AIDS patients are often infected with other disease organisms that may be present in these fluids. These organisms include tuberculosis bacilli in sputum, bacteria such as *Salmonella* and *Shigella, Cryptosporidium* protozoa in feces, and herpesvirus in oral secretions. Therefore, some health care facilities require their employees to use Universal Precautions with all body fluids.

Equipment and Procedures That Contribute to Infection

Surgical procedures and the use of equipment such as catheters and respiratory devices are major contributors to nosocomial infections. The smallest abrasion can provide a site of entry for infectious agents. Infections can arise from a contaminated catheter, inadequate cleaning of the site of catheter insertion, or the movement of organisms from leaky connections. In addition, tubing, joints, containers of fluids, and the fluids themselves also can be contaminated.

All surgical procedures expose internal body parts to air, instruments, surgeons, and other operating room personnel, all of which can be contaminated. These procedures can also allow the patient's own microbiota to enter sites where they can produce infection. For example, bacteria that cause pneumonia can reach the lungs from the pharynx during surgery.

Respiratory devices, including nebulizer jets, that administer oxygen or air and medications to expand passageways in the lungs provide a means for disseminating microorganisms deep into the lungs. Organisms can grow in the reservoir pans of both cold mist and warm steam humidifiers, and the organisms can be dispersed in an aerosol as the machines operate. Therefore, all respiratory equipment should be disinfected or sterilized daily and, if not disposable, should be disinfected before being moved from one patient to the next.

Other devices and procedures account for smaller, but significant, numbers of nosocomial infections. For example, hemodialysis, a procedure for removing wastes from blood, provides a variety of means of introducing microorganisms into the body (Figure 16.21). Devices used to monitor blood pressure in the heart or major vessels or cerebrospinal fluid pressure have tubing extending outside the body. Such devices can be contaminated or can allow introduction of organisms from the patient or from the environment. Improperly cleaned gynecological instruments can transmit disease from one patient to another. Endoscopes, which are introduced through body openings and used to examine the linings of organs such as the bladder, large intestine, stomach, and respiratory passageways, are difficult to sterilize and thus can transfer microorganisms from one patient to another.

Another important factor that contributes to nosocomial infections is the intensive use of antibiotics, especially in hospital settings. How antibiotics contribute to the development of antibiotic-resistant pathogens and how these pathogens contribute to nosocomial infections were discussed in Chapter 14.

Sites of Infection

The sites of nosocomial infections, in order from most to least common, are as follows: urinary tract, surgical wounds, respiratory tract, skin (especially burns), blood (bacteremia), gastrointestinal tract, and central nervous system (Figure 16.22, page 442).

Prevention and Control of Nosocomial Infections

The problem of nosocomial infections is widely recognized, and nearly all hospitals now have infection-con-

MICROBIOLOGIST'S NOTEBOOK

Controlling Nosocomial Infections in a Burn Unit

Drs. Arthur Mason, left, and Albert McManus in the laboratory.

The Institute of Surgical Research, at Brooke Army Medical Center in Fort Sam Houston, Texas, is known worldwide for its pioneering work in burn treatment. Dr. Arthur D. Mason, Jr., a surgeon, is largely responsible for the research on burn infections and ways to combat them. Dr. Albert McManus, chief of the Microbiology Branch, developed a microbiology laboratory that is considered a leader in research of surgical wound infections and in burn care. Colonel Elisabeth Greenfield is the Chief Nurse of the Institute's Burn Center.

Dr. Mason: The treatment of burn victims has changed a great deal in the past half century. When I first entered the field in 1954, two of the major advances in burn care had recently been made. We had learned that skin grafting could be used to replace the skin destroyed by the burn and that shock could be averted by the administration of fluids to replace those lost to the burn.

The advent of effective fluid replacement sharply decreased early loss of life due to burn shock. Thus, a whole new population of burn patients emerged–severely injured people who would not previously have lived for more than a day or two. But, although these patients now lived longer, many still died. In most of these cases, burn wound infection and bacteremia preceded death. Nevertheless, bacteremia was not generally considered to be the cause of death.

Dr. McManus: At the time, bacterial infection of burns was thought to be significant only in so far as it interfered with healing of the burned wound. Organisms such as *Pseudomonas aeruginosa* and other gram-negative bacteria often were found in the blood and wounds of burn patients. Because these organisms were not recognized as causes of infections in other surgical situations, they were considered nonpathogens. Moreover, the first genera-

tions of antibiotics available, such as penicillin, were effective mostly against streptococci and staphylococci—the traditional agents causing wound infections. These antibiotics were of little use against *Pseudomonas*, so treating badly burned patients with these drugs usually made no difference in mortality. Such deaths were considered caused by the burn itself. This fact masked the true role of infection with gram-negative organisms as an often fatal complication in burn patients.

Dr. Mason: During the 1950s we at the Institute set out on a systematic examination of the process of infection in burns. By the end of the decade we had developed an animal model of *Pseudomonas* sepsis in which contamination of nonlethal burn wounds with *P. aeruginosa* consistently caused death. In this way we were able to convince the medical community that often it wasn't the burn itself that proved lethal, but rather the generalized bacteremia that developed—especially with gram-negative organisms.

The next step naturally was to explore possible antimicrobial therapies. We soon found that antibiotics in the bloodstream failed in human infections because they did not penetrate the dead tissue of burns, where there was no longer any blood supply. Even newer antibiotics, such as polymyxin B, that had high activity in the laboratory against strains taken from burned patients did not reach burn wounds in high enough concentrations to be effective. Dead burned tissue was the site of contamination and growth of organisms; by the time the organisms penetrated to the circulatory system, bacterial numbers had become too large for effective antibiotic action.

The logical alternative approach was to try and deliver a drug directly to the surface of the burn site. After much experimentation, we eventually found that sulfamylon, an antibacterial agent used by the German army for wounds during World War II, was

highly effective in controlling bacterial growth in burn wounds. The results were remarkable: Bacteremia was much less frequent, and mortality decreased sharply.

But this Utopian result—a single, simple, effective therapy—didn't last; bacteria are much too versatile for that. In 1969 a strain of *Providencia stuartii* that was resistant to sulfamylon appeared in our patients. Within a year or two we again had a serious infection problem. This challenge was ultimately met with another topical agent, but it was clear that the topical approach alone was not a complete answer. It fell to Dr. McManus to develop a more complete method for infection control in these patients.

Dr. McManus: Clearly, microbial adaptation in response to sulfamylon had occurred. These strains were probably being maintained somewhere in the ward environment and being passed to new patients after admission. In addition to organisms already resistant and present in the burn ward, with each transfer of a patient from another hospital to our burn unit there was a chance that other types of resistant nosocomial pathogens would be added to the burn unit flora.

Most microbiology cultures were taken to identify an organism that had produced the clinical symptoms of infection. This type of information was very useful for describing what infections were occurring, but it helped very little in trying to find the sources and mechanisms for perpetuating resistant organisms on the burn unit. We really did not know how many patients were colonized with these same organisms but were not cultured because they were not sick. Without knowing how commonly cross-contamination occurred in such patients, it was essentially impossible to stop the spread. I therefore proposed a culture system that would provide a wider sample of data, reflecting the colonizing as well as infecting organisms in patients of all burn sizes, at admission and throughout their hospitalization.

This Microbial Surveillance System was based on the premise that colonization of the burn wound and other damaged sites occurs soon after injury. Thus, if the patient survives the initial shock, fluid loss, and the burn itself, colonization will occur from the environment or from the patient's own preinjury flora. Because colonization precedes infection, the information would be readily available to help identify an infecting organism at the time of a clinical change if it occurred. So on admission or during transport by our burn flight team to the burn unit, the patient is cultured at the likely places for infection. Samples of wound, sputum, and urine are taken. These sites are sampled three times per week for the duration of care. In addition, the stool is sampled weekly and cultured on selective media to detect resistance to antibiotics that may be used. Organisms are identified, and antibiotic susceptibility tests are completed on potential problem organisms, such as *Pseudomonas aeruginosa* and *Staphylococcus aureus*. The information is entered into a computerized database available in the burn unit. The culture results are updated daily. Microbial colonization can thus be followed for each patient as well as for the entire patient population on the ward. The population data are used by Colonel Greenfield and her staff to maintain and improve patient isolation and infection control.

Col. Greenfield: As Dr. McManus stated, the data provided by the Microbial Surveillance System are extremely useful to us in many respects. They play an important part in our assigning of nursing care teams to specific groups of patients to reduce the potential for cross-contamination. Patients with specific microbial considerations, such as resistant organisms or increased susceptibility to infection, are further isolated and identified as requiring "cohort" isolation. An individual assigned to a cohorted patient may not provide care to any other patient that day. Ideally, we schedule days off for the individuals caring for such a patient before assigning them to other patients.

Microbial surveillance data are also an excellent Quality Improvement mechanism. Daily examination of the

data allows for early identification of patient cross-contamination and modification of infection control techniques. These data serve as an objective monitor for many of our infection control policies, such as dress code, handwashing, and traffic control.

Finally, microbial surveillance is a basis for monitoring and revising existing infection control policies or for creating new ones. Any consistent pattern of cross-contamination cues an evaluation of current policies.

Dr. McManus: Of course, knowing what organisms are present or are coming doesn't always guarantee that we have the weapons to deal with them successfully. More than a decade ago, before we had the patient isolation rooms, the Institute was called to evacuate three American merchant seamen who had been burned in a fire in their ship's engine room and were admitted to a hospital in a developing country. They were infected with a strain of *Providencia stuartii*. We had done surveillance cultures on the airplane, so we knew the patients had the organism prior to admission at our hospital. Within a few days, all three developed serious infections, and despite the use of infection-control measures, the strain quickly spread to other patients. In addition to being highly transmissible, the organism proved to be resistant not only to many of the antibiotics we could use but also to the disinfectants available for cleaning the unit. The strain harbored a plasmid that was loaded with resistance genes. Within a few weeks, other gram-negative organisms were isolated with essentially the

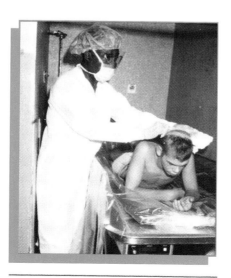

A nurse showering a patient in the Burn Center. Care of burn victims requires much contact between patient and medical staff; cross-contamination of patients is a danger.

same antibiotic resistance pattern. Not only were we having an outbreak of *Providencia* infections but antibiotic resistance was being spread to other organisms by a broad-host-range plasmid. ∞ (Chapter 8, p. 207).

We finally eliminated this troublesome strain and its plasmid by directing all new admissions into a new single-bed-isolation burn unit. The surveillance system documented that the resistance problem was not transmitted to the new ward. We have not had an outbreak of such an organism since we have been in the single-bed-isolation facility. Not only have endemics of resistant organisms stopped, but we have also documented major reductions in the incidence and outcome of infections. *Pseudomonas aeruginosa*, which was the major burn wound pathogen during the 1960s and 1970s, has become a rare problem with only two documented wound invasions in more than 2000 burn admissions. Mortality associated with infection has also significantly declined.

Col. Greenfield: Single-bed isolation was a giant step forward in reducing the incidence of infections in this institution. A comprehensive infection control program, under the direction of an Infection Control Committee, reinforces these statistics. In addition to staffing and isolation guidelines, the infection control program provides for strict surveillance of traffic policies and dress codes for both staff and visitors, environmental surveillance, equipment selection, and employee health. All staff assigned to the Institution, either permanently or temporarily, undergo occupational health screening and provide proof of hepatitis B vaccination. Additionally, they are interviewed by the Infection Control Nurse, who is responsible for educating and training all assigned personnel in infection control policies and procedures.

Dr. Mason: It is gratifying to us that many hospitals and clinicians from all over the world make use of our research. Indeed, many institutions use the Institute as a model. But the most important thing is that at the start of my career most patients who sustained serious burns over 50 percent of their body died. Today, for the first time ever, young adults with burns over 80 percent of their body have an even chance of recovery. And our work has contributed to that change.

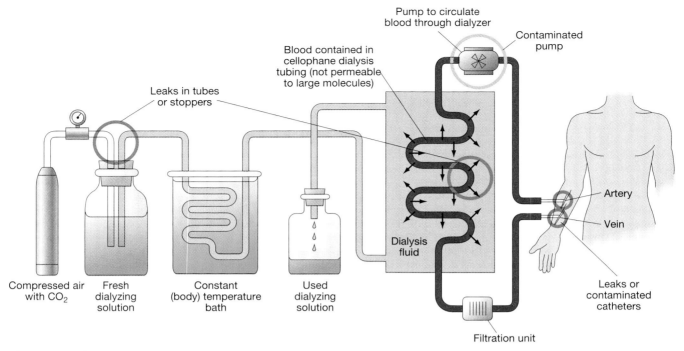

FIGURE 16.21 Possible sites of contamination in hemodialysis equipment.

Labels in figure:
- Pump to circulate blood through dialyzer
- Contaminated pump
- Blood contained in cellophane dialysis tubing (not permeable to large molecules)
- Leaks in tubes or stoppers
- Artery
- Vein
- Dialysis fluid
- Compressed air with CO_2
- Fresh dialyzing solution
- Constant (body) temperature bath
- Used dialyzing solution
- Leaks or contaminated catheters
- Filtration unit

trol programs. In fact, to maintain accreditation by the American Hospital Association, hospitals must have programs that include surveillance of nosocomial infections in both patients and staff, a microbiology laboratory, isolation procedures, accepted procedures for the use of catheters and other instruments, general sanitation procedures, and a nosocomial disease education program for staff members. Most hospitals have an infection-control specialist to manage such a program.

Several techniques are available to prevent the introduction and spread of nosocomial infections. Hand washing is the single most important technique. Physi-

FIGURE 16.22 Relative frequencies of sites of nosocomial infections.

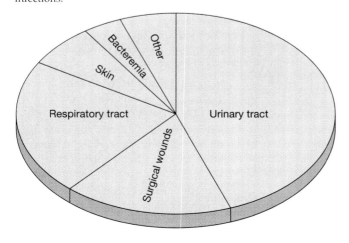

Pie chart labels: Other, Bacteremia, Skin, Respiratory tract, Urinary tract, Surgical wounds

cians, nurses, and other staff members who wash hands thoroughly with soap and water between patient contacts can greatly reduce the risk of spreading diseases among patients. Scrupulous care in obtaining sterile equipment and maintaining its sterility while in use are also important. In addition, the use of gloves when handling infectious materials such as dressings and bedpans and when drawing blood prevents the spread of infections. And, as mentioned earlier, it is important to prevent insect infestations, as flies, ants, or roaches can easily spread an infectious agent.

Other techniques are needed to reduce the development of antibiotic-resistant pathogens. Routine use of antimicrobial agents to prevent infections has turned out to be a misguided effort because it contributes to the development of resistant organisms. Therefore, some hospitals maintain surveillance of antibiotic use. Antibiotics are given for known infections but are given prophylactically (as preventive measures) only in special situations. Prophylactic antibiotics are justified in surgical procedures such as those involving the intestinal tract and repair of traumatic injuries, in which the surgical field is invariably contaminated with potential pathogens. They also are justified in immunosuppressed and excessively debilitated patients, whose natural defense mechanisms may fail.

If all the known techniques for preventing nosocomial infections were practiced rigorously, the incidence of such infections probably could be reduced to half the present level.

THE MENINGITIS BELT

Like many other diseases discussed in Chapter 16, meningococcal meningitis, which is caused by *Neisseria meningitidis*, occurs in roughly 5- to 12-year cycles. During the 1960s, an estimated 3 million cases occurred in China. In mid-April of 1988, patients were entering the hospital in N'Djamena, the capital city of the north central African nation Chad, at a rate of 250 per day. Out in the countryside, thousands of others who were unable to reach medical facilities suffered and died uncounted. In 1989, 40,000 cases were diagnosed in an Ethiopian epidemic; yet the United States does not have such epidemics. Looking specifically at Africa, what epidemiological factors cause meningitis epidemics to sweep across a broad belt of central Africa (Figure 16.23) in these cycles?

In considering such multi-year cycles, epidemiologists also must consider the seasonal cycles in Africa. Each year, whether an epidemic year or one of the years in between, meningitis outbreaks occur only during the dry-season months of January through June. Then, as soon as the rains begin, the number of new cases drops to zero—only to climb again at the start of the dry season. However, it can be shown that the causative meningococci are transmitted year-round. How can the seasonality of meningitis outbreaks be explained?

1. *Environmental factors* The environmental factors of low humidity and heat during the dry season may cause nasal and throat membranes to dry out, thus making them more likely to allow meningococci to enter the bloodstream.

2. *Other diseases* The dry season is also the season of more colds, influenza, and other upper respira-

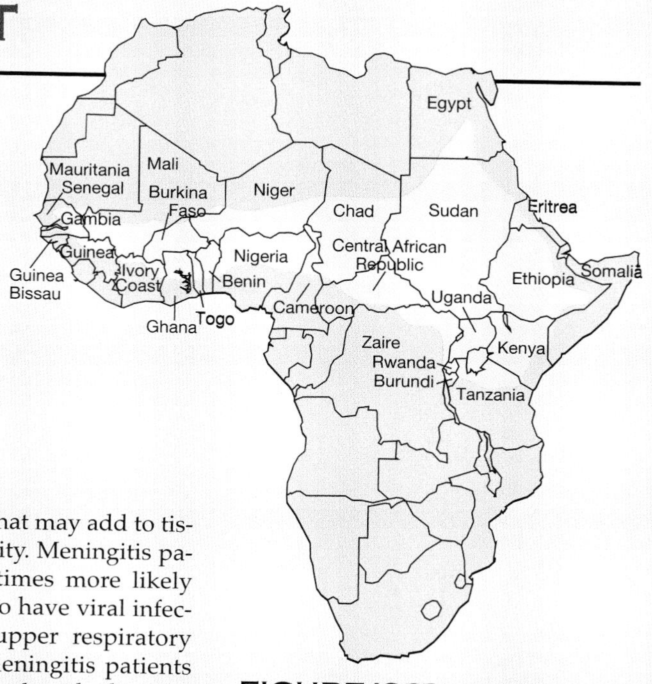

FIGURE 16.23 The meningitis belt across central Africa.

tory diseases that may add to tissue vulnerability. Meningitis patients are 23 times more likely than average to have viral infections of the upper respiratory tract. Many meningitis patients are also infected with the bacterium *Mycoplasma hominis*.

3. *Herd immunity* Why doesn't the seasonal swing erupt into epidemic every year? During an epidemic, most people in a population will develop antibodies against the prevalent strain of causative bacteria. Each year, however, new children are born, and they lack such immunity. Eventually the herd immunity drops so low as to allow another epidemic to occur.

4. *Strain virulence* During the years between epidemics, mutant strains will arise. If one of these new strains is more virulent than previous strains, it could initiate a new epidemic. Studies of epidemics have shown that usually only one strain of meningococci is responsible for any one epidemic. So virulence of strain is an important factor.

5. *Timing* The timing of entry of the mutant strain into the population is also crucial. If the mutant strain enters during the dry sea-

son, those who are exposed to it are likely to develop meningitis. But if a new strain enters during the rainy season, people will become exposed to it without developing the disease. Instead, they will develop antibodies against the new strain and will be immune to it when the following dry season begins.

Clearly the epidemiology of meningitis epidemics is not a simple matter. Several factors are at work, and each may have a greater or lesser effect on an outbreak of a given epidemic. Unfortunately, the available meningitis vaccine confers only short-term immunity; it does not last from one epidemic to the next. However, it can be used during an epidemic to halt its spread.

EPIMIOLOGY

EPIDEMIOLOGY

What Is Epidemiology?

- **Epidemiology** is the study of factors and mechanisms in the spread of diseases in a population.

- In describing infectious diseases, **epidemiologists** use **incidence rate** to refer to the number of new cases in a specific period, **prevalence rate** to refer to the number of people infected at any one time, **morbidity rate** to indicate the number of cases as a proportion of the population, and **mortality rate** to indicate the number of deaths as a proportion of the population.

Diseases in Populations

- In a **sporadic disease,** several isolated cases appear in a population. In an **endemic** disease, many cases appear in a population, but the harm to patients is not sufficient to create a public health problem. In an **epidemic** disease, many cases appear in a population, and patients are sufficiently harmed to create a public health problem. Diseases that spread by **common-source outbreaks** originate with a single contaminated substance, such as a water supply. In **propagated epidemics,** diseases spread by person-to-person contact. A **pandemic** disease is an epidemic disease that has spread over an exceptionally wide geographic area or several geographic areas.

Epidemiologic Studies

- The purpose of **epidemiologic studies** is to learn more about the spread of diseases in populations and how to control those diseases.

- The methods of epidemiologic studies are **descriptive, analytical** (retrospective and prospective), and **experimental.**

Reservoirs of Infection

- **Reservoirs of infection** include humans, other animals, and nonliving sources from which infectious diseases can be transmitted.

- Among human reservoirs, **carriers** often transmit diseases. They are intermittent carriers if they periodically release pathogens.

- Diseases in animal reservoirs can be transmitted by direct contact with animals or by vectors. Diseases that can be naturally transmitted from animals to humans are called **zoonoses.**

- Diseases in nonliving reservoirs are transmitted by water, soil, or wastes.

Portals of Entry

- **Portals of entry** include skin, mucous membranes that line various body systems, tissues, and the placenta.

Portals of Exit

- **Portals of exit** include the nose, ear, mouth, skin, and openings from which products of the digestive, urinary,

and reproductive systems are released. Organisms usually are in body fluids or feces.

Modes of Transmission of Diseases

- Transmission can be by contact, vehicle, or vector. **Direct contact transmission** includes person-to-person **horizontal transmission** and **vertical transmission** from parent to offspring. **Indirect contact transmission** occurs through **fomites** (inanimate objects) and by droplets. **Vehicles** of transmission include water, air, and food. **Vectors** of transmission are usually arthropods, which can transmit disease mechanically or biologically.

- Transmission by carriers, transmission of STDs, and transmission of zoonoses pose special epidemiologic problems.

Disease Cycles

- Some diseases occur in cycles—there are just a few cases for several years, and then many cases suddenly appear.

Herd Immunity

- **Herd immunity,** or group immunity, refers to immunity enjoyed by a large proportion of a population that reduces disease transmission among nonimmune individuals.

- A drop in herd immunity can lead to the sudden appearance of cases of a cyclic disease.

Control of Disease Transmission

- Methods used to control communicable diseases include isolation, quarantine, immunization, and vector control.

- **Isolation** procedures are summarized in Table 16.2. **Quarantine** is rarely used but can prevent exposed individuals from infecting others. Active immunization prevents many infections. Vector control is effective where vectors can be identified and eradicated.

Public Health Organizations

- Public health organizations exist at city, county, state, federal, and world levels. They help establish and maintain health standards, cooperate in the control of infectious diseases, collect and disseminate information, and assist with professional and public education.

Notifiable Diseases

- **Notifiable diseases** are listed in Table 16.3.

NOSOCOMIAL INFECTIONS

NOSOCOMIAL INFECTIONS

- A **nosocomial infection** is an infection acquired in a hospital or other medical facility.

Epidemiology of Nosocomial Infections

- Nosocomial infections can be **exogenous** (caused by external organisms) or **endogenous** (caused by opportunists in normal microbiota). About half are caused by only four types of pathogens, of which many strains are antibiotic resistant.

■ Host susceptibility is an important factor in the development of nosocomial infections.

■ Medical equipment and procedures, including surgery, are often responsible for infections.

■ Modes of nosocomial infections are illustrated in Figure 16.20.

Prevention and Control of Nosocomial Infections

■ Most hospitals have an extensive infection control program. Hand washing, use of gloves, scrupulous attention to maintaining sanitary conditions and sterility where possible, and surveillance of antibiotic use and other hospital procedures help reduce infections.

■ Nosocomial infections could be reduced by half if all known procedures were carefully followed in all medical facilities at all times.

KEY TERMS

aerosol (p. 428)
analytical study (p. 422)
carrier (p. 422)
common-source outbreak (p. 419)
compromised host (p. 438)
contact transmission (p. 426)
descriptive study (p. 421)
direct contact transmission (p. 426)
direct fecal–oral transmission (p. 426)
droplet nucleus (p. 428)
droplet transmission (p. 428)
endemic (p. 418)

endogenous infection (p. 436)
epidemic (p. 418)
epidemiologic study (p. 420)
epidemiologist (p. 416)
epidemiology (p. 416)
etiology (p. 416)
exogenous infection (p. 436)
experimental study (p. 422)
fomite (p. 428)
herd immunity (p. 430)
horizontal transmission (p. 426)
inapparent infection (p. 422)

incidence rate (p. 417)
index case (p. 422)
indirect contact transmission (p. 428)
indirect fecal–oral transmission (p. 428)
isolation (p. 432)
morbidity rate (p. 417)
mortality rate (p. 417)
nosocomial infection (p. 436)
notifiable disease (p. 435)
pandemic (p. 418)
placebo (p. 422)
portal of entry (p. 424)
portal of exit (p. 426)
prevalence rate (p. 417)

propagated epidemic (p. 419)
quarantine (p. 432)
reservoir of infection (p. 422)
sporadic disease (p. 418)
subclinical infection (p. 422)
Universal Precautions (p. 438)
vector (p. 429)
vehicle (p. 428)
vertical transmission (p. 426)
zoonosis (p. 423)

QUESTIONS FOR REVIEW

A Epidemiology

1. What is epidemiology?

2. Match the following:

___ persons in a population who become clinically ill during a specified period of time

___ the total number of sick individuals in a population at a particular time

___ the colonization and growth of an infectious agent in a host

___ the number of new cases of a disease identified in a population during a defined period of time

___ the number of deaths within a population during a specified period of time

a. incidence rate
b. morbidity rate
c. mortality rate
d. prevalence rate
e. none of the above

3. What kind of data could be used to determine when an epidemic is beginning to draw to a close?

B Classification of Diseases

4. Identify the infectious disease pattern in each of the following maps as sporadic, endemic, epidemic, or pandemic.

Number of reported cases of tetanus, 1993

(a)

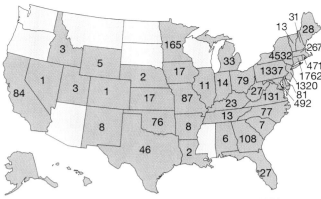

Number of reported cases of Lyme disease, 1994

(b)

5. How does a common-source outbreak differ from a propagated epidemic?

C **Epidemiologic Studies**

6. What is the main purpose of epidemiologic studies?

7. How do methods of epidemiologic studies differ?

D **Reservoirs of Infection**

8. Describe the properties of the various kinds of reservoirs of disease.

9. *Fill in the blanks:*

Pathogen transmission can occur by several modes. Non-living objects such as dishes, doorknobs, and handkerchiefs represent _____. Such objects are usually passed between individuals by _____ transmission. Dried mucus is an example of _____ because these particles can be inhaled directly. Another example of contact transmission is _____ transmission, in which bacteria can be passed from parent to offspring.

Disease-causing microbes can also be transmitted by _____ , which are forms of life that carry the infectious agent. Most such life forms are _____. If the pathogen completes part of its life cycle in the living object, transmission is said to be by _____. However, transmission is said to be by _____ if the pathogen is carried passively on the living object.

E **Portals of Entry and Exit; Modes of Transmission**

10. List the portals of entry and the portals of exit for pathogens.

11. Define and give examples of direct and indirect transmission of pathogens.

12. How do water, air, and food transmit disease?

13. How do vectors and fomites differ?

F **Disease Cycles; Herd Immunity**

14. How can changes in herd immunity contribute to an outbreak of a cyclic disease?

G **Methods of Disease Control**

15. Match the following:

_____ "Healthy" human or animal carriers are separated from the general population.

_____ A patient exhibiting a communicable disease is prevented from having contact with people in the general population.

_____ This mechanism prevents the spread of a disease during its incubation period.

_____ Malaria in the United States was controlled by means of this mechanism.

_____ A patient with HIV disease would be subject to this control mechanism.

a. immunization
b. isolation
c. quarantine
d. vector control
e. none of the above

H **Public Health Organizations; Reporting of Notifiable Diseases**

16. What are the major activities of public health organizations at various governmental levels?

17. Why is reporting of diseases important, and why are there different categories of notifiable diseases?

I **Nosocomial Infections**

18. What is a nosocomial infection?

19. What organisms most often cause nosocomial infections?

20. How does host susceptibility relate to nosocomial infections?

21. Give at least three examples of how medical procedures and sites of nosocomial infections are correlated.

J **Prevention and Control of Nosocomial Infections**

22. What methods are used to control nosocomial infections?

PROBLEMS FOR INVESTIGATION

1. Suppose that the health department of city A mounts a successful campaign to get children immunized against measles. Only 100 out of 10,000 children fail to receive the vaccine. Now suppose that city B, the same size as city A, has not carried out a successful measles vaccination program. Of the 10,000 children in city B, 5000 had measles when the disease last struck the population.

   a. If a child with measles moves to city A, what is the chance that child will encounter a susceptible child?

   b. If a child with measles moves to city B, what is the chance that child will encounter a susceptible child?

   c. Comparing the two scenarios, which city has the higher herd immunity, and in which city is an infected child more likely to transmit the disease to a susceptible child?

2. Why do medical authorities believe it is impossible to prevent all nosocomial infections?

3. Plastic refreezeable ice cubes, filled with water and produced in Hong Kong, were linked to cases of salmonellosis in the United States. What is the most likely explanation? What precautions should be taken to prevent similar cases?

4. What are Universal Precautions? Why are they important?

5. A 7-year-old girl developed pertussis (whooping cough); her 3- and 2-year-old siblings subsequently came down with the disease. Four other children in their apartment building were the next to develop pertussis. The local health department attempted to find the index case for this outbreak. Exhaustive study revealed no cases of pertussis among her schoolmates or any other known contacts. Her father had a deep cough and had taken a few days off from work. The mother was healthy, and an older brother had a runny nose.

   a. Who is a likely index case?

   b. Why did the local health department know about these cases?

   (The answer to this question appears in Appendix F.)

SOME INTERESTING READING

Ayliffe, G. A. J., B. J. Collins, and L. J. Taylor. 1990. *Hospital-acquired infections: Principles and prevention,* 2d ed. Boston: John Wright-PSG.

Barker, D. J. P., and G. Rose. 1990. *Epidemiology in medical practice,* 4th ed. New York: Churchill Livingston.

Benenson, A. S., ed. 1990. *Control of communicable diseases in man.* Washington, D.C.: American Public Health Association.

Brownlee, S. 1995. "The disease busters." *U.S. News & World Report* 118(12):48–58.

Bruce, N. G. 1991. "Epidemiology and the new public health: Implications for training." *Social Science & Medicine* 32(1):103–6.

"Diphtheria epidemic—new independent states of the former Soviet Union, 1990–1994." 1995. *Morbidity and Mortality Weekly Report* 44(10):177–81.

Eagan, J. 1991. "Measles: An infection control nightmare." *RN* 54(6):26–29.

Ellerbrock, T. V., T. J. Bush, M. E. Chamberland, and M. J. Oxtoby. 1991. "Epidemiology of women with AIDS in the United States, 1981 through 1990." *Journal of the American Medical Association* 265(22):2971–75.

Emori, T. G. 1993. "Overview of nosocomial infections, including the role of the microbiology laboratory." *Clinical Microbiological Reviews* 6(4):428–42.

Fleming, D. 1992. "OSHA issues rules for controlling disease exposure in the workplace." *ASM News* 58(3):127–29.

Harkness, G. A. 1995. *Epidemiology in nursing practice.* St. Louis: Mosby-Year Book, Inc.

Mann, J. 1991. "How AIDS has changed epidemiology." *New Scientist* 129(1755):16.

Mylotte, J. M., M. S. Niederman, and W. R. Summer. 1993. "Staying on top of hospital infections." *Patient Care* 27(3):116–45.

Pickering, L. K., and R. R. Reves. 1990. "Occupational risks for child-care providers and teachers." *Journal of the American Medical Association* 263(15):2096–97.

Pugliese, G., P. Lynch, and M. M. Jackson. 1991. *Universal precautions: Policies, procedures and resources.* Chicago: American Hospital Publications.

Ranger, T., and P. Slack, eds. 1992. *Epidemics and ideas: Essays on the historical perception of pestilence.* Cambridge: Cambridge University Press.

Roueché, B. 1991. *The medical detectives.* New York: Truman Talley Books/Plume.

Stolley, P. D., and T. Lasky. 1995. *Investigating disease patterns: The science of epidemiology.* New York: Scientific American Library.

Strobel, G., and S. Dickman. 1995. "Prime suspects lined up in MS mystery." *New Scientist* 146(1971):16.

Swartz, M. N. 1994. "Hospital-acquired infections: Diseases with increasingly limited therapies." *Proceedings of the National Academy of Sciences* 91(7):2420–27.

Valanis, B. 1992. *Epidemiology in nursing and health care.* Norwalk, CT: Appleton and Lange.

Wuethrich, B. 1995. "Playing chicken with an epidemic." *Science* 267(5204):1594.

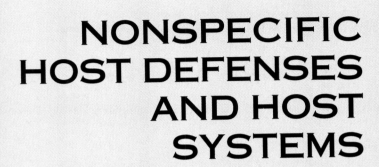

CHAPTER 17

NONSPECIFIC HOST DEFENSES AND HOST SYSTEMS

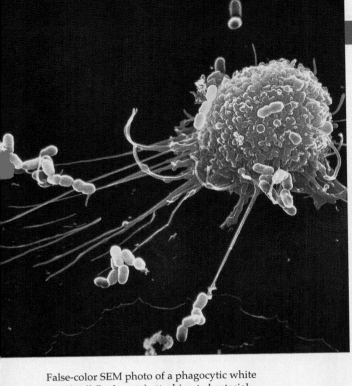

False-color SEM photo of a phagocytic white blood cell (leukocyte) attaching to bacterial cells (*Escherichia coli*). The bacteria will soon be ingested and destroyed.

W̲e can look at infectious disease as a battle between the power of infectious agents to invade and damage the body and the body's powers to resist such invasions. In Chapters 15 and 16 we considered how infectious agents enter and damage the body and how they leave the body and spread through populations. In the next three chapters we consider how the body resists invasion by infectious agents.

We begin this chapter by distinguishing between specific and nonspecific defenses. Then we look at the nonspecific defense mechanisms in more detail to see how they function in protecting the body against infectious agents. We end the chapter with a brief review of the structure of body systems. This survey emphasizes sites where infections often occur and the nature of the defenses available in each system.

NONSPECIFIC AND SPECIFIC HOST DEFENSES

With potential pathogens ever present, why do we rarely succumb to them in illness or death? The answer is that our bodies have defenses for resisting the attack of many dangerous organisms. Only when our resistance fails do we become susceptible to infection by pathogens.

Host defenses that produce resistance can be specific or nonspecific. **Specific defenses** respond to particular agents called *antigens*. Viruses and pathogenic bacteria are examples of such antigens. Specific defenses then respond to these antigens by producing protein *antibodies*. The human body is capable of making thousands of different antibodies, each effective against a particular antigen. Specific responses also involve the activation of the *lymphocytes,* specific cells of the body's immune system. These antibody and cellular responses are more effective against succeeding invasions by the same pathogen than against initial invasions. Chapter 18 focuses on these specific defenses of the immune system.

In the case of many threats to an individual's well being, specific defenses do not need to be called on because the body is adequately protected by its **nonspecific defenses**—those that act against any type of invading agent. Often such defenses perform their function before specific body defense mechanisms are activated. Nonspecific defenses include the following:

1. *Physical barriers,* such as the skin and mucous membranes and the chemicals they secrete.
2. *Chemical barriers,* antimicrobial substances in body fluids such as saliva, mucus, and gastric juices.

A How do nonspecific and specific host defenses differ?

B What are the physical barrier defenses?

C What kinds of cells form cellular barrier defenses?

D What are the stages of phagocytosis, and what kinds of cells are involved?

E What is inflammation, and what do the steps in acute inflammatory processes accomplish?

F What are the causes and the effects of chronic inflammation?

G How does fever function as a nonspecific defense mechanism?

H What roles do interferons play in molecular defense mechanisms?

I How do the molecular defense mechanisms involving complement work?

J What is the acute phase response?

K What are the important body structures and infection sites of the skin, the eyes and ears, and the respiratory, digestive, cardiovascular, lymphatic, nervous, and urogenital systems?

3. *Cellular defenses,* consisting of certain cells that engulf invading microorganisms.

4. *Inflammation,* the reddening, swelling, and temperature increases in tissues at sites of infection.

5. *Fever,* the elevation of body temperature to kill invading agents and/or inactivate their toxic products.

6. *Molecular defenses,* such as interferon and complement, that destroy or impede invading microbes.

The physical and certain chemical barriers operate to prevent pathogens from entering the body. The other nonspecific defenses (cellular defenses, inflammation, fever, and molecular defenses) act to destroy pathogens or inactivate their toxic products that have gained entry or to prevent the pathogens from damaging additional tissues. The nonspecific defenses serve as the body's *first lines of defense* against pathogens. The specific defenses represent the *second lines of defense.* Let's look at each of the nonspecific defenses now; we will discuss the specific defenses in Chapter 18.

PHYSICAL BARRIERS

The skin and mucous membranes protect your body and internal organs from injury and infectious agents.

These two physical barriers are made of cells that line the body surfaces and secrete chemicals, making the surfaces hard to penetrate and inhospitable to pathogens. The **skin,** for example, not only is exposed directly to microorganisms and toxic substances but also is subject to objects that touch, abrade, and tear it. Sunlight, heat, cold, and chemicals can damage the skin. Cuts, scratches, insect and animal bites, burns, and other wounds can disrupt the continuity of the skin and make it vulnerable to infection.

Besides the skin, a **mucous membrane,** or *mucosa,* covers those tissues and organs of the body cavity that are exposed to the exterior. Mucous membranes, therefore, are another physical barrier that makes it difficult for pathogens to invade internal body systems. Let's briefly look at these two physical barriers in more detail.

The Skin

The skin is the largest single organ of the body. The surface consists of a thin **epidermis** and a thicker, underlying layer, the **dermis** (Figure 17.1). The epidermis has several layers of dead *epithelial* (ep-i-the'le-al) cells that function as an excellent barrier against injury and infection of deeper layers of the body. These cells contain

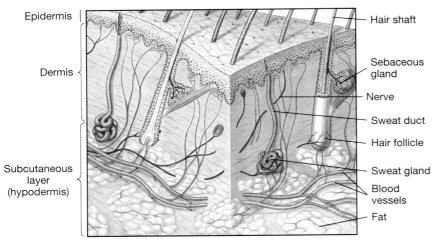

FIGURE 17.1 The skin, consisting of the epidermis and underlying dermis, forms a mechanical barrier to infection. Chemicals secreted onto the skin also retard pathogen attack and infection.

a waterproofing protein called **keratin,** which prevents water-soluble substances from entering the body. In addition, thick skin on the palms of the hands and soles of the feet reduces the likelihood that the skin barrier will be broken. *Calluses* represent thickened areas subject to constant wear. Thus, the intact skin surface prevents pathogenic organisms and other foreign substances from entering the body.

Not only is the skin a physical barrier to infection, its surface is inhabited by a variety of normal microbiota. As described in Chapter 15, the metabolic products secreted by these microbes make it difficult for pathogens to establish a foothold for growth. ∞ (p. 396) For example, the unsaturated fatty acids from *Staphylococcus epidermidis* and *Propionibacterium acnes* are especially toxic to gram-negative bacteria. Lactic acid from lactobacilli also contributes to maintaining an acidic skin pH of 3 to 5, which is inhibitory to many pathogens.

The skin generates antimicrobial substances that make it an even more inhospitable environment. Embedded in the dermis, the durable part of the skin, are sebaceous (se-ba'shus) glands and sweat glands. Most **sebaceous glands** produce an oily secretion called **sebum** (se'bum), which consists mainly of organic acids and other lipids. Although these secretions provide some nutrients for the normal microbiota, the acidic nature of the sebum helps maintain an acidic skin pH that discourages pathogen growth. The quantity of sebum secretions increases at puberty, contributing to the development of acne in some persons. **Sweat** (sudoriferous) **glands** are distributed over the body and empty a watery secretion through pores in the skin. In the armpits and groin, these glands secrete organic substances in the sweat that lower the skin

pH, again inhibiting the growth of most pathogens. The high salt concentration in sweat also inhibits many microorganisms. Sweat is odorless but contains nutrients useful to the growth of resident microbiota. Body odor results from some metabolic byproducts of normal resident bacteria.

Mucous Membranes

Mucous membranes line those tissues and organs that open to the exterior of the body, especially those of the respiratory, digestive, and urogenital systems. Like the skin, mucous membranes have a thin epidermal layer and a deeper connective tissue layer (Figure 17.2). The cells of the epithelial layer block the penetration of microbes and secrete **mucus,** a thick but watery secretion of glycoproteins and electrolytes. Mucus, produced by *goblet cells* and certain glands, forms a protective layer over the epidermal cells, preventing drying and cracking of the mucous membrane. Mucus also traps pathogens before they can establish an infection.

CELLULAR DEFENSES

Although the physical defense barriers do an excellent job of keeping microbes out of our bodies, we constantly suffer minor breaches of the physical defense barriers. A paper cut, the cracking of dry skin, or even brushing our teeth may temporarily breach the physical defenses

FIGURE 17.2 Mucous membranes, much like the skin, consist of epithelial cells that line tissues. The secretion of mucus by goblet cells (special epithelial cells) and glands traps and clears away many potential pathogens and other foreign material.

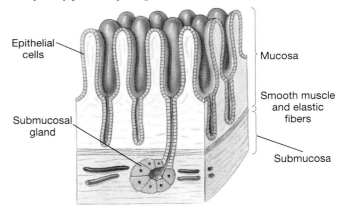

and allow some microbes to enter the blood or connective tissue. However, we survive these daily attacks because ever-present cellular defenses can kill invading microbes or remove them from the blood or tissues.

When the skin is broken by any kind of trauma, microorganisms from the environment may enter the wound. Blood flowing out of the wound helps remove the microorganisms. Subsequent constriction of ruptured blood vessels and the clotting of blood help seal off the injured area until more permanent repair can occur. Still, if microorganisms enter blood through cuts in the skin or abrasions in mucous membranes, cellular defense mechanisms come into play.

Defensive Cells

Cellular defense mechanisms use special-purpose cells found in the blood and other tissues of the body. Blood consists of about 60 percent liquid called **plasma** and 40 percent **formed elements** (cells and cell fragments). Formed elements include **erythrocytes** (red blood cells), **platelets,** and **leukocytes** (white blood cells) (Figure 17.3 and Table 17.1). All are derived from *pluripotent stem cells,* cells that form a continuous supply of blood cells, in the bone marrow. Platelets, which are short-lived fragments of large cells called *megakaryocytes,* are important components of the blood-clotting mechanism.

FIGURE 17.3 Formed (cellular) elements of the blood are derived from pluripotent stem cells (cells that form an endless supply of blood cells) in the bone marrow. The myeloid stem cells differentiate into several kinds of leukocytes, called granulocytes and agranulocytes. Lymphoid stem cells differentiate into B lymphocytes (B cells), T lymphocytes (T cells), and natural killer cells (NK cells).

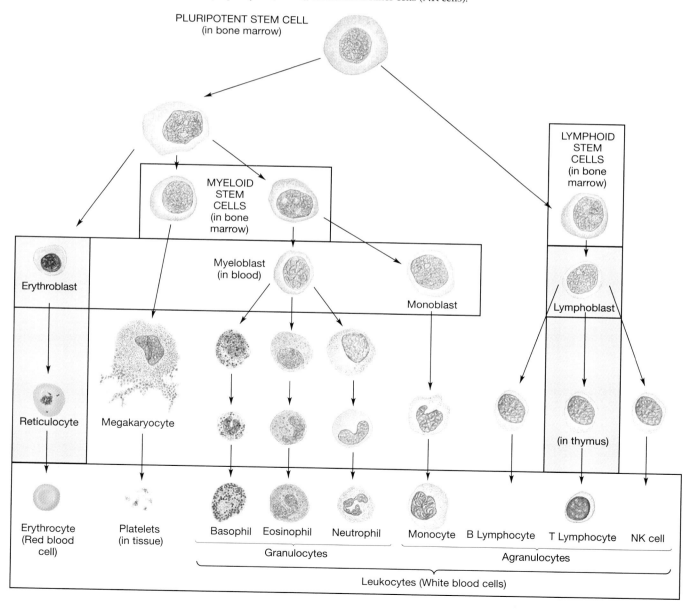

TABLE 17.1   **Formed Elements of the Blood in Healthy Adults**

Element	Normal Numbers (per Microliter*)	Life Span	Functions
Erythrocytes		120 days	Transport oxygen gas from lungs to tissues; transport carbon dioxide gas from tissues to lungs
Adult male	4.6 to 6.2 million		
Adult female	4.2 to 5.4 million		
Newborn	5.0 to 5.1 million		
Leukocytes	5000 to 9000	Hours to days	
Granulocytes			
Neutrophils	50–70%		Phagocytic; contain oxidative chemicals to kill internalized microbes
Eosinophils	1–5%		Release defensive chemicals to damage parasites (worms); phagocytic
Basophils	0.1%		Release histamine and other chemicals during inflammation; responsible for allergic symptoms
Agranulocytes			
Monocytes	2–8%		In tissues, develop into macrophages, which are phagocytic
Lymphocytes	20–40%		Essential to specific host immune defenses; antibody production
Platelets	250,000 to 300,000	5–9 days	Blood clotting

*1 microliter (μl) = 1 mm^3 = 1/1,000,000 l.

Leukocytes are defensive cells that are important to both specific and nonspecific host defenses. These cells are divided into two groups—granulocytes and agranulocytes—according to their cell characteristics and staining patterns with specific dyes.

GRANULOCYTES   **Granulocytes** have granular cytoplasm and an irregularly shaped, lobed nucleus. They are derived from *myeloid stem cells* in the bone marrow (*myelos* is Greek for "marrow"). Granulocytes include basophils, mast cells eosinophils, and neutrophils, which are distinguished from one another by the shape of their cell nuclei and by their staining reactions with specific dyes. **Basophils** release *histamine,* a chemical that helps initiate the inflammatory response. **Mast cells,** which are prevalent in connective tissue and alongside blood vessels, also release histamine and are associated with allergies. **Eosinophils** (e-o-sin′o-fils) are present in large numbers during allergic reactions (Chapter 19) and worm infections. These cells may also detoxify foreign substances and help turn off inflammatory reactions. **Neutrophils,** also called *polymorphonuclear leukocytes* (PMNLs), guard the skin and mucous membranes against infection. These cells respond quickly wherever tissue injury has occurred.

AGRANULOCYTES   **Agranulocytes** lack granular cytoplasm and have round nuclei. These cells include monocytes and lymphocytes. **Monocytes** are derived

from myeloid stem cells, whereas **lymphocytes** are derived from *lymphoid stem cells,* again in the bone marrow. The lymphocytes contribute to specific host immunity. They circulate in the blood and are found in large numbers in the lymph nodes, spleen, thymus, and tonsils.

Neutrophils and monocytes are exceedingly important components of nonspecific host defenses. They are phagocytic cells, or *phagocytes.*

Phagocytes

Phagocytes are cells that literally eat (*phago,* Greek for "eating"; *cyte,* Greek for "cell") or engulf other materials. They patrol, or circulate through the body, destroying dead cells and cellular debris that must be removed constantly from the body as cells die and are replaced. Phagocytes also guard the skin and mucous membranes against invasion by microorganisms. Being present in many tissues, these cells first attack microbes and other foreign material at portals of entry, such as wounds in skin or mucous membranes. If some microbes escape destruction at the portal of entry and enter deeper tissues, phagocytes circulating in blood or lymph mount a second attack on them.

The neutrophils are released from the bone marrow continuously to maintain a stable circulating population. An adult has about 50 billion circulating neutrophils at all times. If an infection occurs, they are

TABLE 17.2	Names of Fixed Macrophages in Various Tissues	
Name of Macrophage	**Tissue**
Alveolar macrophage (dust cell)	Lung
Histiocyte	Connective tissue
Kupffer cell	Liver
Microglial cell	Neural tissue
Osteoclast	Bone
Sinusoidal lining cell	Spleen

usually first on the scene because they migrate quickly to the site of infection. Being avid phagocytes, they are best at inactivating bacteria and other small particles. They are not capable of cell division and are "programmed" to die after only 1 or 2 days.

The monocytes migrate from the bone marrow into the blood. When these cells move from blood into tissues, they go through a series of cellular changes, maturing into macrophages. **Macrophages** are "big eaters" (*macro,* Greek for "big") that destroy not only microorganisms but also larger particles, such as debris left from neutrophils that have died after ingesting bacteria. Although macrophages take longer than neutrophils to reach an infection site, they arrive in larger numbers than neutrophils.

Macrophages can be fixed or wandering. *Fixed macrophages* remain stationary in tissues and are given different names, depending on the tissue in which they reside (Table 17.2). *Wandering macrophages,* like the neutrophils, circulate in the blood, moving into tissues when microbes and other foreign material are present (Figure 17.4). Unlike neutrophils, macrophages can live for months or years. As we will see in Chapter 18, besides having a nonspecific role in host defenses, macrophages also are critical to specific host defenses.

FIGURE 17.4 False-color SEM photo of a macrophage moving over a surface (2150×). The macrophage has spread out from its normal spherical shape and is using its ruffly cytoplasm to move itself and to engulf particles. Macrophages clear the lungs of dust, pollen, bacteria, and some components of tobacco smoke.

The Process of Phagocytosis

Phagocytes digest and generally destroy invading microbes and foreign particles by a process called **phagocytosis** (Chapter 4) or by a combination of immune reactions and phagocytosis ∞ (p. 101) If an infection occurs, neutrophils and macrophages use this four-step process to destroy the invading microorganisms. The phagocytic cells must (1) find, (2) adhere to, (3) ingest, and (4) digest the microorganisms.

Chemotaxis

Phagocytes first must find the invading microorganisms. Both the infectious agents and the damaged tissues release specific chemical substances to which phagocytes are attracted. In addition, basophils and mast cells release histamine, and phagocytes already at the infection site release chemicals called **cytokines** (si'to-kinz). These chemicals are a diverse group of soluble proteins that have specific roles in host defenses, including the attraction of additional phagocytes to the site of the infection. Phagocytes make their way to this site by **chemotaxis,** the movement of cells toward a chemical stimulus. ∞ (Chapter 4, p. 87) We will discuss cytokines in more depth in Chapter 18.

Some pathogens can escape phagocytes by interfering with chemotaxis. For example, most strains of the bacterium that causes gonorrhea (*Neisseria gonorrhoeae*) remain in the urogenital tract, but some strains escape local cellular defenses and enter the blood. Microbiologists believe the invasive strains fail to release the chemical attractants that bring phagocytes to the infection site.

Adherence and Ingestion

Following chemotaxis and the arrival of phagocytes at the infection site, the infectious agents become attached to the plasma membranes of phagocytic cells (see the chapter-opening photograph). The ability of the phagocyte cell membrane to bind to specific molecules on the surface of the microbe is called **adherence.**

A fundamental requirement for many pathogenic bacteria is to escape phagocytosis. The most common means by which bacteria avoid this defense mechanism is an *antiphagocytic capsule.* The capsules present on bacteria responsible for pneumococcal pneumonia (*Streptococcus pneumoniae*) and childhood meningitis (*Haemophilus influenzae*) make adherence difficult for phagocytes. The cell walls of the bacterium responsible for rheumatic fever (*Streptococcus pyogenes*) contain molecules of *M protein,* which interferes with adherence.

To overcome such resistance to adherence, the host's nonspecific defenses can make microbes more susceptible to phagocytosis. If microbes are first coated with antibodies, or with proteins of the *complement system* (to be discussed later in this chapter), phagocytes have a much easier time binding to the microbes. Because both these mechanisms represent molecular defenses, we will discuss them later in this chapter.

Once captured, phagocytes rapidly ingest (engulf) the microbe. The cell membrane of the phagocyte forms fingerlike extensions, called *pseudopodia,* that surround the microbe (Figure 17.5a). These pseudopodia then fuse, enclosing the microbe within a cytoplasmic vacuole called a **phagosome** (Figure 17.5b).

Digestion

Phagocytic cells have several mechanisms for digesting and destroying ingested microbes. One mechanism uses the *lysosomes* found in the phagocyte's cytoplasm. ∞ (Chapter 4, p. 101) These organelles, which contain digestive enzymes and small proteins called *defensins,* fuse with the phagosome membrane, forming a **phagolysosome** (Figure 17.5b). In this way the digestive enzymes and defensins are released into the phagolysosome. The defensins eat holes in the microbes, allowing lysosomal enzymes to digest almost any biological molecule they contact. Thus, lysosomal enzymes rapidly (within 20 minutes) destroy the mi-

crobes, breaking them into small molecules (amino acids, sugars, fatty acids) that the phagocyte can use as building blocks for its own metabolic and energy needs.

Macrophages can also use other metabolic products to kill ingested microbes. These phagocytic cells use oxygen to form hydrogen peroxide (H_2O_2), nitric oxide (NO), superoxide ions (O_2^-), and hypochlorite ions (OCl^-). (Hypochlorite is the ingredient in household bleach that accounts for its antimicrobial action.) All these molecules are effective in damaging plasma membranes of the ingested pathogens.

Once the microbes have been destroyed, there may be some indigestible material left over. Such material remains in the phagolysosome, which now is called a *residual body.* The phagocyte transports the residual body to the plasma membrane, where the waste is excreted (Figure 17.5b).

Just as some microbes interfere with chemotaxis and others avoid adherence, some microbes have developed mechanisms to prevent their destruction within a phagolysosome. In fact, a few pathogens even multiply within phagocytes. Some microbes resist digestion by phagocytes in one of three ways:

FIGURE 17.5 (a) Phagocytosis of two bacterial cells by a neutrophil. Extensions of cytoplasm, called pseudopodia, surround the bacteria. Fusion of the pseudopodia forms a cytoplasmic vacuole, called a phagosome, containing the bacteria. (b) Phagocytes find their way to a site of infection by means of chemotaxis. Phagocytes, including macrophages and neutrophils, have proteins in their plasma membranes to which a bacterium adheres. The bacterium is then ingested into the cytoplasm of the phagocyte as a phagosome, which fuses with lysosomes to form a phagolysosome. The bacterium is digested, and any undigested material within the residual body is excreted from the cell.

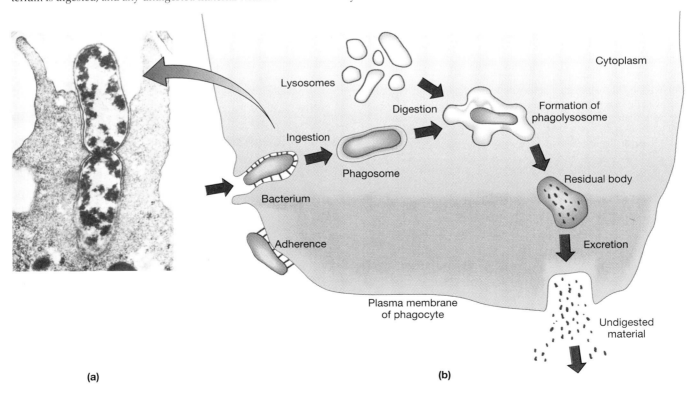

(a)

(b)

1. Some bacteria, such as those that cause the plague (*Yersinia pestis*), produce capsules that are not vulnerable to destruction by macrophages. If these bacteria are engulfed by macrophages, their capsule protects them from lysosomal digestion, allowing the bacteria to multiply, even within a macrophage.

2. Other bacteria—such as those that cause Hansen's disease, or leprosy (*Mycobacterium leprae*), and tuberculosis (*M. tuberculosis*)—and the protozoan that causes leishmaniasis (*Leishmania* species) can resist digestion by phagocytes. In the case of *Mycobacterium,* each engulfed bacillus resides in a membrane-bound, fluid-filled compartment called a *parasitophorous vacuole* (PV). No lysosomal enzyme activity is associated with the PVs. As the bacilli reproduce, new PVs arise. For *Leishmania* infections, each PV contains several protozoan cells. Although the lysosomal enzymes are active in these PVs, microbiologists do not understand how the pathogens resist digestion.

3. Still other microbes produce toxins that kill phagocytes by causing the release of the phagocyte's own lysosomal enzymes into its cytoplasm. Examples of such toxins are **leukocidin,** released by staphylococci, and **streptolysin,** released by streptococci.

Thus, some pathogens survive phagocytosis and can even be spread throughout the body in the phagocytes that attempt to destroy them. Because macrophages can live for months, they can provide pathogens with a long-term, stable environment in which they can multiply out of the reach of other host defense mechanisms.

Extracellular Killing

The phagocytic process described previously represents *intracellular killing*—that is, the microbe is degraded within a defensive cell. However, other microbes, such as viruses and parasitic worms, are destroyed without being ingested by a defensive cell; they are destroyed *extracellularly.*

Neutrophils and macrophages are too small to engulf a large parasite such as a worm (helminth). Therefore, another leukocyte, the eosinophil, takes the leading role in defending the body. Although eosinophils can be phagocytic, they are best suited for excreting toxic enzymes such as *major basic protein* (MBP) that can damage or perforate a worm's body. Once such parasites are destroyed, macrophages can engulf the parasite fragments.

Viruses must get inside cells to multiply. ∞ (Chapter 11, p. 270) Therefore, host defenses must eliminate such infectious agents before they can reproduce in the cells they have infected. The leukocytes responsible for killing intracellular viruses are **natural killer** (NK) **cells.** NK cells are a type of lymphocyte whose activity is greatly increased by exposure to interferons and cytokines. Although the exact mechanism of recognition is not known, NK cells probably recognize specific glycoproteins on the cell surface of virus-infected cells.

Such recognition does not lead to phagocytosis; rather, the NK cells secrete cytotoxic proteins that trigger the death of the infected cell.

INFLAMMATION

Do you remember the last time you cut yourself? If the cut was not too serious, the bleeding soon stopped. You washed the cut and put on a bandage. A few hours later the area around the cut became warm, red, swollen, and perhaps even painful. It had become *inflamed.*

Characteristics of Inflammation

Inflammation is the body's defensive response to tissue damage from microbial infection. It is also a response to mechanical injury (cuts and abrasions), heat and electricity (burns), ultraviolet light (sunburn), chemicals (phenols, acids, and alkalis), and allergies. But whatever the cause of inflammation, it is characterized by *cardinal signs* or symptoms: (1) an increase in temperature, (2) redness, (3) swelling, and (4) pain at the infected or injured site. What happens in the inflammatory process, and why?

The Acute Inflammatory Process

The duration of inflammation can be either acute (short-term) or chronic (long-term). In **acute inflammation,** the battle between microbes (or other agents of inflammation) and host defenses usually is won by the host. In an infection, acute inflammation functions to (1) kill invading microbes, (2) clear away tissue debris, and (3) repair injured tissue. Let's look at acute inflammation more closely. Figure 17.6 illustrates the steps described next.

When cells are damaged, the chemical substance histamine is released from basophils and mast cells. **Histamine** diffuses into nearby capillaries and venules, causing the walls of these vessels to dilate (**vasodilation**) and become more permeable. Dilation increases the amount of blood flowing to the damaged area, and around skin wounds it causes the skin to become red and warm to the touch. Because the vessel walls are more permeable, fluids leave the blood and accumulate around the injured cells, causing **edema** (swelling). The blood delivers clotting factors, nutrients, and other substances to the injured area and removes wastes and some excess fluids. It also brings macrophages, which release cytokines. Some cytokines attract other phagocytes, and another cytokine called *tumor necrosis factor alpha* (TNF-α) additionally causes vasodilation and edema.

All kinds of tissue injury—burns, cuts, infections, insect bites, allergies—cause histamine release. In con-

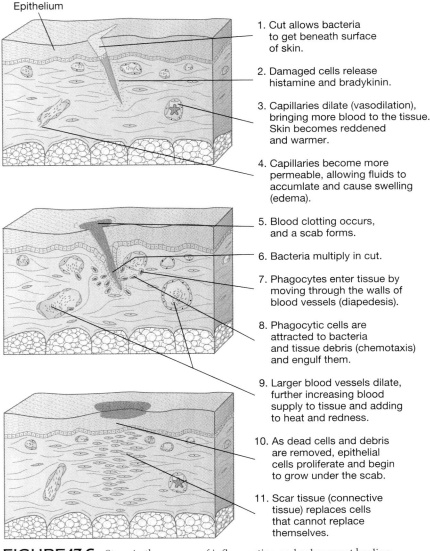

Epithelium

1. Cut allows bacteria to get beneath surface of skin.

2. Damaged cells release histamine and bradykinin.

3. Capillaries dilate (vasodilation), bringing more blood to the tissue. Skin becomes reddened and warmer.

4. Capillaries become more permeable, allowing fluids to accumlate and cause swelling (edema).

5. Blood clotting occurs, and a scab forms.

6. Bacteria multiply in cut.

7. Phagocytes enter tissue by moving through the walls of blood vessels (diapedesis).

8. Phagocytic cells are attracted to bacteria and tissue debris (chemotaxis) and engulf them.

9. Larger blood vessels dilate, further increasing blood supply to tissue and adding to heat and redness.

10. As dead cells and debris are removed, epithelial cells proliferate and begin to grow under the scab.

11. Scar tissue (connective tissue) replaces cells that cannot replace themselves.

FIGURE 17.6 Steps in the process of inflammation and subsequent healing.

junction with its effects on blood vessels, histamine also causes the red, watery eyes and runny nose of hay fever and the breathing difficulties in certain allergies. The drugs called **antihistamines** alleviate such symptoms by blocking the released histamine from reaching its receptors on target organs.

The fluid that enters the injured tissue carries the chemical components of the blood-clotting mechanism. If the injury has caused bleeding, platelets and clotting factors, such as fibrin, stop the bleeding by forming a blood clot in the injured blood vessel. Because clotting takes place near the injury, it greatly reduces fluid movement around damaged cells and walls off the injured area from the rest of the body. Pain associated with tissue injury is thought to be due to the production of **bradykinin,** a small peptide, at the injured site. How bradykinin stimulates pain receptors in the skin is un-

known, but cellular regulators called **prostaglandins** seem to intensify bradykinin's effect.

Inflamed tissues also stimulate **leukocytosis,** an increase in the number of leukocytes in the blood. To do this, the damaged cells release cytokines that trigger the production and infiltration of more leukocytes. Within an hour after the inflammatory process begins, phagocytes start to arrive at the injured or infected site. For example, neutrophils pass out of the blood by squeezing between endothelial cells lining the vessel walls. This process, called **diapedesis** (di-a-pe-de′sis), allows neutrophils to congregate in tissue fluids at the injured region.

As we discussed earlier, when phagocytes reach an infected area, they attempt to engulf the invading microbes by phagocytosis. In that process many of the phagocytes themselves die. The accumulation of dead phagocytes, injured or damaged cells, the remains of ingested organisms, and other tissue debris forms the white or yellow fluid called **pus.** Many bacteria, such as *Streptococcus pyogenes,* cause pus formation because of their ability to produce leukocidins that destroy phagocytes. Viruses lack this activity and do not cause pus formation. Pus continues to form until the infection or tissue damage has been brought under control. An accumulation of pus in a cavity hollowed out by tissue damage is called an **abscess.** Boils and pimples are common kinds of abscesses.

Although the inflammatory process is usually beneficial, it can sometimes be harmful. For example, inflammation can cause swelling of the membranes (meninges) surrounding the brain or spinal cord, leading to brain damage. Swelling, which delivers phagocytes to injured tissue, can also interfere with breathing if it constricts the airways in the lung. Moreover, vasodilation delivers more oxygen and nutrients to injured tissues. Ordinarily this is of greater benefit to host cells than to pathogens, but sometimes it helps the pathogens thrive as well. Even though rapid clotting and the walling off of an injured area prevents pathogens from spreading, it can also prevent natural defenses and antibiotics from reaching the pathogens. Boils must be lanced before therapeutic drugs can reach them. Attempting to suppress the inflammatory process also can be harmful. Such attempts can allow boils to form when natural defenses might otherwise destroy the bacteria.

In summary, cellular defense mechanisms usually prevent an infection from spreading or from getting worse. However, sometimes these nonspecific defense mechanisms are overwhelmed by sheer numbers of microbes or are inhibited by virulence factors that the microbes possess. The pathogens can then invade other parts of the body. For bacterial infections, medical intervention with antibiotics may inhibit microbial growth in injured tissue and reduce the chance of an infection's spreading. Despite such measures, however, infections do spread. In Chapter 18 we will describe the mechanisms by which various lymphocytes act as agents of specific host immune defenses that help overcome an initial infection and prevent future infections by the same microbe.

Repair and Regeneration

During the entire inflammatory reaction, the healing process is also underway. Once the inflammatory reaction has subsided and most of the debris has been cleared away, healing accelerates. Capillaries grow into the blood clot, and **fibroblasts,** connective tissue cells, replace the destroyed tissue as the clot dissolves. The fragile, reddish, grainy tissue seen at the cut site consists of capillaries and fibroblasts called **granulation tissue.** As granulation tissue accumulates fibroblasts and fibers, it replaces nerve and muscle tissues that cannot be regenerated. New epidermis replaces the part destroyed. In the digestive tract and other organs lined with epithelium, an injured lining can similarly be replaced. Although scar tissue is not as elastic as the original tissue, it does provide a strong durable "patch" that allows the remaining normal tissue to function.

Several factors affect the healing process. The tissues of young people heal more rapidly than those of older people. The reason is that the cells of the young divide more quickly, their bodies are generally in a better nutritional state, and their blood circulation is more efficient. As you might guess from the many contributions of blood to healing, good circulation is extremely important. Certain vitamins also are important in the healing process. Vitamin A is essential for the division of epithelial cells, and vitamin C is essential for the production of collagen and other components of connective tissue. Vitamin K is required for blood clotting, and vitamin E also may promote healing and reduce the amount of scar tissue formed.

Chronic Inflammation

Sometimes an acute inflammation becomes a **chronic inflammation,** in which neither the agent of inflammation nor the host is a decisive winner of the battle.

Rather, the agent causing the inflammation continues to produce tissue damage as the phagocytic cells and other host defenses attempt to destroy or at least confine the region of inflammation. In the process, pus may be formed continuously. Such chronic inflammation can persist for years.

Because the cause of inflammation is not destroyed, host defenses attempt to limit or confine the agent so that it cannot spread to surrounding tissue. For example, **granulomatous inflammation** results in granulomas. A **granuloma** is a pocket of tissue that surrounds and walls off the inflammatory agent. The central region of a granuloma contains epithelial cells and macrophages; the latter may fuse to form giant, multinucleate cells. Collagen fibers, which help wall off the inflammatory agent, and lymphocytes surround the core. Granulomas associated with specific disease are sometimes given special names—for example, *gummas* (syphilis), *lepromas* (Hansen's disease), and *tubercles* (tuberculosis).

Tubercles usually contain necrotic (dead) tissue in the central region of the granuloma. As long as necrotic tissue is present, the inflammatory response will persist. If only a small quantity of necrotic tissue is present, the lesions sometimes become hardened as calcium is deposited in them. Calcified lesions are common in tuberculosis patients. When an anti-inflammatory drug such as cortisone is given, the organisms isolated in tubercles may be liberated, and signs and symptoms of tuberculosis reappear (secondary tuberculosis).

FEVER

A rise in temperature in infected or injured tissue is one sign of a local inflammatory reaction. **Fever,** a systemic increase in body temperature, often accompanies inflammation. Fever was first studied in 1868, when the German physician Carl Wunderlich devised a method to measure body temperature. He placed a foot-long thermometer in the armpit of his patients and left it in place for 30 minutes! Using this cumbersome technique, he could record human body temperatures during *febrile* (feverish) illnesses.

Normal body temperature is about 37°C (98.6°F), although individual variations in normal temperature within the range 36.1° to 37.5°C are not uncommon. Fever is defined clinically as an oral temperature above 37.8°C (100.5°F) or rectal temperature of 38.4°C (101.5°F). Fever accompanying infectious diseases rarely exceeds 40°C (104.5°F); if it reaches 43°C (109.4°F), death usually results.

Body temperature is maintained within a narrow range by a temperature-regulating center in the *hypothalamus,* a part of the brain. Fever occurs when the temperature established for this mechanism is reset and raised to a higher temperature. Fever can be caused by many pathogens, by certain immunological processes (such as reactions to vaccines), and by nearly any kind of tissue injury, even heart attacks. Most often, fever is caused by a substance called a **pyrogen** (*pyro*, Greek for "fire"). ∞ (Chapter 15, p. 410) **Exogenous pyrogens** include exotoxins and endotoxins from infectious agents. These toxins cause fever by stimulating the release of an **endogenous pyrogen** from macrophages. The endogenous pyrogen is yet another cytokine, called *interleukin-1* (IL-1), that circulates via the blood to the hypothalamus, where it causes certain neurons to secrete prostaglandins. The prostaglandins then reset the hypothalamus thermostat at a higher temperature, which then causes the body temperature to begin rising within 20 minutes. In such situations, body temperature is still regulated, but the body's "thermostat" is reset at a higher temperature. (The sensation of chills that sometimes accompanies a fever was described in Chapter 15.) ∞ (p. 409)

Fever has several beneficial roles. (1) Fever raises the body temperature above the optimum temperature for growth of many pathogens. This slows their rate of growth, reducing the number of microorganisms to be combated. (2) Fever can heighten the level of immune responses by increasing the rate of chemical reactions in the body. This results in a faster rate at which the body's defense mechanisms attack pathogens, shortening the course of the infection. (3) Fever makes a patient feel ill. In this condition the patient is more likely to rest, preventing further damage to the body and allowing energy to be used to fight the infection.

In an infection, cells also release **leukocyte-endogenous mediator** (LEM). Besides helping to elevate body temperature, LEM decreases the amount of iron absorbed from the digestive tract and increases the rate at which it is moved to iron storage deposits. Thus, LEM lowers the plasma iron concentration. Without adequate iron, growth of microorganisms is slowed. ∞ (Chapter 6, p. 149)

Our current knowledge of the importance of fever has changed the clinical approach to this symptom. In the past, *antipyretics*—fever-reducing drugs such as aspirin—were given almost routinely to reduce fever caused by infections. For the beneficial effects cited above, many physicians now recommend allowing fevers to run their course. Evidence shows that medication can delay recovery. However, if a fever goes above 40°C or if the patient has a disorder that might be worsened by fever, antipyretics are still used. In fact, untreated extreme fever increases the metabolic rate by 20 percent, makes the heart work harder, increases water loss, alters electrolyte concentrations, and can cause convulsions, especially in children. Thus, patients with severe heart disease or fluid and electrolyte imbalances, as well as children subject to convulsions, usually receive antipyretics.

MOLECULAR DEFENSES

Along with cellular defenses, inflammation, and fever, molecular defenses represent another formidable nonspecific defense barrier. These molecular defenses involve the actions of *interferon* and *complement*.

Interferon

As early as the 1930s, scientists observed that infection by one virus prevented for a time infection by another virus. Then, in 1957, a small, soluble protein was discovered that was responsible for this viral interference. This protein, called **interferon** (in-ter-fer'on), "interfered" with virion replication in other cells. Such a molecule suggested to virologists that they might have the "magic bullet" for viral infections, similar to the antibiotics used to treat bacterial infections. As we will see, such hope has dwindled somewhat.

Efforts to purify interferon led to the discovery that many different subtypes of interferon exist in different animal species, and that those produced by one species may be ineffective in other species. For example, interferon produced in a chicken is useful in protecting other chicken cells from viral infection. But chicken interferon is of no use in preventing viral infections in mice or in humans. Different interferons also exist in different tissues of the same animal. In humans there are three groups of interferons, called alpha (α), beta (β), and gamma (γ) (Table 17.3). Analysis of the protein structure and function show α-*interferon* and β-*interferon* to be similar, so they are placed together as *type I interferons*. *Gamma-interferon* is different structurally and functionally and represents the only known *type II interferon*.

Many researchers have tried to determine how these interferons act. The synthesis of α-interferon and β-interferon occurs after a virus infects a cell (Figure 17.7). These interferons do not interfere directly with viral replication. Rather, after viral infection the cell synthesizes and secretes minute amounts of interferon. The interferon then diffuses to adjacent, uninfected cells and binds to their surfaces. Binding stimulates those cells to transcribe specific genes into mRNA molecules, which are then translated to produce many new proteins, most of them enzymes. Together these enzymes are called **antiviral proteins** (AVPs). Although viruses still infect cells possessing the AVPs, many of the proteins interfere with virus replication.

The AVPs are especially effective against RNA viruses. Recall from Chapter 11 that all RNA viruses must either produce dsRNA (Reoviridae) or go through a dsRNA stage during replication of (−) sense or (+) sense RNA. ∞ (p. 275) Two of the AVPs digest mRNA and limit translation of viral mRNA. The result is that the AVPs prevent the formation of new viral nucleic acid and capsid proteins. The infected cell that initially produced the interferon is thus surrounded by

Class	Cell Source	Subtypes	Stimulated By	Effects
Type I				
Alpha-interferon (INF-α)	Leukocytes	20	Viruses	Production of antiviral proteins in neighboring cells
Beta-interferon (INF-β)	Fibroblasts	1	Viruses	Same as INF-α
Type II				
Gamma-interferon (INF-γ)	T lymphocytes and NK cells	1	Viruses and other antigens	Activates tumor destruction and killing of infected cells

cells that can resist the replication of viruses, limiting viral spread.

Gamma-interferon also can block virus replication by AVP synthesis. However, lymphocytes and NK cells do not have to be infected with a virus to synthesize γ-interferon. Rather, it is produced in uninfected lymphocytes and NK cells that are sensitive to specific foreign antigens (viruses, bacteria, tumor cells) present in the body. The exact role of γ-interferon is unclear, but it is known to enhance the activities of lymphocytes, NK cells, and macrophages—the cells needed to attack microbes and tumors. Gamma-interferon (along with tumor necrosis factor-α, or TNF-α) also helps infected macrophages rid themselves of pathogens. For example, we mentioned earlier that macrophages can become infected with *Mycobacterium* bacilli. Such infected macrophages can be activated by γ-interferon and TNF-α, which bind to infected macrophages. New bactericidal activity is thereby triggered within the macrophage, usually leading to death of the bacteria and the restoration of normal macrophage function.

Therapeutic Uses of Interferon

Besides having the ability to block virus replication, interferons can also stimulate specific immune defenses. Therefore, interferons provide a potential therapy for viral infections and tumors. Unfortunately, infected animal cells produce very small quantities of interferons. However, today *recombinant interferon* (rINF) can be produced more cheaply and abundantly by using recombinant DNA techniques. ∞ (Chapter 8, p. 215) Manufacture of recombinant interferon starts with the isolation and copying of the interferon gene and its insertion into plasmids. When recombinant plasmids are mixed with appropriate bacterial or yeast cells, some cells will take up a the gene-containing plasmid and thereby acquire the human interferon gene. By growing these bacterial or yeast cells in very large vats and extracting the interferon that they produce, pharmaceutical companies can produce relatively significant quantities of recombinant interferon.

The ability to produce recombinant interferons spurred research on therapeutic applications for these proteins. In 1986 α-interferon was approved by the FDA for treating hairy cell leukemia, a very rare blood can-

FIGURE 17.7   The mechanism by which interferons α and β act.

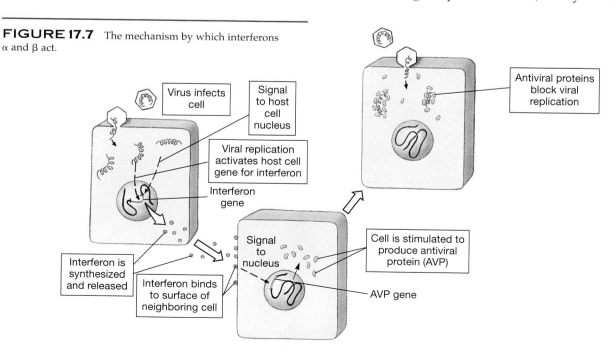

Virus infects cell

Signal to host cell nucleus

Viral replication activates host cell gene for interferon

Interferon gene

Interferon is synthesized and released

Signal to nucleus

Interferon binds to surface of neighboring cell

Cell is stimulated to produce antiviral protein (AVP)

AVP gene

Antiviral proteins block viral replication

BIOTECHNOLOGY

THE NEWEST INTERFERON–TAU

In many mammals the corpus luteum in the ovaries must secrete progesterone to prepare the uterus for embryo implantation. To prevent degradation of the corpus luteum, many embryos secrete the protein *trophoblastin*. In 1993 the amino acid sequence of trophoblastin was determined. Scientists soon realized that the sequence was very similar to that of α-interferon. The sequences were so close that trophoblastin is now considered a type I interferon, called τ (tau)-*interferon*. Interestingly, τ-interferon is as effective as α-interferon in inhibiting viral reproduction. Although humans use a protein other than trophoblastin to maintain the corpus luteum during early pregnancy, we all contain τ-interferon. Scientists have yet to discover which other functions, besides inhibition of viral reproduction, human τ-interferon may have in the body.

Tau-interferon is similar to α-interferon in that both bind to the same sites on adjacent cells to stimulate antiviral protein activity. However, the τ-protein can also bind to a site on adjacent cells that is not recognized by alpha. Therapeutically, τ-interferon has great potential. For example, in cultures of animal cells, τ-interferon inhibits reverse transcriptase, the enzyme needed by HIV to copy itself from an RNA to a DNA form after cell infection. Thus, τ-interferon may be of value in the treatment or prevention of AIDS. This interferon also inhibits the division of cultured tumor cells and so represents a potential new drug in cancer chemotherapy. Perhaps most importantly, τ-interferon does not have the side effects associated with α-interferon. Over the next few years, it should be very interesting to see whether this type I interferon retains these same properties in clinical treatment trials.

cer. Since then, interferons have been approved for treatment of several other viral diseases, including genital warts and cancer. However, in most cases interferon is a treatment, not a cure. Patients must remain on the drug throughout their lives. With hairy cell leukemia, for example, removal of the drug results in a recurrence of the disease in 90 percent of the patients. For hepatitis C virus infection, treatment must be given three times a week for 6 months. Even so, if the patient is taken off treatment, hepatitis C disease will reappear after 6 months in 70 percent of the cases. As of 1994 several clinical trials using interferons for other diseases were in progress.

Other studies have looked at the value of interferons to treat cancer. Tests on one form of bone cancer show that after most of the cancerous tissue is removed by surgery or destroyed by radiation, interferon therapy will reduce the incidence of metastasis (spread).

How interferon stops metastasis is not known. Perhaps it interferes with viral replication. Interferon therapy could also prevent growth of the cancer cells through their destruction by macrophages and NK cells.

The therapeutic use of interferons has some drawbacks. When recombinant interferon is injected, it does not remain stable for very long in the body. This makes delivery of the interferons to the site of infection difficult. Injection of interferon (especially α-interferon) also has side effects, including fatigue, nausea, headache, vomiting, weight loss, and nervous system disorders. Whereas fever normally increases interferon production, which helps the body fight viral infections, the injection of interferon *produces* fever as a side effect. High doses can cause toxicity to the liver, kidneys, heart, and bone marrow.

Moreover, some microbes have developed resistance to interferons. Although some DNA viruses, such as the poxviruses, stimulate interferon synthesis, the human adenoviruses have resistance mechanisms to combat antiviral protein activity. In addition, the hepatitis B virus often fails to stimulate adequate interferon production in infected cells.

The therapeutic usefulness of interferon is clearly not the viral magic bullet that was originally envisioned. Nevertheless, interferons are being used to treat life-threatening viral infections and cancers.

Complement

Complement, or the **complement system,** refers to a set of more than 20 large regulatory proteins that play a key role in host defense. They are produced by the liver and circulate in plasma in an inactive form. These proteins account for about 10 percent (by weight) of all plasma proteins. When complement was discovered, it was believed to be a single substance that "complemented," or completed, certain immunological reactions. Although complement can be activated by immune reactions, its effects are nonspecific—it exerts the same defensive effects regardless of which microorganism has invaded the body.

The general functions of the complement system are to (1) enhance phagocytosis by phagocytes, (2) lyse microorganisms, bacteria, and enveloped viruses directly, and (3) generate peptide fragments that regulate inflammation and immune responses. Furthermore, complement goes to work as soon as an invading microbe is detected; the system makes up an effective host defense long before specific host immune defenses are mobilized.

The complement system works as a cascade. A **cascade** is a set of reactions that amplify some effect—that is, more product is formed in the second reaction than in the first, still more in the third, and so on. Of the proteins so far identified in the complement system, 13 participate in the cascade itself, and 7 activate or inhibit reactions in the cascade.

Complement Function

Two pathways have been identified in the sequence of reactions carried out by the complement system. They are called the **classical pathway** and the **alternative pathway,** or *properdin pathway* (Figure 17.8a). The classical pathway begins when antibodies bind to antigens such as microbes and involves complement proteins C1, C4, and C2 (*C* stands for complement). The alternative pathway is activated by contact between complement proteins and polysaccharides at the pathogen surface. Complement proteins called factor B, factor D, and factor P (*properdin*) replace C1, C4, and C2 in the initial steps. However, the components of both pathways activate reactions involving C3 through C9. Consequently, the effects of the complement systems are the same regardless of the pathway by which C3 is produced. However, the alternative pathway is activated even earlier in an infection than is the classical pathway.

The contributions of the complement system to nonspecific defenses depend on C3, a key protein in the system. Once C3 is formed, it immediately splits into C3a and C3b, which then participate in three kinds

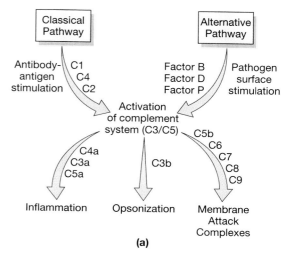

FIGURE 17.8 (a) Classical and alternative pathways of the complement cascade. Although the two pathways are initiated in different ways, they combine to activate the complement system. (b) Activation of the classical complement pathway. In this cascade each complement protein activates the next one in the pathway. The action of C3b is critical for opsonization and, along with C5b, for formation of membrane attack complexes. C4a, C3a, and C5a also are important to inflammation and phagocyte chemotaxis. (IgG is a class of antibodies that we will discuss in Chapter 18.)

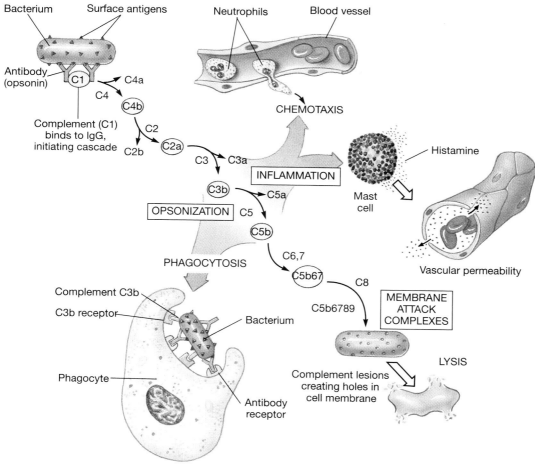

(b)

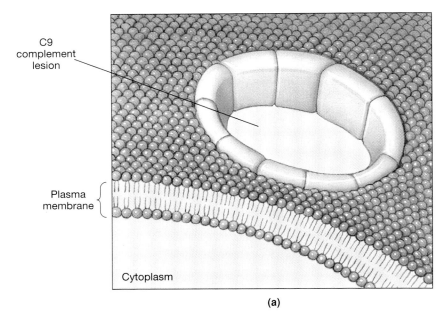

C9
complement
lesion

Plasma
membrane

Cytoplasm

(a)

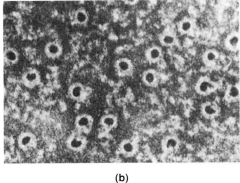

(b)

FIGURE 17.9 (a) Complement lyses a bacterial cell by creating a membrane attack complex (lesion) consisting of 10 to 15 molecules of C9. These protein molecules form a hole in the cell membrane through which the cytoplasmic contents leak out. (b) An EM photo showing the holes formed in red blood cell membranes by C9.

of molecular defenses: opsonization, inflammation, and membrane attack complexes (Figure 17.8b).

OPSONIZATION Earlier, we mentioned that some bacteria with capsules or surface proteins (M proteins) can prevent phagocytes from adhering to them. The complement system can counteract these defenses, making possible a more efficient elimination of such bacteria. First, special antibodies called **opsonins** bind to and coat the surface of the infectious agent. C1 binds to these antibodies, initiating the cascade. C1 causes the cleavage of C4 into C4a and C4b. C4b and C1 then cause C2 to split into C2a and C2b. The C4bC2a complex in turn leads to the splitting of C3 into C3a and C3b. C3b then binds to the surface of the microbe. Complement receptors on the plasma membrane of phagocytes recognize the C3b molecules; this recognition stimulates phagocytosis. This process, initiated by opsonins, is called **opsonization,** or *immune adherence.*

INFLAMMATION The complement system is also potent in initiating and enhancing inflammation. C3a, C4a, and C5a enhance the acute inflammatory reaction by stimulating chemotaxis and thus phagocytosis. These three complement proteins also adhere to the membranes of basophils and mast cells, causing them to release histamine and other substances that increase the permeability of blood vessels.

MEMBRANE ATTACK COMPLEXES Another defense triggered by C3b is cell lysis. By a process called **immune cytolysis,** complement proteins produce lesions in the cell membranes of microorganisms and other types of cells. These lesions cause cellular contents to leak out. To cause immune cytolysis, C3b initiates the splitting of C5 into C5a and C5b. C5b then binds C6 and

C7, forming a C5bC6C7 complex. This protein complex is hydrophobic ∞ (Chapter 4, p. 82) and inserts into the microbial cell membrane. C8 then binds to C5b in the membrane. Each C5bC6C7C8 complex causes the assembly in the cell membrane of up to 15 C9 molecules (Figure 17.9). By extending all the way through the cell membrane, these proteins form a pore and constitute the **membrane attack complex** (MAC). The MAC is responsible for the direct lysis of invading microorganisms. Importantly, host plasma membranes contain proteins that protect against MAC lysis. These proteins prevent damage by preventing the binding of activated complement proteins to host cells. The MAC forms the basis of *complement fixation,* a laboratory test used to detect antibodies against any one of many microbial antigens. That test is described in Chapter 19.

A great advantage of the complement system to host defenses is that, once it is activated, the reaction cascade occurs rapidly. A very small quantity of an activating substance (microbe) can activate a few molecules of C1. They, in turn, activate large quantities of C3; one C4b2a molecule can split 1000 molecules of C3 into C3a and C3b. Thus, sufficient quantities of C3b are quickly available to cause opsonization and inflammation and to produce membrane attack complexes.

Unfortunately, complement activity can be impaired by the absence of one or more of its protein components. Impaired complement activity makes the host more vulnerable to various diseases (Table 17.4), most of which are acquired or congenital. Acquired diseases result from temporary depletion of a complement protein; they subside when cells again synthesize the protein. Congenital complement deficiencies are due to genetic defects that prevent the synthesis of one or more complement components.

TABLE 17.4	Disease States Related to Complement Deficiencies	
Disease State	Complement Deficiencies	
Severe recurrent infections	C3	
Recurrent infections of lesser severity	C1, C2, C5	
Systemic lupus erythematosus (a bodywide immunologic disease)	C1, C2, C4, C5, C8	
Glomerulonephritis (an immunological disease of the kidneys)	C1, C8	
Gonococcal infections	C6, C8	
Meningococcal infections	C6	

The most significant effect of complement deficiencies is the lack of resistance to infection. Deficiencies in several complement components have been observed. The greatest degree of impaired complement function occurs with a deficiency of C3—which is not surprising, because C3 is the key component in the system. In individuals with C3 deficiencies, chemotaxis, opsonization, and cell lysis are all impaired. Such individuals are especially subject to infection by pyogenic bacteria. A deficiency in MAC components (C5–C9) is associated with recurrent infections, especially by *Neisseria* species. Complement deficiencies are less important in defenses against viruses, although some viruses, such as the Epstein–Barr virus, use complement receptors to invade cells.

Acute Phase Response

Observations of acutely ill patients have led to the characterization of the **acute phase response,** a response to acute illness that involves increased production of specific blood proteins called **acute phase proteins.** In an acute phase response, pathogen ingestion by macrophages stimulates the synthesis and secretion of several cytokines. One, called *interleukin-6* (IL-6), travels through the blood and causes the liver to synthesize and secrete the acute phase proteins into the blood. Thus, acute phase proteins form a nonspecific host defense mechanism distinct from both the inflammatory response and host-specific immune defenses. This mechanism appears to recognize foreign substances before the immune system defenses do and acts early in the inflammatory process, before antibodies are produced.

The best understood acute phase proteins are *C-reactive protein* (CRP) and *mannose-*

binding protein (MBP). All humans studied thus far have the capacity to produce CRP and MBP. CRP recognizes and binds to phospholipids, and MBP to mannose sugars, in cell membranes of many bacteria and the plasma membranes of fungi. Once bound, these acute phase proteins act like an opsonin: They activate the complement system and immune cytolysis and stimulate phagocyte chemotaxis. If we knew how to enhance CRP and MBP activity, effective therapies could be developed to combat many bacterial and fungal infections.

In summary, the nonspecific defense mechanisms operate regardless of the nature of the invading agent. They constitute the body's first line of defense against pathogens, whereas the specific defense mechanisms (Chapter 18) constitute the second line of defense. Figure 17.10 reviews the major categories of nonspecific defenses.

FIGURE 17.10 A summary of the body's nonspecific defenses.

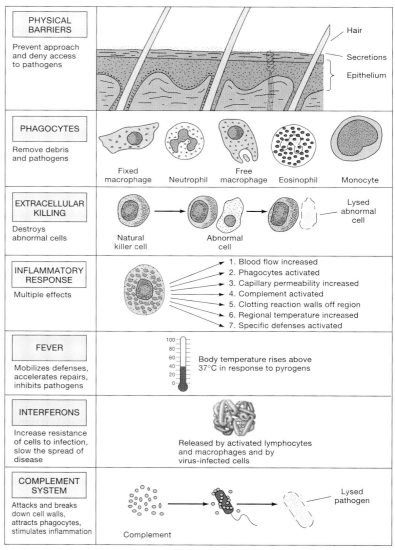

SYSTEM STRUCTURE AND SITES OF INFECTION

We will now look at the major body (host) systems in relation to infection and disease. We emphasize the sites at which infections often occur and the nature of the defenses available in each system.

The Skin

We have already discussed the skin as a physical barrier to nonspecific host defenses (Figure 17.1). The outer epidermis is attached to the thicker, inner dermis. Beneath the dermis is the subcutaneous layer consisting of fat and loose connective tissues. Fibers from the dermis extend down into the subcutaneous layer and anchor the skin to it. The subcutaneous layer attaches to the underlying body tissues and organs. Hair, fingernails and toenails, and glands are called *skin derivatives* because they develop from the epidermis.

The epidermis has several layers of cells. Cells next to the dermis divide throughout the life of the individual and migrate toward the surface, where old cells are sloughed off. Thus, the epidermis is renewed every 15 to 30 days. Lacking blood vessels, it is nourished by nutrients that diffuse from blood vessels in the dermis.

The dermis—the durable part of the skin—contains blood vessels and sensory receptors. Epidermal structures, such as hair follicles, sebaceous (oil) glands, and sweat glands, are embedded in the dermis. As mentioned earlier, these glands produce sebum and sweat, respectively, that lower the skin pH, inhibiting the growth of potential pathogens.

The Eyes and Ears

Being exposed to the atmosphere, the eyes contact millions of microbes every day. The eyes have several external protective structures to lessen the chances of an infection. In fact, these structures work so well that deep eye infections are exceedingly rare. Each eye is protected physically by eyelids, eyelashes, the **conjunctiva** (a mucous membrane covering the inner surface of each eyelid and the anterior region of each eye), and the tough **cornea** (the transparent part of the eyeball exposed to the environment) (Figure 17.11). Eyelashes and eyelids help prevent foreign objects from reaching the cornea.

The eyes also produce antimicrobial substances. Each eye has a **lacrimal gland,** which secretes a lacrimal fluid (tears) that continuously flushes the cornea, keeping it moist as the liquid carries away any microorganisms present. *Lacrimal canals* drain tears from the surface of the eyeball through the *nasolacrimal duct* and into the *nasal cavity*. Tears contain *lysozyme,* an enzyme that breaks down bacterial cell walls. It is especially effective in killing gram-positive organisms, which have a

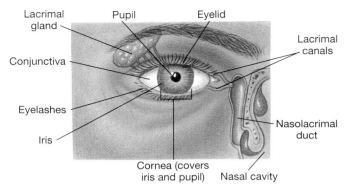

FIGURE 17.11 Structure of the eye. Tears produced over the eye surface contain antibacterial molecules that inhibit potential infections and drain into the nasal cavity. Most eye infections involve the eyelids, conjunctiva, or cornea.

thick peptidoglycan layer (Figure 17.12) ∞ (Chapter 4, p. 78) Lysozyme has no effect on viruses. Tears, like other body secretions, also contain specific defensive chemicals. Glandular cells in the conjunctiva add a mucous substance to the tears. This substance may help trap microorganisms and eliminate them from the eye.

Like the eyes, the ears are exposed to the environment and thus are subject to microbial attack. The ears contain physical protective structures to prevent infection. Each ear is divided into the outer ear, the middle ear, and the inner ear (Figure 17.13). The *outer ear* has a flaplike **pinna,** or auricle (commonly called the ear), covered with skin, and an **auditory canal,** which is is lined with skin that has many small hairs and many ceruminous glands. The **ceruminous** (se-ru'mi-nus) **glands** are modified sebaceous glands that secrete **cerumen** (earwax). Both hairs and wax help trap microorganisms and other foreign objects and prevent them from entering the auditory canal. Nevertheless, this canal can become infected with fungi.

FIGURE 17.12 Lysis of *Micrococcus luteus* bacteria by lysozyme enzymes obtained from tears (two upper disks) and from egg white (two middle disks). The two lowermost disks are controls that have been dipped in sterile distilled water.

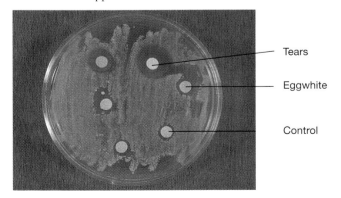

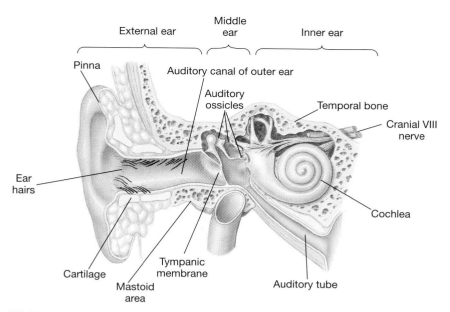

External ear   Middle ear   Inner ear

Pinna
Auditory canal of outer ear
Auditory ossicles
Temporal bone
Cranial VIII nerve
Ear hairs
Cochlea
Cartilage
Tympanic membrane
Mastoid area
Auditory tube

FIGURE 17.13 Structure of the ear. The outer ear is protected from foreign material by ear hairs and earwax. Nevertheless, some ear infections involve the outer ear. Infections of the middle ear usually arise from infections moving up the auditory (eustachian) tube.

The **tympanic membrane,** or *eardrum,* separates the outer and middle ears. The *middle ear* is a small, air-filled cavity containing the small bones called *ossicles* that transmit sound waves from the tympanic membrane to the inner ear. If the tympanic membrane is intact, most middle ear infections arise from microorganisms that move up the *auditory (eustachian) tube* from the nasopharynx. The *inner ear* converts sound waves into nerve impulses carried by the vestibulocochlear cranial nerve (VIII).

Outer and middle ear infections are relatively common—especially in children, because their auditory tubes are shorter and wider, facilitating the transmission of microbes. Because of the relative inaccessibility of the inner ear to pathogens, inner ear infections are rare. However, if such an infection does occur, it can easily spread to the **mastoid area,** a bony projection of the skull located behind and below the auditory canal. There is only a thin bony separation between mastoid and brain. Should the mastoid area become infected, therefore—a condition known as *mastoiditis*—the results can be quite serious, as there is danger that the infection will reach the brain.

The Respiratory System

The **respiratory system** consists of the **upper respiratory tract**—consisting of the nasal cavity, pharynx (far´ingks), larynx (lar´ingks), trachea (tra´ke-a), and bronchi (brong´ki)—and the **lower respiratory tract**—composed of the lungs (Figure 17.14a). This entire system is lined with moist epithelium. However, in the upper respiratory tract this epithelium contains mucus-secreting cells and is covered with cilia.

The Upper Respiratory Tract

Because the respiratory system moves oxygen from the atmosphere to the blood and removes carbon dioxide from the blood to the atmosphere, every breath of air contains millions of microorganisms and other suspended particles. Some microorganisms and particles breathed in are removed by hairs and mucus as the air passes through the **nasal cavity.** If microbes enter the **nasal sinuses,** which are hollow cavities lined with mucous membrane within the skull, sinus infections can occur. Air with any remaining microbes and particles then passes through the **pharynx** (throat), a common passageway for the respiratory and digestive systems. The auditory tube connects the pharynx with the middle ear.

From the pharynx, air and any remaining microbes pass through a series of rigid-walled tubes, the larynx and the trachea. The **larynx** (voice box) contains the vocal cords, which produce sound when they vibrate. The *epiglottis* is a flap of tissue that prevents food and fluids from entering the larynx. The **trachea** (windpipe) branches into primary **bronchi** (singular: *bronchus*), all of which are lined with cilia.

The upper respiratory tract contains a variety of normal microbiota that help prevent infection by pathogens that may be inhaled. In addition, mucus from the membranes that line the nasal cavity and pharynx traps microorganisms and most particles of debris, preventing them from passing beyond the pharynx. Mucus also contains lysozyme. Coughing and sneezing mechanically agitate mucus, increasing exposure of microorganisms to mucus and helping to expel them.

The beating of cilia generally serves to move a cell through its environment. ∞ (Chapter 4, p. 96) In the nasal cavity and bronchi, however, cilia extend from the epithelial cells. Because these cells are anchored in place, they function instead to trap and move microbes and particles near the cell surfaces. There mucus with debris trapped in it is moved up into the pharynx. This mechanism, the **mucociliary escalator,** allows materials in the bronchi to be lifted to the pharynx and to be spit out or swallowed. Nevertheless, the mucous membranes of the upper respiratory system are common sites of infection.

Such infections often spread to the sinuses, the middle ear, and even to the lower respiratory tract.

The Lower Respiratory Tract

From the bronchi, air passes into the lungs. Secondary bronchi divide into smaller **bronchioles,** forming a branching structure known as the *bronchial tree*. This complex branching arrangement greatly increases the surface area exposed to oxygen flowing into, and car-bon dioxide flowing out of, the lungs. Air passing through the terminal bronchioles then enters the **respiratory bronchioles** (Figure 17.14b). These microscopic channels end in a series of saclike **alveoli,** (singular: *alveolus*) which form clusters (nodules). It is in the alveoli that gas exchange occurs. There oxygen diffuses into the blood, and carbon dioxide from the blood diffuses into the alveoli. The bronchial tree, alveoli, blood vessels, and lymphatic vessels form the bulk of the lungs. The surfaces of the lungs and the cavities

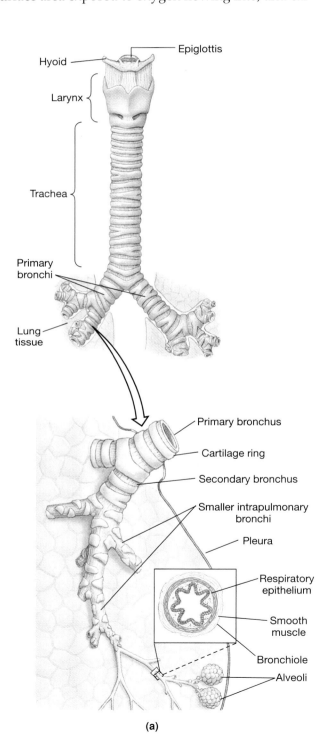

(a)

APPLICATIONS

ALCOHOLICS AND PNEUMONIA

The opening to the larynx is covered by the epiglottis, a flap of tissue that prevents food and fluids from entering the larynx. When the epiglottis fails to operate properly, fluids can be drawn or sucked into the lungs. Such failure in alcoholics often leads to pneumonia. When alcoholics are "sleeping off" drunken episodes, the epiglottis gapes and allows pharyngeal secretions to be drawn into the lungs with breathing. If pneumonia-causing organisms are present in those secretions, they can gain access to and infect the lungs. Once in the lungs, it is unlikely that the microbes will be destroyed because alcoholics have "lazy leukocytes." In other words, their macrophages are less efficient in destroying invading microbes than are those of nonalcoholics.

FIGURE 17.14 Structure of the respiratory system. (a) Much of the upper respiratory tract contains resident microbiota. Parts of the tract are covered with mucus and cilia that trap and move airborne material out. The lower respiratory tract lacks host microbiota and is subject to more severe and dangerous infections. (b) Bronchioles terminate in alveoli. Any microbes entering an alveolus can be destroyed by alveolar macrophages (not shown).

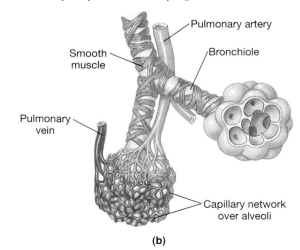

(b)

they occupy are covered by a membrane called the **pleura,** which secretes *serous fluid,* a watery fluid that lubricates some tissues.

Mucus from the lining of the bronchial tree also traps foreign materials that have passed beyond the pharynx. If the upper respiratory tract defense mechanisms fail and microorganisms get into bronchi and bronchioles, alveolar macrophages help remove them. Only when the numbers of organisms exceed the capacity of the phagocytes to destroy them, or when an infection reaches the lungs by way of blood or lymph vessels, does a lower respiratory infection occur.

The Digestive System

The **digestive system** consists of an elongated tube, or tract, that extends from the mouth to the rectum and includes accessory organs that help in digestion of food materials (Figure 17.15). The digestive tract consists of various organs—mouth, pharynx, esophagus, stomach, and intestines— whereas the accessory organs include the teeth, salivary glands, liver, gallbladder, and pancreas. The digestive system has five major functions:

1. *Movement* of food materials through the tract.
2. *Secretion* of digestive juices and mucus to break down food materials.
3. *Digestion,* the process of breaking large food molecules into small molecules that can be passed from the digestive tract into the bloodstream.
4. *Absorption* of digested food substances into the blood or lymph.
5. *Elimination* of indigestible food materials and intestinal microbiota.

Although food contains many microorganisms, most are killed by various defense mechanisms in the digestive tract. Throughout the digestive tract, **mucin** (myu'sin), a glycoprotein in mucus, coats bacteria and prevents their attaching to surfaces.

The Mouth

The mouth, or *oral cavity,* is lined with a mucous membrane and contains the tongue, teeth, and salivary glands. The mouth is a portal of entry for microorganisms; a normal mouth con-

tains more resident microbes than there are people on earth.

Each tooth has a *crown* covered with **enamel** above the gum and a *root* covered with **cementum** below the gum (Figure 17.16). Under these coverings is a porous substance called *dentin,* a central *pulp cavity,* and the *root canals,* where blood vessels and nerves are located. Each tooth is held in a tooth socket by fibers running from the cementum to the bone of the socket. Although enamel is the hardest substance in the body, it can be attacked by microbial-produced acids and enzymes. Microbes also can infect the gums, form pockets of infection between teeth and gums, and spread to the underlying bone.

The salivary glands secrete saliva, containing both antibodies that can coat bacteria and lysozyme that kills some bacteria. The salivary glands themselves, however, are subject to infection.

FIGURE 17.15 Structure of the digestive system. Normal microbiota of the digestive system prevent easy growth of pathogens, and stomach acids and enzymes destroy most disease agents. Vomiting and diarrhea are two expulsion mechanisms used to rid the system of toxic materials, including bacterial toxins.

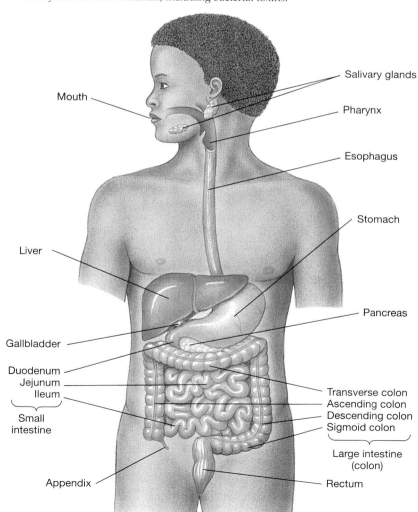

Labels: Mouth, Liver, Gallbladder, Duodenum, Jejunum, Ileum, Small intestine, Appendix, Salivary glands, Pharynx, Esophagus, Stomach, Pancreas, Transverse colon, Ascending colon, Descending colon, Sigmoid colon, Large intestine (colon), Rectum

of the nonspecific and specific defense mechanisms against infection and disease.

Lymphatic Circulation

The process of draining excess fluid from the spaces between cells starts with the *lymphatic capillaries* found

---

FIGURE 17.18 Structure of the lymphatic system. The lymphatic system filters out microbes from the fluids surrounding cells. In so doing, it is subject to infections that overrun the ability of the system to destroy the microbes. Lymphocytes are defensive cells commonly found in the lymphatic system.

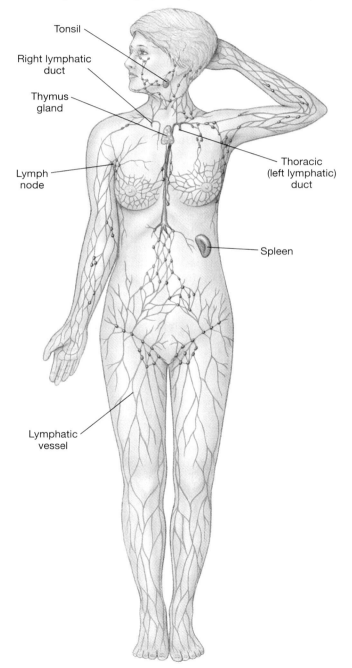

Tonsil

Right lymphatic duct

Thymus gland

Lymph node

Lymphatic vessel

Thoracic (left lymphatic) duct

Spleen

---

throughout the body. These capillaries, which are slightly larger in diameter than blood capillaries, collect the excess fluid and plasma proteins that leak from the blood into the spaces between cells. Once in the lymphatic capillaries, this fluid is called **lymph.** Lymphatic capillaries join to form larger **lymphatic vessels.** As fluid moves through the vessels, it passes through **lymph nodes.** Finally, the lymph is returned to the venous blood via the *right* and *left lymphatic ducts,* which drain the fluids into the right and left subclavian veins. There is no mechanism to move or pump lymphatic fluid. Hence, the flow of lymph depends on skeletal muscle contractions, which squeeze the vessels, forcing the lymph toward the lymphatic ducts. Throughout the lymphatic system, there are one-way valves to prevent backflow of lymph.

Lymphoid Organs

Specific organs of the lymphatic system are essential in the body's defense against infectious agents and cancers. These organs include the lymph nodes, thymus, and spleen. Although all lymphatic organs contain numerous lymphocytes, these cells originate in bone marrow and are released into blood and lymph. They live from weeks to years, becoming dispersed to various lymphatic organs or remaining in the blood and lymph. In humans most lymphocytes are either *B lymphocytes* (B cells) or *T lymphocytes* (T cells). B cells differentiate in the bone marrow itself and migrate to the lymph nodes and spleen. Immature T cells from the bone marrow migrate to the thymus, where they mature; they then migrate to the lymph nodes or spleen. We will discuss these cells in more depth in Chapter 18.

At intervals along the lymphatic vessels, lymph flows through lymph nodes distributed throughout the body. They are most numerous in the thoracic (chest) region, neck, armpits, and groin. The lymph nodes filter out foreign material in the lymph. Most foreign agents passing through a node are trapped and destroyed by the defensive cells present.

Lymph nodes occur in small groups, each group covered in a network of connective tissue fibers called a **capsule** (Figure 17.19). Lymph moves through a lymph node in one direction. Lymph first enters **sinuses,** wide passageways lined with phagocytic cells, in the outer cortex of the lymph node. The *outer cortex* houses large aggregations of B lymphocytes. The lymph then passes through the *deep cortex,* where T lymphocytes exist. The lymph moves through the inner region of a lymph node, the *medulla,* which contains B lymphocytes, macrophages, and plasma cells. Finally, lymph moves through sinuses in the medulla and leaves the lymph node.

This filtration of the lymph is important when an infection has occurred. For example, if a bacterial infection occurs, the bacteria are carried to the lymph nodes. As the lymph passes through the nodes, a majority of the bacteria are removed. Macrophages and other phagocytic cells in the nodes bind to and phagocytize the bacterial cells, thereby initiating a specific immune response (Chapter 18).

The **thymus gland** is a multilobed lymphatic organ located beneath the sternum (breastbone) (Figure 17.18). It is present at birth, grows until puberty, then atrophies (shrinks) and is replaced by fat and connective tissue by adulthood. Around the time of birth, the thymus begins to process lymphocytes and releases them into the blood as T cells. T cells play several roles in immunity: They regulate the development of B cells into antibody-producing cells, and subpopulations of T cells can kill virus-infected cells directly. The thymus also produces at least seven hormones, including cytokines, that stimulate production of T cells in the lymph nodes and spleen.

The **spleen,** located in the upper left quadrant of the abdominal cavity, is the largest of the lymphatic organs (Figure 17.18). Anatomically, the spleen is similar to the lymph nodes. It is encapsulated, lobed, and well supplied with blood and lymphatic vessels. Although it does not filter material, its sinusoids contain many phagocytes that engulf and digest worn-out erythro-

cytes and microorganisms. It also contains B cells and T cells.

Other Lymphoid Tissues

Earlier, we mentioned the lymphoid masses found in the ileum of the small intestine. These Peyer's patches are **lymphoid nodules,** unencapsulated areas filled with lymphocytes. Collectively, the tissues of lymphoid nodules are referred to as **gut-associated lymphatic tissue** (GALT), which are the main sites of antibody production. Similar nodules are found in the respiratory system, urinary tract, and appendix.

The **tonsils** represent another site for the aggregation of lymphocytes. Although these tissues are not essential for fighting infections, they do contribute to immune defenses, as they contain B cells and T cells.

Although lymphatic tissues contain cells that phagocytize microorganisms, if these cells encounter more pathogens than they can destroy, the lymphatic tissues can become sites of infection. Thus, swollen lymph nodes and tonsillitis are common signs of many infectious diseases.

In summary, lymphoid tissues contribute to nonspecific defenses by phagocytizing microorganisms and other foreign material. They contribute to specific immunity through the activities of their B and T cells, which we will discuss in Chapter 18.

FIGURE 17.19 Structure of a lymph node. Lymph nodes are centers for removing microbes. These tissues contain phagocytes and lymphocytes. Swollen lymph nodes are usually an indication of a serious infection.

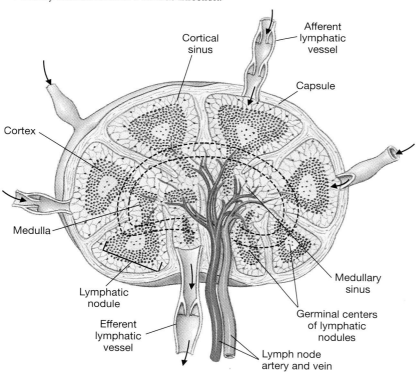

Cortical sinus

Afferent lymphatic vessel

Capsule

Cortex

Medulla

Lymphatic nodule

Efferent lymphatic vessel

Medullary sinus

Germinal centers of lymphatic nodules

Lymph node artery and vein

The Nervous System

As you read this sentence, millions of neural signals allow you to understand words while you maintain an upright posture and carry on various internal processes, such as breathing. The ability of the nervous system to control body functions at the same time depends on many *neurons,* or nerve cells, working together. Structurally, the **nervous system** has two components: the central and peripheral nervous systems (Figure 17.20). The **central nervous system** (CNS), which consists of the brain and spinal cord, receives signals from and issues commands via the **peripheral nervous system** (PNS), composed of nerves that supply all parts of the body. **Nerves** of the peripheral nervous system consist of nerve fibers that transmit sensory information and motor responses. Aggregations of these cell bodies in the PNS are called **ganglia.** The brain and spinal cord are protected by membranes called **meninges** (me-nin'jez), which are sheets of connective tissues. Hollow chambers in the brain and spinal cord and spaces between meninges are filled with *cerebrospinal fluid*.

Like the cardiovascular system, the nervous system is ordinarily sterile and has no normal microbiota.

FIGURE 17.20 Structure of the nervous system. The nervous system normally is free of resident microbes. Infections typically affect the meninges (which cover the brain and spinal cord) or the sensory nerve ganglia.

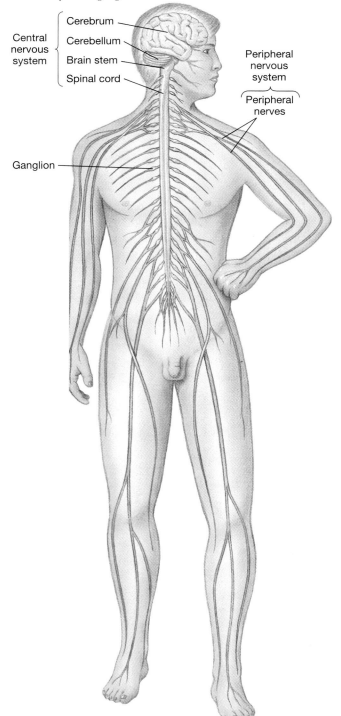

Central nervous system
- Cerebrum
- Cerebellum
- Brain stem
- Spinal cord

Peripheral nervous system
- Peripheral nerves

Ganglion

Bacterial and viral pathogens, however, can enter cerebrospinal fluid from the blood. The CNS is well protected by bone and meninges from invasion by pathogens. In addition, phagocytes of the nervous system called *microglial cells* can destroy invaders that reach the brain and spinal cord. The brain itself has special thick-walled capillaries without pores in their walls. These capillaries form the **blood–brain barrier,** which limits entry of substances into brain cells. Although it does protect the cells from microorganisms and toxic substances, the blood–brain barrier prevents the cells from receiving medications that easily reach other cells.

The Urogenital System

The **urogenital system** includes the organs of both the urinary system and the reproductive system. Thus, in considering the urogenital system we will look at the structures and sites of infection in the urinary system, the female reproductive system, and the male reproductive system.

The Urinary System

The **urinary system** consists of paired kidneys and ureters, the urinary bladder, and the urethra (Figure 17.21a). The **kidneys** regulate the composition of body fluids and remove nitrogenous and certain other wastes from the body. To accomplish this, each kidney contains about 1 million functional units called **nephrons** (Figure 17.21b). In a nephron the fluid part of the blood is filtered from the **glomerulus,** a coiled cluster of capillaries, to the kidney tubules. Starting in the *renal cortex,* nephrons remove solutes, including wastes, and water from the blood. As these materials pass through the tubules known as the *renal medulla,* essential materials such as water, salts, and sugars pass back into the blood. **Urine,** the wastes remaining in the kidney tubules, pass through collecting ducts to the **ureter** of each kidney. The ureters carry urine, which is normally free of microbes, to the **urinary bladder,** where it is stored until released through the **urethra** during micturition (urination). **Urinalysis,** the laboratory analysis of urine specimens, can reveal imbalances in pH or water concentration, the presence of substances such as glucose or proteins, and other conditions associated with infections, metabolic disorders, and other diseases.

The Female Reproductive System

The **female reproductive system** consists of the ovaries, uterine tubes, uterus, vagina, and external genitalia (Figure 17.22, page 474). The paired **ovaries** contain cellular aggregations called **ovarian follicles,** each containing an *ovum* (plural: *ova*), or egg, and surrounding epithelial tissue. During a woman's reproductive years, an ovum capable of being fertilized is released once each month. The **uterine tubes** receive ova and convey them to the uterus. Fertilization usu-

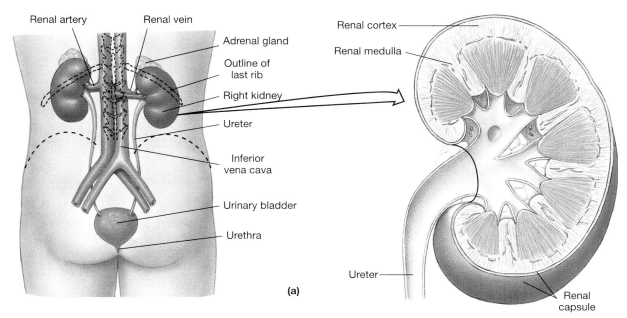

FIGURE 17.21 (a) Rear view of the urinary system, including a blowup of the kidney. Infections of the urinary system often occur near the opening of the urethra, which serves as a portal of entry. The flow of urine out of the system tends to prevent many potential infections. (b) Structure of a nephron.

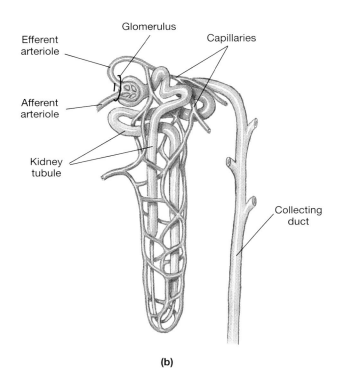

sensitive *clitoris*, two pairs of *labia* (skin folds), and the mucous-secreting **Bartholin glands.** Because they nourish offspring, **mammary glands** are considered part of the female reproductive system. These modified sweat glands develop at puberty and contain gland cells embedded in fat that produce milk and ducts that carry the milk to the nipple.

The Male Reproductive System

The **male reproductive system** consists of the testes, ducts, specific glands, and the penis (Figure 17.23). The **testes** (singular, *testis*) secrete the hormone testosterone into the bloodstream and produce sperm, which are conveyed through a series of ducts to the urethra. Secretions from **seminal vesicles** and the **prostate gland** mix with sperm to form **semen.** Other glands secrete mucus that lubricates the urethra. The **penis** is used to deliver semen into the female reproductive tract during sexual intercourse.

Sites of normal microbiota in the urogenital system also are common sites of infections, partly because of their proximity to the rectum and partly because they provide warm, moist conditions suitable for microbial growth. The normal microbiota of the urogenital system are found mainly at or near the external opening of the urethra of both sexes and the vagina of females. Many infectious diseases of the urogenital system are sexually transmitted, including genital herpes, gonorrhea, syphilis, and nongonoccocal urethritis. Other parts of the urogenital system generally remain

ally occurs in the uterine tubes. The **uterus** is a pear-shaped organ in which a fertilized ovum develops. It is lined with a mucous membrane called the **endometrium,** the outer portion of which is sloughed during menstruation. The **vagina,** also lined with mucous membrane, extends from the **cervix** (an opening at the narrow lower portion of the uterus) to the outside of the body. It allows passage of menstrual flow, receives sperm during intercourse, and forms part of the birth canal. The female external genitalia include the sexually

FIGURE 17.22 Structure of the female reproductive system. The portal of entry for the reproductive tract (vagina) is separate from the urinary tract (urethra). Chemical defenses, including cervical mucus, have antibacterial actions. In addition, secretions moving from the uterine (fallopian) tubes through the uterus and cervix to the vagina make microbial infections less likely.

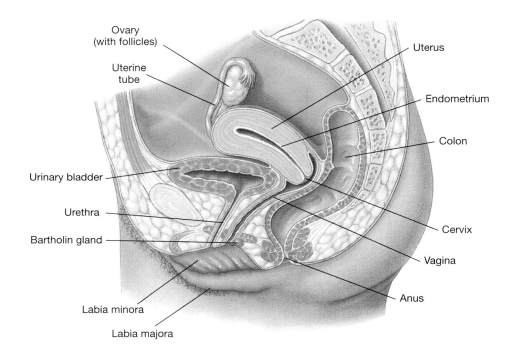

Ovary (with follicles)
Uterine tube
Urinary bladder
Urethra
Bartholin gland
Labia minora
Labia majora
Uterus
Endometrium
Colon
Cervix
Vagina
Anus

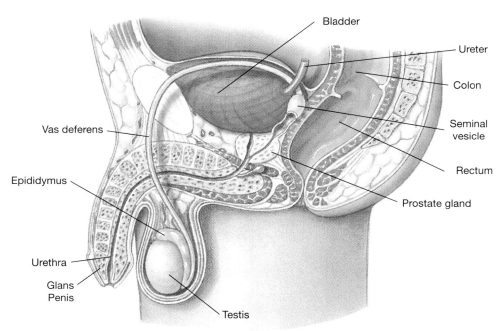

Bladder
Ureter
Colon
Seminal vesicle
Rectum
Prostate gland
Vas deferens
Epididymus
Urethra
Glans Penis
Testis

FIGURE 17.23 Structure of the male reproductive system. Because males have a single portal of entry for the reproductive and urinary systems (the urethra), infections of either system can occur. But chemical secretions containing lysozyme and spermine provide a flow that makes colonization by pathogens more difficult.

sterile, but a breakdown in defense mechanisms can lead to infection.

Defense mechanisms in the urogenital system are numerous. Normal microbiota compete with opportunists and pathogens for nutrients and space and prevent them from causing disease. Urinary *sphincters* (muscles that close openings) act as mechanical barriers to microbes and also help prevent the backflow of urine. The flow of urine through the urethra and of mucus through both the urethra and vagina help wash away microbes. The low pH within both the urethra and, during reproductive years, the vagina prevents invasion by pathogens. Semen contains lysozyme and spermine, which help destroy invading pathogens.

Despite its defenses, the urogenital tract is poorly protected against sexually transmitted diseases. The organisms that cause gonorrhea, syphilis, and nongonoccocal urethritis tolerate acidic conditions and successfully compete with natural microbiota of the urogenital tract.

DEVELOPMENT OF THE IMMUNE SYSTEM

Can all organisms defend themselves against attacks by infectious microbes? For vertebrates, the answer is yes. As we saw in this chapter, they have nonspecific defense mechanisms, and as we will see in Chapter 18, they also have well-developed specific immune defenses.

Defenses against infection are not limited to animals. Other biological kingdoms of also have host defense mechanisms, usually of a chemical nature. Plants, for example, produce chemical defenses that can wall off areas damaged or infected by bacteria or fungi. In fact, an important determinant of how well a given strain of plant can resist infection after pruning or damage is its chemical and physical defensive abilities. Many fungi are plant pathogens, getting their nutrients by parasitizing certain tissues within the plant. To infect a plant, the fungus must penetrate the plant cell. ∞ (Chapter 12, p. 311) During infection the plant cells produce enzymes that release carbohydrate molecules from the fungal cell walls. These fragments of fungal wall, called *elicitors,* trigger an immunological-like response by the plant. Elicitors cause the plant to produce lipidlike chemicals called *phytoalexins.* Phytoalexins inhibit fungal growth by restricting the infection to a small portion of the plant tissue (Figure 17.24). Plant biotechnologists are trying to "breed" this response into other types of plants that are sensitive to fungal invasion.

Invertebrates also have nonspecific defenses for fending off invaders. Phagocytosis is important to invertebrates in obtaining food, but it is also necessary for preventing sedentary organisms permanently fixed to a surface, and living where space is limited, from being overgrown by neighbors. So, phagocytosis is used to defend one's territory. In animals lacking a cardiovascular system, amoebocytes wander through the body, engulfing foreign matter and damaged or aged cells. When your white blood cells phagocytize a bacterium, they are using an ancient mechanism preserved and transformed from simpler life forms.

Opsonization is also observed in invertebrates, made possible by complement-like components of body fluids. For example, fluids in the body cavity of sea urchins share many characteristics with human complement proteins. In fact, complement proteins,

---

FIGURE 17.24 Experimentally damaged areas of tree trunk are walled off in trees that survive attack, thus keeping infection from spreading throughout the entire tree.

like phagocytosis, probably were derived from these early versions in invertebrates. Secretion of antimicrobial enzymes is another means of defense present even in simple protozoa. Thus, nonspecific defense processes, such as phagocytosis and opsonization, are often called a *primitive characteristic* because most animals have these ancient mechanisms.

Almost all invertebrates also can reject grafts of foreign tissue. Vertebrates reject such grafts more vigorously on a second encounter, but invertebrates do not; in fact, the second rejection may be slower than the first. Because invertebrates lack these memory responses, the presence of such specific immune defenses in vertebrates is considered an *advanced characteristic.* These defenses include the B cells, T cells, and antibodies.

Although immune defenses involving the production of specific antibodies are found in all types of fish, the swiftest and most complex immune responses are found in mammals and birds. Birds have a saclike structure, the *bursa of Fabricius,* that is not present in mammals and probably represents a higher state of evolution of the immune system. In chickens, immature B cells in the bone marrow migrate to the bursa of Fabricius. There they are stimulated to mature rapidly and are capable of recognizing foreign substances. In mammals, B cells originate and mature more slowly in the bone marrow. Thus, immune system development culminates in the two-part system of B cells and T cells. In Chapter 18 we will investigate this achievement of specific host defenses.

CHAPTER SUMMARY

NONSPECIFIC AND SPECIFIC HOST DEFENSES

- **Nonspecific defenses** operate regardless of the kind of invading agent; they form a first line of defense that is often effective even before specific defenses are activated.
- **Specific defenses** respond to particular invading agents; provided by the immune system, they form a second line of defense against pathogens.

PHYSICAL BARRIERS

The Skin

- **Skin** consists of an outer **epidermis** and an inner **dermis.**
- Skin infections occur anywhere the skin is broken, in ducts of **sebaceous** and **sweat** (sudoriferous) **glands,** in hair follicles, and sometimes on unbroken skin.

- Chemicals secreted by the skin include acidic and salty secretions, waterproofing **keratin,** and **sebum.**

Mucous Membranes

- **Mucous membranes** consist of a thin layer of cells that secrete **mucus.**

CELLULAR DEFENSES

Defensive Cells

- **Formed elements,** found in blood but derived from bone marrow, provide a cellular defense barrier to infection.
- Defensive cells include **granulocytes—basophils, mast cells, eosinophils,** and **neutrophils—**and **agranulocytes—monocytes** and **lymphocytes.**

Phagocytes

- A **phagocyte** is a cell that ingests and digests foreign substances.
- Phagocytic cells include neutrophils in the blood and in injured tissues, monocytes in the blood, and fixed and wandering **macrophages.**

The Process of Phagocytosis

- The process of **phagocytosis** occurs as follows: (1) Invading microorganisms are located by **chemotaxis,** which is aided by the release of **cytokines** by phagocytes. (2) Ingestion occurs as the phagocyte surrounds and ingests a microbe or other foreign substance into a **phagosome.** (3) Digestion occurs as lysosomes surround a vacuole and release their enzymes into it, forming a **phagolysosome.** Enzymes and defensins break down the contents of the phagolysosome and produce substances toxic to microbes.
- Some microbes resist phagocytosis by producing capsules or specific proteins, preventing release of lysosomal enzymes, and by producing toxins (**leukocidin** and **streptolysin**).

Extracellular Killing

- Eosinophils defend against parasitic worm infections by secreting cytotoxic enzymes.
- **Natural killer** (NK) **cells** secrete products that kill virus-infected cells and certain cancer cells.

INFLAMMATION

Characteristics of Inflammation

- **Inflammation** is the body's response to tissue damage. It is characterized by localized increased temperature, redness, swelling, and pain.

The Acute Inflammatory Process

- **Acute inflammation** is initiated by **histamine** released by damaged tissues, which dilates and increases permeability of blood vessels (**vasodilation**). Activation of cytokines also contributes to initiation of inflammation.
- Dilation of blood vessels accounts for redness and increased tissue temperature; increased permeability accounts for **edema** (swelling).
- Tissue injury also initiates the blood-clotting mechanism.
- **Bradykinin** stimulates pain receptors; **prostaglandins** intensify its effect.
- Inflamed tissues also stimulate an increase in the number of leukocytes in the blood (**leukocytosis**) by releasing cytokines that trigger leukocyte production. Neutrophils and macrophages migrate from the blood to the site of injury (**diapedesis**).
- Leukocytes and macrophages phagocytize microbes and tissue debris.

Repair and Regeneration

- Repair and regeneration occur as capillaries grow into the site of injury and fibroblasts replace the dissolving blood clot. The resulting **granulation tissue** is strengthened by connective tissue fibers (from **fibroblasts**) and the overgrowth of epithelial cells.

Chronic Inflammation

- **Chronic inflammation** is a persistent inflammation in which the inflammatory agent continues to cause tissue injury as host defenses fail to overcome the agent completely.
- **Granulomatous inflammation** is a chronic inflammation in which monocytes, lymphocytes, and macrophages surround necrotic tissue to form a **granuloma.**

FEVER

- **Fever** is an increase in body temperature caused by **pyrogens,** which increase the setting (thermostat) of the temperature-regulating center in the hypothalamus.
- **Exogenous pyrogens** (usually pathogens and their toxins) come from outside the body and stimulate a cytokine that acts as an **endogenous pyrogen.**
- Fever and the chemicals associated with it augment the immune response and inhibit the growth of microorganisms by lowering plasma iron concentrations. Fever also increases the rate of chemical reactions, raises the temperature above the optimum growth rate for some pathogens, and makes the patient feel ill (thereby lowering activity).
- Antipyretics are recommended only for high fevers and for patients with disorders that would be exacerbated by fever.

MOLECULAR DEFENSES

Interferon

- **Interferons** are proteins that act nonspecifically to cause cell killing or to stimulate cells to produce **antiviral proteins.**
- Interferon can be made by recombinant DNA technology and has proved to be therapeutic for certain malignancies; other therapeutic applications are being studied.

Complement

- **Complement** refers to a set of blood proteins that, when activated, produce a **cascade** of protein reactions. The **complement system** can be activated by the **classical pathway** or the **alternative pathway.**
- Action of the complement system is rapid and nonspecific. It promotes opsonization, inflammation, and immune cytolysis through the formation of **membrane attack complexes.** In **opsonization,** invading agents are coated with **opsonins** (antibodies) and C3b complement protein, making the invaders recognizable to phagocytes. In **immune cytolysis,** complement proteins produce lesions on invaders' plasma membranes that cause cell lysis.
- Deficiencies in complement reduce resistance to infection.

Acute Phase Response

- Acutely ill patients increase production of certain blood proteins (**acute phase proteins**). These substances are distinct from those involved in the inflammatory response and act quickly, before antibodies can be made. Such proteins initiate or accelerate inflammation, activate complement, and stimulate chemotaxis of phagocytes.

SYSTEM STRUCTURE AND SITES OF INFECTION

The Skin

- The skin is a physical barrier that secretes antimicrobial chemicals.

The Eyes and Ears

- Protective structures associated with the eyes include the eyelids, eyelashes, **conjunctiva, cornea, and lacrimal glands.**
- Protective structures associated with the ears include ear hairs and **ceruminous glands** in the ear canal.
- Eye infections occur on eyelids, conjunctiva, and cornea; ear infections occur in the outer and middle ears. Eyelashes, eyelids, and hairs and wax in the external ear are mechanical barriers; **lysozyme** in tears is microbicidal.

The Respiratory System

- The **respiratory system** consists of the **nasal cavity, pharynx, larynx, trachea, bronchi, bronchioles,** and **alveoli.**
- Nonspecific defenses include mucus and cilia, which work together in the **mucociliary escalator,** and phagocytic cells.

The Digestive System

- The **digestive system** consists of the mouth, pharynx, esophagus, stomach, and intestines and accessory organs such as the salivary glands, liver, and pancreas. Infections can involve all organs, accessory glands, and areas in and around the teeth.
- Nonspecific defenses include mucus, stomach acid, Kupffer cells in liver **sinusoids,** patches of submucosal lymphatic tissue, and competition from normal microbiota.
- Vomiting and diarrhea are expulsion defense mechanisms for removing toxins.

The Cardiovascular System

- The **cardiovascular system** consists of the heart, an extensive system of blood vessels, and blood.
- Although the cardiovascular system is normally sterile, pathogens can be transported in blood, multiply in blood, and infect the heart valves.
- Nonspecific defenses include the cleansing effect of blood flow out of wounds, constriction of injured blood vessels, blood clotting, phagocytic actions of certain classes of leukocytes and release of histamine, which initiates the inflammatory response.

The Lymphatic System

- The **lymphatic system** consists of a network of **lymphatic vessels, lymph nodes** and **lymphoid nodules,** the **thymus gland,** the **spleen,** and **lymph.**
- All lymphatic tissues that filter blood and lymph are susceptible to infection by pathogens they filter when the pathogens overwhelm defenses.
- Nonspecific defenses consist of the actions of phagocytic cells.

The Nervous System

- The **nervous system** consists of the brain and spinal cord, the peripheral nerves, **meninges,** and cerebrospinal fluid.
- The nervous system is normally sterile, but pathogens can attack meninges, invade nerve endings, or even damage brain tissue.
- Nonspecific defenses include macrophages and the **blood–brain barrier.**

The Urogenital System

- The **urogenital system** includes the **urinary system** and the reproductive system. The urinary system consists of the **kidneys, ureters, urinary bladder,** and **urethra.** The **female reproductive system** consists of the **ovaries, uterine tubes, uterus, vagina,** external genitalia, and **mammary glands.** The **male reproductive system** consists of the **testes,** a system of ducts, glands, and the **penis.**
- Urogenital infections are most common near the openings of the vagina and urethra.
- Nonspecific defenses include urinary sphincters, cleansing action of the outflow of urine, acidity of mucous membranes, and competition from normal microbiota.

KEY TERMS

abscess (*p. 456*)
acute inflammation (*p. 455*)
acute phase protein (*p. 463*)
acute phase response (*p. 463*)
adherence (*p. 453*)
agranulocyte (*p. 452*)
alternative pathway (*p. 461*)
alveolus (*p. 466*)
antihistamine (*p. 456*)
antiviral protein (*p. 458*)
auditory canal (*p. 464*)
Bartholin gland (*p. 473*)

basophil (*p. 452*)
blood–brain barrier (*p. 472*)
bradykinin (*p. 456*)
bronchiole (*p. 466*)
bronchus (*p. 465*)
capsule (*p. 470*)
cardiovascular system
  (*p. 468*)
cascade (*p. 460*)
cementum (*p. 467*)
central nervous system
  (*p. 471*)

cerumen (*p. 464*)
ceruminous gland (*p. 464*)
cervix (*p. 473*)
chemotaxis (*p. 453*)
chronic inflammation
  (*p. 457*)
classical pathway (*p. 461*)
complement (*p. 460*)
complement system (*p. 460*)
conjunctiva (*p. 464*)
cornea (*p. 464*)
cytokine (*p. 453*)

dermis (*p. 449*)
diapedesis (*p. 456*)
digestive system (*p. 467*)
edema (*p. 455*)
electrolyte (*p. 469*)
enamel (*p. 467*)
endogenous pyrogen (*p. 458*)
endometrium (*p. 473*)
eosinophil (*p. 452*)
epidermis (*p. 449*)
erythrocyte (*p. 451*)
exogenous pyrogen (*p. 458*)

QUESTIONS FOR REVIEW

A **Nonspecific vs. Specific Host Defenses**

1. What are the differences between specific and nonspecific defenses?

B **Physical Barrier Defenses**

2. How do the skin and mucous membranes function as a physical defense barrier?

3. What chemicals are secreted from the physical barriers that inhibit pathogen growth and spread?

C **Cellular Barrier Defenses**

4. Match the following:

_____ Its major role is involved with parasite attack.

_____ Macrophages are derived from this cell type.

_____ Red blood cells mature into this cell type.

_____ This granulocyte-type cell has phagocytic properties.

_____ This cell type is involved with antibody production.

   a. basophils

   b. eosinophils

   c. neutrophils

   d. monocytes

   e. B lymphocytes

   f. T lymphocytes

   g. none of the above

5. How do granulocytes differ from agranulocytes?

D **Phagocytosis**

6. Define phagocyte and phagocytosis.

7. Which kinds of phagocytes are found in blood and in tissues?

8. In the following diagram, identify the major steps in the phagocytic process. Describe what happens in each step.

Lysosomes (c) Cytoplasm
(b)
(a)
(d)
Plasma membrane of phagocyte
(e)

9. How do microbes resist phagocytosis?

E **Inflammation; Acute Inflammatory Processes**

10. What is inflammation, and what are its characteristics?

11. How is the acute inflammatory process initiated?

12. How is pain related to the inflammatory response?

13. What are the first events in repair and regeneration?

14. How is healing completed?

F Chronic Inflammation

15. List the distinguishing characteristics of chronic inflammation and granulomatous inflammation.

G Fever

16. What is fever?

17. Describe the roles of exogenous and endogenous pyrogens in fever.

18. What are the benefits of fever?

19. When are antipyretics recommended?

H Interferons

20. What are the interferons, and how do they work?

21. What is the therapeutic value of interferon?

I Complement

22. What is the complement system, and how is it activated?

23. Describe the defensive values of complement.

24. Match the following:

—— This is the result of the assembly of many C9 complement proteins.

—— C3a, C4a, and C5a enhance this effect.

—— This is an example of immune cytolysis.

—— C3b bound to microbes causes this.

—— This process results from the binding of C3b receptors to C3b-coated bacteria.

a. phagocytosis
b. opsonization
c. cell lysis
d. inflammation
e. none of the above

25. What are the effects of complement deficiencies?

J Acute Phase Responses

26. What are the effects of C-reactive protein and mannose-binding protein?

K Important Structures and Sites of Infection

27. List the major components of each of the human body systems.

28. What sites in each body system are most vulnerable to infection?

PROBLEMS FOR INVESTIGATION

1. Recall the last time you suffered from an infectious disease. In which ways do you think your body's nonspecific defenses failed you? Be specific.

2. Patients with cystic fibrosis (a genetic disorder) produce thick secretions that do not drain easily from the respiratory passages. The buildup of such secretions leads to inflammation and the replacement of damaged cells with connective tissue that blocks those respiratory passages. Which types of infections would you expect to find frequently in these patients, and why?

3. Make flow charts to show how nonspecific defenses combat a pathogen that enters the body through (a) a small cut, (b) the upper respiratory tract, (c) the digestive tract, and (d) the urogenital tract.

4. List some ways nonspecific defenses combat a pathogen that has evaded all the defenses considered in Problem 3.

SOME INTERESTING READING

Demling, R. H. 1985. "Burns." *The New England Journal of Medicine* 313 (22):1389–98.

Edelson, R. L., and J. M. Fink. 1985. "The immunologic function of skin." *Scientific American* 252(June):46–53.

Figueroa, J. E., and P. Densen. 1991. "Infectious diseases associated with complement deficiencies." *Clinical Microbiological Reviews* 4(3):359–95.

Graham, N. M. H., et al. 1990. "Adverse effects of aspirin, acetaminophen, and ibuprofen on immune function, viral shedding, and clinical status in rhinovirus-infected volunteers." *Journal of Infectious Diseases* 162(6):1277–82.

Isberg, R. R., and G. Tran Van Nheiu. 1995. "The mechanism of phagocytic uptake promoted by invasin-integrin interaction." *Trends in Cell Biology* 5(3):120–24.

Johnson, H. M., et al. 1994. "How interferons fight disease." *Scientific American* 270(May):68–75.

Kotwal, G. J., et al. 1990. "Inhibition of the complement cascade by the major secretory protein of vaccinia virus." *Science* 250(4982):827–30.

MacKowiac, P. A., ed. 1991. *Fever: Basic mechanisms and management*. New York: Raven Press.

Martini, F. H. 1995. *Fundamentals of anatomy and physiology*, 3d ed. Englewood Cliffs, NJ: Prentice Hall.

Muller, U., et al. "Functional role of type I and type II interferons in antiviral defense." *Science* 264(5167):1918–21.

Pascual, M., and L. E. French. 1995. "Complement in human diseases: Looking towards the 21st century." *Immunology Today* 16(2)58–61.

Russell, D. G. 1995. "*Mycobacterium* and *Leishmania*: Stowaways in the endosomal network." *Trends in Cell Biology* 5(3):125–29.

Thomas, D. O. 1995. "Fever in children: Friend or foe." *RN* 58(4):42–47.

Vines, G. 1995. "Get under your skin." *New Scientist* (Inside Science #78) 145(1960):1–4.

Weissman, G. 1991. "Aspirin." *Scientific American* 264(January):84–90.

Youmans, G. P., P. Y. Paterson, and H. M. Sommers. 1985. *The biologic and clinical basis of infectious diseases*, 3d ed. Philadelphia: W. B. Saunders.

18

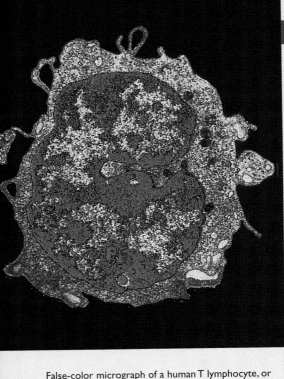

False-color micrograph of a human T lymphocyte, or T cell. T cells are one of several types of cells involved in specific immune responses.

IMMUNOLOGY I: BASIC PRINCIPLES OF SPECIFIC IMMUNITY AND IMMUNIZATION

When you were very young you probably received a variety of immunizations against diphtheria, tetanus, whooping cough, polio, and possibly measles, German measles, and mumps as well. Your parents or grandparents, however, probably became immune to both kinds of measles and to mumps by acquiring and then recovering from these diseases. Either being immunized or acquiring an infectious disease may confer specific immunity to the organism that causes that disease. As we saw in the last chapter, nonspecific host defenses protect the host against infections in a general way. This chapter will show how specific host defenses and immunization protect the host against particular infectious agents.

IMMUNOLOGY AND IMMUNITY

Immunology is the study of **immunity**—the ability of an organism to recognize and defend itself against agents that are "foreign" to it. Many such agents are infectious. Immunology, a mixture of fundamental and applied science, grew out of the study of infectious diseases and the body's responses to them. It has grown to include such fields as allergy, immunochemistry, im-

munopathology, immunogenetics, tumor immunology, serology, and transplantation. Here we will consider specific host defenses to disease.

The word *immune* literally means "free from burden." A number of organs, tissues, cells, cellular products, and mechanisms contribute to a host's **immune system,** or defenses. This system functions to provide effective barriers against penetration by disease agents, to inhibit or destroy invading organisms that do gain access to tissues, and to neutralize or eliminate toxic substances that result, for example, from a disease process. **Susceptibility,** the opposite of immunity, is the vulnerability of the host to harm by infectious and other types of agents.

As we saw in Chapter 17, the host has an arsenal of nonspecific defense weapons (physical, cellular, and molecular) against invading infectious organisms. These defenses are generally available no matter what type of organism invades. ∞ (Chapter 17, p. 448) Activation of these defenses does not require prior exposure to any specific agent.

Besides these nonspecific defenses, vertebrates have additional defenses for mounting a highly specific attack on particular infectious agents or foreign substances. These defenses include various types of cells (especially lymphocytes), cellular secretions (antibodies, cytokines, and interferons), and tissues (such as bone marrow, lymph nodes, and the liver).

TYPES OF IMMUNITY

Immunity to infectious agents can result from *innate immunity, acquired immunity,* or both.

Innate Immunity

Innate immunity is an invariable, hereditary response—an inborn defense. It is independent of previous exposures to disease agents and foreign substances. Instead, innate immunity depends on the nonspecific mechanisms, molecular defenses, and activities of phagocytic cells, such as those that we discussed in Chapter 17.

Innate immunity also includes *species immunity,* by which all individuals of a species are born with resistance to an infectious agent that causes disease in another species. Humans, for example, are immune to most infectious agents that cause disease in pets and other domesticated animals. Canine distemper is a highly contagious and often lethal viral disease in in-

fected puppies; humans do not have the particular cell surface receptors that are recognized by the distemper virus. The bacterium *Mycobacterium avium-intracellulare* causes tuberculosis in birds, but it rarely causes respiratory disease in humans with a functional immune system. (It does infect many persons with AIDS.) Similarly, animals show an innate immunity to many human disease agents. A number of factors, such as age, sex, nutritional status, and general state of health, affect innate immunity.

Acquired Immunity

In contrast to innate immunity, **acquired immunity** is resistance to disease or to foreign substances that develops during an individual's lifetime by some manner other than heredity. It can be either active or passive. **Active immunity** is created when an individual's own immune system takes an active role by producing specific immune defensive cells or protective molecules called *antibodies,* or *immunoglobulins,* upon exposure to an infectious agent or its toxins—whether the exposure is natural or artificial (as in the case of a vaccine). This form of immunity can last for weeks, months, or years. Conversely, **passive immunity** is established when ready-made antibodies or defensive cells are introduced into the body. This form of protection is passive because the individual's own immune system does not make antibodies or defensive cells against the disease agent or its toxins. Passive immunity is only temporary; the antibodies or cells transferred and the immunity they confer persist for only a few weeks to a few months before they are destroyed by the host's immune system. Let's take a closer look at these forms of immunity.

Active Immunity

In *naturally acquired active immunity*, the host's immune system takes an active role in preventing disease when exposed to an infectious agent by producing antibodies and defensive T lymphocytes to fight the infection. The immune system also initiates specific mechanisms that protect against future invasions by the same type of infectious agent. Naturally acquired immunity can last for weeks, months, years, or a lifetime, depending on how long the immune defenses persist.

An individual's response to immunization is an example of active immunity to a specific infectious agent or its toxins. Immunity in this case too is active because the person's immune system produces antibodies and/or specific immune defensive cells. Such immunity is said to be *artificially acquired active immunity*. A vaccine may contain live but weakened (attenuated) organisms, dead organisms, or their inactivated toxins (toxoids). Although injecting vaccine into people is not a natural process, immunizations do protect against future infections.

Passive Immunity

In *naturally acquired passive immunity*, antibodies are transferred in some manner from one person to an-

other. A fetus, for instance, can receive a natural, passive type of immunity in the form of antibodies that cross the placenta from its mother. After childbirth, antibodies can be transferred from mother to newborn via **colostrum** (ko-los'trum), the protein-rich fluid secreted by the mammary glands just after childbirth, before the appearance of true breast milk.

When antibodies are received from another host, *artificially acquired passive immunity* results. For example, a person who is bitten by a rattlesnake may receive a snake antivenin injection. Antivenins are antibodies produced in another animal, such as horses or rabbits.

Relationships among the various types of immunity are shown in Figure 18.1. Most of this chapter discusses acquired immunity and how it can generate specific defenses to particular pathogens and/or foreign substances.

CHARACTERISTICS OF THE IMMUNE SYSTEM

Properties of Antigens (Immunogens)

A host's immune responses are triggered by exposure to foreign substances, which in most cases are large protein molecules. A chemical compound that stimulates an immune response is called an **antigen,** or an **immunogen.** Although there is a functional distinction between these two terms, they are generally used interchangeably.

Antigens are large molecules. Many are protein in nature, but some are polysaccharides, and others are combinations such as glycoproteins (carbohydrate and protein) or nucleoproteins (nucleic acid and protein). A few glycolipids (carbohydrate and lipid) are also **immunogenic**—that is, they are potent stimulators of an-

FIGURE 18.1 The various types of immunity. Nonspecific immunity is largely innate or inborn, whereas specific immunity is acquired.

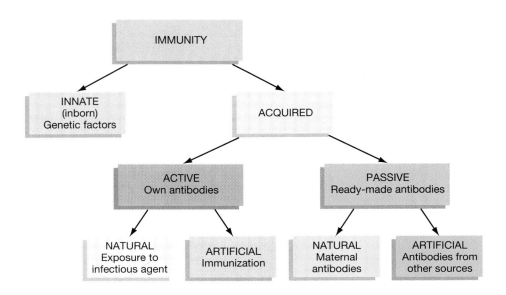

tibody production and defense cell activity. Many antigens are complex molecules with high molecular weights.

A host's immune response is not directed toward the entire antigen molecule, however, but rather to specific chemical groups called **epitopes,** or **antigenic determinants,** on the molecule (Figure 18.2a). Epitopes are responsible for the immunogenic properties of the antigen molecule.

Antigens are found on the surfaces of all cells—including bacteria, other microorganisms, and human cells—and viruses. Bacteria, for instance, can have antigens on capsules, cell walls, attachment pili (fimbriae), and flagella. In fact, most microorganisms have several different antigens, and many copies of each antigen, on their surfaces (Figure 18.2b).

Antigens determine important individual characteristics. For instance, human antigens on the surface of red blood cells determine blood types. Antigens on other cells determine whether a tissue or organ transplanted from a donor will be rejected by the recipient. Knowledge of how the human body responds to different antigens is important in making effective vaccines.

In some instances a small molecule called a **hapten** (hap'ten) can act as an antigen if it binds to a larger protein molecule. Haptens act as epitopes on the surfaces of proteins. They may bind to body proteins and provoke an immune response, even though neither the hapten nor the body protein can do so alone. For example, penicillin molecules represent haptens when they bind to protein molecules and elicit a specific immune response in the form of an allergic reaction.

Properties of Antibodies (Immunoglobulins)

One of the most significant immune system responses is the production of antibodies to antigens carried on pathogens and foreign substances. An **antibody**

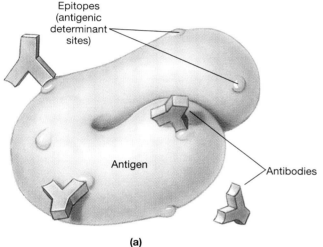

(a)

FIGURE 18.2 (a) A typical antigen–antibody reaction. Antibodies bind to specific chemical groups or structures, called epitopes, or antigenic determinants. (b) A gram-negative bacterial pathogen may have several antigens, or immunogens (for example, for flagella, pili, and cell wall), each with particular epitopes. Large, complex protein molecules may have several different antigenic determinants.

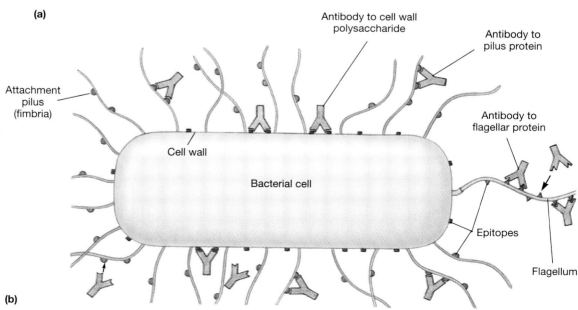

(b)

then is an "anti-antigen" protein produced in response to a particular epitope on the surface of a pathogen or other foreign agent. Many proteins are highly immunogenic.

As Figure 18.2b shows, antibodies do not bind to whole antigens; rather, each kind of antibody binds to a specific epitope. Nor do antibodies destroy the pathogens; rather, antibodies, as we will see, help other cells or molecules inactivate or destroy the infectious agents or their products.

Antibody Structure

Antibodies are a class of proteins called **immunoglobulins** (Ig) found in the blood serum and other body fluids surrounding tissues. Specific kinds of immunoglobulins also are present in the plasma membrane of particular lymphocytes.

Structurally, each antibody molecule of the type most abundant in serum is built from four polypeptide chains—two identical **light** (L) **chains** and two larger, identical **heavy** (H) **chains** (Figure 18.3). The four polypeptides are held together by disulfide bonds. Each light and heavy chain has a *variable region* and a *constant region*. The outer ends of the variable regions of each chain form an **antigen-binding site.** The shape of these regions determines the antigen, or more correctly, the epitope, to which the antibody will bind. Because epitopes differ from one another, the variable regions of immunoglobulin molecules also differ in their amino acid sequences. Each of the millions of different immunoglobulins has its own unique antigen-binding sites. Thus, each pair of light and heavy chains include two **Fab** (fragment antigen-binding) **fragments.**

The chemical structure of the tail portion of constant regions determines the class to which an im-

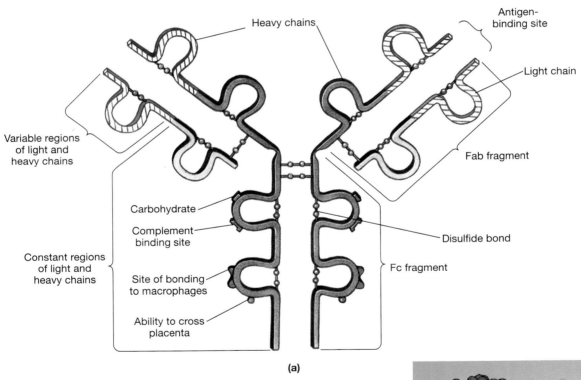

(a)

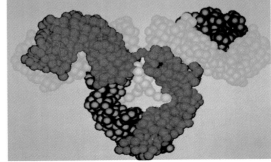

(b)

FIGURE 18.3 (a) The basic structure of the most abundant antibody (immunoglobulin) molecule in serum contains two heavy and two light chains, joined by disulfide bonds to form a Y shape. The upper ends of the Y, consisting of variable regions in both the light and heavy chains, differ from antibody to antibody. These variable regions form the two antigen-binding sites (part of the Fab fragment), which are responsible for the specificity of the antibody. The remaining part of the molecule consists of constant regions that are similar in all antibodies of a particular class. The Fc fragment determines the role each antibody plays in the body's immune responses. (b) A computer model of antibody structure. The two light chains are depicted in green, one heavy chain in red and the other heavy chain in blue.

munoglobulin belongs. These regions, called the **Fc** (fragment crystallizable) **fragments,** may contain sites to which certain immune defense cells of the host will bind. The antigen-binding sites attach to a disease agent's surface epitopes, coating the organism. The free ends of the immunoglobulin molecules (Fc fragments) serve to connect the coated organism to a macrophage or neutrophil and thereby promote phagocytosis.

Classes of Immunoglobulins

On the basis of the characteristics of their heavy chains, five classes of immunoglobulins have been identified in humans and other vertebrates (Table 18.1). The constant regions of all immunoglobulin molecules of the same class have similar amino acid sequences. The five classes are IgG (immunoglobulin class G), IgA, IgM, IgE, and IgD. Each class has different antibody activity and specialized functions.

IgG The main class of antibodies found in the blood is **IgG,** which accounts for as much as 75 percent of the serum immunoglobulin and 20 percent of all plasma proteins. It has the typical Y-shaped structure with two Fab regions, each containing an antigen-binding site. The Fc region attaches to receptors on phagocytic cells, allowing for efficient phagocytosis of antigen and antibody. An additional function of IgG is the activation of complement, which further helps phagocytes destroy the infectious agent. ∞ (Chapter 17, p. 460)

Among the immunoglobulin classes, IgG is the only one that can cross the placenta to provide antibody protection for the fetus. IgG also is the only immunoglobulin class whose Fc region is recognized by phagocytes and *natural killer* (NK) *cells.* ∞ (Chapter 17, p. 455) For example, IgG coating of a virus-infected cell permits NK recognition and destruction of the infected cell.

IgM The first immunoglobulin type produced in an immune response is the **IgM** class, which represents 5 to 10 percent of the immunoglobulin in normal human serum. The secreted form of IgM is a large molecule consisting of five Y-shaped units connected by their Fc regions to a J (joining) chain. Thus, it has 10 peripheral, but identical, antigen-binding sites. IgM in the form of a single Y-shaped unit is part of the B cell plasma membrane. IgM aids in phagocytosis and plays a minor role in inhibiting microbial adhesion to host tissues. It is the first immunoglobulin formed by a fetus. Because of its size, IgM is unable to cross the placenta.

IgA The **IgA** class represents over 10 percent of the total immunoglobulin. IgA occurs in small amounts in blood serum, but a form called *secretory IgA* is the main immunoglobulin in external body secretions such as tears, colostrum, saliva, and urine. It can remain attached to cells lining the respiratory, gastrointestinal, and urogenital tracts, but it is usually secreted across the mucosal surfaces of these systems.

| TABLE 18.1 | Properties of Immunoglobulins |

Property	IgG	IgM	IgA	IgE	IgD
Number of units	1	5	1 or 2	1	1
Activation of complement	Yes	Yes, strongly	Yes, by alternative pathway	No	No
Crosses placenta	Yes	No	No	No	No
Binds to phagocytes	Yes	No	No	No	No
Binds to lymphocytes	Yes	Yes	Yes	Yes	No
Binds to mast cells and basophils	No	No	No	Yes	No
Half-life (days) in serum	21	10	6	2	3
Percentage of total blood antibodies in serum	75	10	10	0.005	0.2
Location	Serum, extra-vascular, and across placenta	Serum and B cell membrane	Transport across epithelium	Serum and extra-vascular	B cell membrane

Although membrane-bound IgA consists of a typical single Y-shaped molecule, secretory IgA consists of two Y-shaped molecules held together at their Fc regions by a J chain (see Table 18.1). Secretory IgA molecules have a **secretory piece** that helps protect them from degradation and aids in the antibodies' secretion. The main function of IgA is to inhibit microorganisms from binding or adhering to tissues. It has a very minor role in activating complement. IgA does not cross the placenta, but it is present in colostrum.

IgE   The **IgE** (also called *reagin*) class has an extremely low concentration in serum. IgE molecules are found mainly in body fluids and beneath the skin and mucosa. The Fc section of IgE has a special affinity for receptors on the plasma membranes of basophils in the blood and of mast cells in tissues. IgE binds to these cells by its Fc region, leaving the antigen-binding sites free to bind antigens. This immunoglobulin class plays a key role in combating helminths and a damaging role in the development of allergies to such agents as drugs, pollens, and certain foods. Asthma and hay fever are common allergic diseases discussed in Chapter 19. Thus, levels of IgE will be elevated both in patients with allergies and in those harboring helminths.

IgD   The **IgD** class is found in the serum and makes up less than 1 percent of the total immunoglobulin. IgD is concentrated along with IgM on the plasma membranes of human B lymphocytes. The main function of IgD presumably is to serve as an antigen receptor in the early stages of the immune response.

In discussing concentrations of antigens and antibodies, immunologists often refer to titers. A **titer** (ti'ter) is the quantity (concentration) of a substance present in a specific volume of body fluid. For example, during an infection, an individual's antibody titer (the concentration of antibody in the serum) normally increases. An increasing antibody titer serves as an indication of an immune response by the body.

CELLS AND TISSUES OF THE IMMUNE SYSTEM

Special cells and tissues are as critical to a host's immune response as are antibodies. Specific responses involve lymphocytes—*B lymphocytes, T lymphocytes,* and *natural killer cells*—and *macrophages.* Lymphocytes develop from **lymphoid stem cells** in the bone marrow. Other leukocytes (white blood cells), erythrocytes (red blood cells), and platelets develop from *myeloid stem cells;* macrophages develop from monocytes. ∞ (Chapter 17, p. 453)

Dual Nature of the Immune System

Lymphocytes give rise to two major types of immune responses—*humoral immunity* and *cell-mediated immunity.* However, the presence of a foreign substance in the body often triggers both kinds of responses.

Humoral (hu'mor-al) **immunity** (also known as *antibody-mediated immunity*) is carried out by antibodies circulating in the blood. When stimulated by an antigen, B lymphocytes initiate a process that leads to the release of antibodies. Humoral immunity is most effective in defending the body against bacterial toxins, bacteria, and viruses before these agents enter cells.

Cell-mediated immunity is carried out by certain T lymphocytes. It occurs at the cellular level, especially in situations in which antigens are embedded in plasma membranes or are inside host cells and thus inaccessible to antibodies. Cell-mediated immunity is most effective in clearing the body of virus-infected cells, but it also may participate in defending against fungi, helminths, cancer, and foreign tissues, such as transplanted organs.

B Lymphocytes

B lymphocytes, commonly called **B cells,** mature in tissue called bursal-equivalent tissue. Differentiation of B cells originally was studied in birds. In these animals, B cells mature in an organ called the *bursa of Fabricius* (Fab-ris'e-us). However, no site equivalent to the bursa of Fabricius exists in mammals. Instead, in humans for example, B cells develop from lymphoid stem cells in the liver of the fetus; after birth these cells develop from lymphoid stem cells in the bone marrow (Figure 18.4). (The *B* in B cell refers to bursa). Thus, B cells can be replaced as needed by new ones from the bone marrow. B cell development involves many steps, culminating in the formation of mature B cells containing IgM and IgD immunoglobulins on their surface.

Following maturation, mature B cells migrate to the lymphoid tissues—lymph nodes, spleen, tonsils, adenoids, and gut-associated lymphoid tissues (GALT). ∞ (Chapter 17, p. 471) If a foreign antigen invades the body, the antigen's binding to the immunoglobulins on B cells triggers the cells to differentiate and mature into **plasma cells,** which secrete antibody. Which class of immunoglobulin is secreted depends on the chemical stimulus received by the B cell. B cells account for up to 15 percent of the lymphocytes circulating in the blood.

T Lymphocytes

Other lymphoid stem cells migrate from the bone marrow to the thymus, where they undergo specialization into **T lymphocytes,** more commonly called **T cells** (*T* for thymus). ∞ (Chapter 17, p. 452) In adults, when the thymus becomes less active than it

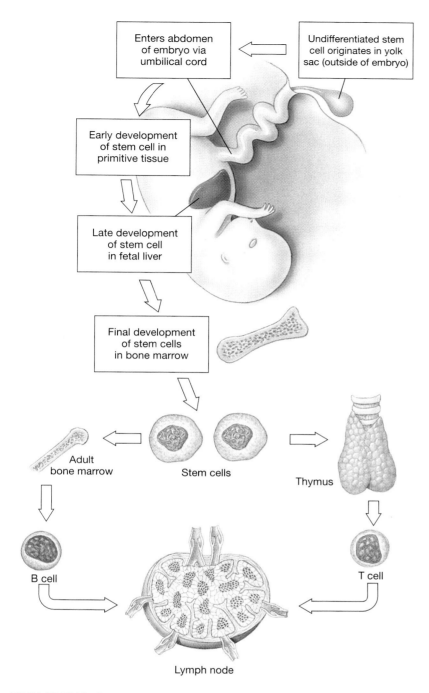

The diagram shows the following flow:

Undifferentiated stem cell originates in yolk sac (outside of embryo) → Enters abdomen of embryo via umbilical cord → Early development of stem cell in primitive tissue → Late development of stem cell in fetal liver → Final development of stem cells in bone marrow → Stem cells

Stem cells → Adult bone marrow → B cell → Lymph node

Stem cells → Thymus → T cell → Lymph node

FIGURE 18.4 Differentiation of stem cells into B cells and T cells occurs in the bone marrow and thymus, respectively. The mature lymphocytes then migrate to lymphoid tissues, such as the lymph nodes.

macrophages. T_C cells kill cells infected with viruses or other microorganisms. T cells are also involved with various types of allergic reactions (Chapter 19).

The characteristics of B cells, T cells, and macrophages, and how each relates to the two types of immune responses, are summarized in Table 18.2. Before we can consider these cells and their actions in greater depth, we must first discuss the various types of surface proteins that occur on lymphocytes.

Lymphocyte Surface Proteins

B cells and T cells physically resemble each other but can be distinguished by epitopes called **cluster of differentiation (CD) markers** located on their plasma membranes. Each marker is designated by a *CD number*, such as CD1, CD2, CD3, and so on.

Another group of membrane proteins on lymphocytes act as *receptors* that recognize epitopes. On B cells, these receptor proteins are IgM or IgD. Such receptors remain attached to the plasma membrane by their Fc region, with the two antigen-binding sites projecting outward (Figure 18.5a). T cells also have membrane receptors. Unlike B cell-associated immunoglobulins, the T cell protein receptors have but one antigen-binding site (Figure 18.5b).

A third group of cell surface proteins are the **major histocompatibility complex** (MHC) molecules. There are two classes of MHC molecules; they differ in the structure of their membrane polypeptides. In mammals, a cluster of genes located on a single chromosome encode for class I and class II molecules. Class I MHC molecules are located on most body cells. Class II MHC molecules are found only on lymphocyte-type cells and macrophages. MHC molecules are recognized by different classes of T cells and are essential to immune recognition reactions, such as those involved in tissue rejection or acceptance. T cell immune responses occur only when a foreign antigen and the MHC antigen are *presented* (displayed) on a cell surface.

Table 18.3 summarizes the surface proteins of the immune cells. We shall return to the critical role of these proteins later.

is in childhood, the number of T cells does not decline. This suggests that mature T cells live for a prolonged period and do not require constant replacement. Like B cells, mature T cells are found in all lymphoid tissues, and they account for about 75 percent of the lymphocytes circulating in the blood.

T cells differentiate further into one of two classes: **helper T (T_H) cells** and **cytotoxic T (T_C) cells.** T_H cells help stimulate other immune cells, such as B cells and

Characteristic	B cells	T cells	Macrophages
Site of production	Bone marrow	Thymus or under thymic hormones	
Type of immunity	Humoral	Cell-mediated; assist in humoral	Humoral and cell-mediated
Subpopulations	Plasma cells and memory cells	Cytotoxic, helper, and memory cells	Fixed and wandering
Presence of surface antibodies	Yes	No	No
Presence of foreign surface antigens	Yes	No	Yes
Presence of receptors for antigens	Yes	Yes	No
Life span	Memory cells: years; others: days or weeks	Years	Years
Secretory product	Antibodies	Cytokines	Cytokines
Distribution (% leukocytes)			
Peripheral blood	15–30	55–75	2–12
Lymph nodes	20	75	5
Bone marrow	75	10	10–15
Thymus	10	75	10

Helper T (T_H) Cells

Helper T cells were so named because they provide chemical signals that are necessary for B cells to differentiate into antibody-producing cells and that extend cell-mediated immune responses. When activated, T_H cells produce **cytokines,** which are soluble chemicals that regulate the activities of T cells, B cells, monocytes, and other cells of the immune system. ∞ (Chapter 17, p. 453) Cytokines are cellular products that include proteins or protein-like molecules, lymphokines and interleukins, that act as chemical mediators. **Lymphokines** are produced by antigen-sensitized lymphocytes. **Interleukins** (IL) are produced by lymphocytes, macrophages, and mast cells. Cytokines provide communication and regulatory functions between different components involved in the immune responses. Interleukins especially are known for their regulation of the growth and differentiation of lymphocytes and of blood-forming stem cells.

Two types of T_H cells are known: T_H1 cells (*inflammatory T cells*) and T_H2 cells. They are distinguished by the type of cytokines they produce. Also, T_H1 cells are involved with cell-mediated immune responses, whereas T_H2 cells are associated with humoral immunity.

Cytotoxic T (T_C) Cells

Cytotoxic T cells destroy certain tumor cells and host cells that have been infected with viruses or with intracellular bacterial pathogens. When a virus has evaded humoral immunity—entering and establishing itself inside a host cell, certain peptide fragments are displayed on the infected cell's plasma membrane along with MHC class I molecules. T_C cells can recognize such peptide fragments and respond by lysing the antigen-bearing cell. Lysis occurs by the release of **cytotoxins,** exotoxins (proteins) that kill the infected

---

FIGURE 18.5 Receptors that recognize antigens or antigen fragments are located on B and T cells. (a) B cell receptors are immunoglobulins of the IgD and IgM classes. Each has two antigen-binding sites. (b) Receptors on T cells are not immunoglobulins but plasma membrane proteins with a single antigen-binding site.

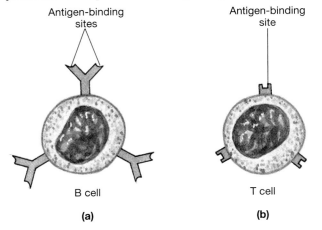

Antigen-binding sites

Antigen-binding site

B cell

T cell

(a)

(b)

TABLE 18.3   **Lymphocyte Surface Proteins**

	Surface Protein		
Lymphocyte	Unique CD	Antigen or Peptide Receptor	MHC
B cell	19, 20, 21	IgD, IgM	Classes I and II
T_H cell	4	T cell receptor	Classes I and II
T_C cell	8	T cell receptor	Classes I and II
NK	16, 56	None of the above	Classes I and II

CD = cluster of differentiation; MHC = major histocompatibility complex cell. The binding of T_C cells to a virus-infected cell triggers the release of the specific cytotoxin **perforin,** which bores holes in the target (infected) cell's plasma membrane, causing essential molecules to leak out of the cell. This process is similar to the action of membrane attack complexes of complement. Another type of cytotoxin, called **granzyme,** enters an infected cell through the pores formed by perforin. Together, these cytotoxins produce 10 to 20 nm-wide pores in the target plasma membrane and thereby kill the cell by osmotic shock. T_C cells are also important in graft rejection and in the lysis of cancer cells that present tumor-specific antigens.

Immunity is important in regulating and terminating the immunological activities of both B and T cells. Prolonged immune responses can interfere with other immunological activities. Some immunologists have suggested that T_H or T_C cells called **suppressor T** (T_S) cells may play a role, but how this may occur is not known. In individuals with AIDS, the proportion of T_S cells to T_H cells is substantially higher than in non-HIV-infected persons because unlike T_H cells, T_S cells are not attacked by HIV. This imbalance, along with the destruction of T_H cells, may contribute to the immunosuppression that is typical of AIDS.

Natural Killer Cells

The natural killer (NK) cells make up a discrete lymphocyte lineage distinct from that of B and T cells. ∞ (Chapter 17, p. 455) NK cells do not have membrane surface markers, which are typical of B and T cells. Defined initially by their ability to kill certain tumor cells and virus-infected cells, NK cells are large lymphocytes. They mature from lymphoid stem cells in the bone marrow and can comprise up to 15 percent of blood lymphocytes.

NK cells kill foreign cells, virus-infected cells, and mutated (tumor) cells in ways similar to those described for T_C cells. NK cells have the same set of cytotoxins that are found in T_C cells—that is, NK cells can use perforins and granzymes to perforate and kill cells by osmotic shock. They also are capable of binding to cells, triggering programmed cell death.

NK cells, however, have a unique killing strategy different from the strategies of T_C cells. If any cell's surface "lacks" a critical "self peptide fragment" associated with MHC class I molecules, NK cells will attack and kill the cell. Foreign cells and virus-infected cells lack this critical fragment, and tumor cells may lack the MHC class I molecule entirely. In all these cases, NK cells will destroy the cells.

Macrophages

When a macrophage (Chapter 17) encounters certain stimuli, it undergoes a process called *activation,* in which its metabolic rate, movement, and phagocytic activity rapidly increase. ∞ (p. 453) In addition, macrophages secrete numerous cytokines, which influence the growth and activities of other cell types.

Activated macrophages are among the most important types of **antigen-presenting cells** (APCs)—cells that process and present (display) foreign peptide fragments in a form that can be recognized by lymphocytes, which can then generate an immune response.

Antigen Processing

During an infection, the immune system's lymphocytes are alerted by the *processing* of the foreign antigens by the APCs. There are two such antigen processing pathways. In the case of a virus-infected cell, virus particles are degraded to peptide fragments. These fragments are then carried to the endoplasmic reticulum, where they associate with MHC class I molecules, forming *MHC class I:peptide complexes.* These complexes then move through the Golgi complex and on to the plasma membrane. ∞ (Chapter 4, p. 95) There they are available for presentation to T_C cells.

In the case of macrophage phagocytosis of a bacterial cell, the bacterium is brought into the cell and digested into peptide fragments in the phagolysosome. ∞ (Chapter 17, p. 454) MHC class II molecules from the endoplasmic reticulum arrive in this compartment, where they bind with a peptide fragment. The MHC class II:peptide complexes then move to the plasma membrane, where they are available for presentation to T_H cells. In both cases, lymphocyte stimulation and triggering of immune responses can take place only when the MHC molecules present the peptide fragments on the cell surface of the APC.

Properties of Immune Responses

Both humoral and cell-mediated responses have four common attributes: (1) recognition of self versus nonself, (2) specificity, (3) heterogeneity, and (4) immunological memory.

Recognition of Self versus Nonself

For the immune system to respond to foreign substances, it must distinguish between host tissues and substances that are foreign to the host. Immunologists refer to normal host substances as **self** and to foreign substances as **nonself.** How does the immune system distinguish self from nonself?

CLONAL SELECTION THEORY The **clonal** (klo'nal) **selection theory,** first proposed by Sir Frank Macfarlane Burnet in the 1950s, explains one way the immune system might distinguish self from nonself (Figure 18.6).

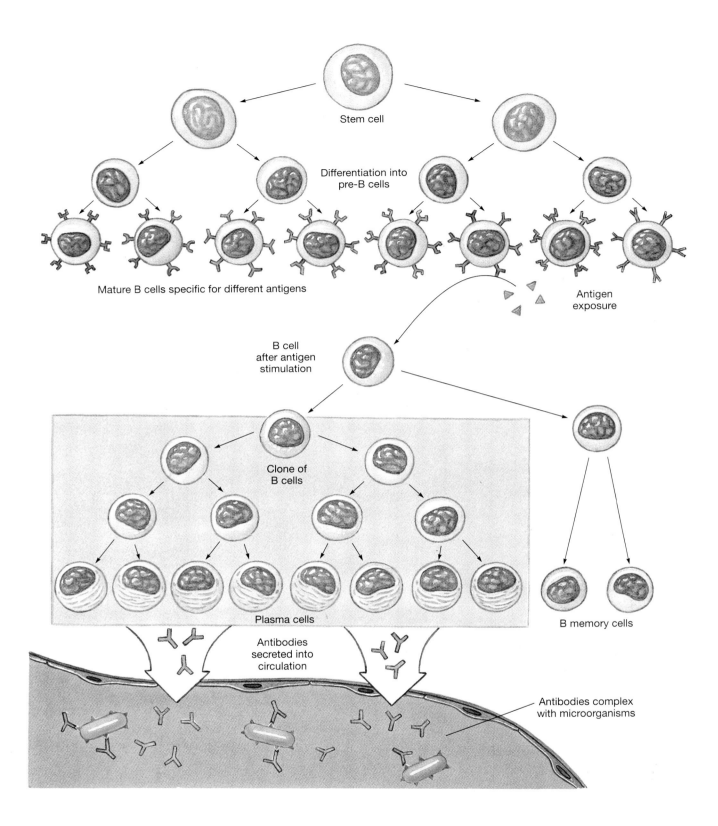

Stem cell

Differentiation into
pre-B cells

Mature B cells specific for different antigens

Antigen
exposure

B cell
after antigen
stimulation

Clone of
B cells

Plasma cells

Antibodies
secreted into
circulation

B memory cells

Antibodies complex
with microorganisms

FIGURE 18.6 According to the clonal selection theory, one of many B cells responds to a particular antigen and begins to divide, thereby producing a large population of identical B cells (a clone). All cells of such a clone produce the same antibody against the original epitope. B memory cells are also produced.

According to this hypothesis, embryos contain many different lymphocytes, each genetically programmed to recognize a particular epitope and make antibodies to destroy it. If a B lymphocyte encounters and recognizes that antigen after development is complete, the lymphocyte divides repeatedly. It produces a **clone,** a group of identical progeny cells that make the same antibody. If, during embryological development, the B lymphocyte encounters its programmed antigen as part of a normal host substance (self), the lymphocyte is destroyed or inactivated. This mechanism removes lymphocytes that can destroy host tissues and thereby creates **tolerance** for—that is, an inability to recognize—self. It also selects for the survival of B lymphocytes that will protect the host from foreign antigens. Because the clonal selection of B cells responds to antigens without the aid of T_H cells, these antigens are called *T-independent antigens.*

CLONAL DELETION In 1953 Peter Medawar showed that the lack of self recognition—tolerance—is acquired during lymphocyte development in the bone marrow. Immunologists now know that immature lymphocytes having receptors for self antigens are removed from the immune system by clonal deletion. **Clonal deletion** is a process in which the binding of lymphocytes to self antigens triggers a condensation and disintegration of these cells. By clonal deletion, lymphocytes carrying receptors to self antigens are eliminated, so no immune response to host components develops (Figure 18.7).

Tolerance can also be produced by irradiation during cancer treatment or by the administration of immunosuppressant drugs to prevent rejection of transplanted organs. Irradiation and drug treatment inhibit the host's immune system, which is then unable to detect and respond to foreign antigens in transplanted organs. But the host's immune system also fails to respond to infectious organisms.

Vertebrates, including humans, have developed an elaborate recognition system for distinguishing self from nonself. This recognition also involves the MHC antigens and is discussed more thoroughly in Chapter 19.

Specificity

By the time the human immune system matures at age 2 to 3, it can recognize millions of foreign substances as nonself. Furthermore, it reacts in a different way to each foreign substance. This property of the immune system is called **specificity.** Due to specificity, each reaction is directed toward a specific foreign antigen, and the response to one antigen generally has no effect on other antigens. However, **cross-reactions,** re-

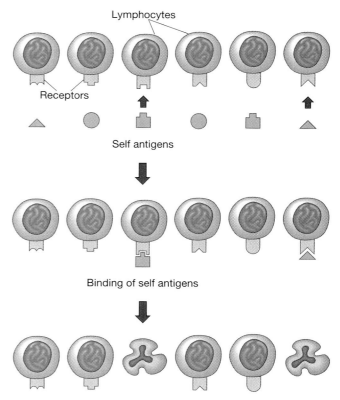

FIGURE 18.7 Clonal deletion is a process that removes those lymphocytes that have receptors for self antigens. When lymphocytes bind to self antigens, clonal deletion occurs; that is, those lymphocytes die as a result of condensation and disintegration of cell nuclei. Lymphocytes lacking self receptors survive.

actions of a particular antibody with very similar antigens, can occur. For example, certain microorganisms, such as *Treponema pallidum* (the bacterium that causes syphilis), have the same haptens as some epitopes on human cells, such as heart muscle cells, although the carrier molecules are quite different. Antibodies against this particular hapten can therefore react with these otherwise vastly different cells. Cross-reactions also occur between strains of microorganisms. Suppose that three strains of pneumococci can cause pneumonia and that each produces a particular capsular antigen, A, B, or C. A person who has recovered from an infection with strain A has anti-A antibodies. This person may also have some resistance to strains B and C because anti-A antibodies cross (that is, they react) with antigens B and C.

Heterogeneity

The ability of the immune system to respond in a specific way allows it to attack particular antigens. But in a lifetime, the human body encounters hundreds of different foreign antigens. The property of **heterogeneity** (versatility, or diversity) refers to the ability of the immune system to produce many different kinds of antibodies, each of which reacts with a different epitope. When a bacterium or other foreign agent has more than one kind of epitope, the immune system may make a different antibody against each. And the immune system is even capable of producing antibodies against molecules it has never before encountered.

Immunological Memory

In addition to its ability to respond in a specific way to a heterogeneous assortment of antigens, the immune system also has the property of **immunological memory**—that is, it can recognize substances it has previously encountered. Immunological memory allows the immune system to respond rapidly to defend the body against an antigen to which it has previously reacted. In addition to producing antibodies during its first reaction to the antigen, the immune system also makes **memory cells,** which stand ready for years or decades to initiate antibody production or lymphocyte production quickly. Consequently, the immune system responds to second and subsequent exposures to an antigen much more rapidly than to the first exposure. This prompt response due to "recall" by memory cells is called an **anamnestic** (secondary) **response.**

The attributes of specific immunity are summarized in Table 18.4. With these attributes in mind,

TABLE 18.4	Main Features of Specific Immunity
Feature	**Description**
Recognition of self versus nonself	The ability of the immune system to tolerate host tissues while recognizing and destroying foreign substances. Due to the destruction (deletion) of clones of lymphocytes during embryonic development.
Specificity	The ability of the immune system to react in a different and particular way to each foreign substance.
Heterogeneity	The ability of the immune system to respond in a specific way to a great variety of different foreign antigens.
Memory	The ability of the immune system to recognize and respond quickly to foreign substances to which it was previously exposed.

APPLICATIONS

HOW B CELLS BUILD DIVERSE ANTIBODIES

How can B cells make antibodies to almost any foreign antigen or foreign substance with which they come in contact? The key to such heterogeneity lies in the immunoglobulin genes within each B cell. When B cells are formed in the bone marrow, each cell randomly pieces together different segments of its antibody genes.

In the embryo, the relatively few gene segments that code for the constant region of each light and heavy chain are not adjacent to the hundreds of gene segments that code for the variable regions. Let's look at how a light chain is built.

Light chains are formed when the DNA that separates a particular variable (V) segment from a constant (C) segment is removed, and the two gene segments are joined by a junction (J) segment. The now-joined segments form one continuous DNA sequence that represents the functional light-chain gene. Heavy chains are formed in a similar manner. Following transcription and translation, light-chain polypeptides are produced. These can be combined with heavy-chain polypeptides to form the functional antibody molecule. Thus, the heterogeneity of antibody-binding sites comes from the random combinations of variable gene segments that join with constant gene segments to form the light and heavy chains.

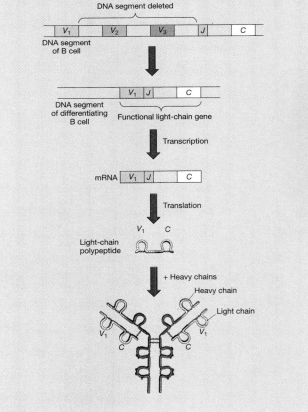

we will now look in more detail at the two kinds of specific immunity—humoral and cell-mediated immunities.

HUMORAL IMMUNITY

Humoral (hyu′mor-al) *immunity* is most effective in defending the body against bacteria, bacterial toxins, and viruses that have not entered cells. **Humoral immune responses** are carried out by antibodies circulating in the blood and tissues. Two types of antigens are recognized: 1) **T-independent** (TI) **antigens,** which do not require the help of T cells to stimulate B cell antibody production, and 2) **T-dependent** (TD) **antigens,** which depend on a collaboration between T_H2 cells and B cells to initiate antibody production. Most common antigens, particularly proteins, are TD antigens.

Following **antigen challenge**—that is, exposure to a foreign antigen—there is a lag period of 4 to 6 days before antibodies can be detected in blood serum. This period reflects the events involved in antigen recognition, antigen processing, and the specific activation of immune system cells. APCs process the antigen so that peptide fragments can be presented to T_H2 cells. The T_H2 cells then bind with B cells that are presenting the same antigen fragment. This binding triggers the production of several interleukins—namely, IL-4, IL-5, IL-6, and IL-10—which stimulate B cell divisions. Such divisions represent the clonal selection of B cells and lead to the formation of antibody-producing plasma cells.

The humoral immune response involves a learning system that responds to antigenic exposure by forming large quantities of specific antibody. A critical aspect of this response is that the antibody titer increases with time and/or with additional antigen exposures.

Primary and Secondary Responses

In humoral immunity the **primary response** to an antigen occurs when the antigen is first recognized by host B cells. After recognizing the antigen, B cells divide to form plasma cells, which begin to synthesize antibodies. Within a few days (a *lag period*), antibodies begin to appear in the blood plasma or serum; the antibodies increase in concentration over a period of 1 to 10 weeks. The first antibodies formed are IgM, which can attack foreign substances directly. As IgM production *declines,* IgG production accelerates and eventually reaches a *plateau* before it too declines. The concentrations of both IgM and IgG can become so low as to be undetectable in plasma samples. However, memory cells persist in lymphoid tissues. These cells retain their ability to recognize a particular antigen. They can survive without dividing for many months to many years.

When an antigen recognized by memory cells enters the blood, a **secondary response** occurs. Memory B cells may serve as the principal APCs in secondary humoral responses. The presence of memory cells makes the secondary response much faster than the primary response. Some memory cells divide rapidly, producing plasma cells, and others remain as memory cells. Plasma cells quickly synthesize and release large quantities of antibodies. In the secondary response, as in the primary response, IgM is produced before IgG. However, IgM is produced in smaller quantities over a shorter period, and IgG is produced sooner and in much larger quantities, than in the primary response. Thus, the secondary response is characterized by a rapid increase in antibodies, most of which are IgG. The primary and secondary responses are compared in Figure 18.8 and in Table 18.5.

How Humoral Responses Eliminate Foreign Antigens

Cytokines, which are polypeptides produced by immune system cells, regulate which antibody will be produced by plasma cells. Under the influence of such T cell factors, the progeny of B cells involved in the early IgM response change or switch over to form IgG, IgA, or IgE molecules. As a consequence of this immunoglobulin class switching, a number of plasma cells produce different antibody types, but all the progeny of a particular clone produce antibodies with the same

APPLICATIONS

HUMORAL IMMUNE RESPONSES: WHAT'S IN A NAME?

The term *humoral* in humoral immune response is derived from the word *humor* (from *umor*, Latin for "liquid"). It originally referred to the four basic body fluids, or "humors"—blood, phlegm, yellow bile, and black bile—which ancient physicians believed must be present in proper proportions for an individual to enjoy good health. If any of these fluids were out of balance, a person was said to be "in bad humor" and likely to be diseased. Because this type of acquired immunity involves antibodies that circulate in the blood fluid, "humoral immune response" seemed logical.

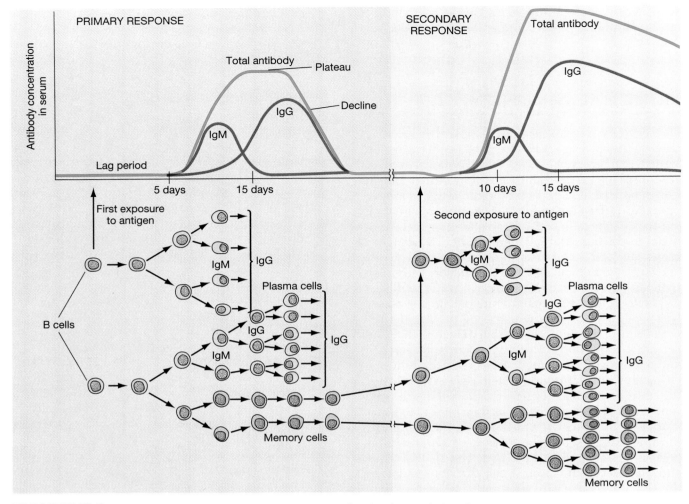

FIGURE 18.8 Primary and secondary responses to an antigen, showing the correlation of antibody concentrations with the activities of B cells. Cytokines trigger the class switching from IgM to IgG.

antigenic specificity. In other words, once a plasma cell is produced, it secretes antibodies with the same antigen-binding properties as the antigen-receptor molecules on the surface of its parent B cell.

TABLE 18.5	**Comparison of Primary and Secondary Immune Responses**	
Feature	Primary Immune Response	Secondary Immune Response
Length of lag period	5 days	1–2 days
Rate of antibody increase	Slow	Fast
Total antibody plateau	Low	High
Antigen sensitivity	Low	High
Antibody affinity	Low	High
Antibody class	IgM and IgG	Mainly IgG

Antibodies provide immunity to infection and disease in three ways—by neutralization, opsonization, or activation of complement. For example, antibodies can neutralize the infectivity of a virus, the harmful effects of an exotoxin, or the ability of a bacterium to colonize a host's tissue. Such **neutralization** is generally brought about by reaction between the antibody and an epitope that is required for attachment of the organism or toxin to a target host cell (Figure 18.9a). IgA and IgG are particularly significant in neutralizing activities.

Microbes that escape neutralization can still be eliminated by *opsonization*. ∞ (Chapter 17, p. 462) In this process, an antibody coats a foreign invader by binding to the invader's epitopes with its own antigen-binding sites. The constant regions of the antibody's heavy chain are then available to attach to specific receptors on macrophages, thus aiding in phagocytosis and the ultimate destruction of the invader (Figure 18.9b).

Immune complexes, which are aggregates of IgM or IgG antibodies and antigen, activate the classical pathway of the *complement cascade,* as described in Chapter 17. ∞ (p. 460) Complement components can intensify a wide range of nonspecific host defenses, including the lysis of many gram-negative bacteria with which antibody has reacted (Figure 18.9c). Similar events occur with blood and tissue cells that carry surface antigens recognized as foreign by the host's immune system.

We have now considered the major characteristics of the humoral immune responses—how B cells are activated, how antibodies are produced, and how they function to eliminate pathogens or toxins. These processes are summarized in Figure 18.10.

MONOCLONAL ANTIBODIES

Generally, when a population of B cells is exposed to an antigen, many B cells will divide and develop into clones, each making a different antibody. Such mixtures are called *polyclonal antibodies* because different antibodies bind to different epitopes on the antigen. Conversely, **monoclonal antibodies** are produced in the laboratory from a clone of cultured cells, all making the same specific antibody to one specific epitope.

FIGURE 18.9 Antibodies produced by humoral immune responses eliminate foreign agents in three ways: (a) neutralization of pathogens and toxins by IgA or IgG, (b) opsonization of bacteria by IgG, and (c) cell lysis initiated by IgM or IgG immune complexes allows for the formation of membrane attack complexes involving complement proteins.

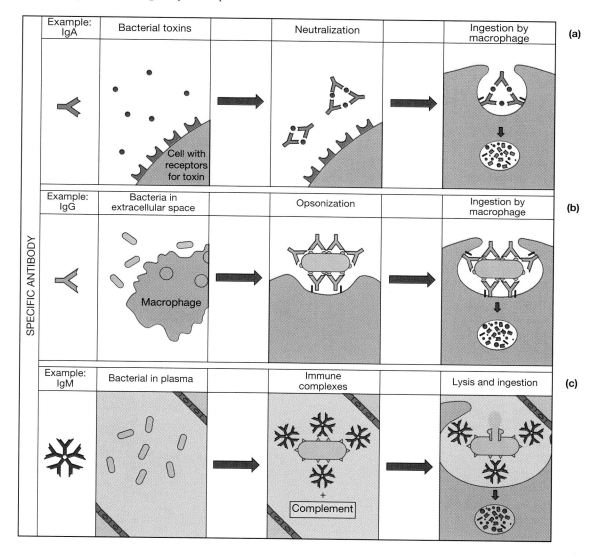

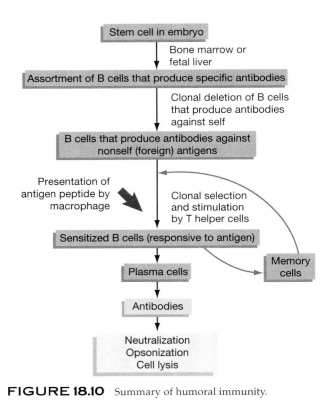

```
Stem cell in embryo
        │ Bone marrow or
        │ fetal liver
        ▼
Assortment of B cells that produce specific antibodies
        │ Clonal deletion of B cells
        │ that produce antibodies
        │ against self
        ▼
B cells that produce antibodies against
        nonself (foreign) antigens

Presentation of        Clonal selection
antigen peptide by     and stimulation
macrophage             by T helper cells
        ▼
Sensitized B cells (responsive to antigen)
        │                    ───► Memory
        ▼                          cells
   Plasma cells
        │
        ▼
    Antibodies
        │
        ▼
  Neutralization
  Opsonization
  Cell lysis
```

FIGURE 18.10 Summary of humoral immunity.

The method for producing monoclonal antibodies was devised by G. Köhler and C. Milstein in the mid-1970s. (They shared the 1984 Nobel Prize in medicine for their work.) After a mouse is injected with the antigen of interest, immune responses in the mouse's spleen activate B cells that recognize the antigen (Figure 18.11). Some of these B cells are mixed with myeloma cells (malignant B cells). The myeloma cells are used because they will divide indefinitely. When the two cell types are mixed, they can be made to fuse with one another to make a cell called a *hybridoma*. Only hybridoma cells grow in the growth medium. The unfused spleen cells cannot continue to divide, and the unfused myeloma cells starve to death.

A hybridoma, being part myeloma, can divide repeatedly. Like its spleen cell parent, it secretes large quantities of a highly specific single (hence "monoclonal") antibody. Which antibody a given hybridoma produces is determined by the antigen to which the spleen cells were sensitized before their progeny were mixed with myeloma cells. Many different hybridomas can be produced by repeated applications of this technique. If one specific antibody is wanted, tests must be used to find which hybridomas are synthesizing that antibody; those clones are then selected and cultured.

Although monoclonal antibodies were first produced in 1975 as research tools, scientists quickly rec-ognized their practical uses, especially in diagnostic tests and in therapy. Several diagnostic procedures that use monoclonal antibodies are now available. Generally, these procedures are quicker and more accurate than previously used procedures. For example, a monoclonal antibody can be used to detect pregnancy only 10 days after conception. Other monoclonal antibodies allow rapid diagnosis of hepatitis, influenza, herpes, streptococcal, and chlamydial infections. Diagnostic tests for other infectious diseases and allergies are rapidly being developed. Progress is also being made in using monoclonal antibodies to diagnose various cancers. In fact, monoclonal antibodies can detect malignant cells before those cells have multiplied to produce a tumor.

Methods of using monoclonal antibodies in therapy are under development. One idea being tested is to attach a specific drug or toxin to the Fc region of monoclonal antibodies, forming *immunotoxins*. Immunotoxins given to a patient should carry the toxic substance directly to the cells bearing the appropriate epitope. The hope is that these therapeutic monoclonal antibodies will damage infected or malignant cells without damaging normal cells. Unfortunately, early trials suggest that the immunotoxins can cause liver damage.

In another approach, monoclonal antibodies against tumor antigens should reduce or destroy the tumor. Patients often display allergic reactions to myeloma proteins that accompany the antibodies. In cancer treatment trials, some tumors targeted by monoclonal antibodies initially did regress but soon reappeared in a mutated form. Oncologists are now trying combination therapy, in which tumors are treated with two or three different monoclonal antibodies simultaneously. The reasoning is that no tumor can mutate fast enough to become resistant to three different antibodies simultaneously.

BIOTECHNOLOGY

DEODORANT MONOCLONAL ANTIBODIES

A unique commercial use of monoclonal antibodies has been patented by the Gillette Company. The company has devised a deodorant that contains monoclonal antibodies that attack *Staphylococcus haemolyticus*, the organism that causes perspiration-associated body odor. These antibodies block a bacterial enzyme necessary for the production of the unpleasant-smelling compounds.

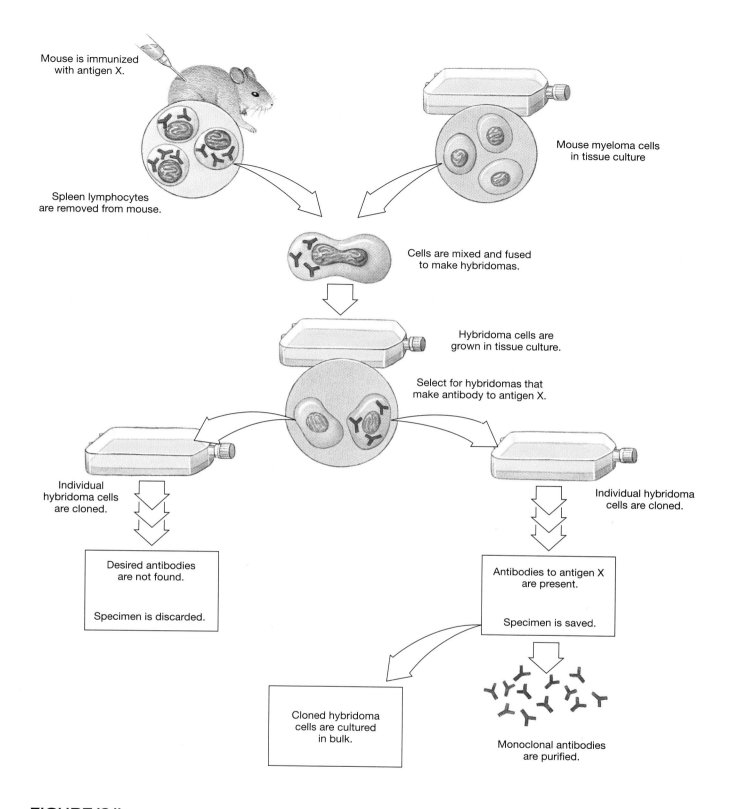

Mouse is immunized with antigen X.

Spleen lymphocytes are removed from mouse.

Mouse myeloma cells in tissue culture

Cells are mixed and fused to make hybridomas.

Hybridoma cells are grown in tissue culture.

Select for hybridomas that make antibody to antigen X.

Individual hybridoma cells are cloned.

Individual hybridoma cells are cloned.

Desired antibodies are not found.

Specimen is discarded.

Antibodies to antigen X are present.

Specimen is saved.

Cloned hybridoma cells are cultured in bulk.

Monoclonal antibodies are purified.

FIGURE 18.11 Production of monoclonal antibodies. Only the hybridoma cells grown in culture will survive, because any unfused spleen cells cannot divide, and any unfused mouse myeloma cells cannot get the nutrients they need to grow.

CELL-MEDIATED IMMUNITY

Humoral immunity, which we just learned involves B cells and T_H2 cells and antibodies, is efficient in removing pathogens and foreign substances from body fluids. However, if a pathogen infects body cells, antibodies are ineffective because they cannot penetrate cells. **Cell-mediated immune responses,** the second type of immune response, then come into play. Cell-mediated immunity involves the direct actions of T cells to clear the body of viruses, bacteria, and other pathogens that have invaded host cells. T-cell-mediated responses also account for rejection of tumor cells and immunological responses to transplanted tissues.

T cells, as noted earlier, either mature in the thymus or develop under the influence of thymic hormones. Unlike B cells, T cells do not make antibodies. However, they do have surface membrane receptors that can recognize peptide fragments that are bound to MHC molecules. The cell-mediated immune response involves the differentiation and actions of different types of T cells and the production of chemical mediators, or cytokines.

The Cell-Mediated Immune Reactions

The cell-mediated immune reactions typically begin with the processing of an antigen by APCs. The major APCs are the macrophages and dendritic cells. *Dendritic cells* are found in lymphoid tissues where T cells are located. Antigen processing involves the ingestion and degradation of the pathogen within an APC. However, during the degradation process, some antigen fragments (peptides) become associated with MHC molecules, either MHC I or MHC II. These MHC:peptide complexes are then placed (presented) on the membrane surface of the APC.

As we described for humoral immune responses, when immature T cells become exposed to antigens (antigen challenge), the cells mature into one of several types of T cells (Figure 18.12). During differentiation, some T memory cells also are formed. As in humoral immunity, the persistence of memory cells in cell-mediated immunity allows the body to recognize antigens to which T cells have previously reacted and to mount more rapid subsequent responses. As it does for B cells, clonal deletion in the thymus removes those T cells that bear receptors for self antigens.

The reactions of cell-mediated immunity are summarized in Figure 18.13. Refer to the figure as you read about the functions of different kinds of T cells.

T Cell Involvement

When a macrophage or dendritic cell presents MHC class II:peptide complex on its surface, only T cells with the proper receptor bind to the complex (Figure 18.13a). Binding stimulates macrophages and the T cells to secrete interleukin-1 (IL-1) and interleukin-2 (IL-2), respectively, which then causes these cells to divide and differentiate into activated T cells. These T cells can differentiate further into either T_H1 cells (inflammatory T cells) or T_H2 cells. What determines which kind of cell is produced is not clearly understood by immunologists. Cytokines produced by the infectious agent and the density of MHC class II:peptide on the APC surface probably are important. However, the consequences of this differentiation are significant.

T_H1 (INFLAMMATORY T) CELLS If the T cells mature as T_H1 cells, they trigger a cell-mediated response characterized by the elimination of immune cells (macrophages) infected with viral, bacterial, or other pathogens. Some bacteria—such as those that cause tuberculosis (*Mycobacterium tuberculosis*), Hansen's dis-

FIGURE 18.12   After T cells are challenged by antigens, the cells differentiate into one of several types of functioning T cells.

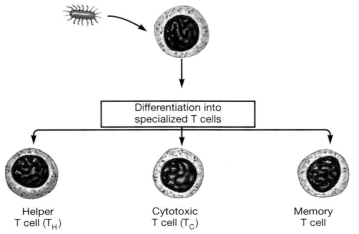

Helper
T cell (T_H)

Cytotoxic
T cell (T_C)

Memory
T cell

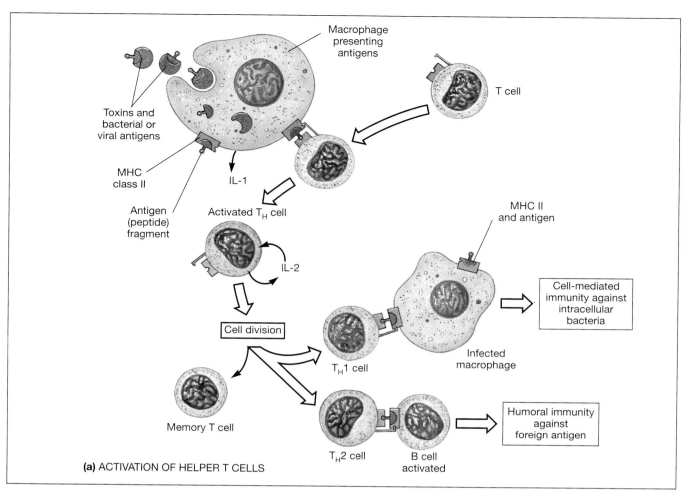

(a) ACTIVATION OF HELPER T CELLS

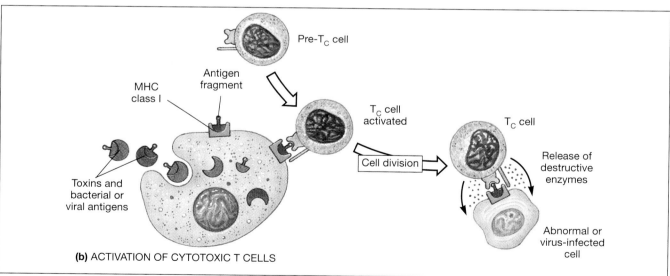

(b) ACTIVATION OF CYTOTOXIC T CELLS

FIGURE 18.13 The reactions in cell-mediated immunity. (a) The macrophage has processed an antigen and inserted an antigen (peptide) fragment into its plasma membrane as an MHC class II molecule. T_H cells have receptors that recognize the peptide fragment on MHC class II. Binding causes the T_H cells to become activated. The activated T cells then differentiate into either T_H1 cells or T_H2 cells. T_H1 cells activate infected macrophages to destroy internal bacterial infections. T_H2 cells activate B cells (humoral immune responses) by binding to MHC class II:peptide presented by the B cells. (b) Presenting the same peptide fragment on MHC class I to T_C cells activates these cells to attack infected cells, especially abnormal or virus-infected cells.

ease (leprosy; *M. leprae*), and brucellosis (*Brucella* species)—and some fungi—the fungi responsible for pneumocystic pneumonia (*Pneumocystis carinii*)—can continue to grow in cytoplasmic vesicles even after they have been engulfed by macrophages. T_H1 cells combat such infections by releasing γ-interferon. This cytokine causes macrophages to become sensitized to other cytokines, especially tumor necrosis factor-α (TNF-α). ∞ (Chapter 17, p. 455) TNF-α increases production of toxic hydrogen peroxide, nitric oxide, and activity of lysosomal enzymes that attack the phagocytized organisms; it thereby accelerates the inflammatory response. If T_H1 cells were not produced, the microbes would continue to survive in macrophages. For example, with tuberculoid leprosy, in which infected areas of the skin are walled off in granulomas, would result. ∞ (Chapter 17, p. 457)

T_H2 CELLS If the activated T cells mature as T_H2 cells, they help activate B cells to differentiate into plasma cells and secrete antibodies. By processing antigen and presenting peptide fragments on MHC class II molecules, T_H2 cells bind to the MHC class II:peptide complex and release interleukins that trigger B cell differentiation.

T_C CELLS APCs also have the ability to place peptide fragments from antigens onto MHC class I molecules (Figure 18.13b). In this situation, the MHC class I:peptide complex is recognized by T_C cells. Activation of these lymphocytes leads to the destruction of virus-infected cells, tumor cells, or foreign cells (in the form of transplants) through the killing mechanisms described earlier in this chapter.

We have now completed the discussion of cell-mediated immunity—how it is initiated and how its effects are produced. Comparing Figure 18.13 with Figure 18.14 and studying Table 18.2 will serve to highlight similarities and differences between humoral and cell-mediated immunities and to provide an overview of acquired immunity.

FIGURE 18.14 Summary of cell-mediated immunity. (CD stands for "cluster of differentiation.")

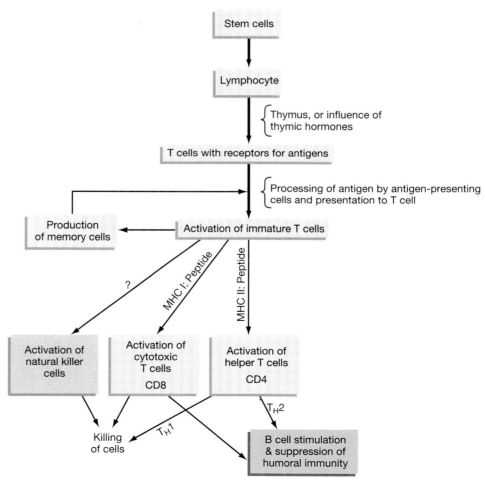

CLOSE-UP

IMMUNE CELL INFECTIONS

An agent that infects T cells is devastating because the agent destroys the very cells that might have combated the infection. HIV is just such an agent. HIV invades T_H cells, dendritic cells, and macrophages (APCs), the very heart of the immune response. Infection and destruction of these cells prevent humoral and cell-mediated responses from occurring as they should. A lack of T_H1 cells and T_H2 cells impairs both humoral and cell-mediated immune responses, including the destruction of cancer (malignant) cells. Thus, because of extensive destruction of APCs and T_H cells by HIV, persons with AIDS are susceptible to a host of opportunistic microorganisms and to various malignancies.

FACTORS THAT MODIFY IMMUNE RESPONSES

The host defenses of young, healthy adults living in an unpolluted environment can prevent nearly all infectious diseases. However, a variety of disorders, injuries, medical treatments, environmental factors, and even age can affect resistance. An individual with reduced resistance is called a **compromised host.**

With regard to age, the very young and the elderly generally are the most susceptible to infections. The very young are susceptible because their immune system is not fully developed; the elderly are susceptible to infections and malignancies because the immune system (especially cell-mediated responses) is among the first systems to decline in function during the aging process. It makes sense to take precautions against unnecessarily exposing infants and elderly people to infectious agents. It also is wise to obtain recommended immunizations during infancy and early childhood.

Emotional or psychological stress can depress the immune system's function. Stress probably affects the endocrine system, which in turn affects the immune system through changes in hormone levels. Certain hormones can depress the inflammatory response, inhibit cytokine secretion, and reduce the number of phagocytic cells and their activity in the body.

An adequate diet, especially sufficient protein and vitamin intake, is essential for maintaining healthy, intact skin and mucous membranes and phagocytic activity. It is also important for lymphocyte production and antibody synthesis. The poor nutrition and poor inflammatory responses of alcoholics and intravenous drug users contribute to lowering their resistance to infection. In the elderly, an inadequate diet can further weaken a declining immunological response.

Regular moderate exercise, such as 45 minutes of brisk walking five times per week, can produce a 20 percent increase in antibody level, which occurs during the exercise and for about 1 hour afterward. NK cell activity is also increased. However, excessive exercise, such as running more than about 12.5 km (20 miles) per week, can depress the immune system. Marathoners who run at their fastest pace for 3 hours experience a drop in NK cell activity of more than 30 percent for about 6 hours. Long-distance runners are more vulnerable to infection, especially of the upper respiratory tract. Studies show that for 12 to 24 hours after a race, these runners experience six times the rate of illness compared with trained runners who did not race.

Traumatic injuries lower resistance and provide access to tissues for microbes. Tissue repair in an inflammatory response competes with other immune processes because both require extensive protein synthesis. When normal nonspecific defense mechanisms that flush microbes from the body—such as tears, urinary excretions, and mucous secretions—are impaired by injuries, pathogens have easier access to tissues. Antibiotics also destroy commensals that sometimes protect the host by competing with pathogens. Impaired defenses and excessive use of antibiotics may allow opportunistic infections to develop.

Finally, environmental factors, including pollution and exposure to radiation, lower resistance to infection. Air pollutants, including tobacco smoke, damage respiratory membranes and reduce their ability to remove foreign substances. They also depress the activities of phagocytes. Excessive exposure to ultraviolet (UV) radiation not only damages cells but also depresses the immune system. These factors can be compounded by induced and inherited immunological disorders. Genetic defects in an individual's immune system itself can result in the absence of B cells, T cells, or both. How these disorders impair immunity is discussed in Chapter 19.

IMMUNIZATION

Throughout the world nearly 3.5 million children, most of them under 5 years of age, die each year of infectious diseases for which immunization is available. About 2 million die of measles, 800,000 of tetanus, and 600,000 of pertussis (whooping cough). Another 4 million die of various kinds of diarrhea. Most of these deaths occur in developing nations. These statistics dramatize three important points about immunization: First, effective immunization has the potential to prevent many deaths. Second, effective vaccines are not yet available for some infectious diseases, such as certain diarrheas in children. Finally, much greater effort is needed to make immunizations available in developing nations.

Active Immunization

To develop active immunity, as noted earlier, the immune system must recognize and neutralize or destroy infectious agents whenever they are encountered. **Active immunization** is the process of inducing active immunity by administering vaccines or toxoids. A **vaccine** is a preparation that contains microorganisms or their parts to which the immune system responds. Vaccines can contain live but attenuated (weakened) organisms, dead organisms, or parts of organisms. A **toxoid** is an inactivated toxin that is no longer harmful but retains its immunogenic activity.

Principles of Active Immunization

Whatever the nature of the immunizing preparation, the mechanism of active immunization is essentially the same. When the vaccine or toxoid is administered, the immune system recognizes it as foreign and produces antibodies and memory cells. This immune response is the same as the one that occurs when an individual acquires a disease. However, with immunization the disease does not occur, either because live pathogens are not used or because they are weakened enough to have lost their virulence. In other words, vaccines retain important immunogenic properties (epitopes) of pathogens but lack the ability to cause disease. Similarly, toxoids are immunogenic but do not cause toxic effects.

An important factor in the longevity of immunity from active immunization is the nature of the immunizing preparation. Vaccines made with live organisms tend to confer longer-lasting immunity than do those made with dead organisms, parts of organisms, or toxoids. The reason is that live organisms continue to reproduce for a short period of time, thereby providing more exposure to their epitopes. For example, measles (both rubella and rubeola) and mumps vaccines and the oral polio vaccine, which contain live viruses, usually confer long-lasting immunity. Intramuscular polio vaccine, which contains killed viruses, and typhoid fever vaccine, which contains dead bacteria, confer immunity that lasts 3 to 5 years. (Supplementary doses are needed to maintain immunity.) Tetanus and diphtheria toxoids confer immunity for about 10 years.

Because immunity is not always long-lasting, "booster shots" are often needed to maintain immunity. As we have noted, the first dose of a vaccine or toxoid stimulates a primary immune response analogous to that during a disease process. Subsequent doses stimulate secondary responses analogous to that following exposure to an organism to which immunity has already developed. Thus, booster shots not only boost immunity by greatly increasing the number of antibodies but also expand the memory cell population available to prevent disease the next time around.

The route of administration of a vaccine can influence the level of immunity. Compared with injecting vaccines into muscle, immunity is more effective when oral vaccines are used against gastrointestinal infections and when nasal aerosols are used against respiratory infections. It makes sense to deliver the antigenic stimulus to the body site affected by the disease.

Vaccines, especially those containing live organisms, must be stored properly to retain their effectiveness. Some require storage at cold temperatures; in tropical nations, serious failures of measles immunization have resulted from inadequate refrigeration. To prevent inactivation, other vaccines must be used within a certain number of hours or days after the vial is opened.

In general, active immunization cannot be used to prevent a disease after a person has been exposed. This is because the time required for immunity to develop is greater than the incubation period of the disease. Rabies immunization is an exception to this rule. Because rabies typically has a long incubation period (2 to 8 weeks), active immunization can be used with some hope that immunity will develop before the rabies virus reaches the brain. The farther the virus must travel to reach the brain, the greater the chance of effective immunization.

Several vaccines have been licensed for general use in the United States. Their properties are summarized in Table 18.6. Many are used by people with special needs, such as for foreign travel or for experimental purposes. Properties of some special-use vaccines are given in Table 18.7 on page 504.

Recommended Immunizations

Three vaccine preparations that immunize against seven diseases are recommended in the United States for routine immunization of normal infants and children: the DTP, oral polio, and MMR vaccines. The *DTP vaccine* contains diphtheria toxoid, tetanus toxoid, and killed *Bordetella pertussis* (whooping cough) bacteria. Although most children tolerate DTP vaccine, a few experience severe complications. The *oral polio vaccine* usually used in the United States contains three different types of live polioviruses. Other vaccines with antigen combinations are available for ad-

Disease	Nature of Material	Route of Administration[a]	Use and Comments	Duration of Effectiveness
Bacterial meningitis	Polysaccharide–protein conjugate	IM	2, 4, 6, and 15 months; 75% effective	14–34 years
Cholera	Killed bacteria	SC, IM, ID	2 doses a week or more apart, 50% effective, may be required for travel	6 months
Diphtheria	Toxoid	IM	3 doses 4 weeks apart plus boosters, 90% effective	10 years
Hepatitis B	Viral antigen	IM	2 doses 4 weeks apart, booster in 6 months	About 5 years
Influenza (viral)	Inactivated virus	IM	1 or 2 doses, depending on type of virus, recommended for high-risk patients and medical personnel; 75% effective	1–3 years
Lobar pneumonia	Polysaccharide	SC, IM	1 dose before chemotherapy	5 years or more
Measles (rubeola)	Live virus	SC	1 dose at 15 months, revaccination around age 12; 95% effective, may prevent disease if given within 48 hours of exposure	Lifelong
Meningococcal meningitis	Polysaccharide	SC	1 dose, recommended during epidemics and for high-risk patients	Lifelong if given after age 2
Mumps	Live virus	SC	1 dose given after age 1, 95% effective	Lifelong
Pertussis (whooping cough)	Killed bacteria	IM	Same as diphtheria; not given to children subject to seizures	10 years
Plague	Killed bacteria	IM	3 doses 4 weeks apart, for travel to some parts of the world	6 months
Poliomyelitis	Live virus	O	2 doses 6 weeks apart plus booster, 95% effective	Lifelong
Rabies	Killed virus	IM	2 doses 1 week apart with third dose in 2 weeks; 80% effective; used after probable exposure	2 years
Rubella	Live virus	SC	1 dose at 15 months, some recommend second dose at 12 years	Lifelong
Smallpox	Live vaccinia virus	ID	1 dose; 90% effective; used only by laboratory workers exposed to poxviruses	3 years
Tetanus	Toxoid	IM	3 doses 4 weeks apart plus boosters	10 years
Tuberculosis	Attenuated bacteria	ID, SC	1 dose for inadequately treated patients and high-risk groups	Lifelong?
Typhoid fever	Live virus	O	2 doses 4 weeks apart; 70% effective; recommended for travel, epidemics, and carriers	3 years
Yellow fever	Live virus	SC	1 dose, recommended for travel to endemic areas	10 years

[a]SC—subcutaneous, IM—intramuscular, ID—intradermal, O—oral

ministration orally or intramuscularly. *MMR vaccine* contains live rubeola (measles) virus, paramyxovirus (mumps), and rubella (German measles) virus. This vaccine can be used to immunize against all three diseases simultaneously, or separate vaccines can be given for each disease.

The recommended age for immunizations varies. DTP and polio vaccines can be administered effectively to babies as young as 2 months of age. However, MMR vaccine is not recommended before 15 months of age. When it is given to younger infants, the level of immunity that results usually is not sufficient to protect

TABLE 18.7	Selected Examples of Materials for Special Immunization and Experimentation	
Infectious Agent	Nature of Material	Uses
Adenovirus	Live virus	Military recruits
Bacillus anthracis	Antigen extract	Handlers of animals and hides
Campylobacter	Attenuated bacteria	Experimentation
Vibrio cholerae	Experimental oral administration to obtain more effective immunization	Toxoids of *Escherichia coli* and *Vibrio cholerae*
Cytomegalovirus	Live virus	Experimentation, may produce latent infections
Equine encephalitis virus	Live/inactivated viruses	Laboratory workers and experimentation

against infection, probably because of immaturity of the immune system. Vaccines against *Haemophilus influenzae* type b (Hib) first became available in 1985 but did not work well in children under 2 years of age. Several FDA-approved Hib vaccines that work well in children are in use. These vaccines could prevent bacterial meningitis, which kills many children and leaves adult survivors mentally disabled, deaf, or otherwise neurologically damaged. The recommended immunization schedule for healthy infants and children in the United States is given in Table 18.8.

Immunization recommendations differ between developed and developing nations. Most developed countries use largely the same immunizations as those recommended in the United States. However, the World Health Organization (WHO) recommends immunization at earlier ages in developing nations to combat serious contagious diseases: oral polio vaccines at birth, DTP and polio vaccines at 6, 10, and 14 weeks, and measles vaccine at 9 months.

Hazards of Vaccines

The greatest breakthroughs in disease prevention have come from immunizations. Use of vaccines has not only prevented a number of infectious diseases but also has eradicated several diseases in certain populations. However, vaccines also pose hazards that must be considered in deciding whether they should be administered to entire populations or to certain individuals. The prevalence and severity of diseases also must be weighed in such decisions.

Active immunization can cause mild fever, malaise (general discomfort), and soreness at the injection site. Thus, individuals already suffering from fever and malaise should not receive immunization because a worsening of their condition might erroneously be attributed to the vaccine. More importantly, the individual's immune system, overburdened by the existing infection, may be unable to mount an adequate response to the vaccine. Allergic reactions sometimes follow the use of influenza and other vaccines that contain egg protein. However, reactions occur in only a few vaccine recipients and are generally much less severe than the disease. An exceedingly small number of vaccine recipients die or suffer permanent damage from vaccines (see the box "The Polio Vaccine Controversy" box in Chapter 24, page 690).

Live vaccines may pose particular hazards to pregnant women, patients with immunological deficien-

TABLE 18.8	Recommended Immunizations for Normal Infants and Children in the United States	
Disease	Vaccine	Dosage Schedule
Diphtheria ⎤	Toxoid	Ages 2, 4, and 6 months, $1\frac{1}{2}$ and 4 to 6 years
Tetanus ⎬ DTP	Toxoid	Same as diphtheria; administered in DTP vaccine
Pertussis ⎦	Killed bacteria	Same as diphtheria; administered in DTP vaccine
Tetanus ⎤ Td	Toxoid	Age 14–16 years and every 10 years thereafter
Diphtheria ⎦	Toxoid	Age 14–16 years and every 10 years thereafter
Poliomyelitis	Live viruses Types I, II, and III	Ages 2 and 4 months, $1\frac{1}{2}$ and 4 to 6 years
Haemophilus influenzae type b (Hib) infection	Polysaccharide–protein conjugate	Ages 2, 4, 6, and 15 months
Measles ⎤	Live virus	Ages 15 months and booster at school age, 11–12 years
Mumps ⎬ MMR	Live virus	Ages 15 months, 11–12 years
Rubella ⎦	Live virus	Ages 15 months, 11–12 years
Hepatitis B	Viral antigen fragments	Birth, ages 2, and 6 months

cies, and patients receiving immunosuppressant therapy, such as radiation or corticosteroid drugs. In pregnant women, live viruses sometimes cross the placenta and infect the fetus, whose immune system is immature. In immunodeficient or immunosuppressed patients, some attenuated viruses may establish a disease process. Therefore, individuals who test positive for the HIV should not receive live-virus vaccines. Such situations present a problem for routine immunization of infants, some of whom, unknown to health care personnel, could have become infected with HIV before birth. It also means that U.S. military and State Department employees and their accompanying families who are being sent to foreign posts must be tested for HIV infection to determine whether they can safely receive the required live-virus vaccines.

Passive Immunization

To induce passive immunity, ready-made antibodies are introduced into an unprotected individual. Because antibodies are found in the serum portion of the blood, these products are often called **antisera.** Although passive immunity is produced quickly, it is only temporary. It lasts only as long as there is a sufficiently high titer of circulating antibodies in the body. **Passive immunization** is established by administering a preparation such as gamma globulin, hyperimmune serum, or an antitoxin that contains large numbers of ready-made antibodies. However, the specificity and degree of this form of immunization depend on the antibody type and concentration used.

Immune serum globulin, formerly called **gamma globulin,** consists of pooled gamma globulin fractions (the portion of serum containing antibodies) from many individuals. This kind of gamma globulin typically contains sufficient antibodies to provide passive immunity to a number of common diseases, such as mumps, measles, and hepatitis A.

If the donors are specially selected, gamma globulins that have high titers of specific kinds of antibodies can be prepared. Such preparations are often called **hyperimmune sera,** or *convalescent sera.* For example, gamma globulin from persons recovering from mumps or from recent recipients of mumps vaccine contains especially high titers of antimumps antibodies. Similar sera can be collected from donors with high titers of antibodies to other diseases. Hyperimmune sera can also be manufactured by introducing particular antigens—for instance, tetanus toxin—into another animal, such as a horse, and subsequently collecting the antibodies from the animal's serum.

Antitoxins are antibodies against specific toxins, such as those that cause botulism, diphtheria, or tetanus. Passive immunization against tetanus toxins

TABLE 18.9	Properties of Materials Available for Passive Immunization
Material	**Uses**
Human gamma globulin	To prevent recurrent infections in patients with deficiencies in humoral immunity and to prevent or lessen disease symptoms after exposure of nonimmune persons to measles or hepatitis A
Specific gamma globulins	Varicella-zoster immune globulin
	To prevent chickenpox in high-risk children; must be given within 4 days of exposure
Hepatitis-B immune globulin	To prevent hepatitis B after exposure (via blood or needles) and to prevent spread of the disease from mothers to newborns
Mumps immune globulin	May prevent orchitis (inflammation of the testes) in adult males exposed to mumps
Pertussis immune globulin	To reduce severity of disease and mortality in children under 3 or in debilitated children
Rabies immune globulin	To prevent rabies after a bite from a possibly rabid animal; applied to the bite if possible and administered intramuscularly
Tetanus immune globulin	To prevent tetanus after injury in nonimmune patients
Vaccinia immune globulin	To halt progress of disease in immunodeficient patients who develop a progressive vaccinia (cowpox) infection

can also be achieved by using tetanus immune globulin, a gamma globulin that contains antibodies against tetanus toxin. The properties of currently available materials used to produce passive immunity are summarized in Table 18.9.

Passive immunization gives immediate immunity to a nonimmune person who is exposed to a disease, or it at least lessens the severity of the disease process. Usually, a vaccine is given after the passive immunizing preparation to provide active immunity. Before the advent of antibiotics, passive immunization was frequently used to prevent or lessen the severity of several kinds of pneumonia and a variety of other infectious diseases. With respect to infectious diseases, the most common current use of passive immunity is to protect people with contaminated wounds against tetanus toxin. Although the incidence of exposure to diphtheria and botulism is lower than that of tetanus, passive immunization can also be used against these diseases.

Passive immunization is also used to counteract the effects of snake and spider bites and to prevent damage to fetuses from certain immunological reactions. Antivenins, or antibodies to the venom of certain poisonous snakes and the black widow spider, are given as emergency treatments. They react with venom molecules that have not already bound to tissues. Thus, the sooner after a bite the antivenin is administered, the more effective it is in counteracting the effects of the venom.

A fetal immunological reaction can occur when a mother with Rh-negative blood carries her second Rh-positive fetus. As is explained in more detail in Chapter 19, the mother becomes sensitized to the Rh-positive red blood cells when her first Rh-positive child is born. The mother's immune system normally makes anti-Rh antibodies, which will harm the second Rh-positive child. To prevent this from happening, anti-Rh antibodies are given to the mother within 72 hours after the birth of the first child and after subsequent births. Like the antivenins, the anti-Rh antibodies bind to Rh-positive red blood cells, so the cells are destroyed before the mother's immune system can make antibodies to them.

As with active immunization, passive immunization may pose some hazards. The most common hazards are allergic reactions. For instance, some antitoxins contain proteins from other animals by virtue of their production in eggs or horses. Such antitoxins are particularly likely to cause allergic reactions, especially when the individual receives them for the second time. Preparations of human origin are safer, at least with respect to the risk of allergic reactions.

Another problem, or at least a temporary disadvantage, of passive immunity is that administering ready-made antibodies can interfere with an individual's ability to produce his or her own antibodies for a limited time. One way this might occur is if ready-made antibodies bind to antigens and prevent them from stimulating the individual's immune system. Thus, maternal antibodies, although they protect newborns from some infections, may interfere with an infant's immune system, preventing the infant from making antibodies.

Future of Immunization

Immunologists continue to search for new vaccines. Effective vaccines should meet five criteria. (1) Vaccines developed should be protective against the disease for which they were designed. (2) The vaccines must be safe and not have adverse side effects. (3) Protection should be sustained, providing long-term protection from infection. (4) Vaccines should generate neutralizing antibodies or protective T cells to vaccine antigens. (5) Vaccines should be practical in terms of stability and use.

Whole-cell killed vaccines (first-generation vaccines) sometimes produce unwanted side effects due to extraneous cellular materials. Therefore, efforts are made to identify and obtain cellular subunits that contain only the purified immunogenic portion of a microorganism that will produce immunity. *Subunit vaccines* (second-generation vaccines) are safer than *attenuated vaccines,* which use live organisms that are treated to eliminate their virulence. There is always the possibility that the organisms may revert to a virulent state. Many researchers consider attenuated vaccines too risky in the hunt for an AIDS vaccine, but they are functional for some other diseases, as is the BCG vaccine for tuberculosis. In general, live organisms produce higher and longer-lasting immunity than do nonliving organisms. *Recombinant DNA vaccines* (third-generation vaccines) are being produced by inserting the genes for specific antigens into the genomes of nonvirulent organisms. The hepatitis B viral antigens have been cloned in yeast cells, extracted and purified before use, to form a safe and very effective vaccine. Rabies virus antigen has been inserted into vaccinia (cowpox) virus and is being tested as a means for controlling rabies in wild animal populations such as raccoons.

Many vaccine researchers believe that by the year 2025, most Americans will be immunized routinely against some 30 diseases, including AIDS; genital herpes; influenza; hepatitis A, B, C, and E; and chickenpox and shingles. These immunizations will be administered in three stages: infancy and early childhood, for childhood diseases; prior to puberty, for certain sexually transmitted diseases; and adulthood, for influenza.

IMMUNITY TO VARIOUS KINDS OF PATHOGENS

This chapter has focused on the basic principles of immunity that apply to all pathogens. However, you may find it helpful to note the ways in which the immune system responds to various kinds of pathogens.

Bacteria

As we saw in Chapter 17 ∞ (p. 448), nonspecific defenses such as skin, mucous membranes, and gastric secretions prevent many bacteria from entering host tissues. When bacteria do infect a host, certain immune responses alter the invading organisms so that they can be phagocytized. Once plasma cells produce specific antibodies, the antibodies can interfere with any of several steps in bacterial invasion. They can attach to pili and capsules, preventing bacterial attachment to cell surfaces. Antibodies can work with complement to opsonize bacteria for later phagocytosis or lysis by other cells of the immune system. Or they can neutralize bacterial toxins or inactivate bacterial enzymes.

Viruses

Viruses infect by invading cells—usually first attacking cells that line body passages. Then they directly invade target organs such as the lungs or travel in the blood (viremia) to target organs or organ systems such as the liver or nervous system. Polioviruses invade cells that line the digestive tract, but they also can enter nerve endings.

Immune responses can combat viral infections at any of these locations. Interferons, secretory IgA, and some IgG antibodies act at the surfaces lining cells and prevent or minimize entry of viruses. IgG and IgM act in the blood to neutralize viruses directly or to promote their destruction by complement. Finally, cytotoxins and cellular immunity via T_C cells and NK cells are especially important in clearing the body of cells infected with viruses. The mechanisms by which the immune system combats viral infections are summarized in Figure 18.15.

Besides specific immune responses to a viral infection, many nonspecific responses can limit infection. Fever is an important defense against viruses. Several viruses, such as influenza, parainfluenza, and rhinoviruses, are temperature-sensitive. They replicate in the lining cells of the respiratory tract, which normally has a temperature between 33 and 35°C—lower than the normal body temperature of 37°C because the cells are cooled as atmospheric air moves over their moist

FIGURE 18.15 How the immune system combats viruses.

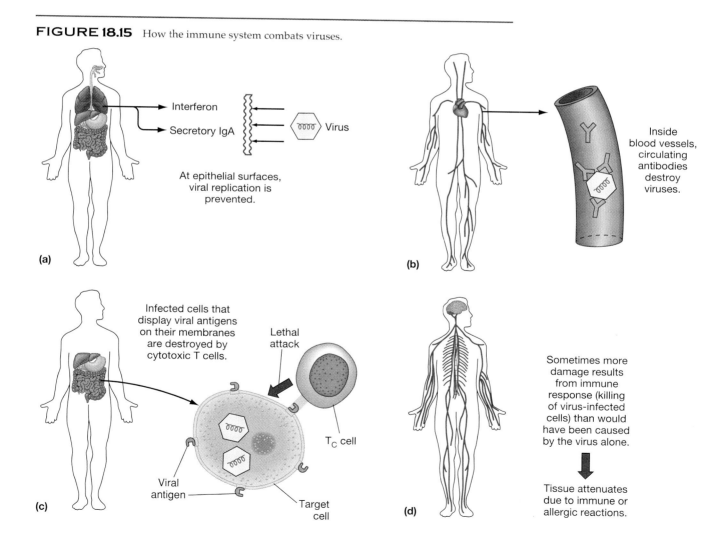

surfaces. When a person has a fever of even 1° to 2°C, the ability of the virus to replicate is reduced. Another benefit of fever in resisting viral infection is that temperature increases cause an increase in interferon production. ∞ (Chapter 17, p. 458)

Fungi

Certain fungal infections progress through a tissue as fungal cells invade and destroy one cell after another. Immunity to fungi is poorly understood, but it appears to be mainly cell-mediated. Fungal skin infections probably are combated by IgA antibodies and T_H cells, which release certain cytokines that activate macrophages. The macrophages, in turn, engulf and digest fungi. Commensal fungi apparently are kept in their place by cell-mediated responses. Supportive evidence comes from studies of individuals with impaired T cell functions. Such persons are extremely likely to become infected with opportunistic fungi, such as *Candida albicans*.

Protozoa and Helminths

Protozoa and helminths are quite dissimilar in size and complexity, but many use similar methods of invading the body. Host defenses against them also are similar, except that allergic reactions to helminths can be severe enough to cause more damage to host cells than to the disease agent. Antigens on the surface of the roundworm *Ascaris* are potent inducers of allergic reactions. Individuals with an *Ascaris* allergy can absorb enough antigens through the skin to cause a severe allergic reaction, even just by coming into contact with fluids in which the worms have been preserved.

Parasitic protozoa and helminths interact with their hosts in ways that do not endanger host survival and thereby ensure parasite survival. These pathogens cause chronic, debilitating diseases that usually are not immediately life-threatening. Most of these parasites are relatively large and difficult to phagocytize. Large helminths, such as heartworms in dogs, can block blood vessels and cause sudden death. When attacked by phagocytes, some helminths release toxins that significantly damage the host. Medical intervention also can be hazardous. Giving drugs to kill some helminths causes them to release large quantities of toxic decay products. Thus, once some worm infections have been

acquired, coexistence with the parasites may be the best course of action.

Many parasites have complex life cycles with more than one host, and some infect animals that serve as reservoirs for human infection. ∞ (Chapter 12, p. 302) Thus, they stand ready to take advantage of appropriate conditions in various hosts. Each life cycle stage of some parasites can have several surface antigens. Although hosts may produce antibodies against such antigens, the ability of certain parasites to change them provides a way to thwart host defenses. For example, the malaria parasite induces host antibody formation before it invades host cells. While the protozoan multiplies inside cells, it produces different antigens, so the original antibodies formed by the host are not effective against the new parasite progeny when they are released.

The host's immune system also combats parasitic protozoa and helminths by cell-mediated processes. Although T_C cells usually are not effective against such parasites, some T cells release cytokines, such as IL-3, that activate macrophages. These macrophages can attack malarial parasites and several kinds of worms, including blood flukes. Other cytokines, such as IL-5, enhance the ability of eosinophils to combat worm infections.

Parasitic protozoa and helminths have available to them an assortment of mechanisms that thwart immune responses. These mechanisms include the following:

1. Some protozoa protect themselves by invading cells, forming protective cases called cysts at certain life cycle stages, or otherwise becoming inaccessible to host defenses.

2. Some protozoa avoid immune recognition by changing their surface antigens (antigenic variation) regularly or with each reproductive cycle.

3. Some parasites suppress the host's immune responses by releasing toxins that damage lymphocytes, enzymes that inactivate IgG, or soluble antigens that thwart the immune system in a variety of ways (Figure 18.16).

4. Some intracellular protozoa suppress the action of phagocytes by inhibiting fusion of lysosomes with vacuoles, by resisting digestion by lysosomal enzymes, or by impairing oxidative metabolism.

Given the various methods protozoa and helminths have to evade and thwart immune responses of hosts, it is not surprising that they cause chronic debilitating infections.

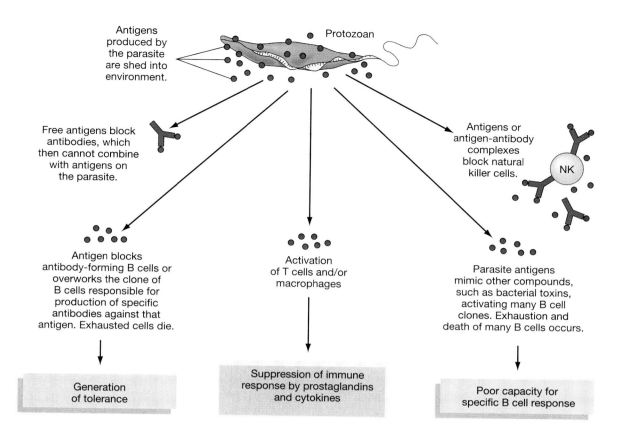

Antigens produced by the parasite are shed into environment.

Protozoan

Free antigens block antibodies, which then cannot combine with antigens on the parasite.

Antigens or antigen-antibody complexes block natural killer cells.

NK

Antigen blocks antibody-forming B cells or overworks the clone of B cells responsible for production of specific antibodies against that antigen. Exhausted cells die.

Activation of T cells and/or macrophages

Parasite antigens mimic other compounds, such as bacterial toxins, activating many B cell clones. Exhaustion and death of many B cells occurs.

Generation of tolerance

Suppression of immune response by prostaglandins and cytokines

Poor capacity for specific B cell response

FIGURE 18.16 How antigens of parasitic protozoans thwart the immune system.

CANCER AND IMMUNOLOGY

Cancer is one of the leading causes of death in developed countries, and its incidence as a cause of death in developing nations is on the rise. Cancer cells can arise by mutation of a normal cell in response to chemical carcinogens, radiation, or viruses; by expression of previously repressed human oncogenes (tumor-producing genes) within cells ∞ (Chapter 11, p. 294); or by a combination of these factors.

Regardless of the means by which they are produced, many cancer cells have certain plasma membrane antigens not found in normal cells. These antigens are an ideal target for destruction by the immune system. According to the theory of *immune surveillance*, T_C cells and NK cells recognize and destroy these abnormal cells before they develop into cancers (Figure 18.17). If this theory is correct, each of us has within our body many different potentially malignant cells, but we develop cancer only if the T_C cells or NK cells fail to identify and destroy the mutated cells. Unfortunately, most tumors and cancers show little regard for control by the immune system.

FIGURE 18.17 Colorized SEM photo of a small T lymphocyte attacking two large tumor cells (5700×).

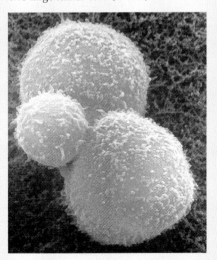

One way T_C cells might fail to recognize malignant cells is by the actions of antigens on the malignant cells themselves. The antigens stimulate the formation of antibodies, which bind to the antigens without damaging the malignant cells. Although this binding does not occur before the antigens sensitize T_C cells, it does block T_C cells from attacking the malignant cells. Other tumors can avoid immune system attacks by lacking abnormal antigens or by losing MHC class I molecules through mutation. Such cells escape immune detection by T_C cells but, as described in this chapter, are still attacked by NK cells.

However, the notion that cancer cells can be destroyed by immune reactions has led researchers to work on developing immunotoxins and cancer vaccines. As discussed earlier in this chapter, an immunotoxin is a monoclonal antibody with an anticancer drug, a microbial toxin such as diphtheria toxin, or a radioactive substance attached to it. The antibody is designed to bind to a specific cancer cell antigen; the attached substance is selected for its ability to destroy the cancer cell. Such immunotoxins are expected to seek and destroy specific cells—the cancer cells that have the appropriate antigen. Cancer vaccines containing one or more cancer cell antigens are also being developed. Some of these vaccines induce immunity to specific cancer cells. Others contain antigens frequently found on certain kinds of cancer cells.

A major difficulty in making immunotoxins and cancer vaccines is that many different kinds of antigens are found on cancer cells from various patients. To be effective against an existing cancer, an immunotoxin or vaccine must be specific for the antigens on the cells of that specific cancer. Similarly, to immunize against common forms of cancer, vaccines need to generate antibodies or T_C cells that will destroy cancer cells should they develop. Another difficulty is that some malignant cells produce immunosuppressive cytokines, such as transforming growth factor-α (TGF-α), that stimulate tumor growth and may interfere with T_H1 cell activity. Immunotoxins and vaccines need to be powerful enough to overcome this inhibition. Finally, some antigens are found on both malignant and normal cells. Great care must be taken to develop immunotoxins and vaccines that will react only with malignant cells.

In October 1991 the first attempt was made to eliminate cancer by immunizing patients against their own tumors. Cells from a malignant melanoma tumor were removed, genes for tumor necrosis (killing) factor were added to the cancer cells' genomes, and the cells were reinjected (infused) into the patient. There, it is hoped, the genetically altered cells will secrete enough of this immune system factor to change the regulation of the immune system so as to overcome its tolerance of the tumor and cause it to begin attacking the cancerous cells. To date, there have been few successes.

Meanwhile, immunization already offers us the ability to prevent about 80 percent of liver cancer cases. Liver cancer is among the most prevalent kinds of cancer. One form of liver cancer, primary hepatocellular carcinoma, accounts for 80 to 90 percent of all cases. It now appears that this form is associated with infection by hepatitis B virus early in life and especially with being a carrier of the virus. Not only has hepatitis B viral DNA been found in chromosomes of cancerous liver cells but also the intact virus has been isolated from cancerous liver cells. The incidence of liver cancer is highest in regions of Africa and Asia where the incidence of hepatitis B infections also is high. Immunization against the hepatitis B virus would protect recipients of the vaccine against infection, prevent them from becoming carriers, and thus protect them against liver cancer.

IMMUNOLOGY AND IMMUNITY

- **Immunology** is the study of **immunity,** which refers to the capacity to recognize and defend against infectious agents and other foreign substances.

- The **immune system** is the defense system that provides the host with specific immunity to particular infectious agents.

- **Susceptibility** is vulnerability to infectious agents.

TYPES OF IMMUNITY

Innate Immunity

- **Innate immunity** provides a nonspecific, hereditary defense.

Acquired Immunity

- **Acquired immunity** provides a specific, nonhereditary defense.

Active Immunity

- In **active immunity,** an individual's own immune system makes antibodies.

Passive Immuniy

- In **passive immunity,** ready-made antibodies are introduced into the body.

- Types of immunity are illustrated in Figure 18.1.

CHARACTERISTICS OF THE IMMUNE SYSTEM

Properties of Antigens (Immunogens)

- An **antigen,** or **immunogen,** is a foreign substance that can elicit a specific immune response. Most antigens are proteins, but some are polysaccharides, nucleoproteins, or glycoproteins.

- Each antigen has several **epitopes,** or **antigenic determinants.**

Properties of Antibodies (Immunoglobulins)

- An **antibody,** or **immunoglobulin,** is a protein produced in response to the presence of an antigen. Antibodies bind to epitopes on the antigen.

- Antibodies consist of **heavy chains** and **light chains,** each pair of which has a **variable region** forming an **antigen-binding site.** Properties of particular kinds of immunoglobulins are summarized in Table 18.1.

CELLS AND TISSUES OF THE IMMUNE SYSTEM

- Specific immune responses involve lymphocytes and macrophages. Lymphocytes develop from **lymphoid stem cells** in the bone marrow.

Dual Nature of the Immune System

- The dual roles of the immune system consist of **humoral immunity,** which is carried out mainly by B cells and plasma cells, and **cell-mediated immunity,** which is carried out mainly by certain T cells.

B Lymphocytes

- **B lymphocytes (B cells)** migrate from the bone marrow to lymphoid tissues and, when exposed to a foreign invader, mature into **plasma cells.**

T Lymphocytes

- Lymphoid stem cells that migrate from the bone marrow to the thymus form **T lymphocytes (T cells).** There are two classes of T cells: **helper T cells** (T_H) and **cytotoxic T cells** (T_C).

- B cells and T cells can be distinguished by epitopes called **cluster of differentiation** (CD) **markers** on their plasma membranes.

- Activated T_H cells produce **cytokines,** chemicals that regulate T cells, B cells, and other immune system cells. Cytokines include **lymphokines** and **interleukins.**

- Two types of T_H cells are known: T_H1 (inflammatory T) and T_H2 cells, distinguishable by the types of cytokines they produce.

- T_C cells kill virus-infected cells and tumor cells by releasing the lethal protein **perforin.**

Natural Killer Cells

- Natural killer (NK) cells are large lymphocytes that kill tumor cells, virus-infected cells, and foreign cells (in transplants).

Macrophages

- Most antigens are processed by **antigen-presenting cells** (APCs)—including macrophages—which present (display) an antigen's peptide fragment so that lymphocytes can recognize them.

- Immune reactions begin with the processing of an antigen by an APC. During processing, the APC combines a foreign antigen fragment (peptide) with its own **major histocompatibility complex** (MHC) molecules and then inserts this complex into its own plasma membrane.

Properties of Immune Responses

- Immune responses distinguish **self** from **nonself.**

- According to the **clonal selection theory,** B cells recognize specific epitopes on antigens according to the particular antibody present on the B cell plasma membrane. When a B cell detects an antigen with which it can react, it binds to the antigen, engulfs it, processes it, displays a peptide fragment as MHC class II to T_H2 cells, and divides many times. A **clone** of genetically identical B cells, which differentiate into many plasma cells and some **memory cells,** is produced.

- **Specificity** refers to the ability of immune responses to respond to and distinguish among different antigens and epitopes.

- **Heterogeneity** refers to the ability of immune responses to produce many different antibodies and cell substances on the basis of the different antigens they encounter.
- **Immunological memory** refers to the ability of T and B cells to recognize substances to which the immune system has previously responded.

HUMORAL IMMUNITY

- In **humoral immune responses,** an antigen stimulates mature B cells to divide and differentiate into plasma cells, which release specific antibodies against a specific epitope.

Primary and Secondary Responses

- **Primary responses** are the immune system's first encounter with foreign antigens.
- Memory cells remain in lymphoid tissue, ready to respond to subsequent exposure to the same antigen.
- **Secondary responses** bring fast and efficient destruction antigens recognized by B and T memory cells. Primary and secondary responses are summarized in Table 18.5.

How Humoral Responses Eliminate Foreign Antigens

- Plasma cells synthesize and release large numbers of antibodies.
- Humoral immunity is most effective against bacteria and some viruses. These microbes are inactivated by **neutralization,** phagocytized after opsonization, or lysed by complement.

MONOCLONAL ANTIBODIES

- **Monoclonal antibodies** are antibodies produced in the laboratory from a clone of cultured cells that make one specific antibody to one specific epitope.
- Specific monoclonal antibodies can be used in some diagnostic tests, and methods to use them in treating infectious diseases and cancer are being developed.

CELL-MEDIATED IMMUNITY

- Cell-mediated immunity concerns the direct actions of certain T cells that defend the body against viral infections and reject tumors and transplanted tissues.
- T cells do not make antibodies but instead have membrane receptors for antigens; these receptors bind to MHC molecules displaying foreign peptide fragments.
- Cell-mediated immune responses involve differentiation and activation of several kinds of T cells and the secretion of cytokines.

The Cell-Mediated Immune Reactions

- Processed antigens on MHC class II molecules bind with T cell receptors. Next, IL-1 secreted from macrophages and IL-2 secreted from T cells activate the T cells, which can then differentiate into T_H1 cells and T_H2 cells.
- Certain pathogenic bacteria can grow in macrophages after they have been phagocytized. T_H1 cells can release γ-interferon, a cytokine that causes such infected macrophages to become resensitized to other cytokines.
- AIDS destroys T_H cells, thereby impairing both humoral and cell-mediated immunity.

FACTORS THAT MODIFY IMMUNE RESPONSES

- Host defenses in healthy adults in an unpolluted environment prevent nearly all infectious diseases. Individuals with reduced resistance are called **compromised hosts.**
- Factors that reduce host resistance include very young or old age, stress, seasonal patterns, poor nutrition, traumatic injury, pollution, and radiation. Complement deficiencies, immunosuppressants, infections such as HIV, and genetic defects impair immune system function.

IMMUNIZATION

Active Immunization

- **Active immunization** induces the same response as the one that occurs during a disease. Immunization challenges the immune system to develop specific defenses and memory cells.
- Active immunization is conferred by **vaccines** and **toxoids.** Vaccines can be made from live, attenuated organisms, dead organisms, or parts of organisms. Toxoids are made by inactivating toxins.
- The benefits of active immunization against life-threatening diseases nearly always outweigh the hazards. Reactions to vaccines can cause serious side effects, but their incidence is lower than the incidences of the diseases themselves.

Passive Immunization

- **Passive immunization** occurs by the same mechanism as natural passive transfer of antibodies.
- Passive immunity is conferred by **antisera** such as **immune serum globulin (gamma globulin), hyperimmune sera,** and **antitoxins.**
- The benefits of passive immunization are limited to providing only temporary protection; the side effects are mainly allergic in nature.

Future of Immunization

- Subunit vaccines produce fewer side effects than whole-cell killed vaccines and offer greater safety than do attenuated vaccines.
- Recombinant DNA vaccines contain genes for antigens of pathogens inserted into nonpathogenic organisms' genomes and are very safe.

IMMUNITY TO VARIOUS KINDS OF PATHOGENS

Bacteria

- Antibodies produced by plasma cells are the chief immunological defense against bacterial antigens. Most immune responses to bacteria serve to promote phagocytosis of the invading cells.

Viruses

- Viral infection is combated by nonspecific defenses, interferon, and antibodies. In addition, the T_C cells of cell-mediated responses and NK cells are important in destroying virus-infected cells.

Fungi

- Immune responses to fungi involve IgA antibodies and are primarily cell-mediated.

Protozoa and Helminths

- Immune responses to parasitic protozoa and helminths are largely cell-mediated. T cells release cytokines that activate macrophages and attract other leukocytes. Allergic reactions to helminths can be damaging to the host.

KEY TERMS

acquired immunity (p. 481)
active immunity (p. 481)
active immunization (p. 502)
anamnestic response (p. 492)
antibody (p. 483)
antigen (p. 482)
antigen-binding site (p. 484)
antigen challenge (p. 493)
antigenic determinant (p. 483)
antigen-presenting cell (p. 489)
antiserum (p. 505)
antitoxin (p. 505)
B cell (p. 486)
B lymphocyte (p. 486)
cell-mediated immune response (p. 498)
cell-mediated immunity (p. 486)
clonal deletion (p. 491)
clonal selection theory (p. 489)

clone (p. 491)
cluster of differentiation marker (p. 487)
colostrum (p. 482)
compromised host (p. 501)
cross-reaction (p. 491)
cytokine (p. 488)
cytotoxic T cell (p. 487)
cytotoxin (p. 488)
epitope (p. 483)
Fab fragment (p. 484)
Fc fragment (p. 485)
gamma globulin (p. 505)
granzyme (p. 489)
hapten (p. 483)
heavy chain (p. 484)
helper T cell (p. 487)
heterogeneity (p. 492)
humoral immune response (p. 493)
humoral immunity (p. 486)
hyperimmune serum (p. 505)
IgA (p. 485)
IgD (p. 486)

IgE (p. 486)
IgG (p. 485)
IgM (p. 485)
immune complex (p. 495)
immune serum globulin (p. 505)
immune system (p. 480)
immunity (p. 480)
immunogen (p. 482)
immunogenic (p. 482)
immunoglobulin (p. 484)
immunological memory (p. 492)
immunology (p. 480)
innate immunity (p. 481)
interleukin (p. 488)
light chain (p. 484)
lymphoid stem cell (p. 486)
lymphokine (p. 488)
major histocompatibility complex (p. 487)
memory cell (p. 492)
monoclonal antibody (p. 495)
neutralization (p. 494)

nonself (p. 489)
passive immunity (p. 481)
passive immunization (p. 505)
perforin (p. 489)
plasma cell (p. 486)
primary response (p. 493)
secondary response (p. 493)
secretory piece (p. 486)
self (p. 489)
specificity (p. 491)
suppressor T cell (p. 489)
susceptibility (p. 480)
T cell (p. 486)
T-dependent antigen (p. 493)
T-independent antigen (p. 493)
titer (p. 486)
T lymphocyte (p. 486)
tolerance (p. 491)
toxoid (p. 502)
vaccine (p. 502)

QUESTIONS FOR REVIEW

A Immunology, Immunity, and Susceptibility

1. What is immunology, and how do immunity and susceptibility differ?

B Innate, Acquired, Active, and Passive Immunities

2. What are the differences between innate and acquired immunity? Indicate with an X the type of acquired immunity (if any) that would develop or be conferred by the following examples:

Example	Active Acquired Immunity	Passive Acquired Immunity
Receiving an antivenin shot		
Getting immunized for tetanus		
Getting sick with flu		
Having chickenpox		

C Antigens and Antibodies

3. List the properties of an antigen.

4. Match the antibody class with its characteristic. More than one characteristic may apply to one antibody class.

   _____ Accounts for 20 percent of total antibody

   _____ A single Y-shaped antibody that triggers complement activity

   _____ Present in body secretions, such as sweat and saliva

   _____ Soluble antibody found only in the circulatory system

   _____ Secreted from helper T cells

   a. IgA
   b. IgD
   c. IgE
   d. IgG
   e. IgM
   f. none of the above

D Immune System Cells and Tissues

5. How do lymphoid stem cells become B cells and T cells?

6. How can lymphocytes be distinguished?

E Self, Specificity, Heterogeneity, and Immunological Memory

7. How is the immune system able to destroy nonself and avoid harming self?

8. What is the difference between specificity and heterogeneity in the immune system?

9. What is the meaning of immunological memory?

F Humoral Immunity

10. Describe the clonal selection theory.

11. What happens when a B cell responds to a foreign antigen?

12. How do the functions of plasma cells and memory cells differ?

13. What is clonal deletion, and how is it related to self and nonself recognition?

14. How do humoral immune responses eliminate bacteria, viruses, and toxins?

G Primary and Secondary Responses

15. How do the primary and secondary responses to the same antigen differ, and what is the significance of those differences?

H Cell-Mediated Immunity

16. Identify the properties of cell-mediated immunity that distinguish it from humoral immunity.

17. What role do macrophages and dendritic cells play in cell-mediated immunity?

18. How is each kind of T cell activated, and what are the main functions of each?

19. How do T_H1 cells bring about macrophage activation?

20. How are T_H cells involved in humoral immunity?

21. How do T_C cells kill virus-infected cells?

22. Match the cell type with the process or activity characteristic of that cell type. Answer choices can be used more than once or not at all.

   _____ Activated by antigen presenting cells (APC) such as macrophages

   _____ Forms myeloid stem cells

   _____ Involved directly in destroying viruses in the blood

   _____ Produces a protein hormone that stimulates T_H cell division

   _____ Aids B cell division

   _____ A granulocyte

   _____ Produces histamines

   a. B cell
   b. helper T cell
   c. cytotoxic T cell
   d. macrophage
   e. plasma cell
   f. memory cell
   g. none of the above

I Modifying Immune Responses

23. What factors contribute to lowered resistance to infection?

24. What factors contribute to impaired function of the immune system?

J Immunization

25. How is active immunization accomplished?

26. How do vaccines and toxoids differ?

27. List the major differences among vaccines.

28. What are the major benefits and hazards of active immunization?

29. How is passive immunization accomplished?

30. How do gamma globulin, hyperimmune serum, and antitoxins differ?

31. List the major benefits and hazards of passive immunization.

PROBLEMS FOR INVESTIGATION

1. How would you respond to parents who wish to avoid (a) all vaccines for their infant or (b) pertussis vaccine for their infant?

2. Explain how a person may be a compromised host on one day, week, or month but not on the next.

3. If you were born without T cells, would you have normal B cell functioning? Why or why not?

4. Identify each of the following types of immunity as innate or acquired, and state whether each is gained actively or passively.

   a. People do not get canine distemper

   b. After a case of polio

   c. After polio vaccine

   d. After an injection of gamma globulin

   e. After ingestion of colostrum

5. Select a bacterium, a virus, and a protozoan that infect the human body. Explain how the body uses innate and acquired immune responses to prevent infection. Explain also how immunity develops if an infection occurs.

6. The parents of a 2-month-old infant delighted in taking him with them on their frequent excursions to shopping malls. One of the grandmothers suggested that they leave him home, as it was flu season and the malls were full of coughing and sneezing people. The mother immediately responded that she was breast-feeding the baby, and with all those antibodies from her, he was surely protected against any diseases in the mall. Later, as she sat at the pediatrician's office holding her sick baby on her lap, she expressed disbelief that he could have caught the flu. If you were the doctor, what fallacies in her beliefs could you have explained to her? (The answer to this question appears in Appendix F.)

SOME INTERESTING READING

Ada, G. L., and G. J. V. Nossal. 1987. "The clonal-selection theory." *Scientific American* 257, no. 2 (August):62–69.

Akbar, A. N., and M. Salmon. 1995. "Selection and survival of activated lymphocytes." *Scientific American* 2(March/April):48–57.

Beardsley, T. 1995. "Better than a cure." *Scientific American* 272 (January):88–95.

Boon, T. 1993. "Teaching the immune system to fight cancer." *Scientific American* 268(March):82–89.

Christensen, D. 1994. "A shot in time." *Science News* 145(22):344–45.

Coghlan, A. 1995. "Grow your own vaccine." *New Scientist* 145(1961):23.

Doherty, P. C. 1995. "Immune memory to viruses." *ASM News* 61(2):68–71.

Engelhard, V. H. 1994. "How cells process antigens." *Scientific American* 271, no. 2 (August):54–61.

Ezzell, C. 1995. "Cancer 'vaccines': An idea whose time has come?" *The Journal of NIH Research* 7(1):46–49.

Golde, D. W. 1991. "The stem cell." *Scientific American* 265, no. 6 (December):86–93.

Gray, D. 1993. "Immunological memory." *Annual Review of Immunology* 11:49–77.

Greene, W. C. 1993. "AIDS and the immune system." *Scientific American* 269, no. 3 (September):98–105.

Janeway, C. A., Jr. 1993. "How the immune system recognizes invaders." *Scientific American* 269, no. 3 (September):72–79.

Janeway, C. A., Jr., and P. Travers. 1994. *Immunobiology: The immune system in health and disease.* New York: Garland.

Kärre, K. 1995. "Express yourself or die: Peptides, MHC molecules, and NK cells." *Science* 267(5200):978–79.

Liu, C.-C., C. M. Walsh, and J. D.-E. Young. 1995. "Perforin: Structure and function." *Immunology Today* 16(4):194–201.

Marx, J. 1995. "The T cell receptor begins to reveal its many facets." *Science* 267(5197):459–60.

Nossal, G. J. V. 1993. "Life, death and the immune system." *Scientific American* 269, no. 3 (September):52–62.

Ozanne, G., and M.-A. d'Halewyn. 1992. "Secondary immune response in a vaccinated population during a large measles epidemic." *Journal of Clinical Microbiology* 30(7):1778–82.

Paul, W. E. 1993. "Infectious diseases and the immune system." *Scientific American* 269, no. 3 (September):90–97.

Rabinovich, N. R., et al. 1994. "Vaccine technologies: View to the future." *Science* 265(5177):1401–4.

Roitt, I. M., J. Brostoff, and D. K. Male. 1993. *Immunology,* 3d edition. St. Louis: Mosby.

Saul, H. 1995. "Flu vaccines wanted: Dead or alive." *New Scientist* 145(1965):26–31.

Service, R. F. 1994. "Triggering the first line of defense." *Science* 265(5178):1522–24.

Smyth, M. J., and J. A. Trapani. 1995. "Granzymes: Exogenous proteinases that induce target cell apoptosis." *Immunology Today* 16(4):202–6.

Weissman, I. L., and M. D. Cooper. 1993. "How the immune system develops." *Scientific American* 269, no. 3 (September):64–71.

Wigzell, H. 1993. "The immune system as a therapeutic agent." *Scientific American* 269, no. 3 (September):126–34.

Wolinsky, H. 1993. "Supershots to the rescue." *American Health* 12(1):10–11.

CHAPTER 19

IMMUNOLOGY II: IMMUNOLOGICAL DISORDERS AND TESTS

Don't forget the chapter on AIDS.

AIDS is a fact of life. So make sure your children get all the facts.
If you need information or assistance, call 1-800-235-2331.

AIDS ACTION
COMMITTEE

As HIV infection rates soar worldwide, AIDS education becomes more imperative. The rates are climbing most rapidly among teenagers, via heterosexual transmission.

I n Chapter 18 we emphasized how specific immune responses defend the body against harmful substances. However, such responses are not always beneficial. Sometimes the humoral or cell-mediated responses react in ways that are physiologically unpleasant or even life-threatening. Perhaps you or someone you know gets a runny nose and watery eyes every time hay fever season rolls around. Maybe you know of people who have other allergies, who have had adverse reactions to a blood transfusion, or who suffer from more severe immunological disorders—even the dreaded AIDS.

In this chapter we will learn more about the immune system by examining the ways the immune system goes awry and reacts inappropriately or inadequately. We also will look at some of the laboratory and clinical methods used to detect and measure immune reactions.

OVERVIEW OF IMMUNOLOGICAL DISORDERS

An **immunological disorder** is a condition that results from an inappropriate or inadequate immune re-

sponse. Most inappropriate responses involve some type of hypersensitivity, whereas inadequate responses are due to an immunodeficiency.

Hypersensitivity

In **hypersensitivity,** or **allergy,** the immune system reacts in an exaggerated or inappropriate way to a foreign substance. Such responses can be thought of as "too much of a good thing"—the immune system responds to a harmless foreign agent by doing harm instead of protecting the body. Although allergy is another name for hypersensitivity, many disorders that people call "allergies" are not due to immunological reactions. These disorders include toxic responses to drugs, digestive upsets from nonallergic responses to foods, and emotional disturbances.

There are four types of hypersensitivity: (1) immediate hypersensitivity (Type I), (2) cytotoxic hypersensitivity (Type II), (3) immune complex hypersensitivity (Type III), (4) and cell-mediated, or delayed, hypersensitivity (Type IV). The type that develops depends on which components of the immune response are involved and on how quickly the reaction develops. **Immediate hypersensitivity,** or *anaphylaxis,* results from a prior exposure to a foreign substance called an *allergen,* an antigen that evokes a hypersensitivity response. Allergies to pollen, foods, and insect stings are examples of immediate hypersensitivity. **Cytotoxic hypersensitivity** is elicited by antigens on cells, especially red blood cells, that the immune system recognizes as foreign. This reaction occurs when a patient receives the wrong blood type during a transfusion. **Immune complex hypersensitivity** is elicited by antigens in vaccines, on microorganisms, or on a person's own cells. Large antigen–antibody complexes form, precipitate on blood vessel walls, and cause tissue injury within hours. **Cell-mediated,** or **delayed, hypersensitivity** is triggered by exposure to foreign substances from the environment (such as poison ivy), infectious disease agents, transplanted tissues, and the body's own tissues and cells. *Delayed hypersensitivity T cells* react with the foreign cells or substances, causing in some cases extensive tissue destruction.

Autoimmune disorders represent a form of hypersensitivity. In such disorders, the body's immune system responds to its own tissues as if they were foreign. Antibodies or T cells attack self-antigens.

Immunodeficiency

In **immunodeficiency** the immune system responds inadequately to an antigen, either because of inborn or acquired defects in B cells or T cells. The weak responses in immunodeficiency disorders are "too little of a good thing"—they leave the individual susceptible to infections, which can be severe and even life-threatening.

Let us now look more closely at the immunological disorders of hypersensitivity and immunodeficiency.

IMMEDIATE (TYPE I) HYPERSENSITIVITY

Immediate (type I) hypersensitivity, or *anaphylactic hypersensitivity,* typically produces an immediate response upon exposure to an allergy-inducing antigen (also known as an *allergen.* Nonallergic persons do not respond to such antigens. **Anaphylaxis** (an"a-fi-lak'sis) is an immediate, exaggerated allergic reaction to antigens. The term *anaphylaxis* refers to detrimental effects to the host caused by an inappropriate immune response. These effects are the opposite of *prophylaxis,* the preventive effects generated by an immune response.

Anaphylaxis is the harmful result of IgE antibodies made in response to allergens. It can be local or generalized (systemic). Localized anaphylaxis appears as reddening of the skin, watery eyes, hives, asthma, and digestive disturbances. Generalized anaphylaxis appears as a systemic life-threatening reaction such as airway constriction or *anaphylactic shock,* a generalized condition resulting from a sudden extreme drop in blood pressure.

Allergens

Immediate hypersensitivity results from two or more exposures to an allergen. An **allergen** is an ordinarily harmless foreign substance (typically a protein or a chemical bound to a protein) that can cause an exaggerated immunological response. The first exposure to the allergen produces no visible signs or symptoms. Allergens include airborne substances such as pollen, household dust, molds, and dander—tiny particles from hair, feathers, or skin. Household dust commonly contains nearly microscopic mites and their fecal pellets. Other allergens include venoms from insect stings, antibiotics, certain foods, sulfites, and foreign substances found in vaccines and in diagnostic or therapeutic materials. Allergens can be introduced into the body by inhalation, ingestion, or injection (Table 19.1).

Mechanism of Immediate Hypersensitivity

The typical sequence of events involved with the mechanism of type I hypersensitivity is *sensitization,* which involves the *production of IgE antibodies (antiallergens), allergen-IgE reactions,* and *local and systemic effects* of those reactions. Such reactions occur only in individuals who have previously been exposed to an allergen. In **sensitization,** the initial exposure to an allergen, B cells are activated (Figure 19.1a). The B

cells differentiate into plasma cells, which produce IgE antibodies against the specific allergen (Figure 19.1b). The IgE antibodies attach by their Fc tails to the surface of mast cells in the respiratory and gastrointestinal tracts and to basophils in the blood (Figure 19.1c). Attachment leaves the antigen (allergen) binding sites of the IgE antibodies free to react with the same allergen upon future exposure. This sequence of sensitization steps does not occur in all people. Why some people become sensitized to normally innocuous substances whereas others do not is poorly understood.

The sensitized mast cells and basophils now are primed to produce a massive chemical response to a second exposure from the same allergen. Although the *sensitizing* (first) *dose* of an allergen can be fairly high, the *triggering* (subsequent) *dose* that causes the hypersensitive symptoms can be quite small. When a second or subsequent encounter occurs with the same allergen, the allergen attaches to sensitized mast cells and basophils, cross-linking the IgE antibodies (Figure 19.1d). Cross-linking causes **degranulation,** the rapid release of *preformed mediators* (chemical substances that induce allergic responses) from cytoplasmic granules in mast cells and basophils (Figure 19.1e). **Histamine** is the main preformed mediator in humans. It dilates capillaries, thereby making them more permeable. It also causes bronchial smooth muscle to contract, increases mucus secretion, and stimulates nerve endings that cause pain and itching.

Prostaglandins and **leukotrienes** are *reaction mediators* (chemical substances that control responses) that are also synthesized and released from mast cells after degranulation has occurred. Prostaglandin D_2 is a cellular messenger molecule produced in mast cells and basophils that also causes constriction of bronchial smooth muscle. *Slow-reacting substance of anaphylaxis* (SRS-A) is another mediator that causes a slow, long-lasting airway constriction in animals. SRS-A consists of three leukotriene mediators. These leukotrienes are 100 to 1000 times as potent as histamines and prostaglandin D_2 in causing prolonged airway constriction. Like hist-

TABLE 19.1	Common Allergens	
Ingested	**Inhaled**	**Injected**
Animal proteins, especially from milk and eggs	Cocaine	Antibiotics, especially cephalosporins and penicillins
Aspirin	Dander	Heroin
Fruits	Dust (household)	Hormones (adrenocorticotropic hormone and animal insulin)
Grains	Face powder	Insect venoms (from bees, hornets, wasps, and yellow jackets)
Hormone preparations	Insecticides	Snake venoms (from vipers and cobras)
Nuts	Mites and their feces	Spider venoms, especially from black widow and brown recluse
Penicillin	Pollen (from grass, trees, and weeds)	
Seafood	Spores (fungal and bacterial)	

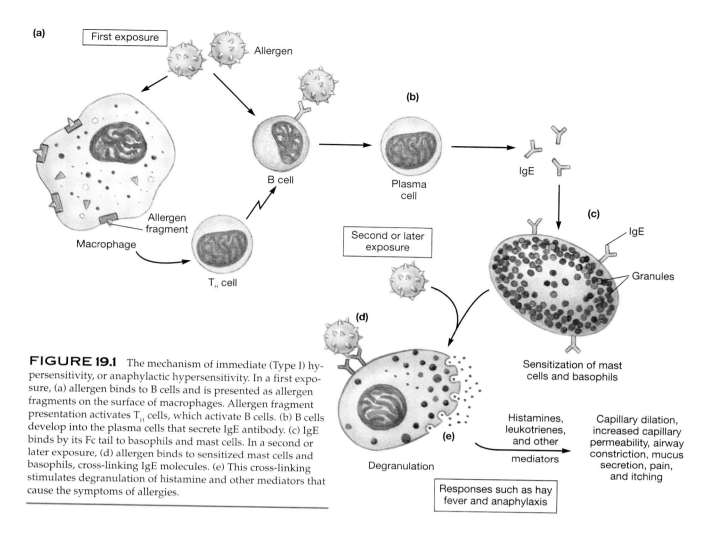

FIGURE 19.1 The mechanism of immediate (Type I) hypersensitivity, or anaphylactic hypersensitivity. In a first exposure, (a) allergen binds to B cells and is presented as allergen fragments on the surface of macrophages. Allergen fragment presentation activates T$_H$ cells, which activate B cells. (b) B cells develop into the plasma cells that secrete IgE antibody. (c) IgE binds by its Fc tail to basophils and mast cells. In a second or later exposure, (d) allergen binds to sensitized mast cells and basophils, cross-linking IgE molecules. (e) This cross-linking stimulates degranulation of histamine and other mediators that cause the symptoms of allergies.

amine, leukotrienes and prostaglandin D$_2$ dilate and increase the permeability of capillaries, increase thick mucus secretion, and stimulate nerve endings that cause pain and itching. Preformed and reaction mediators and their effects are summarized in Table 19.2.

It is a second or later allergen exposure of a sensitized person that produces allergic signs and symptoms. Such hypersensitive responses by mast cells can be triggered by nonallergic factors, too. Emotional stress and temperature extremes often cause mediator release without the involvement of any IgE or allergen.

Localized Anaphylaxis

Atopy (at'o-pe), which literally means "out of place," refers to localized allergic reactions. In atopic individuals, immune reactions occur first at the site where the allergen enters the body. If the allergen enters the skin, it causes a *wheal and flare reaction,* characterized by redness, swelling, and itching (Figure 19.2). If the allergen is inhaled, mucous membranes of the respiratory tract become inflamed, and the patient has a runny nose and watery eyes. If the allergen is ingested, mucous membranes of the digestive tract become inflamed, and the

TABLE 19.2	Mediators of Immediate Hypersensitivity and Their Effects
Mediator	**Effects**
Preformed mediators	
Histamine	Vascular dilation and increased capillary permeability, bronchial smooth muscle contraction, edema of mucosal tissues, secretion of mucus, and itching
Neutrophil and eosinophil chemotactic factors	Attraction of neutrophils, eosinophils, and other leukocytes to the site of an allergic reaction
Reaction mediators	
Leukotrienes (SRS-A)	Prolonged bronchial smooth muscle contraction, increased capillary permeability, edema of mucosal tissues, and secretion of mucus
Prostaglandin D$_2$	Formation of minute blood clots, bronchial smooth muscle contraction, and capillary dilation

patient may have abdominal pain and diarrhea. Some ingested allergens, such as foods and drugs, also cause skin rashes.

Hay fever, or *seasonal allergic rhinitis,* is a common kind of atopy. More than 20 million Americans suffer from the typical signs and symptoms of watery eyes, sneezing, nasal congestion, and sometimes shortness of breath. First described in 1819 as resulting from exposure to newly mown hay, hay fever is now known to result from exposure to airborne pollen—tree pollens in spring, grass pollens in summer, and ragweed pollen (Figure 19.3) in fall. Some plants, such as goldenrod and roses, have long been blamed for hay fever but are innocent because their pollens are too heavy to be airborne for any great distance. It is the far less conspicuous green ragweed flowers that cause much of the misery for hay fever sufferers. On occasion, severe allergic rhinitis (in-flammation of the nasal surfaces) can progress to sinus infections, middle ear problems, and temporary hearing loss. Although they share many symptoms, hay fever can be distinguished from the common cold by the increased numbers of eosinophils in nasal secretions. Finding elevated numbers of eosinophils in blood also suggests allergy (or instead infection with helminths).

Generalized Anaphylaxis

Some anaphylactic reactions are generalized, severe, and immediately life-threatening. In a sensitized person, a generalized reaction begins with sudden reddening of skin, intense itching, and hives, especially over the face, chest, and palms of hands. The disorder

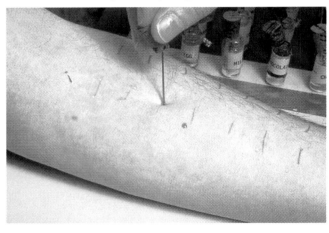
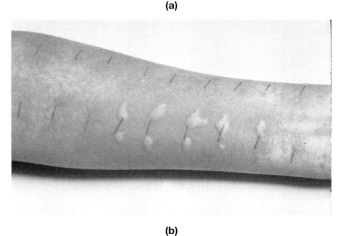

FIGURE 19.2 (a) In one type of allergy testing, possible allergens are placed on prongs and introduced under the patient's skin. (b) If an individual is hypersensitive, a wheal (white raised area) and flare (reddened area) soon become visible on the skin. Such testing is usually done on the extremities to keep any hypersensitive reaction away from major organs.

(a)

(b)

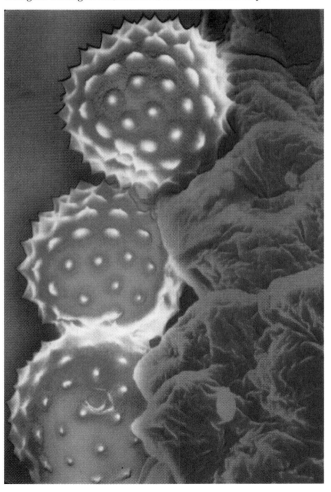

FIGURE 19.3 False-color SEM photo of ragweed (*Ambrosia*) pollen, one of several pollen causes of hay fever. In early spring the culprits are primarily tree pollens such as oak, elm, birch (especially in Europe), and box elder. In late spring and early summer, grass pollens plus those of some broad-leaved plants are most likely to be involved. In late summer and early autumn, the chief allergens are ragweeds, saltbush, and Russian thistle pollens.

THE ORIGINS OF ALLERGIES

Allergies account for almost 10 percent of all visits to physicians' offices in the United States. Why does the immune system often react violently to nonharmful substances? And why are some people allergic, whereas others are not? Research has shown that some people who lack the ability to produce IgE antibodies are prone to lung and sinus infections. Also, people who lack the ability to produce IgG or IgM antibodies often produce IgE antibodies to bacterial infections. These observations suggest that the IgE antibody may play a necessary role in immunity, besides causing allergies.

IgE is known to help fight infections by parasitic helminths. ∞ (Chapter 18, p. 508) IgE antibodies also may protect against ectoparasites (ticks, chiggers, fleas). The American biologist Margie Profet believes that IgE antibodies also are a backup system to protect against the ingestion of toxins.

In the book *Why we get sick: The new science of Darwinian medicine*, the authors, physician Randolph Nesse and evolutionist George Williams, suggest that many allergies that exist today were not common 150 years ago. They say that hay fever was almost nonexistent in England in the early 1800s and rare in Japan even as recently as 1950. Yet today, about 10 percent of the Japanese have hay fever. Nesse and Williams suggest that modern comforts may be to blame. Well-insulated houses with thick carpeting are excellent breeding grounds for dust mites and catchalls for allergens such as pollen and mold spores. In fact, studies show that infants raised in homes relatively clean of allergens develop far fewer allergies than do infants raised in the average home. However, these ideas still do not explain fully why some people have allergies whereas others in the same household or family do not.

Because about 15 percent of Americans suffer from allergies, it stands to reason that any one person would have about a 15 percent chance of being allergic to some allergen. In atopic families, if one parent is atopic, a child has a 25 percent chance of having an allergy. If both parents are atopic, the chance jumps to about 50 percent. Thus, some allergies can run in families and may, in part, have a genetic basis. Profet, Nesse, and Williams also suggest that if someone is simultaneously exposed, for example, to a plant toxin and an allergen, the immune system may respond to the toxin by producing IgE antibodies. In this attack, the immune system sees the allergen as "part of the toxin" and so reacts to it as well. Immune cells remain sensitized, and future exposures of the allergen alone will trigger an IgE response even though the toxin is not present.

Unfortunately, no matter what the causes or origins of allergies, we must suffer through them with only our allergy medications to help us along.

---

can then progress to respiratory anaphylaxis or anaphylactic shock.

In **respiratory anaphylaxis** the airways become severely constricted and filled with mucus secretions, and the allergic individual may die of suffocation. More than 4000 Americans die each year from respiratory anaphylaxis. Another 15 million Americans suffer from **asthma,** which is often caused by inhaled or ingested allergens, emotional stress, aspirin, or cold, dry air. Asthma can also be caused by hypersensitivity to endogenous microorganisms. For example, some patients become sensitized to *Moraxella catarrhalis,* a normal bacterial resident of respiratory mucous membranes.

In **anaphylactic shock** blood vessels suddenly dilate and become more permeable, causing an abrupt and life-threatening drop in blood pressure. Insect bites and stings are a common cause of anaphylactic shock in people sensitized to insect venoms (Figure 19.4a).

Generalized anaphylaxis must be treated immediately. Unless epinephrine (adrenaline) is administered immediately, death can occur. Epinephrine acts by relaxing smooth muscle of respiratory passageways and constricting blood vessels. People who are sensitized to insect venom often carry an emergency anaphylactic kit (Figure 19.4b). The kit contains a tourniquet, benadryl (antihistamine) tablets, and a syringe containing two doses of epinephrine. Having such a kit on hand could easily mean the difference between life and death because of the rapid onset of life-threatening symptoms in patients who already have had anaphylactic reactions.

Genetic Factors in Allergy

In the United States, 40 million people have some kind of allergy. In many cases, genetic factors are thought to contribute to the development of allergies. Although different family members typically have different allergies (one person may suffer from asthma and another from dust allergies), all will have high levels of IgE antibodies. At least 60 percent of children with atopy have a family history of asthma or hay fever, and half these children later develop other allergies. Thus, allergy probably has a genetic basis, possibly in properties of membranes or the performance of various cells involved in immune responses, such as phagocytes. Normal membranes screen out all but the tiniest microorganisms and

(a)

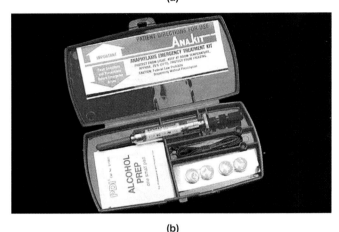

(b)

FIGURE 19.4 (a) Honeybee sting to the face has swollen an eye shut but has not caused the airway to close, as happens in more severe reactions. (b) Anaphylactic kit for emergency use, showing syringe loaded with epinephrine. Many individuals with severe insect-sting allergies carry such kits with them at all times. They are available only by prescription.

virtually all potential allergens. Membranes of allergic individuals, however, are more permeable to larger particles such as pollen grains. Even when allergens pass through membranes, phagocytic cells usually engulf them in normal individuals but sometimes fail to do so completely in allergic individuals.

Treatment of Allergies

One approach to dealing with allergies is to avoid contact with the specific allergen. People with food allergies should not eat a food to which they have had a hypersensitivity reaction. Whatever the allergen, a subsequent exposure generally will trigger degranulation by mast cells (Figure 19.5a). **Desensitization** (hyposensitization) is the only currently available treatment intended to cure an allergy. If denatured allergen is injected subcutaneously ("allergy shots"), it may induce a state of tolerance, preventing the activation of those B cells that mature into IgE-secreting plasma cells (Figure 19.5b). Also, by receiving such injections with gradually increasing doses of the allergen, the patient may produce IgG antibodies, called **blocking antibodies,** against the allergen. Upon reexposure to an allergen, the blocking antibodies combine with the allergen before the allergen has a chance to react with IgE, so mast cells do not release mediators. Thus, increases in IgG and decreases in IgE may work together to make the patient less sensitive to the allergen.

Desensitization has been very successful against insect venoms and drug allergies such as allergies to penicillin. Unfortunately, desensitization does not alleviate the signs and symptoms in many allergies, such as hay fever. In addition, the treatment itself can cause anaphylactic shock because the injections contain the very substance to which the patient is allergic. Patients must remain in the physician's office for 20 to 30 minutes after the injection so that emergency treatment will be available if a generalized anaphylactic reaction occurs.

Other allergy treatments alleviate allergy symptoms but do not cure the disorder. Antihistamines counteract the swelling and redness due to histamine but are not effective against SRS-A or asthmatic conditions, which involve constriction of the airways, whereas antiinflammatory agents such as corticosteroids suppress the inflammatory response. Better therapeutic methods of treating allergies are greatly needed. As we learn more about the properties of leukotrienes and IgE antibodies, perhaps this need can be met.

CYTOTOXIC (TYPE II) HYPERSENSITIVITY

In cytotoxic (Type II) hypersensitivity, specific antibodies react with cell surface antigens on host cells, leading to phagocytosis, killer cell activity, or complement-mediated lysis. The cells to which the antibodies are attached, as well as surrounding tissues, are damaged because of the resulting inflammatory response. Antigens that initiate cytotoxic hypersensitivity typically enter the body in mismatched blood transfusions or during delivery of an Rh-positive infant to an Rh-negative mother.

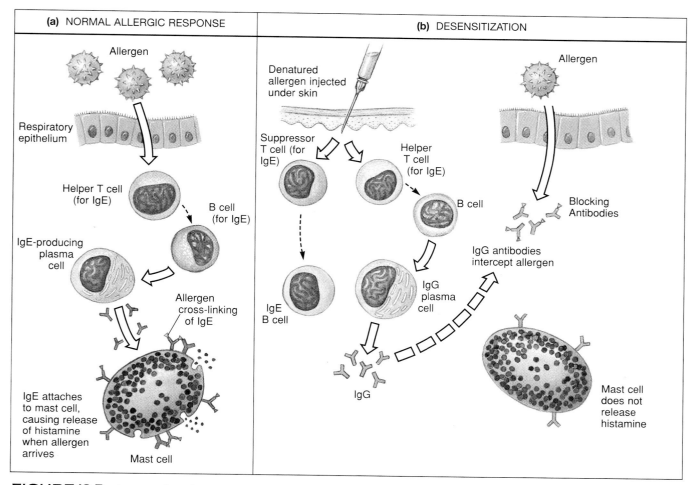

FIGURE 19.5 A proposed mechanism of action for desensitization allergy shots (hyposensitization). (a) In a normal allergic response, natural exposure to an allergen causes helper T cells to stimulate those B cells that mature into plasma cells to make IgE antibodies. After binding to mast cells, a second allergen exposure causes degranulation. (b) Desensitization involves the injection of denatured allergen. Such shots may lead to tolerance, preventing B cells from maturing into plasma cells to make IgE antibodies. Exposure to the allergen also may activate those B cells that mature into the plasma cells that make IgG (blocking) antibody. Such IgG antibodies can bind to incoming allergen before it reaches the IgE molecules attached to mast cells. Complexing of allergen with these attached IgE molecules would cause the mast cells to degranulate and release histamine, so blocking this step is the key to preventing allergic responses.

Mechanism of Cytotoxic Reactions

When an antigen on a plasma membrane is first recognized as foreign, B cells become sensitized and stand ready for antibody production upon a subsequent antigen exposure. During subsequent exposures with the surface antigen, antibodies bind to the antigen and activate complement. Phagocytic cells, such as macrophages and neutrophils, are attracted to the site. The mechanisms of type II hypersensitivity appear to be responsible for the tissue damage in cases of rheumatic fever following a streptococcal infection, in certain viral diseases, in transfusion reactions, and in hemolytic disease of the newborn (mother-infant Rh incompatibility).

Examples of Cytotoxic Reactions

Cytotoxic reactions typical of type II hypersensitivities are exemplified by mismatched blood transfusions and by hemolytic disease of the newborn.

Transfusion Reactions

Normal human red blood cells have genetically determined surface antigens (blood group systems) that form the basis for the different blood types. A **transfusion reaction** can occur when matching antigens and antibodies are present in the patient's blood at the same time. Such reactions can be triggered by any of the

TABLE 19.3	Properties of the ABO Blood Group System	
Blood Type	Antigens on Erythrocytes	Antibodies in Serum
A	A	anti-B
B	B	anti-A
AB	A and B	neither anti-A nor anti-B
O	neither A nor B	anti-A and anti-B

blood group antigens. We will focus on antigens A and B, which determine the **ABO blood group system.** As Table 19.3 shows, four blood types—A, B, AB, and O—are named according to whether red blood cells have antigen A, antigen B, both A and B antigens, or neither antigen. Normally, a person's serum has no IgM antibodies against the antigens present on his or her own red blood cells. However, if a sensitized patient receives red blood cells with a different blood cell antigen during a blood transfusion, IgM antibodies cause a Type II

hypersensitivity reaction against the foreign antigen. The foreign red blood cells are agglutinated (clumped), complement is activated, and hemolysis (rupture of blood cells) occurs within the blood vessels (Figure 19.6). Symptoms of a transfusion reaction include fever, low blood pressure, back and chest pain, nausea, and vomiting. Transfusion reactions can usually be prevented by careful cross-matching of donor and recipient blood group antigens so that the correct blood type can be selected for transfusion (Figure 19.7).

Transfusion reactions to other erythrocyte antigens also occur. However they usually are less serious than reactions to foreign A or B antigens because the antigen molecules are less numerous.

Hemolytic Disease of the Newborn

Another example of a cytotoxic reaction is **hemolytic disease of the newborn,** or *erythroblastosis fetalis.* In addition to the ABO blood group, red blood cells can have **Rh antigens,** so named because the antigens were discovered first in Rhesus monkeys. Blood with Rh antigens on red blood cells is designated Rh-positive; red blood cells lacking Rh antigens are designated Rh-negative. Anti-Rh antibodies normally are not present in the serum of either Rh-positive or Rh-negative blood. Consequently, sensitization is necessary for an Rh antigen–antibody reaction.

FIGURE 19.6 The mechanisms of cytotoxic (Type II) hypersensitivity. Mismatched red blood cell antigen usually is bound to IgM. Complement is activated and results in either subsequent phagocytosis or lysis of the red blood cells.

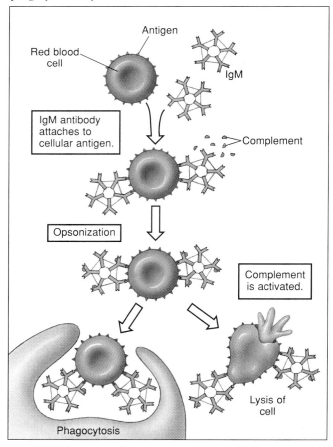

FIGURE 19.7 Careful blood typing and matching of donor and recipient blood prevents most transfusion reactions. Persons with type AB blood can safely receive a transfusion of any of the four major blood types. Persons with type O blood can safely donate blood to recipients of any blood type. (Can you explain why? Refer to Table 19.3.)

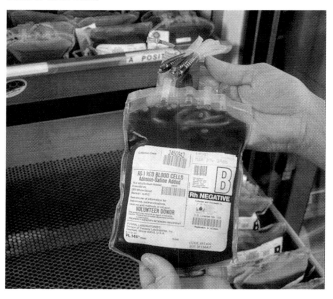

Sensitization typically occurs when an Rh-negative woman carries an Rh-positive fetus, which inherited this blood type from its father. The fetal Rh antigen rarely enters the mother's circulation during pregnancy but can leak across the placenta during delivery, miscarriage, or abortion (Figure 19.8a). The Rh-negative mother's immune system then becomes sensitized to the Rh antigen and can produce anti-Rh antibodies if it again encounters the Rh antigen.

Because sensitization usually occurs at delivery, the first Rh-positive child of an Rh-negative mother rarely suffers from hemolytic disease. But when a sensitized Rh-negative mother carries a second or subsequent Rh-positive fetus, the mother's anti-Rh antibodies cross the placenta and cause a Type II hypersensitivity reaction in the fetus (Figure 19.8b). If this occurs, fetal red blood cells agglutinate, complement is activated, and the red

blood cells are destroyed. The result is hemolytic disease of the newborn. The baby is born with an enlarged liver and spleen caused by efforts of these organs to eliminate damaged red blood cells (Figure 19.8c,d). Such babies are jaundiced due to excessive bilirubin—a product of the breakdown of red blood cells—in their blood.

Hemolytic disease of the newborn can be prevented by giving Rh-negative mothers intramuscular injections of anti-Rh IgG antibodies (Rhogam) within 72 hours after delivery. The antibodies presumably bind to Rh antigens on the fetal red blood cells that have leaked into the mother's blood. These anti-Rh antibodies destroy the fetal red blood cells before they can act to sensitize her immune system. It is essential to treat all Rh-negative women after delivery, miscarriage, or abortion in case the fetus may have been Rh-

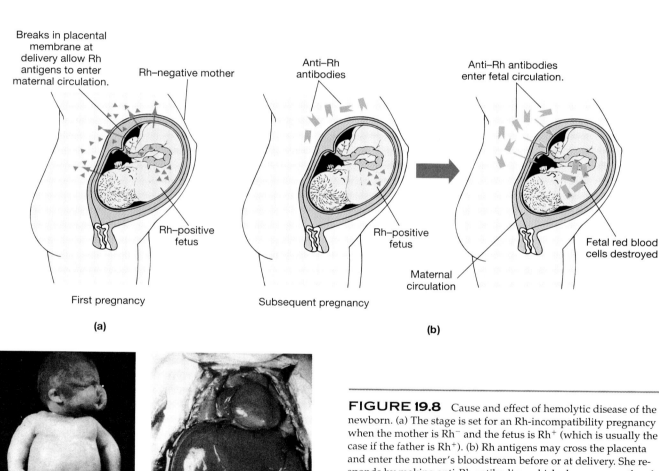

(a)

(b)

(c)   (d)

FIGURE 19.8 Cause and effect of hemolytic disease of the newborn. (a) The stage is set for an Rh-incompatibility pregnancy when the mother is Rh⁻ and the fetus is Rh⁺ (which is usually the case if the father is Rh⁺). (b) Rh antigens may cross the placenta and enter the mother's bloodstream before or at delivery. She responds by making anti-Rh antibodies, which also can cross the placenta. Even if antibody production is not stimulated until delivery, the resulting antibodies will persist in the mother's circulation and attack the red blood cells of any subsequent Rh⁺ fetus. To prevent this situation, Rhogam (anti-Rh antibody) is injected into the mother early in the pregnancy, immediately after delivery, and in cases of miscarriage or abortion. Rhogam reduces exposure to the antigen and thus lessens anti-Rh antibody production. (c) Child affected by hemolytic disease caused by Rh incompatibility. (d) The liver is greatly enlarged.

positive. Today, anti-Rh antibodies are often adminis-tered to Rh-negative women during pregnancy as well. Such treatment at 3 and 5 months prevents sensitiza-tion to the fetus in case fetal antigens leak into the mother's circulation, which can result from hard coughing or sneezing. Before anti-Rh antibody came to be given preventively, hemolytic disease of the new-born occurred in about 0.5 percent of all pregnancies; 12 percent of these terminated in stillbirths.

Examples of Immune Complex Disorders

We will illustrate immune complex disorders with two phenomena, the systemic serum sickness and the lo-calized Arthus reaction. Other disorders that involve immune complexes, such as rheumatoid arthritis and systemic lupus erythematosus, are discussed later in connection with autoimmune diseases because the an-

IMMUNE COMPLEX (TYPE III) HYPERSENSITIVITY

Immune complex (Type III) hypersen-sitivity results from the formation of antigen–antibody complexes. Under normal circumstances these large im-mune complexes are engulfed and de-stroyed by phagocytic cells. Hyper-sensitivity occurs when antigen–antibody complexes persist or are con-tinuously formed.

Mechanism of Immune Complex Disorders

Like anaphylactic and cytotoxic reac-tions, immune complex disorders also are initiated after sensitization. Upon subsequent exposure to the sensitizing antigen, specific IgG antibodies com-bine with the antigen in the blood to form an *immune complex* and activate complement (Figure 19.9). Normally, large immune complexes are removed by phagocytosis in the liver and spleen. However, immune complexes are often quite small and fail to bind tightly to Kupffer cells in the liver, thereby escaping elimination from the blood, and are deposited in organs, tis-sues, or joints. Such antigen-antibody complexes and complement, in turn, cause basophils and mast cells to re-lease histamine and other mediators of allergic reactions, with the effects de-scribed earlier. Phagocytes attracted chemotactically to these sites of activ-ity release hydrolytic enzymes, caus-ing tissue damage that is acute but can become chronic if the antigen remains for long periods of time.

FIGURE 19.9 The mechanism of immune complex (Type III) hypersensitivity. Immune complexes are formed when antigen is introduced into a previously sensitized individual. When the resulting immune complex is deposited, it activates complement, producing fever, itching, rash or hemorrhagic areas, joint pain, and acute inflammation. On a systemic basis, this can cause serum sickness.

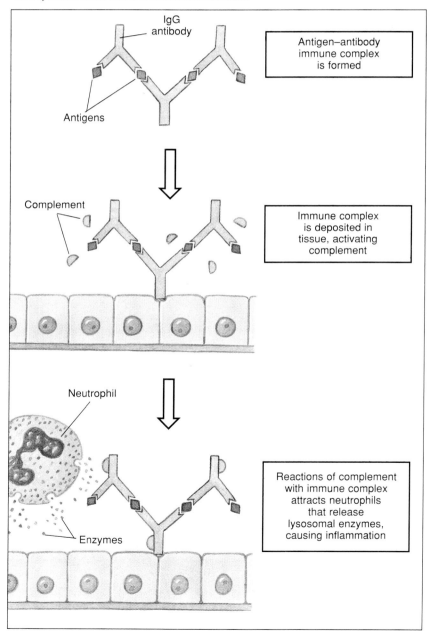

tibodies involved in these disorders react with the patient's own tissues. Acute glomerulonephritis following certain streptococcal infections is another immune complex disease, in which the glomeruli of the kidneys can be severely damaged (Chapter 25).

Serum sickness was frequently seen in the pre-antibiotic era when large doses of antitoxin sera were used to immunize people passively against infectious diseases such as diphtheria. Diphtheria toxin given to horses caused them to make antibodies against the toxin. A patient then received the horse serum, which contained not only antidiphtheria toxin antibody but also horse proteins. A sensitized patient's immune system would make sufficient antibodies against these horse proteins to form immune complexes consisting of human antibody reacting against horse serum protein, upon second exposure. These immune complexes, which are removed slowly by phagocytic cells, would attach to the glomeruli of the kidneys. The filtration capacity of the glomeruli was thereby impaired, causing proteins and blood cells to be excreted in the urine. Immune complexes are also deposited in joints and in skin blood vessels.

Patients with serum sickness usually have fever, enlarged lymph nodes, decreased numbers of circulating leukocytes, and swelling at the injection site.

Most patients recover from serum sickness as the complexes eventually are cleared from the blood and tissue repair occurs in the glomeruli. However, the disorder became chronic in many diphtheria patients because they received horse serum daily over the course of the disease.

Today, serum sickness is rare and usually is due to second exposure to a foreign substance in a biological product, such as horse serum in a vaccine preparation. An advantage of vaccines made by genetic engineering is that they do not contain such foreign substances. When the use of any biological product is contemplated, the patient should be tested for sensitivity first. A small quantity of the product should be given intradermally (within the skin) or intravenously. A wheal and flare reaction or a drop of 20 points or more in blood pressure following intravenous injection indicate hypersensitivity, and the product should not be given.

The **Arthus reaction** is a local reaction seen in the skin after subcutaneous (under the skin) or intradermal injection of an antigenic substance. It occurs in patients who already have large quantities of antibodies (mainly IgG) to the antigen. In 4 to 10 hours, edema and hemorrhage develop around the injection site as immune complexes and complement trigger cell damage and platelet aggregation (Figure 19.10). In severe reactions,

FIGURE 19.10 The mechanism that produces an Arthus reaction and hemorrhagic areas. In severe cases, (a) injection of horse protein antigens leads to (b) immune complex formation. (c) In association with complement, (d) neutrophils release lysosomal enzymes that damage the blood vessel wall. (e) Immune complexes also trigger platelet aggregation that can obstruct blood flow. (f) Complement also attracts more neutrophils to the site and (g) causes mast cell degranulation. (h) Finally, platelet and complement trigger endothelial retraction. Tissue death can occur if cells are cut off from blood vessel flow.

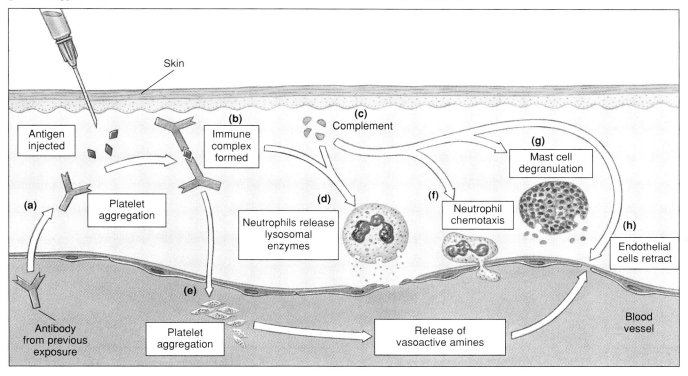

FIGURE 19.11 Arthus reaction in patient showing extensive area of hemorrhagic damage that will result in tissue necrosis.

senting cells that present antigen fragments to T_H1 (inflammatory T) cells. ∞ (Chapter 18, p. 498) When APCs again present the same antigen during a second, later exposure, the sensitized T_H1 cells release various cytokines, including γ-interferon and migration inhibiting factor (MIF). Gamma-interferon stimulates macrophages to ingest the antigens. If the antigens are on microorganisms, the macrophages usually, but not always, kill the microorganisms. MIF prevents migration of macrophages, so they remain localized at the site of the hypersensitivity reaction. Other cytokines are presumed to cause the hypersensitivity reaction itself. Such reactions account for patches of raw, reddened skin in eczema, swelling, and granulomatous lesions. These processes are summarized in Figure 19.12.

tiny clots obstruct blood vessels, and cells normally nourished by the blocked vessels die (Figure 19.11).

CELL-MEDIATED (TYPE IV) HYPERSENSITIVITY

Cell-mediated (Type IV) hypersensitivity is also called *delayed hypersensitivity* because reactions take more than 12 hours to develop. These reactions are mediated by T cells—specifically, a type of T_H1 cell (sometimes called a **delayed hypersensitivity T [T_{DH}] cell**)—not by antibodies.

Mechanism of Cell-Mediated Reactions

Cell-mediated hypersensitivity occurs as follows. On first exposure, antigen molecules bind to antigen-pre-

Examples of Cell-Mediated Disorders

Three common examples of delayed hypersensitivity—contact dermatitis, tuberculin hypersensitivity, and granulomatous hypersensitivity—illustrate the diversity of cell-mediated reactions.

Contact dermatitis occurs in sensitized individuals on second or subsequent exposure to allergens such as oils from poison ivy, rubber, certain metals, dyes, soaps, cosmetics, some plastics, topical medications, and other substances (Table 19.4). Unlike Type I hypersensitivities, Type IV hypersensitivities do not appear to run in families. Molecules too small to cause immune reactions pass through the skin, where they become antigenic by binding to normal proteins on Langerhans cells of the epidermis. These cells, which carry MHC class II antigens, migrate to the lymph nodes, where they act as antigens presenting cells to T_H1 cells. Within 4 to 8 hours after the next exposure, a hypersensitivity reaction begins, and eczema occurs within 48 hours.

FIGURE 19.12 The mechanism of cell-mediated, or delayed (Type IV), hypersensitivity. This type of reaction is mediated by T cells rather than by B cells as in types I, II, and III. T cells that have become sensitized to a particular antigen release cytokines on subsequent contact with the same antigen fragment. These cytokines cause inflammatory reactions that attract macrophages to the site. By degranulation, APCs release mediators that add to the inflammatory response. Contact dermatitis and poison ivy rash are examples of cell-mediated hypersensitivity.

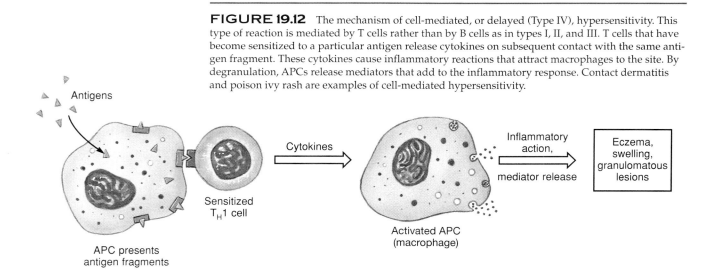

Antigens

APC presents antigen fragments

Sensitized T_H1 cell

Cytokines

Activated APC (macrophage)

Inflammatory action, mediator release

Eczema, swelling, granulomatous lesions

TABLE 19.4	Selected Contact Allergens
Allergen	**Common Sources of Contact**
Benzocaine	Topical anesthetic
Chromium	Jewelry, watches, chrome-tanned leather, cement
Formaldehyde	Facial tissues, nail hardeners, synthetic fabrics
Latex	Surgical or examination gloves
Nickel	Jewelry, watches, objects made of stainless steel and white gold
Mercaptobenzothiazole	Rubber goods
Methapyrilene	Topical antihistamine
Merthiolate	Topical antiseptic
Neomycin	Topical antibiotic
Oleoresin	Oil from poison ivy and similar plants

(a)

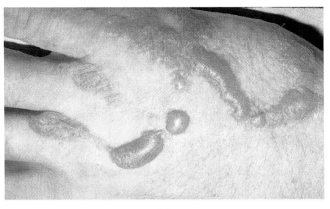

(b)

FIGURE 19.13 (a) Poison ivy (*Toxicodendron radicans*), showing the leaves with their characteristic three leaflets. Poison ivy vines also contain the irritating oil urushiol, so it is important to be able to recognize them in winter, when leaves may not be present. (b) Poison ivy dermatitis, showing fluid-filled vesicles.

Urushiol (u'ru-she-ol), an oil from the poison ivy plant, is a major cause of contact dermatitis in the United States (Figure 19.13). Most people get poison ivy from direct contact with leaves or other plant parts, but some get it by inhaling smoke from burning brush that contains poison ivy plants. Poison ivy is particularly severe when oil droplets come in contact with respiratory membranes. Sensitivity to poison ivy can develop at any age, even among people who have come in contact with it for years without reacting to it. One way to minimize a reaction to poison ivy is to wash exposed areas thoroughly with strong soap or detergent within minutes of contact, before much oil has penetrated the skin and bound chemically to skin cells. Once a person is sensitized, an exceedingly small quantity of oil will elicit a reaction upon the next exposure. Scratching lesions does not spread the oil, but it can lead to infections. Cashews and mangoes contain substances chemically similar to urushiol, and some people display delayed hypersensitivity (including digestive disturbances) to these substances.

Tuberculin hypersensitivity occurs in sensitized individuals when they are exposed to *tuberculin,* an antigenic lipoprotein from the tubercle bacillus *Mycobacterium tuberculosis.* Similar antigens from the bacterium that causes leprosy (*Mycobacterium leprae*) and the protozoan that causes leishmaniasis (*Leishmania tropica*) produce similar reactions in sensitized individuals. The antigen activates T_H1 cells, which in turn release cytokines that cause large numbers of lymphocytes, monocytes, and macrophages to infiltrate the dermis. The normally soft tissues of the skin then form a raised, hard, red region called an **induration** (Figure 19.14). In a **tuberculin skin test,** a purified protein derivative (PPD) from *Mycobacterium tuberculosis* is injected subcutaneously. If a person has been exposed to the bacterium

APPLICATIONS

POISON IVY? BUT IT DOESN'T GROW HERE!

Contact dermatitis similar to poison ivy was seen in American military personnel in Japan after World War II. Lesions appeared on elbows, forearms, and in a horseshoe shape on the buttocks and thighs. Knowing that poison ivy did not grow in Japan, the medical staff was puzzled. The puzzle was solved when scientists discovered that oils from a Japanese plant containing a small quantity of urushiol were used to manufacture lacquer. Although the quantity of urushiol in lacquer was not sufficient to sensitize Japanese people, it did elicit an allergic response in Americans previously sensitized to poison ivy. Resting the arms on countertops explained lesions on elbows and forearms; contact with toilet seats explained the horseshoe-shaped lesions.

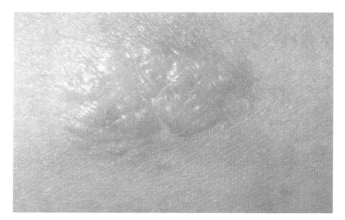

FIGURE 19.14 A positive tuberculin skin test reaction. The raised area of induration should be observed and measured after 48 to 72 hours. A positive reaction will measure 5 mm or more; 2 mm or smaller is negative; 3 and 4 mm are considered doubtful.

or had received the BCG vaccine, an induration will form within 48 hours. The diameter and elevation of the induration indicate whether further tests are needed.

Granulomatous hypersensitivity, the most serious of the cell-mediated hypersensitivities, usually occurs when macrophages have engulfed pathogens but have failed to kill them. Inside the macrophages the protected pathogens survive and sometimes continue to divide. T_H1 cells sensitized to an antigen of the pathogen elicit the hypersensitivity reaction, attracting several cell types to the skin or lung. A granuloma in the skin (leproma) or lung (tubercle) develops. This kind of hypersensitivity is the most delayed of all, appearing 4 weeks or more after exposure to the antigen. Such persistent and chronic antigenic stimuli are also typical of the bacterial disease listeriosis as well as many fungal and helminthic infections.

Characteristics of the four types of hypersensitivity are summarized in Table 19.5.

AUTOIMMUNE DISORDERS

Autoimmune disorders occur when individuals become hypersensitive to specific antigens on cells or tissues of their own bodies, despite mechanisms that ordinarily create tolerance to those self antigens. The antigens elicit an immune response in which **autoantibodies,** antibodies against one's own tissues, are produced. An autoimmune response can be T-cell-mediated as well. These disorders are characterized by cell destruction in various types of hypersensitivity reactions. Although autoimmune disorders arise from a response to a self antigen, they range over a wide spectrum—from those that affect a single organ or tissue (organ-specific) to those that are systemic, affecting many organs and tissues (Table 19.6).

Autoimmunization

Autoimmunization is the process by which hypersensitivity to "self" develops. Such a response is usually sustained and long-lasting and can cause long-term tissue damage. Immunologists are beginning to understand this process better. Several different mechanisms of autoimmunity probably exist:

1. *Genetic factors* may predispose a person toward autoimmune disorders. For example, the children of a parent who has autoantibodies to a single organ are likely to develop autoantibodies to the same or to a different single organ. As we shall see later, individuals who have genes for certain histocompatibility antigens are at greater than normal risk of developing particular autoimmune disorders.

TABLE 19.5	Characteristics of the Types of Hypersensitivity			
	Type I	**Type II**	**Type III**	**Type IV**
Characteristic	Immediate	Cytotoxic	Immune Complex	Cell-Mediated
Main mediators	IgE	IgG, IgM	IgG, IgM	T cells
Other mediators	Mast cells, basophils, histamine, prostaglandins, leukotrienes	Complement	Complement, inflammatory factors, eosinophils, neutrophils	Lymphokines, macrophages
Antigen	Soluble or particulate	On cell surfaces	Soluble or particulate	On cell surfaces
Reaction time	Seconds to 30 minutes	Variable, usually hours	3 to 8 hours	24 hours to 4 or more weeks
Nature of reaction	Local wheal and flare, airway restriction, anaphylactic shock	Clumping of erythrocytes, cell destruction	Acute inflammation effects	Cell-mediated cell destruction
Therapy	Desensitization, antihistamines, steroids	Steroids	Steroids	Steroids

TABLE 19.6   The Spectrum of Autoimmune Disorders

Disorder	Organ(s) or Tissues Affected	Autoantibody Target
Organ-Specific Disorders		
Addison's disease	Adrenal glands	Adrenal gland proteins
Autoimmune hemolytic anemia	Erythrocytes	Red blood cell membrane proteins
Glomerulonephritis	Kidneys	Streptococcal cross-reactivity with kidney
Graves' disease	Thyroid gland	TSH receptor
Hashimoto's thyroiditis	Thyroid gland	Thyroglobulin
Idiopathic thrombocytopenic purpura	Blood platelets	Platelet glycoproteins
Juvenile diabetes	Pancreas	Beta cells and insulin
Myasthenia gravis	Skeletal muscles	Acetylcholine receptor
Pernicious anemia	Stomach	Vitamin B_{12} binding site
Postvaccine/postinfection encephalomyelitis	Myelin	Measles cross-reactivity with myelin
Premature menopause	Ovaries	Corpus luteum
Rheumatic fever	Heart	Streptococcal cross-reactivity with heart
Spontaneous male infertility	Testes	Spermatozoa
Ulcerative colitis	Colon	Colon cells
Systemic (Disseminated) Disorders		
Goodpasture's syndrome	Basement membranes	Basement membrane
Polymyositis/dermatomyositis	Muscles and skin	Cell nuclei
Rheumatoid arthritis	Joints	Cell nuclei, gamma globulins
Scleroderma	Connective tissues	Nucleoli
Sjögren's syndrome	Lacrimal and salivary glands	Cell nuclei
Systemic lupus erythematosus	Many tissues	Cell nuclei, histones

2. In addition to predisposing genetic factors, **antigenic,** or *molecular,* **mimicry** can occur. T_H cells might attack tissue antigens that are similar to antigens of some pathogens. Some children who suffer rheumatic fever (caused by *Streptococcus pyogenes*) develop rheumatic heart disease later in life. For some reason the immune system of such individuals "sees" the heart valve tissue as similar to certain streptococcal antigens and attacks the heart valves.

3. The thymus is critical to the normal development of T cells. Aside from the T_H cells that recognize nonself antigens, T_H cells that recognize self antigens can exist if *clonal deletion* fails to remove these self-reactive T cells. ∞ (Chapter 18, p. 491) If they survive and proliferate, they can attack self antigens and trigger B cell activity with antibody production. Antigens hidden in tissues and lacking contact with B or T cells during immune system development or clonal deletion could be released through physical injury. These antigens then will be perceived as foreign by the immune system (see the box "Sympathetic Blindness").

4. Mutations might give rise to aberrant proteins to which B cells react, producing plasma cells that make autoantibodies.

5. Viral components inserted into host cell membranes might act as antigens, or virus–antibody complexes might be deposited in tissues.

6. The sympathetic nervous system, which along with the parasympathetic system controls internal body functions, helps regulate the immune system.

Examples of Autoimmune Disorders

Autoimmune disorders usually are chronic inflammatory disorders with symptoms that can alternately worsen and lessen. They affect about 6 percent of all

APPLICATIONS

SYMPATHETIC BLINDNESS

Lens proteins are normally confined within the lens capsule of the eye. Because they are never exposed to lymphocytes during development, the immune system never acquires tolerance for them. Sometimes an eye injury allows these proteins to leak into the bloodstream, where they elicit an immune response. The antibodies formed in this way then attack proteins in the undamaged eye. Because circulation (which transports antibodies) tends to be better in that eye, the immune response in the healthy eye can be more intense than that in the injured eye. This phenomenon sometimes leads to sympathetic blindness, or loss of sight in the uninjured eye.

humans and can affect one organ or many. We now look at three examples to illustrate the diversity of such disorders.

Myasthenia Gravis

Myasthenia gravis is an autoimmune disease that afflicts approximately 25,000 Americans. It affects primarily women in their 20s and 30s and men in their 40s and 50s. The disease usually affects skeletal muscles of the limbs and those involved in eye movements, speech, and swallowing. The principle symptoms of the disease are progressive weakness and muscle fatigue.

For muscles to contract normally, neurons secrete the neurohormone acetylcholine across the gap (junction) between neuron and muscle (Figure 19.15a).

FIGURE 19.15 Myasthenia gravis involves a loss of acetylcholine receptors from the neuromuscular junction. (a) A normally functioning neuromuscular junction has many acetylcholine receptors that bind acetylcholine. (b) Myasthenic patients have significantly fewer receptors for acetylcholine.

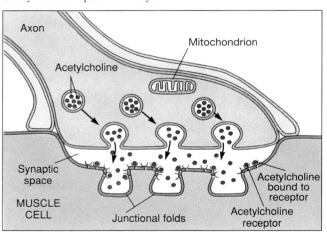

(a)

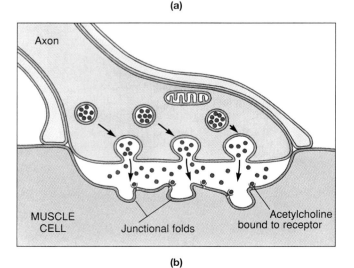

(b)

When acetylcholine receptors on the muscle cells bind acetylcholine, contraction of the muscle occurs. Evidence suggests that in patients with myasthenia gravis, muscle contraction is prevented by IgG autoantibodies that either block the acetylcholine receptor or cause a reduction in the number of acetylcholine receptors (Figure 19.15b). In fact, most myasthenic patients have only 30 to 50 percent as many receptors as do unaffected persons.

Although myasthenia gravis is one of the best-understood autoimmune diseases, the reason why autoantibodies are formed is not well understood. One possibility is that autoantibodies are triggered by an immune response to an infectious virus or bacterium that has antigens mimicking a part of the acetylcholine receptor. Myasthenia gravis was once considered a fatal or disabling disease. Today myasthenic patients can be treated with drugs or immunosuppressive steroids so that they can live full lives. However, no means of preventing autoantibodies from forming or of removing them once they have formed is available.

Rheumatoid Arthritis

In contrast with myasthenia gravis, which affects single-organs, **rheumatoid arthritis** (RA) affects mainly the joints of the hands and feet, although it can extend to other tissues. Of all forms of arthritis, RA is most likely to lead to crippling disabilities and to develop early in life (between the ages of 30 and 40). It is one of the most common autoimmune diseases, affecting about 2 million Americans. It is two to three times more prevalent in women than in men.

RA is characterized by inflammation and destruction of cartilage in the joints, often causing deformities in the fingers (Figure 19.16). Despite continued research, the cause of RA is unknown. Some researchers believe that an infectious microbe (mycoplasma or virus) is the

FIGURE 19.16 Joint inflammation is typical in people suffering from rheumatoid arthritis. In many cases, joint inflammation and destruction are so severe that they result in misshapen digits. In this γ-ray photograph, swollen joints appear as bright spots.

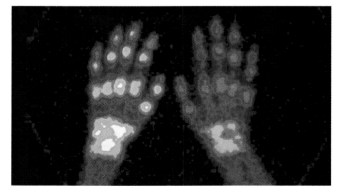

cause, leading to antigenic mimicry and ultimately to an attack of self antigens. Others believe that a self antigen recognized by the immune system as foreign is involved. Whatever the stimulus, inflammation activates specific cells in the joint and attracts lymphocytes. T_H1 cells recognize a self antigen in the joint and trigger the activation of B cells, which differentiate into plasma cells and secrete IgG antibody. These T_H1 cells also release several cytokines that researchers believe trigger resident phagocytic cells to release degrading enzymes from lysosomes. ∞ (Chapter 4, p. 101) These enzymes attack the cartilage, causing inflammation in the joints. RA patients also have circulating IgM antibodies that are specific for their own IgG antibodies. These autoantibodies, called **rheumatoid factors,** are used as a diagnostic test for RA. However, there is no evidence that these factors are involved with the formation of immune complexes. All these enzymes, factors, and cells increase the inflammatory response, leading to swollen, painful joints.

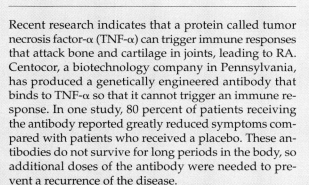

BIOTECHNOLOGY

NEW TREATMENTS FOR RHEUMATOID ARTHRITIS

Recent research indicates that a protein called tumor necrosis factor-α (TNF-α) can trigger immune responses that attack bone and cartilage in joints, leading to RA. Centocor, a biotechnology company in Pennsylvania, has produced a genetically engineered antibody that binds to TNF-α so that it cannot trigger an immune response. In one study, 80 percent of patients receiving the antibody reported greatly reduced symptoms compared with patients who received a placebo. These antibodies do not survive for long periods in the body, so additional doses of the antibody were needed to prevent a recurrence of the disease.

The major component of joint cartilage is type II collagen. In animal models used to study RA, oral administration of type II collagen suppressed T cell–mediated responses and reversed RA symptoms. In a human trial, 28 RA patients received chicken type II collagen for 3 months, while 31 (the control group) received a placebo. After the 3 months, there was a significant reduction in swollen and painful joints in the groups given chicken collagen as compared with the control group. In fact, four of the collagen-treated patients had complete remission of RA. Although there were no side effects or evidence of sensitization to the oral collagen treatment, optimum dose and long-term effects of oral administration need to be determined. In fact, similar oral therapies are under investigation for several other autoimmune disorders.

Although no cure exists for RA, treatment can alleviate symptoms. Hydrocortisone lessens inflammation and reduces joint damage, but long-term use weakens bones and causes such undesirable side effects as a reduction of normal immune responses. Aspirin decreases inflammation and reduces pain with fewer side effects. Physical therapy is used to keep joints movable. In severe cases, surgical replacement of damaged joints can restore movement.

Systemic Lupus Erythematosus

About 200,000 Americans suffer from **systemic lupus erythematosus** (SLE), a systemic autoimmune disease. The name is derived from the reddened skin rash (erythematose) that resembles a wolf's mask (*lupus* is Latin for "wolf"). The butterfly-shaped rash appears over the nose and cheeks of about 30 percent of SLE patients (Figure 19.17). SLE occurs 10 to 20 times as often in women as in men, with 80 percent of cases occurring in women during their reproductive years.

In SLE, autoantibodies (IgG, IgM, IgA) are made primarily against components of DNA but can also be made against blood cells, neurons, and other tissues. As the normal dying process of cells (skin, intestinal, kidney) occurs, anti-DNA antibodies attack the remnants of these cells. Immune complexes are deposited between the dermis and epidermis and in blood vessels, joints, glomeruli of the kidneys, and the central nervous system. They cause inflammation and interfere with normal functions at these sites.

Inflammation of blood vessels, heart valves, and joints are common effects of interference. Arthritis is the most common clinical characteristic of SLE; a patchy skin rash on the upper trunk and extremities is a common skin manifestation. This rash often is precipitated by exposure to sunlight. Most SLE patients eventually die from kidney failure as glomeruli fail to

FIGURE 19.17 The characteristic butterfly-shaped rash of systemic lupus erythematosus.

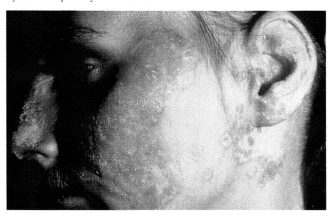

remove wastes from the blood. Among individuals with SLE, men tend to have a nonsystemic *discoid* form of the disease. It produces disk-shaped skin lesions and is less serious than the systemic form in its side effects.

SLE cannot be cured. Treatment depends on individual disease characteristics. It can include antipyretics to control fever, corticosteroids to reduce inflammation, and immunosuppressant drugs to prevent or decrease further autoimmune reactions.

TRANSPLANTATION

Transplantation is the transfer of tissue, called **graft tissue,** from one site to another. An **autograft** involves the grafting of tissue from one part of the body to another—for example, the use of skin from a patient's chest to help repair burn damage on a leg. A graft between genetically identical individuals (identical twins in humans, or members of highly inbred animal strains) is called an **isograft** (*iso,* Greek for "equal"). A graft between two people who are not genetically identical is termed an **allograft** (*allo,* Greek for "different"). Most organ transplants fall into this category. A transplant between individuals of different animal species is known as a **xenograft** (*xeno,* Greek for "foreign").

Early transplantation experiments involved the grafting of skin from one animal to another of the same species. The grafts appeared healthy at first but in a few days to a few weeks became inflamed and fell off. First thought to be due to infection, this reaction, called **transplant rejection,** is now known to be due to the destruction of the grafted tissue by the recipient's (that is, the host's) immune system. This process, which depends on T cells, also accounts for rejection of most organ transplants in humans. Transplants recognized as nonself are rejected.

A much less common transplantation effect is **graft-versus-host** (GVH) **disease,** in which the transplanted tissue contains immunocompetent cells that launch primarily a cell-mediated response against the recipient's tissues. This response occurs most often when immunodeficient patients receive bone marrow transplants. GVH was more common before immunosuppressive drugs were introduced to block the immunological responses.

Histocompatibility Antigens

All human cells, and those of many other mammals, have a set of self antigens called **histocompatibility antigens** (*histo,* Latin for "tissue"). The genes producing these molecules are called the **major histocompatibility complex** (MHC). Only identical twins have exactly the same MHC molecules, but all family members have a mixture of similar and different MHC mole-

cules. These antigens are located on the surfaces of cells, including the cells of kidneys, hearts, and other commonly transplanted organs. If donor and recipient histocompatibility antigens are different, as they likely would be when donors and recipients are not related, recipient T cells recognize these cells as foreign and destroy the donor tissue.

To try to prevent allograft rejection, it is necessary to determine if the graft has histocompatibility antigens not found in the recipient. Like red blood cell antigens, histocompatibility antigens can be identified by laboratory tests so that donor and recipient tissues can be as closely matched as possible. Such tests are one of several methods of *tissue typing,* or testing the compatibility of donor and recipient tissues. Because the first studies in humans involved antibody reaction with leukocytes, the MHC molecules on human cells are called **human leukocyte antigens** (HLAs). Human HLAs are determined by a set of genes located on chromosome 6. They are designated A, B, C, DR, DQ, and DP. The information in each gene specifies a particular antigen. For example, HLA-B is so highly variable that there are alleles for 51 different antigens. Overall, about 120 different antigens are recognized in humans and produced from the HLA genes.

For tissue typing, it would be impossible to do a complete typing of all HLAs present in an individual. However, the HLA-DR antigens are known to generate the strongest rejection reactions. Therefore, the tissues of prospective transplant recipients are typed as to which of the 20 HLA-DR antigens are present. When a donor organ becomes available, it also is typed. It is transplanted into the recipient whose antigens most nearly match. This procedure reduces the chances of rejection. Identical twins are the best match for allografts because all their HLA antigens are the same, but siblings of the same parents can have some matching HLA antigens in common.

Transplant Rejection

Like other immune reactions, transplant rejection displays specificity and memory. ∞ (Chapter 18, pp. 491–492) Rejection usually is associated with mismatched HLA-DR antigens. Certain cells that present antigens to phagocytes increase the likelihood of rejection. The fact that HLA-DR antigens are found on T cells and macrophages that carry out rejection reactions may explain why these antigens are so important in graft rejection.

T cells are responsible for rejection of grafts of solid tissue, such as a kidney, heart, skin, or other organs. In animal experiments, allografts are retained by animals that lack T cells and are rejected by those lacking B cells. More specifically, T_H2 cells lead to rejection (Figure 19.18). These cells help stimulate cytotoxic T

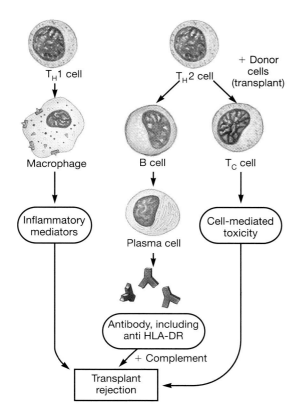

FIGURE 19.18 Transplant rejection is a combination of both cell-mediated and humoral immune reactions. T_H1 (inflammatory T) cells activate macrophages, which produce inflammatory mediators. T_H2 cells trigger both T_C and B cell activation. B cell activation leads to the production of plasma cells that synthesize antibodies, including anti–HLA-DR. Inflammatory mediators, T_C cell–mediated toxicity, and antibodies, along with complement, bring about transplant rejection.

(T_C) cells that reject the transplant through cell-mediated cytotoxicity. The T_H2 cells can also activate B cells to produce plasma cells and antibodies that cause rejection through lytic damage. Macrophages that are activated by T_H1 cells secrete inflammatory mediators and cause cytotoxic damage to the transplant. Natural killer cells can also act in transplant rejection.

The time required for rejection to occur varies from minutes to months. *Hyperacute rejection,* which is a cytotoxic hypersensitivity reaction, occurs when the recipient is already sensitized at the time the graft is done. For example, in kidney transplants, in which the graft is immediately supplied with host blood, extensive tissue destruction occurs within minutes to hours. (Corneal transplants, however, are not rejected because the cornea lacks blood vessels, and antibodies cannot reach them.) *Accelerated rejection* takes several days because it requires cells to reach the graft. *Acute rejection* occurs in days to weeks, requiring T cell sensitization after transplantation. Rejection that begins months to years after transplantation represents a *chronic rejection.*

This slow process is typical of cardiac and kidney transplants in which an interaction between the immune system and the transplant leads to eventual dysfunction of the transplant.

Immunosuppression

When a patient is facing an organ transplant, HLA antigens are probably not a complete match. Therefore, it is important to prevent immune reactions that would destroy the organ. The minimizing of immune reactions is called **immunosuppression.** Ideally, immunosuppression should be as specific as possible—it should cause the immune system to tolerate only the antigens in transplanted tissue and allow the immune system to continue to respond to infectious agents.

In practice, radiation or cytotoxic drugs, both of which impair immune responses, are used to minimize rejection reactions. *Radiation* (X-rays) of lymphoid tissues suppresses the immune system, preventing rejection. Radiation also destroys other lymphoid functions, including the ability of the immune system to recognize infectious microbes. **Cytotoxic drugs,** such as azathioprine and methotrexate, damage many kinds of cells. But because they interfere with DNA synthesis, these drugs cause most damage to rapidly dividing cells. Because B cells and T cells divide rapidly after sensitization, the drugs exert a somewhat selective effect on the immune system.

BIOTECHNOLOGY

DISGUISING TISSUES AIDS IN TRANSPLANTS

Transplanted tissues are rejected because the recipient's immune system recognizes them as foreign. However, scientists have succeeded in disguising foreign cells by covering HLAs that act as antigens and trigger rejection. Ordinarily the binding of antibodies to antigens on cells starts a process that leads to death of the cells. The researchers modified the antibodies such that although the antibodies fit tightly to foreign cell surface proteins, they did not destroy the cells. Human pancreatic cells with HLA proteins thus "covered" were transplanted into mice. The mouse immune system ignored the human cells, allowing them to live and produce insulin for more than 6 months. Further research may eventually provide transplants to treat diabetics. An advantage of using disguised cells is that immunosuppressants now used with transplants will be unnecessary. These drugs, sometimes taken for the rest of the patients' life, leave patients highly vulnerable to infections and often cause other undesirable side effects, including a predisposition to cancer.

Radiation and cytotoxic drugs impair T cell responses to infections. In contrast, the fungus-derived peptide cyclosporine A (CsA) suppresses, but does not kill, T cells, and it does not affect B cells. It is particularly useful in preventing transplant rejection: It allows T cells to regain function after the drug is stopped, and it does not reduce resistance to infections provided by B cells. The use of immunosuppressive drugs, especially CsA, has greatly increased the success rate of organ transplants. However, CsA may increase the patient's risk of developing cancer.

DRUG REACTIONS

Most drug molecules are too small to act as allergens. If a drug combines with a protein, however, the protein-drug complex sometimes can induce hypersensitivity. All four types of hypersensitivity have been observed in drug reactions.

Type I hypersensitivity can be caused by various drugs. Most reactions are localized, but generalized anaphylactic reactions sometimes occur, especially when drugs are given by injection. Orally administered drugs are less likely to cause hypersensitivity reactions because they are absorbed more slowly. Hypersensitivity reactions require prior sensitization and depend on the production of IgE antibodies. Although penicillin is one of the safest drugs in use, 5 to 10 percent of patients receiving it repeatedly become sensitized. Of the sensitized patients, about 1 percent develop generalized anaphylactic reactions, which account for about 300 deaths per year in the United States.

Type II hypersensitivity (Figure 19.19) can occur when the drug binds to a plasma membrane directly; when it binds to a plasma protein, and the complex binds to a plasma membrane; or when it alters a plasma membrane in such a way that cellular antigens trigger autoantibody production. All such reactions involve IgG or IgM and complement. Their targets—erythrocytes, leukocytes, or platelets—are destroyed by complement-dependent cell lysis. Many antibiotics, sulfonamides, quinidine, and methyldopa elicit Type II reactions.

Type III hypersensitivity appears as serum sickness and can be caused by any drug that participates in the formation of immune complexes. Symptoms appear several days after administration, when sufficient quantities of immune complexes have accumulated to activate the complement system. A few patients sensitized to penicillin develop serum sickness.

Type IV hypersensitivity usually occurs as contact dermatitis after topical application of drugs. Antibiotics, antihistamines, local anesthetics, and additives such as lanolin are frequent agents of Type IV reactions. Medical personnel who handle drugs sometimes develop Type IV hypersensitivities.

IMMUNODEFICIENCY DISEASES

Immunodeficiency diseases arise from an absence of active lymphocytes or phagocytes, the presence of defective lymphocytes or phagocytes, or the destruction

FIGURE 19.19 Drug reactions based on Type II hypersensitivity. A drug (or one of the products of its metabolism in the body) may bind to the plasma membrane of a blood cell, bind to a blood (plasma) protein to form a complex that binds to a plasma membrane, or alter a plasma membrane protein. Autoantibodies—IgG or IgM—are produced and then bind to the complex and activate complement to lyse the cell.

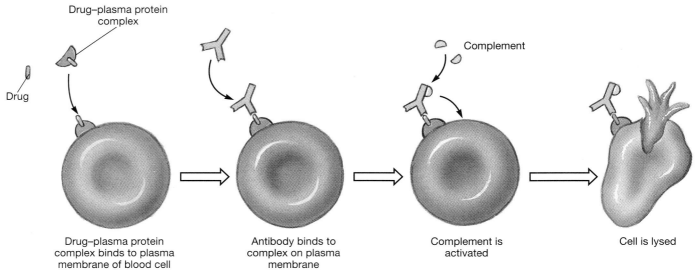

Drug–plasma protein complex binds to plasma membrane of blood cell

Antibody binds to complex on plasma membrane

Complement is activated

Cell is lysed

of lymphocytes. Such diseases invariably lead to impaired and inadequate immunity. **Primary immunodeficiency diseases** are caused by genetic defects in embryological development, such as failure of the thymus gland or Peyer's patches to develop normally. The result is a lack of T cells or B cells, or defective T and B cells. **Secondary immunodeficiency diseases** can be caused by (1) infectious agents, such as those responsible for leprosy, tuberculosis, measles, and AIDS; (2) malignancies, such as Hodgkin's disease or multiple myeloma; or (3) immunosuppressants, some chemotherapeutic drugs, certain antibiotics, and radiation. Such agents damage T cells or B cells after the cells have developed normally.

Primary Immunodeficiency Diseases

Agammaglobulinemia, the first immunodeficiency disease to be understood, is a B cell deficiency. It occurs primarily in male infants in which B cells, and therefore antibodies, are absent. After maternal antibodies are lost by about 9 months of age, affected infants develop severe infections because they cannot produce IgM, IgA, IgD, and IgE antibodies and produce only small amounts of IgG. Agammaglobulinemia is treated with massive doses of immune serum (gamma) globulin to replace missing antibodies and with antibiotics to prevent infections.

DiGeorge syndrome results from a deficiency of T cells, probably caused by a congenital defect in the thymus during embryological development. Cell-mediated immunity is impaired, so viral diseases pose a greater-than-usual threat. Although B cells are normal, their activation requires T_H cell activation. ∞

(Chapter 18, p. 488) Therefore, humoral immunity also is affected, because there are no functional T_H2 cells.

Severe combined immunodeficiency (SCID) is particularly debilitating because both B and T cells are absent. SCID can have several genetic origins. For example, stem cells in the bone marrow that normally give rise to lymphocytes fail to develop properly because of a defective gene for IL-2, for the enzyme adenosine deaminase (ADA), or for MHC molecules. An infant who inherits this condition is doomed to die within the first few years of life unless he or she is kept in a germ-free environment until satisfactory treatment can be devised. (See the box "The Boy in the Bubble.")

Bone marrow transplants can be effective in SCID patients if a compatible donor (usually a brother or sister) is found. If the transplant is not compatible, the transplanted lymphocytes respond immunologically to antigens in recipient's tissues. This is another example of GVH disease, and it can be lethal.

Gene therapy, which attempts to replace a defective gene with a functional, therapeutic copy of the gene, has been used to treat SCID—with spectacular results. A few children have had their bone marrow cells removed and "infected" with a reproductively deficient retrovirus carrying the missing gene for ADA (which is essential to cell maturation). The cells were then returned to their bodies. Although the patients who received the infected bone marrow transplant must periodically have additional transplants, all are living normal lives!

Secondary Immunodeficiency Diseases

Immunodeficiency diseases are not always inherited; sometimes they are *acquired* as a result of infections,

APPLICATIONS

THE BOY IN THE BUBBLE

David, a child with SCID, had to be isolated from all sources of infectious agents because he lacked both B cells and T cells. He lived in a series of specially designed germ-free "bubbles" for most of his life. One such bubble was a self-contained, sterile space suit. At age 12 he received a bone marrow transplant intended to provide him with immune functions. He was watched carefully for signs of GVH disease after the transplant. In about 5 months he developed symptoms similar to those of GVH disease and soon died. On autopsy it was discovered that he died not of GVH disease but of a malignancy caused by the Epstein–Barr virus, which had contaminated the marrow transplant. Researchers concluded that the lack of immune surveillance of tumor cells proved to be the immunodeficiency most responsible for his death.

David, a boy born without an immune system, at age 6 in his self-contained, sterile suit, a mobile isolation system designed for him by NASA. Other equipment included a pushcart with a battery-powered motor and a seat.

malignancies, autoimmune diseases, or other conditions. For example, congenital rubella infections can decrease T cell function and antibody production to the extent that infants fail to respond to vaccines. Once patients develop immunodeficiencies, they may suffer from chronic or frequent recurrent infections.

Among malignant diseases that produce immunodeficiencies, those of lymphoid tissues suppress T cell function, and those of bone marrow suppress both T cell function and antibody production. Autoimmune diseases, some kidney disorders, severe burns, malnutrition or starvation, and anesthesia also can cause temporary or permanent immunodeficiencies.

Acquired Immune Deficiency Syndrome (AIDS)

Certainly the most well known secondary immunodeficiency is **acquired immune deficiency syndrome** (AIDS), an infectious disease caused by the **human immunodeficiency virus** (HIV). The virus gradually but relentlessly destroys the patient's immune system. The lack of a functional immune system leaves the body open to a variety of malignancies and opportunistic infections, most of which are rarely seen among people who are not suffering from advanced HIV disease or AIDS. These complications—either alone or in combination—eventually prove fatal (Table 19.7).

HIV specifically targets and damages T_H cells, macrophages, dendritic cells, and Langerhans cells. All these immune cells bear on their surface an antigen called the CD4 marker, to which HIV binds. Infection by HIV is indicated by the presence of anti-HIV antibodies. The infectious cycle for the virus was described in Chapter 11. ∞ (p. 285)

In January 1995 AIDS researchers provided new clues as to how HIV eventually defeats the immune system. After a person is infected with HIV, an enormous battle ensues between HIV and the immune system. As immune cells destroy virus particles, more virus particles are produced. As virus-infected cells are destroyed by T_C cells, more immune cells replace those killed. Each day as HIV disease progresses, an estimated 1 billion virus particles are produced and destroyed as 2 billion immune cells are replaced! Although these early battles result in a draw, the virus eventually wins the war. As the years pass, it becomes more difficult to replace immune cells such as T_H. In addition, because HIV can mutate quickly, producing virus particles that progressively become more variable in structure, the immune system simply cannot keep up the fight.

Without activated T_H cells and macrophages, the immune system cannot "see" infectious microbes. Because T_H cells are greatly diminished in number, B cells are not stimulated to form plasma cells, which produce antibodies to combat infections. (The anti-HIV antibodies detected early in the course of infection are made before T_H cell populations become too depleted to stimulate B cells.) Similarly, cytokines are produced in amounts insufficient to activate macrophages and T_C cells. A drop in T_H cell count can be used to predict the onset of disease symptoms. A normal T_H count is

TABLE 19.7	Infections Frequently Found in AIDS Patients
Pathogen	**Disease**
Bacteria	
Mycobacterium tuberculosis	Tuberculosis
Mycobacterium avium-intracellulare	Disseminated tuberculosis
Legionella pneumophila	Pneumonia
Salmonella species	Gastrointestinal disease
Viruses	
Herpes simplex	Skin and mucous membrane lesions, pneumonia
Cytomegalovirus	Encephalitis, pneumonia, gastroenteritis, fevers
Epstein–Barr	Oral hairy leukoplakia, possibly lymphoma
Varicella-zoster	Chickenpox, shingles
Fungi	
Pneumocystis carinii	*Pneumocystis carinii* pneumonia
Candida albicans	Mucous membrane and esophagus infections (thrush)
Cryptococcus neoformans	Meningitis, kidney disease
Histoplasma capsulatum	Pneumonia, disseminated infections, fevers
Other opportunistic fungi	Varies with opportunist
Protozoa	
Toxoplasma gondii	Encephalitis
Cryptosporidium species	Severe diarrhea

800 to 1200 per μl of blood. If the count remains above 400, 8 percent of patients will develop AIDS within 18 months. With a T_H count of 200, 33 percent progress to AIDS; if the count is below 100, 58 percent will develop AIDS within 18 months.

Progrexssion of HIV Disease and AIDS

The sequence of events in HIV disease has now been established in some detail. The progression depends heavily on how much of the virus a person is exposed to (the *viral burden*) and how often the exposure is repeated. The CDC classification is based on the absence or presence of certain signs and symptoms and includes laboratory test results. Thus, persons categorized as being in groups 1 to 3 have HIV disease. Persons in group 4 have been diagnosed as having AIDS (Figure 19.20). Although diseases such as hairy leukoplakia (a white lesion appearing on the tongue) are diagnostic of most AIDS patients, other opportunistic diseases or cancers vary among AIDS patients. For ex-

ample, latent viral infections such as those caused by herpes simplex and cytomegalovirus, normally kept in check by the immune system, flare up and create a variety of disease symptoms. Severe diarrhea can be caused by opportunistic pathogens, including several species of *Cryptosporidium* (Chapter 22); encephalitis can be caused by *Toxoplasma gondii* (Chapter 24); and yeast infections can be caused by *Candida albicans* (Chapter 20). Pneumonia produced by the fungus *Pneumocystis carinii* is common to 80 percent of AIDS patients prior to their death. If patients do not first die of another opportunistic infection, about 50 percent will develop respiratory disease—caused either by *Mycobacterium tuberculosis* or by *Mycobacterium avium-intracellulare*. Overall, 88 percent of AIDS deaths result from an opportunistic infection.

Most AIDS patients develop malignancies not commonly found in the general population. A malignancy called **Kaposi's sarcoma** causes blood vessels to grow into tangled masses that are filled with blood and easily ruptured. In the skin and viscera, this sarcoma shows up as prominent pink or purplish spots (Figure 19.21).

FIGURE 19.20 CDC classification of HIV disease and AIDS. The progression of an HIV infection to AIDS is shown along with diseases and dysfunctions that can occur.

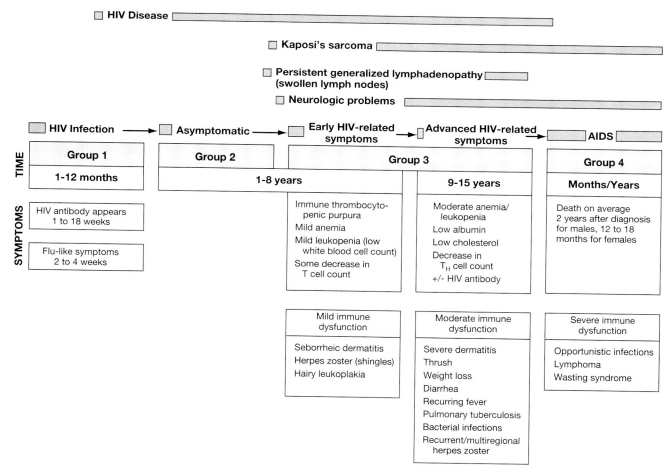

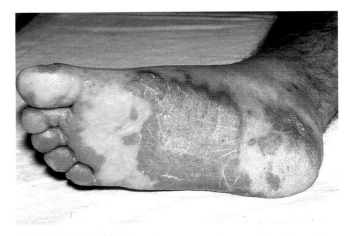

FIGURE 19.21 Kaposi's sarcoma, a tumor of blood vessels, seen in AIDS patients as dark purple areas.

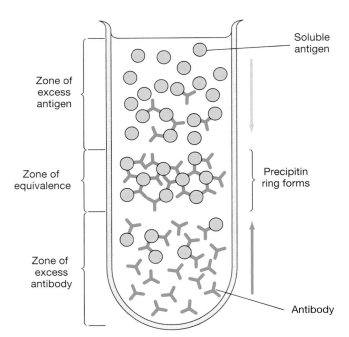

FIGURE 19.22 The precipitin test for antibodies. When IgG or IgM antibodies (which are soluble) react with soluble antigens, they quickly form small complexes, which over time combine into larger lattices that precipitate from the solution. Such precipitation occurs, however, only when there is an appropriate ratio of antigen to antibody. In the test, antibodies are placed in the bottom of a narrow tube. Soluble antigen is added, and the two are allowed to diffuse toward each other. Where the necessary concentration ratio is achieved (the zone of equivalence), precipitation takes place, visible as a hazy "precipitin ring" in the tube.

It can spread to the digestive tract, lungs, liver, spleen, and lymph nodes. However, there have been no reported cases of deaths among AIDS patients from Kaposi's sarcoma.

About 30 percent of individuals newly infected with HIV will progress to group 4 within 5 years. Within 15 years, 90 percent of HIV-infected patients will develop AIDS. Some of the drugs currently being used or tested against AIDS appear capable of prolonging the life of AIDS patients. (See the Essay "Acquired Immune Deficiency Syndrome" at the end of this chapter.)

IMMUNOLOGICAL TESTS

In Chapter 18 we considered how certain immunological reactions—agglutination, cell lysis by complement and by IgM antibodies, and neutralization of viruses and toxins—kill pathogens. Now we will consider how those and other reactions are used as laboratory tests to detect and quantify antigens and antibodies. Such laboratory tests make up the branch of immunology called **serology,** so named because many of the tests are performed on serum samples. Today, some laboratory tests also use monoclonal antibodies derived from the culture fluid of animal tissue grown in culture. ∞ (Chapter 18, p. 495) The tests and reactions described here represent a broad but incomplete sampling of laboratory and clinical tests. The selection of the "right" test depends on the nature of the pathogen or disease being analyzed.

The Precipitin Test

Historically, one of the first serologic tests to be developed was the **precipitin test** (Figure 19.22), which can

be used to detect antibodies or antigens. This test is based on a **precipitation reaction** in which antibodies called *precipitins* react with antigens, diffuse toward each other, and form a visible precipitate. During such reactions antigen–antibody complexes form within seconds. Latticelike networks of these complexes, which are visually opaque, form minutes to hours later.

Many modifications have been made to the basic precipitin test to increase its sensitivity to detect specific antigen–antibody complexes. **Immunodiffusion tests** are based on the same principle as the precipitin test, but they are carried out in a thin layer of agar that has solidified on a glass slide. Immunodiffusion tests are used to determine if more than one antigen—and therefore more than one antibody—is present in a serum specimen. Small wells are made in the solidified agar, and the antigens and antibodies are placed in separate wells. Antigen–antibody complexes appear as detectable precipitation lines (after staining) in the agar between the wells. After diffusion occurs, one or more bands of precipitation can be detected, each representing a different antigen–antibody complex (Fig-

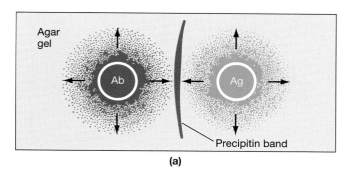

(a)

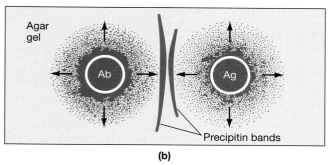

(b)

FIGURE 19.23 In immunodiffusion, a modification of the precipitin reaction, wells are made in an agar gel and filled with test solutions of antigen (Ag) and antibody (Ab). (a) A single antibody and antigen diffuse outward from the wells, meet, react with each other, and precipitate. They form a line called a precipitin band, which is visualized by staining. (b) Two different antigen–antibody complexes diffuse at different rates, producing separate bands.

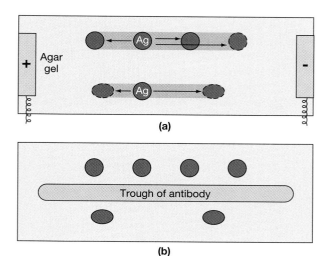

(a)

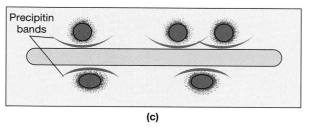

(b)

(c)

FIGURE 19.24 Immunoelectrophoresis. (a) Antigens placed in an agar gel are separated by means of an electric current. Positively charged molecules are drawn toward the negative pole; negatively charged ones move toward the positive pole. (b) A trough is cut into the agar between the wells and filled with antibody. (c) Curved precipitin bands form where antigens and antibodies diffuse to meet and react.

ure 19.23). The bands can be made more visible by washing the agar surface and applying a stain that colors the antigen–antibody complexes. An advantage of immunodiffusion tests is that in a single test medium several antigens can be reacted with one kind of antibody or several kinds of antibodies with one antigen.

When serum samples contain several antigens, **immunoelectrophoresis** can be used to detect separate antigen–antibody complexes. The antigens are placed in a well on an agar-coated slide. An electric current then is passed through the gel. This process is called **electrophoresis.** During electrophoresis different antigen molecules migrate at different rates, depending on the size and electric charges of the molecules. Following electrophoresis, the antibody is placed in a trough made along one or both sides of the slide, and diffusion is allowed to occur. The results of immunoelectrophoresis are similar to those obtained in other immunodiffusion tests—precipitin bands form wherever matching antigen and antibody precipitate (Figure 19.24). Thus, the advantage of immunoelectrophoresis is the ability to separate several antigens that might be present in a serum sample.

Radial immunodiffusion provides a quantitative measure of antigen or antibody concentrations. In this

test, antibody is added to molten agar, and the solution is allowed to solidify as a thin layer on a glass slide. Antigen samples of different concentrations are placed in wells made in the agar slide. After diffusion the antigen concentration is determined by measuring the diameter of the ring of precipitation around the antigen (Figure 19.25). Similarly, antibody concentrations can be determined by placing antibody samples of different concentrations in wells in a gel containing the antigen.

Agglutination Reactions

When antibodies react with antigens on cells, they can cause **agglutination,** or clumping together of the cells. One application of **agglutination reactions** is to determine whether the quantity of antibodies against a particular infectious agent in a patient's blood is increasing. The quantity of antibodies is called the **antibody titer.** It is reported as the reciprocal of the greatest serum dilution in which agglutination occurs. For example, an antiserum that agglutinated an agent's

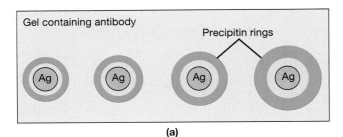

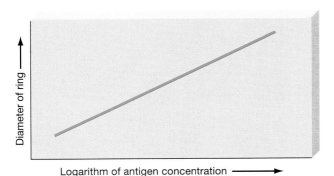

FIGURE 19.25 In radial immunodiffusion, (a) wells cut into antibody-containing agar sheets are filled with antigen. The antigen diffuses outward, complexing with antibody as it goes. When the ratio of antigen to antibody is optimal, the complexes precipitate in a ring. (b) The diameter of the ring is proportional to the logarithm of the antigen concentration, which can be determined by reference to a standard curve. The concentration of an antibody can also be determined by placing it in a well cut into an antigen-containing agar sheet and comparing the size of the resulting ring with a standard curve for that antibody.

FIGURE 19.26 Hemagglutination test, used for matching blood types, is based on the agglutination reaction. In A, red blood cells were mixed with serum having antibodies to the antigens on the cells. The complex of cells and antibodies clumped together. In B, the serum added did not contain antibodies that recognize the blood type antigens on the cells, so no clumping occurred.

antigen at a dilution of 1:256 but not at 1:512 would be reported as an antibody titer of 256. An increase in the antibody titer over time indicates that the patient's immune system is attacking the agent. Diagnosis of the disease agent is possible when it can be shown that the patient's serum had no antibodies against the agent before the onset of disease or that a rise in titer occurred during the course of the disease. This production of antibodies in the serum resulting from infection (or immunization) is called **seroconversion.** The **tube agglutination test** measures antibody titers by comparing various dilutions of the patient's serum against the same known quantity of the antigen (cells).

Hemagglutination, or agglutination of red blood cells, is similar to the agglutination tests except the antigens are on the surface of red blood cells. Hemagglutination is used in blood typing (Figure 19.26). In addition, hemagglutination tests can be used to detect viruses, such as those that cause measles and influenza. Such viruses bind to and cross-link red blood cells, causing **viral hemagglutination**. This process is inhibited by adding antibodies to the viruses. Because these antiviral antibodies bind to the viruses, the

viruses cannot agglutinate red blood cells. Such inhibition is the basis of the **hemagglutination inhibition test,** which can be used to diagnose measles, influenza, and other viral diseases.

Today, these types of agglutination tests usually are performed in plastic *microtiter plates*. These plates contain 96 separate wells, so many tests can be done simultaneously. In the plate shown in Figure 19.27, antibody dilutions are added to the wells of the plate. Then, equal concentrations of red blood cells are added to each well. If sufficient antibody is present to agglutinate the red blood cells, the antibody–cell complexes sink into a diffuse layer at the bottom of the well. If the antibody titer is too low, the red blood cells settle and form a red "button" at the bottom of the well.

Earlier we saw that severe hemolytic disease of the newborn results from an Rh factor incompatibility between mother (Rh negative) and fetus (Rh positive). In a hemagglutination test, although anti-Rh antibodies will bind to Rh antigen on red blood cells, there are not sufficient Rh antigens to cause clumping with anti-Rh antibodies (Figure 19.28a). The **Coomb's antiglobulin test** is designed to detect such antibodies. If red blood cells coated with anti-Rh antibody are treated with an antibody that recognizes the anti-Rh antibody, the antibody–antibody–cell complexes will agglutinate (Figure 19.28b). Thus, if a patient's serum contains Rh antibodies or if the red blood cells are Rh positive, agglutination will occur.

The body's natural defenses use complement to bind to antigen–antibody complexes, helping to destroy pathogens. This same ability is used in the laboratory or clinic to detect very small quantities of antibodies. The **complement fixation test** is a multistep procedure that begins with the inactivation of complement from a patient's serum by heating. The serum is

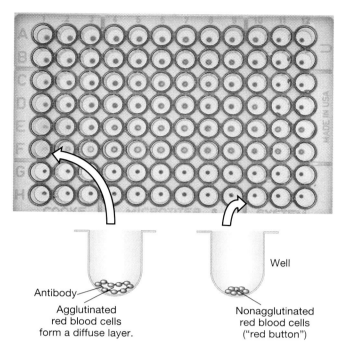

FIGURE 19.27 Hemagglutination tests often are carried out in microtiter plates. These plates contain 96 plastic wells, so many tests can be done simultaneously. When hemagglutination occurs, antibody reacts with antigen. All wells contain red blood cells. Positive hemagglutination results in a diffuse clumping of red blood cells (as in row F, column 1). Negative hemagglutination is seen as a clump of red blood cells ("red button") at the bottom of the wells (as in rows G and H).

In the figure, labels read: Antibody, Agglutinated red blood cells form a diffuse layer., Well, Nonagglutinated red blood cells ("red button").

then diluted, and known quantities of nonhuman complement and the test antigen are added separately (Figure 19.29a). The antigen is specific to the antibody being sought. This mixture is incubated to allow the antigen to react with any antibody present. Next, an indicator system typically consisting of sheep red blood cells and antibody against those cells are added. If the antibody to the test antigen was present in the patient's serum, the antigen–antibody reaction will have fixed (combined with) the complement. Hence the blood cells will not be lysed, and the test will be positive, forming a red button of undamaged cells (Figure 19.29b). But if the antibody was not present, free complement remaining in the mixture will be fixed by the indicator system, resulting in the lysis of the cells. The test will be negative. The complement fixation test is used to diagnose such bacterial diseases as pertussis and gonorrhea and fungal diseases, including histoplasmosis. The Wassermann test for syphilis uses complement fixation.

Neutralization reactions can be used to detect bacterial toxins and antibodies to viruses. Immunity to diphtheria, which depends on the presence of diphtheria antitoxins (antibodies to diphtheria toxin) can be detected by the **Schick test.** In this test a person is inoculated with a small quantity of diphtheria toxin. If the person is immune to the disease, diphtheria antitoxins (circulating in the blood) will neutralize the toxin, and no adverse reaction will occur. If the person is not immune and the antitoxin is not present, the toxin will cause tissue damage, detected as a swollen reddened area at the injection site after 48 hours.

Viral neutralization occurs when antibodies bind to viruses and neutralize them, or prevent them from infecting cells. In the laboratory or clinic, a patient's serum and a test virus are added to a cell culture or a chick embryo. If the serum contains antibodies to the virus, these antibodies will neutralize the virus and prevent the cells of the culture or the embryo from becoming infected.

FIGURE 19.28 The Coomb's antiglobulin test. (a) Anti-Rh antibodies are allowed to react with red blood cells. If Rh antigens are present on the blood cells, there are not enough of them to produce a hemagglutination reaction. (b) Therefore, anti-human antibodies prepared in rabbits are reacted with the red blood cell–antibody complexes. If Rh antigens are present on the red blood cells, hemagglutination will occur. A person with these red blood cells would be Rh-positive.

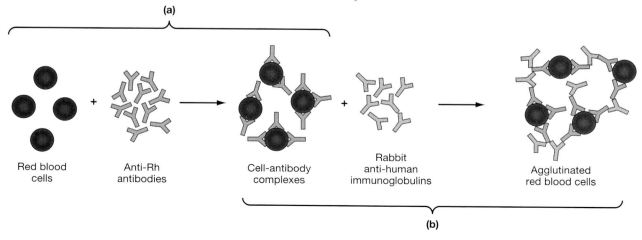

(a)

Red blood cells + Anti-Rh antibodies → Cell-antibody complexes + Rabbit anti-human immunoglobulins → Agglutinated red blood cells

(b)

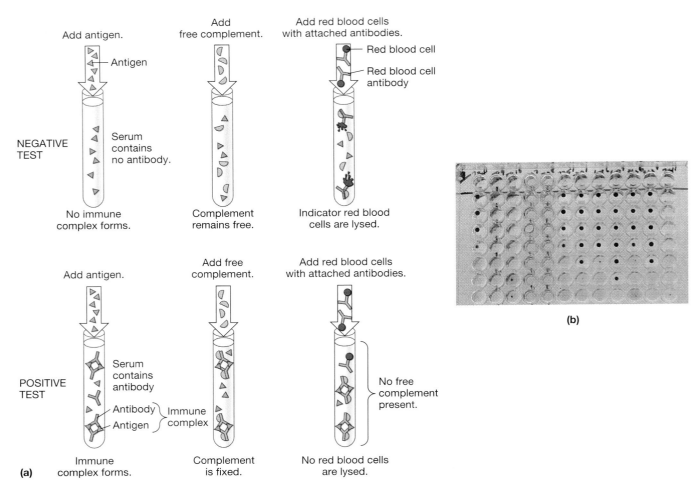

Add antigen.

Antigen

NEGATIVE TEST

Serum contains no antibody.

No immune complex forms.

Add free complement.

Complement remains free.

Add red blood cells with attached antibodies.

Red blood cell

Red blood cell antibody

Indicator red blood cells are lysed.

Add antigen.

POSITIVE TEST

Serum contains antibody

Antibody
Antigen
Immune complex

Immune complex forms.

(a)

Add free complement.

Complement is fixed.

Add red blood cells with attached antibodies.

No free complement present.

No red blood cells are lysed.

(b)

FIGURE 19.29 (a) The complement fixation test for antibodies. In the first step, the serum to be tested is diluted, and antigen to the antibody being sought is added. If the antibody is present, it reacts with the antigen and forms immune complexes. In the second step, free complement is added. If immune complexes have formed, the complement will interact with them and be fixed; if no immune complexes have formed, the complement will remain free. In the third step, sheep red blood cells with bound sheep red blood cell antibody molecules are added. If free complement is present, it will lyse the red blood cells. This is a negative test result—which indicates that there was no antibody in the original serum. If all the complement has already been fixed by the earlier immune complex, the red blood cells will not be lysed. This is a positive test result—it indicates that antibody was present in the original serum specimen. (b) In a positive test, complement has been fixed, and the red blood cells will not be lysed. Instead, they form a characteristic red button in the bottom of the well.

Tagged Antibody Tests

The most sensitive immunological tests used to detect antibodies or antigens use antibodies that have a "molecular tag" that is easy to detect even at very low concentrations. In fact, the concentrations are so low that precipitation or agglutination does not occur.

Immunofluorescence makes use of antibodies (usually IgG) to which fluorescent dye molecules are bound (tagged) at the tail (Fc) ends of the antibodies. For example, IgG antibodies tagged with fluorescein isothiocyanate glow a bright yellow green when exposed to ultraviolet (UV) light. Fluorescent tagged antibodies can be used to detect antigens, other antibodies, or comple-

ment at their locations on cells or within tissues (Figure 19.30). Because such cells or tissue samples can be examined with a fluorescence microscope, this technique is particularly useful in the research laboratory to locate cellular antigens and autoantibodies. A fluorescent tagged antibody that detects another antibody is known as an *anti-antibody;* one that detects complement is an *anti-complement antibody.* Immunofluorescence is helpful in diagnosing syphilis, gonorrhea, HIV infection, and Legionnaires' disease, to name a few.

Radioimmunoassay (RIA) also can be used to detect very small quantities (nanograms) of antigens and antibodies. To measure an antibody in a test specimen by RIA, a known antigen is placed in a saline (salt) so-

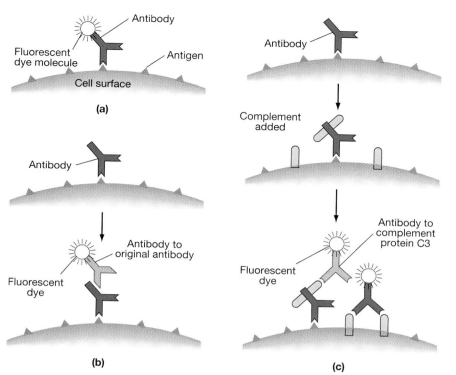

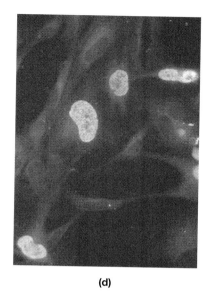

(a)

(b)

(c)

(d)

FIGURE 19.30 Immunofluorescence. Fluorescein is a fluorescent dye molecule that can be complexed with other molecules. When viewed with ultraviolet light (UV) under a fluorescence microscope, it will fluoresce, revealing the presence of the "tagged" molecule. To detect directly the presence of a specific antigen in a tissue, (a) a solution of fluorescein-tagged antibody to that antigen is prepared, added to cells or to a thin section of tissue, incubated, and then washed. Any dye-tagged antibody that has complexed with antigen in tissue will fluoresce when viewed by fluorescence microscopy. (b) In indirect testing the antibody to the antigen being sought is not itself tagged. Instead, its presence is detected by means of a fluorescein-tagged antibody (anti-antibody) to the original antibody. (c) Complement (protein C3) can be added to the tissue section along with the antibody, and a fluorescein-tagged antibody to one of the complement proteins can then be used to detect the presence of antigen–antibody complexes or complement attached to cells. (d) Immunofluorescent staining of influenza virus–infected lung cells. The cells were treated with fluorescent dye–tagged antibodies against the specific virus. The nucleus of infected cells is bright yellow green, indicating the location of the viral antigen.

lution and incubated in plastic well plates (Figure 19.31a). Some antigen molecules attach to the plastic; those that do not attach are washed away. The antibody being measured is added and allowed to bind with the antigen (Figure 19.31b). Then a radioactively tagged anti-antibody is applied (Figure 19.31c). After an incubation period, the excess unbound antibody is removed by washing. Radioactive material remaining in the well is measured with a radiation counter to determine the concentration of antibody in the test specimen.

Enzyme-linked immunosorbent assay (ELISA) is a modification of RIA in which the anti-antibody, instead of being radioactive, has an enzyme tag attached to it (Figure 19.32a). After the antibody being measured has reacted with the antigen, the anti-antibody–enzyme complex is added (Figure 19.32b). Finally, a substrate that the enzyme converts to a colored product is

applied (Figure 19.32c). The amount of colored product is proportional to the concentration of the antibody. RIA and ELISA are among the most widely used tests for antibodies or for antigens.

An important application of ELISA is the detection of HIV antibodies. The test was developed to screen the nation's blood supply and protect recipients of blood products from infection. ELISA is a sensitive test—so sensitive that the American Red Cross Blood Services reports that false-positive results (color product detected when HIV antibodies are not present) occur at a rate of about 0.2 percent.

To confirm the presence of an HIV infection, a more expensive test, called **Western blotting,** can be performed on samples from an individual who has a positive ELISA test. In Western blotting, HIV proteins are isolated from the individual and are first separated

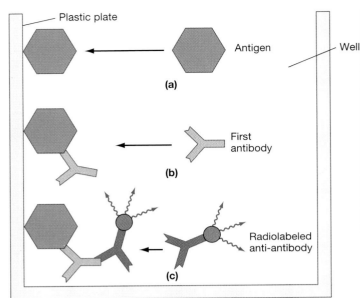

FIGURE 19.31 Radioimmunoassay (RIA) is used to detect very small quantities of antibody. (a) Antigen first is bound to a well of a plastic plate. After excess unbound antigen is washed out, the solution being tested for antibody is added and allowed to react. (b) If antibody is present, it reacts with the antigen. (c) After any unbound antibody is washed out, a radioactively labeled second antibody (anti-antibody) specific to the first antibody is added. The amount of bound radioactive anti-antibody present is measured; it is proportional to the concentration of the antibody in the original solution.

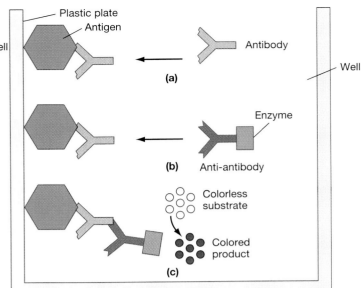

FIGURE 19.32 Enzyme-linked immunosorbent assay (ELISA) is a modification of RIA. (a) As in RIA, antigen bound to a well of a plastic plate reacts with the antibody being detected. (b) In one form of the ELISA, an anti-antibody is then added. However, rather than being radioactively labeled as in RIA, this antibody has a covalently attached enzyme. (c) A substrate specific to the enzyme is then added. If this enzyme has bound to the original antibody, it can catalyze a reaction, converting the colorless substrate into a color product. Such ELISA tests are done routinely as an initial test to detect HIV in blood samples.

in a gel by an electric current, similar to the procedure used in immunoelectrophoresis. The separated proteins are then transferred ("blotted") to cellulose filter paper. Next, serum from the individual is added to the blot. If HIV antibodies are present, they will react with the separated HIV proteins. Such anti-

gen–antibody complexes can be visualized by the addition of an enzyme-labeled anti-human antibody. When an enzyme substrate is added, colored bands appear on the paper (Figure 19.33). Thus, Western blotting can determine the exact viral antigens to which the HIV antibodies are specific.

FIGURE 19.33 Western blotting test for HIV antigens in blood. (a) HIV antigens in a gel are separated by an electric current, forming bands of separate antigens. (b) The antigen bands are transferred (blotted) to cellulose paper. (c) HIV antibodies tagged with dye are added. Any antibodies recognizing specific HIV antigens attach to that antigen and form a visible band.

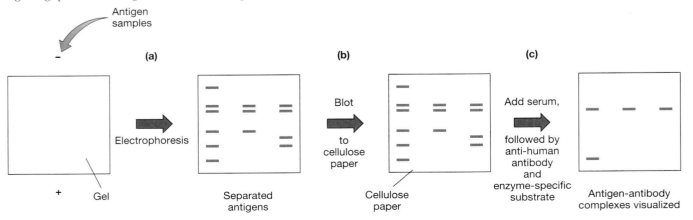

ACQUIRED IMMUNE DEFICIENCY SYNDROME (AIDS)

A IDS can be caused by at least two different types of human immunodeficiency viruses (HIV), designated HIV-1 and HIV-2. Most cases of AIDS in the United States, Canada, and Europe are caused by HIV-1. HIV-2, which is most common in certain parts of West Africa, may be less virulent. Both forms are tested for in screening the U.S. blood supply.

Epidemiology of AIDS

AIDS has been called the epidemic of the century; certainly, few diseases have had such a dramatic impact. Since 1981 an estimated 1.5 million Americans have been infected with HIV. (The actual number is likely to be higher.) Amazingly, only about 25 percent of the 1.5 million individuals know they are infected! Between 1981 and the end of 1994, about 442,000 Americans had AIDS and 260,000 died from it. By mid-1997 an estimated 600,000 Americans will have AIDS and more than 350,000 will have died from it. In 1994 alone, some 78,000 new AIDS cases were reported, with an estimated 37,000 deaths. AIDS has become the seventh leading cause of death among 1- to 4-year-olds, sixth among 15- to 24-year-olds, and *first* among 25- to 44-year-olds.

The AIDS pandemic is a global phenomenon. WHO officials estimate that by the end of 1995, the worldwide number of people infected with HIV will exceed 20 million (Figure 19.34). But in many developing nations, AIDS cases also are probably undiagnosed or unreported. The region most seriously affected is Sub-Saharan Africa, where over 10 million people are infected. The virus is spreading fastest in South and Southeast Asia (India, Myanmar, Thailand), where 2.5 million people are estimated to be infected. Another 2 million people in Latin America and the Caribbean are believed to be infected.

The Origin of AIDS

The identification of SIV, a virus similar to HIV in certain species of African monkeys (Chapter 11), has led many scientists to suggest that HIV evolved in nonhuman primates. ∞ (p. 295)

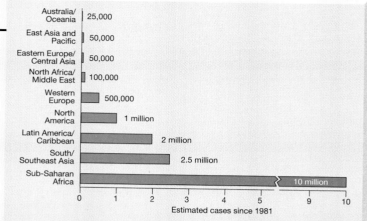

FIGURE 19.34 World Health Organization estimates of the number of people infected with HIV since 1981, by region. *Source:* American Association for World Health.

Although recent studies of the genetic information in these two viruses shows that they are only distantly related, there may be other retroviruses more closely related to HIV. Perhaps mutations in one such strain gave rise to the predecessor of HIV.

Early evidence for the origin of HIV comes from studies of human blood stored in England and Zaire since 1959, in which HIV-1 antibodies were found. The virus may have existed in relatively isolated regions, perhaps in Central Africa, for decades. Migration of rural people to rapidly growing cities, where population density was much higher and sexual contact was more casual and more frequent, could have brought about a great increase in the number of infected individuals. The expansion of international travel in recent years could rapidly spread the virus to many other parts of the world. The virus probably made multiple entries into the United States before becoming established. The oldest documented case of AIDS was attributed to a 25-year-old British sailor who died in 1959 of what was then called a "previously unknown viral disease." But in 1995 new evidence suggested that this person may not have had AIDS. So for now the oldest documented HIV case dates back to 1976 in a male from Zaire, Africa.

Who Gets AIDS, and How

All available evidence suggests that it is virtually impossible to become in-

fected with HIV through casual contact. Rather, a person becomes infected with the AIDS virus only through intimate contact with the body fluids of an infected individual and by transmission from infected mother to fetus. The virus is most commonly transmitted through blood, semen, and vaginal secretions. It seems likely that, however it is transmitted, the virus must make contact with a break or abrasion in the skin or with mucous membranes to cause infection. For this reason, all practices that lead to an exchange of body fluids carry a risk of HIV infection. These include:

1. *Sexual contact with an infected individual.* All forms of sexual intercourse—heterosexual and homosexual, active and passive, vaginal, anal, and oral—carry the risk of HIV infection. Condoms can reduce but not eliminate transmission because they have a significant failure rate (17 to 50 percent in various studies), usually the result of improper use, and not all types of condoms are equally effective in blocking HIV. Natural skin condoms allow passage of the virus; latex condoms are much safer.

2. *The sharing of unsterilized needles by intravenous drug users.*

3. *Receipt of a blood transfusion or blood products contaminated with HIV.* HIV infection by blood transfusion caused many AIDS cases in the early 1980s. Many of the in-

fected persons were hemophiliacs, who receive injections of blood products to let their own blood clot properly. Transfusions are much less of a viral threat today, thanks to the testing of donated blood for HIV antibodies and to recombinant DNA technology. And, popular fears to the contrary, it is not possible to acquire the HIV virus by donating blood because new, sterile needles are used.

4. *Passage from an infected mother to an infant.* Between 7 and 65 percent of infants carried by HIV-positive women become HIV infected. HIV transmission may be possible while the fetus is in the uterus, during delivery, and through breast feeding. Preliminary studies indicate that more newborns are infected during delivery than before birth.

Health care personnel treating AIDS patients or HIV-infected patients are at risk of becoming infected. The CDC has recommended the following precautions to minimize that risk (see also the Universal Precautions in Chapter 16): ∞ (p. 438)

1. *Wear gloves, masks, protective eyewear, and gowns* for touching blood, body fluids, mucous membranes, or skin lesions of patients and for procedures that might release droplets of body fluids. Discard these items, and wash hands immediately and thoroughly after each patient. Like other medical personnel, dentists and their technicians should consider blood, saliva, and gingival fluids of all patients to be potentially infective and use these procedures to prevent contact with such fluids.

2. *Avoid injury from needles* and other sharp objects, and discard them in puncture-proof containers.

3. *Use mouthpieces, resuscitation bags, or other ventilator devices* for emergency resuscitation.

4. Workers with skin lesions should *avoid direct patient care and handling of contaminated equipment.*

Keep in mind that almost any infectious disease (such as tuberculosis) that an HIV disease patient or AIDS patient may have contracted poses a danger to health care personnel. Of course, this problem is not limited to HIV infection; other types of infections can be a threat to health care workers.

Treatment of AIDS

There is no cure for HIV disease or for AIDS. This is not surprising, for very few antiviral agents are known, and retroviruses have proved particularly hard to attack. Nevertheless, several drugs are being used to prolong the lives of HIV-infected patients and to alleviate their symptoms. Zidovudine (formerly called AZT), the first drug licensed by the FDA for treatment of AIDS, became available in 1987. This drug, as well as a few other *nucleoside analogues,* interfere with HIV replication. In 1995 the cytokine interleukin-2 (IL-2) was shown to trigger new T_H cell production, replenishing those cells destroyed by HIV. In addition, there are drugs to slow the progression of opportunistic infections. For example, acyclovir is used for treating herpes infections, trimethoprim–sulfamethoxazole for *Pneumocystis* pneumonia, pyrimethamine for toxoplasmosis, and fluconazole for *Candida* infections. Kaposi's sarcoma is treated with α-interferon.

Also in 1995, drugs called *protease inhibitors* were shown to be powerful viral replication inhibitors. Use of these drugs helps restore immune system function in patients in the early stages of HIV disease. In fact, these drugs may be up to 20 times more powerful than zidovudine. However, these drugs are not a cure; they only may slow disease progression.

What about an AIDS Vaccine?

The outlook for an AIDS vaccine is not promising. Scientists expect a period of trial and error lasting beyond the year 2000, because research with HIV presents unusual problems. For one thing, HIV has a high mutation rate due to the imprecise operation of its reverse transcriptase. Thus, even if a successful vaccine against one strain of the virus is produced, another strain might not be affected by the vaccine, and new strains would likely develop. Outbreaks of influenza every few years are caused by similarly high mutation rates in the influenza virus.

Developing a vaccine poses more problems. Attenuated viruses cannot be used in a vaccine because they contain DNA that can be incorporated into the host's genome, possibly later giving rise to AIDS. Whole inactivated viruses, which have been used successfully in polio and influenza vaccines, are inappropriate for an AIDS vaccine. Such vaccines would, for HIV and other lentiviruses, predispose the host to severe infections. Recombinant viruses, such as HIV antigens on a vaccinia virus, are unacceptable because the host might develop disseminated vaccinia. Also, immunocompromised hosts may develop a variety of severe complications. Thus, many problems must be solved before we can have a safe and effective vaccine.

The Social Perspective: Economic, Legal, and Ethical Problems

AIDS and HIV disease will have an increasingly significant economic impact in the coming years. The lifetime cost of medical care for an AIDS patient in the United States can reach $100,000 or more. On the basis of CDC estimates, the total cost for care of all U.S. AIDS patients in 1994 was almost $9 billion. If the care of AIDS patients is a burden in the United States, where average income exceeds $12,000 per person per year, imagine the catastrophic burden in many developing countries, where average annual income is less than $200 per person.

U.S. laws protect the confidentiality of medical information, including AIDS test results. The laws also provide equal opportunity with respect to employment, housing, and education, but many AIDS patients continue to encounter various kinds of discrimination. Protecting the rights of uninfected citizens and those of health professionals who treat AIDS patients also must be considered. Other legal issues concern the responsibility of HIV-infected individuals not to transmit the disease, and the liabilities of distributors of blood products.

Many ethical issues are related to the legal issues. A major question is how the epidemic can be curtailed without infringing on individual freedoms. Another question weighs the moral obligation of health professionals to care for all patients against the risk of acquiring a fatal disease. Still other questions relate to allocation of scarce medical resources.

Overview Of Immunological Disorders

- An **immunological disorder** results from an exaggerated or an inadequate immune response.

Hypersensitivity

- **Hypersensitivity,** or **allergy,** is due to an exaggerated reaction to an antigen.

Immunodeficiency

- **Immunodeficiency** is due to an inadequate immune response.

IMMEDIATE (TYPE I) HYPERSENSITIVITY

- **Anaphylaxis** is due to harmful effects caused by an immediate exaggerated immune response.

Allergens

- An **allergen** is an ordinarily innocuous substance that can trigger a harmful immunological response in an allergen-sensitized person; common allergens are listed in Table 19.1.

Mechanism of Immediate Hypersensitivity

- The mechanism of **immediate hypersensitivity** is summarized in Figure 19.1; mediators of immediate hypersensitivity are summarized in Table 19.2.

- In the **sensitization** process the affected individual produces IgE antibodies that attach to mast cells and basophils. A second or later exposure to the same allergen will trigger cellular **degranulation** with the release of **histamine** and other preformed mediators. The synthesis of reaction mediators also is responsible for symptoms of immediate hypersensitivity.

Localized Anaphylaxis

- **Atopy** is a localized reaction to an allergen in which histamine and other mediators elicit wheal and flare reactions and other signs and symptoms of allergy.

Generalized Anaphylaxis

- **Generalized anaphylaxis** is a systemic reaction in which the airways are constricted (**respiratory anaphylaxis**) or blood pressure is greatly decreased (**anaphylactic shock**).

Genetic Factors in Allergy

- Genetic factors are thought to contribute to the development of allergies.

Treatment of Allergies

- Allergy is treated by **desensitization** (hyposensitization), as shown in Figure 19.5; symptoms are alleviated with antihistamines.

CYTOTOXIC (TYPE II) HYPERSENSITIVITY

Mechanism of Cytotoxic Reactions

- The mechanism of **cytotoxic hypersensitivity** is summarized in Figure 19.6.

Examples of Cytotoxic Reactions

- **Transfusion reactions** result in red blood cell **hemolysis** due to the production of antibodies to antigens on transfused red blood cells.

- **Hemolytic disease of the newborn** results when anti-Rh antibodies from a sensitized mother react with Rh antigens in an Rh-positive fetus.

IMMUNE COMPLEX (TYPE III) HYPERSENSITIVITY

Mechanism of Immune Complex Disorders

- The mechanism of **immune complex hypersensitivity** is summarized in Figure 19.9.

Examples of Immune Complex Disorders

- **Serum sickness** occurs when foreign (sometimes animal) antigens in sera combine with antibody to form immune complexes, which are deposited in various tissues.

- The **Arthus reaction** is a local immune response to an antigenic substance (usually an injected substance) that causes edema and hemorrhage.

CELL-MEDIATED (TYPE IV) HYPERSENSITIVITY

Mechanism of Cell-Mediated Reactions

- The mechanism of **cell-mediated (delayed) hypersensitivity** is summarized in Figure 19.12.

Examples of Cell-Mediated Disorders

- **Contact dermatitis** occurs after second contact with poison ivy, metals, or another substance and usually appears as eczema.

- Other kinds of cell-mediated hypersensitivity include **tuberculin hypersensitivity** and **granulomatous hypersensitivity.**

- Characteristics of the four types of hypersensitivities are summarized in Table 19.5.

AUTOIMMUNE DISORDERS

- **Autoimmune disorders** arise from a hypersensitivity to self antigens on cells and in tissues. Such diseases often produce **autoantibodies** to these self antigens. Autoimmune diseases can also be cell-mediated.

Autoimmunization

- **Autoimmunization** occurs when the immune system responds to a body component as if it were foreign. Genetic factors and **antigenic mimicry** are among the mechanisms that can cause autoimmune disease.

■ Tissue damage from autoimmune disorders can be caused by cytotoxic, immune complex, and cell-mediated hypersensitivity reactions.

Examples of Autoimmune Disorders

■ Examples of autoimmune diseases include **myasthenia gravis, rheumatoid arthritis,** and **systemic lupus erythematosus.**

TRANSPLANTATION

■ **Transplantation** involves the transfer of **graft tissue** from one site to another on the same individual (**autograft**), between genetically identical individuals (**isograft**), from one individual to another nonidentical individual (**allograft**), or between different species (**xenograft**).

Histocompatibility Antigens

■ Genetically determined **histocompatibility antigens** are found on the surface membranes of all cells. Some are correlated with increased risk of certain diseases.

■ **Human leukocyte antigens** (HLAs) in graft tissue are a main cause of **transplant rejection,** but immunocompetent cells in bone marrow or other types of grafts sometimes destroy host tissue, as in **graft-versus-host (GVH) disease.**

Transplant Rejection

■ **Transplant rejection** is due to the presence of foreign HLAs. **Hyperacute rejection** is a Type II hypersensitivity, occurring in an already sensitized recipient. **Accelerated rejection** reactions are mainly cell-mediated (Type IV) hypersensitivities. **Chronic rejection** occurs over several months to years.

Immunosuppression

■ **Immunosuppression** is a lowering of the responsiveness of the immune system to materials it recognizes as foreign. Immunosuppression is produced by **radiation** and by **cytotoxic drugs.** It minimizes transplant rejection but also can reduce the host's immune response to infectious agents.

DRUG REACTIONS

■ All four types of hypersensitivity reactions have been observed in immunological drug reactions.

IMMUNODEFICIENCY DISEASES

■ **Immunodeficiency diseases** arise from an absence of formed lymphocytes and other components of the immune system (**primary immunodeficiency**) or from the destruction of already formed lymphocytes (**secondary immunodeficiency**).

Primary Immunodeficiency Diseases

■ B cell deficiency, or **agammaglobulinemia,** leads to a lack of humoral immunity.

■ T cell deficiency, such as **DiGeorge syndrome,** leads to a lack of cell-mediated and humoral immunity.

■ Deficiencies of both B and T cells, or **severe combined immunodeficiency disease** (SCID), leads to a lack of both humoral and cell-mediated immunity.

Secondary Immunodeficiency Diseases

■ Secondary immunodeficiencies are acquired through infections, malignancies, or autoimmune disorders.

■ **Acquired immune deficiency syndrome** (AIDS) is caused by the **human immunodeficiency virus (HIV).** HIV destroys T_H cells and eventually impairs all immune functions.

■ HIV-infected individuals progress through a series of stages that lead to AIDS. Group 4 individuals suffer from opportunistic infections and malignancies such as **Kaposi's sarcoma**.

IMMUNOLOGICAL TESTS

■ **Serology** is the use of laboratory tests to detect antigens and antibodies.

The Precipitin Test

■ The **precipitin test** can be used to detect antibodies. Modifications of this test include **immunodiffusion, immunoelectrophoresis,** and **radial immunodiffusion.** All these tests rely on the formation of antigen–antibody complexes that precipitate from solutions or in agar gels.

Agglutination Reactions

■ **Agglutination reactions** depend on the clumping of antigen and antibody combinations. Such tests can be used to determine one's **antibody titer** or if **seroconversion** has occurred.

■ **Hemagglutination** includes the clumping of red blood cells by viruses and the binding of antibodies to specific antigen-bearing red blood cells. The **hemagglutination inhibition test** can be used to diagnose for measles or diseases caused by other viruses. **Coomb's antiglobulin test** is used to detect anti-Rh antibodies.

■ The **complement fixation test** indirectly detects antibodies in serum to antigens by determining whether complement combines (is fixed) with antigen–antibody complexes.

■ **Neutralization reactions** can detect bacterial toxins and antibodies to certain viruses.

Tagged Antibody Tests

■ **Immunofluorescence** allows detection of products of immune reactions on cells or within tissues.

■ Assays for antigens and antibodies can be performed with radioactive antibodies (**radioimmunoassay**) or antibodies containing enzymes coupled to them (**enzyme-linked immunosorbent assay, ELISA**). **Western blotting** detects specific antibodies to specific antigens.

ABO blood group system (p. 524)

acquired immune deficiency syndrome (p. 538)

agammaglobulinemia (p. 537)

agglutination (p. 541)

agglutination reaction (p. 541)

allergen (p. 518)

allergy (p. 516)

allograft (p. 534)

anaphylactic shock (p. 521)

anaphylaxis (p. 517)

antibody titer (p. 541)

antigenic mimicry (p. 531)

Arthus reaction (p. 527)

asthma (p. 521)

atopy (p. 519)

autoantibody (p. 530)

autograft (p. 534)

autoimmune disorder (p. 530)

autoimmunization (p. 530)

blocking antibody (p. 522)

cell-mediated hypersensitivity (p. 517)

complement fixation test (p. 542)

contact dermatitis (p. 528)

Coomb's antiglobulin test (p. 542)

cytotoxic drug (p. 535)

cytotoxic hypersensitivity (p. 517)

degranulation (p. 518)

delayed hypersensitivity (p. 517)

delayed hypersensitivity (T_{DH}) cell (p. 528)

desensitization (p. 522)

DiGeorge syndrome (p. 537)

electrophoresis (p. 541)

enzyme-linked immunoabsorbent assay (ELISA) (p. 545)

graft tissue (p. 534)

graft-versus-host disease (p. 534)

granulomatous hypersensitivity (p. 530)

hemagglutination (p. 542)

hemagglutination inhibition test (p. 542)

hemolytic disease of the newborn (p. 524)

histamine (p. 518)

histocompatibility antigen (p. 534)

human immunodeficiency virus (p. 538)

human leukocyte antigen (p. 534)

hypersensitivity (p. 516)

immediate hypersensitivity (p. 517)

immune complex hypersensitivity (p. 517)

immunodeficiency (p. 517)

immunodeficiency disease (p. 536)

immunodiffusion test (p. 540)

immunoelectrophoresis (p. 541)

immunofluorescence (p. 544)

immunological disorder (p. 516)

immunosuppression (p. 535)

induration (p. 529)

isograft (p. 534)

Kaposi's sarcoma (p. 539)

leukotriene (p. 518)

major histocompatibility complex (p. 534)

myasthenia gravis (p. 532)

neutralization reaction (p. 543)

precipitation reaction (p. 540)

precipitin test (p. 540)

primary immunodeficiency disease (p. 537)

prostaglandin (p. 518)

radial immunodiffusion (p. 541)

radioimmunoassay (p. 544)

respiratory anaphylaxis (p. 521)

Rh antigen (p. 524)

rheumatoid arthritis (p. 532)

rheumatoid factor (p. 533)

Schick test (p. 543)

secondary immunodeficiency disease (p. 537)

sensitization (p. 518)

seroconversion (p. 542)

serology (p. 540)

serum sickness (p. 527)

severe combined immunodeficiency (p. 537)

systemic lupus erythematosus (p. 533)

transfusion reaction (p. 523)

transplantation (p. 534)

transplant rejection (p. 534)

tube agglutination test (p. 542)

tuberculin hypersensitivity (p. 529)

tuberculin skin test (p. 529)

viral hemagglutination (p. 542)

viral neutralization (p. 543)

Western blotting (p. 545)

xenograft (p. 534)

QUESTIONS FOR REVIEW

A Types of Hypersensitivities

1. List the characteristics of an immunological disorder.

2. What is a hypersensitivity, and what are the four types of hypersensitivity?

3. What is immunodeficiency, and how does it differ from hypersensitivity?

B Immediate Hypersensitivities

4. What is an allergen, and what properties make substances likely to be treated as antigens?

5. List examples of common allergens.

6. Explain the mechanism of an immediate hypersensitivity reaction.

7. What substances mediate immediate hypersensitivity reactions, and how do they contribute to the reaction?

8. What events occur in atopy, and what makes them happen?

9. What events occur in generalized anaphylaxis, and what makes them happen?

10. How might allergies be treated or cured?

C Cytotoxic Hypersensitivities

11. Explain the mechanism of a cytotoxic hypersensitivity reaction.

12. What triggers a transfusion reaction, and what are the effects of such a reaction?

13. How can cytotoxic hypersensitivity reactions be prevented?

D Immune Complex Disorders

14. What triggers an immune complex hypersensitivity reaction, and what are its effects?

15. What causes an Arthus reaction, and how can it be prevented?

E Cell-Mediated Hypersensitivities

16. Explain the mechanism of cell-mediated hypersensitivity.

17. Why are cell-mediated reactions called delayed hypersensitivity?

18. How does a tuberculin reaction differ from granulomatous hypersensitivity?

19. In the following table, identify the type(s) of hypersensitivity—if any—with which each disease or statement is associated.

	Type of Hypersensitivity			
	I	II	III	IV
Hemolytic disease of the newborn				
Rheumatoid arthritis				
Contact dermatitis				
Animal dander				
Counteracted by epinephrine injection				
Serum sickness				

F Autoimmune Disorders

20. What is autoimmunity?

21. What kind of immunological defect leads to autoimmunity?

22. For each autoimmune disorder described, tell how tissue damage can be related to a particular kind of hypersensitivity.

G Organ Transplant Rejection

23. What are HLA antigens?

24. How are HLAs related to transplant rejection and disease risk?

25. What is immunosuppression?

26. Under what circumstances would immunosuppressants be used, and what are the hazards of these drugs?

H Drug Reactions and Hypersensitivity

27. How are immunological drug reactions related to hypersensitivity?

I Immunodeficiency Diseases

28. How are immunodeficiencies related to cells of the immune system?

29. Aside from AIDS infections, how can immunodeficiencies be acquired?

30. Explain how HIV causes AIDS, and list the sequence of events leading to AIDS.

31. Match the following:
_____ a lack of both B and T cells
_____ congenital defect in the thymus produces few T cells
_____ patients lack functional acetylcholine receptors
_____ inflammation and destruction of cartilage in joints
_____ produces a butterfly-shaped rash on nose and cheeks

a. systemic lupus erythematosus
b. rheumatoid arthritis
c. agammaglob-ulinemia
d. myasthenia gravis
e. SCID
f. none of the above

J Immunologic Tests

32. Briefly describe the mechanism of each of the immunological tests discussed in the text.

33. Diagram _____ represents an immunodiffusion test in which only one antigen–antibody complex is present; in diagram _____, two different antigen–antibody complexes are present.

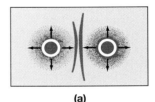 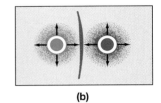

(a)   (b)

K AIDS

34. Where is AIDS spreading the fastest, and why is its incidence rising?

35. What is thought to be the origin of HIV?

36. How is HIV transmitted, and what treatments are available for AIDS?

37. What problems are associated with making an AIDS vaccine?

38. Describe some of the economic, legal, and ethical problems associated with AIDS.

PROBLEMS FOR INVESTIGATION

1. From members of your family (or another family), obtain as much information as possible about allergies that have been diagnosed over several generations. Use this information to determine whether the allergies are inherited or caused by some other factor.

2. How would you approach the following legal and ethical problems created by AIDS:

   a. Should a child with AIDS be allowed to attend school?

   b. Should a person with AIDS be allowed to hold a job as long as the illness permits?

   c. Should the government be allowed to make testing for HIV antibodies mandatory?

   d. Are there special circumstances in which AIDS testing should be mandatory?

   e. Who should have access to the results of a test for AIDS?

   f. Should AIDS be a reportable disease?

   g. Does a medical worker have the right to refuse to treat an AIDS patient?

(continued)

h. Should insurance companies be allowed to refuse to provide health or life insurance to an AIDS patient?

i. Must all hospitals accept AIDS patients?

j. Is the use of a large portion of medical resources to care for an AIDS patient justified if it means denying care to indigent mothers and their children?

k. What are the implications to the U.S. health care system of having 90 percent of all hospital beds occupied by AIDS patients?

l. What are the implications to a country, such as some in Africa, in which cities may have HIV infection rates of 30 percent?

3. Explain the positive and negative aspects of the use of immunosuppressive drugs for organ transplants or autoimmune diseases

4. Some people suffer hypersensitivity reactions when they wear jewelry. What are they reacting to, and are all such people reacting to the same thing? What type of hypersensitivity is this? What, if anything, can be done?

5. A 53-year-old woman who had smoked since the age of 17 decided to give up smoking. Shortly after stopping, she was hospitalized with an autoimmune form of ulcerative colitis. Stress caused her to begin smoking again. Her colitis rapidly disappeared. When she stopped smoking, her colitis returned. Over the years, a pattern emerged: no colitis while smoking; return of colitis upon cessation of smoking. What could account for this pattern? (The answer to this question appears in Appendix F.)

6. As a California teenager neared death from myelogenous leukemia, her parents, unable to find a compatible donor to replace her cancerous bone marrow, decided to conceive another child. They hoped that it would have the type needed to save the older sister. Their gamble paid off. Before the transplant, 19-year-old Anissa received chemotherapy to destroy her own bone marrow. Healthy marrow taken from the hip of her 13-month-old-sister, Marissa, replaced it. This was not the first family to have conceived children to provide donor tissue for a relative, but it was the first to announce it publicly. Immediate controversy arose as to the morality of such actions. How do you feel about this?

SOME INTERESTING READING

Benditt, J., ed. 1993. "AIDS: The unanswered questions." Special Feature Issue, *Science* 260(5112):1253–93.

Couroux, P., B. C. Schieven, and Z. Hussain. 1993. "*Pneumocystis carinii*." *ASM News* 59(4):179–81.

Drachman, D. B. 1994. "Myasthenia gravis." *The New England Journal of Medicine* 330(25):1797–1810.

Elliott, M. J., et al. 1994. "Randomised double-blind comparison of chimeric monoclonal antibody to tumor necrosis factor α (cA2) versus placebo in rheumatoid arthritis." *The Lancet* 344(8930):1105–10.

Emini, E. A. 1995. "Hurdles in the path to an HIV-1 vaccine." *Scientific American: Science and Medicine* 2, no. 3 (May/June):38–47.

Fackelmann, K. A. 1995. "HIV's infectious nature." *Science News* 147(2):22.

Greene, W. C. 1993. "AIDS and the immune system." *Scientific American* 269, no. 3 (September):98–105.

Hingley, A. T. 1993. "Food allergies: When eating is risky." *FDA Consumer* 27(12):27–31.

Hirsch, M. S., and R. T. D'Aquila. 1993. "Therapy for human immunodeficiency virus infection." *The New England Journal of Medicine* 328(23):1686–95.

Horner, W. E., et al. 1995. "Fungal allergens." *Clinical Microbiology Reviews* 8, no. 2 (April):161–79.

Lichtenstein, L. M. 1993. "Allergy and the immune system." *Scientific American* 269, no. 3 (September):116–24.

Minkoff, H. L., and J. A. DeHovitz. 1991. "Care of women infected with the human immunodeficiency virus." *Journal of the American Medical Association* 266(16):2253–58.

Nicoll, A., and P. Brown. 1994. "HIV: Beyond reasonable doubt." *New Scientist* 14(1908):24–28.

Ness, R. M., and G. C. Williams. 1994. *Why we get sick: The new science of Darwinian medicine*. New York: Times Books.

Pantaleo, G., C. Graziosi, and A. S. Fauci. 1993. "The immunopathogenesis of human immunodeficiency virus infection." *The New England Journal of Medicine* 328 (5):327–35.

Profet, Margie. 1991. "The function of allergy: Immunological defense against toxins." *Quarterly Review of Biology* 66(1):23–62.

Rennie, J. 1991. "Graft without corruption." *Scientific American* 265, no. 3 (September):18–22.

Rosen, F. S., M. D. Cooper, and R. J. P. Wedgwood. 1995. "The primary immunodeficiencies." *The New England Journal of Medicine* 333(7):431–37.

Schochetman, G., and J. R. George, eds. 1994. *AIDS testing*. New York: Springer-Verlag.

Schwartz, R. H. 1993. "T cell anergy." *Scientific American* 269, no. 2 (August):62–71.

Seachrist, L. 1995. "Food for healing: Oral tolerance therapy aims to neutralize autoimmune diseases." *Science News* 148(10):158–59.

Steinman, L. 1993. "Autoimmune disease." *Scientific American* 269, no. 3 (September):106–14.

Stine, G. J. 1995. *AIDS update: 1994–1995*. Englewood Cliffs, NJ: Prentice Hall.

Tarantola, D., and J. Mann. 1993. "Coming to terms with the AIDS pandemic." *Issues in Science and Technology* 9(3):41–48.

Trentham, D. E., et al. 1993. "Effects of oral administration of type II collagen on rheumatoid arthritis." *Science* 261(5129):1727–29.

Ungvarski, P. J., and J. Schmidt. 1992. "AIDS patients under attack." *RN* 55(11):36–44.

Zurlinden, J., and R. Verheggen. 1994. "HIV vaccines, a report from the front." *RN* 57(1):36–40.

DISEASES OF THE SKIN AND EYES; WOUNDS AND BITES

Tumorlike growths that develop in cases of molluscum contagiosum, which is caused by a virus.

H aving considered the general principles of microbiology in the first five units of this text, we are now ready to apply those principles to understanding infectious diseases in humans. As you study diseases, keep in mind the information on normal microbiota and disease processes in Chapter 15, on epidemiology in Chapter 16, and on host systems in Chapter 17. We begin our study of human diseases with diseases of the skin.

DISEASES OF THE SKIN

Your skin covers your body and accounts for 15 percent of your body weight. It provides an effective barrier to invasion by most microbes except when it is damaged. ∞ (Chapter 17, p. 449) Very few microbes can penetrate unbroken skin; however, mucous membranes are more easily entered. Here we will consider the kinds of skin diseases that occur when the skin surface fails to prevent microbial invasion.

Bacterial Skin Diseases

Many bacteria are found among the normal microbiota of the skin. Ordinarily they are kept from invading tis-

sues by an intact skin surface and nonspecific defense mechanisms of the skin. Bacterial and other skin infections usually arise from a failure of these defenses. They are generally diagnosed by appearance and by clinical history.

Staphylococcal Infections

FOLLICULITIS AND OTHER SKIN LESIONS Everyone has had a pimple at some time; most likely it was caused by *Staphylococcus aureus,* the most pathogenic of the staphylococci. Staphylococcal skin infections are exceedingly common because the organisms are nearly always present on the skin. Strains of staphylococci colonize the skin and upper respiratory tract of infants within 24 hours of birth. (Half of all adults and virtually all children are nasal carriers of *S. aureus.*) Infection occurs when these organisms invade the skin through a hair follicle, producing **folliculitis** (fol-lik"u-li'tis), also referred to as **pimples** or **pustules.** An infection at the base of an eyelash is called a **sty.** A larger, deeper, pus-filled infection is an **abscess;** an exterior abscess is known as a **furuncle** (fu'rung-kl), or **boil** (Figure 20.1). An estimated 1.5 million Americans have such infections annually. Further spread of infection, particularly on the neck and upper

THIS CHAPTER FOCUSES ON THE FOLLOWING QUESTIONS:

A What kinds of pathogens cause skin diseases?

B What are the important epidemiologic and clinical aspects of skin diseases?

C What kinds of pathogens cause eye diseases?

D What are the important epidemiologic and clinical aspects of eye diseases?

E What kinds of pathogens infect wounds and bites?

F What are the important epidemiologic and clinical aspects of wound and bite infections?

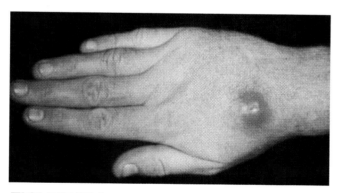

FIGURE 20.1 A furuncle, or deep, pus-filled lesion, caused by *Staphylococcus aureus*.

well as by nasal droplets and fomites. Staphylococci usually cause infection in older patients only when foreign bodies such as catheters or splinters are present. Although 5 million organisms must be injected into the skin to cause infection, only 100 are needed if they are soaked into a suture and tied into the skin.

SCALDED SKIN SYNDROME **Scalded skin syndrome** is caused by certain exotoxin-producing strains of *S. aureus*. Two different exotoxins, both called *exfoliatins*, are known. The genes for one are carried on the bacterial chromosome, whereas the genes for the other are located on a plasmid. An individual staphylococcal strain may carry genes for one, both, or neither of these exotoxins. The exotoxins are called exfoliatins because they travel through the bloodstream to sites far from the inital infection, causing the upper skin layers to separate and peel off in leaflike (foliar) sheets.

Scalded skin syndrome is most common in infants but can occur in adults as well. It begins with a slight reddened area, often around the mouth. Within 24 to 48 hours, it spreads to form large, soft, easily ruptured vesicles over the whole body. The skin over vesicles and adjacent reddened areas peels, leaving large, wet, scalded-looking areas (Figure 20.2). The lesions dry and scale, and the skin returns to normal in 7 to 10

back, creates a massive lesion called a **carbuncle** (kar'bung-kl). Encapsulation of abscesses prevents them from shedding organisms into the blood, but it also prevents circulating antibiotics from reaching the abscesses in effective quantities. Thus, in addition to antibiotic treatment, it is usually necessary to lance and drain abscesses surgically.

Staphylococcal infections are easily transmitted. Asymptomatic carriers, hospital personnel, and hospital visitors often spread staphylococci via the skin as

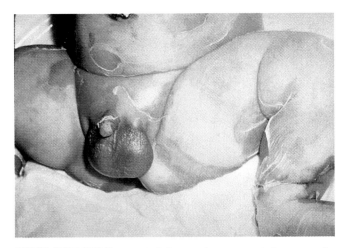

FIGURE 20.2 Scalded skin syndrome in an infant, caused by *Staphylococcus aureus*. Reddened areas of skin peel off, leaving wet, scalded-looking areas.

days. High fever is generally present. Bacteremia is common and can lead to septicemia and death within 36 hours. Exotoxins are highly antigenic. ∞ (Chapter 15, Table 15.4) They stimulate the production of antibodies that prevent recurrence of this syndrome—but only if reinfection occurs by the same *S. aureus* strain.

Streptococcal Infections

SCARLET FEVER **Scarlet fever,** sometimes called scarlatina, is caused by *Streptococcus pyogenes,* which also causes the familiar strep throat. Strains of the organism that cause scarlet fever have been infected by a temperate phage that enables them to produce an erythrogenic ("red-producing") toxin that causes the scarlet fever rash. Patients who have previously been exposed to the toxin, and so have antibodies that can neutralize it, can develop strep throat without the scarlet fever rash. However, such people can still spread scarlet fever to others. Three different erythrogenic toxins have been identified. A person can develop scarlet fever once from each toxin.

In the United States most scarlet fever organisms currently have low virulence. In past decades, when strains were more virulent, it was a much-feared killer. However, even low-virulence strains can cause serious complications, such as glomerulonephritis or rheumatic fever. The use of penicillin has appreciably lowered the mortality rate. Convalescent carriers can shed infective organisms from the nasopharynx for weeks or months after recovery. Fomites also are an important source of streptococcal infections.

ERYSIPELAS **Erysipelas** (er"i-sip'e-las; from the Greek *erythros,* for red, and *pella,* for skin), also called

St. Anthony's fire, has been known for over 2000 years and is caused by hemolytic streptococci. Before antibiotics became available, erysipelas often occurred after wounds and surgery and sometimes after very minor abrasions. Mortality was high. Today, it rarely occurs, and mortality is low. The disease begins as a small, bright, raised, rubbery lesion at the site of entry. Lesions spread as streptococci grow at lesion margins, producing toxic products and enzymes such as hyaluronidase. ∞ (Chapter 15, p. 401) Lesions are so sharply defined that they appear to be painted on. The organisms spread through lymphatics and can cause septicemia, abscesses, pneumonia, endocarditis, arthritis, and death if untreated. Curiously, erysipelas tends to recur at old sites. Instead of developing immunity, patients acquire greater susceptibility to future attacks.

Pyoderma and Impetigo

Pyoderma, a pus-producing skin infection, is caused by staphylococci, streptococci, and corynebacteria, singly or in combination. **Impetigo** (im-pe-ti'go), a highly contagious pyoderma, is caused by staphylococci, streptococci, or both (Figure 20.3). Fluid from early pustules usually contains streptococci, whereas fluid from later lesions contains both. Streptococcal strains that cause skin infections usually differ from those that cause strep throat. Impetigo occurs almost exclusively in children; why adults are not susceptible is unknown. Easily transmitted on hands, toys, and furniture, impetigo can rapidly spread through a daycare center. Impetigo rarely produces fever, and it is easily treated with penicillin. Lesions usually heal without scarring, but skin can be discolored for several weeks, and pigment can be permanently lost.

FIGURE 20.3 Impetigo, a highly contagious infection, is caused by staphylococci, streptococci, or both together.

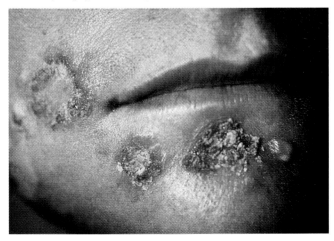

PUBLIC HEALTH

FLESH-EATING BACTERIA

In June 1994, newspapers and television sounded an alert after about a dozen people (mostly in England) died from infection with a strain of *Streptococcus* that destroys flesh at the fearful rate of several inches per hour. This deadly strain, known as "strep A," probably acquired the tissue-destroying toxin genes via infection by a virus. Once the tissue dies, the bacteria live on the remains of the dead flesh, churning out more toxin. The toxin diffuses out of the dead area into the circulatory system, where it kills more tissue and causes the condition *necrotizing fascitis*. Antibiotics are of little value in killing the bacteria because the circulatory system, which would carry the drugs to the bacteria, does not function in dead tissue. Thus the bacteria remain safe inside the destroyed flesh but continue to produce toxins that spread and kill more cells. The dead tissue must be surgically removed, even if it means amputation of a limb.

Fortunately, only a few cases of this "flesh-eating" *Streptococcus* occurred in the United States. Outbreaks elsewhere soon ended as well. Approximately every ten years a different strain of *Streptococcus* gains dominance; perhaps the flesh-eating strain is nearing the end of its ten-year cycle.

Acne

Acne (acne vulgaris) affects more than 80 percent of teenagers and many adults, too. It is most often the result of male sex hormones that stimulate sebaceous glands to increase in size and secrete more sebum. Acne occurs in both males and females because the hormones are produced by the adrenal glands as well as by the testes. Microorganisms feed on sebum, and ducts of the glands and surrounding tissues become inflamed. "Blackheads" are a mild form of acne in which hair follicles and sebaceous glands become plugged with sebum and keratin. In more severe cases (cystic acne) the plugged ducts become inflamed, rupture, and release secretions. Bacteria, especially *Propionibacterium acnes,* infect the area and cause more inflammation, more tissue destruction, and scarring. Such lesions can be widely distributed over the body, and some become encysted in connective tissue.

Acne is treated with frequent cleansing of the skin and topical ointments to reduce the risk of infection. Sometimes acne sufferers are advised to avoid fatty foods, but a connection between diet and acne is not well established. Dermatologists often prescribe oral antibiotics such as tetracycline in low doses to control bacterial infections in the lesions. However, continuous use of antibiotics depletes natural intestinal flora and can contribute to the development of antibiotic-resistant strains of bacteria. ∞ (Chapter 14, p. 363) The drug Accutane, derived from a molecule related to vitamin A, is now used to treat severe and persistent acne. It seems to inhibit sebum production for several months after treatment is stopped, but it can cause serious side effects, such as intestinal bleeding. In most cases acne disappears or decreases in severity as the body adjusts to the hormonal changes of puberty and as the functioning of the sebaceous glands stabilizes.

Burn Infections

Severe burns destroy much of the body's protective covering and provide ideal conditions for infection. Burn infections, which are usually nosocomial, account for 80 percent of deaths among burn patients. *Pseudomonas aeruginosa* is the prime cause of life-threatening burn infections, but *Serratia marcescens* and species of *Providencia* also often infect burns. Many strains of these gram-negative bacilli are antibiotic-resistant. (See the Microbiologist's Notebook "Controlling Infection in a Burn Unit.") ∞ (p. 440)

The thick crust or scab that forms over a severe burn is called **eschar** (es'kar). Bacteria growing in or on eschar pose no great threat, but those growing beneath it cause severe local infections and can move into the blood. Delivering antibiotics to infections under eschar is difficult because the eschar lacks blood vessels. Topical antimicrobials can seep through the eschar; removal of some of the eschar by a surgical scraping technique called **débridement** (da-bred-maw' or de-bred'ment) helps them to reach infection sites.

Prevention of burn infections is difficult even when patients are isolated within hospital burn units. Having lost skin, patients have lost the benefit of the movement of white blood cells to sites of infection in the skin. Such patients also have fluid and electrolyte deficiencies because of seepage from burned tissue. Finally, their appetites are depressed at a time when extensive tissue repair increases metabolic needs.

Burn infections are as difficult to diagnose as they are to treat. Early signs of infection can consist of only a mild loss of appetite or increased fatigue. Definitive diagnosis is made by finding more than 10,000 bacteria per gram of eschar. *Pseudomonas aeruginosa* infection should be suspected when a greenish discoloration appears at the burn site and cultures have a grapelike odor. This bacterium produces tissue-killing toxins that erode skin. It is extremely resistant to antimicrobial drugs and has been found growing in surgical scrub solutions.

APPLICATIONS

WHAT GOES ROUND IN WHIRLPOOL BATHS?

Whirlpools baths can be quite dangerous if not properly cleaned and disinfected daily. Warm water, plus skin flakes, oils, and secretions makes a perfect breeding ground for microbes. The aerosol from a bubbling whirlpool can contain millions of bacteria per cubic meter of air. *Pseudomonas aeruginosa,* a frequent colonizer, is very difficult to eliminate—it can be back in the water just hours after disinfection. Often it grows attached to the sides of the whirlpool tub or inside the pipes connected to it. Disinfection may kill all the organisms in the water, but when the same bacteria attach to a surface and form a biofilm, they are much more resistant to the disinfecting chemicals. A person in the bath who vigorously rubbed his or her eyes could cause a corneal abrasion and thereby become susceptible to bacterial eye infection. *Pseudomonas* infection can cause severe scarring of the cornea and distortion of vision, or even blindness. Once such an infection is started, you can lose an eye almost overnight. Inhalations of aerosols can cause *Pseudomonas* pneumonia, which has a very poor prognosis. Many people break out in a body rash in whirlpools due to a hypersensitivity reaction to being immersed in water filled with *Pseudomonas.* And *Pseudomonas* isn't the only microbe to worry about in or on whirlpools, as two elderly widows discovered. They developed herpes lesions on the backs of their thighs after sitting on the moist rim of a hot tub where a previous occupant with a herpes outbreak had been seated.

Viral Skin Diseases

Rubella

NATURE OF THE DISEASE **Rubella,** or **German measles,** is the mildest of several human viral diseases that cause **exanthema** (ex-an-the'mah), or skin rash. A rash, the main symptom of rubella, appears first on the trunk 16 to 21 days after infection, but the virus (a togavirus) spreads in the blood and other tissues before the rash appears. Infected adult women often suffer from temporary arthritis and arthralgia (joint pain) from dissemination of the virus to joint membranes. These complications are seen less frequently in adult men.

Congenital rubella syndrome results from infection of a developing embryo across the placenta. When a woman becomes infected with rubella during the first 8 weeks of pregnancy, severe damage to the embryo's organ systems is likely because they are developing. After the eighteenth week of pregnancy, damage is rare. The spread of rubella viruses in the infant kills many cells, persistently infects other cells, reduces the rate of cell division, and causes chromosomal abnormalities. Many infants are stillborn, and those that survive may suffer from deafness, heart abnormalities, liver disorders, and low birth weight.

INCIDENCE AND TRANSMISSION Prior to the development of a rubella vaccine, nearly all humans caught rubella at some time, but many cases were not detected. Half the cases in young children and up to 90 percent of those in young adults are not recognized. In the United States the incidence of rubella was nearly 30 cases per 100,000 people in 1969, when a vaccine was licensed. Rubella, including congenital rubella, has now been nearly eliminated (Figure 20.4).

Transmission is mainly by nasal secretions shortly before, and for about a week following, the appearance of a rash. Many infected individuals do not have a rash and transmit the virus without knowing it. Rubella is highly contagious, especially by direct contact among children aged 5 to 14. Infants infected before birth are rubella carriers; they excrete viruses and expose the hospital staff and visitors, including pregnant women, to the disease.

DIAGNOSIS Rubella can be positively diagnosed by a variety of laboratory tests. Determination of a four-fold rise in rubella-specific IgM antibody levels is particularly useful in identifying newborn carriers and assessing immunity of pregnant women exposed to rubella.

IMMUNITY AND PREVENTION The currently available rubella vaccine, part of a combined attenuated viral vaccine preparation (MMR), is the only means of preventing rubella. Children should receive it both to protect themselves from measles and to protect infants, too young to receive the vaccine, from being exposed to a source of infection. High levels of immunity (Chapter 16) must be maintained, or outbreaks will occur. ∞ (p. 430) The vaccine produces lower antibody levels than does infection, and immunity probably is not as long-lasting as that produced by infection. Also, viruses appear in the nasopharynx a few weeks after immunization, and some individuals experience mild symptoms of the disease. To prevent infection of fetuses, a second immunization is recommended for females before they become sexually active. If a woman is pregnant when immunized, viruses from the vaccine may be able to infect the fetus. To prevent transmission of the virus from child to mother to fetus, caution must be exercised in immunizing young children whose mothers are pregnant.

Measles

NATURE OF THE DISEASE **Measles,** or **rubeola,** is a febrile (accompanied by fever) disease with a rash

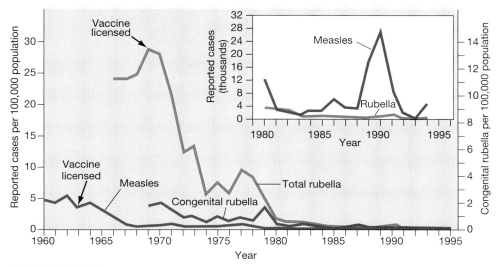

FIGURE 20.4 Incidence rates of rubella (German measles) cases and measles (rubeola) cases dropped rapidly in the United States after vaccines were licensed during the 1960s. Failure to vaccinate, however, has led to a recent increase of cases of measles.

caused by the rubeola virus. This paramyxovirus invades lymphatic tissue and blood. After the virus enters the body through the nose, mouth, or conjunctiva, symptoms appear in 9 to 11 days in children and in 21 days in adults. **Koplik's spots,** red spots with central bluish specks (Figure 20.5), appear on the upper lip and cheek mucosa 2 or 3 days before other symptoms such as fever, conjunctivitis, and cough. These symptoms persist for 3 or 4 days and sometimes progressively worsen. They are followed by a rash, which spreads during a 3- to 4-day period from the forehead to the upper extremities, trunk, and lower extremities and disappears in the same order several days later. The rash is caused by the reaction of T cells with virus-infected cells in small blood vessels. Without such re-

actions, as in patients with defective cell-mediated immunity, no rash appears, but the virus is free to invade other organs. ∞ (Chapter 19, p. 530) If the virus invades the lungs, kidneys, or brain, the common childhood disease often is fatal.

The most common complications of measles are upper respiratory and middle ear infections. **Measles encephalitis,** a more serious complication, occurs in only 1 or 2 patients per 1000 but has a 30 percent mortality rate and leaves one-third of survivors with permanent brain damage. Another complication, **subacute sclerosing panencephalitis** (SSPE), occurs in only 1 in 200,000 cases and is nearly always fatal. It is due to the persistence of measles viruses in brain tissue and causes death of nerve cells, with progressive mental deterioration and muscle rigidity. SSPE manifests itself 6 to 8 years after measles, usually in children who had measles before age 3. In poorly nourished children, measles causes intestinal inflammation with extensive protein loss and shedding of viruses in stools. More than 15 percent of children infected with measles in developing countries die of the disease or its complications.

INCIDENCE AND TRANSMISSION The measles virus is highly contagious; its portal of entry is the respiratory tract. A susceptible person has a 99 percent chance of infection if exposed directly to someone who is releasing the virus while coughing and sneezing. Before widespread immunization, most children had measles before they were 10 years old. In populations such as that of the United States, where periodic epidemics have occurred and most children are well nourished, the disease is serious but rarely fatal. In populations lacking immunity from periodic epidemics or

FIGURE 20.5 Measles (rubeola), showing the typical rash and Koplik's spots in the mouth.

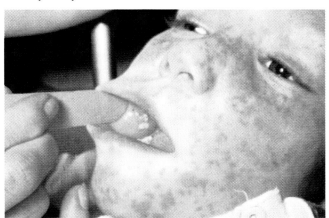

in malnourished children, measles is a killer. In 1875, when measles was first introduced into Fiji, 30 percent of the population died of the disease.

DIAGNOSIS AND TREATMENT   Measles is diagnosed by its symptoms. Treatment is limited to alleviating symptoms and dealing with complications. Secondary bacterial infections can be effectively treated with antibiotics.

IMMUNITY AND PREVENTION   In a nonimmunized population, epidemics of numerous cases over a 3- or 4-week period generally occur every 2 to 5 years over wide regions. Measles vaccine now prevents such epidemics in many developed countries. In the United States, mandatory measles vaccination has greatly reduced the incidence of the disease (Figure 20.4). Because the MMR vaccine containing attenuated viruses of measles, mumps, and rubella is generally used, the incidence of all three diseases has decreased simultaneously. ∞ (Chapter 18, p. 503) Eradication of a disease requires that about 90 percent of the population be immune. Religious groups that reject immunization, immigrants who don't understand its importance, and apathetic parents who don't have their children immunized make eradication difficult. Compounding these problems have been federal government cutbacks in funding used to purchase vaccine for clinics. Hence some poor children in the United States are going without vaccine.

Immunity acquired from having measles is lifelong. Recent epidemics among college students immunized as children suggests that immunity from the vaccine may not be lifelong or that the children were immunized before 15 months of age. A few cases of SSPE have been attributed to the vaccine, but the incidence of SSPE is much lower than it was prior to the use of the vaccine.

Chickenpox and Shingles

ONE VIRUS, TWO DISEASES   The **varicella-zoster virus** (VZV), a herpesvirus, causes both **chickenpox** (varicella) and **shingles** (zoster). The same virus can be isolated from lesions of either disease. Chickenpox is a highly contagious disease that causes skin lesions and usually occurs in children. There are probably over 3 million cases per year in the United States alone. Shingles is a sporadic disease that appears most frequently in older and immunocompromised individuals. A congenital varicella syndrome also is known to occur in about 5 percent of the U.S. population.

In chickenpox the virus enters the upper respiratory tract and conjunctiva and replicates at the site of entry. New viruses are carried in blood to various tissues, where they replicate several more times. Release

PUBLIC HEALTH

MEASLES

Humans are the only known hosts of measles, but distemper in dogs is caused by a virus with antigens similar to those of the measles virus. Measles can spread only if a continuous chain exists between infected and susceptible people. That is because the virus doesn't display latency and cannot survive for long outside the body, no other organism serves as a reservoir, and immunity from infection is lifelong. It has been estimated that a world population of at least 300,000 was required to perpetuate the virus, a population size not reached until about 2500 B.C. Measles probably appeared as a human pathogen after that time—possibly as a mutant of a virus like the one that causes distemper.

of these viruses causes fever and malaise. In 14 to 16 days after exposure, small, irregular, rose-colored skin lesions appear. The fluid in them becomes cloudy, and they dry and crust over in a few days. The lesions appear in cyclic crops over 2 to 4 days as the viruses go through cycles of replication. They start on the scalp and trunk and spread to the face and limbs, sometimes to the mouth, throat, and vagina, and occasionally to the respiratory and gastrointestinal tracts.

Chickenpox, although sometimes thought of as a mild childhood disease, can be fatal. The viruses invade and damage cells that line small blood vessels and lymphatics. Circulating blood clots and hemorrhages from damaged blood vessels are common. Death from varicella pneumonia is due to extensive blood vessel damage in the lungs and the accumulation of erythrocytes and leukocytes in alveoli. Cells in the liver, spleen, and other organs also die because of damage to blood vessels within them.

In shingles, painful lesions like those of chickenpox usually are confined to a single region supplied by a particular nerve (Figure 20.6a). Such eruptions arise from latent viruses acquired during a prior case of chickenpox. During the latent period, these viruses reside in ganglia in the cranium and near the spine. When reactivated, the viruses spread from a ganglion along the pathway of its associated nerve(s). Pain and burning and prickling of the skin occur before lesions appear. The viruses damage nerve endings, cause intense inflammation, and produce clusters of skin lesions indistinguishable from chickenpox lesions (Figure 20.6b). Symptoms range from mild itching to continuous, severe pain and can include headache, fever, and malaise. Lesions often appear on the trunk in a girdlelike pattern (*zoster,* girdle) but can infect the

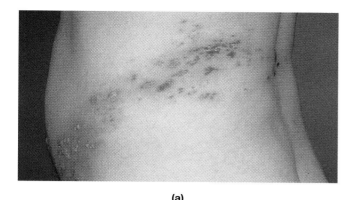
(a)

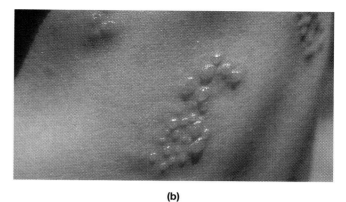
(b)

FIGURE 20.6 Shingles lesions usually result from varicella-zoster virus infections acquired during exposure to childhood chickenpox. The virus can remain latent within the body for many years before being reactivated in adulthood. (a) Vesicles commonly form a belt around the chest or hips, following the pathway of a nerve. (b) The small, yellow vesicles dry up and heal by scabbing, but they can be excruciatingly painful and itchy.

face and eyes. Shingles is most severe in individuals with malignancies or immune disorders. In such patients, lesions may cover wide areas of the skin and sometimes spread to internal organs, where they can be fatal.

The latent virus is activated when cell-mediated immunity drops below a critical minimal level, as can occur in lymphatic cancers, spinal cord trauma, heavy-metal poisoning, or immunosuppression. In other cases, no cause of viral reactivation can be identified. Release of newly replicated viruses increases antibody production, but the antibodies may fail to stop viral replication. Recovery from shingles usually is complete. Second and third cases do occur, and they depend on the degree of development of cell-mediated immunity and local interferon production. Chronic shingles is found in the immunocompromised, including AIDS patients. New vesicles constantly erupt, whereas old ones fail to heal, which can be very debilitating.

INCIDENCE AND TRANSMISSION Chickenpox is endemic in industrialized societies in the temperate zone, and its incidence is highest in March and April. The primary infection usually occurs between the ages of 5 and 9. In general, chickenpox in adults who have not had it as children is more severe than in children. Shingles also is age-related, with most cases appearing in people over 45 years of age.

Infection can be spread by respiratory secretions and contact with moist lesions but not from crusted lesions. Children experiencing a mild case, with only a few lesions and no other symptoms, often spread the disease. In rare cases, adults with partial immunity can contract shingles from exposure to children with chickenpox. Susceptible children, however, can easily

contract chickenpox from exposure to adults with shingles.

DIAGNOSIS AND TREATMENT Chickenpox is usually diagnosed by a history of exposure and the nature of lesions, although a rapid laboratory test has recently become available. Shingles may be impossible to distinguish from other herpes lesions without laboratory tests. Treatment is limited to relieving symptoms, but aspirin should not be given to children because of the risk of Reye's syndrome. Acyclovir, once hoped to be useful in early stages of the disease to reduce its severity, has not proven successful. Antiviral agents are being tested on infections in immunosuppressed patients and those with disseminated disease.

IMMUNITY AND PREVENTION Having chickenpox as a child confers lifelong immunity in most cases. Recurrences (in the form of shingles) are seen only in individuals with low concentrations of VZV antibodies and waning cell-mediated immunity. Chickenpox was the last of the common communicable childhood diseases for which no vaccine was widely available—until 1995, when a vaccine was finally licensed for use in the United States. An attenuated varicella vaccine, developed in Japan, prevents chickenpox when administered as late as 72 hours after exposure. It was used overseas for all children. The U.S. Food and Drug Administration approved the vaccine for general use in 1995. Because the mechanism of latency and reactivation is not well understood, however, immunologists had some reservations about administering vaccines containing viruses that can become latent. Cases have been found in which shingles has been caused by reactivation of the virus in the vaccine. Nevertheless, the vaccine is considered relatively safe and effective.

Other Pox Diseases

Smallpox, a formerly worldwide and serious disease, has now been eradicated (see the Essay at the end of this chapter). Other poxviruses cause a variety of diseases.

COWPOX **Cowpox,** caused by the vaccinia virus, causes lesions (similar to a smallpox vaccination) at abrasion sites, inflammation of lymph nodes, and fever. Vaccinia viruses can also cause a progressive disease, with numerous lesions and symptoms more like those of smallpox, that can be treated with the drug methisazone. Cattle appear to transmit the disease to humans.

Not only was the vaccinia virus used by Jenner to immunize against smallpox but also it was the first animal virus to be obtained in sufficient quantity for chemical and physical analysis. ∞ (Chapter 1, p. 13) The modern vaccinia virus may have become attenuated by several centuries of passage on calves' skin, a procedure used in making smallpox vaccine.

MOLLUSCUM CONTAGIOSUM The **molluscum contagiosum** virus is unusual for several reasons. First, it differs immunologically from both orthopoxviruses and parapoxviruses, the two major groups of poxviruses. Second, it elicits only a slight immune response. Third, although infected cells cease to synthesize DNA, the virus induces neighboring uninfected cells to divide rapidly. Thus, this virus may be intermediate between viruses that cause specific diseases and those that induce tumors. Molluscum contagiosum affects only humans and is distributed worldwide. It causes pearly white to light pink painless tumorlike growths scattered over the skin (see the chapter-opening photograph). The disease usually affects children and young adults, and it can persist for years. It is acquired by personal contact or from items such as gym equipment and swimming pools. Treatment generally involves the removal of growths by chemicals or by localized freezing.

Warts

Human **warts,** or **papillomas,** are caused by **human papillomaviruses** (HPV). HPV specifically attack skin and mucous membranes. Warts grow freely in many sites in the body—the skin, the genital and respiratory tracts, and the oral cavity. Viral infection lasts a lifetime. Even when warts disappear or are removed, the virus remains in surrounding tissue, and warts may recur or form malignant tumors.

NATURE OF WARTS Warts vary in appearance, area of occurrence, and pathogenicity. Some are barely visible and self-limiting—that is, they do not grow or spread—and others, such as laryngeal warts, are larger but benign. A few warts are malignant. **Genital warts,** also known as condylomata acuminata (kon"di-lo-ma'ta a-kum"i-na'ta) (Figure 20.7a), for example, sometimes become malignant, and some of the viral strains that produce them are strongly associated with cervical cancer. (Women who have had genital warts removed are advised to have Pap tests twice a year.) Warts of all types grow larger than normal in people with AIDS or other immunodeficiencies.

TRANSMISSION OF WARTS Papillomaviruses are transmitted by direct contact, usually between humans, or by fomites. **Dermal warts** (Figure 20.7b) form when the virus enters the skin or mucous membranes through abrasions. Genital warts are sexually transmitted, and *juvenile onset laryngeal warts* are acquired during passage through an infected birth canal. The incubation period varies from 1 week to 1 month for der-

---

FIGURE 20.7 Warts are caused by human papillomaviruses. (a) Genital warts around the vagina, a probable cause of cervical cancer, and (b) dermal warts.

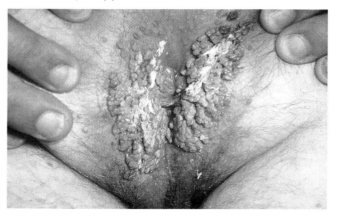

(a)

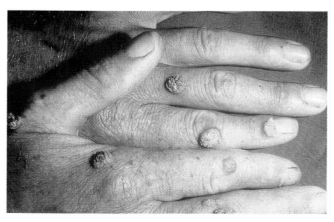

(b)

APPLICATIONS

MIND OVER BODY

According to Lewis Thomas, "Warts are wonderful structures. Fully grown, nothing in the body has so much the look of toughness and permanence as a wart, yet, inexplicably and often very abruptly, they come to the end of their lives and vanish without a trace." Thomas goes on to relate how warts "can be ordered off the skin by hypnotic suggestion." In a controlled study, 14 patients with multiple warts were hypnotized and the suggestion made that all warts on one side of the body go away. Within a few weeks, this happened in 9 patients, including one who "got mixed up and destroyed the warts on the wrong side."

mal warts and 8 to 20 months for genital warts. Genital warts and warts acquired at birth will be discussed in Chapter 25.

DERMAL WARTS Epithelial cells become infected and proliferate to form dermal warts, which have distinct boundaries and remain above the basement membrane between the epidermis and the dermis. Children and young adults are more likely than older people to have dermal warts. Only a few warts are present at any one time, and most regress in less than 2 years. Removing one wart often causes regression of all others. In spontaneous regression, all warts usually disappear at the same time. Regression probably is an immunologic phenomenon.

DIAGNOSIS AND TREATMENT Warts can be distinguished by immunological tests and microscopic examination of tissues. Enzyme immunoassay and immunofluorescent antibody tests can detect about three-fourths of the cases in which viruses are found microscopically. These tests sometimes fail because certain papillomas, especially genital and laryngeal papillomas and those progressing toward malignancy, produce only small quantities of antigens.

Available treatments for various kinds of warts are not entirely satisfactory. The most widely used treatment, cryotherapy, involves the freezing of tissue with liquid carbon dioxide or liquid nitrogen and excision of the infected tissue. Caustic chemical agents, such as podophylin, salicylic acid, and glutaraldehyde; surgery; antimetabolites such as 5-fluorouracil; and interferon to block viruses also are used to get rid of warts. Recurrences are still common.

Fungal Skin Diseases

The fungi that invade keratinized tissue are called **dermatophytes** (dur'ma'to-fitz), and fungal skin diseases are called **dermatomycoses** (dur"ma-to-mi-ko'ses). These diseases can be caused by any one of several organisms, mainly from three genera: *Epidermophyton*, *Microsporum*, and *Trichophyton* (Table 20.1). They cause various types of ringworm and attack the skin, nails, and hair.

Fungi that invade subcutaneous tissues live freely in soil or on decaying vegetation and can be found in bird droppings and as airborne spores. These fungi enter tissue through a wound and sometimes spread to lymph vessels. Subcutaneous fungal infections usually spread slowly and insidiously; response to treatment is likewise slow.

Ringworm

Ringworm occurs in several forms (including athlete's foot, discussed next) and is highly contagious. For example, ringworm of the scalp is easily acquired at hairstyling establishments if strict sanitary practices are not followed. Ringworm involves the skin, hair, and nails, and most forms are named according to where they are found. **Tinea corporis** (body ringworm) causes ringlike lesions with a central scaly area. (The shape of these lesions is what originally gave rise to the misleading name "ringworm.") **Tinea cruris** (groin ringworm, or "jock itch") occurs in skin folds in the pubic

TABLE 20.1	Dermatomycoses and Common Organisms Found in Lesions				
Dermatomycosis	*Epidermophyton floccosum*	*Microsporum canis*	*Trichophyton mentagrophytes*	*Trichophyton rubrum*	*Trichophyton tonsurans*
Body ringworm		X	X	X	X
Nail ringworm	X			X	
Groin ringworm	X		X	X	
Scalp ringworm		X	X		X
Beard ringworm			X	X	
Athlete's foot	X		X	X	

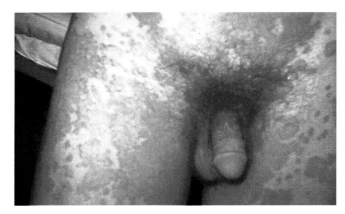

FIGURE 20.8 Tinea cruris, ringworm of the groin ("jock itch"), is caused by the fungus *Trichophyton.*

region (Figure 20.8). **Tinea unguium** (ringworm of the nails) causes hardening and discoloration of fingernails and toenails. In **tinea capitis** (scalp ringworm), hyphae grow down into hair follicles and often leave circular patterns of baldness. **Tinea barbae** (barber's itch) causes similar lesions in the beard.

None of the dermatomycoses result in severe diseases, and lesions usually do not invade other tissues; but they are unsightly, itchy, and persistent. The causative agents grow well at skin temperature, which is slightly below body temperature. Tissue damage caused by dermatophytes can allow secondary bacterial infections to develop.

DIAGNOSIS AND TREATMENT Diagnosis of ringworm can be made by microscopic examination of scrapings from lesions, but observation of the skin itself is often sufficient. Although fungi in tissues generally do not form spores, those in laboratory cultures often do, and workers must be especially careful not to become infected from escaping spores. Large numbers of people were infected when spores from a mistakenly opened Petri plate in a university laboratory were dispersed by the building's ventilation system.

Treatment generally consists of removing all dead epithelial tissues and applying a topical antifungal ointment. If lesions are widespread or difficult to treat topically, as when they infect nailbeds, griseofulvin is administered orally. Prevention of ringworm requires avoiding contaminated objects and spores.

ATHLETE'S FOOT In **athlete's foot,** or **tinea pedis,** hyphae invade the skin between the toes and cause dry, scaly lesions. Fluid-filled lesions develop on moist, sweaty feet. Subsequently, the skin cracks and peels, and a secondary bacterial infection leads to itchy, soggy, white areas between the toes. Athlete's foot, a form of ringworm, results from an ecological imbal-

ance between normal flora and host defenses. The fungi that cause athlete's foot are widespread in the environment; they infect tissues when body defenses fail to repel them. Prevention of athlete's foot depends on maintaining healthy, clean, dry feet that can resist the opportunism of fungi.

Subcutaneous Fungal Infections

SPOROTRICHOSIS **Sporotrichosis** is caused by *Sporothrix schenckii,* which usually enters the body from plants, especially sphagnum moss and rose and barberry thorns. The disease can also be acquired from other humans, dogs, cats, horses, and rodents. It is most common in the midwestern United States, especially in the Mississippi valley. A lesion appears first as a nodular mass at the site of a minor wound. The mass ulcerates, becomes chronic, granulomatous, and pus-filled, and can spread easily to lymphatic vessels. In rare instances it disseminates to internal organs, especially the lungs. Diagnosis is made by culturing specimens of pus or tissue taken from lesions. The cutaneous and lymphatic forms can be treated with potassium iodide; disseminated infections require amphotericin B. People who work with plants or soil should cover injured skin areas to protect themselves from exposure to contaminated materials.

BLASTOMYCOSIS North American **blastomycosis** (Figure 20.9), caused by the fungus *Blastomyces dermatitidis,* is most common in soil of the central and southeastern United States. The fungus enters the body through the lungs or wounds, where it causes disfiguring granulomatous, pus-producing lesions and multiple abscesses in skin and subcutaneous tissue. This condition, called **blastomycetic dermatitis,** for unknown reasons affects chiefly males in their thirties and forties. In some cases, the fungus travels in the blood and invades internal organs, causing **systemic blasto-**

FIGURE 20.9 Blastomycosis lesion on the face.

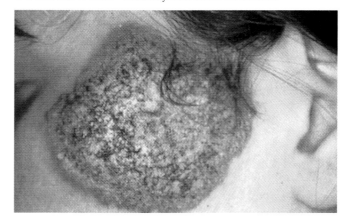

mycosis. The lungs can be infected directly by the inhalation of spores, and organisms can travel from the lungs to infect other tissues. The disease usually causes relatively mild respiratory symptoms, fever, and general malaise. It can be diagnosed by finding budding yeast cells in sputum or pus, and is treated with amphotericin B or the less toxic hydroxystilbamidine.

Opportunistic Fungal Infections

Certain fungi, such as some yeasts and black molds, can invade the tissues of humans with impaired resistance. Two yeasts, *Candida albicans* and *Cryptococcus neoformans,* and certain molds, such as *Aspergillus fumigatus, A. niger,* and the zygomycetes *Mucor* and *Rhizopus,* account for many opportunistic fungal diseases.

CANDIDIASIS *Candida albicans,* an oval, budding yeast, is present among the normal flora of the digestive and urogenital tracts of humans. In debilitated individuals it can cause **candidiasis** (kan"di-di'uh-sis), or **moniliasis,** in one or several tissues. Superficial candidiasis appears as **thrush** (Figure 20.10a), milky patches of inflammation on oral mucous membranes, especially in infants, diabetics, debilitated patients, and those receiving prolonged antibiotic therapy. ∞ (Chapter 14, p. 363) The disease appears as **vaginitis** when vaginal secretions contain large amounts of sugar, as occurs during pregnancy, when oral contraceptives are used, when diabetes is poorly controlled, or when women wear tight synthetic undergarments, which promote warmth and moisture and thus favor yeast growth. Some strains of *Candida* may be sexually transmitted. Cannery workers whose hands are in water for long periods sometimes develop skin and nail lesions (Figure 20.10b). *Candida* can invade the lungs, kidneys, and heart or be carried in the blood, where it causes a severe toxic reaction. Candidiasis, the most common nosocomial fungal infection, is seen in patients with diseases such as tuberculosis, leukemia, and AIDS.

Finding budding cells in lesions, sputum, or exudates confirms the diagnosis of candidiasis. Various antifungal drugs are used to treat it. *Candida* is ubiquitous; infections can be prevented in large part by preventing debilitating conditions.

ASPERGILLOSIS **Aspergillosis** in humans can be caused by various species of *Aspergillus,* but especially *A. fumigatus.* These fungi grow on decaying vegetation. *Aspergillus* initially invades wounds, burns, the cornea, or the external ear, where it thrives in earwax and can ulcerate the eardrum. In immunosuppressed patients, it can cause severe pneumonia. Diagnosis is made by finding characteristic hyphal fragments in tissue biopsies. Antifungal agents are only modestly successful in treatment. Because the mold is so ubiquitous that exposure is inevitable, prevention depends mainly on host defenses.

ZYGOMYCOSES Certain zygomycetes of the genera *Mucor* and *Rhizopus* can infect susceptible humans, such as untreated diabetics, and cause **zygomycoses.** Once established, the fungus invades the lungs, central nervous system, and tissues of the eye orbit and can be rapidly fatal, presumably because it evades body defenses. It can be diagnosed by finding broad hyphae in the lumens and walls of blood vessels. Antifungal drugs may or may not be effective in treating zygomycoses.

FIGURE 20.10 (a) *Candida* infections of the oral cavity (thrush), seen as white patches, are a common complication of AIDS, diabetes, and prolonged antibiotic therapy. (b) *Candida* infections of the nails are very difficult to eradicate.

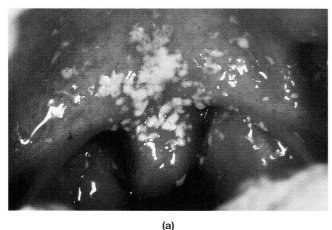

(a)

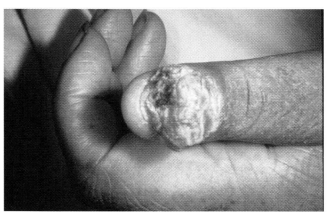

(b)

Other Skin Diseases

Madura foot, or **maduromycosis,** occurs mainly in the tropics. It is caused by a variety of soil organisms, including fungi of the genus *Madurella* and filamentous actinomycetes, such as *Actinomadura, Nocardia, Streptomyces,* and *Actinomyces.* Organisms enter the body through breaks in the skin, especially in people who do not wear shoes. Initial pus-filled lesions spread and form connected lesions that eventually become chronic and granulomatous. Untreated, the organisms invade muscle and bone, and the foot becomes massively enlarged (Figure 20.11). Madura foot is diagnosed by finding white, yellow, red, or black granules of intertwined hyphae in pus. So-called sulfur granules in pus are yellow hyphae and not sulfur. Unless antibiotic therapy begins early in the infection and is sufficiently prolonged, amputation may be necessary. Keeping soil particles out of wounds prevents this disease.

Skin reactions to cercariae (larvae) of several helminth species of schistosomes cause **swimmer's itch.** These cercariae parasitize birds, domestic animals, and primates and can burrow into human skin. Immune reactions cause reddening and itching and usually prevent cercariae from reaching the blood and causing schistosomiasis. ∞ (Chapter 12, p. 317) Swimmer's itch occurs throughout the United States but is especially common in the Great Lakes area.

Dracunculiasis (drah-kun"kew-li'a-sis) is caused by a parasitic helminth called a guinea worm. This disease is discussed in the Microbiologist's Notebook "Carter Foundation Hopes for Eradication of Guinea Worm from World by 1997." ∞ (Chapter 12, p. 320)

Skin diseases are summarized in Table 20.2.

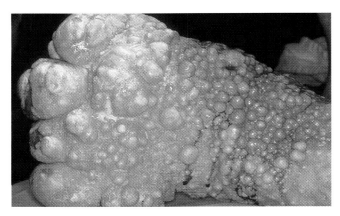

FIGURE 20.11 Madura foot can be caused by true fungi or by actinomycetes, funguslike bacteria. Lesions filled with pus and pathogens may distort the foot to the point at which amputation is necessary.

DISEASES OF THE EYES

Like skin diseases, eye diseases frequently result from pathogens that enter from the environment, and they cause damage to an exterior portion of our bodies. For that reason we discuss them here. If you need to, refer to Chapter 17 for an overview of the eye and its structure. ∞ (p. 464)

Bacterial Eye Diseases

Ophthalmia Neonatorum

Ophthalmia neonatorum, or conjunctivitis of the newborn, is a pyogenic (pus-forming) infection of the eyes caused by organisms such as *Neisseria gonorrhoeae* and *Chlamydia trachomatis.* Organisms present in the birth canal enter the eyes as a baby is born. The resulting infection can cause **keratitis,** an inflammation of the cornea, which can progress to perforation and destruction of the cornea and blindness. Early in the 1900s, 20 to 40 percent of children in European institutions for the blind had suffered from this disease. In developing countries the disease is still prevalent, but the true incidence is unknown because high infant mortality makes it difficult to estimate the number of original infections. Although infections are most common in newborns (Figure 20.12, p. 568), adults can transfer organisms by hands or by fomites from genitals to their eyes.

Penicillin was once the treatment of choice, but resistant strains have developed, and tetracycline is now used. Tetracycline has the advantage of also being effective against chlamydias and other organisms, with which mothers may also be infected. Preventive mea-

APPLICATIONS

ALGAL INFECTIONS

Most algae manufacture their own food and are not parasitic, but some strains of the alga *Prototheca* have lost their chlorophyll and survive by parasitizing other organisms. Found in water and moist soil, they enter the body through skin wounds. By 1987, 45 cases of protothecosis had been reported, 2 from cleaning home aquariums. Protothecosis was first observed on the foot of a rice farmer, and most subsequent cases have occurred on legs or hands. In immunodeficient patients the parasite can invade the digestive tract or peritoneal cavity. A few skin infections have responded to oral potassium iodide or intravenous amphotericin B and tetracycline therapy, but no satisfactory treatment has been found for others.

TABLE 20.2   **Summary of Skin Diseases**

Disease	Agent	Characteristics
Bacterial Skin Diseases		
Folliculitis	*Staphylococcus aureus*	Skin abscess; encapsulated, so not reached by antibiotics
Scalded skin syndrome	*S. aureus*	Vesicular lesions over entire skin surface, fever; most common in infants
Scarlet fever	*Streptococcus pyogenes*	Sore throat, fever, rash caused by toxin; can lead to rheumatic fever and other complications
Erysipelas	*S. pyogenes*	Skin lesions spread to systemic infection; rare today, but common and fatal before antibiotics were available
Pyoderma and impetigo	Staphylococci, streptococci	Skin lesions, usually in children; easily spread by hands and fomites
Acne	*Propionibacterium acnes*	Skin lesions caused by excess of male sex hormones; infection is secondary; common in teenagers
Burn infections	*Pseudomonas aeruginosa* and other bacteria	Growth of bacteria under eschar, often a nosocomial infection; difficult to diagnose and treat; causative agents typically antibiotic-resistant
Viral Skin Diseases		
Rubella	Rubella virus	Mild disease with maculopapular exanthema (discolored, pimply rash); infection early in pregnancy can lead to congenital rubella; vaccine has greatly reduced incidence
Measles	Rubeola virus	Severe disease with fever, conjunctivitis, cough, and rash; encephalitis is a complication; disease occurs mainly in children; vaccine has greatly reduced incidence
Chickenpox	Varicella-zoster virus	Generalized macular (discolored) skin lesions
Shingles	Varicella-zoster virus	Pain and skin lesions, usually on trunk; occurs in adults with diminished immunity; susceptible children exposed to cases of shingles can develop chickenpox
Smallpox	Smallpox virus	Eradicated by immunization as a human disease
Other pox diseases	Other poxviruses	Clear or bluish vesicles on skin surfaces; human infections are rare
Warts	Human papillomaviruses	Dermal warts are self-limiting; malignant warts occur in immunologic deficiencies
Fungal Skin Diseases		
Dermatomycoses	Dermatophytes	Dry, scaly lesions on various parts of the skin; difficult to treat
Sporotrichosis	*Sporothrix schenckii*	Granulomatous, pus-filled lesions; sometimes disseminates to lungs and other organs
Blastomycosis	*Blastomyces dermatitidis*	Granulomatous, pus-filled lesions that develop in lungs and wounds; sometimes disseminates to other organs
Candidiasis	*Candida albicans*	Patchy inflammation of mucous membranes of the mouth (thrush) or vagina (vaginitis); disseminated nosocomial infections occur in immunodeficient patients
Aspergillosis	*Aspergillus* species	Wound infection in immunodeficient patients; also infects burns, cornea, and external ear
Zygomycosis	*Mucor* and *Rhizopus* species	Occurs mainly with untreated diabetes; begins in blood vessels and can rapidly disseminate
Other Skin Diseases		
Madura foot	Various soil fungi and actinomycetes	Initial lesions spread and become chronic and granulomatous; can require amputation
Swimmer's itch	Cercariae of schistosomes	Itching due to cercariae burrowing into skin; immunological reaction prevents their spread
Dracunculiasis	*Dracunculus medinensis*	Larvae ingested in crustaceans in contaminated water migrate to skin and emerge through lesion; juveniles cause severe allergic reactions

sures have almost eradicated ophthalmia neonatorum in developed countries. A few drops of 1 percent silver nitrate solution in the eyes immediately after birth kills gonococci, but it is not effective against chlamydias and can irritate the eyes. Antibiotics such as penicillin, tetracycline, and erythromycin are effective against most known bacterial causes of ophthalmia neonatorum.

Bacterial Conjunctivitis

Bacterial conjunctivitis, or pinkeye, is an inflammation of the conjunctiva caused by organisms such as *Staphylococcus aureus, Streptococcus pneumoniae, Neisseria gonorrhoeae, Pseudomonas* species, and *Haemophilus influenzae* biogroup *aegyptius*. The BPF clone of the latter causes Brazilian purpuric fever, in which young children ini-

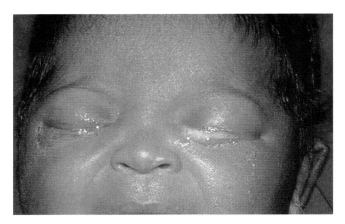

FIGURE 20.12 Gonococcal infection of the eye. Most often this condition is seen in newborns, who acquire the infection as they pass through an infected birth canal. Adults, however, can also transfer bacteria to the eye from the genitals.

tially have conjunctivitus and then develop potentially fatal septicemia with extensive hemorrhaging and symptoms of meningitis. Bacterial conjunctivitis is extremely contagious, especially among children, and can spread rapidly through schools and day-care centers. Children rub itchy, runny eyes and transfer organisms to their playmates. In warm weather, gnats attracted to the moisture of tears may pick up organisms on their feet and transfer them to other persons' eyes. Topically applied sulfonamide ointment is an effective treatment. Children should not return to school until their infection is completely eliminated.

Trachoma

Trachoma (Figure 20.13), from the Greek word meaning "pebbled," or "rough," is caused by specific strains

FIGURE 20.13 Trachoma, caused by *Chlamydia trachomatis*, is the leading cause of preventable blindness in the world, affecting some 500 million people. Note the pebbled appearance of the vastly swollen conjunctiva.

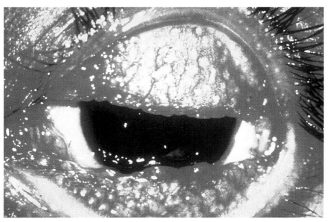

of *Chlamydia trachomatis*. The disease is marked by severly swollen conjunctiva with a pebbled appearance. Scarring of the eyelids causes the eyelashes to point inward, which leads to scarring and destruction of the cornea and eventually to blindness. Trachoma is the leading cause of preventable blindness worldwide; 500 million people have trachoma, and over 20 million are already blind from it. Although uncommon in the United States, except among Native Americans in the Southwest, the disease is widespread in parts of Asia, Africa, and South America, sometimes affecting 90 percent of the population. Flies are important mechanical vectors, and close mother-child contact facilitates human transfer.

PUBLIC HEALTH

SAVE YOUR EYES

When soft contact lenses became popular in the 1970s, ophthalmologists began seeing fungal and bacterial eye infections among their wearers. Fungi are not part of the normal flora of the eye. Some contact lens users developed fungal keratitis (corneal ulcers) because the fungus grew in their hydrogel lens and then attacked the eye surface. Fungi of the genus *Fusarium* are usually the cause. People who complain that they can't get the spots off their contact lenses may not realize that those little spots are fungal colonies growing in the lens material itself. This problem typically results from poor hygiene and improper disinfection.

Fungal eye infections are a special problem for two reasons. First, we don't have an extensive arsenal of effective drugs against fungi, as we do against bacteria. Second, fungal infections are slow to develop, so people may not pay much attention to them at first.

Conversely, bacterial eye infections, such as those caused by *Pseudomonas*, can develop so rapidly that a patient can suffer serious eye damage within 24 hours. Careless use of mascara is one common way of getting a corneal ulcer, usually caused by *Pseudomonas*. Within 24 to 48 hours after initial use of health and beauty aids such as mascara, the material in the container is well colonized by *Pseudomonas* and other organisms. Scratching an eye with a contaminated mascara applicator introduces organisms into the cornea, allowing an infection to start. The individual often suffers severe pain and redness within 3 to 4 hours after the injury and seeks treatment, unlike the victim of the slow, insidious attack of fungi.

Cosmetics, especially mascara, should be discarded after 3 to 6 months of use. Preservatives in them do not last forever, and bacteria multiply rapidly. The big "one-year-supply" size jar of a cosmetic may not be a bargain after all.

TRY IT

WHAT GROWS IN YOUR HEALTH AND BEAUTY AIDS?

Are the shampoos, contact lens fluids, mascara, cold cream, and aerosol spray cans in your house sterile? Or are you smearing and spraying microorganisms on yourself and into your environment? Try culturing brand new bottles of products as well as ones that have been open for varying times. A product may have passed the quality control tests at the factory, but after months on a store shelf, the antimicrobial "preservatives" that were originally added may no longer be effective. Single-dose containers such as the little foil packets of shampoo in a hotel room are not required to have the preservatives needed in products that will have repeated reuse with reinoculation of organisms into the product each time. With aerosol cans, check both the liquid inside and the nozzle. The product may be sterile, but a contaminated nozzle can spray organisms into the air. Does the storage case for your contact lenses have a biofilm of organisms on its inner sides? If you make up your own shampoo or contact lens fluids from concentrates, how do they compare with store-bought versions in terms of sterility?

Viral Eye Diseases

Epidemic Keratoconjunctivitis

Epidemic keratoconjunctivitis (EKC), caused by an adenovirus, is sometimes called *shipyard eye;* workers are frequently infected from dust particles in the environment. After 8 to 10 days' incubation, the conjunctiva become inflamed, and eyelid edema, pain, tearing, and sensitivity to light follow. Within 2 days the infection spreads to the corneal epithelium and sometimes to deeper corneal tissue. Clouding of the cornea can last up to 2 years. EKC can be nosocomially acquired in eye clinics and ophthalmologists' offices.

Acute Hemorrhagic Conjunctivitis

Another viral eye disease, acute hemorrhagic conjunctivitis (AHC), caused by an enterovirus, appeared in 1969 in Ghana. Serological studies showed it was not prevalent anywhere in the world before then. First seen in the United States in 1981, the disease occurs chiefly in warm, humid climates under conditions of crowding and poor hygiene. AHC causes severe eye pain, abnormal sensitivity to light, blurred vision, hemorrhage under conjunctival membranes, and sometimes transient inflammation of the cornea. Onset is sudden, and

recovery usually is complete in 10 days. A rare complication is a paralysis resembling poliomyelitis.

Parasitic Eye Diseases

Onchocerciasis–River Blindness

The filarial (threadlike) larvae of the roundworm *Onchocerca volvulus* cause onchocerciasis (ong" ko-ser-si'a-sis), or river blindness, in many parts of Africa and Central America (Figure 20.14). Adult worms and microfilariae (small larvae) accumulate in the skin in nodules. ∞ (Chapter 12, p. 325) When a blackfly bites an infected individual, it ingests microfilariae, which mature in the fly and move to its mouthparts. When the fly bites again, infectious microfilariae enter the new host's skin and invade various tissues, including the eyes. In many small villages where the people depend on water from blackfly-infested rivers, nearly all inhabitants who live past middle age are blind.

Adult worms cause skin nodules that can abscess. Microfilariae cause skin depigmentation and severe dermatitis via immunological responses to live filariae or toxins from dead ones. The worst tissue damage occurs as the worms invade the cornea and other parts of the eye. Over several years they cause eye blood vessels to become fibrous, and total blindness ensues by about age 40.

Onchocerciasis can be diagnosed by finding microfilariae in thin skin samples or adult worms visible through the skin. The drug ivermectin kills adult worms quickly and microfilariae over several weeks. Drugs are available to kill microfilariae rapidly, but toxins from very many dying microfilariae can cause ana-

FIGURE 20.14 Onchocerciasis, or river blindness, is caused by the roundworm *Onchocerca volvulus,* whose microfilarial stages are transmitted by bite of the blackfly. In some African villages, nearly all adults are blind and must be led by children, the only ones in the village who can still see, although they are already infected with the worms that will eventually blind them also.

phylactic shock. Onchocerciasis could be prevented by eliminating blackflies, which congregate along rivers. DDT destroys some flies, but it allows resistant ones to survive, and it accumulates in the environment.

The small size of the pygmies of Uganda is due to onchocerciasis infections. When mothers are infected by *O. volvulus* (and most of them are), the parasite damages the pituitary gland of their fetuses. The result is dwarfism from a deficiency of growth hormone. A forest-dwelling strain of *O. volvulus* does not cause blindness but leads to severe itching as well as psychological distress in more than 9 million infected Africans. Treatment with ivermectin is expected to begin in 1996.

Loaiasis

The eye worm *Loa loa,* a filarial worm endemic to African rain forests, is transmitted to humans by deer flies. Adult worms live in subcutaneous tissues and eyes (Figure 20.15); microfilariae appear in peripheral blood in the daytime and concentrate in the lungs at night. Deer flies feed during the day and acquire microfilariae from infected humans. The worms develop in the flies, migrate to their mouthparts, and are transferred to other humans when the flies feed again. Microfilariae migrate through subcutaneous tissue, leaving a trail of inflammation, and often settle in the cornea and conjunctiva. Although they usually do not cause blindness, the shock of finding a worm over an inch long in one's eye is undoubtedly a

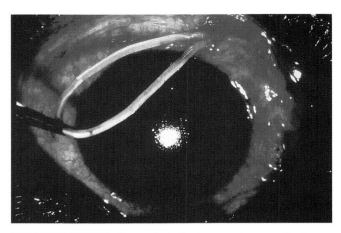

FIGURE 20.15 Removal of a *Loa loa* worm from a patient's eye.

traumatic experience! In fact, the name *Loa loa* is believed to be derived from a voodoo term meaning the rising of a devil within you.

Loaiasis (lo'ah-i"a-sis) is diagnosed by finding microfilariae in blood or by finding worms in the skin or eyes. It is treated by excising adult worms or by the use of suramin or other drugs to eradicate microfilariae. Control could be achieved by eradicating deer flies, but this is an exceedingly difficult task.

Eye diseases are summarized in Table 20.3.

TABLE 20.3	Summary of Eye Diseases	
Disease	**Agent**	**Characteristics**
Bacterial Eye Diseases		
Ophthalmia neonatorum corneal lesions and can lead to	*Neisseria gonorrhoeae*	Infection acquired during passage through birth canal causes corneal lesions and can lead to blindness; silver nitrate or antibiotics have nearly eradicated it in developed countries
Bacterial conjunctivitis	*Haemophilus influenzae* biogroup *aegyptius*, *Staphylococcus aureus*, *Streptococcus pneumoniae*, *Neisseria gonorrhoeae*, *Pseudomonas* species	Highly contagious inflammation of the conjunctiva in young children
Trachoma	*Chlamydia trachomatis*	Infection and destruction of cornea and conjunctiva; cause of preventable blindness
Keratitis	Bacteria, viruses, and fungi	Ulceration of the cornea; occurs mainly in immunodeficient and debilitated patients
Viral Eye Diseases		
Epidemic keratoconjunctivitis	Adenovirus	Inflammation of the conjunctiva that spreads to the cornea; transmitted in dust particles; also nosocomial
Acute hemorrhagic conjunctivitis	Enterovirus	Severe pain and hemorrhage under conjunctiva; highly contagious under crowded, unsanitary conditions
Parasitic Eye Diseases		
Onchocerciasis	*Onchocerca volvulus*	Microfilariae enter skin through blackfly bite and invade eyes and other tissues; causes dermatitis and blindness; occurs in tropics
Loaiasis	*Loa loa*	Microfilariae enter skin through deer fly bite; cause inflammation of conjunctiva and cornea

PUBLIC HEALTH

ACANTHAMOEBA KERATITIS

In the mid-1980s, 24 cases of *Acanthamoeba* keratitis were reported to the CDC, 20 of which were in users of contact lenses. Since then, over 100 cases have been reported. The amoeba that causes this condition is common in fresh water and soil and is found in brackish water, sea water, and hot tubs as well as contaminated contact lens cleaning solution. It can also travel via airborne dust. The protozoan infection causes severe eye pain and destruction of corneal epithelium. Some patients have been successfully treated with ketoconazole or miconazole, but because of extensive damage to the cornea, others have required corneal transplants. Contamination of water sources by this amoeba is responsible for the closing of the historic baths in Bath, England, where a girl died from an amoebic brain infection.

FIGURE 20.16 Gas gangrene has blackened the toes of an infected foot.

WOUNDS AND BITES

Intact skin protects against most infectious disease agents, but we have seen that some are able to penetrate unbroken skin and that mucous membranes are not always effective barriers. In other cases, wounds and bites break the protective barrier of skin, allowing organisms to cause disease. (We discuss rabies, probably the most well known bite-related disease, in Chapter 24 because it is so closely associated with the nervous system.)

Wound Infections

Gas Gangrene

Associated with deep wounds, **gas gangrene** often is a mixed infection caused by two or more species of *Clostridium*, especially *C. perfringens* (found in 80 to 90 percent of cases), *C. novyi*, and *C. septicum*. Spores of these obligately anaerobic organisms are introduced by injuries or surgery into tissues where circulation is impaired and dead and anaerobic tissue is present. In body regions where oxygen concentration is low, spores germinate, multiply, and produce toxins and enzymes such as collagenases, proteases, and lipases that kill other host cells and extend the anaerobic environment.

The onset of gas gangrene occurs suddenly 12 to 48 hours after injury. As the organisms grow and ferment muscle carbohydrates, they produce gas, mainly hydrogen, and gas bubbles distort and destroy tissue (Figure 20.16). Such tissue is called **crepitant** (rattling) **tissue.** The bubbles audibly "snap, crackle, and pop"

when the patient is repositioned. Foul odor is such a prominent feature of gas gangrene that medical personnel can diagnose it without even entering the patient's room. High fever, shock, massive tissue destruction, and blackening of skin accompany this rapidly spreading disease. If left untreated, death occurs swiftly. Diagnosis usually is made on the basis of clinical findings, and treatment is begun before laboratory test results are available. Penicillin is administered, and dead tissue is removed or limbs are amputated. Gas gangrene often follows illegal abortions performed under unsanitary conditions; it usually necessitates a hysterectomy. *C. perfringens* is sometimes present in bile and can occasionally cause gas gangrene of abdominal muscles following gall bladder or bile duct surgery, especially if bile is spilled.

The use of *hyperbaric* (high-pressure) oxygen chambers to treat gas gangrene is somewhat controversial. Patients are placed in chambers containing 100 percent oxygen at 3 atmospheres pressure for 90 minutes two or three times a day. The actual mechanism by which high-pressure oxygen aids recovery is not known, but presumably it kills or inhibits obligate anaerobes. Gas gangrene can be prevented by adequate cleansing of wounds, delaying closing of wounds, and providing drainage when they are closed. No vaccine is available.

Other Anaerobic Infections

In addition to clostridia, certain non-spore-forming anaerobes are associated with some infections. *Bacteroides* and *Fusobacterium* species are normally present in the digestive tract, and *Bacteroides* accounts for nearly half of human fecal mass. *Fusobacterium* sometimes causes oral infections. If introduced into the abdominal cavity, genital region, or deep wounds by surgery or human bites, it causes infections there, too.

Such infections are difficult to treat because the organisms are resistant to many antibiotics. Abscesses caused by anaerobic organisms must be drained surgically, and appropriate antibiotics used as supportive therapy.

Cat Scratch Fever

Cats are mechanical vectors of **cat scratch fever,** a disease caused by two different organisms: *Afipia felis,* a gram-negative bacillus with a single flagellum, and, more commonly, the rickettsia *Bartonella (Rochalimaea) henselae.* The organisms are found mainly in capillary walls or in microabscesses. A cat scratch lesion resembling Kaposi's sarcoma has been found in AIDS patients. Another cat scratch lesion found in AIDS patients, as well as in other immunocompromised patients, is bacillary peliosis, in which blood-filled cavities form in bone marrow, liver, and spleen. Presumably cats acquire the organisms from the environment and carry them on claws and in the mouth. When cats scratch, bite, or lick, they transmit the organisms to humans. After 3 to 10 days, a pustule appears at the site of entry, and the patient has a mild fever, headache, sore throat, swollen glands, and conjunctivitis for a few weeks. Diagnosis is based on clinical findings and a history of contact with cats. Treatment with the antibiotics tetracycline or doxycycline is effective against *Bartonella (Rochalimaea)*-caused cases, whereas *Afipia*-caused cases generally do not respond, making only symptomatic relief possible. No vaccine is available, and the only means of prevention is to avoid contact with cats.

Rat Bite Fever

One type of **rat bite fever** is caused by *Streptobacillus moniliformis,* which is present in the nose and throat of about half of all wild and laboratory rats. However, only about 10 percent of people bitten by rats develop rat bite fever. Most cases result from bites of wild rats; half are reported in children under 12 living in overcrowded, unsanitary conditions. It can also result from bites and scratches of mice, squirrels, dogs, and cats.

Rat bite fever begins as a localized inflammation at a bite site that heals promptly. In 1 to 3 days, headache begins and new lesions appear elsewhere, especially on palms and soles. Fever is intermittent. By the distribution and appearance of the rash, the disease sometimes is mistaken for Rocky Mountain spotted fever. An arthritis that develops can permanently damage joints.

Another form of rat bite fever, **spirillar fever,** is caused by *Spirillum minor.* First described in Japan as *sodoku* and now known to exist worldwide, this disease is still poorly understood. The initial bite heals easily, but 7 to 21 days later it flares up and occasionally forms an open ulcer. Chills, fever, and inflamed lymph nodes accompany a red or dark purple rash that spreads out from the wound site. After 3 to 5 days symptoms subside, but they can return after a few days, weeks, months, or even years.

Diagnosis of both forms of rat bite fever is made by dark-field examination of *exudates* (oozing fluids). The disease is treated with streptomycin or penicillin; without treatment the mortality rate is about 10 percent. Technicians bitten by rodents should disinfect the bite site, seek medical treatment, and be alert for rat bite fever symptoms.

Arthropod Bites and Diseases

A variety of arthropods, including ticks, mites, and insects, cause human disease either directly or as vectors of pathogens. Many arthropods are ectoparasites that feed on blood. Their bites can be painful, and their toxins can lead to the development of anaphylactic shock. Here we consider the direct effects of bite injuries caused by arthropods; the diseases they transmit are discussed in other chapters.

Tick Paralysis

As ectoparasites, ticks attach themselves to the skin of a host, where they cause both local and systemic effects. The local effect is a mild inflammation at the bite site. Systemic effects usually occur when any of several species of hard-bodied ticks attach to the back of the neck, near the base of the skull, and feed for several days. This allows deep diffusion of anticoagulants and toxins secreted into the bite via the tick's saliva. The anticoagulant prevents the host's blood from clotting while the tick feeds on it. The toxins can cause **tick paralysis,** especially in children. Although the exact chemical composition of these toxins is not known, they appear to be produced by the ovaries of the ticks. They cause fever and paralysis, which affects first the limbs and eventually respiration, speech, and swallowing. Removal of the ticks when symptoms first appear prevents permanent damage. Failure to remove ticks can lead to death by cardiac or respiratory arrest. Both humans and livestock can be affected.

Chigger Dermatitis

The term *mites* refers not to a particular species but to an assortment of species. As ectoparasites, adult mites attach to a host long enough to obtain a blood meal and then usually drop off. Chiggers, the larvae of certain species of *Trombicula* mites, burrow into the skin and release proteolytic enzymes that cause host tissue at the bite site to harden into a tube. The chiggers then insert mouthparts into the tube and feed on blood. They cause itching and inflammation in most people and can cause a violent allergic reaction called **chigger dermatitis** in sensitive individuals. Attempts to "smother" a chigger in its tube by painting nail polish, for example, over the itchy bite are misguided. Chiggers are especially prevalent along the southeastern coast of the United States. Unlike some South American chiggers, those found in the United States cease feeding and drop to the ground hours before the itching begins.

Scabies and House Dust Allergy

Scabies, or **sarcoptic mange,** is caused by the itch mite *Sarcoptes scabiei.* By the time intense itching appears, the lesions usually are quite widespread. Scratching lesions and causing them to bleed provides an opportunity for secondary bacterial infections. Scabies is spread by close human contact and can be transmitted by sexual activities. Outbreaks are especially problematic in hospitals and nursing homes. Disinfection of linens and strict isolation are necessary to prevent spread of the infestation.

House dust mites are ubiquitous and present another problem. We all inhale airborne mites or their excrement. Although this phenomenon is unappealing, it causes disease only in individuals with an allergy to house dust. Two other mites are human commensals: *Demodex folliculorum* lives in hair follicles, and *D. brevis* in sebaceous glands. The incidence of these mites in humans increases with age from 20 percent in young adults to nearly 100 percent in the elderly.

Flea Bites

The sand flea, *Tunga penetrans,* is also sometimes called a chigger because it burrows into the skin and lays its eggs. Sand fleas cause extreme itching, inflammation, and pain and provide sites for secondary infections, including infection with tetanus spores. Surgical removal and sterilization of the wound is used to treat sand flea infections. Fleas in homes and on pets are controlled by insecticides with residual effects for days to weeks after application. Wearing shoes and avoiding sandy beaches helps protect against sand fleas. Although all fleas encountered by humans are a nuisance, only those that carry infectious agents are a public health hazard.

Pediculosis

Common expressions such as "nit-picking," "going over with a fine-toothed comb," "lousy," and "nitty-gritty" attest to association of lice with humans over the ages. To stay alive, lice must remain on their hosts for all but very short periods of their life. They glue their eggs (nits) to fibers of clothing and hair. Two varieties of the louse *Pediculus humanus* parasitize hu-

mans. One lives mainly on the body and on clothing in temperate climates where clothing usually is worn; the other lives on hair in any climate. **Pediculosis,** or lice infestation, results in reddened areas at bites, dermatitis, and itching. A lymph exudate from bites provides an ideal medium for secondary fungal infections, especially in the hair. The crab louse, *Phthirus pubis* (Figure 12.19b), clings to skin more tightly than does the body louse and causes intense itching at bites, especially in the pubic area. It is transmitted between humans by close physical contact, but the louse itself is not known to transmit other diseases. Insecticides can be used to eradicate lice, but sanitary conditions and good personal hygiene must be maintained to prevent their return.

Other Insect Bites

Blackflies inflict vicious wounds, and sensitive individuals who are bitten get **blackfly fever,** characterized by an inflammatory reaction, nausea, and headache. Bloodsucking flies, such as the tsetse fly and the deer fly, are related to horse flies. All inflict painful bites and sometimes cause anemia in domestic animals.

Myiasis (mi-i'a-sis) is an infection caused by maggots (fly larvae). Wild and domestic animals are susceptible to myiasis in wounds. Human myiasis can occur when larvae of the botfly, a greenish, metallic-looking fly, penetrate mucous membranes and small wounds. The larvae of more than 20 species of flies, including the common housefly, can inhabit the human intestine. One such species is the botfly, aptly named *Gastrophilus intestinalis.* The Congo floor maggot, the only known bloodsucking larva, sucks human blood. In the daytime it hides in soil and debris or beneath dirt floors; at night it seeks blood meals from hosts that are usually asleep.

Screwworm fly larvae enter cattle via wounds or openings such as the ears (Figure 20.17). The maggots tunnel beneath the animals' skin, causing great pain and opening avenues for infection. One steer may have over 100 maggots tunneling in its head, and the poor beast will run frantically, barely stopping to eat or drink. Ranchers lost billions of dollars before the screwworm was eradicated in North America (see the box "Screwworm Eradication"). It continues to plague other parts of the world today.

Many mosquitoes, bedbugs, and bloodsucking insects feed on humans and leave painful, itchy bites. Bedbugs hide in crevices during the day and come out at night to feed on their sleeping victims. Inflammation at bites results from an allergic reaction to the bug's saliva. Large numbers of bites can lead to anemia, especially in children. Mosquitoes and many other insects can be eradicated by the application of insecti-

FIGURE 20.17 A screwworm fly larva (*Cochliomyia hominivorax*), or maggot.

cides that have long-term residual effects. They can be kept out of dwellings by the use of tight housing construction and solid roofs instead of thatch, and by household cleanliness.

Wound and bite infections are summarized in Table 20.4.

CLOSE-UP

SCREWWORM ERADICATION

An ingenious method of controlling screwworm maggots, which do great harm to cattle, was developed in the 1930s by Edward Knipling of the U.S. Department of Agriculture. He captured and irradiated male screwworm flies with radioactive cobalt, making them incapable of fertilizing eggs. If females, which mate only once, mate with sterile males, they lay infertile eggs, so far fewer new flies are produced. Irradiated males must be released frequently to maintain a high percentage of sterile males in the population. Another control method uses insects, especially wasp larvae, that are parasitic on other insects. Farmers sometimes purposely release parasitic wasps to infect crop-destroying insects and thereby reduce crop damage. Both the use of sterilized males and release of parasitic insects utilize natural habits of insects; neither contributes to environmental pollution, as most pesticides do.

TABLE 20.4	Summary of Wound and Bite Infections	
Disease	**Agent**	**Characteristics**
Wound Infections		
Gas gangrene	*Clostridium perfringens* and other species	Deep wound infections with gas production in anaerobic tissue; tissue necrosis and death can result if not treated promptly
Cat scratch fever	*Afipia felis* and *Bartonella* (*Rochalimaea*) *henselae*	Pustules at scratch site, fever, and conjunctivitis
Rat bite fever	*Streptobacillus moniliformis*	Inflammation at bite site, dissemination of lesions, intermittent fever
Spirillar fever	*Spirillum minor*	Inflammation at rat bite site heals; later reinflammation, fever, and rash
Athropod Bites and Infections		
Tick paralysis	Various ticks	Toxins introduced with tick bite cause fever and ascending motor paralysis
Chigger dermatitis	*Trombicula* mites	Larvae burrow into skin and cause itching and inflammation; can cause violent allergic reactions
Scabies	*Sarcoptes scabiei*	Widespread lesions with intense itching; other mites cause house dust allergy when their feces are inhaled by allergic individuals
Flea bites	*Tunga penetrans*	Itching and inflammation from adult females in skin
Pediculosis	*Pediculus humanus*	Inflammation at louse bite sites and itching; *Phthirus pubis* (pubic louse) found in pubic areas
Other Insect Bites		
Blackfly fever	Blackfly	Bites cause severe inflammatory reaction in sensitive individuals
Myiasis	Fly larvae	Maggots infect wounds in animals; congo floor maggot sucks human blood; screwworms injure cattle
Mosquito and other bites	Mosquitoes, some flies, bedbugs	Painful, itchy bites; several insects serve as disease vectors

SMALLPOX–THE ERADICATION OF A HUMAN SCOURGE

In 1980 the World Health Organization (WHO) officially proclaimed that smallpox had been eradicated worldwide. This proclamation marked the end of centuries of sickness and death from smallpox. Thus, health professionals will no longer encounter patients infected with smallpox. However, the disease and its consequences are of historical interest, and the methods used to eradicate it provide an important lesson in the control of infectious diseases.

HISTORY OF THE DISEASE

Smallpox first appeared sometime after 10,000 B.C. in a small agricultural settlement in Asia or Africa. The mummy of Ramses V, who died in Egypt in 1160 B.C., has smallpox scars on the face, neck, shoulders, and arms. Smallpox ravaged villages in India and China for centuries, and an epidemic occurred in Syria in A.D. 302. The Persian physician Rhazez clearly described the disease in A.D. 900.

Crusaders brought smallpox to Europe in the twelfth century. The Spaniards carried the virus to the West Indies in 1507, and Cortez's army introduced it into Mexico in 1520. In each instance, the disease was introduced into a population that lacked immunity, and its spread was rampant. As many as 3.5 million Native Americans may have died from smallpox, and by the eighteenth century more than half the inhabitants of Boston had become infected. Slave traders introduced smallpox into Central Africa in the sixteenth and seventeenth centuries, and immigrants from India brought it to South Africa in 1713. The disease reached Australia in 1789.

HISTORY OF IMMUNIZATION

The idea of immunization itself originated in Asia with the technique of variolation to prevent smallpox. In variolation, threads saturated with fluid from smallpox lesions were introduced into a scratch or dangled in the sleeve of a nonimmune person. This practice immunized some people, but it also started epidemics because live, virulent virus was used. Lady Mary Wortley Montague, wife of the British Ambassador to Turkey, introduced the practice to England in 1717. Despite high morbidity (illness) and even mortality (death) from this technique, General George Washington ordered all his troops variolated in 1777. By 1792, 97 percent of the population of Boston had been variolated.

Smallpox immunization was greatly improved by the English physician Edward Jenner, who noticed that milkmaids with cowpox scars never became infected with smallpox. It is now known that immunity to the less severe cowpox confers immunity to smallpox as well. In 1796 Jenner inoculated an 8-year-old boy with cowpox and 6 weeks later inoculated him with smallpox. As Jenner had expected, the boy remained healthy. In the United States in 1799, the physician Benjamin Waterhouse introduced the cowpox vaccine by vaccinating his own children and then exposing them to smallpox.

Although it was well known by then that smallpox could be prevented by vaccination with cowpoxvirus, in 1947 an unimmunized traveler from Mexico infected 12 people in New York City, and 2 of them died. The last 8 cases of smallpox in the United States occurred in the Rio Grande Valley in 1949.

Smallpox was still endemic in 33 countries in 1967 when the WHO established its immunization campaign; in 1977 only a single natural case of smallpox occurred worldwide (Figure 20.18). When smallpox viruses escaped via air ducts from a laboratory in Birmingham, England, in 1978, an unimmunized medical photographer became infected and died. Her mother

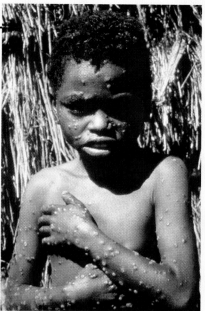

FIGURE 20.18 One of the last cases of smallpox. Lesions are most numerous on the face and arms. As the disease progresses, the blisterlike lesions become opaque and eventually form crusts that fall off. Scarring is common.

suffered a mild case of the disease and recovered. The director of the laboratory from which the virus escaped committed suicide. After this disaster, smallpox stock cultures have been maintained in only two maximum containment laboratories—one in the Atlanta division of the CDC and the other in the Research Institute for Viral Preparations Moscow. However, smallpox surveillance continues and is especially rigorous in West Africa from Zaire to Sierra Leone because of the high incidence there of a similar disease, monkeypox (Figure 20.19).

THE DISEASE; DIAGNOSIS AND IMMUNITY

The smallpox virus enters the throat and respiratory tract. During a 12-day incubation period, it infects phagocytic cells and later blood cells. The infection spreads to skin cells, causing pus-filled vesicles. Acute systemic symptoms begin with fever, backache, and headache. Vesicles appear first in the mouth and throat. They then rapidly spread to the face, forearms, hands, and finally the trunk and legs. Vesicles become opaque and pustular and encrust within about 2 weeks. Death is most likely to occur 10 to 16 days after onset of the first symptoms.

FIGURE 20.19 A child with the skin lesions of monkeypox.

Scrapings from lesions are used to differentiate smallpox from leukemia, chickenpox, and syphilis. In a nonimmune population, smallpox is highly contagious; in a largely immunized population, it spreads very slowly. But now that smallpox has been eradicated worldwide, complications of vaccination are far more life-threatening than the disease itself.

THE FINAL DESTRUCTION
Public health authorities around the world have concluded that the major danger from smallpox now exists in the chance of another accidental escape from a laboratory. Therefore, it was proposed that in December 1993, officials in both Atlanta and Moscow would place all smallpox stocks into a metal container that would be autoclaved at 130°C for 45 minutes. Smallpox would presumably have been eradicated from our planet.

Some scientists, however, have expressed concern that valuable knowledge would be lost with the destruction of this very interesting and potentially useful virus. It is one of the largest viruses known and has a very large double-stranded DNA genome that, unlike most viruses, is able to replicate in the cytoplasm rather than in the nucleus of a host cell. To ensure that we will be able to study this virus

in the future, laboratories in the United States and Russia are currently sequencing the bases in the genomes of at least four strains of the virus. More work remains to be done, and the December 1993 date for the destruction was abandoned, as was a subsequent June 1995 date. It remains to be seen if this virus will ever be destroyed.

Scientists are concerned that merely knowing the genetic sequence of an organism is not enough. The organism must be seen in action, with all its parts interacting, before we can fully understand how its genome *functions*. These researchers feel that it is necessary to keep some smallpox virus intact. Other researchers disagree, maintaining that the ability of the smallpox virus to rapidly infect vast quantities of tissue makes it an ideal candidate for carrying desired genes into humans as part of genetic engineering techniques. First, however, the harmful parts of the viral genome must be identified and removed—leaving a partially empty virus "shell" into which desirable genes can be loaded for transfer.

In the unlikely event that smallpox should someday recur, sufficient vaccine to immunize 300 million people is stockpiled. Fortunately, as with rabies, vaccination even after exposure can prevent development of the disease.

CHAPTER SUMMARY

- The agents and characteristics of the diseases discussed in this chapter are summarized in Tables 20.2, 20.3, and 20.4. Information in those tables is not repeated in this summary.

DISEASES OF THE SKIN

Bacterial Skin Diseases

- Bacterial skin diseases are usually transmitted by direct contact, droplets, or fomites.

- Most such diseases (including **scalded skin syndrome, scarlet fever, erysipelas,** and **pyoderma**) can be treated with penicillin or other antibiotics. Treating **acne** with antibiotics may contribute to the development of antibiotic-resistant organisms. Burn infections often are nosocomial and caused by antibiotic-resistant organisms.

Viral Skin Diseases

- **Rubella (German measles), measles (rubeola),** and **chickenpox** viruses are usually transmitted via nasal secretions. Pox diseases (including **smallpox, cowpox,** and **molluscum contagiosum**) and **warts (papillomas)** are usually transmitted by direct contact.

- Treatment of viral skin diseases can usually relieve some symptoms but not cure the disease; warts can be excised or treated chemically.

Fungal Skin Diseases

- Fungal skin diseases are called **dermatomycoses.** They include various types of **ringworm,** including **athlete's foot (tinea pedis); sporotrichosis; blastomycosis; candidiasis (moniliasis); aspergillosis;** and **zygomycoses.**

- Subcutaneous fungal infections often persist despite treatment with topical fungicides. Systemic infections are even more difficult to eradicate.
- Because of the difficulty in treating fungal skin infections, prevention by maintaining healthy skin and avoiding contamination of wounds is especially important.

Other Skin Diseases

- Other skin diseases include those caused by both fungi and bacteria (such as **Madura foot**) and by parasitic helminths (**swimmer's itch** and **dracunculiasis**).
- Table 20.2 summarizes diseases of the skin.

DISEASES OF THE EYES

Bacterial Eye Diseases

- Bacterial eye diseases include **ophthalmia neonatorum, bacterial conjunctivitis** (pinkeye), and **trachoma.** They are transmitted by direct contact, by fomites, by insect vectors, and (in certain infections) by contamination of infants during delivery.
- Most bacterial eye diseases are treated with antibiotics and prevented by good sanitation.

Viral Eye Diseases

- Viral eye diseases, such as **epidemic keratoconjunctivitis** and **acute hemorrhagic conjunctivitis,** are transmitted by dust particles or direct contact. No effective treatments are available, but good sanitation can help prevent such diseases.

Parasitic Eye Diseases

- Controlling insect vectors reduces transmission of parasitic helminthic eye diseases, which include **onchocerciasis (river blindness)** and **loaiasis.**
- Table 20.3 summarizes diseases of the eyes.

WOUNDS AND BITES

Wound Infections

- **Gas gangrene** and other anaerobic infections can be prevented by careful cleaning and draining of deep wounds; they are treated with penicillin, antitoxins, and sometimes with hyperbaric oxygen.

Other Anaerobic Infections

- **Cat scratch** and **rat bite fevers** can be prevented by avoiding injuries from cats and rats, respectively. Laboratory workers should be aware of dangers of infections from rat bites.

Arthropod Bites and Diseases

- Arthropod bites cause diseases such as **tick paralysis, chigger dermatitis, scabies (sarcoptic mange), pediculosis,** and **myiasis.** Such diseases can be prevented by good sanitation and hygiene and protecting the skin from bites.
- Table 20.4 summarizes infections that result from wounds and bites.

KEY TERMS

abscess (p. 554)
acne (p. 557)
acute hemorrhagic conjunctivitis (p. 569)
aspergillosis (p. 565)
athlete's foot (p. 564)
bacterial conjunctivitis (p. 567)
blackfly fever (p. 573)
blastomycetic dermatitis (p. 564)
blastomycosis (p. 564)
boil (p. 554)
candidiasis (p. 565)
carbuncle (p. 555)
cat scratch fever (p. 572)
chickenpox (p. 560)
chigger dermatitis (p. 573)
congenital rubella syndrome (p. 558)
cowpox (p. 562)
crepitant tissue (p. 571)
débridement (p. 557)
dermal wart (p. 562)
dermatomycosis (p. 563)

dermatophyte (p. 563)
dracunculiasis (p. 566)
epidemic keratoconjunctivitis (p. 569)
erysipelas (p. 556)
eschar (p. 557)
exanthema (p. 558)
folliculitis (p. 554)
furuncle (p. 554)
gas gangrene (p. 571)
genital wart (p. 562)
German measles (p. 558)
human papillomavirus (p. 562)
impetigo (p. 556)
keratitis (p. 566)
Koplik's spot (p. 559)
loaiasis (p. 570)
Madura foot (p. 566)
maduromycosis (p. 566)
measles (p. 558)
measles encephalitis (p. 559)
molluscum contagiosum (p. 562)

moniliasis (p. 565)
myiasis (p. 573)
onchocerciasis (p. 569)
ophthalmia neonatorum (p. 566)
papilloma (p. 562)
pediculosis (p. 573)
pimple (p. 554)
pustule (p. 554)
pyoderma (p. 556)
rat bite fever (p. 572)
ringworm (p. 563)
river blindness (p. 569)
rubella (p. 558)
rubeola (p. 558)
sarcoptic mange (p. 573)
scabies (p. 573)
scalded skin syndrome (p. 555)
scarlet fever (p. 556)
shingles (p. 560)
smallpox (p. 562)
spirillar fever (p. 572)
sporotrichosis (p. 564)
St. Anthony's fire (p. 556)

sty (p. 554)
subacute sclerosing panencephalitis (SSPE) (p. 559)
swimmer's itch (p. 566)
systemic blastomycosis (p. 564)
thrush (p. 565)
tick paralysis (p. 572)
tinea barbae (p. 564)
tinea capitis (p. 564)
tinea corporis (p. 563)
tinea cruris (p. 563)
tinea pedis (p. 564)
tinea unguium (p. 564)
trachoma (p. 568)
vaginitis (p. 565)
varicella-zoster virus (p. 560)
wart (p. 562)
zygomycosis (p. 565)

A, B Skin Diseases

1. What causes folliculitis, furuncles, and carbuncles, and how are they treated?

2. How does scarlet fever differ from strep throat, and how can a person have it more than once?

3. What are the causes and effects of erysipelas and scalded skin syndrome?

4. How are pyoderma and impetigo spread?

5. Match the following:

   ____ scarlet fever
   ____ scalded skin syndrome
   ____ impetigo
   ____ erysipelas
   ____ carbuncle
   ____ acne

   a. *Streptococcus*
   b. *Staphylococcus*
   c. both (a) and (b)
   d. neither (a) nor (b)

6. How are microorganisms involved in acne?

7. What organisms are frequently found in burn infections?

8. What are the major problems in diagnosis, treatment, and prevention of burn infections?

9. In what ways do rubella and measles (rubeola) differ? Name some possible complications of these diseases.

10. What are the similarities between chickenpox and shingles?

11. What pox diseases are seen in humans?

12. How are warts acquired, and what can be done to get rid of them?

13. Match the following:

    ____ German measles
    ____ measles
    ____ chickenpox
    ____ Koplik's spots
    ____ warts
    ____ shingles
    ____ cowpox
    ____ SSPE

    a. varicella
    b. vaccinia virus
    c. rubeola
    d. papillomavirus
    e. zoster
    f. rubella

14. What organisms cause dermatomycoses, and how are they treated?

15. Name the type of ringworm:

    a. tinea _____ (body ringworm)
    b. tinea _____ (ringworm of the nails)
    c. tinea _____ (barber's itch)
    d. tinea _____ (jock itch, or groin ringworm)
    e. tinea _____ (scalp ringworm)

f. What type of ringworm is likely to cause the lesions shown in the accompanying photograph?

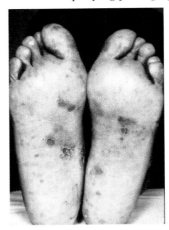

16. How can systemic fungal diseases be treated and prevented?

17. Under what conditions do opportunistic fungal infections occur, and how can they be prevented?

18. What are the causes of Madura foot, and how can it be prevented?

19. How are swimmer's itch and dracunculiasis similar, and how are they different?

20. Match the following:

    ____ trachoma
    ____ river blindness
    ____ thrush
    ____ swimmer's itch
    ____ eyeworm usually *not* causing blindness
    ____ gas gangrene
    ____ sodoku
    ____ pediculosis
    ____ myiasis

    a. *Clostridium*
    b. *Spirillum*
    c. maggots
    d. schistosomes
    e. *Chlamydia*
    f. lice
    g. *Loa loa*
    h. *Candida*
    i. onchocerciasis

C, D Eye Diseases

21. How does conjunctivitis differ from keratitis?

22. Which bacteria can infect the eyes, and how do they reach the eyes?

23. What causes ophthalmia neonatorum, and under what circumstances does it occur?

24. How is bacterial conjunctivitis spread, and how is it treated?

25. What causes trachoma, and how is it prevented?

26. How does bacterial conjunctivitis differ from trachoma?

27. What kinds of eye diseases are caused by viruses, and how can they be prevented?

28. How are parasitic eye diseases acquired, and how can they be prevented?

E,F Wound and Bite Infections

29. What causes gas gangrene? How does the disease progress, and how is it treated and prevented?

30. What other anaerobic organisms cause wound infections?

31. What is cat scratch fever?

32. What is rat bite fever, and why is it a hazard to laboratory workers?

33. What are the major effects of arthropod bites and infections? How can they be prevented?

PROBLEMS FOR INVESTIGATION

1. Study the problem of blindness resulting from infectious diseases, and devise a scheme for reducing its incidence.

2. How can pathogens such as *Acanthamoeba* be detected in the eyewash stations in your science lab?

3. Why does *Candida albicans* sometimes cause disease when a person takes large amounts of antibiotics?

4. List and describe the skin diseases seen in AIDS patients.

5. Prepare a report on how the burn unit in your area hospital attempts to prevent burn infection. What is their success rate?

6. A patient who recently arrived from Africa came to the emergency room late one night complaining of vision problems. Imagine the surprise of the resident on duty who, upon peering through the pupil of the patient's eye, saw a slender, 2.5-cm-long worm moving vigorously inside the eye. Slides made from peripheral blood showed numerous microfilariae. What was the diagnosis? (The answer to this question appears in Appendix F.)

SOME INTERESTING READING

Centers for Disease Control. 1986. "*Acanthamoeba* keratitis associated with contact lenses—United States." *Morbidity and Mortality Weekly Report* 35, no. 25 (June 27):405.

Croen, K. D., and S. E. Straus. 1991. "Varicella-zoster virus latency." *Annual Review of Microbiology* 45:265–82.

Crum, C. P., et al. 1991. "Pathobiology of papillomavirus-related cervical diseases." *Clinical Microbiology Reviews* 4:270–85.

Edwards, J. E., Jr. 1991. "Invasive *Candida* infection: Evolution of a fungal pathogen." *New England Journal of Medicine* 324, no. 15 (April 11):1060–62.

Ellner, P. D., and H. C. Neu. 1992. *Understanding infectious disease.* St. Louis: Mosby.

Fenner, F. 1982. "A successful eradication campaign: Global eradication of smallpox." *Reviews of Infectious Disease* 4(5):916–22.

Fleiszig, S. M. J., and N. Efron. 1992. "Microbial flora in eyes of current and former contact lens wearers." *Journal of Clinical Microbiology* 30(5):1156–61.

Gradon, J. D., J. G. Timpone, and S. M. Schnittman. 1992. "Emergence of unusual opportunistic pathogens in AIDS: A review." *Clinical Infectious Disease* 15(1):134–57.

Hart, C. A. 1992. *Color atlas of pediatric infectious disease.* St. Louis: Mosby.

Kilvington, S., et al. 1990. "Laboratory investigation of *Acanthamoeba* keratitis." *Journal of Clinical Microbiology* 28 (12):2722–25.

Kwon-Chung, K. J., and J. E. Bennet. 1992. *Medical mycology.* Philadelphia: Lea and Febiger.

McNeil, M. M., and J. M. Brown. 1994. "*The medically important aerobic actinomycetes: Epidemiology and microbiology.*" Clinical Microbiology Reviews 7, no. 3 (July):357–417.

New, A. R. 1993. "Controlling 'yeast' infections." *FDA Consumer* 27 (December):10–13.

Noble, W. C., ed. 1992. *The skin microflora and microbial skin disease.* New York: Cambridge University Press.

Radentz, W. H. 1991. "Fungal skin infections associated with animal contact." *American Family Physician* 43(April): 1253–56.

Stehlin, I. B. 1995. "First vaccine for chickenpox." *FDA Consumer* 29 (September):6–10.

Strickland, G. T. 1991. *Hunter's tropical medicine,* 7th ed. Philadelphia: W. B. Saunders.

Weitzman, I., and R. C. Summerbell. 1995. "The dermatophytes." *Clinical Microbiology Reviews* 8, no. 2 (April):240–59.

Wood, D. L., and P. A. Brunell. 1995. "Measles control in the United States: Problems of the past and challenges for the future." *Clinical Microbiology Reviews* 8, no. 2 (April):260–67.

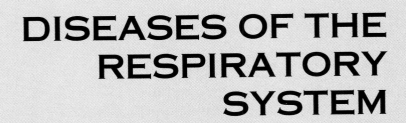

CHAPTER 21

DISEASES OF THE RESPIRATORY SYSTEM

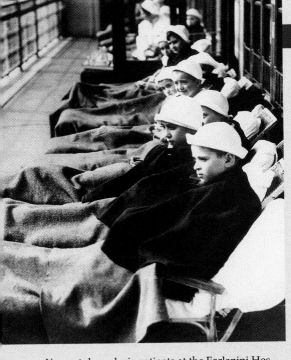

Young tuberculosis patients at the Forlanini Hospital in Rome in 1938. In the United States, the disease was so common that sanitariums were built in most large cities to care for TB patients alone. The disease is on the rise once again.

T he human respiratory system can be infected by various bacteria, viruses, fungi, and at least one helminth. Whether respiratory infections become established depends on host–microbe relationships (Chapter 15) and the condition of the respiratory system and its nonspecific defenses. ∞ (Chapter 17, p. 465) Respiratory infections are divided into the upper respiratory infections, including ear infections, and the lower respiratory infections.

DISEASES OF THE UPPER RESPIRATORY TRACT

Bacterial Upper Respiratory Diseases

Bacterial infections of the *upper respiratory tract* (see Figure 17.14a) are exceedingly common. ∞ (p. 466) They are easily acquired through the inhalation of droplet nuclei from infected persons or carriers, especially in winter, when people are crowded indoors in poorly ventilated areas.

Pharyngitis and Related Infections

Pharyngitis, or sore throat, is an infection of the pharynx. It is frequently caused by a virus but is sometimes bacterial in origin. **Laryngitis** is an infection of the larynx, often with loss of voice. **Epiglottitis,** an infection of the epiglottis, can close the airway and cause suffocation. *Croup,* a viral infection found in children, also involves the larynx and the epiglottis. Difficult breathing, spasms of the larynx, and membrane formation may occur. When an infection reaches the sinus cavities, bronchi, or tonsils, the conditions are referred to as **sinusitis, bronchitis,** and **tonsilitis,** respectively. When an infection spreads to the lungs, it is no longer an upper respiratory disease and is called *pneumonia.*

STREPTOCOCCAL PHARYNGITIS Less than 10 percent of cases of pharyngitis are caused by the group A β-hemolytic *Streptococcus pyogenes.* This infection, familiarly known as *strep throat,* is most common in children 5 to 15 years old. It is acquired by inhaling droplet nuclei from active cases or healthy carriers. Dogs and other family pets also can be carriers. The detection and elimination of carrier states is often difficult in re-

THIS CHAPTER FOCUSES ON THE FOLLOWING QUESTIONS:

A Which bacteria cause upper respiratory infections, and what are the important epidemiologic and clinical aspects of these diseases?

B Which viruses cause upper respiratory infections, and what are the important epidemiologic and clinical aspects of these diseases?

C Which bacteria cause lower respiratory infections, and what are the important epidemiologic and clinical aspects of these diseases?

D Which viruses cause lower respiratory infections, and what are the important epidemiologic and clinical aspects of these diseases?

E Which fungi cause lower respiratory infections, and what are the important epidemiologic and clinical aspects of these diseases?

F Which parasites cause respiratory infections, and what are the important epidemiologic and clinical aspects of these diseases?

current or cluster outbreaks. Contaminated food, milk, and water also can spread the disease, so it is important that infected persons and carriers not handle food. In strep throat, the throat typically becomes inflamed, and the adenoids and lymph nodes in the neck swell. The tonsils become quite tender and develop white, pus-filled lesions (Figure 21.1). Onset is usually abrupt, with chills, headache, acute throat soreness, especially upon swallowing, and often nausea and vomiting. Fever is generally high. The absence of cough and nasal discharge helps distinguish strep throat from the common cold. Strains of *S. pyogenes* infected with a specific toxin-producing temperate phage cause scarlet fever along with pharyngitis. ∞ (Chapter 20, p. 556)

Diagnosis is made by positive throat culture. A rapid enzyme-labeled antibody screening test using a throat swab can be completed in a physician's office within minutes. Immediate treatment is important. If treatment is delayed, *S. pyogenes* can interact with the immune system and give rise to rheumatic fever (Chapter 23), which occurs in 3 percent of untreated cases. For this reason treatment with penicillin or one of its derivatives is often begun even before culture results are available.

LARYNGITIS AND EPIGLOTTITIS Laryngitis can be caused by bacteria such as *Haemophilus influenzae* and *Streptococcus pneumoniae,* by viruses alone, or by a combination of bacteria and viruses. Acute epiglottitis is almost invariably caused by *H. influenzae.* Inflammation of the tissues rapidly closes the airway, causing difficulty in breathing or even death.

SINUSITIS More than half the cases of sinusitis are caused by *Streptococcus pneumoniae* or by *Haemophilus influenzae,* but some cases are caused by *Staphylococcus aureus* or *Streptococcus pyogenes.* Swelling of sinus cavity linings slows or prevents drainage and leads to pressure and severe pain. When drainage is impeded, mucus accumulates and fosters bacterial growth. Secretions consisting of mucus, bacteria, and phagocytic cells then collect in the sinuses. Chronic sinusitis can permanently damage sinus linings and cause *polyps* (smooth, pendulous growths) to form. Applying moist heat over the sinuses, instilling drops of vasoconstrictors, such as ephedrine, into the nasal passages, humidifying the air, and holding the head in a position to promote drainage help alleviate symptoms. Treatment with antibiotics such as penicillin and medica-

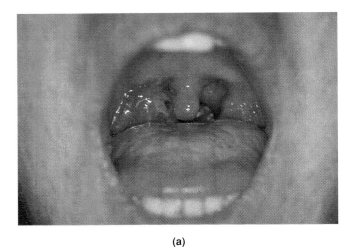

(a)

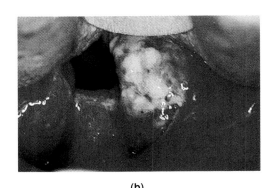

(b)

FIGURE 21.1 Strep throat, a common form of pharyngitis, showing (a) enlarged and reddened adenoids at the sides of the throat and (b) white, pus-filled lesions on tonsils.

tions to lessen discomfort may be necessary. Swimmers and divers often suffer from sinusitis if water is forced into the sinuses. Such people can prevent water from entering their sinuses by using nose clips or by exhaling as they submerge and continuing to exhale while their head is underwater.

BRONCHITIS Bronchitis involves the bronchi and bronchioles but does not extend into the alveoli. About 15 percent of the general population has chronic bronchitis. It is most common in older people and is linked to smoking, air pollution, inhalation of coal dust, cotton lint, and other particles, and heredity. Patients cough up sputum containing mucus, organisms, and phagocytic cells. Common causative agents include *Streptococcus pneumoniae, Mycoplasma pneumoniae,* and various species of *Haemophilus, Streptococcus,* and *Staphylococcus.* Infections can spread to the alveoli of the lungs and cause pneumonia. Diagnosis is made from sputum cultures. By the time an infected individual seeks medical attention, respiratory membranes may have been permanently damaged. Antibiotic treatment can halt further deterioration but cannot reverse the damage already done. Eventually severe shortness of breath develops.

Diphtheria

Although fewer than 10 cases per year are now seen in the United States, **diphtheria** was once a feared killer. A century ago, 30 to 50 percent of patients died; most deaths occurred by suffocation in children under age 4. Diphtheria is still a problem today in the former Soviet Union, where the disease exists in epidemic proportions.

Sequelae, or adverse signs that follow a disease, are common in diphtheria. Myocarditis, an inflammation of the heart muscle, and polyneuritis, an inflammation of several nerves, account for deaths even after apparent recovery. Significant cardiac abnormalities occur in 20 percent of patients. Neurologic problems, including paralysis, can occur after particularly severe cases.

CAUSATIVE AGENT Diphtheria is caused by strains of *Corynebacterium diphtheriae* infected with a prophage that carries an exotoxin-producing gene. ∞ (Chapter 11, p. 284) Club-shaped cells of this gram-positive rod grow side by side in palisades (like the upright logs used to fortify old forts). These cells contain metachromatic granules of phosphates, which turn reddish when stained with methylene blue (Figure 21.2). **Diphtheroids,** are corynebacteria that are found in or on such body sites as the nose, throat, nasopharynx, urinary tract, and skin. These organisms differ from *C. diphtheria* by not being toxin-producers. On inoculation into a guinea pig, toxin-producing *C. diphtheriae* strains cause disease, but diphtheroids do not. Some laboratories also use a gel diffusion test to detect toxin-producing strains.

To produce toxin, the bacterium must be infected by an appropriate strain of bacteriophage in the lysogenic

FIGURE 21.2 *Corynebacterium diphtheriae,* stained with methylene blue, showing metachromatic granules of phosphates that have stained a deep reddish blue.

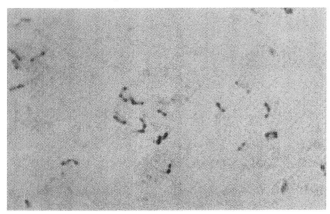

prophage state. ∞ (Chapter 11, p. 284) In other words, the bacteriophage DNA must be integrated into the bacterial chromosome, where its toxin-producing genes are expressed. "Curing" the bacterium of its phage infection eliminates its ability to produce toxin. The phage-infected cell also fails to produce toxin unless the patient's blood iron concentration drops to a critically low level.

THE DISEASE   Humans are the only natural hosts for *C. diphtheriae*. Diphtheria usually is spread by droplets of respiratory secretions. Infection usually begins in the pharynx 2 to 4 days after exposure. The organism, damaged epithelial cells, fibrin, and blood cells combine to form a **pseudomembrane** (Figure 21.3). Although it is not a true membrane, removing it leaves a raw, bleeding surface that is soon covered by another pseudomembrane. The pseudomembrane can block the airway and cause suffocation. The organisms very rarely invade deeper tissues or spread to other sites, but the extremely potent toxin spreads throughout the body and kills cells by interfering with protein synthesis. The heart, kidneys, and nervous system are most susceptible.

Diphtheria organisms are sometimes found in unusual sites. They can invade the nasal cavities, which have relatively few blood vessels. There they cause a milder disease because less toxin enters the blood. They also can invade the skin in *cutaneous diphtheria* and cause tissue injury. A small amount of toxin is generally absorbed by the blood. Cutaneous diphtheria is a tropical disease associated with poor hygiene; it is rare in the United States.

TREATMENT AND PREVENTION   Diphtheria is treated by administering both antitoxin to counteract toxin and antibiotics such as penicillin or erythromycin

FIGURE 21.3   Pseudomembrane of diphtheria. This is not a true membrane; rather, it adheres to the underlying tissue. If torn off, it will leave a raw, bloody surface and will eventually re-form. However, it must be removed when it blocks the airway.

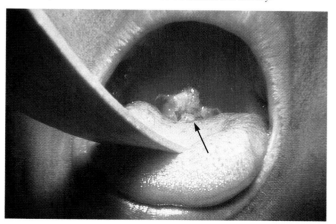

to kill the organisms. The disease can be prevented with DTP (*diphtheria, tetanus, pertussis*) vaccine, which contains diphtheria toxoid, tetanus toxoid, and killed *Bordetella pertussis* and is administered in a series of shots beginning at age 2 months. Boosters are needed throughout childhood and adulthood. In the past, when diphtheria was endemic in the United States, most of the adult population maintained, or even acquired, immunity by continual exposure to small quantities of the bacterium. Such exposures acted as natural booster shots. Now that diphtheria is no longer endemic, immunity depends on the vaccine. Because diphtheria is not well controlled elsewhere in the world, especially in the former Soviet Union, vaccination must be continued in the United States to prevent epidemics should the disease spread from other places. During diphtheria outbreaks, quarantine is a necessary precaution.

Ear Infections

Ear infections commonly occur as **otitis media** in the middle ear and as **otitis externa** in the external auditory canal. Because otitis media usually produces a puslike exudate, it often is referred to as otitis media

CLOSE-UP

DIPHTHERIA AND THE "DESERT FOX"

Erwin Rommel, the German general known as the "Desert Fox," led the North African tank warfare during World War II. Newsreels show him with a handkerchief pressed to his nose—not because of desert dust but because of nasal diphtheria. This chronic disease required that he be flown back to Berlin periodically for treatment, leaving his troops without his brilliant and charismatic leadership. Battles lost and men disheartened because of his absence might have given the Allies the edge they needed to defeat the Nazis in North Africa.

While Rommel was present in Berlin, he joined other generals in a plot to assassinate Hitler. The plot failed, and Rommel's role was discovered. As Gestapo agents were taking him to headquarters, he reportedly died of a heart attack in the car. Whether Rommel was executed, committed suicide, or knowing the game was up, actually succumbed to a pounding heart on the way to Berlin is not known. A public trial of a celebrated war hero for treason would not have been politically expedient. But his heart, weakened by years of exposure to diphtheria toxin, could easily have given out. Had Rommel never contracted diphtheria, who knows how the outcome of World War II might have changed?

with effusion (OME). *Streptococcus pneumoniae, S. pyogenes,* and *Haemophilus influenzae* account for about half of acute cases. Organisms such as species of *Proteus, Klebsiella,* and *Pseudomonas* are often responsible for chronic cases. Otitis externa usually is caused by *Staphylococcus aureus* or *Pseudomonas aeruginosa.* Pseudomonad infections are common in swimmers because the organisms are highly resistant to chlorine.

OME infections arise from the passage of pharyngeal organisms through the Eustachian tube. Fever and earache, which arise from pressure the pus creates in the middle ear, usually are present, but some cases are asymptomatic. The disease is treated with antibiotics, usually penicillin. It is important to continue treatment until all organisms are eradicated to prevent complications. Even after successful therapy, sterile fluid can remain in the middle ear, impairing vibration of the tiny bones of hearing there and decreasing sound transmission. Tubes are sometimes inserted to prevent fluid accumulation and repeated middle ear infections (Figure 21.4). If the impairment occurs during speech development, as it often does, speech can be adversely affected. Some children thought to be inattentive or defiant really cannot hear. As children grow older, the Eu-

stachian tube changes shape and develops an angle that prevents most organisms from reaching the middle ear—much to the relief of the children and their parents.

Viral Upper Respiratory Diseases

The Common Cold

The common cold, or **coryza,** probably causes more misery and loss of work hours than any other infectious disease, but it is not life-threatening. Although exact statistics are not available on the number of infections per year, Americans lose more than 200 million days of work and school per year because of colds. Economically, colds are a bane to employers, who must deal with lost work time, and a boon to manufacturers and retailers of cold remedies. Cold viruses are ubiquitous and present year round, but most infections occur in early fall or early spring. After an incubation period of 2 to 4 days, signs and symptoms such as sneezing, inflammation of mucous membranes, excessive mucus secretion, and airway obstruction appear. Sore throat, malaise, headache, cough, and tracheobronchitis occasionally occur. Pus and blood may be present in nasal secretions. The illness lasts about 1 week. The severity of symptoms is directly correlated with the quantity of viruses released from infected epithelial cells of the upper respiratory tract. Viral cold infections can predispose an individual to secondary bacterial infections such as pneumonia, sinusitis, bronchitis, and otitis media. These infections add time, misery, and additional cost to the original cold. More serious diseases such as whooping cough and respiratory syncytial virus infections can be mistaken early on for colds, so correct diagnosis and subsequent treatment are delayed.

CAUSATIVE AGENTS Different cold viruses predominate during different seasons. In fall and spring, rhinoviruses are in the majority. Parainfluenza virus is present all year but peaks in late summer. In mid-December the coronaviruses appear. Adenoviruses are present at a low level all year. About 200 different viruses can cause colds.

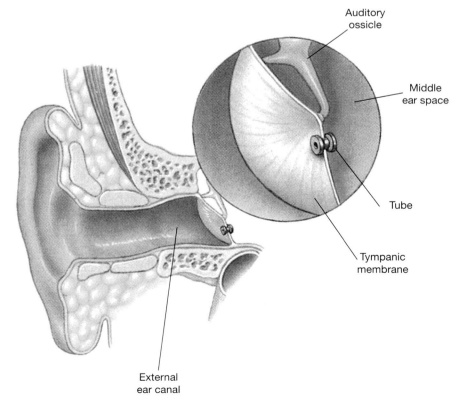

FIGURE 21.4   Tubes to reduce middle ear infections are placed through the tympanic membrane (eardrum) to promote drainage. This can be done in the physician's office. The phlange on the rear end of the tube is designed to hold it in place.

Auditory ossicle

Middle ear space

Tube

Tympanic membrane

External ear canal

Rhinoviruses are the most common cause of colds, accounting for about half of all cases. They are resistant to antibiotics, chemotherapeutics, and disinfectants but are quickly inactivated by acidic conditions. They grow best at 33° to 34°C and thus replicate in the epithelium of the upper respiratory tract, where air movements lower the tissue temperature. Rhinoviruses can be isolated from nasal secretions and throat washings and identified by their sensitivity to low pH and resistance to ether and chloroform, a unique combination of traits among viruses. At least 113 different rhinoviruses have been identified, all with different antigens. Natural immunity is short-lived, and no effective vaccine has been developed. Even if a person becomes immune to some rhinoviruses, there are always others to cause another cold.

The second most common cause of colds is another group of ubiquitous viruses, the **coronaviruses.** Coronaviruses have clublike projections that give them a halo (*corona*, Latin for "halo" or "crown"). The projections are responsible for attaching the virus to host cells as well as stimulating the immune system to produce antibodies against the virus. In addition to causing colds, rhinoviruses also cause acute respiratory distress, in some cases mild pneumonia, and acute gastroenteritis. They infect the epithelium of both the respiratory and digestive tracts. In the digestive tract they reduce absorptive capacity and cause diarrhea, dehydration, and electrolyte imbalances.

TRANSMISSION Cold viruses are spread by fomites and by close contact with infected persons. Blowing the nose and handling used tissues contaminates the fingers, so anything touched becomes contaminated. Conscientiously using tissues once, discarding them immediately in covered containers, and thoroughly washing hands after each nose wipe can significantly reduce the spread of rhinoviruses.

DIAGNOSIS AND TREATMENT Most people diagnose and treat their own colds with remedies that alleviate some symptoms. Over-the-counter antihistamines are reasonably effective in counteracting inflammatory reactions as the body attempts to defend itself against the viruses. Experiments with certain kinds of human interferon have shown potential in blocking or limiting rhinovirus infections, but they must be given with another agent that helps the interferon reach epithelial cells beneath a thick blanket of mucus. A quantity of alpha interferon sufficient to block infection causes irritation of membranes and bleeding. Different combinations of interferons and other factors are being explored to control rhinovirus infections.

Parainfluenza

Parainfluenza is characterized by rhinitis (nasal inflammation), pharyngitis, bronchitis, and sometimes pneumonia, mainly in children. **Parainfluenza viruses** (paramyxoviruses; Chapter 11) initially attack the mucous membranes of the nose and throat. ∞ (p. 278) In very mild cases, symptoms may be inapparent. When symptoms do appear, the first symptoms are cough and hoarseness for 2 or 3 days, harsh breathing sounds, and a red throat. Symptoms can progress to a barking cough and a high-pitched, noisy respiration called *stridor* (stri'dor). Recovery is usually quick—within a few days.

Of the four parainfluenza viruses capable of infecting humans, two can cause croup. **Croup** is defined as any acute obstruction of the larynx and can be caused by a variety of infectious agents, including parainfluenza viruses. Both the larynx and the epiglottis become swollen and inflamed; the high-pitched barking cough of croup results from partial closure of these structures. Increased humidity from a cool-mist vaporizer or from a hot shower helps relieve croup symptoms.

By age 10, most children, whether or not they have had recognizable illness, have antibodies to all four parainfluenza viruses. Thus, the incidence of infection is very high, although the incidence of clinically ap-

---

APPLICATIONS

MANY VERSUS FEW VERSIONS OF A VIRUS

Why have scientists been able to develop a vaccine against polio but still have no success against the common cold? There are only three distinct forms (serotypes) of wild poliovirus. Once a separate vaccine was made against each poliovirus serotype, the job was done. However, there are at least 113 different rhinoviruses, the most common cause of colds. We would need 113 vaccines to attack just the rhinoviruses, and then more for the coronaviruses and adenoviruses that also can cause colds.

Is it just good luck that there are only three, and not 103, types of poliovirus? No, it probably has to do with the route of entry of these pathogens. Polioviruses must be able to survive the acidic conditions of the stomach in order to establish an infection. Mutations in their capsid structure almost always leave them vulnerable to low (acidic) pH. Hence, mutated forms do not succeed. Rhinoviruses are spared this harrowing route. Capsid changes due to mutations allow them to evade the defenses of the upper respiratory tract easily. So some mutated forms do succeed. The result? There are far too many versions of cold viruses for vaccine researchers to keep up with.

parent disease is much lower. Epidemics and smaller outbreaks of parainfluenza infections occur primarily in fall but also sometimes in early spring after the "flu" season. The viruses are spread by direct contact or by large droplets. The causative viruses can be inactivated by drying, increased temperature, and most disinfectants, so they do not remain long on surfaces or in the environment. Resistance to infection comes from secretory IgA that defends mucous membranes against infection, and not from blood-borne IgG. ∞ (Chapter 18, p. 486) Reinfection with parainfluenza viruses is rare, so secretory immunoglobulins must create effective immunity. However, efforts to make a vaccine for parainfluenza viruses have not been successful.

Diseases of the upper respiratory tract are summarized in Table 21.1.

DISEASES OF THE LOWER RESPIRATORY TRACT

Bacterial Lower Respiratory Diseases

Among the bacterial diseases of the lower respiratory tract are two of the great killer infections of history: pneumonia and tuberculosis. The advent of antibiotic therapy brought these diseases under control to a considerable extent. Both are making comebacks today as a result of the spread of AIDS and of treatment with immunosuppressive drugs for transplant patients and with anti-inflammatory agents for autoimmune disorders, such as rheumatoid arthritis and multiple sclerosis. Lowered resistance, overcrowding, chronic diseases, aging, and other immunosuppressive factors also contribute to the severity of the problem.

Whooping Cough

Whooping cough, also called **pertussis,** is a highly contagious disease known only in humans. The word pertussis means "violent cough," and the Chinese call it the "cough of 100 days." Although distributed worldwide, strains found in the United States are less virulent than most. However, jet travel could bring virulent strains from North Africa or other parts of the world at any time. Whooping cough is a major health problem in developing nations, where lack of immunization allows 80 percent of those exposed to contract the disease. In the United States, concern and negative publicity regarding vaccine safety discouraged some parents from having very young children immunized. As a result, incidence of the disease more than doubled during the 1980s and reached nearly 6600 in 1993 (Figure 21.5a). The disease tends to occur sporadically, especially in infants or young children; 50 percent of cases occur in the first year of life (Figure 21.5b).

Before the development of the vaccine, nearly every child got whooping cough. Adults who contract it today either were not vaccinated, or their immunity has declined. Partial immunity lessens the severity of the disease. Many adults may fail to display the characteristic "whooping" sound and so are misdiagnosed. Such people serve as a reservoir to spread the infection.

TABLE 21.1	Summary of Diseases of the Upper Respiratory Tract	
Disease	**Agent(s)**	**Characteristics**
Bacterial Upper Respiratory Diseases		
Pharyngitis	*Streptococcus pyogenes*	Inflammation of the throat; fever without cough or nasal discharge
Laryngitis and epiglottitis	*Haemophilus influenzae, Streptococcus pneumoniae*	Inflammation of the larynx and epiglottis, often with loss of voice
Sinusitis	*H. influenzae, S. pneumoniae, S. pyogenes, Staphylococcus aureus*	Inflammation of sinus cavities, sometimes with severe pain
Bronchitis	*Streptococcus pneumoniae, Mycoplasma pneumoniae,* and others	Inflammation of bronchi and bronchioles with a mucopurulent (mucus- and pus-filled) cough; shortness of breath in chronic cases
Diphtheria	*Corynebacterium diphtheriae*	Inflammation of the pharynx with pseudomembrane and systemic effects of toxin
Otitis externa	*Staphylococcus aureus, Pseudomonas aeruginosa*	Inflammation of external ear canal; common in swimmers
Otitis media	*Streptococcus pneumoniae, S. pyogenes, Haemophilus influenzae*	Pus-filled infection of the middle ear with pressure and pain
Viral Upper Respiratory Diseases		
Common cold	Rhinoviruses, coronaviruses	Sore throat, malaise, headache, and cough
Parainfluenza	Parainfluenza viruses	Nasal inflammation, pharyngitis, bronchitis, croup, sometimes pneumonia

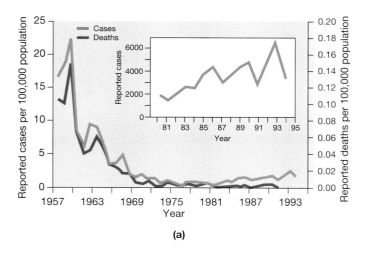

(a)

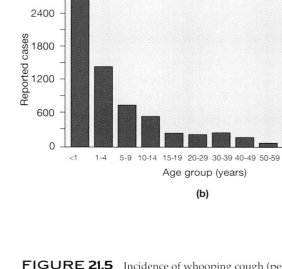

(b)

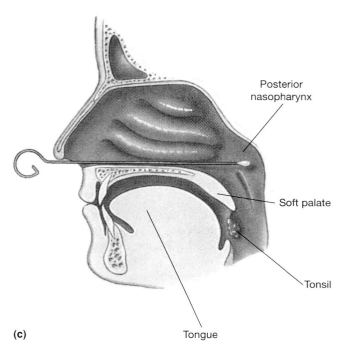

Posterior nasopharynx

Soft palate

Tonsil

(c)

Tongue

FIGURE 21.5 Incidence of whooping cough (pertussis) in the United States (a) by year and (b) by age, 1993. (c) The technique of culturing for whooping cough. A swab attached to a thin, flexible wire is inserted through the nostrils, and the patient is asked to cough several times.

CAUSATIVE AGENT *Bordetella pertussis*, a small, aerobic, encapsulated gram-negative coccobacillus first isolated in 1906, is the usual causative agent of whooping cough. Only about 5 percent of the cases are due to *Bordetella parapertussis* and *B. bronchiseptica*, which usually produce a milder disease. *B. bronchiseptica* is a normal resident of canine respiratory tracts, where it sometimes causes "kennel cough."

Susceptible people become infected by inhaling respiratory droplets. The organisms colonize cilia lining the respiratory tract. Only active cases of whooping cough are known to shed organisms; carriers of the disease are unknown. *Bordetella pertussis* does not invade tissues or enter the blood, but it does produce several substances that contribute to its virulence. It produces an endotoxin, an exotoxin, and *hemagglutinins*

(hem"ah-gloo'ti-ninz), surface antigens that help it attach to the cilia of epithelial cells in the upper respiratory tract. There, the toxins destroy the ciliated epithelial cells. Those cells are then sloughed, leaving a surface of nonciliated cells, which in turn allows mucus to accumulate within the respiratory airway.

THE DISEASE After an incubation period of 7 to 10 days, the disease progresses through three stages: *catarrhal*, *paroxysmal*, and *convalescent*. The **catarrhal stage** is characterized by fever, sneezing, vomiting, and a mild, dry, persistent cough. A week or two later the **paroxysmal** (par-oks-iz'mal), or intensifying, **stage** begins as mucus and masses of bacteria fill the airway and immobilize the cilia. Strong, sticky, ropelike strings of mucus in the airway elicit violent, paroxysmal coughing. Failure to keep the airway open leads to **cyanosis** (si-an-o'sis), or a bluing of the skin, because too little oxygen gets to the blood. Keeping the airway open is especially difficult in infants; airway blockage accounts for the high death rate in patients under 1 year of age. Straining to draw in air gives the characteristic loud "whooping" sound. Coughing sieges occur several times a day and cause exhaustion. Sometimes coughing is so severe that it causes hemorrhage, convulsions, and rib fractures. Vomiting that usually follows a coughing siege leads to dehydration, nutrient deficiency, and electrolyte imbalance—all especially dangerous in infants.

After a paroxysmal stage, usually of 1 to 6 weeks duration but sometimes longer, the patient enters the **convalescent stage.** Milder coughing can continue for several months before it eventually subsides. Secondary infections with other organisms are common at this stage.

DIAGNOSIS AND TREATMENT   Whooping cough is diagnosed by obtaining organisms from the posterior nasal passages (Figure 21.5c) and culturing them on freshly made charcoal–blood agar medium, which has largely replaced the classic potato–blood–glycerol agar (Bordet-Gengou) medium. Once typical colonies appear, fluorescent antibody stain can be used for identification. Whooping cough treatment includes antitoxin given early in the disease to combat toxin, and erythromycin to shorten the paroxysmal, or second, stage. ∞ (Chapter 18, p. 505) The antibiotic cannot entirely eliminate that stage, but it does reduce the number of viable organisms being shed. Supportive measures such as suctioning, rehydration, oxygen therapy, and attention to nutrition and electrolyte balance are very important, as is appropriate treatment of secondary infections.

PREVENTION   Pertussis vaccine has saved many lives, but it is not entirely safe. In the United States it lowered the number of cases from 227,319 in 1938 to

PUBLIC HEALTH

WHOOPING COUGH

In years gone by, it was not uncommon to see nannies in Manhattan's Central Park wearing white uniforms covered with blood from holding infants who suffered paroxysms of coughing. The infants also produced plentiful aerosol droplets containing infectious organisms. Very few younger Americans have seen a case of whooping cough. Ask an older family member or friend what they remember about it or how it was when they had it.

PUBLIC HEALTH

DTP VACCINE LIABILITY

Deaths and injuries from pertussis and other vaccines have resulted in numerous lawsuits against the manufacturers. After several years of deliberations, the National Vaccine Compensation Act was passed by Congress and went into effect January 1, 1988. This law establishes a trust fund and a minimum compensation of $250,000 per affected child. It requires parents who accept compensation to sign away their right to pursue further court action. Surcharges of $4.56 on each DTP injection and $4.44 on each MMR (mumps, measles, rubella) injection go into the trust fund. Unfortunate side effects of the surcharges are that some parents cannot afford to have their children immunized and that public health clinics cannot afford to immunize all the children of eligible low-income parents.

3590 in 1994, but each year vaccine reactions cause 5 to 20 children to die and another 50 to suffer permanent brain damage. Yet the number of children hurt by the vaccine is far smaller than the number who would die without it. Vaccination begins with the DTP series at 2 months of age, as soon as the immune system is capable of responding to antigens in the vaccine. Early immunization is important because pertussis antibodies do not cross the placenta, so infants have no passive immunity to whooping cough. Immunization of unvaccinated children over 7 years of age or adults is not recommended, due to the risks associated with the vaccine. Recombinant DNA technology is being used for the development of acellular vaccines. Because they contain only bacterial proteins, such preparations could reduce or even eliminate the side effects associated with current preparations.

Recovery from the disease confers immunity, but not for a lifetime. Second cases are known, especially in adulthood, but they are usually milder than the first. The United States is fortunate right now in having low-virulence strains of *B. pertussis.* African strains are far more virulent and would probably cause massive outbreaks if they were to become established here.

Classic Pneumonia

In the United States, more deaths result from pneumonia than from any other infectious disease. **Pneumonia,** an inflammation of lung tissue, can be caused by bacteria, viruses, fungi, certain helminths, chemicals, radiation, and some allergies. Infectious forms of the disease develop when pathogens that are able to evade upper respiratory defenses are inhaled. Such organisms initiate the disease process by colonizing the upper respiratory tract and then entering the lower respiratory tract accidentally during a deep breath or suppressed cough, or by means of a large amount of mucus. Several bacteria are known causes of pneumonia, including *Streptococcus pneumoniae,* also known as the pneumococcus; *Staphylococcus aureus; Klebsiella pneumoniae;* and *Chlamydia pneumoniae,* a recently recognized cause of human respiratory infections that is responsible for about 10 percent of community-acquired cases of pneumonia. *Pseudomonas aeruginosa* is another respiratory pathogen. It causes pneumonia in immunocompromised persons and hospitalized patients. Once the organism gains access to damaged respiratory epithelium, it attaches by means of its pili and multiplies (Figure 21.6).

CLASSIFICATION OF PNEUMONIAS   Pneumonias are classified by site of infection as lobar or bronchial. **Lobar pneumonia** affects one or more of the five major lobes of the lungs. It is a serious primary disease that affects about 500,000 people in the United States annually. *Streptococcus pneumoniae* is the most common cause. Fibrin deposits are characteristic of lobar pneumonia. When they solidify, they cause **consolidation,** or blockage of air spaces. **Pleurisy,** inflammation of the pleural membranes that causes painful breathing, often accompanies lobar pneumonia.

   Bronchial pneumonia begins in the bronchi and can spread into the surrounding tissues toward the alveoli in a patchy manner. Although also most likely to be caused by pneumococci, bronchial pneumonia differs from lobar pneumonia in two ways: (1) It often appears as a secondary infection after a primary infection such as a viral influenza or heart disease, or an-

other lung disease, and (2) the plentiful fibrin deposits are not seen in lobar pneumonia. Bronchial pneumonia can follow exposure to chemicals or aspiration of vomitus or other fluids. Infants sometimes aspirate amniotic fluid during birth, especially during delivery by Caesarean section. Bronchial pneumonia is common in elderly and debilitated patients.

TRANSMISSION   Both lobar and bronchial pneumonia are transmitted by respiratory droplets and, in the winter, by carriers—including health care workers—who have had contact with pneumonia patients. As much as 60 percent of a population can be carriers in closely associated groups, such as personnel on military bases or children in preschools.

THE DISEASE   After a few days of mild upper respiratory symptoms, the onset of pneumococcal pneumonia is sudden. The infected person suffers violent chills and high fever (up to 106°F). Chest pain, cough, and sputum containing blood, mucus, and pus follow. Fever may end 5 to 10 days after onset when untreated or within 24 hours after antibiotics are given.

   Pneumococcal pneumonia is the only infectious disease in the top 10 causes of death in the United States. With immediate and adequate antibiotic therapy, the mortality rate is 5 percent; without treatment, it is 30 percent. This compares favorably with *Klebsiella* pneumonia, which despite best treatment still has a mortality rate of 50 percent. *Klebsiella* causes an extremely severe pneumonia that can lead to chronic ulcerative lesions in the lungs and extensive destruction of lung tissue. Failure to seek medical attention promptly keeps the overall mortality rate for pneumonia in the United States at 25 percent. Conditions that predispose toward pneumonia include old age, chilling, drugs, anesthesia, alcoholism, and a variety of disease states.

DIAGNOSIS, TREATMENT, AND PREVENTION
Diagnosis of pneumonia is based on clinical observations, X-rays, or sputum culture. *Klebsiella* pneumonia is usually treated with cephalosporins. Penicillin is the drug of choice for treatment of pneumococcal pneumonia. Following recovery, immunity exists for a few months only against the particular serotype that caused the infection. Thus, a patient can develop case after case of pneumonia from infection with other serotypes or other species. More than 85 serotypes are recognized. Of these, only 23 are responsible for 85 percent of pneumococcal disease in the United States. Artificial immunity can be induced with the polyvalent vaccine Pneumovax. This vaccine contains 23 serotype antigens, so it protects against most pneumococcal pneumonias in all age groups except children under 2 years of age. Immunization is especially recommended for the elderly and at-risk populations.

FIGURE 21.6   Pneumonia-causing bacteria; SEM photo showing *Pseudomonas aeruginosa* (see arrow) attaching to human respiratory epithelium.

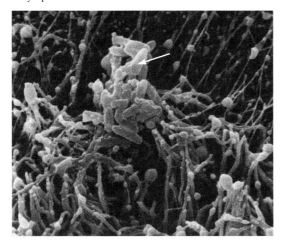

Mycoplasma Pneumonia

One of the tiniest bacterial pathogens known, *Mycoplasma pneumoniae* ordinarily causes mild, and sometimes inapparent, upper respiratory tract infections. In 3 to 10 percent of infections it causes **primary atypical pneumonia,** or *Mycoplasma* pneumonia, usually a mild pneumonia with an insidious onset. The disease is said to be atypical because the symptoms are different from those of classic pneumonia. Patients often remain ambulatory, so the disease is sometimes called **walking pneumonia.** The mortality rate is less than 0.1 percent. It is unusual in preschool children and is most common among young people 5 to 19 years old. *Mycoplasma pneumoniae* may cause up to 20 percent of all noninfluenza community-acquired pneumonias.

Transmission is by respiratory secretions in droplet form, and onset of symptoms follows an incubation period of 12 to 14 days. Fever lasts 8 to 10 days and gradually declines along with the cough and chest pain. Unusual features of *Mycoplasma* pneumonia are that alveoli decrease in size by inward swelling of the alveolar walls and do not fill with fluid.

Diagnosis can be made by isolating *M. pneumoniae* from sputum or from a nasopharyngeal swab. This process takes 2 to 3 weeks due to the slow growth rate of the organisms (Figure 21.7). Serologic tests such as indirect immunofluorescence, latex agglutination, and ELISA are also of value during the early stages of the disease. Commercially available DNA probes are also used and produce results similar to those of culture. However, treatment is usually based on clinical symptoms.

Erythromycin and tetracycline are the drugs of choice. Penicillin has no effect because *Mycoplasma* lack cell walls (the site of penicillin's antibacterial action). Even untreated cases have a favorable prognosis. No vaccine is currently available, so prevention requires avoidance of contact with infected persons and their secretions.

Legionnaires' Disease

In 1976 many war veterans attending a convention in Philadelphia became victims of a mysterious ailment that became known as **Legionnaires' disease.** After 29 deaths and much frantic investigation, the previously unidentified causative organism, *Legionella pneumophila*, was finally isolated (Figure 21.8). Researchers at CDC later found antibodies to *L. pneumophila* in frozen blood samples from unidentified outbreaks that occurred decades ago. One wonders how the organism was overlooked for such a long time. However, it is sufficiently different from previously classified bacteria that a new genus had to be created for it.

Legionella pneumophila is a weakly gram-negative, strictly aerobic bacillus with fastidious nutritional requirements. It does not ferment sugars and has an obscure life cycle. More than 20 *Legionella* species have

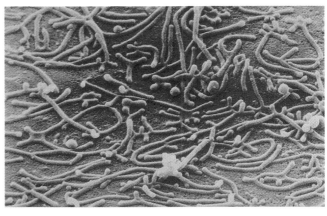

(a)

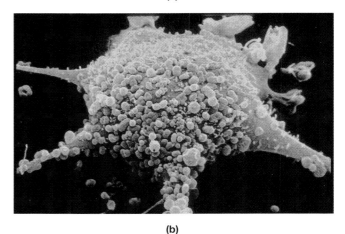

(b)

FIGURE 21.7 (a) *Mycoplasma* cells (151,000×), which lack a cell wall, assume irregular shapes. (b) A cell infected by numerous *Mycoplasma* cells.

been identified. Most are free-living in soil or water and do not ordinarily cause disease. However, some strains live as intracellular parasites of amoebas of many species such as *Acanthamoeba, Naegleria, Hartmanella,* and *Echinamoeba.* Some of these amoebas col-

FIGURE 21.8 Color-enhanced TEM photo of *Legionella pneumophila* (25,000×), the cause of Legionnaires' disease.

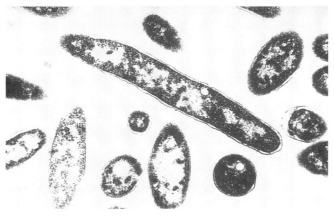

onize cooling towers, shower heads, and other wet areas. Legionellosis is transmitted when organisms growing in soil or water become airborne and enter the patient's lungs as an aerosol. Person-to-person transmission has never been documented. Air conditioners, ornamental fountains, produce sprayers in grocery stores, humidifiers, and vaporizers in patient rooms have been implicated in the spread of the disease. Such devices should be regularly disinfected. Once inhaled, the *Legionella* organisms are taken up into ameoboid phagocytes by a spirally progressing form of phagocytosis. They thrive inside the acidic conditions of the phagolysosome (Chapter 17), multiplying and eventually rupturing the cell. ∞ (p. 454)

After an incubation period of 2 to 10 days, Legionnaires' disease appears with fever, chills, headache, diarrhea, vomiting, fluid in the lungs, pain in the chest and abdomen, and, less frequently, profuse sweating and mental disorders. When death occurs, it is usually due to shock and kidney failure. In nonpneumonic legionellosis, after about 48 hours' incubation, the patient suffers 2 to 5 days of flulike symptoms without infiltration of the lungs.

Pontiac fever is a mild legionellosis named for an outbreak in 1968 affecting 144 people—95 percent of the employees of the Pontiac, Michigan, Health Department. None died; all recovered within 3 to 4 days.

Direct fluorescent antibody tests, ELISA, and commercially available genetic probes are used to diagnose *Legionella* infections. Erythromycin is used to treat them; most other antibiotics have no effect. Because it is difficult to distinguish Legionnaires' disease from other pneumonias, erythromycin is typically given along with a second antibiotic such as penicillin to treat any pneumonia-like disease.

Control of *Legionella*-associated infections depends on maintaining adequate chlorine levels in all potable water sources, cooling towers, and other reservoirs of potable water when not in use. Periodic cleaning of surfaces in air conditioners, humidifiers, and similar equipment is valuable in reducing the incidence of outbreaks, especially in hospitals.

Tuberculosis

Tuberculosis (TB, formerly called *consumption*) has plagued humankind since ancient times, as indicated by skeletal damage in Egyptian mummies and earlier human remains. It remains a massive global health problem today. One and a half billion people (six times the U.S. population) have tuberculosis. Each year 600,000 die, and 3 to 5 million new cases arise.

INCIDENCE IN THE UNITED STATES The incidence of tuberculosis in the general U.S. population decreased by nearly 6 percent per year from 1953, when uniform national reporting began, to 1986. Tem-

porary increases in the 1980s were caused by Asian and Haitian refugees, many of whom became infected in crowded, unsanitary camps and escape boats. Disproportionately large numbers of resident nonwhite groups—blacks, Eskimos, Native Americans, and Hispanics—suffer from tuberculosis, sometimes in epidemic proportions (Figure 21.9). In 1986 the incidence increased by 2.6 percent. Many new cases occurred among AIDS patients; others appeared to represent reactivations of old infections triggered by immunodeficiency, crowding, stress, and use of anti-inflammatory drugs. In 1993 the number of new cases reported per month rose from 778 in January to 5130 in December. The increase continues and people are afraid that tuberculosis may return and grow into an even bigger and more frightening plague than AIDS. Tuberculosis is an airborne disease to which anyone could be exposed, whereas behavior modification can lessen much of one's chances of exposure to the HIV virus.

CAUSATIVE AGENTS The causative agents of tuberculosis are members of the genus *Mycobacterium,* with *M. tuberculosis* causing the vast majority of cases (Table 21.2). *Mycobacterium tuberculosis* was discovered by Robert Koch in 1882, when the disease was called the "White Plague" of Europe. Certain other agents, referred to as atypical mycobacteria, also cause tuberculosis, especially *M. avium-intracellulare* complex (MAC) in AIDS patients. All mycobacteria are straight or slightly curved rods that stain acid-fast, as illustrated in Figure 3.27. ∞ (p. 67)

Certain properties of mycobacteria are closely associated with their role in tuberculosis. Waxes and long-chain mycolic acids in mycobacterial cell walls make mycobacteria difficult to Gram stain, contribute to environmental survival of these organisms, and protect them from some host defenses. Being obligate aerobes sensitive to slight decreases in oxygen concentration, mycobacteria grow best in the apical, or upper portions of the lungs, which are the most highly oxygenated. Pathogenic mycobacteria have an exceedingly long generation time (12 to 18 hours, compared with 20 to 30 minutes for most bacteria), which accounts for the long time (up to 8 weeks) it takes to produce a visible colony on laboratory media. Mycobacteria are highly resistant to drying and can remain viable for 6 to 8 months in dried sputum, a property that contributes to public health problems. They are, however, quite sensitive to direct sunlight.

THE DISEASE Tuberculosis is acquired by the inhalation of droplet nuclei of respiratory secretions or particles of dry sputum containing tubercle bacilli. Young children and elderly people are particularly at risk, so screening school, day-care, and nursing home workers for tuberculosis is important. After organisms are inhaled, they multiply very slowly *inside* white

FIGURE 21.9 Incidence of tuberculosis in the United States (a) by state and (b,c) by age and ethnic background, 1993.

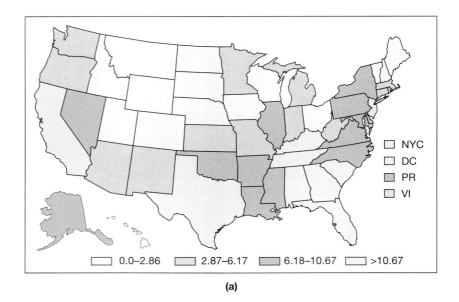

(a)

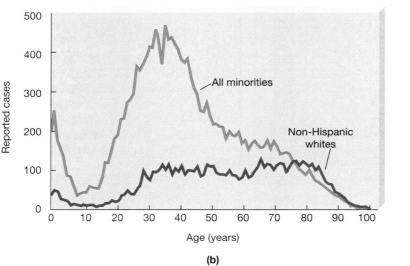

(b)

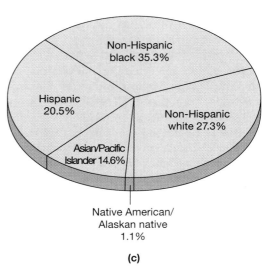

(c)

TABLE 21.2	Mycobacteria that Cause Human Disease
Species	**Disease**
Mycobacterium tuberculosis	Tuberculosis
M. avium-intracellulare (MAC complex)	Tuberculosis-like disease in humans, transmitted from birds and swine
M. bovis	Tuberculosis, transmitted from cattle; could be transmitted from nonhuman primates
M. fortuitum complex	Wound infections
M. kansasii	Tuberculosis-like disease
M. leprae	Hansen's disease (leprosy)
M. marinum	Cutaneous lesions in humans, tuberculosis in fish
M. ulcerans	Ulcerative lesions

blood cells that have phagocytized them. They elicit a host response that includes neutrophil infiltration and fluid accumulation within the alveoli of the lungs. The organisms eventually rupture and destroy the neutrophils. Later, macrophages and lymphocytes move into the area. Alveolar macrophages also phagocytize the living tubercle bacilli, which are again able to multiply within and destroy their new hosts. Rupture of the dead phagocytes releases infective organisms. No toxins are produced. As additional cells are infected, an acute inflammatory response occurs. A large quantity of fluid is released, especially in lung tissue, where it produces pneumonia-like symptoms. Lesions sometimes heal, but more often they produce massive tissue necrosis or solidify to become chronic granulomas, or **tubercles.** Tubercles consist of central accumulations of enlarged macrophages, multinucleate Langerhans giant cells containing tubercle bacilli, peripheral lymphocytes, macrophages, and newly formed connective tissue. The central portion of the tubercle undergoes destruction, typically giving it a cheesy, or **caseous,** appearance. Some organisms gain access to the lymphatic and circulatory systems. About 3 to 4 weeks after exposure, delayed hypersensitivity and cell-mediated immunity develop. On occasion, however, the host fails to respond immunologically. Uncontrolled multiplication of tubercle bacilli then takes place in the lungs, resulting in the formation of numerous tubercles. Organisms are subsequently spread by the circulatory system to other body tissues and organisms. This condition is called **miliary tuberculosis,** named after the small lesions that form that resemble millet seeds. Lesions can be walled off from the rest of the lung by encapsulation when the host has sufficient resistance. Lesions near blood vessels can perforate the vessels and cause hemorrhage, which leads to bloody sputum, a major symptom of tuberculosis.

Granulomas can keep viable organisms walled off for decades. When the immune system becomes impaired by age or other infections, granulomas can open and the disease can be reactivated. People with AIDS often have tuberculosis, either from new infections or reactivation of old ones. Except for initial infections among immigrants, AIDS patients, and residents of certain large cities such as New York and Washington, D.C., most tuberculosis cases in the United States are reactivations rather than first-time infections.

Tuberculosis of the bones can cause extensive erosion, especially in the spine (Figure 21.10). The urogenital tract, meninges, lymphatic system, and peritoneum also are prone to develop extrapulmonary tuberculosis. In **disseminated tuberculosis,** now frequently seen in AIDS patients, infected cells become

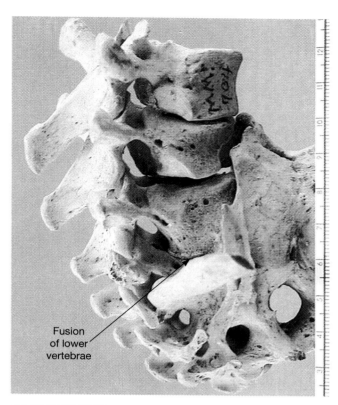

FIGURE 21.10 Tuberculosis of the spinal column. Note that several of the lower vertebrae have fused due to infection damage.

casts: As the tubercle bacillus multiplies in a cell, the cell's organelles and membranes are destroyed, and a clump of microorganisms in the shape, or cast, of the former cell remains. When this process occurs in the intestine, a replica of the intestine consisting of live *Mycobacterium* can be seen.

Although primary tuberculosis is nowadays most likely to affect the lungs, it can also affect the digestive tract. Before milk pasteurization became widespread, primary digestive tract infections were often seen, especially in children. Most were caused by *Mycobacterium bovis* transmitted in raw milk or other dairy products. By law, dairy cattle are now regularly screened for tuberculosis, and infected animals are destroyed. However, some cows escape testing, so drinking raw, unpasteurized milk is a health risk. Today tuberculosis of the digestive tract usually occurs as a secondary infection site, as when pathogen-laden sputum coughed up from a primary lung infection is swallowed.

DIAGNOSIS, TREATMENT, AND PREVENTION

Tuberculosis can be diagnosed by sputum culture, but because the organisms grow very slowly, cultures must be kept for at least 8 weeks before they are declared negative. Genetic probes are also of value in identification of mycobacteria. Chest X-rays fail to reveal lesions outside the lungs and detect only relatively large ones within them. Therefore, screening is now done by

CLOSE-UP

REACTIVATED TB

When cortisone is used to treat the inflamed arthritic joints of aging patients, the drug also exerts an anti-inflammatory action on granulomatous lesions of tuberculosis. By counteracting factors that keep the lesions walled off, the drug can reactivate the tuberculosis organisms. Most elderly tuberculosis patients today were exposed to mycobacteria in their youth, and many developed granulomatous lesions, although their disease was never diagnosed. Bringing relief to their aching joints also can activate quiescent tuberculosis. Those who care for children can unknowingly infect the children. Primary tuberculosis has thus become a problem in day-care centers.

MICROBIOLOGIST'S NOTEBOOK
Multidrug-Resistant Tuberculosis

An interview with:

Paula I. Fujiwara, M.D., MPH—Director of Epidemiology and Surveillance, Assistant Director for Medical Affairs, Bureau of Tuberculosis Control, Department of Health, New York City; Medical Officer, Division of Tuberculosis Elimination, Centers for Disease Control and Prevention, Atlanta, GA

With clinical comments by:

Wilfrid Desir, M.D.—Assistant Professor of Emergency Medicine, New York Medical College at Metropolitan Hospital; Clinical Instructor in Medicine, Columbia University College of Physicians & Surgeons at Harlem Hospital, New York City

Q: How many patients are treated by your department?

Dr. Fujiwara: We treat about 40 percent of the people with tuberculosis in New York City. In 1994 there were 2995 cases in all. There's a difference between tuberculosis infection and tuberculosis disease. This difference is not appreciated by the general public. If you're infected with tuberculosis, your chance of coming down with the actual disease varies based on your immune status. The person who's not immunocompromised has a *lifetime* risk of about 10 percent, but the immunocompromised person has a risk of 7 to 8 percent *a year.* There are causes of immunosuppression besides HIV, such as cancer and chronic steroid administration, but the one that has contributed to the resurgence of tuberculosis has been HIV.

Dr. Desir: I provide direct care to patients in the emergency room. I also teach interns and residents and supervise their work. We see about 100, 120 adult patients per day in the emergency room. That includes medical, surgical, and orthopedic patients. I see tuberculosis in people of all ages. I see chronic cases, new cases, people with HIV, as well as people who have no previous history of any condition but are found on a chest X-ray to have tuberculosis. A significant number of patients don't exhibit any symptoms. Tuberculosis can be silent—although it is a devastating illness.

Q: What is multidrug-resistant tuberculosis?

Dr. Fujiwara: Multidrug-resistant tuberculosis is caused by an organism that's resistant to one or more of the medications used to treat *Mycobacterium tuberculosis*—at least to isoniazid and to rifampin. Although there is tuberculosis that's resistant to one medication, multidrug-resistant tuberculosis is the most serious type of resistance, because isoniazid and rifampin are the two most important medications in short-course tuberculosis chemotherapy. Drug resistance essentially triples the length of time we have to treat.

If you give one drug, such as streptomycin (the first antituberculosis medication—it's still used, by the way), you'll kill off all the organisms that are susceptible to streptomycin. But then you'll leave the streptomycin-resistant strains there. They'll grow and overtake the population. So the basic theory and concept of tuberculosis treatment is to use at least two medications to treat the organism. Then the odds of the patient's developing drug resistance are much less, because the chances of a tuberculosis organism being naturally resistant to two medications are extremely small.

Q: How many types of drug-resistant tuberculosis are there?

Dr. Fujiwara: There are two types, but they're intertwined; the distinctions are blurred. One type occurs if you've developed tuberculosis and you're treated with one medication, the wrong medication—then you select out for the organism that was resistant. This is called acquired resistance. The second type occurs if the person with the resistant organism subsequently coughs and this resistant organism is inhaled by another person—then that person is infected from the start with drug-resistant tuberculosis. This is called primary resistance. So there are two ways of developing resistance to antituberculosis medication.

Q: How did drug resistance develop in the first place?

Dr. Fujiwara: Drug resistance developed because it takes a long time to treat tuberculosis, and people don't take their medicines for the whole period of time. If a person says, "I won't take this pill because I don't like the side effects," or "I have only enough money to buy one kind and not the other kind," then he or she can develop drug resistance.

Q: If they're sick, don't people realize they must take their medication?

Dr. Fujiwara: It's not hard to understand. Let's say you have a bacterial infection in the lung, and your doctor tells you to take your medicines for 10 days. You start to feel good after 5 days, and you don't take the medicine anymore. With tuberculosis, you rapidly feel much better after a week or two of treatment. The incentive to continue taking medicines for a minimum of 6 months is pretty low. Also, some of our patients have very disorganized lives; they have other priorities. If you're homeless you may not have a place to keep your medicines. Or medicines that are given to you can be stolen. This is something that we *know* happens.

The other thing about not taking medications is that the medicines can be very expensive. Tuberculosis medications in many areas across the United States and the world are free. For example, tuberculosis treatment is free in New York City—but people don't know that. One woman was being yelled at by her husband because the medications cost too much. He told her, "Don't take all of them every day. Just take some every day. Take this one today, and take that one tomorrow so they'll last longer." That's probably the worst thing that one could do!

Dr. Paula Fujiwara.

So there are many reasons—lack of access to care, lack of money—why people find it difficult to take their medications. But the main issue is that it's hard to take medications after you feel well and you don't think you need them.

Q: How can health care workers be protected against tuberculosis?

Dr. Fujiwara: The most important protection for a health worker is to make sure the patient is taking treatment. Aside from that, there is what we call the *hierarchy of controls:* 1) administrative, 2) environmental, and 3) personal protective devices. There are administrative methods to prevent transmission; if patients are coughing, then you should consider isolating them. You want to make sure that protocols and procedures are in place to ensure that people who are suspected of having tuberculosis are rapidly identified and evaluated. The other issue is environmental protection—adequate ventilation, use of ultraviolet lights, ... And the third step is the use of personal protective devices—for example, masks called particulate respirators—a plain surgical mask doesn't work.

Dr. Desir: There is a high level of suspicion of tuberculosis in our patient population. We're very aggressive in terms of tuberculosis. In fact, there is a questionnaire, specifically aimed at tuberculosis, that we use on every patient. The triage nurse asks about fever, night sweats, cough, and weight loss. Sometimes we apologize for asking those kinds of questions. We explain that it's nothing personal, and it doesn't mean that the person *has* tuberculosis. Some people are ashamed to mention anything connected with tuberculosis. They see their condition as a social stigma.

Our suspicions are aroused by people with severe weight loss or with a prior history of tuberculosis. Some-

times these individuals had previous admissions and were discharged, and they now come back sick again, so you know who they are. There is also a high level of suspicion for patients who come as referrals from alcohol treatment centers, for patients from drug detoxification programs, or for people who are known to be HIV positive. You assume that such individuals have tuberculosis.

If there is a definite suspicion of tuberculosis, the patient is given a mask and placed in an isolation room where there is a special air filter—even before a chest X-ray is taken. Everyone who comes into that room, including the family, must wear gloves, a gown, and a particulate respirator. Warning signs are posted on the door to the isolation room. There the patients wait until a bed is available—2 hours, 4 hours, sometimes even a day or two.

If tuberculosis is confirmed, one of the first things we do is talk to the family and advise them to be checked. This includes children, grandparents, and anyone else who might be staying with the family . . . or any other close contact. We refer them to a public health station, or they may register in the emergency room, to have a skin test and/or a chest X-ray. The health stations run by the New York City Department of Health are quite efficient in terms of detection and follow-up of tuberculosis.

Patients are usually started on four-drug therapy, such as isoniazid, rifampin, pyrazinamide, and methambutol. But with drug-resistant and HIV-associated tuberculosis on the rise, patients will be started on five medications at the same time. Patients who are not compliant with medication are treated by Directly Observed Therapy (DOT).

Q: What is Directly Observed Therapy?

Dr. Fujiwara: We have responsible health workers, called Public Health Advisors, who directly observe every person taking medication. There have been many studies on adherence to tuberculosis treatment, and there's no relation to education, economic status . . . any factor like that. So we advocate that every patient should be treated using DOT. The Department of Health in New York City, via the Bureau of Tuberculosis Control, runs one of the largest DOT programs in the United States.

Q: How do you get people to come in?

Dr. Fujiwara: We provide enablers and incentives for people to

come in. For many people it's an issue of not having enough money to come to the clinic, so we provide bus fare. We'll also provide food for the people in the clinics. Some of them don't have enough food at home, so they know if they come to the clinic, they'll eat. We also have social workers on site to do eligibility for Medicaid and housing. But for people who cannot, or will not, come in to us, we'll go out to them. We provide DOT in the patient's home . . . workplace . . . on the street corner . . . we've even gone into crack dens. We'll do anything to have the person under DOT.

Q: This must be done every day?

Dr. Fujiwara: In the beginning. For strains that are susceptible to the standard medication, after a certain period of time you can treat twice a week instead of every day. In drug-susceptible tuberculosis (which is the majority), all the medications can be taken at once. One time a day. It's more effective that way. In the case of multidrug-resistant tuberculosis, most medication can be taken all at once, if the person can tolerate it.

Q: Can tuberculosis be cured?

Dr. Fujiwara: Yes. The hard work is getting people to take their medicine. It's both simple and hard. It's simple because the concept is so simple; it's hard because it's difficult to do over a period of time. But there's not going to be a new drug or some magic bullet that's going to make tuberculosis go away. It's just really hard work—following the patient until he or she is fully cured . . . using DOT whenever possible. It's not a question of a better pill. It's a question of being steady, consistent, with your eye on the goal of treatment until cure. You have to have a good, strong control program looking after the aspects of the disease in order to get it under control.

Dr. Wilfrid Desir.

skin tests rather than by X-ray examinations. In a skin test, a small quantity of *purified protein derivative* (PPD), a protein from tuberculosis bacilli, is injected intracutaneously and examined 48 to 72 hours later for an *induration*, or a raised, not necessarily reddened bump. Induration is a delayed hypersensitivity reaction to the PPD. A positive skin test indicates previous exposure and some degree of immune response. It does not indicate that the person now has or has ever had tuberculosis—the immune response may have prevented infection or eliminated or walled off organisms. In fact, health care workers, teachers, and others who have frequent skin tests may eventually develop a positive response due to sensitization to the test material itself.

Treatment is with isoniazid and rifampin for at least 1 year. Many strains of *Mycobacterium*, however (particularly atypical species that are often found in AIDS patients but are rare in healthy individuals), are now resistant to isoniazid. Such strains must be treated with a "second-line" or sometimes even a "third-line" drug. If treatment is successful, sputum cultures become negative for mycobacteria within 3 weeks. Before the development of isoniazid, patients were often sent to tuberculosis sanatoriums, where cold, fresh air was believed to have curative effects. Patients were bundled up and left outdoors for most of the day. In reality, the rest and proper nutrition were what really helped some patients survive.

Tuberculosis can be prevented by vaccination with attenuated organisms in the vaccine BCG, or bacillus of Calmette and Guérin. The French bacteriologists Albert Calmette and Camille Guérin developed the vaccine at the Pasteur Institute in Paris in the early 1900s. BCG immunization is widely practiced in other parts of the world where the prevalence of tuberculosis is high. ∞ (Chapter 18, p. 506) The vaccine's use in the United States is limited to high-risk individuals, such as health care workers regularly exposed to tuberculosis patients, or to personnel specializing in pulmonary and chest diseases.

Psittacosis and Ornithosis

In the early 1900s, **psittacosis** (sit"ah-ko'sis), or *parrot fever,* a respiratory disease associated with psittacine birds, such as parrots and parakeets, was discovered. Today at least 130 different species of birds, including ducks, chickens, and turkeys, are found to carry a form of this disease. Because most of these birds are not psittacines, their disease is referred to as **ornithosis** (from *ornis,* Greek for "bird"). Wild and domestic birds can be infected, often without showing symptoms. However, stresses such as overcrowding, chilling, or shipping to pet shops can activate the disease. Both forms of the disease are caused by *Chlamydia psittaci* and are spread by direct contact, infectious nasal droplets, and feces. Organisms can be found in every organ of an infected bird. The birds have diarrhea and a mucopurulent discharge from the nose and mouth.

Humans usually acquire the disease from birds. Poultry-plant workers are especially susceptible. Human-to-human transmission also has been documented, and medical personnel caring for patients can contract the disease. Organisms are inhaled and spread systemically to the lungs and reticuloendothelial system, especially Küpffer's cells. Most cases of the disease are mild and self-limiting, but some patients develop a serious pneumonia. After an incubation time of 1 to 2 weeks, onset of symptoms is sudden, with sore throat, coughing, difficulty in breathing, headache, fever, and chills. This disease is difficult to distinguish from other pneumonias on a clinical basis. Definitive diagnosis is made by inoculation into tissue culture—but with great care, because the organism is so highly infective.

With tetracycline treatment, the mortality rate from ornithosis pneumonia is about 5 percent; in untreated cases, it is around 20 percent. No vaccine is available, so prevention involves strict quarantine regulations for birds entering the United States.

Q Fever

Q fever was first described in Queensland, Australia. The *Q* stands not for Queensland but for "query," because determining which organism caused it remained a question for a long time. Q fever is now known to be caused by *Coxiella burnetii*, an organism included among the rickettsias. Unlike other rickettsias, this organism survives long periods outside cells and can be transmitted aerially as well as by ticks. For a long time it was a mystery how Q fever organisms can survive

CLOSE-UP

FASHIONABLE TUBERCULOSIS

In the late nineteenth century, highly intellectual and creative individuals were thought to be especially susceptible to tuberculosis. The merits of poets, artists, and musicians were validated by their deaths in unheated attics. Death from tuberculosis was a prominent theme in opera—the heroines of both *La Traviata* and *La Boheme* die of it. Much art of the period shows pale, lethargic women reclining across sofas, with the bright pink cheeks that marked tuberculosis. Indeed, tuberculosis became a fashionable disease and set the standards for beauty. The buxom, healthy, energetic woman was considered less beautiful than the pale, emaciated, languorous woman.

PARROT SMUGGLING AND ORNITHOSIS

Unlike parakeets, many parrots do not breed easily in captivity and so are imported from jungles. The cost of capture, shipping to the United States, and holding in quarantine to detect such diseases as ornithosis is high, so a parrot may cost $3000. To increase profits unscrupulous dealers have stolen and smuggled so many parrots that the parrot population has been decimated in many jungle areas. And most birds don't survive the conditions of smuggling. They are drugged, wrapped so that they cannot flap, and tied underneath automobile fenders, where they are exposed to the dust of unpaved roads and exhaust fumes. Thousands of parrots die this way each year. A few birds survive and reach pet shops without having been quarantined. If they are infected when they reach the hands of a legitimate pet dealer, they can infect the entire stock. All the birds must then be destroyed.

Bird being held at a U.S. Department of Agriculture quarantine station before being allowed to enter the country. The cage's air supply is being filtered to prevent the spread of any airborne diseases that the animal may have.

---

7 to 9 months in wool, 6 months in dried blood, and over 2 years in water or skim milk. The mystery now appears to have been solved. *C. burnetii* has two forms, called large- and small-cell variants (Figure 21.11). Electron microscope studies suggest that a type of endospore may be formed at one end of the large-cell variant, thereby giving resistance uncharacteristic of other rickettsias.

Once *C. burnetii* cells with endospore-like bodies inside them have been inhaled, they are phagocytized by host cells. *Coxiella* grow in phagolysosomes of host cells. There they display the unusual property of responding to the acidic conditions by increasing their metabolism and multiplying rapidly until they almost fill infected cells. Eventually the cells rupture, releasing new groups of pathogenic bacteria. These pathogens, when spread by the bloodstream, then infect other body cells. There appears to be no exit from the human body.

Coxiella burnetii exist all over the world, especially in cattle- and sheep-raising areas. Wild animals and domestic sheep and cattle are the normal hosts of *C. burnetii*. The bacteria are transmitted via tick bites, feces, and genital secretions of infected animals. Humans become infected by inhaling aerosol droplets from infected domestic animals, which usually do not appear ill. Farmers become infected while attending a cow giving birth or miscarrying if the placenta is laden with organisms. (In Canada recently a dozen card players all contracted Q fever after an infected pet cat had a litter of kittens in the same room.) Slaughterhouse and tannery workers become infected by inhaling dried tick feces from the hides of animals. Because

of its resistance to drying, *Coxiella* can remain viable in the environment for long periods of time. Transmission to humans also occurs by the ingestion of milk from contaminated animals. In areas such as Los Angeles, an estimated 10 percent of the cows shed organisms into their milk. Rates are lower in some other areas. Flash pasteurization eliminates this hazard. ∞ (Chapter 13, p. 346) In fact, flash pasteurization was

---

FIGURE 21.11 False-color TEM photo of small- and large-cell forms of *Coxiella burnetii* (40,700×) growing in a placental cell. The cell walls of the large forms have less peptidoglycan with no cross-links—hence the diversity of cell shapes.

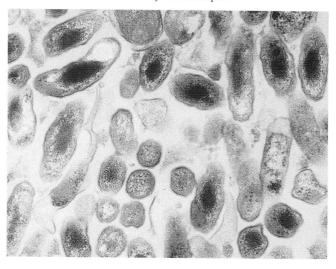

designed to kill Q fever organisms without giving the milk a "cooked" flavor.

Symptoms of Q fever—chills, fever, headache, malaise, and severe sweats—are very similar to those of primary atypical pneumonia. The incubation period is 18 to 20 days. Diagnosis is made by serologic testing or by direct immunofluorescent antibody staining. Treatment is with antibiotics such as tetracycline or fluoroquinolone. Lifelong immunity generally follows an attack of Q fever; a vaccine is available for workers with occupational exposure.

Untreated or inadequately treated cases of Q fever can go into long periods of remission, sometimes lasting years. Cortisone treatment can reactivate it. In chronic cases, endocarditis and heart valve infections are sometimes seen. Endocarditis is invariably fatal, as the organisms do not respond to antibiotic therapy. Special strains appear to be genetically adapted to infect the heart and are very resistant, causing the most serious disease. Other genetic strains cause milder forms of disease.

Nocardiosis

Nocardiosis is characterized by tissue lesions and abscesses. It is caused by the aerobic, acid-fast staining filamentous bacterium *Nocardia asteroides* (Figure 21.12). The organisms are found in soil and water. Disease is usually initiated by inhalation of *N. asteroides*. The bacterium was first identified in 1888 as the causative agent of a disease in cattle known as *farcy*. In humans the primary infection site is the lungs in about three-fourths of all cases, but the disease can also originate in the skin and in other organs. The disease usually occurs in immunosuppressed patients. Human-to-human transmission is rare. Mortality rates as high as 85 percent have been reported. However, with early diagnosis and aggressive treatment with sulfonamides in combination with trimethoprim, mortality can be

---

FIGURE 21.12 *Nocardia asteroides* (magnified 1000×), seen as purple filaments under acid-fast staining.

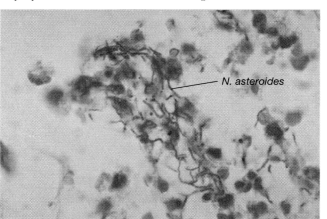

N. asteroides

kept below 50 percent. Diagnosis is based on finding the organism in sputum or other specimens and on culture results.

Viral Lower Respiratory Diseases

Influenza

Influenza is the last remaining great plague from the past. Since Hippocrates described an influenza-like outbreak in 412 B.C., repeated influenza epidemics and pandemics have been recorded, even in recent times (Figure 21.13). One of the greatest killers of all time was the pandemic of swine flu (also known as Spanish flu) of 1918–1919 when 20 to 40 million people died. (See the Essay at the end of this chapter.) The causative virus was unusually virulent, and the crowded, unsanitary conditions created by World War I increased the spread of the virus.

CAUSATIVE AGENTS Influenza is caused by **orthomyxoviruses.** These RNA viruses have an envelope surface antigen hemagglutinin that is responsible for their infectivity. It attaches specifically to a receptor on erythrocytes and other host cells. Some influenza viruses have an enzyme called *neuraminidase* (nur"am-in'i-das), which helps the virus penetrate the mucus layer protecting the respiratory epithelium (Figure 21.14). The enzyme also plays a role in the budding of new virus particles from infected cells. These properties allow identification of influenza viruses in the laboratory.

On the basis of their nucleoprotein antigens, three major influenza virus serotypes are recognized: types A, B, and C. Influenza viruses have a tendency to undergo **antigenic variations** (*changeability*), or mutations that affect viral antigens. Thus, immunity developed through infection with one influenza virus is often insufficient to prevent infection by a variant.

Influenza virus type A, first isolated in 1933, causes epidemics and pandemics at irregular intervals because it periodically undergoes antigenic variation. Secondary bacterial pneumonias are the most important and serious complications of influenza virus A infections. Various species of birds and mammals are infected with strains of influenza A and may serve as reservoirs of infection.

Influenza B viruses also undergo antigenic changes, but less extensively and at a slower rate than do influenza A viruses. Epidemics caused by influenza B viruses are limited geographically and tend to center around schools and other institutions. These viruses are found only in humans.

Influenza C viruses are structurally different from the A and B viruses. Infections with C viruses are rarely recognized. When the disease is recognized, it is typically limited to the children in a single family or single classroom. The low infectivity of C viruses, which lack

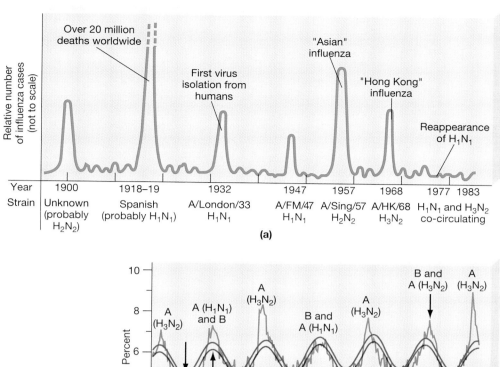

(a)

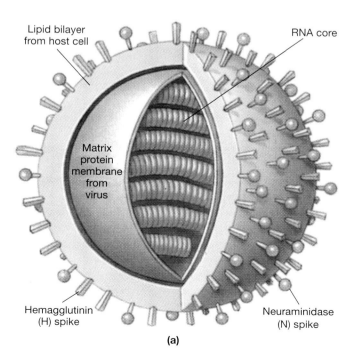

(b)

FIGURE 21.13 Incidence of influenza in the United States (a) since 1900 and (b) recently (as a proportion of all deaths in 121 cities). Influenza varieties are labeled by type and by hemagglutinin (H) and neuraminidase (N) numbers. (See the box "Flu Vaccines" on p. 601.)

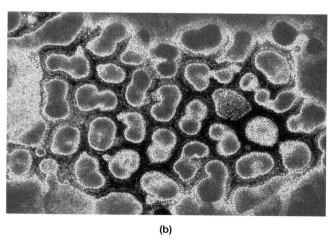

(a)

(b)

FIGURE 21.14 (a) The influenza virus, showing hemagglutinin and neuraminidase spikes on its outer surface and an RNA core. (b) False-color TEM photo of influenza virus type A.

neuraminidase, suggests that this enzyme may enhance infectivity in viruses that have it.

ANTIGENIC VARIATION Antigenic variation in influenza viruses occurs by two processes, *antigenic drift* and *antigenic shift*. **Antigenic drift** results from mutations in genes that code for hemagglutinin and neuraminidase. Such mutations change the configuration of the part of the antigen molecule that stimulates the production of, and combines with, specific antibodies. Thus, antibodies formed against the parental hemagglutinin or neuraminidase are less effective in inhibiting the mutated forms of these viral components of viral offspring. If antigenic drift occurs naturally and is sufficient, an influenza epidemic will follow. **Antigenic shift** results from gene reassortment, possibly after two different viruses infect the same cell and exchange large segments of their genome. It represents more dramatic changes; the viral strains that emerge are significantly different antigenically from previously known strains. Thus, antibodies formed against one type of hemagglutinin are not protective against another type. Antigenic shift generally precedes a major pandemic. It is fortunate that antigenic shift is rare because the strains it produces cause severe pandemics. Six have occurred in the last century—in 1890, 1900, 1918, 1957, 1968, and 1977.

Immune responses vary according to antigenic changes. It is as if a contest occurs between the virus's ability to elude the immune system and the immune system's ability to recognize and inactivate the virus. The antigenic change in most viruses is small, and existing antibodies prevent the viruses from causing infections. The viruses cause severe disease in a few people who lack antibodies, mild disease in others who have some effective antibodies, and no disease in others. The diverse assortment of antibodies that hosts acquire from exposure to several different viral strains tends to protect them from new strains. For a virus to cause an epidemic, it must have undergone sufficient antigenic change to elude most available antibodies.

Influenza viruses replicate in the cytoplasm of nucleated host cells but not in erythrocytes, which lack a nucleus. The effects of influenza viruses on cells they infect are noteworthy. The viruses acquire their envelopes by budding through host cell membranes and can do so without immediately killing the host cell. ∞ (Chapter 11, p. 288) A host cell typically produces thousands of viruses per minute over many hours before the cell's macromolecules are exhausted and the cell dies.

THE DISEASE Influenza, like many other viral respiratory infections, is a relatively superficial infection. The viruses enter the body through inhalation of virus-containing droplets or indirect contact with infectious respiratory secretions. Immediate invasion of the oropharyngeal epithelial lining results. The invading viruses multiply and spread quickly to other portions of the respiratory tract, including mucus-secreting and ciliated epithelial cells. First the cilia are destroyed; then the cells are damaged. Severely damaged cells die and are sloughed. Although host cells begin regeneration immediately, it takes 10 days or more to restore an intact ciliated epithelium. Loss of the mucociliary escalator, ordinarily a major host defense, allows bacterial invasion and enhanced adherence of bacteria to virus-infected cells. ∞ (Chapter 17, p. 465) Impaired phagocytosis and accumulation of fluid in the lungs add to the risk of secondary bacterial infections, especially pneumonia. Death can result from influenza alone, secondary bacterial infection alone, or a combination. Although antibiotics reduce the risk of death from bacterial infections, they have no effect on viral infections. ∞ (Chapter 14, p. 381)

In a nonimmune person, disease signs and symptoms begin to appear 36 to 48 hours after infection. Fever, malaise, and muscle soreness are the most common symptoms; cough, nasal discharge, sore throat, and gastroenteritis (usually only in children) also occur frequently. The fever usually lasts about 3 days. As systemic symptoms decrease, respiratory symptoms increase. The severity of the disease is directly proportional to the quantity of viruses released from cells. Release of viruses starts as the first malaise appears and peaks about a day before maximum fever and interferon production occur. Viral shedding usually ceases within 8 days of initial exposure. Although the acute phase of the illness is over in about a week, fatigue, cough, and weakness may persist over several weeks.

INCIDENCE AND TRANSMISSION In northern temperate zones, influenza appears in late November or early December and disappears in April. In most years the greatest number of cases are seen between January and mid-March. In any one region influenza has a 5- to 7-week period of high prevalence. Indoor crowding, poor air circulation, and dry air expedite the spread of the virus. Schools furnish almost ideal conditions for influenza transmission and usually are the focal points of outbreaks. People often claim to have the "flu," however, when they are actually suffering from another condition such as an allergy or a bad cold (Table 21.3). Gastrointestinal problems, often called "stomach flu," are probably not flu at all but are most likely due to another viral infection.

DIAGNOSIS AND TREATMENT The best specimens for isolation of viruses are throat swabs taken as early in the illness as possible. The viruses can be cultured in embryonated chick eggs and various cell lines and identified by hemagglutination inhibition and immunofluorescent antibody tests.

Moderately effective treatment is now becoming available for influenza. The drug amantadine blocks

Symptoms	Cold	Influenza	Pneumonia
Fever	Rare	Characteristic high (100.4–104°F) sudden onset, lasts 3 to 4 days	May or may not be high
Headache	Occasional	Prominent	Occasional
General aches and pains	Slight	Usual; often quite severe	Occasionally quite severe
Fatigue and weakness	Quite mild	Extreme; can last up to a month	May occur depending on type
Exhaustion	Never	May occur early and prominent	May occur depending on type
Runny, stuffy nose	Common	Sometimes	Not characteristic
Sneezing	Usual	Sometimes	Not characteristic
Sore throat	Common	Sometimes	Not characteristic
Chest discomfort, cough	Mild to moderate; hacking cough	Can become severe	Frequent and may be severe
Complications	Sinus and ear infections	Bronchitis, pneumonia; can be life-threatening	Widespread infections of other organs; can be life-threatening, especially in elderly and debilitated persons

SOURCE: American Lung Association.

influenza A virus replication, probably by interfering with uncoating. ∞ (Chapter 14, p. 381) If given soon after the onset of symptoms, amantadine is effective in shortening the disease's course and reducing the severity. It is useful for short-term protection of selected individuals but impractical to use for the duration of an influenza epidemic unless patients are especially compromised. Furthermore, it prevents users from developing antibodies to the virus and has some unpleasant side effects. Rimantidine is more effective and less toxic. ∞ (Chapter 14, p. 381) Ribavirin has been demonstrated to inactivate both A and B viruses.

IMMUNITY AND PREVENTION Because of influenza viruses' ability to mutate, annual immunization is recommended, especially for high-risk persons, such as those with chronic conditions. The killed virus vaccine provides effective protection against influenza, and annual immunization increases the diversity of the recipient's antibodies. Even less frequent immunization can provide some degree of protection. Although immunized individuals sometimes become infected with flu viruses, they generally have a milder disease of shorter duration than do nonimmunized individuals. Furthermore, they shed fewer viruses over a shorter duration of time. Thus, immunization of some people tends to reduce the spread of the disease to other people.

A human gene that confers resistance to influenza was identified at the University of California in Santa Barbara. This gene is turned on by interferon. It produces a specific protein, called the Mx protein, that prevents the virus from making viral RNA and proteins.

Respiratory Syncytial Virus Infection

The **respiratory syncytial virus** (RSV) is the most common and most dangerous and costly cause of lower respiratory tract infections in children under 1 year and especially in male infants 1 to 6 months old. The virus derives its name from the fact that it causes cells in cultures to fuse their plasma membranes, lose independent identity, and become multinucleate masses, or **syncytia** (sin-sish'ya; singular: *syncytium*). The disease, a form of **viral pneumonia**, begins with a 3- or 4-day period of fever as the virus infects the respiratory tract. Fever is followed by hyperventilation, hyperinflation, and infiltration of the lungs with fluid. If an infant with RSV is placed on a rapidly pulsing respirator, the infant can shed enough virus to infect an entire intensive care nursery. Major outbreaks of this nature generally result in fatalities. Viruses are released for weeks to months after infection, and reinfection is common. In

PUBLIC HEALTH

FLU VACCINES

Each year around June, the CDC selects the strains to be used as vaccine for the coming flu season according to which strains are most prevalent and what alterations they have undergone. Strains are named for the geographic location where they were first isolated, a number assigned by the laboratory that isolated them, the year of isolation, and—for type A viruses—the numbers of their envelope antigens, hemagglutinin (H) and neuraminidase (N). The vaccine for the 1994–95 season contained three strains: influenza A/Texas/36/91-like (H_1N_1), A/Shangdong/9/93-like (H_3N_2), and influenza B/Panama/45/90-like antigens. Tested by inoculation into laboratory animals, the vaccines are thoroughly evaluated for safety and effectiveness before they are given to humans.

older children and adults, RSV affects mainly the upper respiratory tract, and adults who carry the virus in their nose can be the source of nursery infections.

The incidence and transmission of RSV is quite similar to that for parainfluenza viruses, and simultaneous infection by both kinds of viruses is not uncommon. The virus can be identified by enzyme immunoassay performed on nasal secretions or by a direct immunofluorescent test on cells from such secretions. Ribavirin shortens the duration of the disease. There is no vaccine. The best means of prevention is to reduce an infant's exposure to crowds and other sources of infection.

Hantavirus Pulmonary Syndrome

In May and June 1993, 24 cases of severe respiratory illness among residents of the Four Corners area of the southwestern United States were reported. Healthy adults suddenly became ill and died within a few hours. Fear raced through Arizona, Colorado, New Mexico, and parts of the Navajo reservations where deaths had occurred. Investigation identified a previously unreported hantavirus as the cause of the disease, later named **hantavirus pulmonary syndrome** (HPS). The virus was identified by PCR techniques and RNA sequencing (Chapter 7)and was shown to be distinct from all other known hantaviruses. ∞ (p. 190) Hantavirus illness had been known elsewhere in the world for years (Chapter 23), but with renal and hemorrhagic symptoms—not pulmonary illness. The new virus was tentatively named Muerto Canyon virus, after the area from which it was isolated. After complaints from residents of the area, however, the name Sin Nombre ("No-Name") virus was proposed.

The fatality rate (60 percent) has been more than 10 times higher than for other hantaviruses. During 1994, other cases of HPS were found throughout the United States—in Florida, Louisiana, Indiana, and Rhode Island. In each case, a different, new hantavirus was found to be the cause. Each virus was carried by rodents that appeared healthy but shed virus in urine, feces, and saliva. Muerto Canyon virus is carried by the deer mouse, *Peromyscus maniculatus*. The Florida variety of the virus is carried by the cotton rat, *Sigmodon hispidus;* the vector for the Louisiana strain has not yet been identified. Virus from dried excreta becomes airborne and is inhaled. Exposure to rodent-contaminated areas, both indoors and outdoors, should be avoided. Ribavirin has been somewhat useful in treatment, but further studies are needed.

Acute Respiratory Disease

Acute respiratory disease (ARD) is a lower respiratory tract viral disease that ranges from mild to severe. Respiratory symptoms such as sore throat, cough, and other cold symptoms, fever, headache, and malaise are

PUBLIC HEALTH

THE MEDICINE MEN KNEW

In May 1993 public health officials questioned everyone for clues to the mysterious outbreak of respiratory illness that eventually was recognized as hantavirus pulmonary syndrome. Navajo medicine men, who are careful observers of nature, pointed out that there was an exceptional crop of pinyon pine nuts that year. Almost every tree on the Navajo reservation had borne nuts for the entire year, whereas normally just a few trees do so each year, and then only for a few weeks. Rodents were burying the biggest nuts from this bumper crop in burrows. People hunting pine nuts dug up these storehouses and were exposed to rodent feces and urine containing hantavirus. CDC investigators were able to tie the changes in the nut crop to the outbreak.

frequently seen. Severe viral pneumonia that lasts about 10 days sometimes occurs in ARD.

Cases of viral ARD have been seen in military training facilities in the United States and in Europe. Epidemics usually start 3 to 6 weeks after the start of training. The epidemic nature of the disease has been attributed to crowding of people from different geographic areas under stressful conditions. However, epidemics have not been observed in colleges and other institutions where conditions are similar, so other factors may be involved in such epidemics.

Adenoviruses cause about 5 percent of cases of ARD in children under 5 years old. Usually the symptoms are mild and nonspecific—stuffy nose, cough, and nasal discharge. More severe symptoms such as tonsillitis, pharyngitis, bronchitis, bronchiolitis, croup, and often conjunctivitis and abdominal pain may also appear. Adenovirus pneumonia accounts for about 10 percent of all childhood pneumonia and is occasionally fatal.

Fungal Respiratory Diseases

Compared with bacteria and viruses, fungi are much less frequent causes of respiratory diseases. Most fungal infections are seen in immunodeficient and debilitated patients. Blastomycosis, usually a skin disease, can also cause mild respiratory symptoms. ∞ (Chapter 20, p. 564)

Coccidioidomycosis

The soil fungus *Coccidioides immitis* causes **coccidioidomycosis** (kok"sid-i-oy'do-mi-ko"sis), or San Joaquin valley fever (Figure 21.15). This organism is found mainly in warm, arid regions of the southwestern United States and Mexico. Infection usually occurs

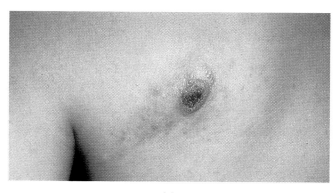

(a)

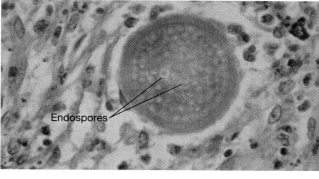

Endospores

(b)

FIGURE 21.15 (a) A skin lesion of coccidioidomycosis on the back. (b) Tissue section from the cerebrospinal fluid of a coccidioidomycosis patient, showing a spherule containing many fungal endospores of *Coccidioides immitis.*

as a result of inhaling dust particles laden with fungal spores. Dogs, cattle, sheep, and wild rodents deposit spores in feces, and the spores easily become airborne. Susceptible individuals have become infected merely from standing on the platform at a train station for a short period of time to get a breath of "fresh air." Travelers passing through dust storms in the Southwest often become infected and return to states where the disease is not expected. Recent increases in disease incidence in California are due in part to earthquake activity, which has produced excess dust. When spores are inhaled by susceptible individuals, an influenza-like illness results. Coccidioidomycosis is always highly infectious, and it can be self-limiting or progressive. In fewer than 1 percent of these individuals, dissemination to the meninges or the bones occurs within 1 year of the initial infection. Dissemination is much more common in blacks than in whites.

Large spherules containing numerous small endospores (reproductive spores that are not ordinarily infectious—not to be confused with bacterial endospores) can be found in sputum, pus, spinal fluid, or biopsied tissue (Figure 21.15b). The fungus forms cottony, white mycelia (colonies) in culture. Various immunological tests are available to aid in diagnosis,

but some give false positive results in individuals who have antibodies against other fungal diseases. A skin test is also available to determine previous exposure to the fungus. The test is performed in the same manner as the tuberculosis skin test and measures the delayed hypersensitivity to antigens of *C. immitis.* Individuals exhibiting a positive skin test are immune to a second attack of the disease.

Coccidioidomycosis in its common form is an acute, self-limiting infection that generally does not require treatment. However, in cases of disseminated disease, therapy is needed. Amphotericin B is the most effective agent available. Newer polyenes, such as the azoles, show promise. If treatment fails, the disease is often fatal. Prevention is difficult, but reducing dust in endemic regions may be helpful. A vaccine is being developed for coccidioidomycosis.

Histoplasmosis

The soil fungus *Histoplasma capsulatum* causes **histoplasmosis,** or *Darling's disease.* The disease is endemic to the central and eastern United States but is found globally in major river valleys. The highest incidence is seen in the Mississippi and Ohio valleys, where 80 percent of the population shows immunological evidence of exposure to the fungus. *H. capsulatum* thrives in soil mixed with feces and especially in chicken houses and in caves containing bat guano (feces). Cave dust is responsible for so many cases that the disease is sometimes referred to as cave sickness.

The fungus enters the body by the inhalation of conidia (fungal spores). ∞ (Chapter 12, p. 313) These conidia are engulfed, but not killed, by macrophages and thus travel in macrophages throughout the body. Person-to-person spread does not occur. Although most infections do not produce disease symptoms, granulomatous lesions occur in the lungs and spleen of susceptible individuals. Inhaling large numbers of conidia can cause pulmonary infection resembling pneumonia. In a few individuals, especially the very young, very old, or those receiving immunosuppressants, *H. capsulatum* can spread to the spleen, liver, and lymph nodes. Anemia, high fever, and an enlarged spleen and liver are common in disseminated histoplasmosis, and death often results.

Diagnosis is made by microscopic identification of small ovoid cells of the organism inside infected human cells. When cultured at body temperature, the fungi look like budding yeasts; at room temperature, they form mycelia with hyphae and spores. An intradermal skin test is also available to determine previous exposure to *H. capsulatum.*

Supportive therapy is used for pulmonary histoplasmosis, and amphotericin B is sometimes effective in treating the disseminated disease. Humans can very easily become infected in environments such as chicken houses and caves, where the air is laden with spores. Some employers hire only individuals with positive histoplasmosis skin tests, and therefore immunity to histoplasmosis, to work in high-risk environments. Spraying infected soil and fecal deposits with 3 percent formaldehyde destroys some of the spores.

Cryptococcosis

Cryptococcosis is caused by *Filobasidiella* (formerly *Cryptococcus*) *neoformans,* a budding, encapsulated yeast. Organisms typically enter the body through the skin, nose, or mouth. Birds carry the fungi on their feet and beak. Although birds do not suffer from cryptococcosis, they do disseminate the opportunistic yeast. These organisms thrive on the nitrogenous waste creatinine, which is present in high concentration in bird feces.

Cryptococcosis is usually characterized by mild symptoms of respiratory infection, but it can become systemic if large quantities of the spores are inhaled by debilitated patients. *Filobasidiella neoformans* often spreads to the meninges, which become thickened and matted, and the organism can invade brain tissue. As with other fungal opportunists, cryptococcosis is increasing in incidence among AIDS patients.

Observing the organism in body fluids confirms the diagnosis. A latex agglutination test is also used to detect the presence of capsular material in body fluids. Flucytosine and amphotericin B can be used in combination to treat systemic disease. Because *F. neoformans* thrives in pigeon droppings, some degree of prevention can be achieved by reducing pigeon populations and decontaminating droppings with alkali.

Pneumocystis Pneumonia

Pneumocystis carinii (Figure 21.16) was long thought to be a protozoan of the sporozoan group. It is now thought to be an opportunistic fungus. It invades cells of the lungs and causes alveolar septa to thicken and the epithelium to rupture. Then the parasites and a foamy exudate from cells collect in the alveoli. The disease, called *Pneumocystis* **pneumonia,** occurs in infants, the elderly, and the immunocompromised. The marked increase in the incidence of this disease in recent years is due mainly to its ability to infect persons with AIDS. The fungus can spread to other organs and cause extrapulmonary infections.

Diagnosis is made by finding organisms in biopsied lung tissue or bronchial lavage (washings from the bronchial tubes). Treatment of *Pneumocystis* infection generally involves the administration of a combination of trimethoprim and sulfamethoxazole or pentamidine. HIV-infected persons and AIDS patients are treated with these drugs as a preventive measure.

Aspergillosis

Aspergillosis (Chapter 20), called *farmer's lung disease* when it occurs in the lungs, is usually caused by *Aspergillus fumigatus* or *A. flavus.* ∞ (p. 565) Other species of *Aspergillus* and even other genera of fungi also can cause farmer's lung disease. Fungal spores inhaled from piles of rotting vegetation or compost may cause clinical allergy, such as asthma, or may produce invasive infection in the lower respiratory tract. Masses of fungal mycelium may grow large enough to be visible on X-rays as a *fungus ball*, or *aspergilloma*. These masses can obstruct gas exchange and cause death by asphyxiation. *Aspergillus* growing in the lungs may also serve as antigens that trigger chronic asthma. Amphotericin B is the drug of choice for invasive infections. Immunosuppressed, immunodeficient (AIDS), and diabetic patients are at higher than normal risk. Fungus balls are very difficult to treat, and surgery is sometimes needed to remove them.

Parasitic Respiratory Diseases

The lung fluke *Paragonimus westermani* is found in many parts of Asia and the South Pacific (Figure 21.17). The fluke's life cycle starts when egg-laden feces are

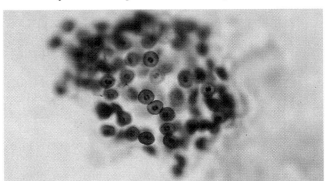

FIGURE 21.16 *Pneumocystis carinii* in sputum. The organism is a frequent cause of pneumonia in AIDS patients.

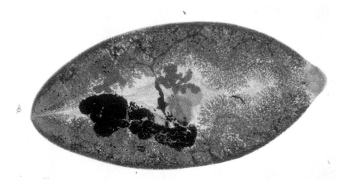

FIGURE 21.17 The lung fluke *Paragonimus westermani* (magnified 13×): an adult worm stained to show internal structures.

released into water; the eggs hatch and invade first a snail and then a crab or crayfish, in which the last larval form, or metacercariae, develop. ∞ (Chapter 12, p. 317) When a human eats an infected shellfish, the metacercariae leave the crab or crayfish as it is digested in the human's small intestine. The larvae then bore through the intestine and embed in the abdominal wall temporarily. They soon leave it and penetrate the diaphragm and membranes around the lungs to reach the bronchioles. The larvae mature into adults and lay eggs in the bronchioles. When the host coughs, the eggs move into the pharynx, are swallowed, and exit the body in the feces. Infected humans have a chronic cough, bloody sputum, and difficulty breathing. Diagnosis can be made by finding eggs in sputum or by any one of several immunologic tests. The drug praziquantel is effective in treating lung fluke infections. Infections can be avoided by cooking shellfish before eating them.

Diseases of the lower respiratory tract are summarized in Table 21.4.

TABLE 21.4	Summary of Diseases of the Lower Respiratory Tract	
Disease	Agent(s)	Characteristics
Bacterial lower respiratory diseases		
Whooping cough	Bordetella pertussis	Catarrhal stage with fever, sneezing, vomiting, and mild cough; paroxysmal stage with ropy mucus and violent cough; convalescent stage with mild cough
Classic pneumonia	Streptococcus pneumoniae, Staphylococcus aureus, Klebsiella pneumoniae	Inflammation of bronchi or alveoli of lungs with fluid accumulation and fever
Mycoplasma pneumonia	Mycoplasma pneumoniae	Mild inflammation of bronchi or alveoli
Legionnaires' disease	Legionella pneumophila	Inflammation of the lungs, fever, chills, headache, diarrhea, vomiting, and fluid in lungs
Tuberculosis	Mycobacterium tuberculosis	Tubercles in lungs and sometimes in other tissues; organisms can persist in walled-off lesions and be reactivated
Ornithosis	Chlamydia psittaci	Pneumonia-like disease transmitted to humans by birds
Q fever	Coxiella burnetii	Disease similar to Mycoplasma pneumonia but transmitted by ticks, aerosols, and fomites
Nocardiosis	Nocardia asteroides	Pneumonia-like disease seen in immunodeficient patients
Viral lower respiratory diseases		
Influenza	Influenza viruses	Viruses subject to antigenic variation, with new strains causing epidemics; inflammation of oropharyngeal membranes, fever, malaise, muscle pain, cough, nasal discharge, and gastroenteritis
Respiratory syncytial virus infection	Respiratory syncytial virus	Febrile disease of the respiratory tract; can cause viral pneumonia
Hantavirus pulmonary syndrome	Hantaviruses	Fever, kidney abnormalities; in severe cases shock, bleeding, and pulmonary edema occur
Acute respiratory disease	Adenoviruses	Mild cough and nasal discharge; can cause viral pneumonia
Fungal respiratory diseases		
Coccidioidomycosis	Coccidioides immitis	Influenza-like illness; dissemination to meninges and bones can occur
Histoplasmosis	Histoplasma capsulatum	Granulomatous lesions in lungs and spleen in susceptible individuals; can cause pneumonia
Cryptococcosis	Filobasidiella (Cryptococcus) neoformans	Usually a mild pulmonary disease; pneumonia and dissemination to meninges can occur
Blastomycosis	Blastomyces dermatitidis	Usually a skin disease (see Chapter 20); sometimes disseminated to lungs or acquired directly by inhalation; mild respiratory symptoms
Pneumocystis pneumonia	Pneumocystis carinii	Rupture of alveolar septa, foamy sputum; occurs mainly in immunodeficient patients
Aspergillosis	Aspergillus species	Allergic asthmatic response to inhalation of spores or invasive infection of lung; fungal balls can cause asphyxiation
Parasitic respiratory diseases		
Lung fluke infection	Paragonimus westermani	Larvae mature in bronchioles and cause chronic cough, bloody sputum, and difficulty breathing

THE GREAT FLU PANDEMIC OF 1918

Where could he be? He should have been home an hour ago! In normal times, this wouldn't have been so alarming—but those weren't normal times. Was his body one of those piled up like cordwood on a street corner downtown?

Such fears raced through the minds of frantic families during the great Spanish flu pandemic of 1918. Seemingly healthy people dropped dead without warning. Undertakers couldn't meet the demands of a pandemic that killed half a million Americans in only 10 months. (The U.S. Civil War claimed about that many lives over 4 years.) Philadelphia alone recorded 528 bodies piled up, awaiting burial, in just one day. Numerous others lay undiscovered. Rescue teams went door to door, seeking victims who were too sick to call out for aid. Wagons made regular rounds, stopping at corners to pick up bodies that families, neighbors, storekeepers, or passers-by had carried there. No wonder families worried when someone was late getting home. Who knew which of the 25 million infected Americans would die?

Unlike many diseases, influenza took its heaviest toll among young, healthy individuals. Some died of high fever; others succumbed to secondary bacterial infections, chiefly pneumonia. The immediate cause of death for many patients was lung damage. Tremendous pressure built up until the lungs degenerated completely (on autopsy, they typically had the consistency of pudding). If the crushing pressure was relieved, the patient had a chance of survival. One physician called to the bedside of a teenage girl could do little but comfort the parents. Suddenly a stream of blood shot from her nose and soaked the trio. Projectile hemorrhaging had reduced the pressure in her lungs, and she lived. But people who survived cases with high fever often developed Parkinson's disease. We now associate Parkinson's disease with the elderly. But after the 1918 pandemic, it affected all ages, including many small children.

The cause of this grim experience was something too tiny for the researchers of 1918 to see—the swine flu virus. What they did see with light microscopes were secondary bacterial invaders. (One bacillus was named *Haemophilus influenzae* in the mistaken belief that it was the cause of the pandemic.) Indeed, in one New York City family, all six members were found to be infected with different species of bacteria. No wonder the medical world was confused. Attempts at vaccine production proved futile. Few people even guessed at the existence of the virus.

Where had this virus come from? A new type of influenza had appeared in France in April 1918 as American and European troops fought during the final months of World War I. From France the virus traveled to Spain. That summer it spread throughout Europe, China, and West Africa. By then, the English-speaking countries had dubbed it the "Spanish lady." The Lady entered Boston in August and crossed the United States within a month. Around the world she went, causing a pandemic. By the time it had subsided in 1920, more than 25 million had died. The country hardest hit was India, where death wiped out the entire population increase of a decade. In some parts of the world, nearly half the population died. There is some evidence that the virus took nonhuman victims as well—baboons in South Africa, for example.

It is not certain what brought on this deadly kind of flu. The influenza virus, with a long history as a human pathogen, is one of the most rapidly mutating viruses known. But the natural mutation process may have had human help. Mustard gas, which was widely used in trench warfare in France, is a powerful alkylating agent—a known mutagen. Could it have interacted with the virus to produce the new, virulent strain? No one can be sure.

The term *swine flu* was not used until years after the pandemic was over. Influenza was unknown in pigs before the Spanish flu reached the United States in 1918. Thereafter, swine were found to suffer from a complex disease involving both an influenza virus and a bacterium. The virus was believed to be the one that caused the Spanish flu. Today, the virus that causes influenza in swine is not the one that causes "swine flu" in humans. The name swine flu may not have been a mistake, however, for the rapidly evolving virus strains are likely to have mutated quite a bit since 1918.

Is the virus that caused the great pandemic still a threat? In 1976, the U.S. government mounted a massive immunization program against swine flu. Flu viruses almost always march westward around the globe. By looking at the viruses that have caused problems to the east, U.S. public health officials decide which viral strains to incorporate into next year's vaccine. So when swine flu was detected in Asia in 1975, it sent chills up and down medical spines. Not only could mil-

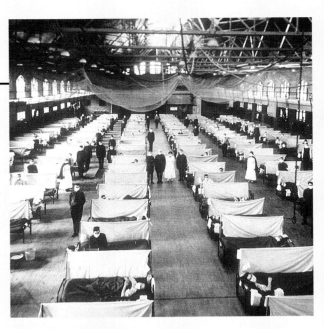

FIGURE 21.18 During the height of the great flu pandemic of 1918, the gymnasium of Iowa State University was temporarily converted into a hospital ward.

lions of deaths occur—there were also the sequelae. How many Parkinson's cases would be survivors of a 1976 pandemic?

Nervously the government exhorted all to be immunized. Then came a disturbing development. Some recipients of swine flu vaccine had subsequently developed *Guillain-Barré syndrome,* resulting in complete paralysis. (Although most persons recover completely, a small percentage die of the paralysis.) Which would be worse: swine flu or paralysis? Lines at vaccine clinics shortened. With fewer people immunized, what would happen when the swine flu arrived? The medical community waited. But swine flu never came. Was it a reprieve or just a postponement? No one can say.

CHAPTER SUMMARY

The agents and characteristics of the diseases discussed in this chapter are summarized in Tables 21.1 and 21.4. Information in these tables is not repeated in this summary.

DISEASES OF THE UPPER RESPIRATORY TRACT

Bacterial Upper Respiratory Diseases

- Infections related to **pharyngitis,** or sore throat, include **laryngitis, epiglottitis, sinusitis,** and **bronchitis.** These diseases usually are transmitted by respiratory droplets. If severe enough, they can be treated with penicillin or other antibiotics.

- **Diphtheria,** no longer common in the United States, occurs only in humans. Both the organism and its toxin, produced by genes from a lysogenic prophage, contribute to disease signs and symptoms. A major disease sign is the formation of a **pseudomembrane** that can block the airway. Diphtheria is spread by respiratory droplets, treated with antitoxin and antibiotics such as penicillin, and prevented by DTP vaccine.

- Ear infections occur in the middle (**otitis medea**) and outer (**otitis externa**) ear. Organisms reach the middle ear via the Eustachian tube and usually can be eradicated by penicillin.

Viral Upper Respiratory Diseases

- The common cold, **coryza,** is transmitted by fomites and aerosols. Treatment is limited to alleviating symptoms; no vaccine is available.

- **Parainfluenza** infections range from inapparent disease to severe **croup.** Most children develop antibodies to **parainfluenza viruses** by age 10.

DISEASES OF THE LOWER RESPIRATORY TRACT

Bacterial Lower Respiratory Diseases

- **Whooping cough,** or **pertussis,** is distributed worldwide, but immunization has decreased its incidence. It is transmitted by respiratory droplets and treated with antitoxin and erythromycin. Vaccine prevents the disease but may also cause some complications and deaths.

- Classic **pneumonia** can be **lobar** or **bronchial;** it is transmitted by respiratory droplets and carriers. *Klebsiella* pneumonia is more severe than pneumococcal pneumonia. Penicillin is the drug of choice for pneumococcal pneumonia, and vaccine is recommended for high-risk populations. *My-coplasma* pneumonia is transmitted by respiratory droplets and is treated with erythromycin or tetracycline.

- **Legionnaires' disease** is transmitted via aerosols from contaminated water and treated with erythromycin.

- **Tuberculosis** (TB) has been a major health problem worldwide for centuries, and its incidence in the United States is increasing. It is transmitted by respiratory droplets. Inactive but viable organisms can persist for years walled off in **tubercles.** Treatment with isoniazid is effective except for resistant organisms, which must be treated with "second- or third-line" drugs. A vaccine is available worldwide, but its use in the United States is limited to high-risk individuals.

- **Ornithosis** is transmitted to humans from infected birds. It is usually mild but can cause a serious pneumonia. The causative organism is dangerous to handle in the lab.

- **Q fever** is transmitted by ticks, aerosol droplets, and fomites. It is treated with tetracycline, and a vaccine is available for workers with occupational exposure.

- **Nocardiosis,** marked by lesions and abscesses, infects primarily the lungs but can infect the skin or other organs.

Viral Lower Respiratory Diseases

- The viruses that cause **influenza** display **antigenic variation** (mutation that affects antigens). The disease occurs mainly from December through April. It is transmitted in crowded, poorly ventilated conditions and diagnosed by immunological tests. Vaccine can prevent the disease, but its effectiveness can be lessened by antigenic variation.

- **Respiratory syncytial virus** (RSV) infections and **acute respiratory disease** are important acute respiratory tract illnesses. RSV infections tend to be severe, especially in young children.

- **Hantavirus pulmonary syndrome,** a severe infection, has a high mortality. The hantavirus infects primarily rodents and is found in their excreta.

Fungal Respiratory Diseases

- Opportunistic fungal infections occur mainly in immunocompromised and debilitated patients. These infections usually are transmitted by spores, and some can be treated with amphotericin B.

- In the United States, **coccidioidomycosis** occurs in warm, arid regions; **histoplasmosis** is endemic in eastern states, and **cryptococcosis** occurs wherever there are infected birds, especially pigeons.

■ *Pneumocystis* **pneumonia,** an opportunistic fungal infection, is a common cause of death among AIDS patients.

■ **Aspergillosis** is contracted by the inhalation of fungal spores and can involve the growth of large masses of fungal mycelia (fungus balls) that are visible on X-rays.

Parasitic Respiratory Diseases

■ Lung fluke infections occur mainly in Asia and the South Pacific where infected shellfish are eaten. The disease can be treated with drugs and prevented by cooking shellfish adequately.

KEY TERMS

acute respiratory disease (p. 602)
antigenic drift (p. 600)
antigenic shift (p. 600)
antigenic variation (p. 598)
aspergillosis (p. 604)
bronchial pneumonia (p. 589)
bronchitis (p. 580)
caseous (p. 593)
catarrhal stage (p. 587)
coccidioidomycosis (p. 602)
consolidation (p. 589)
convalescent stage (p. 588)
coronavirus (p. 585)
coryza (p. 584)

croup (p. 585)
cryptococcosis (p. 604)
cyanosis (p. 588)
diphtheria (p. 582)
diphtheroid (p. 582)
disseminated tuberculosis (p. 593)
epiglottitis (p. 580)
hantavirus pulmonary syndrome (p. 602)
histoplasmosis (p. 603)
influenza (p. 598)
laryngitis (p. 580)
Legionnaires' disease (p. 590)
lobar pneumonia (p. 589)

miliary tuberculosis (p. 593)
nocardiosis (p. 598)
ornithosis (p. 596)
orthomyxovirus (p. 598)
otitis externa (p. 583)
otitis media (p. 583)
parainfluenza (p. 585)
parainfluenza virus (p. 585)
paroxysmal stage (p. 587)
pertussis (p. 586)
pharyngitis (p. 580)
pleurisy (p. 589)
Pneumocystis pneumonia (p. 604)
pneumonia (p. 588)
Pontiac fever (p. 591)

primary atypical pneumonia (p. 590)
pseudomembrane (p. 583)
psittacosis (p. 596)
Q fever (p. 596)
respiratory syncytial virus (p. 601)
rhinovirus (p. 585)
sinusitis (p. 580)
syncytium (p. 601)
tonsilitis (p. 580)
tubercle (p. 592)
tuberculosis (p. 591)
viral pneumonia (p. 601)
walking pneumonia (p. 590)
whooping cough (p. 586)

QUESTIONS FOR REVIEW

A **Bacterial Upper Respiratory Infections**

1. How is strep throat diagnosed and treated?
2. Compare and contrast laryngitis, sinusitis, bronchitis, and pharyngitis.
3. What causes diphtheria, and how does it damage tissue?
4. How is diphtheria diagnosed, treated, and prevented?
5. What are the causes and effects of ear infections?

B **Viral Upper Respiratory Infections**

6. What agents cause colds, and how are they transmitted?
7. Why are colds difficult to treat and to prevent?
8. How are parainfluenza and croup related?

C **Bacterial Lower Respiratory Infections**

9. *Fill in the blanks:* _____ is the agent that causes whooping cough, which is also known as _____. This disease is a world health problem because _____. It can be transmitted by _____ and diagnosed by obtaining _____. The disease is treated by _____, and _____ is used to prevent it.
10. List the characteristics of each stage of whooping cough, and describe the contribution of toxin to each.
11. Contrast lobar pneumonia with bronchial pneumonia.
12. List the characteristics of pneumococcal pneumonia, and contrast it with other typical bacterial pneumonias.

13. How is classical pneumonia diagnosed, treated, and prevented?
14. List the characteristics of primary atypical pneumonia.
15. List the characteristics of Legionnaires' disease, and describe its treatment.
16. What causes tuberculosis? Why is it a global problem?
17. How do mycobacteria cause tissue damage, and what tissues can they affect?
18. How is tuberculosis transmitted, and what populations are most susceptible?
19. How is tuberculosis diagnosed, treated, and prevented?
20. Describe ornithosis and its diagnosis, treatment, and prevention.
21. Describe Q fever and its diagnosis, treatment, and prevention.
22. What is nocardiosis, and how is it treated?

D **Viral Lower Respiratory Infections**

23. Describe influenza and the viruses that cause it.
24. How is influenza diagnosed, treated, and prevented?
25. What kinds of viruses can cause viral pneumonia?

E **Fungal Lower Respiratory Infections**

26. Under what circumstances do fungal respiratory infections arise?

27. Compare and contrast coccidioidomycosis, histoplasmosis, and cryptococcosis.

28. What is *Pneumocystis* pneumonia, and how is it caused?

F Parasitic Respiratory Infections

29. How do people become infected with lung flukes, and what are their effects?

30. Match the following:

____ diphtheria	**a.** *Coxiella*
____ tuberculosis	**b.** *Paragonimus*
____ Q fever	**c.** orthomyxoviruses
____ ornithosis	**d.** *Pneumocystis*
____ lung fluke infection	**e.** *Mycobacterium*
____ whooping cough	**f.** *Corynebacterium*
____ valley fever	**g.** *Chlamydia*
____ influenza	**h.** *Coccidioides*
____ Pontiac fever	**i.** *Bordetella*
____ pseudomembrane	**j.** *Legionella*

31. Using the accompanying diagram, name six diseases each of the upper and lower respiratory systems. Indicate the specific part of the respiratory system each is likely to affect; identify the infectious agent, and describe each as bacterial, viral, fungal, or parasitic.

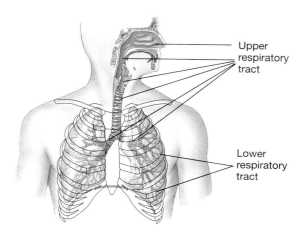

Upper respiratory tract

Lower respiratory tract

PROBLEMS FOR INVESTIGATION

1. Devise a plan for reducing the incidence of upper respiratory infections.

2. A patient has *Mycoplasma* pneumonia. Should the physician prescribe penicillin? Why or why not?

3. What would be the disease consequences of having lungs that are constructed in one large unit rather than in lobes? Why?

4. Prepare a report on current worldwide efforts to control tuberculosis.

5. In recent issues of *Morbidity and Mortality Weekly Reports*, look up the CDC statistics for notifiable respiratory diseases in your state. Compare these with other parts of the country.

6. A 34-year-old man was hospitalized with respiratory symptoms of pneumonia. Culture of his sputum revealed infection by *Pneumocystis carinii*. What other condition(s) would you also suspect he has? (The answer to this question appears in Appendix F.)

SOME INTERESTING READING

Bone, R. C. 1991. "Chlamydial pneumonia and asthma: A potentially important relationship." *Journal of the American Medical Association* 266, no. 2 (July 10):265.

Bloom, B. R., ed. 1994. *Tuberculosis: Pathogenesis, protection, and control.* Washington, D.C.: ASM Press.

Breiman, R. F., et al. 1994. "Emergence of drug-resistant pneumococcal infections in the United States." *Journal of the American Medical Association* 271 (June 15):1831–35.

Chazen, G. 1987. "Nocardia." *Infection Control* 8(6):260.

Collier, R. 1974. *The plague of the Spanish lady.* New York: Atheneum.

Cousins, D. V., et al. 1992. "Use of polymerase chain reaction for rapid diagnosis of tuberculosis." *Journal of Clinical Microbiology* 30(1):255–58.

Edelstein, P. H. 1985. "Environmental aspects of *Legionella*." *ASM News* 51(9):460–67.

Fincher, J. 1989. "America's deadly rendezvous with the 'Spanish lady'." *Smithsonian* 19 (January):130–45.

Gilligan, P. H. 1991. "Microbiology of airway disease in patients with cystic fibrosis." *Clinical Microbiology Reviews* 4(1):35–51.

Grady, D. 1993. "Death at the corners." *Discover* 14(December):82–91.

Granstrom, M., et al. 1991. "Specific immunoglobulin for treatment of whooping cough." *Lancet* 338, no. 8777 (November 16):1230–33.

Middleton, D. B. 1991. "An approach to pediatric upper respiratory infections." *American Family Physician* 44, no. 5 (November):33–40.

Radetsky, P. 1989. "Taming the wily rhinovirus." *Discover* 10(April):38–43.

Reimer, L. G. 1993. "Q fever." *Clinical Microbiology Reviews* 6(3):193–98.

Ryan, F. 1993. *The forgotten plague: How the battle against tuberculosis was won and lost,* 3d ed. Boston: Little, Brown.

Salyers, A. A., and D. D. Whitt. 1994. *Bacterial pathogenesis: A molecular approach.* Washington, D.C.: ASM Press.

Schlossberg, D., ed. 1994. *Tuberculosis.* Boston: Springer-Verlag.

Sepkowitz, K. A., et al. 1995. "Tuberculosis in the AIDS era." *Clinical Microbiology Reviews* 8:180–99.

Snider, D. E., et al. 1994. "Multi-drug-resistant tuberculosis." *Scientific American: Science and Medicine* 1, no. 3 (May/June):16–25.

Stevens, D. A. 1995. "Coccidioidomycosis." *The New England Journal of Medicine* 332, no. 16 (April 20):1077–82.

ORAL AND GASTROINTESTINAL DISEASES

A girl paddles through a canal full of excrement and refuse in Belen, Peru. The water, used for cooking, bathing, swimming, and fishing, breeds cholera, which is now affecting large areas of South America without proper water and sewage treatment.

We all have a wealth of firsthand experience with diseases of the oral cavity and gastrointestinal tract. We've had plaque removed from our teeth and cavities filled. We've had episodes of nausea, vomiting and diarrhea—usually of short duration—after we ate contaminated food or drank impure water. Fortunately, for most of us these problems have been minor inconveniences rather than serious illnesses. In other parts of the world, however, especially where water and sewage treatment are inadequate, gastrointestinal infections are major problems. In this chapter we will consider diseases of the oral cavity and intestinal tract, both minor and serious.

Like the foregoing chapters dealing with diseases of body systems, this chapter presumes a knowledge of disease processes, normal and opportunistic microbiota (flora) (Chapter 15), epidemiology (Chapter 16), host systems and host defenses (Chapter 17), and immunity (Chapters 18 and 19). It also presupposes familiarity with the characteristics of bacteria, viruses, and eukaryotic microorganisms and helminths (Chapters 9 through 12). The anatomy and physiology of the digestive system, especially of the teeth, should be reviewed from Chapter 17. ∞ (p. 467)

DISEASES OF THE ORAL CAVITY

Although we would like to believe otherwise, our mouths are microbial breeding grounds. About 2 billion organisms live in a healthy human mouth. We pick up microbes in many ways—from food, kissing, and dirty fingers, to name a few—and the microbes thrive.

Bacterial Diseases of the Oral Cavity

Dental Plaque

Dental plaque is a continuously formed coating of microorganisms and organic matter on tooth surfaces. Plaque formation, although not a disease itself, is the first step in tooth decay and gum disease. Scrupulous and frequent cleaning of teeth minimizes but does not entirely prevent plaque formation. Plaque begins to form within 24 hours after cleaning. Unless it is removed regularly, plaque can become so securely attached to teeth that it can no longer be removed by home methods. Professional cleaning removes plaque,

THIS CHAPTER FOCUSES ON THE FOLLOWING QUESTIONS:

A What kinds of pathogens cause diseases of the oral cavity, and what are the important epidemiologic and clinical aspects of these diseases?

B Which bacteria cause gastrointestinal diseases, and what are the important epidemiologic and clinical aspects of such diseases?

C Which viruses cause gastrointestinal diseases, and what are the important epidemiologic and clinical aspects of such diseases?

D Which protozoa cause gastrointestinal diseases, and what are the important epidemiologic and clinical aspects of such diseases?

E Which fungi cause gastrointestinal diseases, and what are the important epidemiologic and clinical aspects of such diseases?

F Which helminths cause gastrointestinal diseases, and what are the important epidemiologic and clinical aspects of such diseases?

but it begins to form again even before you get home from the dentist's office.

Plaque formation begins as positively charged proteins in saliva adhere to negatively charged enamel surfaces and form a *pellicle* (film) over the tooth surface. Cocci such as *Streptococcus mutans* and some filamentous bacteria among the normal oral flora microbiota attach to the newly formed pellicle. These organisms may hydrolyze sucrose (table sugar) to glucose and fructose, both of which can be metabolized for energy production and growth. Plaque-forming bacteria such as *S. mutans* can also convert sucrose to polymers of glucose units (such as the polysaccharide dextran) that serve as bridges holding together cells in plaque. Plaque consists of up to 30 different genera of bacteria and their products, such as dextran, saliva proteins, and minerals (Figure 22.1). If plaque is not removed thoroughly and regularly, streptococci, lactobacilli, and other acid-producing bacteria may accumulate within it in layers 300 to 500 cells thick. These organisms metabolize fructose and other sugars that diffuse into the plaque and set the stage for tooth decay. Plaque that accumulates near the gumline also offers protection to bacteria in the gingival (gum) crevices between the teeth and the gums. These bacteria include species of *Actinomyces, Veillonella, Fusobacterium,* and sometimes spirochetes as well as the streptococci just noted. If plaque is allowed to accumulate, some crevices change into anaerobic pockets full of bacteria that can irritate the gums or destroy the bone in which the teeth are set. In some people, this gumline plaque mineralizes into *calculus* (*tartar*) that can also irritate the gums and contributes to inflammation and bleeding.

Dental Caries

THE DISEASE PROCESS Dental caries, or tooth decay, is the chemical dissolution of enamel and deeper parts of teeth. Taking its name from the Latin *cariosus,* which means rotten, caries is the most common infectious disease in developed countries where the diet contains relatively large amounts of refined sugar. Unchecked, the decay can proceed through enamel, into dentin, into the pulp cavity, and eventually can cause an abscess in the bone that supports the tooth. Sugars easily diffuse through plaque to bacteria embedded in it, but acids produced by bacterial fermentation fail to diffuse out. The acids gradually dissolve

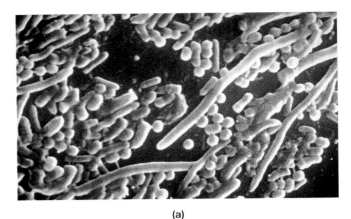

(a)

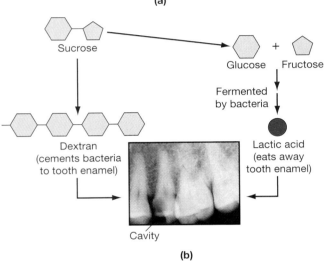

(b)

FIGURE 22.1 Dental plaque. (a) A variety of organisms that accumulate in plaque deposits. (b) Production of sticky dextran from sucrose enables bacteria to adhere to tooth surfaces, where the lactic acid produced in this process eats away tooth enamel and forms cavities.

enamel, after which protein-digesting enzymes break down any remaining material.

The combination of sucrose and the action of *S. mutans* on it account for much tooth decay. Consequently, the more sucrose you eat and the more frequently you eat it, the greater your risk of dental caries. Saliva helps rinse sugars from the mouth, but its rinsing efficiency varies with flow rate and mouth shape. Sugar accumulates in poorly rinsed areas. Although starchy foods are only partially digested in the mouth, those that are sticky can adhere to tooth surfaces and remain on the teeth long enough for bacterial action to contribute to tooth decay. Sugar alcohols, such as sorbitol and xylitol used in "sugar-free" chewing gums, do not contribute to tooth decay because many bacteria cannot metabolize them. Bacteria that can utilize sorbitol, such as *S. mutans,* metabolize it slowly. They release acid at a rate that allows buffers in saliva to neutralize it, preventing a drop in pH in areas of plaque.

TREATMENT AND PREVENTION Dental caries are treated by removing decay and filling the cavity with *resins* (plastic materials) or with *amalgam* (a mixture of silver and other metals). Dental caries can be prevented or their incidence greatly reduced by limiting the intake of sugary and sticky foods, brushing regularly with plaque-removing toothpaste, and flossing between teeth. Vaccines to prevent dental caries are being developed against strains of *S. mutans,* which are most prevalent at decay sites. An injectable vaccine that elicits circulating IgG production has been successful in monkeys. An oral vaccine that causes secretory IgA production has been successful in rats. Neither vaccine has yet been adequately tested in humans. Stress may be a factor in the effectiveness of a vaccine because it appears to depress the immune system. This was noted when dental students were found to secrete less salivary IgA while taking examinations than while on summer vacation. ∞ (Chapter 18, p. 485)

The use of **fluoride** has been the most significant factor in reducing tooth decay. It works by hardening the surface enamel of teeth. Fluoride reduces the solubility of tooth enamel by inhibiting demineralization and also enhances remineralization. To understand how fluoride affects enamel, imagine the tooth surface as consisting of the ends of rods packed together like a fistful of pencils (Figure 22.2). No matter how tightly the rods are packed, channels exist between them. Acids produced by bacteria in plaque seep through channels and dissolve the enamel rods. As channels are enlarged, more acid enters and dissolves more enamel. Eventually, sufficient enamel is eroded to form cavities, in which dental caries can form. Fluoride fills the spaces between rods with a hard mineralized material that strengthens the tooth surface and prevents acid penetration. Fluoride may also inhibit certain enzymes that produce phosphates needed by bacteria to capture energy from nutrients. Without the phosphates, the bacteria die.

Numerous studies have shown fluoride to be safe and effective in the prevention of tooth decay. Added to city water supplies in a 1-part-per-million concentration, fluoride reduces tooth decay in children by as much as 60 percent. Because caries occur mainly in childhood and adolescence, a successful program to prevent childhood tooth decay is of major importance. Yet nearly half the U.S. population obtains water from wells or from nonfluoridated community supplies. In some communities that lack fluoridated water, children receive fluoride tablets or gels at school. Ingestion of fluoride by children during tooth formation strengthens the entire tooth, whereas topical application affects only the surface of the tooth.

Although fluoride is most beneficial before age 20, it provides some benefits at any age, even when water is fluoridated and fluoride toothpastes and mouth-

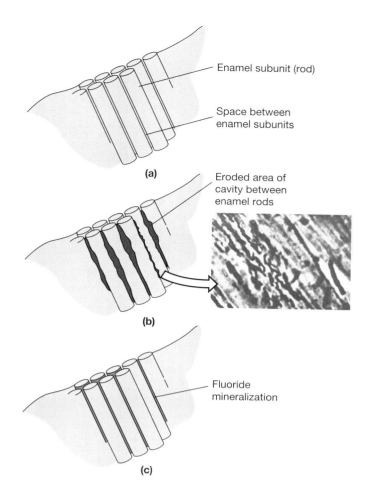

FIGURE 22.2 The effect of fluoride on tooth decay: (a) normal tooth structure; (b) dental caries forming as a result of bacterially produced acid seeping down between rods of enamel; (c) mineralization by application of fluoride, which fills in the spaces between enamel rods, thereby preventing acid from seeping in.

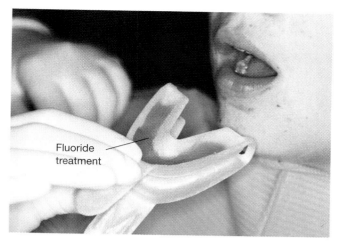

FIGURE 22.3 Visits to the dentist can be minimized by proper oral hygiene and the use of fluoride-containing toothpaste or oral rinse. Fluoride can also be applied in the dentist's office.

washes are used. Fluoride gel treatments should be given every 6 months from about age 4 to adulthood. Each time teeth are cleaned, a little surface is removed, and remineralization with topical fluoride helps restore the surface (Figure 22.3).

Another means of preventing dental caries is sealing of the teeth with a type of resin that mechanically bonds to the tooth structure. Because the grinding surfaces of teeth have tiny pits and crevices that fill with plaque that cannot be removed with a toothbrush, these surfaces are more susceptible to decay than are smooth sides. Applying sealant to teeth provides nearly complete protection against decay as long as the sealant remains, which may be 10 years or longer. Applied at about age 6 or 7 after secondary grinding teeth have erupted, sealants will last through most of the decay-prone years. In the sealing process, teeth are first thoroughly cleaned and etched with acid; then the sealant is applied. Tiny cavities cease to grow in sealed teeth and eventually become sterile as the decay organisms die. Sealant can be applied only to teeth without fillings, but it can be used on adults as well as children.

Periodontal Disease

THE DISEASE PROCESS When bacteria become trapped in gingival crevices, they cause both tooth decay and gum inflammation. **Periodontal disease** is a combination of gum inflammation and erosion of periodontal ligaments and the alveolar bone which supports teeth (Figure 22.4). The disease is a chronic, generally painless infection that affects over 80 percent of teenagers and adults and is the major cause of tooth loss.

Plaque formation is the initiating event in periodontal disease. Organisms in gingival crevices are believed to produce endotoxins and acids that in turn produce an inflammatory response. This response breaks down the epithelial cells of the gums, and new groups of organisms replace previous crevice inhabitants as the disease progresses. If the process is not arrested, the gums recede and can become necrotic, and the teeth loosen as surrounding bone and ligaments are eroded and weakened.

In its mildest form, periodontal disease is called **gingivitis** (jin-jiv-i'tis), which affects only the gums. In its most severe form, gingivitis is called **acute necrotizing ulcerative gingivitis** (ANUG), or trench mouth. The disease got its name because it was common among soldiers in World War I under stress "in the trenches." It is now common in young people, and stress seems to be an important factor in its development. ANUG responds to antibiotics, but more advanced forms of periodontal disease usually do not.

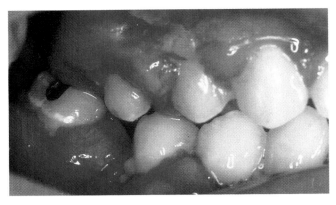

(a)

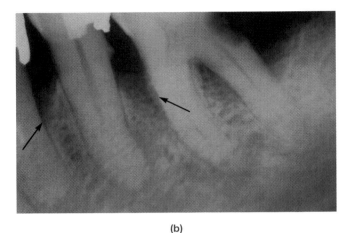

(b)

FIGURE 22.4 Advanced periodontal disease, showing (a) severe gum inflammation, which can lead to (b) loss of bone (see arrows) surrounding roots of teeth, causing loosening and eventual loss of teeth.

Unchecked, periodontal disease can lead to chronic **periodontitis,** which affects the bone and tissue that support the teeth as well as the gums. The loss of bone due to periodontitis is not reversible.

Periodontal disease occurs when plaque is allowed to accumulate, resulting in overgrowth of potentially virulent bacteria such as *Porphyromonas gingivalis, Actinobacillus actinomycetemcomitans, Prevotella intermedia, Bacteroides forsythus, Fusobacterium nucleatum,* and other gram-negative rods. Many of the organisms living in the mouth, especially, spirochetes, have yet to be isolated, cultured, and identified. Certain other bacteria, such as *Streptococcus mitis, S. sanguis, Veillonella* species, and some *Actinomyces* species, may help control virulent bacterial populations. If plaque is not removed by brushing or flossing, calcium is deposited on plaque surfaces, forming a very rough, hard crust called **tartar,** or *calculus.* Tartar strongly binds to teeth surfaces. As more plaque accumulates, tartar forms a thick layer. The calculus contributes to the conditions leading to

bleeding of gums. As the condition worsens, pockets of inflammation form in the gum crevices. Measuring the depth of these pockets serves as an indication of the extent of periodontal disease.

A 1982 study of organisms associated with periodontal disease, which was done by Paul H. Keyes, illustrates a method of studying a complex microbial environment. Keyes obtained plaque from individuals with healthy gums, mildly inflamed gums, and severely diseased gums. When he examined them by phase-contrast microscopy, he found distinctly different populations of microorganisms. Plaque from healthy gums contains nonmotile filamentous bacteria, a few colonies of cocci, fewer than five leukocytes per microscopic field, and no amoebas or spirochetes. Plaque from mildly inflamed gums shows a greater variety of microbes. Nonmotile organisms form dense masses surrounded by clusters of motile spirochetes and rapidly spinning bacilli too numerous to count. Some spirochetes and leukocytes are present, but amoebas and trichomonads are absent. Plaque from severely diseased gums contains dense mats of nonmotile filaments, cocci, and numerous motile organisms. Extending from the dense mats are brushlike aggregations of spirochetes and flexible rods that move in rippling waves and migrate from one surface to another. Amoebas always are present, trichomonads sometimes are seen, and leukocytes are numerous.

More recent studies suggest that of 300 species of bacteria in the mouth, *Porphyromonas gingivalis* may be a specific cause of some cases of periodontal disease. Researchers at the University of Texas have succeeded in causing a burst of periodontal disease in monkeys given doses of the bacterium. The researchers have also had some success in treating the monkeys with rifampin.

Plaque organisms do not appear to penetrate gum tissue. Their effects are caused largely by the secretion of destructive enzymes, inflammation, and allergic reactions to bacterial products spreading from the plaque.

TREATMENT AND PREVENTION The treatment of chronic periodontal disease is a somewhat controversial subject in dentistry today, and patients vary in their responses to different treatments. Treatments include antimicrobial mouth rinses, brushing with a mixture of bicarbonate of soda and hydrogen peroxide, surgery to eliminate pockets, and antibiotic therapy in unresponsive or rapidly progressing cases. Chronic periodontal disease can be prevented or its onset delayed by daily thorough cleaning of teeth and, most important, frequent professional removal of plaque from the pockets. Once organisms erode gums and form pockets between the teeth and gums, infections must be kept under control.

Viral Diseases of the Oral Cavity

Mumps

Mumps is caused by a paramyxovirus somewhat similar to the measles (rubeola) virus. The virus is transmitted by saliva or as aerosol droplets, enters via the oral cavity or via the respiratory tract, and invades cells of the oropharynx. After initially replicating in the upper respiratory tract, the virus travels in the blood to the salivary glands and sometimes to other glands and organs, such as the testes and the meninges. Swelling of the parotid glands appears 14 to 21 days after initial infection and can persist as long as 7 days. Viruses are released and the infected person is contagious for 7 days before the glands swell and for up to 9 days after swelling subsides. Mumps viruses are excreted in urine for up to 2 weeks after the onset of symptoms.

Humans are the only known hosts for the mumps virus. The virus is found worldwide, occurring especially in the spring. Infections are most common in children 6 to 10 years of age. Although up to 85 percent of an exposed susceptible population will be infected by mumps, 20 to 40 percent will have no symptoms. When the disease appears in postpubertal males, 20 to 30 percent develop **orchitis,** inflammation of the testes. Such infections are capable of causing sterility but rarely do so. Other complications of mumps, regardless of the infected person's age or sex, include meningoencephalitis, eye and ear infections, and inflammation of glands other than the parotid, such as the ovaries and pancreas (where it may be a cause of ju-

venile onset diabetes). An effective vaccine for mumps contains viruses weakened by passage through a sequence of embryonated egg inoculations. The vaccine usually is administered in combination with measles and rubella vaccines—collectively known as the MMR vaccine. The incidence of mumps has decreased dramatically since the vaccine was licensed and put to use in 1967 (Figure 22.5). The vaccine is recommended for all persons who were born after 1957 and have no immunity to the disease. Infants are immunized after 15 months of age.

Other Diseases

Thrush, an oral infection by the yeast *Candida albicans,* was discussed in Chapter 20 and illustrated in Figure 20.10a. ∞ (p. 565) Herpes simplex virus infections, a cause of fever blisters and cold sores on the lips and in the mouth, are discussed in Chapter 25. Diseases of the oral cavity are summarized in Table 22.1.

GASTROINTESTINAL DISEASES CAUSED BY BACTERIA

Bacterial Food Poisoning

Food poisoning is caused by ingesting food contaminated with preformed toxins. It also can be caused by the ingestion of foods contaminated with pesticides, heavy metals, or other toxic substances. In food poisoning caused by microbial toxins, organisms that can

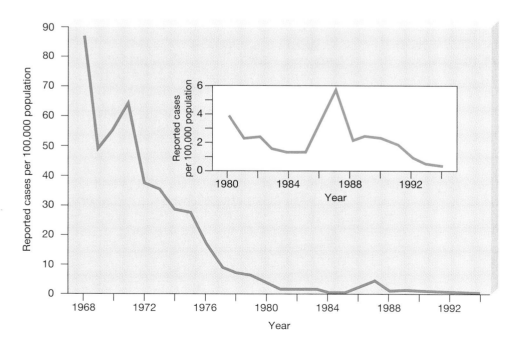

FIGURE 22.5 The incidence of mumps has declined sharply since a vaccine became available in 1967.

TABLE 22.1 Summary of Oral Diseases

Disease	Agent(s)	Characteristics
Bacterial diseases of the oral cavity		
Dental caries	*Streptococcus mutans* and other species	Erosion of tooth enamel and other structures by acids from microbial metabolism
Periodontal disease	Various bacteria, *Porphyromonas gingivalis*	Inflammation and destruction of gums, loosening of teeth, erosion of bone
Viral disease of the oral cavity		
Mumps	Paramyxovirus	Inflammation and swelling of salivary glands and sometimes testes, epididymis, and other tissues

continue to produce toxin may also be ingested with the toxins. However, tissue damage is due to action of the toxin, so most cases of microbial food poisoning are intoxications rather than infections. Because the toxin is preformed, the onset of symptoms in intoxication is more rapid than in infection. Food poisoning can be prevented by following proper food-handling procedures, described in Chapter 27.

Bacteria that produce toxins responsible for food poisoning include *Staphylococcus aureus, Clostridium perfringens, C. botulinum,* and *Bacillus cereus.* Staphylococci usually enter food by way of infected food handlers. Other organisms that cause food poisoning are ubiquitous soil organisms found in water, feces, sewage, and nearly all foods. We will consider the nature of the food poisoning intoxication caused by each of these organisms.

Staphylococcal Enterotoxicosis

Certain strains of *Staphylococcus aureus* cause food poisoning, or **enterotoxicosis** (en"ter-o-tox-i-ko'sis), by releasing certain **enterotoxins**—enterotoxins A or D—which are exotoxins that inflame the intestinal lining and inhibit water adsorption from the intestine. Because the organisms are relatively resistant to heat and drying, foods easily become contaminated with them from food handlers or from the environment. The organisms multiply and release toxin in uncooked or inadequately cooked foods, especially if the foods are unrefrigerated. Nearly any food can be contaminated with *S. aureus,* but those with a starch or cream base are most likely candidates. Cream pies, dairy products, poultry products, and picnic foods such as potato salad are common culprits. Contamination is difficult to detect because it produces no change in food's appearance, taste, or odor. Unlike most exotoxins, the toxin is heat-stable and withstands boiling for 30 minutes. Hence cooking foods can kill the organisms but does not destroy the toxin. ∞ (Chapter 15, p. 404)

When food contaminated with *S. aureus* enterotoxin enters the intestine, the toxin acts directly. The organisms usually continue to produce toxin (but do not multiply). When it comes in contact with the mucosa, the toxin causes tissue damage only after it has entered the blood and has circulated back to the intestine. Symptoms such as abdominal pain, nausea, vomiting, and **diarrhea** (excessive frequency and looseness of bowel movements), but usually not fever, appear 1 to 6 hours after ingestion of contaminated food. The time required for symptoms to appear depends on how long it takes for absorption of a sufficient quantity of toxin to produce them. A food loaded with toxin elicits symptoms quickly; one with only a little toxin takes longer. Once elicited, symptoms usually last about 8 hours. For otherwise healthy adults, no treatment is required because the disease is self-limiting. It can be severe in infants, the elderly, and debilitated patients. Recovery from food poisoning does not confer immunity, as sufficient antibody is not produced. The best way of preventing food poisoning by *S. aureus* is to use sanitary food-handling procedures.

Other Kinds of Food Poisoning

An enterotoxin from *Clostridium perfringens* also causes food poisoning. ∞ (Chapter 15, p. 404) The enterotoxin, which is released only during sporulation, is produced under anaerobic conditions, as when undercooked meats and gravies are kept warm for a while. The main symptom is diarrhea. Compared with *S. aureus* food poisoning, *C. perfringens* food poisoning takes longer to appear—8 to 24 hours after ingestion—and lasts longer (about 24 hours). It too is self-limiting and can be prevented by sanitary food handling.

Although *C. perfringens* enterotoxin causes food poisoning, the organism itself can infect tissues. Upon introduction of spores into a wound, germination to vegetative cells can result in gas gangrene (Chapter 20) and anaerobic cellulitis (inflammation of connective tissue). ∞ (p. 571) Vegetative cells multiply and release toxins and enzymes that diffuse into surrounding healthy tissue, causing cellular destruction and death.

Botulism, which is caused by a neurotoxin of *Clostridium botulinum,* is acquired when toxin-contaminated food is eaten. Although it is a kind of food poisoning, it has little effect on the digestive system. Its effects on the nervous system are discussed in Chapter 24.

Bacillus cereus secretes a toxin that acts as an emetic—that is, it induces vomiting. The incubation period and signs and symptoms resemble those of staphylococcal food poisoning. Symptoms occur less than 12 hours after ingestion and last only a short time. Such toxins are often found in contaminated rice or meat dishes. *Bacillus cereus* exists as a saprophyte in water and soil.

Food poisoning by *Pseudomonas cocovenenans,* which occurs in Polynesia, is named **bongkrek disease,** after its association with the native coconut delicacy bongkrek. The bacterium produces a potent, and often fatal, toxin that is frequently found in dishes prepared with coconut.

Bacterial Enteritis and Enteric Fevers

For most of us diarrhea is an unpleasant inconvenience, but it can be deadly. In 1900 in New York City the death rate among infants from diseases grouped together as diarrhea was 5603 per 100,000. Although today the death rate is less than 60 per 100,000, the disease is still life-threatening, and death from diarrhea can occur in hours in infants.

Enteritis is an inflammation of the intestine. **Bacterial enteritis** is an intestinal infection, not an intoxication, as is food poisoning. The causative bacteria actually invade and damage the intestinal mucosa or deeper tissues. Enteritis that affects chiefly the small intestine usually causes diarrhea. When the large intestine is affected, the result is often called **dysentery,** a severe diarrhea that often contains large quantities of mucus and sometimes blood or even pus. Some pathogens spread through the body from the intestinal mucosa and cause systemic infections, as is the case with typhoid fever. Such infections are called **enteric fevers.**

Salmonellosis

Salmonellosis (sal"mo-nel-o'sis) is a common enteritis caused by some members of the genus *Salmonella.* The annual reported incidence of salmonellosis (excluding typhoid fever) in the United States climbed from 20,000 cases in the early 1970s to more than 65,000 cases in 1986 but has dropped to about 42,000 cases in 1993. Many more cases go unreported: some health experts estimate the true prevalence to exceed 2 million. The classification of the salmonellas has recently been revised. The number of species within the genus *Sal-*

monella has been reduced to three: *Salmonella typhi, S. choleraesuis* (usually a swine pathogen), and *S. enteritidis.* About 2000 strains of salmonellas have been identified by their surface antigens and grouped into **serovars** (se'ro-varz). Prior to 1972 each was named as a separate species. Most of the 2000 strains have been consolidated into the species *S. enteritidis.* The former *S. typhimurium* is now considered to be a serovar (strain) of *S. enteritidis.* However, most investigators still refer to *S. typhimurium* rather than more properly to *S. enteritidis* serovar *typhimurium.* Identification of serovars is sometimes useful in tracing the source of disease outbreaks.

Salmonella species other than *S. typhi,* the cause of typhoid fever, can be found in the gastrointestinal tracts of many animals, including poultry, wild birds, and rodents. In these hosts the bacteria either cause obvious disease or are carried without producing harmful effects. It is now known that chicken eggs can become infected if the hens laying them are infected. Salmonellas have also been traced to contaminated water and to food contaminated by carriers. The practice of confining poultry and pigs to small spaces while being reared for market makes feeding and caring for the animals more efficient, but it also facilitates the transmission of *Salmonella* and other pathogens among them.

Salmonella infection is generally associated with the ingestion of improperly prepared, previously contaminated food. Meat and dairy products are most likely candidates. Foods containing uncooked eggs can also be a source.

Signs and symptoms of salmonellosis include abdominal pain, fever, and diarrhea with blood and mucus. They appear 8 to 48 hours after ingestion of organisms and are associated with the organisms' invading the mucosa of both the small and large intestines. Fever probably is caused by endotoxins, toxins that are released from a cell only when it is lysed. In otherwise healthy adults, salmonellosis lasts 1 to 4 days and is self-limiting. Antibiotics usually are not given because they tend to induce carrier states and to contribute to the development of antibiotic-resistant strains. Infants and elderly or debilitated patients often have more severe and prolonged symptoms. In such cases, antibiotics may be prescribed.

Other serovars of *Salmonella* also cause disease. Because of their ability to invade intestinal tissue and enter the blood, *S. typhimurium* and *S. paratyphi* cause a somewhat more serious condition called **enterocolitis,** or enteric fever. Symptoms and bacteremia appear after an incubation period of 1 to 10 days. Enteric symptoms such as fever and chills can last 1 to 3 weeks. Chronic infections of the gallbladder and other tissues are not uncommon. A carrier state becomes established when, after the patient's recovery, organisms from

PUBLIC HEALTH

BABY CHICKS AND TURTLES

At one time baby chicks and ducks were frequently purchased to delight small children at Easter. In many localities they are no longer available. Although the little birds often were badly mistreated by children, the ban on their sale probably resulted from local health regulations aimed at preventing salmonellosis. If you or your children do handle chicks, consider all their droppings to be infectious, and practice careful hand washing.

Similar bans have been passed against the sale of baby turtles, which also carry *Salmonella*. If you handle turtles, follow the same precautions recommended for handling chicks. Washing the turtle's bowl in the kitchen sink can be hazardous to your health; if you do so, be sure to clean the sink thoroughly. Better yet, whenever possible clean the bowl outdoors.

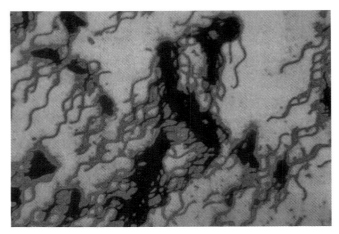

FIGURE 22.6 *Salmonella typhi* (magnified 2000×), the causative agent of typhoid fever. Note the flagella.

chronically infected tissues continue to be excreted with feces. Broad-spectrum antibiotics rid carriers of organisms but can activate the disease in some carriers by upsetting the balance of intestinal microbiota.

Prevention of salmonellosis and enterocolitis depends on the maintenance of sanitary water and food supplies and eradication of organisms from carriers. The organisms cannot be entirely eradicated because poultry and other animals serve as reservoirs, and no effective vaccine is available.

Typhoid Fever

Typhoid fever, one of the most serious of the epidemic enteric infections, is caused by *Salmonella typhi* (Figure 22.6). The disease is rare in places where good sanitation is practiced but is more common where faulty water and sewage systems exist. Uncooked shellfish, raw fruit, or raw vegetables have often served as sources of the pathogen in outbreaks in the United States. In 1942 the total number of U.S. cases was 5000. For most of the last 30 years, however, fewer than 500 cases per year have been reported in the United States. The organisms enter the body in food or water and invade the mucosa of the upper small intestine. From there they invade lymphoid tissues and are phagocytized and disseminated. The organisms multiply in the phagocytes, emerge, and continue to multiply intracellularly.

Bacteremia and septicemia occur at the same time as symptoms appear. During the first week, the patient suffers from headache, malaise, and fever, probably due to an endotoxin. During the second week, the pa-

tient's condition worsens. The organisms invade many tissues, including the intestinal mucosa, and are excreted in the stools. *Salmonella typhi* thrive and multiply in bile; organisms from the gallbladder reinfect the intestinal mucosa and lymphoid tissue such as Peyer's patches. Characteristic "rose spots" often appear on the trunk and abdomen for a few days. Abdominal distention and tenderness and enlargement of the spleen are common complaints, but diarrhea usually is absent. Unlike most other infections, in which leukocytes increase in number, leukocytes decrease in number in typhoid fever. Some patients become delirious and can suffer complications such as internal hemorrhage, perforation of the bowel, and pneumonia. Chloramphenicol is the antibiotic of choice in treating typhoid fever, but some strains of *S. typhi* are resistant to it.

By the fourth week, symptoms subside, convalescence begins, and immunity develops. Cell-mediated immunity provides protection against future infection. Antibodies to surface antigens of the bacterium also are produced, but these are of more use in laboratory diagnosis than in protecting the patient against infection. The Widal test detects antibodies and is used to confirm the diagnosis of typhoid fever. A live, attenuated oral typhoid vaccine is now available. It stimulates cell-mediated as well as humoral immunity, including the production of secretory antibody (IgA), which aids in prevention of mucosal invasion. ∞ (Chapter 18, p. 485) After the initial series of doses, a booster dose is needed every 3 years. Means of protection against typhoid fever include good sanitation and sewage disposal and proper hygiene.

Shigellosis

Shigellosis (shig″el-o′sis), or **bacillary dysentery,** can be caused by several serovars (se′-ro-var), of *Shigella*

THE SAGA OF TYPHOID MARY

1901—Mary Mallon is working as a cook for a family. A visitor with typhoid fever arrives, and Mary develops typhoid fever. One month later the family laundress falls ill with typhoid.

1902—Mary changes employers. Two weeks later the laundress develops typhoid and is soon followed by six more members of the household. Mary quickly leaves.

1903—Now working as a cook in a household in Ithaca, New York, Mary is believed to have started a waterborne typhoid outbreak that spreads widely, killing 1300 people. Mary quickly leaves.

1904—Mary moves on to a household on Long Island; within three weeks, four fellow servants develop typhoid.

1906—Moving again, Mary arrives, and six people fall ill with typhoid, all within 1 week. One girl dies. Two weeks later, Mary—having been suspected as the cause—flees to another family. After another 2 weeks, the family's laundress develops typhoid fever.

1907—Using a false name, Mary starts work as a cook for a family in New York City. Two months later, two household members develop typhoid; one dies. Mary leaves soon after the disease breaks out, but this time the New York City Health Department tracks her down. In March 1907 she refuses to cooperate with Health Department workers who want to test her stool for the presence of typhoid bacilli that would mark this otherwise healthy woman as a carrier of typhoid. Police arrive and forcibly remove her to a hospital on an island in the East River. Tests reveal typhoid bacilli in her stool specimens; presumably they are being shed from her bacilli-colonized gallbladder. No antibiotics are available at this time—nor will they be until the late 1940s. Surgical re-

moval of Mary's gallbladder is the only way to end her carrier state. Not a believer in the germ theory of disease, Mary is certain that she will die under the surgeon's knife, all for foolish nonsense, and refuses to undergo the surgery. She is sentenced to involuntary isolation in the hospital until such time as she is no longer a menace to public health.

1907–1910—Mary's stools are cultured every few days. Some days there are no typhoid bacilli; other days huge numbers appear. Meanwhile, Mary has attained public celebrity status—imprisoned against her will, threatened by the surgeon's knife!

1910—Public sympathy leads to Mary's release. She promises never again to work as a cook.

1915—Twenty-five cases of typhoid occur at the Sloane Hospital for Women in New York City. Eight die—mostly doctors and nurses. The hospital cook steps out "for a few minutes," never to return. She does not use the name Mary Mallon. But when the Health Department catches up with her, it is indeed Mary again, insisting on her right to be employed as a cook—the only job she likes.

It is back to the island hospital again for Mary. On arrival at the quarantine facilities, she offers to work in the kitchen—an offer authorities refuse. Mary steadfastly refuses a gallbladder operation and forcefully insists that she will work as a cook if she gets out. What to do? Mary remains quarantined there for the rest of her life.

1938—Mary Mallon dies at age 70 from a stroke.

Had you been a Health Department official, a judge, a lawyer, or someone else associated with Mary, what would you have done? Before antibiotics and vaccines were available, quarantine was a common practice.

(Figure 22.7). They include *S. dysenteriae* (serovar A), *S. flexneri* (serovar B), *S. boydii* (serovar C), and *S. sonnei* (serovar D). The latter causes 80 percent of the cases in the United States. Shigellosis was first described in the fourth century B.C.. Although less invasive than salmonellosis, it spreads rapidly in overcrowded conditions with poor sanitation. Humans and other higher primates such as chimpanzees and gorillas are the only reservoirs of infection, but the organisms can persist in foods for up to a month. *Shigella* also creates a hazard for zoo workers. ∞ (Chapter 14, p. 370)

The pathogens are spread by contaminated food, fingers, flies, feces, and fomites. Playing, bathing, and washing clothes in contaminated water play significant roles in transmitting *Shigella*. In areas of good san-

FIGURE 22.7 False-color TEM photo of *Shigella* (41,250×), causative agent of shigellosis (bacillary dysentery).

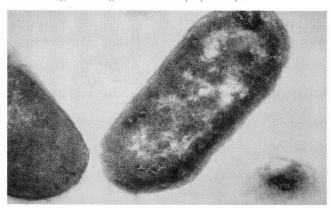

itation, infection is usually acquired by poor hand-washing.

Children aged 1 to 10 are most susceptible to *Shigella,* which accounts for 15 percent of the infant diarrhea cases in the United States. For the past several decades, the total number of annual cases of shigellosis in the United States has ranged between 15,000 and 20,000. Many recent outbreaks have been recorded in day-care centers. The ingestion of just 10 organisms can be sufficient to cause infection. Thus, even slight lapses in hygiene can allow easy spread of the disease. In developing nations it is a major cause of infant mortality, with one-third of all infant deaths due to dehydration from shigellosis and other enteric pathogens. Essential fluid replacement treatment often is not available in developing countries.

Once contaminated food or water is ingested, *Shigella* survive the acidity of the stomach, pass through the small intestine, and attach to portions of it and the large intestine. The pathogens invade host cells and induce the cells to create special filaments for invading adjacent host cells. After an incubation period of 1 to 4 days, abdominal cramps, fever, and profuse diarrhea with blood and mucus suddenly appear. The severity of signs and symptoms ranges from most to least serious in the same order as the serovar designations A to D. In the most serious cases, diarrhea can cause dangerous protein deficiencies—called *kwashiorkor* (kwash-e-or'kor)—and vitamin B$_{12}$ deficiencies, which, together with loss of electrolytes, can result in neurological damage. Along with the fever-eliciting endotoxin found in all serovars, *S. dysenteriae* also produces Shiga toxin which acts as a neurotoxin. The neurotoxin activity is believed to be responsible for the severity and relatively high fatality rate of *S. dysenteriae* disease due to convulsions and coma. All serovars cause ulceration and bleeding of the intestinal lining and sometimes deeper intestinal layers. Symptoms persist for 2 to 7 days and usually are self-limiting, but they can cause severe dehydration and fluid and electrolyte imbalances.

Specific diagnosis of shigellosis can be difficult because the organisms are very sensitive to acids in feces. The organisms die if fecal specimens are not maintained in a buffered transport medium. Viable specimens can be obtained by swabbing a bowel lesion during internal examination of the bowel. Stool specimens are inoculated into selective media. Distinctive colonies usually appear after 24 hours of incubation.

Treatment is necessary in children and debilitated patients. Restoring fluid and electrolytes is essential to recovery. Usually hydration fluids, with or without antibiotics, are used for this purpose. A combination of antibiotics—ampicillin, tetracycline, and nalidixic acid—is used. Nalidixic acid is a synthetic agent that inhibits bacterial DNA synthesis, especially in gram-

negative intestinal pathogens. Prevention is difficult because many people have inapparent infections. A carrier state, usually lasting less than 1 month, exists and accounts for many new cases by fecal–oral transmission. Any breakdown in sanitation can lead to transmission by way of feces, fingers, and flies. Immunity following recovery from shigellosis is transient. From limited experience in recent years, oral vaccines containing certain strains of *S. flexneri* and *S. sonnei* seem to be safe and effective.

Asiatic Cholera

Asiatic cholera, so named because of its high incidence in Asia, can affect people anywhere sanitation is poor and fecal contamination of water occurs. Worldwide, more than 100,000 cases are reported annually. In the United States, fewer than 10 cases are reported per year. Some of those are transmitted to humans from contaminated shellfish in Gulf Coast states. In endemic regions, such as parts of Asia, 5 to 15 percent of patients die; when a seasonal epidemic occurs, as many as 75 percent of patients die.

The causative organism, *Vibrio cholerae* (Figure 22.8), can survive outside the body in cool alkaline water, especially if organic and/or fecal matter is present. When ingested, it invades the intestinal mucosa, multiplies, and releases a potent enterotoxin. The enterotoxin, known as *choleragen,* binds to epithelial cells of the small intestine and makes plasma membranes highly permeable to water. This action results in a significant secretion of fluids and chloride ions and the inhibition of sodium absorption. At this time the intestinal lining becomes shredded, causing numerous small, white flecks resembling rice grains to be passed in the feces. Infected individuals experience severe nausea, vomiting, abdominal pain, and diarrhea. Stools rapidly become clear, containing numerous mucus plugs, thus giving rise to

FIGURE 22.8 False-color SEM photo of *Vibrio cholerae* (3000×), causative agent of asiatic cholera. The bacterium is slightly curved and has a single polar flagellum.

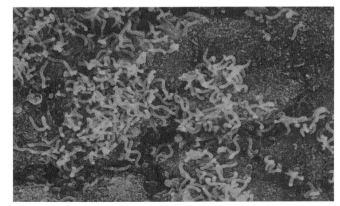

CLOSE-UP

CHOLERA

"The speed with which cholera killed was profoundly alarming, since perfectly healthy people could never feel safe from sudden death when the infection was anywhere near. In addition, the symptoms were peculiarly horrible: radical dehydration meant that a victim shrank into a wizened caricature of his former self within a few hours, while ruptured capillaries discolored the skin, turning it black and blue. The effect was to make mortality uniquely visible: patterns of bodily decay were exacerbated and accelerated, as in a timelapse motion picture, to remind all who saw it of death's ugly horror and utter inevitability."

—William H. McNeill, 1976

APPLICATIONS

ORAL REHYDRATION

Oral rehydration therapy, introduced relatively recently, is so simple and effective that one wonders why it was not discovered earlier. The formula for the solution used is:

$\frac{1}{2}$ teaspoon salt (NaCl)

$\frac{1}{4}$ teaspoon sodium bicarbonate (NaHCO$_3$)

$\frac{1}{4}$ teaspoon potassium chloride (KCl)

4 tablespoons sugar (sucrose)

Dissolve in 1 liter water.

The ingredients are extremely inexpensive and easy to obtain, even in developing countries. The solution can be prepared in the field by people with little or no training. If uncontaminated water is used, the solution is almost certain to be safe. In severe cases of cholera, however, oral rehydration must be used as a supplement to, not as a substitute for, intravenous rehydration.

the term "rice-water stools." As many as 22 liters of fluids and electrolytes can be lost per day, so all patients, regardless of age, are subject to severe dehydration. Special "cholera cots" made of canvas with a hole cut beneath the buttocks have been used. A bucket marked in liters is placed below the hole to measure the fluid lost so that it can be replaced. Most deaths are attributable to shock due to greatly reduced blood volume.

Fluid and electrolyte replacement is the most effective treatment for cholera. During the 1971 epidemic in India and Pakistan, medical personnel were able to save large numbers of victims of the disease largely because of the availability of replacement therapy. Treatment with tetracycline reduces the duration of the symptoms but does not eliminate the organism or the toxin. Recovery from the disease confers only temporary immunity. Many recovered patients remain in a carrier state and can infect others and reinfect themselves. The only available vaccine is not very effective and is not widely used. However, a toxoid-type vaccine is being tested. It also may be possible to develop a vaccine that makes use of secreted IgA against the cholera organism.

A strain of *V. cholerae* known as the El Tor biotype (named after the quarantine camp where it was first isolated) causes a form of cholera that is slower and more insidious in its onset than the classical form of the disease. Cholera outbreaks have often resulted in the imposition of quarantine and the halting of shipping from port cities. In the past, therefore, countries in South and Central America that were economically dependent on their exports would sometimes report that cholera was not present, only "El Tor." The two forms of the disease, however, are essentially the same.

(See the Essay at the end of this chapter for a discussion of recent outbreaks of cholera in Peru and Rwanda.)

Vibriosis

An enteritis called **vibriosis** (vib-re-o'sis) is caused largely by *Vibrio parahaemolyticus*. Although the disease is most common in Japan, where raw fish is considered a delicacy, the organism is widely distributed in marine environments. In the United States, infections usually are acquired from contaminated fish and shellfish that have not been thoroughly cooked. Most outbreaks in the United States occur at outdoor festivities where crab and shrimp are served without thorough cooking and proper refrigeration. Some outbreaks have been traced to the eating of contaminated raw oysters. *Vibrio parahaemolyticus* also can infect skin wounds of people exposed to contaminated water. Once inside the intestine, the organisms colonize the mucosa and release an enterotoxin. Symptoms of nausea, vomiting, diarrhea, and abdominal pain appear about 12 hours after contaminated food or water has been ingested, and they last 2 to 5 days. The disease usually is not treated, and no vaccine is available.

Traveler's Diarrhea

Among the 250 million people who travel internationally each year, it has been estimated that over 100 million suffer from a self-limiting mild to severe diarrhea. Of these travelers, 30 percent are confined to bed, and another 40 percent are forced to curtail their activities. This disorder, officially called **traveler's diarrhea,** also has been called "Delhi belly," "Montezuma's revenge," and some even less attractive names.

The most common causes of traveler's diarrhea are pathogenic strains of *Escherichia coli,* which account for 40 to 70 percent of all cases. Strains differ with geographic location, so travelers are likely to be exposed to new strains. Some strains of *E. coli* are normal inhabitants of the human digestive tract, and only certain strains are capable of causing enteritis. **Enteroinvasive strains** have a plasmid with a gene coding for a particular surface antigen called K antigen, which enables the strains to attach to and invade mucosal cells. **Enterotoxigenic strains** contain a plasmid that enables them to make an enterotoxin. They attach to the mucosa by attachment pili or fimbriae. These organisms also cause numerous cases of infant diarrhea. Other causes of traveler's diarrhea include bacteria such as *Shigella, Salmonella, Campylobacter,* rotaviruses, and protozoa such as *Giardia* and *Entamoeba.* Jet lag and other stresses of travel do not cause diarrhea, but they can lower resistance to infection. Travelers can experience symptoms of diarrhea even when no pathogens are present. Such cases are due to unusual kinds and amounts of substances dissolved in water, which lead to gastrointestinal upsets.

Symptoms of traveler's diarrhea vary from mild to severe and include nausea, vomiting, diarrhea, bloating, malaise, and abdominal pain. A typical case causes 4 to 5 loose stools per day for 3 or 4 days. Fluid loss is greater with invasive strains than with toxigenic strains. The disease is especially hazardous in infants, who are subject to severe dehydration. Bottle-fed infants are much more likely to become infected than breast-fed infants. Before their normal microbiota are sufficiently established to compete with the pathogens, newborns are especially at risk of acquiring pathogenic strains from hospital workers. After infancy, children most often acquire *E. coli* infections during travel in foreign countries, especially those with poor sanitation.

Travelers sometimes medicate themselves with antibiotics before and during a stay in a foreign country. This treatment is not recommended because antibiotics usually are not effective and because such use contributes to the development of antibiotic-resistant strains. A better practice is to keep antidiarrhea medicine available and to use it only after symptoms appear.

Traveler's diarrhea can persist for months or years as a postinfectious irritable bowel syndrome. It can also cause lactose intolerance by damaging intestinal lining cells that normally produce the enzyme lactase, which digests lactose (milk sugar). During bouts of diarrhea, patients should eliminate from their diet dairy products and other foods containing lactose. After a

few weeks, small quantities of such foods can be reintroduced and gradually increased as long as they are well tolerated. In some cases normal lactose tolerance is lost forever. *Escherichia coli* and other bacteria, the protozoan *Giardia,* and certain helminths frequently cause lactose intolerance.

Escherichia coli has significance far beyond its ability to cause diarrhea. It is an important indicator organism because it is always present in water contaminated with fecal material. *Escherichia coli* is usually more numerous than other organisms and is easier to isolate. Finding *E. coli* in water indicates that any pathogens found in feces might also be present. Deadly outbreaks of toxin-producing *E. coli* strain 0157:H7 in 1992 and since have been attributed to undercooked hamburgers served at fast-food restaurants. Several cases involved the kidneys and resulted in the condition known as *hemolytic-uremic syndrome.*

Escherichia coli is also an extremely versatile opportunistic pathogen—it can infect any part of the body subject to fecal contamination, including the urinary and reproductive tracts and the abdominal cavity after perforation of the bowel. It is present in many bacteremias, causes septicemias, and can infect the gallbladder, meninges, surgical wounds, skin lesions, and lungs, especially in debilitated and immunodeficient patients.

Other Kinds of Bacterial Enteritis

Certain strains of *Campylobacter jejuni* and *C. fetus* can be found in food and water and are becoming increasingly associated with human gastroenteritis, especially in infants and in elderly or debilitated patients. Some strains of *C. fetus* cause infectious abortions in several kinds of domestic animals. Although these organisms apparently do not multiply in foods, they are transmitted passively in undercooked chicken, unpasteurized milk, and poultry held in unchlorinated water during processing. Improper cooking of contaminated poultry can lead to enteritis. In some areas *Campylobacter* surpasses *Salmonella* as the major cause of foodborne enteric disease. Many health departments have begun testing for *Campylobacter* in food-handling areas and on equipment.

Campylobacter infections cause copious diarrhea, foul-smelling feces, fever, and abdominal pain. They also cause arthritis in 2 to 10 percent of infected children but rarely in adults. Because large quantities of fluid can be lost, dehydration and fluid and electrolyte imbalances are common among the populations most affected. The disease is treated with fluid and electrolyte replacement and sometimes with tetracycline and erythromycin.

Yersiniosis (yer-sin"e-o'sis), a severe enteritis, is caused by *Yersinia enterocolitica.* The disease is most common in western Europe, but some cases are seen in the United States. This free-living organism is found mainly in marine environments, but it can survive in many places. Infection can be acquired from water, milk, seafoods, fruits, and vegetables, even when they are refrigerated, because the organism grows more rapidly at refrigerator temperatures than at body temperature. The organism is most easily identified when cultured at 25°C, a temperature at which the organisms are motile and easily distinguished from other bacteria. Yersiniosis symptoms, which are related to release of an enterotoxin, are similar to other kinds of enteritis, but abdominal pain usually is more severe, and white blood cells increase in number. Yersiniosis sometimes is misdiagnosed as appendicitis because of the similarity of symptoms.

A new source of *Yersinia* infection has caused the CDC to issue a warning to preparers of the Southern delicacy chitterlings, or chitlins (pork intestines). Until the preparers have very carefully washed their hands,

they should avoid touching children or anything used by children. Children should not be allowed to handle raw chitterlings. Fifteen children in Atlanta recently became ill, mainly from contact with preparers, whereas the adults remained well.

Bacterial Infections of the Stomach, Esophagus, and Intestines

Peptic Ulcer and Chronic Gastritis

Recent studies have revealed a bacterial cause of peptic ulcers and chronic gastritis and a probable cofactor of stomach cancer (Figure 22.9 and book cover). The organism, *Helicobacter pylori* (formerly called *Campylobacter pylori*), was first cultured in 1982 from gastric biopsy tissues by Barry Marshall and J. Robin Warren. It is able to survive the very acidic conditions of the stomach by generating ammonia from urea. The am-

FIGURE 22.9 (a) Healthy stomach lining is (b) invaded by *Helicobacter pylori*, a spiral-shaped bacterium. Recent research indicates that the organism is the cause of (c) chronic gastritis and (d) peptic ulcers and (e) may be involved in stomach cancer. Antibiotic treatment can lead to permanent cure of ulcers if reinfection does not occur.

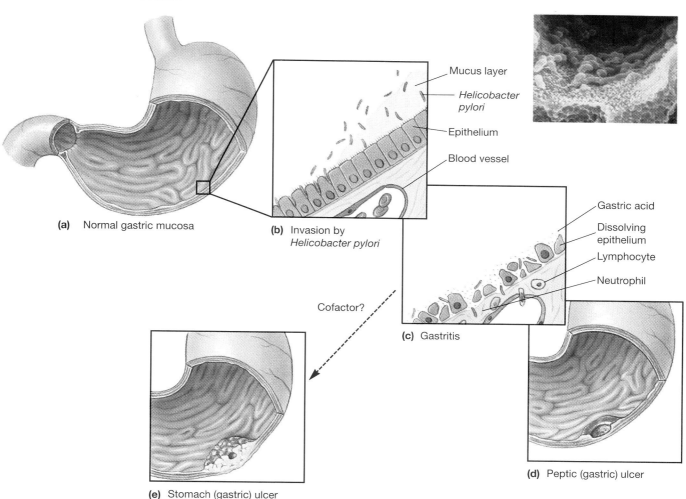

(a) Normal gastric mucosa

(b) Invasion by *Helicobacter pylori*

Mucus layer
Helicobacter pylori
Epithelium
Blood vessel

Gastric acid
Dissolving epithelium
Lymphocyte
Neutrophil

(c) Gastritis

Cofactor?

(d) Peptic (gastric) ulcer

(e) Stomach (gastric) ulcer

monia is believed to neutralize gastric acidity around the *Helicobacter* cells, thereby allowing the organisms to survive and reproduce. They colonize and multiply in the gastric mucosa directly above the epithelial cell layer of the stomach.

Peptic ulcers are lesions of the mucous membranes lining the esophagus, stomach, or duodenum. The lesions are caused by the sloughing away of dead inflammatory tissue and exposure to acid; they eventually result in an excavation into the surface of the organ. Four million Americans suffer from ulcers each year. Approximately 10 percent of the population will suffer from an ulcer at some point in their lives. Ulcers are responsible for about 46,000 operations and 14,000 deaths per year.

Chronic gastritis (stomach inflammation) may be so mild as to cause no noticeable signs or symptoms, or it can produce pain and indigestion. It is observed in 70 to 95 percent of people who have peptic ulcers. Severe gastritis may lead to ulceration.

Helicobacter pylori is present in 95 percent of patients with duodenal ulcers and in 70 percent of those with gastric ulcers. In developed countries, it is uncommon for children to be infected, but the rate of infection increases by about 1 percent per year of age above 20, eventually reaching an average of 20 to 30 percent among U.S. adults. In developing countries, it is commonly found in children and reaches levels of 80 percent of the adult population. In Latin America, stomach cancer rates are the highest in the world, and so is the rate of *H. pylori* infection. Interestingly, U.S. Hispanics also have a 75 percent infection rate. In Japan, rates of stomach cancer and of *H. pylori* infection are dropping together. It may be that just having chronic inflammation of the stomach for a long period of time—which occurs when a person suffers *H. pylori* infection early in life—predisposes toward stomach cancer. Therefore scientists are unwilling to call *H. pylori* a definite cause of stomach cancer but feel it may be only a cofactor because most people infected with *H. pylori* never develop stomach cancer. However, among the various kinds of stomach cancer, the most common type (gastric intestinal) has an 89 percent correlation with *H. pylori* infection.

No one is certain yet of the route of infection or the portal of exit. It is of great importance, however, if we find that we can prevent or cure ulcers by eliminating *H. pylori* infection. Currently available drugs such as Tagament only control, but cannot cure, ulcers. Experimental treatments with antibiotics have reported cure rates as high as 80 to 90 percent, but these drugs must be used very carefully. Patients treated only with drugs that suppress stomach acid had a relapse rate of 75–95 percent within 2 years. Researchers expect to have better tests, drugs, and treatment plans worked out in a few years.

Approaches to *H. pylori* identification include direct detection of the organism in gastric biopsies, testing for the presence of the enzyme urease in biopsy tissue, and culturing specimens on special media. Serological tests for specific antibody against *H. pylori* have been developed and are undergoing evaluation.

Pseudomembranous Colitis

Pseudomembranous colitis, a condition characterized by the formation of a membrane covering on the mucosal surface of the colon, is caused by *Clostridium difficile*. The signs and symptoms, diagnosis, and treatment for *C. difficile* disease were discussed in the Essay at the end of Chapter 14. ∞ (p. 387)

Bacterial Infections of the Gallbladder and Biliary Tract

The gallbladder and its associated ducts can also be sites of bacterial infection. The liver produces bile, which is stored in the gallbladder and slowly released into the small intestine. Organisms with lipid-containing envelopes are destroyed by the action of bile, which breaks down lipids. Most enteric viruses, poliovirus, and hepatitis A virus (discussed in the next section) lack lipid envelopes and are protected from bile activity. Organisms such as the typhoid bacillus

APPLICATIONS

FUNGAL PRODUCT ALLEVIATES FLATULENCE

Troubled by intestinal gas? Well, everyone has it. The average person releases about a liter per day. This gas, called *flatus*, is produced by colon bacteria as they metabolize compounds that you have been unable to digest. Complex sugars from foods such as beans, peas, cabbage, brussels sprouts, oat bran, bananas, apples, and high-fiber cereals are broken down by some bacteria, releasing hydrogen gas plus other gases and compounds as byproducts. Other intestinal bacteria consume the hydrogen. The balance between these two groups of bacteria partially explains why some people produce more gas than others.

The commercial product Beano, produced by and collected from the mold *Aspergillus niger*, contains the sugar-digesting enzyme that people lack. The enzyme, an α-galactosidase, breaks the alpha linkages in the complex sugars, digesting them before they reach the bacteria in the colon, thus reducing the amount of flatulence.

are so resistant to the action of bile that they can actually grow in the gallbladder itself. They are shed from the gallbladder into the intestine and are eliminated in feces. People carrying these organisms in the gallbladder are asymptomatic. (See the box "The Saga of Typhoid Mary," p. 619.)

Gallstones, formed from crystals of cholesterol and calcium salts, can block the bile ducts, decreasing the flow of bile and predisposing the individual to inflammation of the gallbladder (*cholecystitis*) or biliary ducts (*cholangitis*). Distension of the gallbladder due to the accumulation of bile fosters the entry of microbes—most commonly *E. coli*—into the bloodstream, probably through minute tears in the gallbladder wall. Infection of the gallbladder can also ascend into the liver.

Diagnosis of gallbladder infections is based on clinical findings of recurring pains (*biliary colic*), nausea, vomiting, chills, fever, and often jaundice due to the absorption of blocked bile into the bloodstream. Despite prompt treatment with antibiotics, gallbladder infections cannot be cured unless the causative obstruction is removed, either by surgery or by the spontaneous passage of the gallstone.

Gastrointestinal diseases caused by bacteria are summarized in Table 22.2.

TABLE 22.2	Summary of Gastrointestinal Diseases Caused by Bacteria	
Disease	**Agent(s)**	**Characteristics**
Bacterial food poisoning		
Staphylococcal enterotoxicosis	*Staphylococcus aureus*	Heat-stable enterotoxin causes tissue damage, abdominal pain, nausea, vomiting
Other kinds of food poisoning	*Clostridium perfringens, C. botulinum, Bacillus cereus*	Diarrhea and sometimes intestinal infection and gas gangrene
Bacterial enteritis and enteric fevers		
Salmonellosis	*Salmonella typhimurium, S. enteritidis*	Abdominal pain, fever, diarrhea with blood and mucus from toxin; enterocolitis from invasion of organisms; chronic infections and carrier states occur
Typhoid fever	*Salmonella typhi*	Organisms invade mucosa and lymphatics, multiply in phagocytes and other tissues; high fever and "rose spots"; carrier state and life-threatening complications can occur
Shigellosis	*Shigella* strains	Organisms cause intestinal lesions and release toxins; symptoms include cramps, fever, profuse diarrhea with blood and mucus
Asiatic cholera	*Vibrio cholerae*	Organisms invade intestinal lining, release potent toxin that increases lining permeability; symptoms include nausea, vomiting, copious diarrhea, fluid imbalances
Vibriosis	*Vibrio parahaemolyticus*	Organisms colonize mucosa and release toxin; nausea, vomiting, diarrhea are self-limiting
Traveler's diarrhea	Pathogenic strains of *Escherichia coli*, other bacteria (also viruses and protozoa)	Organisms can invade mucosa and/or produce toxin, cause nausea, vomiting, diarrhea, bloating, malaise, abdominal pain; self-limiting except for postinfection complications; dehydration and death in infants
Other kinds of bacterial enteritis	*Campylobacter jejuni, C. fetus, Yersinia enterocolitica*	*Campylobacter* species cause enteritis in infants and debilitated patients; *Yersinia* releases a toxin that causes enteritis with pain resembling appendicitis
Bacterial infections of the upper gastrointestinal tract		
Ulcers, stomach cancer	*Helicobacter pylori*	Definite association with ulcers; probable cofactor in stomach cancer
Cholecystitis, cholangitis	Usually *E. coli*	Blockage of bile ducts by gallstones causes inflammation of gallbladder and bile ducts; accumulation of bile can cause infection to spread to bloodstream or liver
Pseudomembranous colitis	*Clostridium difficile*	Formation of a pseudomembrane on the mucosal surface of the colon; occurs in patients whose gastrointestinal microbiota have been altered by the use of antibiotics

GASTROINTESTINAL DISEASES CAUSED BY OTHER PATHOGENS

Viral Gastrointestinal Diseases

Viral Enteritis

Rotavirus infection is a major cause of **viral enteritis** among infants and young children. Rotaviruses are transmitted by the fecal–oral route, replicate in the intestine, damage the intestinal epithelium, and cause a watery diarrhea within 48 hours. Rotavirus infection is a major cause of infant morbidity and mortality in developing countries. There 3 to 5 billion cases occur annually, with 5 to 10 million deaths in children under age 5. Rotavirus infections account for a third of childhood deaths in some countries. Rotavirus infections also occur in infants and young children in developed countries, where the infections are often nosocomial. Special care should be used with hospitalized children to prevent such infections. The number of cases rises dramatically during the winter in the United States. This timing helps distinguish it from bacterial diarrheas.

Rotaviruses are reoviruses that replicate their double-stranded RNA in such great numbers that they can easily be found by electron microscopy in fecal suspensions (Figure 22.10); no effort to concentrate the viruses is required. Immunoelectron microscopy (immunologic techniques combined with electron microscopy) can be used to identify rotaviruses when antibodies from a patient's serum are reacted with virions in diagnostic specimens. In addition, many hospitals now use ELISA tests that detect rotaviruses in stool specimens. Although there is no specific treatment, restoring fluid and electrolyte balance is crucial and should be prompt.

Rotavirus infection in humans is important in medical practice in the United States. Antibodies to the virus

FIGURE 22.10 Rotaviruses (625,000×), in fecal suspension, resemble little wheels and are a cause of diarrhea.

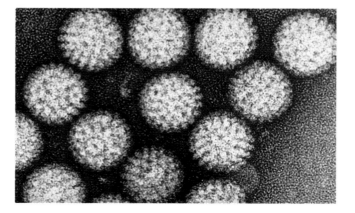

have been identified in as many as 90 percent of groups of children tested, even though the infection may not have been identified at the time it occurred. Much more research is needed to determine how immunity is produced and how the disease might be controlled.

Enteritis can also be caused by viruses other than rotaviruses. Species of *Enterovirus* such as the echoviruses (*enteric cytopathic human orphan viruses*) can cause mild gastrointestinal symptoms and damage intestinal cells. They also sometimes infect other tissues, and certain species can cause meningoencephalitis (inflammation of the brain and meninges).

Patients who have received bone marrow transplants are especially susceptible to infection with rotaviruses, enteroviruses, and the bacterium *Clostridium difficile*. As many as 55 percent of such patients succumb to these infections.

The *Norwalk virus* (named for a 1968 outbreak in Norwalk, Ohio) is responsible for nearly half of all U.S. outbreaks of acute infectious nonbacterial enteritis. Norwalk virus infection affects older children and adults more often than it affects preschoolers or infants. Outbreaks occur throughout the year and are common at schools, camps, and nursing homes and on cruise ships. The infection is the second most common cause of illness (after respiratory disease) among U.S. families and occurs worldwide. It is characterized by 1 to 2 days of diarrhea, vomiting, or both. Immunity does not follow an attack, which makes development of a vaccine unlikely. Careful sanitary practices are the best means of prevention.

Hepatitis

Hepatitis, an inflammation of the liver, usually is caused by viruses (Figure 22.11 and Table 22.3). It also can be caused by an amoeba and various toxic chemicals. The most common viral hepatitis is **hepatitis A,** formerly called **infectious hepatitis.** It is caused by the hepatitis A virus (HAV), a single-stranded RNA virus usually transmitted by the fecal–oral route. **Hepatitis B,** formerly called **serum hepatitis,** is caused by the hepatitis B virus (HBV), a double-stranded DNA virus usually transmitted via blood. A third type of hepatitis is transmitted parenterally (by blood) and is probably caused by at least two viral agents. This type is diagnosed in the absence of HAV and HBV as **hepatitis C** (HCV), formerly called *non-A, non-B (NANB) hepatitis.* A fourth type of hepatitis, transmitted by the fecal–oral route and formerly called *non-A, non-B, non-C hepatitis,* has been separated out as **hepatitis E** (HEV). An especially severe form of the disease **hepatitis D,** or **delta hepatitis,** is caused by the presence of both hepatitis D virus (HDV) and HBV. However, HDV alone does not cause disease; it cannot infect without HBV.

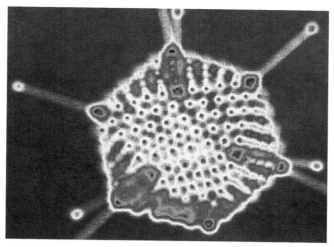

FIGURE 22.11 Computer graphic of the hepatitis A virus (magnified 412,000×), showing capsomeres and penton spikes projecting from the corners of the virion.

Hepatitis A

Hepatitis A occurs most often in children and young adults, especially in autumn and winter. It can occur in epidemics if a population is subjected to water or food, especially shellfish, contaminated with HAV. Outbreaks due to contaminated food in fast-food restaurants have been on the rise.

Hepatitis A has an incubation period of 15 to 40 days and begins as an acute febrile illness. After entering the body through the mouth, the viruses replicate in the gastrointestinal tract and spread through the blood to the liver, spleen, and kidneys. Jaundice, a yellowing of the skin common in hepatitis, is caused by impaired liver function: The liver fails to rid the body of a yellow substance called **bilirubin,** which is a product of the breakdown of hemoglobin from red blood cells. Other symptoms of hepatitis are malaise, nausea, diarrhea, abdominal pain, and lack of appetite for a period of 2 days to 3 weeks. Chronic infections are rare, and recovery usually is complete. Immunological tests are available to detect hepatitis A viruses

TABLE 22.3	Comparison of Types of Viral Hepatitis				
Characteristic	**Hepatitis A**	**Hepatitis B**	**Hepatitis C**	**Hepatitis D**	**Hepatitis E**
Alternate names	Infectious hepatitis; epidemic hepatitis; short-term hepatitis	Serum hepatitis	Parenterally transmitted non-A, non-B, posttransfusion hepatitis	Delta hepatitis	Enterically transmitted non-A, non-B, non-C hepatitis
Agent	HAV RNA virus, Picornaviridae	HBV DNA virus, Hepadnaviridae	HCV At least two unclassified RNA viruses; ? Flavivirus, ? Togavirus	HDV Defective RNA virus; has hepatitis B capsid	HEV Unclassified RNA virus; ? Calcivirus
Transmission	Fecal–oral	Blood and other body fluids; crosses placenta with high frequency	Blood and blood products; occasionally crosses placenta	Blood, must coinfect or superinfect with hepatitis B; can cross the placenta	Fecal–oral; more common in adults than in children
Incubation period	15–40 days; average, 28 days	45–180 days; average, 90 days	Short, 2–4 weeks; long, 8–12 weeks	2–12 weeks	2–6 weeks
Severity of disease	Self-limiting; usually mild, rarely severe	Subclinical to severe; most recover completely	Subclinical to severe; most resolve spontaneously	Severe; high mortality rate	Moderate but high mortality in pregnant women
Carrier state	No	Yes, is associated with 80% of liver cancer	Yes, possible association with liver cancer	Yes	No
Chronic liver disease	No	Yes	Yes	Yes	No
Vaccines	Yes	Yes	No	No	No

and host antibodies against them. There is no treatment for hepatitis other than alleviating symptoms. A vaccine for hepatitis A is now available. Gamma globulin injections are also used to provide temporary immunity.

Hepatitis B

Hepatitis B occurs in people of all ages with about the same incidence throughout the year. It can be transmitted by intravenous or percutaneous (into the skin) injections, by anal/oral sexual practices (common among homosexual males), contact with other virus-containing body secretions (including semen and breast milk), and by contaminated needles among intravenous drug users. Health care workers who have routine contact with patients' body fluids (especially blood) have a higher incidence of the disease than does the general community. Transmission via contaminated semen in artificial insemination has been documented.

Hepatitis B has an incubation period of 45 to 180 days, with an average of 90 days. The virus replicates in cells of the liver, lymphoid tissues, and blood-forming tissues. It can persist in the blood for years, thus creating a carrier state. The onset of symptoms is insidious, and fever is uncommon. Otherwise the symptoms are similar to those of hepatitis A, except that chronic active hepatitis B frequently destroys liver cells.

Immunological methods are available to detect hepatitis B virus and the host's antibodies. Treatment relieves some symptoms but does not cure the disease. An effective vaccine is available, and government regulations require that it be provided by employers for health care workers who may have contact with blood or body fluids that might contain HBV. The vaccine is safe. It is produced by recombinant DNA technology—the insertion of the appropriate hepatitis B virus genes into a plasmid. The plasmid is then inserted into yeast cells, thereby avoiding all contact with human cells. This yeast-produced vaccine has been given to over 2 million people in the United States and is 95 percent effective. When given to hepatitis B-infected pregnant women, it reduces from 90 to 23 percent the number of infants who become carriers. This number can be reduced to 5 percent when gamma globulin is given along with the vaccine. Some 40 percent of carriers die of liver disease, and in some parts of the world nearly 90 percent of mothers are infected. Health officials are currently urging that all infants be vaccinated at birth. But such vaccination is expensive and is probably beyond the means of many developing countries, where it is most important.

The hepatitis B virus is unusually stable and resists drying and irradiation. Its double-stranded, circular DNA has a gap in one strand that may help it insert into liver cell DNA. Insertion of viral DNA into liver cell DNA, in turn, may contribute to liver cell carcinoma, a kind of cancer that occurs much more fre-

quently in people who have had hepatitis B than in the general population.

Hepatitis C

Hepatitis C virus has been recovered from parenterally transmitted non-A, non-B hepatitis cases. Because hepatitis C disease has two different incubation periods—2 to 4 weeks and 8 to 12 weeks—some researchers believe that there may be two different causative agents. Hepatitis C can be distinguished from other kinds of hepatitis by the high blood concentration of a liver enzyme, alanine transferase. Various enzymes are released into the blood from damaged liver cells in all types of hepatitis, but this particular enzyme is unusually elevated in HCV hepatitis. Although the disease usually is mild or even inapparent, the infection can be severe in compromised individuals and becomes chronic in about half those infected. No vaccine is available, and immunity does not follow infection.

Hepatitis E

Transmitted through fecally contaminated water supplies, hepatitis E has caused large outbreaks in Asia and Africa. It is more common in adults than in children. The mortality rate is low (1 percent), except in pregnant women, for whom it is about 20 percent. No vaccine is available, and immunity does not follow infection.

Hepatitis D

Hepatitis D has an incubation period of 2 to 12 weeks. That period is shorter when HBV carriers are superinfected with HDV than when individuals are infected with both viruses at the same time. HDV alone fails to cause disease because it requires HBV antigens for replication. HDV and HBV together can result in death.

The hepatitis virus story has not ended. Recently three more viruses have been discovered that are associated with hepatitis-like disease. If these viruses turn out to be causative agents for true hepatitis cases, it has been proposed that they be named hepatitis F, G, and H viruses. Some investigators believe that there may be dozens more viruses that can cause little-known forms of hepatitis.

Protozoan Gastrointestinal Diseases

Giardiasis

The flagellated protozoan *Giardia intestinalis* (Figure 22.12a) was first observed by Leeuwenhoek in 1681 when he was studying organisms in his own stools. It is, however, a far older organism. Examination of

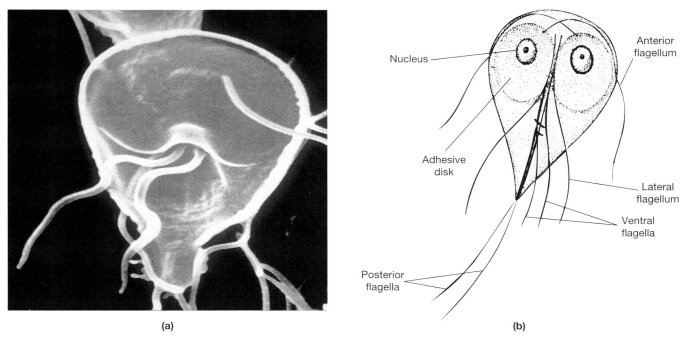

FIGURE 22.12 (a) *Giardia intestinalis* (magnified 35,600×), a protozoan parasite that causes diarrhea. (b) The structure of *G. intestinalis*.

Giardia's DNA has shown it to be the most primitive DNA of any eukaryote, very similar to that of older prokaryotes.

Giardia infects the small intestine of humans, especially children, and causes a disorder called **giardiasis** (je"ar-di'a-sis). After cysts ingested from fecal material pass through the stomach and small intestine, motile trophozoites are released in the colon. ∞ (Chapter 12, p. 307) The parasite has an adhesive disk by which it attaches to the bowel wall (Figure 22.12b). It feeds mainly on mucus, forming cysts that are deposited in the mucus and passed intermittently in mucous stools. Symptoms of giardiasis include inflammation of the bowel, diarrhea, dehydration, and weight loss. Nutritional deficiencies are common in infected children because the parasites can occupy much of the intestinal absorptive area. Fat absorption is greatly reduced, and deficiencies of fat-soluble vitamins are common. Diarrhea is copious and frothy from bacterial action on unabsorbed fats, but it is not bloody because the parasitic protozoa usually do not invade cells. Some infected persons experience severe joint inflammation and an itchy rash even before they have diarrhea. This *reactive arthritis of Giardia* fails to respond to anti-inflammatory drugs ordinarily used to treat arthritis. The arthritis disappears along with the diarrhea when antiprotozoan drugs are given.

Giardiasis is transmitted through food, water, and hands contaminated with fecal matter. It is occasionally transmitted from wild animals and is thus called "beaver fever" by backpackers and hunters in the western United States. Contaminated water supplies in Aspen, Colorado, Leningrad, Russia, and probably many other places have caused large numbers of cases. In some child-care centers, up to 70 percent of the children are infected with the pathogen. *Giardia* cysts are not killed by ordinary sewage treatment and chlorination. Some localities in Pennsylvania are being forced to drink bottled water until they can afford to add sand filters to their treatment plants to trap *Giardia* cysts.

Diagnosis is made by microscopic examination and finding cysts of the protozoan in stools. *Giardia* is found in the mucous sheets that line the intestinal surface. These sheets are released every few days. Samples such as those taken from a bedpan should include some mucus. Because of the intermittent passing of cyst-containing mucus, daily stool samples for several days are needed to increase the likelihood of a positive diagnosis. Trophozoites are sometimes found in watery stools. Immunofluorescent antibody techniques and ELISA procedures are also used for diagnosis.

Quinacrine (Atabrine), furazolidine, and metronidazole (Flagyl) are used to treat giardiasis. The disease can be prevented by maintaining pure water supplies uncontaminated by human or animal wastes.

Amoebic Dysentery and Chronic Amebiasis

Amebiasis (am-e-bi'a-sis) is caused by *Entamoeba histolytica* (Figure 22.13), a major pathogenic amoeba. It was originally isolated from intestinal ulcers of a pa-

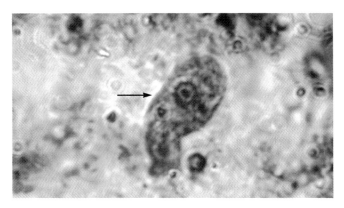

FIGURE 22.13 *Entamoeba histolytica* trophozoite (active amoebic form; 1550×).

tient who had succumbed to severe diarrhea, used to infect a dog in which the same disease appeared, and recovered from the dog, thereby demonstrating Koch's postulates. ∞ (Chapter 15, p. 397) Amebiasis can appear as a severe acute disease called **amoebic dysentery** or as **chronic amebiasis,** which can suddenly revert to the acute stage. Approximately 400 million people are infected worldwide, most with chronic amebiasis. The proportion of the infected population varies from 1 percent in Canada to 5 percent in the United States to 40 percent in tropical areas.

Humans become infected with the parasite by ingesting cysts in food or water contaminated with fecal matter. After ingestion and passage through the stomach and small intestine, cysts rupture and release amoeboid trophozoites in the colon. Trophozoites reproduce asexually within the colon. There they feed on bacteria that make up the normal microbiota of the large intestine. They may cause very little trouble to the host, or they may invade the intestinal mucosa, where they can live indefinitely. Once they have invaded the intestinal mucosa, the parasites multiply and cause significant ulceration. Sometimes their proteolytic enzymes digest deep into, or even through, the bowel wall. Thus, the protozoa sometimes enter blood vessels and travel to other tissues, or they allow bacteria in fecal material to enter the body cavity and cause peritonitis. Patients with amebiasis have abdominal tenderness, 30 or more bowel movements per day, and dehydration from excessive fluid loss. If the parasites invade liver and lung tissue, they can cause abscesses. Bacterial infection of lesions in any tissues can occur.

Because fecal material becomes dehydrated as it passes through the colon, *E. histolytica* trophozoites tend to encyst there. The cysts are passed with the feces. Cysts can survive up to 30 days in a cool, moist environment and are not killed by normal chlorine concentrations in water. The most common means of transmission is through the fecal–oral route. Flies and cockroaches can also be mechanical vectors. Infection can be acquired through sexual practices in which there is ingestion of fecal material.

Amoebic infections can be diagnosed by finding trophozoites or cysts in stools, but several stool samples on consecutive days may be needed to find them. Immunofluorescent antibody and ELISA procedures are also available for diagnosis. Metronidazole is widely used in treatment, even though it has been found to be mutagenic in bacteria and carcinogenic in rats. Antibiotics also are used to prevent or cure secondary bacterial infections. Such infections can be prevented by sanitary handling of water and food.

Balantidiasis

Balantidium coli (Figure 22.14a,b) is the only ciliated protozoan that causes human disease. It is distributed

FIGURE 22.14 (a) *Balantidium coli*, in a fecal smear, is a very large ciliated protozoan that causes diarrhea. It is the only ciliate known to infect humans. (b) Structure of *B. coli.*

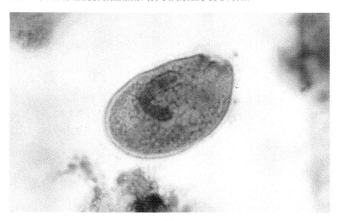

(a)

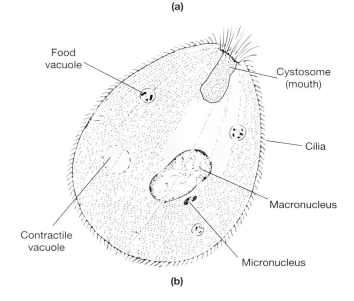

(b)

worldwide, particularly in the tropics, but human infection is rare except in the Philippines. *Balantidium coli* is transmitted by cysts in fecal matter. After being ingested, cysts rupture and release trophozoites that invade the walls of the large intestine, causing a dysentery known as **balantidiasis** (bal"an-tid-i'a-sis). Symptoms of the disease are similar to those of amoebic dysentery; as with amoebic dysentery, perforation of the intestine from balantidiasis can lead to fatal peritonitis.

Diagnosis is made by finding trophozoites or cysts in fecal specimens. Tetracycline or metronidazole are used to treat the disease, but some people remain carriers even after treatment. As with other organisms transmitted through fecal matter, infection can be prevented by good sanitation. Pigs serve as a reservoir of infection, so contact with their feces must be avoided.

Cryptosporidiosis

Protozoans of the genus *Cryptosporidium* commonly cause opportunistic infections worldwide, probably by fecal–oral transmission from kittens and puppies. These organisms live in or under the membrane of cells lining the digestive and respiratory systems. After being swallowed, cysts burst in the intestine, releasing protozoa that invade intestinal cells or migrate to other tissues. In 1993 an outbreak of watery diarrhea affected some 403,000 Milwaukee residents; the outbreak was attributed to *Cryptosporidium* infection of one of the city's water treatment plants. In immunocompetent individuals, the disease is self-limiting, but in immunosuppressed patients, it causes severe diarrhea—up to 25 bowel movements per day with a loss of as much as 17 liters of fluid. The majority of severe cases of **cryptosporidiosis** (krip"to-spor-id-e-o'sis) have been seen in AIDS patients. No effective treatment has been found.

Effects of Fungal Toxins

Fungi produce a large number of toxins, and most come from members of the genera *Aspergillus* and *Penicillium*. Their various effects on humans include loss of muscle coordination, tremors, and weight loss. Some are carcinogenic.

Aspergillus flavus and other aspergilli produce poisonous substances called **aflatoxins** (af"lah-tox'inz). Aflatoxins are the most potent carcinogens yet discovered. Although the effects of the toxins on humans are not fully understood, their presence in foodstuffs may cause cancer of the liver. The toxins reach humans in food made from mold-infested grain and peanuts.

Claviceps purpurea is a fungus that grows parasitically on rye and wheat (Figure 22.15). **Ergot** (er'got) is the name of the vegetative structure *C. purpurea* forms as well as the name of the toxin and the disease it

CLOSE-UP

"HOLD THE AFLATOXIN!"

Aflatoxin can be present in peanut butter if any moldy peanuts are used to make it. In one study, 7 percent of peanut butter samples tested contained aflatoxin. Similar toxins can be present in jellies. Even when the moldy top layer of jelly is discarded, some toxins may have diffused down into the jelly. If, in addition, moldy bread is used, a potentially highly carcinogenic sandwich can be concocted from these simple ingredients.

causes in grains. Although most strains of these grains grown in the United States are genetically resistant to the ergot fungus, many strains grown in other parts of the world are not. Ergot causes a variety of effects on humans. When the fungus is harvested with rye and incorporated into foodstuffs, it can cause **ergot poisoning,** or *ergotism*—hallucinations, high fever, convulsions, gangrene of the limbs, and ultimately death.

FIGURE 22.15 (a) Micrograph of a cross-section through the parasitic fungus *Claviceps purpurea* (magnified 65×), which produces ergot. (b) The structure of ergotamine, a toxic alkaloid produced by the ergot-producing mold.

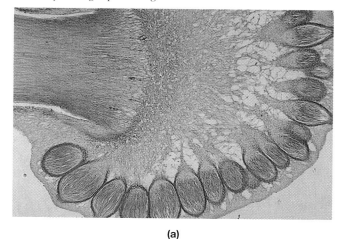

(a)

(b)

CLOSE-UP

ERGOT POISONING

Ergot poisoning has been implicated in the witchcraft trials of Massachusetts and Connecticut in 1692. The now-known symptoms of ergot poisoning are quite like the visions and fits reported in individuals hanged as witches. *Claviceps purpurea* thrives on rye, which was commonly used in bread making in colonial times. Bread pans from this period contain evidence of ergot-contamination. Weather records for 1690 to 1692 show that New England summers were especially cool and moist and that winters also were especially cold—ideal for the growth of *C. purpurea*. All the conditions necessary for ergot poisoning among the population were present.

The same substance that causes ergot poisoning in large quantities can be used therapeutically in small quantities. Drugs derived from ergot are used in carefully measured doses to control bleeding in childbirth, induce abortions, treat migraine headaches, and lower high blood pressure.

Mushroom toxins are found mainly in various species of *Amanita,* which are widely distributed around the world (Figure 22.16). The mushroom tox-ins phallotoxin and amatoxin act on liver cells. They cause vomiting, diarrhea, and jaundice. Ingestion of a sufficient quantity of the toxin can be lethal or can so damage the liver that an organ transplant is necessary.

Helminth Gastrointestinal Diseases

A wide variety of helminths can parasitize the human intestinal tract, and some also invade other tissues. Although most are prevalent only in tropical regions, several are endemic to the United States. Health care workers in the United States should be alert to the possibility that patients may have acquired such parasites while traveling in tropical areas.

Fluke Infections

Foodborne fluke infections affect 40 million people worldwide. The sheep liver fluke *Fasciola hepatica* (Figure 22.17) is the most thoroughly studied of all flukes. It is found in humans in South America, Cuba, northern Africa, and some parts of Europe. Its intermediate host is a snail. Cercaria (larval forms) develop in the snail and upon their release mature in water and encyst as metacercaria on water vegetation. When humans eat such vegetation, especially watercress, the metacercaria are released in the large intestine, bore through the intestinal wall, and migrate to the liver. There they feed on blood, block bile ducts, and cause inflammation. Sometimes they migrate to the eyes, brain, or lungs. Adult flukes can be found in the bile ducts and the gallbladder. Infections can be diagnosed by finding eggs in stool specimens. Infected humans can be treated with bithionol and other antihelminthic agents. Infection can be prevented by not eating water vegetation unless it has been cooked.

FIGURE 22.16 Deadly toxin–producing mushrooms. *Amanita* mushrooms kill by means of a toxin that inhibits RNA polymerase. (a) *Amanita muscaria,* which was formerly used as an insecticide. It was sprinkled with sugar and set out to attract flies, which died after nibbling on the insecticide-laced sugar. (b) *Amanita virosa,* commonly called the "destroying angel."

(a)　　　　　　　(b)

FIGURE 22.17 *Fasciola hepatica,* the sheep liver fluke (magnified 2.5×), stained to reveal its internal structures.

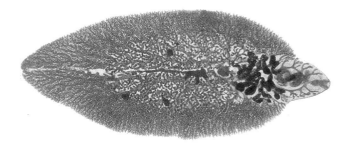

The Chinese liver fluke *Clonorchis sinensis* is widely distributed in Asia. As much as 80 percent of the population in rural areas is infected; some travelers and users of raw imported products also become infected. The life cycle is similar to *Fasciola*'s except that a second intermediate host, typically a fish but sometimes a crustacean, is required. Metacercaria excyst (emerge from a cyst) in the duodenum and migrate to the liver, and adult flukes take up residence in bile ducts. They destroy bile duct epithelium, block ducts, and sometimes perforate and damage the liver. The incidence of liver cancer is unusually high in areas where fluke infections are high, but it is unknown whether the fluke is responsible. Finding eggs in feces confirms the diagnosis, but there is no effective treatment. The parasite could be eradicated by cooking fish and crustaceans. However, the cultural habit of eating raw fish and shellfish and the lack of fuel for cooking maintain infections.

Yet another fluke, *Fasciolopsis buski*, is common in pigs and humans in the Orient. It lives in the small intestine and causes chronic diarrhea and inflammation. If several flukes are present, they can cause obstruction, abscesses, and **verminous intoxication,** an allergic reaction to toxins in the flukes' metabolic wastes. Antihelminthic drugs can be used to rid the body of flukes. Human infection can be prevented by controlling snails, avoiding uncooked vegetation, and stopping the use of human excreta for fertilizer.

Tapeworm Infections

Human tapeworm infections can be caused by several species, most of which have worldwide distribution. Illustrations of the tapeworm life cycle were provided in Chapter 12. ∞ (p. 318) Humans are most often infected by eating uncooked or poorly cooked, contaminated pork or beef. Tapeworm infections can also occur through contact with infected dogs or from eating infected raw fish.

The pork tapeworm *Taenia solium* (Figure 22.18a) reaches a length of 2 to 7 m, and the beef tapeworm *T. saginata* reaches a length of 5 to 25 m. These worms usually enter the body as larvae in raw or poorly cooked meat, especially pork. Viable larvae can develop into adult tapeworms. When adult tapeworms develop in the intestine, they absorb large quantities of nutrients and lead to malnutrition even when the person has an adequate diet. Long, ribbonlike worms may tangle up into a mass that blocks passage of materials through the intestine.

When humans ingest tapeworm eggs (or ova) instead of larvae (Figure 22.18b), the egg coverings disintegrate in the small intestine, and the released larvae penetrate the intestinal wall and enter the blood. In 60 to 70 days, a larva migrates to various tissues and develops into a **cysticercus,** or bladder worm. ∞ (Chapter 12, p. 318) The bladder worm consists of an oval white sac with the tapeworm head invaginated into it.

FIGURE 22.18 Tapeworms. (a) False-color SEM photo of the scolex, or head, of *Taenia solium,* showing suckers and hooks, which aid in attaching to the intestinal lining of its host. (b) Proglottids, or body segments, of *T. pisiformis* (2×). Small new ones grow behind the scolex; they increase in size as they age and move backward away from the scolex. The last row of proglottids is filled with mature eggs. (c) Protoscolices of *Echinococcus granulosus.* Each round structure, the size of a grain of sand, contains the head of a tapeworm and can grow into a complete worm. As many as millions of such structures are found in a fluid-filled hydatid cyst, which can contain up to 15 liters of fluid. When even a more moderately sized cyst forms in the brain, it can do extensive damage.

(a)

(b)

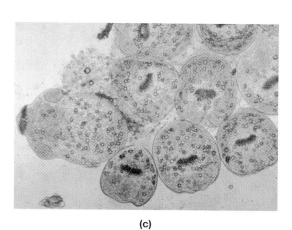

(c)

In humans, cysticerci are frequently found in the brain, where they can reach a diameter of 6 cm. When pigs ingest tapeworm eggs, the cysticerci migrate to muscles and encyst. The pigs' defenses cause the encysted larvae to be surrounded by calcium deposits. In humans, calcification fails to stop the growth of cysticerci; if vital organs are involved, patients may suffer paralysis and convulsions. When a cysticercus dies, it releases toxins and usually causes a severe, or even fatal, allergic response. Human tapeworm infections can be prevented by the sanitary disposal of human wastes and by thorough cooking of meats and fish. Even freezing meats at −5°C for at least a week appears to kill the parasites.

Humans ingest eggs of *Echinococcus granulosus* tapeworms through contact with infected dogs, especially when the dogs lick the faces of small children. Eggs of this tapeworm are especially likely to produce cysts, called **hydatid cysts** (Figure 22.18c), in vital tissues such as the liver, lungs, and brain. ∞ (Chapter 12, p. 319) The cysts, which can contain hundreds of tiny immature worm heads and often reach the size of a grapefruit or larger, exert pressure on organs. If a cyst ruptures, as may happen during an attempt at surgical removal, it can release all these infective units. A ruptured cyst can also cause severe allergic reactions such as anaphylactic shock.

Humans can ingest *Hymenolepis nana* tapeworm eggs in cereals or other foods that contain parts of infected insects. Such tapeworms in the intestine, especially in children, can cause diarrhea, abdominal pain, and convulsions.

The broad fish tapeworm, *Diphyllobothrium latum,* is common in fish-eating carnivores. It reaches humans through ingestion of raw or poorly cooked, contaminated fish in sushi and other dishes. Fish tapeworm infections are common in Scandinavia, Russia, and the Baltic—approaching 100 percent infestation of the population in some areas—and have occurred in the Great Lakes area of the United States. The worm requires as intermediate hosts both a small crustacean and a fish to complete its life cycle. When humans ingest infected fish tissue, the worms coiled up in the fish muscle reach and mature in the intestine. Adult worms attach to the intestine and begin producing eggs. While attached, the parasites absorb large quantities of vitamin B_{12} and impair the victim's ability to absorb the vitamin. Vitamin B_{12} deficiency, or pernicious anemia, from tapeworm infections is especially high in Finland.

Tapeworm infections are diagnosed by finding eggs or proglottids in feces. Infections can be treated with niclosamide and other antihelminthic agents. Prompt diagnosis and treatment are important to remove worms before they invade tissues beyond the intestine. Human infections could be completely prevented by avoiding raw meats and fish and infected dogs.

Trichinosis

Trichinosis (trik"in-o'sis) is caused by the small roundworm *Trichinella spiralis*, also sometimes called *trichina worm*. This parasite, unlike most, is more common in temperate than in tropical climates. (Almost all U.S. adults have antibodies to *Trichinella* and therefore carry around a few worms. But a few will not cause symptoms.) The parasite usually enters the digestive tract as encysted larvae (Figure 22.19) in poorly cooked pork, but infections have been traced to venison and meat from other game animals and to horse meat in France. In the intestine, cysts release larvae that develop into adults. The adults mate, the males die, and the females produce living larvae before they too die. The larvae migrate through blood and lymph vessels to the liver, heart, lungs, and other tissues. When they reach skeletal muscles, especially eye, tongue, and chewing muscles, they form cysts.

These parasites cause tissue damage as adults and as migrating and encysted larvae. The adult females penetrate the intestinal mucosa and release toxic wastes that produce symptoms similar to those of food poisoning. Wandering larvae damage blood vessels and any tissues they enter. Death can result from heart failure, kidney failure, respiratory disorders, or reactions to toxins. Encysted larvae cause muscle pain. Trichinosis is difficult to diagnose, but muscle biopsies and immunological tests are sometimes positive. Treatment is directed toward relieving symptoms because the disease cannot be cured. It can be prevented by eating only thoroughly cooked meat. Freezing does not necessarily kill encysted larvae, and microwave cooking is safe only if the internal temperature of the meat reaches 77°C. Microwave cooking depends on geometry. Because pieces of meat are irregularly shaped,

FIGURE 22.19 *Trichinella spiralis* curled up as a cyst, embedded in striated muscle fibers (magnified 1000×).

they must be rotated during cooking to heat evenly. Studies of pork experimentally infected with *Trichinella* show survival of some worms whenever cooking is done without rotation.

Hookworm Infections

Hookworm is most often caused by one of two species of small roundworms—*Ancylostoma duodenale* and *Necator americanus*. Although these parasites have a complex life cycle, that cycle can occur in a single host, and the host is often a human. Eggs in feces quickly hatch in moist soil. There the eggs release free-living larvae that feed on bacteria and organic debris, grow, molt, and become mature parasitic larvae. If these larvae reach the skin, typically of the feet or legs, they burrow through it to reach blood vessels that carry them to the heart and lungs. The larvae then penetrate lung tissue, and some are coughed up and swallowed. In the intestine the larvae burrow into villi and mature into adult worms. The adult worms mate and start the cycle over again.

As hookworm larvae burrow through the skin, host inflammatory reactions kill many of them, although bacterial infection of penetration sites causes **ground itch.** In the lungs the parasites cause many tiny hemorrhages, but they cause the greatest damage to the lining of the entire small intestine. They feed on blood and cause abdominal pain, loss of appetite, and protein and iron deficiencies, so people infected with hookworms often appear to be lazy. These effects are especially debilitating in people whose diets are barely adequate without the burden of worm infections.

Diagnosis is made by finding eggs or worms in feces, but samples must be concentrated to find them. Tetrachloroethylene is effective against *Necator* infections. It is inexpensive and easy to administer in mass treatment efforts. Bephenium hydroxynaphthalate (Alcopar), mebendazole, and several other drugs, although more expensive, kill both species of hookworms. Dietary supplements should be provided for all hookworm patients. Hookworm is preventable through sanitary disposal of human wastes, but stopping the use of human excreta as fertilizer and getting uneducated people to use latrines can be difficult. Plantation workers often defecate repeatedly in areas near fields where they work. If infected, such individuals serve as a continuous source of larvae to themselves and to others.

In 1991, cases of human hookworm infection by the dog hookworm *Ancylostoma caninum* (Figure 22.20) were linked with intestinal problems such as diarrhea, abdominal pain, and weight loss. Larvae of other species of hookworm for which humans are not the normal host sometimes penetrate the skin and cause *cutaneous larva migrans,* or *creeping eruption.* Severe skin

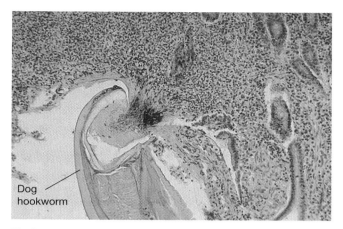

FIGURE 22.20 Dog hookworm, *Ancylostoma caninum* (250×), about 1 cm long, attached to intestinal epithelium, where it sucks blood. It has taken a large piece of intestine into its mouth. (Heavy infestations may result in anemia.)

inflammation results from body defenses that prevent further migration of the parasites. Such infections often are acquired from infected cats and dogs and can be treated with thiabendazole.

Ascariasis

Ascaris lumbricoides (Figure 22.21) is a large roundworm 25 to 35 cm long that causes **ascariasis** (as″kar-i′a-sis). People become infected by ingesting food or water contaminated with *Ascaris* eggs. Once in the intestine, the eggs hatch, and larvae penetrate the intestinal wall and enter lymph vessels and venules. Although the larvae can invade and cause immunological reactions in al-

FIGURE 22.21 *Ascaris lumbricoides,* a large roundworm. Females may reach a length of 35 cm and can produce several hundred eggs per day. The eggs are passed out in feces and can survive in soil for months or even years.

PUBLIC HEALTH

SUSHI

In recent years, many Americans have fallen in love with sushi, the Japanese delicacy made of raw fish. If you are one of them, have you ever worried about your chances of ingesting a live worm with that tasty snack? Have you wondered what the consequences might be if you did?

In Japan, several hundred people each year find out. Typically they consume sushi at dinner and awake in the small hours of the morning with such agonizing pain that they are rushed to the hospital. There, examination of the stomach with a fiber-optic gastroscope shows *Anisakis* larvae penetrating the gastric or duodenal mucosa. In some cases, the damage by this roundworm is so severe that a portion of the stomach wall must be removed. In more fortunate cases, the worm larvae can be removed by forceps attached to the gastroscope. This infection is called *anisakiasis* (an"is-a-ki'a-sis). There is no treatment for it other than removal of the larvae.

The Dutch have had their problems with anisakiasis too, thanks to the relatively recent introduction of a new dish, raw salted herring, which immediately became very popular. The worm is killed by freezing, and Dutch law now requires that all herring served this way must be frozen first.

What are the odds that a sushi dinner will give you anisakiasis in the United States? Only a half dozen or so cases have been reported in the United States; curiously, these cases tended to involve problems somewhat higher up in the digestive tract. A California man had eaten raw white sea bass 10 days before he felt a peculiar sensation at the back of his throat. Coughing, he reached into his mouth and pulled out a lively 7.5-cm-long worm. The other U.S. victims have similarly coughed up and removed their own worms—one as quickly as 4 hours after a meal of cod fillet. Still, considering the number of people in the United States who eat raw or undercooked fish, the total number of diagnosed cases of anisakiasis is so low that there is not much to worry about—so far!

Ascaris worms cause three kinds of damage:

1. Larvae burrowing through lung tissue cause *Ascaris* pneumonitis, which involves hemorrhage, edema, and blockage of alveoli from worms, dead leukocytes, and tissue debris. If secondary bacterial pneumonia develops, it can be fatal.

2. Adult worms cause malnutrition, but because they feed mainly on the contents of the intestine, they do little damage to the mucosa. They also release toxic wastes that elicit allergic reactions. If sufficiently numerous, they cause intestinal blockage and sometimes perforation. The peritonitis that follows perforation is nearly always fatal.

3. Wandering worms cause abscesses in the liver and other organs and sometimes traumatize victims by crawling from body openings.

Diagnosis is made by finding eggs or worms in feces. The adult worms can be eradicated from the body by piperazine, mebendazole, and several other drugs, but no treatment is available to rid the body of larvae. Infection is fully preventable by good sanitation and personal hygiene.

Toxocara species, another type of roundworm, ordinarily parasitize cats and dogs. It has been estimated that in the United States, 98 percent of puppies, including those from good kennels, are infected. Infection rates may be almost as high in dogs and cats of any age. *Visceral larva migrans* is the migration of larvae of these parasites in human tissues such as liver, lung, and brain, where they cause tissue damage and allergic reactions. The risk of such human infections can be minimized by worming pets periodically, disposing of pet wastes carefully, and keeping children's sandboxes inaccessible to pets.

Trichuriasis

Trichuriasis (trik"u-ri'a-sis) is caused by the **whipworm**, *Trichuris trichiura*, which is distributed nearly worldwide. It is estimated that nearly 300 million people are infected, some of them in the southeastern United States. For humans to be infected, human feces must be deposited on warm, moist soil in shady areas. Small children, who put dirty hands in their mouth, are particularly susceptible to infection. Eggs deposited with feces contain partially developed embryos. When the eggs are swallowed, they hatch. Juveniles crawl into enzyme-secreting glands of the intestine called crypts of Lieberkühn, where they develop. They return to the intestinal lumen (central space), where they reach full maturity within 3 months of initial infection.

Adult whipworms damage the intestinal mucosa and feed on blood. They cause chronic bleeding, anemia, malnutrition, allergic reactions to toxins, and susceptibility to secondary bacterial infection. Infections

most any tissue, most move through the respiratory tract to the pharynx and are swallowed. Larvae move to the small intestine, mature, and begin to produce eggs. Eggs are especially resistant to acids and can develop in 2 percent formalin. They also resist drying, and people can be infected by airborne eggs. In some areas of the southern United States where the soil never freezes, 20 to 60 percent of children are infected. Mebendazole and pyrantel are effective in treatment.

can result in rectal bleeding in children. Infection is diagnosed by finding worm eggs in stools. The drug mebendazole is effective in ridding the intestine of parasites. Sanitary disposal of wastes is essential to prevent reinfection.

Strongyloidiasis

Strongyloidiasis (stron″jil-oi-di′a-sis) is caused by *Strongyloides stercoralis* (Figure 22.22). This parasite is unusual in that females produce eggs by *parthenogenesis*—that is, without fertilization by a male. Adult females attach to the small intestine, burrow into underlying layers, and release eggs containing noninfective larvae. Many eggs hatch in the intestine and are passed with the feces. In soil, larvae can become free-living adults or develop into infective larvae and penetrate the skin of new hosts. Infective larvae, which penetrate the skin, are carried by blood to the lungs. There they bore their way to the trachea, travel to the pharynx, and are swallowed. When they arrive in the small intestine, they develop into adults and restart the life cycle.

The *Strongyloides* larvae cause itching, swelling, and bleeding at penetration sites, which often become infected with bacteria. Migrating larvae cause immunological reactions in the host, but the reactions usually do not stop the parasites. Coughing and burning sensations in the chest occur in lung infections, and burning and ulceration occur in intestinal infections. Secondary bacterial infection in any tissue can lead to septicemia. The diarrhea and fluid loss associated with intestinal parasites is severe and difficult to control even with electrolyte therapy. Consequently, patients often die of complications such as heart failure or paralysis of respiratory muscles.

Strongyloides infection in humans usually occurs when larvae are encountered in contaminated soil or water. Diagnosis is difficult because larvae can be found in fecal smears only in massive infections. Work is in progress to develop a reliable immunological test. The drugs thiabendazole and cambendazole have the greatest effect on the parasites with the fewest undesirable side effects.

Pinworm Infections

Pinworm infections are caused by the small roundworm *Enterobius vermicularis*. Like the hookworm, this parasite can complete its life cycle without an alternate host. Adult pinworms attach to the epithelium of the large intestine and mate, and the females produce eggs. Egg-laden females migrate toward the anus during the night, release their eggs on the exterior of the anus, and then crawl back in. These eggs are easily transmitted to other people by bedclothes, by debris under fingernails of those who scratch the itchy area around the anus, and even by inhalation of airborne eggs. Ingested eggs hatch in the small intestine and release larvae that mature and reproduce in the large intestine.

Although pinworm infection usually is not debilitating, it does cause considerable discomfort and can interfere with adequate rest and nutrition, especially in children. Infection with large numbers of the worms can cause the rectum to protrude from the body. Pinworm infections are diagnosed by finding eggs around the anus; at night or right after the host awakens, the pinworms can be picked up with the sticky side of cellophane tape affixed to a wooden tongue depressor. If one member of a family has pinworms, all are presumed infected and are treated with piperazine or another antihelminthic agent. These agents are generally inexpensive and nontoxic. Bed linens, clothing, and towels should be washed and the house thoroughly cleaned at the time of treatment. The treatment and cleaning are repeated in 10 days. Despite these efforts, reinfection in families is very likely.

Gastrointestinal diseases caused by pathogens other than bacteria are summarized in Table 22.4.

FIGURE 22.22 *Strongyloides stercoralis* (magnified 250×), a roundworm parasite of the small intestine.

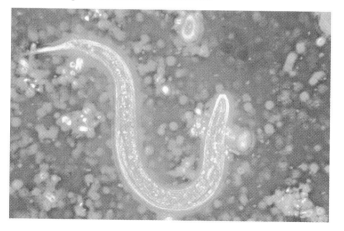

Disease	Agent(s)	Characteristics
Viral gastrointestinal diseases		
Viral enteritis	Rotaviruses	Viral replication destroys intestinal epithelium; causes diarrhea and dehydration that can be fatal under age 5
Hepatitis A	Hepatitis A virus	Viral replication in intestinal and other cells causes malaise, nausea, diarrhea, abdominal pain, lack of appetite, fever, and jaundice; usually self-limiting; vaccine available
Hepatitis B	Hepatitis B virus	Viral replication and symptoms similar to hepatitis A, except onset is insidious and fever usually absent; chronic infections and carrier state occur; transmitted parenterally; vaccine available
Hepatitis C	Unknown, but may involve two agents	Mild symptoms, but disease often chronic
Hepatitis D	Hepatitis D and B viruses	Can result in fatal hepatitis
Hepatitis E	Hepatitis E virus	Moderate disease, but high mortality in pregnant women
Protozoan gastrointestinal diseases		
Giardiasis	*Giardia intestinalis*	Parasite attaches to intestinal wall; feeds on mucus and causes inflammation, diarrhea, dehydration, nutritional deficiencies, and sometimes reactive arthritis
Amoebic dysentery	*Entamoeba histolytica*	Parasites ulcerate mucosa and cause severe acute diarrhea, abdominal tenderness, and dehydration; parasites can live indefinitely in intestine, causing latent amebiasis
Balantidiasis	*Balantidium coli*	Organisms invade walls of intestine; cause dysentery, and sometimes perforation and peritonitis
Cryptosporidiosis	*Cryptosporidium* species	Organisms live in or under mucosal cells and cause severe diarrhea in immunocompromised patients
Helminth gastrointestinal diseases		
Fluke infections	*Fasciola hepatica, Clonorchis sinensis, F. buski*	Organisms excyst in intestine and migrate to liver, block ducts, and damage tissues; *F. buski* causes intestinal obstruction, abscesses, and verminous intoxication
Tapeworm infections	*Taenia solium, T. saginata, Echinococcus granulosis, Hymenolepis nana, Diphyllobothrium latum*	Most infections caused by ingesting encysted larvae, which mature and erode intestinal mucosa; ingestion of eggs allows larval forms to develop in humans as cysticerci or hydatid cysts that can damage the brain and other vital organs
Trichinosis	*Trichinella spiralis*	Larvae excyst in intestine, mature, and produce new larvae, which migrate to various tissues and encyst in muscles, where they cause pain; larvae cause tissue damage, and adults release toxins
Hookworm infections	*Ancylostoma duodenale, Necator americanus*	Larvae penetrate skin and migrate via blood vessels to heart and lungs; coughed-up larvae enter digestive tract and burrow into villi; mature worms feed on blood, cause abdominal pain, loss of appetite, and protein and iron deficiencies
Ascariasis	*Ascaris lumbricoides*	Eggs hatch in the intestine, and larvae cause immunological reactions in many tissues; adults cause malnutrition by feeding on nutrients in gut; wandering worms cause abscesses
Trichuriasis (whipworm)	*Trichuris trichiura*	Eggs hatch in the intestine, and juveniles invade crypts of Lieberkühn and mature; adults damage intestinal mucosa, causing chronic bleeding, anemia, malnutrition, and allergic reactions to toxins
Strongyloidiasis	*Strongyloides stercoralis*	Filariae penetrate the skin, bore to trachea, climb to pharynx and are swallowed; mature worms bore into intestine and release embryos; cause inflammation, bleeding, and immunologic reactions at various sites, severe diarrhea and fluid loss
Pinworm infections	*Enterobius vermicularis*	Ingested eggs give rise to mature worms in the intestine; infection interferes with nutrition, especially in children

CHOLERA AROUND THE WORLD– AND ON OUR DOORSTEP

n 1892 the great German hygienist Max von Pettenkofer, in the presence of witnesses, raised a broth culture of *Vibrio cholerae* to his lips and swallowed approximately 1 billion organisms. His aim was to disprove Robert Koch's studies, done in India, showing that cholera is a contagious disease spread by polluted water. Von Pettenkofer firmly believed that cholera was neither contagious nor related to drinking water. He thought the disease was due instead to interactions of microbes and soil, with "soil factors" playing the greatest role. Control of the disease would depend on the removal of these "soil factors." Within a few days, von Pettenkofer developed a mild but genuine case of cholera, and *V. cholerae* organisms were recovered from his stool. However, he insisted that his was *not* a real case of cholera. So, a few days later, his assistant repeated the experiment, fell severely ill with cholera, but recovered. The culture used in these experiments was a weakly virulent strain supplied by the bacteriologist Georg Gaffky, who had guessed at what von Pettenkofer intended to do and did not want the 74-year-old man to die.

In February 1991 Peruvians saw a similar scene broadcast on their tele-visions. The president of the country, Alberto Fujimori, and his wife were filmed eating raw fish and proclaiming that this popular dish was safe to eat. At that time 45,000 Peruvians had become cholera victims, and at least 193 had died from the disease (Figures 22.23a,b). To slow the epidemic, the Peruvian Health Ministry declared that raw fish was likely to be contaminated and therefore unsafe to eat. Physicians worldwide criticized President Fujimori for his foolhardy defense of the Peruvian fisheries industry. Fujimori neglected to make it clear to his viewers that the fish he and his wife ate had been caught far out to sea, away from the sewage-contaminated coastal and river waters where poorer Peruvians caught their fish. Ironically, the Peruvian minister of fisheries, Felix Canal, attempted the same demonstration and promptly came down with cholera.

In April 1991 eight people in New Jersey contracted cholera by eating crab meat illegally imported from Ecuador. And in September 1991, Alabama state health officials closed an oyster reef near Dauphin Island, in the Gulf of Mexico, that they said was teeming with the deadly strain of *V. cholerae* that was causing the outbreak in South America. Thankfully, health officials swiftly caught and managed these two outbreaks. But cholera can quickly veer out of control, causing untold suffering and death.

Cholera had existed on the Indian subcontinent for centuries before the first Europeans arrived. Portuguese explorers described it early in the sixteenth century, but it did not invade other areas until 1817. The world is now gripped by the seventh cholera pandemic to have swept the globe. The United States experienced the second, third, and fourth of these, beginning in 1832, 1849, and 1866, respectively. The fourth pandemic left more than 50,000 Americans dead, with outbreaks continuing until about 1878. In 1887 and 1892 boatloads of infected passengers arrived in New York City as part of the fifth pandemic. However, a drastic shift in knowledge and attitudes in the United States prevented its spread. During the 1832 outbreak, cholera was viewed as a punishment from God. Indeed, health authorities resisted pleas to clean up streets and water supplies; to do so, they felt, would be to oppose God's will. Pigs were allowed to wander the New York streets, and they left feces and garbage in their wake. In contrast,

---

FIGURE 22.23 (a) Contaminated water in Chimbote, Peru, is used for washing, drinking, and bathing. Cholera organisms from excrement of victims also enter this water and spread rapidly through the population. (b) Peruvian boys playing in refuse piles hold a puppy while pigs in the background consume edible garbage and add their own urine and feces to the piles—an ideal situation for spread of disease. (c) A Rwandan refugee fills his water jug and women wash clothes in the same lake in which the body of a cholera victim decomposes in 1994.

(a)   (b)   (c)

by 1866 most intelligent physicians realized that cholera was a contagious disease spread by filth and ignorance. During that pandemic, physicians and city officials learned a great deal about control measures. Implementing these measures prevented future pandemics from spreading extensively into the United States.

The current pandemic, caused by the El Tor strain of *V. cholerae*, began in Indonesia in 1958. From there it spread to Macao, Hong Kong, and the Philippines and reached eastern Europe by 1965. It is still rampant in parts of Africa and Asia. In 1989 sub-Saharan Africa accounted for three-quarters of all cases in the world. In the summer of 1994, tribal conflicts in the central African nation of Rwanda claimed more than 500,000 lives, and 3 million Rwandans fled to surrounding nations, including Zaire, Burundi, and Tanzania. There, the refugees set up primitive camps while they waited for the slaughter to subside. With no running water, no sanitation facilities, and little food, the refugees succumbed to cholera, dysentery, and malaria before worldwide relief efforts came to their aid. About 50,000 Rwandan refugees contracted cholera in the camps, and thousands died due to lack of treatment (Figure 22.23c). With proper

treatment, cholera victims can recover, but treatment requires clean water for victims to drink to replace lost fluids. In late July 1994 the United States and Canada joined the relief efforts by sending physicians, food, clean water, and portable latrines to the region.

Curbing the disease is a different matter where sanitation remains primitive over the long term. Health officials fear that cholera will establish itself as an endemic disease in South America and will bring death for decades to poor areas that lack proper sewage and water treatment systems. Carried by ship from Africa, cholera reached a port city north of Lima, Peru, in January 1991. It spread through the untreated drinking waters and via infected people and contaminated food to Lima. Within 3 months, some 175,000 people had sickened, and 1258 had died. Because only about one-fourth of infected people show symptoms, vast numbers of Peruvians must have been infected. Meanwhile, cholera spread into nearby countries, including Colombia, Ecuador, Chile, Guatemala, Brazil, Mexico—and even the United States. Because the current vaccine protects only half of those who receive it, and then only for 6 months, the World Health Organization is not recommending mass vaccination campaigns.

Many South American countries have instituted educational programs about cholera transmission and offer practical solutions to the sanitation problems. For instance, the Peruvian government now uses television and radio to broadcast information about boiling water, proper handling of food, and the necessity of disposing of garbage and sewage away from drinking water supplies. The government has distributed millions of chlorination tablets to its citizens. In addition, physicians and nurses take special courses about the treatment of cholera, and food handlers are trained in workshops about how to handle food safely. As a result of this massive educational effort, many areas of Peru have witnessed a decrease in the number of cholera cases.

Overcoming ingrained habits, however, is not easy. One man in Peru said he would not boil his water because boiled water doesn't quench his thirst. Children continue to play in sewage outfalls, and the livestock that roam the streets in many Peruvian cities are an ongoing threat to the health of its citizens. The fight against cholera is also a battle against ignorance, but with proper education, perhaps this fight can be won.

CHAPTER SUMMARY

- The agents and characteristics of the diseases discussed in this chapter are summarized in Tables 22.1, 22.2, and 22.4. Information in those tables is not repeated here.

DISEASES OF THE ORAL CAVITY

Bacterial Diseases of the Oral Cavity

- **Dental plaque** is a continuously formed coating on teeth that consists of microorganisms in organic matter.
- **Dental caries** (tooth decay) is the chemical breakdown of enamel and deeper parts of the teeth. Caries can be prevented by good oral hygiene, use of **fluoride,** and application of sealants.
- **Periodontal disease** involves the gums, periodontal ligaments, and alveolar bone and can be prevented by prevention of plaque buildup. It is treated with plaque removal techniques, mouth rinses, peroxide—sodium bicarbonate, surgery, or antibiotics.

Viral Diseases of the Oral Cavity

- **Mumps** is transmitted by saliva or as an aerosol and occurs worldwide, mainly in children. It can be prevented by a vaccine.

GASTROINTESTINAL DISEASES CAUSED BY BACTERIA

Bacterial Food Poisoning

- **Food poisoning** is caused by ingesting food containing preformed toxins. It can be prevented by sanitary handling and proper cooking and refrigeration of foods. Bacterial toxins in foods cause intoxications.
- Staphylococcal **enterotoxicosis** usually results from eating contaminated, poorly refrigerated foods, especially dairy and poultry products.
- Other kinds of food poisoning usually arise from eating undercooked, contaminated meats and gravies and rice.

Bacterial Enteritis and Enteric Fevers

- **Enteritis** is an inflammation of the intestine. **Enteric fever** is a systemic disease caused by pathogens that invade other tissues. All enteritises and enteric fevers are transmitted via the fecal–oral route and can be prevented by good sanitation.
- **Salmonellosis** is a self-limiting form of enteritis that is treated with antibiotics only in high-risk patients.
- **Typhoid fever** is an enteric infection caused by the bacterium *Salmonella typhi*. The disease is treated with chloramphenicol. Cell-mediated immunity follows infection. The available vaccine is limited in its effectiveness.
- **Shigellosis,** or **bacillary dysentery,** is treated with antibiotics, and recovery does not produce immunity.
- **Asiatic cholera,** common in Asia and other regions with poor sanitation, is treated with fluid and electrolyte replacement and tetracycline. Recovery can produce a long-lasting immunity, but the duration is unknown. Vaccines are not effective in cases of massive exposures.
- **Vibriosis** is a mild disease common where raw seafood is eaten.
- **Traveler's diarrhea** occurs in more than 1 million travelers each year. It is usually self-limiting but can have complications.
- *Escherichia coli* is an indicator of fecal contamination and an opportunistic pathogen.

Bacterial Infections of the Stomach, Esophagus, and Intestines

- *Helicobacter pylori* is considered to be the cause of peptic ulcers and chronic gastritis and a probable cofactor of stomach cancer.

Bacterial Infections of the Gallbladder and Biliary Tract

- Bile destroys most organisms that have lipid envelopes. *Salmonella typhi* is resistant to bile; it can live in the gallbladder and be shed in feces without causing symptoms. Gallstones blocking bile ducts can cause infections of the gallbladder and ducts, usually by *E. coli*. Infection can spread to the bloodstream or ascend to the liver.

GASTROINTESTINAL DISEASES CAUSED BY OTHER PATHOGENS

Viral Gastrointestinal Diseases

- Most viral gastrointestinal diseases arise from the ingestion of contaminated water or food and are transmitted via the fecal–oral route, but hepatitis B, C, and D are transmitted parenterally from contaminated blood and other body fluids.
- **Rotavirus** infections kill many children in developing countries.
- Treatment for viral **hepatitis** relieves symptoms. Gamma globulin gives temporary immunity to **hepatitis A.** Vaccines are available for both hepatitis A and **hepatitis B.**

Protozoan Gastrointestinal Diseases

- Protozoan gastrointestinal diseases arise from contaminated food and water via the fecal–oral route and can be prevented by good sanitation.
- Diagnosis usually is made by finding protozoan cysts in fecal matter. Most such diseases can be treated with antiprotozoan agents.
- **Giardiasis** is especially common in children. **Amoebic dysentery, chronic amebiasis,** and **balantidiasis** occur worldwide but mainly in tropical areas. **Cryptosporidiosis** occurs mainly in immunocompromised patients, but recent outbreaks have involved persons with normal immune systems.

Effects of Fungal Toxins

- **Aflatoxins** are potent carcinogens produced by fungi of the genus *Aspergillus;* humans ingest them from moldy grain and peanuts.
- **Ergot poisoning** comes from eating grains contaminated with *Claviceps purpurea.* Small quantities of **ergot** toxin can be used medicinally.
- Mushroom toxins, which are associated mainly with species of *Amanita,* cause vomiting, diarrhea, jaundice, and hallucinations. In sufficient quantities, they are fatal.

Helminth Gastrointestinal Diseases

- Helminth gastrointestinal diseases are acquired mainly in tropical regions and include several kinds of fluke, roundworm, and tapeworm infections.
- Helminths that infect humans often have complex life cycles in which some animals, fish, vegetation, snails, and crustaceans may also serve as hosts.
- Most such diseases can be diagnosed by finding eggs in fecal specimens. They can be prevented by good sanitation—avoidance of contaminated water and soil and thoroughly cooking foods that might be contaminated.

KEY TERMS

QUESTIONS FOR REVIEW

A Diseases of the Oral Cavity

1. What is dental plaque, and how does it form?

2. What is dental caries, how does it form, and how is it treated?

3. What is periodontal disease, and how is it treated?

4. What causes mumps, and what complications can arise from the disease?

5. Could mumps be eradicated? If so, how?

B Bacterial Gastrointestinal Diseases

6. True or False: Food poisoning is not considered an infection. Explain.

7. How is food poisoning acquired, and how can it be prevented?

8. List the characteristics of staphylococcal enterotoxicosis.

9. How do other kinds of food poisoning differ from that caused by staphylococci?

10. Compare and contrast enteritis with enteric fever.

11. How are enteric infections transmitted, and how can they be prevented?

12. List the characteristics of salmonellosis.

13. What is the cause of typhoid fever, and how does the disease progress?

14. How is damage caused by *Salmonella typhi* related to disease signs and symptoms?

15. How is typhoid fever diagnosed, treated, and prevented?

16. How do shigellosis and cholera differ with respect to cause, nature of the illness, and treatment?

17. What causes traveler's diarrhea, and how should travelers prepare to deal with the disease?

C Viral Gastrointestinal Diseases

18. Which viruses cause gastroenteritis?

19. Childhood deaths from viral enteritis can be reduced by _____.

20. Compare and contrast the types of hepatitis with respect to cause, nature of the illness, and prevention.

D Protozoan Gastrointestinal Diseases

21. Explain how the subject of each of the accompanying diagrams interacts with each of the others in causing giardiasis. What are the symptoms of the disease, and how can it be prevented and treated?

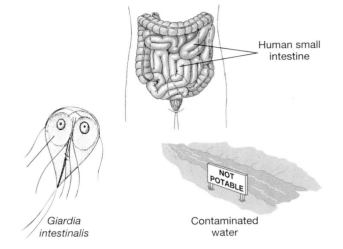

Human small intestine

Giardia intestinalis

Contaminated water

NOT POTABLE

22. Compare and contrast latent amebiasis with amoebic dysentery.

23. What causes balantidiasis, and how can it be prevented and treated?

24. Under what circumstances does cryptosporidiosis occur?

E Fungal Gastrointestinal Diseases

25. Which fungi produce toxins, and how do these toxins affect humans?

26. Humans can avoid fungal toxins by _____.

F Helminth Gastrointestinal Diseases

27. Summarize the variations in the life cycles of flukes, and explain how the variations affect the ways humans can become infected.

28. How do adult tapeworms and bladder worms differ, and how do they affect humans?

29. List the causes and effects of trichinosis infections.

30. How can humans avoid helminth infections?

31. Why would improving food and water sanitation not prevent hookworm infections?

32. List the distinguishing characteristics of ascariasis, trichuriasis, and strongyloidiasis.

33. What are the consequences of pinworm infections, and why are they most likely in children?

34. Match the following:

_____ dental caries
_____ mumps
_____ typhoid fever
_____ bacillary dysentery
_____ cholera
_____ peptic ulcer
_____ viral enteritis
_____ flagellated protozoan
_____ ciliated protozoan
_____ aflatoxins
_____ live in intestinal mucus
_____ ergotism
_____ sheep liver fluke
_____ hydatid cysts

a. *Salmonella*
b. *Balantidium*
c. *Ancylostoma*
d. *Aspergillus*
e. *Enterobius*
f. *Fasciola*
g. *Echinococcus*
h. *Trichuris*
i. *Shigella*
j. paramyxovirus
k. *Helicobacter*
l. *Giardia*
m. *Streptococcus mutans*
n. *Vibrio*

_____ scolex
_____ hookworm
_____ whipworm
_____ pinworm
_____ tapeworm
_____ pseudomembranous colitis
_____ flatworm
_____ transmitted by poultry and eggs

o. *Claviceps*
p. *Clostridium difficile*
q. rotavirus

(continued)

35. Match the following:

_____ enterically transmitted
_____ bloodborne
_____ have vaccine(s) for
_____ defective RNA virus
_____ associated with 80% of liver cancer

a. hepatitis A
b. hepatitis B
c. hepatitis C
d. hepatitis D
e. hepatitis E

PROBLEMS FOR INVESTIGATION

1. Read from other sources about *Clonorchis sinensis*. Explain why it is unlikely that Chinese liver fluke will ever become a threat in the United States, even though large numbers of persons in rural populations elsewhere in the world are infested with *C. sinensis*. Give an example of a parasite that could easily become established in the United States. How does this organism differ from *C. sinensis?*

2. Select an enteric disease that is common in the United States, and devise a public health program to eradicate it.

3. Select an enteric disease that is common in the tropics, and devise a public health program to eradicate it.

4. Survey the current literature to determine which vaccines are under development to prevent enteric diseases.

5. Prepare a report on the various kinds of hepatitis vaccines, their methods of manufacture, and the costs and sequence of doses required. In what ways are the latest versions safer than the earlier ones?

6. A veterans' hospital in California reported an outbreak of 26 cases of hepatitis B infection in a single ward. All the affected patients had blood drawn by a hand-held finger-stick instrument. Nurses always changed the lancet, which penetrated the skin, after each use. They sometimes failed to change the disposable prong that is held against the finger to position the device properly. The prong does not penetrate the skin. What is the probable means by which the virus spread from person to person?

(The answer to this question appears in Appendix F.)

SOME INTERESTING READING

Boyce, T. G., D. L. Swerdlow, and P. M. Griffin. 1995. "*Escherichia coli* 0157:H7 and the hemolytic-uremic syndrome." *The New England Journal of Medicine* 333, no. 6 (August 10):364–68.

Cover, T. L., and M. J. Blaser. 1995. "*Helicobacter pylori:* A bacterial cause of gastritis, peptic ulcer disease, and gastric cancer." *ASM News* 61(1):21–26.

Culthbert, J. A. 1994. "Hepatitis C: Progress and problems." *Clinical Microbiology Reviews* 7:505–32.

Fang, G. 1993. "Intestinal *Escherichia coli* infections." *Current Opinions in Infectious Disease* 6:48.

Genco, R., et al., eds. 1994. *Molecular pathogenesis of periodontal disease.* Washington, D.C.: ASM Press.

Hill, D. R. 1993. "Salmonellosis and typhoid fever." *Current Opinions in Infectious Disease* 6:48.

Kaper, J. B., et al. 1995 "Cholera." *Clinical Microbiology Reviews* 8, no. 1: 48–86.

Luna, A., and A. D. Walling. 1988. "Cysticercosis." *American Family Physician* 37 (January):105.

MacKenzie, W. R., et al. 1994. "A massive outbreak in Milwaukee of *Cryptosporidium* infection transmitted through the public water supply." *The New England Journal of Medicine* 331, no. 3 (July 21):161–67.

Matossian, M. K. 1982. "Ergot and the Salem witchcraft affair." *American Scientist* 70 (July–August):355–57.

Mbithi, J. N., et al. 1992. "Survival of hepatitis A virus on human hands and its transfer on contact with animate and inanimate surfaces." *Journal of Clinical Microbiology* 30 (April):757–63.

Raloff, J. 1995. "Defending us from our dirty mouths." *Science News* 147, no. 11 (March 18):166.

Rippey, S. R. 1994. "Infectious diseases associated with molluscan shelfish consumption." *Clinical Microbiology Reviews* 7:419–25.

Stroh, M. 1992. "In the mouths of babes: No more cavities?" *Science News* 141, no. 5 (February 1):70.

Wachsmuth, I. K., et al. 1994. *Vibrio cholerae and cholera: Molecular to global perspectives.* Washington, D.C: ASM Press.

Wachsmuth, I. K., et al. 1993. "The molecular epidemiology of cholera in Latin America." *Journal of Infectious Diseases* 167 (March):621–26.

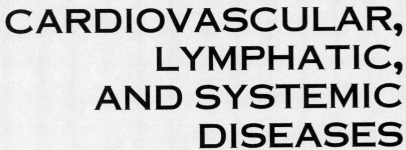

CARDIOVASCULAR, LYMPHATIC, AND SYSTEMIC DISEASES

This colorized fourteenth-century engraving, *The Plague of Florence*, depicts the devastation that resulted when the plague hit Europe.

D iseases of the cardiovascular and lymphatic systems frequently affect several other systems because the infectious agents are easily disseminated through blood and lymph. ∞ (Chapter 17, p. 469) Therefore, diseases that usually affect multiple systems are included in this chapter. As with other chapters on the diseases of organ systems, familiarity with the properties of various groups of organisms and with disease processes, host systems, and host defenses will be helpful in understanding the diseases described here. ∞ (Chapter 15)

CARDIOVASCULAR AND LYMPHATIC DISEASES

Bacterial Septicemias and Related Diseases

Ordinarily blood is sterile. When organisms enter the blood from a wound or an infection, *bacteremia*, in which bacteria circulate without multiplying, can occur. Immune defenses ordinarily eliminate these organisms. If they are not eliminated, bacteremia may progress to **septicemia**, or *blood poisoning*, as the organisms rapidly multiply. ∞ (Chapter 15, p. 407)

Septicemias

Before antibiotics, septicemia was often fatal; even with antibiotics, it is still not easy to treat. Gram-positive organisms such as *Staphylococcus aureus* and *Streptococcus pneumoniae* once commonly caused most septicemias. Today, broad-spectrum antibiotics have made septicemias from these organisms less frequent. However, other bacterial species—including *Pseudomonas aeruginosa, Bacteroides fragilis,* and species of *Klebsiella, Proteus, Enterobacter,* and *Serratia* have come into play. These organisms cause **septic shock,** a life-threatening septicemia accompanied by low blood pressure and the collapse of blood vessels. Probably one-third of all septicemias are now gram-negative septic shock, and 10 percent of all septicemias are caused by multiple organisms. Endotoxins produced by these organisms are directly responsible for shock. Antibiotics often worsen the situation; when they kill organisms, the disintegrating organisms release larger quantities of endotoxin, causing more damage to the host's blood vessels, and blood pressure drops even further. A few

644

THIS CHAPTER FOCUSES ON THE FOLLOWING QUESTIONS:

A Which pathogens cause bacterial septicemias and related diseases, and what are the important epidemiologic and clinical aspects of these diseases?

B Which helminths cause diseases of the blood and lymph, and what are the important epidemiologic and clinical aspects of these diseases?

C Which bacterial pathogens cause systemic diseases, and what are the important epidemiologic and clinical aspects of these diseases?

D Which rickettsia cause systemic diseases, and what are the important epidemiologic and clinical aspects of these diseases?

E Which viruses cause systemic diseases, and what are the important epidemiologic and clinical aspects of these diseases?

F Which protozoa cause systemic diseases, and what are the important epidemiologic and clinical aspects of these diseases?

cases of septic shock are caused by organisms that produce exotoxins.

Symptoms of septicemia include fever, shock, and **lymphangitis** (lim-fan-ji'tis), or red streaks due to inflamed lymphatic vessels beneath the skin (Figure 23.1). One-third of all septicemia cases are nosocomial and appear within 24 hours after an invasive medical procedure has been performed. The transition from bacteremia to septicemia can be sudden or gradual. Therefore, hospital patients who have undergone invasive procedures should be carefully watched for signs of septicemia. Septicemia has a mortality of 50 to 70 percent and accounts for about 100,000 deaths per year in the United States alone.

Diagnosis of septicemia is made by culturing of blood, catheter tips, urine, or other sources of infection. In treating septicemias, blood pressure must be elevated and stabilized; then the infectious organisms must be eliminated by appropriate antibiotic therapy.

Puerperal Fever

Puerperal (pu-er'per-al) **fever,** also called *puerperal sepsis* or **childbed fever,** was a common cause of death before antibiotics became available. ∞ (Chapter 1,

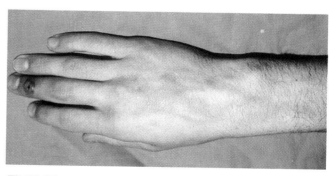

FIGURE 23.1 Lymphangitis from an infected burn. The reddened streaks in the arm indicate the spread of organisms through lymph vessels, a symptom of septicemia. The old-fashioned name for such an infection was blood poisoning.

p. 12) It is caused by β-hemolytic streptococci (*Streptococcus pyogenes*), which are normal vaginal and respiratory microbiota. They can also be introduced during delivery by medical personnel. Streptococci pass through irritated uterine surfaces and invade the blood, giving rise to septicemia. Signs and symptoms of the disease are chills, fever, pelvic distention and tenderness, and a bloody vaginal discharge. Streptococci can be isolated from blood cultures to diagnose puerperal fever. Penicillin is effective except against resistant organisms, and mortality is low with prompt treatment, but recovery usually takes many weeks, and relapses are common.

Rheumatic Fever

Rheumatic fever is a multisystem disorder following infection by β-hemolytic *Streptococcus pyogenes*. That rheumatic fever can follow such infections has been known for decades, but the mechanisms by which it occurs are not yet completely understood. Some form of genetic predisposition is suspected because a certain HLA antigen is present in 75 percent of rheumatic fever patients but in only 12 percent of the general population.

Most rheumatic fever patients are between the ages of 5 and 15. Onset of the disease usually occurs 2 to 3 weeks after a strep throat, but it can occur within 1 week or as late as 5 weeks after the initial infection. Strep throat symptoms have disappeared by the time the rheumatic fever symptoms begin. Classic signs and symptoms include fever, arthritis, and a rash. Evidence of damage to the mitral valve of the heart confirms a specific diagnosis of rheumatic fever. Weeks or months later subcutaneous nodules appear, especially near the elbows. Approximately 3 percent of untreated strep throat cases progress to rheumatic fever. Culturing of streptococci and serological tests are helpful in diagnosis, as is a previous history of streptococcal infection.

Heart damage in rheumatic fever results from immunological events. Certain streptococcal strains have an antigen that is very similar to heart cell antigens. Antibodies that bind to one antigen will bind to the other; that is, the antibodies are *cross-reactive*. In the immune reaction, lymphocytes probably become sensitized to the antigen and attack heart tissue as well as the streptococci. The resulting heart damage can be fatal. Antibiotic therapy will not reverse existing damage, but it can prevent further damage and can be used to prevent recurrences. Once rheumatic fever occurs, mitral valve deformities contribute to eddies in blood flow that predispose toward bacterial colonization of heart valve surfaces. (This condition, *bacterial endocarditis*, is discussed next.) Individuals at risk of rheumatic fever should receive a prophylactic antibiotic—usually penicillin—before dental work or other invasive procedures to prevent possible streptococcal infection.

Prompt treatment of β-hemolytic *S. pyogenes* infections with antibiotics before cross-reactive antibodies can form is the only practical way to prevent rheumatic fever. No effective vaccine exists; vaccines produced so far elicit damaging antibodies, but this problem might someday be solved. Anti-inflammatory drugs such as steroids and aspirin can lessen scarring of heart tissue.

Bacterial Endocarditis

Bacterial endocarditis (en′-do-kar-di″tis), or *infective endocarditis*, is a life-threatening infection and inflammation of the lining and valves of the heart. It can be subacute or acute. Two out of three patients have the subacute type, which manifests itself as fever, malaise, bacteremia, and regurgitating heart murmur usually lasting 2 weeks or more. It occurs primarily in people over age 45 who have a history of valvular disease from rheumatic fever or congenital defects. Many microbes, including fungi, can cause endocarditis, but most cases are due to bacteria, especially strains of *Streptococcus* or *Staphylococcus,* many of them normal residents of the mouth or throat. Acute endocarditis is a rapidly progressive disease that destroys heart valves and causes death in a few days.

In bacterial endocarditis, organisms from another body site of infection are transported to the heart. A **vegetation** develops, in which exposed collagen fibers on damaged valvular surfaces trigger fibrin deposition (Figure 23.2). Transient bacteria attach to fibrin and form a bacteria–fibrin mass. Vegetations deform heart valves, decrease their flexibility, and prevent them from closing completely. Blood flows backward from ventricles into atria when the ventricles contract, decreasing the pumping efficiency of the heart. Congestive heart failure, an accumulation of fluids around the heart, is the most common complication and direct cause of death from bacterial endocarditis.

Bacterial endocarditis is diagnosed from blood cultures and is treated with penicillin or other antibiotics, depending on the susceptibilities of causative organisms. In some cases surgical valve replacement is necessary. Untreated endocarditis results in death. Antibiotics cure about half of all patients; surgery cures another quarter, and one-quarter die. Deaths are most frequent among intravenous drug abusers and others with compromising conditions.

Myocarditis (mi′o-kar-di″tis), an inflammation of the heart muscle (myocardium), and **pericarditis** (per′e-kar-di″tis), an inflammation of the protective membrane around the heart (pericardial sac), also can be caused by microbial infections. Although most such infections are viral, *Staphylococcus aureus* causes 40 percent of all pericarditis cases. Untreated cases have a mortality rate of nearly 100 percent, whereas cases receiving appropriate treatment have a mortality rate between 20 and 40 percent.

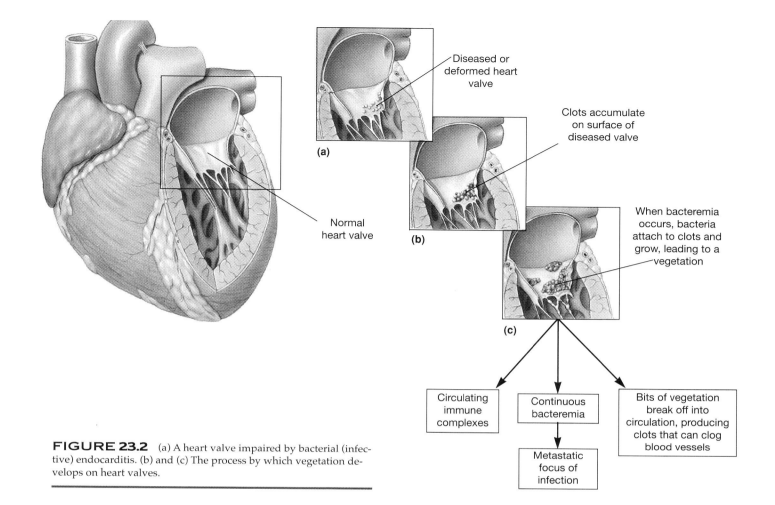

FIGURE 23.2 (a) A heart valve impaired by bacterial (infective) endocarditis. (b) and (c) The process by which vegetation develops on heart valves.

Labels in figure:

(a) Normal heart valve / Diseased or deformed heart valve

(b) Clots accumulate on surface of diseased valve

(c) When bacteremia occurs, bacteria attach to clots and grow, leading to a vegetation

Circulating immune complexes

Continuous bacteremia → Metastatic focus of infection

Bits of vegetation break off into circulation, producing clots that can clog blood vessels

Helminthic Diseases of the Blood and Lymph

Schistosomiasis

Three species of blood flukes of the genus *Schistosoma* cause **schistosomiasis** (shis"to-so-mi'a-sis), and each requires a particular snail intermediate host to complete its life cycle (see Figure 12.15). ∞ (p. 317) The first of these helminths was identified by the German parasitologist Theodor Bilharz in the 1850s, so the disease is also called *bilharzia* (bil-har'ze-a). Bilharzia has been known since biblical times; some think Joshua's curse on Jericho was the placing of blood flukes in communal wells. In fact, the Egypt of the pharaohs was called "land of the menstruating men" by ancient writers because the prevalence of flukes made bloody urine so common. Eggs of these parasites have been found in the bladder walls of Egyptian mummies.

WHO reports some 200 million cases of schistosomiasis worldwide. Incidence of the disease has increased significantly in Egypt since the building of the Aswan Dam in 1960 because collection of water behind the dam created exceptionally favorable conditions for snail hosts. *Schistosoma japonicum* is found in Asia, *S. haematobium* in Africa, and *S. mansoni* (Figure 23.3) in Africa, South America, and the Caribbean. The latter two species probably reached South America during slave trading days, but only *S. mansoni* found an appropriate snail host there. New species have been reported in Vietnam (for instance, *S. mekongi*) and in other geographic areas.

Humans become infected by free-swimming cercariae that have emerged from their snail hosts (Figure 23.3b). Cercariae penetrate the skin when humans wade in snail-infested waters, migrate to blood vessels, and are carried to the lungs and liver. The flukes mature and migrate to veins between the intestine and liver or sometimes the urinary bladder, where they mate and produce eggs. Adults have a special ability to coat themselves with host antigens, thereby evading the host's immune system. Some eggs become trapped in the tissues and cause inflammation; others penetrate the intestinal wall and are excreted in the feces. Cer-

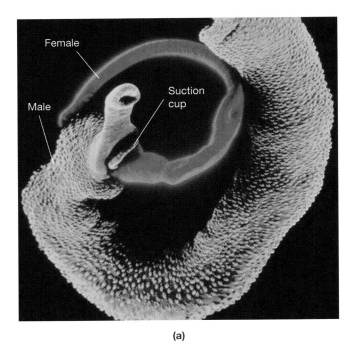

(a)

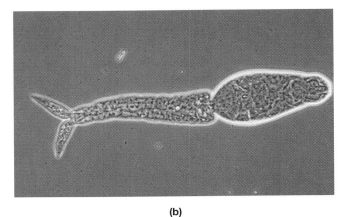

(b)

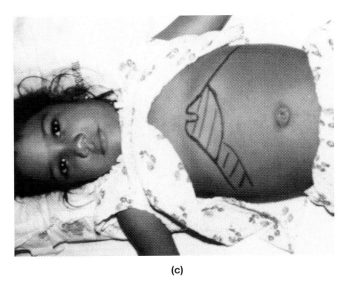

(c)

FIGURE 23.3 Schistosomes, or blood flukes, are trematodes that cause schistosomiasis in Africa, South America, and the Caribbean. (a) A mating pair of schistosomes. The larger, male worm holds the smaller, female worm in a groove of his body. The male attaches itself to the wall of a human blood vessel by means of the suction cup just below its head. The worms, thinner than a cotton fiber, are barely visible to the unaided eye. A mating pair can live in the human body for up to 10 years. (b) The cercarial stage (160×) leaves snails and penetrates the skin of humans wading or swimming in water. (c) A child suffering from schistosomiasis. The tracing delineates her enlarged liver, typical of this disease.

cariae cause dermatitis at penetration sites and tissue damage during migration. Metacercariae and adults migrate to and invade the liver (Figure 23.3c), where they cause cirrhosis. These flukes also can invade and damage other organs.

Schistosome eggs are highly antigenic, and allergic reactions to them are responsible for much of the damage caused by blood flukes. If the eggs are released near the spinal cord, the resulting inflammation can cause neurological disorders. The eggs most often damage blood vessels, but which vessels are damaged depends on the species. *Schistosoma japonicum* severely damages blood vessels in the small intestine; *S. mansoni*, those of the large intestine; and *S. haematobium*, those of the urinary bladder. Symptoms of schistosome bladder infections include pain on urination, bladder inflammation, and bloody urine.

Diagnosis can be made by finding eggs in feces or urine, but eggs may not be present in chronic cases. Intradermal injection of schistosome antigen and measurement of the area of the wheal or a complement fixation test are good immunological methods of diagnosis. ∞ (Chapter 19, p. 542) Until recently, toxic antimony compounds were used to treat the disease.

Several new drugs, especially praziquantal, seem to be quite effective and less toxic. Some investigators, however, believe that praziquantal-resistant schistosomes are on the increase.

Schistosomiasis could be prevented entirely if human wastes were not emptied into rivers or if humans never waded in snail-infested waters. The practice of wading in the local river to wash the body after defecation or urination is an important means of transmission wherever the infection occurs. Chemical molluscicides have been used to reduce snail populations, but it is difficult to determinine the proper concentrations under varying river conditions. Recent experiments using predatory snails to destroy cercariae-carrying snails have shown great promise. Finally, work toward developing a vaccine is underway. One new vaccine in testing does not prevent infection but protects against damage caused by schistosomes.

Filariasis

Filariasis (fil″ar-i′a-sis) can be caused by several different roundworms. *Wuchereria bancrofti* (see Figure 12.18) is a common cause of this tropical disease, for which WHO reports over 100 million cases worldwide. ∞ (p. 322) Adult worms are found in the lymph glands and ducts of humans. Females release embryos called *microfilariae*. They are present in peripheral blood vessels during the night and retreat to deep vessels, especially those of the lungs, during the day. Mosquitoes also are essential hosts in the life cycle of this parasite, and several species of night feeders among the genera *Culex, Aedes,* and *Anopheles* serve as hosts. When a mosquito bites an infected person, it ingests microfilariae that develop into larvae and migrate to the mosquito's mouthparts. When the mosquito bites again, the larvae are infected and can infect another person. They enter the blood, develop, and reproduce in the lymph glands and ducts, thereby completing the life cycle. Adult worms are responsible for inflammation in lymph ducts, fever, and eventual blockage of lymph channels in affected areas. Repeated infections over a period of years can lead to **elephantiasis** (el″ef-an-ti′a-sis), gross enlargement of limbs, the scrotum, and sometimes other body parts due to the blockage of lymph vessels and subsequent accumulation of fluid and connective tissue in those vessels (Figure 23.4a).

Filariasis is diagnosed by finding microfilariae in thick blood smears (Figure 23.4b) made from blood samples taken at night or by an intradermal test. The drugs diethylcarbamizine (Hetrazan) and metronidazole are effective in treating the disease. Swollen limbs are wrapped in pressure bandages to force lymph from them; if distortion is not too great, nearly normal size can be regained. To control the disease, it would be necessary to treat all infected individuals and to eradicate the species of mosquitoes that carry the parasite. Limited progress has been made.

Cardiovascular and lymphatic diseases are summarized in Table 23.1.

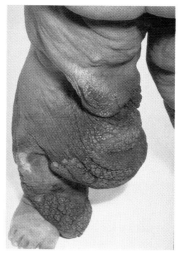

(a)

FIGURE 23.4 (a) Elephantiasis of the leg, caused by the roundworm *Wuchereria bancrofti.* Swelling results from blockage of the lymphatic system by adult worms. Another common site of elephantiasis is the scrotum. (b) The microfilaria stage of the life cycle (2900×) is transmitted to humans by the bite of mosquitoes.

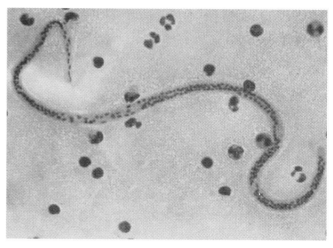

(b)

TABLE 23.1	Summary of Cardiovascular and Lymphatic Diseases	
Disease	**Agent(s)**	**Characteristics**
Bacterial septicemias and related diseases		
Septicemia	Various bacterial species	Septic shock due to endotoxins of causative agent(s), fever, lymphangitis
Puerperal fever	*Streptococcus pyogenes*	Organisms from uterus invade blood and cause septicemia, pelvic distention, bloody discharge
Rheumatic fever	*Streptococcus pyogenes*	Fever, arthritis, rash, mitral valve damage due to immunological reaction
Bacterial endocarditis	*Staphylococcus* or *Streptococcus* strains	Inflammation and vegetation of heart valves and lining, fever, malaise, bacteremia, heart murmur; congestive heart failure that can cause death
Parasitic diseases of the blood and lymph		
Schistosomiasis	*Schistosoma haematobium, S. mansoni, S. japonicum*	Dermatitis from cercaria, cirrhosis of liver from eggs, allergic reactions to eggs, tissue damage in intestine and urinary bladder
Filariasis	*Wuchereria bancrofti*	Inflammation and blockage of lymph ducts, leading to elephantiasis, fever

CLOSE-UP

HEARTWORM IN DOGS

Heartworm infections are common in dogs, and one such infection has been reported in a Maryland farmer. The heartworm is a roundworm (nematode), *Dirofilaria immitis*, transmitted by mosquitoes. Upon biting a dog, an infected mosquito introduces microscopic worm larvae into the dog's blood. The larvae migrate to the skin, mature to about 8 cm in length, and move to the dog's heart. The worms complete maturation in the heart, mate, and release microfilariae that become infective larvae only in a mosquito. Adult worms 15 to 30 cm long accumulate in the right side of the heart and also are found in the lungs and liver. The dog becomes weak, tires easily, coughs, and has difficulty breathing. Ultimately it dies of heart damage and circulatory failure.

Ivermectin, which has proved so valuable in fighting the roundworm that causes river blindness, was originally developed for veterinary use. ∞ (Chapter 20, p. 569) When given in monthly doses, it is highly effective against heartworm larvae. However, it can be administered only to dogs that do not already have adult heartworms. If adult worms are present, the drug kills them, but the worms are more dangerous in death than in life. Disintegration of worm remains releases toxic substances and debris that can clog blood vessels while further weakening the wall of the heart where they had resided.

A dog heart that has been opened to show the presence of many adult heartworms, *Dirofilaria immitis*.

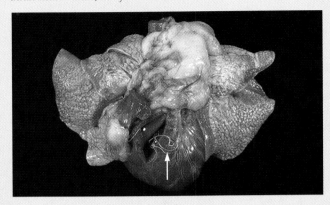

SYSTEMIC DISEASES

Bacterial Systemic Diseases

Anthrax

Anthrax is a zoonosis that affects mostly plant-eating animals, especially sheep, goats, and cattle. Meat-eating animals can acquire the disease by eating infected flesh or by inhaling anthrax spores but the disease is not spread from live animal to animal. Each year, many thousands of animals worldwide have anthrax, but only 20,000 to 100,000 humans develop it, mostly in Africa, Asia, and Haiti. An anthrax outbreak in Russia in 1979 killed at least 64 people. Health authorities in the United States have made vigorous efforts to eradicate anthrax and to prevent its import from other countries. Only five human cases have occurred in the United States since 1980, and no more than six per year have been reported since 1970.

THE DISEASE   The causative agent of anthrax, *Bacillus anthracis,* was discovered in 1877 by Robert Koch. The bacillus is a large, gram-positive, facultatively anaerobic, endospore-forming rod. The endospores form only under aerobic conditions and are not found in tissues or circulating blood. Veterinarians and farmers should be very careful to avoid contaminating the soil or other materials, as endospores can remain viable over 50 years.

Most human anthrax results from contact with endospores during occupational exposure on farms or in industries that handle wool, hides, meat, or bones. Respiratory anthrax, or "woolsorters' disease," was such a problem in nineteenth-century England that legislation was enacted to protect textile workers from this occupational disease.

Of human anthrax cases, 90 percent are cutaneous, 5 percent are respiratory, and 5 percent are intestinal. Cutaneous anthrax has a mortality rate of 10 to 20 percent when untreated, but only 1 percent with adequate treatment. Respiratory anthrax is always fatal, regardless of treatment. Clots form inside pulmonary capillaries and lymph nodes, causing swelling that obstructs airways. Intestinal anthrax has a fatality rate of 25 to 50 percent. In addition, regardless of the initial site of infection, septicemia leads to meningitis (which is almost always fatal in 1 to 6 days) in about 5 percent of patients.

Cutaneous anthrax develops 2 to 5 days after endospores enter epithelial layers of the skin. Lesions 1 to 3 cm in diameter develop at the site of entry (Figure 23.5). Eventually, the center of the lesion becomes black and necrotic; finally, it heals but leaves a scar. Recovery probably confers some—but not total—immunity.

Symptoms of intestinal anthrax closely mimic those of food poisoning. ∞ (Chapter 22, p. 615) An

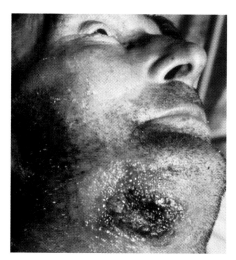

FIGURE 23.5 A case of cutaneous anthrax. Spores introduced into the skin through abrasions or cuts germinate. The newly emerged organisms multiply, release exotoxin locally, and invade adjacent tissue, where they produce extensive damage.

outbreak in the United States resulted from consumption of imported cheese made from unpasteurized goat's milk. The cheese was served at a wine and cheese party to a group of physicians. Some of them traced the source of the infection to a particular cheese factory in France, which was closed as a result.

DIAGNOSIS, TREATMENT, AND PREVENTION
Anthrax is diagnosed by culturing blood or examining smears from cutaneous lesions of patients with a history of possible exposure. The disease is treated with penicillin or tetracycline. A vaccine is available and can be given annually to workers with occupational exposure to anthrax, but industries still should maintain dust-free environments and provide respirators to prevent endospore inhalation. Worker education and on-site employee health services for prompt detection of infections also are important. Unimmunized visitors must be kept away from work areas. Clothing worn by workers should be sterilized and cleaned at the facil-

PUBLIC HEALTH

PRAIRIE HERITAGE: DEADLY SPORES

For centuries, vast herds of bison roamed the Great Plains of the United States. Periodically they suffered outbreaks of anthrax. Their dead carcasses rotted, leaving anthrax spores in the soil. Then came the great cattle drives of the 1800s, from Texas to Canada. Carcasses were left along the trails, adding more anthrax spores to the soil. By the 1930s ranchers realized that the bodies of anthrax-infected animals must be dealt with so as to prevent soil contamination. Some ranchers still remember the huge bonfires of burning carcasses in the 1930s and 1940s. But by then the soil was contaminated with anthrax spores—which remain viable even today.

In dry years, wind blows away the topsoil, exposing spores from the older, deeper layers. In wet years, or along river and creek beds, erosion washes away topsoil, likewise exposing and spreading spores. From Texas to Canada, yearly anthrax immunization of cattle is largely routine. But after 10 or 15 years without an outbreak, ranchers may get a false sense of security and discontinue use of the vaccine. That is what happened in South Dakota during the summer of 1993.

In August 1993 an unvaccinated herd of beef cattle in southeastern South Dakota began to sicken. On August 13, after three animals had died, a veterinarian was summoned. His diagnosis was pulmonary emphysema. On August 15, a second veterinarian arrived and confirmed that diagnosis. The rancher was told that it was safe to send the rest of the herd to slaughter. Soon 19 animals had died, 9 others went to slaughter immediately, and the rest were shipped to the stockyards to be sold.

At the slaughterhouses the animals' livers did not look healthy to inspectors. The rancher was notified of the pos-

sible diagnosis of anthrax. Laboratory testing on August 17 confirmed the presence of anthrax. An immediate order of quarantine was issued the following morning.

The slaughtered animals' remains were retrieved, and the meat plants and storage lockers had to be disinfected by steaming. Fortunately, none of the animals at the stockyards had been sold yet. All were shipped back to the ranch from which they had come and were held in quarantine. There they received anthrax vaccine plus antibiotics, and a booster shot 2 weeks later. The manure and bedding from the areas they had occupied in the stockyards had to be collected, hauled away, and sterilized.

The dead animals had been sent to a rendering plant, where carcasses or waste parts such as bones and fat are boiled down. Products from rendering plants may eventually be used in food products for humans or other animals. Thus, these 19 carcasses were a danger. Had the animals been milk cows, their milk could also have been contaminated.

State health inspectors and others who had handled the herd were given tetracycline to prevent the development of anthrax infection. Happily, no human cases occurred. But by then 32 head of cattle had died. The last 13 carcasses were burned. Thirty days after the last death, the rest of the herd was released from quarantine. Nearby ranchers who had neglected to vaccinate their herds swiftly administered the vaccine.

South Dakota and the other Great Plains states have periodic outbreaks. It is part of life there. Health authorities must be ever vigilant against the prairie heritage of spores from the distant past.

ity to prevent family members from becoming infected by handling it.

Animal immunization is an important means of prevention. Farmers must avoid using bone meal contaminated with anthrax spores and must dispose of infected animals by burying them in deep, lime-lined pits. Lime prevents earthworms from bringing anthrax endospores to the surface; incineration is used but must be done properly to avoid the spread of bits of contaminated carcass and spores by the wind. Veterinarians must be especially careful when working with infected animals or giving vaccinations because accidental vaccination of humans with a vaccine intended for animals can cause anthrax.

Plague

From 1937 through 1974, fewer than 10 cases per year of plague, a zoonosis, were reported in the United States, with no cases reported in some years. Then 20 cases were reported in 1975 and 40 in 1983, mostly in rural settings in the Rocky Mountain states (Figure 23.6). Plague remains an endemic disease in certain other areas of the world, but the number of cases worldwide is much smaller than in the great pandemics mentioned in Chapter 1. ∞ (p. 7)

THE DISEASE The causative agent of plague, *Yersinia pestis*, is a short gram-negative rod. In *sylvatic plaque*, organisms infect wild rodents, especially rats, and are transmitted by animal-to-animal contact and occasionally to humans by flea bites. As infected rats die of plague, their body temperature drops, and their blood coagulates; hungry fleas jump to nearby sources of warmth and liquid blood. The new host usually is another rat, but in crowded, rat-infested living quarters or when contact is made with the carcass of a

FIGURE 23.6 (a) Incidence of plague in the United States, 1955–1994. (b) Regions of endemic plague.

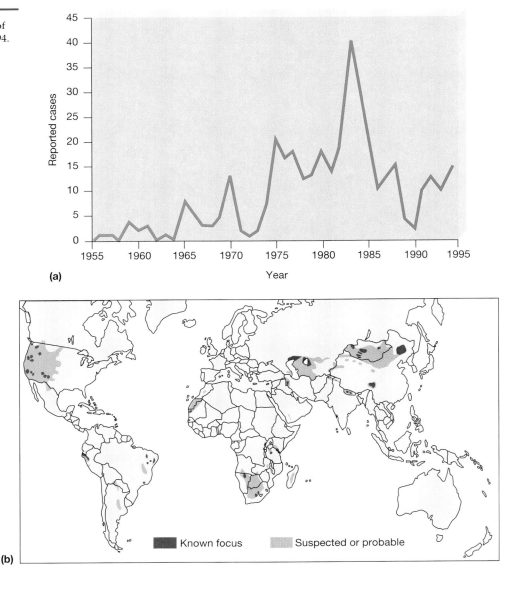

(a)

(b)

FIGURE 23.7 Drawing from a fourteenth-century Flemish manuscript, showing a physician lancing a plague-caused bubo. A second patient awaits the lancing of his underarm bubo.

PUBLIC HEALTH

PLAGUE IN THE UNITED STATES

Plague transmitted by rats in cities is sometimes referred to as *urban plague*, in contrast to *sylvatic plague* carried by wild rodents in rural areas. Sylvatic plague is endemic in 15 western states, where it is found in gophers, chipmunks, pack rats, prairie dogs, and ground squirrels. Plague is thought to have persisted for years among similar wild rodent populations of the Central Asian steppe before breaking out to decimate Europe in 1346. So far only sporadic cases have occurred in the American West, often among hunters and Native Americans on reservations. Some years ago a pet-store owner in Los Angeles trapped some desert rodents with the intention of selling them. Unfortunately for him—but perhaps fortunately for the community—he died of plague before he could sell any of the dangerous animals as pets.

plague-infected animal, the next host can easily be a human.

The flea itself suffers from plague infection. The plague organisms ingested from a sick rat multiply and block the flea's digestive tract until food (blood meals) cannot pass through it. The flea gets hungrier, bites more ferociously, and infects new victims as it dispenses plague organisms with each bite. Eventually the flea dies, but this is little consolation to a new human victim, who has a 50 to 60 percent chance of dying if not treated.

Once inside the host, plague bacilli multiply and travel in lymphatics to lymph nodes, where they cause hemorrhages and immense lymph node enlargements called **buboes** (singular: *bubo*), especially in the groin and armpit (Figure 23.7). Buboes are characteristic of **bubonic plague** and appear after an incubation time of 2 to 7 days. Hemorrhages turn the skin black—hence the name Black Death. Deaths from bubonic plague can be prevented with adequate timely antibiotic treatment. If organisms move from the lymphatics into the circulatory system, **septicemic plague** develops. It is characterized by hemorrhage and necrosis in all parts of the body, meningitis, and pneumonia. This form of plague is invariably fatal despite the best modern care. **Pneumonic plague** occurs with lung involvement and can be spread when aerosol droplets from a coughing patient are inhaled. It, too, has a mortality rate approaching 100 percent despite excellent care. Medical personnel working with patients are more likely to acquire pneumonic than bubonic plague.

DIAGNOSIS, TREATMENT, AND PREVENTION
Plague can be diagnosed by fluorescent antibody tests or by the identification of *Y. pestis* from stained smears of sputum or from fluid aspirated from lymph nodes. It is treated with streptomycin, tetracycline, or both.

Fortunately no drug-resistant strains have yet appeared.

Recovery from a case of plague confers lifetime immunity. A vaccine is available to protect medical personnel and the families of victims. However, during treatment of plague patients in the Vietnam War, it was discovered that workers, even those protected by immunization, can become pharyngeal carriers of plague organisms for a short period of time. Plague can be prevented by immunizing those who travel to endemic regions, controlling rat populations, and maintaining surveillance for infections in wild rodent populations. CDC surveys have found plague only among wild rural (usually desert) rodents, which are not likely to come in contact with urban populations. However, plague has moved eastward from the California coast, where it first arrived on a ship from China that docked at San Francisco in 1899. If the disease spreads to urban rats, many of which are resistant to rat poisons, the risk to humans will increase.

Tularemia

THE DISEASE **Tularemia,** caused by *Francisella tularensis,* is a zoonosis found in more than 100 mammals—especially cottontail rabbits, muskrats, and rodents—and arthropod vectors, such as ticks and deer flies. In ticks the pathogen is incorporated in eggs as they leave the ovaries in **transovarian transmission**—infection of eggs before they are fertilized—and thereby passes from one generation to the next. Although in about half of human cases the vector is never identified, tularemia is most often associated with cottontail rabbits; the number of cases reported always rises significantly during rabbit-hunting season.

Francisella tularensis, a small gram-negative coccobacillus with worldwide distribution, was first isolated in 1911 from Tulare County, California, for which the species was named. The genus was named after Edward Francis, who did much of the early work on this organism. The annual incidence of tularemia in the United States (Figure 23.8) has dropped from more than 2000 cases in 1939 to fewer than 200 cases in recent years. The disease is an occupational hazard for taxidermists.

Tularemia can be acquired in three ways. First, organisms usually enter through minor cuts, abrasions, or bites. Second, organisms can be inhaled, especially from aerosols formed during the skinning of infected animals. Third, organisms can be acquired by consumption of contaminated water or meat, resulting in an intestinal form of the disease. Swimming in a river near a colony of infected river rats and eating undercooked infected rabbit have been reported as sources of infection. Freezing, even for years, does not destroy the organisms.

Entry through the skin results in the **ulceroglandular** form of disease. After a 48-hour incubation period, symptoms begin with abrupt high fever of 40° to 41° C (104° to 106°F) with chills and shaking. If untreated, fever, severe headache, and buboes can last a month. An ulcer sometimes forms at the site of entry. The handling of animals and skins is most likely to cause ulcers on hands, whereas a bite by an arthropod vector is most likely to cause buboes (lymph node lesions) in the groin or armpit. The patient is initially disabled for 1 to 2 months and can have frequent relapses. The mortality is about 5 percent if left untreated. In the days of American pioneers, rabbit was a major part of the diet, and tularemia also must have been a major feature of pioneer life. Until the 1960s, when biological warfare was banned, *F. tularensis* was one organism studied by government scientists for possible use in biological warfare.

Bacteremia from lesions can lead to **typhoidal tularemia,** a septicemia that resembles typhoid fever. Touching the eyes with contaminated hands can lead to conjunctivitis, but this happens in very few cases. Inhalation of organisms or their spread from blood produces a patchy bronchial pneumonia that leads to lung tissue necrosis and 30 percent mortality.

DIAGNOSIS, TREATMENT, AND PREVENTION Diagnosis from blood cultures is difficult; the highly infectious organisms are hard to grow on ordinary laboratory media. Only 50 organisms are sufficient to produce a human infection, regardless of the route of administration. Laboratory infections are easily acquired, so culturing of *Francisella tularensis* should be attempted only by very experienced personnel in isolation laboratories equipped with special air-flow hoods and other safety devices. Animal inoculations should be avoided. Agglutination tests are the standard methods of diagnosis ∞ (Chapter 19, p. 541) Streptomycin is the drug of choice for all forms of tularemia.

Prevention by eliminating the organism from wild reservoir populations is impractical. Therefore, one should avoid handling sickly animals, wear gloves when handling or skinning wild game, and, in tick-infested areas, wear protective clothing and frequently search clothing and skin for ticks. A vaccine exists, but it is not always protective and must be readministered every 3 to 5 years.

Brucellosis

THE DISEASE Brucellosis, also called **undulant fever** or **Malta fever,** is a zoonosis highly infective for humans. It is caused by several species of *Brucella*. *Brucella melitensis* was isolated in 1887 on the Mediterranean isle of Malta by Sir David Bruce. *Brucella* are small gram-negative bacilli, each of which has a preferred host: *B. abortus,* cattle; *B. melitensis,* sheep and goats; *B. suis,* swine; and *B. canis,* dogs. In addition to preferred hosts, each species can infect several other hosts, including humans. The incidence of brucellosis in the United States has dropped sharply from more

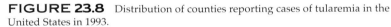

FIGURE 23.8 Distribution of counties reporting cases of tularemia in the United States in 1993.

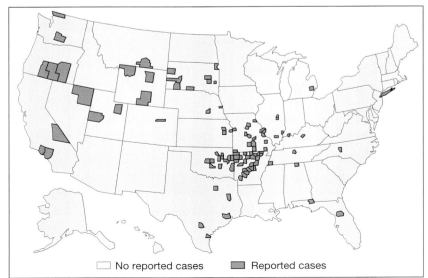

☐ No reported cases   ◼ Reported cases

than 6000 cases per year at the end of World War II to fewer than 200 cases per year since 1978.

Brucella enter hosts through the digestive tract via contaminated dairy products and animal feed, the respiratory tract via aerosols, or the skin via contact with infected animals on farms or in slaughterhouses. Inside the host these facultative intracellular parasites multiply and move through the lymphatic system into the blood, where they cause an acute bacteremia within 1 to 6 weeks. Uncontrolled infections form granulomas (Chapter 17) in the reticuloendothelial system. ∞ (p. 457) Brucellosis has a gradual onset with a daily fever cycle—high in the afternoon and low at night after profuse sweating. These episodes are caused by the release of bacteria from granulomas into the bloodstream. The spleen, lymph nodes, and liver can be enlarged, and jaundice can be present, but symptoms may be too mild to diagnose. The initial acute phase lasts from several weeks to 6 months. Recovery usually is spontaneous, but chronic aches and nervousness can develop. If attributed to a psychiatric disturbance, the disease may not be diagnosed or properly treated.

DIAGNOSIS, TREATMENT, AND PREVENTION

Brucellosis is diagnosed by serologic tests and treated with tetracycline. Streptomycin, gentamicin, or rifampin may be added in severe cases. Treatment must be prolonged because the organisms are well protected from antibiotics in the blood. When death does occur, it is usually from endocarditis. Brucellosis can be prevented by pasteurizing dairy products, immunizing animal herds, and providing education and protective clothing for workers with occupational exposure. No human vaccine is available at present.

Cattle ranchers in the vicinity of Yellowstone, Montana, face a danger from diseased bison who wander outside the park boundaries. Over 50 percent of the Yellowstone Park bison herd are believed to be infected with brucellosis. They can spread it to cattle when they mingle with them, especially during hard winters, when food may be more abundant on ranchers' lands. The state of Montana has spent millions of dollars in the last decade to eradicate brucellosis from its herds.

Relapsing Fever

THE DISEASE **Relapsing fever** is an acute arthropodborne disease characterized by alternating fever and nonfever periods. About a dozen species of the genus *Borrelia* are the causative agents (Figure 23.9). Two different vectors transmit relapsing fever: soft ticks of the genus *Ornithodoros,* and human body and head lice of the genus *Pediculus.* Fifteen species of soft ticks are known to transmit relapsing fever. Louseborne infections are called **epidemic relapsing fever,** whereas tickborne infections are called **endemic relapsing fever.**

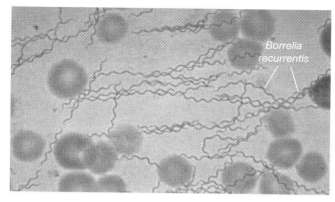
(a)

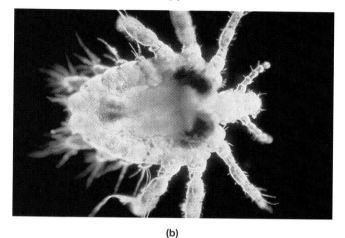
(b)

FIGURE 23.9 Relapsing fever is caused by various *Borrelia* species, such as (a) *Borrelia recurrentis* (among blood cells; magnified 1250×), carried by ticks and by (b) lice such as the body louse.

Relapsing fever appeared in ancient Greece; since then outbreaks have occurred during times of war, poverty, and in situations of extreme overcrowding that occur, for example, after natural disasters. During epidemics, mortality can reach 30 percent of untreated cases.

Borrelia are large, spiral, gram-negative bacteria; their coarser, more irregular spirals and ease of staining distinguish them from *Treponema* (syphilis) spirochetes.

Transmission of relapsing fever varies with the vector involved. Lice must be crushed and their body contents scratched into the skin to transmit the disease. Each louse acquires the organisms by biting an infected host. Ticks transmit disease organisms in their salivary secretions when biting and by transovarian transmission. The ticks can survive for up to 5 years without a meal and still contain live infectious *Borrelia.* Eradication of these vectors is impossible. The ticks feed mainly at night for only about half an hour; victims such as campers or occupants of houses infested with tick-carrying rodents never realize they have been bitten. Lack of this important piece of clinical history makes diagnosis difficult.

Much of what we know about relapsing fever was learned before the antibiotic era, when syphilis was treated by inducing high fever. Patients were purposely infected with relapsing fever or with malaria; the high fever killed the syphilis organisms and left the patient with a more manageable disease. After an incubation period of 3 to 5 days, relapsing fever begins with a sudden onset of chills and high fever. Fever persists for 3 to 7 days and ends by crisis. About 16 percent of patients never experience a relapse, but most, after 7 to 10 days, have fever for 2 or 3 days. Typically, more fevers followed by intervals of relief occur—hence the name relapsing fever. The disease is particularly dangerous in pregnant women because the organisms can cross the placenta and infect the fetus.

Relapses are explained by changes in the organisms' antigens. During a febrile period, the body's immune response destroys most of the organisms. The few that remain have surface antigens the host's immune system fails to recognize. These organisms multiply during a relief period until they are numerous enough to cause a relapse. Each relapse represents a new population of organisms that have evaded the host's defense mechanisms.

DIAGNOSIS, TREATMENT, AND PREVENTION Diagnosis is made by identifying the organisms in stained blood smears prepared during a rising fever phase. Tetracycline or chloramphenicol are used to treat relapsing fever. Immunity following recovery is usually short-lived. No vaccine is available, so prevention primarily is directed toward tick and louse control as well as educating the public.

Lyme Disease

THE DISEASE Alteration of ecosystems can give rise to new human diseases or to increased incidence and recognition of previously unidentified diseases (*emerging diseases*), as has been demonstrated in the case of **Lyme disease.** More Virginia white-tailed deer, which thrive along borders between forests and clearings and are a major reservoir of the disease agent, now inhabit the United States than when the Pilgrims landed. Settlers clearing fields created suitable habitats; as hunting of deer for food has declined, their populations have increased to record levels. With this increase has come Lyme disease, first described in 1974 by Allen Steere and his colleagues at Yale University and named after the Connecticut town where the earliest recognized cases oc-

curred. The disease has now been identified on three continents and in more than 46 states in the United States. It is common along with deer from Cape Cod to Virginia and in Minnesota (Figure 23.10).

In 1982 Willy Burgdorfer of the National Institutes of Health laboratory in Montana isolated and described the causative organism, *Borrelia burgdorferi,* a previously unknown spirochete. By 1985 Lyme disease was the most commonly reported tickborne disease in this country. It is carried by *Ixodes dammini,* the deer tick, which feeds on deer and small mammals such as mice. The tick takes three blood meals during its 2-year life cycle (Figure 23.11a). It ingests contaminated blood in one meal and transmits disease during a subsequent one. Dogs, horses, and cows as well as humans can be infected. As the tick spreads to areas of high human density, the risk of tickborne infections increases.

DIAGNOSIS, TREATMENT, AND PREVENTION Symptoms of Lyme disease vary, but most patients develop flulike symptoms shortly after being bitten by an infected tick. A bull's-eye rash at the site of the bite is also seen in about half of all cases (Figure 23.11b). Weeks or months later, other symptoms occur. Arthritis is the most common symptom, but loss of insulating myelin from nerve cells can cause symptoms resembling Alzheimer's disease and multiple sclerosis. Myocarditis occurs in some patients. Because most patients do not seek medical attention early, Lyme disease usually is diagnosed by clinical symptoms after arthritis appears. An antibody test is available. Lyme disease is treated with antibiotics such as doxycycline and amoxicillin, which are more effective the earlier

FIGURE 23.10 Distribution of Lyme disease in the United States in 1994. What appears to be a dramatic spread of the disease may really be better diagnosis due to heightened awareness. However, the disease is also spreading in some areas.

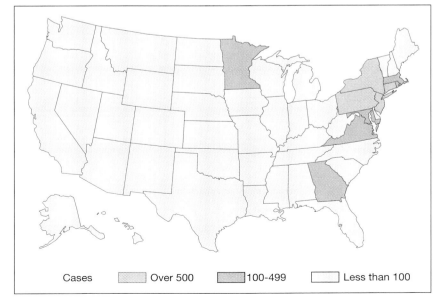

Cases ▢ Over 500 ▢ 100-499 ▢ Less than 100

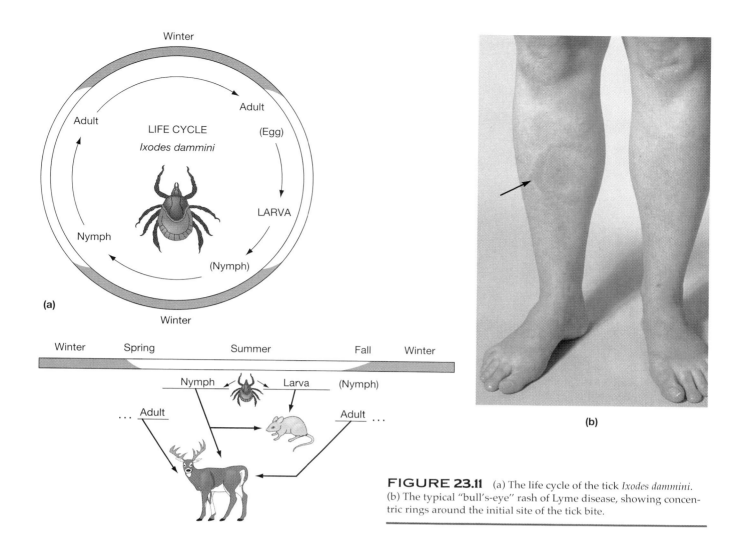

(a)

FIGURE 23.11 (a) The life cycle of the tick *Ixodes dammini*. (b) The typical "bull's-eye" rash of Lyme disease, showing concentric rings around the initial site of the tick bite.

they are administered in the course of the disease. Unfortunately, the earlier stages are often misdiagnosed. Furthermore, undiagnosed subclinical infections are common in endemic areas.

Although no vaccine for humans currently exists, a vaccine to protect dogs from Lyme disease is now available. Dogs are six times more likely than humans to develop the disease. Cats are less likely to develop Lyme disease than dogs, possibly because of their continual grooming. Other animals such as cattle are also susceptible, as a Wisconsin dairy farmer discovered when three-quarters of his 60 dairy cows went lame from Lyme arthritic enlargements of leg joints, some nearly as large as basketballs. Only about 10 percent of infected people develop crippling Lyme arthritis. Researchers have found a genetic predisposition in those who do, with 89 percent having HLA DR 4 or RLA DR 2 tissue antigens. ∞ (Chapter 19, p. 534) Antibiotics given in the first 6 weeks after infection generally prevent this chronic arthritis. After a certain point in the course of the infection, however, antibiotics are no longer able to prevent the development of arthritis in susceptible individuals.

Because no vaccine for humans is available, Lyme disease can be controlled only by avoiding tick bites. Bites are difficult to avoid, however, because the tick—especially the immature form, which often carries the spirochete—is very tiny. A drastic attempt to control the disease was tried on Great Island, Massachusetts, where an entire 52-member deer herd was killed. Putting antitick eartags on deer and mice, a difficult job, has not been very effective. Investigators have also tried saturating cotton balls with insecticide and leaving them around for the mice to take back to their nests, thereby killing off many of their ticks. Now, a European wasp that parasitizes the ticks is being released in tick-infected areas. This biological control method has given excellent results, almost exterminating the tick in some areas.

Bacterial systemic diseases are summarized in Table 23.2.

TABLE 23.2   **Summary of Bacterial Systemic Diseases**

Disease	Agent(s)	Characteristics
Anthrax	*Bacillus anthracis*	Cutaneous lesions become necrotic; respiratory infections always fatal; intestinal infections similar to food poisoning
Plague	*Yersinia pestis*	Bubonic plague causes buboes, or enlarged lymph nodes, and hemorrhages turn skin black; septicemic plague occurs when organisms invade blood and cause hemorrhage and necrosis in many tissues; pneumonic plague from inhalation of organisms causes pneumonia
Tularemia	*Francisella tularensis*	Ulceroglandular form causes high fever, headache, buboes; bacteremia leads to typhoidal tularemia; inhalation leads to bronchopneumonia; relapses often occur
Brucellosis	*Brucella* species	Gradual onset of symptoms, cyclic fever, enlarged lymph nodes and liver, jaundice
Relapsing fever	*Borrelia recurrentis*	Several days of high fever, respites, and shorter periods of fever due to changes in organisms' antigens; can cross placenta
Lyme disease	*Borrelia burgdorferi*	Rash and flulike symptoms at onset, later arthritis and nerve and heart damage; can cross placenta

Rickettsial and Related Systemic Diseases

Rickettsias are named after Howard T. Ricketts, who identified them as the causative agents of typhus and Rocky Mountain spotted fevers. Both he and another investigator, Baron von Prowazek, died of laboratory infections from these highly infectious organisms. Rickettsias are small, gram-negative coccobacilli and are microscopic obligate intracellular parasites. ∞ (Chapter 10, p. 256) Those that cause typhus fever grow in the cytoplasm of infected cells; those that cause spotted fever grow in both the nucleus and the cytoplasm. Rickettsias can be cultured only in cells. Embryonated eggs also are often used, but they should be handled only in specially equipped isolation laboratories. Laboratory-derived rickettsial infections are common.

Rickettsial diseases have several properties in common. The organisms invade and damage the cells of blood vessel linings and cause the linings to leak. This leakage causes skin lesions and especially **petechiae** (pe-te′ke-e; singular: *petechia*), pinpoint-size hemorrhages most common in skin folds. It also causes necrosis in organs such as the brain and heart. Although each disease produces a particular kind of skin rash, all rickettsial diseases cause fever, headache, extreme weakness, and liver and spleen enlargement. Headaches can be very painful, causing the patient to appear confused.

Except for Q fever (Chapter 21), rickettsial diseases are transmitted among vertebrate hosts by an arthropod vector. ∞ (p. 596) Humans often are accidental hosts of zoonoses, but they are the sole hosts of epidemic typhus fever and trench fever. Because of the hazards of culturing organisms, rickettsial diseases usually are diagnosed by clinical findings and serologic tests. The diseases are treated with tetracycline or chloramphenicol; even these antibiotics merely inhibit,

but do not kill, rickettsias. Therapy must be prolonged until the body's defenses can overcome the infection. Often rickettsias are not totally eliminated; they can remain latent in lymph nodes and have been known to reactivate 20 years after initial infection.

Typhus Fever

Typhus fever occurs in a variety of forms, including *epidemic, endemic (murine),* and *scrub typhus. Brill–Zinsser disease* is a recurrent form of endemic typhus.

Epidemic typhus, also called *classic, European,* or *louseborne typhus,* is caused by *Rickettsia prowazekii.* The disease is most frequently seen during wars and other conditions of overcrowding and poor sanitation. In 1812 epidemic typhus helped drive Napoleon from Russia; more recently, during World War I, it infected over 30 million Russians and killed 3 million. The history of warfare is filled with instances in which typhus was the "commanding general." These and other examples of the effects of epidemic typhus on human affairs are well documented in Hans Zinsser's *Rats, lice, and history.* Only after the discovery of the pesticide DDT during World War II were typhus epidemics halted.

Epidemic typhus is transmitted by human body lice. After a louse feeds on an infected person, rickettsias multiply in its digestive tract and are shed in its feces. When a louse bites, it also defecates; infected lice deposit organisms next to a bite and themselves die of typhus in a few weeks. As victims scratch bites, they inoculate organisms into the wound. Lice become infected by biting an infected human. They abandon dead bodies or people with high fevers, moving to and infecting new hosts.

After about 12 days' incubation, onset of fever and headache is abrupt and is followed 6 or 7 days later by

CLOSE-UP

MICROBES AND WAR

"Soldiers have rarely won wars. They more often mop up after the barrage of epidemics. And typhus, with its brothers and sisters—plague, cholera, typhoid, dysentery—has decided more campaigns than Caesar, Hannibal, Napoleon, and all the . . . generals of history. The epidemics get the blame for defeat, the generals the credit for victory. It ought to be the other way 'round. . . ."

—Hans Zinsser, 1935

a rash on the trunk that spreads to the extremities but rarely affects palms or soles. Antibiotic therapy should be started immediately. Without treatment the disease typically lasts up to 3 weeks, and mortality ranges between 3 and 40 percent. The disease can be prevented by eradicating lice with insecticides and by maintaining hygienic living conditions. A vaccine is available. Recovery generally gives lifetime immunity, except when Brill–Zinsser disease occurs.

Brill–Zinsser disease, or *recrudescent typhus,* is a recurrence of a typhus infection. It is named for Nathan Brill and Hans Zinsser, who studied it in the 1930s among New York City's Eastern European immigrants. Compared with first infections, this disease has milder symptoms, is shorter in duration, and often does not cause a skin rash. It is caused by reactivation of latent organisms harbored in lymph nodes, sometimes for years. Lice feeding on patients with Brill–Zinsser disease can transmit the organism and cause initial typhus infections in susceptible individuals. The disease can be prevented by preventing typhus itself and can be reduced in incidence by adequate antibiotic therapy of those already infected. Despite rigorous treatment, some victims of typhus still develop Brill–Zinsser disease later. It can be distinguished from epidemic typhus by the type of antibodies formed shortly after onset. Epidemic typhus first elicits IgM and then IgG antibodies, whereas Brill–Zinsser disease, being a secondary response, elicits primarily IgG antibodies. ∞ (Chapter 18, p. 493).

Endemic typhus, or **murine typhus** (so named for its association with rats—*murine* refers to rats and mice) is a fleaborne typhus caused by *Rickettsia typhi*. It occurs in isolated pockets around the world, including southeastern and Gulf Coast states in the United States, especially Texas. Its incidence in the United States is fewer than 100 cases per year (Figure 23.12). Fleas from infected rats defecate while biting, infecting the humans they bite. The host rubs organisms into the bite wound or transfers them to mucous membranes, another portal of entry. After 10 to 14 days' incubation, onset of fever, chills, and a crushing headache is abrupt, followed by a rash in 3 to 5 days. The disease is self-limiting and lasts about 2 weeks if untreated. Mortality is about 2 percent.

Scrub typhus, or *tsutsugamushi* (soot"soo-ga-moosh'e) *disease,* is caused by *Rickettsia tsutsugamushi.* "Tsutsugamushi" is Japanese for "bad little bug." The "bug" that transmits this disease is a mite that feeds on rats in Japan, Australia, and parts of Southeast Asia. Mites drop off rodent hosts and infect humans with their bites. Scrub typhus was a problem during World War II and the Vietnam War when soldiers crawled on their bellies through low scrub vegetation to avoid snipers. After 10 to 12 days' incubation, scrub typhus begins abruptly with fever, chills, and headache. Many patients develop sloughing lesions at the bite sites and later a generalized spotty rash. In untreated cases, the fatality rate can reach 50 percent, but with prompt antibiotic treatment, fatalities are rare. Although no vaccine is available, infections can be prevented by controlling mite populations.

Rocky Mountain Spotted Fever

Rocky Mountain spotted fever was first recognized around 1900 in Rocky

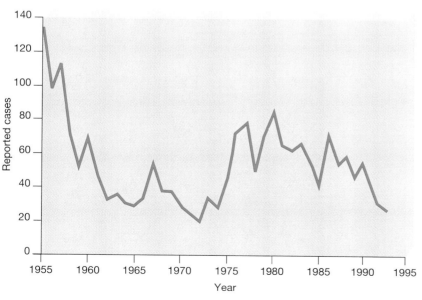

FIGURE 23.12 Incidence of endemic (murine) typhus in the United States, 1955–1993.

Mountain states such as Idaho and Montana. However, its geographic distribution suggests that "Appalachian spotted fever" would be a more appropriate name (Figure 23.13). The incidence has stayed below 1000 cases per year since the 1930s but might increase as more people spend time in tick-infested areas. This disease is caused by *Rickettsia rickettsii* and is transmitted by ticks of the genus *Dermacentor* (Figure 23.14a). After 3 to 4 days' incubation, onset of fever, headache, and weakness is abrupt, followed in 2 to 4 days by a rash (Figure 23.14b). The rash begins on ankles and wrists, is prominent on palms and soles, and progresses toward the trunk—just the reverse of the progression in typhus. Spots are caused by blood leaking out of damaged blood vessels beneath the skin surface; they coalesce as blood leaks from many damage sites. Blood vessels in organs throughout the body are similarly damaged.

Strains vary considerably in virulence; likewise, mortality varies from 5 percent to 80 percent, with an average of 20 percent in untreated cases. Prompt antibiotic treatment keeps the mortality rate to between 5 and 10 percent. Rocky Mountain spotted fever can be prevented by wearing protective clothing and by vigilantly inspecting clothing and skin during visits to tick-infested areas. Inspecting

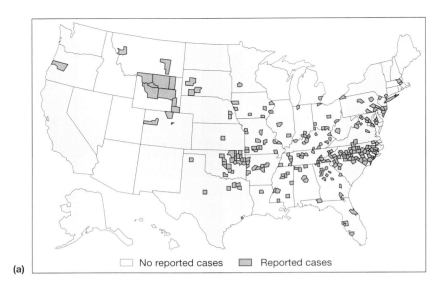

(a)

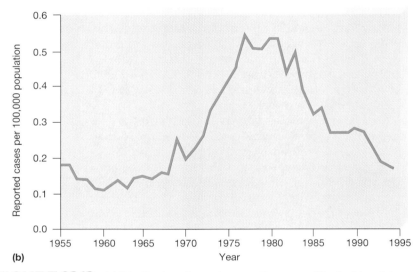

(b)

FIGURE 23.13 (a) Distribution of counties reporting cases of Rocky Mountain spotted fever in the United States in 1993. (b) Incidence of the disease in the United States, 1955–1994.

FIGURE 23.14 (a) Vectors of Rocky Mountain spotted fever include the dog tick *Dermacentor variabilis* (magnified 8×). (b) Rash of the disease on a patient's feet.

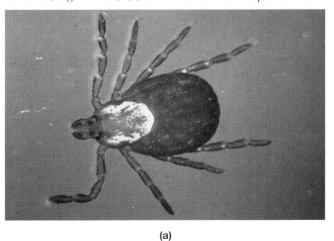

(a)

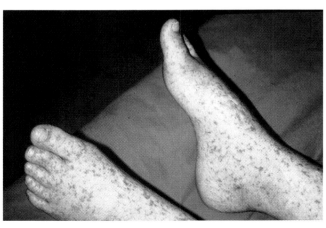

(b)

children's hair is especially important. The only vaccine is not completely effective.

Worldwide, there are many other types of spotted fevers. Each is caused by a specific rickettsial species and is typically named for its location, such as "Siberian spotted fever."

Rickettsialpox

Rickettsialpox, caused by *Rickettsia akari,* was first discovered in New York City in 1946 and is now known to occur in Russia and Korea as well. It is carried by mites found on house mice. The disease is relatively mild, and its lesions resemble those of chickenpox. Because of misdiagnosis as chickenpox or other diseases, incidence and mortality data are unreliable, but no fatalities have been reported. It can be prevented by controlling rodents.

Trench Fever

Trench fever, also called *shinbone fever,* resembles epidemic typhus in that it is transmitted among humans by lice and is prevalent during wars and under unsanitary conditions. Stress probably is a predisposing factor. The causative agent is the bacterium *Bartonella (Rochalimaea) quintana,* classified with rickettsias even though it is not an obligate intracellular parasite. It can be cultured in artificial media and has worldwide distribution, but it rarely causes disease.

Trench fever was first seen in World War I among soldiers living in trenches and wearing the same clothing day after day. British "trench coats" were developed to protect troops who were without shelter during bad weather, but they were not entirely effective. Soldiers and their clothing, including the trench coats, became infested with body lice; exhausted soldiers crowded together in filth fell victim to the disease. After World War I the disease disappeared and did not reappear until World War II. Cases of trench fever have recently been found among the urban poor. The symptoms of trench fever are a 5-day fever and severe leg pain, but many soldiers have reported recurrent symptoms, including mental confusion and depression, as long as 19 years after infection. No vaccine is available, and prevention depends on control of lice.

Bartonellosis

Bartonellosis (bar″to-nel-o′sis) is caused by *Bartonella bacilliformis,* named for the Peruvian physician A. L. Barton, who first described it in 1901. The disease occurs in two forms: **Oroya fever,** or *Carrion's disease,* an acute fatal fever with severe anemia, and **verruga peruana** (ver-oo′gah per-oo-a′nah), a chronic, nonfatal skin disease (Figure 23.15). Both are found only on the

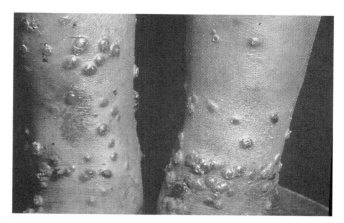

FIGURE 23.15 Lesions of verruga peruana.

western slopes of the Andes in Peru, Ecuador, and Colombia—the habitat of the sandfly *Phlebotomus,* which transmits the organism. In 1885 Daniel Carrion, a Peruvian medical student, inoculated himself with material from a wartlike verruga peruana lesion to show a connection between it and Oroya fever. His death 39 days later from Oroya fever clearly demonstrated the connection.

After being transmitted to a human host by the bite of an infected sandfly, *B. bacilliformis* enters the blood and multiplies during an incubation period of a few weeks to 4 months. Little is known of the epidemiology of bartonellosis, but humans appear to be the sole reservoir. Oroya fever is a severe, febrile, hemolytic anemia. Verruga peruana causes only skin lesions that persist for 1 month to 2 years but usually last for about 6 months. Lesions heal spontaneously but can recur. Oroya fever probably develops in people with no immunity, and verruga peruana occurs in those with partial immunity. Penicillin, tetracyline, or streptomycin can cure Oroya fever but not verruga peruana. No vaccine is available, and prevention depends on the control of sandflies.

Ehrlichiosis

Some recently identified human pathogens related to the rickettsias belong to the genera *Ehrlichia* and *Bartonella.* They are gram-negative coccobacilli and are microscopic obligate intracellular parasites.

Ehrlichiosis, originally recognized as a disease in dogs, has now been found in humans. *Ehrlichia canis* and *E. chaeffeensis,* the causative agents, are spread by dog ticks. Clinically, the human form of the disease resembles other rickettsial diseases. Typical symptoms include fever, headache, hepatitis, and muscle pain. Blood abnormalities, such as a reduction of white blood cells, are common. The absence of a rash distin-

TABLE 23.3　Summary of Rickettsial Diseases

Disease	Causative Organism	Geographic Area of Prevalance	Arthropod Vector Reservoir	Vertebrate Reservoir
Typhus group				
Epidemic (classic, European) typhus	*Rickettsia prowazekii*	Worldwide	Louse	Human
Brill–Zinsser disease (recrudescent typhus)	*R. prowazekii*	Worldwide	(Recurring infection)	Human
Endemic (murine) typhus	*R. typhi*	Worldwide, small scattered foci	Flea	Rodents
Scrub typhus group				
Scrub typhus (tsutsugamushi disease)	*R. tsutsugamushi*	Japan, Southeast Asia	Mite	Rat
Spotted fever group				
Rocky Mountain spotted fever	*R. rickettsii*	Western Hemisphere	Tick	Rodents, dogs
Rickettsialpox	*R. akari*	United States, Korea, Russia	Mite	House mouse
Trench fever	*Bartonella (Rochalimaea) quintana*	Worldwide, but disease only during wars	Louse	Human
Bartonellosis	*Bartonella bacilliformis*	Western slopes of the Andes	Sandfly	Humans only known host
Ehrlichiosis	*Ehrlichia canis, E. chaffeensis*	Southeastern and south central United States	Tick	Dogs, humans

guishes ehrlichiosis from Rocky Mountain spotted fever. Diagnosis requires special serological tests, but finding typical erhlichial intracytoplasmic inclusions in white blood cells may suggest the presence of *Ehrlichia*.

The characteristics of rickettsial diseases are summarized in Table 23.3.

Bacillary Angiomatosis

Bacillary angiomatosis is caused by *Bartonella henselae*, another rickettsial organism. The disease involves small blood vessels of the skin and internal organs. It is seen in persons with AIDS and other immunocompromised patients. Molecular probes and related techniques are used in diagnosis. The organism is also the cause of cat scratch fever. ∞ (Chapter 20, p. 572)

Viral Systemic Diseases

Dengue Fever

Dengue (den'ga or den'ge) **fever,** first characterized in 1780 by the American physician Benjamin Rush in Philadelphia, has also been called *breakbone fever* because of the severe bone and joint pain it causes. Other symptoms include high fever, headache, loss of appetite, nausea, weakness, and, in some cases, a rash.

The disease is self-limiting and runs its course in about 10 days. Four distinct immunological types of the dengue virus—an arbovirus of the family Flaviviridae—have been identified, and two of them have been correlated with disease symptoms. A first dengue fever infection produces the symptoms just noted. A second dengue fever infection with a different immunologic type of the virus causes a hemorrhagic form of the disease. The hemorrhage occurs while the virus is replicating in circulating lymphocytes. It involves an immune response made possible by prior infection. Other symptoms are rapid breathing and low blood pressure that may progress to shock. The shock is reversible if treatment is initiated promptly. Serologic tests are available to diagnose dengue fever, and a vaccine against one immunological type of the virus appears to confer immunity.

Dengue fever is distributed worldwide in tropical areas, causing 500,000 to 3 million cases per year, with occasional episodes in the subtropics. It is endemic in 101 countries. Its main vector is *Aedes aegypti*, although in some areas the Asian tiger mosquito *A. albopictus* can be important. South America is currently experiencing serious outbreaks of dengue fever. Since 1985 health officials have been concerned about the arrival and spread of *A. albopictus* in the United States. The mosquito apparently arrived in used tire casings imported from Asia. In 1980 the first outbreak of dengue fever in the United States in about 40 years occurred. Since then, all four serotypes have arrived in the United States. The rapid spread and aggressive biting habits of *A. albopictus* could bring dengue fever to areas that previously had been safe. Mosquito control is the primary method of preventing the disease, as vaccines are not available for all dengue viral serotypes.

Yellow Fever

Yellow fever was first studied by Carlos Finlay and Walter Reed in the early 1900s when infection of workers threatened to disrupt the construction of the Panama Canal. Although adequate techniques to identify the causative flavivirus (Chapter 11), an arbovirus, were not available, Finley and Reed identified the mosquito vector, *Aedes aegypti*, and instituted control measures that prevented transmission of the disease. ∞ (p. 277) The disease is now limited to tropical areas of Central and South America and Africa. Incidence is greatest in remote jungle areas, where monkeys serve as reservoirs of infection and carrier mosquitoes bite both monkeys and people. In recent decades the number of reported cases each year has varied from 12 to 304, but yellow fever incidence has been vastly underreported. The actual number of cases worldwide is probably 200,000 annually, 30,000 of whom die each year.

Yellow fever causes fever, nausea, and vomiting, which coincide with viremia. Liver damage from viral replication in liver cells causes the jaundice for which the disease is named. The disease is of short duration: In less than a week, the patient has either died or is recovering. In most instances the fatality rate is about 5 percent, but in some epidemics it reaches 30 percent. Two strains of yellow fever viruses are used to produce vaccines. The Dakar strain is scratched into the skin, whereas the 17D strain is administered subcutaneously. Both are effective in establishing immunity.

Infectious Mononucleosis

In 1962 Dennis Burkitt suggested that a virus caused the lymphoid malignancy now called *Burkitt's lymphoma,* found in children in East Africa. This herpesvirus, now called **Epstein–Barr virus** (EBV) (Figure 23.16), is known to cause both Burkitt's lymphoma, most cases of **infectious mononucleosis,** and oral hairy leukoplakia, a disease found among AIDS cases (see Table 11.3). ∞ (Chapter 11, p. 280) EBV infects primarily human B lymphocytes. It replicates like most other herpesviruses and derives its envelope from the inner nuclear membrane of the host cell. The virus has an unusually large number of genes—more than 50 different proteins are produced by complete expression of the DNA in EBV.

EBV enters the body through the oropharynx. The virus first infects epithelial cells and eventually B cells.

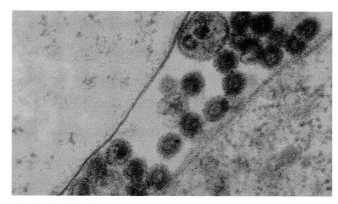

FIGURE 23.16 False-color TEM photo of the Epstein–Barr virus (200,000×), which causes infectious mononucleosis and, along with other factors, leads to Burkitt's lymphoma.

It establishes a persistent infection in which viruses are shed for months to years. The viruses invade such sites as lungs, bone marrow, and lymphoid organs, where they infect certain kinds of mature B lymphocytes. They penetrate B lymphocytes over a 12-hour period, and EBV replication begins within 6 hours after penetration. Viral DNA replicates much faster than cellular DNA. Viral DNA can exist as circular plasmids or can become integrated into the cellular DNA.

EBV exerts three significant effects on lymphocytes:

1. The virus acts on antibody-producing cells and elicits EBV antibodies.

2. Infection and transformation of viral genes to the cellular DNA (or presence of the viral chromosome as a free plasmid in the host cell) are complex events that occur primarily in B lymphocytes, which have receptors for EBV. The cells produce a variety of antigens, some of which are recognized by T cells. This recognition causes the T cells to proliferate, which accounts for the excess of lymphocytes seen in infectious mononucleosis.

3. Other antigens are induced on the surface of some infected B cells. They appear to play a role in some B cell and T cell interactions and may account for some symptoms of infectious mononucleosis.

The proliferation of EBV-infected lymphocytes is limited by cytotoxic T cells and cells that make humoral antibodies and complement. ∞ (Chapter 18, pp. 488–489) If these defenses fail to limit lymphocyte proliferation, uncontrolled B cell proliferation can lead to B cell cancer or Burkitt's lymphoma.

Infectious mononucleosis is an acute disease that affects many systems. Lymphatic tissues become inflamed, some liver cells become necrotic, and monocytes accumulate in liver sinusoids. In some cases myocarditis and glomerulonephritis are seen. The incubation period for the disease is from 30 to 50 days. Mild symptoms—headache, fatigue, and malaise—occur during the first 3 to 5 days of the disease and worsen as the disease progresses. About 80 percent of patients have a sore throat during the first week. The spleen is enlarged, and cells in lymphoid tissues in the oropharynx multiply. The tonsils are coated with a gray exudate, and the soft palate may be covered with petechiae. Secondary infection with β-hemolytic streptococci frequently occurs. Although the disease causes great discomfort and requires several weeks of recuperation, fatalities are rare and usually result from underlying immunological defects.

Diagnosis of EBV infections is complicated by the fact that the disease resembles cytomegalovirus infections, toxoplasmosis, and acute leukemia. The distinguishing symptoms of EBV infections are the concurrent sore throat, multiplication of lymphocytes, and the presence of antibodies against the antigens on sheep and human erythrocytes. Infectious mononucleosis is treated with bed rest and antibiotics for secondary infections. Ampicillin is not used because it causes a rash in infectious mononucleosis patients. The presence of IgG antibodies to viral capsid protein indicates a past infection, and their numbers furnish an index of immunity. Increasing numbers of IgM antibodies to the protein are evidence of a current infection. No vaccine is available.

In developing countries almost all adults have antibodies to EBV by age 1. Exposure to the virus in infancy produces mild symptoms or no symptoms and confers immunity to later infection. Where living standards are higher, a more severe disease is seen later in life. The incidence of infectious mononucleosis in the United States is highest among relatively affluent teenagers and young adults; 10 to 15 percent become infected. The affected age group and the large inoculum required to transmit the disease may be responsible for its being called the "kissing disease." Actually, infectious mononucleosis is not highly contagious. About 15 percent of people who have recovered from the disease are shedding low levels of virus in their saliva at any one time. Those who have recently been infected will shed the virus continuously for about 18 months. About half of immunosuppressed individuals constantly shed large amounts of the virus.

BURKITT'S LYMPHOMA   **Burkitt's lymphoma,** a tumor of the jaw, is seen mainly in children (Figure 23.17). It occurs about 6 years after the primary infection with EBV, but genetic and environmental factors play a role in the development of the disease. The tumor frequently arises from a single cell. The immune system of affected individuals appears to be normal, but it is incapable of eliminating the tumor cells.

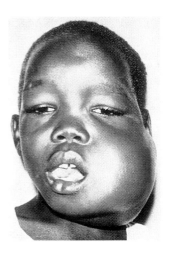

FIGURE 23.17
Burkitt's lymphoma, a form of cancer of the jaw that results from infection with the Epstein–Barr virus, is usually seen only in African children.

Burkitt's lymphoma is found mainly in regions of Africa where malaria is endemic, and infection with malarial parasites may enhance growth of the virus or interfere with the immune response.

OTHER EFFECTS   Another tumor associated with EBV, which also shows a distinct geographic localization, is *nasopharyngeal carcinoma,* found most often in China and rarely in the Western Hemisphere. Individuals with immunological defects are especially susceptible to developing lymphomas, presumably because they lack the necessary immune mechanisms to eliminate the malignant cells. Cyclosporine A, used to depress immunity in organ transplant patients, enhances lymphoid cell growth in donor organs. Organs from donors who have had EBV infections may contain EBV; immunosuppression from any cause, including AIDS, can release EBV and lead to mononucleosis or malignancy. Recipients of EBV-free organs will probably have latent EBV infections of their own, which can then proliferate due to the use of immunosuppresive drugs to prevent organ rejection. Blood transfusions can also transmit EBV.

CHRONIC FATIGUE SYNDROME   Since 1985, researchers have been attempting to determine whether a *chronic EBV syndrome* exists. Patients report fever and persistent fatigue along with a variety of other nonspecific symptoms similar to those of infectious mononucleosis. Some, but not all, have EBV antibodies. Others have had measles or herpesvirus infections. Because a direct relationship between the symptoms and previous EBV infection has not been established, the illness has been renamed **chronic fatigue syndrome.** More study is needed to determine whether the syndrome is associated with one or more previous viral illnesses or whether it might be a psychological disorder. Some recent studies point to possible immune system defects. Herpes human virus number 6

(HHV6), the recently discovered cause of the childhood disease roseola, is also a candidate for causative agent of chronic fatigue syndrome.

Other Viral Infections

FILOVIRUS FEVERS **Filoviruses,** or filamentous viruses, display unusual variability in shape. Some are branched, others are fishhook- or U-shaped, and still others are circular. They contain negative-sense RNA in a helical capsid and vary in length from 130 to 4000 nm. ∞ (Chapter 11, p. 278) Two filoviruses have been associated with human disease. The **Ebola virus** caused outbreaks of hemorrhagic fever first in 1976, with a mortality of 88 percent in Zaire and 51 percent in Sudan. Nearly a fifth of the population of rural areas of Central Africa have antibodies to Ebola. Transmission is person to person. A 1995 outbreak of Ebola virus in Zaire became world news; more than 200 cases were documented, with a mortality rate of about 75 percent. (See the box "Ebola Virus Scare.") The *Marburg virus* was first recognized in Germany when technicians preparing monkey kidney cell cultures died of a hem-

orrhagic disease. Nosocomial Marburg virus infections have since been encountered with a mortality of about 25 percent. Hemorrhage into skin, mucous membranes, and internal organs, death of cells of the liver, lymph tissue, kidneys, and gonads, and brain edema also have been observed. The virus has been isolated directly from monkeys and from laboratory inoculation of guinea pigs.

BUNYAVIRUS FEVERS Infections caused by bunyaviruses begin suddenly with fever and chills, headache, and muscle aches. Although usually not fatal and without permanent effects, the infections are temporarily incapacitating. When encephalitis occurs, it progresses slowly, either because the viruses replicate slowly in neural tissues or because certain viruses capable of replicating in neural tissue are selected. Rats, bats, and animals with hooves serve as reservoirs of infection. Tropical and temperate forest mosquitoes are vectors.

The *LaCrosse bunyavirus* has been identified in the northeastern and north-central United States. It causes a mild disease in adults but can cause seizures, con-

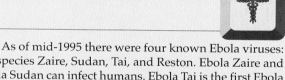

PUBLIC HEALTH

EBOLA VIRUS SCARE

In April and May of 1995, an outbreak of the Ebola virus occurred in Zaire, Africa. The outbreak was traced to a forest worker who had become infected in December 1994. People infected with the virus develop symptoms of hemorrhagic fever, including fever and muscle aches. Most patients experience respiratory and kidney problems, abdominal pain, sore throat, and severe bleeding as well. Because the blood of Ebola patients cannot clot properly, bleeding occurs from needle injection sites, the digestive system, internal organs, and the skin. The virus is spread by direct contact with body fluids. Thus, outbreaks can be limited through isolation of infected individuals, sterilization of needles and syringes, proper disposal of hospital waste and corpses, and use of masks, gowns, boots, and gloves by hospital personnel—who are at special risk. Thousands of dollars worth of such supplies, as well as medicines and blood plasma, have been donated from around the world to limit the spread of the latest Ebola outbreak.

The reservoir for the Ebola virus is unknown. The first known outbreaks occurred in 1976 near the Ebola River (for which the virus was named) in Zaire and in western Sudan. That first Zaire outbreak resulted from Ebola-contaminated needles and syringes that were reused without sterilization. In fact, most of the outbreaks, including the 1995 one, were the result of poor medical conditions in hospitals.

As of mid-1995 there were four known Ebola viruses: subspecies Zaire, Sudan, Tai, and Reston. Ebola Zaire and Ebola Sudan can infect humans. Ebola Tai is the first Ebola virus known to be passed to humans from other animals—naturally infected chimpanzees of the Tai Forest in the Ivory Coast. Because infections of chimps, like humans, can lead to death, the original source of the virus is not the chimps. Ebola Reston, isolated from infected monkeys delivered to Reston, Virginia, in 1989, appears to be spread only among monkeys, but in this case aerosol transmission was documented.

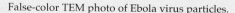

False-color TEM photo of Ebola virus particles.

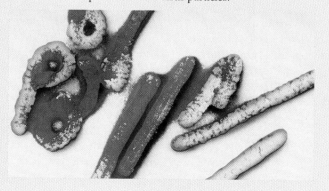

vulsions, mental confusion, and paralysis in children. The *California encephalitis virus,* a bunyavirus initially isolated from mosquitoes in the San Joaquin Valley, has similar effects on humans.

Bunyaviruses called **phleboviruses** (because they are carried by the sandfly *Phlebotomus papatasii*) have been recovered from human infections. The **Rift Valley fever** virus causes epidemics and has unpredictable virulence. It causes sudden vomiting, joint pain, and slowing of the heart rate. One epidemic of Rift Valley fever in 1975 in central Africa infected thousands but left only 4 people dead. Only 2 years later the disease appeared in Egypt, where 200,000 cases and 598 deaths were recorded.

The *hantaviruses* are associated with hemorrhagic fevers, including Korean or epidemic hemorrhagic fever (see the Essay at the end of Chapter 23, "Combating Hantaviruses"), and kidney disease. They are distributed widely over Eurasia and find reservoirs in rodents. They cause capillary leakage, hemorrhage, and cell death in the pituitary gland, heart, and kidneys. Kidney damage can be severe, and low blood pressure can proceed quickly to shock, from which one-third of patients die. Even more die if bleeding occurs in the gastrointestinal tract and the central nervous system or if fluid accumulates in the lungs. The major source of such infections is contact with infected rodents or their excreta. Some rodents shed viruses in feces and saliva for 30 days and into their urine for a year. Humans under 10 or over 60 years are rarely infected, perhaps because they are less likely to come in contact with infected materials.

ARENAVIRUS FEVERS Like bunyaviruses, arenaviruses cause hemorrhagic fevers. Of these **Lassa fever** is perhaps the most widely known. It is an African disease that begins with pharyngeal lesions and proceeds to severe liver damage. The prognosis is poor in 20 to 30 percent of cases in which hemorrhage from mucous membranes occurs. Several other arenavirus infections, including **Bolivian hemorrhagic fever,** have been identified in humans, especially in Africa and South America.

Bolivian and other *South American hemorrhagic fevers* are multisystem diseases with insidious onset and progressive effects. The viruses attack lymphatic tissues and bone marrow and cause vascular damage, bleeding, and shock. However, death, which occurs in about 15 percent of cases, usually results from damage to the central nervous system. How the viruses affect the nervous system is not known.

COLORADO TICK FEVER **Colorado tick fever** is caused by an **orbivirus** (a member of the Reoviridae) that is transmitted to humans by dog ticks from reservoir animals such as squirrels and chipmunks. High

CLOSE-UP

TRADING ONE DISEASE FOR ANOTHER

Bolivian hemorrhagic fever is of interest because of its relationship to efforts to eliminate malaria through improved sanitation. Prior to these efforts, hemorrhagic fever had existed in Bolivia for years without causing a major problem. This situation changed when a campaign was initiated to control malaria by eradicating mosquitoes. Large quantities of pesticides were used, which killed many village cats. In the absence of the cats, the rodent population multiplied and invaded thatch-roofed human dwellings. The rodents carried arenaviruses, which caused the incidence of hemorrhagic fever to rise to epidemic levels.

levels of viremia with infection of immature erythrocytes occur. The patient suffers from headache, backache, and fever, but recovery usually is complete.

PARVOVIRUS INFECTIONS Two parvovirus infections are now known to affect cats and dogs, respectively. **Feline panleukopenia virus** (FPV) causes severe disease with fever, decreased numbers of white blood cells, and enteritis in cats. The virus replicates in blood-forming and lymphoid tissues and secondarily invades the intestinal mucosa. In 1978 a new virus, the **canine parvovirus,** appeared and infected dogs in widespread geographic areas. This virus first surfaced in North America, Europe, and Australia and rapidly spread worldwide. It causes severe vomiting and diarrhea in dogs of all ages and sudden death with myocarditis in puppies under 3 months old. When the canine parvovirus first appeared in an area, the death rate often exceeded 80 percent. Vaccines are now available for both feline panleukopenia and canine parvovirus.

APLASTIC CRISIS A member of the genus *Erythrovirus* called B19 has been identified as the probable cause of **aplastic crisis**—a period in which erythrocyte production ceases—in sickle cell anemia. The virus appears to replicate in rapidly dividing cells of the bone marrow. Afflicted children soon are in acute distress. Although normal erythrocytes remain functional in the blood for about 120 days, sickled cells survive only 10 to 15 days. Under such circumstances severe destruction of erythrocytes occurs. A child generally experiences only one such crisis, possibly because immunity to the virus develops. Another bit of evidence that aplastic crisis is infectious is that, although it occurs only in sickle cell anemia patients, it does occur in 3- to 5-year cycles within particular communities.

FIFTH DISEASE   The erythrovirus B19 destroys the stem cells that give rise to red blood cells. This is not a major problem in healthy adults or children, but it is a serious danger to those who have chronic hemolytic anemias, such as sickle cell anemia, and who therefore have difficulty maintaining normal levels of red blood cells. It is also a danger to the fetus if a pregnant woman contracts the virus. The virus can be transmitted across the placenta and cause the fetus to develop fatal anemia. The virus does not, however, cause birth defects. Immunodeficient patients cannot control replication of the virus and may develop chronic anemia. After B19 was recognized as the cause of aplastic crisis in sickle cell anemia, it was also found to cause **fifth disease** (*erythema infectiosum*) in normal children. It is common in children of ages 5 to 14 and often goes unnoticed. Infected children have a bright red rash on the cheeks that may spread to the trunk and extremities. A low-grade fever may accompany the rash. Often the infection is totally asymptomatic. The virus appears to be spread via the respiratory route. The disease is self-limiting and confers lifetime immunity. The name "fifth disease" comes from a nineteenth-century list of childhood rash diseases. The first disease on this list is scarlet fever; the second is rubeola; the third, rubella; the fourth, epidemic pseudoscarlatina (a type of sepsis); and the fifth, erythema infectiosum.

COXSACKIE VIRUS INFECTIONS   *Coxsackie viruses* have an affinity for the pericardium and myocardium. They can cause meningoencephalitis, diarrhea, rashes, pharyngitis, and liver disease. Epidemic muscle pain, diabetes, and inflammation of the pancreas, heart muscle, and the pericardium are associated with coxsackie B viruses. Coxsackie viruses are highly infectious and readily spread among family members and in institutions. Most infections probably arise from fecal–oral transmission, but because the viruses can be isolated from nasal secretions, infection also may occur by respiratory fomites. Coxsackie virus infections during pregnancy can cause congenital defects, but their incidence is much lower than that for rubella virus infections, and aborting the fetus is usually not recommended. No effective means are available for treatment, immunization, or prevention.

Protozoan Systemic Diseases

Leishmaniasis

Three species of protozoa of the genus *Leishmania* can cause **leishmaniasis** (lish"man-i′a-sis) in humans (Figure 23.18a). The protozoa are transmitted by sandflies. When an infected sandfly bites, the parasites enter the blood of the host and are phagocytized by

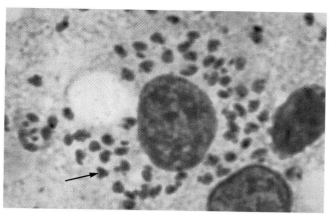

(a)

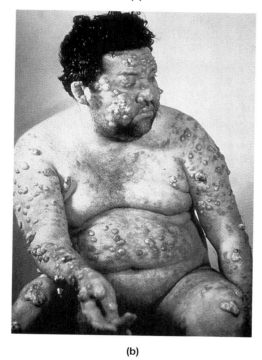

(b)

FIGURE 23.18   (a) *Leishmania donovani,* seen as small dots among cells. (b) A patient suffering from leishmaniasis, probably caused by *L. braziliensis.*

macrophages. ∞ (Chapter 17, p. 453) The parasites multiply inside the macrophages, and new parasites are released when the macrophages rupture. Leishmaniasis is endemic to most tropical and subtropical countries, where appropriate species of sandfly vectors are available. Rodents can be reservoir hosts of the disease. Some 12 million cases worldwide are reported by WHO.

THE DISEASE   *Leishmania donovani* causes **kala azar** (kah′lah ah-zar′; Hindi for "black poison"), or *visceral leishmaniasis.* Symptoms include high irregular fever, progressive weakness, wasting, and protrusion of the abdomen due to extensive liver and spleen enlarge-ment. Extensive damage to the immune system results when parasites destroy large numbers of phagocytic cells. If untreated, the disease is usually fatal in 2 to 3 years, and it can be fatal in 6 months in patients with impaired immunity and secondary infections.

Other leishmanias are more localized in their ef-fects and are rarely fatal. *L. tropica* causes a cutaneous lesion, sometimes called an *oriental sore,* at the site of a sandfly bite. *L. braziliensis* causes skin and mucous membrane lesions and sometimes nasal and oral polyps (Figure 23.18b). Some parents, observing that people who had oriental sores rarely got kala azar, have purposely infected children with oriental sores on inconspicuous parts of the body to protect them from the more serious disease.

DIAGNOSIS, TREATMENT, AND PREVENTION

Diagnosis is made by identifying the protozoa in blood smears in kala azar and from scrapings of skin and mu-cous membrane lesions. Antimony compounds are used to treat both kala azar and skin and mucous mem-brane lesions. However, such drugs are very toxic. Pre-vention depends mainly on controlling sandfly breed-ing and elimination of rodent reservoir infections.

Malaria

Several species of the protozoan *Plasmodium* are capa-ble of causing **malaria,** one of the most severe of all par-asitic diseases. Malaria is one of the world's greatest

PUBLIC HEALTH

DESERT STORM LEISHMANIASIS

Several dozen cases of leishmaniasis have been con-firmed in U.S. military personnel who served in the Per-sian Gulf War early in 1991. The disease is fatal in 90 percent of untreated cases. However, currently avail-able treatment lowers the mortality rate to 10 percent. Health officials have banned blood transfusions and organ transplants from Gulf War veterans to prevent the spread of the disease.

public health problems. It is endemic in most tropical areas (Figure 23.19). The number of clinical cases has been estimated at up to 500 million cases worldwide, with 1.5 million to 3 million deaths—of which more than 1 million occur in children under age 5. Nearly all adults in Africa and India have been infected, and annual eco-nomic losses due to malaria exceed $1 billion in Africa alone. At one time malaria was thought to have been eradicated from the United States, but military person-nel, travelers, and immigrants have carried the disease with them to the United States from endemic areas.

Members of the genus *Plasmodium* are amoeboid, intracellular protozoa that infect erythrocytes and other tissues. They are transmitted to humans through the

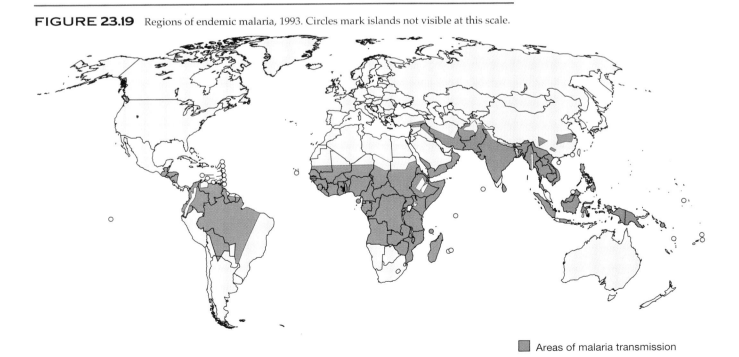

FIGURE 23.19   Regions of endemic malaria, 1993. Circles mark islands not visible at this scale.

▨ Areas of malaria transmission

bite of *Anopheles* mosquitoes. *Plasmodium* species have a complex life cycle (see Figure 12.4). ∞ (p. 307) At least four species, *P. vivax, P. malariae, P. ovale,* and *P. falciparum,* infect humans. They can be identified by their effects on, and in some cases by their appearance in, red blood cells and by the nature of the disease the protozoa cause.

Individuals who carry the allele (Chapter 7) for hemoglobin S, the abnormal gene associated with sickle cell anemia, have a selective and protective advantage against the lethal effects of *P. falciparum*-caused malaria. ∞ (p. 165) Individuals with two such alleles on corresponding chromosomes have the lethal sickle cell disease, because the abnormal hemoglobin S gives up oxygen, causing the red blood cells to deform and become nonfunctional. ∞ (Chapter 15, p. 399) Individuals carrying only one sickle cell gene do not develop the disease and are protected from the severe effects of the form known as falciparum malaria. A high percentage of West African blacks are protected from various malarial infections.

THE DISEASE In the pathogenesis of malaria, sporozoites (Chapter 12) enter the blood from the bite of an infected female mosquito. ∞ (p. 306) (Male mosquitoes are not equipped to take a blood meal.) The parasites disappear from the blood within an hour and invade cells of the liver and other organs. In about a week they begin releasing merozoites, which invade and reproduce in red blood cells as trophozoites (Figure 23.20). At intervals of 48 to 72 hours, depending on the infecting species of *Plasmodium,* blood cells rupture in a characteristic pattern and release more merozoites,

which infect other red blood cells. The release of merozoites soon becomes synchronized and corresponds to intervals of high fever. Some merozoites become gametocytes, which can undergo sexual reproduction in mosquitoes, should they feed on the patient's blood. Even after the initial disease has subsided, patients are subject to relapses as dormant protozoa become activated, emerge from the liver, and initiate a new cycle of disease. Relapses do not occur after infections caused by *P. falciparum* because this species does not remain in the liver.

Of the four species of malarial parasites, *P. falciparum* causes the most severe disease because it agglutinates red blood cells and obstructs blood vessels. Such obstruction causes tissue **ischemia,** or reduced blood flow with oxygen and nutrient deficiency and waste accumulation. This species can also cause malignant malaria—an especially virulent, rapidly fatal disease—and a condition called **blackwater fever.** In blackwater fever, large numbers of erythrocytes are lysed, probably because of the host's autoimmune reaction to the parasites. Products of hemoglobin breakdown cause jaundice and kidney damage. Pigments from hemoglobin blacken the urine and give the disease its name.

DIAGNOSIS, TREATMENT, AND PREVENTION
The main means of diagnosing malaria is by identifying the protozoa in red blood cells. The species of *Plasmodium* responsible for a particular infection can be identified by the distinctive appearance of erythrocytes invaded by the parasite. Chloroquine (Aralen) is the drug of choice for all forms of malaria in the acute stage. A serious problem in the treatment of malaria is that some strains, especially strains of *P. falciparum,* have become resistant to chloroquine. Drugs have recently been found that can be administered with chloroquine to overcome such resistance. This strategy has been tested in monkeys but not yet in humans. A traveler who will enter a malarial region can take chloroquine prophylactically for a week before entry, during the stay, and for 6 weeks after leaving the area. The drug suppresses clinical symptoms of malaria, but it does not necessarily prevent infection. Disease caused by *P. vivax* or *P. ovale* can appear months or years after a person leaves a malarial area, even when the suppressant drug has been taken. Primaquine is the drug of choice for eliminating parasites from the liver and other tissues if they have been infected.

Attempts to destroy mosquitoes that carry malaria have been an important component of malaria control efforts. In the early 1960s the pesticide DDT was used successfully to eradicate malaria-carrying mosquitoes from the United States. (But it was soon banned in the United States due to its toxicity.) DDT and other in-

FIGURE 23.20 The "ring" stage of the malarial parasite *Plasmodium falciparum,* seen as dark circular structures within red blood cells. At this stage, merozoites have become trophozoites upon invading the host's red blood cells.

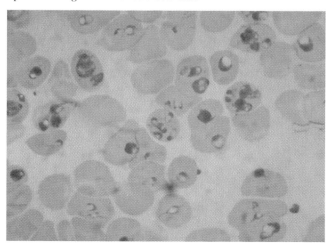

secticides also have been tried in other areas, especially in Africa. Unfortunately, the region in which such mosquitoes thrive in Africa is so vast that no insecticide spraying program has been effective. Some of the mosquitoes, particularly those that carry *P. falciparum*, have now become resistant to DDT and probably to other pesticides. Thus, the use of insecticides has made malaria more deadly by increasing the proportion of mosquitoes that carry the more virulent parasite.

Researchers at the CDC and their colleagues at other institutions have recently developed a strain of *Anopheles gambiae* that is highly resistant to infection with malaria parasites. The researchers believe that the resistance is due to a relatively simple genetic change and that it might be possible to induce such resistance in natural vector populations. If large numbers of *A. gambiae*, the most important malaria vector in Africa, could be made resistant, transmission of malaria could be greatly reduced.

Another control effort is directed toward developing a malaria vaccine. One problem has been to identify the stage of the parasite responsible for triggering the immune response in humans. An antigen on the sporozoite has now been identified, and the gene for this antigen can be cloned using recombinant DNA technology. Consequently, the antigen can be manufactured and used to make a vaccine. Thus, great strides have been made toward the development of an effective vaccine; the vaccine will hopefully be available in the near future. (See the Chapter 12 Essay, "The War Against Malaria: Missteps and Milestones." ∞ [p. 327]) But even when a vaccine becomes available, administering it to the vast numbers of people living where malaria is endemic will be a monumental task. Mechanisms to distribute the vaccine and ways of gaining the people's cooperation will be needed. The cost of the vaccine in the quantities required will be another major problem.

Toxoplasmosis

Toxoplasma gondii (Figure 23.21) is a widely distributed protozoan that infects many warm-blooded animals—both domestic and wild. It is an intracellular parasite and can invade many tissues. Humans usually become infected through contact with the feces of domestic cats that forage for natural foods, especially infected rodents (Figure 23.22). Cats that are kept inside and are fed on canned or cooked foods are unlikely to acquire the parasites. Another common means of transmission is the consumption of raw or undercooked contaminated meat. The French, who consume large amounts of steak tartare (raw ground beef), have the highest incidence of infection in the world.

THE DISEASE *Toxoplasma gondii* causes only mild lymph node inflammation in most humans. In the ma-

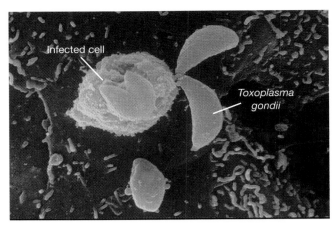

FIGURE 23.21 Crescent-shaped protozoans, *Toxoplasma gondii*, leaving an infected cell in which they have multiplied (SEM photo, magnified 3600X). This organism can be a danger to immunocompromised patients and to pregnant women, in whom it can cause serious birth defects and miscarriages.

jority of cases, the infection is chronic, asymptomatic, and self-limiting. However, *T. gondii* can cause serious **toxoplasmosis,** especially in developing fetuses, newborn infants, and sometimes in young children. The organism can be transferred across the placenta of an infected mother to the fetus. There it causes serious congenital defects, including the accumulation of cerebrospinal fluid, abnormally small head, blindness, mental retardation, and disorders of movement. It can also be responsible for stillbirths and spontaneous abortions. If infection occurs after birth, the symptoms are similar to, but less severe than, those seen in fetuses. In patients with severe immunosuppression, such as AIDS patients, the disease can appear as encephalitis and can also cause dermatologic problems.

DIAGNOSIS, TREATMENT, AND PREVENTION
Toxoplasmosis can be diagnosed by finding the parasites in the blood, cerebrospinal fluid, or tissues, by animal inoculation with subsequent isolation of the organisms, or by indirect immunofluorescence tests. ∞ (Chapter 19, p. 544) Pyrimethamine and trisulfapyridine are used in combination to treat toxoplasmosis, but no treatment can reverse permanent damage from prenatal infection. To prevent this disease, pregnant women should avoid contact with raw meat and cat feces, and cats should be kept out of sandboxes where children play, especially if a child might carry the organism to a pregnant woman.

Babesiosis

Several species of the sporozoan *Babesia* can cause **babesiosis** (ba-be″se-o′sis). Cattle are affected with babesiosis caused by the tickborne protozoan *Babesia*

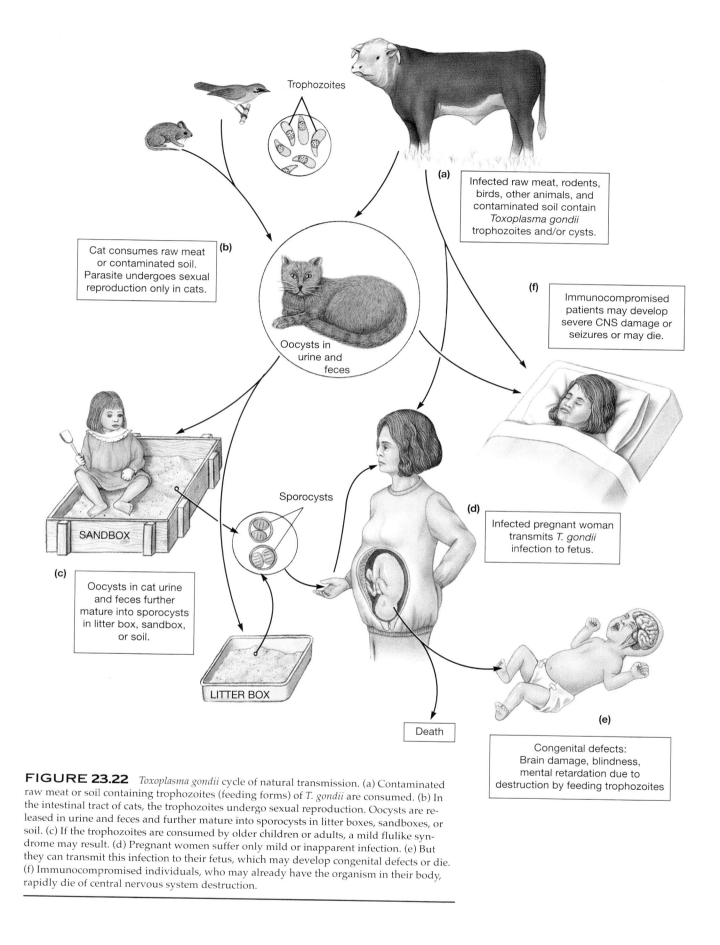

FIGURE 23.22 *Toxoplasma gondii* cycle of natural transmission. (a) Contaminated raw meat or soil containing trophozoites (feeding forms) of *T. gondii* are consumed. (b) In the intestinal tract of cats, the trophozoites undergo sexual reproduction. Oocysts are released in urine and feces and further mature into sporocysts in litter boxes, sandboxes, or soil. (c) If the trophozoites are consumed by older children or adults, a mild flulike syndrome may result. (d) Pregnant women suffer only mild or inapparent infection. (e) But they can transmit this infection to their fetus, which may develop congenital defects or die. (f) Immunocompromised individuals, who may already have the organism in their body, rapidly die of central nervous system destruction.

Labels within the figure:

Trophozoites

(a) Infected raw meat, rodents, birds, other animals, and contaminated soil contain *Toxoplasma gondii* trophozoites and/or cysts.

(b) Cat consumes raw meat or contaminated soil. Parasite undergoes sexual reproduction only in cats.

Oocysts in urine and feces

(f) Immunocompromised patients may develop severe CNS damage or seizures or may die.

Sporocysts

(d) Infected pregnant woman transmits *T. gondii* infection to fetus.

SANDBOX

(c) Oocysts in cat urine and feces further mature into sporocysts in litter box, sandbox, or soil.

LITTER BOX

Death

(e) Congenital defects: Brain damage, blindness, mental retardation due to destruction by feeding trophozoites *bigemina*, but *B. microti* is most often associated with human infections. The parasites enter the blood via bites of infected ticks and invade and multiply in red blood cells.

THE DISEASE   Although many cases are asymptomatic, when symptoms appear, they usually begin with sudden high fever, headache, and muscle pain. Anemia and jaundice can occur as red blood cells are destroyed. The symptoms last for several weeks and are followed by a prolonged carrier state. If babesiosis occurs in a person who has undergone spleen removal,

it is usually fatal in 5 to 8 days. This is because the lack of a spleen impairs the body's ability to break down defective red blood cells.

DIAGNOSIS, TREATMENT, AND PREVENTION
Diagnosis is made from blood smears, but the parasite can be confused with *Plasmodium falciparum*. Chloroquine is the drug of choice for treatment, and avoiding tick bites is the best means of protection.

The properties of nonbacterial systemic diseases are summarized in Table 23.4.

TABLE 23.4	Viral and Protozoan Systemic Diseases	
Disease	**Agent(s)**	**Characteristics**
Viral systemic diseases		
Dengue fever	Dengue fever virus	Severe bone and joint pain, high fever, headache, loss of appetite, weakness, sometimes a rash
Yellow fever	Yellow fever virus	Fever, anorexia, nausea, vomiting, liver damage, jaundice
Infectious mononucleosis	Epstein–Barr virus	Headache, fatigue, malaise, usually sore throat, secondary streptococcal infections common
Other viral fevers	Filoviruses, bunyaviruses, phleboviruses, arenaviruses, orbivirus, and coxsackie viruses	Some cause hemorrhagic fevers; some cause encephalitis, joint pain, slow heart rate, infection of erythrocytes, diarrhea, rashes, sore throat, liver disease, meningitis, inflammation of the heart and the sac around it
Protozoan systemic diseases		
Kala azar	*Leishmania donovani*	Visceral leishmaniasis with irregular fever, weakness, wasting, enlarged liver and spleen
Localized leishmaniasis	*L. tropica, L. braziliensis*	Oriental sore and skin and mucous membrane lesions
Malaria	*Plasmodium* species	Periods of high fever associated with release of parasites from red blood cells; relapses can occur; one species can cause malignant malaria and blackwater fever
Toxoplasmosis	*Toxoplasma gondii*	Mild lymph node inflammation in adults; can cross placenta and cause serious damage to nervous system of fetus; also causes damage in small children and immunosuppressed patients
Babesiosis	*Babesia microti*	High fever, headache, muscle pain, anemia, and jaundice; fatal in patients whose spleens have been removed

COMBATING HANTAVIRUSES

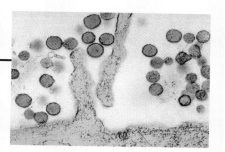

FIGURE 23.23 Color-enhanced TEM photo of a hantavirus.

For the second year in a row, the winter of 1993 brought mild, wet weather to the high desert and mountains of Four Corners, where Arizona, Colorado, New Mexico, and Utah meet. Plants thrived there, and along with them came insects. The populations of the pale gray mouse and white deer mouse, feeding on piñon pine nuts and grasshoppers, skyrocketed. These events would probably have gone unnoticed but for one consequence: People in the Four Corners area suddenly began dying.

The unknown disease began with fever and muscle aches. Within days it progressed to respiratory collapse as the lungs filled with fluid leaking from capillaries. Medical officers reported the first 3 cases on May 14, 1993, and 5 more within 3 days. By June 7 a total of 24 cases had surfaced, a few identified retrospectively to December 1992. Half these patients had died—some within 24 hours of first being seen by a physician. Those who survived the pulmonary crisis appeared to recover fully in a few days. Most patients had no history of debilitating disease and were not immunocompromised. Physicians and epidemiologists ruled out plague, anthrax, influenza—even toxins such as chemical warfare agents potentially escaping from a nearby army depot. A coalition of agencies, including the University of New Mexico Health Sciences Center and the CDC, launched an all-out effort to identify the cause of the dangerous new disease.

Similarities to a disease known as hemorrhagic fever with renal syndrome (HFRS), which is common in Asia, provided a clue. HFRS—characterized by initial high fever followed by shock, hemorrhage and usually kidney failure—came to the attention of Western medicine in 1951, during the Korean War. Not until 1978 did researchers isolate the causative agent of HFRS, a virus carried by the Asian striped field mouse (*Apodemus agrarius*). Researchers called the agent Hantaan virus after the Hantaan River in Korea, where it was first isolated.

Other strains of the Hantaan virus later turned up in Europe, and a new genus of viruses, *hantavirus*, was named (Figure 23.23). Hantaviruses belong to the family Bunyaviridae. (Chapter 11, p. 279) Whereas other bunyaviruses move from host to host via insect vectors, hantaviruses appear to spread from their normal host—rodents—to humans without a vector.

HFRS and the Four Corners mystery disease had two major differences. HFRS ultimately attacks the kidneys; the new disease attacked the lungs. And HFRS kills only a few percent of those diagnosed; the new disease killed more than half, and with frightening rapidity. The CDC had several hantavirus antigens available from long-term army research stemming from the Korean War. In immunological tests, several patients revealed a low-level antibody response to some of these antigens. The conclusion: An unknown hantavirus caused the disease.

Researchers next used the polymerase chain reaction (PCR) technique (Chapter 7, p. 190) to analyze the new virus. Portions of the genome resembled but were not identical with previously known hantaviruses. Now sure that the culprit was a hantavirus, epidemiologists named the emerging disease **hantavirus pulmonary syndrome** (HPS). The new virus became known by several names, including "FC hantavirus," for Four Corners. (The ICTV will decide the name in 1996.) Since the Four Corners outbreak, other HPS cases have turned up in 21 states and in Canada. As of April 1995, agencies had reported 106 cases in the United States, with 55 fatalities.

With the knowledge that other hantaviruses replicate in rodents, epidemiologists soon fingered the deer mouse (*Peromyscus maniculatus*) in the Four Corners area. About a third of trapped deer mice harbored the new virus. Through PCR, researchers identified portions of the virus genome in preserved deer mouse specimens collected as early as 1975 and in preserved human tissues. At least five rodent species other than deer mice carry hantavirus in the United States, but only two are known to cause disease in humans. Two young men died of HPS acquired from white-footed mice in Suffolk County, New York. A nonfatal case in Florida appears to have come from a cotton rat. But by far the most common cause is contact with droppings of the deer mouse. The virus causes no symptoms in the mouse and appears to be transmitted to humans by airborne dust containing dried droppings.

Like all viruses, the FC hantavirus is difficult to treat. During the 1993 outbreak in the U.S. Southwest, physicians tested the broad-spectrum antiviral drug ribavirin (Virazole) on HPS patients. Ribavirin had proved effective against HFRS—thanks to the work of the U.S. Army Medical Research Institute of Infectious Disease—and is used in China, which has about 250,000 HFRS cases each year. Physicians hoped the drug would work in HPS patients. Thus far tests show no evidence of efficacy against the FC hantavirus, but more drug trials are planned.

The problem with ribavirin comes from the 24-hour lag between administration and the time the drug reaches effective blood levels. Because HPS progresses so rapidly, 24 hours can mean the difference between life and death. A test for hantavirus antibodies—developed at breakneck speed the summer after the initial 1993 outbreak—might identify patients in the early stages of the disease, while time remains for ribavirin to work. Efforts are under way to make the diagnostic test available in rural clinics that typically see patients early in the course of the disease.

An entirely different treatment may prove most useful in the long run. Many of the symptoms of HPS (and of HFRS as well) appear immunologic in origin rather than a direct result of viral action. The viruses cause little or no abnormality in cells in the laboratory or in cells lining lung capillaries of patients. Signs of immunologic response include readily detectable hantavirus-specific antibodies, tissue infiltration by lymphocytes, and the simultaneous appearance of immature immune cells and pulmonary symptoms. Given these hints, researchers hope to find a way to short-circuit the immune response that appears to kill so many HPS patients.

- The agents and characteristics of the diseases discussed in this chapter are summarized in Tables 23.1 through 23.4. Information in those tables is not repeated in this summary.

CARDIOVASCULAR AND LYMPHATIC DISEASES

Bacterial Septicemias and Related Diseases

- **Septicemia,** or blood poisoning, involves multiplication of bacteria in the blood. Septicemias and related diseases are diagnosed by culturing appropriate samples and are treated with antibiotics.

- **Rheumatic fever** and **bacterial endocarditis** occur most often in patients who have had previous streptococcal infections.

Helminthic Diseases of the Blood and Lymph

- **Schistosomiasis** is acquired from schistosome larvae that penetrate the skin and is diagnosed by finding eggs in feces. It is treated with praziquantel and could be prevented by the eradication of infected snails or avoidance of snail-infested water.

- **Filariasis** is transmitted by mosquitoes and is diagnosed by finding microfilariae in blood. It is treated with diethylcarbamizine or metronidazole and could be prevented if infected mosquitoes could be eradicated.

SYSTEMIC DISEASES

Bacterial Systemic Diseases

- **Anthrax** is acquired through contact with *Bacillus anthracis* spores from infected domestic animals or their hides. It is diagnosed by culturing blood or from smears from lesions and is treated with penicillin or tetracycline. The disease can be prevented by immunizing animals and humans with occupational exposure and by careful burial of infected animals.

- Plague has occurred in periodic epidemics since the Middle Ages, remains endemic in some regions, and is increasing in incidence in the United States. The form that is transmitted by fleas from infected rats is known as **bubonic plague,** which forms enlarged lymph nodes, called **buboes.** If the disease progresses to the circulatory system, it is called **septicemic plague.** Lung involvement results in **pneumonic plague,** which is contagious and transmitted by aerosols. Plague is diagnosed by stained smears and antibody tests and treated with streptomycin or tetracycline. It can be prevented by controlling rat populations and immunizing people who enter endemic areas.

- **Tularemia** is transmitted through the skin, inhaled, or ingested. It can be diagnosed by agglutination tests and treated with streptomycin. It is prevented by avoiding infected mammals and arthropods; the vaccine is short-lasting and not fully protective.

- **Brucellosis** is transmitted to humans through the skin from domestic animals by the ingestion of contaminated dairy products and by inhalation or ingestion and is diagnosed by serological tests. It is treated with prolonged antibiotic therapy and can be prevented by avoiding infected animals and contaminated fomites.

- **Relapsing fever** is transmitted by lice and ticks and can be diagnosed from blood smears. It is treated with tetracycline or chloramphenicol and can be prevented by avoiding or controlling ticks and lice.

- **Lyme disease** is transmitted by ticks from infected deer and other animals and is diagnosed by clinical signs and serological tests. It is treated with antibiotics and can be prevented by avoiding tick bites.

Rickettsial and Related Systemic Diseases

- **Typhus fever** occurs in several forms. **Epidemic typhus,** transmitted by human body lice, usually occurs in unsanitary, overcrowded conditions. It has a high mortality rate unless treated with antibiotics.

- **Brill–Zinsser disease,** or recrudescent typhus, is a recurrence of a latent typhus infection. **Endemic,** or **murine, typhus** is carried by fleas, and **scrub typhus,** by mites from infected rats.

- **Rocky Mountain spotted fever,** carried by ticks, damages blood vessels. Strains of the causative rickettsia vary in virulence, and the mortality of the untreated infection can be high.

- **Rickettsialpox** is carried by mites that live on house mice. **Trench fever,** transmitted by lice, is prevalent under unsanitary conditions, most often among persons under stress. **Bartonellosis,** transmitted by sandflies, occurs in two forms: **Oroya fever,** an acute fever that causes life-threatening anemia, and **verruga peruana,** a self-limiting skin rash.

- Recently identified human pathogens resembling the rickettsias include *Ehrlichia canis* and *E. chaeffeenis,* which cause **ehrlichiosis,** and *Bartonella henselae,* which causes **bacillary angiomatosis.**

Viral Systemic Diseases

- **Dengue fever,** an arbovirus disease, can be diagnosed by serologic tests; a vaccine is available against one immunological type of dengue fever virus.

- **Yellow fever,** another arbovirus disease, is diagnosed by symptoms and can be prevented with vaccine.

- **Infectious mononucleosis** is caused by the **Epstein-Barr virus,** diagnosed by symptoms, and treated symptomatically and with antibiotics for secondary infections. In developing countries babies have mild symptoms and produce antibodies by age 1, but in developed countries patients are teenagers or young adults, who have a much more serious disease.

- **Chronic fatigue syndrome** has been linked with the Epstein–Barr virus, which is responsible for infectious mononucleosis, **Burkitt's lymphoma,** nasopharyngeal carcinoma, and oral hairy leukoplakia.

- Other viral infections include **filovirus** fevers (such as **Ebola virus** infection), bunyavirus fevers (such as **Rift Valley fever**), arenavirus fevers (such as **Lassa fever** and **Bolivian hemorrhagic fever**), **Colorado tick fever, feline panleukopenia virus** and **canine parvovirus** infections, and **fifth disease.**

Protozoan Systemic Diseases

- **Leishmaniasis** occurs in tropical and subtropical countries where appropriate species of sandfly vectors are available. It is diagnosed from blood smears or scrapings from lesions and is treated with antimony. The disease could be prevented by controlling sandfly breeding and eliminating rodent reservoir infections.

- **Malaria** is one of the world's greatest public health problem; it kills a million people annually, most of them children. Cases in the United States come from people who have been in endemic areas. Malaria is transmitted by female *Anopheles* mosquitoes and is diagnosed by identifying protozoa in blood smears. Active disease is treated with chloroquine (except for resistant strains), and latent parasites are eliminated with primaquine. Research is in progress to find a way to control mosquitoes and to develop an effective vaccine.

- **Toxoplasmosis** is usually transmitted from feces of cats that have consumed infected rodents and by human ingestion of contaminated raw or undercooked meat. It is diagnosed by finding parasites in body fluids or tissues and is treated with pyrimetamine and trisulfapyridine. The disease can be prevented by avoiding contaminated materials.

- **Babesiosis** is transmitted by ticks and is diagnosed from blood smears. It is treated with chloroquine and can be prevented by avoiding contact with ticks.

KEY TERMS

anthrax   (*p. 650*)
aplastic crisis   (*p. 666*)
babesiosis   (*p. 670*)
bacillary angiomatosis
   (*p. 662*)
bacterial endocarditis
   (*p. 646*)
bartonellosis   (*p. 661*)
blackwater fever   (*p. 669*)
Bolivian hemorrhagic fever
   (*p. 666*)
Brill–Zinsser disease
   (*p. 659*)
brucellosis   (*p. 654*)
bubo   (*p. 653*)
bubonic plague   (*p. 653*)
Burkitt's lymphoma
   (*p. 644*)
canine parvovirus   (*p. 666*)
childbed fever   (*p. 645*)
chronic fatigue syndrome
   (*p. 664*)

Colorado tick fever   (*p. 666*)
dengue fever   (*p. 662*)
Ebola virus   (*p. 665*)
ehrlichiosis   (*p. 661*)
elephantiasis   (*p. 649*)
endemic relapsing fever
   (*p. 655*)
endemic typhus   (*p. 659*)
epidemic relapsing fever
   (*p. 655*)
epidemic typhus   (*p. 658*)
Epstein–Barr virus   (*p. 663*)
feline panleukopenia virus
   (*p. 666*)
fifth disease   (*p. 667*)
filariasis   (*p. 648*)
filovirus   (*p. 665*)
hantavirus pulmonary syndrome   (*p. 673*)
infectious mononucleosis
   (*p. 663*)
ischemia   (*p. 669*)

kala azar   (*p. 667*)
Lassa fever   (*p. 666*)
leishmaniasis   (*p. 667*)
Lyme disease   (*p. 656*)
lymphangitis   (*p. 645*)
malaria   (*p. 668*)
Malta fever   (*p. 654*)
murine typhus   (*p. 659*)
myocarditis   (*p. 646*)
orbivirus   (*p. 666*)
Oroya fever   (*p. 661*)
pericarditis   (*p. 646*)
petechia   (*p. 658*)
phlebovirus   (*p. 666*)
pneumonic plague   (*p. 653*)
puerperal fever   (*p. 645*)
relapsing fever   (*p. 655*)
rheumatic fever   (*p. 646*)
rickettsialpox   (*p. 661*)
Rift Valley fever   (*p. 666*)
Rocky Mountain spotted
   fever   (*p. 659*)

schistosomiasis   (*p. 647*)
scrub typhus   (*p. 659*)
septicemia   (*p. 644*)
septicemic plague   (*p. 653*)
septic shock   (*p. 644*)
toxoplasmosis   (*p. 670*)
transovarian transmission
   (*p. 653*)
trench fever   (*p. 661*)
tularemia   (*p. 653*)
typhoidal tularemia
   (*p. 654*)
typhus fever   (*p. 658*)
ulceroglandular   (*p. 654*)
undulant fever   (*p. 654*)
vegetation   (*p. 646*)
verruga peruana   (*p. 661*)
yellow fever   (*p. 663*)

A **Bacterial Septicemias and Related Diseases**

1. How does septicemia differ from bacteremia?

2. List the causative agents and symptoms of septicemia.

3. Identify the main characteristics of puerperal fever.

4. What are the cause and nature of tissue damage in rheumatic fever?

5. How can rheumatic fever be diagnosed, treated, and prevented?

6. What is bacterial endocarditis, and why is it life-threatening?

B **Helminthic Diseases of the Blood and Lymph**

7. Complete the following table to describe how each disease is acquired, what its effects are, and how it is diagnosed, treated, and prevented.

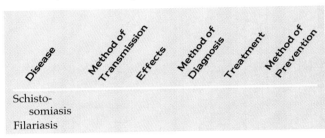

Disease	Method of Transmission	Effects	Method of Diagnosis	Treatment	Method of Prevention
Schisto-somiasis					
Filariasis					

C **Bacterial Systemic Diseases**

8. What is the cause of anthrax, and how do humans acquire it?

9. What are the forms of anthrax, and how do their effects, treatment, and prevention differ? What are their pathogenic effects?

10. Complete the following table to describe the cause of each disease, what its effects are, and how it is diagnosed, treated, and prevented.

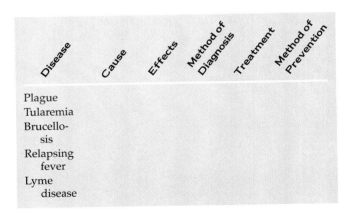

Disease	Cause	Effects	Method of Diagnosis	Treatment	Method of Prevention
Plague					
Tularemia					
Brucello-sis					
Relapsing fever					
Lyme disease					

11. Where is plague found today?

12. How do humans acquire tularemia?

13. How do humans acquire brucellosis?

14. Why is Lyme disease often misdiagnosed?

D **Rickettsial Systemic Diseases**

15. List the general characteristics of the rickettsias and rickettsial diseases.

16. How are rickettsias transmitted, and how are diseases they cause diagnosed, treated, and prevented?

17. Compare the various kinds of typhus fever.

18. How does Rocky Mountain spotted fever differ from other rickettsial diseases?

19. What are the main features of rickettsialpox?

20. Compare trench fever with typhus fever.

21. In what forms is bartonellosis seen? What causes it, and how can it be treated and prevented?

E **Viral Systemic Diseases**

22. Compare and contrast dengue fever with yellow fever. These diseases might be eradicated by ——————.

23. List the special characteristics of the Epstein–Barr virus. What diseases does it cause, and how are they diagnosed and treated?

24. What viruses other than EBV cause systemic infections in humans?

F **Parasitic Protozoan Systemic Diseases**

25. List the different kinds of leishmaniases, and describe how can they be diagnosed, treated, and prevented.

26. Why is malaria one of the world's greatest public health problems?

27. How is malaria diagnosed and treated, and how might it be controlled?

28. How do people get toxoplasmosis, and how can it be prevented?

29. How do people get babesiosis, and how can it be prevented?

30. Match the following:

—— tularemia	**a.** *Brucella*
—— plague	**b.** *Borrelia*
—— Lyme disease	**c.** *Plasmodium*
—— puerperal fever	**d.** *Leishmania*
—— anthrax	**e.** Epstein–Barr virus
—— undulant fever	**f.** *Yersinia*
—— malaria	**g.** *Streptococcus*
—— relapsing fever	**h.** *Francisella*
—— infectious mononucleosis	**i.** *Bacillus*
—— kala azar	
—— Burkitt's lymphoma	
—— rheumatic fever	

PROBLEMS FOR INVESTIGATION

1. Using your present knowledge of microbiology and any information you can find in other sources, hypothesize why Lyme disease and the organism that causes it were not recognized earlier.

2. Discuss the similarities and differences associated with arthropodborne infections. Are such infections more likely to be systemic than limited to particular tissues? Defend your answer.

3. Why do you suppose there have been no major outbreaks of bubonic plague in this century? Do you think the United States could ever have a plague epidemic? Why or why not?

4. Suppose that you are a park employee or a health department official. Devise a way to prevent people from getting arthropodborne infections. What information would you give to a hiker about to go into a tick-infested area?

5. What practical advice would you give to a pregnant woman regarding toxoplasmosis?

6. What prophylactic measures should you follow if you travel in a malarial zone? What could you do to lessen the number of mosquito bites you get?

7. A 47-year-old biologist had been studying small wild mammals in a rural area outside La Paz, Bolivia. She developed chills, sudden high fever, swollen lumps in her right armpit, headache, loss of appetite, and aching back and hip muscles. She immediately returned to Washington, D.C., where her disease was diagnosed, and she was hospitalized in an isolation room. After 6 days of streptomycin therapy, she was discharged. Make a diagnosis. (The answer to this question appears in Appendix F.)

SOME INTERESTING READING

"AAP issues policy on treatment of Lyme disease in children." 1991. *American Family Physician* 44, no. 1 (July):308–11.

Boyles, S. 1990. "Allogenic bone marrow transplantation in adults with Burkitt's lymphoma or acute lymphoblastic leukemia in first complete remission." *NCI Cancer Weekly* (June 4):24–26.

CDC. 1995. "Outbreak of Ebola viral hemorrhagic fever—Zaire, 1995." *Morbidity and Mortality Weekly Report* 44, no. 19 (May 19):381–82.

Defoe, D. 1721. (Reprinted 1960). *A journal of the plague year*. New York: Signet Classics.

Duchin, J. S., et al. 1994. "Hantavirus pulmonary syndrome: A clinical description of 17 patients with a newly recognized disease." *The New England Journal of Medicine* 330, no. 14 (April 17):949–55.

Fishbein, D. B., and D. T. Dennis. 1995. "Tick-borne diseases—a growing risk." *The New England Journal of Medicine* 333, no. 7 (August 17):452–53.

Gutin, J. A. C. 1993. "The infection unto death." *Discover* 14(November):114–20.

Kantor, F. S. 1994. "Disarming lyme disease." *Scientific American* 271, no. 3 (September):34–39.

Knoop, F. C., et al. 1993. "Leishmaniases of the New World: Current concepts and implications for future research." *Clinical Microbiology Reviews* 6, no. 3 (July):230–50.

Lanciotti, R. S., et al. 1994. "Molecular evolution and epidemiology of dengue—3 viruses." *Journal of General Virology* 75(1):65–75.

Levy, D. L. 1995. "Hantavirus pulmonary syndrome: Outbreak of a new disease caused by a new virus." *Postgraduate Medicine* 97 (March):127–39.

Magnarelli, L. A., et al. 1991. "Rickettsiae and *Borrelia burgdorferi* in ixodid ticks." *Journal of Clinical Microbiology* 29(12):2798–2804.

"Malaria in travelers returning to the United States." 1991. *American Family Physician* 44, no. 6 (December):2183.

McEvedy, C. 1988. "The bubonic plague." *Scientific American* 258 (February):118–23.

Melby, P. C., et al. 1992. "Cutaneous leishmaniasis: Review of 59 cases seen at the National Institutes of Health." *Clinical Infectious Disease* 15:924–37.

Meldrum, S. C., et al. 1992. "Human babesiosis in New York state: An epidemiological description of 136 cases." *Clinical Infectious Disease* 15:1019–23.

Preston, R. 1994. *The hot zone*. New York: Random House.

Relman, D. A. 1995. "Has trench fever returned?" *The New England Journal of Medicine* 332, no. 7 (February 16):463–64.

Rybak, L. P. 1990. "Deafness associated with Lassa fever." *Journal of the American Medical Association* 264, no. 16 (October 24/31):2119.

Sexton, D. J., et al. 1991. "Spotted fever group rickettsial infections in Australia." *Reviews of Infectious Diseases* 13, no. 5 (September/October):876–86.

Strickland, G. T., and S. L. Hoffman. 1994. "Strategies for the control of malaria." *Scientific American: Science and Medicine* 1, no. 4 (July/August):24–33.

White, N. J. 1994. "Tough test for malaria vaccine." *Lancet* 344 (October 29):1172–74.

Young, S. 1991. "Dragonflies help to defeat dengue fever." *New Scientist* 130, no. 1766 (April 27):26.

Zinsser, H. 1935. *Rats, lice, and history*. Boston: Little, Brown.

CHAPTER 24

DISEASES OF THE NERVOUS SYSTEM

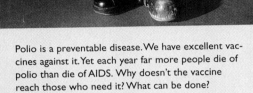

Polio is a preventable disease. We have excellent vaccines against it. Yet each year far more people die of polio than die of AIDS. Why doesn't the vaccine reach those who need it? What can be done?

As with cardiovascular and lymphatic diseases, the diseases of the nervous system also often affect other systems. It is assumed that the reader is familiar with disease processes (Chapter 15) and with host systems and host defenses (Chapters 17 and 18).

DISEASES OF THE BRAIN AND MENINGES

Bacterial Diseases of the Brain and Meninges

Bacterial Meningitis

Bacterial meningitis is an inflammation of the *meninges,* the membranes that cover the brain and spinal cord. ∞ (Chapter 17, p. 471) This life-threatening disease can be caused by several kinds of bacteria, each of which has a prevalence that can be correlated with the age of the host (Table 24.1). Meningitis causes *necrosis* (death of tissues in a given area), clogging of blood vessels, increased pressure within the skull from edema, decreased cerebrospinal fluid flow, and impaired central nervous sys-

TABLE 24.1	Types of Bacterial Meningitis	
Age	Most Frequent Causative Agents	Comments
Newborn (0–2 months)	*Escherichia coli,* other Enterobacteriaceae, *Streptococcus* species	Average mortality about 50%; incidence 40–50/ 100,000 live births; maternal transmission
Preschool (2 months– 5 years)	*Haemophilus influenzae,* type b; *Neisseria meningitidis*	Maximum incidence 6–8 months; overall incidence 180/ 100,000 children
Youth and young adult (5– 40 years)	*Neisseria meningitidis, Streptococcus pneumoniae*	Sporadic or epidemic
Mature adult (over 40 years)	*Streptococcus pneumoniae, Staphylococcus* species	Sporadic

tem function. The early symptoms are headache, fever, and chills. In rare instances, seizures develop. Onset can be insidious or fulminating. Death occurs from shock

A Which pathogens cause bacterial diseases of the brain and meninges, and what are the important epidemiologic and clinical aspects of these diseases?

B Which pathogens cause viral diseases of the brain and meninges, and what are the important epidemiologic and clinical aspects of these diseases?

C Which pathogens cause bacterial nerve diseases, and what are the important epidemiologic and clinical aspects of these diseases?

D Which pathogens cause viral nerve diseases, and what are the important epidemiologic and clinical aspects of these diseases?

E Which pathogens cause parasitic diseases of the nervous system, and what are the important epidemiologic and clinical aspects of these diseases?

and other serious complications within hours of the appearance of symptoms.

Most cases of meningitis are acute, but some are chronic. Acute meningitis is acquired from carriers or endogenous organisms. Organisms gain access to the meninges directly during surgery or trauma, or spread in blood to them from other infections such as pneumonia and otitis media. Host defenses in the arachnoid layer, one of the meninges, ordinarily combat bacteremia. If organisms overwhelm the defenses, however, meningitis results. Chronic meningitis occurs with extension of underlying diseases such as syphilis or tuberculosis. The bacteria causing these underlying diseases are slow growers; in such cases, onset of typical meningitis symptoms is insidious, occurring over a period of weeks.

Meningitis is diagnosed by culturing cerebrospinal fluid. The fluid usually is turbid—sometimes so thick with pus that it is difficult to remove with a syringe. Antibiotic treatment varies with the causative organism. If tubercular meningitis is suspected, isoniazid therapy is called for immediately. Treatment for tubercular meningitis lasts a year of more, whereas some very rare cases of fungal meningitis may require several years of treatment.

MENINGOCOCCAL MENINGITIS The bacterium *Neisseria meningitidis* has caused 2000–3000 cases of meningitis per year for the past decade in the United States. Mortality is about 85 percent when the disease is untreated but only 1 percent with the best treatment. The 15 percent mortality rate in the United States probably reflects delay in seeking treatment. This disease was the leading cause of death from infectious disease among U.S. armed forces during World War II.

In meningococcal meningitis the organisms colonize the nasopharynx, spread to the blood, and make their way to the meninges, where they grow rapidly (Figure 24.1). In a complication called Waterhouse–Friderichsen syndrome, the meningococci invade all parts of the body, and death occurs within hours from endotoxin shock. The immediate cause of death is usually clotting of blood followed by massive hemorrhage in the adrenal glands (located above the kidneys), leading to fatal deficiency of essential adrenal hormones. A lesser degree of hemorrhaging is sometimes seen in meningitis patients who develop a petechial skin rash.

Penicillin is the drug of choice for treatment; the prevalence of resistant strains makes sulfonamides no longer useful. Third-generation cephalosporins and ampicillin are also used now. A vaccine is available, but

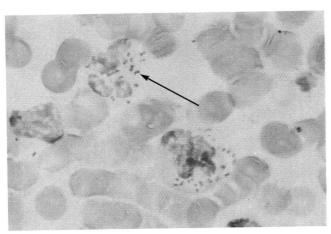

FIGURE 24.1 Meningococci infection of the meninges is a cause of meningitis. These meningococci have been phagocytized by white blood cells in a cerebrospinal fluid specimen.

it is not effective against the most common type B meningococci. The risk of this disease can be decreased by prevention of overtiring and overcrowding. The number of inches required between bunks in military barracks is based on experience with meningococcal outbreaks.

In some closed environments, such as military bases, dormitories, and day-care centers, 90 percent of the population can be carriers of meningococci, yet only 1 per 1000 carriers develops the disease. Among members of patient households, 80 to 90 percent are carriers, compared with only 5 to 30 percent in the general public. Antibiotics can eliminate the carrier state.

HAEMOPHILUS MENINGITIS   About two-thirds of bacterial meningitis cases during the first year of life are caused by *Haemophilus influenzae* type B (Hib). Among children, 30 to 50 percent are carriers of this organism; among adults, the figure is only 3 percent. Humans are exposed to *H. influenzae* early in life and rapidly acquire immunity, so the disease is rare in adults. Only 10 percent of children between ages 3 and 6 lack antibodies, and all those over age 6 have antibodies. Without treatment this disease is nearly always fatal. Even with treatment, one-third die. Of those who recover, 30 to 50 percent have serious mental retardation, and 5 percent are permanently institutionalized because of damage to the central nervous system. *Haemophilus* meningitis is the leading cause of mental retardation in the United States and worldwide. Hib vaccines are available, which all children age 5 and under should receive. ∞ (Chapter 18, p. 502) These vaccines have dramatically reduced incidence of the disease.

STREPTOCOCCUS MENINGITIS   Among adults *Streptococcus pneumoniae* is the most common cause of

meningitis. Organisms generally spread via the blood from lung, sinus, mastoid, or ear infections. Mortality is 40 percent.

Listeriosis

Another kind of meningitis, **listeriosis,** is caused by *Listeria monocytogenes*, a small, gram-positive bacillus that is widely distributed in nature. Foodborne transmission by improperly processed milk, cheese, meat, and vegetables is the most common source of infection. It is sometimes acquired as a zoonosis and is particularly threatening to those with impaired immune systems. Although not an especially significant human disease for several decades, listeriosis is now a leading cause of infection in kidney transplant patients. In pregnant women the bacillus can cross the placenta, infect the fetus, and cause abortion, stillbirth, or neonatal death. Listeriosis is responsible for many cases of fetal damage.

Brain Abscesses

Microorganisms that cause **brain abscesses** reach the brain from head wounds or via blood from another site. As would be expected with wounds, multispecies infections are common, and anaerobes are as likely to be responsible as are aerobes. Most such abscesses occur in patients under age 40, but two age periods—birth to 20 years and 50 to 70 years—show peak incidences. The infection gradually grows in mass and compresses the brain. The masses can be detected by CAT scans or X-rays, and causative agents can be identified by serologic tests and by culturing cerebrospinal

APPLICATIONS

REACHING THE BRAIN

The blood–brain barrier places some limits on the use of antibiotics to treat diseases of the central nervous system. ∞ (Chapter 17, p. 472) Beta-lactam antibiotics such as penicillin, for example, do not penetrate the blood-brain barrier easily, although they pass through inflamed meninges more easily than through noninflammed meninges. When administered orally or intravenously, the average concentration of antibiotics in the cerebrospinal fluid generally reaches only about 15 percent of their concentration in plasma. Some other antibiotics, however, such as chloramphenicol and tetracycline, are lipid-soluble and so diffuse easily across the blood–brain barrier. Unfortunately, fewer antibodies and complement pass from the blood into brain tissue than into other tissues. These factors often combine to make treatment of brain infections very difficult.

fluid. In very early stages, antibiotic treatment can be sufficient, but later surgical drainage or removal of abscesses is usually necessary. Abscesses in areas of the brain that control the heart or other vital organs cannot be treated surgically. Without treatment, half the patients die, but with the current best treatment, only 5 to 10 percent die.

Viral Diseases of the Brain and Meninges

Viral Meningitis

Unlike bacterial meningitis, which is always fatal if untreated, **viral meningitis** is usually self-limiting and nonfatal. Enteroviruses account for approximately 40 percent of viral meningitis cases, and mumps virus for 15 percent. The causative virus in 30 percent of cases remains unidentified.

Rabies

Rabies was described by Democritus in the fifth century B.C. and by Aristotle in the fourth century B.C. Pasteur made real progress in understanding the disease when he found evidence of the infectious agent in saliva, the central nervous system, and peripheral nerves. He attenuated the agent and proved that a suspension of it could be used to prevent rabies. In 1903 the Italian physician Adelchi Negri discovered inclusion bodies ∞ (Chapter 15, p. 405) called *Negri bodies*, or clusters of viruses, in neurons (Figure 24.2). Negri bodies were used to diagnose rabies for more than 50 years until an immunofluorescent antibody test (IFAT) was developed in 1958. Still used today, the IFAT is so sensitive that after an animal suspected of having rabies bites a human, it can be killed immediately and its brain examined for rabies antigens. Prior to the availability of IFAT, such animals had to be held for 30 days, or until symptoms of rabies developed, be-

FIGURE 24.2 Negri bodies of rabies in the cerebellum of a human brain.

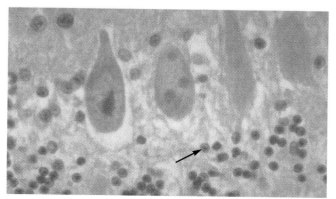

fore a search for Negri bodies could be made. Most biting animals are not rabid, so the IFAT has saved many people both the anguish of wondering if they have been exposed to rabies and the discomfort and risk of treatment.

The virus that causes rabies has a worldwide distribution. It infects all mammals exposed to it, so the possibilities for reservoir infections are almost limitless. Different types of rabies are found in different regions of the world. In almost all of Asia, Africa, Mexico, and Central and South America, rabies is endemic in dogs. In Canada, the United States, and Western Europe, wildlife rabies predominates, and dog rabies is controlled. The WHO lists 60 countries, including England, Australia, Japan, Sweden, and Spain, as rabies-free. This success is due to animal vaccination and quarantine programs.

Identifying rabid animals can be a problem. About half of all rabid dogs release viruses into saliva 3 to 6 days before they show symptoms of rabies. In contrast, 90 percent of rabid cats have viruses in their saliva about 1 day before they become symptomatic. Any change in the behavior of an animal can be a warning that it might be rabid. A "friendly" wild animal approaching people or a gentle family animal snapping without provocation sometimes indicates impending rabies symptoms. Small wild animals and some domestic ones, too, will bite if suddenly grabbed and held. Biting is their only defense in such a stressful situation. Much unnecessary anguish could be avoided if people, especially children, were taught to appreciate unfamiliar animals without touching them.

Animals vary in their susceptibility to rabies, and susceptibility is directly correlated with the role of the animals in maintaining reservoir infections. Foxes, coyotes, skunks, raccoons, and bats are highly susceptible. No infected fox has ever been known to survive, whereas 10 to 20 percent of dogs and 30 to 40 percent of mongooses do survive. Bats are particularly dangerous because they are asymptomatic and shed viruses into their feces, urine, and saliva. Two explorers of bat-infested caves in Texas died of rabies. Dogs, cats, cattle, horses, and sheep are less susceptible. Whether a person becomes infected depends mainly on whether an animal is shedding viruses in its saliva at the time of the bite. Even animals later proved to have rabies may not have been shedding viruses at the time they inflicted bites.

THE DISEASE Rabies is caused by the **rabies virus,** an RNA-containing rhabdovirus. ∞ (Figure 11.3c, p. 277) After entering the body through an animal bite or other break in the skin, the rabies virus first replicates in injured tissue for 1 to 4 days and then migrates to nerves, where it replicates slowly until it reaches the spinal cord. It progresses rapidly up the spinal cord to the brain by the flow of cytoplasm through axons. The time from infection to the appearance of symptoms varies from 13 days to 2 years but is usually between 20 and 60 days. The length of time required for symptoms to appear is proportional to the distance between the wound and the brain and is affected by the accessibility of nerve fibers. Thus, a bite on the face, which is well supplied with nerves and close to the brain, produces symptoms much more quickly than does a bite on the leg. The rabies virus has a predilection for nervous tissue but also infects salivary glands and the respiratory tract lining. There are even documented cases of transmission of rabies virus via corneal transplants.

It is the typically lengthy incubation period of rabies that allows postexposure immunization to be possible. Ordinarily there is sufficient time for the bitten individual to be vaccinated and to respond by making enough protective antibodies to prevent onset of the disease. Once symptoms have occurred, it is too late to vaccinate, and death usually follows quickly.

In humans the first symptoms are headache, fever, nausea, and partial paralysis near the bite site. These symptoms persist for 2 to 10 days and then worsen until the acute neurological phase of the disease ensues. The patient's gait becomes uncoordinated as paralysis becomes more general. Hydrophobia (fear of water) occurs as throat muscles undergo painful spasms, especially during swallowing. Aerophobia (fear of moving air) occurs because the skin is hypersensitive to any sensations. Confusion, hyperactivity, and hallucinations also occur. Within 10 to 14 days of the onset of symptoms, the patient typically goes into a coma and dies. Of all human patients who have suffered from clinical rabies, only two have been known to survive and make a complete recovery. Both had some degree of protection from an earlier immunization, and they were known to have been bitten by a rabid animal, so postexposure immunological treatment was started immediately.

CLOSE-UP

RABIES IN DOGS

Rabies symptoms in a dog begin with the dog acting as if it has a sore throat or has something caught in its throat. As the disease progresses, the dog may become groggy or paralyzed (dumb rabies) or agitated and aggressive, biting anything that disturbs it (furious rabies). Throat muscle spasms, difficulty in swallowing, and drooling also are indicative of rabies. Eventually the dog becomes apathetic and stuporous and slides into a final coma.

DIAGNOSIS, TREATMENT, AND PREVENTION A sample from a brain or skin biopsy can be stained by the IFAT to identify rabies virus antigen before the patient dies. Finding the antigen confirms the diagnosis, but failure to find it does not rule out rabies. Sometimes a diagnosis can be made before death by testing cerebrospinal fluid or serum for neutralizing antibodies, which increase 10 to 12 days after the symptoms appear.

The bite of a rabid animal is treated by first thoroughly cleaning it with soap and flushing it with large amounts of water. Hyperimmune rabies serum is introduced into and around the wound in hope of neutralizing viruses before they reach the nervous system, where they are beyond reach of the antibodies. Interferon also can be applied to the wound. A series of injections of vaccine is given to induce the production of neutralizing antibodies. The biting animal should be located and confined for examination by IFAT.

The best means of preventing rabies is to immunize pets, and such immunization is required in many countries. Attempts to reduce rabies in raccoons have been made by using vaccine in small sponges covered with food bait. When the raccoons eat the bait, they also ingest enough vaccine to prevent them from acquiring rabies. Preliminary studies conducted in the United States show promising results. When raccoons from a bait-treated wildlife area were trapped, 15 out of 16 survived a challenge dose of rabies virus. All 16 caught in the nonbaited control area died of the same challenge doses. It is not certain whether the animal that died in the experimental group ever ate any bait. It is estimated that a distribution of one bait pellet per acre, at a cost of $1 per pellet, could prove sufficient to reduce rabies dramatically in wild raccoons.

Rabies immunization is recommended for veterinarians and their staffs, hunters who may have contact with wild animals, and technicians who work with the virus. The first vaccine, and for many years the only one available, was developed by Pasteur. This vaccine contained viruses modified by 50 transfers, involving drying of infected spinal cords, from one rabbit to another. It was used in all cases of suspected rabies because the disease might develop while the patient waited for test results on the biting animal. The vaccine was given subcutaneously (into the skin) in 14 or more daily injections into the abdomen, where the thick tissue slows the rate of absorption. These injections had unpleasant to serious side effects, including severe abdominal pain, fatigue, fever, and sometimes rabies infection. A relatively new vaccine produced from viruses grown in human diploid fibroblast cultures elicits high levels of neutralizing antibodies with only a few injections and minimal side effects. The current vaccine is given intramuscularly on days 0, 3, 7, 14, and 28. In addition, hyperimmune globulin is placed deep in the wound and infiltrated around the wound.

Encephalitis

THE DISEASE Encephalitis is an inflammation of the brain caused by a variety of togaviruses or by a flavivirus. We will consider the following four diseases, each of which is caused by a different virus: (1) **eastern equine encephalitis** (EEE), seen most often in the eastern United States; (2) **western equine encephalitis** (WEE), seen in the western United States; ∞ (Figure 16.6, p. 420) (3) **Venezuelan equine encephalitis** (VEE), seen in Florida, Texas, Mexico, and much of South America; and (4) **St. Louis encephalitis** (SLE), seen from east to west in the central United States. ∞ (Figure 16.3, p. 418) The equine varieties, caused by togaviruses, are so named because they infect horses more often than humans. The life cycles of these viruses generally involve transmission from a mosquito to a bird, back to a mosquito, and then to a horse, human, or other mammal, and finally back to a mosquito. The St. Louis variety, caused by a flavivirus, is so named because the first epidemic was identified in St. Louis in 1933. That virus appears to be transmitted mostly between English sparrows, mosquitoes, and humans.

The viruses, which are introduced into the body through bites of infected mosquitoes, first multiply in the skin and spread to lymph nodes. Viremia involving especially large numbers of viruses follows. In a few infections, the viruses invade the central nervous system, where they cause shrinkage and lysis of neurons. WEE appears every summer, and about a third of the cases occur in children under 1 year of age. Fever and headache are common symptoms, and convulsions sometimes occur. EEE is a much more serious disease; it causes a severe necrotizing infection of the brain. The disease is fatal in 50 to 80 percent of cases, and survivors often suffer permanent brain damage. Fortunately, because swamp birds are the major reservoir for the virus and swamp mosquitoes the major vectors, very few humans become infected. VEE is mainly a disease of horses; when it occurs in humans, it resembles influenza.

SLE occurs in late summer epidemics about every 10 years and causes the most severe symptoms in elderly patients. The illness starts with malaise, fever, and chills as consequences of viremia. Other common symptoms include anorexia, myalgia (muscle pain), sore throat, and drowsiness. In addition, certain patients have symptoms of a urinary tract infection, neurological disorders, altered states of consciousness, and convulsions. Complications can include secondary bacterial infections, blood clots in the lungs, and gastrointestinal hemorrhage. Most patients escape the complications and recover fully.

DIAGNOSIS, TREATMENT, AND PREVENTION Encephalitis sometimes can be diagnosed by isolating the causative agent from cell cultures or mice inoculated with blood or spinal fluid. Cultures can be negative when the disease is present because the viremic phase of the disease usually is over before the patient seeks medical attention. Serological methods can be used to identify antibodies at any time during and following the illness. Treatment only alleviates symptoms. Vaccines are available for immunizing horses, but they are rarely used in humans because of the danger of inducing a virulent form of the disease. Prevention by eradicating mosquito vectors is a more appropriate means of decreasing the already low incidence of encephalitis among humans.

Other Viral Diseases of the Brain and Meninges

HERPES MENINGOENCEPHALITIS Herpes simplex virus, which usually is responsible for cold sores, also can cause **herpes meningoencephalitis.** This disease often follows a generalized herpes infection in a newborn infant, child, or adult. The virus reaches the brain by ascending from the trigeminal ganglion. The disease has a rapid onset with fever and chills, headache, convulsions, and altered reflexes. In the middle-aged or elderly, meningoencephalitis causes confusion, loss of speech, hallucinations, and sometimes seizures. Most patients die in 8 to 10 days; survivors usually display neurological damage.

POLYOMAVIRUS INFECTIONS Polyomaviruses enter the body through the respiratory or gastrointestinal tract. Initial replication takes place in the cells the virus first enters. The viremia that follows allows the viruses to reach target organs, particularly the kidneys, lungs, and brain. Polyomaviruses, which are papovaviruses, were first recognized in the 1960s as viral particles in the enlarged nuclei of oligodendrocytes. These are the cells that produce myelin, the lipoprotein that coats nerve fibers in the central nervous system. Infected oligodendrocytes are observed to surround areas that lack myelin in the brains of patients dying from **progressive multifocal leukoencephalopathy.**

One cause of progressive multifocal leukoencephalopathy is now known to be the JC virus, a polyomavirus named with the initials of a victim from whom it was isolated. Onset is insidious, with vision and speech impairment being the first signs. Typical symptoms of viral infections, such as fever and headache, are absent. Mental deterioration, limb paralysis, and blindness follow. Diagnosis is difficult because cerebrospinal fluid remains normal, and only nonspecific changes are seen in electroencephalograms. The JC virus infects and kills oligodendrocytes but does not affect neurons. On occasion a young patient develops this disease as a complication of a primary infection, but most cases result from reactivated latent viruses from childhood infections.

Other polyomaviruses have been isolated from various patients. The BK virus was isolated from the urine of a kidney-transplant patient. In one instance a 16-year-old boy with an immunodeficiency had BK viremia and developed kidney inflammation with viruses present in kidney cells. Irreversible kidney failure resulted. The BK virus also has been associated with, but not isolated from, respiratory illness and cystitis, an inflammation of the urinary bladder.

Half the children in the United States have antibodies to the JC virus by age 14; antibodies to the BK virus are found by age 4. Although JC and BK viruses apparently persist in most humans for years without causing disease, they sometimes reappear as complications of chronic diseases, immunodeficiencies, and disorders in which lymphocytes proliferate. Pregnancy, diabetes, organ transplantation, antitumor therapy, and immunodeficiency diseases, including AIDS, are among the conditions that can reactivate polyomaviruses. Many kidney-transplant patients, for example, excrete either BK or JC viruses, but the shedding of these viruses rarely has serious consequences. However, unchecked viral multiplication, which is particularly likely with T cell deficiencies, can sometimes occur and cause clinically apparent disease. Both JC and BK viruses are oncogenic in laboratory animals and possibly also in humans. No diagnostic tests for polyomaviruses are available for routine use; nor is any treatment available for the infections, even if they can be recognized.

PUBLIC HEALTH

AMOEBIC INVADERS

Soil amoebas of the genera *Naegleria* and *Acanthamoeba* are opportunistic human pathogens. All such amoebas are associated with meningoencephalitis. *Naegleria fowleri* usually is seen in swimmers. The amoebas are thought to enter nasal passages and make their way along nerves to the meninges. *Acanthamoeba polyphaga* causes ulceration of eyes or skin. If it invades the central nervous system, death occurs a few weeks after the onset of neurological symptoms. The major source of human infection is contaminated water in hot tubs. While the tubs are covered, the amoebas accumulate on the water surface. Removing the lid disperses cysts among the bathers as they enter the tub.

OTHER DISEASES OF THE NERVOUS SYSTEM

Bacterial Nerve Diseases

Hansen's Disease

Hansen's disease, the currently preferred name for **leprosy,** has been known since biblical times, when many things, even houses, were said to have "leprosy." Many so-called leprosy cases were other skin diseases such as fungal and viral infections, and the houses probably had fungi growing on their walls.

The disease is vastly underreported, but an estimated 15 million cases exist worldwide today, mainly in Asia, Africa, and South America (Figure 24.3). Hansen's disease also occurs in the United States, where a peak number of 361 cases were reported in 1985, mostly among immigrants from countries where Hansen's disease is endemic. Infected people sometimes show no symptoms when they enter the United States. No reliable test is available to disclose all these subclinical cases, although the **lepromin skin test,** similar to the tuberculin skin test for tuberculosis, detects some of them. Health care workers should watch for Hansen's disease among immigrants.

THE DISEASE   The acid-fast bacillus *Mycobacterium leprae* is found in all cases of Hansen's disease. Although *M. leprae* was the first bacterium to be recognized as a human pathogen, demonstrating that it fulfills Koch's postulates has been slow because the organism is difficult to grow in the laboratory. For one thing, it reproduces very slowly, having a 12-day division cycle. Recently developed methods for growing *M. leprae* in such diverse hosts as nine-banded armadillos, chimpanzees, mangabey monkeys, and mice have made the bacterium available for research. It may now be possible to develop a vaccine. PCR techniques allow *M. leprae* to be identified in samples of skin or nasal tissues, thus simplifying diagnosis.

Clinical forms of Hansen's disease vary along a spectrum from tuberculoid to lepromatous. In the **tuberculoid,** or anesthetic, form (Figure 24.4a), areas of skin lose pigment and sensation. In the **lepromatous** (lep-ro'mat-us), or nodular, form (Figure 24.4b), a granulomatous response (Chapter 17) causes enlarged, disfiguring skin lesions called **lepromas.** ∞ (p. 457) Incubation time averages 2 to 5 years for the tuberculoid form and 9 to 12 years for the lepromatous form.

Mycobacterium leprae is the only bacterium known to destroy peripheral nerve tissue; it also destroys skin and mucous membranes. The organism has a predilection for cooler parts of the human body, such as the nose, ears, and fingers, but large numbers of organisms are seen throughout the body, except for the central nervous system. Continuous bacteremia of 1000 or-

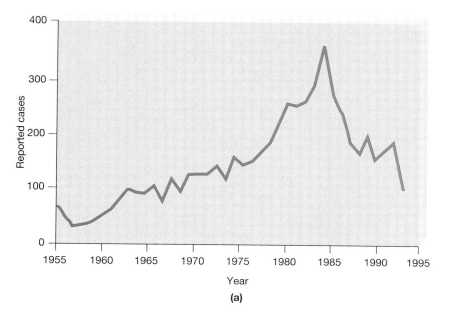

FIGURE 24.3 Incidence of Hansen's disease (leprosy) (a) in the United States, 1955–1994, and (b) worldwide.

FIGURE 24.3 Incidence of Hansen's disease (leprosy) (a) in the United States, 1955–1994, and (b) worldwide.

ganisms per milliliter of blood has been demonstrated in lepromatous cases. Large numbers of bacilli are shed in respiratory secretions and in pus discharged from lesions. Although Hansen's disease is not highly contagious, shedding of organisms probably transmits the disease to those with extensive close contact with patients, such as children of infected parents.

As Hansen's disease progresses, it deforms hands and feet (Figure 24.5). Severe lepromatous disease erodes bone: Fingers and toes become needle-like, pits develop in the skull, nasal bones are destroyed, and teeth fall from the jawbone. Surgery sometimes can restore the use of extremely crippled hands and feet. The National Hansen's Disease Center, a U.S. Public Health Service facility in Carville, Louisiana, has pioneered development of these special surgical techniques. In the past decade, it was recognized that changes in the feet of diabetic patients can be helped by these same surgical techniques. Thus, Carville now has an active teaching program for surgeons who will use knowledge gained from one of the

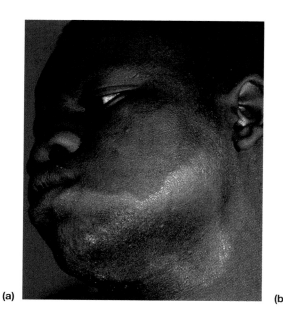
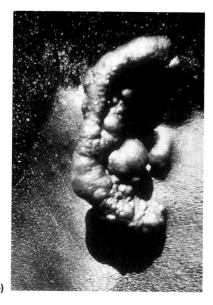

FIGURE 24.4 Hansen's disease appears in two extremes with gradations and combinations between them. (a) In the tuberculoid, or anesthetic, form, areas of skin lose pigment and sensitivity. A pin may be stuck into these "anesthetized" areas and not be felt because of destruction of nerves and nerve endings. (b) The nodular form is characterized by disfiguring granulomas called lepromas.

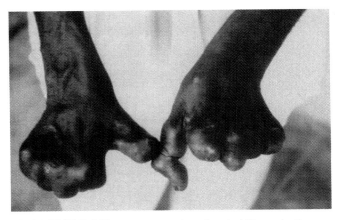

FIGURE 24.5 Deformed "claw" hand of Hansen's disease. This deformity can be treated surgically in its early stages, thereby preventing crippling.

most shunned diseases to help thousands of victims of diabetes.

Examination of ancient skeletons has provided insights into the epidemiology of Hansen's disease in past centuries. It is clear that the disease traveled from the Old World to the New World and that in the past, even allowing for misdiagnosis, its incidence in Europe was much higher than it is today. Genetic factors may predispose toward resistance to Hansen's disease; as susceptible individuals have died, resistant ones have made up a greater percentage of the population.

DIAGNOSIS, TREATMENT, AND PREVENTION
Hansen's disease is diagnosed by PCR or by finding the organism in acid-fast-stained smears and in scrapings from lesions or biopsies. The disease is treated with dapsone and rifampin, but dapsone-resistant strains are beginning to appear. Treatment greatly reduces nodules of lepromatous disease, but it cannot restore lost tissue. Until recently, victims of Hansen's disease were isolated in special hospitals called leprosariums. Now the disease can be arrested, and the people can live nearly normal lives without infecting others in their community. (They must still sleep in separate bedrooms and use only their own linens and utensils, however, and they cannot live in a household where children are present.)

Immune responses in Hansen's disease are cell mediated and vary from strong to weak. Strong responses and distinctly positive skin tests are seen in patients with the less serious tuberculoid disease. Weak responses and negative skin tests are seen in patients with rapidly progressing lepromatous disease. However, test results may change over time from positive to negative and vice versa as immune response rises and falls. Lepromatous patients usually have adequate cell-mediated response to other antigens, so their lack of immunity is not due to generalized T cell absence or dysfunction. The absence of T cells in "nude" mice, which are hairless and lack a thymus, makes them suitable organisms for growing large quantities of organisms.

Vaccine is not available for Hansen's disease. Even if a vaccine became available now, its effectiveness would take years to determine because of the disease's long incubation period. Avoiding exposure and receiving prophylactic chemotherapy after exposure are the only means of prevention.

Tetanus

Tetanus is caused by the obligately anaerobic, gram-positive, spore-forming rod *Clostridium tetani* (Figure 24.6). The organism can be cultured in the laboratory only under strict anaerobic conditions. *Clostridium tetani* endospores are exceedingly resistant to drying, disinfectants, and heat. Boiling for 20 minutes does not kill them, and they can survive for years if not exposed to sunlight. ∞ (Chapter 4, p. 92) Spores are found in all soils but especially in those enriched with manure. The organisms are part of the normal bowel microbiota of horses and cattle and about 25 percent of humans. Therefore, handling bedpans, dirty diapers, or other objects contaminated with feces can transmit organisms to persons who have any breaks in their skin.

Since the development of tetanus vaccine in 1933, the incidence of tetanus in the United States has

FIGURE 24.6 (a) TEM photo of *Clostridium tetani* bacillus with large, dark terminal endospore (105,800×). (b) A freeze-etch TEM preparation, showing the rounded endospore inside a *Clostridium* bacillus (107,250×).

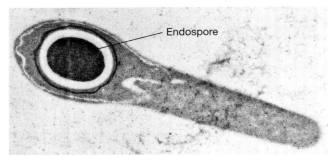

Endospore

(a)

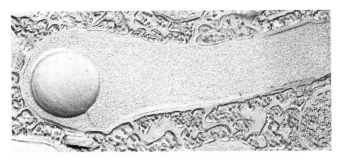

(b)

steadily dropped, with the annual number of cases remaining below 100 since 1975. The highest incidence is in older people, especially women. Vaccine was not available during their childhoods, and they did not receive it during military service, as men did. They remain susceptible to tetanus spores as they enjoy gardening in their retirement years. The elderly should be immunized for their protection.

THE DISEASE   To cause tetanus, spores must be deposited deep in tissues, where oxygen is unavailable. This occurs in deep cuts and puncture wounds. Stepping on a rusty nail has a reputation for leading to tetanus, but it is tetanus endospores, not rust, that cause the disease—a shiny new nail can be just as dangerous if the spores are present. Making puncture wounds bleed helps flush tetanus spores and other organisms from them. Once inside the host, the noninvasive tetanus organisms stay at the wound site and release a powerful exotoxin; tetanus is a toxin-mediated disease. After 4 to 10 days' incubation, symptoms begin, with generalized muscle stiffness followed by spasms that affect every muscle. An arched back and clenched fists and jaws (hence the term *lockjaw*), are classic symptoms (Figure 24.7). Spasms can be violent enough to break bones. Eventually, respiratory muscles become paralyzed, heart function is disturbed, and, with rare exceptions, the patient dies. Survivors experience a period of sore muscles but suffer no further sequelae. Before vaccine was available, many soldiers died from tetanus. On battlefields strewn with horses and manure, contamination of wounds with tetanus spores was inevitable. War-related cases were virtually eliminated by vaccinating soldiers; only 12 cases occurred during World War II.

TREATMENT AND PREVENTION   Tetanus toxoid vaccine given prior to injuries protects against the toxin. Antitoxin and antibiotics are given to nonimmunized patients when injuries are treated. Because antitoxin must be administered to inactivate the toxin before the immune system has time to become sensitized to it, infection treated in this way confers no immunity. Patients should receive toxoid immunization after they recover.

Tetanus neonatorum is acquired through the raw stump of the umbilical cord. In some societies, contaminated knives are used to cut umbilical cords after a baby is delivered, and mud is smeared on the cut end. In parts of some developing nations, 10 percent of deaths within a month of birth are due to neonatal tetanus.

Botulism

Botulism derives its name from the Latin word *botulus*, which means "sausage." It was coined at a time when the disease was often acquired from eating sausages. Botulism is caused by *Clostridium botulinum*, a spore-forming obligate anaerobe that releases a potent exotoxin (neurotoxin). (See the box "Clinical Use of Botulinum Toxin" in Chapter 15.) ∞ (p. 404) The disease occurs in three forms: foodborne, infant, and wound. Foodborne botulism accounts for 90 percent of cases and is caused by ingestion of toxin, usually from improperly home-canned nonacid foods, especially green beans and green peppers (Figure 24.8). Thus, foodborne botulism is an intoxication; the organisms do not infect tissues. Infant botulism and wound botulism involve both infection and intoxication because the organisms grow in tissues and produce toxin.

Endospores of *C. botulinum* are more heat resistant than those of any other anaerobe; they withstand several hours at 100°C and 10 minutes at 120°C. They also are very resistant to freezing (down to −190°C) and irradiation. Found in most soils in the Northern Hemisphere, these endospores remain viable for long periods of time and enable the organism to withstand aerobic conditions. Endospores will germinate only in anaerobic conditions.

The ability of *C. botulinum* to form toxin depends on infection with a bacteriophage. This phage carries the information for botulism toxin production. If infected with an appropriate bacteriophage, *C. botulinum* produces one of eight different toxins, of which only four cause human disease. The other toxins cause disease in various

FIGURE 24.7   A soldier dying of tetanus, an all too common cause of death in the days of cavalry troops. The extreme contraction of all muscles, from those of the face to those of the toes, is a classic symptom of the disease.

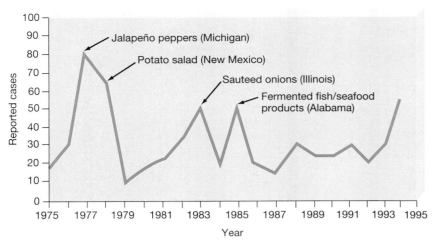

FIGURE 24.8 Incidence of foodborne botulism in the United States, 1975–1994, and the foods that caused certain outbreaks.

other animals. If a strain of the bacillus is "cured" of its phage infection, it no longer produces toxin. If later infected with a different phage, the bacillus will produce another toxin.

Botulism toxin is the most potent toxin known—even more toxic than *Shigella* and tetanus toxins. As little as 0.000005 μg can kill a mouse! One ounce could kill the entire U.S. population. Originally thought to be an exotoxin, it is now known to be produced inside the cytoplasm and released only upon death and autolysis of the cell. It is activated by proteolytic enzymes, possibly including trypsin in the host's intestine. The toxin is colorless, odorless, and tasteless; people have died from a single taste of affected food. If endospores are not destroyed, they germinate in food during storage under anaerobic conditions and can release large quantities of toxin. Whereas endospores are highly heat resistant, the toxin can be inactivated by only a few minutes' boiling. Boiling home-canned foods vigorously before serving would eliminate most foodborne botulism.

THE DISEASE Botulism is a neuroparalytic disease with sudden onset and rapidly progressing paralysis. It ends in death from respiratory arrest if not treated promptly. The toxin acts at junctions between neurons and muscle cells and prevents the release of acetylcholine, the chemical that neurons release to cause muscle cells to contract. The toxin thus paralyzes muscles in a relaxed state—starting with small eye muscles, progressing to the larynx and pharynx, and on to the respiratory muscles. This causes double vision, difficulty in speaking and swallowing, and difficulty in breathing. It causes no fever but can cause gastrointestinal disturbances. Although the toxin is an antigen, people who have recovered do not have antibodies, so the amount of antigen required to elicit antibodies must be greater than the lethal dose.

DIAGNOSIS AND TREATMENT Diagnosis is based on clinical symptoms and history, with confirmation later by demonstrating toxin in serum, feces, or food remains. Although the confirmation test takes 24 to 96 hours, treatment with a polyvalent antitoxin is started immediately. A polyvalent antitoxin is used to ensure effectiveness against all toxins that affect humans. Help in maintaining respiration is important and may be continued for up to 2 months. Antibiotics are of no use because foodborne botulism is due to preformed toxin and not to growth of organisms. With proper treatment, the mortality rate is less than 10 percent.

Infant botulism was first recognized in 1976, and incidence has ranged between 30 and 100 cases per year, mostly in California. The disease is associated with feeding honey to infants. Studies in California have shown that 10 percent of the jars of honey sold there contain botulism endospores. Endospores germinate and grow in the immature digestive tract of infants, probably because they lack appropriate competing microbiota. As toxin is absorbed, the infant becomes lethargic and loses the ability to suck and swallow, so the disease is often called "floppy baby" syndrome. Infant botulism usually occurs in infants under 6 months of age and rarely after 12 months. If parents would not give honey to children under 1 year, most cases could be prevented. The prognosis is excellent, and death is rare, but the child usually must remain hospitalized for several months.

CLOSE-UP

BOTULISM IN WATERFOWL

Botulism is an important animal disease, especially during ecological disturbances. Wetland birds are particularly at risk. It is a leading cause of death among ducks in the western United States. After storms uproot marsh vegetation, decomposition uses up available oxygen, so small aquatic invertebrates die. Botulism endospores in their digestive tract or in mud germinate and produce toxin. Ducks eat the dead invertebrates, die, and fill with toxin. Flies lay eggs on dead ducks; the eggs hatch into toxin-filled maggots. Healthy ducks eat the maggots, die, and provide a site for new fly eggs. The cycle is repeated until thousands of ducks have died and no more ducks are left at that site.

Wound botulism is the least common form of botulism; no more than one case per year has been seen in the United States since 1942. It occurs in deep, crushing wounds. Tissue damage impairs circulation and creates anaerobic conditions, so endospores germinate, multiply, and produce toxin. Toxin enters the blood and is distributed throughout the body. It reaches junctions between neurons and muscle cells about a week after injury and causes progressive paralysis. The mortality rate is about 25 percent.

Viral Nerve Diseases

Poliomyelitis

Poliomyelitis is a very ancient disease; its effects are clearly depicted in Egyptian wall paintings thousands of years old. As recently as the early 1950s, it was a dreaded disease in the United States, with nearly 58,000 cases reported in the peak year of 1952. The coming of summer struck terror in the minds of parents, and the diagnosis of a case of paralytic polio in the community was cause for outright panic. In the United States today, those most likely to become infected are members of religious groups who are opposed to immunization, and illegal aliens who are unprotected by a vaccine.

THE DISEASE Poliomyelitis is caused by three strains of polioviruses (picornaviruses) that have an affinity for motor neurons of the spinal cord and brain. Although most poliovirus infections are inapparent or mild and nonparalytic, the virus reaches the central nervous system in 1 to 2 percent of cases. High fever, back pain, and muscle spasms result. In less than 1 percent of cases, these symptoms are accompanied by partial or complete paralysis of muscles in a relaxed state. The nature and degree of paralysis depend on which neurons in the spinal cord and brain are infected and how severely they are damaged, or if they are lysed. Any paralysis remaining after several months is permanent. The very old and the very young are likely to suffer paralysis as a result of poliovirus infection. Malnutrition, physical exhaustion, corticosteroids, radiation, and pregnancy can increase the severity of the disease.

Poliovirus infections in small children in impoverished areas may go undetected, whereas teenagers and young adults in affluent areas sometimes acquire severe, paralytic poliovirus infections. Good sanitation in the affluent areas reduces exposure and, therefore, natural immunity to the viruses.

DIAGNOSIS, TREATMENT, AND PREVENTION Diagnosis of poliomyelitis is made by isolating the virus from pharyngeal swabs or feces, culturing it, and noting its cytopathic effects. Methods also are available for identifying antibodies to the virus in serum.

Treatment alleviates symptoms, but patients with paralyzed breathing muscles must forever live in an "iron lung" (Figure 24.9).

Before vaccine became available in 1955, only nonspecific public health measures were available to prevent the spread of poliomyelitis. Schools, swimming pools, and other places where crowds, especially crowds of children, gathered were closed. Large quantities of insecticides were sprayed in the mistaken belief that insect bites somehow played a role in transmission of the disease. Transmission is now known to occur both by the fecal–oral route and from pharyngeal secretions, thus explaining the dangers of fecally contaminated swimming pools in summer. During the first few years that vaccine was available, it could not be made in sufficient quantities to immunize the whole population. Clinics were set up to immunize pregnant women and young children.

FIGURE 24.9 During polio epidemics (before 1955) in the United States, (a) row after row of iron lungs was filled with patients (b) such as this 2-year-old girl. Patients sometimes remained in them for years, or until their death.

(a)

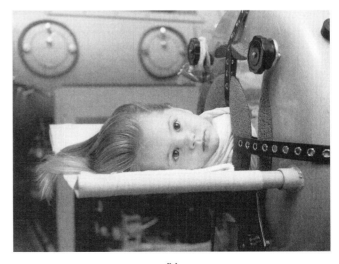

(b)

VACCINES In 1955 the injectable Salk polio vaccine became available. It contained viruses that were inactivated by formalin at neutral pH but that still retained their antigenic properties. Unfortunately, before the technique was perfected, some batches of vaccine still contained infectious viruses. More than 200 cases of polio and 10 deaths resulted. In 1963 the oral Sabin vaccine, which contained attenuated live viruses, was introduced. In addition to ease of administration, this vaccine has the advantages of longer-lasting immunity and prevention of fecal–oral transmission by eliminating viruses in the gastrointestinal tract, where they multiply. Vaccine use has reduced polio incidence in the United States from about 29,000 cases in 1955 to 20 cases in 1969 in unimmunized and immunosuppressed individuals (Figure 24.10). In October 1995 CDC recommended a combination of polio vaccines to reduce the incidence of vaccine-related polio. (See the box below.) Infants

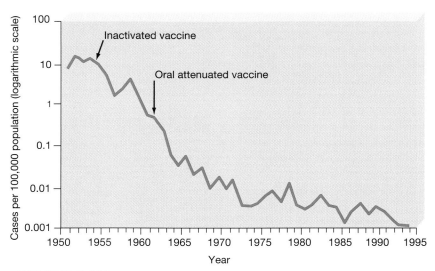

FIGURE 24.10 Incidence of polio in the United States, 1951–1994. The drastic drop since vaccines became available (first Salk, then Sabin) has not happened in South and Central America. Thus, many people still suffer from this preventable disease. In fact, more people die each year from vaccine-preventable diseases than die from AIDS.

would receive shots of the inactivated vaccine at 2 and 4 months, then oral doses of the attenuated vaccine at 6 months and up to 18 months. The oral vaccine is

---

PUBLIC HEALTH

THE POLIO VACCINE CONTROVERSY

In the United States, the attenuated-virus vaccine developed by Albert Sabin is almost universally used. Some other countries, however, have returned to the inactivated-virus vaccine developed by Jonas Salk. Which is preferable?

Both vaccines have advantages and disadvantages. The inactivated-virus vaccine can be incorporated with other pediatric vaccines, and at the appropriate dose it can be given to immunodeficient patients. However, it does not produce immunity in all recipients, and recipients may require booster shots. Moreover, virulent viruses are used to make the vaccine; a tragedy could occur if they are not completely inactivated.

The attenuated-virus vaccine induces immunity similar to that produced by a natural infection in that the vaccine causes antibodies to be produced in the gut (secretory IgA) as well as in the bloodstream. This reduces the possibility that people who are themselves immune can serve as reservoirs of infection by carrying the virus in their intestines and passing it on to others. Immunity develops quickly and may be lifelong. Oral administration requires less skill and is more acceptable than injection to many people. Finally, the oral vaccine remains potent without refrigeration for a longer time than does the injectable vaccine.

Unfortunately, the live virus used in the oral vaccine does occasionally mutate, and some mutants are virulent. The viruses multiply in gut tissue of the vaccinated individual and are shed in their feces. Thus, changing the dia-

per of a recently vaccinated child can pose a hazard, especially to family members or day-care workers and immunodeficient patients, depending on the degree of hygiene practiced. About six cases of polio are caused by the vaccine each year in the United States, and about that many are caused by close contact with these immunized children. Liability costs are now included in the price of the oral vaccine to cover awards that may have to be made to victims of vaccine-induced polio. In the United States, these costs have made the oral vaccine more expensive than the injected vaccine, although the oral vaccine is cheaper to make and administer. In warm countries with endemic poliomyelitis and other viral infections, repeated administration has often failed to induce immunity, probably because immune responses being mounted against other viruses prevent an adequate immune response to the vaccine organisms. Finally, the attenuated-virus vaccine, like all vaccines that use live organisms, cannot be administered to immunosuppressed or immunocompromised persons.

Polio has been virtually eliminated in Israel through the use of a combination of inactivated-virus and attenuated-virus vaccines. This procedure shows promise for controlling polio in developing countries with warm climates, fecally contaminated water, and endemic polio. However, the cost of obtaining and administering both vaccines will delay implementation.

---

needed because it is effective against "wild" polio virus that, although eradicated from the United States, could be brought in from other areas by travelers.

Postpolio syndrome is a condition in which people who survived polio, years before, suffer weakening or paralysis of muscles, which requires them again to use crutches and braces. It is neither infectious nor a recurrence of the disease. It is believed to be due to overuse of compensating muscles that have labored too hard for too many years and now cannot function properly.

Parasitic Diseases of the Nervous System

African Sleeping Sickness

African sleeping sickness, or **trypanosomiasis** (tripan"o-so-mi'a-sis), is a disease of equatorial Africa caused by protozoan blood parasites of the genus *Trypanosoma*. Although 100 or more species of this parasite infect various vertebrates and invertebrates, two species, *T. brucei gambiense* and *T. brucei rhodesiense,* cause disease in humans. Typically, trypanosomes have an undulating membrane and a flagellum (Figure 24.11a), but in some stages they are shorter and lack a flagellum. For humans to get African sleeping sickness, they must be bitten by an infected tsetse (tset'se) fly (Figure 24.11b). When a tsetse fly bites, it injects infectious trypanosomes, sometimes hundreds in a single bite, into the blood of its victim. The flies serve as vectors and as hosts for part of the life cycle of trypanosomes. Although transmission usually is from one human to another via the fly, game animals serve as natural reservoirs for *T. brucei rhodesiense.*

THE DISEASE African sleeping sickness is a progressive disease characterized according to the tissue in which the parasites congregate during stages that affect first the blood, then the lymph nodes, and finally the central nervous system. Although the parasites do not actually invade cells, they can damage every tissue and organ in the body.

The bites of tsetse flies cause a local inflammatory reaction. After an incubation period of 2 to 23 days, fever appears initially for about a week while parasites are in the blood and at irregular intervals as the parasites are released from lymph nodes. Patients are able to work through the first and second stages, but they suffer from various symptoms—shortness of breath, cardiac pain, disturbed vision, anemia, and weakness—that become increasingly severe. Invasion of the nervous system by the parasites causes headache, apathy, tremors, and an uncoordinated, shuffling gait. Pain and stiffness in the neck and paralysis occur as the disease progresses. Eventually, the patient cannot be roused to eat, becomes emaciated, has convulsions, sleeps continuously, goes into a profound coma, and dies.

Infection with *T. brucei gambiense* produces a slowly progressive, chronic disease that, if untreated, lasts several years before central nervous system symptoms intensify and death occurs. Infection with *T. brucei rhodesiense* produces a more rapidly progressive disease; it is often fatal within a few months, before central nervous system damage is apparent.

DIAGNOSIS, TREATMENT, AND PREVENTION Diagnosis of African sleeping sickness is made by finding the trypanosome parasite in the blood. Until recently, arsenic-containing drugs were used to treat the disease, but they cause eye damage, and the parasites quickly become tolerant to them. Pentamidine, suramin, and melarsoprol are now used, usually in that order. If pentamidine, the least toxic drug, fails to combat the infection, more toxic ones are tried. Results of treatment with any drug are generally more successful if treatment is begun before central nervous system involvement occurs. A combination of Berenil and nitroimidazole is being used to treat the disease after it has progressed to the central nervous system.

Preventing human infection is nearly impossible because of the wide range of tsetse flies (4.5 million square miles) and possibly because of the reservoir of infection in large game animals. Some control has been achieved by clearing brush where flies congregate and

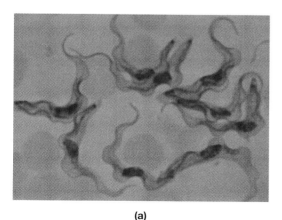

(a)

(b)

FIGURE 24.11 (a) *Trypanosoma brucei gambiense* (magnified 1855×), seen in a blood sample, causes African sleeping sickness. (b) Trypanosomes are spread by the bite of the tsetse fly.

NAGANA AND HDL

In up to 25 percent of Africa, it is impossible to raise live-stock because of the tsetse-fly-borne disease *nagana*. The causative agent, the flagellated parasite *Trypanosoma brucei brucei*, probably originated in Africa. It lives in the blood of large native grazing animals without harming them, pre-sumably because a long association has allowed host and parasite to adapt to each other. But when humans migrated into the areas inhabited by tsetse flies, the domesticated livestock the humans brought with them were not able to survive infection with *T. brucei brucei*. When bitten by tsetse flies, humans are also infected by the parasite but do not become ill; the parasites rapidly disintegrate in human blood.

Which component of human blood protects against *T. brucei brucei*? The surprising answer involves a protein never before recognized as providing protection against in-fectious disease. The high-density lipoprotein (HDL) frac-tion of blood serum, known for its role in removing cho-lesterol from the blood, contains a small subfraction that binds to hemoglobin released from dying red blood cells. This binding allows the iron in hemoglobin to be recycled within the body. When a tsetse fly bite releases *T. brucei bru-cei* into the human bloodstream, the protein–hemoglobin complex is taken in by the trypanosome and eventually en-ters the parasite's lysosomes. The acidic pH inside the lyso-somes causes the protein–hemoglobin to undergo enzy-matic changes that result in the release of free radicals. These extremely reactive molecules destroy the membranes of the lysosomes, whose contents—lysozyme enzymes—are ex-pelled into the cytoplasm. There the enzymes digest the par-asite from within.

Only humans, some apes, and Old World monkeys have the gene for this *trypanosome lytic factor* (TLF). How-ever, it is hoped that genetic engineering techniques can be used to transfer the gene into livestock, thereby producing trypanosome-resistant breeds of cattle, for example.

---

by applying aerial pesticides. Another significant means of reducing the tsetse fly population is to release irradiated male flies, which fail to produce viable sperm. The eggs of females that mate with them fail to develop.

Trypanosomes that cause African sleeping sickness have a special means of evading the host's defenses. The intermittent fever of the disease is directly corre-lated with increasing numbers of parasites in the blood. The unusual thing about these parasites is that each time parasites appear in the blood, they have a glyco-protein coat different from that of previously released parasites. By the time the immune system has devel-oped antibodies to a trypanosome surface antigen, the trypanosome has a different surface antigen. This abil-ity to alter antigens has thwarted efforts to produce a vaccine for African sleeping sickness. When try-panosomes first enter a human, they seem to be able to make only about 15 antigens; later they can make 100 or more. Researchers hope to produce a vaccine that will confer immunity against any antigen the try-panosome might have when it enters the human body. Such a vaccine would allow the body to attack the try-panosome before an infection becomes established.

Should a vaccine be developed, a means to mount a costly, massive vaccination program would be needed. Because available vaccines, such as those for measles and mumps, have not been administered to many of the children in the region endemic to African sleeping sickness, the prospect for mass immunization is not bright. More than 20,000 people per year are likely to continue to die of the disease.

Chagas' Disease

Chagas' (Chah'gahs) **disease,** named for the Brazilian physician Carlos Chagas, who first described it in 1909, is caused by *Trypanosoma cruzi*. The disease occurs spo-radically in the southern United States and is endemic in Mexico, Central America, and much of South Amer-ica. It affects more than 18 million people (Figure 24.12). *T. cruzi* looks like the trypanosomes that cause African sleeping sickness. It is transmitted by several kinds of reduviid bugs (Chapter 12) that are hosts for the sexual phase of the trypanosome's life cycle. ∞ (p. 326) Each species of bug occupies a particular region, so the bugs that transmit Chagas' disease in Mexico are different from those that transmit it in South America. Reduviid bugs often bite near the eyes; they defecate as they bite, depositing infectious parasites on the skin. Humans al-most automatically rub such a bug bite and thereby transfer parasites to the eyes or the bite wound.

THE DISEASE Chagas' disease begins with subcu-taneous inflammation around the bug bite. After 1 to 2 weeks, the parasites have made their way to lymph nodes, where they repeatedly divide and form aggre-gates called **pseudocysts.** Wherever the pseudocysts rupture, they cause inflammation and tissue necrosis. These parasites enter cells either by invasion or by

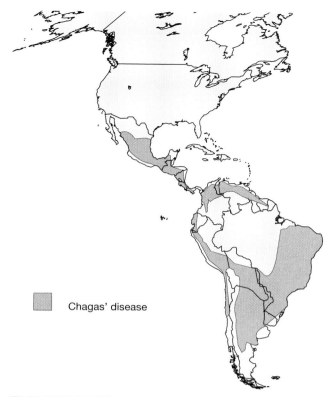

FIGURE 24.12 Distribution of Chagas' disease, 1994.

phagocytosis and can damage lymphatic tissues, all kinds of muscle, and especially supporting tissues around nerve ganglia. Destruction of nerve ganglia in the heart accounts for nearly three-fourths of deaths from heart disease among young adults in endemic areas.

Chagas' disease appears in an acute form and a chronic form. The acute disease, which is most common in children under 2, is characterized by severe anemia, muscle pain, and nervous disorders. In especially virulent acute disease, death can occur in 3 to 4 weeks, but many patients recover after several months of less virulent disease. The chronic disease, which is seen mainly in adults, probably arises from a childhood infection. It is a mild disease and is sometimes asymptomatic but often causes enlargement of various organs. Insidious damage to nerves can cause several severe effects. In the digestive tract, it slows or stops muscle contractions; in the heart, it can cause irregular heartbeat and accumulation of fluid around the heart; in the central nervous system, it can cause paralysis by destroying motor centers. *T. cruzi* also crosses the placenta, so chronically infected mothers often give birth to infants with severe acute disease.

DIAGNOSIS, TREATMENT, AND PREVENTION

Several diagnostic techniques are available for Chagas' disease. The parasites can be found in the blood during fever in acute cases. Small animals such as guinea pigs and mice can be inoculated with blood from patients and observed for disease symptoms. This technique is known as *xenodiagnosis. Xenos* is Greek for "strange" or "foreign"; in this context it refers to the use of an organism different from a human. Chronic Chagas' disease sometimes can be detected by allowing patients to be bitten by uninfected laboratory-reared reduviid bugs and examining the bugs in 2 to 4 weeks for trypanosomes, which develop in the bug's intestine.

Despite several ways to diagnose the disease, no effective treatment is available. Drugs used to treat other trypanosome infections are of no use because they fail to reach the parasites inside cells. Work is underway to develop new drugs and a vaccine, but until these are available, control of the reduviid vectors is the only means of reducing misery from this disease. Treating homes with insecticides offers some protection, but the bugs crawl into crevices in walls and thatched roofs and are difficult to eradicate.

The agents and characteristics of the diseases discussed in this chapter are summarized in Table 24.2.

TABLE 24.2   Summary of Diseases of the Nervous System

Disease	Agent(s)	Characteristics
Bacterial diseases of the brain and meninges		
Bacterial meningitis	See Table 24.1	Tissue necrosis, brain edema, headache, fever, occasionally seizures
Listeriosis	*Listeria monocytogenes*	A kind of meningitis seen in fetuses and immunodeficient patients
Brain abscesses	Various anaerobes	Infection that grows in mass and compresses brain
Viral diseases of the brain and meninges		
Rabies	Rabies virus	Invades nerves and brain; headache, fever, nausea, partial paralysis, coma, and death ensue unless patient has immunity
Encephalitis	Several encephalitis viruses	Shrinkage and lysis of neurons of the central nervous system; headache, fever, and sometimes brain necrosis and convulsions
Herpes meningoencephalitis	Herpesvirus	Fever, headache, meningeal irritation, convulsions, altered reflexes
Progressive multifocal leukoencephalopathy	Polyomavirus, JC virus	Infects oligodendrocytes in areas of brain lacking myelin; mental deterioration, limb paralysis, and blindness
Bacterial nerve diseases		
Hansen's disease	*Mycobacterium leprae*	Range of symptoms from loss of skin pigment and sensation to lepromas and erosion of skin and bone
Tetanus	*Clostridium tetani*	Toxin-mediated disease; muscle stiffness, spasms, paralysis of respiratory muscles, heart damage, and usually death
Botulism	*Clostridium botulinum*	Preformed toxin from food prevents release of acetylcholine; paralysis and death result unless treated promptly; in infants and wounds, endospores germinate and produce toxin
Viral nerve diseases		
Poliomyelitis	Several types of polioviruses	Fever, back pain, muscle spasms, partial or complete flaccid paralysis from destruction of motor neurons
Parasitic diseases of the nervous system		
African sleeping sickness	*Trypanosoma brucei gambiense,* *T. brucei rhodesiense*	Fever, weakness, anemia, tremors, shuffling gait, apathy; as parasites invade nervous system, emaciation, convulsions, and coma ensue
Chagas' disease	*Trypanosoma cruzi*	Subcutaneous inflammation, damage to lymphatic tissues, muscle, and nerve ganglia; muscle pain and paralysis of intestinal, heart, and skeletal muscle

MYSTERIOUS BRAIN INFECTIONS

Since the 1920s, when *Creutzfeldt–Jakob disease* (CJD) was first investigated, several degenerative diseases of the nervous system have been identified. Although the existence of prions is not universally accepted, many researchers believe that such diseases are associated with prions. ∞ (Chapter 11, p. 292) These diseases are referred to collectively as *transmissible spongiform encephalopathies* because in damaging neurons, they give brain tissue a "spongy" appearance (Figure 24.13). They include *kuru*, CJD, and a special form of CJD called *Gerstmann–Strassler disease* in humans; *scrapie* in sheep and goats; *transmissible mink encephalopathy; chronic wasting disease* of elk and mule deer; and *"mad cow"* disease (*bovine spongiform encephalopathy*) in British dairy cattle. In addition, in 1991, 29 cases of transmissible spongiform encephalopathies were reported in British cats, and 2 were reported in ostriches in the Berlin Zoo.

A prominent feature of all prion-associated diseases is the lack of any inflammatory response, which is a hallmark of other infectious diseases. There is, however, an increase in the size of *astrocytes*—the cells that regulate the passage of materials from the blood to neurons—throughout the central nervous system. These cells apparently produce large quantities of a filamentous protein called *amyloid*, which is characteristic of a variety of degenerative diseases of the nervous system.

Like other infectious agents, prions are transmissible, although disease symptoms may not appear until years after initial infection. Kuru, which occurred mainly in New Guinea, is transmitted through small breaks in the skin. Why this disease appeared mainly in women was puzzling until it was discovered that New Guinean women prepared the bodies of the dead for cannibalistic consumption and smeared their own bodies with the raw flesh of the corpses. Prions in diseased tissues entered the blood of the women or the children playing at their feet, traveled to the brain, and eventually caused kuru. Men and adolescent boys live apart from the women and small children and so were spared infection with kuru.

D. Carleton Gajdusek, a U.S. investigator at the National Institutes of Health (NIH), won a Nobel Prize in 1976 for his work on kuru. He found that 1 to 15 years following inoculation with the kuru prion, onset of symptoms began with headaches, minor loss of coordination, and a tendency to giggle at inappropriate times. Three months later, victims needed crutches to walk or stand; a month after that, victims could not move except for spasms. At that point, the swallowing muscles no longer functioned, and malnutrition became a serious problem. Relatives would prechew food and massage it down the esophagus of victims (Figure 24.14), but within 1 year, the patients died.

Some cases of kuru have been traced to inadvertent inoculation with prions present in corneal transplant tissue from patients having undiagnosed CJD, to dwarfed children who received human growth hormone injections made from cadaver pituitaries, and to silver electrodes implanted in the brain during surgical procedures following use in a CJD patient. The same electrodes, located 17

FIGURE 24.13 (a) A section through the cerebral cortex of a normal human brain (magnified 370×) reveals a solid structure, whereas (b) a section through the brain of a patient with Creutzfeldt–Jakob disease (370×) shows many holes. It is clear why CJD is referred to as a subacute *spongiform* encephalopathy.

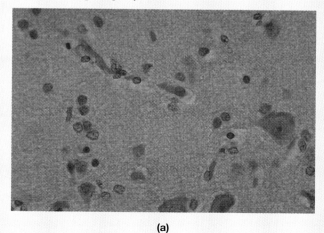

(a)

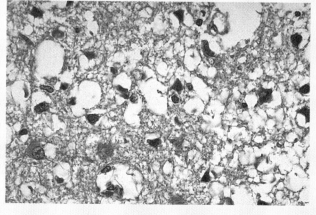

(b)

FIGURE 24.14 Kuru victims have reached the point where they must be fed prechewed food. Since cannibalistic rites have been stopped in New Guinea, the disease has disappeared there.

FIGURE 24.15 Sheep infected with scrapie will rub or scrape—sometimes until they are bloodied—against fences, poles, or trees. This fatal disease has no cure.

months later after numerous supposed sterilizations, were implanted into a chimpanzee's brain, where they caused CJD. These findings have caused some newer methods of instrument sterilization to be abandoned. Autoclaving at 121°C at 15 psi of pressure for 1 hour destroys infectivity, something that years of storage in formaldehyde fails to do.

In one experiment, Gajdusek and Paul Brown (also at NIH) buried scrapie-infected hamster brains in soil and left them undisturbed for 3 years. When the brain material was dug up, it was still infectious. Gajdusek and Brown suspect that infected material may retain its deadliness for more than a decade under such circumstances. Thus, sheep that graze in fields where contaminated carcasses have been placed have become infected with scrapie (Figure 24.15). Marking of burial sites and reexamination of the accepted technique of adding corrosive quicklime to the bodies need to be done before eradication can be achieved.

In most cases of CJD, no source of prions has been identified. In the most thoroughly studied instance of Gerstmann–Strassler syndrome, CJD developed in every generation of a family line for over 100 years. In this form of CJD, a genetic mechanism seems to facilitate prion infection, but how this mechanism works is not clear. Perhaps the presence of a particular gene renders patients susceptible to infection by prions from outside the body, or perhaps the gene activates synthesis of prions within the body.

Mad cow disease reached a peak rate in the early 1990s, with 400–500 deaths of British cattle per week, but it has now diminished. It is thought to have existed in Britain since at least the late 1960s, and the number of cases began to mushroom in early 1987. At that time the method of rendering (boiling down) animal remains for livestock feed was changed so as to omit a solvent extraction step and to increase the number of sheep heads used (including the brains). There had been considerable debate as to whether prions can cross species lines in nature. But prions have clearly been transmitted from one species to another in laboratory trials. The British government has prohibited the addition of beef brains to hamburger, formerly a common practice. The British have also changed their butchering methods so that knife or saw blades do not pass through the spinal cord, thus preventing blades from contaminating edible parts. Many countries, including the United States, have banned the import of British beef, cattle, and beef products.

The possibility that slow-onset prions play a role in other neurodegenerative diseases such as Alzheimer's and Parkinson's diseases is being investigated. Alzheimer's disease, first described by Alois Alzheimer in 1907, now affects more than 2 million people in the United States—more than 5 percent of the population over age 65. Amyloid proteins arranged in structures called *neurofibrillary tangles,* or *plaques,* have been found at autopsy in the brains of Alzheimer's patients, but the holes found in kuru and other spongiform encephalopathies are missing. Whether prions play a role in the deposition of the proteins is not yet known. So far researchers have not been able to transmit Alzheimer's disease to laboratory animals, as has been done with kuru, CJD, and scrapie. Whether this is due to nonsusceptibility of the animals or the absence of an infectious agent also is not yet known. However, bits of β-amyloid protein injected into rat brains have recently produced a disease similar to Alzheimer's in these rats. Genetic factors may be involved, at least in Alzheimer's disease and CJD, because multiple cases of both diseases have been seen in some families. Much remains to be learned about prions, neurodegenerative diseases, and the relationship between them.

- The agents and characteristics of the diseases discussed in this chapter are summarized in Table 24.2. Information in that table is not repeated in this summary.

DISEASES OF THE BRAIN AND MENINGES

Bacterial Diseases of the Brain and Meninges

- **Bacterial meningitis,** acquired from carriers or endogenous organisms, can usually be treated with penicillin. A vaccine is available to protect children from *Haemophilus* meningitis.

- **Listeriosis** can be transmitted by improperly processed dairy products and can cross the placenta. It is a great threat to immunodeficient patients.

- **Brain abscesses** arise from wounds or as secondary infections. Antibiotics are effective if given early in an infection, and surgery can be used later unless the abscess is in a vital area of the brain.

Viral Diseases of the Brain and Meninges

- **Viral meningitis** is usually self-limiting and nonfatal.

- **Rabies** has a worldwide distribution except for a few rabies-free countries. It is difficult to control because of the large number of small mammals that serve as reservoirs.

- Rabies is diagnosed by IFAT and treated by thoroughly cleaning bite wounds and injecting hyperimmune rabies serum and giving vaccine. It can be prevented by immunizing pets and people at risk and by avoidance of contact with wild animals by those who are not immunized.

- **Encephalitis** is transmitted by mosquitoes, often from horses, and can sometimes be diagnosed by culturing blood or spinal fluid in cells or mice. Vaccine is available for horses.

- **Herpes meningoencephalitis** often follows a generalized herpes infection.

OTHER DISEASES OF THE NERVOUS SYSTEM

Bacterial Nerve Diseases

- **Hansen's disease (leprosy)** affects millions of people worldwide. Many patients are asymptomatic for years after infection, and no diagnostic test is available to identify such patients. Dapsone and rifampin are used to treat Hansen's disease, but no vaccine is available.

- **Tetanus** typically follows introduction of spores into deep wounds. It is treated with antitoxin and antibiotics. If all people received the available vaccine, tetanus could be prevented.

- **Botulism** is acquired by ingesting foods that contain a potent preformed neurotoxin and is treated with polyvalent antitoxin. In **infant botulism** and **wound botulism,** spores germinate and produce toxin.

Viral Nerve Disease

- Prior to the development of vaccines in the 1950s, **poliomyelitis** was a common and dreaded disease. It can be diagnosed from cultures and by immunological methods. Treatment merely alleviates symptoms and includes the use of an iron lung to maintain breathing when respiratory muscles are paralyzed.

- Both an injectable and an oral polio vaccine are available; each has advantages and disadvantages. Worldwide use of vaccine could eradicate poliomyelitis.

Parasitic Diseases of the Nervous System

- **African sleeping sickness** occurs in equatorial Africa and is transmitted by the tsetse fly. It is diagnosed by finding parasites in the blood and can be treated with pentamidine and other drugs.

- **Chagas' disease** occurs from the southern United States to all but the southernmost part of South America and is transmitted by various species of bugs. It is diagnosed by finding parasites in the blood and by xenodiagnosis. No effective treatment is available.

KEY TERMS

African sleeping sickness (p. 691)
bacterial meningitis (p. 678)
botulism (p. 687)
brain abscess (p. 680)
Chagas' disease (p. 692)
eastern equine encephalitis (p. 683)
encephalitis (p. 683)

Hansen's disease (p. 684)
herpes meningoencephalitis (p. 683)
infant botulism (p. 688)
leproma (p. 684)
lepromatous (p. 684)
lepromin skin test (p. 684)
leprosy (p. 684)
listeriosis (p. 680)
poliomyelitis (p. 689)

progressive multifocal leukoencephalopathy (p. 683)
pseudocyst (p. 692)
rabies (p. 681)
rabies virus (p. 681)
St. Louis encephalitis (p. 683)
tetanus (p. 686)
tetanus neonatorum (p. 687)

trypanosomiasis (p. 691)
tuberculoid (p. 684)
Venezuelan equine encephalitis (p. 683)
viral meningitis (p. 681)
western equine encephalitis (p. 683)
wound botulism (p. 689)

A Bacterial Diseases of the Brain and Meninges

1. What is meningitis, and why is it life threatening?
2. Which bacteria can cause meningitis, and what is distinctive about the disease caused by each organism?
3. How is meningitis diagnosed, treated, and prevented?
4. List the special characteristics of listeriosis.
5. Under what circumstances do brain abscesses occur, and how are they treated?

B Viral Diseases of the Brain and Meninges

6. List the characteristics of rabies. What treatment is available?
7. Why is rabies difficult to control, and what control measures are available?
8. Humans acquire encephalitis by _____ . The season of the year during which it is most common is _____ .
9. How can encephalitis be diagnosed, treated, and prevented?
10. Under what conditions does herpes meningoencephalitis occur, and what is its prognosis?

C Bacterial Nerve Diseases

11. What changes have occurred in the incidence of Hansen's disease over the years?
12. Contrast the tuberculoid and lepromatous forms of Hansen's disease.
13. What is the cause of tetanus, and how do humans acquire it?
14. What are the effects and prognosis for tetanus?
15. What is tetanus neonatorum?
16. How do the forms of botulism differ, and how do humans acquire them?

17. Complete the following table.

Disease	Method of Diagnosis	Treatment	Method of Prevention
Hansen's disease			
Tetanus			
Botulism			

D Viral Nerve Diseases

18. What are the effects of poliomyelitis?
19. How is poliomyelitis diagnosed and treated?
20. True or False: Vaccine has not altered the incidence of poliomyelitis. Explain.
21. List the advantages and disadvantages of injectable and oral polio vaccines.

E Parasitic Diseases of the Nervous System

22. What agents cause African sleeping sickness, and how does the disease vary according to the causative agent?
23. What efforts have been made to control African sleeping sickness?
24. What causes Chagas' disease, and what are its effects?
25. Complete the following table to describe where each disease occurs and how it is transmitted.

Disease	Location of Occurrence	Mode of Transmission
African sleeping sickness		
Chagas' disease		

PROBLEMS FOR INVESTIGATION

1. Propose a mechanism that allows a prion to cause a disease.
2. Survey the problems associated with controlling one of the neurological diseases discussed in this chapter, and propose a better way to control the disease.
3. Read about the U.S. Public Health Service leprosarium at Carville, Louisiana, from its establishment in 1894 to its current research projects.
4. Review the current literature to determine what progress has been made in understanding mysterious brain infections.

Centers for Disease Control and Prevention. 1994. "Progress toward global eradication of poliomyelitis, 1988–1993." *Morbidity and Mortality Weekly Reports* 43, no. 27 (July 15):499–503.

Cohn, J. P. 1989. "Leprosy: Out of the Dark Ages." *FDA Consumer* 23 (September):24–27.

Farrar, W. E. 1994. *Areas of infections of the nervous system.* St. Louis: Mosby.

Fox, J. L. 1990. "Rabies vaccine field test undertaken." *ASM News* 56(11):579–83.

Fraser, H., et al. 1992. "Transmission of bovine spongiform encephalopathy and scrapie to mice." *Journal of General Virology* 73(8):1891–97.

Gray, L. D., and D. P. Fedorko. 1992. "Laboratory diagnosis of bacterial meningitis." *Clinical Microbiology Reviews* 5(2):130–45.

Kinnunen, L., T. Pöyry, and T. Hovi. 1992. "Genetic diversity and rapid evolution of poliovirus in human hosts." *Current Topics in Microbiology and Immunology* 176:49–61.

Kirchhoff, L. V. 1993. "American trypanosomiasis (Chagas' disease)—a tropical disease now in the United States." *The New England Journal of Medicine* 329, no. 9 (August 26):639–44.

Koprowski, H. 1995. "Visit to an ancient curse." *Scientific American: Science and Medicine* 2, no. 3 (May/June):48–57.

Marx, J. L. 1987. "Leukemia virus linked to nerve disease." *Science* 236 (May 29):1059–61.

Moore, P. S., and C. V. Broome. 1994. "Cerebrospinal meningitis epidemics." *Scientific American* 271, no. 5 (November):38–45.

Prusiner, S. B. 1993. "Biology of prion diseases." *Journal of Acquired Immune Deficiency Syndromes* 6:663–65.

Sanford, J. P. 1995. "Tetanus—forgotten but not gone." *The New England Journal of Medicine* 332, no. 12 (March 23):812–13.

Smith, A. B., J. D. Esko, and S. L. Hajduk. 1995. "Killing of trypanosomes by the human haptoglobin-related protein." *Science* 268(April 14):284–86.

Tappero, J. W., et al. 1995. "Reduction in the incidence of human listeriosis in the United States." *Journal of the American Medical Association* 273, no. 14 (April 12):1118–22.

Townsend, G. C., and W. M. Scheld. 1995. "Microbe–endothelium interactions in blood–brain barrier permeability during bacterial meningitis." *ASM News* 61(6):294–98.

Tuomanen, E. 1993. "Breaching the blood–brain barrier." *Scientific American* 268 (February):80–84.

Whitley, R. J. 1990. "Viral encephalitis." *The New England Journal of Medicine* 323, no. 4 (July 26):242–50.

Wispelwey, B., et al. 1987. "Brain abscesses." *Clinical Neuropharmacology* 10(6):483.

UROGENITAL AND SEXUALLY TRANSMITTED DISEASES

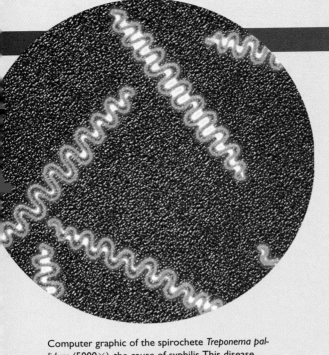

Computer graphic of the spirochete *Treponema pallidum* (5000×), the cause of syphilis. This disease, which is usually transmitted sexually, was far more deadly 500 years ago than it is today, but it can still severely damage body systems.

I n this chapter we conclude our discussion of diseases with diseases of the urogenital system, including sexually transmitted diseases. As with the other chapters on diseases, keep in mind what you have learned about normal microbiota ∞ (Chapter 15, p. 396), disease processes ∞ (Chapter 15, p. 400), and body defenses (Chapters 17, 18, and 19). In studying urogenital diseases, it is important to recall that the urinary and reproductive (genital) systems are closely associated. Infections in one system easily spread to the other.

UROGENITAL DISEASES USUALLY NOT TRANSMITTED SEXUALLY

Bacterial Urogenital Diseases

Urinary Tract Infections

Urinary tract infections (UTIs) are among the most common of all infections seen in clinical practice. Second only to respiratory infections, they account for 3 million office visits per year in the United States. UTIs

cause **urethritis** (u-re-thri′tis), or inflammation of the urethra, and **cystitis** (sis-ti′tis), or inflammation of the bladder. Because infection easily spreads from the urethra to the bladder, most infections are properly named **urethrocystitis** (u-re″thro-sis-ti′tis). Infectious agents reach the bladder more easily through the short (4 cm) female urethra than through the longer (20 cm) male urethra. Thus, women are affected some 40 to 50 times as often as men; by the age of 30, 20 percent of all women have acquired a UTI. In males the prostate gland is closely associated with the urethra and bladder, so **prostatitis** (pros-ta-ti′tis), or inflammation of the prostate gland, often accompanies UTIs.

Each year one out of five women develops **dysuria** (dis-yur′-e-ah), or pain and burning on urination indicative of urethral infection. One-fourth of these infections will develop into chronic cystitis, which will plague the unlucky victims intermittently for years. Symptoms of cystitis include continued dysuria, frequent and urgent urination, and sometimes pus in the urine. Elderly women are prone to UTIs, and up to 12 percent of some groups suffer chronically (Table 25.1).

A major cause of UTIs is incomplete emptying of the bladder during urination. Retained urine serves as a reservoir for microbial growth, thereby fostering in-

THIS CHAPTER FOCUSES ON THE FOLLOWING QUESTIONS:

A What bacteria cause urogenital diseases that are not usually sexually transmitted, and what are the important epidemiologic and clinical aspects of these diseases?

B What parasites cause urogenital diseases that are not usually sexually transmitted, and what are the important epidemiologic and clinical aspects of these diseases?

C What bacteria cause sexually transmitted urogenital diseases, and what are the important epidemiologic and clinical aspects of these diseases?

D What viruses cause sexually transmitted urogenital diseases, and what are the important epidemiologic and clinical aspects of these diseases?

TABLE 25.1	Urinary Tract Infections by Age and Sex	
Age	**Female**	**Male**
First 4 months of life	0.7%	1.3%
Four months to 5 years	4.5%	0.5%
Five years to 60 years	1–4%	<0.1%
Over 60 years	12%	1–4%
Very elderly	up to 30%	up to 30%

fection. Any factor that interferes with the flow of urine and the complete emptying of the bladder can therefore predispose an individual to UTIs. The bladder is sometimes compressed by a "sagging" uterus or by the expansion of the uterus in pregnancy. Pregnancy can also cause decreased flow of urine through the ureters. Even the ring of a diaphragm can exert enough pressure on the bladder or ureters to interfere with voiding. In men the prostate tends to enlarge with age and constrict the urethra. Finally, the problem may be behavioral rather than mechanical: Some people simply do not visit the bathroom often enough. It is important for both females and males to empty the bladder frequently and completely. People with various kinds of

paralysis who cannot void completely tend to have frequent UTIs.

UTIs originating in one area often spread throughout the urinary tract by "ascending" or "descending." Infections usually begin in the lower urethra and can ascend to cause inflammation of the kidneys, or **pyelonephritis** (pi"e-lo-ne-fri'tis). Less often, infections begin in the kidneys and descend to the urethra. Although UTIs do not occur more frequently during pregnancy, they usually are more serious because they are likely to ascend. Among pregnant women who have bacteria present in the urine, 40 percent develop pyelonephritis if the infection is not treated promptly. Descending UTIs originate outside the urinary tract. Organisms that enter the bloodstream from a focal infection, such as an abscessed tooth, can be filtered by the kidneys and cause a single infection or a chronic infection. For this reason it is often suggested that persons with chronic or frequent UTIs visit their dentist to see if some undiagnosed gum infection might be the source.

Escherichia coli is the causative agent in 80 percent of UTIs, but other enteric bacteria from feces can also cause such infections. Poor hygiene, such as wiping from back to front with toilet tissue, especially in females, can introduce fecal organisms into the urethra. It is important to teach children good toilet habits and the reasons for them. When *Chlamydia* or *Ureaplasma* are responsible for infection, they usually are sexually transmitted and cause nongonococcal urethritis, which is discussed later.

Between 35 and 40 percent of all nosocomial infections are UTIs. Outpatients have a 1 percent chance of developing a UTI following a single catheterization, whereas hospitalized patients have a 10 percent chance. A great many patients with an indwelling catheter develop a UTI within the first week of use as organisms from skin, the lower urethra, or the catheter colonize the urethra. People with permanent catheters because of paralysis fight a never-ending battle against UTIs. *Staphylococcus epidermidis* often causes infections in patients with indwelling catheters. *Pseudomonas aeruginosa* commonly causes infections that follow the use of instruments to examine the urinary tract. However, *E. coli* causes nearly half of nosocomial UTIs, and *Proteus mirabilis* causes about 13 percent of them.

UTIs are diagnosed by identifying organisms in urine cultures (Figure 25.1). Normal urine in the bladder is sterile, but urine is inevitably contaminated by bacteria as it passes through the lower part of the urethra. Even a clean-catch, midstream urine specimen will contain 10,000 to 100,000 organisms per milliliter. Low numbers of organisms do not necessarily rule out infection; in pyelonephritis and acute prostatitis, organisms sometimes enter the urine only in small numbers. Generally, however, only a single type of pathogen will be present at any one time. Finding several species in urine almost always means that the specimen was contaminated and the test should be repeated.

UTIs are treated with antibiotics, such as amoxicillin, trimethoprim, and quinolones, or with sulfonamides, according to susceptibilities of their causative agents. Prompt treatment helps prevent the spread of infection. UTIs can be prevented by good personal hygiene and frequent, complete emptying of the bladder.

PUBLIC HEALTH

WHO IS PRONE TO UTIS?

A simple blood test may soon be used to reveal whether you are prone to UTIs and might benefit from aggressive preventive treatment such as taking antibiotics. Researchers at the Memorial Sloan–Kettering Cancer Center in New York City have found that women with one of two specific types of Lewis blood groups have nearly a fourfold higher risk of developing UTIs. The correlation was evident as early as age 2. The Lewis factors in blood are tested for in the same fashion as are A, B, O blood types, but using different antisera. Such testing is available at blood banks across the United States. The correlation between UTIs and the Lewis blood groups has yet to be fully explained. However, it is speculated that cells lining the urinary tract of people with the two Lewis blood types are more receptive to bacterial attachment, thus leading to infection.

Prostatitis

The symptoms of prostatitis are urgent and frequent urination, low fever, back pain, and sometimes muscle and joint pain. Most men have had at least one prostate infection by age 40. *Escherichia coli* is the cause of 80 percent of the cases, but it is still uncertain how the bacteria reach the prostate. Four routes of infection are possible: (1) by ascent through the urethra, (2) by backflow of contaminated urine, (3) by passage of fecal organisms from the rectum through lymphatics and to the prostate, and (4) by descent of bloodborne organisms. Although uncommon, chronic prostatitis is a major cause of persistent UTIs in males, and it can cause infertility. Acute prostatitis usually responds well to appropriate antibiotic therapy without leaving sequelae.

Pyelonephritis

Pyelonephritis, an inflammation of the kidney, is usually caused by the backup of urine and consequent ascent of microorganisms. Urine backup can be caused by a number of factors, including lower urinary tract blockage or anatomical defects. Young children in particular often have imperfectly formed urinary tract valves that do not prevent urine from backing up. *Escherichia coli* causes 90 percent of outpatient cases and 36 percent of those in hospitalized patients, but yeasts

FIGURE 25.1 Analysis of a urine sample. In this test, the color produced by a diagnostic reagent determines the presence of substances such as protein and glucose—which can indicate bacterial growth as in a urinary tract infection.

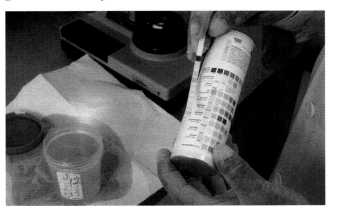

such as *Candida* are occasionally responsible for the infection.

Pyelonephritis—and any other UTI—can be asymptomatic. When present, the symptoms are indistinguishable from those of cystitis, except that sometimes chills and fever occur. Dilute urine is another common finding, leading to frequent urination and **nocturia** (nok-tu're-ah), or nighttime urination. Patients must be carefully evaluated to identify underlying, predisposing conditions such as kidney stones or other blockages, which need to be relieved. Pyelonephritis is more difficult to treat than lower UTIs, but nitrofurantoin, sulfonamides, trimethoprim, ampicillin, and quinolones usually are effective. During kidney failure or impaired kidney function, drugs must be used with care to prevent toxic accumulations.

Glomerulonephritis

Glomerulonephritis (glom-er"u-lo-nef-ri'tis), or *Bright's disease,* causes inflammation and damage to the glomeruli of the kidneys. It is an immune complex disease that sometimes follows a streptococcal or viral infection. Antigen–antibody complexes are filtered out in the kidney, where they cause inflammation of glomerular capillaries (Figure 25.2). The inflamed vessels leak blood and protein into the urine. Because of the risk of glomerulonephritis, organisms from throat and other infections that might be streptococcal (Chapter 21) should be cultured and appropriate antibiotics given. ∞ (p. 581) Although most people recover from glomerulonephritis, some have permanent residual kidney damage, and a few die.

FIGURE 25.2 Glomerulonephritis occurs when (a) immune complexes are deposited on the filtering surfaces of the glomerulus, which is located inside the Bowman's capsule in the cortex of the kidney. (b) Basement membranes of cells thicken. (c) Podocytes (foot cells) in the glomerulus, which normally act as filters, fuse their slits, and (d) leukocytes are attracted into the tissues. Kidney function is diminished, sometimes to the point of kidney failure.

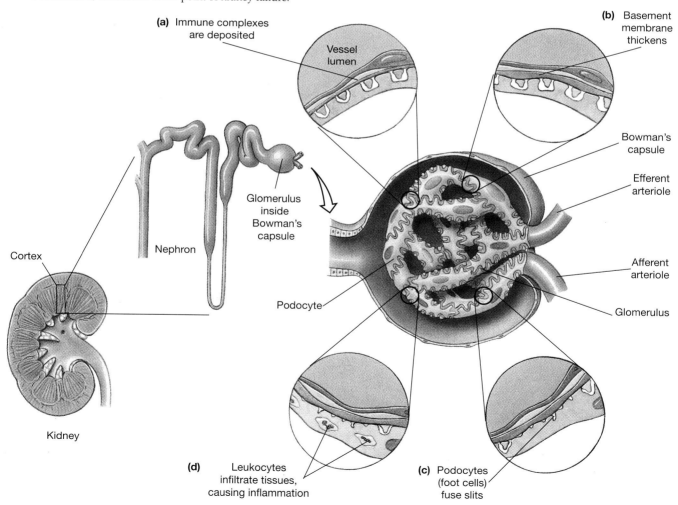

(a) Immune complexes are deposited

Vessel lumen

(b) Basement membrane thickens

Bowman's capsule

Efferent arteriole

Afferent arteriole

Glomerulus

Nephron

Glomerulus inside Bowman's capsule

Cortex

Podocyte

Kidney

(d) Leukocytes infiltrate tissues, causing inflammation

(c) Podocytes (foot cells) fuse slits

Leptospirosis

Leptospirosis is caused by the spirochete *Leptospira interrogans* (Figure 25.3). It is a zoonosis usually acquired by humans through contact with contaminated urine, directly or in water or soil. Dogs, cats, and many wild mammals carry these spirochetes (recall the "Dog Germs" box in Chapter 16). ∞ (p. 430) In some parts of the world, more than 50 percent of the rats are carriers of *Leptospira*. The bacteria live within the convoluted tubules of the kidneys and are shed into the urine. Often rain washes them from streets and the soil into natural bodies of water. They die quickly in brackish or acidic water but can survive for 3 months or longer in neutral or slightly basic water. Ponds or rivers bordering pastures where animals graze are especially likely to be contaminated.

The organisms enter the body through mucous membranes of the eyes, nose, or mouth or through skin abrasions. Parents of small children often become infected from contact with pets. The adults frequently clean up after the pets and become infected with leptospires. Dogs, being ignorant of the niceties of toilet training, often account for a cluster infection that affects all who shared a swimming or wading pool with the dogs.

After an incubation period of 10 to 12 days, leptospirosis usually occurs as a febrile and otherwise nonspecific illness. In most cases, an uneventful recovery takes place in 2 to 3 weeks. However, 5 to 30 percent of the untreated cases result in death. A particularly virulent form of the infection, *Weil's syndrome*, is characterized by jaundice and significant liver damage.

Diagnosis can be made by direct microscopic examination of blood, but it is often not even considered because of the nonspecific symptoms and low incidence of leptospirosis. Many cases are asymptomatic, so it is difficult to know the true incidence of the disease. Generally, fewer than 100 cases per year are reported in the United States.

Leptospires are susceptible to almost any antibiotic if it is given in the first 2 or 3 days of the disease, but not after the fourth day. Patients have long-term immunity to the particular strain of *Leptospira* that infected them but not to any other strains. Leptospirosis can be prevented by vaccinating pets and avoiding swimming in or contact with contaminated water.

Bacterial Vaginitis

Vaginitis, or vaginal infection, usually is caused by opportunistic organisms that multiply when the normal vaginal microbiota are disturbed by antibiotics or other factors. Predisposing conditions include diabetes mellitus, pregnancy, use of contraceptive pills, menopause, and conditions that result in imbalances of estrogen and progesterone, all of which change the pH and sugar concentration in the vagina. Several organisms account for a share each of vaginitis cases. The bacterium *Gardnerella vaginalis*, in combination with anaerobic bacteria, accounts for about one-third of the cases. *Mobiluncus* is seen in a few infections; it may be a separate organism or a clinical variant of *G. vaginalis*. The protozoan *Trichomonas vaginalis*, which can be opportunistic but usually is transmitted sexually, accounts for about one-fifth of the cases. The fungus *Candida albicans* causes most other cases.

GARDNERELLA VAGINALIS This tiny gram-negative bacillus or coccobacillus is present in the normal urogenital tract in 20 to 40 percent of healthy women. The normal vaginal pH is 3.8 to 4.4 in women of reproductive age and near neutral in young girls and elderly women. When the vaginal pH reaches 5 to 6, *Gardnerella vaginalis* interacts with anaerobic bacteria such as *Bacteroides* and *Peptostreptococcus* to cause vaginitis. None of these organisms alone produces disease. Because different anaerobes interact with *Gardnerella*, this kind of vaginitis sometimes is called *nonspecific vaginitis*. *Gardnerella* vaginitis produces a frothy, fishy-smelling vaginal discharge. The discharge, although usually small in volume, contains millions of organisms. Males occasionally get **balantitis** (bal-an-ti'-tis), an infection of the penis that corresponds to female vaginitis. Lesions appear on the penis after sexual contact with a woman who has vaginitis.

Diagnosis can be made when wet mounts of the discharge exhibit "clue cells," vaginal epithelial cells covered with many tiny rods or coccobacilli (Figure 25.4). Metronidazole (Flagyl) suppresses vaginitis by eradicating the anaerobes necessary for continuance of the disease but allows normal lactobacilli to repopulate the vagina. This effect supports the notion that *Gardnerella* vaginitis requires an association with anaer-

FIGURE 25.3 False-color TEM photo of *Leptospira interrogans* (11,000×), a spirochete ordinarily found in animals. The spirochete sometimes causes liver infections and jaundice in humans.

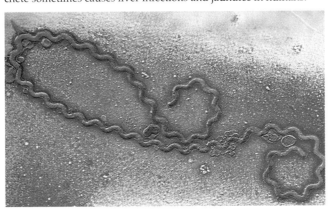

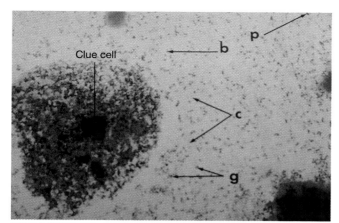

FIGURE 25.4 Clue cell of *Gardnerella* infection in a vaginal smear (g: *Gardnerella*; b, c, p: various bacteria of the normal microbiota). Note the many bacilli adhering to the surface of the clue cell.

obes. Ampicillin and tetracycline are also sometimes used in treatment. Unflavored (plain) live-culture yogurt used as a douche effectively replaces lactobacilli of normal microbiota killed by antibiotic treatment.

Toxic Shock Syndrome

Infection with certain toxigenic strains of *Staphylococcus aureus* can produce **toxic shock syndrome** (TSS). Prior to 1977 only 2 to 5 cases were reported each year, but then the incidence rose suddenly, reaching a peak of about 320 cases in September 1980. Since 1986 fewer than 25 cases per year have been reported. The sudden rise was associated with new superabsorbent, but abrasive, tampons, which were left in the vagina for longer than usual periods of time. The tampons caused small tears in the vaginal wall and provided appropriate conditions for bacteria to multiply.

Between 5 and 15 percent of women have *S. aureus* among their vaginal microbiota, but only a small fraction of these strains cause TSS. Most cases occur in menstruating women, but men, children, and postmenopausal women with focal infections of *S. aureus* occasionally develop TSS. Organisms enter the blood or grow in accumulated menstrual flow in tampons. They produce *exotoxin C,* which enhances the effects of an endotoxin; how these toxins exert their effects is not clearly understood. Clinical manifestations include fever, low blood pressure (shock), and a red rash, particularly on the trunk, that later peels. Immediate treatment with nafcillin has held the mortality rate to 3 percent. Deaths, when they occur, usually are due to shock. Recurrence is a frequent possibility, especially during subsequent menstrual cycles. Antibiotics can be given prophylactically to prevent recurrences, but women who have had TSS can reduce their risk of recurrence by ceasing to use tampons.

CLOSE-UP

MASTITIS

The female breast, a part of the reproductive system, sometimes becomes infected. Mastitis (breast infection), usually caused by *Staphylococcus aureus*, is most likely to occur in nursing mothers. Appropriate antibiotic treatment generally gives prompt results with no sequelae. The disease is not limited to humans. In rabbits, for example, mastitis (known as "blue breasts") is common in nursing does and may be treated with penicillin.

Although infrequent changing of superabsorbent tampons accounts for most cases of TSS, the contraceptive sponge now causes some. Males with boils, furuncles, or other staphylococcal infections also sometimes get TSS.

Parasitic Urogenital Diseases

Trichomoniasis

Although **trichomoniasis** is transmitted primarily by sexual intercourse, it is discussed here because of its similarity to other kinds of vaginitis and because children (and rarely adults) can be infected from contaminated linens and toilet seats. At least three species of protozoa of the genus *Trichomonas* can parasitize humans, but only *T. vaginalis* causes trichomoniasis. The others are commensals; *T. hominis* is found in the intestine, and *T. tenax* in the mouth. *T. vaginalis* is a large flagellate with four anterior flagella and an undulating membrane (Figure 25.5). It infects urogenital tract surfaces in both males and females and feeds on bac-

FIGURE 25.5 *Trichomonas vaginalis,* showing characteristic undulating membranes and flagella.

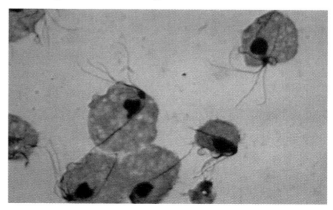

PUBLIC HEALTH

VAGINAL YEAST INFECTIONS

In addition to the protozoan *Trichomonas*, fungi and viruses also can cause urogenital infections. In fungal vaginitis, most often caused by *Candida albicans*, lesions are raised gray or white patches surrounded by red areas, and the discharge is scanty, thick, and curdlike. *Candida albicans* is a relatively common organism, being found among the natural microbiota of 20 percent of nonpregnant women and 30 percent of pregnant ones. *Candida* causes opportunistic infections, especially in patients with AIDS or uncontrolled diabetes, or in those being treated with tetracycline. Diabetes should be regulated and tetracycline therapy stopped so that *Candida* vaginitis can be effectively treated with Nystatin or imidazole. Other opportunistic fungi that can cause urogenital infections include *Aspergillus, Cryptococcus,* and *Histoplasma.* ∞ (Chapter 20, p. 565)

teria and cell secretions. Because the optimum pH for the organism is 5.5 to 6.0, it infects the vagina only when vaginal secretions have an abnormal pH. The symptoms of trichomoniasis are intense itching and a copious white discharge, especially in females.

Trichomoniasis is diagnosed by microscopic examination of smears of vaginal or urethral secretions and treated with metronidazole (Flagyl) and restoration of normal vaginal pH in women. Flagyl cannot be used during pregnancy because it causes abortions, but it is important to get rid of the infection before delivery to prevent infecting the infant. A vinegar douche usually is effective.

The effects of different species of *Trichomonas* illustrate the extreme variation in the degree of damage parasites can do. *T. hominis* and *T. tenax* are considered commensals. However, *T. foetus* causes serious genital infections in cattle. It is the leading cause of spontaneous abortion in cows and, in fact, is responsible for losses to U.S. cattle breeders of nearly $1 million per year.

SEXUALLY TRANSMITTED DISEASES

Sexually transmitted diseases (STDs) have become an increasingly serious public health problem in recent years, in part because of changing sexual behaviors. In addition, some causative agents are becoming resistant to antibiotics, and no vaccines have been developed to control any STDs. Consequently, the only means of preventing sexually transmitted diseases is to avoid exposure to them.

Acquired Immune Deficiency Syndrome (AIDS)

Although AIDS is in many cases an STD, it is not transmitted exclusively by sexual contact. We discussed it in Chapter 19 with disorders of the immune system. ∞ (p. 538)

Bacterial Sexually Transmitted Diseases

Gonorrhea

The term **gonorrhea** means "flow of seed" and was coined in A.D. 130 by the Greek physician Galen, who mistook pus for semen. By the thirteenth century, the venereal (from Venus, goddess of love in Roman mythology) transfer of this disease was known. But it was not until the mid-nineteenth century that gonorrhea was recognized as a specific disease; until then it was thought to be an early symptom of syphilis. The causative organism, *Neisseria gonorrhoeae,* was first described by the German physician Albert Neisser in 1879. It is a gram-negative, spherical or oval diplococcus with flattened adjacent sides and resembles a pair of coffee beans facing each other (Figure 25.6).

Drying kills the organisms in 1 to 2 hours, but they can survive for several hours on fomites. Cases in which *Neisseria* survived improper laundering in the hospital have been documented; in dried masses of pus, the bacteria can survive for 6 to 7 weeks! So although these organisms are ordinarily thought of as being very fragile, some are quite robust.

The infectivity of *Neisseria* is related in several ways to its pili. Attachment pili (fimbriae) enable gonococci to attach to epithelial cells that line the urinary tract so that they are not swept out with the passage of urine. (In fact, sloughing of epithelial cells is one of the body's defenses against such infections.) These bacteria also use their pili to attach to sperm; conceivably, swimming sperm could carry gonococci into the upper part of the reproductive tract. Strains lacking pili usually are nonvirulent. The search for a vaccine against gonococcal pili is under way, but success has not yet been achieved.

Gonococci produce an endotoxin that damages the mucosa in fallopian tubes and releases enzymes such as proteases and phospholipases that may be important in pathogenesis. They also produce an extracellular protease that cleaves IgA, the immunoglobulin present in secretions. ∞ (Chapter 18, p. 485) Gonococci adhere to

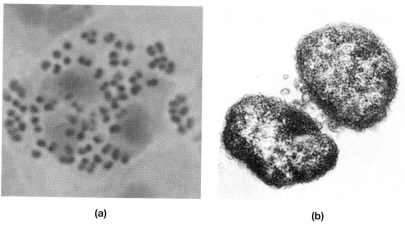

(a)   (b)

FIGURE 25.6 Diplococci of *Neisseria gonorrhoeae* are seen (a) as small, dark purple dots inside the cytoplasm of leukocytes in a urethral smear (1900×), and (b) magnified 40,000× by TEM, showing the internal structure of the pair.

neutrophils and are phagocytized by them. ∞ (Chapter 17, p. 452) Phagocytosis kills some of the bacteria, but survivors multiply inside PMNLs. A typical wet mount of a urethral discharge will show PMNLs with diplococci in their cytoplasm (Figure 25.6). Gonococci also obtain iron for their own metabolic needs from the iron transport protein transferrin.

THE DISEASE Humans are the only natural hosts for gonococci. Gonorrhea is transmitted by carriers who either have no symptoms or have ignored them. As many as 40 percent of males and 60 to 80 percent of females remain asymptomatic after infection and can act as carriers for 5 to 15 years. Very few organisms are required to establish an infection; half of a group of males given a urethral inoculum of only 1000 organisms developed gonorrhea. Following a single sexual exposure to an infected individual, about one-third of males become infected. After repeated exposures to the same infected person, three-quarters of males eventually become infected. Of males who develop symptoms, 95 percent have pus dripping from the urethra within 14 days, and many develop symptoms sooner (Figure 25.7a). The typical incubation period is 2 to 7 days.

Both the contraceptive pill and IUDs have contributed to the epidemic of gonorrhea now seen in the Western world (Figure 25.8). Their introduction in the 1960s led to increased sexual freedom and decreased use of condoms and spermicides. Women using the contraceptive pill have a 98 percent chance of being infected upon exposure because the pill alters vaginal conditions in favor of gonococcal growth. Although condoms and spermicides are less reliable contraceptives, they do offer some protection from gonorrhea. Gonorrhea spreads into the endometrial cavity and fal-

lopian tubes two to nine times faster in women who use IUDs than in those who do not use them. The exposure of blood vessels during menstruation allows bacteria to enter the circulatory system and thus facilitates the development of bacteremia.

Although gonorrhea is thought of as a venereal disease (VD), it also affects other parts of the body (Figure 25.7b). Pharyngeal infections, which develop in 5 percent of those exposed by oral sex, are most common in women and homosexual men. The patient may develop a sore throat, but most pharyngeal cases are asymptomatic. The infected tissues act as a focal source for bacteremia. Anorectal infection, especially common in homosexual men, occurs in the last 5 to 10 cm of the rectum. It can be either asymptomatic or painful, with constipation, pus, and rectal bleeding. Women with vaginal gonorrhea also often have anorectal in-

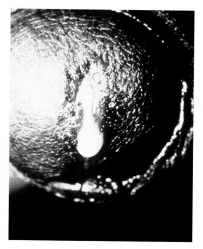

(a)

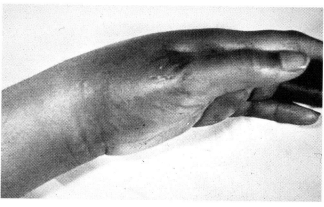

(b)

FIGURE 25.7
Symptoms of gonorrhea sometimes include (a) in males, a urethral "drip" of pus and (b) an infrequent complication, gonococcally caused arthritis.

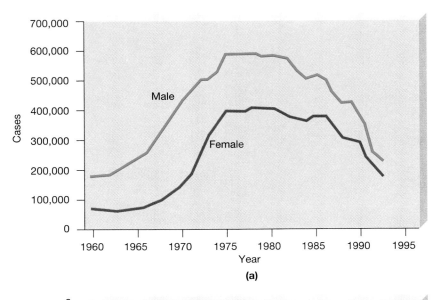

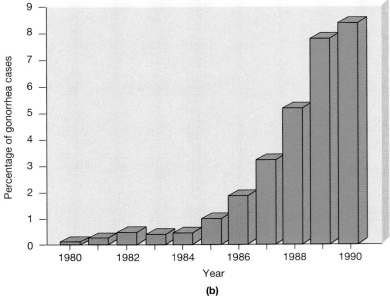

FIGURE 25.8 (a) Incidence of gonorrhea, by sex, in the United States, 1960–1993. The sharp upswing in cases of gonorrhea follows the increased use of alternative methods of birth control in place of condoms, beginning in the 1960s. (b) The percentage of cases of gonorrhea caused by penicillin-resistant strains, United States, 1980–1990.

fection. Such infection can occur in the absence of anorectal intercourse due to contamination of the anal opening by organism-laden vaginal discharges. Medical personnel doing internal pelvic examinations can avoid spreading the infection by using a different gloved finger for anal examination than was inserted into the vagina.

The urethra is the most common site of gonorrhea infections in males. The most common site of infection in females is the cervix, followed by the urethra, anal canal, and pharynx. Skene's glands in the urethra and Bartholin's glands near the vaginal outlet also can be infected in females. As many as half of infected females

develop **pelvic inflammatory disease** (PID), in which the infection spreads throughout the pelvic cavity. Studies done in Sweden showed that sterility often follows PID because of tubal occlusion by scarring. The rate of sterility increases with the number of infections—13 percent after one infection, 35 percent after two infections, and 75 percent after three infections.

Disseminated infections, which occur in 1 to 3 percent of cases, produce bacteremia, fever, joint pain, and skin lesions that can be pustular, hemorrhagic, or necrotic. When organisms reach the joints, they can cause arthritis. Gonococcal arthritis is now the most common joint infection in people 16 to 50 years old.

Another complication of gonorrhea is infection of the lymphatics that drain the pelvis. Scarring produces tight, inflexible tissue that immobilizes pelvic organs into the condition known as "frozen pelvis."

Transfer of organisms by contaminated hands, or fomites such as towels, can result in eye infections. If untreated, severe scarring of the cornea and blindness can result. Newborns can acquire ophthalmia neonatorum during passage through the birth canal of an infected mother. ∞ (Chapter 20, p. 566) Pus that accumulates behind swollen eyelids can spurt forth with great pressure and infect health care workers. Such infections have occurred despite immediate treatment.

Age affects the site and type of gonococcal infection. During the first year of life, infection usually results from accidental contamination of the eye or vagina by an adult. Between 1 year and puberty, gonorrhea usually occurs as vulvovaginitis in girls who have been sexually molested. A girl's vaginal epithelium has less keratin and is more susceptible to infection than a woman's. The girls exhibit pain on urination, vulvar and perianal soreness, discomfort on defecation, and a yellowish-green discharge from vaginal and urethral openings. Both boys and girls can develop a pus-filled anal discharge indicative of anorectal gonorrhea as a result of sexual abuse. Failure of a hospital to launder sheets properly led to an outbreak of vulvovaginitis in female pediatric patients who acquired it from sitting on the sheets.

DIAGNOSIS, TREATMENT, AND PREVENTION

Diagnosis is made by identifying *N. gonorrhoeae* in laboratory cultures. Being a rather fastidious organism, it requires high humidity and ambient carbon dioxide for growth. Temperatures of 35° to 37°C and pH of 7.2 to 7.6 are optimal. Many species of *Neisseria* exist. A positive result on a screening test may not be due to *N. gonorrhoeae*; confirmatory tests should always be done. Samples from patients suspected of having *Neisseria* infections are inoculated from a swab directly into a special medium for transport and incubation in a laboratory. ∞ (Chapter 6, p. 157)

Sulfonamides were the first agents found to treat gonococcal infections. As some strains became resistant to sulfonamides, penicillin became available. At first, penicillin G was effective in very small amounts, but now much larger doses must be used. Some strains of gonococci are entirely resistant to penicillin because they produce the enzyme β-lactamase, which enables them to break down this antibiotic. CDC now recommends a single intramuscular (250 mg) dose of the cephalosporin ceftriaxone. Increases in resistance have made it very difficult to combat gonorrhea. Organisms typically acquire antibiotic resistance by means of their sex pili. One wonders if it will always

PUBLIC HEALTH

DRUG-RESISTANT GONORRHEA

The relatively new quinolone antibiotics have not yet been employed to treat gonorrhea in the United States. However, they have been so used in the Philippines, with the result that quinolone-resistant strains of *Neisseria gonorrhoeae* have already emerged there. If these strains make their way to the United States, the quinolones will be obsolete for treating gonorrhea before they can be used generally. Their use is currently restricted to special cases.

be possible to find another antibiotic to combat resistant strains.

Gonorrhea patients often have other STDs. A 7-day course of treatment with oral doxycycline has the advantage of killing *Chlamydia*, which may concurrently be present. Forty-five percent of patients having gonorrhea also test positive for *Chlamydia*. Follow-up cultures should be done 7 to 15 days after completion of treatment to be sure the infection is cured. Additional follow-up cultures are also advised at 6 weeks, as 15 percent of women who are negative at 7 to 14 days are positive again at 6 weeks—possibly because of reinfection. All sexual partners must be treated.

The best means of preventing gonorrhea is to avoid sexual contact with infected individuals. No vaccine is available.

Syphilis

Syphilis is caused by the spirochete *Treponema pallidum*, an active motile organism with fastidious growth requirements. ∞ (Figure 3.25b, p. 65) The evolution of this organism has paralleled human evolution (see the Essay at the end of this chapter). *Treponema pallidum* eluded discovery until it was finally stained in 1905. Today in the United States, syphilis is much less common than gonorrhea. However, the incidence of syphilis has, on average, been on the rise since 1960, although it has been declining steadily since 1990 (Figure 25.9).

The disease is ordinarily transmitted by sexual means, but it can be passed in body fluids such as saliva. It thereby creates a hazard for dentists, dental hygienists, and those who kiss. It is not transmitted in food, water, or air or by arthropod vectors. Humans are its only reservoir. Donated blood need not be screened for syphilis because any spirochetes present are killed when the blood is refrigerated.

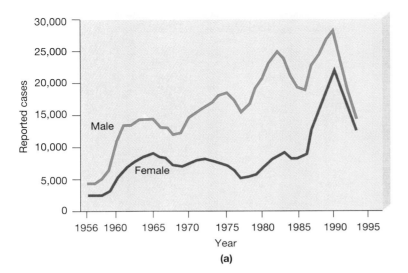

(a)

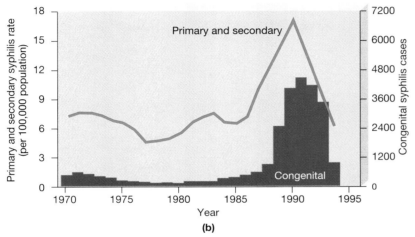

(b)

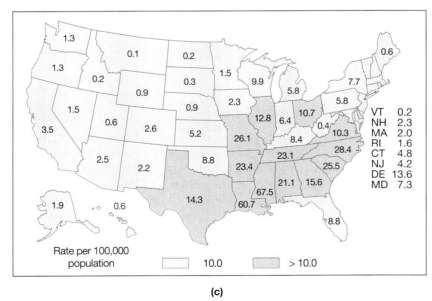

(c)

FIGURE 25.9 (a) Incidence of syphilis infections, by sex, United States, 1956–1993. (b) Incidence of syphilis among women and congenital syphilis through 1994. As more women contract syphilis, transmission to the fetus becomes more common, leading to a parallel increase in incidence of congenital syphilis. (c) Incidence of syphilis by state, 1993.

THE DISEASE A typical case of syphilis progresses as follows:

1. Incubation stage: Over a period of 2 to 6 weeks after entering the body, the organisms multiply and spread throughout the body.

2. Primary stage: An inflammatory response at the original entry site causes formation of a **chancre** (shang'ker), a hard, painless, nondischarging lesion about 1/2 inch in diameter. One or more primary chancres usually develop on the genitals but can develop on lips or hands (Figure 25.10). In females, chancres on the cervix or another internal location sometimes escape detection. The patient often is embarrassed to seek medical attention for a lesion in a genital location and hopes that it will just "go away." And go away it does after about 4 to 6 weeks, without leaving any scarring. The patient thinks all is well, but the disease has merely entered the next stage.

3. Primary latent period: All external signs of the disease disappear, but blood tests diagnostic for syphilis are positive.

4. Secondary stage: Symptoms appear, disappear, and reappear over a period of about 5 years, during which the patient is contagious. These symptoms include a copper-colored rash, particularly on the palms of the hands and the soles of the feet (Figure 25.11a), and various pustular rashes and skin eruptions. Painful, whitish mucous patches swarming with spirochetes appear on the tongue, cheeks, and gums. Kissing spreads the spirochetes to others. These lesions heal uneventfully, and the patient again thinks all is well. But the disease now has entered the next stage.

5. Secondary latent stage: Again all symptoms disappear, and blood tests can be negative. This stage can persist for life or for a highly variable period, or may never occur. Symptoms can reoccur at any time during latency. In some patients syphilis does not progress beyond this stage, but in many patients it progresses to the tertiary stage.

6. Tertiary stage: Permanent damage occurs throughout various systems of the body. A wide assortment of symptoms can appear; syphilis has been called the "great imitator" because its symptoms can mimic those of so many other

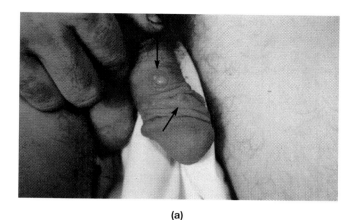

(a)

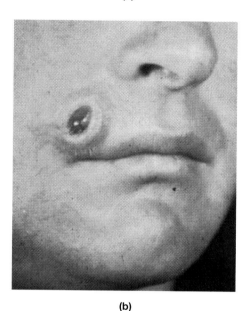

(b)

FIGURE 25.10 Primary chancres of syphilis on (a) a genital site (penis) and (b) on an extragenital site (the face).

diseases. Most involve the cardiovascular and nervous systems. Blood vessels and heart valves are damaged. In long-standing cases, calcium deposits in heart valves can be so extensive as to be visible on a chest X-ray. Neurological damage, called **neurosyphilis,** can include thickening of the meninges; ataxia (a-tax'e-ah), or an unsteady gait or inability to walk; and paresis (par'e-sis), or paralysis and insanity. These symptoms often are due to formation of granulomatous inflammations called **gummas** (gum'ahz). ∞ (Chapter 17, p. 457) Internal gummas typically destroy neural tissue, whereas external gummas destroy skin tissue (Figure 25.11b). Mental illness accompanies neural damage. In the preantibiotic era, as many as half the beds in mental hospitals were occupied by patients with tertiary syphilis.

DIAGNOSIS, TREATMENT, AND PREVENTION Diagnostic tests include fluorescent antibody and treponemal immobilization tests. Actively motile organisms can be observed under the dark-field microscope while specific antibody against *Treponema pallidum* is added. Immobilization of organisms by the antibody is 98 percent confirmatory for syphilis. Other blood tests such as VDRL (Venereal Disease Research Laboratory) and Wasserman tests have a high frequency of false positives, as they are based on the detection of tissue damage. A severe case of influenza, a myocardial infarction, or an autoimmune disease can cause sufficient damage to elicit a false positive reaction. Therefore, screening tests such as the VDRL must always be followed up by a confirmatory test, such as the fluorescent antibody test.

Syphilis usually is treated with penicillin, but tetracycline and erythromycin also are effective. The longer the patient has had syphilis, the more important are continued treatment and testing to ensure that organisms have been eradicated. No vaccine is available, and recovery does not confer immunity.

FIGURE 25.11 Signs of (a) secondary syphilis, a typical papular rash, and (b) of tertiary syphilis, a gumma.

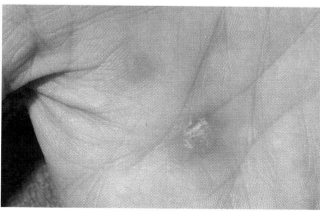

(a)

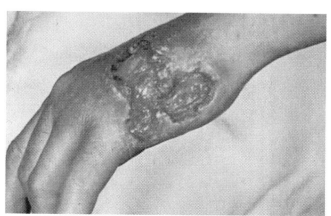

(b)

CONGENITAL SYPHILIS Congenital syphilis occurs when treponemes cross the placenta from mother to baby. At birth or shortly thereafter, the infant may show such signs as notched incisors, or Hutchinson's teeth (Figure 25.12a), a perforated palate, saber shins (in which the shin bone projects sharply on the front of the leg) (Figure 25.12b), an aged-looking face with saddle-nose (a flat, saddle-shaped nose) (Figure 25.12c), and a nasal discharge. Women should be tested for syphilis prior to becoming pregnant, and as part of their prenatal testing if they are already pregnant, to prevent such congenital infections.

Chancroid

Chancroid (shang'kroid), called soft chancre to distinguish it from the hard, painless chancre of syphilis, is caused by *Haemophilus ducreyi*. Named for Augusto Ducrey, the Italian dermatologist who first observed it in skin lesions in 1889, the organism is a small gram-negative rod that occurs in strands.

Relatively rare in the United States, with approximately 800 cases reported per year, chancroid is seen most frequently in developing countries of Africa, the Caribbean, and Southeast Asia. Its worldwide incidence is believed to be greater than that of either gonorrhea or syphilis. Most cases in the United States occur in immigrants and a few in military personnel. In 1982 CDC reported a sudden increase in the number of cases in southern California—from an average of 29 per year to over 400 per year. Of these, 90 percent were in Hispanic men, most of whom had recently immigrated from Mexico.

THE DISEASE Chancroid begins with the appearance of soft, painful lesions called chancres, which bleed easily, on the genitals 3 to 5 days after sexual exposure. They often occur on the labia and clitoris in females and on the penis in males. However, the infection can be present without apparent lesions, the only symptom being a burning sensation after urination. Chancres also can occur on the tongue and lips. Regardless of their location, chancres are extremely infective. Medical personnel sometimes acquire lesions on the hands merely from contact with chancres. In about one-third of the patients, chancroid spreads to the groin, where it forms enlarged masses of lymphatic tissue called *buboes*. These appear about 1 week after infection, swell to great size, and can break through the skin, discharging pus to the surface (Figure 25.13).

DIAGNOSIS AND TREATMENT Chancroid is diagnosed by identifying the organism in scrapings from a lesion or in fluids from a bubo. Patients with chancroid also often have syphilis and other STDs. Thus, a patient with a positive diagnosis for one STD should be tested for other STDs. Untreated lesions can persist for months.

FIGURE 25.12 Signs of congenital syphilis. (a) Hutchinson's teeth, showing notched central incisors. (b) Saber shin, displaying arching of the anterior edge of the tibia. (c) Saddle-nose, which causes a snuffling type of breathing.

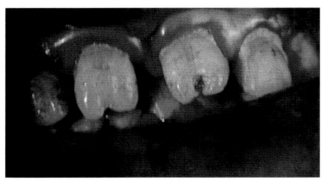

(a)

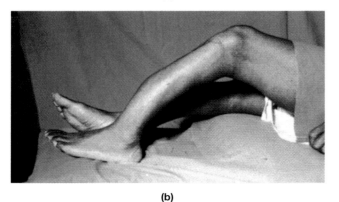

(b)

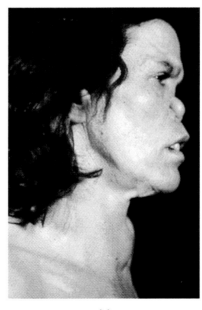

(c)

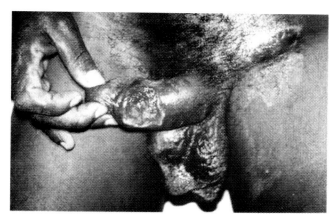

FIGURE 25.13 Lesions of chancroid on the penis, showing a draining buboe in the adjacent groin area. Chancroid is caused by *Haemophilus ducreyi*.

Often the disease resolves on its own. Infection does not confer permanent immunity, and the disease can be acquired again and again. It is treated with antibiotics such as tetracycline, erythromycin, sulfanilamide, or a combination of trimethoprim and sulfamethoxazole. With treatment lesions heal rapidly, but they often leave deep scars with much tissue destruction.

Nongonococcal Urethritis

As the name suggests, **nongonococcal urethritis** (NGU) is a gonorrhea-like STD caused by organisms other than gonococci. Most cases are caused by *Chlamydia trachomatis*, but some are caused by mycoplasmas. The prevalence of chlamydial infections is greater than that of any other STDs and is increasing dramatically.

CHLAMYDIAL INFECTIONS *Chlamydia trachomatis* is a tiny spherical bacterium with a complex intracellular life cycle. ∞ (Chapter 10, p. 257) In addition to causing NGU, subspecies strains of *C. trachomatis* cause a wide range of disorders, including conjunctivitis and lymphogranuloma venereum (discussed later). The CDC has estimated that 3 to 5 million Americans contract *Chlamydia* NGU each year. The subspecies that causes inclusion conjunctivitis in newborns also causes 30 to 50 percent of NGU cases in males as well as 30 to 50 percent of vulvovaginitis cases and half of cervicitis (inflammation of the cervix) cases in women. The large numbers of humans who are infected and who have organisms in their body secretions makes transmission of chlamydial infections especially easy.

Following an incubation period of 1 to 3 weeks, symptoms of NGU similar to but milder than gonorrhea begin. A scanty watery urethral discharge is observed, especially after passage of first morning urine. Sometimes this is accompanied by tingling sensations in the penis. Many chlamydial NGU infections are asymptomatic, and fortunately most chlamydial STDs produce no significant complications or aftereffects. However, inflammation of the epididymis (the tube through which sperm pass from the testis) can lead to sterility.

Pelvic inflammatory disease, which can be caused by more than 20 different infectious agents, is a common complication of NGU as well as of gonorrhea. Chlamydial PID increases the risk of sterility and ectopic pregnancy, a pregnancy in which the embryo begins development outside the uterus (for example, in a fallopian tube or the peritoneal cavity). Surveys of pregnant women reveal that 11 percent have *Chlamydia* in the cervix; postpartum fever is common among infected women. Their infants can develop neonatal chlamydial pneumonia, a rarely fatal disease that accounts for 30 percent of all pneumonias in children under 6 months of age. Because the condition typically does not develop until 2 or 3 months after delivery, its connection with cervical chlamydial infection is often overlooked.

Chlamydial infections are difficult to control. Infants can become infected while passing through the birth canal of an infected mother. Silver nitrate used to prevent gonorrheal ophthalmia neonatorum does not protect against *Chlamydia*, but erythromycin protects against both. In venereal disease clinics, penicillin used to treat syphilis and gonorrhea fails to eliminate chlamydial infections. The recent rise in chlamydial infections could be stopped with tetracycline and sulfa drugs if all sexual partners were treated.

Adult **inclusion conjunctivitis** can result from self-inoculation with *C. trachomatis* from genitals via fingers or towels; it is especially common in sexually active young adults. It closely resembles trachoma. ∞ (Chapter 20, p. 568) Prior to the widespread use of chlorine in pools, it was called "swimming pool conjunctivitis." Reinfection can occur, but whether by inadequate immune response or by infection with another of eight different infectious strains is unknown.

PUBLIC HEALTH

CHLAMYDIAL INFECTIONS

Before 1983 only 200 laboratories in the United States could culture chlamydias. Newly developed diagnostic tests such as enzyme immunoassays and monoclonal antibody tests are revealing a previously unsuspected epidemic of chlamydial infections. Thus, chlamydial infection is now known to be one of the most prevalent diseases in American society. In Britain, where chlamydial infection is a reportable disease, statistics confirm similar high prevalences.

Each year about 75,000 infants acquire chlamydial **inclusion blennorrhea** (blen-or-e'ah), a name derived from the Greek for "mucus flow." This usually benign conjunctivitis begins as a pus-filled mucous discharge 7 to 12 days after delivery and subsides with erythromycin treatment or spontaneously after a few weeks or months. If it persists, it becomes indistinguishable from childhood trachoma and can lead to blindness.

MYCOPLASMAL INFECTIONS NGU also can be caused by *Mycoplasma hominis*, which is frequently among the normal urogenital microbiota, especially in women. Mycoplasmas, which have no cell wall, apparently infect by fusing their cell membranes with the host cell membranes. Such infections are very common; more than half of normal adults have antibodies to *M. hominis*. Although most often associated with NGU, *M. hominis* sometimes causes PID in women and opportunistic urethritis in men. Mycoplasmas on the cervix during pregnancy probably colonize the placenta and cause spontaneous abortions, premature births, and low birth rate. They also foster ectopic pregnancies.

Yet another causative agent of NGU is *Ureaplasma urealyticum*, formerly called T-strain (T for tiny) mycoplasma. Between 1 and 2.5 million people in the United States are infected. One of the smallest bacteria known to cause human disease, its name is based on the fact that it requires a 10 percent urea medium to grow. Of patients seen in venereal disease clinics, 50 to 80 percent carry *U. urealyticum* in addition to other sexually transmitted pathogens. The organism accounts for more than half of all infections that make couples infertile. Low sperm counts and poor sperm mobility have been observed in males; the organisms bind tightly to sperm and can be transmitted by them to sexual partners. They are a major cause of fetal death, recurrent miscarriage, prematurity, and low birth weight—itself a leading cause of neonatal death.

Diagnosis is made by culture from urethral and vaginal discharges and from placental surfaces. When both members of an infected couple are treated, pregnancy is achieved in 60 percent of cases, compared with a success rate of only 5 percent in untreated couples infected with this organism. Because mycoplasmas lack a cell wall, penicillin is not effective against them. Therefore, people being treated with penicillin for other STDs will not be cured of NGU. Tetracycline is most often used because it also controls *Chlamydia*. The 15 percent of *Mycoplasma* strains resistant to tetracycline can be treated with erythromycin and spectinomycin.

Lymphogranuloma Venereum

Another STD, **lymphogranuloma venereum** (LGV), is common in tropical and subtropical regions. Although only about 250 cases per year occur in the United States, mostly in southeastern states, as many as 10,000 cases are treated annually in a single Ethiopian clinic. This nonreportable disease is about 20 times more common in males than in females.

THE DISEASE The causative agent of LGV was identified in 1940 as a highly invasive strain of *Chlamydia trachomatis*. Within 7 to 12 days after contact with the organism, lesions appear at the site of infection—usually the genitals, but sometimes the oral cavity. In most cases the lesions rupture and heal without scarring. Other symptoms include fever, malaise, headache, nausea and vomiting, and a skin rash. One week to 2 months later, the organisms invade the lymphatic system and cause regional lymph nodes to become enlarged, painful, and pus-filled buboes (Figure 25.14a). Victims have been known to use a razor blade to open the buboes to gain relief (Figure 25.14b), but aspiration with a sterile needle is a safer treatment. Lymph node inflammation occasionally obstructs and scars lymph vessels, causing edema of genital skin and elephantiasis (massive enlargement) of the external genitalia in both males and females. Rectal infections often occur in homosexual men. In women, lymph from the vagina drains toward

FIGURE 25.14 Bilateral buboes of lymphogranuloma venereum, caused by *Chlamydia trachomatis*, (a) early in development and (b) after reaching such size that the lesion was opened for drainage.

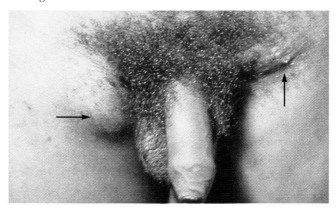

(a)

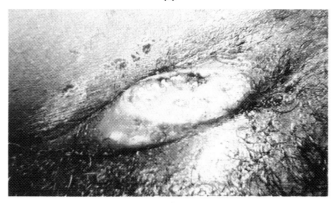

(b)

the rectum, and lymph nodes in the walls of the rectum become chronically enlarged in 25 percent of the cases. This causes rectal blockage that usually requires surgery. Untreated cases produce a bloody, pus-filled anal discharge and can eventually lead to rectal perforation. Organisms transferred from the hands to the eyes can cause conjunctivitis. In rare instances the disease progresses to meningitis, arthritis, and pericarditis.

Spontaneous "cures" sometimes represent latent infections. Latency can be prolonged, as is evidenced by males who infect sex partners many years after their own initial infection. The genital tracts and rectums of chronically infected, but at times asymptomatic, persons serve as reservoirs of infection.

DIAGNOSIS AND TREATMENT   LGV is diagnosed by finding chlamydias as inclusions stained with iodine in pus from lymph nodes. Serological tests are available, but they frequently give false-positive results. Tetracycline is the drug of choice for treating LGV, but erythromycin and sulfamethoxasole also are effective. Enlarged lymph nodes can take 4 to 6 weeks to subside even after successful antibiotic therapy.

Granuloma Inguinale

Granuloma inguinale, or donovanosis, is caused by the small gram-negative encapsulated rod *Calymmatobacterium granulomatis,* which is related to *Klebsiella.* The disease is uncommon in the United States, with about 50 cases reported per year, mostly among male homosexuals. It is common in India, western coastal Africa, South Pacific islands, and some South American countries, from which it is occasionally brought to the United States. The epidemiology of granuloma inguinale is not completely understood. Some cases appear to be sexually transmitted and others not. Even in genital cases, infectivity is low and many partners of infected individuals do not get it.

Granuloma inguinale appears as irregularly shaped, painless ulcers on or around the genitals 9 to 50 days after intercourse. Fever is absent. Ulcers can be spread to other body regions by contaminated fingers. As ulcers heal, skin pigmentation is lost. Without treatment, tissue damage can be extensive. Diagnosis from scrapings of lesions is confirmed by finding large mononuclear cells called **Donovan bodies** (Figure 25.15). Antibiotics such as ampicillin, tetracycline, erythromycin, and gentamicin offer effective treatment.

Viral Sexually Transmitted Diseases

Herpesvirus Infections

Two closely related herpesviruses cause disease in humans (Chapter 11). ∞ (p. 279) **Herpes simplex virus**

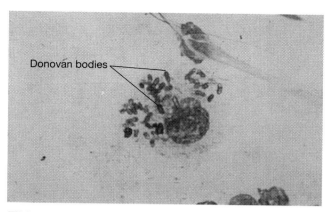

FIGURE 25.15   Scrapings of lesions of granuloma inguinale, caused by *Calymmatobacterium granulomatis,* reveal encapsulated bacterial cells inside the much larger macrophages. These bacterial cells resemble closed safety pins and are called Donovan bodies.

type 1 (HSV-1) typically causes fever blisters (cold sores), and **herpes simplex virus type 2** (HSV-2), sometimes called *herpes hominis virus,* typically causes **genital herpes.** Both HSV-1 and HSV-2 have an incubation period of 4 to 10 days, cause the same kind of lesions, and have been isolated from skin and mucous membranes of oral and genital lesions. Among oral infections, 90 percent are caused by HSV-1 and 10 percent by HSV-2. Among genital infections, 85 percent are caused by HSV-2 and 15 percent by HSV-1. The presence of HSV-1 in the genital region and HSV-2 in the oral region is most commonly due to the practice of oral sex. Genital herpes is by far the most common and most severe of the herpes simplex viral infections.

Initial HSV-1 or HSV-2 infection can be asymptomatic, especially in children, or cause localized lesions with or without symptoms of acute infection. Most adults have antibodies to herpesviruses, but only 10 to 15 percent have experienced symptoms. In both HSV-1 and HSV-2 infections, vesicles form under keratinized cells and fill with fluid from virus-damaged cells, particles of cell debris, and inflammatory cells. Vesicles are painful, but they heal completely in 2 to 3 weeks without scarring unless there is a secondary bacterial infection. Adjacent lymph nodes enlarge and are sometimes tender.

Latency ∞ (Chapter 11, p. 279) is a hallmark of herpes infections. More than 80 percent of the adult population worldwide harbors these viruses, but only a small proportion experience recurrent infections. Within 2 weeks of an active infection, the viruses travel via sensory neurons to ganglia (Figure 25.16). Within the ganglia they replicate slowly or not at all. They can reactivate spontaneously or be activated by fever, ultraviolet radiation, stress, hormone imbalance, menstrual bleeding, a change in the immune system, or trauma. The infection can spread to and kill cells in the adrenal glands, liver, spleen, and lungs. In fatal herpes

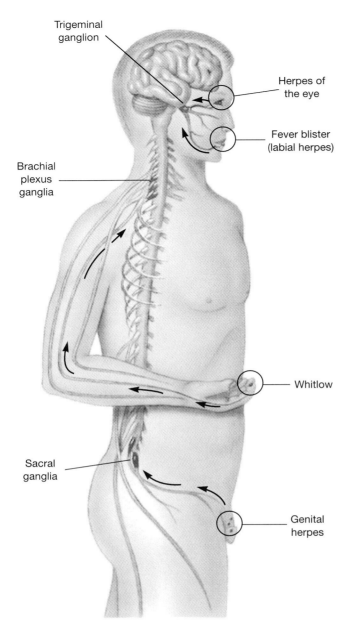

FIGURE 25.16 Herpes simplex viruses are permanent "houseguests" once acquired. After an occurrence of a lesion, they migrate to ganglia, from which they travel back to the site of the original lesion at the time of the next outbreak. An infrequent exception to this pattern sometimes occurs at the trigeminal ganglion, at which three branches of the trigeminal (fifth cranial) nerve join. Virus from a lesion on the lower lip could migrate out along the ophthalmic branch on a subsequent episode, causing the eye to become affected, or even migrate backward into the brain, causing meningitis and brain damage. Why this sometimes happens is not known; fortunately, it is very rare.

Labels on figure:
- Trigeminal ganglion
- Herpes of the eye
- Fever blister (labial herpes)
- Brachial plexus ganglia
- Whitlow
- Sacral ganglia
- Genital herpes

encephalitis, soft, discolored lesions appear in both gray and white matter of the brain.

After reactivation, the virus moves along the nerve axon to the epithelial cells. There it replicates, causing recurrent lesions. These lesions, which always recur in exactly the same place as the original infection, are smaller, shed fewer viruses, contain more inflammatory cells, and heal more rapidly than primary lesions. Successive recurrences usually become milder until they finally cease. While the virus is in a neuron, neither humoral nor cellular immunity can combat it. Once the virus reaches target epithelial cells and starts to replicate, antibodies can neutralize the viruses, and T cells can eliminate virus-infected cells. These immunological processes that make it difficult to isolate viruses from vesicular fluid in recurrent lesions also reduce their severity and duration. Recurrences can be limited to one or two episodes or can appear periodically for the life of the patient but typically occur five to seven times. Even if recurrences cease, the viruses remain latent in ganglia, and long-dormant viruses can be reactivated by severe stress, trauma, or impaired immune function (as for example in AIDS patients).

"HSV shedders," people who shed viruses while remaining asymptomatic, pose a significant problem. As many as 1 in 200 women have been shown to shed viruses even when they have no observable lesions and in some cases no knowledge of ever having had a herpes infection. Such women pose a serious threat to any infant they might bear.

GENITAL HERPES Genital herpes infections usually are acquired after the onset of sexual activity because the virus is transmitted mainly by sexual contact. However, the virus can survive for short periods of time in moist areas such as hot tubs. Changes in sexual practices, especially an increase in oral sex, have increased the incidence of HSV-1 in genital lesions and HSV-2 in oral lesions. More than 20 million Americans now have genital herpes, and half a million new cases are seen each year.

In females the vesicles appear on the mucous membranes of the labia, vagina, and cervix. Ulcerations sometimes spread over the vulva and can even appear on the thighs. In males tiny vesicles appear on the penis and foreskin and are accompanied by urethritis and a watery discharge (Figure 25.17). The prostate gland and seminal vesicles also can be affected. Both sexes experience intense pain and itching at the sites of lesions and swelling of lymph nodes in the groin.

A person infected with herpesvirus is contagious any time viruses are being shed. Shedding always occurs when active lesions are present and usually starts a few days before lesions appear. It can occur continuously even when lesions are not present. Therefore, abstention from sexual contacts when lesions are present does not always prevent spread of the disease. In recent years promiscuous sexual practices and ignorance of, or lack of concern about, transmitting the disease have greatly increased the number of cases of genital herpes. This incurable disease has now become one of the most common STDs.

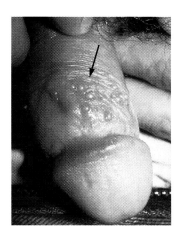

FIGURE 25.17 Penile herpes lesions.

Women infected with genital herpes may be subject to three other serious problems. First, the incidence of miscarriages among women with genital herpes is higher than that for uninfected women. Second, when infected women become pregnant, the infant must be delivered by Caesarean section. Finally, infected women have an increased risk of developing a kind of cancer called cervical carcinoma. Among women with cervical carcinoma, 80 percent have antibodies to HSV-2; they have been infected and probably still harbor the viruses. Most have also had multiple sexual partners. The exact roles of HSV-2, wart viruses (discussed later), or other factors associated with sexual activity in the development of the malignancy are not yet known. Until more information is available, avoiding sexual contact with virus-infected individuals and with multiple partners seems prudent.

NEONATAL HERPES **Neonatal herpes** (Figure 25.18) can appear at birth or up to 3 weeks after birth. Babies most often become infected by delivery through a birth canal contaminated with HSV-2, but they can also become infected through contaminated equipment

FIGURE 25.18 Neonatal herpes can be acquired when the baby passes through the birth canal of an infected mother. This could be avoided by Caesarean section delivery. Sometimes lesions are so extensive that they cover almost the entire skin surface. Such children either suffer profound brain damage or do not survive.

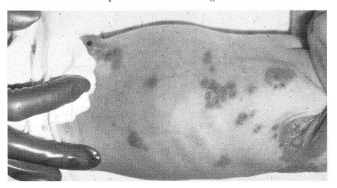

and hospital procedures. In rare instances, infants are infected *in utero*. Because neonates are highly susceptible to HSV infections, they should not be cared for by individuals with such infections. Infected mothers who must care for their infants must scrupulously follow sanitary procedures.

At diagnosis, two-thirds of infected infants have skin vesicles; the others already have disseminated infection with neural or visceral lesions. Skin infections disseminate in 70 percent of infected infants. Infants with disseminated infections display poor appetite, vomiting, diarrhea, respiratory difficulties, and hypoactivity. Some also have neurological symptoms, jaundice, and eye disorders. Neonates with disseminated infections deteriorate rapidly and usually die within 10 days. The few that survive usually have central nervous system and eye damage. Occasionally infants have only a few vesicles, but latent viruses may later cause significant damage. Early diagnosis and treatment of neonatal herpesvirus infections are essential for survival and to reduce the likelihood of neurological damage.

OTHER HERPES SIMPLEX INFECTIONS A variety of manifestations of herpesvirus infections have been seen. Most are thought to be caused by HSV-1, but HSV-2 may be responsible for some. They include gingivostomatitis, herpes labialis, keratoconjunctivitis, herpes meningoencephalitis, herpes pneumonia, eczema herpeticum, traumatic herpes, herpes gladiatorium, and whitlows. HSV-1 is spread by contaminated secretions and through contact with lesions and usually is acquired in childhood from relatives, nurses, or other children. The incidence of the virus is especially high within families, hospitals, and other institutions. Most humans are infected with HSV-1 in the first 18 months of life. Many of these infections are inapparent, and the first apparent lesions result from reactivations.

Gingivostomatitis (jin″gi-vo-sto″mah-ti′tis), lesions of the mucous membranes of the mouth, is most common in children 1 to 3 years old. After an incubation period of 2 to 20 days, small vesicles appear around the mouth over a 7-day period. Each vesicle accumulates fluid, scabs over, and heals in 2 to 3 weeks. Recurrent lesions typically occur in the form of **herpes labialis** (la′be-ah-lis), or fever blisters on the lips. If the eyes instead of the mouth are the site of initial infection, vesicles appear on the cornea and eyelids, causing **keratoconjunctivitis** (ker″a-to-kon-junk-ti-vi′tis).

The most serious manifestation of HSV infection is *herpes meningoencephalitis,* which can follow a generalized herpes infection in a newborn, child, or adult. How the virus crosses the blood–brain barrier to enter the central nervous system has recently been discovered. The virus remains latent in the ganglion of the trigeminal nerve, which supplies parts of the face and mouth, until some unknown factor reactivates it. Then, instead of moving outward to epithelial sites such as the tongue or lip, it ascends along the nerve into the brain. The disease has a rapid onset, with fever, headache, meningeal irritation, convulsions, and altered reflexes. In the middle-aged and elderly, meningoencephalitis can appear without any preceding symptoms. The patient experiences increasing confusion, hallucinations, and sometimes seizures. Most patients die in 8 to 10 days; survivors usually have permanent neurological damage.

Herpes pneumonia is rare. It usually is seen only in burn patients, alcoholics, and patients with AIDS or other immunodeficiencies. **Eczema herpeticum** is a generalized eruption caused by entry of the virus through the skin. **Traumatic herpes** occurs when the virus enters traumatized skin in the area of a burn or other injury. **Herpes gladiatorium** occurs in skin injuries of wrestlers. A **whitlow** (Figure 25.19) is a herpetic lesion on a finger that can result from exposure to oral, ocular, and probably genital herpes lesions. Dental technicians, nurses, and other medical personnel should therefore use rubber gloves when treating patients with herpes lesions. A person with a whitlow can spread the infection to the mouth, eyes, and genital areas or to other persons.

DIAGNOSIS, TREATMENT, AND PROGNOSIS
Herpesviruses are most easily isolated from vesicular fluid and cells from the base of a lesion and can be grown in the laboratory in a variety of cell types. The cytopathic effect of the viruses appears in the culture quickly. The time to obtain a diagnosis has been greatly decreased by rapid immunological tests. Speedy diagnosis is especially important for women in labor or near delivery, as a Caesarean delivery can prevent exposure of the infant to the virus.

In recent years several drugs have been used with varying degrees of success in treating HSV-1 and HSV-2 infections. Trifluorothymidine is effective in treating ocular herpes. Although acyclovir fails to prevent recurrences reliably, it does prevent the spread

CLOSE-UP

CORTISONE SUPPRESSION OF THE IMMUNE SYSTEM

Cortisone, which is used to treat many inflammatory lesions, should not be used to treat herpes lesions, especially ocular lesions. Because cortisone suppresses the immune system, it allows viral multiplication to increase and to cause more extensive cellular injury. The damage can cause perforation of the cornea.

of lesions, decrease viral shedding, and shorten healing time. Best results are obtained with acyclovir when it is used at the beginning of a primary infection. The use of acyclovir, Ara-A, and vidarabine in various combinations early in the infection has increased survival rates in herpes meningoencephalitis and neonatal herpes. The prognosis for alleviating genital herpes symptoms is reasonably good. Most patients have few if any recurrences if they take acyclovir daily but may have a recurrence if they stop taking it. No treatment can eradicate latent viruses.

IMMUNITY AND PREVENTION Although immunological knowledge of herpesviruses is sufficient to make a vaccine, such a vaccine is not yet available. Even if a vaccine were available and administered to all who lack HSV antibodies, it would take years to eradicate the disease. Large numbers of people already harbor latent viruses, so a vaccine to inactivate the viral genetic information responsible for *latency* is needed. A vaccine that elicits antibodies like those from natural infections would probably not be effective; such antibodies apparently have no effect on latent viruses and cannot alone prevent recurrences.

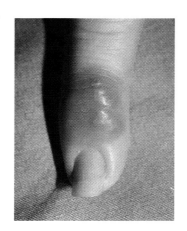

FIGURE 25.19 Herpes virus infections include herpetic whitlow, a very painful infection of the finger, which can be spread to other areas of the body, for example, by rubbing one's eyes. Health care workers should protect themselves by wearing latex gloves.

PUBLIC HEALTH

HERPES SINGLES CLUBS

Some herpes victims join "herpes singles clubs" in the hope of being able to enjoy sexual activity with other members without inflicting the disease on uninfected persons. Although limiting one's choice of partners in this way may prevent the spread of herpes infections to "outsiders," it can also give herpes sufferers a false sense of security. Because several different strains of both HSV-1 and HSV-2 exist, people can infect each other with different strains of HSV. Each new strain can cause a new primary infection and very painful lesions.

The best available means of preventing HSV infections is to avoid contact with individuals with HSV-1 or HSV-2 lesions. If all infected individuals were to refrain from sexual activity, especially when they have active lesions, some new cases of genital herpes could be prevented. If pregnant women would advise their obstetricians of herpes infections, their infants could be protected from exposure during delivery. Even these precautions will not prevent the spread of genital herpes. Most infected individuals shed viruses a few days before lesions appear, and some shed viruses continuously without having any lesions.

Genital Warts

Condylomas (kon-dil-o'mahz), or **genital warts,** caused by the human papillomavirus, most often occur in the sexually promiscuous young adult population. The incidence of genital warts has increased rapidly in recent years to the extent that such warts are now among the most prevalent STDs. Two-thirds of the sexual partners of infected individuals also develop warts. The warts can be papillary or flat. In males the warts appear on the penis, anus, and perineum; in females they are on the vagina, cervix, perineum, and anus.

Like dermal warts (Chapter 20), genital warts cause irritation and sometimes intense itching. They can persist or regress spontaneously. ∞ (p. 562) Genital warts often become infected with bacteria, and those that persist for many years can transform into malignant growths. Some victims suffer psychological damage from the presence of warts. Warts temporarily increase in number and size during pregnancy but decrease after delivery. Infants can be infected during delivery. The number of warts increases in immunosuppressed patients and in those with AIDS and other immunological deficiencies. (Treatment of warts was discussed in Chapter 20. ∞ (p. 563)

A special hazard of genital warts is their epidemiologic association with cervical carcinoma. Lesions previously diagnosed as cervical dysplasia (displa'se-ah), or abnormal cell growth, are now more often properly recognized as condylomas even though both can be found side by side on the cervix. Distinguishing these conditions from cervical carcinoma is even more difficult. Examining lesions for viral capsid antigens and viral nucleic acids can be helpful. Such examinations suggest that two strains of human papillomaviruses, HPV-8 and HPV-16, are most likely to be found in cervical carcinomas. The extent to which they contribute to such malignancies is not clear, but the virus has been found in 90 percent of patients with cervical carcinoma.

Laryngeal Papillomas

Laryngeal papillomas (pap-il-lo'maz) are benign growths that can be dangerous if they block the airway.

Hoarseness, voice changes, and respiratory distress occur when the airway becomes obstructed. Children are more likely to have laryngeal papillomas than adults. Surgical removal, sometimes every 2 to 4 weeks, is the only treatment for these obstructive growths. Also, there is danger of spreading the virus to the lungs during surgery. Laryngeal papillomas are usually caused by HPV-6 and HPV-11, which are thought to infect infants during birth to women with active genital warts.

Cytomegalovirus Infections

The **cytomegaloviruses** (CMVs) constitute a widespread and diverse group of herpesviruses. In general, each strain of CMV is capable of infecting a single species. The majority of human CMV infections occur in older children and adults and go unnoticed because they do not produce clinically apparent symptoms. An estimated 80 percent of U.S. adults carry the virus. When there are symptoms, they include malaise, myalgia, protracted fever, abnormal liver function, and lymph node inflammation without swelling. Symptoms are more severe in patients with AIDS and other immunodeficiencies.

Initially the virus can be recovered from the oropharynx; viremia with many virus-infected neutrophils can last for months. The viruses replicate and have low pathogenicity but are excreted intermittently over many months, during which time they can infect others. Viruses are shed in all body fluids—saliva, blood, semen, breast milk—but they are found most often and in the largest quantities in urine even a year or more after infection. In a symptomatic primary attack of CMV, cell-mediated immunity is depressed, and the ratio of T helper to T suppressor cells is reversed. However, the immune system returns to normal during convalescence. In subclinical cases cell-mediated immunity remains active. Large quantities of long-lasting antibodies are produced in response to CMV infection, but they do not prevent viral shedding. The virus is usually spread through long, close contact with children who are shedding the virus, but it can be spread by blood transfusions, organ transplants, and sexual intercourse.

FETAL AND INFANT CMV INFECTIONS In fetuses and infants, CMV infections can be life threatening because the virus disseminates widely to various organs. Fetuses become infected by viruses that cross the placenta from infected mothers. Unlike rubella infection, maternal CMV infections are rarely detected, so the risk to the fetus is unknown. Maternal antibodies also can cross the placenta and inactivate small quantities of virus. Maternal hormones suppress CMV, but their effect diminishes as the pregnancy progresses. Both fetuses and neonates are dependent on maternal defenses against CMV because their own immune systems are too immature to mount a successful defense. Tests for maternal antibody levels are available. Young children are the most likely source of infection to non-immune women.

In severe CMV infections in which the fetus has been infected with large numbers of viruses (about 4000 cases per year in the United States), intrauterine growth retardation and severe brain damage can occur (Figure 25.20). Many babies have CMV-infected cells in the inner ear and hearing loss due to nerve damage. Some have jaundice with liver damage, and others have impaired vision. Less severe infections (another 4500–6000 cases per year) cause damage to certain brain areas and mild central nervous system disorders with or without damage to hearing or sight. Mortality can be as high as 30 percent.

CMV infections contracted after birth generally cause fewer permanent defects than those contracted before birth. However, these infections can cause severe illness. Among babies infected with CMV, 5 percent have typical generalized *cytomegalic inclusion disease* (CID); another 5 percent have atypical, less generalized infections; the rest have subclinical but chronic infections. Many infants with CID infections have significant mental or sensory disorders. These include an abnormally small brain accompanied by intellectual deficits and inflammation of the eyes with impaired vision. In subclinical cases the prognosis is better, but 10 percent will have deafness or other sensory problems. Some will experience intellectual and behavioral disorders later. Babies who acquire CMV infections from blood transfusions have a gray pallor and symptoms like those of CID. Pneumonia, respiratory deterioration, and death may follow because the

infants receiving transfusions usually are premature and debilitated before transfusions are given. Moreover, CMV infections sometimes are not diagnosed in the presence of many other disorders. Finally, CMV sometimes causes pneumonia in infants 1 to 6 months of age in the presence of other infections, such as *Chlamydia trachomatis* and *Ureaplasma urealyticum*.

DISSEMINATED CMV In severe disseminated disease, the virus can be present in many organs. When the kidneys become infected, immune complexes are deposited in the glomeruli, but renal dysfunction is rare except in renal transplants. Subclinical hepatitis occurs in both children and adults. Lung infections are common in infants and immunosuppressed adults. Brain involvement is rare except in fetuses.

CMV is most virulent when present as a primary infection in immunosuppressed patients. Between 1 and 4 months after an organ transplant, the patient develops symptoms like those of infectious mononucleosis, with prolonged fever and enlargement of the liver and spleen. Pneumonia is frequent and severe in bone marrow transplant patients, and mortality can be as high as 40 percent. Hepatitis is usually reversible, but damage to the eyes is likely to be progressive and irreversible.

DIAGNOSIS, TREATMENT, AND PREVENTION Definitive diagnosis of CMV infections requires identification of the virus from clinical specimens, a procedure that can take up to 6 weeks. New, faster techniques use monoclonal antibodies to detect viral antigens. It is anticipated that nucleic acid probes, which make use of the base sequences unique to CMV nucleic acids, will allow identification of the virus in clinical specimens in less than 24 hours.

No effective treatment for CMV infections in infants is available. The prognosis is poor when infections occur in fetuses because the damage is done be-

FIGURE 25.20 Baby with birth defects due to congenitally acquired cytomegalovirus. The virus, sometimes called salivary gland virus, can cause grave damage to persons who have impaired immune systems.

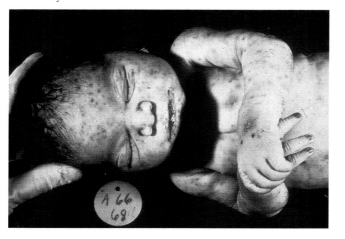

CLOSE-UP

CMV AND CANCER

Because CMV stimulates cellular RNA and DNA synthesis, it also may play a role in the development of cancer. Antigens and other CMV components have been found in human tumors including prostate and cervical cancers and Kaposi's sarcoma. (A newly discovered human herpesvirus is also suspected of playing a role in development of Kaposi's sarcoma.) Whether CMV contributes to the cancer or is merely a passenger in the malignant cells is unknown.

fore a diagnosis is made in the neonate. Interferon and hyperimmune gamma globulin given before and after organ transplantation reduce the incidence and severity of CMV infections in transplant patients. When blood transfusions are required, donor and recipient should be matched for CMV antigens to minimize introduction of CMV to which the recipient is not immune. Also, donors serologically negative for CMV should be used with infants, especially premature infants, to avoid the possibility of transmitting CMV through a transfusion. No effective vaccine exists.

The agents and characteristics of the diseases discussed in this chapter are summarized in Table 25.2.

TABLE 25.2	Summary of Urogenital and Sexually Transmitted Diseases	
Disease	**Agent(s)**	**Characteristics**
Bacterial urogenital diseases		
Urinary tract infections	*Escherichia coli, Proteus mirabilis,* and other bacterial species	Dysuria; sometimes leads to chronic cystitis; often ascend or descend in urinary tract
Prostatitis	*E. coli* and other bacterial species	Dysuria, urgent and frequent urination, low fever, back pain; can cause infertility
Pyelonephritis	*E. coli* and other bacterial species, sometimes the yeast *Candida*	Inflammation of pelvis of kidney, often caused by urinary tract blockage; dysuria, nocturia, sometimes fever
Glomerulonephritis	Streptococcal or viral infections from other sites	Deposition of immune complexes causes inflammation of glomeruli; can cause permanent kidney damage
Leptospirosis	*Leptospira interrogans*	Fever, nonspecific symptoms; can lead to Weil's syndrome with jaundice and liver damage
Bacterial vaginitis	*Gardnerella vaginalis* with anaerobes	Frothy, fishy-smelling discharge, pain and inflammation
Toxic shock syndrome	*Staphylococcus aureus*	Toxins reach blood and cause fever, rash, and shock that can lead to death
Parasitic urogenital disease		
Trichomoniasis	*Trichomonas vaginalis*	Intense itching, copious white discharge
Bacterial sexually transmitted diseases		
Gonorrhea	*Neisseria gonorrhoeae*	Infectious organisms release endotoxin that damages mucosa; pus-filled discharge; can cause PID and infect other systems
Syphilis	*Treponema pallidum*	Chancre develops in primary stage; mucous membrane lesions and rash occur in secondary stage; permanent cardiovascular and neurological damage often occur in tertiary stage
Chancroid	*Haemophilus ducreyi*	Painful, bleeding lesions on genitals; often enlarged lymphatic buboes
Nongonococcal urethritis	*Chlamydia trachomatis* and mycoplasmas	Scanty watery urethral discharge, inflammation, sometimes sterility; can cause neonatal infections and fetal death
Lymphogranuloma venereum	*Chlamydia trachomatis*	Genital lesions, fever, malaise, headache, nausea, vomiting, skin rash; lymph nodes become pus-filled buboes
Granuloma inguinale	*Calymmatobacterium granulomatis*	Painful ulcers on genitals and other sites; loss of skin pigmentation as ulcers heal
Viral sexually transmitted diseases		
Herpes simplex infections	Herpes simplex viruses	Fever blisters usually caused by HSV-1, genital herpes usually caused by HSV-2 (both are latent viruses); recurrent painful vesicular lesions, neonatal herpes and a variety of other manifestations
Genital warts	Human papillomaviruses	Warts on genitals, vagina, and cervix; irritation and sometimes intense itching; can contribute to cervical carcinoma
Cytomegalovirus infections	Cytomegaloviruses	Often asymptomatic but severe in fetuses, neonates, and immunodeficient patients; malaise, myalgia, fever, inflamed lymph nodes, neural damage and death in fetuses and neonates

HOMO SAPIENS AND TREPONEMA PALLIDUM–AN EVOLUTIONARY PARTNERSHIP

Historically, four different diseases—syphilis, pinta, yaws, and bejel—have been caused by the same pathogen, *Treponema pallidum*. DNA studies show the causative organisms to be extremely similar for all four diseases. Only recently has the causative agent of pinta evolved significantly from *T. pallidum*; it has been given a separate species name, *T. carateum*. Nevertheless, the organisms cause different diseases. Both C. J. Hackett and Ellis Hudson suggest that the treponemes have evolved in accordance with changing lifestyles of humans. The following discussion offers an explanation for that evolution.

The first pathogen infected some wild animal and occasionally infected people as a zoonosis. As such, it caused the set of symptoms known as *pinta*, a mild skin disease that does not reach deeper tissues. Its red, slate blue, and white lesions are disfiguring but not life threatening (Figure 25.21). Found in the tropics, it is passed from person to person through direct skin contact. Warm skin, moist with perspiration, provides an ideal environment for the treponeme.

Around 10,000 B.C., probably in Africa, the pathogen mutated and gave rise to *yaws*, a disease that thrived in the somewhat more densely settled human villages. This disease still has its highest incidences in tropical Africa and in Latin America. Typical symptoms are deep oozing skin lesions that can penetrate underlying tissues and erode bone but that do not reach the viscera (Figure 25.22). Most victims are children 4 to 10 years old.

By 7000 B.C. people carrying the treponeme moved northward. The climate was cooler and drier, required more clothing, and made transmission via skin contact more difficult. The pathogen survived best in warm, moist areas such as the mouth, groin, and armpits. Transfer to a new host was most likely via oral contact such as kissing or by sharing bits of food or dishes. Today the disease is known as nonvenereal syphilis, or *bejel*. The strain responsible for bejel is still more invasive; it causes granulomatous lesions like those of syphilis and attacks tissues of the skeletal, cardiovascular, and nervous systems. It is often transferred by young children, because older ones are generally past the infectious stage.

Bejel is almost always acquired prior to puberty and is not sexually transmitted. It is never transmitted congenitally, presumably because the mother's infection becomes nontransmissible before she is old enough to become pregnant. Bejel epidemics occurred during the fifteenth century in Africa and in the seventeenth and eighteenth centuries in Scotland and Norway. Today the worldwide incidence of bejel is higher than that of venereal syphilis.

As standards of living improved and people covered more of their bodies, the surest means of transmitting the treponeme was via sexual activity and warm, moist reproductive tracts. The pathogen, again evolving in accordance with changing human lifestyles, came to cause venereal syphilis, in which the primary chancre appears to be analogous to yaws lesions. The disease is more severe than bejel because it penetrates organs to a greater and more devastating extent.

There is further evidence for the evolution of treponemes with human culture. First, none of the four diseases can be found in the same geographic area as any other. Second, infection with yaws prevents concurrent syphilis. Third, people with yaws who move to cooler climates lose their yaws symptoms and develop those of bejel instead. Evidently, the treponeme is extremely adaptive to its environment. Indeed, syphilis itself seems to have evolved considerably in the past five centuries. Accounts that date from the 1490s describe the disease as

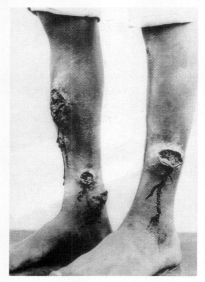

FIGURE 25.22 Lesions and bone damage caused by yaws. The bone damage can be similar to that of saber shin produced by syphilis (Figure 25.12b).

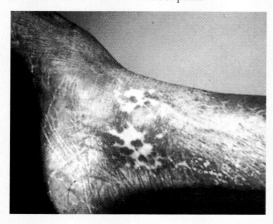

FIGURE 25.21 The rash of pinta.

killing its victims in a matter of weeks. Today it is far less virulent; the incubation period alone is longer than that, and people with syphilis may survive for decades.

If the theory of parallel evolution of pathogen and humans is correct, historians who blamed Columbus's sailors for bringing the organism back from the New World were mistaken. The fact that the disease appeared in Europe at about the same time as Columbus's voyages is probably just a coincidence. The treponeme most likely arrived in Europe with slaves from Africa before the discovery of America, eventually evolving to the point at which it could cause venereal syphilis. Evidence for this hypothesis has been found in Native-American burial grounds in New England. The remains of individuals who died before the arrival of settlers from the Old World show no trace of syphilis. Signs of syphilis appear first in the skeletons of young native women dating from the time of early European settlement.

CHAPTER SUMMARY

- The agents and characteristics of the **sexually transmitted diseases** (STDs) discussed in this chapter are summarized in Table 25.2. Information in that table is not repeated in this summary.

UROGENITAL DISEASES USUALLY NOT TRANSMITTED SEXUALLY

Bacterial Urogenital Diseases

- **Urinary tract infections** (UTIs), which are exceedingly common, can ascend or descend to spread through the urogenital system. They are diagnosed from urine cultures, treated with various antibiotics, and can be prevented by good personal hygiene and complete emptying of the bladder.
- **Prostatitis** can result from the spread of a urinary tract infection and usually can be treated with antibiotics.
- **Pyelonephritis** also can result from the spread of a urinary tract infection; it can be difficult to treat because renal failure allows drugs to accumulate to toxic levels.
- **Glomerulonephritis** usually results from throat and other infections in which immune complexes deposit in glomeruli. Prompt treatment of other infections minimizes the risk of glomerulonephritis.
- **Leptospirosis** is acquired through contact with contaminated urine and is diagnosed by microscopic examination of blood. It can be treated with antibiotics and prevented by vaccinating pets, which are usually responsible for transmitting the disease to humans.
- Bacterial and other forms of **vaginitis** occur when normal vaginal microbiota are disturbed. *Gardnerella vaginitis* is diagnosed by the presence of clue cells. It is most often treated with metronidazole to eradicate anaerobes and allow restoration of normal microbiota.
- **Toxic shock syndrome** most often arises from the use of superabsorbent tampons. It should be treated promptly with nafcillin and can be prevented by avoiding such tampons.

Parasitic Urogenital Diseases

- **Trichomoniasis** usually is transmitted sexually, is diagnosed from smears of secretions, and is treated with metronidazole.

SEXUALLY TRANSMITTED DISEASES

Bacterial Sexually Transmitted Diseases

- **Gonorrhea** is diagnosed by finding organisms in cultures and is treated with penicillin. With penicillin-resistant organisms, another antibiotic is used.
- **Syphilis** is diagnosed by immunological tests and is treated with tetracycline or erythromycin.
- **Chancroid** occurs mainly in developing countries. It is diagnosed by finding organisms from lesions or buboes and is treated with tetracycline and other antibiotics.
- The incidence of **nongonococcal urethritis** is increasing dramatically. It can be diagnosed by culturing samples from discharges and is treated with erythromycin, tetracycline, or other antibiotics, depending on the susceptibility of the causative agent.
- **Lymphogranuloma venereum** occurs mainly in tropical and subtropical regions. It is diagnosed by finding chlamydias as inclusions in pus and is treated with tetracycline or another antibiotic.
- **Granuloma inguinale** is common in India and several other countries. It is diagnosed by finding **Donovan bodies** in scrapings from lesions and can be treated with a variety of antibiotics.
- All the preceding diseases are usually transmitted by sexual contact and can be prevented by avoiding such contacts. No vaccines are available.

Viral Sexually Transmitted Diseases

- **Genital herpes** is the most severe of herpes simplex virus infections. It is diagnosed by finding viruses in vesicular fluids or by immunological tests and can be treated, but not cured, with acyclovir and other antiviral agents.
- **Genital warts** must be distinguished from cervical dysplasia and carcinoma; treatment of warts was discussed in Chapter 20.
- **Cytomegalovirus** infections are of greatest significance in fetuses, newborns, and the immunosuppressed. Rapid diagnosis can be made by monoclonal antibody tests, but no effective treatment is available.

26

ENVIRONMENTAL MICROBIOLOGY

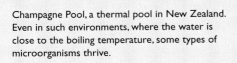

Champagne Pool, a thermal pool in New Zealand. Even in such environments, where the water is close to the boiling temperature, some types of microorganisms thrive.

Microorganisms are found in every environment—in the air we breathe, the food we eat, the soil where food is grown, and the water we drink. Many microbes benefit humans, whereas only a few cause human disease. One might consider environmental microbiology to be of little significance in the health sciences, but that is not the case. Humans, being one of many organisms in an environment, affect and are affected by both living and nonliving components of an environment, including its microorganisms. To control disease, health scientists need to know how to control microorganisms in air, food, soil, and water. To do that they need to understand the roles of microorganisms in the environment.

FUNDAMENTALS OF ECOLOGY

Ecology is the study of the relationships among organisms and their environment. These relationships include interactions of organisms with physical features—the **abiotic factors**—of the environment and interactions of organisms with one another—the **biotic factors**—of the environment. An **ecosystem** comprises all the organisms in a given area together with the surrounding abiotic and biotic factors.

The Nature of Ecosystems

Ecosystems are organized into various biological levels. The **biosphere** is the region of the earth inhabited by living organisms. It consists of the *hydrosphere* (earth's water supply), the *lithosphere* (the soil and rock that include the earth's crust), and the *atmosphere* (the gaseous envelope surrounding earth). A wide diversity of organisms exists within the biosphere. A terrestrial ecosystem, such as a desert, tundra, grassland, or tropical rain forest, is characterized by a particular climate, soil type, and organisms. The hydrosphere is divided into freshwater and marine ecosystems.

The organisms within an ecosystem live in communities. An ecological **community** consists of all the kinds of organisms that are present in a given environment. Microorganisms can be categorized as indigenous or nonindigenous to an environment. **Indigenous,** or *native,* **organisms** are always found in a given environment. They generally are able to adapt to normal seasonal changes or changes in the quantity of available nutrients in the environment. For example, *Spirillum volutans* is indigenous to stagnant water, various species of *Streptomycetes* are indigenous to soil, and *Escherichia coli* is indigenous to the human digestive tract. Regardless of variations in the environment (except for cataclysmic changes), an environment will always support the life of its indigenous organisms. **Nonindigenous organisms** are temporary inhabitants of an environment. They become numerous when growth conditions are favorable for them and disappear when conditions become unfavorable.

Communities are made up of *populations,* groups of organisms of the same species. In general, communities composed of many populations of organisms are more stable than those composed of only a few populations—that is, of only a few different species. The various species create a system of "checks and balances" such that the numbers of each species remain relatively constant.

The basic unit of the population is the individual organism. Organisms occupy a particular habitat and niche. The *habitat* is the physical location of the organism. Microorganisms are so small that they often occupy a **microenvironment,** a habitat in which the oxygen, nutrients, and light are stable, including the environment immediately surrounding the microbe. A particle of soil could be the microenvironment of a bacterium. This environment is more important to the bacterium than the more extensive *macroenvironment*. An organism's *niche* is the role it plays in the ecosystem—that is, its use of biotic and abiotic factors in its environment. Microbes may be *producers, consumers,* or *decomposers.* We will discuss each of these groups in the next section.

Energy Flow in Ecosystems

Energy is essential to life, and radiant energy from the sun is the ultimate source of energy for nearly all organisms in any ecosystem. (The chemolithotrophic bacteria that extract energy from inorganic compounds are an exception.) ∞ (Chapter 5, p. 129) Organisms called **producers** (autotrophs) capture energy from the sun. They use this energy and various nutrients from soil or water to synthesize the substances they need to grow and to support their other activities. Energy stored in the bodies of producers is transferred through an ecosystem when **consumers** (heterotrophs) obtain nutrients by eating the producers or other consumers. **Decomposers** obtain energy by digesting dead bodies or wastes of producers and consumers. The decom-

CLOSE-UP

BACTERIA MAKE THE WORLD GO 'ROUND

The bacteria particularly . . . are still more important than we. Omnipresent in infinite varieties, . . . they release the carbon and nitrogen held in the dead bodies of plants and animals which would—without bacteria and yeasts—remain locked up forever in useless combinations, removed forever as further sources of energy and synthesis. Incessantly busy in swamp and field, these minute benefactors release the frozen elements and return them to the common stock, so that they may pass through other cycles as parts of other living bodies. . . . Without the bacteria to maintain the continuities of the cycles of carbon and nitrogen between plants and animals, all life would eventually cease. . . . Without them, the physical world would become a storehouse of well-preserved specimens of its past flora and fauna . . . useless for the nourishment of the bodies of posterity. . . .

—Hans Zinsser, 1935

posers release substances that producers can use as nutrients. The flow of energy and nutrients in an ecosystem is summarized in Figure 26.1.

Microorganisms can be producers, consumers, or decomposers in ecosystems. Producers include photosynthetic organisms among bacteria, cyanobacteria, protists, and eukaryotic algae. Although green plants are the primary producers on land, microorganisms fill this role in the ocean. Consumers include heterotrophic bacteria, protists, and microscopic fungi. (To the extent that viruses divert a cell's energy to the synthesis of new viruses, they too act as consumers.) Many microorganisms act as decomposers. In fact, they play a greater role in the decomposition of organic substances than larger organisms do.

BIOGEOCHEMICAL CYCLES

As they carry on essential life processes, living organisms incorporate water molecules and carbon, nitrogen, and other elements from their environment into their bodies. Without decomposers to ensure the flow of nutrients through ecosystems, much matter would

FIGURE 26.1 The flow of energy (E) and nutrients (N) in ecosystems. Energy flows through the system (it is obtained continuously from the sun), whereas nutrients in the environment must be recycled for new life to continue.

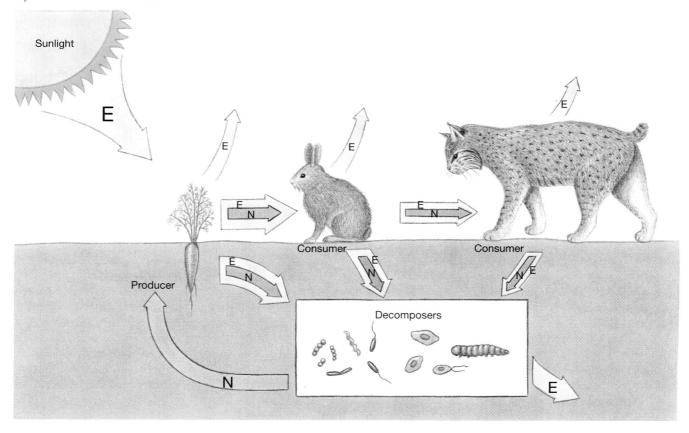

soon be incorporated into bodies and wastes, and life would soon become extinct. Although the supply of energy from sunlight is continuously renewed, the supply of water and the chemical elements that serve as nutrients is fixed. These materials must be recycled continuously to make them available to living organisms. The mechanisms by which such recycling occurs are referred to collectively as **biogeochemical cycles;** *bio* refers to living things, and *geo* refers to the earth, the environment of the living things.

The Water Cycle

The **water cycle,** or **hydrologic cycle** (Figure 26.2), recycles water. Water reaches the earth's surface as precipitation from the atmosphere. It enters living organisms during photosynthesis and by ingestion. It leaves them as a byproduct of respiration, in wastes, and by evaporation from the surfaces of living things, such as by *transpiration* (water loss from pores in plant leaves). Like all other living things, microorganisms use water in metabolism, but they also live in water or very moist environments. Many form spores or cysts that help them survive periods of drought, but vegetative cells must have water.

The Carbon Cycle

In the **carbon cycle** (Figure 26.3), carbon from atmospheric carbon dioxide (CO_2) enters producers during photosynthesis or chemosynthesis. Consumers obtain carbon compounds by eating producers, other consumers, or the remains of either. Carbon dioxide is returned to the atmosphere by respiration and by the actions of decomposers on the dead bodies and wastes of other organisms. Carbon compounds can be deposited in peat, coal, and oil and released from them during burning. A small but significant quantity of carbon dioxide in the atmosphere comes from volcanic activity and from the weathering of rocks, many of which contain the carbonate ion, CO_3^{2-}. The oceans and carbonate rocks are the largest reservoirs of carbon, but recycling of carbon through these reservoirs is very slow.

As we have seen in earlier chapters, all microorganisms require some carbon source to maintain life. Most carbon entering living things comes from carbon dioxide dissolved in bodies of water or in the atmosphere. Even the carbon in sugars and starches ingested by consumers is derived from carbon dioxide. Because the atmosphere contains only a limited quantity of car-

FIGURE 26.2 The water cycle.

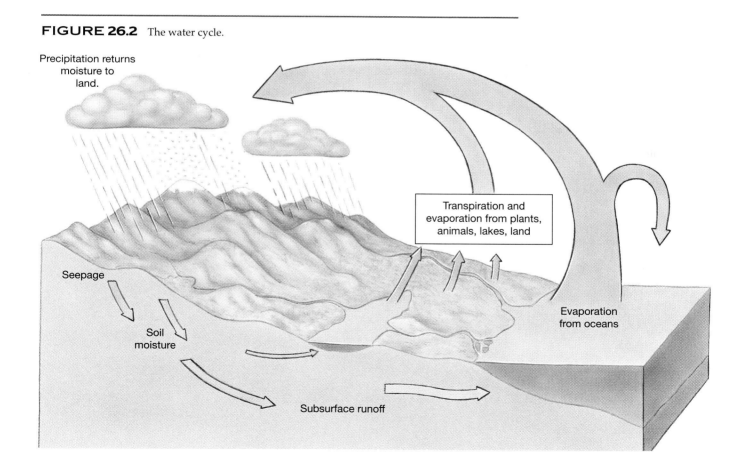

Precipitation returns moisture to land.

Seepage

Soil moisture

Subsurface runoff

Transpiration and evaporation from plants, animals, lakes, land

Evaporation from oceans

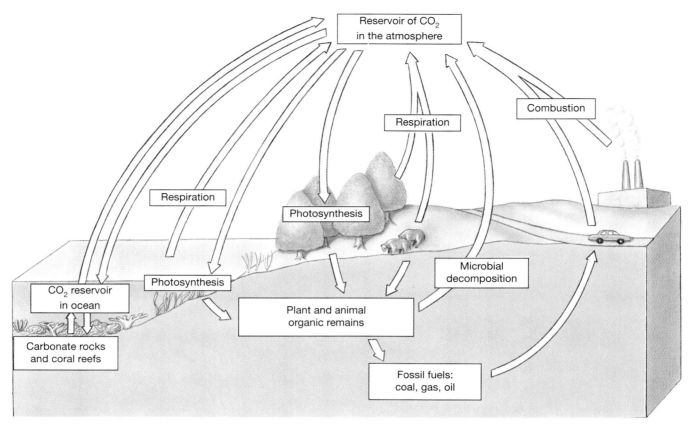

FIGURE 26.3 The carbon cycle.

THE GREENHOUSE EFFECT

Atmospheric carbon dioxide and water vapor form a blanket over the earth's surface, creating a "greenhouse effect." These gases allow the sun's radiation to penetrate the atmosphere, thereby reaching the earth's surface and warming it and the atmosphere. But they trap much of the infrared (heat) radiation produced by the warm surface, reflecting it back to earth. Solar energy is thus captured within the "greenhouse." The overall effect is to reduce temperature variation and to raise the temperature near the earth's surface. Over the past century, human activities have released relatively large quantities of carbon dioxide from coal and oil combustion. In fact, the concentration of CO_2 in the atmosphere has increased by more than 10 percent since 1958. Yet, forests, which absorb carbon dioxide and replenish the earth's oxygen through photosynthesis, have been cut down and burned at an ever-increasing rate. Some scientists think that these processes are causing a general warming trend, increasing average temperatures around the world and possibly changing the balance of organisms in ecosystems. This trend would make now-temperate regions too warm to grow wheat and other food crops, create droughts in other areas,

and make new deserts. It might even melt polar ice, raising the level of oceans and flooding coastal cities.

Global warming might lead to an increased incidence of infectious diseases as range of the vectors (mosquitos, snails, flies, and such) carrying the diseases expand. Several scenarios have been developed by computer-based modeling. According to one model, a global temperature rise of just 3°C over the next century could result in 50 million to 80 million new malaria cases per year. Other diseases, including schistosomiasis, African trypanosomiasis, and dengue and yellow fevers, are also likely to spread with global warming.

Marine photosynthetic plankton (*phytoplankton*) utilize CO_2 and are thought to affect Earth's temperature. (See the box "Microbial Determinants of Weather," p. 743.) In June 1995 researchers reported that iron added in regular doses to part of the Pacific Ocean promoted extensive phytoplankton growth. The plankton absorbed huge amounts of CO_2 from the air. Scientists are quick to warn, however, against the prospect of seeding the oceans with iron to reduce global CO_2 levels. Iron inputs would alter the food chain and could increase levels of other greenhouse gases.

bon dioxide (0.03 percent), recycling is essential for maintaining a continuous supply of atmospheric carbon dioxide.

The Nitrogen Cycle and Nitrogen Bacteria

In the **nitrogen cycle** (Figure 26.4), nitrogen moves from the atmosphere through various organisms and back into the atmosphere. This cyclic flow depends not only on decomposers but also on various nitrogen bacteria.

Decomposers use several enzymes to break down proteins in dead organisms and their wastes, releasing nitrogen in much the same way they release carbon. Proteinases break large protein molecules into smaller molecules. Peptidases break peptide bonds to release amino acids. Deaminases remove amino groups from amino acids and release ammonia. Eventually, free nitrogen gas finds its way back into the atmosphere. Many soil microorganisms produce one or more of these enzymes. Clostridia, actinomycetes, and many fungi produce extracellular proteinases that initiate protein decomposition.

Nitrogen bacteria fall into one of three categories according to the roles they play in the nitrogen cycle: 1) nitrogen-fixing bacteria, 2) nitrifying bacteria, and 3) denitrifying bacteria.

Nitrogen-Fixing Bacteria

Nitrogen fixation is the reduction of atmospheric nitrogen gas (N_2) to ammonia (NH_3). Organisms that can fix nitrogen are essential for maintaining a supply of physiologically usable nitrogen on earth. About 255 million metric tons of nitrogen is fixed annually—70 percent of it by nitrogen-fixing bacteria. Bacteria and cyanobacteria fix nitrogen in many different environments—from Antarctica to hot springs, acid bogs to salt flats, flooded lands to deserts, in marine and fresh water, and even in the gut of some organisms.

The energy for nitrogen fixation can come from fermentation, aerobic respiration, or photosynthesis. The

FIGURE 26.4 The nitrogen cycle.

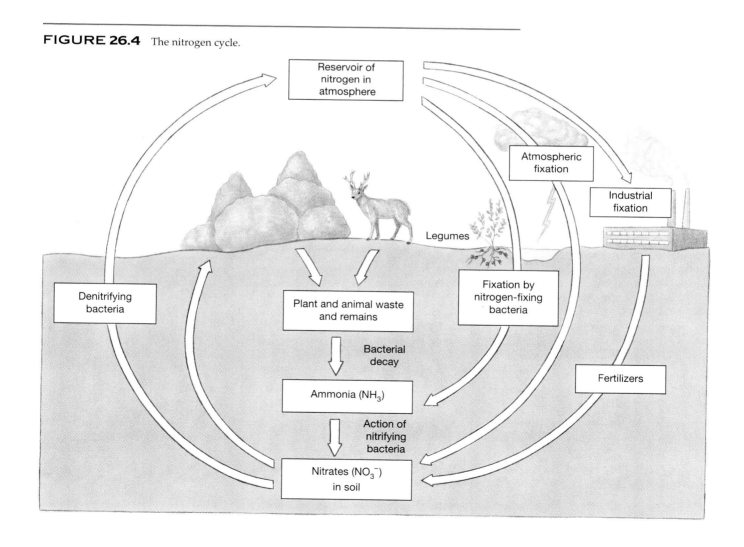

various organisms that fix nitrogen live independently, in loose associations, or in intimate symbiosis. Regardless of the environment or the associations of the organisms, nitrogen-fixing bacteria must have a functional nitrogen-fixing enzyme called **nitrogenase,** a reducing agent that supplies hydrogen as well as energy from ATP. In aerobic environments, nitrogen fixers must also have a mechanism to protect the oxygen-sensitive nitrogenase from inactivation.

Free-living aerobic nitrogen fixers include several species of the genus *Azotobacter* and some methylotrophic bacteria. Cyanobacteria are also free-living nitrogen fixers. *Azotobacter* is found in soils and is a versatile heterotroph whose growth is limited by the amount of organic carbon available. *Methylotrophic bacteria* can fix nitrogen when provided with methane, methanol, or hydrogen from various substrates. Cyanobacteria can fix nitrogen by using hydrogen from hydrogen sulfide, so they increase the availability of nitrogen in sulfurous environments.

Nitrogen-fixing facultative anaerobes include species of *Klebsiella, Enterobacter, Citrobacter,* and *Bacillus.* In addition, a number of obligate anaerobes, including photosynthetic Rhodospirillaceae and bacteria of the genera *Clostridium, Desulfovibrio,* and *Desulfotomaculum,* also fix nitrogen. Several species of *Klebsiella* capable of fixing nitrogen are found in *rhizomes* (subsurface stems) of legumes such as peas and beans and in the intestines of humans and other animals. Nitrogen fixation has been observed in about 12 percent of *Klebsiella pneumoniae* organisms from patients. Various species of *Clostridium* are found in soils and muds. They use a variety of organic substances for energy and withstand unfavorable conditions as spores. They tolerate a range of pH from 4.5 to 8.5 but fix nitrogen best at pH 5.5 to 6.5. *Desulfovibrio* and

Desulfotomaculum, anaerobic sulfate reducers that live in mud and soil sediments, fix nitrogen at pH 7 to 8.

Some nitrogen fixers are found in symbiotic association with other organisms, which provide them with organic carbon sources. For example, the cyanobacterium *Anabaena* (Figure 26.5a) is found in pores of the leaves of *Azolla,* a small water fern found in many parts of the world. The nitrogen-fixing *Anabaena* supplies the necessary nitrogen; the fern supplies substrates for energy capture in ATP and reductants for nitrogen fixation. Together, they fix 100 kg of nitrogen per hectare (about two football fields in area) per year and are used as "green manure" in rice cultivation in southeast Asia.

Rhizobium is the primary symbiotic nitrogen fixer. It lives in *nodules,* growths extending from the roots of certain plants, usually legumes (Figure 26.5b and c). In this symbiotic relationship, the plant benefits by receiving nitrogen in a usable form, and the bacteria benefit by receiving nutrients needed for growth. When paired with legumes, *Rhizobium* can fix 150 to 200 kg of nitrogen per hectare of land per year. In the absence of legumes, the bacteria fix only about 3.5 kg of nitrogen per hectare per year—less than 2 percent of that fixed in the symbiotic relationship. Farmers often mix nitrogen-fixing bacteria with seed peas and beans before planting to ensure that nitrogen fixation will be adequate for their crops to thrive.

The mechanism by which *Rhizobium* establishes a symbiotic relationship with a legume has been the subject of many research studies. *Rhizobium* multiplies in the vicinity of legume roots, probably under the influence of root secretions. As the rhizobia increase in numbers, they release enzymes that digest cellulose and the substance that cements cellulose fibers together in the root cell walls. The rhizobia then change from their free-living rod shape and become spherical, flagellated

---

FIGURE 26.5   (a) SEM photo of the beadlike, filamentous *Anabaena azollae,* a nitrogen-fixing cyanobacterium that lives in a mutualistic relationship with the water fern *Azolla.* (b) Nodules on the roots of a bean plant, resulting from invasion of the plant by nitrogen-fixing bacteria. (c) A cross section through a root of a leguminous plant shows densely packed *Rhizobium* bacteria inside the nodule.

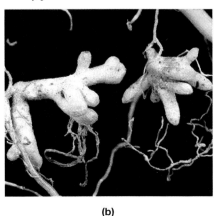
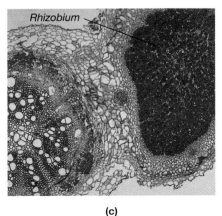

(a) (b) (c)

TRY IT

NITROGEN-FIXING BACTERIA IN ROOT NODULES

It's easy to find and visualize nitrogen-fixing bacteria in root nodules. To do this, first dig up the root system from a leguminous plant such as beans or peas. Carefully wash away the dirt to expose the nodules. Crush a nodule on a clean glass slide. Add a small drop of water, and use the nodule pieces or an inoculating loop to spread the fluid out into a thin film. Remove the nodule pieces. Allow the slide to air-dry, and then quickly pass it through a Bunsen burner flame three or four times to heat-fix the organisms. Flood it with methylene blue stain for 1 minute. Rinse the slide with water, dry it, and examine it under the oil immersion lens. The bacilli you see are likely to belong to the genus *Rhizobium*.

cells called **swarmer cells.** These cells are thought to produce indoleacetic acid, a plant growth hormone that causes curling of root hairs. The swarmer cells then invade the root hairs and form hyphalike networks, killing some root cells and proliferating in others. Swarmer cells become large, irregularly shaped cells called **bacteroids,** which are tightly packed into root cells, probably under the influence of chemical substances in the plant cells. Accumulations of bacteroids in adjacent root cells form nodules on the plant roots.

Bacteroids contain the enzyme nitrogenase, which catalyzes the following reaction:

$$N_2 \ + \ 3\,H_2 \ \xrightarrow{\textit{Rhizobium}} \ 2\,NH_3$$

nitrogen gas   hydrogen gas   ammonia

However, this enzyme is inactivated by oxygen, so nitrogen fixation can occur only when oxygen is prevented from reaching the enzyme. Nitrogenase is protected from oxygen by a kind of hemoglobin, a red pigment that binds oxygen. This particular hemoglobin is synthesized only in root nodules containing bacteroids, because part of the genetic information for its synthesis is in the bacteroids and part is in the plant cells. Synthesis of nitrogenase is repressed in the presence of excess ammonium (NH_4^+) and derepressed in the presence of free nitrogen. Thus, fixation occurs only when "fixed" nitrogen is in short supply and free nitrogen is available.

Rhizobium species vary in their capacity to invade particular legumes and in their capacity to fix nitrogen once they have invaded. Some species cannot invade any legumes, whereas others invade only certain legumes. This invasive specificity is determined genetically (probably by a single gene or a group of closely related genes) and can be altered by genetic transformation. ∞ (Chapter 8, p. 197) Such a transformation might enable a species of *Rhizobium* to invade a group of legumes it previously could not colonize.

Although symbiotic nitrogen fixation occurs mainly in association between rhizobia and legumes, other such associations are known. The alder tree, which grows in soils low in nitrogen, has root nodules similar to those formed by rhizobia. These nodules contain nitrogen-fixing actinomycetes of the genus *Frankia*.

Nitrifying Bacteria

Nitrification is the process by which ammonia or ammonium ions are oxidized to nitrites or nitrates. Carried out by autotrophic bacteria, nitrification is an important part of the nitrogen cycle because it supplies plants with *nitrate* (NO_3^-), the form of nitrogen most usable in plant metabolism. Nitrification occurs in two steps. In each step, nitrogen is oxidized and energy is captured by the bacteria that carry out the reaction. Various species of *Nitrosomonas* (Figure 26.6) and related genera, gram-negative rod-shaped bacteria, produce *nitrite* (NO_2^-), the reduced form of nitrates:

$$NH_4^+ \ + \ 1\tfrac{1}{2}O_2 \ \xrightarrow{\textit{Nitrosomonas}}$$

ammonium ion   oxygen gas

$$NO_2^- \ + \ H_2O \ + \ 2\,H^+ \ + \ energy$$

nitrite   water   hydrogen ions

FIGURE 26.6 TEM photo (magnified 17,750×) of a microcolony of the nitrifying bacterium *Nitrosomonas*.

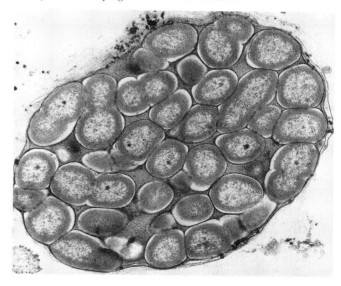

Species of *Nitrobacter* and related genera, also gram-negative rods, produce nitrates:

$$NO_2^- + \tfrac{1}{2}O_2 \xrightarrow{\text{\textit{Nitrobacter}}} NO_3^- + \text{energy}$$

nitrite   oxygen   nitrate

These bacteria use energy derived from the preceding reactions to reduce carbon dioxide in autotrophic metabolism. Because oxygen is required for the nitrification reactions, the reactions occur only in oxygenated water and soils. Furthermore, because nitrite is toxic to plants, it is essential that these reactions be carried out in sequence to provide nitrates and to prevent excessive accumulation of nitrites in soil.

Denitrifying Bacteria

Denitrification is the process by which nitrates are reduced to nitrous oxide (N_2O) or nitrogen gas (unbalanced equation):

$$NO_3^- \longrightarrow NO_2^- \longrightarrow N_2O \longrightarrow N_2$$

nitrate   nitrite   nitrous   nitrogen
                          oxide   gas

Although this process does not occur to any significant degree in well-oxygenated soils, it does occur in oxygen-depleted, waterlogged soils. Most denitrification is performed by *Pseudomonas* species, but it can also be accomplished by *Thiobacillus denitrificans, Micrococcus denitrificans,* and several species of *Serratia* and *Achromobacter.* These bacteria, although usually aerobic, use nitrate instead of oxygen as a hydrogen acceptor under anaerobic conditions. Another process that reduces soil nitrate is the reduction of nitrate to ammonia. Several anaerobic bacteria carry out this process (called *dissimilative nitrate reduction*) in a complex reaction that can be summarized as follows (unbalanced equation):

$$NO_3^- \longrightarrow H_2 \longrightarrow NH_3 \longrightarrow N_2O$$

nitrate   hydrogen   ammonia   nitrous
            gas   oxide

Although the latter process reduces the quantity of available nitrate, it retains the nitrogen in the soil in other forms.

Denitrification is a wasteful process because it removes nitrates from the soil and interferes with plant growth. It is responsible for significant losses of nitrogen from fertilizer applied to soils. Another unfortunate effect of denitrification is the production of nitrous oxide, which is converted to nitric oxide (NO) in the atmosphere. Nitric oxide, in turn, reacts with ozone (O_3) in the upper atmosphere. Ozone provides a barrier between living things on earth and the sun's ultraviolet

radiation. If enough ozone is destroyed, it no longer serves as an effective screen, and living things can be exposed to excessive ultraviolet radiation, which can cause cancer and mutations. ∞ (Chapter 7, p. 184)

The Sulfur Cycle and Sulfur Bacteria

The **sulfur cycle** (Figure 26.7), which involves the movement of sulfur through an ecosystem, resembles the nitrogen cycle in several respects. Sulfhydryl (—SH) groups in proteins of dead organisms are converted to hydrogen sulfide (H_2S) by a variety of microorganisms. This process is analogous to the release of ammonia from proteins in the nitrogen cycle. Hydrogen sulfide is toxic to living things and thus must be oxidized rapidly. Oxidation to elemental sulfur is followed by oxidation to sulfate (SO_4^{2-}), which is the form of sulfur most usable by both microorganisms and plants. This process is analogous to nitrification. The sulfur cycle is of special importance in aquatic environments, where sulfate is a common ion, especially in ocean water.

The various sulfur bacteria can be categorized according to their roles in the sulfur cycle. These roles include sulfate reduction, sulfur reduction, and sulfur oxidation.

Sulfate-Reducing Bacteria

Sulfate reduction is the reduction of sulfate (SO_4^{2-}) to hydrogen sulfide (H_2S). The sulfate-reducing bacteria are among the oldest life forms, probably more than 3 billion years old. They include the closely related genera *Desulfovibrio, Desulfomonas,* and *Desulfotomaculum.* In these bacteria, sulfate is the final electron acceptor in anaerobic oxidation, as oxygen is the final electron acceptor in aerobic oxidation. By reducing sulfate, these bacteria produce large amounts of hydrogen sulfide. However, for this process to occur, energy from ATP is required to phosphorylate the sulfate and convert it to ADP—SO_4. ADP—SO_4 then can act as an electron acceptor and compete successfully for substrates.

Sulfate-reducing bacteria (Figure 26.8) are strict anaerobes. They are widely distributed and predominate in nearly all anaerobic environments. As discussed in Chapter 6, sulfate-reducing bacteria can be psychrophilic, mesophilic, thermophilic, or halophilic. ∞ (p. 145) The variety of organic carbon sources they can metabolize is limited. Most use lactate, pyruvate, fumarate, malate, or ethanol, and some can use glucose and citrate. End products of such metabolism are typically acetate and carbon dioxide. Some sulfate reducers use products derived from the anaerobic degradation of plant material by other organisms. *Desulfovibrio, Desulfomonas,* and *Desulfotomaculum* oxidize fatty acids and a variety of other organic acids.

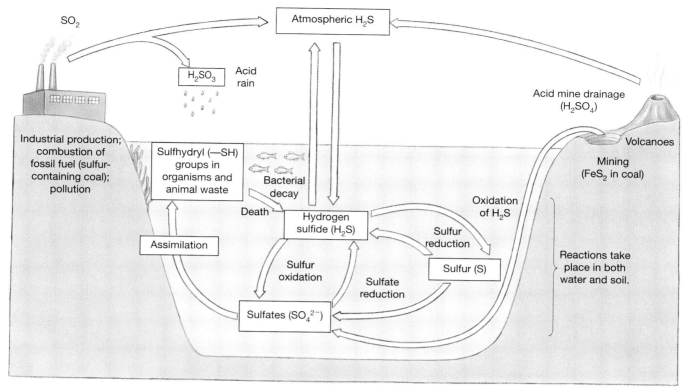

FIGURE 26.7 The sulfur cycle.

Sulfur-Reducing Bacteria

Sulfur reduction is the reduction of sulfur to hydrogen sulfide. Like sulfate-reducing bacteria, sulfur-reducing bacteria are anaerobes. They can use intracellular or extracellular sulfur as an electron acceptor in fermentation. The sulfur can be in elemental form or in disulfide bonds of organic molecules. This process provides energy for the organisms when sunlight is not available for photosynthesis.

FIGURE 26.8 The sulfate-reducing bacterium *Desulfotomaculum acetoxidans,* showing spherical spores and gas vacuoles (phase contrast, magnified 5000×).

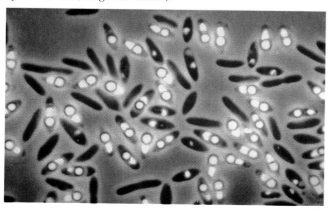

Sulfur-Oxidizing Bacteria

Sulfur oxidation is the oxidation of various forms of sulfur to sulfate. *Thiobacillus* and similar bacteria oxidize hydrogen sulfide, ferrous sulfide, or elemental sulfur to sulfuric acid (H_2SO_4). When this acid ionizes, it greatly decreases the pH of the environment, sometimes lowering the pH to 1 or 2. Sulfur-oxidizing organisms are responsible for oxidizing ferrous sulfide in coal-mining wastes, and the acid they produce is extremely toxic to fish and other organisms in streams fed by such wastes.

Other Biogeochemical Cycles

In addition to water, carbon, nitrogen, and sulfur—whose cycles we have examined—phosphorus and other elements also move through ecosystems cyclically. Any element (including trace elements) that appears in the cells of living organisms must be recycled to extract it from dead organisms and make it available to living ones. We conclude our discussion of biogeochemical cycles with a brief description of the phosphorus cycle.

The **phosphorus cycle** (Figure 26.9) involves the movement of phosphorus among inorganic and organic forms. Soil microorganisms are active in the phosphorus cycle in at least two important ways:

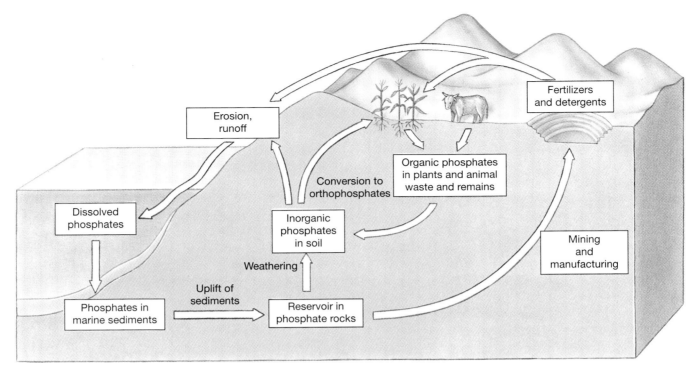

FIGURE 26.9 The phosphorus cycle.

CLOSE-UP

INDICATORS OF A HEALTHY ENVIRONMENT

At the 1995 annual meeting of the North American Benthological Society, researchers discussed new techniques for determining the integrity and stability of ecological environments. Many of these researchers use bacterial diversity as a measure of the health of an ecosystem.

Bacteria (and fungi) regulate most ecosystem dynamics—such as the biogeochemical cycles—by producing materials that are necessary for the growth of other organisms in a specific environment. Therefore, if a habitat is found to contain a healthy and diverse community of microbes, researchers can say that it is, in fact, a healthy environment. If there is instead a lack of microbial diversity, then nutrients and other abiotic factors are insufficient, and the ecosystem can be said to have limited functions. In such cases, waste products are not broken down or recycled efficiently.

J. Vaun McArthur, an ecologist at the Savannah River Ecology Laboratory in Georgia, and his colleagues are using techniques of molecular biology to learn more about the diversity of microbial communities that inhabit the streams and swamps of the 310-mile² Savannah River Site near Aiken, South Carolina. These ecologists would like to determine, for example, the distribution and abundance of

bacteria along any stream in the world. With new molecular techniques, the researchers are able to make observations on individual microbes and species associations, to judge what microbes are present, and to understand the functions within microbial communities.

According to McArthur, Upper Three Runs Creek, a relatively uncontaminated stream that runs through the Savannah River Site, has the highest diversity of aquatic invertebrates in North America—and possibly in the world. Upper Three Runs is home to a number of invertebrate species found nowhere else and to many others that are rare. McArthur and his colleagues also found that in the study area, invertebrate species diversity and productivity are significantly higher in the streams than in the adjacent floodplain swamp. Because the stream has more oxygen gas and water flow than does the swamp, decomposition and recycling by microorganisms produces more organic matter in the stream. With more nutrients available, invertebrate populations are larger and more diverse there. The bottom line: The more ecologists understand about ecosystem dynamics, the more we come to appreciate the essential role of microbes in our environment.

(1) They break down organic phosphates from decomposing organisms to inorganic phosphates, and (2) they convert inorganic phosphates to orthophosphate (PO_4^{3-}), a water-soluble nutrient used by both plants and microorganisms. These functions are particularly important because phosphorus is often the limiting nutrient in many environments.

AIR

Having discussed briefly some fundamentals of ecology and of biogeochemical cycles, we now explore different kinds of environments and the microorganisms they contain. We will begin with the air around us and then consider soil and water.

Microorganisms Found in Air

Microorganisms do not grow in air, in part because it lacks the nutrients needed for metabolism and growth. However, spores are carried in air, and vegetative cells can be carried on dust particles and water droplets in air. The kinds and numbers of airborne microorganisms vary tremendously in different environments. Large numbers of many different kinds of microorganisms are present in indoor air where humans are crowded together and building ventilation is poor. Small numbers have been detected at altitudes of 3000 m.

Among the organisms found in air, mold spores are by far the most numerous, and the predominant genus is usually *Cladosporium*. Bacteria commonly found in air include both aerobic spore formers such as *Bacillus subtilis* and non–spore formers such as *Micrococcus* and *Sarcina*. Algae, protozoa, yeasts, and viruses also have been isolated from air. While coughing, sneezing, or even talking, infected humans can expel pathogens along with water droplets. Health care workers should handle wastes from patients carefully to avoid producing aerosols (tiny droplets that remain suspended for some time) of pathogens. ∞ (Chapter 16, p. 428)

Determining the Microbial Content of Air

Airborne microbes can be detected by collecting those that happen to fall onto an agar plate or in a liquid medium. A special centrifuge-like air-sampling instrument provides a better measure of airborne microbes (Figure 26.10a).

Methods for Controlling Microorganisms in Air

Chemical agents, radiation, filtration, and laminar airflow all can be used to control microorganisms in air. Certain chemical agents, such as triethylene glycol, resorcinol, and lactic acid dispersed as aerosols, kill many if not all microorganisms in room air. These agents are highly bactericidal, remain suspended long enough to act at normal room temperature and humidity, are nontoxic to humans, and do not damage or discolor objects in the room.

Ultraviolet radiation has little penetrating power and is bactericidal only when rays make direct contact with microorganisms in air. ∞ (Chapter 13, p. 348) Thus, ultraviolet lamps must be carefully placed to ensure treatment of all air in a room. They are most useful in maintaining sterile conditions in rooms only sporadically occupied by humans. They can be turned off while technicians are performing sterile techniques and turned on again when the room is not in use. Humans entering a room while ultraviolet lights are on must wear pro-

FIGURE 26.10 (a) An air-sampling device used to measure airborne bacteria and fungi. (b) Technicians are protected from the airborne spread of organisms by the use of a laminar flow hood, which suctions air away from the opening and filters it before expelling it.

(a)

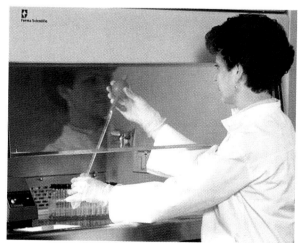

(b)

SICK BUILDINGS

The Environmental Protection Agency (EPA) estimates that 30 million to 75 million U.S. workers are at risk of getting sick from the buildings in which they work. Since the late 1970s, indoor air specialists from the National Institute for Occupational Safety and Health (NIOSH) have investigated more than 1000 instances of building-related illness. Some of the ailments can be fatal. Sick building syndrome (SBS) is a collection of symptoms including headache, dizziness, nausea, burning eyes, nosebleeds, and respiratory and skin problems. Symptoms vary among individuals.

How do you know when a building—instead of a cold, allergies, or another cause—is to blame for SBS symptoms? If your symptoms get worse the more time you spend in a building and improve when you are away from it, take a close look at the building. The ventilation system has been implicated in many cases of SBS. In many new "tight" buildings, occupants are dependent on a central ventilation system for the air they breathe. The system pulls in air from outside, runs it through filters, warms or cools it, and then delivers it throughout the building by a series of ducts. Once the air has circulated, return ducts channel some of it outside. Dirt, dust, insects, mold, mildew, particles of building materials, and other detritus are found on the system's filters. Whatever gets into the ventilation system gets into workers' lungs as well. New filters and proper cleaning of the system's equipment often correct ventilation problems.

Another possible cause of SBS is too little air. In the 1930s the American Society of Heating, Refrigeration, and Air Conditioning Engineers set 15 ft^3 of fresh outdoor air per person per minute as the ventilation standard. The current standard is 20 ft^3—but it assumes that no more than seven people will occupy a 1000-ft^2 area. With more people, more air is needed. The "fleece factor," or pollution from "manmade mineral fibers" (MMMF), may also cause SBS. New carpets, padded partitions, curtains, and upholstery are treated with volatile compounds that are released into the air. Copy machines, paper shredders, laser printers, and blueprint copiers release vapors and particles that may be irritating. Finally, biological contaminants such as molds, fungi, and bacteria thrive in building ventilation systems and are spread as aerosols. In recent years, bacteria that collect in ventilation systems have proved fatal—for example, when *Legionella* species have caused Legionnaires' disease. (Chapter 21, p. 590)

If you suspect that you are in a sick building, document your symptoms, check the ventilation system, inspect the air vents for dirt, check for exhaust vents around copy machines, and so on. Ask the maintenance engineers how much fresh outdoor air is circulating in cubic feet per minute per person. Once you have targeted hazards, convince someone to do something about them.

tective clothing and special glasses to protect the eyes from burns. Ultraviolet lamps can also be installed in air ducts to reduce the number of microorganisms entering a room through its ventilation system.

Air filtration involves passing air through fibrous materials, such as cotton or fiberglass. It is useful in industrial processes in which sterile air must be bubbled through large fermentation vats. Cellulose acetate filters can be installed in a *laminar airflow system* (Figure 26.10b) to remove microorganisms that may have escaped into the air underneath the laminar flow hood. The air is suctioned away from the opening, filtered, and then returned to the room.

SOIL

We might think of the soil we walk on as an inert substance, but nothing could be further from the truth. Soil, in fact, is teeming with microscopic and small macroscopic organisms, and it receives animal wastes and organic matter from dead organisms. Microorganisms act as decomposers to break down this organic matter into simple nutrients that can be used by plants and by the microbes themselves. (Animals, of course, obtain their nutrients from plants or from other animals.) Soil microorganisms are thus extremely important in recycling substances in ecosystems.

Components of Soil

Soil is divided into a number of layers, or *soil horizons* (Figure 26.11). These horizons include the *topsoil, subsoil,* and *parent material,* which overlie bedrock. Soil contains inorganic and organic components. The inorganic components are rocks, minerals, water, and gases. The organic components are **humus** (nonliving organic matter) and living organisms. Soils differ greatly in the relative proportions of these components. Topsoil, the surface layer of soil, contains the greatest number of microorganisms because it is well supplied with oxygen and nutrients. Lower layers of soil (subsoil and parent material) and soils depleted of oxygen and nutrients contain fewer organisms.

The most abundant inorganic components of soil are pulverized rocks and minerals. *Weathering* of rocks—mechanical or chemical breakdown—releases minerals, and the elements that make up those minerals, into the soil. The most abundant elements in many soils are silicon, aluminum, and iron. Small quantities of other elements, such as calcium, potassium, magnesium, sodium, phosphorus, nitrogen, and sulfur, also are present in soil.

In addition to rocks and minerals, soil contains water and the gases carbon dioxide, oxygen, and nitrogen. Water molecules adhere to soil particles or are interspersed among them. The amount of water in soil

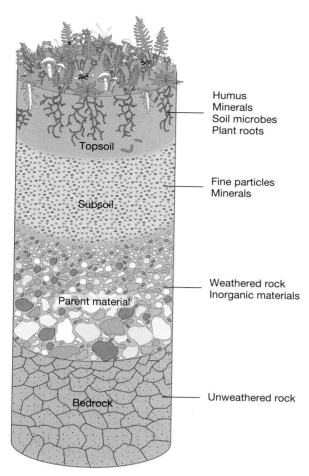

FIGURE 26.11 Soil horizons—a vertical segment of the main soil layers.

is highly variable and depends on the climate, the quantity of recent rainfall, and the drainage of the soil. Gases are dispersed among soil particles or dissolved in water. The concentration of gases in the soil also varies with the metabolic activity of soil organisms. Compared with atmospheric air, soil is lower in oxygen and higher in carbon dioxide.

Humus is constantly changing as organisms die and as decomposers degrade complex molecules to simpler ones. Soils differ greatly in the amount of humus they contain. Most soils are 2 to 10 percent humus, but a peat bog can be 95 percent humus.

In addition to microorganisms, soil also contains the root systems of many plants, a variety of invertebrate animals (nematodes, earthworms, snails, slugs, insects, centipedes, millipedes, and spiders), and a few burrowing reptiles and mammals. Despite the great diversity of soil organisms, microorganisms are by far the most numerous both in total numbers and in the number of species present. As we have seen, microorganisms convert humus to nutrients usable by other organisms. They also make great demands on soil nu-

trients. When nutrients are abundant, the number of microbes increases rapidly. As nutrients are depleted, the number of microbes decreases accordingly.

Microorganisms in Soil

All the major groups of microorganisms—bacteria, fungi, algae, and protists, as well as viruses—are present in soil, but bacteria are more numerous than all other kinds of microorganisms (Figure 26.12). Among the bacteria in soil are autotrophs, heterotrophs, aerobes, anaerobes, and, depending on the soil temperature, mesophiles and thermophiles. In addition to nitrogen-fixing, nitrifying, and denitrifying bacteria, soil contains bacteria that digest special substances such as cellulose, protein, pectin, butyric acid, and urea.

Soil fungi are mostly molds. Both mycelia and spores are present mainly in topsoil, the aerobic surface layer of the soil. Fungi serve two functions in soil. They decompose plant tissues such as cellulose and lignin, and their mycelia form networks around soil particles, giving the soil a crumbly texture. ∞ (Chapter 12, p. 308) In addition to molds, yeasts are abundant in soils in which grapes and other fruits are growing.

Small numbers of cyanobacteria, algae, protists, and viruses are found in most soils. Algae are found only on the soil surface, where they can carry on photosynthesis. In desert and other barren soils, algae contribute significantly to the accumulation of organic matter in the soil. Protists, mostly amoebas and flagellated protozoa, also are found in many soils. They feed on bacteria and may help control bacterial populations. Soil viruses infect mostly bacteria, but a few infect plants (see the box "Plant Viruses" in Chapter 11) or animals. ∞ (p. 274)

FIGURE 26.12 The relative proportions of various kinds of organisms found in soil. "Other organisms" include such things as algae, protists, and viruses.

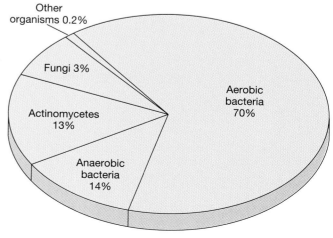

Factors Affecting Soil Microorganisms

Soil microorganisms, like all other organisms, interact with their environment. Their growth is influenced both by abiotic factors and by other organisms. The microorganisms, in turn, affect the physical characteristics of soil and other organisms in the soil.

Abiotic factors in soil, as in any other environment, include moisture, oxygen concentration, pH, and temperature. The moisture and oxygen content of soil are closely related. Spaces among soil particles ordinarily contain both water and oxygen, and aerobic organisms thrive there. However, in waterlogged soils all the spaces are filled with water, so only anaerobic bacteria grow there.

Soil pH, which can vary from 2 to 9, is an important factor in determining which microorganisms will be present. Most soil bacteria have an optimum pH between 6 and 8, but some molds can grow at almost any soil pH. Molds thrive in highly acidic soils partly because of reduced competition from bacteria for available nutrients. Lime neutralizes acidic soil and increases the bacterial population. Fertilizer containing ammonium salts has two effects on soil: (1) It provides a nitrogen source for plants; and (2) when it is metabolized by certain bacteria, the bacteria release nitric acid, which decreases the soil pH and increases the mold population.

Soil temperature varies seasonally from below freezing to as high as 60°C at soil surfaces exposed to intense summer sunlight. Mesophilic and thermophilic bacteria are quite numerous in warm to hot soils, whereas cold-tolerant mesophiles (and not true psychrophiles) are present in cold soils. Most soil molds are mesophilic and are found mainly in soils of moderate temperature.

Exceedingly wide variations exist in physical characteristics of soil and in the numbers and kinds of organisms it contains—even in soil samples taken only a few centimeters apart. Such observations led ecologists to develop the concept of microenvironments. The interactions among organisms and between the organisms and their environment can be quite different in different microenvironments, no matter how close together they are.

Importance of Decomposers in Soil

Soil microorganisms that are decomposers are important in the carbon cycle because of their ability to decompose organic matter. The decomposition of complex organic substances from dead organisms is a stepwise process that requires the action of several kinds of microorganisms. The organic substances include cellulose, lignins, and pectins in the cell walls of plants, glycogen from animal tissues, and proteins and fats from both plants and animals. Cellulose is degraded by bacteria, especially those of the genus *Cytophaga*, and by various fungi. Lignins and pectins are partially digested by fungi, and the products of fungal action are further digested by bacteria. Protozoa and nematodes also can play a role in the degradation of lignins and pectins. Proteins are degraded to individual amino acids mainly by fungi, actinomycetes, and clostridia.

Under the anaerobic conditions of waterlogged soils in marshes and swamps, methane is the main carbon-containing product. It is produced by three genera of strictly anaerobic bacteria—*Methanococcus, Methanobacterium,* and *Methanosarcina.* In addition to degrading carbon compounds to methane, they also obtain energy by oxidizing hydrogen gas:

$$4\,H_2 \;+\; CO_2 \;\longrightarrow\; CH_4 \;+\; 2\,H_2O$$

hydrogen   carbon   methane   water
gas   dioxide

In one way or another, organic substances are metabolized to carbon dioxide, water, and other small molecules. In fact, for each naturally occurring organic compound, there is one or more organisms that can decompose it. Thus, carbon is continuously recycled. However, certain organic compounds manufactured by humans resist the actions of microorganisms. Accumulations of these synthetic substances create environmental hazards.

Nitrogen enters the soil through the decomposition of proteins from dead organisms and through the actions of nitrogen-fixing organisms. In addition to protein decomposition, which introduces nitrogen into the soil, gaseous nitrogen is fixed both by free-living microorganisms and by symbiotic microorganisms associated with the roots of legumes, as previously described.

Soil Pathogens

Soil pathogens are primarily plant pathogens, many of which have been discussed in earlier chapters. A few soil pathogens can affect humans and other animals. The main human pathogens found in soil belong to the genus *Clostridium.* ∞ (Chapter 24, p. 686) All are anaerobic spore formers. *Clostridium tetani* causes tetanus and can be introduced easily into a puncture wound. *Clostridium botulinum* causes botulism. Its spores, found on many edible plants, can survive in incompletely processed foods to produce a deadly toxin. *Clostridium perfringens* causes gas gangrene in poorly cleaned wounds. Grazing animals can contract anthrax from spores of *Bacillus anthracis* in the soil. In fact, most soil organisms that infect warmblooded animals exist

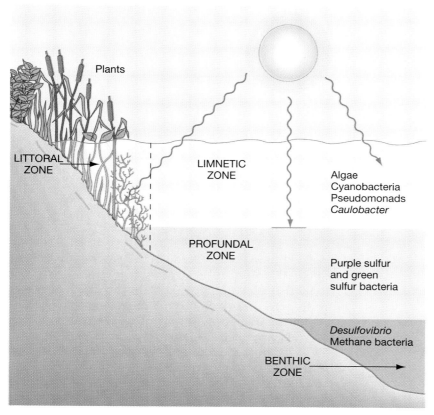

FIGURE 26.13 Zonation in a typical lake or pond, with organisms commonly found in each zone.

water near the shore where light penetrates to the bottom. The *limnetic zone* comprises the sunlit water away from the shore; resident microorganisms include algae and cyanobacteria. Between the limnetic zone and the lake sediment is the *profundal zone*. When organisms in the limnetic zone die, they sink to the profundal zone, where they provide nutrients for other organisms. The sediment, or *benthic zone*, is composed of organic debris and mud (Figure 26.13).

In aquatic environments, water temperatures vary from 0°C to nearly 100°C. Most microorganisms thrive in water at moderate temperatures. However, some thermophilic bacteria have been found in geysers at a water temperature above 90°C, and psychrophilic fungi and bacteria have been found in water at 0°C. The pH of fresh water varies from 2 to 9. Although most microorganisms grow best in waters of nearly neutral pH, a few have been found in both extremely acidic and extremely alkaline waters. Most natural waters are rich in nutrients, but the quantities of various nutrients in different bodies of water vary considerably. Sometimes nutrients become so abundant that a "bloom," or sudden proliferation of organisms in a body of water, occurs (Figure 26.14).

In aquatic environments, oxygen can be the limiting factor in the growth of microorganisms. Because of

as spores because soil temperatures are usually too low to maintain vegetative cells of these pathogens.

WATER

All aquatic ecosystems—fresh water, ocean water, even rainwater—contain microorganisms as well as inorganic substances. Most of the organisms in such environments have been considered in earlier chapters. Here we will consider properties of the environments, interactions of microorganisms with their environment, transmission of human pathogens in water, and methods for maintaining safe water supplies.

Freshwater Environments

Freshwater environments include surface water, such as lakes, ponds, rivers, and streams, and groundwater that runs through underground strata of rock. Although groundwater contains few microorganisms, surface water contains large numbers of many different kinds of microorganisms.

Ponds and lakes are divided vertically into zones. The shoreline, or *littoral zone*, is an area of shallow

FIGURE 26.14 An overabundance of nutrients has allowed an algal bloom to develop on this pond. Bubbles of oxygen can be seen collecting underneath some of the algae. However, algal blooms often lead to increased biological oxygen demand (BOD) as they die off and decompose, using up oxygen needed by other pond organisms such as fish.

oxygen's low solubility in water, its concentration never exceeds 0.007 g per 100 g of water. When water contains large quantities of organic matter, decomposers rapidly deplete the oxygen supply as they oxidize the organic matter. Oxygen depletion is much more likely in standing water in lakes and ponds than in running water in rivers and streams because the movement of running water causes it to be continuously oxygenated.

Yet another factor affecting microorganisms in aquatic environments is the depth to which the sunlight penetrates the water. Of minor importance in fresh water except for deep lakes, it is very important in the ocean. Photosynthetic organisms are limited to locations with adequate sunlight.

Which organisms are present depends on the temperature and pH of the water, dissolved minerals, the depth of sunlight penetration, and the quantity of nutrients in the water. Aerobic bacteria are found where oxygen supplies are adequate, whereas anaerobic bacteria are found in oxygen-depleted waters. Eukaryotic algae, cyanobacteria, and sulfur bacteria are limited to water that receives adequate sunlight. *Desulfovibrio* and methane bacteria are found in lake sediments. Protists are found in many different freshwater environments.

Marine Environments

The marine, or ocean, environment covers about 70 percent of the earth's surface and is therefore larger than all other environments combined. Compared with fresh water, the ocean is much less variable in both temperature and pH. Except in the vicinity of volcanic vents in the sea floor, where water reaches 250°C ∞ (Chapter 9, p. 232), ocean water ranges from 30°C to 40°C at the surface near the equator to 0°C in polar regions and in the lowest depths. At any one location and depth, the temperature is nearly constant. The pH of ocean water ranges from nearly neutral to slightly alkaline (pH 6.5 to 8.3). This pH range is suitable for growth of many microorganisms, and sufficient carbon dioxide dissolves in the water to support photosynthetic organisms.

Ocean water, which is about seven times as salty as fresh water, displays a remarkably constant concentration of dissolved salts—3.3 to 3.7 g per 100 g of water. Thus, organisms that live in marine environments must be able to tolerate high salinity but need not tolerate variations in salinity.

Marine environments display a far greater range of hydrostatic pressure than do fresh waters. Hydrostatic pressure increases with depth at a rate of approximately 1 atm per 10 m, so the pressure at a depth of 1000 m is 100 times that at the surface. A few microorganisms, including archaeobacteria, have been

FIGURE 26.15 Nutrients flow into the ocean with the influx of water from rivers. The light-colored area of water is silt-bearing, nutrient-rich river water.

isolated from Pacific Ocean trenches at depths greater than 1000 m!

Other factors that vary with the depth of ocean water are penetration of sunlight and oxygen concentration. Sunlight of sufficient intensity to support photosynthesis penetrates ocean water to depths of only 50 to 125 m, depending on the season, latitude, and transparency of the water. Oxygen diffuses into surface water and is released by photosynthetic organisms in sunlit water. However, deep water lacks both sunlight and oxygen.

Nutrient concentrations vary in ocean water, depending on depth and proximity to the shore (Figure 26.15). Nutrients are most plentiful near the mouths of rivers and near the shore, where runoff from land enriches the water. However, ocean water is generally lower in phosphates and nitrates than fresh water. Nutrients in the waters of the open ocean are relatively dilute. Photosynthetic organisms near the surface serve as food for the heterotrophic organisms at the same or deeper levels. Decomposers are usually found in bottom sediments, where they release nutrients from dead organisms.

Large numbers of many different kinds of microorganisms live in the ocean. Yet, much remains to be learned about the exact numbers and the particular species living in various parts of the ocean. The primary producers of the ocean are photosynthetic microorganisms called *phytoplankton*. They are motile and contain oil droplets or other devices for buoyancy, allowing them to remain in sunlit waters. Phytoplankton include cyanobacteria, diatoms, dinoflagellates, chlamydomonads (kla"mi-do-mo'nadz), and a variety of other protists and eukaryotic algae (Figure 26.16).

Many of the consumers of the ocean are heterotrophic bacteria. Which species inhabit a given region is determined by the temperature and pH of the water and by the available nutrients. Members of the genera *Pseudomonas, Vibrio, Achromobacter,* and *Flavobacterium* are common in ocean water. Protozoa,

APPLICATIONS

MICROBIAL DETERMINANTS OF WEATHER

Oceanic phytoplankton absorb large amounts of carbon dioxide as they perform photosynthesis. Atmospheric carbon dioxide is one of the greenhouse gases that trap heat and raise global temperature. Seasonal fluctuations of phytoplankton populations affect the earth's temperature more than we had previously realized. Recent research has revealed that massive populations of marine viruses may be at the root of these processes. Earlier, the ocean was thought to contain relatively few viruses. Viral infections destroy phytoplankton and thus may help determine earth's weather.

FIGURE 26.16 Producers of the ocean are called phytoplankton. They include organisms such as (a) the marine diatom *Isthmia nervosa* (magnified 625×) and (b) these diatoms (100×) collected in a plankton tow from Rhode Island Sound.

(a)

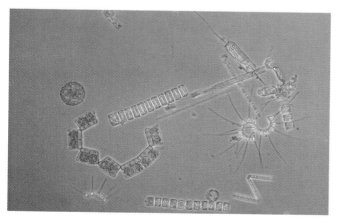

(b)

especially radiolarians and foraminiferans, and a variety of fungi also feed on producers in the open ocean. However, most of these organisms inhabit an aquatic zone beneath the region of intense sunlight.

Between the stratum in which these consumers live and the bottom of the ocean is a relatively uninhabited stratum. In the sediments at the bottom of the ocean, the number of microorganisms again increases. The bottom dwellers are generally strict or facultative anaerobic decomposers. Many of them contribute significantly to the maintenance of biogeochemical cycles and produce such substances as ammonium ion, hydrogen sulfate, and nitrogen gas.

Water Pollution

Water is considered polluted if a substance or condition is present that renders the water useless for a particular purpose. Thus, the concept of water pollution is a relative one, depending on both the nature of the pollutants and the intended uses of the water. For example, human drinking water is considered polluted if it contains pathogens or toxic substances. Water that is too polluted for drinking may be safe for swimming; water that is too polluted for swimming may be acceptable for boating, industrial uses, or generating electrical power. The EPA has established drinking water standards and methods for water testing. Water that is fit for human consumption is termed **potable water.**

Pollutants

The major types of water pollutants are organic wastes such as sewage and animal manures, industrial wastes, oil, radioactive substances, sediments from soil erosion, and heat. Organic wastes suspended in water are usually decomposed by microorganisms provided that the water contains sufficient oxygen for oxidation of the substances. The oxygen required for such decomposition is called **biological oxygen demand** (BOD). When BOD is high, the water can be depleted of oxygen rapidly. Anaerobes increase in number while populations of aerobic decomposers dwindle, leaving behind large quantities of organic wastes. Organic wastes also can contain pathogenic organisms from among the bacteria, viruses, and protozoa.

Industrial wastes contain metals, minerals, inorganic and organic compounds, and some synthetic chemicals. The metals, minerals, and other inorganic substances can alter the pH and osmotic pressure of the water, and some are toxic to humans and other organisms. Synthetic chemicals can persist in water because most decomposers lack enzymes to degrade them. Oil is another important water pollutant. Radioactive substances released into water persist as a hazard to living organisms until they have undergone natural radioactive decay.

Probing the Microbial Foundation of Deep-Sea Communities

Douglas Nelson at work in the laboratory.

I'll never forget my first deep-sea submersible dive. After we'd come up and gotten out, I stood on the deck of the support ship and repeated to myself at least 20 times, "I can't believe that's down there!" We were surrounded by miles and miles of bare, flat ocean as far as I could see. But far beneath us, in the midst of an equally bare ocean bottom, were rich biological islands surrounding the roiling plumes of hydrothermal vents. You'd never suspect such incredibly abundant biological communities would be possible under those conditions. And the entire community depends on chemoautotrophic sulfur bacteria. Some are free-living. Others are endosymbionts (Chapter 4)—they live inside and nourish various invertebrates such as clams, tube worms, and mussels. ∞ (p. 102) I study these relationships and the bacterial metabolism.

My name is Douglas Nelson, and I'm a microbiologist at the University of California, Davis. Although someone predicted from my third-grade class photo that I'd become a scientist, I didn't set out to become a microbiologist. With a bachelor's degree in chemistry, I turned to ecology in graduate school. I got interested in microbiology when I was a teaching assistant in a limnology course. That's the branch of ecology concerned with bodies of fresh water. The metabolism of bacteria in warm springs intrigued me—and still does. Now I study the same types of metabolism in the bacteria that are the basis of the communities found at hydrothermal (deep-sea) vents.

Deep-sea vents occur where there are cracks in the ocean floor. Cold seawater trickles down through the cracks, picks up heat and minerals from the earth's interior, and then pours forth continuously through the vents. Vent water can be extremely hot—up to 400°C—or it can be only a few degrees warmer than normal deep-ocean water (which is in the 2 to 4°C range), but it's rich in reduced-sulfur

compounds. That's why, even though vents occur deep in the ocean where life is not generally abundant, they support unique communities of unusual organisms.

The discovery of these vent communities would have been impossible without a specially designed submersible craft such as *Alvin*. Sponsored by the Woods Hole Oceanographic Institution (WHOI) of Woods Hole, Massachusetts, *Alvin* is about 2 m across inside. That's just big enough to hold a pilot and two observers, who are generally scientists with experiments underway. The pressure outside is about 4000 lb/in^2 (about 250 atm)—the weight of a Volvo station wagon pressing on every square inch of submersible surface—so *Alvin*'s viewports are small and several inches thick. It takes over an hour to descend to the depth of the vents, 2500 m, and as long to go back up. Specialized arms and other tools on *Alvin* gather various samples and measure temperature or the concentrations of various chemicals as they change in time. *Alvin* has made nearly 3000 dives without incident, but in case of emergency, built-in explosive bolts can blow off its arms and other outside equipment to make it buoyant or to free it from entanglement.

Many kinds of analyses must await transport of samples to laboratories on the mother vessel. Some scientists collect sediment cores or rock cores. I use a "slurp gun," a sort of underwater vacuum cleaner made with a large jar that catches the slurped material. A very fine filter over the exit traps mats of bacteria inside the jar. We don't try to maintain pressure for most bacterial or animal samples. Animal collections come up in nonpressurized containers called clam buckets or worm coffins. Any pressure-retaining sample vessel loaded at the vents could be dangerous at surface pressures. It would be like having a bomb on board.

Host organisms seem to withstand the pressure transition—at least,

they survive the ascent—but they don't live very long once at surface pressures. If they are repressurized (in Plexiglas vessels to allow observation), the animals survive for 1 to 2 weeks and can be experimentally manipulated. The bacterial symbionts of those animals are apparently not pressure sensitive. The symbionts are equally active at vent pressures and at surface pressures, at least for 4 to 5 hours, the longest tried so far. Few if any truly barophilic (pressure-loving) bacteria have been isolated from around hydrothermal vents. Even *E. coli* can grow well at up to several hundred atmospheres of pressure. The only bacteria that *require* high pressure, obligate barophiles, come from great depths where pressures are 900 to 1000 atm—four times the pressure at the vents.

When *Alvin* returns from a dive, as many as 25 scientists are waiting to get their hands on the materials brought up. The scientists who've just spent 10 hours in the sub have to decide whether they're going to grab a quick bite or skip dinner and begin the evening's research. Many analyses must be begun immediately, for any number of reasons. For example, the hydrogen sulfide content of vent waters is altered by biological and nonbiological oxidations, and prolonged stress of host animals may diminish the viability of symbionts.

Microbiologists studying the symbionts that live in tube worms may have to wait until the blood and various tissues have been sampled before obtaining the trophosome—a spongy tissue that fills the trunk of tube worms and is packed (at 10^{10} to 10^{11} cells per milliliter) with endosymbiotic bacteria. These worms (from the phylum Vestimentifera) have no mouth, gut, or anus. They're nourished entirely by their bacterial endosymbionts.

Temperatures at vents that support dense animal communities—from 10 to 25°C—are only slightly warmer than normal deep-ocean water. The bacteria living inside tube worms are mesophilic, preferring temperatures of 25 to 35°C, whereas bacteria living inside mussels are psychrophiles and show little or no metabolic activity at 22°C or above. ∞ (Chapter 6, p. 145) Mussels occupy peripheral regions away from the warm vent waters.

Interestingly, microbial sulfur metabolism and symbioses exploiting it aren't unique to hot springs and deep-sea vents. They occur elsewhere too—for instance, in many marine sediments. What's important is the combination of hydrogen sulfide and oxygen. In many relatively shallow near-shore marine waters (200 m or less), anaerobic sulfate-reducing bacteria decompose at least half the organic matter (such as dead seaweed) that reaches the sediments. These bacteria use the abundant sulfate anions from seawater as a terminal electron acceptor (Chapter 5) while they oxidize a wide variety of fermentation end products and produce copious amounts of the waste product hydrogen sulfide. ∞ (p. 129)

The diffusion of oxygen into these sediments can't keep up with oxygen use in the top few millimeters by aerobic bacteria. Below this level, anaerobic conditions develop that favor the production of hydrogen sulfide. Host organisms such as clams build burrows there and pump oxygen-rich water down into the sediments. There the water contacts bacterially produced hydrogen sulfide, and bacterial symbionts of the hosts have access to these key substrates. *Solemya velum*, a small clam that lives in sediments a few kilometers from WHOI, is involved in one such symbiotic association. Only after the general principle of symbiosis based on chemoautotrophic sulfur bacteria had been discovered—in large part by WHOI scientists traveling great distances to deep-sea vents—did the researchers find examples in their own backyard.

Thiosulfate is another partially reduced sulfur compound that is not abundant in hydrothermal vent waters but is the preferred substrate for certain vent symbionts. It accumulates in the blood of vent organisms and thus is made available to symbionts living in these organisms. It represents a partial oxidation product of hydrogen sulfide; because it is less toxic than hydrogen sulfide, it is presumably more readily stored by the host for future use by symbionts.

In addition to endosymbionts, numerous free-living bacteria also live at the vents. These include filamentous bacteria, such as very large representatives of the genus *Beggiatoa*. Individual cells can have a diameter of 150 μm and form chains several centimeters long. The conditions in the subsurface region of many vents can approximate a chemostat (Chapter 6), with factors such as temperature and hydrogen sulfide concentration maintained within a narrow range. (p. 139) Natural selection should favor bacteria that grow most rapidly, allowing a population to live in subsurface chambers. Slower-growing strains wash out into cold ambient seawater that will not provide sufficient nutrients.

The vents don't last forever, so how do the various organisms get from one vent to another? There can be hundreds of kilometers of cold, sulfide-depleted territory between vents. How do the organisms survive en route? One theory is that high-temperature, free-living archaeobacteria exist everywhere beneath the earth's crust in a continuum from vent to vent. Some symbiotic bacteria may require a stage in their life cycle that remains viable in a nongrowing state for long periods of time in cold water. Such bacteria may drift extensively before they reach a new site where conditions again select for them to grow rapidly and become established. It has also been proposed that dead whales act as waystations for certain bivalve-symbiont associations; the whales' tremendous organic biomass could fuel hydrogen sulfide production for many years. We do know from DNA studies that all symbionts of a given host species of tube worm at one vent site are the same; between vents the differences are trivial.

Somehow the bacteria get to the vent, as do their invertebrate hosts. But how do the bacteria colonize their hosts? There are three possibilities. (1) The symbionts may pass vertically, from one generation to the next, within a host's eggs. This passage has been demonstrated with gene probes for a number of bivalve symbionts. However, PCR analysis (Chapter 7) strongly suggests that eggs of tube worms do not contain symbionts. ∞ (p. 190) (2) Symbionts may pass horizontally from adults to larval or juvenile stages. (3) The bacteria could be acquired by reinfection from the environment, possibly by intake from the surroundings. This third route requires a free-living form of the symbiont. We know that juveniles of some species of tube worms have transient digestive tracts that actually have to pass through the brain to get to the outside, where symbionts could be acquired.

An area ripe for future studies centers on the question, How do symbionts feed their hosts? In some associations there is evidence that symbionts are digested in lysosomal vacuoles. In other associations very limited data suggest that the symbionts may overproduce a few key organic compounds that "leak" out and feed the host.

This is just a partial list of research topics that interest me. You could say it's quite a shift for someone who started out in chemistry. But I advise students to follow what interests them; don't choose on the basis of established career niches today. Research topics can be trendy; by the time you finish your education, you may be only one of a whole parade of people following a particular trend. If you follow what interests you, you'll be happier, and you may end up leading a parade!

A thriving deep-sea vent community surrounding a hydrothermal vent.

Bare ocean floor, devoid of life, immediately adjacent to the rich vent community.

Soil particles, sand, and minerals from soil erosion enter water from agricultural, mining, and construction activities. Nitrates, phosphates, and other nutrients enter the water from detergents, fertilizers, and animal manures. Abundant nutrient enrichment of water, called **eutrophication,** leads to excessive growth of algae and other plants. Eventually the plants become so dense that sunlight cannot penetrate the water. Many algae and other plants die, leaving large quantities of dead organic matter in the water. The high BOD of this organic matter leads to oxygen depletion and the persistence of undecomposed matter in the water.

Even heat can act as a water pollutant when large quantities of heated water are released into rivers, lakes, or oceans. Increasing the temperature of water decreases the solubility of oxygen in it. The altered temperature and decreased oxygen supply significantly change the ecological balance of the aquatic environment.

The effects of water pollution are summarized in Table 26.1.

TABLE 26.1	Effects of Water Pollution	
Pollutant	**Effects**	**Comment**
Organic wastes (sewage, decaying plants, animal manures, wastes from food processing plants, oil refineries, and leather, paper, and textile plants)	Increase biological oxygen demand of water	If adequate oxygen is available, these substances can be degraded by microorganisms usually present in water. If oxygen becomes depleted, decomposition is limited to what can be done by anaerobic decomposers. Water plants may be killed, and animals may be killed or caused to migrate.
Pathogenic organisms	Cause disease in humans who drink the water	Most bacteria are well controlled in public drinking water, but certain viruses especially those that cause hepatitis, still cause human disease. More effective means of removing viruses during purification are needed.
Inorganic chemicals and minerals	Increase the salinity and acidity of water and render it toxic	Such chemicals should be removed during waste treatment. Heavy metals such as mercury, which are toxic to humans, should be prevented from entering water supplies.
Synthetic organic chemicals (herbicides, pesticides, detergents, plastics, wastes from industrial processes)	Can cause birth defects, cancer, neurological damage, and other illness	Because these substances are not biodegradable, chemical or physical means must be used to remove them during waste treatment. Many such substances become magnified (increased in concentration) as they are passed along food chains.
Plant nutrients	Cause excessive and sometimes uncontrolled growth of aquatic plants (eutrophication); impart undesirable odors and tastes to drinking water	Removal of excess phosphates and nitrates from water during waste treatment is costly and difficult.
Sediments from land erosion	Cause silting of waterways and destruction of hydroelectric equipment near dams; reduce light reaching plants in water and oxygen content of water	
Radioactive wastes	Can cause cancer, birth defects, radiation sickness when in large doses	Effects can be magnified through food chains. Because such wastes are difficult to remove from water, preventing them from reaching water is exceedingly important.
Heated water	Reduces oxygen solubility in water; alters habitats and kinds of organisms present; encourages growth of some aquatic life, but can decrease growth of desired organisms such as fish	

Pathogens in Water

Human pathogens in water supplies usually come from contamination of the water with human feces. When water is contaminated with fecal material, any pathogens that leave the body through the feces— many bacteria and viruses and some protozoa—can be present. The most common pathogens transmitted in water are listed in Table 26.2. Water usually is tested for fecal contamination by isolating *Escherichia coli* from a water sample. *Escherichia coli* is called an **indicator organism** because, as *E. coli* is a natural inhabitant of the human digestive tract, its presence in water indicates that the water is contaminated with fecal material.

Water Purification

Purification Procedures

Purification procedures for human drinking water are determined by the degree of purity of the water at its source. Water from deep wells or from reservoirs fed by clean mountain streams requires very little treatment to make it safe to drink. In contrast, water from rivers that contain industrial and animal wastes and even sewage from upstream towns requires extensive treatment before it is safe to drink. Such water is first allowed to stand in a holding reservoir until some of the particulate matter settles out. Then alum (aluminum potassium sulfate) is added to cause **flocculation,** or precipitation of suspended colloids (Chapter 2) such as clay. ∞ (p. 33) Many microorganisms are also removed from the water by flocculation.

Following flocculation treatment, water is filtered. **Filtration,** the passage of water through beds of sand, removes nearly all the remaining microorganisms. Charcoal can be used instead of sand for filtration; it has the advantage of removing organic chemicals that are not removed by sand. Finally, water is chlorinated.

PUBLIC HEALTH

IS OUR DRINKING WATER SAFE?

In the spring of 1993, the worst outbreak of waterborne disease in U.S. history occurred in Milwaukee. More than 400,000 cases of gastrointestinal illness and more than 100 deaths resulted from parasitic contamination of the drinking water supply. The city's water treatment facilities were not using the proper screens to exclude the waterborne parasite *Cryptosporidium parvum* from release into the drinking water. A more recent outbreak in Las Vegas has been linked to the deaths of 35 HIV-infected individuals.

U.S. residents drink about 420 million l of water daily, and they expect a clean, safe stream of water when they turn on the faucet or water fountain. On the basis of CDC estimates, 940,000 Americans become ill each year from contaminated water, and 900 die. Contaminants include disease-producing microorganisms such as *C. parvum* and chemicals such as lead, nitrates, arsenic, and radon. Scientists have found that even chlorine reacts with decaying organic matter in water to form carcinogenic chemical byproducts. Chief sources of water pollution include municipal and industrial wastes in landfills, residential septic systems, active and abandoned oil and gas wells, active and inactive coal and mineral mines, active and inactive underground storage tanks at gas stations, and a vast quantity of pesticides, fertilizers, and improperly disposed of motor oil.

Although 70 percent of the earth's surface is covered by water, 97 percent of it is in oceans and seas, and 2 percent is frozen in glaciers and polar ice. Only about 1 percent is available for human use. The Safe Drinking Water Act was signed into law in 1974, but chemical poisoning and microbial contamination still exist and in some areas have increased due to failure to upgrade water purification systems. To avoid crises such as the Milwaukee outbreak and to preserve the purity of our 1 percent, the EPA is expected to enact new regulations in December 1995 that will require water-treatment plants serving more than 10,000 people to test its source water for *Cryptosporidium* cysts. If these initial tests reveal contamination, the facility must also test the treated water to determine if the cysts were removed during treatment. The cysts are *not* killed by the chlorine treatment used by most water treatment plants. In June 1995 the CDC released a statement advising immunocompromised individuals to consider alternatives to untreated tap water. In people with impaired immune systems, infection with *Cryptosporidium* can cause severe, potentially fatal disease.

TABLE 26.2	Human Pathogens Transmitted in Water

Organisms	Diseases Caused
Salmonella typhi	Typhoid fever
Other *Salmonella* species	Salmonellosis (gastroenteritis)
Shigella species	Shigellosis (bacillary dysentery)
Vibrio cholerae	Asiatic cholera
Vibrio parahaemolyticus	Gastroenteritis
Escherichia coli	Gastroenteritis
Yersinia enterocolitica	Gastroenteritis
Campylobacter fetus	Gastroenteritis
Legionella pneumophilia	Legionnaire's disease (pneumonia)
Hepatitis A virus	Hepatitis
Poliovirus	Poliomyelitis
Giardia intestinalis	Giardiasis
Balantidium coli	Balantidiasis
Entamoeba histolytica	Amoebic dysentery
Cryptosporidium parvum	Cryptosporidiosis (gastroenteritis)

Chlorination, the addition of chlorine to water, readily kills bacteria but is less effective in destroying viruses and cysts of pathogenic protozoa. The amount of chlorine required to destroy most of the microorganisms is increased by the presence of organic matter in the water. Chlorine may combine with some organic molecules to form carcinogenic substances. Although current water chlorination procedures have not been proved to increase the risk of cancer in humans, the long-term effects of the interaction of chlorine and organic compounds are difficult to assess.

Tests for Water Purity

Water purity is usually tested by looking for **coliform bacteria.** The coliform bacteria, which include *E. coli,* are gram-negative, non-spore-forming, aerobic or facultative anaerobic bacteria that ferment lactose and produce acid and gas. Most municipal water supplies are regularly tested for the presence of coliform bacteria. The presence of a significant number of coliforms is evidence that the water may not be safe for drinking. Three methods of testing for coliform bacteria currently in use are the multiple-tube fermentation method, the membrane filter method, and the ONPG and MUG test.

The **multiple-tube fermentation method** (Figure 26.17) involves three stages of testing: the presumptive test, the confirmed test, and the completed test. In the **presumptive test,** a water sample is used to inoculate lactose broth tubes. Each tube receives a water volume of 10 ml, 1 ml, or 0.1 ml. The tubes are incubated at 35°C and observed after 24 and 48 hours for evidence of gas production. Gas production provides presumptive evidence that coliform bacteria are present. In using the presumptive test, microbiologists analyzing water samples determine approximate numbers of organisms by means of the most probable number method (Chapter 6). (p. 143)

Because certain noncoliform bacteria also produce gas, additional testing is necessary to confirm the presence of coliforms. In the **confirmed test,** samples from the highest dilution showing gas production are streaked onto eosin–methylene blue (EMB) agar plates. EMB prevents growth of gram-positive organisms. Coliforms

(which are gram-negative) produce acid; under acidic conditions the eosin and methylene blue dyes are absorbed by the organisms of a colony. Thus, after 24-hour incubation, coliform colonies have dark centers and may also have a metallic greenish sheen. Observing such colonies confirms the presence of coliforms. In the **completed test,** organisms from dark colonies are used to inoculate lactose broth and agar slants. The production of acid and gas in the lactose broth and the microscopic identification of gram-negative, non-spore-forming rods from slants constitute a positive completed test.

In the **membrane filter method** (Figure 26.18), a 100-ml water sample is drawn through a sterile membrane filter that has pores about 0.45 μm in diameter. This membrane, which traps bacteria on its surface, is then incubated on the surface of a sterile absorbent pad previously saturated with an appropriate growth medium. After incubation, colonies form on the filter where bacteria were trapped during filtration. The presence of more than one colony per 100 ml of water

FIGURE 26.17 The multiple-tube fermentation test for water purity.

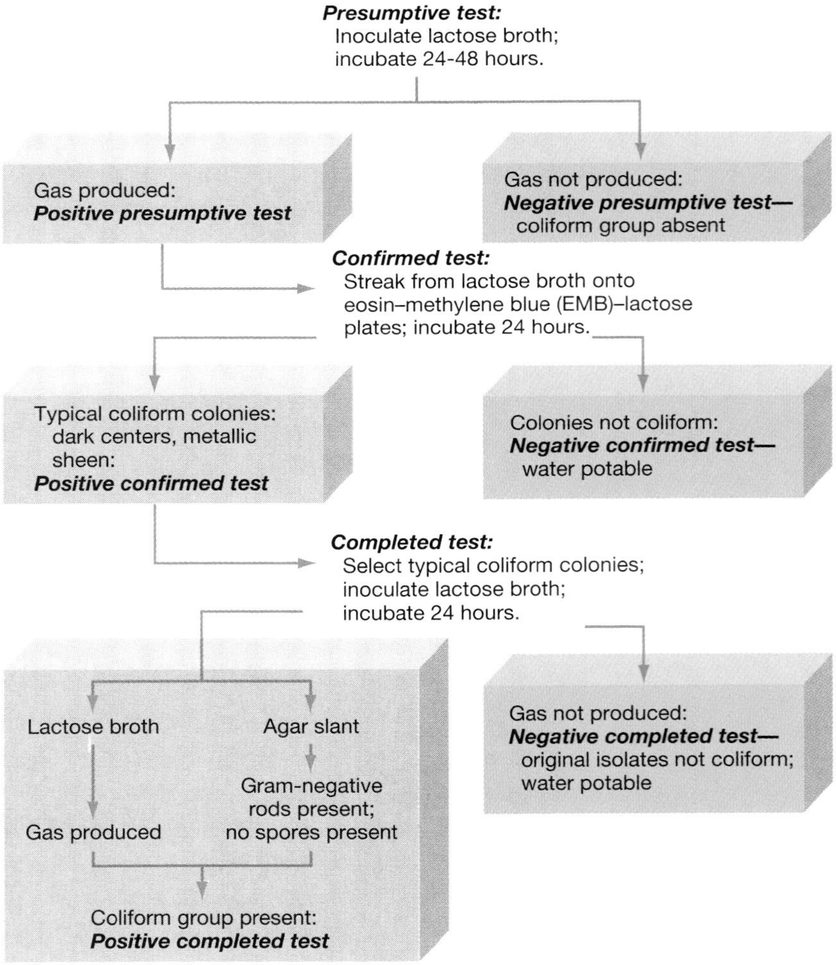

Presumptive test:
Inoculate lactose broth; incubate 24-48 hours.

Gas produced:
Positive presumptive test

Gas not produced:
Negative presumptive test— coliform group absent

Confirmed test:
Streak from lactose broth onto eosin–methylene blue (EMB)–lactose plates; incubate 24 hours.

Typical coliform colonies: dark centers, metallic sheen:
Positive confirmed test

Colonies not coliform:
Negative confirmed test— water potable

Completed test:
Select typical coliform colonies; inoculate lactose broth; incubate 24 hours.

Lactose broth

Agar slant

Gram-negative rods present; no spores present

Gas produced

Gas not produced:
Negative completed test— original isolates not coliform; water potable

Coliform group present:
Positive completed test

(a)

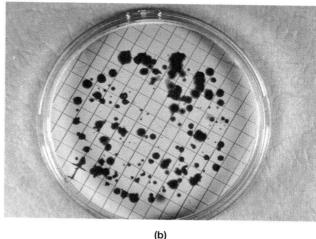

(b)

FIGURE 26.18 The membrane filter test for water purity involves (a) filtering water samples and trapping organisms on filter pads, which are then incubated. (b) After incubation, the coliform colonies that appear are counted.

indicates that the water may be unsafe for human consumption. If colonies are observed, additional tests can be performed to identify them specifically. The membrane filter method is much faster, and permits larger volumes of water to be tested, than the multiple-tube fermentation method.

The **ONPG and MUG test** relies on the ability of coliform bacteria to secrete enzymes that convert a substrate (Chapter 5) into a product that can be detected by a color change. ∞ (p. 111) In this test, a water sample is inoculated into a nutrient-broth tube containing the substrates ONPG (o-nitrophenyl-β-D-galactopyranoside) and MUG (4-methylumbelliferyl-β-D-glucuronide). After a suitable incubation period, if coliforms are present, they will have secreted the enzymes β-galactosidase and β-glucuronidase and other substances. Beta-galactosidase hydrolyzes OPNG into a yellow-colored product, whereas β-glucuronidase hydrolyzes MUG into a product that fluoresces blue when illuminated with ultraviolet light (Figure 26.19). This test can be used in conjunction with the other tests for water purity.

In addition to contamination with coliforms, drinking water can also contain other organisms sometimes referred to as "nuisance" organisms. Although these organisms do not produce human disease, they can affect the taste, color, or odor of the water. Some also can form insoluble precipitates inside water pipes. Nuisance organisms include sulfur bacteria, iron bacteria, slime-forming bacteria, and algae. Among the sulfur bacteria are *Desulfovibrio,* which produces hydrogen sulfide, and *Thiobacillus*, which produces sulfuric acid that can corrode pipes. Iron bacteria deposit insoluble iron compounds that can obstruct water flow in pipes. Eukaryotic algae, diatoms, and cyanobacteria repro-

duce rapidly in water exposed to sunlight (as in reservoirs). They can become so numerous that they clog filters used in the purification process. Identifying these various nuisance organisms in water is a tedious task because different tests are required for each kind of organism. None of the tests are performed routinely but may be done when citizens complain of objectionable tastes, odors, or colors in water.

Water can also be contaminated with various organic and inorganic substances, especially when water supplies are drawn from rivers that have received upstream effluents of industrial wastes. Although it is at least theoretically possible to detect these substances

FIGURE 26.19 The ONPG and MUG test. A positive ONPG produces a yellow color; a positive MUG gives off a blue fluorescence on ultraviolet illumination. A sample without coliforms remains clear.

through chemical analyses, such tests are rarely performed.

SEWAGE TREATMENT

Sewage is used water and the wastes it contains. It is about 99.9 percent water and about 0.1 percent solid or dissolved wastes. These wastes include household wastes (human feces, detergents, grease, and anything else people put down the drain or garbage disposal unit), industrial wastes (acids and other chemical wastes and organic matter from food-processing plants), and wastes carried by rainwater that enters sewers.

Sewage treatment is a relatively modern practice. Until recent years, many large U.S. cities dumped untreated sewage into rivers and oceans; many cities along the Mediterranean Sea still do! When small amounts of sewage are dumped into fast-flowing, well-oxygenated rivers, the natural activities of decomposers in the river purify the water. However, large amounts of sewage overload the purification capacity of rivers. Cities downstream are then forced to take their water supplies from rivers that contain wastes of the cities upstream. Fortunately, most U.S. cities now have some form of sewage treatment.

Complete sewage treatment consists of three steps: primary, secondary, and tertiary treatment (Figure 26.20). In **primary treatment,** physical means are used

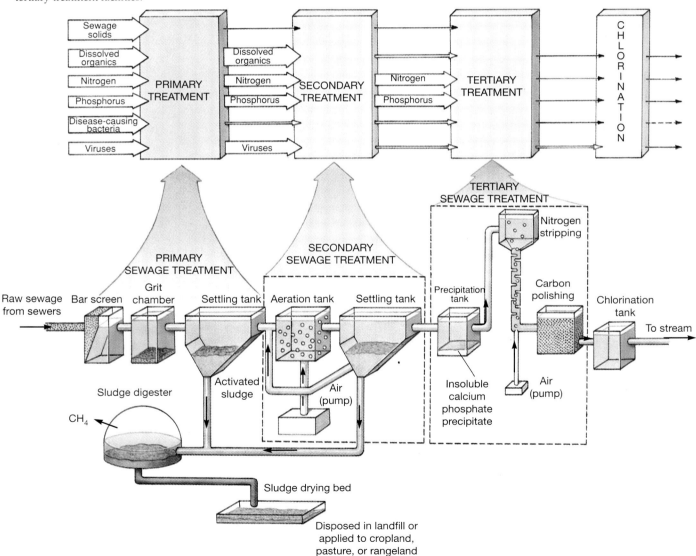

FIGURE 26.20 An overview of a sewage treatment plant, showing primary, secondary, and tertiary treatment facilities.

to remove solid wastes from sewage. In **secondary treatment,** biological means (actions of decomposers) are used to remove solid wastes that remain after primary treatment. In **tertiary treatment,** chemical and physical means are used to produce an effluent of water pure enough to drink. Let us look at each of these processes in more detail.

Primary Treatment

As raw sewage enters a sewage treatment plant, several physical processes are used to remove wastes in primary treatment. Screens remove large pieces of floating debris, and skimmers remove oily substances. Water is then directed through a series of sedimentation tanks, where small particles settle out. The solid matter removed by these procedures accounts for about half the total solid matter in sewage. Flocculating substances can be used to increase the amount of solids that settle out and thus the proportion of solids removed by primary treatment. Sludge is removed from the sedimentation tanks intermittently or continuously, depending on the design of the treatment plant.

Secondary Treatment

The effluent from primary treatment flows into secondary treatment systems. These systems are of two types: trickling filter systems and activated sludge systems. Both systems make use of the decomposing activity of aerobic microorganisms. The BOD is high in secondary treatment systems, so the systems provide for continuous oxygenation of the wastewater.

In a **trickling filter system** (Figure 26.21), sewage is sprayed over a bed of rocks about 2 m deep. The individual rocks are 5 to 10 cm in diameter and are

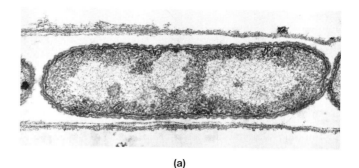

(a)

(b)

FIGURE 26.22 Two sheathed bacteria used in trickling filters are (a) *Sphaerotilus* (29,300×) and (b) *Beggiatoa* (400×).

coated with a slimy film of aerobic organisms such as *Sphaerotilus* and *Beggiatoa* (Figure 26.22). Spraying oxygenates the sewage so that the aerobes can decompose organic matter in it. Such a system is less efficient but less subject to operational problems than an activated sludge system. It removes about 80 percent of the organic matter in the water.

In an **activated sludge system,** the effluent from primary treatment is constantly agitated, aerated, and added to solid material remaining from earlier water treatment. This **sludge** contains large numbers of aerobic organisms that digest organic matter in wastewater. However, filamentous bacteria multiply rapidly in such systems and cause some of the sludge to float on the surface of the water instead of settling out. This phenomenon, called **bulking,** allows the floating matter to contaminate the effluent. The sheathed bacterium *Sphaerotilus* (Figure 26.22a), which sometimes proliferates rapidly on decaying leaves in small streams and causes a bloom, can interfere with the operation of sewage systems in this way. Its filaments clog filters and create floating clumps of undigested organic matter.

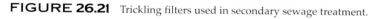

FIGURE 26.21 Trickling filters used in secondary sewage treatment.

FIGURE 26.23 Injection of municipal sewage sludge into farmland is designed to return nutrients to the soil. But if treatment is not complete, it may also add pathogens to the soil.

to nitrogen gas. Finally, chlorine is used to destroy any remaining organisms. Water that has received tertiary treatment can be released into any body of water without danger of causing eutrophication. Such water is pure enough to be recycled into a domestic water supply. However, the chlorine-containing effluent, when released into streams and lakes, can react to produce carcinogenic compounds that may enter the food chain or be ingested directly by humans in their drinking water. It would be safer to remove the chlorine before releasing the effluent, but this is rarely done today, although the cost is not great. Ultraviolet lights are now replacing chlorination as the final treatment of effluent. ∞ (Chapter 13, p. 348) They destroy microbes without adding carcinogens to our streams and waters.

Septic Tanks

About 50 million rural families in the United States do not have access to city sewer connections or their treatment facilities. These homes rely on backyard **septic tank** systems (Figure 26.24). Homeowners must be careful not to flush or put materials such as poisons and grease down the drain, as these might kill the beneficial microbes in the septic tank that decompose sludge solids that accumulate there. This would necessitate immediate pumping of the tank by a vehicle known as the "honey wagon" to prevent sewage from backing up into the house. Even with normal operation, it is occasionally necessary to pump the sludge from the tank and haul it to a sewage treatment plant.

Soluble components of the sewage continue out of the septic tank into the drainage (leaching) field. There they seep through perforated pipe, past a gravel bed and into the soil, which filters out bacteria and some viruses and binds phosphate. Soil bacteria decompose organic materials. Drainage fields must be placed where they will not allow seepage into wells, a difficult problem on hills or in densely populated areas. Drainage fields cannot be used where the water table is too high or the soil is insufficiently permeable, such as in rocky areas. After 10 years or more, the average drainage field clogs up and can no longer be used.

Sludge from both primary and secondary treatments can be pumped into **sludge digesters.** Here, oxygen is virtually excluded, and anaerobic bacteria partially digest the sludge to simple organic molecules and the gases carbon dioxide and methane. The methane can be used for heating the digester and providing for other power needs of the treatment plant. Undigested matter can be dried and used as a soil conditioner or as landfill (Figure 26.23).

Tertiary Treatment

The effluent from secondary treatment contains only 5 to 20 percent of the original quantity of organic matter and can be discharged into flowing rivers without causing serious problems. However, this effluent can contain large quantities of phosphates and nitrates, which can increase the growth rate of plants in the river. Tertiary treatment is an extremely costly process that involves physical and chemical methods. Fine sand and charcoal are used in filtration. Various flocculating chemicals precipitate phosphates and particulate matter. Denitrifying bacteria convert nitrates

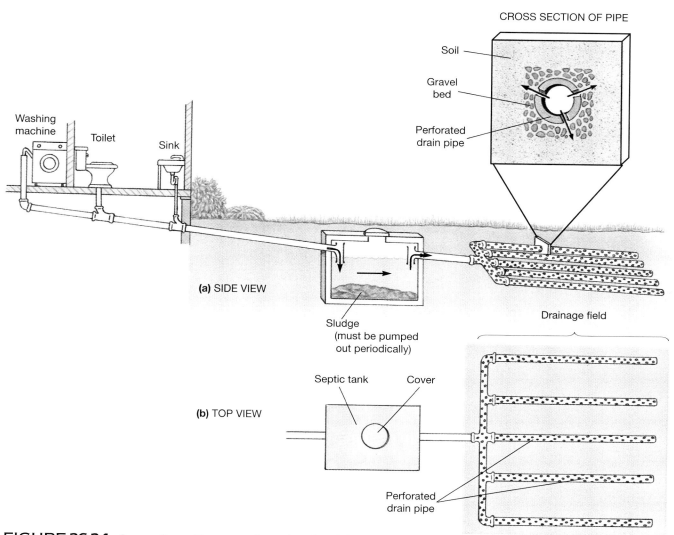

CROSS SECTION OF PIPE

Soil

Gravel bed

Perforated drain pipe

Washing machine

Toilet

Sink

(a) SIDE VIEW

Sludge (must be pumped out periodically)

Drainage field

Septic tank

Cover

(b) TOP VIEW

Perforated drain pipe

FIGURE 26.24 Sewage disposal by means of a septic tank and drainage field system. (a) Solid materials settle out as sludge, which undergoes microbial decay, while soluble materials continue to the drainage field. There (b) they seep out into the soil, which filters out bacteria and some viruses and binds phosphate. Soil bacteria decompose organic materials.

ESSAY

PUTTING MICROBES TO WORK TO CLEAN OUR ENVIRONMENT

In 1991 manufacture of the East German auto the Trabant ceased. The two-cylinder, two-stroke engine couldn't compete with the Mercedes and BMWs of what was then West Germany. Imagine the pain, after perhaps 10 years of saving to buy one, of finding that what should have been a status symbol was now a costly dinosaur. Customers seeking to trade in their Trabants on a bigger car not only got no trade-in allowance but were charged $300 more to dispose of their old Trabant. Space in landfills is scarce and costly in Germany.

Biotechnology may soon have a solution to the problem by means of microbial *biodegradation,* or decomposition of humanmade materials by microbial activity. Trabants are made mostly of plastic. They are so light that two men can lift one into a dumpster. This plastic has a great deal of agricultural waste (cellulose) from the former Soviet Union incorporated into it, which should make it easily biodegradable by microbes. Scientists are searching the soil of landfills for microbes that

are degrading the plastic of Trabants already buried there. These organisms could be used to disintegrate the glut of unwanted Trabants more rapidly (Figure 26.25). One Berlin biotechnology company claims to have developed a bacterium that will eat a Trabant in 20 days, leaving only a small pile of compost.

Plastics such as soda bottles and disposable diapers are an ever-increasing component of already overcrowded landfills. Some companies are producing biodegradable plastics meant to be broken down by microbes over time. In fact, the manufacture of those plastics depends on bacteria. Poly-β-hydroxyalkanes (PHAs) are biodegradable polyesters (polymers formed with ester bonds; Chapter 2) made by bacteria such as *Alcaligenes eutrophus.* ∞ (p. 37) These bacteria store PHAs in granules as a reserve of carbon. When carbon is readily available in the form of glucose, *A. eutrophus* produces large quantities of PHAs to store it for lean times. When carbon is scarce, the bacterium uses its

own stores of carbon in the granules. Thus, this one bacterium can be used to manufacture PHAs—which are then formed into bottles, plastic bags, and such—as well as to degrade those products once their useful days have passed!

Although old Trabants and plastics are a nuisance, they aren't harmful, unlike contaminated soil and water. Dangerous chemical contaminants must be removed from soil and water. **Bioremediation** is a process that uses naturally occurring or genetically engineered microorganisms such as yeast, fungi, and bacteria to transform harmful substances into less toxic or nontoxic compounds. Microorganisms break down a variety of organic compounds in nature to obtain nutrients, carbon, and energy for growth and survival. Bioremediation promotes the growth of microorganisms to degrade contaminants by utilizing those contaminants as carbon and energy sources.

Bioremediation has been used since the late 1970s to degrade petro-

FIGURE 26.25 End of a dream: Dozens of discarded Trabant cars are stacked up in a Berlin junkyard. Scientists are looking for microbes that will eat these cars, which are made mostly of plastic. East Germans once saved for years to afford one of these "dream" cars.

leum products and hydrocarbons. (See the box "Microbial Cleanup" in Chapter 5). ∞ (p. 126) In March 1989, the supertanker *Exxon Valdez* ran aground in Prince William Sound, Alaska. The leaking tanker flooded the shoreline with about 42 million liters (11 million gal) of crude oil. The gravel and sand beaches were saturated with oil more than half a meter thick. A massive cleanup program was organized. At first conventional cleanup techniques—such as booms, high-pressure hot-water sprays, skimmers, and manual scrubbers—were used. The shore remained black and gooey, as these techniques were unable to remove all the oil from beneath rocks and within beach sediments.

Scientists from the EPA decided to use bioremediation to enhance the cleanup. They sprayed the beach with fertilizer (nutrients) in hopes of stimulating the growth of the native microorganisms and promoting the use of the oil wastes as the microbes' carbon source. The areas sprayed with fertilizer were soon nearly clean of oil to a depth of about one-third meter, while untreated areas were still coated with sticky oil. Studies of one treated beach show that 60 percent of total hydrocarbons and 45 percent of polycyclic aromatic hydrocarbons (PAHs), which are potentially toxic, were degraded by bacteria within 3 months. The *Exxon Valdez* cleanup was a bioremediation success story.

Another recent bioremediation success story involves the natural oil-detoxification actions of microbes. In 1995, scientists studying the Persian Gulf region, where lakes of oil had been left after pipelines and oil wells were damaged in the 1991 Gulf War, concluded that a natural recovery of the region's plants was occurring. Samir Radwan and his colleagues in Kuwait found that the roots of wildflowers in the oil-soaked desert were healthy and oil-free. Cultures made of bacteria and fungi from the sand revealed several types of known oil-eating microorganisms, such as the bacterium *Arthrobacter*. The researchers believe that they may have found a cheap, safe, and natural method for cleaning up oil spills on land—cultivation of plants whose roots recruit microbial oil-eaters.

The applications for bioremediation are rapidly expanding. It has already proved to be a successful means of decomposing wastes in landfills. It may be suitable for cleaning up soil and groundwater contaminated by leaks from underground tanks storing petroleum, heating oil, and other materials. The wood-preservative industry also appears to be a promising area for bioremediation. Each year the United States uses 450,000 tons of creosote, an oily liquid that is distilled from coal tar and used as a wood preservative. Creosote sometimes leaks from its holding tanks and seeps into the soil and underlying groundwater. The white-rot fungus *Phanerochaete chrysosporium* has been shown to degrade pentachlorophenol, the main contaminant at wood-preserving sites. This superfungus can also de-

stroy other toxic compounds in the soil, including dioxins, polychlorinated biphenyls (PCBs), and PAHs.

At present only naturally occurring microbes are used in bioremediation. Scientists are experimenting with genetic engineering techniques, however, to develop microorganisms for use at hazardous waste sites. Before these bioengineered organisms can be used in the field, the EPA requires that they undergo a safety review in accordance with the Toxic Substance Control Act to evaluate any possible risks to human health or to the environment.

There are both advantages and disadvantages to the use of bioremediation. On the advantageous side, bioremediation is an ecologically sound, "natural" process; it destroys target chemicals at the contamination site instead of transferring contaminants from one site to another; and the process is usually less expensive than other methods used for cleanup of hazardous wastes. On the disadvantageous side, cleanup using bioremediation often takes longer than other remedial methods such as excavation or incineration, and bioremediation techniques are not yet refined for sites with mixtures of contaminants. More research is needed to perfect this technology. Nevertheless, bioremediation holds enormous promise for the future. As scientists develop more practical uses for bioremediation, this technology will become more important in cleaning up and protecting the environment.

FUNDAMENTALS OF ECOLOGY

- **Ecology** is the study of relationships among organisms and their environment.

The Nature of Ecosystems

- An **ecosystem** includes all the **biotic** and **abiotic factors** of an environment. It will always support life of **indigenous organisms** and sometimes supports **nonindigenous** ones. All the living organisms in an **ecosystem** make up a **community.**

Energy Flow in Ecosystems

- Energy in an ecosystem flows from the sun to **producers** to **consumers. Decomposers** obtain energy from digesting dead bodies and wastes of other organisms. The nutrients they release can be recycled.

BIOGEOCHEMICAL CYCLES

- Although energy is continuously available, nutrients must be recycled from dead organisms and waste to make them available for other organisms.

The Water Cycle

- The **water cycle** is summarized in Figure 26.2.

The Carbon Cycle

- The **carbon cycle** is summarized in Figure 26.3.

The Nitrogen Cycle and Nitrogen Bacteria

- The **nitrogen cycle** is summarized in Figure 26.4.
- **Nitrogen fixation** is the reduction of atmospheric nitrogen to ammonia. It is accomplished by certain free-living aerobes and anaerobes but mostly by *Rhizobium.*
- *Rhizobium* cells accumulate around legume roots and change to **swarmer cells,** which invade root cells and become bacteroids. **Nitrogenase** in **bacteroids** catalyzes nitrogen-fixation reactions.
- **Nitrification** is the conversion of ammonia to nitrites and nitrates; nitrites are formed by *Nitrosomonas,* and nitrates by *Nitrobacter.*
- **Denitrification** is the conversion of nitrates to nitrous oxide and nitrogen gas. It is accomplished by a variety of organisms, especially in waterlogged soils.

The Sulfur Cycle and Sulfur Bacteria

- The **sulfur cycle** is summarized in Figure 26.7.
- Various bacterial species carry out **sulfate reduction, sulfur reduction,** or **sulfur oxidation.**

Other Biogeochemical Cycles

- The **phosphorus cycle** is summarized in Figure 26.9.
- All elements found in living organisms must be recycled.

AIR

Microorganisms Found in Air

- Microorganisms are transmitted by air but do not grow in air. Air is analyzed by exposing agar plates to it and by drawing air over the surface of agar or into a liquid medium and then studying the organisms found.

Methods for Controlling Microorganisms in Air

- Microorganisms in air are controlled by chemical agents, radiation, filtration, and unidirectional airflow.

SOIL

Components of Soil

- Soil contains inorganic rocks, minerals, water, gases, organic matter (**humus**), and many microorganisms.

Microorganisms in Soil

- Microbes of all major taxonomic groups are found in soil.
- Physical factors that affect soil microorganisms include moisture, oxygen concentration, pH, and temperature.
- Organisms also alter the characteristics of their environment as they use nutrients and release wastes.
- Soil microorganisms are important as decomposers in the carbon cycle and in all phases of the nitrogen cycle.

Soil Pathogens

- Soil pathogens affect mainly plants and insects.
- Various species of *Clostridium* are important human pathogens found in soil.

WATER

Freshwater Environments

- Freshwater environments are characterized by low salinity and variability in temperature, pH, and oxygen concentration.
- Microorganisms from all major taxonomic groups are found in freshwater environments. Bacteria—aerobes where oxygen is plentiful, and anaerobes where oxygen is depleted—are especially abundant.

Marine Environments

- Marine environments are characterized by high salinity, smaller variability in temperature, pH, and oxygen concentration. As depth increases, pressure increases and sunlight penetration decreases. Microorganisms from all major taxonomic groups are also found in marine environments. Photosynthetic organisms are found nearest the surface, heterotrophs in surface and lower strata, and decomposers in bottom sediments.

Water Pollution

- Water is polluted if a substance or condition is present that renders the water useless for a particular purpose.

- Table 26.1 summarizes the effects of water pollution.
- Many human pathogens can be transmitted in water.
- Such pathogens are listed in Table 26.2.

Water Purification

- Water purification involves **flocculation** of suspended matter, **filtration,** and **chlorination.**
- Tests for water purity are designed to detect **coliform bacteria;** they include the **multiple-tube fermentation** and **membrane filter methods** and the **ONPG and MUG test.**

SEWAGE TREATMENT

- **Sewage** is used water and the wastes it contains.

Primary Treatment

- **Primary treatment** is the physical removal of solid wastes.

Secondary Treatment

- **Secondary treatment** is the removal of organic matter by the action of aerobic bacteria.

Tertiary Treatment

- **Tertiary treatment** is the removal of most organic matter, nitrates, phosphates, and any surviving microorganism by physical and chemical means.

Water Purification

- Soil bacteria decompose soluble components of sewage in the drainage field of a **septic tank** system.

Septic Tanks

- Soil bacteria decompose soluble components of sewage in the drainage field of a **septic tank** system.

KEY TERMS

abiotic factor *(p. 726)*
activated sludge system *(p. 751)*
bacteroid *(p. 733)*
biogeochemical cycle *(p. 729)*
biological oxygen demand *(p. 743)*
bioremediation *(p. 754)*
biosphere *(p. 726)*
biotic factor *(p. 726)*
bulking *(p. 751)*
carbon cycle *(p. 729)*
chlorination *(p. 748)*
coliform bacterium *(p. 748)*
community *(p. 726)*
completed test *(p. 748)*

confirmed test *(p. 748)*
consumer *(p. 727)*
decomposer *(p. 727)*
denitrification *(p. 734)*
ecology *(p. 726)*
ecosystem *(p. 726)*
eutrophication *(p. 746)*
filtration *(p. 747)*
flocculation *(p. 747)*
humus *(p. 738)*
hydrologic cycle *(p. 729)*
indicator organism *(p. 747)*
indigenous organism *(p. 726)*
membrane filter method *(p. 748)*
microenvironment *(p. 727)*

multiple-tube fermentation method *(p. 748)*
nitrification *(p. 733)*
nitrogen cycle *(p. 731)*
nitrogen fixation *(p. 731)*
nitrogenase *(p. 732)*
nonindigenous organism *(p. 727)*
ONPG and MUG test *(p. 749)*
phosphorus cycle *(p. 735)*
potable water *(p. 743)*
presumptive test *(p. 748)*
primary treatment *(p. 750)*
producer *(p. 727)*
secondary treatment *(p. 751)*

septic tank *(p. 752)*
sewage *(p. 750)*
sludge *(p. 751)*
sludge digester *(p. 752)*
sulfate reduction *(p. 734)*
sulfur cycle *(p. 734)*
sulfur oxidation *(p. 735)*
sulfur reduction *(p. 735)*
swarmer cell *(p. 733)*
tertiary treatment *(p. 751)*
trickling filter system *(p. 751)*
water cycle *(p. 729)*

QUESTIONS FOR REVIEW

A Ecology and Energy Flow

1. What is ecology?
2. Describe some important properties of an ecosystem.
3. Use the following diagram to explain how energy flows through an ecosystem. In doing so, identify parts (a) through (d).

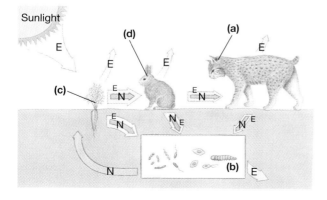

B Recycling; Water and Carbon Cycles

4. Why is recycling of substances essential in an ecosystem?
5. Summarize the water cycle.
6. Summarize the carbon cycle.

C Other Biogeochemical Cycles

7. Summarize the nitrogen cycle.
8. What is nitrogen fixation, and how does *Rhizobium* contribute to it?
9. What is nitrification? Explain how that process is accomplished.
10. What is denitrification? Explain how that process is accomplished.
11. Summarize the sulfur cycle and the roles of microorganisms in it.

12. Match the correct process to each of the following bacteria:

_____ *Thiobacillus*   **a.** nitrogen fixation

_____ *Desulfovibrio*   **b.** nitrification

_____ *Pseudomonas* species   **c.** denitrification

_____ *Nitrosomonas*   **d.** sulfur reduction

_____ *Rhizobium*   **e.** sulfate reduction

_____ *Azotobacter*   **f.** sulfur oxidation

_____ *Serratia*

_____ *Nitrobacter*

_____ *Desulfomonas*

13. Briefly summarize the phosphorus cycle.
14. Why are all elements found in living things recycled?

D Microorganisms in Air
15. What kinds of microorganisms are found in air?
16. How is air analyzed for microorganisms?
17. How are microorganisms in air controlled?

E Microorganisms in Soil
18. List the typical components of soil.
19. What are the properties of soil components?
20. What kinds of microorganisms are found in soil?
21. In what ways do soil microorganisms interact?
22. How do soil microorganisms participate in biogeochemical cycles?
23. What human pathogens are found in soil?

F Freshwater and Marine Environments
24. List the main characteristics of freshwater environments.
25. What kinds of microorganisms are found in fresh water?
26. List the main characteristics of marine environments.
27. What kinds of microorganisms are found in ocean water?

G Water Pollution and Waterborne Pathogens
28. What is water pollution?
29. How does water pollution affect microorganisms and humans?

30. _____ is considered to be an indicator organism for fecal contamination in water.
31. **a.** List four waterborne bacteria that cause gastroenteritis.
    b. List other human pathogens that are transmissible in water.

H Water Purification and Testing
32. What processes are used to purify water?
33. What tests are used to determine water purity?

I Sewage
34. What is sewage?
35. Match each of the following processes with the type of sewage treatment used:

_____ An activated sludge system agitates and aerates the effluent.   **a.** primary treatment

_____ Denitrifying bacteria convert nitrates to nitrogen gas.   **b.** secondary treatment

_____ Screens remove large pieces of debris.   **c.** tertiary treatment

_____ Decomposition by aerobic organisms is used.

_____ Flocculating substances are used to increase solids that settle out.

_____ Fine sand and charcoal filters are employed.

_____ Sewage is spread over a bed of rocks in a trickling filter system.

_____ Sludge digesters utilize anaerobic bacteria.

_____ Chlorination is the final step.

_____ Skimmers are used to remove oily substances.

_____ Viruses and dissolved organics can be removed.

_____ Phosphates are precipitated and removed.

PROBLEMS FOR INVESTIGATION

1. Determine the nature of air or water pollution in your locale. Explain how such pollution affects microorganisms and humans. If possible, suggest ways to reduce its effects.

2. How would you protect workers in sewage disposal plants from being infected with waterborne diseases and from transmitting such diseases to people outside the treatment plants?

3. Propose a method of controlling photosynthetic bacteria in water supplies.

4. Visit a local water purification plant or a sewage treatment plant. Determine which microbiological techniques are used in that plant and the purposes for which they are used.

5. Obtain a sample of your drinking water, and perform the multiple-tube fermentation method to check water purity.

6. Read about alternative water treatment methods that can be used by campers or in times of emergency.

7. What roles do microbes play in compost piles that recycle nutrients?

SOME INTERESTING READING

American Public Health Association. 1992. *Standard methods for the examination of water and waste water*. Washington, D.C.

Atlas, R. M., and R. Bartha. 1993. *Microbial ecology: Fundamentals and applications*, 3d ed. Redwood City, CA: Benjamin Cummings.

Bragg, J., et al. 1994. "Effectiveness of bioremediation for the *Exxon Valdez* oil spill." *Nature* 368(6470):413–18.

Canby, T. Y., and C. O'Read. 1993. "Bacteria." *National Geographic* 184(2):36–62.

Chollar, S. 1990. "The poison eaters: Some cultivated bacteria are finally getting down to business, gobbling up hazardous pollutants." *Discover* 11(April):76–78.

Endo, S. 1994. "An ocean of resources." *Science World* 50(11):14–18.

Folger, T. 1995. "Biosphere III." *Discover* 16(January):66.

Freckman, D. W. 1995. "Soil biodiversity: A new sense of hitting pay dirt." *ASM News* 61(6):280–1.

MacKenzie, W. R., et al. 1994. "A massive outbreak in Milwaukee of *Cryptosporidium* infection transmitted through the public water supply." *The New England Journal of Medicine* 331, no. 3 (July 21):161–67.

McCabe, A. 1990. "The potential significance of microbial activity in radioactive waste disposal." *Experientia* 46(8):779–87.

Mitchell, R. 1992. *Environmental microbiology*. New York: Wiley-Liss.

Nebel, B. J., and R. T. Wright. 1996. *Environmental science: The way the world works*, 5th ed. Englewood Cliffs, NJ: Prentice Hall.

"Oil-fighting bacteria discovered." 1994. *Science Teacher* 61(6):16.

Olson, B. H. 1991. "Tracking and using genes in the environment." *Environmental Science & Technology* 25(April):604–11.

Stetter, K. O. 1995. "Microbial life in hyperthermal environments." *ASM News* 61(6):280–1.

Stone, R. 1995. "If the mercury soars, so may health hazards." *Science* 267(5200):957–58.

Stover, D. 1992. "Toxic avengers." *Popular Science* 24(1):70–76.

Swannell, R., and I. Head. 1994. "Bioremediation comes of age." *Nature* 368(6470):396–97.

Tate, N. 1994. *The sick-building syndrome: How indoor air pollution is poisoning your life—and what you can do*. New York: New Horizon.

CHAPTER 27

Aging of gouda cheese. The action of microorganisms
ripens cheese. Humans have taken advantage of such
uses of microbes since the dawn of civilization.

APPLIED
MICROBIOLOGY

T
hroughout history, humans have employed microorganisms. We have long used microorganisms in making food and medicine, and today
other industrial applications of microbes abound. And
because biotechnology is advancing rapidly, the future
of applied microbiology looks very bright. As we consider the many applications of microbiology—from
food production to mining—we must also discuss food
spoilage and methods of preservation.

MICROORGANISMS FOUND
IN FOOD

Anything that people eat or drink can be used as food
by microorganisms, too (Figure 27.1). Most substances
consumed by humans are derived from plants, which,
of course, grow in soil, or from animals, which live in
contact with soil and so have soil organisms on them.
Although soil organisms usually are not human
pathogens, many can cause food spoilage. The handling of foods from harvest or slaughter to human consumption provides many more opportunities for the
foods to become contaminated with microorganisms.

FIGURE 27.1 Microorganisms utilize the same foods that
humans do, as can be seen from these strawberries covered with
"gray mold."

Unsanitary practices by food handlers and unsanitary
working conditions frequently lead to contamination
of foods with pathogens. Improper storage and preparation procedures at home and especially in restaurants
can lead to further contamination with pathogens. Improper refrigeration of prepared foods is a major source
of food poisoning.

THIS CHAPTER FOCUSES ON THE FOLLOWING QUESTIONS:

A What kinds of microorganisms are found in different categories of food?

B How are diseases transmitted in food?

C How can food spoilage and disease transmission be prevented, and what standards relate to these problems?

D How are microorganisms used as food or in the making of food products?

E How are microbes used in the manufacture of beer, wine, and spirits?

F How can microbes be used in industry, and what problems are associated with their use?

G What is the role of microorganisms in the manufacture of simple organic compounds, antibiotics, enzymes, and other biologically useful substances?

H How are microorganisms used in mining?

I How are microorganisms used in waste disposal?

Grains

When properly harvested, the various kinds of edible grains, such as rye and wheat, are dry. Because of the lack of moisture, few microorganisms thrive on them. However, if stored under moist conditions, grains can easily become contaminated with molds and other microorganisms. Insects, birds, and rodents also transmit microbial contaminants to grains.

Raw grains contaminated with the mold *Claviceps purpurea*, known as *ergot*, cause ergot poisoning, or ergotism. Compounds produced by this fungus are hallucinogenic and can alter behavior or even be deadly if eaten. Certain species of *Aspergillus* infecting peanuts and some grains produce *aflatoxins*. These toxic compounds have proved to be potent mutagens and carcinogens. ⊂⊃ (Chapter 22, p. 631)

Most grains are used in making breads and cereals. Humans have made breads for thousands of years, and some loaves of bread on display in the British Museum are 4000 years old. Although the first instance of leavening bread with yeast probably occurred by accident, special strains of *Saccharomyces cerevisiae* that produce large amounts of carbon dioxide are now added to bread dough to make

it rise. We will describe this process in more detail later.

Like raw grain, bread is susceptible to contamination and spoilage by various molds. *Rhizopus nigricans* is the most common bread mold, but several species of *Penicillium, Aspergillus,* and *Monilia* also thrive in bread. Contamination of bread with *M. sitophila*, a pink bread mold, is especially dreaded by bakers because it is almost impossible to eliminate from a bakery once it has become established. Rye bread is particularly likely to become contaminated with *Bacillus* species, which hydrolyze proteins and starch and give the bread a stringy texture. If not killed by baking, *Bacillus* spores germinate and cause rapid and extensive damage to freshly baked bread.

Fruits and Vegetables

Millions of commensal bacteria, especially *Pseudomonas fluorescens,* are found on the surfaces of fruits and vegetables. These foods also easily become contaminated with organisms from soil, animals, air, irrigation water, and equipment used to pick, transport, store, or process them. Pathogens such as *Salmonella, Shigella,*

Entamoeba histolytica, Ascaris, and a variety of viruses can be transmitted on the surfaces of fruits and vegetables. However, the skins of most plant foods contain waxes and release antimicrobial substances, both of which tend to prevent microbial invasion of internal tissues.

Certain vegetables are particularly vulnerable to microbial attack and spoilage. Leafy vegetables and potatoes are susceptible to bacterial soft rot by *Erwinia carotovora.* The fungus *Phytophthora infestans* caused the Irish potato famine of 1846. Biotechnology is being called upon to help prevent such crop damage. Attempts are under way, for example, to insert into soybeans a gene for disease resistance, found in wild mustard plants, that could yield pathogen-resistant soybean crops. Similarly, a gene that protects plants against the bacterium *Pseudomonas syringae* may soon be introduced into corn, bean, and tomato plants.

Fruits are likewise susceptible to spoilage by microbial action. Tomatoes, cucumbers, and melons can be damaged by the fungus *Fusarium,* which causes soft rot and cracking of tomato skins. Fruit flies pick up the fungus from infected tomatoes and transmit it to healthy ones as they deposit eggs in surface cracks. Other insects pierce tomatoes to feed on them and at the same time introduce *Rhizopus,* which breaks down pectin and can turn a tomato into a bag of water. Fresh fruit juices, because of their high sugar and acid content, provide an excellent medium for growth of molds, yeasts, and bacteria of the genera *Leuconostoc* and *Lactobacillus.* Grapes and berries are damaged by a wide variety of fungi, and large numbers of stone fruits, such as peaches, can be destroyed overnight by brown rot due to *Monilia fructicola. Penicillium expansum,* which grows on apples, produces the toxin patulin, which can easily contaminate cider. Other *Penicillium* species produce blue and green mold on citrus fruits.

Meats and Poultry

Meat animals arrive at slaughterhouses with numerous and varied microorganisms in the gut and feces, on hides and hoofs, and sometimes in tissues. At least 70 pathogens have been identified among these microbes. Nearly all animal carcasses in slaughterhouses in the United States are inspected by a veterinarian or a trained inspector, and those found to be diseased are condemned and discarded (Figure 27.2). Inspection, however, cannot guarantee that meat is parasite-free. The inspector examines the carcass surface and the heart (where helminths are often concentrated) but cannot look inside each cut of meat. Some of the most common diseases identified in slaughterhouses are abscesses, pneumonia, septicemia, enteritis, toxemia, nephritis, and pericarditis. Lymphadenitis, an inflam-

FIGURE 27.2 Meat inspection at the slaughterhouse helps provide, but cannot guarantee, a safe supply of meat for consumption.

mation of the lymph nodes, is especially common in sheep and lambs.

Even after animals are slaughtered and the carcasses are hung in refrigerated rooms to age, microorganisms sometimes spoil the meat. Several molds grow on refrigerated meats, and *Cladosporium herbarum* can grow on frozen meats. Mycelia of *Rhizopus* and *Mucor* produce a fluffy, white growth referred to as "whiskers" on the surfaces of hanging carcasses. The bacterium *Pseudomonas mephitica* releases hydrogen sulfide and causes green discoloration on refrigerated meat under low oxygen conditions. Several species of *Clostridium* cause putrefaction called **bone stink** deep in the tissues of large carcasses.

Ground meats sometimes contain helminth eggs and always contain large numbers of lactobacilli and molds. In most areas butcher shops are required to maintain two separate meat-grinding machines—one for pork and one for other meats. This practice is encouraged because it is difficult to clean a machine thoroughly enough to eliminate any possibility of transmitting raw pork bits, which may carry the helminth *Trichinella spiralis,* the cause of trichinosis, to other meats. ∞ (Chapter 12, p. 319) Lactobacilli in ground meats produce acids that retard the growth of enteric pathogens. Nevertheless, ground meats are subject to spoilage even when refrigerated and should be frozen if not to be used within a day or two. All meats, but especially ground meats, should be cooked thoroughly to kill pathogens.

More than 20 bacterial genera have been found on dressed poultry, and mishandling of poultry in restaurants accounts for many foodborne infections. Nearly half these infections have been traced to *Salmonella* and a fourth each to *Clostridium perfringens* and *Staphylococcus aureus.* Freezing fails to rid poultry of *Salmonella.*

BIG OR LITTLE END UP WHEN STORING EGGS?

Does it make a difference whether you refrigerate eggs with the big end or the little end up? Big end up is the recommendation of food scientists. Why? The object is to keep the egg yolk and any embryo residing on it as close to the center of the egg as possible. This maximizes the distance an invading microbe would have to swim from the shell to the yolk. The egg white is filled with chemical hazards for the bacteria. Lysozymes attack the bacterial cell walls, killing many bacteria by rupturing their walls. Nutrients, vitamins, and metal ions of iron, copper, and zinc are wrapped tightly by proteins and other substances in the egg white and so are unavailable to the bacteria. Deprived of nutrients and chewed up by lysozymes, bacteria do not survive the swim; the embryo remains safe.

But why big end up? The natural tendency of the lipid-rich yolk is to rise, just as oil floats to the top of water. Bird eggs have two ropy cords (look for them next time you crack a raw egg) called chalaza, which act much like the ropes of a hammock to suspend the yolk from the shell lining. The larger chalaza is at the small end of the egg and can hold the yolk down better, preventing it from rising too close to the edge of the protective egg white.

Pseudomonads and several other gram-negative bacteria are common contaminants of poultry, where they cause slime and offensive odors.

You might suppose that eggs, with their hard shells, would be free of microbial contamination. Most are, but the shells are porous, and pseudomonads and some other bacteria, as well as fungi such as *Penicillium*, *Cladosporium,* and *Sporotrichum*, grow on eggshells. These microbes can pass through pores in the shells to infect the inside of eggs. *Salmonella* also survives on eggshells and can enter broken eggs or be deposited with bits of shell in foods. Hens infected with *S. pullorum* lay infected eggs. The CDC reports that 1 out of every 10,000 eggs has *Salmonella* inside the shell. Any pathogens on eggshells or in eggs can be transmitted to humans unless eggs and foods that contain them are cooked thoroughly.

Fish and Shellfish

Fresh fish abound with microorganisms. Several species of enteric bacteria and clostridia, enteroviruses, and parasitic worms are commonly found on fresh fish.

POULTRY AND EGGS

Salmonella species are found in many animal reservoirs, especially poultry. Recent investigations of *Salmonella* in the poultry industry were not reassuring, and consumers can assume that poultry products will be heavily contaminated. What steps can you take to maneuver defensively in the sea of *Salmonella*?

First, assume that the shells of all eggs are contaminated, and wash your hands after handling them. (But do not wash the eggs—washing removes a protective surface coating that helps prevent microbes from entering the eggs.) *Salmonella* are part of the intestinal microbiota of poultry, and eggs surely come in contact with feces, feathers, and contaminated surfaces. Cracked eggs cannot be sold for human consumption, but in some areas they can be sold as pet food. When organisms enter an egg, they find a superb, nutrient-rich culture medium in which they can rapidly multiply.

Second, when baking, control your urge to lick raw batter if you've added any eggs to it. If the batter contains eggs, even powdered eggs in mixes, it might contain *Salmonella*. Pieces of shell dropped into batter, even when promptly retrieved, can inoculate it with organisms. Also, raw eggs can be contaminated internally if laid by an infected hen or if the eggs have undetected cracks or have been immersed in water.

Third, handle raw poultry carefully. Poultry is naturally contaminated with fecal organisms, and contamination is exacerbated by the industry's practice of holding poultry in water baths during certain phases of the plucking and evisceration processes. The water becomes a *Salmonella* broth, and the surfaces of the birds become uniformly contaminated. Thus, use poultry promptly and discard contaminated wrappings and juices carefully. If you have laid poultry on a countertop or cutting board, scrub those surfaces thoroughly with hot soapy water before placing other foods on them. And don't touch other foods until you have washed your hands thoroughly.

Contaminated eggs being destroyed.

Many of these organisms survive shipment of fish packed in crushed ice, especially if the fish are packed too tightly together or are pressed against the slats of contaminated crates.

Shellfish, such as oysters and clams, carry many of the same organisms as fish. Raw oysters typically carry *Salmonella typhimurium* and sometimes *Vibrio cholerae*. Clams are especially likely sources of human infection because they are filter feeders—that is, they obtain food by filtering water and extracting microbes. If clams are exposed to increased levels of sewage, red tides, and other sources of large numbers of pathogens or toxin producers, clamming may be prohibited until the numbers of organisms decrease. Scallops are less likely to transmit diseases to humans because only the muscular part of the organism, and not its digestive tract, is eaten.

Among crustaceans, shrimp are extremely likely to be contaminated. Some studies have shown that over half the breaded shrimp on the market contain in excess of 1 million bacteria (and more than 5000 coliforms) per gram. Such high bacterial counts probably result from growth of bacteria during processing before the shrimp are frozen. Lobsters and crabs are even more perishable than shrimp and can carry a variety of enteric pathogens. Along the U.S. Gulf Coast, improperly cooked crabs have transmitted cholera. Crabs also carry *Clostridium botulinum* ∞ (Chapter 24, p. 687) and the pathogenic fungi *Cryptococcus* and *Candida*. Keep in mind, however, that the mere presence of microorganisms in seafoods—or in any other foods—does not necessarily mean that the foods are spoiled or contaminated with pathogens. In fact, *Lactobacillus bulgaricus*, which produces hydrogen peroxide, can be used to inhibit growth of other organisms found on seafood.

Milk

Modern mechanized milking and milk handling has greatly reduced the microbial content of raw milk (Figure 27.3). However, breeding dairy cattle for increased milk production has resulted in exceptionally large udders and teats that easily admit bacteria. The first few milliliters of milk drawn from such cows can contain as many as 15,000 bacteria per milliliter, whereas the last milk drawn is free of microorganisms. Most microorganisms in freshly drawn milk are *Staphylococcus epidermidis* and *Micrococcus*, but *Pseudomonas, Flavobacterium, Erwinia,* and some fungi also can be present.

Microorganisms have many opportunities to enter milk before it is consumed. Hand milking, as opposed to mechanical milking, allows organisms from the body of the cow to enter the milk. These include *Escherichia coli*, which give milk a fecal flavor, and *Acinetobacter johnsoni* (formerly known as *Alcaligenes viscolactis*), which is especially abundant during summer months and causes a viscous slime to form in milk. Storing, transporting, and processing milk allow for contamination with any organisms in containers and for growth of those already present. Infectious organisms in milk usually come either from infected cows or from unsanitary practices of milk handlers. Diseased cattle can transmit *Mycobacterium bovis* and *Brucella* species to their milk. Dairy herds are tested for tuberculosis (with infected animals removed from the herd) and vaccinated against brucellosis (undulant fever); hence the risk of transmitting these diseases to humans is small. *Staphylococcus aureus, Salmonella* species, and other enteric bacteria can enter milk through unsanitary handling. Bacteria and molds will grow even in dried milk, which is made from pasteurized liquid

PUBLIC HEALTH

SEAFOOD

Seafood is highly perishable and should always be as fresh as possible when eaten. If you buy a lobster to cook, make sure that it is alive when you bring it home. If it is dead when you are ready to cook it, don't. Similarly, make sure that the shells of clams or oysters are tightly closed—not only when you buy them but also, if you are serving them uncooked, when you are ready to serve them. If you find a shell open when you are about to serve a raw clam or oyster, don't serve it. Better yet, remember that eating raw seafood is a calculated risk.

FIGURE 27.3 A merry-go-round milking parlor in New Jersey. Mechanized procedures for milking and milk handling have greatly reduced the microbial content of raw milk.

milk but is unsterilized, if the water content of the powder reaches 10%.

Some microorganisms, such as certain *Pseudomonas* species and some soil organisms, grow in refrigerated milk. These organisms are psychrophilic; although they normally grow at higher temperatures, they can grow at 5°C (refrigerator temperature). They also survive the concentration of chlorine normally used to purify drinking water.

Organisms that sour milk include *Streptococcus lactis* and species of *Lactobacillus*. When these microbes release enough lactic acid to bring the pH below 4.8, the milk proteins coagulate, and the milk is said to have soured. Souring of milk does not mean that the milk is unsafe for human consumption, but it does greatly alter the taste and appearance of the milk.

Various microorganisms that are involved in the spoilage of food are summarized in Table 27.1.

Other Edible Substances

Humans consume sugar, spices, condiments, tea, coffee, and cocoa—all of which are subject to microbial contamination—in addition to major nutrients. Fresh,

PUBLIC HEALTH

CANDY

Candy is generally safe from microbial contamination because its high sugar concentration creates an osmotic pressure too high for most organisms to survive. Occasionally, the cream filling in a batch of chocolates becomes contaminated with *Salmonella* or *Clostridium*. Chocolate lovers have been surprised to discover that the gas produced by clostridia has caused their favorite chocolates to explode.

dry, refined sugar is sterile, but raw sugar-cane juice supports growth of fungi such as *Aspergillus, Saccharomyces,* and *Candida* and several species of bacteria, including *Bacillus* and *Micrococcus*. Most are removed by filtration, and the remainder are killed by heat during evaporation of the juice. Foods to which sugar is added are especially susceptible to spoilage because the sugar is an excellent nutrient for many organisms. *Bacillus stearothermophilus,* a facultative anaerobe that

TABLE 27.1	Microorganisms Involved in Food Spoilage	
Food	**Organism**	**Type of Spoilage**
Grains		
Bread	*Rhizopus nigricans*	Bread mold
	Penicillium species	
	Aspergillus species	
	Monilia sitophilia	
	Bacillus species	Protein and starch hydrolysis—stringy texture
Fruits and Vegetables		
Leafy vegetables	*Erwinia carotovora*	Bacterial soft rot
Potatoes	*Phytophthora infestans*	Potato blight
Tomatoes, melons	*Fusarium* species	Soft rot; cracking skin
Grapes, berries, "stone" fruit	*Monilia fructicola*	Brown rot
Apples	*Penicillium expansum*	Patulin, a toxic cider contaminant
Meat		
	Rhizopus species	"Whiskers," white fluffy growth
	Mucor species	
	Pseudomonas	Green discoloration on refrigerated meat
	Clostridium species	Bone stink
Poultry		
	Pseudomonads	Sliminess
	Penicillium	Eggshell contamination
	Cladosporium species	
	Sporotrichum species	
Milk		
	Acinetobacter johnsoni	Sliminess
	Streptococcus lactis	Sour milk
	Lactobacillus species	

grows best at 55° to 60°C, can multiply rapidly during processing of foods. Another thermophilic anaerobe, *Clostridium thermosaccharolyticum,* is often responsible for the gas production that causes cans to bulge. Conversly, sugar in high concentration acts as a preservative. The high sugar concentration in such foods as jellies, jams, candies, and candied fruits creates sufficient osmotic pressure to inhibit microbial growth.

Maple trees are tapped, and sap is collected from them, in the early spring. The sap becomes increasingly contaminated as the weather becomes warmer. Organisms that thrive in maple sugar sap include species of *Leuconostoc, Pseudomonas,* and *Enterobacter.* Although these organisms may consume large quantities of sugar, they are killed as the sap is evaporated to syrup or sugar.

Honey can contain toxins if it is made of nectars from such plants as *Rhododendron* or *Datura.* It can also contain spores of *Clostridium botulinum.* Although the spores do not germinate in the honey, in infants they can germinate after the honey is ingested. Their toxins can cause "floppy baby syndrome."

Spices have been used in food preservation and embalming for centuries and have thereby gained a reputation as antimicrobials. This reputation is undeserved; spices more often mask odors of putrefaction than prevent spoilage. Leeuwenhoek, who first observed bacteria in spices, reported that water containing whole peppercorns teemed with them. Among the many microorganisms found in spices (Table 27.2), most are not pathogens. The small amounts of spices used in cooking are probably not a health hazard.

Condiments such as salad dressings, catsup, pickles, and mustard usually are markedly acidic. Although low pH prevents growth of many microorganisms, some molds are able to grow in these foods if they are not refrigerated.

Americans consume huge quantities of carbonated beverages and coffee and lesser quantities of tea and

CLOSE-UP

TEA OR COFFEE FOR BRITAIN

When the coffee plantations of Ceylon (now Sri Lanka) were destroyed in the 1860s by coffee rust, the fields were replanted with tea. The British, who had depended on Ceylon for their coffee supplies, were forced to switch to tea. Thus, the lowly fungus played a major role in changing the British into a nation of tea drinkers. The Irish, two decades earlier, had not been so lucky. When a fungal disease destroyed the potato crop in the late 1840s, no replacement capable of feeding the population was available. A million people starved to death, and over a million emigrated.

Could such catastrophes occur today? The answer, unfortunately, is probably yes. Most of the food crops grown today are potentially vulnerable to diseases and pests because they are monotypes, or pure strains, lacking the genetic diversity that is typical of wild plants. An outbreak of plant disease could therefore spread very rapidly through an entire crop. Agricultural scientists are using genetic engineering techniques to develop strains that are resistant to diseases and insects—rust-resistant strains of coffee, for example (as well as varieties that are naturally low in caffeine). However, there is always the danger that such strains will turn out to be vulnerable to a new species of microorganism—either a mutation or one to which previously cultivated varieties were not susceptible. This happened to the U.S. corn crop in the 1970s and resulted in large economic losses.

cocoa. The automated equipment in modern manufacturing plants can prepare carbonated beverages aseptically, but syrups can become contaminated with molds if there is a mechanical failure. Syrups sold in bulk to restaurants also can become contaminated. Freshly har-

TABLE 27.2	Numbers of Bacteria in Spices				
	Number of Microorganisms per Gram of Dry Sample				
Type of Spice	Total Aerobes	Coliforms	Yeasts and Molds	Aerobic Spores	Anaerobic Spores
Bay leaves	520,000	0	3,300	9,200	<2
Cloves	3,000	0	18,005	<2	<2
Curry	>7,500,000	0	70	>240,000	>240,000
Marjoram	370,000	0	18,000	54,000	>24,000
Paprika	>5,500,000	600	2,300	>240,000	>620
Pepper	>2,000,000	0	15	>240,000	>24,000
Sage	6,800	0	10	7	>001,700
Thyme	1,900,000	0	11,000	160,000	>24,000
Turmeric	1,300,000	50	70	>110,000	>240,430

Source: Adapted from Karlson and Gunderson, 1965, *Food Technology* 1986, with permission from the Institute of Food Technologies.

vested coffee beans are subject to contamination by various molds and by microorganisms transmitted to them by insects. The coffee rust *Hemileia vastatrix,* a fungus, has devastated coffee plantations in Asia and is now a serious problem in South America. Tea leaves allowed to become moist are susceptible to contamination by *Aspergillus* and *Penicillium* molds, which impart an unpleasant aroma to tea brewed from them.

Microorganisms are useful in preparing coffee and cacao beans for marketing. The bacterium *Erwinia dissolvens* is used to digest pectin in outer coverings of coffee beans. Other bacteria are used to dissolve the covering of cacao beans, the beans from which cocoa and chocolate are made. These beans are subsequently treated with fermenting bacteria before they are roasted.

PREVENTING DISEASE TRANSMISSION AND FOOD SPOILAGE

Diseases acquired from food are due mainly to the direct effects of microorganisms or their toxins (Table 27.3), but they can result from microbial action on food substances. Industrialization has increased the spread of foodborne pathogens. Large processing plants provide opportunities for contamination of great quantities of food unless sanitation is strictly practiced. In institutions that feed large numbers of people, contaminated food will cause many cases of disease. The increased popularity of convenience foods, especially fast foods, has also raised the risk of infection.

In addition to the enteric diseases described in Chapter 20, several other diseases can be transmitted in food. *Klebsiella pneumoniae* is commonly found in the human digestive tract. Although it is thought of mainly as a pathogen of the respiratory system, it can cause diarrhea in infants, abscesses, and nosocomial wound

and urinary tract infections. Tuberculosis can be transmitted by fomites in food, unpasteurized milk and cheese, and meats from infected animals.

Several diseases can be transmitted to humans who eat infected meat. They include anthrax, brucellosis, Q fever, and listeriosis. Because of meat inspection procedures, animal and meat handlers are exposed to these diseases much more often than are consumers. Adirondack disease, caused by *Yersinia enterocolitica,* is thought to be transmitted by eating infected meat, but it also may be transmitted through milk and water. The disease has many forms, including mild to severe gastroenteritis, arthritis, glomerulonephritis, a fatal typhoidlike septicemia, and fatal ileitis, an inflammation of the small intestine that can be mistaken for appendicitis. People have also become infected with *Erysipelothrix rhusiopathiae* by eating infected pork. The resulting disease—called erysipelas (er-i-sip'e-las) in animals and erysipeloid in humans—infects swine, sheep, and turkeys. It is most likely to infect farmers and packing plant workers through skin wounds. The organism can infect skin, joints, and the respiratory tract.

Viruses are frequently transmitted through food. Enteroviruses are often spread during unsanitary handling of food, especially by asymptomatic food handlers. Droplet infection of food can transmit echovirus and coxsackie respiratory infections. Poliomyelitis viruses can be transmitted through milk and other foods. And the viruses responsible for hepatitis A can be transmitted through shellfish from contaminated waters. The virus that causes lymphocytic choriomeningitis, a flulike illness, can be spread to foods by mice.

Milk is an ideal medium for the growth of many pathogens. In addition to toxin producers and pathogens found in other foods, milk can contain organisms from cows. These organisms include *Mycobacterium bovis, Brucella* species, *Listeria monocytogenes,* and *Coxiella burnetii.* Spore-producing organisms

TABLE 27.3	Pathogenic Organisms Transmitted in Food and Milk	
Organism	Disease	Vector
Staphylococcus aureus	Food poisoning	Infected food handlers, unrefrigerated foods, milk from infected cows
Clostridium perfringens	Food poisoning	Unrefrigerated foods
Bacillus cereus	Food poisoning	Unrefrigerated foods
Clostridium botulinum	Botulism	Inadequately processed canned goods
Salmonella species	Salmonellosis	Infected food handlers, poor sanitation, contaminated seafood
Shigella species	Shigellosis	Infected food handlers, poor sanitation
Enteropathogenic *Escherichia coli*	Traveler's diarrhea and other diseases	Infected food handlers (sometimes asymptomatic), poor sanitation, contaminated meat
Campylobacter	Gastroenteritis	Undercooked chicken and raw milk
Vibrio cholerae	Cholera	Poor sanitation
Vibrio parahaemolyticus	Asian food poisoning	Undercooked fish and shellfish

such as *Bacillus anthracis* enter milk from infected cows or from soil. But milk also contains certain antibacterial substances, including lysozyme, agglutinins, leukocytes, and lactenin. Lactenin is a combination of thiocyanate, lactoperoxidase, and hydrogen peroxide. It is also present in human milk and other body secretions and may help prevent enteric infections in newborns. Fermented milk contains bacteria such as *Leuconostoc cremoris* that kill pathogens. A crucial factor in preventing spoilage and disease transmission in food and milk is cleanliness in handling. Other common-sense practices—prompt use of fresh foods, careful refrigeration, and prompt and adequate processing of foods to be stored—also help control disease transmission and spoilage.

Food Preservation

Many procedures used to preserve foods are based on practices begun early in human civilization. The ability to maintain a stable year-round food supply was essential in allowing humans to abandon a nomadic lifestyle, which followed the food supply, for a more settled village lifestyle. These practices were most likely based on simple observations like the following: Grain that was kept dry did not turn moldy. Dried and salted foods remained edible over a long period of time. And milk allowed to sour or made into cheese was usable over a much longer period of time than was fresh milk. Modern methods of food and milk preservation still utilize some of these early methods, but they also use heat and cold and other specialized procedures.

Many of the methods of food preservation have been described under antimicrobial physical agents in Chapter 10. They include canning using moist heat; refrigeration, freezing; lyophilization and drying; and the use of radiation. A number of chemical food additives also are used to retard spoilage.

Canning

The most common method of food preservation is **canning**—the use of moist heat under pressure. This method, analogous to laboratory autoclaving, is used to preserve fruits, vegetables, and meats in metal cans or glass jars (Figure 27.4). If properly carried out, canning destroys all harmful spoilage microorganisms, including most heat-resistant endospores, prevents spoilage, and averts any hazard of disease transmission. Foods so treated may remain edible for years.

Some thermophilic anaerobic endospores, such as those of *Bacillus stearothermophilus*, may remain alive even after commercial canning. Hence canned food should not be stored in a hot environment, such as a car trunk. At elevated temperatures, the endospores can germinate, grow, and cause spoilage. Usually this

FIGURE 27.4 Commercial canning equipment (being used here to can tomatoes) destroys all microbes and their spores, thereby preventing food spoilage and disease transmission.

produces gas, which causes the ends of cans to bulge such that the ends can be pressed up and down. Such spoilage also usually produces acid, which gives a sour flavor. Such changes are called **thermophilic anaerobic spoilage.** In some cases, however, spoilage due to growth of such spores does not cause cans to bulge with gas; such spoilage is called **flat sour spoilage.** Cans can also bulge due to **mesophilic spoilage,** which occurs when canning procedures were improperly followed or the seal has been broken. This type of spoilage can take place at room temperature, unlike flat sour and thermophilic spoilage, which occur only in properly processed and sealed cans that have been stored at high temperatures.

Because of the danger of botulism and other kinds of spoilage in improperly processed canned foods, people who do home canning should carefully follow instructions in a good, up-to-date canning handbook. The U.S. Department of Agriculture (USDA) now recommends that all low-acid foods be processed by pressure cooking. The USDA also recommends that jellies and jams, formerly packed hot and sealed with wax, be processed in a boiling water bath and sealed with lids like other canned goods to prevent accumulation of mold toxins. (See the box "Home Canning" in Chapter 13.) ∞ (p. 345) *Any can that has bulging ends— whether produced at home or commercially—should be discarded.*

Although many condiments are heat processed, some of their properties help preserve them. Sugar in jellies and jams increases osmotic pressure and retards growth of microorganisms. (Saccharin does not have this effect, so artificially sweetened products may require more stringent precautions to prevent spoilage than do those with a high sugar content.) Likewise, the high acidity of pickles and other sour foods helps prevent microbial growth.

Refrigeration and Freezing

Refrigeration at temperatures slightly above freezing (about 4°C) is suitable for preserving foods only for a few days. It does not prevent growth of psychrophilic organisms that can cause food poisoning. Freezing, another common method of food preservation, involves storage of foods at temperatures below freezing (about −10°C in most home freezers). All kinds of foods can be preserved by freezing for several months and some for much longer periods of time. An advantage of freezing is that it preserves the natural flavor of foods better than does canning. However, freezing has two disadvantages: (1) It causes some foods, especially fruits and watery vegetables, to become somewhat soft upon thawing and detracts from their appearance. (2) Although freezing prevents growth of most microorganisms, it does not destroy them. As soon as food begins to thaw, microorganisms begin to grow. In fact, freezing and thawing of foods actually promotes growth of microorganisms. Ice crystals puncture cell and plasma membranes and cell walls, allowing nutrients to escape from foods. These nutrients are then readily available to support microbial growth. Consequently, it is important never to thaw and refreeze foods.

Drying and Lyophilization

Drying (dessication, or dehydration) is one of the oldest methods used for food preservation. A certain level of water activity is necessary for microbial growth. Ideally, if more than 90 percent of the water is removed, the food can be stored in that state. Drying stops microbial growth but does not kill all microorganisms in or on food. Foods can be dehydrated by natural means, such as sun drying, or by artificial means by passing heated air over the food with controlled humidity. Addition of salt, high concentrations of sugar, or chemical preservatives that alter osmotic pressure and reduce water content are often included in the drying process.

Currently, **lyophilization** (freeze-drying) is employed in the food industry almost exclusively for the preparation of instant coffee and dry yeast for breadmaking. (The technique is also used to preserve bacterial cultures—see Chapter 13.) ⨾ (p. 347) Lyophilization involves drying frozen food in a vacuum. The

process yields higher quality foods than are produced by ordinary drying methods.

Irradiation

Irradiation of food as a method of preservation is still new and quite controversial due to public concern about the hazards of radiation. There are two categories of radiation used to control microorganisms in food: nonionizing and ionizing.

Ultraviolet (UV) radiation, a form of *nonionizing radiation,* is limited by poor penetrating power. The radiation wavelength and exposure time also determine the effectiveness of this method of preservation. UV radiation is effective as a sanitizing agent for food-processing equipment and other surfaces. Microwaves, another form of nonionizing radiation, are useful for food preparation, cooking, and processing but not for preservation. Microwaves do not kill microorganisms directly, but the heat generated during cooking can be microbicidal. Uneven heating can interfere with this antimicrobial activity. Therefore, frequent rotation of food is necessary during microwaving.

Ionizing radiation, such as gamma rays, has great penetrating ability and is microbicidal. This type of radiation can be utilized either before or after packaging, depending on the foodstuff. Gamma radiation from cobalt-60 or cesium-137 has been used in Japan and some European countries for food preservation for a number of years. The U.S. Food and Drug Administration (FDA) has declared such irradiation safe for preserving certain foods. (See the essay "The Great Irradiation Debate: Are Gamma Rays Really Safe for Food?" at the end of this chapter.) Radiation has successfully been used to control microbial growth on fresh fish during transport to market and to kill insects in spices. It has also proved effective in reducing spoilage of fresh fruits and vegetables. Most recently the USDA proposed rules providing for the irradiation of fresh poultry. The public has much skepticism about irradiated foods. It should be emphasized that irradiation kills microorganisms but does not cause the food itself to become radioactive.

Chemical Additives

A large number of chemical compounds are added to various foods to kill microorganisms or retard their growth. We will describe a few examples and their uses.

Organic acids, some of which occur naturally in some foods, lower the pH of foods enough to prevent growth of human pathogens and toxin-producing bacteria. Acids such as benzoic, sorbic, and propionic inhibit growth of yeast and other fungi in margarine, fruit juices, breads, and other baked goods.

PUBLIC HEALTH

COMBATING *SALMONELLA* IN POULTRY

Salmonella infection in poultry has been a serious source of food spoilage and causes food poisoning in as many as 4 million people in the United States each year. Microbiologists have found that the sugar D-mannose blocks the ability of *Salmonella* bacteria to adhere to chicken intestinal tissue. Birds fed 2.5 percent solutions of D-mannose the first 10 days of their lives resisted an oral dose of 100 million *S. typhimurium* bacteria on their third day of life. At the age of 50 days, when broilers are usually sent to market, only one bird had become colonized with *Salmonella*—99 percent fewer than among those that had received only plain water. D-mannose has no side effects, is a natural sugar, and is metabolized as an energy source by the bird as part of its regular food.

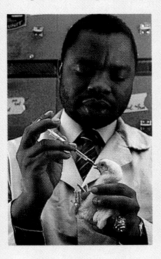

A U.S. Department of Agriculture scientist tests the ability of mannose to prevent *Salmonella* infection in chicks.

Alkylating agents such as ethylene oxide and propylene oxide are used only in nuts and spices. Sulfur dioxide, which is most effective at an acidic pH, is used in the United States only to bleach dried fruits and to eliminate bacteria and undesired yeasts in wineries. Ozone, a highly reactive form of oxygen, is used to kill coliform bacteria in shellfish and to treat water used in beverages. It has the advantage of leaving no residue. The disadvantages of ozone are that it tends to give foods a rancid taste by oxidizing fats, and it can damage molecules, especially lung lysozyme, if inhaled.

Sodium chloride, perhaps one of the first food additives, increases osmotic pressure in foods, keeping most microorganisms from growing. Salt used to cure meats is especially useful in preventing growth of clostridia deep in tissues, although fungi eventually grow on the surfaces of salted foods. Salt dehydrates bacteria and makes it difficult for them to take in water and nutrients. A recent discovery suggests that salt, in addition to increasing osmotic pressure, may create electric charges on the surface of meats, preventing bacteria from adhering to the surfaces.

Still other chemical additives have special uses. Halogen compounds, such as sodium hypochlorite, disinfect water and food surfaces. Gaseous chlorine prevents growth of microorganisms on food-processing equipment. Nitrates and nitrites suppress microbial growth in meats, especially ground meats and cold cuts. During cooking, however, they can be converted to nitrosamines, which are carcinogenic and toxic to the liver. We continue to use nitrates and nitrites because we have no good alternatives, particularly for sausage. The name botulism comes from the Latin word for sausage, so common was food poisoning from sausage in the days before the use of nitrites. Another reason nitrites are used is to keep colors bright, especially the red color of fresh meat. Carbon dioxide kills microorganisms, except for some fungi, in carbonated beverages. It also retards maturation of fruits and decreases spoilage during shipment. Finally, quaternary ammonium compounds (quats) can be used to sanitize many objects—utensils, cows' udders, fresh vegetables, and the surfaces of eggshells, which they do not penetrate.

Antibiotics

Antibiotics are added to foods and milk in some countries. In the United States only the anticlostridial agent *nisin*, a bacteriocin produced naturally during fermentation of milk by *Streptococcus lactis*, may be used. Antibiotic use in foods and milk is prohibited for the following good reasons:

- The antibiotics might be relied on instead of good sanitation.
- Pathogenic microorganisms might develop resistance to the antibiotics, so treatment of the diseases they cause would become difficult or impossible.
- Humans might be sensitized to the antibiotics and subsequently suffer allergic reactions.
- The antibiotics might interfere with the activities of microorganisms essential for fermenting milk and making cheese.

Pasteurization of Milk

Prevention of spoilage and disease transmission in milk begins with the maintenance of health in both dairy animals and in milk handlers. In the past, bovine tuberculosis was sometimes transmitted to humans from cows' milk. Many children were infected early in life and usually died by age 15. Establishment of mandatory testing of dairy herds for tuberculosis every 3 years (the time it takes an infection to progress to a transmissible stage) has greatly decreased incidence of the disease in the United States.

Milk is collected under clean, but not sterile, conditions and is usually subjected to pasteurization. Two methods of pasteurization are currently in use: (1) In **high-temperature short-time (HTST) pasteurization,** or **flash pasteurization,** milk is heated to 71.6°C for at least 15 seconds. (2) In **low-temperature long-time (LTLT) pasteurization,** or the **holding method,** milk is heated to 62.9°C for at least 30 minutes. Both methods destroy vegetative cells of pathogens likely to be found in milk and decrease the numbers of organisms that can cause souring. Following pasteurization, milk is quickly cooled and refrigerated in sealed containers until it is used.

Milk can be preserved and kept safe to drink by means other than pasteurization. In Europe and increasingly in parts of the United States, milk may be sterilized by **ultra-high temperature treatment (UHT)** rather than simple pasteurization. UHT milk is heated to 87.8°C for 3 seconds. Such milk can be kept in sealed paper containers—called aseptic packaging—and stay unrefrigerated for about 6 months. Milk can also be preserved as canned condensed milk, which also is sterilized. It is reconstituted by adding an equal volume of water. Although sterilized milk is completely free of microorganisms, the heat treatment necessary to render it sterile alters its flavor.

Various chemical additives are sometimes used in milk. Addition of hydrogen peroxide to milk reduces the temperature required to destroy most pathogens. However, it fails to kill mycobacteria, and its use in milk has been banned in the United States for this reason.

Standards for Food and Milk Production

Because food and milk production are carefully regulated by federal, state, and local laws, consumers are better protected in the United States than in many other countries. Despite regulations, some hazards remain because of the use of food additives or heat treatment. The FDA regulates inspection of meat and poultry, accurate labeling, and quality standards for products being shipped across state lines. Other similar regulations are imposed by many state and local agencies within their own jurisdictions. And the USDA is now proposing that meat and poultry be inspected by microscope—a much stricter testing requirement than has ever been in place (Figure 27.5).

Many food producers maintain quality checks on their own products. In canning factories, for example,

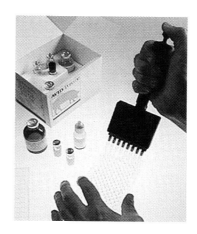

FIGURE 27.5 This new blood test kit can detect trichinosis infection in pigs.

counts of microorganisms in samples of foods are made during processing in an effort to minimize the number of organisms in the foods. Milk, because it is an extremely good growth medium for microorganisms, is subjected to several tests (Table 27.4). The use of these tests virtually guarantees high-quality milk to consumers.

CLOSE-UP

RAW MILK

The interest in natural foods has led some consumers to insist on obtaining raw milk. Although federal regulations prevent shipment of such milk across state lines, local processors, especially in California, have acquiesced to consumer demands and sell it within the state. The CDC has objected strenuously to this practice because of its potential for disease transmission. Unfortunately, no way has yet been found to prevent in-state sale of raw milk—and its accompanying hazards.

The issue of raw-milk safety is not limited to the United States. In France, unpasteurized milk is used to make cheeses with flavors unique to small regions. Hence, when the European Union, headquartered in Brussels, Belgium, proposed a ban on the use of raw milk in cheeses, the French objected. Such a ban, they argued, would lead to a dull sameness of flavor of all cheeses because, in addition to killing harmful bacteria, pasteurization kills good bacteria that give special aroma and flavors. Some cheese producers insisted that heating the milk to pasteurization temperatures would ruin the quality of their cheese by changing its texture. The extra processing would also add to the expense of production. Recently, the president of the European Union announced that this was all a misunderstanding and that the Union had intended to draw up different standards for raw-milk products.

TABLE 27.5	Fermented Milk Beverages
Characteristics	**Beverages and Countries Where Made**
Less than 1% lactic acid	Sour cream, cultured buttermilk (United States)
	Filmjolk (Finland)
2–3% lactic acid	Yogurt (United States); called leben in Egypt, matzoon in Armenia, naja in Bulgaria, and dahi in India
	Tarho (Hungary)
	Kos (Albania)
	Fru-fru (Switzerland)
	Kaimac (Yugoslavia)
	Acidophilus milk (United States)
Alcoholic (1–3%)	Koumiss, kefir, and araka (former Soviet Union)
	Fuli and puma (Finland)
	Taette (Norway)
	Lang (Sweden)

it is similar to buttermilk except that it is more acidic and lacks the flavor imparted by the leuconostocs. The Balkan product kefir is made from the milk of cows, goats, or sheep, and the fermentation is usually carried out in goatskin bags. The Russian product koumiss is made from mare's milk. In kefir and koumiss, *Streptococcus lactis, L. bulgaricus,* and yeasts are responsible for the production of lactic acid, alcohol, and other products. These products are usually made by continuous fermentation—fresh milk is added as fermented product is removed.

Cheeses

The first step in making almost any cheese is to add lactic acid bacteria and either **rennin** (an enzyme from calves' stomachs) or bacterial enzymes to milk. The bacteria sour the milk, and the enzymes coagulate the

TRY IT

RODS IN YOGURT

Yogurt is a nutrient-rich culture of bacteria. Commercially produced yogurt labeled "with active cultures" is teeming with bacteria. Place a tiny dab of yogurt on a microscope slide, and stir in a drop of water. Add a coverslip, and examine the sample under high power with a compound microscope. You will see masses of rod-shaped cells. These cells are *Lactobacillus bulgaricus,* one of the organisms used to ferment milk to yogurt.

APPLICATIONS

COMBATING LACTOSE INTOLERANCE

Certain strains of *Lactobacillus acidophilus* metabolize cholesterol in milk fat. Acidophilus milk made by appropriate strains might be enjoyed by individuals who must restrict their cholesterol intake. *Lactobacillus acidophilus* also adds the enzyme lactase to milk. Lactase is needed to digest the lactose present in milk. Such milk might be tolerated by lactose-intolerant people, who cannot drink milk because they lack this enzyme.

milk protein casein. The solid portion—the **curd**—is used to make cheese, and the liquid portion—the **whey**—is a waste product of this process (Figure 27.7). Sometimes lactic acid is extracted from whey. In separating the curd and the whey, different amounts of moisture are removed according to the kind of cheese being made. For soft cheeses, the whey is simply allowed to drain from the curd; for harder cheeses, heat and pressure are used to extract more moisture. Nearly all cheeses are salted. Salting helps remove water, prevents growth of undesired microorganisms, and contributes to the flavor of the cheese.

A few cheeses, such as cream cheese, cottage cheese, and ricotta, are unripened, but most cheeses are ripened. Ripening involves the action of microorganisms on the curd after it is pressed into a particular form. Soft cheeses are ripened by the action of microorganisms that occur naturally or are inoculated onto the surface of the pressed curd. Because the enzymes that ripen the cheese must diffuse from the surface into the center of the cheese, soft cheeses are relatively small. In contrast, hard cheeses are ripened by the action of microorganisms distributed through the curd. Because microbial action does not depend on diffusion, these cheeses can be quite large (Figure 27.8). Cheese is ripened by microorganisms in a cool, moist environment. Many modern factories have environmentally controlled rooms, but some still use natural caves similar to those where cheeses were first ripened.

Cheeses can be classified by their consistency (soft to hard), by the kind of microorganisms involved in the ripening process, and by the length of time required for ripening (Table 27.6, page 776). The ripening period is shorter for soft cheeses (1 to 5 months) than for hard cheeses (2 to 16 months).

Several microbial actions, such as decomposition of the curd and fermentation, occur during the ripening of cheeses. Prior to ripening, the curd consists of protein, lactose, and if the cheese is made from whole

(a)

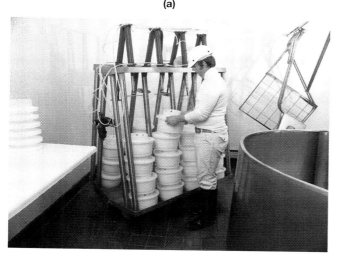

(b)

(c)

FIGURE 27.7 Making gouda cheese. (a) Lactobacilli and the enzyme rennin are added to pasteurized milk. The bacteria sour the milk, and the rennin coagulates the milk protein casein. (b) The milk curdles into the solid curd and the liquid whey. Curds are drained and placed into "hoops," which are then squeezed in a press. (c) The pressed cheeses are removed from their hoops and floated in a tank of brine (salt solution). The high salt concentration extracts even more moisture from the cheese by osmosis (recall Chapter 4) and thus hardens it. ⊂⊃ (p. 98) The salt also adds flavor to the cheese and prevents the growth of undesired microorganisms.

FIGURE 27.8 Ripening of gouda cheese involves the action of microorganisms on the curd after it is pressed into its form. A plastic coating is added to the cheese to keep it from drying out. The cheese is then placed on shelves in the aging room, where the microorganisms that are distributed throughout the curd will ripen it. Because gouda is a hard cheese, it must ripen for a relatively long period—3 months to a year—depending on the size of the wheel.

milk, fat. As microorganisms act on the curd, they first break down the lactose to lactic acid and other products, such as alcohols and volatile acids. Proteolytic enzymes break down protein, more extensively in soft than in hard cheese. *Brevibacterium linens* and the mold *Penicillium camemberti* are especially adept at releasing proteolytic enzymes. Lipase, particularly in *Penicillium roqueforti*, releases short-chain fatty acids such as butyric, caproic, and caprylic acids. These acids and their

TABLE 27.6 Classification of Ripened Cheeses

Consistency and Ripening Period	Examples	Organisms Associated with Ripening
Soft (1–2 months)	Limburger	*Streptococcus lactis, S. cremoris, Brevibacterium linens*
Soft (2–5 months)	Brie and Camembert	*S. lactis, S. cremoris, Penicillium camemberti,* and *P. candidum*
Semihard (1–8 months)	Muenster and Brick	*S. lactis, S. cremoris, Brevibacterium linens*
Semihard (2–12 months)	Roquefort and Blue	*S. lactis, S. cremoris,* and *P. roqueforti* or *P. glaucum*
Hard (3–12 months)	Cheddar and Colby	*S. lactis, S. cremoris, S. durans,* and *Lactobacillus casei*
	Edam and Gouda	*S. lactis* and *S. cremoris*
	Gruyere and Swiss	*S. lactis, S. thermophilus, S. helveticus, Propionibacterium shermani,* or *L. bulgaricus* and *P. freudenreichii*
Hard (12–16 months)	Parmesan	*S. lactis, S. cremoris, S. thermophilus,* and *L. bulgaricus*

oxidation products contribute significantly to the flavor of the cheeses. The effects of fermentation in cheese ripening are most easily seen in Swiss cheese. Bacteria of the genus *Propionibacterium* ferment lactic acid and produce propionic acid, acetic acid, and carbon dioxide. The acids flavor the cheese, and the carbon dioxide, which becomes trapped in the curd, produces the characteristic holes in the cheese.

Other Products

A great variety of fermented foods and food products are made throughout the world; here we consider only a few. (Beers, wines, and spirits are discussed later.)

VINEGAR Vinegar is made from ethyl alcohol by the acetic acid bacterium, *Acetobacter aceti* (Figure 27.9), which oxidize the alcohol to acetic acid. Commercially produced vinegar contains about 4 percent acetic acid. Cider vinegar is made from alcohol in fermented apple cider; wine vinegar is made from alcohol in wine.

SAUERKRAUT Sauerkraut was first made in Europe in the sixteenth century. Bacteria naturally present on

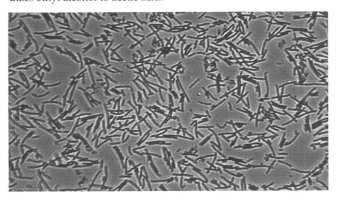

FIGURE 27.9 Manufacture of vinegar makes use of the bacterium *Acetobacter aceti* (magnified 400×). This organism oxidizes ethyl alcohol to acetic acid.

cabbage leaves act on shredded cabbage placed in layers in large crocks. Enough dry salt is placed between the layers to make a 2 percent to 3 percent salt solution as the salt draws water out of the cabbage. The cabbage is firmly packed and weighted down to create an anaerobic environment. Although many organisms cannot tolerate such an environment, anaerobic, halophilic species of *Lactobacillus* and *Leuconostoc* can carry out fermentation under these conditions. These microbes produce lactic acid, acetic acid, carbon dioxide, alcohol, and small amounts of other substances. After 2 to 4 weeks at room temperature, the cabbage is changed by fermentation to sauerkraut, which can be refrigerated until eaten or preserved by canning.

PICKLES Pickles are made by essentially the same process as that used to make sauerkraut. Fresh cucumbers, either whole, sliced, or ground, are packed in brine and allowed to ferment for a few days to a few weeks. *Leuconostoc mesenteroides* is the major fermenting organism in low-salt brines (less than 5 percent salt), whereas species of *Pediococcus* are more active in high-salt brines (5 percent or higher salt concentration). A problem in pickle-making is to prevent a yeast film from forming on the surface of vats. Direct sunlight and ultraviolet light help solve this problem. After fermentation, vinegar and spices are added to sour pickles; sugar also is added to sweet pickles. Most pickles are pasteurized. Some pickles, such as sweet pickles and pickle relish, are made without fermentation. After the pickles have been soaked in brine for a few hours, they are seasoned with vinegar and spices, heat-processed, and sealed. The entire operation can be done in a day in any kitchen.

OLIVES Olives are treated with lye to hydrolyze oleuropein, a very bitter phenolic glucoside they contain that gives an undesirable flavor. Green olives are fermented in a 5 percent to 8 percent salt solution by *Leuconostoc mesenteroides* and *Lactobacillus plantarum*. They are packed in water and pasteurized. Ripe olives are picked when reddish in color but not fully ripe. They are oxidized with tannins to blacken them, fermented in dilute (less than 5 percent) salt solution, packed in water in cans, and processed at 116°C for 60 minutes.

POI Poi, a common food in the South Pacific, is made from ground roots of the taro plant. A paste of ground roots is allowed to ferment through the action of a succession of naturally occurring organisms. Pseudomonads and coliforms begin the fermentation process, and lactobacilli later take over. Yeasts add alcohol to the mixture.

SOY SAUCE Soy sauce is made in a stepwise process (Figure 27.10). A salted mixture of crushed soybeans and wheat is treated with the mold *Aspergillus oryzae* to break down starch into fermentable glucose. The product, called *koji*, is mixed with an equal quantity of salt solution to make a mixture called moromi. Moromi is fermented for 8 to 12 months at low temperature and with occasional stirring. The fermenting organisms are mainly the bacterium *Pediococcus soyae* and the yeasts *Saccharomyces rouxii* and *Torulopsis* species. Lactic acid, other acids, and alcohol are produced. Upon completion of fermentation, the liquid and solid parts of the moromi are separated. The liquid part is bottled as soy sauce, and the solid part is sometimes used as animal food.

SOY PRODUCTS Other soy products include miso, tofu, and sufu. *Miso* is a fermented paste of soybeans made like soy sauce. *Tofu* is a soft curd of soybeans. It is made from soybeans ground to make a milk, which is boiled to inactivate enzymes. The curd is precipitated with calcium or magnesium sulfate and pressed into a soft mass similar in consistency to soft cheese. *Sufu* is made by action of a fungus on soy curd. Cubes of curd are dipped in a mixture of salt and citric acid, inoculated with *Mucor*, and incubated until coated with mycelia. The fungus-coated cubes are aged 6 weeks in a rice-wine brine.

BEER, WINE, AND SPIRITS

Beer and wine are made by fermenting sugary juices; spirits, such as whiskey, gin, and rum, are made by fermenting juices and distilling the fermented product. **Distillation** separates alcohol and other volatile substances from solid and nonvolatile substances. Strains of *Saccharomyces* are the fermenters for all alcoholic beverages. Many different strains have been developed, each having distinctive characteristics. Both the organisms and how they are used are carefully guarded brewers' secrets.

(a)

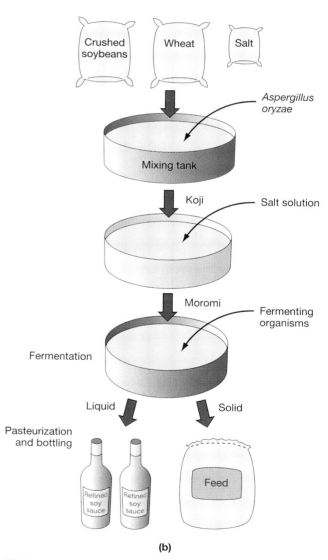

(b)

FIGURE 27.10 (a) Fermentation of soy and wheat is carried out in large steel tanks. (b) Schematic diagram of the fermentation of soy.

To make beer, cereal grains (usually barley) are **malted** (partially germinated) to increase the concentration of starch-digesting enzymes that provide the sugar for fermentation (Figure 27.11). Malted grain is

crushed and mixed with hot water (about 65°C), producing **mash.** After a few hours, a liquid extract called **wort** is separated from the mix. *Hops* (flower cones from the hop plant) are added to the wort for flavoring, and the mixture is boiled to stop enzyme action and precipitate proteins. A strain of *Saccharomyces* is added. Fermentation produces ethyl alcohol, carbon dioxide, and other substances, including amyl and isoamyl alcohols and acetic and butyric acids, which add to the flavor of the beer. After fermentation, the yeast is removed, and the beer is filtered, pasteurized, and bottled.

Most wine is made from juice extracted from grapes (Figure 27.12), although it can be made from any fruit—and even from nuts or dandelion blossoms. Juice is treated with sulfur dioxide to kill any wild yeasts that may already be present. Sugar and a strain of *Saccharomyces* are then added, and fermentation proceeds. Although ethyl alcohol is the main product of fermentation, other products similar to those in beer add to the flavor of the wine. In both beer and wine, the particular characteristics of the juice and the yeast strain determine the flavor of the final product. When fermentation is completed, liquid wine is siphoned to separate it from yeast sediment, and if necessary, cleared with agents such as charcoal to remove suspended particles. Finally, it is bottled and aged in a cool place.

Spirits are made from the fermentation of a variety of foods, including malted barley (Scotch whiskey), rye (rye whiskey, gin), corn (bourbon), wine or fruit juice (brandy), potatoes (vodka), and molasses (rum). After fermentation, distilling separates alcohol and other volatile substances that impart flavor from the solid and nonvolatile substances. Because of distillation, the alcohol content of spirits ranges from 40 to 50 percent—much higher than the typical 12 percent for wine and 6 percent for beer. (Wines do not contain more alcohol because when the alcohol concentration reaches 12 to 15 percent, it poisons the yeasts carrying out the fermentation. To produce fortified wines such as sherry and cognac, extra alcohol is added after fermentation.)

INDUSTRIAL AND PHARMACEUTICAL MICROBIOLOGY

Industrial microbiology deals with the use of microorganisms to assist in the manufacture of useful products or to dispose of waste products. **Pharmaceutical microbiology** is a special branch of industrial microbiology concerned with the manufacture of products used in treating or preventing disease. Today many industrial and pharmaceutical processes make use of genetic engineering, as we saw in Chapter 8. ∞ (pp. 210–220)

FIGURE 27.11 Schematic diagram of the fermentation of beer.

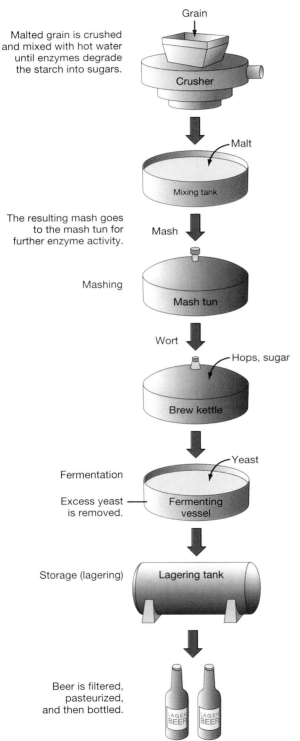

Malted grain is crushed and mixed with hot water until enzymes degrade the starch into sugars.

Grain

Crusher

Malt

Mixing tank

The resulting mash goes to the mash tun for further enzyme activity.

Mash

Mashing

Mash tun

Wort

Hops, sugar

Brew kettle

Yeast

Fermentation

Excess yeast is removed.

Fermenting vessel

Storage (lagering)

Lagering tank

Beer is filtered, pasteurized, and then bottled.

LAGER BEER

(a)   (b)   (c)

FIGURE 27.12 Fermentation of wine begins with (a) the rapid bubbling of the grape juice and yeast mixture, which is (b) stored in two-story-high stainless steel fermentation vats until the fermentation process is complete. (c) The wine is then transferred to wooden barrels, where it is aged, sometimes for many years. During this time, the flavor mellows and develops fully.

Industrial microbiology, albeit in primitive form, had its beginnings more than 8000 years ago when the Babylonians fermented grain to make beer. However, little was understood about fermentation until Pasteur studied the process in the nineteenth century. Over the next several decades, various researchers studied fermentation and its products, but their findings were largely ignored until shortages of materials for making explosives in World War I created a use for them. Both glycerol and acetone are used to make explosives and other materials. The Germans developed a process for making glycerol; the British used acetone–butanol fermentation by *Clostridium acetobutylicum* to make acetone. An important sidelight of the acetone–butanol fermentation was the development of techniques to maintain pure cultures in industrial fermentation vats.

The serendipitous discovery by Fleming in 1928 that *Penicillium notatum* kills *Staphylococcus aureus* was the beginning of the antibiotic industry. ∞ (Chapter 14, p. 358) Today the manufacture of antibiotics is an immense component of pharmaceutical microbiology. Concurrent with the development of antibiotics was the development and industrial production of a variety of vaccines. The manufacture of antibiotics, vaccines, and many other pharmaceuticals all require pure-culture technology.

In recent years genetic engineering has been used to cause cells to synthesize products they otherwise would not make or to increase the yield of products they normally make. Genetic engineering may make it possible to program organisms to carry out specific industrial and pharmaceutical processes. Such processes might well be more efficient and more profitable than any others now available.

Today hundreds of different substances are manufactured with the aid of microorganisms. Various species of yeasts, molds, bacteria, and actinomycetes are used in manufacturing processes. The organisms themselves are sometimes useful, as when they can serve as a source of protein. Animal feed consisting of microorganisms is called **single-cell protein (SCP).** Single-cell protein is an important high-yield, relatively inexpensive source of protein-rich food. More often the valuable substance is a product of microbial metabolism.

Useful Metabolic Processes

The production of complex molecules and metabolic end products in amounts that are commercially profitable requires the manipulation of microbial processes. In nature, microbes have regulatory mechanisms, such as induction and repression, that cause them to make

---

BIOTECHNOLOGY

VITAMINS FROM WASTES

Fermentation wastes from breweries contain significant quantities of vitamins, proteins, and carbohydrates. Releasing these wastes into streams is prohibited in some areas because it enriches the waters, causing overgrowth of algae. Some breweries dispose of these wastes by drying them and marketing them as supplements to animal feed.

Yeasts also can be grown specifically for the vitamins they produce. Certain *Candida* species can make and release into the medium up to 0.1 mg of riboflavin per gram of dry weight of yeast. This process has limited commercial value because the organisms are poisoned by traces of iron in the medium, and iron is nearly always present from equipment. Higher fungi, such as some that parasitize coffee and other plants, are not harmed by traces of iron and can be used commercially to produce riboflavin.

Saccharomyces uvarum produces ergosterol, a sterol that can be converted to vitamin D by ultraviolet radiation. This process is commercially feasible when carbon sources are adequate and vats are aerated.

INDUSTRIAL AND PHARMACEUTICAL MICROBIOLOGY **779**

substances only in the amounts needed. ∞ (Chapter 7, p. 178) An important task in industry is to manipulate regulatory mechanisms so that the organisms continue producing large quantities of useful substances for humans. Industrial microbiologists accomplish this in several ways: (1) by altering nutrients available to the microbes, (2) by altering environmental conditions, (3) by isolating mutant microbes that produce excesses of a substance because of a defective regulatory mechanism, and (4) by using genetic engineering to program organisms to display particular synthetic capabilities. In some instances these efforts have been remarkably successful. The industrial strain of the mold *Ashbya gossypii* produces 20,000 times as much of the vitamin riboflavin as it uses. Industrial strains of *Propionibacterium shermanii* and *Pseudomonas denitrificans* produce 50,000 times as much cobalamin (vitamin B_{12}) as they use.

Problems of Industrial Microbiology

It is one thing to cause a microbe to carry out a useful process in a test tube. It is quite another to adapt the process so that it will be profitable in a large-scale industrial setting. In the past, most industrial processes have been carried out in large fermentation vats. Many new processes simply do not work in large vats, so smaller vats must be used (Figure 27.13). Moreover, many of today's industrial microorganisms have been so extensively modified by selection of mutants or by genetic engineering that their products, which are useful to humans, may be useless or even toxic to the organisms.

(a)

(b)

FIGURE 27.13 Some fermentations do not progress satisfactorily (a) in large vats (at a Philadelphia beer-brewing plant) and therefore must be conducted (b) in small-scale containers.

CLOSE-UP

"EXTREMOZYMES"

Microbes have been discovered in places where, until recently, we did not believe organisms could survive. These microorganisms, dubbed "extremophiles," are found in desiccating salt marshes, in arctic ice, deep in oil wells, and just above steaming heat vents at the bottom of the ocean—the planet's harshest environments. Scientists are most interested in the enzymes that regulate the metabolism of these organisms. Enzymes from ordinary microorganisms are too sensitive to be useful in most industrial processes, because they would be denatured by the high temperatures or destroyed by toxic solvents. The enzymes these hardy organisms produce—called "extremozymes"—can survive extreme heat, cold, pressure, and salinity. Thus, these enzymes would prove useful in environments, such as industrial solvents, that would be deadly to most organisms.

Isolating and purifying the product, with or without killing the organisms, often presents technical difficulties. When the product remains within the cell, the plasma membrane must be disrupted to obtain the product. Membranes can be disrupted by spraying a medium containing the organisms through a nozzle under high pressure or by putting the organisms in alcohol or salt solutions that cause the product to form a precipitate. Product molecules from disrupted cells sometimes can be collected on resin beads of appropriate size and electric charge. Secreted products can be collected relatively easily, sometimes without killing the organisms. This can be done by using a **continuous reactor,** in which fresh medium is introduced on one side and medium containing the product is withdrawn on the other side (Figure 27.14). Continuous reactors, which operate under strictly controlled temperature and pH conditions, are used in many types of industrial fermentations.

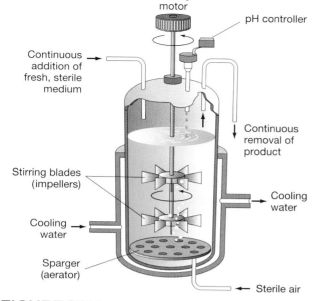

FIGURE 27.14 A continuous reactor. Fresh culture medium is introduced on one side, and medium containing product is withdrawn at the other side in a continuous process.

USEFUL ORGANIC PRODUCTS

Simple Organic Compounds

Simple organic compounds such as solvents and organic acids can be manufactured with the aid of microorganisms. The solvents include ethanol (ethyl alcohol), butanol, acetone, and glycerol. The acids include acetic, lactic, and citric acids. Although microorganisms are not now regularly used to make these substances, microbial synthesis will become economically feasible as the cost of raw materials from petroleum increases, especially if organisms can be genetically programmed to increase their productivity.

Ethanol, an industrial chemical with annual sales of about $300 million, is used not only as a solvent but also in the manufacture of antifreeze, dyes, detergents, adhesives, pesticides, explosives, cosmetics, and pharmaceuticals. In addition, ethanol is used as a fuel, either alone or mixed with gasoline. Microbes are employed to produce ethanol for industrial purposes in much the same way they are used to make alcoholic beverages, but microbiologists are developing new methods. In one method, cellulose is extracted from wood, digested to sugars, and fermented by thermophilic clostridia. Because these organisms act at high temperatures, their metabolic rates are higher than those of other fermenters, and they produce alcohol at a more rapid rate. Also, because the effluent is already heated, less energy is needed to distill and

purify the product. In another method, the bacterium *Zymomonas mobilis,* which ferments sugar twice as fast as yeast can, is used (Figure 27.15).

The yeast *Pachysolen tannophilus* produces relatively large quantities of alcohol from five-carbon sugars. This is significant because grain products contain both five-carbon and six-carbon sugars. Thus, making alcohol with yeasts that use only six-carbon sugars fails to extract a significant amount of energy from the grain. Producing alcohol for fuel is currently almost a break-even activity; the energy obtained from the six-carbon sugars in the alcohol is approximately equal to the energy required to produce it. Extracting energy from both five- and six-carbon sugars would be much more economical.

Clostridium acetobutylicum acting on starch or *C. saccharoacetobutylicum* acting on sugar produces both butanol and acetone. Butanol is used in the manufacture of brake fluid, resins, and gasoline additives. Acetone is used mainly as a solvent.

Glycerol is made by adding sodium sulfite to a yeast fermentation. The sodium sulfite shifts the metabolic pathway so that glycerol, and not alcohol, is the main product. Glycerol is used as a lubricant and as an emollient in a variety of foods, toothpaste, cosmetics, and paper.

Of the organic acids made by microorganisms, acetic acid (vinegar) is used in the greatest volume in the manufacture of rubber, plastics, fabrics, insecticides, photographic materials, dyes, and pharmaceuticals. Acetic acid bacteria oxidize ethanol to make vinegar. But thermophilic bacteria can make it from cellulose, and *Acetobacterium woodii* and *Clostridium aceticum* can make it from hydrogen and carbon diox-

FIGURE 27.15 Equipment for the industrial production of ethanol, using *Zymomonas mobilis* as a catalyst.

ide. Other acids made by microorganisms include lactic and citric acids. *Lactobacillus delbrueckii* metabolizes glucose to lactic acid, which is used to acidify foods, to make synthetic textiles and plastics, and in electroplating. The mold *Aspergillus niger* makes citric acid with great efficiency when molasses is used as the fermentation substrate. Citric acid is widely used to acidify and improve the flavor of foods.

Antibiotics

The antibiotic industry came into being in the 1940s with the manufacture of penicillin (Figure 27.16a), and about 100 antibiotics have been manufactured in quantity since then. The market value of antibiotics worldwide now exceeds $5 billion annually.

Industrial microbiologists work diligently to find ways to cause organisms to make their particular antibiotic in large quantities. An effective way is to induce mutations and then screen the progeny to find

strains that produce more antibiotic than the parent strain. Such methods, combined with improved fermentation procedures, have been quite successful. For example, a strain of *Penicillium chrysogenum* that once produced 60 mg of penicillin per liter of culture now produces 20 g/l. Genetic engineering may lead to more effective methods.

The reason only 2 percent of all known antibiotics have been marketed is that many antibiotics are either too toxic to use therapeutically or of no greater benefit than antibiotics already in use. Because pathogens continually develop resistance to antibiotics, industrial microbiologists not only seek new antibiotics but also try to make available antibiotics more effective. They look for ways to increase potency, improve therapeutic properties, and make antibiotics more resistant to inactivation by microorganisms. These efforts have led to the development of *semisynthetic antibiotics* (Figure 27.16b), which are made partly by microorganisms and partly by chemists. ∞ (Chapter 14, p. 357) This collaboration is illustrated by the way chemists have modified β-lactam antibiotics (penicillins and cephalosporins), which some microorganisms destroy with the enzyme β-lactamase. Binding a molecule such as clavulanic acid to the β-lactam ring prevents the enzyme from inactivating the antibiotic. The antibiotic augmentin, for example, is a semisynthetic antibiotic consisting of amoxicillin and clavulanic acid.

Enzymes

All enzymes used in industrial processes are synthesized by living organisms. With a few exceptions, such as the extraction of the meat tenderizer papain from papaya fruit, industrial enzymes are made by microorganisms. Enzymes are extracted from microorganisms rather than being synthesized in the laboratory because laboratory synthesis is as yet too complicated to be practical. Enzymes, like other proteins, consist of complex chains of amino acids in specific sequences. Organisms use genetic information to synthesize enzymes easily, but laboratory chemists find this a tedious and expensive process. Enzymes are especially useful in industrial processes because of their specificity. They act on a certain substrate and yield a certain product, thus minimizing problems of product purification.

The methods industrial microbiologists use to produce enzymes include screening for mutants, and gene manipulation. Screening for mutants that produce large quantities of an enzyme has been an effective technique. Gene manipulation has been somewhat less effective. Compared with synthetic pathways for antibiotics, those for enzymes are simpler, and fewer genes are involved. Thus, gene manipulation shows great promise for increasing the yield of some enzymes

FIGURE 27.16 (a) The amber-colored droplets seen on the surface of this culture of the mold *Penicillium notatum* are the antibiotic penicillin, which it has produced. (b) Synthetic and semisynthetic antibiotics are now made in the laboratory for use alongside those produced by microorganisms.

(a)

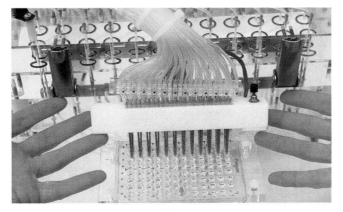

(b)

BIOTECHNOLOGY

MOWING THE GRASS CAN BE REWARDING

A Danish enzyme researcher returned from his summer vacation to find a compost heap of grass cuttings left by the brother-in-law who had thoughtfully mowed his lawn. Recognizing this pile as a potential treasure-trove, the researcher took samples from its center and isolated fungi that had been growing there. One fungus caused milk to coagulate. Laboratory analysis revealed an enzyme now trade-named Rennilase. This enzyme is a substitute for rennet, a mixture containing the enzyme rennin, formerly collected from the stomachs of lambs and calves at the slaughterhouse. Rennilase is used to coagulate milk in making cheese. One ounce can curdle 50 gallons of milk into curds and whey.

In the United States over $100 million is spent each year on rennet, which is of uncertain purity and whose cost has risen sharply in recent years. Therefore the dairy industry welcomes industrially produced rennin. The FDA has approved a version of rennin that was bioengineered by inserting the cow gene for rennin into *Escherichia coli*. The altered *E. coli cells* are grown in large fermentation tanks, and the enzyme is collected.

and making industrial production of others feasible. Only about 200 of some 2000 known enzymes are now produced commercially, so there is room for significant progress in this area.

Of the commercially available enzymes, proteases and amylases are produced in greatest quantities. Proteases, which degrade proteins and are added to detergents to increase cleaning power, are made industrially by molds of the genus *Aspergillus* and bacteria of the genus *Bacillus*. Amylases, which degrade starches into sugars, also are made by *Aspergillus* species. Other useful degradative enzymes are lipase from the yeast *Saccharomycopsis* and lactase from the mold *Trichoderma* and the yeast *Kluyveromyces*. Another important industrial enzyme is invertase (glucose isomerase) from *Saccharomyces;* it converts glucose to fructose, which is used as a sweetener in many processed foods.

PROTEOLYTIC ENZYMES   Pancreatic enzymes were first used more than 70 years ago to remove blood stains from butchers' aprons without weakening the cloth. These enzymes were tried as laundry aids, but it was found that they were inactivated by soap. In the 1970s proteolytic enzymes from bacteria, which retain their activity in the presence of detergents and hot water, were added to detergents. When this was first tried, workers in the detergent factories developed respiratory problems and skin irritations. The enzymes were removed from the detergents when the illnesses were traced to airborne enzyme molecules. Today, enzymes are incorporated in coated granules that dissolve in the wash instead of in the workers' environment.

Proteolytic enzymes have also been added to drain cleaners, where they are especially useful in degrading hair, which often clogs bathroom drains. A drain cleaner with a lipase should do a good job on kitchen drains.

ENZYMES IN PAPER PRODUCTION   To make high-quality paper, much of the lignin, a coarse material in wood, is removed by expensive chemical means to produce a purer cellulose wood pulp. The fungus *Phanerochaete chrysosporium* secretes enzymes that digest both lignin and cellulose. If the enzymes can be separated and purified, one might be used selectively to digest lignin, leaving cellulose unchanged. If perfected, this process would provide a cheap way to prepare wood pulp for high-quality paper. A byproduct of the research may benefit humans. Another lignin-digesting enzyme has been identified in a strain of *Streptomyces* bacteria. This enzyme modifies lignin into a molecule that enhances antibody production in mice. Might it someday do the same for humans?

Amino Acids

Microbial production of amino acids has become a commercially successful industry. Twenty different amino acids are utilized by animals in the production of proteins. Eight of the 20 are *essential amino acids*—they cannot be synthesized by the animal and therefore must be provided in the diet. Over 30,000 tons of lysine, an essential amino acid, are manufactured by microbial fermentation annually. Lysine is produced by a mutant strain of *Corynebacterium glutamicum* in which the biosynthetic pathway has been altered to promote the production of large quantities of the amino acid. Lysine is added to animal feed as a supplement and sold in health-food stores for human consumption.

Glutamic acid, another product of microbial fermentation, is also produced by a mutant strain of *C. glutamicum*. This mutant contains a high level of the enzyme glutamic acid dehydrogenase, which increases the yield of glutamic acid. The bacterium is grown in a culture medium deficient in biotin, which results in the formation of "leaky" plasma membranes. Such membranes permit the excretion of glutamic acid from the cell. Glutamic acid is used to make the flavor enhancer monosodium glutamate (MSG). About 200,000 tons is produced annually. Other microbially synthesized amino acids include phenylalanine, aspartic acid, and tryptophan.

Other Biological Products

Vitamins, hormones, and single-cell proteins are the major categories of other biologically useful products of industrial microbiology. The ability of microorganisms to make vitamin B_{12} and riboflavin was cited earlier as an example of highly successful amplification of microbial synthesis. Vaccines, of course, are exceedingly useful products (Figure 27.17). The development of hepatitis vaccine was described in Chapter 8. ∞ (p. 218)

Microorganisms also are used in the production of steroid hormones. The process used is called **bioconversion,** a reaction in which one compound is converted to another by enzymes in cells. The first application of bioconversion to hormone synthesis was the use of the mold *Rhizopus nigricans* to hydroxylate progesterone. This microbial step simplified the chemical synthesis of cortisone from bile acids from 37 to 11 steps. It reduced the cost of cortisone from $200 per gram to $6 per gram. (Subsequent improvements in procedures have brought the price below $0.70 per gram.) Other hormones that can now be produced industrially include insulin, human growth hormone, and somatostatin. They are made by recombinant DNA technology, using modified strains of *Escherichia coli.*

Single-cell proteins, as noted earlier, are whole organisms rich in protein. Currently used in animal feed, they may someday feed humans, too. An important advantage of single-cell protein is that it can be made from substances of less value than the protein produced. Certain *Candida* species make protein from paper pulp wastes, *Saccharomycopsis* makes it from petroleum wastes, and the bacterium *Methylophilus* makes it from methane or methanol.

MICROBIOLOGICAL MINING

As the availability of mineral-rich ores decreases, methods are needed to extract minerals from less concentrated sources. This need spawned the new discipline known as **biohydrometallurgy,** the use of microbes to extract metals from ores. Copper and other metals originally were thought to be leached from the wastes of ore-crushing as a result of an inorganic chemical reaction such as those reactions used to extract metals from ores. It was then discovered that this leaching is due to the action of *Thiobacillus ferrooxidans.* This chemolithotrophic acidophilic bacterium lives by oxidizing the sulfur that binds copper, zinc, lead, and uranium into their respective sulfide minerals, with a resultant release of the the pure metal. Copper in low-grade ores is often present as copper sulfide. When acidic water is sprayed on such ore, *T. ferrooxidans* obtains energy as it uses oxygen from the atmosphere to oxidize the sulfur atoms in sulfide ores to sulfate. The bacterium doesn't use the copper; it merely converts it to a water-soluble form that can be retrieved and used by humans (Figure 27.18a).

Other minerals also can be degraded by microbes. *T. ferrooxidans* releases iron from iron sulfide by the same process (Figure 27.18b). Combinations of *T. ferrooxidans* and a similar organism, *T. thiooxidans,* degrade some copper and iron ores more rapidly than either one does alone. Another combination of organisms, *Leptospirillum ferrooxidans* and *T. organoparus,* degrades pyrite (FeS_2) and chalcopyrite ($CuFeS_2$), although neither organism can degrade the minerals alone. Other bacteria can be used to mine uranium, and bacteria may eventually be used to remove arsenic, lead, zinc, cobalt, and gold.

FIGURE 27.17 Hepatitis B vaccine is being produced by recombinant DNA technology. Technicians use a chromatography column to separate key proteins from batches of yeast cells.

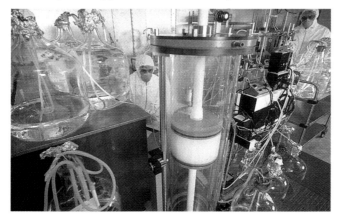

MICROBIOLOGICAL WASTE DISPOSAL

Sewage-treatment plants (Chapter 26) are prime examples of microbiological waste-disposal systems, but sewage disposal is a relatively simple problem compared with the problems associated with disposing of chemical pollutants and toxic wastes. ∞ (p. 750) Some wastes persist in the environment and contaminate water supplies for wildlife and humans. Bioremediation, the use of microorganisms to dispose of some chemical wastes, can help avert a monumental environmental disaster from accumulation of toxic wastes. ∞ (Chapter 26, p. 754)

(a)

(b)

FIGURE 27.18 Mining is made easier in many places by the activity of microorganisms. (a) Microbial leaching (extraction) helps concentrate copper from ore when conventional chemical approaches are not economically feasible. When acid solutions are added to dump surfaces, acidophilic bacteria such as *Thiobacillus ferrooxidans* quickly catalyze the oxidation of copper sulfate. (b) Microorganisms in these ponds reclaim metals from low-grade ores. Ions generated by microbial activity include ferric ions (orange) and ferrous ions (green).

APPLICATIONS

BACTERIOFORM GOLD

Bacterioform gold is gold that has accumulated in or on cell surfaces of bacteria. It most likely originates in nearby gold-containing rocks that break down, releasing the gold into solution in streams and water-logged soil. There, bacteria may act as scavengers of the soluble gold complexes. Budding bacteria that resemble members of the genus *Pedomicrobium* are known to form bacterioform gold. Examination of single cells suggests that the bacterioform structures may not be solid gold but a thin (less than 100 nm) covering built up on the original cells. This process is similar to the encrusting with iron and manganese oxides of other budding bacteria in the genus *Pedomicrobium*. Bacterioform gold has been found in Alaska, China, and South Africa.

Three strains of microorganisms have been found to deactivate Arochlor 1260, one of the most highly toxic of polychlorinated biphenyl (PCB) compounds. Other organisms have been shown to detoxify chemical substances such as cyanide and dioxin and to degrade oil spilled in the ocean. Genetic engineering has been used to develop a bacterium capable of detoxifying the defoliant Agent Orange, and work is under way to modify bacteria to detoxify other toxins.

One of the problems in developing microorganisms that can degrade toxic substances is the limited information available on the genetic characteristics of microorganisms found in wastes. Many researchers in this area focus their efforts on organisms found in wastes because these organisms probably already have degradatory capabilities. The researchers think it should be somewhat easier to modify them to metabolize other wastes than to use better-known organisms that have no known capacity to degrade wastes.

THE GREAT IRRADIATION DEBATE: ARE GAMMA RAYS SAFE FOR FOOD?

What method of food preservation allows you to store onions sprout-free for up to three months, keeps refrigerated strawberries firm and fresh for three weeks, and virtually eliminates *Salmonella* from poultry? It's irradiation, but it may not be coming to a store near you. Despite its approval by the FDA and its endorsement by the World Health Organization, the American Medical Association, and even Julia Child, irradiation is practiced by only one major poultry company in the United States, and only one U.S. company is in the business of irradiating fruits and vegetables. Although advocates praise irradiation as an effective way to reduce bacterial and insect contamination of foods, critics charge that the practice may cause more harm than the organisms it destroys.

Irradiation is not a new technology; for over 20 years, selected food industries have used irradiation to decontaminate foods. In 1963 the FDA approved irradiation for the elimination of mold from wheat and potatoes. In the 1980s the FDA added dried spices, which attract insects, and pork, which carries the parasitic *Trichina* worm, to the list of foods approved for irradiation. In 1986 fruits and vegetables were approved, and poultry soon followed. The FDA is currently considering a proposal to allow irradiation of beef before it is shipped to markets.

The process of irradiating foods is simple. The food is passed on a conveyor belt through a lead-shielded, concrete chamber. Within the chamber, radioactive elements—generally cobalt-60 or cesium-137—bombard the food with gamma rays. The process does not make the food itself radioactive, any more than a dental X-ray makes your teeth radioactive. The gamma rays do, however, kill insects, molds, and bacteria and may slow the ripening of certain fruits and vegetables, extending their shelf life (Figure 27.19).

Proponents of irradiation think its time has come. They point to the frightening statistics gathered from poultry processing plants—that up to 60 percent of all birds are infected with *Salmonella* and that up to 100 percent are infected with *Campylobacter*. Both kinds of bacteria can cause serious gastrointestinal illnesses in humans if the poultry is not refrigerated or cooked properly. And although *Salmonella* and *Campylobacter* food poisonings are usually not fatal, other types are. In 1992 an outbreak of a deadly form of food poisoning caused by *Escherichia coli* strain 0157:H7 killed four children in the Pacific Northwest. The outbreak was traced to contaminated, undercooked beef served at a fast-food restaurant. Since then, 45 outbreaks of this kind of food poisoning have occurred throughout the United States. Supporters of irradiation say that the practice could have prevented many of these cases of food poisoning.

Tests show that irradiation is effective in destroying many unwelcome organisms. It kills 99.5 to 99.9 percent of *Salmonella* and *Campylobacter* bacteria in poultry, as well as fungi, parasites, and nonpathogenic bacteria that can cause food to spoil. And because irradiation kills insects, farmers would not have to use as many pesticides on their crops, reducing our exposure to these chemicals.

Opponents say that irradiation offers a false sense of security. Irradiation can't kill all the harmful bacteria that infect poultry and beef, and consumers who put their trust in irradia-

FIGURE 27.19 Radiation can be used to sterilize food, as these two boxes of strawberries demonstrate. Irradiating food does not make it radioactive; it is safe to eat.

NON · IRRADIATED ·

IRRADIATED · (0.2 M RAD)

STRAWBERRIES -
15 DAYS STORAGE 38°F (4°C)

tion may take short cuts when storing, handling, and cooking food. The food may look and smell fresh, but *Salmonella* and *Campylobacter*—which don't cause food to look or smell spoiled—may still be present in small amounts. In conditions favorable for bacterial growth, such as a warm kitchen in which a chicken is defrosting in the sink, these bacteria can multiply, with disastrous results. And irradiation doesn't touch toxins that many pathogenic bacteria may have already deposited in the food. In many types of food poisoning, the toxins, not the bacteria, are what cause the nasty symptoms. Moreover, the most deadly food pathogen of all, *Clostridium botulinum*, is impervious to radiation.

These aren't the only drawbacks to irradiation, critics contend. Irradiation may create harmful chemicals in food. When gamma rays strike tissues, they disrupt chemical bonds. Living cells, such as bacteria, die, but some chemical bonds re-form in new configurations, creating chemicals that were not present before irradiation. Researchers have found minute amounts of formaldehyde and benzene—both cancer-causing chemicals—in samples of irradiated food. Irradiation supporters counter that many foods, such as eggs, naturally contain these chemicals and that some cooking techniques generate these chemicals in other foods. But along with benzene and formaldehyde, *unique radiolytic products* (URPs) form when the protein, fat, and carbohydrate molecules are rearranged during irradiation. It is not known whether these products are cancer-causing. Irradiation advocates contend that URPs are not new to preservation methods; they've only recently been identified by more sensitive detection techniques.

Another criticism is that irradiation depletes nutrients in food. The amounts of vitamins E, A, and C and the B vitamin thiamin are reduced during the irradiation process, sometimes by as much as 9 percent. Critics charge that the extension of shelf life irradiation promises can't make up for the loss of essential nutrients.

Some people think that irradiated food tastes "funny." One taste tester commented that irradiated poultry has the taste of "burnt hair." The "off" taste may result from the oxidation of fat, which produces byproducts that create less-than-fresh flavors and odors. The same process occurs when food is kept too long in the freezer. Other tasters, however, could detect no difference in taste between irradiated foods and nonirradiated foods.

Are irradiated foods coming to your grocer's shelves soon? It's not likely. Most poultry producers continue to rely on a disinfectant bath to rid their chickens of bacteria. A major spice manufacturer did irradiate its spices for a short time in the 1970s but has since reverted to chemical methods of decontamination. Until the perceived dangers are thoroughly studied and fears put to rest, Americans will continue to shun irradiated food.

CHAPTER SUMMARY

MICROORGANISMS FOUND IN FOOD

- Anything that humans eat is food for microorganisms, too.
- Many organisms in food are commensals; some cause spoilage, and a few cause human disease.

Grains

- Grains stored in moist areas become contaminated with various molds.
- Flour is purposely inoculated with yeasts to make bread.

Fruits and Vegetables

- Fruits and vegetables are subject to soft rot and mold damage.

Meats and Poultry

- Meats and poultry contain many kinds of microorganisms, some of which cause zoonoses. Poultry typically is contaminated with *Salmonella*, *Clostridium perfringens*, and *Staphylococcus aureus*.

Fish and Shellfish

- Seafood can be contaminated with several kinds of bacteria and viruses.

Milk

- Milk can contain organisms from cows, milk handlers, and the environment.

Other Edible Substances

- Sugars support growth of various organisms, which are killed during refining.
- Spices contain large numbers of microorganisms; condiments may support mold growth.
- Syrups for carbonated beverages can become contaminated with molds.
- Tea, coffee, and cocoa are also subject to mold contamination if not kept dry.

PREVENTING DISEASE TRANSMISSION AND FOOD SPOILAGE

- Common diseases transmitted in food and milk are listed in Table 27.3.
- Good sanitation reduces the chance of getting foodborne diseases. The most important factor in preventing spoilage and disease transmission in food and milk is cleanliness in handling.

Food Preservation

- Methods of food preservation include **canning,** refrigeration, freezing, **lyophilization,** drying, ionizing radiation, and use of chemical additives.

Pasteurization of Milk

- Methods of preserving milk include pasteurization and sterilization.

Standards for Food and Milk Production

- Certain standards for food and milk production are maintained by federal, state, and local laws. Tests for milk quality are summarized in Table 27.4.

MICROORGANISMS AS FOOD AND IN FOOD PRODUCTION

Algae, Fungi, and Bacteria As Food

- Rapid human population growth has created a need for new food sources.
- Yeasts can be grown on a variety of wastes and are good sources of cheap protein and vitamins. Equipment to grow them is expensive, and some effort will be needed to persuade humans to consider them acceptable as food.
- Algae can be grown in lakes and on sewage and also are good sources of protein. Problems with growing them include the danger of viral contamination and lack of acceptability as food.

Food Production

- Yeast is used to leaven bread.

- The fermentative capabilities of certain bacteria are used to make dairy products such as buttermilk, sour cream, yogurt, a variety of fermented beverages, and cheeses.
- In cheese-making, the **whey** of milk is discarded, and microorganisms ferment the **curd** and impart flavor and texture to a cheese.
- Other foods produced by microbial fermentation include vinegar, sauerkraut, pickles, olives, poi, soy sauce, and other soy products.

BEER, WINE, AND SPIRITS

- Beer is made from **malted** grains; hops are added, and the mixture is fermented.
- Wine is made by fermenting fruit juices.
- Spirits are made by fermenting various substances and distilling the products.

INDUSTRIAL AND PHARMACEUTICAL MICROBIOLOGY

- **Industrial microbiology** deals with the use of microorganisms to assist in the manufacture of useful products or to dispose of waste products.
- **Pharmaceutical microbiology** deals with use of microorganisms in the manufacture of medically useful products.

Useful Metabolic Processes

- Modifications of microbial processes in industry include altering nutrients available to microbes, altering environmental conditions, isolating mutants that produce excesses of useful products, and modifying the organisms by genetic engineering.

Problems of Industrial Microbiology

- One problem in industrial microbiology is the development of small-scale processes; other problems concern techniques for recovering products.

USEFUL ORGANIC PRODUCTS

Simple Organic Compounds

- Microorganisms can make alcohols, acetone, glycerol, and organic acids. Microbial processes to make such products are likely to become more economically feasible in the future.

Antibiotics

- Antibiotics are derived from *Streptomyces, Penicillium, Cephalosporin,* and *Bacillus* species.
- Many antibiotics have a β-lactam ring; some semisynthetic antibiotics are made by modifying that ring so that microorganisms cannot degrade it.

Enzymes

■ Enzymes extracted from microorganisms include proteases, amylases, lactases, lipases, and invertase. They are used in detergents, drain cleaners, enrichment of food, and manufacture of paper.

Amino Acids

■ Several amino acids, such as lysine and glutamic acid, can be produced microbially.

Other Biological Products

■ Vitamins and hormones are usually made by manipulating organisms so that they produce excessive amounts of these useful products.

■ **Single-cell proteins** consist of whole organisms rich in proteins; they are used mainly as animal feeds.

MICROBIOLOGICAL MINING

■ Microbes are currently used to extract copper from low-grade ores. Other minerals that can be extracted by microorganisms include iron, uranium, arsenic, lead, zinc, cobalt, and nickel.

MICROBIOLOGICAL WASTE DISPOSAL

■ Sewage-treatment plants (Chapter 26) use microorganisms in waste disposal.

■ A few organisms have been found that degrade toxic wastes, and research is under way to identify or develop others.

KEY TERMS

bioconversion   *(p. 784)*
biohydrometallurgy
   (p. 784)
bone stink   *(p. 762)*
canning   *(p. 768)*
continuous reactor   *(p. 780)*
curd   *(p. 774)*
distillation   *(p. 777)*
flash pasteurization   *(p. 771)*

flat sour spoilage   *(p. 768)*
high-temperature short-time
   pasteurization   *(p. 771)*
holding method   *(p. 771)*
industrial microbiology
   (p. 778)
leavening agent   *(p. 773)*
low-temperature long-time
   pasteurization   *(p. 771)*

lyophilization   *(p. 769)*
malted   *(p. 778)*
mash   *(p. 778)*
mesophilic spoilage
   (p. 768)
pharmaceutical micro-
   biology   *(p. 778)*
rennin   *(p. 774)*
single-cell protein   *(p. 779)*

thermophilic anaerobic
   spoilage   *(p. 768)*
ultra-high temperature
   treatment   *(p. 771)*
whey   *(p. 774)*
wort   *(p. 778)*

QUESTIONS FOR REVIEW

A **Microorganisms in Food**

1. What types of microorganisms are found in food and milk?

2. List organisms that cause food spoilage.

B **Disease Transmission in Food**

3. List the major diseases transmitted in food, identify the causative organism(s), and explain how they enter foods.

C **Preventing Food Spoilage and Disease Transmission**

4. True or False: As long as the shells of eggs are not broken, eggs cannot harbor microorganisms.

5. In each accompanying diagram, explain how organisms get into milk, and list the diseases they cause.

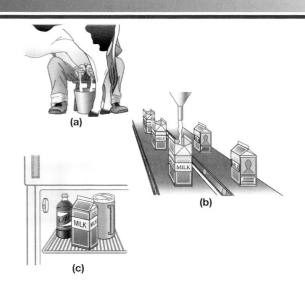

(a)

(b)

(c)

6. The single most important factor in preventing spoilage and disease transmission in food and milk is:

   a. adequate refrigeration   **c.** thorough cooking

   b. cleanliness in handling   **d.** irradiation of foods

7. List the methods of preserving foods, and describe how they are carried out.

8. Compare the holding method of pasteurization, flash pasteurization, and sterilization of milk.

9. How are standards for food and milk production maintained?

D Microorganisms and Food Manufacture

10. How are microorganisms used as food?

11. What problems prevent development of food uses of microorganisms?

12. How are microorganisms used in bread-making?

13. How are microorganisms used in making dairy products?

14. How are microorganisms used in making other foods?

E Microorganisms and Beer, Wine, and Spirits

15. Distinguish among the methods of producing beer, wine, and spirits.

F Microorganisms and Industry

16. List some common industrial uses of microorganisms.

17. What kinds of problems often arise in adapting microbiological processes for industrial use?

G Microorganisms and Biologically Useful Substances

18. What simple organic compounds can microorganisms make?

19. How do natural and semisynthetic antibiotics differ?

20. Why does the search for antibiotics and improvements in antibiotics continue?

21. How are enzymes obtained for industrial use, and how are they used?

22. What other biologically useful products are made in industrial microbiology?

H Microorganisms and Mining

23. What benefits can be derived from mining with microorganisms?

I Microorganisms and Waste Disposal

24. Describe some of the advantages of using microbes in waste disposal.

PROBLEMS FOR INVESTIGATION

1. Even though enzymes are used in cheese production, why are microorganisms also important?

2. What waste materials in your community could be used to grow single-cell proteins?

3. Can you think of a way to use yeast as a new type of food that would be acceptable to people in the United States?

4. Devise methods to test samples of spices for bacterial content; if laboratory facilities are available, carry out your tests.

5. Read about home canning techniques. What precautions are taken in the processing directions?

SOME INTERESTING READING

Anderson, D. M. 1994. "Red tides." *Scientific American* 271, no. 2 (August):62–68.

Baltz, R. H., G. D. Hegeman, and P. L. Skatrud, eds. 1993. *Industrial microorganisms: Basic and molecular applied genetics.* Washington, D.C.: American Society for Microbiology.

Becker, E. W. 1994. *Microalgae: Biotechnology and microbiology.* New York: Cambridge University Press.

Blumenthal, D. 1990. "Food irradiation: Toxic to bacteria, safe for humans." *FDA Consumer* 24(November):11–15.

Bradley, D. 1993. "A giant step for artificial enzymes." *New Scientist* 138(1867):16.

Coghlan, A. 1994. "Cellular 'control box' to switch on genes." *New Scientist* 142(1920):17.

Edlow, J. 1992. "Gut reactions." *American Health* 11(June):66–70.

Erickson, D. 1991. "Industrial immunology: Antibodies may catalyze commercial chemistry." *Scientific American* 265(September):174–75.

Fox, P. F., and J. Law. 1991. "Enzymology of cheese ripening." *Food Biotechnology* 5(3):239–62.

Foulke, J. 1994. "How to outsmart dangerous E. coli strains." *FDA Consumer* 28(1):11.

Glick, B. R., and J. J. Pasternak. 1994. *Molecular biotechnology: Principles and applications of recombinant DNA.* Washington, D.C.: ASM Press.

Kosikowski, F. V. 1985. "Cheese." *Scientific American* 252(May):88–99.

MacDonald, K. L., and M. T. Osterholm. 1993. "The emergence of *Escherichia coli* 0157:H7 infection in the United States." *Journal of the American Medical Association* 269(17):2264–66.

Merson, J. 1992. "Mining with microbes." *New Scientist* 133(1802):17–19.

"Mining with microbes: A labor of bug." 1990. *Science News* 137(15):236.

Papazian, R. 1992. "Food irradiation: A hot issue." *Harvard Health News Letter* 17(10):1–3.

Roman, M. 1994. "Something fishy." *American Health* 13(2):76–79.

Stix, G. 1995. "A recombinant feast: New bioengineered crops move toward market." *Scientific American* 272(March)38–40.

Tombs, M. P. 1990. *Biotechnology in the Food Industry.* Philadelphia: Open University Press.

METRIC SYSTEM MEASUREMENTS, CONVERSIONS, AND MATH TOOLS

METRIC SYSTEM PREFIXES

pico (p) $= 10^{-12}$
nano (n) $= 10^{-9}$
micro (μ) $= 10^{-6}$
milli (m) $= 10^{-3}$
centi (c) $= 10^{-2}$
deci (d) $= 10^{-1}$
kilo (k) $= 10^{3}$

LENGTH

1 kilometer (km) = 0.62 mile
1 meter (m) = 39.37 inches = 3.281 feet
1 meter = 100 centimeters = 1000 millimeters
1 centimeter (cm) = 10 millimeters = 0.394 inch
1 millimeter (mm) = 0.0394 inch
1 micrometer (μm) = 10^{-6} meter
1 nanometer (nm) = 10^{-9} meter
1 angstrom (Å) = 10^{-10} meter

VOLUME

1 liter (l) = 1.057 quarts
1 liter = 1000 milliliters
1 milliliter (ml) = 1 cm^3 = 0.061 cubic inch
1 mm^3 = 10^{-3} cm^3 = 10^{-6} liter

MASS

1 kilogram (kg) = 1000 grams = 2.205 pounds
1 pound = 453.6 g
1 gram (g) = 1000 milligrams = 0.0353 ounce
1 ounce = 28.35 g
1 milligram (mg) = 10^{-3} g
1 microgram (μg) = 10^{-6} g

TEMPERATURE

degrees Fahrenheit (°F) $= \frac{9}{5}$(°C) + 32
degrees Celsius (°C) $= \frac{5}{9}$(°F − 32)

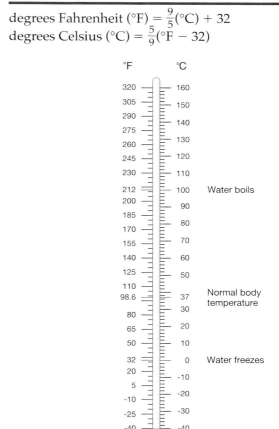

$0°C = 32°F$ (freezing point of water)
$100°C = 212°F$ (boiling point of water)
$37°C = 98.6°F$ (normal body temperature)

EXPONENTIAL (SCIENTIFIC) NOTATION

Numbers that are either very large or very small are usually represented in *exponential* (scientific) *notation* as a number between 1 and 10 multiplied by a power of 10. In this kind of expression, the small raised number to the right of the 10 is the *exponent*.

Number	Exponential Form	Exponent
1,000,000	1×10^6	6
100,000	1×10^5	5
10,000	1×10^4	4
1,000	1×10^3	3
100	1×10^2	2
10	1×10^1	1
1		
0.1	1×10^{-1}	-1
0.01	1×10^{-2}	-2
0.001	1×10^{-3}	-3
0.000 1	1×10^{-4}	-4
0.000 01	1×10^{-5}	-5
0.000 001	1×10^{-6}	-6
0.000 000 1	1×10^{-7}	-7

Numbers greater than 1 have *positive* exponents, which tell how many times a number must be *multiplied* by 10 to obtain the correct value. For example, the expression 5.2×10^3 means that 5.2 must be multiplied by 10 three times:

$$5.2 \times 10^3 = 5.2 \times 10 \times 10 \times 10 = 5.2 \times 1000 = 5200$$

In so doing, we move the decimal point three places to the right:

$$5\ \underset{1\ \ 2\ \ 3}{2\ 0\ 0.}$$

The value of a positive exponent indicates *how many places to the right the decimal point must be moved* to give the correct number in ordinary decimal notation.

Numbers less than 1 have *negative* exponents, which tell how many times a number must be *divided* by 10 (or multiplied by one-tenth) to obtain the correct value. Thus, the expression 3.7×10^{-2} means that 3.7 must be divided by 10 two times:

$$3.7 \times 10^{-2} = \frac{3.7}{10 \times 10} = \frac{3.7}{100} = 0.037$$

In so doing, we move the decimal point two places to the left:

$$0.\ \underset{2\ \ \ 1}{0\ 3}\ 7$$

The value of a negative exponent indicates *how many places to the left the decimal point must be moved* to give the correct number in ordinary decimal notation.

CONVERTING DECIMAL NUMBERS TO EXPONENTIAL NOTATION

To convert a number greater than 1 from decimal notation to exponential notation, first move the decimal point to the *left* until only a single digit is to the left of the decimal point. The *positive* exponent needed for exponential notation is the same as *the number of places the decimal point was moved*.

$$\underset{5\ \ 4\ \ 3\ \ 2\ \ 1}{6\ 3\ 5\ 7\ 8\ 1.} = 6.35781 \times 10^5$$

To convert a number less than 1 from decimal notation to exponential notation, first move the decimal point to the *right* until a single *nonzero* digit is to the left of the decimal point. The *negative* exponent needed for exponential notation is the same as *the number of places the decimal point was moved*.

$$0.\ \underset{1\ \ 2\ \ 3\ \ 4}{0\ 0\ 0\ 4\ 2}\ 6 = 4.26 \times 10^{-4}$$

MULTIPLYING EXPONENTIAL NUMBERS

To multiply two numbers in exponential form, *add* the exponents. For example:

$$\begin{aligned}
(3.5 \times 10^3) \times (4.2 \times 10^4) &= 3.5 \times 4.2 \times 10^{(3+4)} \\
&= 14.7 \times 10^7 \\
&= 1.47 \times 10^8 = 1.5 \times 10^8 \\
&\quad \text{(rounded off)}
\end{aligned}$$

$$\begin{aligned}
(5.2 \times 10^4) \times (4.6 \times 10^{-3}) &= 5.2 \times 4.6 \times 10^{[4+(-3)]} \\
&= 23.92 \times 10^1 \\
&= 2.392 \times 10^2 = 2.4 \times 10^2 \\
&\quad \text{(rounded off)}
\end{aligned}$$

DIVIDING EXPONENTIAL NUMBERS

To divide two numbers in exponential form, *subtract* the exponents. For example:

$$\begin{aligned}
\frac{4.1 \times 10^4}{6.2 \times 10^6} &= \frac{4.1}{6.2} \times 10^{(4-6)} = 0.6613 \times 10^{-2} \\
&= 6.613 \times 10^{-3} = 6.6 \times 10^{-3} \\
&\quad \text{(rounded off)}
\end{aligned}$$

$$\begin{aligned}
\frac{6.6 \times 10^3}{8.4 \times 10^{-2}} &= \frac{6.6}{8.4} \times 10^{[3-(-2)]} = 0.7857 \times 10^5 \\
&= 7.857 \times 10^4 = 7.9 \times 10^4 \\
&\quad \text{(rounded off)}
\end{aligned}$$

pH EQUATION

The most convenient way to express the acidity or alkalinity of a solution is in terms of the *concentration of protons, or hydrogen ions* (H$^+$), present in the solution. Thus, to calculate the pH of a solution, you need to know the H$^+$ concentration—expressed as [H$^+$]—of that solution. Conversely, if you know the pH, you can determine [H$^+$]. A simple logarithmic equation can be used in either case:

$$pH = -\log_{10}[H^+]$$

In words, the pH of a solution is equal to the negative logarithm (to the base 10) of the hydrogen ion concentration, where [H$^+$] is given in moles per liter. For example, water normally has [H$^+$] = 10^{-7} moles/liter. The pH of water then is $-\log_{10}[10^{-7}] = -(-7) = 7$.

Your stomach "juices" have [H$^+$] = 10^{-2} moles/liter. Therefore, the pH of your stomach juices is 2.

To calculate [H$^+$] from the pH equation when you know the pH, consider a structure in an animal cell with pH = 5. For this structure [H$^+$] must be 10^{-5} moles/liter. Antacids have a pH of 9, so for antacids [H$^+$] = 10^{-9} moles/liter.

MOST PROBABLE NUMBER (MPN) TABLE

The following table can be used to determine the MPN of any organism in a test medium. ∞ (Chapter 6, p. 143) One of its most useful applications is in testing water purity. For example, the multiple-tube fermentation method provides an estimate for the MPN of coliforms. ∞ (Chapter 26, p. 748)

MPN Index for Combinations of Positive and Negative Results When Five Tubes Are Used per Dilution (five each of 10 ml, 1 ml, and 0.1 ml)

Number of Tubes with Positive Results				Number of Tubes with Positive Results			
10 ml	1 ml	0.1 ml	MPN Index/100 ml	10 ml	1 ml	0.1 ml	MPN Index/100 ml
0	0	0	<2	4	3	1	33
0	0	1	2	4	4	0	34
0	1	0	2	5	0	0	23
0	2	0	4	5	0	1	30
1	0	0	2	5	0	2	40
1	0	1	4	5	1	0	30
1	1	0	4	5	1	1	50
1	1	1	6	5	1	2	60
1	2	0	6	5	2	0	50
2	0	0	4	5	2	1	70
2	0	1	7	5	2	2	90
2	1	0	7	5	3	0	80
2	1	1	9	5	3	1	110
2	2	0	9	5	3	2	140
2	3	0	12	5	3	3	170
3	0	0	8	5	4	0	130
3	0	1	11	5	4	1	170
3	1	0	11	5	4	2	220
3	1	1	14	5	4	3	280
3	2	0	14	5	4	4	350
3	2	1	17	5	5	0	240
4	0	0	13	5	5	1	300
4	0	1	17	5	5	2	500
4	1	0	17	5	5	3	900
4	1	1	21	5	5	4	1600
4	1	2	26	5	5	5	≥1600
4	2	0	22				
4	2	1	26				
4	3	0	27				

SOURCE: A. E. Greenberg, L. S. Clesceri, and A. D. Eaton, eds. *Standard methods for the examination of water and wastewater,* 18th ed. Washington, D.C.: American Public Health Association, 1992.

CLASSIFICATION OF BACTERIA AND VIRUSES

CLASSIFICATION OF BACTERIA ACCORDING TO *BERGEY'S MANUAL OF SYSTEMATIC BACTERIOLOGY*

The classification of bacteria changes as our understanding of them improves. The classification scheme currently accepted by microbiologists is set forth in the four volumes of *Bergey's Manual of Systematic Bacteriology* (Baltimore: Williams & Wilkins): Volume 1, 1984, R. Krieg; Volume 2, 1986, P. Sneath; Volume 3, 1989, J. Staley; and Volume 4, 1989, S. Williams.

This is a working model only, and changes were made even as the volumes were being published. For example, Section 26 in Volume 4 is an updated, slightly enlarged version of Section 17 in Volume 2. Some organisms present special difficulties and will, therefore, be listed in two different places until further research can determine their appropriate position. Bacteria not yet validated by publication in the *International Journal of Systematic Bacteriology* appear in quotation marks.

KINGDOM PROCARYOTAE

Division I.	**Gracilicutes**	Procaryotes with thinner cell walls, implying a gram-negative type of cell wall
Division II.	**Firmicutes**	Procaryotes with thick and strong skin, indicative of a gram-positive type of cell wall
Division III.	**Tenericutes**	Procaryotes of a pliable, soft nature, indicative of the lack of a rigid cell wall
Division IV.	**Mendosicutes**	Procaryotes having faulty cell walls, suggesting the lack of conventional peptidoglycan-type of cell wall

VOLUME I

SECTION 1: The Spirochetes

Order I.	*Spirochaetales*
Family I.	*Spirochaetaceae*
Genus I.	*Spirochaeta*
Genus II.	*Cristispira*
Genus III.	*Treponema*
Genus IV.	*Borrelia*
Family II.	*Leptospiraceae*
Genus I.	*Leptospira*

Other Organisms
Hindgut spirochetes of termites and *Cryptocercus punctulatus* (wood-eating cockroach)

SECTION 2: Aerobic/Microaerophilic, Motile, Helical/Vibrioid Gram-Negative Bacteria

Genus	*Aquaspirillum*
Genus	*Spirillum*
Genus	*Azospirillum*
Genus	*Oceanospirillum*
Genus	*Campylobacter*
Genus	*Bdellovibrio*
Genus	*Vampirovibrio*

SECTION 3: Nonmotile (or Rarely Motile), Gram-Negative Curved Bacteria

Family I.	*Spirosomaceae*
Genus I.	*Spirosoma*
Genus II.	*Runella*
Genus III.	*Flectobacillus*

Other Genera

Genus	*Ancyclobacter*
Genus	*Meniscus*
Genus	*"Brachyarcus"*
Genus	*"Pelosigma"*

SECTION 4: Gram-Negative Aerobic Rods and Cocci

Family I.	*Pseudomonadaceae*
Genus I.	*Pseudomonas*
Genus II.	*Xanthomonas*
Genus III.	*Frateuria*
Genus IV.	*Zoogloea*
Family II.	*Azotobacteraceae*
Genus I.	*Azotobacter*
Genus II.	*Azomonas*
Family III.	*Rhizobiaceae*
Genus I.	*Rhizobium*
Genus II.	*Bradyrhizobium*
Genus III.	*Agrobacterium*
Genus IV.	*Phyllobacterium*
Family IV.	*Methylococcaceae*
Genus I.	*Methylococcus*
Genus II.	*Methylomonas*
Family V.	*Halobacteriaceae*
Genus I.	*Halobacterium*
Genus II.	*Halococcus*
Family VI.	*Acetobacteraceae*
Genus I.	*Acetobacter*
Genus II.	*Gluconobacter*
Family VII.	*Legionellaceae*
Genus I.	*Legionella*

Family VIII.	*Neisseriaceae*
Genus I.	*Neisseria*
Genus II.	*Moraxella*
Genus III.	*Acinetobacter*
Genus IV.	*Kingella*
Other Genera	
Genus	*Beijerinckia*
Genus	*Derxia*
Genus	*Xanthobacter*
Genus	*Thermus*
Genus	*Thermomicrobium*
Genus	*Halomonas*
Genus	*Alteromonas*
Genus	*Flavobacterium*
Genus	*Alcaligenes*
Genus	*Serpens*
Genus	*Janthinobacterium*
Genus	*Brucella*
Genus	*Bordetella*
Genus	*Francisella*
Genus	*Paracoccus*
Genus	*Lampropedia*
Genus	*Cyclobacterium*

SECTION 5: Facultatively Anaerobic Gram-Negative Rods

Family I.	*Enterobacteriaceae*
Genus I.	*Escherichia*
Genus II.	*Shigella*
Genus III.	*Salmonella*
Genus IV.	*Citrobacter*
Genus V.	*Klebsiella*
Genus VI.	*Enterobacter*
Genus VII.	*Erwinia*
Genus VIII.	*Serratia*
Genus IX.	*Hafnia*
Genus X.	*Edwardsiella*
Genus XI.	*Proteus*
Genus XII.	*Providencia*
Genus XIII.	*Morganella*
Genus XIV.	*Yersinia*
Other Genera of Enterobacteriacae	
Genus	*Obesumbacterium*
Genus	*Xenorhabdus*
Genus	*Kluyvera*
Genus	*Rahnella*
Genus	*Cedecea*
Genus	*Tatumella*
Family II.	*Vibrionaceae*
Genus I.	*Vibrio*
Genus II.	*Photobacterium*
Genus III.	*Aeromonas*
Genus IV.	*Plesiomonas*
Family III.	*Pasteurellaceae*
Genus I.	*Pasteurella*
Genus II.	*Haemophilus*
Genus III.	*Actinobacillus*
Other Genera	
Genus	*Zymomonas*
Genus	*Chromobacterium*
Genus	*Cardiobacterium*
Genus	*Calymmatobacterium*
Genus	*Gardnerella*
Genus	*Eikenella*
Genus	*Streptobacillus*

SECTION 6: Anaerobic Gram-Negative Straight, Curved and Helical Rods

Family I.	*Bacteroidaceae*
Genus I.	*Bacteroides*
Genus II.	*Fusobacterium*
Genus III.	*Leptotrichia*
Genus IV.	*Butyrivibrio*
Genus V.	*Succinimonas*
Genus VI.	*Succinivibrio*
Genus VII.	*Anaerobiospirillum*
Genus VIII.	*Wolinella*
Genus IX.	*Selenomonas*
Genus X.	*Anaerovibrio*
Genus XI.	*Pectinatus*
Genus XII.	*Acetivibrio*
Genus XIII.	*Lachnospira*

SECTION 7: Dissimilatory Sulfate- or Sulfur-Reducing Bacteria

Genus	*Desulfuromonas*
Genus	*Desulfovibrio*
Genus	*Desulfomonas*
Genus	*Desulfococcus*
Genus	*Desulfobacter*
Genus	*Desulfobulbus*
Genus	*Desulfosarcina*

SECTION 8: Anaerobic Gram-Negative Cocci

Family I.	*Veillonellaceae*
Genus I.	*Veillonella*
Genus II.	*Acidaminococcus*
Genus III.	*Megasphaera*

SECTION 9: The Rickettsias and Chlamydias

Order I.	*Rickettsiales*
Family I.	*Rickettsiaceae*
Tribe I.	*Rickettsieae*
Genus I.	*Rickettsia*
Genus II.	*Rochalimaea*
Genus III.	*Coxiella*
Tribe II.	*Ehrlichieae*
Genus IV.	*Ehrlichia*
Genus V.	*Cowdria*
Genus VI.	*Neorickettsia*
Tribe III.	*Wolbachieae*
Genus VII.	*Wolbachia*
Genus VIII.	*Rickettsiella*
Family II.	*Bartonellaceae*
Genus I.	*Bartonella*
Genus II.	*Grahamella*
Family III.	*Anaplasmataceae*
Genus I.	*Anaplasma*
Genus II.	*Aegyptianella*
Genus III.	*Haemobartonella*
Genus IV.	*Eperythrozoon*
Order II.	*Chlamydiales*
Family I.	*Chlamydiaceae*
Genus I.	*Chlamydia*

SECTION 10: The Mycoplasmas

Division *Tenericutes*	
Class I. *Mollicutes*	
Order I.	*Mycoplasmatales*
Family I.	*Mycoplasmataceae*
Genus I.	*Mycoplasma*
Genus II.	*Ureaplasma*
Family II.	*Acholeplasmataceae*

Genus I.	Acholeplasma
Family III.	Spiroplasmataceae
Genus I.	Spiroplasma

Other Genera
| Genus | Anaeroplasma |
| Genus | Thermoplasma |
Mycoplasma-like organisms of plants and invertebrates

SECTION 11: Endosymbionts

A. Endosymbionts of Protozoa
Endosymbionts of ciliates
Endosymbionts of flagellates
Endosymbionts of amoebas
Taxa of endosymbionts:
Genus I.	Holospora
Genus II.	Caedibacter
Genus III.	Pseudocaedibacter
Genus IV.	Lyticum
Genus V.	Tectibacter
B. Endosymbionts of Insects
Blood-sucking insects
Plant sap–sucking insects
Cellulose and stored grain feeders
Insects feeding on complex diets
Taxon of endosymbionts:
| Genus | Blattabacterium |
C. Endosymbionts of Fungi and Invertebrates Other than Arthropods
Fungi
Sponges
Coelenterates
Helminths
Annelids
Marine worms and mollusks

VOLUME 2

SECTION 12: Gram-Positive Cocci

Family I.	Micrococcaceae
Genus I.	Micrococcus
Genus II.	Stomatococcus
Genus III.	Planococcus
Genus IV.	Staphylococcus
Family II.	Deinococcaceae
Genus I.	Deinococcus

Other Genera
| Genus | Streptococcus |
Pyogenic Hemolytic Streptococci
Oral Streptococci
Lactic Acid Streptococci
Anaerobic Streptococci
Other Streptococci
Genus	Enterococcus
Genus	Leuconostoc
Genus	Pediococcus
Genus	Aerococcus
Genus	Gemella
Genus	Peptococcus
Genus	Peptostreptococcus
Genus	Ruminococcus
Genus	Coprococcus
Genus	Sarcina

SECTION 13: Endospore-forming Gram-Positive Rods and Cocci

| Genus | Bacillus |
| Genus | Sporolactobacillus |

Genus	Clostridium
Genus	Desulfotomaculum
Genus	Sporosarcina
Genus	Oscillospira

SECTION 14: Regular, Nonsporing, Gram-Positive Rods

Genus	Lactobacillus
Genus	Listeria
Genus	Erysipelothrix
Genus	Brochothrix
Genus	Renibacterium
Genus	Kurthia
Genus	Caryophanon

SECTION 15: Irregular, Nonsporing, Gram-Positive Rods

| Genus | Corynebacterium |
Plant Pathogenic Species of Corynebacterium
Genus	Gardnerella
Genus	Arcanobacterium
Genus	Arthrobacter
Genus	Brevibacterium
Genus	Curtobacterium
Genus	Caseobacter
Genus	Microbacterium
Genus	Aureobacterium
Genus	Cellulomonas
Genus	Agromyces
Genus	Arachnia
Genus	Rothia
Genus	Propionibacterium
Genus	Eubacterium
Genus	Acetobacterium
Genus	Lachnospira
Genus	Butyrivibrio
Genus	Thermoanaerobacter
Genus	Actinomyces
Genus	Bifidobacterium

SECTION 16: The Mycobacteria

| Family | Mycobacteriaceae |
| Genus | Mycobacterium |

SECTION 17: Nocardioforms
(These organisms are considered again in Section 26.)

Genus	Nocardia
Genus	Rhodococcus
Genus	Nocardioides
Genus	Pseudonocardia
Genus	Oerskovia
Genus	Saccharopolyspora
Genus	Micropolyspora
Genus	Promicromonospora
Genus	Intrasporangium

VOLUME 3

SECTION 18: Anoxygenic Phototrophic Bacteria

I. Purple Bacteria
Family I.	Chromatiaceae
Genus I.	Chromatium
Genus II.	Thiocystis
Genus III.	Thiospirillum
Genus IV.	Thiocapsa
Genus V.	Lamprobacter
Genus VI.	Lamprocystis
Genus VII.	Thiodictyon
Genus VIII.	Amoebobacter

Genus IX. *Thiopedia*
Family II. *Ectothiorhodospiraceae*
Genus *Ectothiorhodospira*
Purple nonsulfur bacteria
Genus *Rhodospirillum*
Genus *Rhodopila*
Genus *Rhodobacter*
Genus *Rhodopseudomonas*
Genus *Rhodomicrobium*
Genus *Rhodocyclus*
II. Green Bacteria
Green sulfur bacteria
Genus *Chlorobium*
Genus *Prosthecochloris*
Genus *Pelodictyon*
Genus *Ancalochloris*
Genus *Chloroherpeton*
Addendum to the green sulfur bacteria
Multicellular filamentous green bacteria
Genus *Chloroflexus*
Genus *Heliothrix*
Genus *"Oscillochloris"*
Genus *Chloronema*
III. Genera Incertae Sedis
Genus *Heliobacterium*
Genus *Erythrobacter*

SECTION 19: Oxygenic Photosynthetic Bacteria

Group I. Cyanobacteria
Preface
Taxa of the Cyanobacteria
Subsection I. Order *Chroococcales*
1. Genus I. *Chamaesiphon*
2. Genus II. *Gloeobacter*
3. *Synechococcus*-group
4. Genus III. *Gloeothece*
5. *Cyanothece*-group
6. *Gloeocapsa*-group
7. *Synechocystis*-group
Subsection II. Order *Pleurocapsales*
1. Genus I. *Dermocarpa*
2. Genus II. *Xenococcus*
3. Genus III. *Dermocarpella*
4. Genus IV. *Myxosarcina*
5. Genus V. *Chroococcidiopsis*
6. *Pleurocapsa*-group
Subsection III. Order *Oscillatoriales*
Genus I. *Spirulina*
Genus II. *Arthrospira*
Genus III. *Oscillatoria*
Genus IV. *Lyngbya*
Genus V. *Pseudanabaena*
Genus VI. *Starria*
Genus VII. *Crinalium*
Genus VIII. *Microcoleus*
Subsection IV. Order *Nostocales*
Family I. *Nostocaceae*
Genus I. *Anabaena*
Genus II. *Aphanizomenon*
Genus III. *Nodularia*
Genus IV. *Cylindrospermum*
Genus V. *Nostoc*
Family II. *Scytonemataceae*
Genus I. *Scytonema*
Family III. *Rivulariaceae*
Genus I. *Calothrix*

Subsection V. Order *Stigonematales*
Genus I. *Chlorogloeopsis*
Genus II. *Fischerella*
Genus III. *Stigonema*
Genus IV. *Geitleria*
Group II. Order *Prochlorales*
Family I. *Prochloraceae*
Genus *Prochloron*
Other taxa
Genus *"Prochlorothrix"*

SECTION 20: Aerobic Chemolithotrophic Bacteria and Associated Organisms

Nitrifying Bacteria
Family *Nitrobacteraceae*
Nitrite-oxidizing bacteria
Genus I. *Nitrobacter*
Genus II. *Nitrospina*
Genus III. *Nitrococcus*
Genus IV. *Nitrospira*
Ammonia-oxidizing bacteria
Genus V. *Nitrosomonas*
Genus VI. *Nitrosococcus*
Genus VII. *Nitrosospira*
Genus VIII. *Nitrosolobus*
Genus IX. *"Nitrosovibrio"*
Colorless Sulfur Bacteria
Genus *Thiobacterium*
Genus *Macromonas*
Genus *Thiospira*
Genus *Thiovulum*
Genus *Thiobacillus*
Genus *Thiomicrospira*
Genus *Thiosphaera*
Genus *Acidiphilium*
Genus *Thermothrix*
Obligately Chemolithotrophic Hydrogen Bacteria
Genus *Hydrogenobacter*
Iron- and Manganese-Oxidizing and/or Depositing Bacteria
Family *"Siderocapsaceae"*
Genus I. *"Siderocapsa"*
Genus II. *"Naumaniella"*
Genus III. *"Siderococcus"*
Genus IV. *"Ochrobium"*
Magnetotactic Bacteria
Genus *Aquaspirillum*
(*A. magnetotacticum*)
Genus *"Bilophococcus"*

SECTION 21: Budding and/or Appendaged Bacteria

Prosthecate Bacteria
Budding bacteria
Buds produced at tip of prostheca
Genus *Hyphomicrobium*
Genus *Hyphomonas*
Genus *Pedomicrobium*
Buds produced on cell surface
Genus *Ancalomicrobium*
Genus *Prosthecomicrobium*
Genus *Labrys*
Genus *Stella*
Bacteria that divide by binary transverse fission
Genus *Caulobacter*
Genus *Asticcacaulis*
Genus *Prosthecobacter*
Nonprosthecate Bacteria
Budding bacteria

Lack peptidoglycan
  Genus   *Planctomyces*
  Genus   *"Isophaera"*
Contain peptidoglycan
  Genus   *Ensifer*
  Genus   *Blastobacter*
  Genus   *Angulomicrobium*
  Genus   *Gemmiger*
Nonbudding, stalked bacteria
  Genus   *Gallionella*
  Genus   *Nevskia*
Other bacteria
  Nonspinate bacteria
    Genus   *Seliberia*
    Genus   *"Metallogenium"*
    Genus   *"Thiodendron"*
  Spinate bacteria

SECTION 22: Sheathed Bacteria

  Genus   *Sphaerotilus*
  Genus   *Leptothrix*
  Genus   *Haliscominobacter*
  Genus   *"Lieskeella"*
  Genus   *"Phragmidiothrix"*
  Genus   *Crenothrix*
  Genus   *"Clonothrix"*

SECTION 23: Nonphotosynthetic, Nonfruiting Gliding Bacteria

  Order I.   *Cytophagales*
    Family I.   *Cytophagaceae*
    Genus I.   *Cytophaga*
    Genus II.   *Capnocytophaga*
    Genus III.   *Flexithrix*
    Genus IV.   *Sporocytophaga*

    Other Genera
    Genus   *Flexibacter*
    Genus   *Microscilla*
    Genus   *Chitinophaga*
    Genus   *Saprospira*
  Order II.   *Lysobacterales*
    Family I.   *Lysobacteraceae*
    Genus I.   *Lysobacter*
  Order III.   *Beggiatoales*
    Family I.   *Beggiatoaceae*
    Genus I.   *Beggiatoa*
    Genus II.   *Thiothrix*
    Genus III.   *Thioploca*
    Genus IV.   *"Thiospirillopsis"*

    Other Families and Genera
    Family   *Simonsiellaceae*
    Genus I.   *Simonsiella*
    Genus II.   *Alysiella*
    Family   *"Pelonemataceae"*
    Genus I.   *"Pelonema"*
    Genus II.   *"Achroonema"*
    Genus III.   *"Peloploca"*
    Genus IV.   *"Desmanthus"*

    Other Genera
    Genus   *Toxothrix*
    Genus   *Leucothrix*
    Genus   *Vitreoscilla*
    Genus   *Desulfonema*
    Genus   *Achromatium*
    Genus   *Agitococcus*
    Genus   *Herpetosiphon*

SECTION 24: Fruiting Gliding Bacteria: The Myxobacteria

  Order *Myxococcales*
    Family I.   *Myxococcaceae*
    Genus   *Myxococcus*
    Family II.   *Archangiaceae*
    Genus   *Archangium*
    Family III.   *Cystobacteraceae*
    Genus I.   *Cystobacter*
    Genus II.   *Melittangium*
    Genus III.   *Stigmatella*
    Family IV.   *Polyangiaceae*
    Genus I.   *Polyangium*
    Genus II.   *Nannocystis*
    Genus III.   *Chondromyces*

SECTION 25: Archaeobacteria

  Group I. Methanogenic Archaeobacteria
    Order I.   *Methanobacteriales*
      Family I.   *Methanobacteriaceae*
      Genus I.   *Methanobacterium*
      Genus II.   *Methanobrevibacter*
      Family II.   *Methanothermaceae*
      Genus   *Methanothermus*
    Order II.   *Methanococcales*
      Family   *Methanococcaceae*
      Genus   *Methanococcus*
    Order III.   *Methanomicrobiales*
      Family I.   *Methanomicrobiaceae*
      Genus I.   *Methanomicrobium*
      Genus II.   *Methanospirillum*
      Genus III.   *Methanogenium*
      Family II.   *Methanosarcinaceae*
      Genus I.   *Methanosarcina*
      Genus II.   *Methanolobus*
      Genus III.   *Methanothrix*
      Genus IV.   *Methanococcoides*
    Other Taxa
      Family   *Methanoplanaceae*
      Genus   *Methanoplanus*
      Other Genus   *Methanosphaera*
  Group II. Archaeobacterial Sulfate Reducers
    Order   *"Archaeoglobales"*
      Family   *"Archaeoglobaceae"*
      Genus   *Archaeglobus*
  Group III. Extremely Halophilic Archaeobacteria
    Order   *Halobacteriales*
      Family   *Halobacteriaceae*
      Genus I.   *Halobacterium*
      Genus II.   *Haloarcula*
      Genus III.   *Haloferax*
      Genus IV.   *Halococcus*
      Genus V.   *Natronobacterium*
      Genus VI.   *Natronococcus*
  Group IV. Cell Wall–less Archaeobacteria
      Genus   *Thermoplasma*
  Group V. Extremely Thermophilic S^0-Metabolizers
    Order I.   *Thermococcales*
      Family   *Thermococcaceae*
      Genus I.   *Thermococcus*
      Genus II.   *Pyrococcus*
    Order II.   *Thermoproteales*
      Family I.   *Thermoproteaceae*
      Genus I.   *Thermoproteus*
      Genus II.   *Thermofilum*
      Family II.   *Desulfurococcaceae*
      Genus   *Desulfurococcus*

Other Bacteria
Genus   *Staphylothermus*
Genus   *Pyrodictium*
Order III.   *Sulfolobales*
Family   *Sulfolobaceae*
Genus I.   *Sulfolobus*
Genus II.   *Acidianus*

VOLUME 4

SECTION 26: Nocardioform Actinomycetes
(This is an updated, slightly enlarged version of Section 17 in Volume 2.)

Genus   *Nocardia*
Genus   *Rhodococcus*
Genus   *Nocardioides*
Genus   *Pseudonocardia*
Genus   *Oerskovia*
Genus   *Saccharopolyspora*
Genus   *Faenia*
Genus   *Promicromonospora*
Genus   *Intrasporangium*
Genus   *Actinopolyspora*
Genus   *Saccharomonospora*

SECTION 27: Actinomycetes with Multilocular Sporangia

Genus   *Geodermatophilus*
Genus   *Dermatophilus*
Genus   *Frankia*

SECTION 28: Actinoplanetes

Genus   *Actinoplanes*
Genus   *Ampullariella*
Genus   *Pilimelia*
Genus   *Dactylosporangium*
Genus   *Micromonospora*

SECTION 29: Streptomycetes and Related Genera

Genus   *Streptomyces*
Genus   *Streptoverticillium*
Genus   *Kineosporia*
Genus   *Sporichthya*

SECTION 30: Maduromycetes

Genus   *Actinomadura*
Genus   *Microbispora*
Genus   *Microtetraspora*
Genus   *Planobispora*
Genus   *Planomonospora*
Genus   *Spirillospora*
Genus   *Streptosporangium*

SECTION 31: Thermomonospora and Related Genera

Genus   *Thermomonospora*
Genus   *Actinosynnema*
Genus   *Nocardiopsis*
Genus   *Streptoalloteichus*

SECTION 32: Thermoactinomycetes

Genus   *Thermoactinomyces*

SECTION 33: Other Genera

Genus   *Glycomyces*
Genus   *Kibdelosporangium*
Genus   *Kitasatosporia*
Genus   *Saccharothrix*
Genus   *Pasteuria*

ADDENDUM

Some novel actinomycete taxa that have been validly published or validated since Volume 4 of *Bergey's Manual of Systematic Bacteriology* went to press.

Placement of certain genera has presented difficulties, as indicated by the following examples:

(a) The genus *Gardnerella*. The organisms of this genus have had a checkered taxonomic history and it is still not entirely clear whether they should be placed in Volume 1 with gram-negative bacteria or in Volume 2 with gram-positive bacteria.

(b) The genus *Butyrvibrio*. Although the cells stain gram-negative, the ultrastructure of the cell wall is of the gram-positive type. It is not clear whether the genus should be placed in Volume 1 or Volume 2.

(c) The genus *Xanthobacter*. The cells stain gram-positive or gram-variable, yet the cell wall structure and composition, as well as nucleic acid hybridization data, indicate that the organisms are of the gram-negative type.

(d) The genus *Chromobacterium*. Although 80 percent of the strains attack glucose fermentatively and grow well anaerobically, the remainder attack glucose oxidatively and grow slowly under anaerobic conditions. It is consequently difficult to assign the organisms definitely to either Section 5 (Facultatively Anaerobic Gram-Negative Rods) or Section 4 (Gram-Negative Aerobic Rods and Cocci). Nucleic acid hybridization studies indicate a relationship to certain genera of aerobic rods.

(e) The genus *Zymomonas*. Although the organisms are facultatively anaerobic (a few obligately anaerobic), they are related genetically, phenotypically and ecologically to the acetic acid bacteria, which are aerobic. Moreover, the occurrence of the Entner-Doudoroff pathway is typical of aerobic bacteria.

(f) The genus *Thermoplasma*. The lack of a cell wall makes this genus compatible with Section 10 (the Mycoplasmas); however, studies of the ribosomal RNA, as well as various phenotypic characteristics, indicate that the genus is related to the archaeobacteria, covered in Volume 3 of the *Manual*.

(g) The genera *Halobacterium* and *Halococcus*. Although these extreme halophiles are compatible with Section 4 (Gram-Negative Aerobic Rods and Cocci), nucleic acid studies and certain phenotypic characteristics indicate the genus is related to the archaeobacteria, covered in Volume 3.

As an interim solution to some of these problems, some taxa are described not only in Volume 1 but in an appropriate subsequent volume as well.

CLASSIFICATION OF VIRUSES

The classification of viruses is undergoing greater changes than is bacterial taxonomy. Most viruses have not even been classified due to a lack of data concerning their reproduction and molecular biology. Estimates suggest that more than 30,000 viruses are being studied in laboratories and reference centers worldwide.

The classification and viral information presented here follows the outline given in Chapter 11 (Tables 11.1 and 11.2). Information also can be found in *Human Virology: A Text for Students of Medicine, Dentistry, and Microbiology* (L. Collier and J. Oxford, 1993, Oxford University Press), and in *Virology* (J. Levy, H. Fraenkel-Conrat, and R. Owens, 2d ed., 1994, Prentice Hall).

The 21 families of viruses listed here are primarily those that infect vertebrates. Thus, these families represent only a small part of the 71 families and more than 3000 viruses recognized in *Virus Taxonomy—Sixth Report of the International Committee on Taxonomy of Viruses* (F. A. Murphy, et al. [eds.], 1995, New York: Springer-Verlag Wien).

1. Family: Picornaviridae

Genera:
- *Enterovirus* (gastrointestinal viruses, poliovirus, coxsackie viruses A and B, echoviruses)
- *Hepatovirus* (hepatitis A virus)
- *Cardiovirus* (encephalomyocarditis virus of mice and other rodents)
- *Rhinovirus* (upper respiratory tract viruses, common cold viruses)
- *Aphthovirus* (foot-and-mouth disease virus)

Naked, polyhedral, positive-sense, ssRNA. Synthesis and maturation take place in the host cell cytoplasm. Viruses are released via cell lysis.

2. Family: Caliciviridae

Genus:
- *Calicivirus* (Norwalk viruses and similar viruses causing gastroenteritis, hepatitis E virus)

Naked, polyhedral, positive-sense, ssRNA. Synthesis and maturation take place in the host cell cytoplasm. Viruses are released via cell lysis.

3. Family: Togaviridae

Genera:
- *Alphavirus* (eastern, western, and Venezuelan equine encephalitis viruses, Semliki forest virus)
- *Rubivirus* (rubella virus)
- *Arterivirus* (equine arteritis virus, simian hemorrhagic fever virus)

Enveloped, polyhedral, positive-sense, ssRNA. Synthesis occurs in the host cell cytoplasm; maturation involves budding of nucleocapsids through the host cell plasma membrane. Viruses are released via cell lysis (*Arterivirus*). Many replicate in arthropods and vertebrates.

4. Family: Flaviviridae

Genera:
- *Flavivirus* (yellow-fever virus, dengue fever virus, St. Louis and Japanese encephalitis viruses, tickborne encephalitis virus)
- *Pestivirus* (bovine diarrhea virus, hog cholera virus)
- Hepatitis C virus

Enveloped, polyhedral, positive-sense, ssRNA. Synthesis occurs in the host cell cytoplasm; maturation involves budding through host cell endoplasmic reticulum and Golgi apparatus membranes. Most replicate in arthropods.

5. Family: Coronaviridae

Genus:
- *Coronavirus* (common cold viruses, avian infectious bronchitis virus, feline infectious peritonitis virus, mouse hepatitis virus)

Enveloped, helical, positive-sense, ssRNA. Synthesis occurs in the host cell cytoplasm; maturation involves budding through membranes of the endoplasmic reticulum and Golgi apparatus. Viruses are released via cell lysis.

6. Family: Rhabdoviridae

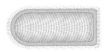

Genera:
- *Vesiculovirus* (vesicular stomatitis-like viruses)
- *Lyssavirus* (rabies and rabieslike viruses)
- [Unnamed] (proposed, for bovine ephemeral feverlike viruses)

[Unnamed] (proposed, for many of the ungrouped rhabdoviruses of mammals, birds, fish, arthropods, and plants)

Enveloped, helical, negative-sense, ssRNA. Synthesis occurs in the host cell nucleus; maturation occurs via budding from the host cell plasma membrane. Many replicate in arthropods.

7. Family: Filoviridae

Genus:
   Filovirus (Marburg and Ebola viruses)

Enveloped; long, filamentous forms, sometimes with branching, and sometimes U-shaped, 6-shaped, or circular; negative-sense, ssRNA. Synthesis occurs in the host cell cytoplasm; maturation involves budding from the host cell plasma membrane. Viruses are released via cell lysis. These viruses are "Biosafety Level 4" pathogens—they must be handled in the laboratory under maximum containment conditions.

8. Family: Paramyxoviridae

Genera:
   Paramyxovirus (parainfluenza viruses 1–4, mumps virus, Newcastle disease virus)
   Morbillivirus (measles and measleslike viruses, canine distemper virus)
   Pneumovirus (respiratory syncytial virus)

Enveloped, helical, negative-sense, ssRNA. Synthesis occurs in the host cell cytoplasm; maturation involves budding through the host cell plasma membrane. Viruses are released via cell lysis. Morbilliviruses can cause persistent infections.

9. Family: Orthomyxoviridae

Genera:
   Influenzavirus A and B (influenza viruses A and B)
   Influenzavirus C (influenza C virus)

Enveloped, helical, negative-sense, ssRNA (eight segments). Synthesis occurs in the host cell nucleus; maturation takes place in the host cell cytoplasm. Viruses are released through budding from the host cell's plasma membrane. These viruses can reassort genes during mixed infections.

10. Family: Bunyaviridae

Genera:
   Bunyavirus (Bunyamwera supergroup)
   Phlebovirus (sandfly fever viruses)
   Nairovirus (Nairobi sheep disease–like viruses)
   Uukuvirus (Uukuniemi-like viruses)
   Hantavirus (hemorrhagic fever viruses, Korean hemorrhagic fever, Sin Nombre hantavirus)

Enveloped, spherical, negative-sense, ssRNA (three segments; Phlebovirus ambisense ssRNA). Synthesis occurs in the host cell cytoplasm; maturation occurs within the Golgi apparatus. Viruses are released via cell lysis. Closely related viruses can reassort genes during mixed infections.

11. Family: Arenaviridae

Genus:
   Arenavirus (Lassa fever virus, lymphocytic choriomeningitis virus, Machupo virus, Junin virus)

Enveloped, helical, ambisense, ssRNA (two segments). Synthesis occurs in the host cell cytoplasm; maturation involves budding from the host cell plasma membrane. Virions contain ribosomes. The human pathogens Lassa, Machupo, and Junin viruses are "Biosafety Level 4" pathogens—they must be handled in the laboratory under maximum containment conditions.

12. Family: Reoviridae

Genera:
   Orthoreovirus (reoviruses 1, 2, and 3)
   Orbivirus (Orungo virus)
   Rotavirus (human rotaviruses)
   Cypovirus (cytoplasmic polyhidrosis viruses)
   Coltivirus (Colorado tick fever virus)
   Plant reovirus 1/3 (plant reoviruses subgroups 1, 2, and 3)

Each genus differs in morphology and physiochemical details. In general, virions are naked, polyhedral, dsRNA (10–12 segments). Synthesis and maturation take place in the host cell cytoplasm. Viruses are released via cell lysis. Virions contain ribosomes.

13. Family: Birnaviridae

Genera:
  Birnavirus (infectious pancreatic necrosis virus of fish and infectious bursal disease virus of fowl)

Naked, polyhedral, dsRNA (two segments). Synthesis and maturation take place in the host cell cytoplasm. Viruses are released via cell lysis.

14. Family: Retroviridae

Genera:
  MLV-related virus (spleen necrosis virus, mouse and feline leukemia viruses)
  Mammalian type-B (mouse mammary tumor virus)
  Type D (squirrel monkey retrovirus)
  ALV-related virus (avian leukemia virus, rous sarcoma virus)
  HTLV-BLV group (human T cell leukemia virus HTLV-I, HTLV-II, bovine leukemia virus)
  Spumavirus (the foamy viruses)
  Lentivirus (human, feline, simian, and bovine immunodeficiency viruses)

Enveloped, spherical, negative-sense, ssRNA (two identical strands). Synthesis occurs in the host cell cytoplasm; maturation involves budding through the host cell plasma membrane. These viruses contain the enzyme reverse transcriptase. The retroviruses (except the Spumavirus and Lentivirus genera) represent the RNA tumor viruses, causing leukemias, carcinomas, and sarcomas).

15. Family: Hepadnaviridae

Genera:
  Orthohepadnavirus (hepatitis B virus)
  Avihepadnavirus (duck hepatitis virus)

Enveloped, polyhedral, partially dsDNA. Synthesis and maturation take place in the host cell nucleus. Surface antigen production occurs in the cytoplasm. Persistence is common and is associated with chronic disease and neoplasia.

16. Family: Parvoviridae

Genera:
  Parvovirus (feline leukopenia virus, canine parvovirus)
  Dependovirus (adeno-associated viruses)
  Densovirus (insect parvoviruses)
  Erythrovirus (human erythrovirus B19)

Naked, polyhedral, negative-sense, ssDNA (*Parvovirus*) or positive-sense and negative-sense, ssDNA (other genera). Synthesis and maturation occur in rapidly dividing host cells, specifically in the host cell nucleus. Viruses are released via cell lysis.

17. Family: Papovaviridae

Genera:
  Papillomavirus (wart viruses, genital condylomas, DNA tumor viruses)
  Polyomavirus (human polyoma-like viruses, SV-40)

Naked, polyhedral, dsDNA. Synthesis and maturation take place in the host cell nucleus. Viruses are released via cell lysis.

18. Family: Adenoviridae

Genera:
  Mastadenovirus (human adenoviruses A–F, infectious canine hepatitis virus)
  Aviadenovirus (avian adenoviruses)

Naked, polyhedral, dsDNA. Synthesis and maturation take place in the host cell nucleus. Viruses are released via cell lysis.

19. Family: Herpesviridae

Subfamily:
  Alphaherpesvirinae
Genera:
  Simplexvirus (herpes simplex viruses 1 and 2)
  Varicellovirus (varicella-zoster virus)

Subfamily:
  Betaherpesvirinae
Genera:
  Cytomegalovirus (human cytomegalovirus)
  Muromegalovirus (murine cytomegalovirus)
Subfamily:
  Gammaherpesvirinae
Genera:
  Lymphocryptovirus (Epstein-Barr viruses)
  Rhadinovirus (saimiri-ateles-like viruses)

Enveloped, polyhedral, dsDNA. Synthesis and maturation occur in the host cell nucleus, with budding through the nuclear envelope. Although most herpesviruses cause persistent infections, virions can be released by rupture of the host cell plasma membrane.

20. Family: Poxviridae

Subfamily:
  Chordopoxvirinae
Genera:
  Orthopoxvirus (vaccinia and variola viruses, cowpox virus)

Parapoxvirus (orf virus, pseudocowpox virus)
Avipoxvirus (fowlpox virus)
Capripoxvirus (sheep pox virus)
Leporipoxvirus (myxoma virus)
Suipoxvirus (swinepox virus)
Yatapoxvirus (yabapox virus and tanapox virus)
Molluscipoxvirus (molluscum contagiosum virus)
Subfamily:
  Entomopoxvirinae
Genus:
  *Entomopoxvirus **A/B/C*** (poxviruses of insects)

External envelope, large, brick-shaped (or ovoid), dsDNA. Synthesis and maturation take place in the portion of the host cell cytoplasm called viroplasm ("viral factories"). Viruses are released via cell lysis.

21. Family: Iridoviridae

Genera:
  Iridovirus (small iridescent insect viruses)
  Chloriridovirus (large iridescent insect viruses)
  Ranavirus (frog viruses)
  Lymphocystivirus (lymphocystis viruses of fish)

Enveloped (missing on some insect viruses), polyhedral, dsDNA. Synthesis occurs in both the host cell nucleus and cytoplasm. Most virions remain cell-associated.

WORD ROOTS COMMONLY ENCOUNTERED IN MICROBIOLOGY

a-, an- — not, without, absence — abiotic, not living; anaerobic, in the absence of air

acantho- — thorn or spinelike — *Acanthamoeba*, an amoeba with spinelike projections

actino- — having rays — *Actinomyces*, a bacterium forming colonies that look like sunbursts

aero- — air — aerobic, in the presence of air

agglutino- — clumping or sticking together — hemagglutination, clumping of blood cells

albo- — white — *Candida albicans*, a white fungus

amphi- — around, doubly, both — Amphitrichous describes flagella found at both ends of a bacterial cell.

ant-, anti- — against, versus — Antibacterial compounds kill bacteria.

archaeo- — ancient — Archaeobacteria are thought to resemble ancient forms of life.

arthro- — joint — arthritis, inflammation of joints

asco- — sac, bag — Ascospores are held in a saclike container, the ascus.

-ase — denotes enzyme — lipase, an enzyme attacking lipids

aureo- — golden — *Staphylococcus aureus* has gold-colored colonies.

auto- — self — Autotrophs, self-feeding organisms.

bacillo- — rod — bacillus, a rod-shaped bacterium

basid- — base, foundation — basidium, fungal cell bearing spores at its end

bio- — life — biology, the study of living things

blast- — bud — blastospore, spore formed by budding

bovi- — cow — *Mycobacterium bovis*, bacterium causing tuberculosis in cattle

brevi- — short — *Lactobacillus brevis*, a bacterium with short rod-shaped cells

butyr- — butter — Butyric acid gives rancid butter its unpleasant odor.

campylo- — curved — *Campylobacter*, a curved bacterium

carcino- — cancer — A carcinogen causes cancer.

caryo-, karyo- — center, kernel — Prokaryotic cells lack a true, discrete nucleus.

caseo- — cheese — caseous, cheeselike lesions

caul- — stalk, stem — *Caulobacter*, a stalked bacterium

ceph-, cephalo- — of the head or brain — encephalitis, inflammation of the brain

chlamydo- — cloaked, hidden — *Chlamydia* are difficult bacteria to detect.

chloro- — green — chlorophyll, a green pigment

chromo- — colored — Metachromatic granules stain various colors within a cell.

chryso- — golden — *Streptomyces chryseus*, a bacterium forming golden colonies

-cide — to kill — Fungicide kills fungi.

co-, con- — with, together — Congenital, existing from birth

cocc- — berry — *Streptococcus*, spherical bacteria in chains

coeno- — shared in common — coenocytic, many nuclei not separated by septa

col-, colo- — colon — coliform bacteria, found in the colon (large intestine)

conidio- — dust — conidia, tiny dustlike spores produced by fungi

coryne- — club — *Corynebacterium diphtheriae*, a club-shaped bacterium

-cul — little, tiny — molecule, a tiny mass

cut-, -cut — skin — cutaneous, of the skin

cyan- — blue — cyanobacteria, formerly called the blue-green algae

cyst-, -cyst — bladder — cystitis, inflammation of the urinary bladder

cyt-, -cyte — cell — leukocyte, white blood cell

de- — lack of, removal — decolorize, to remove color

dermato- — skin — dermatitis, inflammation of the skin

di-, diplo- — two, double — diplococci, pairs of spherical cells

dys- — bad, faulty, painful — dysentery, a disease of the enteric system

ec-, ecto-, ex- — outside, outer — ectoparasite, found on the outside of the body

em-, en- — in, inside — encapsulated, inside a capsule

-emia — of the blood — pyemia, pus in the blood

endo- — inside — endospore, spore found inside a cell

entero- — intestine — enteric, bacteria found in the intestine

epi- — atop, over — epidemic, a disease spreading over an entire population at one time

erythro- — red — lupus erythematosus, disease with a red rash

etio- — cause — etiology, study of the causes of disease

eu- — true, good, normal — eukaryote, cell with a true nucleus

exo- — outside — exotoxin, toxin released outside of a cell

extra- — outside, beyond — extracellular, outside of a cell

fil- — thread — filament, thin chain of cells

flav- — yellow — flavivirus, cause of yellow fever

-fy — to become, make — solidify, to become solid

galacto- | milk galactose, monosaccharide from milk sugar

gamet- | marriage gamete, a reproductive cell, such as egg or sperm

gastro- | stomach gastroenteritis, inflammation of the stomach and intestines

gel- | to stiffen, congeal gelatinous, jellylike

gen-, -gen | to give rise to pathogen, microbe that causes disease

-genesis | origin, development pathogenesis, development of disease

germ, germin- | bud germination, process of growing from a spore

-globulin | protein immunoglobulins, proteins of the immune system

haem-, hem- | blood hemagglutination, clumping of blood cells

halo- | salt halophilic, organisms that thrive in salty environments

hepat- | liver hepatitis, inflammation of the liver

herpes | creeping herpes zoster, or shingles, in which vesicles erupt sequentially along a nerve pathway

hetero- | different, other heterotroph, organism deriving nutrition from other sources

histo- | tissue histology, the study of tissues

homo- | same homologous, having the same structure

hydro- | water hydrologic cycle, water cycle

hyper- | over, above hyperbaric oxygen, higher than atmospheric pressure oxygen

hypo- | under, below hypodermic, going beneath the skin

im-, in- | not insoluble, cannot be dissolved

inter- | between intercellular, between cells

intra- | inside intracellular, inside a cell

io- | violet iodine, element that is purple in gaseous state

iso- | same, equal isotonic, having the same osmotic pressure

-itis | inflammation of meningitis, inflammation of the meninges

kin- | moving kinetic energy, energy of movement

leuko- | white leukocyte, white blood cell

lip-, lipo- | fat, lipid lipoprotein, molecule having both fatty and proteinaceous parts

-logy, -ology | study of microbiology, study of microbes

lopho- | tuft lophotrichous, having a tuft or group of flagella

luc-, luci- | light luciferase, enzyme that catalyzes a light-producing reaction

luteo- | yellow *Micrococcus luteus,* bacterium producing yellow colonies

lys-, -lysis | splitting cytolysis, rupture of a cell

macro- | large macroconidia, large spores

meningo- | membrane meninges, membranes of the brain

meso- | middle mesophile, organism growing best at medium temperatures

micro- | small, tiny microbiology, study of tiny forms of life

mono- | one, single monosaccharide, a single sugar unit

morph- | shape, form pleiomorphic, having many different shapes

multi- | many multicellular, having many cells

mur- | wall muramic acid, a component of cell walls

muri-, mus- | mouse murine, in or of mice

mut-, -mute | to change mutagen, agent that causes genetic change

myc-, -myces | fungus *Actinomyces,* a bacterium that resembles a fungus

myxo- | slime, mucus myxomycetes, slime molds

necro- | dead, corpse necrotizing toxin, causes death of tissue

nema-, -nema | thread *Treponema,* nematode, threadlike organisms

nigr- | black *Rhizopus nigricans,* a black mold

oculo- | eye binocular, microscope with two eyepieces

-oid | like, resembling toxoid, harmless molecule that resembles a toxin

-oma | tumor carcinoma, tumor of epithelial cells

onco- | mass, tumor oncogenes, genes that cause tumors

-osis | condition of brucellosis, condition of being infected with Brucella

pan- | all, universal pandemic, a disease affecting a large part of the world

para- | beside, near, abnormal parainfluenza, a disease resembling influenza

patho- | abnormal pathology, study of abnormal diseased states

peri- | around peritrichous, flagella located all around an organism

phago- | eating phagocytosis, cell eating by engulfing

philo-, -phil, -phile | loving, preferring capnophile, organism needing higher than normal levels of carbon dioxide

-phob, -phobe | hating, fearing hydrophobic, water-repelling

-phore | bearing, carrying electrophoresis, technique in which ions are carried by an electric current

-phyte | plant dermatophyte, fungus that attacks skin

pil- | hair pilus, hairlike tube on bacterial surface

-plast | formed part chloroplast, green body inside plant cell

pod-, -pod | foot podocyte, foot cell of kidney

poly- | many polyribosomes, many ribosomes on the same piece of messenger RNA

post- | afterward, behind post-streptococcal glomerulonephritis, kidney damage following a streptococcal infection

pre-, pro- | before, toward prepubertal, before puberty

pseudo-	false pseudopod, projection resembling a foot, false foot	super-	above, more than superficial mycosis, fungal infection of the surface tissues
psychro-	cold psychrophilic, preferring extreme cold	sym-, syn-	together symbiosis, living together
pyo-	pus pyogenic, producing pus	tact-, -taxis	touch chemotaxis, orientation or movement in response to chemicals
pyro-	fire, heat pyrogen, fever-producing compound	tax-, taxon-	arrangement taxonomy, the classification of organisms
rhin-	nose rhinitis, inflammation of nasal membranes	thermo-	heat thermophile, organism preferring or needing high temperatures
rhizo-	root mycorrhiza, symbiotic growth of fungi and roots	thio-	sulfur *Thiobacillus*, organism that oxidizes hydrogen sulfide to sulfates
rhodo-	red *Rhodospirillum*, a large red spiral bacterium	tox-	poison toxin, a harmful compound
-rrhea	flow diarrhea, abnormal flow of liquid feces	trans-	through, across transduction, movement of genetic information from one cell to another
rubri-	red *Rhodospirillum rubrum*, a large red spiral bacterium	trich-	hair monotrichous, having a single, hair-like flagellum
saccharo-	sugar polysaccharide, many sugar units linked together	-troph	feeding, nutrition phototroph, organism that makes its own food, using energy from light
sapro-	rotten, decaying saprophyte, organism living on dead matter	uni-	one, singular unicellular, composed of one cell
sarco-	flesh sarcoma, tumor made up of muscle or connective tissue	undul-	waving undulant fever, disease in which fever rises and falls
schizo-	to split schizogony, a type of fission in malarial parasites	vac-, vaccin-	cow vaccine, disease-preventing product originally produced by inoculating it onto skin of calves
-scope, -scopy	to see, examine microscopy, use of the microscope to examine small things	vacu-	empty vacuole, empty-appearing structure in cytoplasm
sept-, septo-	partition, wall septum, wall between cells	vesic-	blister, bladder vesicle, small blisterlike lesions
septi-	rotting septic, exhibiting decomposition due to bacteria	vitr-	glass in vitro, grown in laboratory glassware
soma-, -some	body chromosome, colored body (when stained)	xantho-	yellow *Xanthomonas oryzae*, bacterium producing yellow colonies
spiro-	coil spirochete, spiral-shaped bacterium	xeno-	strange, foreign xenograft, graft from a different species
sporo-	spore sporocidal, spore killing	zoo-	animal protozoan, first animal
staphylo-	in bunches, like grapes staphylococci, spherical bacteria growing in clusters	zygo-	yoke, joining zygote, fertilized egg
-stasis, stat-	stopping, not changing bacteriostatic, able to stop the growth of bacteria	-zyme	ferment enzymes, biological catalysts, some of which are involved in fermentation
strepto-	twisted *Streptobacillus*, twisted chains of bacilli		
sub-	under, below subclinical, signs and symptoms not clinically apparent		

SAFETY PRECAUTIONS IN THE HANDLING OF CLINICAL SPECIMENS

Concern for maintaining safe conditions in school and hospital laboratories, in other work settings, and especially during patient interactions has led the federal government to formulate various regulations and recommendations. These are far too extensive to reproduce here in their entirety; several are nearly 200 pages in length. As an introduction to the kinds of safety measures that should be taken, a few of the guidelines set forth in these publications are provided below. Some key references and sources are also listed, with the hope that they will stimulate the interested reader to investigate further.

In 1983 CDC published a document entitled "Guidelines for isolation precautions in hospitals" that contained a section headed "Blood and body fluid precautions." The recommendations in this section specified precautions to be observed regarding contact with the blood or body fluids of any patient known or suspected to be infected with bloodborne pathogens.

In August 1987 CDC published a document entitled "Recommendations for prevention of HIV transmission in health-care settings." In contrast with the 1983 document, the 1987 publication recommended that blood and body fluid precautions be consistently used for *all* patients regardless of their bloodborne infection status. These blood and body fluid precautions as they pertain to all patients are referred to as "Universal Blood and Body Fluid Precautions," or more simply as "Universal Precautions."

Following publication of this document, there were many requests for clarification—for example, as to which bodily fluids these precautions should apply. This led to a CDC publication on June 24, 1988, in *MMWR* entitled "Update: Universal precautions for prevention of transmission of human immunodeficiency virus, hepatitis B virus, and other bloodborne pathogens in health-care settings."

Copies of these two most recent reports (CDC August 1987 and June 1988) are available through the National AIDS Information Clearinghouse, P.O. Box 6003, Rockville, MD 20850.

In these two publications CDC makes the following recommendations regarding sharp instruments:

1. Take care to prevent injuries when using needles, scalpels, and other sharp instruments or devices; when handling sharp instruments after procedures; when cleaning used instruments; and when disposing of used needles. Do not recap used needles by hand; do not remove used needles from disposable syringes by hand; and do not bend, break, or otherwise manipulate used needles by hand. Place used disposable syringes and needles, scalpel blades, and other sharp items in puncture-resistant containers for disposal. Locate the puncture-resistant containers as close to the use area as is practical.

2. Use protective barriers to prevent exposure to blood, body fluids containing visible blood, and other fluids to which universal precautions apply. The type of protective barrier(s) should be appropriate for the procedure being performed and the type of exposure anticipated.

3. Immediately and thoroughly wash hands and other skin surfaces that are contaminated with blood, body fluids containing visible blood, or other body fluids to which universal precautions apply.

Glove use is recommended during phlebotomy (drawing blood samples) but cannot protect against penetrating injuries. Some institutions have relaxed recommendations for using gloves for phlebotomy procedures by skilled phlebotomists in settings where the prevalence of bloodborne pathogens is known to be very low (for example, volunteer blood-donation centers). Such institutions should periodically reevaluate their policy. Gloves should always be available to health care workers who wish to use them for phlebotomy. In addition, the following general guidelines apply:

1. Use gloves for performing phlebotomy when the health care worker has cuts, scratches, or other breaks in his/her skin.

2. Use gloves in situations where the health care worker judges that hand contamination with blood may occur, for example, when performing phlebotomy on an uncooperative patient.

3. Use gloves for performing finger and/or heel sticks on infants and children.

4. Use gloves when persons are receiving training in phlebotomy.

In March 1989 the Environmental Protection Agency (EPA) published a new set of "Standards for the tracking and management of medical waste," in part designed to prevent the deplorable pollution of our nation's beaches by medical wastes (*Federal Register,* March 24, 1989, pp. 12325–95). In May 1989 the Occupational Safety and Health Administration (OSHA) published a new set of rules for "Occupational exposure to bloodborne pathogens" (*Federal Register,* May 30, 1989, pp. 23041–139). Both publications provide very detailed information about procedures that must be followed.

Another useful and quite detailed publication, issued by CDC in February 1989 and reprinted in *MMWR* for June 23, 1989, is entitled "Guidelines for prevention of transmission of human immunodeficiency virus and hepatitis B virus to health-care and public-safety workers."

For those concerned primarily with the teaching laboratory, we recommend the publication "Handling infectious materials in the education setting" (G. Ballman, *American Clinical Laboratory,* July 1989, pp. 10–11). Other publications of interest include *Biosafety in Microbiological and Biomedical Laboratories* (CDC, U.S. Department of Health and Human Services, Public Health Service, 1988), which describes the four biosafety levels classified by CDC and recommended safety precautions for each, and "Labeling of microbial risks" (C. Robinson and T. H. Hatfield, *Journal of College Science Teaching,* May 1995, pp. 407–9), which evaluates that classification system of biosafety levels.

METABOLIC PATHWAYS

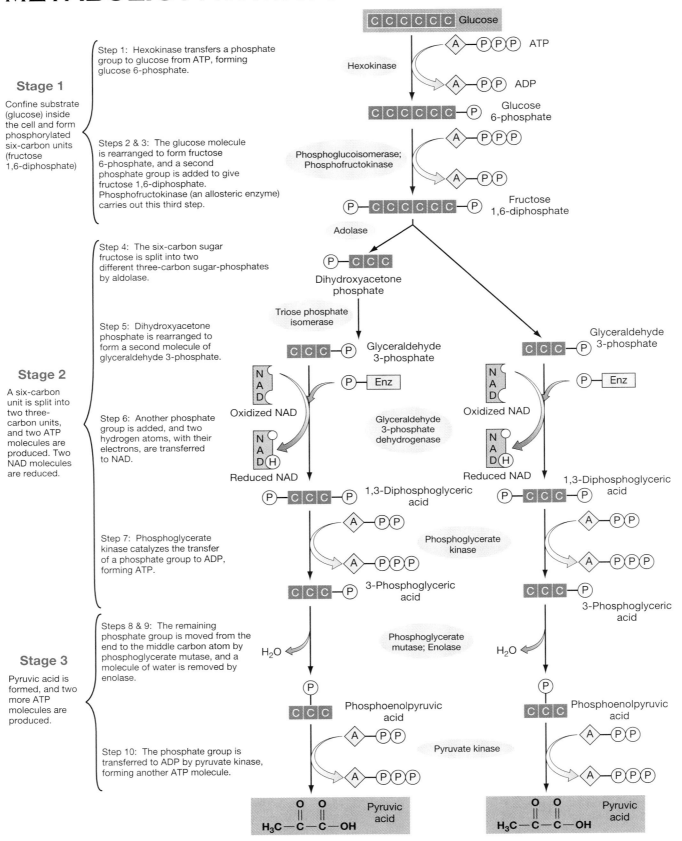

Stage 1

Confine substrate (glucose) inside the cell and form phosphorylated six-carbon units (fructose 1,6-diphosphate)

Step 1: Hexokinase transfers a phosphate group to glucose from ATP, forming glucose 6-phosphate.

Steps 2 & 3: The glucose molecule is rearranged to form fructose 6-phosphate, and a second phosphate group is added to give fructose 1,6-diphosphate. Phosphofructokinase (an allosteric enzyme) carries out this third step.

Stage 2

A six-carbon unit is split into two three-carbon units, and two ATP molecules are produced. Two NAD molecules are reduced.

Step 4: The six-carbon sugar fructose is split into two different three-carbon sugar-phosphates by aldolase.

Step 5: Dihydroxyacetone phosphate is rearranged to form a second molecule of glyceraldehyde 3-phosphate.

Step 6: Another phosphate group is added, and two hydrogen atoms, with their electrons, are transferred to NAD.

Step 7: Phosphoglycerate kinase catalyzes the transfer of a phosphate group to ADP, forming ATP.

Stage 3

Pyruvic acid is formed, and two more ATP molecules are produced.

Steps 8 & 9: The remaining phosphate group is moved from the end to the middle carbon atom by phosphoglycerate mutase, and a molecule of water is removed by enolase.

Step 10: The phosphate group is transferred to ADP by pyruvate kinase, forming another ATP molecule.

FIGURE E.1 Glycolysis (Embden–Meyerhof pathway). Each of the 10 steps of glycolysis is catalyzed by a specific enzyme, which is indicated in a purple oval. (Refer to Chapter 5 for an explanation; Figure 5.11 shows a simplified version of the process. p. 117)

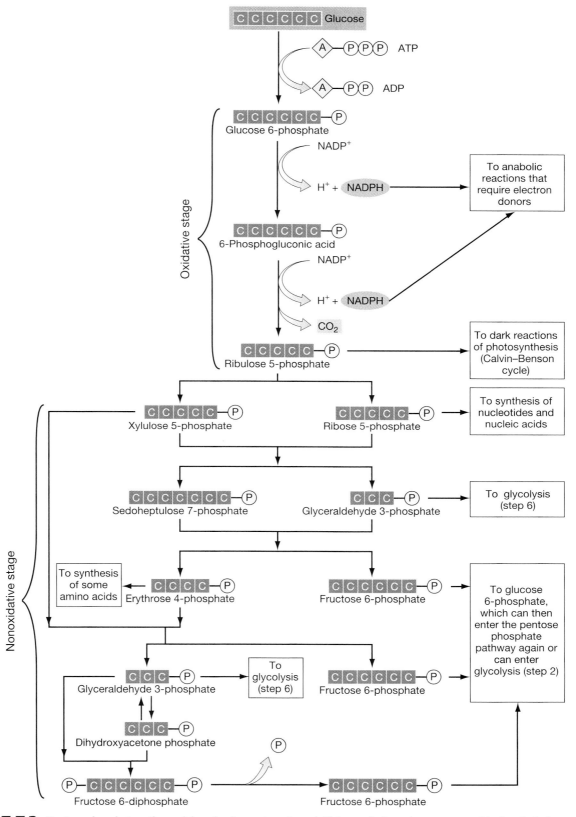

FIGURE E.2 Pentose phosphate pathway (phosphogluconate pathway). This metabolic pathway occurs with glycolysis. It provides an alternative pathway for the breakdown of glucose as well as pentoses (five-carbon sugars). This pathway plays three important roles: (1) It provides intermediate pentoses, especially ribose, that the bacterial cell must use to synthesize nucleic acids. (2) This pathway's intermediates can be used to synthesize some amino acids. (3) The pentose phosphate pathway reduces NADP to NADPH. This coenzyme, like NADH, is an electron carrier and thus is a source of reducing power. The fates of several intermediates are indicated. For clarity, the specific enzymes catalyzing these reactions and the structural formulas of substrates have been omitted. (Refer to Chapter 5 for an explanation of this pathway. ∞ p. 118)

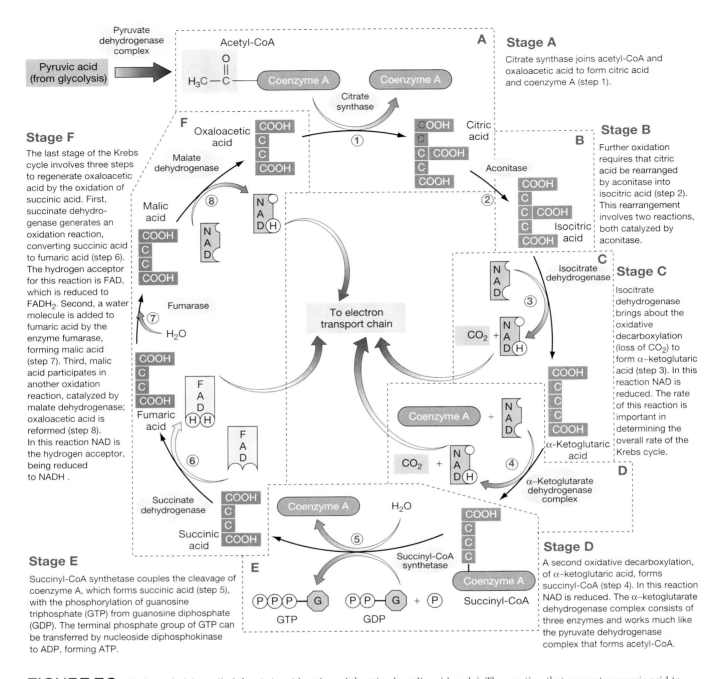

Stage A

Citrate synthase joins acetyl-CoA and oxaloacetic acid to form citric acid and coenzyme A (step 1).

Stage B

Further oxidation requires that citric acid be rearranged by aconitase into isocitric acid (step 2). This rearrangement involves two reactions, both catalyzed by aconitase.

Stage C

Isocitrate dehydrogenase brings about the oxidative decarboxylation (loss of CO_2) to form α–ketoglutaric acid (step 3). In this reaction NAD is reduced. The rate of this reaction is important in determining the overall rate of the Krebs cycle.

Stage D

A second oxidative decarboxylation, of α–ketoglutaric acid, forms succinyl-CoA (step 4). In this reaction NAD is reduced. The α–ketoglutarate dehydrogenase complex consists of three enzymes and works much like the pyruvate dehydrogenase complex that forms acetyl-CoA.

Stage E

Succinyl-CoA synthetase couples the cleavage of coenzyme A, which forms succinic acid (step 5), with the phosphorylation of guanosine triphosphate (GTP) from guanosine diphosphate (GDP). The terminal phosphate group of GTP can be transferred by nucleoside diphosphokinase to ADP, forming ATP.

Stage F

The last stage of the Krebs cycle involves three steps to regenerate oxaloacetic acid by the oxidation of succinic acid. First, succinate dehydrogenase generates an oxidation reaction, converting succinic acid to fumaric acid (step 6). The hydrogen acceptor for this reaction is FAD, which is reduced to $FADH_2$. Second, a water molecule is added to fumaric acid by the enzyme fumarase, forming malic acid (step 7). Third, malic acid participates in another oxidation reaction, catalyzed by malate dehydrogenase; oxaloacetic acid is reformed (step 8). In this reaction NAD is the hydrogen acceptor, being reduced to NADH .

FIGURE E.3 Krebs cycle (also called the citric acid cycle and the tricarboxylic acid cycle). The reaction that converts pyruvic acid to acetyl-CoA precedes the Krebs cycle (see Figure 5.16.). This reaction is catalyzed by a pyruvate dehydrogenase complex, which contains three enzymes. Each of the eight steps of the Krebs cycle is also catalyzed by a specific enzyme, as indicated in a purple oval. (Refer to Chapter 5 for an explanation; Figure 5.17 shows a simplified version of the process.∞ p. 121)

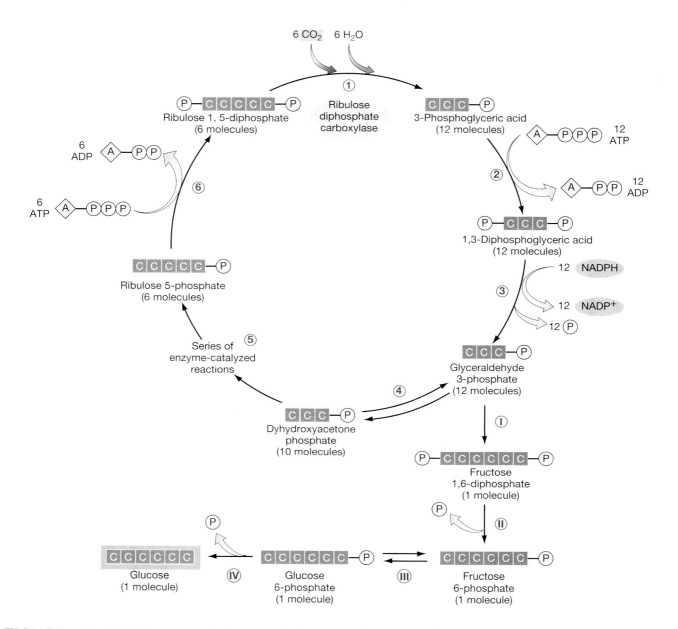

FIGURE E.4 Calvin–Benson cycle (dark reactions of photosynthesis). Each step of the Calvin–Benson cycle is catalyzed by a specific enzyme, which for simplicity is not shown. Steps 1 through 3 produce 12 three-carbon intermediates. These three steps are dependent on photophosphorylation products (ATP and NADPH). Two of every 12 three-carbon molecules undergo chemical reactions (steps 1 through IV) to produce a six-carbon glucose molecule. The other 10 three-carbon molecules are recycled (steps 4 through 6), forming 6 five-carbon molecules. These are phosphorylated by ATP to ribulose-1,5-diphosphate. Each of these five-carbon molecules then combines with a CO_2 molecule, starting the process once again. The enzyme catalyzing this step is *ribulose diphosphate carboxylase*, the most prevalent enzyme in the biological world. (Refer to Chapter 5 for an explanation of the process. ∞ p. 128)

ANSWERS TO SELECTED PROBLEMS FOR INVESTIGATION

Chapter 12

7. Cysticercosis is acquired by ingesting tapeworm eggs shed in human feces, not by ingesting pork. The disease is widely endemic in rural areas of Latin America, Asia, and Africa. It is probable that the girl obtained her infection from one of her neighbor's friends via the fecal-oral route.

Chapter 13

8. **a)** A leak was discovered in the lines supplying ethylene oxide to the ethylene oxide sterilizer.
   b) Ethylene oxide is a carcinogen.

Chapter 14

6. Pneumonia-like diseases should be given dual antibiotic therapy, with one being erythromycin, the only commonly used antibiotic that is effective against Legionnaires' disease.

Chapter 15

7. **a)** An intoxication. Such swift onset of symptoms was due to the presence of preformed toxin in the eggs. Bacteria could not have multiplied that quickly so as to cause an infection.
   b) Endotoxins can withstand prolonged heating. Perhaps the eggs were soft and runny.
   c) No, endotoxins are weak antigens and cannot be converted to toxoids for immunization.
   d) *Salmonella* bacteria are part of the normal microbiota in the chicken's gut and feces. Contamination of the shell is unavoidable. The bacteria entered through a crack and grew inside.

Chapter 16

5. **a)** The father is most likely the index case. Adults with pertussis lack the characteristic "whooping" cough, usually presenting signs and symptoms typical of a heavy chest cold. That means that adults are the unidentified cases who spread the disease to children, in whom it *is* recognized.
   b) Because pertussis is a notifiable disease.

Chapter 18

6. She can give her baby antibodies against only diseases that she has had. Because new strains of flu virus arise frequently, she would not have antibodies against these strains. Also, some kinds of antibodies, such as those against pertussis, do not cross the placenta at all. She also might not secrete as much antibody in her breast milk as she imagines. The bulk of antibodies received by the baby are in colostrum, not in milk.

Chapter 19

5. Autoimmune diseases improve when the immune system is suppressed. Smoking suppresses the immune system. When the woman was smoking, her immune system was sufficiently suppressed for her to remain asymptomatic. Rebound of her immune system after she stopped smoking caused symptoms to return.

Chapter 20

6. *Loa loa* worm, which is endemic to Africa.

Chapter 21

6. An impaired immune system—probably due to HIV infection.

Chapter 22

6. The prong probably became contaminated with blood, which could enter the wound made in a subsequent patient's finger when blood was then drawn.

Chapter 23

7. Bubonic plague. Plague is endemic to Bolivia, and the buboes in the armpit are highly diagnostic. CDC confirmed the diagnosis from samples aspirated from one of the buboes.

Chapter 25

7. Yes. The virus that causes genital warts is associated with 90 percent of all cases of cervical cancer. The virus can cause inapparent infections that last for years. It is transmitted sexually, and the husband may have transmitted it to both wives, or he may have acquired it from the first wife and then passed it on to the second wife. (Moral of the story: Ask a widower what his first wife died of.)

abiotic factor A physical feature of the environment that interacts with organisms.

ABO blood group system One of the blood typing systems that is based on the presence or absence of blood group antigens A and B on red blood cells.

abortive infection Viral infection in which viruses enter a cell but are unable to express all their genes to make infectious progeny.

abscess An accumulation of pus in a cavity hollowed out by tissue damage.

absorption Process in which light rays are neither passed through nor reflected off an object but are retained and either transformed to another form of energy or used in biological processes.

accidental parasite A parasite that invades an organism other than its normal host.

acid A substance that releases hydrogen ions when it is dissolved in water.

acidic dye See **anionic dye**.

acidophile An acid-loving organism that grows best in an environment with a pH of 4.0 to 5.4.

acme During the illness phase of the disease process, the time of most intense signs and symptoms. (Sometimes referred to as **fulminating**.)

acne Skin condition caused by bacterial infection of hair follicles and the ducts of sebaceous glands.

acquired immune deficiency syndrome (AIDS) An infectious disease caused by the human immunodeficiency virus that destroys the individual's immune system.

acquired immunity Immunity obtained in some manner other than by heredity.

acridine derivative A chemical mutagen that can be inserted between bases of the DNA double helix, causing frameshift mutations.

actinomycetes Gram-positive bacteria that tend to form filaments.

activated sludge system Procedure in which the effluent from the primary stage of sewage treatment is agitated, aerated, and added to sludge containing aerobic organisms that digest organic matter.

activation energy The energy required to start a chemical reaction.

active immunity Immunity created when an organism's own immune system produces antibodies or other defenses against an agent recognized as foreign.

active immunization Use of vaccines to control disease by increasing herd immunity through stimulation of the immune response.

active site Area on the surface of an enzyme to which its substrate binds.

active transport Movement of molecules or ions across a membrane against a concentration gradient; requires expenditure of energy from ATP.

acute disease A disease that develops rapidly and runs its course quickly.

acute hemorrhagic conjunctivitis Eye disease caused by an enterovirus.

acute inflammation The relatively short duration of inflammation during which time host defenses destroy invading microbes and repair tissue damage.

acute necrotizing ulcerative gingivitis (ANUG) A severe form of periodontal disease. (Also called trench mouth.)

acute phase protein Protein, such as C-reactive protein or mannose-binding protein, that forms a nonspecific host defense mechanism during an acute phase response.

acute phase response A response to an acute illness that produces specific blood proteins called acute phase proteins.

acute respiratory disease (ARD) Viral disease that occurs in epidemics with cold symptoms as well as fever, headache, and malaise; sometimes causes viral pneumonia.

adenovirus A medium-sized, naked DNA virus that is highly resistant to chemical agents and often causes respiratory infections or diarrhea.

adherence The attachment of a microorganism to a host's cell surface.

adhesin A protein or glycoprotein on attachment pili (fimbriae) or capsules that helps a microorganism attach to a host cell.

adsorption The attachment of the virus to the host cell in the replication process.

aerobe An organism that uses oxygen, including ones that must have oxygen.

aerobic respiration Process in which aerobic organisms gain energy from the catabolism of organic molecules via the Krebs cycle and oxidative phosphorylation.

aerosol A cloud of tiny liquid droplets suspended in air.

aerotolerant anaerobe A bacterium that can survive in the presence of oxygen but does not use oxygen in its metabolism.

aflatoxin Fungal toxin that is a potent carcinogen; found in food made from contaminated grain or peanuts infested with *Aspergillus flavus* and other aspergilli.

African sleeping sickness Disease of equatorial Africa caused by protozoan blood parasites of the genus *Trypanosoma*. (Also called **trypanosomiasis**.)

agammaglobulinemia Primary immunodeficiency disease caused by failure of B cells to develop, resulting in lack of antibodies.

agar A polysaccharide extracted from certain marine algae and used to solidify medium for the growth of microorganisms.

agar plate A plate of nutrient medium solidified with agar.

agglutination reaction A reaction of antibodies with antigens that results in agglutination, the clumping together of cells or other large particles.

agranulocyte A leukocyte (monocyte or lymphocyte) that lacks granules in the cytoplasm and has rounded nuclei.

alcoholic fermentation Fermentation in which pyruvic acid is reduced to ethyl alcohol by electrons from reduced NAD (NADH).

algae (singular: **alga**) Photosynthetic, eukaryotic organisms in the kingdoms Protista and Plantae.

alkaline Condition caused by an abundance of hydroxyl ions (OH^-), resulting in a pH of greater than 7.0. (Also called basic.)

alkaliphile A base- (alkaline) loving organism that grows best in an environment with a pH of 7.0 to 11.5.

alkylating agent A chemical mutagen that can add alkyl groups ($-CH_3$) to DNA bases, altering their shapes and causing errors in base pairing.

allele The form of a gene that occupies the same place (locus) on the DNA molecule as another form but may carry different information for a trait.

allergen An ordinarily innocuous foreign substance that can elicit an adverse immunological response in a sensitized person.

allergy See **hypersensitivity**.

allograft A graft of tissue between two organisms of the same species that are not genetically identical.

allosteric site The site at which a noncompetitive inhibitor binds.

alpha (α) hemolysin A type of enzyme that partially lyses red blood cells, leaving a greenish ring in the blood agar medium around the colonies.

alpha (α) hemolysis Incomplete lysis of red blood cells by bacterial enzymes.

alternative pathway   One of the two sequences of reactions in nonspecific host responses by which proteins of the complement system are activated.

alveolus   A saclike structure arranged in clusters at the ends of respiratory bronchioles, having walls one cell layer thick, where gas exchange occurs.

amantadine   An antiviral agent that prevents penetration by influenza A viruses.

Ames test   Test used to determine whether a particular substance is mutagenic, based on its ability to induce mutations in auxotrophic bacteria.

amino acid   An organic acid containing an amino group and a carboxyl group, composing the building blocks of proteins.

aminoglycoside   An antimicrobial agent that blocks bacterial protein synthesis.

amoebic dysentery   Severe, acute form of amebiasis, caused by *Entamoeba histolytica.*

amoeboid movement   Movement by means of pseudopodia that occurs in cells without walls, such as amoebas and some white blood cells.

amphibolic pathway   A metabolic pathway that can yield either energy or building blocks for synthetic reactions.

amphitrichous   The presence of flagella at both ends of the bacterial cell.

anabolic pathway   A chain of chemical reactions in which energy is used to synthesize biologically important molecules.

anabolism   Chemical reactions in which energy is used to synthesize large molecules from simpler components. (Also called synthesis.)

anaerobe   An organism that does not use oxygen, including some organisms that are killed by exposure to oxygen.

analytical study   An epidemiological study that focuses on establishing cause-and-effect relationships in the occurrence of diseases in populations.

anamnestic response   Prompt immune response due to "recall" by memory cells. (Also called **secondary response.**)

anaphylactic shock   Condition resulting from a sudden extreme drop in blood pressure caused by an allergic reaction.

anaphylaxis   An immediate, exaggerated allergic reaction to antigens, usually leading to detrimental effects.

Ångstrom   (Å)   Unit of measurement equal to 0.0000000001 m, or 10^{-10} m. No longer officially recognized.

Animalia   The kingdom of organisms to which all animals belong.

animal passage   The rapid transfer of a pathogen through animals of a species susceptible to infection by the pathogen.

anion   A negatively charged ion.

anionic dye   An ionic compound, used for staining bacteria, in which the negative ion imparts the color. (Also called **acidic dye.**)

antagonism   The decreased effect when two antibiotics are administered together.

anthrax   A zoonosis caused by *Bacillus anthracis* that exists in cutaneous, respiratory ("woolsorters disease"), or intestinal forms; transmitted by endospores.

antibiosis   The natural production of an antimicrobial agent by a bacterium or fungus.

antibiotic   A chemical substance produced by microorganisms that can inhibit the growth of or destroy other microorganisms.

antibody   A protein produced in response to an antigen that is capable of binding specifically to that antigen. (Also called **immunoglobulin.**)

antibody titer   The quantity of a specific antibody in an individual's blood, often measured by means of agglutination reactions.

anticodon   A three-base sequence in tRNA that is complementary to one of the mRNA codons, forming a link between each codon and the corresponding amino acid.

antigen   A substance that the body identifies as foreign and toward which it mounts an immune response. (Also called **immunogen.**)

antigen binding site   The site on the antibody to which the antigen (epitope) binds.

antigen challenge   Exposure to a foreign antigen.

antigenic determinant   See **epitope.**

antigenic drift   Process of antigenic variation that results from mutations in genes coding for hemagglutinin and neuraminidase.

antigenic mimicry   Self antigen that is similar to an antigen on a pathogen.

antigenic shift   Process of antigenic variation probably caused by a reassortment of viral genes.

antigenic variation   Mutations in influenza viruses that occur by antigenic drift and antigenic shift.

antigen presenting cell   An immunological cell, such as a macrophage, dendritic cell, or B cell, that processes antigen fragments and presents peptide fragments from the antigen on its cell surface.

antihistamine   Drug that alleviates symptoms caused by histamine.

antimetabolite   A substance that prevents a cell from carrying out an important metabolic reaction.

antimicrobial agent   A chemotherapeutic agent used to treat diseases caused by microbes.

antiparallel   The opposite head-to-tail arrangement of the two strands in a DNA double helix.

antiseptic   A chemical agent that can be safely used externally on tissues to destroy microorganisms or to inhibit their growth.

antiserum   (plural: **antisera**) Serum that contains antibodies.

antitoxin   An antibody against a specific toxin.

antiviral protein   A protein induced by interferon that interferes with the replication of viruses.

apicomplexan   A parasitic protozoan such as *Plasmodium,* that generally has a complex life cycle. (Also called sporozoan.)

aplastic crisis   A period during which erythrocyte production ceases.

apoenzyme   The protein portion of an enzyme.

arachnid   An arthropod with two body regions, four pairs of legs, and mouth parts that are used in capturing and tearing apart prey.

archaeobacteria   Prokaryotic organisms lacking peptidoglycan in their cell walls and differing from eubacteria in many ways.

arenavirus   An enveloped RNA virus that causes Lassa fever and certain other hemorrhagic fevers.

arthropod   Makes up the largest group of living organisms, characterized by a jointed chitinous exoskeleton, segmented body, and jointed appendages associated with some or all of the segments.

Arthus reaction   A local reaction seen in the skin after subcutaneous or intradermal injection of an antigenic substance, an immune complex (type III) hypersensitivity.

ascariasis   Disease caused by a large roundworm, *Ascaris lumbricoides,* acquired by ingestion of food or water contaminated with eggs.

Ascomycota   See **sac fungus.**

ascospore   One of the eight sexual spores produced in each ascus of a sac fungus.

ascus   (plural: **asci**) Saclike structure produced by sac fungi during sexual reproduction.

aseptic technique   A set of procedures used to minimize chances that cultures will be contaminated by organisms from the environment.

Asiatic cholera   Severe gastrointestinal disease caused by *Vibrio cholerae;* common in areas of poor sanitation and fecal contamination of water.

aspergillosis   Skin infection caused by various species of *Aspergillus,* which can cause severe pneumonia in immunosuppressed patients. (Also called farmer's lung disease.)

asthma   Respiratory anaphylaxis caused by inhaled or ingested allergens or by hypersensitivity to endogenous microorganisms.

athlete's foot   A form of ringworm in which hyphae invade the skin between the toes, causing dry, scaly lesions. (Also called t**inea pedis.**)

atom   The smallest chemical unit of matter.

atomic number   The number of protons in an atom of a particular element.

atomic weight   The sum of the number of protons and neutrons in an atom.

atopy   Localized allergic reactions that occur first at the site where an allergen enters the body.

atrichous   A bacterial cell without flagella.

attachment pilus   Type of pilus that helps bacteria adhere to surfaces. (Also called **fimbria.**)

attenuation   (1) A genetic control mechanism that terminates transcription of an operon prematurely when the gene products are not needed. (2) The weakening of the disease-producing ability of an organism.

auditory canal   Part of the outer ear lined with skin that contains many small hairs and ceruminous glands.

autoantibody   An antibody against one's own tissues.

autoclave   An instrument for sterilization by means of moist heat under pressure.

autograft   A graft of tissue from one part of the body to another.

autoimmune disorder   An immune disorder in which individuals are hypersensitive to antigens on cells of their own bodies.

autoimmunization   The process by which hypersensitivity to "self" develops; it occurs when the immune system responds to a body component as if it were foreign.

autotroph   An organism that uses carbon dioxide gas to synthesize organic molecules.

autotrophy   "Self-feeding"—the use of CO_2 as a source of carbon atoms for the synthesis of biomolecules.

auxotroph   An organism that requires special substances in its growth medium to maintain growth.

auxotrophic mutant   An organism that has lost the ability to synthesize one or more metabolically important enzymes through mutation.

axial filament   A subsurface filament attached near the ends of the cytoplasmic cylinder of spirochetes that causes the spirochete body to rotate like a corkscrew. (Also called **endoflagellum.**)

babesiosis   A protozoan disease caused by the apicomplexan *Babesia microti* and other species of *Babesia*.

bacillary angiomatosis   A disease of the small blood vessels of the skin and internal organs caused by the rickettsial organism *Bartonella henselae*.

bacillary dysentery   See **shigellosis.**

bacillus   (plural: **bacilli**) A rodlike bacterium.

bacteremia   An infection in which bacteria are transported in the blood but do not multiply in transit.

bacteria   (singular: **bacterium**) All prokaryotic organisms.

bacterial conjunctivitis   A highly contagious inflammation of the conjunctiva caused by various bacterial species. (Also called pinkeye.)

bacterial endocarditis   A life-threatening infection and inflammation of the lining and valves of the heart. (Also called infective endocarditis.)

bacterial enteritis   An intestinal infection caused by bacterial invasion of intestinal mucosa or deeper tissues.

bacterial lawn   A uniform layer of bacteria grown on the agar surface in a petri dish.

bacterial meningitis   An inflammation of the meninges, the membranes that cover the brain and spinal cord, caused by any one of several bacterial species.

bactericidal   Referring to an agent that inhibits the growth of bacteria.

bacteriocin   A protein released by some bacteria that inhibits the growth of other strains of the same or closely related species.

bacteriocinogen   A plasmid that directs production of a bacteriocin.

bacteriophage   A virus that infects bacteria. (Also called **phage.**)

bacteriostatic   Referring to an agent that kills bacteria.

bacteroid   Irregularly shaped cell usually found in tight packets that develop from *Rhizobium* swarmer cells and form nodules in the roots of leguminous plants.

balantidiasis   Type of dysentery caused by the ciliated protozoan *Balantidium coli*.

balantitis   An infection of the penis.

barophile   An organism that lives under high hydrostatic pressure.

Bartholin gland   A mucus-secreting gland of the female external genitalia.

bartonellosis   Rickettsial disease, caused by *Bartonella bacilliformis*, that occurs in two forms. (See **Oroya fever** and **verruga peruana.**)

base   A substance that absorbs hydrogen ions or donates hydroxyl ions.

base analog   A chemical mutagen similar in molecular structure to one of the nitrogenous bases found in DNA that causes point mutations.

basic dye   See **cationic dye.**

Basidiomycota   See **club fungus.**

basidiospore   A sexual spore of the club fungi.

basidium   (plural: **basidia**) A clublike structure in club fungi bearing four external spores on short, slender stalks.

basophil   A leukocyte that migrates into tissues and helps initiate the inflammatory response by secreting histamine.

B cell   See **B lymphocyte.**

benign   Referring to a noncancerous tumor.

beta (β) hemolysin   A type of enzyme that completely lyses red blood cells, leaving a clear ring in the blood agar medium around the colonies.

beta (β) hemolysis   Complete lysis of red blood cells by bacterial enzymes.

beta oxidation   A metabolic pathway that breaks down fatty acids into 2-carbon pieces.

bilirubin   A yellow substance, the product of the breakdown of hemoglobin from red blood cells.

binary fission   Process in which a bacterial cell duplicates its components and divides into two cells.

binocular   Referring to a light microscope having two eyepieces (oculars).

binomial nomenclature   The system of taxonomy developed by Linnaeus in which each organism is assigned a genus and specific epithet.

biochemistry   The branch of organic chemistry that studies the chemical reactions of living systems.

bioconversion   A reaction in which one compound is converted to another by enzymes in cells.

biogeochemical cycle   Mechanism by which water and elements that serve as nutrients are recycled.

biohydrometallurgy   The use of microbes to extract metals from ores.

biological oxygen demand   (BOD) The oxygen required to degrade organic wastes suspended in water.

biological vector   An organism that actively transmits pathogens that complete part of their life cycle within the organism.

bioremediation   A process that uses naturally occurring or genetically engineered microorganisms to transform harmful substances into less toxic or nontoxic compounds.

biosphere   The region of the earth inhabited by living organisms.

biosynthesis   The step of viral replication during which new nucleic acids and viral proteins are made.

biotic factor   An organism in the biosphere.

blackfly fever   Illness resulting from bites by blackflies, characterized by an inflammatory reaction, nausea, and headache.

blackwater fever   Malaria caused by *Plasmodium falciparum* that results in jaundice and kidney damage.

blastomycetic dermatitis   Fungal skin disease caused by *Blastomyces dermatitidis*; characterized by disfiguring, granulomatous, pus-producing lesions.

blastomycosis   Fungal skin disease caused by *Blastomyces dermatitidis* that enters the body through wounds.

blocking antibody   IgG antibody, elicited in allergy patients by increasing doses of allergen, that complexes with allergen before it can react with IgE antibody.

blood agar   Type of medium containing sheep blood, used to identify organisms that cause hemolysis, or breakdown of red blood cells.

blood–brain barrier Formation in the brain of special thick-walled capillaries without pores in their walls that limits entry of substances into brain cells.

B lymphocyte A lymphocyte that is produced in and matures in bursal-equivalent tissue, it gives rise to antibody-producing plasma cells. (Also called **B cell.**)

body tube Microscope part that conveys an image from the objective to the eyepiece.

boil See **furuncle.**

Bolivian hemorrhagic fever A multisystem disease caused by an arenavirus with insidious onset and progressive effects.

bone stink Putrefaction deep in the tissues of large carcasses that is caused by several species of *Clostridium.*

bongkrek disease Type of food poisoning caused by *Pseudomonas cocovenenans,* named for a native Polynesian coconut dish.

botulism Disease caused by *Clostridium botulinum.* The most common form, foodborne botulism, results from ingestion of preformed toxin and is, therefore, an intoxication rather than an infection.

bradykinin Small peptide thought to cause the pain associated with tissue injury.

brain abscess A pus-filled cavity caused by microorganisms reaching the brain from head wounds or via blood from another site.

bread mold A fungus with complex mycelia composed of aseptate hyphae with chitinous cross walls. (Also called **Zygomycota** or conjugation fungus.)

bright-field illumination Illumination produced by the passage of visible light through the condenser of a light microscope.

Brill–Zinsser disease A recurrence of an epidemic typhus infection caused by reactivation of latent organisms harbored in the lymph nodes. (Also called recrudescent typhus.)

broad spectrum Referring to the range of activity of an antimicrobial agent that attacks a wide variety of microorganisms.

bronchial pneumonia Type of pneumonia that begins in the bronchi and can spread through surrounding tissue toward the alveoli.

bronchiole A finer subdivision of the air-conveying bronchi.

bronchitis An infection of the bronchi.

bronchus (plural: **bronchi**) A subdivision of the trachea that conveys air to and from the lungs.

brucellosis A zoonosis highly infective for humans, caused by any of several species of *Brucella.* (Also called **undulant fever** or **Malta fever.**)

bubo Enlargement of infected lymph nodes, especially in the groin and armpit, due to accumulation of pus; characteristic of bubonic plague and other diseases.

bubonic plague A bacterial disease, caused by *Yersinia pestis* and transmitted by flea bites, that spreads in the blood and lymphatic system.

budding Process that occurs in yeast and a few bacteria in which a small new cell develops from the surface of an existing cell.

bulking Phenomenon in which filamentous bacteria multiply, causing sludge to float on the surface of water rather than settling out.

bunyavirus An enveloped RNA virus that causes some forms of respiratory distress and hemorrhagic fever.

Burkitt's lymphoma A tumor of the jaw, seen mainly in African children; caused by the Epstein–Barr virus.

burst size The number of new virions released in the replication process. (Also called **viral yield.**)

burst time The time from absorption to release of phages (in the replication process).

cancer An uncontrolled, invasive growth of abnormal cells.

candidiasis A yeast infection caused by *Candida albicans* that appears as thrush (in mouth) or vaginitis. (Also called **moniliasis.**)

canine parvovirus A parvovirus that causes severe disease in dogs.

canning The use of moist heat under pressure to preserve food.

capillary A blood vessel that branches from an arteriole.

capnophile An organism that prefers carbon dioxide gas for growth.

capsid The protein coating of a virus, which protects the nucleic acid core from the environment and usually determines the shape of the virus.

capsomere A protein aggregate that makes up a viral capsid.

capsule (1) A protective structure outside the cell wall, secreted by the organism. (2) A network of connective fibers covering organs such as the lymph nodes.

carbapenem A bactericidal antibiotic that acts on bacterial cell walls.

carbohydrate A compound composed of carbon, hydrogen, and oxygen that serves as the main source of energy for most living things.

carbon cycle Process by which carbon from atmospheric carbon dioxide enters living and nonliving things and is recycled through them.

carbuncle A massive pus-filled lesion resulting from an infection, particularly of the neck and upper back.

carcinogen A cancer-producing substance.

cardiovascular system Body system that supplies oxygen and nutrients to all parts of the body and removes carbon dioxide and other wastes from them.

carrier An individual who harbors an infectious agent without having observable clinical signs or symptoms.

cascade A set of reactions in which magnification of effect occurs, as in the complement system.

casein hydrolysate A substance derived from milk protein that contains many amino acids; used to enrich certain media.

caseous Characterizing lesions with a "cheesy" appearance that form in lung tissue of patients with tuberculosis.

catabolic pathway A chain of chemical reactions that capture energy by breaking down large molecules into simpler components.

catabolism The chemical breakdown of molecules in which energy is released.

catabolite repression Process by which the presence of a preferred nutrient (often glucose) represses the genes coding for enzymes used to metabolize some alternative nutrient.

catalase An enzyme that converts hydrogen peroxide to water and molecular oxygen.

catarrhal stage Stage of whooping cough characterized by fever, sneezing, vomiting, and a mild, dry, persistent cough.

cation A positively charged ion.

cationic dye An ionic compound, used for staining bacteria, in which the positive ion imparts the color. (Also called **basic dye.**)

cat scratch fever A disease caused by *Afipia felis* or, more commonly, *Bartonella (Rochalimaea) henselae* and transmitted in cat scratches and bites.

cavitation The formation of a cavity inside the cytoplasm of a cell.

cell culture A culture in the form of a monolayer from dispersed cells and continuous cultures of cell suspensions.

cell-mediated (type IV) hypersensitivity Type of allergy elicited by foreign substances from the environment, infectious agents, transplanted tissues, and the body's own malignant cells. (Also called **delayed hypersensitivity.**)

cell-mediated immune response The immune response to an antigen carried out at the cellular level by T cells.

cell-mediated immunity The immune response involving the direct action of T cells to activate B cells or to destroy microbe-infected cells, tumor cells, or transplanted cells (organ transplants).

cell membrane A selectively permeable lipoprotein bilayer that forms the boundary between a bacterial cell's cytoplasm and its environment. (See **plasma membrane.**)

cell strain Dominant cell type resulting from subculturing.

cell theory Theory formulated by Schleiden and Schwann that cells are the fundamental units of all living things.

cell wall Outer layer of most bacterial, algal, fungal, and plant cells that maintains the shape of the cell.

cellular slime mold Funguslike protist consisting of amoeboid, phagocytic cells that aggregate to form a pseudoplasmodium.

cementum The hard, bony covering of the tooth below the gumline.

central nervous system The brain and spinal cord.

cephalosporin An antibacterial agent that inhibits cell wall synthesis.

cercaria A free-swimming fluke larva that emerges from the snail or mollusk host.

cerumen Earwax.

ceruminous gland A modified sebaceous gland that secretes cerumen.

cervix An opening at the narrow lower portion of the uterus.

Chagas' disease Disease caused by *Trypanosoma cruzi* that occurs in the southern United States and is endemic to Mexico; transmitted by several kinds of reduviid bugs.

chancre A hard, painless, nondischarging lesion; a symptom of primary stage syphilis.

chancroid Sexually transmitted disease caused by *Haemophilus ducreyi* that causes soft, painful skin lesions on the genitals, which bleed easily.

chemical bond The interaction of electrons in atoms that form a molecule.

chemical equilibrium A steady state in which there is no net change in the concentrations of substrates or products.

chemically nondefined medium See **complex medium.**

chemiosmosis Process of energy capture in which a proton gradient is created by means of electron transport and then used to drive the synthesis of ATP.

chemoautotroph An autotroph that obtains energy by oxidizing simple inorganic substances such as sulfides and nitrites.

chemoheterotroph A heterotroph that obtains energy from breaking down ready-made organic molecules.

chemolithotroph See **chemoautotroph.**

chemostat A device for maintaining the logarithmic growth of a culture by the continuous addition of fresh medium.

chemotaxis A nonrandom movement of an organism toward or away from a chemical.

chemotherapeutic agent Any chemical substance used to treat disease. (Also called **drug.**)

chemotherapeutic index The maximum tolerable dose of a particular drug per kilogram body weight divided by the mini-

mum dose per kilogram body weight that will cure the disease.

chemotherapy The use of chemical substances to treat various aspects of disease.

chickenpox A highly contagious disease, characterized by skin lesions, caused by the varicella–zoster herpesvirus; usually occurs in children.

chigger dermatitis A violent allergic reaction caused by chiggers, the larvae of *Trombicula* mites.

childbed fever See **puerperal fever.**

chitin A polysaccharide found in the cell walls of most fungi and the exoskeletons of arthropods.

chlamydias Tiny, nonmotile, spherical bacteria; all are obligate intracellular parasites with a complex life cycle.

chloramphenicol A bacteriostatic agent that inhibits protein synthesis.

chlorination The addition of chlorine to water to kill bacteria.

chloroplast A chlorophyll-containing organelle found in eukaryotic cells that carry out photosynthesis.

chloroquine An antiprotozoan agent effective against the malaria parasite.

chocolate agar Type of medium made with heated blood, so named because it turns a chocolate brown color.

chromatin The appearance of chromosomes as fine threads in cells.

chromatophore The internal membranes of photosynthetic bacteria and cyanobacteria.

chromosomal resistance Drug resistance of a microorganism due to a mutation in chromosomal DNA.

chromosome A structure that contains the DNA of organisms.

chromosome mapping The identification of the sequence of genes in a chromosome.

chronic amebiasis Chronic infection caused by the protozoan *Entamoeba histolytica.*

chronic disease A disease that develops more slowly than an acute disease, is usually less severe, and persists for a long, indeterminate period.

chronic fatigue syndrome Disease of uncertain origin similar to mononucleosis with symptoms including persistent fatigue and fever. (Previously called chronic EBV syndrome.)

chronic inflammation A condition in which there is a persistent, indecisive standoff between an inflammatory agent and the phagocytic cells and other host defenses attempting to destroy it.

cilium (plural: **cilia**) A short cellular projection used for movement that beats in coordinated waves.

ciliate A protozoan that moves by means of cilia that cover most of its surface.

classical pathway One of the two sequences of reactions by which proteins of the complement system are activated.

clonal deletion The process in which the binding of lymphocytes to self antigens triggers a genetically programmed destruction of those lymphocytes.

clonal selection theory Theory that explains how exposure to an antigen stimulates a lymphocyte capable of making antibodies against that particular antigen to proliferate, giving rise to a clone of identical antibody-producing cells.

clone A group of genetically identical cells descending from a single parent cell.

club fungus A fungus, including mushrooms, toadstools, rusts, and smuts, that produces spores on basidia. (Also called **Basidiomycota.**)

cluster of differentiation marker An antigen found on the cell surface of B and T cells that can be used to distinguish the cells from one another.

coagulase A bacterially produced enzyme that accelerates the coagulation (clotting) of blood.

coarse adjustment Focusing mechanism of a microscope that rapidly changes the distance between the objective lens and the specimen.

coccidioidomycosis Fungal respiratory disease caused by the soil fungus *Coccidioides immitis.* (Also called valley fever.)

coccus (plural: **cocci**) A spherical bacterium.

codon A sequence of three bases in mRNA that specifies a particular amino acid in the translation process.

coelom The body cavity between the digestive tract and body wall in higher animals.

coenzyme An organic molecule bound to or loosely associated with an enzyme.

cofactor An inorganic ion necessary for the function of an enzyme.

colicin A protein released by some strains of *Escherichia coli* that inhibits growth of other strains of the same organism.

coliform bacterium Gram-negative, non-spore-forming, aerobic or facultatively anaerobic bacterium that ferments lactose and produces acid and gas; significant numbers may indicate water pollution.

colloid A mixture formed by particles too large to form a true solution dispersed in a liquid.

colonization Growth of microorganisms on epithelial surfaces such as skin or mucous membranes.

colony A group of descendants of an original cell.

Colorado tick fever Disease caused by an orbivirus carried by dog ticks, characterized by headache, backache, and fever.

colostrum The protein-rich fluid secreted by the mammary glands just after childbirth, prior to the appearance of breast milk.

commensal An organism that lives in or on another organism without harming it and that benefits from the relationship.

commensalism A symbiotic relationship in which one organism benefits and the other one neither benefits nor is harmed by the relationship.

common-source outbreak An epidemic that arises from contact with contaminated substances.

communicable infectious disease Infectious disease that can be spread from one host to another. (Also called **contagious disease.**)

community All the kinds of organisms present in a given environment.

competence factor A protein released into the medium that facilitates the uptake of DNA into a bacterial cell.

competitive inhibitor A molecule similar in structure to a substrate that competes with that substrate by binding to the active site.

complement A set of more than 20 large regulatory proteins that circulate in plasma and when activated form a nonspecific defense mechanism against many different microorganisms. (Also called **complement system.**)

complement fixation test A complex serologic test used to detect small quantities of antibodies.

complement system See **complement.**

complementary base pairing Hydrogen bonding between adenine and thymine (or uracil) bases or between guanine and cytosine bases.

completed test The final test for coliforms in multiple-tube fermentation in which organisms from colonies grown on eosin methylene blue agar are used to inoculate lactose broth and agar slants.

complex medium A growth medium that contains certain reasonably well defined materials but that varies slightly in chemical composition from batch to batch. (Also called **chemically nondefined medium.**)

complex virus A virus, such as a bacteriophage or poxvirus, that has an envelope or specialized structures.

compound A chemical substance made up of atoms of two or more elements.

compound light microscope A light microscope with more than one lens.

compromised host An individual with reduced resistance, being more susceptible to infection.

condenser Device in a microscope that converges light beams so that they will pass through the specimen.

condyloma See **genital wart.**

confirmed test Second stage of testing for coliforms in multiple-tube fermentation in which samples from the highest di-

lution showing gas production are streaked onto eosin methylene blue agar.

congenital rubella syndrome Complication of German measles causing death or damage to a developing embryo infected by virus that crosses the placenta.

congenital syphilis Syphilis passed to a fetus when treponemes cross the placenta from mother to child before birth.

conidium (plural: **conidia**) A small, asexual, aerial spore organized into chains in some bacteria and fungi.

conjugation (1) The transfer of genetic information from one bacterial cell to another by means of conjugation pili. (2) The exchange of information between two ciliates (protists).

conjugation pilus A type of pilus that attaches two bacteria together and provides a means for the exchange of genetic material. (Also called **sex pilus.**)

conjunctiva Mucous membranes of the eye.

consolidation Blockage of air spaces as a result of fibrin deposits in lobar pneumonia.

constitutive enzyme An enzyme that is synthesized continuously regardless of the nutrients available to the organism.

consumer An organism that obtains nutrients by eating producers or other consumers. (Also called **heterotroph.**)

contact dermatitis Cell-mediated (type IV) hypersensitivity disorder that occurs in sensitized individuals on second exposure of the skin to allergens.

contact transmission A mode of disease transmission effected directly, indirectly, or by droplets.

contagious disease See **communicable infectious disease.**

contamination The presence of microorganisms on inanimate objects or surfaces of the skin and mucous membranes.

continuous cell line Cell culture consisting of cells that can be propagated over many generations.

continuous reactor A device used in industrial and pharmaceutical microbiology to isolate and purify a microbial product often without killing the organisms.

control variable A factor that is prevented from changing during an experiment.

convalescent stage The stage of an infectious disease during which tissues are repaired, healing takes place, and the body regains strength and recovers.

Coomb's antiglobulin test An immunological test designed to detect anti-Rh antibodies.

core The living part of an endospore.

cornea The transparent part of the eyeball exposed to the environment.

coronavirus Virus with clublike projections that causes colds and acute upper respiratory distress.

cortex A laminated layer of peptidoglycan between the membranes of the endospore septum.

corynebacteria Club-shaped, irregular, non-spore-forming, gram-positive rods.

coryza The common cold.

countable number A number of colonies on an agar plate small enough so that one can clearly distinguish and count them (30 to 300 per plate).

covalent bond A bond between atoms created by the sharing of pairs of electrons.

cowpox Disease caused by the vaccinia virus and characterized by lesions, inflammation of lymph nodes, and fever; virus is used to make vaccine against smallpox and monkeypox.

crepitant tissue Distorted tissue caused by gas bubbles in gas gangrene.

crista A fold of the inner mitochondrial membrane.

cross-reaction Immune reaction of a single antibody with different antigens that are similar in structure.

cross-resistance Resistance against two or more similar antimicrobial agents through a common mechanism.

croup Acute obstruction of the larynx that produces a characteristic high-pitched, barking cough.

crustacean A usually aquatic arthropod that has a pair of appendages associated with each body segment.

cryptococcosis Fungal respiratory disease caused by a budding, encapsulated yeast, *Filobasidiella neoformans.*

cryptosporidiosis Disease caused by protozoans of the genus *Cryptosporidium,* common in AIDS patients.

curd The solid portion of milk resulting from bacterial enzyme addition and used to make cheese.

cyanobacteria Photosynthetic, prokaryotic, typically unicellular organisms that are members of kingdom Monera.

cyanosis Bluish skin characteristic of oxygen-poor blood.

cyclic photophosphorylation Pathway in which excited electrons from chlorophyll are used to generate ATP without the splitting of water or reduction of NADP.

cysticercus An oval white sac with a tapeworm head invaginated into it. (Also called bladder worm.)

cystitis Inflammation of the bladder.

cyst A spherical, thick-walled cell that resembles an endospore, formed by certain bacteria.

cytokine One of a diverse group of soluble proteins that have specific roles in host defenses.

cytomegalovirus One of a widespread and diverse group of herpesviruses that often produces no symptoms in normal adults but can severely affect AIDS patients and congenitally infected children.

cytopathic effect (CPE) The visible effect viruses have on cells.

cytoplasm The semifluid substance inside a cell, excluding, in eukaryotes, the cell nucleus.

cytoplasmic streaming Process by which cytoplasm flows from one part of a eukaryotic cell to another.

cytoskeleton A network of protein fibers that support, give rigidity and shape to a eukaryotic cell, and provide for cell movements.

cytotoxic drug A drug that interferes with DNA synthesis, used to suppress the immune system and prevent the rejection of transplants.

cytotoxic T cell (T_c) Lymphocyte that destroys virus-infected cells.

cytotoxic (type II) hypersensitivity Type of allergy elicited by antigens on cells, especially red blood cells, that the immune system treats as foreign.

cytotoxin Toxin produced by cytotoxic T cells that kills infected host cells.

dark-field illumination In light microscopy, the light that is reflected from an object rather than passing through it, resulting in a bright image on a dark background.

dark reactions Part of photosynthesis in which carbon dioxide gas is reduced by electrons from reduced NADP (NADPH) to form various carbohydrate molecules, chiefly glucose. (Also called carbon fixation.)

dark repair Mechanism for repair of damaged DNA by several enzymes that do not require light for activation; they excise defective nucleotide sequences and replace them with DNA complementary to the unaltered DNA strand.

daughter cell One of the two identical products of cell division.

deaminating agent A chemical mutagen that can remove an amino group ($—NH_2$) from a nitrogenous base, causing a point mutation.

death phase See **decline phase.**

débridement A scraping technique used to remove eschar from a burn to reach infection sites.

decimal reduction time (DRT) The length of time needed to kill 90 percent of the organisms in a given population at a specified temperature. (Also called **D value.**)

decline phase (1) The fourth of four major phases of the bacterial growth curve in which cells lose their ability to divide (due to less supportive conditions in the medium) and thus die. (Also called **death phase.**) (2) In the stages of a disease, the period during which the host defenses finally overcome the pathogen and symptoms begin to subside.

decomposer Organism that obtains energy by digesting dead bodies or wastes of producers and consumers.

defined synthetic medium A synthetic medium that contains known specific kinds and amounts of chemical substances.

definitive host An organism that harbors the adult, sexually reproducing form of a parasite.

degranulation Release of histamine and other preformed mediators of allergic reactions by sensitized mast cells and basophils after a second encounter with an allergen.

dehydration synthesis A chemical reaction that builds complex organic molecules.

delayed hypersensitivity See **cell-mediated (type IV) hypersensitivity.**

delayed hypersensitivity T cells Those T cells (inflammatory T_H1) that produce lymphokines in cell-mediated (type IV) hypersensitivity reactions.

deletion The removal of one or more nitrogenous bases from DNA, usually producing a frameshift mutation.

delta hepatitis See **hepatitis D.**

denaturation The disruption of hydrogen bonds and other weak forces that maintain the structure of a globular protein, resulting in the loss of its biological activity.

dengue fever Viral systemic disease that causes severe bone and joint pain. (Also called breakbone fever.)

denitrification The process by which nitrates are reduced to nitrous oxide or nitrogen gas.

dental caries The erosion of enamel and deeper parts of teeth. (Also called tooth decay.)

dental plaque A continuously formed coating of microorganisms and organic matter on tooth enamel.

deoxyribonucleic acid (DNA) Nucleic acid that carries hereditary information from one generation to the next.

dermal wart Wart resulting from the viral infection of epithelial cells.

dermatomycosis A fungal skin disease.

dermatophyte A fungus that invades keratinized tissue of the skin and nails.

dermis The thick inner layer of the skin.

descriptive study An epidemiologic study that notes the number of cases of a disease, which segments of the population are affected, where the cases have occurred, and over what time period.

desensitization Treatment designed to cure allergies by means of injections with gradually increasing doses of allergen.

Deuteromycota See **Fungi Imperfecti.**

diapedesis The process in which leukocytes pass out of blood into inflamed tissues by squeezing between cells of capillary walls.

diarrhea Excessive frequency and looseness of bowel movements.

diatom An alga or plantlike protist that lacks flagella and has a glasslike outer shell.

dichotomous key Taxonomic key used to identify organisms; composed of paired (either-or) statements describing characteristics.

differential medium A growth medium with a constituent that causes an observable change (in color or pH) in the medium when a particular chemical reaction occurs, making it possible to distinguish between organisms.

differential stain Use of two or more dyes to differentiate among bacterial species or to distinguish various structures of an organism—for example, the Gram stain.

diffraction Phenomenon in which light waves, as they pass through a small opening, are broken up into bands of different wavelengths.

DiGeorge syndrome Primary immunodeficiency disease caused by failure of the thymus to develop properly, resulting in a deficiency of T cells.

digestive system The body system that converts ingested food into material suitable for the liberation of energy or for assimilation into body tissues.

dikaryotic Referring to fungal cells within hyphae that have two nuclei, produced by plasmogamy in which the nuclei have not united.

dilution method A method of testing antibiotic sensitivity in which organisms are incubated in a series of tubes containing known quantities of a chemotherapeutic agent.

dimer Two adjacent pyrimidines bonded together in a DNA strand, usually as a result of exposure to ultraviolet rays.

dimorphism The ability of an organism to alter its structure when it changes habitats.

dinoflagellate An alga or plantlike protist, usually with two flagella.

diphtheria A severe upper respiratory disease caused by *Corynebacterium diphtheriae;* can produce subsequent myocarditis and polyneuritis.

diphtheroid Organism found in normal throat cultures that fails to produce exotoxin but is otherwise indistinguishable from diphtheria-causing organisms.

dipicolinic acid Acid found in the core of an endospore that contributes to its heat resistance.

diploid A eukaryotic cell that has paired sets of chromosomes.

diploid fibroblast strain A culture derived from fetal tissues that retains fetal capacity for rapid, repeated cell division.

direct contact transmission Mode of disease transmission requiring person-to-person body contact.

direct fecal-oral transmission Direct contact transmission of disease in which

feline panleukopenia virus A parvovirus that causes severe disease in cats.

female reproductive system The host system consisting of the ovaries, uterine tubes, uterus, vagina, and external genitalia.

fermentation Anaerobic metabolism of the pyruvic acid produced in glycolysis.

fever A body temperature that is abnormally high.

fibroblast A new connective tissue cell that replaces fibrin as a blood clot dissolves, forming granulation tissue.

fifth disease A normal disease in children caused by the *Erythrovirus* called B19; characterized by a bright red rash on the cheeks and a low-grade fever. (Also called erythema infectiosum.)

filariasis Disease of the blood and lymph caused by any of several different roundworms carried by mosquitoes.

filovirus A filamentous virus that displays unusual variability in shape. Two filoviruses, the Ebola virus and the Marburg virus, have been associated with human disease.

filter paper disk method Method of evaluating the antimicrobial properties of a chemical agent using filter paper disks placed on an inoculated agar plate.

filtration (1) A method of estimating the size of bacterial populations in which a known volume of air or water is drawn through a filter with pores too small to allow passage of bacteria. (2) A method of sterilization that uses a membrane filter to separate bacteria from growth media. (3) The filtering of water through beds of sand to remove most of the remaining microorganisms after flocculation in water treatment plants.

fimbria See **attachment pilus.**

fine adjustment Focusing mechanism of a microscope that very slowly changes the distance between the objective lens and the specimen.

five-kingdom system System of classifying organisms into one of five kingdoms: Monera (Prokaryotae), Protista, Fungi, Plantae, and Animalia.

flagellum (plural: **flagella**) A long, thin, helical appendage of certain cells that provides a means of locomotion.

flagellar staining A technique for observing flagella by coating the surfaces of flagella with a dye or a metal such as silver.

flash pasteurization See **high-temperature short-time pasteurization.**

flat sour spoilage Spoilage due to the growth of spores that does not cause cans to bulge with gas.

flatworm A primitive, unsegmented, hermaphroditic often parasitic worm. (Also called Platyhelminthes.)

flavivirus A small, enveloped, (+) sense RNA virus that causes a variety of encephalitides, including yellow fever.

flavoprotein An electron carrier in oxidative phosphorylation.

flocculation The addition of alum to cause precipitation of suspended colloids, such as clay, in the water purification process.

fluctuation test A test to demonstrate that resistance to chemical substances occurs spontaneously rather than being induced.

fluid-mosaic model Current model of membrane structure in which proteins are dispersed in a phospholipid bilayer.

fluke A flatworm with a complex life cycle; can be an internal or external parasite.

fluorescence Emission of light of one color when irradiated with another, shorter wavelength of light.

fluorescence microscopy Use of ultraviolet light in a microscope to excite molecules so that they release light of different colors.

fluorescent antibody staining Procedure in fluorescence microscopy that uses a fluorochrome attached to antibodies to detect the presence of an antigen.

fluoride Chemical that helps in reducing tooth decay by poisoning bacterial enzymes and hardening the surface enamel of teeth.

focal infection An infection confined to a specific area from which pathogens can spread to other areas.

folliculitis Local infection produced when hair follicles are invaded by pathogenic bacteria. (Also called **pimple** or **pustule.**)

fomite A nonliving substance capable of transmitting disease, such as clothing, dishes, or paper money.

food poisoning A gastrointestinal disease caused by ingestion of foods contaminated with preformed toxins or other toxic substances. (Also called **enterotoxicosis.**)

formed elements Cells and cell fragments comprising about 40 percent of the blood.

F pilus A bridge formed from an F^+ cell to an F^- cell for conjugation.

F plasmid Extrachromosomal DNA found in F^+ cells that contains the genetic information for the synthesis of F pili.

F′ plasmid An F plasmid that has been imprecisely separated from the bacterial chromosome so that it carries a fragment of the bacterial chromosome.

frameshift mutation Mutation resulting from the deletion or insertion of one or more bases.

freeze-etching Technique in which water is evaporated under vacuum from the freeze-fractured surface of a specimen before observation with electron microscopy.

freeze-fracturing Technique in which a cell is first frozen and then broken with a knife so that the fracture reveals structures inside the cell when observed by electron microscopy.

fulminating See **acme.**

functional group Part of a molecule that generally participates in chemical reactions as a unit and gives the molecule some of its chemical properties.

fungi (singular: **fungus**) The kingdom of nonphotosynthetic, eukaryotic organisms that absorb nutrients from their environment.

Fungi Imperfecti Group of fungi termed "imperfect" because no sexual stage has been observed in their life cycles. (Also called **Deuteromycota.**)

furuncle A large, deep, pus-filled infection. (Also called a **boil.**)

gamete A male or female reproductive cell.

gametocyte A male or female sex cell.

gamma globulin See **immune serum globulin.**

ganglion An aggregation of neuron cell bodies.

gas gangrene A deep wound infection, destructive of tissue, often caused by a combination of two or more species of *Clostridium.*

gene A linear sequence of DNA nucleotides that form a functional unit within a chromosome.

gene amplification A technique of genetic engineering in which plasmids or bacteriophages carrying a specific gene are induced to reproduce at a rapid rate within host cells.

generalized transduction Type of transduction in which a fragment of DNA from the degraded chromosome of an infected bacterial cell is accidentally incorporated into a new phage particle during viral replication and thereby transferred to another bacterial cell.

generation time Time required for a population of organisms to double in number.

genetic code The one-to-one relationship between each codon and a specific amino acid.

genetic engineering The use of various techniques to purposefully manipulate genetic material in order to alter the characteristics of an organism in a desired way.

genetic fusion A technique of genetic engineering that allows transposition of genes from one location on a chromosome to another location; the coupling of genes from two different operons.

genetic homology The similarity of DNA base sequences among organisms.

genetics The science of heredity, including the structure and regulation of genes and how these genes are passed between generations.

gene transfer Movement of genetic information between organisms by transformation, transduction, or conjugation.

genital herpes See **herpes simplex virus type 2.**

genital wart An often malignant wart associated with sexually transmitted viral disease having a very high association rate with cervical cancer. (Also called **condyloma.**)

genome The genetic information in an organism or virus.

genotype The genetic information contained in the DNA of an organism.

genus A taxon consisting of one or more species; the first name of an organism in the binomial system of nomenclature—for example, *Escherichia* in *Escherichia coli.*

germ theory of disease Theory that microorganisms (germs) can invade other organisms and cause disease.

German measles See **rubella.**

germination The start of the process of development of a spore or an endospore.

giardiasis A gastrointestinal disorder caused by the flagellated protozoan *Giardia intestinalis.*

gingivitis The mildest form of periodontal disease, characterized by inflammation of the gums.

gingivostomatitis Lesions of the mucous membranes of the mouth.

glomerulonephritis Inflammation of and damage to the glomeruli of the kidneys. (Also called Bright's disease.)

glomerulus A coiled cluster of capillaries in the nephron.

glycocalyx Term used to refer to all substances containing polysaccharides found external to the cell wall.

glycolysis An anaerobic metabolic pathway used to break down glucose into pyruvic acid while producing some ATP.

glycoprotein A long, spikelike molecule made of carbohydrate and protein that projects beyond the surface of a cell or viral envelope; some viral glycoproteins attach the virus to receptor sites on host cells, while others aid fusion of viral and cellular membranes.

glycosidic bond A covalent bond between two monosaccharides.

Golgi apparatus An organelle in eukaryotic cells that receives, modifies, and transports substances coming from the endoplasmic reticulum.

gonorrhea A sexually transmitted disease caused by *Neisseria gonorrhoeae.*

graft tissue Tissue that is transplanted from one site to another.

graft-versus-host (GVH) disease Disease in which host antigens elicit an immunological response from graft cells that destroys host tissue.

gram molecular weight See **mole.**

Gram stain A differential stain that uses crystal violet, iodine, alcohol, and safranin to differentiate bacteria. Gram-positive bacteria stain dark purple; gram-negative ones stain pink/red.

granulation tissue Fragile, reddish, grainy tissue made up of capillaries and fibroblasts that appears with the healing of an injury.

granule An inclusion that is not bounded by a membrane and contains compacted substances that do not dissolve in the cytoplasm.

granulocyte A leukocyte (basophil, mast cell, eosinophil, neutrophil) with granular cytoplasm and irregularly shaped, lobed nuclei.

granuloma In a chronic inflammation, a collection of epithelial cells, macrophages, lymphocytes, and collagen fibers.

granuloma inguinale A sexually transmitted disease caused by *Calymmatobacterium granulomatis.* (Also called donovanosis.)

granulomatous hypersensitivity Cell-mediated hypersensitivity reaction that occurs when macrophages have engulfed pathogens but have failed to kill them.

granulomatous inflammation A special kind of chronic inflammation characterized by the presence of granulomas.

granzyme A cytotoxin produced by cytotoxic T cells that help kill infected host cells.

griseofulvin An antifungal agent that interferes with fungal growth.

ground itch Bacterial infection of sites of penetration by hookworms.

group translocation An active transport process in bacteria that chemically modifies a substance so it cannot diffuse out of the cell.

growth curve The different growth periods of a bacterial or phage population.

gumma A granulomatous inflammation, symptomatic of syphilis, that destroys tissue.

gut-associated lymphatic tissue (GALT) Collective name for the tissues of lymphoid nodules, especially those in the digestive, respiratory, and urogenital tracts; main site of antibody production.

halobacteria One of the groups of the archaeobacteria that live in very concentrated salt environments.

halophile A salt-loving organism that requires moderate to large concentrations of salt.

hanging drop A special type of wet mount often used with dark-field illumination to study motility of organisms.

Hansen's disease The preferred name for leprosy; caused by *Mycobacterium leprae*, it exhibits various clinical forms ranging from tuberculoid to lepromatous.

hantavirus pulmonary syndrome (HPS) The "Sin Nombre" hantavirus responsible for severe respiratory illness.

haploid A eukaryotic cell that contains a single, unpaired set of chromosomes.

hapten A small molecule that can act as an antigenic determinant when combined with a larger molecule.

heat fixation Technique in which air-dried smears are passed through an open flame so that organisms are killed, adhere better to the slide, and take up dye more easily.

heavy chain (H chain) Larger of the two identical pairs of chains comprising immunoglobulin molecules.

helminth A worm, with bilateral symmetry; includes the roundworms and flatworms.

helper T cell (T_H) Lymphocytes that stimulate other immune cells, such as B cells and macrophages.

hemagglutination Agglutination (clumping) of red blood cells; used in blood typing.

hemagglutination inhibition test Serologic test used to diagnose measles, influenza, and other viral diseases, based on the ability of antibodies to viruses to prevent viral hemagglutination.

hemoglobin The oxygen-binding compound found in erythrocytes.

hemolysin An enzyme that lyses red blood cells.

hemolysis The lysis of red blood cells.

hemolytic disease of the newborn Disease in which a baby is born with enlarged liver and spleen caused by efforts of these organs to destroy red blood cells damaged by maternal antibodies; mother is Rh-negative and baby, Rh-positive. (Also called erythroblastosis fetalis.)

hepadnavirus A small, enveloped DNA virus with circular DNA; one such virus causes hepatitis B.

hepatitis An inflammation of the liver, usually caused by viruses but sometimes by an amoeba or various toxic chemicals.

hepatitis A Common form of viral hepatitis caused by a single-stranded RNA virus transmitted by the fecal–oral route. (Formerly called **infectious hepatitis.**)

hepatitis B Type of hepatitis caused by a double-stranded DNA virus usually transmitted in blood or semen. (Formerly called **serum hepatitis.**)

hepatitis C Type of hepatitis distinguished by a high level of the liver enzyme alanine transferase; usually mild or inapparent infection but can be severe in compromised individuals. (Formerly called non-A, non-B hepatitis.)

hepatitis D Severe type of hepatitis caused by presence of both hepatitis D and hepatitis B viruses. (Also called **delta hepatitis.**)

hepatitis E Type of hepatitis transmitted through fecally contaminated water supplies.

hepatovirus One of three major groups of picornaviruses that can infect nerve and is responsible for causing hepatitis A.

herd immunity The proportion of individuals in a population who are immune to a particular disease. (Also called group immunity.)

heredity The transmission of genetic traits from an organism to its progeny.

hermaphroditic Having both male and female reproductive systems in one organism.

herpes gladiatorium Herpesvirus infection that occurs in skin injuries of wrestlers; transmitted by contact or on mats.

herpes labialis Fever blisters (cold sores) on lips.

herpes meningoencephalitis A serious disease caused by a herpesvirus that can cause permanent neurological damage or death and that sometimes follows a generalized herpes infection or ascends from the trigeminal ganglion.

herpes pneumonia A rare form of herpes infection seen in burn patients, alcoholics, and AIDS patients.

herpes simplex virus type 1 (HSV-1) A virus that most frequently causes fever blisters (cold sores) and other lesions of the oral cavity, and less often causes genital lesions.

herpes simplex virus type 2 (HSV-2) A virus that typically causes genital herpes, but which can also cause oral lesions. (Sometimes called herpes hominis virus.)

herpesvirus A relatively large, enveloped DNA virus that can remain latent in host cells for long periods of time.

heterogeneity The ability of the immune system to produce many different kinds of antibodies, each specific for a different antigenic determinant.

heterotroph An organism that uses organic compounds to produce biomolecules.

heterotrophy "Other-feeding"—the use of carbon atoms from organic compounds for the synthesis of biomolecules.

high-energy bond A chemical bond that releases energy when hydrolyzed; the energy can be used to transfer the hydrolyzed product to another compound.

high frequency of recombination (Hfr) strain A strain of F^+ bacteria in which the F plasmid is incorporated into the bacterial chromosome.

high-temperature short-time pasteurization (HTST) Process in which milk is heated to 71.6° C for at least 15 seconds. (Also called flash pasteurization.)

histamine Amine released by basophils and tissues in allergic reactions.

histocompatibility antigen An antigen found in the membranes of all human cells that is unique in all individuals except identical twins.

histone A protein that contributes directly to the structure of eukaryotic chromosomes.

histoplasmosis Fungal respiratory disease endemic to the central and eastern United States, caused by the soil fungus *Histoplasma capsulatum.* (Also called Darling's disease.)

holding method See low-temperature long-time pasteurization.

holoenzyme A functional enzyme consisting of an apoenzyme and a coenzyme or cofactor.

homolactic-acid fermentation A pathway in which pyruvic acid is directly converted to lactic acid using electrons from reduced NAD (NADH).

hookworm A disease caused by two species of small roundworms, *Ancylostoma duodenale* and *Necator americanus,* whose larvae burrow through skin of feet, enter the blood vessels, and penetrate lung and intestinal tissue.

horizontal transmission Direct contact transmission of disease in which pathogens are usually passed by handshaking, kissing, contact with sores, or sexual contact.

host Any organism that harbors another organism.

host range The different types of organisms that a microbe can infect.

host specificity The range of different hosts in which a parasite can mature.

human immunodeficiency virus (HIV) One of the retroviruses that is responsible for AIDS.

human leukocyte antigen A lymphocyte antigen used in laboratory tests to determine compatibility of donor and recipient tissues for transplants.

human papillomavirus Virus that attacks skin and mucous membranes, causing papillomas or warts.

humoral immune response A response to foreign antigens carried out by antibodies circulating in the blood.

humoral immunity The immune response most effective in defending the body against bacteria, bacterial toxins, and viruses that have not entered cells.

humus The nonliving organic components of soil.

hyaluronidase A bacterially produced enzyme that digests hyaluronic acid, which helps hold the cells of certain tissues together, thereby making tissues more accessible to microbes. (Also called spreading factor.)

hybridoma A hybrid cell resulting from the fusion of a cancer cell with another cell, usually an antibody-producing white blood cell.

hydatid cyst An enlarged cyst containing many tapeworm heads.

hydrogen bond A relatively weak attraction between a hydrogen atom carrying a partial positive charge and an oxygen or nitrogen atom carrying a partial negative charge.

hydrologic cycle See water cycle.

hydrolysis A chemical reaction that produces simpler products from more complex organic molecules.

hydrophilic Water-loving.

hydrophobic Water-repelling.

hydrostatic pressure Pressure exerted by standing water.

hyperimmune serum A preparation of immune serum globulins having high titers of specific kinds of antibodies. (Also called convalescent serum.)

hyperparasitism The phenomenon of a parasite itself having parasites.

hypersensitivity Disorder in which the immune system reacts inappropriately, usually by responding to an antigen it normally ignores. (Also called allergy.)

hypertonic solution A solution containing a concentration of dissolved material greater than that within a cell.

hypha (plural: hyphae) A long, thread-like structure of cells in fungi or actinomycetes.

hypothesis A tentative explanation for an observed condition or event.

hypotonic solution A solution containing a concentration of dissolved material lower than that within a cell.

IgA Class of antibody found in the blood and secretions.

IgD Class of antibody found on the surface of B cells and rarely secreted.

IgE Class of antibody that binds to receptors on basophils in the blood or mast cells in the tissues; responsible for allergic or immediate (type I) hypersensitivity reactions.

IgG The main class of antibodies found in the blood; produced in largest quantities during secondary response.

IgM The first class of antibody secreted into the blood during the early stages of a primary immune response (a rosette of five immunoglobulin molecules) or found on the surface of B cells (a single immunoglobulin molecule).

illness phase In an infectious disease, the period during which the individual experiences the typical signs and symptoms of the disease.

imidazole An antifungal agent that disrupts fungal plasma membranes.

immediate (type I) hypersensitivity Response to a foreign substance (allergen) resulting from prior exposure to the allergen. (Also called anaphylactic hypersensitivity.)

immersion oil Substance used to avoid refraction at a glass–air interface when examining objects through a microscope.

immune complex An antigen–antibody complex that is normally eliminated by phagocytic cells.

immune complex disorder A disorder caused by antigen–antibody complexes that precipitate in the blood and injure tissues; elicited by antigens in vaccines, on microorganisms, or on a person's own cells. (Also called immune complex (type III) hypersensitivity.)

immune cytolysis Process in which the membrane attack complex of complement produces lesions on cell membranes through which the contents of the bacterial cells leak out.

immune serum globulin A pooled sample of antibody-containing fractions of serum from many individuals. (Also called **gamma globulin.)**

immune system Body system that provides the host organism with specific immunity to infectious agents.

immunity The ability of an organism to recognize and defend itself against infectious agents.

immunocompromised Referring to an individual whose immune defenses are weakened due to fighting another infectious disease, or because of an immunodeficiency disease or an immunosuppressive agent.

immunodeficiency Inborn or acquired defects in lymphocytes (B or T cells).

immunodeficiency disease A disease of impaired immunity, caused by lack of lymphocytes, defective lymphocytes, or destruction of lymphocytes.

immunodiffusion test A serologic test similar to the precipitin test but carried out in an agar gel medium.

immunoelectrophoresis Serologic test in which antigens are first separated by gel electrophoresis and then allowed to react with antibody placed in a trough in the gel.

immunofluorescence Referring to the use of antibodies to which a fluorescent substance is bound to detect antigens, other antibodies, or complement within tissues.

immunogen See **antigen.**

immunogenic Something that is a potent stimulator of antibody production and defense cell activity.

immunoglobulin (Ig) The class of protective proteins produced by the immune system in response to a particular epitope. (Also called **antibody.)**

immunological disorder Disorder that results from an inappropriate or inadequate immune response.

immunological memory The ability of the immune system to recognize substances it has previously encountered.

immunology The study of specific immunity and how the immune system responds to specific infectious agents.

immunosuppression Minimizing of immune reactions using radiation or cytotoxic drugs.

impetigo A highly contagious pyoderma caused by staphylococci, streptococci, or both.

inapparent infection An infection that fails to produce symptoms, either because too few organisms are present or because host defenses effectively combat the pathogens. (Also called **subclinical infection.)**

incidence rate The number of new cases of a particular disease per 100,000 population seen in a specific period of time.

inclusion A granule or vesicle found in the cytoplasm of a bacterial cell.

inclusion blennorrhea A mild chlamydial infection of the eyes in infants.

inclusion body (1) An aggregation of reticulate bodies within chlamydias. (2) A form of cytopathic effect consisting of viral components, masses of viruses, or remnants of viruses.

inclusion conjunctivitis A chlamydial infection that can result from self-inoculation with *Chlamydia trachomatis.*

incubation period In the stages of an infectious disease, the time between infection and the appearance of signs and symptoms.

index case The first case of a disease to be identified.

index of refraction A measure of the amount that light rays bend when passing from one medium to another.

indicator organism An organism such as *Escherichia coli* whose presence indicates the contamination of water by fecal matter.

indigenous organism An organism native to a given environment. (Also called native organism.)

indirect contact transmission Transmission of disease through fomites.

indirect fecal–oral transmission Transmission of disease in which pathogens from feces of one organism infect another organism.

induced mutation A mutation produced by agents called mutagens that increase the mutation rate.

inducer A substance that binds to and inactivates a repressor protein.

inducible enzyme An enzyme coded for by a gene that is sometimes active and sometimes inactive.

induction The stimulation of a temperate phage (prophage) to excise itself from the host chromosome and initiate a lytic cycle of replication.

induration A raised, hard, red region on the skin resulting from tuberculin hypersensitivity.

industrial microbiology Branch of microbiology concerned with the use of microorganisms to assist in the manufacture of useful products or disposal of waste products.

infant botulism Form of botulism in infants associated with ingestion of honey. (Also called "floppy baby" syndrome.)

infection The multiplication of a parasite organism, usually microscopic, within or upon the host's body.

infectious disease Disease caused by infectious agents (bacteria, viruses, fungi, protozoa, and helminths).

infectious hepatitis See **hepatitis A.**

infectious mononucleosis An acute disease that affects many systems, caused by the Epstein–Barr virus.

infestation The presence of helminths (worms) or arthropods in or on a living host.

inflammation The body's defensive response to tissue damage caused by microbial infection.

influenza Viral respiratory infection caused by orthomyxoviruses that appears as epidemics.

initiating segment That part of the F plasmid that is transferred to the recipient cell in conjugation with an Hfr bacterium.

innate immunity Immunity to infection that exists in an organism because of genetically determined characteristics.

insect An arthropod with three body regions, three pairs of legs, and highly specialized mouthparts.

insertion The addition of one or more bases to DNA, usually producing a frameshift mutation.

interferon A small protein often released from virus-infected cells that binds to adjacent uninfected cells, causing them to produce antiviral proteins that interfere with viral replication.

interleukin A cytokine produced by leukocytes.

intermediate host An organism that harbors a sexually immature stage of a parasite.

intoxication The ingestion of a microbial toxin that leads to a disease.

intron Region of a gene (or mRNA) in eukaryotic cells that does not code for a protein. (Also called intervening region.)

invasiveness The ability of a microorganism to take up residence in a host.

ion An electrically charged atom produced when an atom gains or loses one or more electrons.

ionic bond A chemical bond between atoms resulting from attraction of ions with opposite charges.

iris diaphragm Adjustable device in a microscope that controls the amount of light passing through the specimen.

ischemia Reduced blood flow to tissues with oxygen and nutrient deficiency and waste accumulation.

isograft A graft of tissue between genetically identical individuals.

isolation Situation in which a patient with a communicable disease is prevented from contact with the general population.

isomer An alternative form of a molecule having the same molecular formula but different structure.

isoniazid An antimetabolite that is bacteriostatic against the tuberculosis-causing mycobacterium.

isotonic Fluid containing the same concentration of dissolved materials as is in a cell; causes no change in cell volume.

isotope An atom of a particular element that contains a different number of neutrons.

kala azar Visceral leishmaniasis caused by *Leishmania donovani.*

Kaposi's sarcoma A malignancy often found in AIDS patients in which blood vessels grow into tangled masses that are filled with blood and easily ruptured.

karyogamy Process by which nuclei fuse to produce a diploid cell.

keratin A waterproofing protein found in epidermal cells.

keratitis An inflammation of the cornea.

keratoconjunctivitis Condition in which vesicles appear on the cornea and eyelids.

kidney One of a pair of organs responsible for the formation of urine.

Kirby–Bauer method See **disk diffusion method.**

Koch's postulates Four postulates formulated by Robert Koch in the 19th century; used to prove that a particular organism causes a particular disease.

Koplik's spots Red spots with central bluish specks that appear on the upper lip mucosa in early stages of measles.

Krebs cycle A sequence of enzyme-catalyzed chemical reactions that metabolizes 2-carbon units called acetyl groups to CO_2 and H_2O. (Also called tricarboxylic acid cycle and the citric acid cycle.)

Kupffer cells Phagocytic cells that remove foreign matter from the blood as it passes through sinusoids.

lacrimal gland Tear-producing gland of the eye.

lactobacilli Type of regular, nonsporing, gram-positive rods found in many foods; used in production of cheeses, yogurt, sourdough, and other fermented foods.

lagging strand The new strand of DNA formed in short, discontinuous DNA segments during DNA replication.

lag phase First of four major phases of the bacterial growth curve, in which organisms grow in size but do not increase in number.

large intestine The lower area of the intestine that absorbs water and converts undigested food into feces.

laryngeal papilloma Benign growth caused by herpesviruses that can be dangerous if such papillomas block the air-

way; infants are often infected during birth by mothers having genital warts.

laryngitis An infection of the larynx, often with loss of voice.

larynx The voicebox.

Lassa fever Hemorrhagic fever, caused by arenaviruses, that begins with pharyngeal lesions and proceeds to severe liver damage.

latency The ability of a virus to remain in host cells for long periods of time while retaining the ability to replicate.

latent disease A disease characterized by periods of inactivity either before symptoms appear or between attacks.

latent period Period of a bacteriophage growth curve that spans the time from penetration through biosynthesis.

latent viral infection An infection typical of herpesviruses, in which an infection in childhood that is brought under control later in life is reactivated.

leading strand The new strand of DNA formed as a continuous strand during DNA replication.

leavening agent An agent, such as yeast, that produces gas to make dough rise.

legionellas The causative bacterial agent in Legionnaires' disease, *Legionella pneumophila.*

Legionnaires' disease Disease caused by *Legionella pneumophila,* transmitted by airborne bacteria.

leishmaniasis A parasitic systemic disease caused by three species of protozoa of the genus *Leishmania* and transmitted by sandflies.

leproma An enlarged, disfiguring skin lesion that occurs in the lepromatous form of Hansen's disease (leprosy).

lepromatous Referring to the nodular form of Hansen's disease (leprosy) in which a granulomatous response causes enlarged, disfiguring skin lesions called lepromas.

lepromin skin test Test used to detect Hansen's disease (leprosy); similar to the tuberculin test.

leprosy See **Hansen's disease.**

leptospirosis A zoonosis caused by the spirochete *Leptospira interrogans,* which enters the body through mucous membranes or skin abrasions.

leukocidin An exotoxin produced by many bacteria, including the streptococci and staphylococci, that kills phagocytes.

leukocyte A white blood cell.

leukocyte-endogenous mediator A substance that helps raise the body temperature while decreasing iron absorption (increasing iron storage).

leukocytosis An increase in the number of white blood cells (leukocytes) circulating in the blood.

leukostatin An exotoxin that interferes with the ability of leukocytes to engulf microorganisms that release the toxin.

leukotriene A reaction mediator released from mast cells after degranulation that causes prolonged airway constriction, dilation and increased permeability of capillaries, increased thick mucus secretion, and stimulation of nerve endings that cause pain and itching.

L forms Irregularly shaped naturally occurring bacteria with defective cell walls.

ligase An enzyme that joins together DNA segments.

light chain (L chain) Smaller of the two identical pairs of chains constituting immunoglobulin molecules.

light microscopy The use of any type of microscope that uses visible light to make specimens observable.

light reactions The part of photosynthesis in which light energy is used to excite electrons from chlorophyll, which are then used to generate ATP and NADPH.

light repair Repair of DNA dimers by a light-activated enzyme. (Also called **photoreactivation.**)

lipid A Toxic substance found in the cell wall of gram-negative bacteria.

lipid One of a group of complex, water-insoluble compounds.

lipopolysaccharide Part of the outer layer of the cell wall in gram-negative bacteria. (Also called **endotoxin.**)

listeriosis A type of meningitis caused by *Listeria monocytogenes* that is threatening to those with impaired immune systems.

loaiasis Tropical eye disease caused by the filarial worm *Loa loa.*

lobar pneumonia Type of pneumonia that affects one or more of the five major lobes of the lungs.

local infection An infection confined to a specific area of the body.

locus The location of a gene on a chromosome.

log phase Second of four major phases of the bacterial growth curve, in which cells divide at an exponential or logarithmic rate.

logarithmic rate See **exponential rate.**

lophotrichous Having two or more flagella at one or both ends of a bacterial cell.

lower respiratory tract Thin-walled bronchioles and alveoli where gas exchange occurs.

low-temperature long-time pasteurization (LTLT) Procedure in which milk is heated to 62.9° C for at least 30 minutes. (Also called **holding method.**)

luminescence Process in which absorbed light rays are reemitted at longer wavelengths.

Lyme disease Disease caused by *Borrelia burgdorferi,* carried by the deer tick.

lymph The excess fluid and plasma proteins lost through capillary walls that is found in the lymphatic capillaries.

lymphangitis Symptom of septicemia in which red streaks due to inflamed lymphatics appear beneath the skin.

lymphatic system Body system, closely associated with the cardiovascular system, that transports lymph in lymphatic vessels through body tissues and organs; performs important functions in host defenses and specific immunity.

lymphatic vessel Vessel that returns lymph to the blood circulatory system.

lymph node An encapsulated globular structure located along the routes of the lymphatic vessels that helps clear the lymph of microorganisms.

lymphocyte A leukocyte (white blood cell) found in large numbers in lymphoid tissues that contribute to specific immunity.

lymphogranuloma venereum A sexually transmitted disease, caused by *Chlamydia trachomatis,* that attacks the lymphatic system.

lymphoid nodule A small, unencapsulated aggregation of lymphatic tissue that develops in many tissues, especially the digestive, respiratory, and urogenital tracts, collectively called gut-associated lymphatic tissue (GALT); they are the body's main sites of antibody production.

lymphoid stem cell A cell in the bone marrow from which lymphocytes develop.

lymphokine A cytokine secreted by T cells when they encounter an antigen.

lyophilization The drying of a material from the frozen state; freeze-drying.

lysis The destruction of a cell by the rupture of a cell or plasma membrane, resulting in the loss of cytoplasm.

lysogen The combination of a bacterium and a temperate phage.

lysogenic Pertaining to a bacterial cell in the state of lysogeny.

lysogenic conversion The ability of a prophage to prevent additional infections of the same cell by the same type of phage; also the conversion of a non-toxin-producing bacterium into a toxin-producing one by a temperate phage.

lysogeny The ability of temperate bacteriophages to persist in a bacterium by the integration of the viral DNA into the host chromosome and without the replication of new viruses or cell lysis.

lysosome A small membrane-bound organelle in animal cells that contains digestive enzymes.

lytic cycle The sequence of events in which a bacteriophage infects a bacterial cell, replicates, and eventually causes lysis of the cell.

lytic phage See **virulent phage.**

macrolide A large-ring compound, such as erythromycin, that is antibacterial by affecting protein synthesis

macrophage Ravenously phagocytic leukocytes found in tissues.

madura foot Tropical disease caused by a variety of soil organisms (fungi and actinomycetes) that often enter the skin through bare feet. (Also called **maduromycosis.**)

maduromycosis See **madura foot.**

major histocompatibility complex (MHC) A group of cell surface proteins that are essential to immune recognition reactions.

malaria A severe parasitic disease caused by several species of the protozoan *Plasmodium* and transmitted by mosquitoes.

male reproductive system The host system consisting of the testes, ducts, specific glands, and the penis.

malignant Relating to a tumor that is cancerous.

Malta fever See **brucellosis.**

malted Referring to cereal grains that are partially germinated to increase the concentration of starch-digesting enzymes.

mammary gland A modified sweat gland that produces milk and ducts that carry the milk to the nipple.

mash Malted grain that is crushed and mixed with hot water.

mast cell A leukocyte that releases histamine during an allergic response.

mastigophoran A flagellate protozoan such as *Giardia.*

mastoid area Portion of the temporal bone prominent behind the ear opening.

matrix Fluid-filled inner portion of a mitochondrion.

maturation The process by which complete virions are assembled from newly synthesized components in the replication process.

measles A febrile disease with rash caused by the rubeola virus, which invades lymphatic tissue and blood. (Also called **rubeola.**)

measles encephalitis A serious complication of measles that leaves many survivors with permanent brain damage.

mebendazole An antihelminthic agent that blocks glucose uptake by parasitic roundworms.

mechanical stage Attachment to a microscope stage that holds the slide and allows precise control in moving the slide.

mechanical vector A vector in which the parasite does not complete any part of its life cycle during transit.

medium A mixture of nutritional substances on or in which microorganisms grow.

megakaryocyte Large cell normally present in bone marrow that gives rise to platelets.

meiosis Division process in eukaryotic cells that reduces the chromosome number by half.

membrane attack complex A set of proteins in the complement system that lyses invading bacteria by producing lesions in their cell membranes.

membrane filter method Method of testing for coliform bacteria in water in which bacteria are filtered through a membrane and then incubated on the membrane surface in growth medium.

memory cell Long-lived B or T lymphocyte that can carry out an anamnestic or secondary response.

meninges Three layers of membrane that protect the brain and spinal cord.

merozoite A malaria trophozoite found in infected red blood or liver cells.

mesophile An organism that grows best at temperatures between 25° and 40°C, including most bacteria.

mesophilic spoilage Spoilage due to improper canning procedures or because the seal has been broken.

messenger RNA (mRNA) A type of RNA that carries the information from DNA to dictate the arrangement of amino acids in a protein.

metabolic pathway A chain of chemical reactions in which the product of one reaction serves as the substrate for the next.

metabolism The sum of all chemical processes carried out by living organisms.

metacercaria The postcercarial encysted stage in the development of a fluke, prior to transfer to the final host.

metachromasia Property of exhibiting a variety of colors when stained with a simple stain.

metachromatic granule A polyphosphate granule that exhibits metachromasia. (Also called **volutin.**)

metastasize Relating to the spread of malignant tumors to other body tissues.

methanogens One of the groups of the Archaeobacteria that produce methane gas.

metronidazole An antiprotozoan agent effective against *Trichomonas* infections.

microaerophile A bacterium that grows best in the presence of a small amount of free oxygen.

microbe See **microorganism.**

microbial antagonism The ability of normal microbiota to compete with pathogenic organisms and in some instances to effectively combat their growth.

microbial growth Increase in the number of cells, due to cell division.

microbiology The study of microorganisms.

micrococci Aerobes or facultative anaerobes that form irregular clusters by dividing in two or more planes.

microenvironment A habitat in which the oxygen, nutrients, and light are stable, including the environment immediately surrounding the microbe.

microfilament A protein fiber that makes up part of the cytoskeleton in eukaryotic cells.

microfilaria An immature microscopic roundworm larva.

micrometer (μm) Unit of measure equal to 0.000001 m or 10^{-6} m; formerly called a micron (μ).

microorganism Organism studied with a microscope; includes the viruses. (Also called **microbe**.)

microscopy The technology for making very small things visible to the unaided eye.

microtubule A protein tubule that forms the structure of cilia, flagella, and part of the cytoskeleton in eukaryotic cells.

microvillus (plural: **microvilli**) A minute projection from the surface of an animal cell.

miliary tuberculosis Type of tuberculosis that invades all tissues, producing tiny lesions.

minimum bactericidal concentration (MBC) The lowest concentration of an antimicrobial agent that kills microorganisms, as indicated by absence of growth following subculturing in the dilution method.

minimum inhibitory concentration (MIC) The lowest concentration of an antimicrobial agent that prevents growth in the dilution method of determining antibiotic sensitivity.

miracidium Ciliated, free-swimming first-stage fluke larva that emerges from an egg.

mitochondrion An organelle in eukaryotic cells that carries out oxidative reactions that capture energy in ATP.

mitosis Process by which the cell nucleus of a eukaryotic cell divides to form identical daughter nuclei.

mixed infection An infection caused by several species of organisms present at the same time.

mixture Two or more substances combined in any proportion and not chemically bound.

mole The weight of a substance in grams equal to the sum of the atomic weights of the atoms in a molecule of the substance. (Also called **gram molecular weight**.)

molecular mimicry Imitation of the behavior of a normal molecule by an antimetabolite.

molecule Two or more atoms chemically bonded together.

molluscum contagiosum A viral infection characterized by flesh-colored, painless lesions.

Monera The kingdom of prokaryotic organisms that are unicellular and lack a true cell nucleus. (Also called **Prokaryotae**.)

moniliasis See **candidiasis**.

monoclonal antibody A single, pure antibody produced in the laboratory by a clone of cultured hybridoma cells.

monocular Refers to a light microscope having one eyepiece (ocular).

monocyte A ravenously phagocytic leukocyte, called a macrophage after it migrates into tissues.

monolayer A suspension of cells that attach to plastic or glass surfaces as a sheet one cell layer thick.

monosaccharide A simple carbohydrate, consisting of a carbon chain or ring with several alcohol groups and either an aldehyde or ketone group.

monotrichous A bacterial cell with a single flagellum.

morbidity rate The number of persons contracting a specific disease in relation to the total population (cases per 100,000).

mordant A chemical that helps a stain adhere to the cell or cell structure.

mortality rate The number of deaths from a specific disease in relation to the total population.

most probable number (MPN) A statistical method of measuring bacterial growth, used when samples contain too few organisms to give reliable measures by the plate count method.

mother cell A cell that has approximately doubled in size and is about to divide into two daughter cells. (Also called **parent cell**.)

mucin A glycoprotein in mucus that coats bacteria and prevents their attaching to surfaces.

mucociliary escalator Mechanism involving ciliated cells that allows materials in the bronchi, trapped in mucus, to be lifted to the pharynx and spit out or swallowed.

mucous membrane A covering over those tissues and organs of the body cavity that are exposed to the exterior. (Also called mucosa.)

mucus A thick but watery secretion of glycoproteins and electrolytes secreted by the mucous membranes.

multiple-tube fermentation method Three-step method of testing for coliform bacteria in drinking water.

mumps Disease caused by a paramyxovirus that is transmitted by saliva and invades cells of the oropharynx.

murine typhus See **endemic typhus**.

mutagen An agent that increases the rate of mutations.

mutation A permanent alteration in an organism's DNA.

mutualism A symbiotic relationship in which both organisms benefit from the relationship.

myasthenia gravis Autoimmune disease specific to skeletal muscle, especially muscles of the limbs and those involved in eye movements, speech, and swallowing.

mycelium (plural: **mycelia**) In fungi, a mass of long, threadlike structures (hyphae) that branch and intertwine.

mycobacteria Slender, acid-fast rods, often filamentous; include organisms that cause tuberculosis, leprosy, and chronic infections.

mycology The study of fungi.

mycoplasmas Very small bacteria with cell membranes, RNA and DNA, but no cell walls.

mycosis (plural: **mycoses**) A disease caused by a fungus.

myiasis An infestation caused by maggots (fly larvae).

myocarditis An inflammation of the heart muscle.

NAD Nicotinamide adenine dinucleotide, a coenzyme that carries hydrogen atoms and electrons.

naked virus A virus that lacks an envelope.

nanometer (nm) Unit of measure equal to 0.000000001 m or 10^{-9} m; formerly called a millimicron (mμ).

narrow spectrum The range of activity of an antimicrobial agent that attacks only a few kinds of microorganisms.

nasal cavity Part of the upper respiratory tract where air is warmed and particles are removed by hairs as they pass through.

nasal sinus A hollow cavity within the skull that is lined with mucous membrane.

natural killer (NK) **cell** A lymphocyte that can destroy virus-infected cells, malignant tumor cells, and cells of transplanted tissues.

negative (−) sense RNA An RNA strand made up of bases complementary to those of a positive (+) sense RNA.

negative staining Technique of staining the background around a specimen, leaving the specimen clear and unstained.

nematode See **roundworm**.

neonatal herpes Infection in infants, usually with HSV-2, most often acquired during passage through a birth canal contaminated with the virus.

neoplasm A localized tumor.

neoplastic transformation The uncontrollable division of host cells caused by the infection with a DNA tumor virus.

nephron A functional unit of the kidney in which fluid from the blood is filtered.

nerve A bundle of neuron fibers that relays sensory and motor signals throughout the body.

nervous system The body system, comprising the brain, spinal cord, and nerves, that coordinates the body's activities in relation to the environment.

neuron A conducting nerve cell.

neurosyphilis Neurological damage, including thickening of the meninges, ataxia, paralysis, and insanity, that results from syphilis.

neurotoxin A toxin that acts on nervous system tissues.

neutral Referring to a solution with a pH of 7.

neutralization   Inactivation of microbes or their toxins through the formation of antigen-antibody complexes.

neutralization reaction   An immunological test used to detect bacterial toxins and antibodies to viruses.

neutron   An uncharged subatomic particle in the nucleus of an atom.

neutrophile   An organism that grows best in an environment with a pH of 5.4 to 8.5.

neutrophil   A phagocytic leukocyte. (Also called polymorphonuclear leukocyte, PMNL.)

niclosamide   An antihelminthic agent that interferes with carbohydrate metabolism.

nitrification   The process by which ammonia or ammonium ions are oxidized to nitrites or nitrates.

nitrofuran   An antibacterial drug that damages cellular respiratory systems.

nitrogen cycle   Process by which nitrogen moves from the atmosphere through various organisms and back into the atmosphere.

nitrogen fixation   The reduction of atmospheric nitrogen gas to ammonia.

nitrogenase   Enzyme in nitrogen-fixing bacteroids that catalyzes the reaction of nitrogen gas and hydrogen gas to form ammonia.

nocardioforms   Gram-positive, non-motile, pleomorphic, aerobic bacteria, often filamentous and acid-fast; include some skin and respiratory pathogens.

nocardiosis   Respiratory disease characterized by tissue lesions and abscesses; caused by the filamentous bacterium *Nocardia asteroides*.

nocturia   Nighttime urination, often a result of urinary tract infections.

Nomarski microscopy   Differential interference contrast microscopy; utilizes differences in refractive index to visualize structures, producing a nearly three-dimensional image.

noncommunicable infectious disease   Disease caused by infectious agents but not spread from one host to another.

noncompetitive inhibitor   A molecule that attaches to an enzyme at an allosteric site (a site other than the active site), distorting the shape of the active site so that the enzyme can no longer function.

noncyclic photoreduction   The photosynthetic pathway in which excited electrons from chlorophyll are used to generate ATP and reduce NADP with the splitting of water molecules.

nongonococcal urethritis   A gonorrhea-like sexually transmitted disease most often caused by *Chlamydia trachomatis* and mycoplasmas.

nonindigenous organism   An organism temporarily found in a given environment.

noninfectious disease   Disease caused by any factor other than infectious agents.

nonself   Antigens recognized as foreign by an organism.

nonsense codon   A set of three bases in a gene (or mRNA) that does not code for an amino acid. (Also called **terminator codon.**)

nonspecific defenses   Those host defenses against pathogens that operate regardless of the invading agent.

nonsynchronous growth   Natural pattern of growth during the log phase in which every cell in a culture divides at some point during the generation time, but not simultaneously.

normal microbiota   Microorganisms that live on or in the body but do not usually cause disease. (Also called normal flora.)

nosocomial infection   An infection acquired in a hospital or other medical facility.

notifiable disease   A disease that a physician is required to report to public health officials.

nuclear envelope   The double membrane surrounding the cell nucleus in a eukaryotic cell.

nuclear pore   An opening in the nuclear envelope that allows for the transport of materials between nucleus and cytoplasm.

nuclear region   Central location of DNA, RNA, and some proteins in bacteria; not a true nucleus. (Also called **nucleoid.**)

nucleic acids   Long polymers of nucleotides that encode genetic information and direct protein synthesis.

nucleocapsid   The nucleic acid and capsid of a virus.

nucleoid   See **nuclear region.**

nucleolus   (plural: **nucleoli**) Area in the nucleus of a eukaryotic cell that contains RNA and serves as the site for the assembly of ribosomes.

nucleoplasm   The semifluid portion of the cell nucleus in eukaryotic cells that is surrounded by the nuclear envelope.

nucleotide   An organic compound consisting of a nitrogenous base, a five-carbon sugar, and one or more phosphate groups.

numerical aperture   The widest cone of light that can enter a lens.

numerical taxonomy   Comparison of organisms based on quantitative assessment of a large number of characteristics.

nutritional complexity   The number of nutrients an organism must obtain to grow.

nutritional factor   One factor that influences both the kind of organisms found in an environment and their growth.

objective lens   Lens in a microscope closest to the specimen that creates an enlarged image of the object viewed.

obligate aerobe   A bacterium that must have free oxygen to grow.

obligate anaerobe   A bacterium that is killed by free oxygen.

obligate   Requiring a particular environmental condition.

obligate intracellular parasite   An organism or virus that can live or multiply only inside a living host cell.

obligate parasite   A parasite that must spend some or all of its life cycle in or on a host.

obligate psychrophile   An organism that cannot grow at temperatures above 20°C.

obligate thermophile   An organism that can grow only at temperatures above 37°C.

ocular lens   Lens in microscope that further magnifies the image created by the objective lens.

ocular micrometer   A glass disk with an inscribed scale that is placed inside the eyepiece of a microscope; used to measure the actual size of an object being viewed.

Okazaki fragment   One of the short, discontinuous DNA segments formed on the lagging strand during DNA replication.

onchocerciasis   An eye disease caused by the filarial larvae of the nematode *Onchocerca volvulus*, transmitted by blackflies; common in Africa and Central America. (Also known as **river blindness.**)

oncogene   A cancer-causing gene.

ONPG and MUG test   Water purity test that relies on the ability of coliform bacteria to secrete enzymes that convert a substrate into a product that can be detected by a color change.

Oomycota   See **water mold.**

operon   A sequence of closely associated genes that includes both structural genes and regulatory sites that control transcription.

ophthalmia neonatorum   Pyogenic infection of the eyes caused by *Neisseria gonorrhoeae*. (Also known as conjunctivitis of the newborn.)

opportunist   A species of resident or transient microbiota that does not ordinarily cause disease but can do so under certain conditions.

opsonin   An antibody that promotes phagocytosis when bound to the surface of a microorganism.

opsonization   The process by which microorganisms are rendered more attractive to phagocytes by being coated with antibodies (opsonins) and C3b complement protein. (Also called immune adherence.)

optical microscope   See **compound light microscope.**

optimum pH   The pH at which microorganisms grow best.

orbivirus   Type of virus that causes Colorado tick fever.

orchitis   Inflammation of the testes; a symptom of mumps in postpubertal males.

organelle   An internal membrane-bound structure found in eukaryotic cells.

organic chemistry   The study of compounds that contain carbon.

ornithosis Disease with pneumonia-like symptoms, caused by *Chlamydia psittaci* and acquired from birds. (Previously called **psittacosis** and parrot fever.)

Oroya fever One form of bartonellosis; an acute fatal fever with severe anemia. (Also called Caffion's disease.)

orthomyxovirus A medium-sized, enveloped RNA virus that varies in shape from spherical to filamentous and has an affinity for mucus.

osmosis A special type of diffusion in which water molecules move from an area of higher concentration to one of lower concentration across a selectively permeable membrane.

osmotic pressure The pressure required to prevent the net flow of water molecules by osmosis.

otitis externa Infection of the external ear canal.

otitis media Infection of the middle ear.

outer membrane A bilayer membrane, forming part of the cell wall of gram-negative bacteria.

ovarian follicle An aggregation of cells in the ovary containing an ovum.

ovary In the female, one of a pair of glands that produce ovarian follicles, which contain an ovum and hormone-secreting cells.

oxidation The loss of electrons and hydrogen atoms.

oxidative phosphorylation Process in which the energy of electrons is captured in high-energy bonds as phosphate groups combine with ADP to form ATP.

pandemic An epidemic that has become worldwide.

papilloma See **wart.**

papovavirus A small, naked DNA virus that causes both benign and malignant warts in humans; some types cause cervical cancer.

parainfluenza Viral disease characterized by nasal inflammation, pharyngitis, bronchitis, and sometimes pneumonia, mainly in children.

parainfluenza virus Virus that initially attacks the mucous membranes of the nose and throat.

paramyxovirus A medium-sized, enveloped RNA virus that has an affinity for mucus.

parasite An organism that lives in or on, and at the expense of, another organism, the host.

parasitism A symbiotic relationship in which one organism, the parasite, benefits from the relationship, whereas the other organism, the host, is harmed by it.

parasitology The study of parasites.

parfocal For a microscope, remaining in approximate focus when minor focus adjustments are made.

paroxysmal stage Stage of whooping cough in which mucus and masses of bacteria fill the airway, causing violent coughing.

parvovirus A small, naked DNA virus.

passive immunity Immunity created when ready-made antibodies are introduced into, rather than created by, an organism.

passive immunization The process of inducing immunity by introducing ready-made antibodies into a host.

***Pasteurella–Haemophilus* group** Very small gram-negative bacilli and coccobacilli that lack flagella and are nutritionally fastidious.

pasteurization Mild heating to destroy pathogens and other organisms that cause spoilage.

pathogen Any organism capable of causing disease in its host.

pathogenicity The capacity to produce disease.

pediculosis Lice infestation, resulting in reddened areas at bites, dermatitis, and itching.

pellicle (1) A thin layer of bacteria adhering to the air–water interface of a broth culture by their attachment pili. (2) A strengthened plasma membrane of a protozoan cell. (3) Film over the surface of a tooth at the beginning of plaque formation.

pelvic inflammatory disease An infection of the pelvic cavity in females, caused by any of several organisms including *Neisseria gonorrhoeae* and *Chlamydia.*

penetration The entry of the virus (or its nucleic acid) into the host cell in the replication process.

penicillin An antibacterial agent that inhibits cell wall synthesis.

penis Part of the male reproductive system used to deliver semen to the female reproductive tract during sexual intercourse.

peptide bond A covalent bond joining the amino group of one amino acid and the carboxyl group of another amino acid.

peptidoglycan A structural polymer in the bacterial cell wall that forms a supporting net. (Also called murein.)

peptococci Anaerobes that form pairs, tetrads, or irregular clusters; they lack both catalase and the enzymes to ferment lactic acid.

peptone A product of enzyme digestion of proteins that contains many small peptides; a common ingredient of a complex medium.

perforin A cytotoxin produced by cytotoxic T cells that bores holes in the plasma membrane of infected host cells.

pericarditis An inflammation of the protective membrane around the heart.

periodontal disease A combination of gum inflammation, decay of cementum, and erosion of periodontal ligaments and bone that support teeth.

periodontitis A chronic periodontal disease that affects the bone and tissue that support the teeth and gums.

peripheral nervous system All nerves outside the central nervous system.

periplasm Those substances (enzymes, transport proteins) located in the periplasmic space of gram-negative bacteria or in the older cell wall of gram-positive bacteria.

periplasmic enzyme An exoenzyme produced by gram-negative organisms, which acts in the periplasmic space.

periplasmic space The space between the cell membrane and the outer membrane in gram-negative bacteria that is filled with periplasm.

peritrichous Having flagella distributed all over the surface of a bacterial cell.

permanent parasite A parasite that remains in or on a host once it has invaded the host.

permease An enzyme complex involved in active transport through the cell membrane.

peroxisome An organelle filled with enzymes that in animal cells oxidize amino acids and in plant cells oxidize fats.

persistent viral infection The continued production of viruses within the host over many months or years.

pertussis See **whooping cough.**

petechia (plural: **petechiae**) A pinpoint-size hemorrhage, most common in skin folds, that often occurs in rickettsial diseases.

pH A means of expressing the hydrogen-ion concentration, and thus the acidity, of a solution.

phage See **bacteriophage.**

phagocyte A cell that ingests and digests foreign particles.

phagocytosis Ingestion of solids into cells by means of the formation of vacuoles.

phagolysosome A structure resulting from the fusion of lysosomes and a phagosome.

phagosome A vacuole that forms around a microbe within the phagocyte that engulfed it.

pharmaceutical microbiology A special branch of industrial microbiology concerned with the manufacture of products used in treating or preventing disease.

pharyngitis An infection of the pharynx, usually caused by a virus but sometimes bacterial in origin; a sore throat.

pharynx The throat, a common passageway for the respiratory and digestive systems with tubes connecting to the middle ear.

phase-contrast microscopy Use of a light microscope having a condenser that accentuates small differences in the refractive index of various structures within the cell.

phenol coefficient A numerical expression for the effectiveness of a disinfectant relative to that of phenol.

phenotype The specific observable characteristics displayed by an organism.

phlebovirus Bunyavirus that is carried by the sandfly *Phlebotomus papatsii*.

phospholipid A lipid composed of glycerol, two fatty acids, and a polar head group; found in all membranes.

phosphorescence Continued emission of light by an object when light rays no longer strike it.

phosphorus cycle The cyclic movement of phosphorus between inorganic and organic forms.

phosphorylation The addition of a phosphate group to a molecule, often from ATP, generally increasing the molecule's energy.

phosphotransferase system A mechanism that uses energy from phosphoenolpyruvate to move sugar molecules into cells by active transport.

photoautotroph An autotroph that obtains energy from light.

photoheterotroph A heterotroph that obtains energy from light.

photolysis Process in which light energy is used to split water molecules into protons, electrons, and oxygen molecules.

photoreactivation See **light repair.**

photosynthesis The capture of energy from light and use of this energy to manufacture carbohydrates from carbon dioxide.

phototaxis A nonrandom movement of an organism toward or away from light.

phylogenetic Pertaining to evolutionary relationships.

physical factor Factor in the environment, such as temperature, moisture, pressure, or radiation, that influences the kinds of organisms found and their growth.

picornavirus A small, naked RNA virus; different genera are responsible for polio, the common cold, and hepatitis A.

pilus (plural: **pili**) A tiny hollow projection used to attach bacteria to surfaces (attachment pilus) or for conjugation (conjugation pilus).

pimple See **folliculitis.**

pinna Flaplike external structure of the ear.

pinworm A small roundworm, *Enterobius vermicularis*, that causes gastrointestinal disease.

placebo An unmedicated, usually harmless substance given to a recipient as a substitute for or to test the efficacy of a medication or treatment.

Plantae The kingdom of organisms to which all plants belong.

plaque A clean area in a bacterial lawn culture where viruses have lysed cells.

plaque assay A viral assay used to determine viral yield by culturing viruses on a bacterial lawn and counting plaques.

plaque-forming unit A plaque counted on a bacterial lawn that gives only an approximate number of phages present, because a given plaque may have been due to more than one phage.

plasma Liquid portion of the blood, excluding the formed elements.

plasma cell A large lymphocyte differentiated from a B cell that synthesizes and releases antibodies like those on the B cell surface.

plasma membrane A selectively permeable lipoprotein bilayer that forms the boundary between the cytoplasm of a eukaryotic cell and its environment. (See **cell membrane.**)

plasmid A small, circular, independently replicating piece of DNA in a cell that is not part of its chromosome and can be transferred to another cell. (Also called extrachromosomal DNA.)

plasmodial slime mold Funguslike protist consisting of a multinucleate amoeboid mass, or plasmodium, that moves about slowly and phagocytizes dead matter.

plasmodium A multinucleate mass of cytoplasm that forms one of the stages in the life cycle of a plasmodial slime mold.

plasmogamy Sexual reproduction in fungi in which haploid gametes unite and their cytoplasm mingles.

plasmolysis Shrinking of a cell, with separation of the cell membrane from the cell wall, resulting from loss of water in a hypertonic solution.

platelet A short-lived fragment of large cells called megakaryocytes, important component of the blood-clotting mechanism.

pleomorphism Phenomenon in which bacteria vary widely in form, even within a single culture under optimal conditions.

pleura Serous membrane covering the surfaces of the lungs and the cavities they occupy.

pleurisy Inflammation of pleural membranes that causes painful breathing; often accompanies lobar pneumonia.

Pneumocystis **pneumonia** A fungal respiratory disease caused by *Pneumocystis carinii*.

pneumonia An inflammation of lung tissue caused by bacteria, viruses, or fungi.

pneumonic plague Usually fatal form of plague transmitted by aerosol droplets from a coughing patient.

point mutation Mutation in which one base is substituted for another at a specific location in a gene.

polar compound A molecule with an unequal distribution of charge due to an unequal sharing of electrons between atoms.

poliomyelitis Disease caused by any of several strains of polioviruses that attack motor neurons of the spinal cord and brain.

polyacrylamide gel electrophoresis (PAGE) A technique for separating proteins from a cell based on their molecular size.

polyene An antifungal agent that increases membrane permeability.

polymer A long chain of repeating subunits.

polymerase chain reaction (PCR) A technique that rapidly produces a billion or more identical copies of a DNA fragment without needing a cell.

polymyxin An antibacterial agent that disrupts the cell membrane.

polynucleotide A chain of many nucleotides.

polypeptide A chain of many amino acids.

polyribosome A long chain of ribosomes attached at different points along an mRNA molecule. (Also called polysome.)

polysaccharide A carbohydrate formed when many monosaccharides are linked together by glycosidic bonds.

Pontiac fever A mild variety of legionellosis.

porin A protein in the outer membrane of gram-negative bacteria that nonselectively transports polar molecules into the periplasmic space.

portal of entry A site at which microorganisms can gain access to body tissues.

portal of exit A site at which microorganisms can leave the body.

positive chemotaxis Movement of an organism toward a chemical.

positive (+) sense RNA An RNA strand that encodes information for making proteins needed by a virus.

potable water Water that is fit for human consumption.

pour plate A plate containing separate colonies and used to prepare a pure culture.

pour plate method Method used to prepare pure cultures using serial dilutions, each of which is mixed with melted agar and poured into a sterile petri plate.

poxvirus DNA virus that is the largest and most complex of all viruses.

precipitation reaction Immunological test in which antibodies called precipitins react with antigens to form latticelike networks of molecules that precipitate from solution.

precipitin test Immunological test used to detect antibodies that is based on the precipitation reaction.

prediction The expected outcome if a hypothesis is correct.

preserved culture A culture in which organisms are maintained in a dormant state.

presumptive test First stage of testing in multiple-tube fermentation in which gas production in lactose broth provides presumptive evidence that coliform bacteria are present.

prevalence rate The number of people infected with a particular disease at any one time.

primaquine   An antiprotozoan agent that interferes with protein synthesis.

primary atypical pneumonia   A mild form of pneumonia with insidious onset. (Also called mycoplasma pneumonia and **walking pneumonia.**)

primary cell culture   A culture that comes directly from an animal and is not subcultured.

primary immunodeficiency disease   A genetic or developmental defect in which T cells or B cells are lacking or nonfunctional.

primary infection   An initial infection in a previously healthy person.

primary response   Humoral immune response that occurs when an antigen is first recognized by host B cells.

primary structure   The specific sequence of amino acids in a polypeptide chain.

primary treatment   Physical treatment to remove solid wastes from sewage.

prion   An exceedingly small infectious particle alleged to consist of protein without any nucleic acid.

probe   A single-stranded DNA fragment that has a sequence of bases that can be used to identify complementary DNA base sequences.

prodromal phase   In an infectious disease, the short period during which nonspecific symptoms such as malaise and headache sometimes appear.

prodrome   A symptom indicating the onset of a disease.

producer   Organism that captures energy from the sun and synthesizes food. (Also called **autotroph.**)

product   The material resulting from an enzymatic reaction.

productive infection   Viral infection in which viruses enter a cell and produce infectious progeny.

proglottid   One of the segments of a tapeworm, containing the reproductive organs.

progressive multifocal leukoencephalopathy   Disease caused by the JC polyomavirus with symptoms including mental deterioration, limb paralysis, and blindness.

Prokaryotae   See **Monera.**

prokaryote   Microorganism that lacks a cell nucleus and membrane-bound internal structures; all bacteria in the kingdom Monera (Prokaryotae) are prokaryotes.

prokaryotic cell   A cell that lacks a cell nucleus; includes all bacteria.

propagated epidemic   An epidemic that arises from person-to-person contacts.

prophage   The DNA of a lysogenic phage that has integrated into the host cell chromosome.

propionibacteria   Pleomorphic, irregular, nonsporing, gram-positive rods.

prostaglandin   A reaction mediator that acts as a cellular regulator, often intensifying pain.

prostate gland   Gland located at the beginning of the male urethra whose milky fluid discharge forms a component of semen.

prostatitis   Inflammation of the prostate gland.

protein   A polymer of amino acids joined by peptide bonds.

protein profile   A technique for visualizing the proteins contained in a cell; obtained by the use of polyacrylamide gel electrophoresis.

proto-oncogene   A normal gene that can cause cancer in uncontrolled situations; often the normal gene comes under the control of a virus.

protein repressor   Substance produced by host cells that keeps a virus in an inactive state and prevents the infection of the cell by another phage of the same type.

Protista   The kingdom of organisms that are unicellular but contain internal organelles typical of the eukaryotes.

protist   A unicellular eukaryotic organism that is a member of the kingdom Protista.

proton   A positively charged subatomic particle located in the nucleus of an atom.

protoplast   A gram-positive bacterium from which the cell wall has been removed.

protoplast fusion   A technique of genetic engineering in which genetic material is combined by removing the cell walls of two different types of cells and allowing the resulting protoplasts to fuse.

prototroph   A normal, nonmutant organism. (Also called wild type.)

protozoa   (singular: **protozoan**) Single-celled, microscopic, animal-like protists in the kingdom Protista.

provirus   Viral DNA that is incorporated into a host-cell chromosome.

pseudocoelom   A primitive body cavity, typical of nematodes, that lacks the complete lining found in higher animals.

pseudocyst   An aggregate of trypanosome protozoa that forms in lymph nodes in Chagas' disease.

pseudomembrane   A combination of bacilli, damaged epithelial cells, fibrin, and blood cells resulting from infection with diphtheria that can block the airway, causing suffocation.

pseudomonads   Aerobic motile rods with polar flagella.

pseudoplasmodium   A multicellular mass composed of individual cellular slime mold cells that have aggregated.

pseudopodium   A temporary footlike projection of cytoplasm associated with amoeboid movement.

psittacosis   See **ornithosis.**

psychrophile   A cold-loving organism that grows best at temperatures of 15° to 20° C.

puerperal fever   Disease caused by β-hemolytic streptococci, which are normal vaginal and respiratory microbiota that can be introduced during child delivery by medical personnel. (Also called **childbed fever** or puerperal sepsis.)

pure culture   A culture that contains only a single species of organism.

purine   The nucleic acid bases adenine and guanine.

pus   Fluid formed by the accumulation of dead phagocytes, the materials they have ingested, and tissue debris.

pustule   See **folliculitis.**

pyelonephritis   Inflammation of the kidneys.

pyoderma   A pus-producing skin infection caused by staphylococci, streptococci, and corynebacteria, singly or in combination.

pyrimidine   Any of the nucleic acid bases thymine, cytosine, and uracil.

pyrogen   A substance that acts on the hypothalamus to set the body's "thermostat" to a higher-than-normal temperature.

Q fever   Pneumonialike disease caused by *Coxiella burnetii,* a rickettsia that survives long periods outside cells and can be transmitted aerially as well as by ticks.

quarantine   The separation of humans or animals from the general population when they have a communicable disease or have been exposed to one.

quaternary ammonium compound (quat)   A cationic detergent that has four organic groups attached to a nitrogen atom.

quaternary structure   The three-dimensional structure of a protein molecule formed by the association of two or more polypeptide chains.

quinine   An antiprotozoan agent used to treat malaria.

quinolone   A bactericidal agent that inhibits DNA replication.

quinone   A nonprotein, lipid-soluble electron carrier in oxidative phosphorylation. (Also called coenzyme Q.)

R group   An organic chemical group attached to the central carbon atom in an amino acid.

rabies   A viral disease that affects the brain and nervous system with symptoms including hydrophobia and aerophobia; transmitted by animal bites.

rabies virus   An RNA-containing rhabdovirus that is transmitted through animal bites.

rad   A unit of radiation energy absorbed per gram of tissue.

radial immunodiffusion   Serological test used to provide a quantitative measure of antigen or antibody concentration by measuring the diameter of the ring of precipitation around an antigen.

radiation   Light rays, such as X-rays and ultraviolet rays, that can act as mutagens.

radioimmunoassay (RIA) Technique that uses a radioactive anti-antibody to detect very small quantities of antigens or antibodies.

radioisotope Isotope with unstable nuclei that tends to emit subatomic particles and radiation.

rat bite fever A disease caused by *Streptobacillus moniliformis* transmitted by bites from wild and laboratory rats.

reactant Substance that takes part in a chemical (enzymatic) reaction.

recombinant DNA DNA combined from two different species by restriction enzymes and ligases.

recombination The combining of DNA from two different cells, resulting in a recombinant cell.

redia The development stage of the fluke immediately following the sporocyst stage.

reduction The gain of electrons and hydrogen atoms.

reference culture A preserved culture used to maintain an organism with its characteristics as originally defined.

reflection The bouncing of light off an object.

refraction The bending of light as it passes from one medium to another medium of different density.

regulator gene Gene that controls the expression of structural genes of an operon through the synthesis of a repressor protein.

regulatory site The promotor and operator regions of an operon.

relapsing fever Disease caused by various species of *Borrelia,* most commonly by *B. recurrentis;* transmitted by lice.

release The exit from the host cell of new virions, which usually kills the host cell.

rennin An enzyme from calves' stomachs used in cheese manufacture.

reovirus A medium-sized RNA virus that has a double capsid with no envelope; causes upper respiratory and gastrointestinal infections in humans.

replica plating A technique used to transfer colonies from one medium to another.

replication Process by which an organism or structure (especially a DNA molecule) duplicates itself.

replication cycle The series of steps of virus replication in a host cell.

replication fork A site at which the two strands of the DNA double helix separate during replication and new complementary DNA strands form.

reservoir host An infected organism that makes parasites available for transmission to other hosts.

reservoir of infection Site where microorganisms can persist and maintain their ability to infect.

resident microbiota Species of microorganisms that are always present on or in an organism.

resistance The ability of a microorganism to remain unharmed by an antimicrobial agent.

resistance (R) gene A component of a resistance plasmid that confers resistance to a specific antibiotic or to a toxic metal.

resistance (R) plasmid A plasmid that carries genes that provide resistance to various antibiotics or toxic metals. (Also called **R factor.**)

resistance transfer factor (RTF) A component of a resistance plasmid that implements transfer by conjugation of the plasmid.

resolution The ability of an optical device to show two items as separate and discrete entities rather than as a fuzzily overlapped image.

resolving power A numerical measure of the resolution of an optical instrument.

respiratory anaphylaxis Life-threatening allergy in which airways become constricted and filled with mucous secretions.

respiratory bronchiole Microscopic channel in the lower respiratory system that ends in a series of alveoli.

respiratory syncytial virus (RSV) Cause of lower respiratory infections affecting children under 1 year old; causes cells in culture to fuse their plasma membranes and become multinucleate masses (syncytia).

respiratory system Body system that moves oxygen from the atmosphere to the blood and removes carbon dioxide and other wastes from the blood.

restriction endonuclease An enzyme that cuts DNA at precise base sequences.

reticulate body An intracellular stage in the life cycle of chlamydias.

retrovirus An enveloped RNA virus that uses its own reverse transcriptase to transcribe its RNA into DNA in the cytoplasm of the host cell.

reverse transcriptase An enzyme found in retroviruses that copies RNA into DNA.

R factor See **resistance (R) plasmid.**

rhabdovirus A rod-shaped, enveloped RNA virus that infects insects, fish, various other animals, and some plants.

Rh antigen An antigen found on some red blood cells; discovered in the cells of Rhesus monkeys.

rheumatic fever A multisystem disorder following infection by β-hemolytic *Streptococcus pyogenes* that can cause heart damage.

rheumatoid arthritis Autoimmune disease that affects mainly the joints but can extend to other tissues.

rheumatoid factor IgM found in the blood of patients with rheumatoid arthritis, and their relatives.

rhinovirus A virus that replicates in cells of the upper respiratory tract and causes the common cold.

ribonucleic acid (RNA) Nucleic acid that carries information from DNA to sites where proteins are manufactured in cells and that directs and participates in the assembly of proteins.

ribosomal RNA (rRNA) A type of RNA that, together with specific proteins, makes up the ribosomes.

ribosome Site for protein synthesis consisting of RNA and protein, located in the cytoplasm.

rickettsialpox Mild rickettsial disease with symptoms resembling those of chickenpox; caused by *Rickettsia akari* and carried by mites found on house mice.

rickettsias Small, nonmotile, gram-negative organisms; obligate intracellular parasites of mammalian and arthropod cells.

rifamycin An antibacterial agent that inhibits ribonucleic acid (RNA) synthesis.

Rift Valley fever Disease caused by bunyaviruses that occurs in epidemics.

ringworm A highly contagious fungal skin disease that can cause ringlike lesions.

river blindness See **onchocerciasis.**

RNA polymerase An enzyme that binds to one strand of exposed DNA during transcription and catalyzes the synthesis of RNA from the DNA template.

RNA tumor virus Any retrovirus that causes tumors and cancer.

Rocky Mountain spotted fever Disease caused by *Rickettsia rickettsia* and transmitted by ticks.

rotavirus Virus transmitted by the fecal–oral route that replicates in the intestine, causing diarrhea and enteritis.

roundworm A worm with a long, cylindrical, unsegmented body and a heavy cuticle. (Also called **nematode.**)

rubella Viral disease characterized by a skin rash; can cause severe congenital damage. (Also called **German measles.**)

rubeola See **measles.**

rule of octets Principle that an element is chemically stable if it contains eight electrons in its outer shell.

sac fungus A member of a diverse group of fungi that produces saclike asci during sexual reproduction. (Also called **Ascomycota.**)

salmonellosis A common enteritis characterized by abdominal pain, fever, and diarrhea with blood and mucus; caused by *Salmonella* species.

sapremia A condition caused when saprophytes release metabolic products into the blood.

saprophyte An organism that feeds on dead or decaying organic matter.

sarcina A group of eight cocci in a cubical packet.

sarcodine An amoeboid protozoan.

sarcoptic mange See **scabies.**

saturated fatty acid A fatty acid containing only carbon–hydrogen single bonds.

scabies Highly contagious skin disease caused by the itch mite *Sarcoptes scabiei*. (Also called **sarcoptic mange.**)

scalded skin syndrome Infection caused by staphylococci consisting of large, soft vesicles over the whole body.

scanning electron microscope (SEM) Type of electron microscope used to study the surfaces of specimens.

scarlet fever Infection caused by *Streptococcus pyogenes* that produces an erythrogenic toxin. (Sometimes called scarlatina.)

Schaeffer–Fulton spore staining A differential stain used to make endospores easier to visualize.

schistosomiasis Disease of the blood and lymph caused by blood flukes of the genus *Schistosoma*. (Also called bilharzia.)

schizogony Multiple fission, in which one cell gives rise to many cells.

scolex Head end of a tapeworm, with suckers and sometimes hooks that attach to the intestinal wall.

scrub typhus A typhus caused by *Rickettsia tsutsugamushi;* transmitted by mites that feed on rats. (Also called tsutsugamushi disease.)

sebaceous gland Epidermal structure, associated with hair follicles, that secretes an oily substance called sebum.

sebum Oily substance secreted by the sebaceous glands.

secondary immunodeficiency disease Result of damage to T cells or B cells after they have developed normally.

secondary infection Infection that follows a primary infection, especially in patients weakened by the primary infection.

secondary response Humoral immune response that occurs when an antigen is recognized by memory cells; more rapid and stronger than a primary response. (Also called **anamnestic response.**)

secondary structure The folding or coiling of a polypeptide chain into a particular pattern, such as a helix or pleated sheet.

secondary treatment Treatment of sewage by biological means to remove remaining solid wastes after primary treatment.

secretory piece A part of the IgA antibody that protects the immunoglobulin from degradation and helps in the secretion of the antibody

secretory vesicle Small membrane-bound structure that stores substances coming from the Golgi apparatus.

selective medium A medium that encourages growth of some organisms and suppresses growth of others.

selective toxicity The ability of an antimicrobial agent to harm microbes without causing significant damage to the host.

selectively permeable Able to prevent the passage of certain specific molecules and ions while allowing others through.

self Molecules that are not recognized as antigenic or foreign by an organism.

semen The male fluid discharge at the time of ejaculation, containing sperm and various glandular and other secretions.

semiconservative replication Replication in which a new DNA double helix is synthesized from one strand of parent DNA and one strand of new DNA.

seminal vesicle A saclike structure whose secretions form a component of semen.

semisynthetic drug An antimicrobial agent made partly by laboratory synthesis and partly by microorganisms.

sense codon A set of three DNA (or mRNA) bases that code for an amino acid.

sensitization Initial exposure to an antigen, which causes the host to mount an immune response against it.

septicemia An infection caused by rapid multiplication of pathogens in the blood. (Also called blood poisoning.)

septicemic plague Fatal form of plague that occurs when bubonic plague bacteria move from the lymphatics to the circulatory system.

septic shock A life-threatening septicemia with low blood pressure and blood-vessel collapse, caused by endotoxins.

septic tank An underground tank for receiving sewage, where solid materials settle out as sludge, which must be pumped periodically.

septum (plural: **septa**) A cross-wall separating two fungal cells.

sequela (plural: **sequelae**) The aftereffect of a disease, after recovery from it.

serial dilution A method of measurement in which successive 1:10 dilutions are made from the original sample.

seroconversion The identification of a specific antibody in serum as a result of an infection.

serology The branch of immunology dealing with laboratory tests to detect antigens and antibodies.

serovar Strain; a subspecies category.

serum The liquid part of blood after cells and clotting factors have been removed.

serum hepatitis See **hepatitis B.**

serum killing power Test used to determine effectiveness of an antimicrobial agent in which a bacterial suspension is added to the serum of a patient who is receiving an antibiotic, and incubated.

serum sickness Immune complex disorder that occurs when foreign antigens in sera cause immune complexes to be deposited in tissues.

severe combined immunodeficiency (SCID) Primary immunodeficiency disease caused by failure of stem cells to develop properly, resulting in deficiency of both B and T cells.

sewage Used water and the wastes it contains.

sexually transmitted disease (STD) An infectious disease spread by sexual activities.

shadow casting The coating of electron microscopy specimens with a heavy metal, such as gold or palladium, to create a three-dimensional effect.

shigellosis Gastrointestinal disease caused by several strains of *Shigella* that invade intestinal lining cells. (Also called **bacillary dysentery.**)

shinbone fever See **trench fever.**

shingles Sporadic disease caused by reactivation of varicella–zoster herpesvirus that appears most frequently in older and immunocompromised individuals.

sign A disease characteristic that can be observed by examining the patient, such as swelling or redness.

simple diffusion The net movement of particles from a region of higher to one of lower concentration; does not require energy from a cell.

simple stain A single dye used to reveal basic cell shapes and arrangements.

single-cell protein Animal feed consisting of microorganisms.

sinus A large passageway in tissues, lined with phagocytic cells.

sinusitis An infection of the sinus cavities.

sinusoid An enlarged capillary.

skin The largest single organ of the body that presents a physical barrier to infection by microorganisms.

slime layer A thin protective structure loosely bound to the cell wall that protects the cell against drying, helps trap nutrients, and sometimes binds cells together.

slime mold A funguslike protist.

sludge Solid matter remaining from water treatment that contains aerobic organisms that digest organic matter.

sludge digester Large fermentation tank in which sludge is digested by anaerobic bacteria into simple organic molecules, carbon dioxide, and methane gas.

small intestine The upper area of the intestine where digestion is completed.

smallpox A formerly worldwide and serious viral disease that has now been eradicated.

smear A thin layer of liquid specimen spread out on a microscope slide.

solute The substance dissolved in a solvent to form a solution.

solution A mixture of two or more substances in which the molecules are evenly distributed and will not separate out on standing.

solvent The medium in which substances are dissolved to form a solution.

sonication The disruption of cells by sound waves.

specialized transduction Type of transduction in which the bacterial DNA transduced is limited to one or a few genes lying adjacent to a prophage that are accidentally included when the prophage is excised from the bacterial chromosome.

species A group of organisms with many common characteristics; the narrowest taxon.

specific defense A host defense that operates in response to a particular invading pathogen.

specific epithet The second name of an organism in the binomial system of nomenclature, following that of the genus—for example, *coli* in *Escherichia coli.*

specificity (1) The property of an enzyme that allows it to accept only certain substrates and catalyze only one particular reaction. (2) The property of a virus that restricts it to certain specific types of host cells. (3) The ability of the immune system to mount a unique immune response to each antigen it encounters.

spectrum of activity Refers to the range of different microbes against which an antimicrobial agent is effective.

spheroplast A gram-negative bacterium that lacks the cell wall but has not lysed.

spike A glycoprotein projection that extends form the viral capsid or envelope and is used to attach to or to fuse with host cells.

spindle apparatus A system of microtubules in the cytoplasm of a eukaryotic cell that guides the movement of chromosomes during mitosis and meiosis.

spirillar fever A form of rat bite fever, caused by *Spirillum minor,* first described as sodoku in Japan.

spirillum (plural: **spirilla**) A flexible, wavy-shaped bacterium.

spirochetes Corkscrew-shaped motile bacteria.

spleen The largest lymphatic organ; acts as a blood filter.

spontaneous generation The theory that living organisms can arise from nonliving things.

spontaneous mutation A mutation that occurs in the absence of any agent known to cause changes in DNA; usually caused by errors during DNA replication.

sporadic disease A disease that is limited to a small number of isolated cases posing no great threat to a large population.

spore A resistant reproductive structure formed by fungi and actinomycetes; different from a bacterial endospore.

spore coat A keratin-like protein material that is laid down around the cortex of an endospore by the mother cell.

sporocyst Larval form of a fluke that develops in the body of its snail or mollusk host.

sporotrichosis Fungal skin disease caused by *Sporothrix schenckii* that often enters the body from plants.

sporozoite A malaria trophozoite present in the salivary glands of infected mosquitoes.

sporulation The formation of spores, such as endospores.

spread plate method A technique used to prepare pure cultures by placing a diluted sample of cells on the surface of an agar plate and then spreading the sample evenly over the surface.

St. Anthony's fire See **erysipelas.**

St. Louis encephalitis Type of viral encephalitis most often seen in humans in the central United States.

stain A molecule that can bind to a structure and give it color. (Also called dye.)

standard bacterial growth curve A graph plotting the number of bacteria versus time and showing the phases of bacterial growth.

stationary phase The third of four major phases of the bacterial growth curve in which new cells are produced at the same rate that old cells die, leaving the number of live cells constant.

sterility The state in which there are no living organisms in or on a material.

sterilization The killing or removal of all microorganisms in a material or on an object.

steroid A lipid having a four-ring structure; includes cholesterol, steroid hormones, and vitamin D.

stock culture A reserve culture used to store an isolated organism in pure condition for use in the laboratory.

strain A subgroup of a species with one or more characteristics that distinguish it from other subgroups of that species.

streak plate method Method used to prepare pure cultures in which bacteria are lightly spread over the surface of agar plates, resulting in isolated colonies.

streptococci Aerotolerant anaerobes that form pairs, tetrads, or chains by dividing in one or two planes; most lack the enzyme catalase.

streptokinase A bacterially produced enzyme that digests (dissolves) blood clots.

streptolysin Toxin produced by streptococci that kills phagocytes.

streptomycetes Gram-positive, filamentous, sporing, soil-dwelling bacteria; producers of many antibiotics.

streptomycin An antibacterial agent that blocks protein synthesis.

stroma The fluid-filled inner portion of a chloroplast.

strongyloidiasis Parasitic disease caused by the roundworm *Strongyloides stercoralis* and a few closely related species.

structural gene A gene that carries information for the synthesis of a specific polypeptide.

structural protein A protein that contributes to the structure of cells, cell parts, and membranes.

sty An infection at the base of an eyelash.

subacute disease A disease that is intermediate between an acute and a chronic disease.

subacute sclerosing panencephalitis (SSPE) A complication of measles, nearly always fatal, that is due to the persistence of measles viruses in brain tissue.

subclinical infection See **inapparent infection.**

subculturing The process by which cells from an existing culture are transferred to fresh medium in new containers.

substrate (1) The substance on which an enzyme acts. (2) A surface or food source on which a cell can grow or a spore can germinate.

sulfate reduction The reduction of sulfate ions to hydrogen sulfide.

sulfonamide A synthetic, bacteriostatic agent that blocks the synthesis of folic acid. (Also called sulfa drug.)

sulfur cycle The cyclic movement of sulfur through an ecosystem.

sulfur oxidation The oxidation of various forms of sulfur to sulfate.

sulfur reduction The reduction of elemental sulfur to hydrogen sulfide.

superinfection A secondary infection resulting from the removal of normal microbiota, allowing colonization by pathogenic, and often antibiotic-resistant, microbes.

superoxide A highly reactive form of oxygen that kills obligate anaerobes.

superoxide dismutase An enzyme that converts superoxide to molecular oxygen and hydrogen peroxide.

suppressor T cell (T_s) Possibly a type of cytotoxic or helper T cell that inhibits immune responses.

surface tension A phenomenon in which the surface of water behaves like a thin, invisible, elastic membrane.

surfactant A substance that reduces surface tension.

susceptibility The vulnerability of an organism to harm by infectious agents.

swarmer cell Spherical, flagellated *Rhizobium* cell that invades the root hairs of leguminous plants, eventually to form nodules.

sweat gland Epidermal structure that empties a watery secretion through pores in the skin.

swimmer's itch Skin reaction to cercariae of some species of the helminth *Schistosoma.*

symbiosis The living together of two different kinds of organisms.

symptom A disease characteristic that can be observed or felt only by the patient, such as pain or nausea.

synchronous growth Hypothetical pattern of growth during the log phase in which all the cells in a culture divide at the same time.

syncytium (plural: **syncytia**) A multinucleate mass in a cell culture for example, caused by the respiratory syncytial virus.

syndrome A combination of signs and symptoms that occur together.

synergism Referring to an inhibitory effect produced by two antibiotics working together that is greater than either can achieve alone.

synthetic drug An antimicrobial agent synthesized chemically in the laboratory.

synthetic medium A growth medium prepared in the laboratory from materials of precise or reasonably well defined composition.

syphilis A sexually transmitted disease, caused by the spirochete *Treponema pallidum*, characterized by a chancre at the site of entry and often eventual neurological damage.

systemic blastomycosis Disease resulting from invasion by *Blastomyces dermatitidis* of internal organs, especially the lungs.

systemic infection An infection that affects the entire body. (Also called generalized infection.)

systemic lupus erythematosus A widely disseminated, systemic autoimmune disease resulting from production of antibodies against DNA and other body components.

tapeworm Flatworm that lives in the adult stage as a parasite in the small intestine of animals.

tartar Calcium deposition on dental plaque forming a very rough, hard crust.

taxon (plural: **taxa**) A category used in classification, such as species, genus, order, family.

taxonomy The science of classification.

T cell See **T lymphocyte.**

T-dependent antigen Antigen requiring helper T cell (T_H2) activity to activate B cells.

teichoic acid A polymer attached to peptidoglycan in gram-positive cell walls.

temperate phage A bacteriophage that does not cause a virulent infection; rather, its DNA is incorporated into the host cell chromosome, as a prophage, and replicated with the chromosome.

template DNA used as a pattern for the synthesis of a new nucleotide polymer in replication or transcription.

temporary parasite A parasite that feeds on and then leaves its host (such as a biting insect).

teratogen An agent that induces defects during embryonic development.

teratogenesis The induction of defects during embryonic development.

terminator codon A codon that signals the end of the information for a particular protein. (Also called **nonsense codon** or stop codon.)

tertiary structure The folding of a protein molecule into globular shapes.

tertiary treatment Chemical and physical treatment of sewage to produce an effluent of water pure enough to drink.

test A shell made of calcium carbonate and common to some protists.

testis (plural: **testes**) One of a pair of male reproductive glands that produce testosterone and sperm.

tetanus Disease caused by *Clostridium tetani* in which muscle stiffness progresses to eventual paralysis and death. (Also called lockjaw.)

tetanus neonatorum Type of tetanus acquired through the raw stump of the umbilical cord.

tetracycline An antibacterial agent that inhibits protein synthesis.

thallus The body of a fungus.

tetrad Cuboidal groups of four cocci.

theca A tightly affixed, secreted outer layer of dinoflagellates that often contains cellulose.

therapeutic dosage level Level of drug dosage that successfully eliminates a pathogenic organism if maintained over a period of time.

thermal death point The temperature that kills all the bacteria in a 24-hour-old broth culture at neutral pH in 10 minutes.

thermal death time The time required to kill all the bacteria in a particular culture at a specified temperature.

thermoacidophile A member of one of the groups of the archaeobacteria that live in extremely hot, acidic environments.

thermophile A heat-loving organism that grows best at temperatures from 50° to 60° C.

thermophilic anaerobic spoilage Spoilage due to endospore germination and growth in which gas and acid are produced, making cans bulge.

thrush Milky patches of inflammation on oral mucous membranes; a symptom of candidiasis, caused by *Candida albicans*.

thylakoid An internal membrane of chloroplasts that contains chlorophyll.

thymus gland Multilobed lymphatic organ located beneath the sternum that processes lymphocytes into T cells.

tick paralysis A disease characterized by fever and paralysis due to anticoagulants and toxins secreted into a tick's bite via the ectoparasite's saliva.

tincture An alcoholic solution.

T-independent antigen Antigen not requiring helper T cell (T_H2) activity to activate B cells.

tinea barbae Barber's itch; a type of ringworm that causes lesions in the beard.

tinea capitis Scalp ringworm, a form of ringworm in which hyphae grow in hair follicles, often leaving circular patterns of baldness.

tinea corporis Body ringworm, a form of ringworm that causes ringlike lesions with a central scaly area.

tinea cruris Groin ringworm, a form of ringworm that occurs in skin folds in the pubic region. (Also called jock itch.)

tinea pedis See **athlete's foot.**

tinea unguium A form of ringworm that causes hardening and discoloration of fingernails and toenails.

tissue culture Culture made from a single tissue, assuring a reasonably homogenous set of cultures in which to test the effects of a virus or to culture an organism.

titer The quantity of a substance needed to produce a given reaction.

T lymphocyte Thymus-derived cell of the immune system and agent of cellular immune responses. (Also called **T cell.**)

togavirus A small, enveloped RNA virus that multiplies in many mammalian and arthropod cells.

tolerance A state in which antigens no longer elicit an immune response.

tonsil Lymphoid tissue that contributes immune defenses in the form of B cells and T cells.

tonsillitis A bacterial infection of the tonsils.

TORCH series A group of blood tests used to identify teratogenic diseases in pregnant women and newborn infants.

total magnification Obtained by multiplying the magnifying power of the objective lens by the magnifying power of the ocular lens.

toxemia The presence and spread of exotoxins in the blood.

toxic shock syndrome (TSS) Condition caused by infection with certain toxigenic strains of *Staphylococcus aureus;* often associated with the use of superabsorbent but abrasive tampons.

toxin Any substance that is poisonous to other organisms.

toxoid An exotoxin inactivated by chemical treatment but which retains its antigenicity and therefore can be used to immunize against the toxin.

toxoplasmosis Disease caused by the protozoan *Toxoplasma gondii* that can cause congenital defects in newborns.

trace element Minerals, such as copper, iron, zinc, and cobalt ions, that are required in minute amounts for growth.

trachea The windpipe.

trachoma Eye disease caused by *Chlamydia trachomatis* that can result in blindness.

transcription The synthesis of RNA from a DNA template.

transduction The transfer of genetic material from one bacterium to another by a bacteriophage.

transfer RNA (tRNA) Type of RNA that transfers amino acids from the cytoplasm to the ribosomes for placement in a protein molecule.

transformation A change in an organism's characteristics through the transfer of naked DNA.

transfusion reaction Reaction that occurs when matching antigens and antibodies are present in the blood at the same time.

transgenic State of permanently changing an organism's characteristics by integrating foreign DNA (genes) into the organism.

transient microbiota Microorganisms that may be present in or on an organism under certain conditions and for certain lengths of time at sites where resident microbiota are found.

translation The synthesis of protein from information in mRNA.

transmission The passage of light through an object.

transmission electron microscope (TEM) Type of electron microscope used to study internal structures of cells; very thin slices of specimens are used.

transovarian transmission Passing of a pathogen from one generation of ticks to the next as eggs leave the ovaries.

transplantation The moving of tissue from one site to another.

transplant rejection Destruction of grafted tissue or of a transplanted organ by the host immune system.

transposable element A mobile genetic sequence that can move from one plasmid to another plasmid or chromosome.

transposal of virulence A laboratory technique in which a pathogen is passed from its normal host sequentially through many individual members of a new host species, resulting in a lessening or even total loss of its virulence in the original host.

transposition The process whereby certain genetic sequences in bacteria or eukaryotes can move from one location to another.

transposon A mobile genetic sequence that contains the genes for transposition as well as one or more other genes not related to transposition.

traumatic herpes Type of herpes infection in which the virus enters traumatized skin in the area of a burn or other injury.

traveler's diarrhea Gastrointestinal disorder generally caused by pathogenic strains of *Escherichia coli*.

trench fever Rickettsial disease, caused by *Rochalimaea quintana*, resembling epidemic typhus in that it is transmitted by lice and is prevalent during wars and under unsanitary conditions. (Also called **shinbone fever**.)

treponemes Spirochetes belonging to the genus *Treponema*.

triacylglycerol A molecule formed from three fatty acids bonded to glycerol.

trichinosis A disease caused by a small nematode, *Trichinella spiralis*, that enters the digestive tract as encysted larvae in poorly cooked meat, usually pork.

trichocyst Tentacle-like structure on ciliates for catching prey or for attachment.

trichomoniasis A parasitic urogenital disease, transmitted primarily by sexual intercourse, that causes intense itching and a copious white discharge, especially in females.

trichuriasis Parasitic disease caused by the whipworm, *Trichuris trichiura*, that damages intestinal mucosa and causes chronic bleeding.

trickling filter system Procedure in which sewage is spread over a bed of rocks coated with aerobic organisms that decompose the organic matter in it.

trophozoite Vegetative form of a protozoan such as *Plasmodium*.

trypanosomiasis See **African sleeping sickness.**

tube agglutination test Serologic test that measures antibody titers by comparing various dilutions of the patient's serum against known quantities of an antigen.

tubercle A solidified lesion or chronic granuloma that forms in the lungs in patients with tuberculosis.

tuberculin hypersensitivity Cell-mediated hypersensitivity reaction that occurs in sensitized individuals when they are exposed to tuberculin.

tuberculin skin test An immunological test for tuberculosis in which a purified protein derivative from the *Mycobacterium tuberculosis* is injected subcutaneously, resulting in an induration if there was previous exposure to the bacterium.

tuberculoid Referring to the anesthetic form of Hansen's disease (leprosy) in which areas of skin lose pigment and sensation.

tuberculosis Disease caused mainly by *Mycobacterium tuberculosis*.

tularemia Zoonosis caused by *Francisella tularensis*, most often associated with cottontail rabbits.

tumor An uncontrolled division of cells, often caused by viral infection.

turbidity A cloudy appearance in a culture tube indicating the presence of organisms.

tympanic membrane Membrane separating the outer and middle ear. (Also called the eardrum.)

type strain Original reference strain of a bacterial species, descendants of a single isolation in pure culture.

typhoid fever An epidemic enteric infection caused by *Salmonella typhi*; uncommon in areas with good sanitation.

typhoidal tularemia Septicemia that resembles typhoid fever, caused by bacteremia from tularemia lesions.

typhus fever Rickettsial disease that occurs in a variety of forms including epidemic, endemic (murine), and scrub typhus.

tyrocidin An antibacterial agent that disrupts cell membranes.

ulceroglandular Referring to the form of tularemia caused by entry of *Francisella tularensis* through the skin and characterized by ulcers on the skin and enlarged regional lymph nodes.

ultrahigh temperature treatment (UHT) A method of sterilizing milk and dairy products by raising the temperature to 87.8° C for 3 seconds.

uncoating Process in which protein coats of animal viruses that have entered cells are removed by proteolytic enzymes.

undulant fever See **brucellosis.**

Universal Precautions A set of guidelines established by the CDC to reduce the risks of disease transmission in hospital and medical laboratory settings.

unsaturated fatty acid A fatty acid that contains at least one double bond between adjacent carbon atoms.

upper respiratory tract The nasal cavity, pharynx, larynx, trachea, bronchi, and larger bronchioles.

ureter Tube that carries urine from the kidney to the urinary bladder.

urethra Tube through which urine passes from the bladder to the outside during micturition (urination).

urethritis Inflammation of the urethra.

urethrocystitis Common term used to describe urinary tract infections involving the urethra and the bladder.

urinalysis The laboratory analysis of urine specimens.

urinary bladder Storage area for urine.

urinary system Body system that regulates the composition of body fluids and removes nitrogenous and other wastes from the body.

urinary tract infection (UTI) A bacterial urogenital infection that causes urethritis or cystitis.

urine Waste collected in the kidney tubules.

urogenital system Body system that (1) regulates the composition of body fluids and removes certain wastes from the body and (2) enables the body to participate in sexual reproduction.

use-dilution test A method of evaluating the antimicrobial properties of a chemical agent using standard preparations of certain test bacteria.

uterine tube A tube that conveys ova from the ovaries to the uterus. (Also called Fallopian tubes or oviducts.)

uterus The pear-shaped organ in which a fertilized ovum implants and develops.

vaccine A substance that contains an antigen to which the immune system responds.

vacuole A membrane-bound structure that stores materials such as food or gas in the cytoplasm of eukaryotic cells.

vagina The female genital canal, extending from the cervix to the outside of the body.

vaginitis Vaginal infection, often caused by opportunistic organisms that multiply when the normal vaginal microbiota are disturbed by antibiotics or other factors.

variable Anything that can change in an experiment.

varicella–zoster virus A herpesvirus that causes both chickenpox and shingles.

vasodilation Dilation of the capillary and venule walls during an acute inflammation.

vector (1) A self-replicating carrier of DNA; usually a plasmid, bacteriophage, or eukaryotic virus. (2) An organism that transmits a disease-causing organism from one host to another.

vegetation A growth that forms on damaged heart valve surfaces in bacterial endocarditis; exposed collagen fibers elicit fibrin deposits, and transient bacteria attach to the fibrin.

vegetative cell A cell that is actively metabolizing nutrients.

vehicle A nonliving carrier of an infectious agent from its reservoir to a susceptible host.

Venezuelan equine encephalitis Type of viral encephalitis seen in Florida, Texas, Mexico, and South America; infects horses more frequently than humans.

verminous intoxication An allergic reaction to toxins in the metabolic wastes of liver flukes.

verruga peruana One form of bartonellosis; a chronic nonfatal skin disease.

vertical transmission Direct contact transmission of disease in which pathogens are passed from parent to offspring in an egg or sperm, across the placenta, or while traversing the birth canal.

vesicle A membrane-bound inclusion in cells.

vibrio A comma-shaped bacterium.

vibriosis An enteritis caused by *Vibrio parahaemolyticus,* acquired from eating contaminated fish and shellfish that have not been thoroughly cooked.

villus (plural: **villi**) A multicellular projection from the surface of a mucous membrane, functioning in absorption.

viral enteritis Gastrointestinal disease caused by rotaviruses, characterized by diarrhea.

viral hemagglutination Hemagglutination caused by binding of viruses, such as those that cause measles and influenza, to red blood cells.

viral meningitis Usually self-limiting and nonfatal form of meningitis.

viral neutralization The binding of antibodies to viruses, which is used in an immunological test to determine if a patient's serum contains viruses.

viral pneumonia Disease caused by viruses such as respiratory syncytial virus.

viral specificity Refers to the specific types of cells within an organism that a virus can infect.

viral yield See **burst size.**

viremia An infection in which viruses are transported in the blood but do not multiply in transit.

viridans group A group of streptococci that often infect the valves and lining of the heart and cause incomplete (alpha) hemolysis of red blood cells in laboratory cultures.

virion A complete virus particle, including its envelope if it has one.

viroid An infectious RNA particle smaller than a virus and lacking a capsid that causes various plant diseases.

virulence The degree of intensity of the disease produced by a pathogen.

virulence factor A structural or physiological characteristic that helps a pathogen cause infection and disease.

virulent phage A bacteriophage that enters the lytic cycle when it infects a bacterial cell, causing eventual lysis and death of the host cell. (Also called **lytic phage.**)

virus A submicroscopic, parasitic, acellular microorganism composed of a nucleic acid (DNA or RNA) core inside a protein coat.

visceral larva migrans The migration of larvae of *Toxocara* species in human tissues, where they cause damage and allergic reactions.

vitamin A substance required for growth that the organism cannot make.

volutin Polyphosphate granules. (Also called **metachromatic granule.**)

walking pneumonia See **primary atypical pneumonia.**

wandering macrophages Phagocytic cells that circulate in the blood or move into tissues when microbes and other foreign material are present.

wart A growth on the skin and mucous membranes caused by infection with human papillomaviruses. (Also called **papilloma.**)

water cycle Process by which water is recycled through precipitation, ingestion by organisms, respiration, and evaporation. (Also called the **hydrologic cycle.**)

water mold A funguslike protist that produces flagellated asexual spores (zoospores) and large, motile gametes. (Also called **Oomycota.**)

wavelength The distance between successive crests or troughs of a light wave.

Western blotting A technique used to transfer and identify proteins.

western equine encephalitis Type of viral encephalitis seen most often in the western United States; infects horses more frequently than humans.

wet mount Microscopy technique in which a drop of fluid containing the organisms (often living) is placed on a slide.

wetting agent A detergent solution often used with other chemical agents to penetrate fatty substances.

whey The liquid portion (waste product) of milk resulting from bacterial enzyme addition.

whipworm *Trichuris trichiura,* a worm that causes trichuriasis infestation of the intestine.

whitlow A herpetic lesion on a finger that can result from exposure to oral, ocular, and probably genital herpes.

whooping cough A highly contagious respiratory disease caused primarily by *Bordetella pertussis.* (Also called **pertussis.**)

wort The liquid extract from mash.

wound botulism Rare form of botulism that occurs in deep wounds when tissue damage impairs circulation and creates anaerobic conditions in which *Clostridium botulinum* can multiply.

xenograft A graft between individuals of different species.

yeast extract Substance from yeast containing vitamins, coenzymes, and nucleosides; used to enrich media.

yellow fever Viral systemic disease found in tropical areas, carried by the mosquito *Aedes aegypti.*

yersiniosis Severe enteritis caused by *Yersinia enterocolitica.*

Ziehl–Neelsen acid-fast stain A differential stain for organisms that are not decolorized by acid in alcohol, such as the bacteria that cause Hansen's disease (leprosy) and tuberculosis.

zone of inhibition A clear area that appears on agar in the disk diffusion method, indicating where the agent has inhibited growth of the organism.

zoonosis (plural: **zoonoses**) A disease that can be transmitted from animals to humans.

zygomycosis Disease in which certain fungi of the genera *Mucor* and *Rhizopus* invade lungs, the central nervous system, and tissues of the eye orbit.

Zygomycota See **bread mold.**

zygospore In bread molds, a thick-walled, resistant, spore-producing structure enclosing a zygote.

zygote A cell formed by the union of gametes (egg and sperm).

Amphitrichous bacteria, 86
Amphotericin B, 380
  diseases treated with, 380, 564, 565, 566, 603, 604
Ampicillin, 369
Amylases, 151, 783
Anabaena, nitrogen fixer, 732
Anabaena azollae, 732
Anabolic pathway, 110
Anabolism, 31–32
  definition of, 108
  and enzyme repression, 178–180
Anaerobes, 120
Anaerobic metabolism, 116–120
  fermentation, 118–120
  glycolysis, 116–118
Anaerobic transfer chamber, 157
Anal contact and transmisson of diseases, 429
Analytical Profile Index (API), 159
Analytical studies, 422
  prospective studies, 422
  retrospective studies, 422
Anamnestic response, 492
Anaphylaxis
  anaphylactic shock, 362, 521
  generalized, 520–521
  localized, 519–520
  process in, 517–518
  respiratory, 521
  treatment of, 521
Ancylostoma caninum, 635
Ancylostoma duodenale, 635
Anderson, James, 314
Anemia, 380
Anesthetics, 536
Angiomatosis, 257
Angle of refraction, 55
Angstrom, 52
Animal bites
  cat scratch fever, 572
  rat bite fever, 572
  spirillar fever, 572
Animalcules, 8
Animalia, characteristics of, 233
Animals
  animal passage, and virulence, 395
  animal reservoirs for disease, 423–424, 430–431
  disease fighting at zoo, 370–371
  transgenic, 217
  types of disease transmitted by, 423
  viral diseases, 278, 280
Anion, 28
Anionic (acidic) dyes, 65
Anisakiasis, from raw seafood, 636
Anopheles, 302, 326, 669
Anopheles gambiae, 670
Anorectal infection, 708
Anoxygenic phototrophic bacteria, characteristics of, 263–264
Antagonism, of antibiotics, 365–366
Antarctica, 148
Antepar, 383
Antheridia, 313
Anthrax, 259, 650–652
  prevention of, 651–652
  transmission of, 650–651
  treatment of, 651
Antiallergen, 518
Antibiosis, 310, 357
Antibiotic resistance, 363–366
  acquisition of resistance, 363–364
  and antibiotics in animal feeds, 366
  chromosomal resistance, 364
  cross-resistance, 365
  extrachromosomal resistance, 364
  first/second/third line drugs, 365
  genetic resistance, 363–364
  in hospitalized patients, 383, 386
  limiting resistance, methods of, 365–366
  mechanisms of resistance, 364–365
Antibiotics
  aminoglycosides, 374–375
  antibiotic resistance, 363–366
  as antimetabolites, 361–362
  and blood–brain barrier, 680

carbapenems, 374
cell membrane function disruption, 360–361
cell wall synthesis inhibition, 360
cephalosporins, 372–374
chloramphenicol, 375
definition of, 357
discovery of, 18–19
erythromycin, 375
ethambutol, 377
as food additive, 770
functions of, 357
history of, 18–19
ideal agent, characteristics of, 369
isoniazid, 377
lincomycin, 375
making from soil microbes, 372
and microorganisms, 1
nitrofurans, 377
nucleic acid synthesis inhibition, 361
penicillins, 369, 372
phages as alternative to, 283
polymyxins, 374
produced by *Bacillus,* 259
produced by sac fungi, 313
produced by *Streptomyces,* 261
production of, 782
protein synthesis inhibition, 361
quinolones, 377
resistance to, 208
rifamycins, 376–377
selective toxicity, 358–359
semisynthetic, 782
side effects, 362–363
spectrum of activity, 359–360
sulfonamides, 377
tetracyclines, 375
tyrocidins, 374
Antibiotic sensitivity
  automated methods, 368–369
  dark diffusion method, 367
  dilution method, 367–368
  serum killing power, 368
Antibodies, 218, 483–486
  fluorescent antibody staining, 59
  functions of, 484
  immunoglobulins, classes of, 485–486
  structure of, 484–485
Antibody titer, 541–542
Anticoagulants, 376
Anticodons, 173
Anti-complement antibody, 544
Antifungal agents
  amphotericin B, 380
  flucytosine, 381
  griseofulvin, 380–381
  imidazoles, 380
  nystatin, 380
  terbinafine, 381
  tolnaftate, 381
  triazoles, 380
Antigen-binding site, 484
Antigen challenge, 493
Antigenic determinants, 483
Antigenic drift, 600
Antigenic mimicry, 531
Antigenic shift, 600
Antigen-presenting cells, 489
Antigens, 218, 482–483
  antigen processing, 489
  fluorescent antibody staining, 59
  and humoral immunity, 493–495
  properties of, 482–483
  and transplantation, 534
Antihelminthic agents
  ivermectin, 383
  mebendazole, 383
  niclosamide, 383
  piperazine, 383
Antihistamines, 456, 522
Antimetabolites, 361–362
Antimicrobial methods
  chemical agents, 334–342
  in hospitals, 352
  physical methods, 342–351

sterilization/disinfection, terms related to, 334
  See also individual methods
Antimony, 383
Antiparallel, 168
Antiprotozoan agents
  chloroquine, 382
  metronidazole, 382–383
  nifurtimox, 383
  pentamidine isethionate, 383
  primaquine, 382
  pyrimethamine, 383
  quinine, 382
  suramin sodium, 383
Antipyretics, 458
Antisense drug, 363
Antiseptics, meaning of, 332, 334
Antisera, 505
Antitoxins, 505
Antivenins, 506
Antiviral agents
  acyclovir, 381
  amantadine, 381
  and drug-resistant viruses, 382
  ganciclovir, 381
  idoxuridine, 381
  inosiplex, 382
  interferons, 382
  levamisole, 382
  purine, 381
  ribavirin, 381
  rimantadine, 381
  trifluridine, 381
  vidarabine, 381
Antiviral proteins, 458–459
Aperture, numerical, 53
Apicomplexans, characteristics of, 306–307
Aplastic anemia, 375
Aplastic crisis, 666
Apoenzymes, 113
Appendix, 467
Applied microbiology, 5, 760–790
Aquaria, 340, 566
Aquaspririlum, 252
Aquaspririlum magnetotacticum, magnetic properties of, 85
Ara-A, 718
Arachnids, characteristics of, 323, 325
Aralen, 382
Arboviruses, 4, 275
Archaeobacteria
  characteristics of, 264
  halobacteria, 264
  locations for discovery of, 232, 266
  methanogens, 264
  taxonomic problem, 231
  thermoacidophiles, 264
Arenaviruses, 666
  characteristics of, 279
  types of, 666
Aristotle, 681
Armadillo, culturing *Mycobactorium leprae* in, 399
Armillaria bulbosa, 314
Armillaria ostoyae, 314
Arsenicals, 383
Arsenophenylglycine, 18
Arteries, 469
Arterioles, 469
Arthritis, rheumatoid. *See* Rheumatoid arthritis
Arthrobacter, oil-eating ability, 755
Arthropods, 4, 323–326
  arachnids, 323, 325
  characteristics of, 323
  classification of, 323
  crustaceans, 326
  diseases transmitted by, 324–325
  insects, 325–326
  types of, 233
Arthus reaction, 527–528
Artifically acquired immunity, 482
Art works, microbial invasion of, 45
Asbestos, 350
Ascariasis, 635–636
  damage from, 636

Lipids, 37–40
    enzymes for breakdown of, 151
    fatty acids, 37–38
    lipid dissolving agents, 336
    metabolism of, 126
    phospholipids, 38
    solubility of, 37
    steroids, 38–40
Lipopolysaccharide, 78, 80
Lipoproteins, 78
Lister, Joseph, 12, 13, 335
Listeria, 346
Listeria monocytogenes, disease from, 260, 680, 767
Listeriosis, 260, 680
    transmission of, 680
Liver, 467, 468
    drugs toxic to, 372, 375, 376
    inflammation of, 258
    macrophages in, 453
    and mushroom toxins, 632
Liver cancer, 510, 631
Liver flukes, 317
Liver infections, 280
Lobar pneumonia, 589
Local infection, meaning of, 407
Localized anaphylaxis, 519–520
Lockjaw, 687
Locus, 165
Lodestone, 85
Logarithmic growth rate, 138
Logarithmic values, 34
Log phase, microbial growth, 138–139
Lophotrichous bacteria, 86
Louse, diseases from, 325
Lower respiratory tract, 466–467, 586–602
Low-temperature long-time pasteurization, 771
Luciferase, 132
Lucretius, contributions of, 7
Luminescence, 55
Luminescent bacteria, 132, 255
Lumpy jaw, 261
Lung, 465–467. *See also* Respiratory system
    diseases of, 261
    macrophages in, 453
Lung fluke infections, 604–605
Lupus, 533–534
Luria, Salvador, 186
Lwoff, André, 284–285
Lyase, 112
Lyme disease, 252, 324, 656–657
    prevention of, 657
    symptoms of, 656
    transmission of, 656
    treatment of, 656–657
Lymph, 470
Lymphadenitis, 255
Lymphangitis, 645
Lymphatic infections
    filariasis, 648–649
    lymphangitis, 645
Lymphatic system, 469–471
    functions of, 469–470
    lymphatic circulation, 470
    lymphatic tissues, types of, 471
    lymphoid organs, 470–471
Lymphatic vessels, 470
Lymph nodes, 470
Lymphocytes, 452, 486–489
    B lymphocytes, 486
    cytotoxic T cells, 488–489
    development of, 486
    helper T cells, 488
    surface proteins, 487, 488
    T lymphocytes, 486–487
    types of, 470
Lymphocytic choriomeningitis, 767
Lymphogranuloma venereum, 258, 714–715
    diagnosis, 715
    organism in, 714
    progression of, 714–715
    treatment, 715
Lymphoid nodules, 471
Lymphoid stem cells, 486

Lymphokines, 488
Lyophilization, 158, 228
    food, 347, 769
Lysine, microbial, 783
Lysis, 201
Lysogen, 284
Lysogenic, 201
Lysogenic conversion, 284
Lysogeny, 201, 284–285
Lysosomal enzymes, 454–455
Lysosomes, 95
Lysozyme, 18, 82, 282, 464, 465
Lytic cycle, 201, 282

MacConkey agar, 155, 156
Macfarlane Burnet, Frank, 489
MacLeod, Colin, 198
    contributions of, 19
Macrolide, 376
Macrophages, 453, 489
    activation, 489
    effect of HIV on, 538
    fixed, 453
    immune function of, 489
    names of, in tissues, 453
    wandering, 453
"Mad cow" disease, 695–696
Madura foot, 566
Madurella, 566
Maduromycosis, 566
Maggots, diseases from, 573
"Magic bullet," 18
Magnetosomes, 85
Magnetotactic bacteria, 85, 265
Magnetotaxis, 85
Magnification, total, 58
Major histocompatibility complex (MHC), 487
Major basic protein (MBP), 455
Malachite green, 68
Malaria, 74, 303, 306–307, 326, 668–670
    causes of, 327, 668–669
    diagnosis of, 669
    prevention of, 327, 328, 433
    progression of, 669
    resurgence of, 327–328
    treatment/prevention of, 328, 382, 669–670
Male cells, 204
Male reproductive system
    components of, 473–474
    infections of, 473, 474
Malignant, 293
Malignant cells, 290, 496
Malignant diseases, 412, 538
Mallon, Mary, 619
Malnutrition, 538, 559
Malta fever. *See* Brucellosis
Maltase, 112, 151
Malted grain, 778
Mammary glands, 473
Mannitol, 36
Mannitol-fermentation test, 120
D-Mannose, 770
Marble, bacterial action on, 45
Marburg virus, 665
Margulis, Lynn, 102, 228, 248
Marine environments, 742–743
Marshall, Barry, 623
Mascara, 569
Mash, 778
Mason, Dr. Arthur D., Jr., 440
Mast cells, 452
Mastigophorans, characteristics of, 306
Mastitis, 259, 705
Mastoid area, 465
Materia Medica (Dioscorides), 15
Matrix, 92
Maturation, viral replication, 282, 288
MBC. *See* Minimum bactericidal concentration
McArthur, J. Vaun, 736
McCarty, Maclyn, 198
    contributions of, 19
McClintock, Barbara, 209
    contributions of, 19
McManus, Dr. Albert, 440
McNeill, William H., 419

Measles, 558–560
    complications of, 559
    transmission of, 559
    vaccine for, 560
Measles encephalitis, 559
Meat
    inspection of, 762
    organisms of, 762–763, 767
Mebendazole, 383
Mechanical stage, of microscope, 57–58
Mechanical vector, 302
Medawar, Peter, 272, 491
Media. *See* Culture media
Medical products, from recombinant DNA, 215, 218
Medicine, and bacterial enzymes, 401–402
Medium, function of, 138
Mefloquine, 382
Megakaryocytes, 451
Meiosis, 92
Melarsoprol, 691
Melioidosis, 253
Membrane attack complex, 462
Membrane filter method, 748–749
Membrane filters, pore sizes, 350
Membranes. *See also* Cell membrane; Plasma
        membrane
    in cell walls, 78
    transport across, 96–101, 130–131
Membrane transport, 130–131
    active transport, 130–131
    phosphotransferase system, 131
Memory, in immune system, 492–493
Memory cells, 492
Mendel, Gregor, 19
Meninges, 471–472
Meningitis, 4, 253, 255, 409, 678–680, 694
Meningococcal meningitis, 679–680
    progression of, 679
    treatment of, 679–680
Meningoencephalitis, herpes, 683, 717–718
Mental disease, nature of, 400
Mental retardation from meningitis, 680
Mercury, for treating syphilis, 15
Mercury compounds, as antimicrobial agents, 339
Merozoites, 307
Merthiolate, 339
Meselson, Matthew, 168
Mesenteric lymphadenitis, 255
Mesophiles, 145
Mesophilic spoilage, 768
Mesosomes, 84
Messenger RNA (mRNA), 171–175
    properties of, 173
Metabolic pathway, 110, 118
Metabolism
    aerobic metabolism, 120–125
    anaerobic metabolism, 116–120
    chemoautotrophy, 129
    definition of, 108
    energy capture in, 109–110
    and enzymes, 111–116
    of fats, 126
    metabolic pathway, 110, 118
    photoautotrophy, 127–129
    photoheterotrophy, 129
    of proteins, 126–127
Metabolism regulation, 175–180
    enzyme induction, 178
    enzyme repression, 178–180
    feedback inhibition, 175, 177
Metacercariae, 317
Metachromasia, 85
Metachromatic granules, 85
Metals
    denaturing effect of, 336
    heavy, as antimicrobial agents, 339–340, 342
    for treating disease, 15
Metarhizium anisopliae, 311
Metastasize, 293
Metchnikoff, Elie, contributions of, 13–14
Methadone, 377
Methane, 36, 740, 752
Methane bacteria, 36, 264, 742
Methanobacterium, 740
Methanococcus, 740

Methanogens
   characteristics of, 264
   organisms producing methane, 740
Methanosarcina, 740
Methicillin, 369
Methotrexate, 535
Methylene blue, 65
   as antimicrobial agent, 342, 349
   for staining, 327, 582
Methylophilus, 784
Methylotrophic bacteria, nitrogen fixers, 732
Metric units, A1
   related to microscopy, 52, 53
Metronidazole, 382–383
   diseases treated with, 383, 629, 630, 631, 706
   side effects, 383
MHC. *See* Major histocompatibility complex
Mice, 290
Miconazole, 380
Micrasterias, 3
Microaerophiles, 147
   culture of, 156
Microbe Hunters, 14
Microbes, 2–4
   algae, 3
   bacteria, 3
   fungi, 3
   protozoa, 4
   viruses, 3
Microbial antagonism, 397
Microbial competition, 394
Microbial ecology, study of, 5
Microbial genetics, study of, 5
Microbial growth
   adaptation to limited nutrients, 151
   carbon sources for, 148–149
   chemostat, 139
   colonies, 140
   control of. *See* Antimicrobial methods
   cultures, 153–158
   decline phase, 140
   and enzymes, 150–151
   exponential/logarithmic growth, 138
   generation time, 138, 139
   lag phase, 138
   and light, 148
   log phase, 138–139
   and moisture, 147
   nitrogen sources for, 149
   and nutritional complexity, 150
   and osmotic pressure, 148
   and oxygen, 146–147
   and pH, 144–145
   sporulation, 151–153
   standard bacterial growth curve, 138
   stationary phase, 139–140
   and sulfur/phosphorus, 149
   synchronous and nonsynchronous growth, 138–139
   and temperature, 145–146
   and trace elements, 149
   and vitamins, 149–150
Microbial growth measures, 140–145. *See also* individual types
Microbial metabolism, study of, 5
Microbial taxonomy, study of, 5
Microbiologists
   household products research, case example, 16–17
   investigative methods of, 22
   role of, 4–6
Microbiology
   fields of study in, 5
   history of. *See* History of microbiology
   reasons for study of, 1–2
   scope of study of, 2–6
Micrococci
   in air, 737
   characteristics of, 259
Microccus denitrificans, 734
Microcystis aeruginosa, 63
Microenvironment, nature of, 727
Microfilaments, 95
Microfilariae, 322

Microglial cells, 472
Micrographs, 61
Micrometer, 52
Micromonospora, conidia formation, 153
Micromonospora purpurea, 265
Microorganisms, 0. *See also* Algae; Bacteria; Fungi; Viruses; Protozoa
Microscopes, 64
   development of, 50–52
   light, parts of, 57–58
Microscopy
   atomic force microscopy, 69
   electron microscopy, 60–63
   light in, 52–53, 55–57
   light microscopy, 57–60
   metric units used, 52
   microscope, invention of, 8, 50–52
   Nomarski, 60, 64
   numerical aperture, 53–54
   resolution, 52–53
   resolving power, 53
   scanning electron microscopy, 63
   scanning tunneling microscopy, 69–70
   smears, 64–65
   staining, 65–68
   transmission electron microscopy, 61–63
   wet mounts, 64
Microsporum, 563
Microtox Acute Toxicity Test, 132
Microtubules, 95
Microvilli, 468
Microwave radiation, as antimicrobial method, 349
Migration inhibiting factor (MIF), 528
Miliary tuberculosis, 593
Milk
   milk products, production of, 773–774
   organisms of, 764–765, 767–768
   pasteurization, 10–11, 346, 771
   raw milk, 771
   tests for quality of, 772
Milstein, C., 496
Minerals, 155, 264
Minimum bacterial concentration, 368
Minimum inhibitory concentration, 368
Mining, biohydrometallurgy, 784
Minocin, 375
Minocycline, 375
Miracidia, 317
Miscarriages, and , genital herpes, 252, 717
Mitchell, Peter, 124
Mites, 424, 572–573
   types of, 573, 574
Mitochondria, 92–93, 94
Mitosis, 92
Mixed-acid fermentation, 118
Mixed infections, 408
Mixotricha paradoxa, 102
Mixture, formation of, 33
MMR vaccine, 503, 560
Mobiluncus, disease from, 704
Moist heat sterilization, 343–344, 346
Mokolo virus, 278
Molds, 3. *See also* Bread molds; Fungi; Water molds
   as allegens, 518
   in food, 760–761
   infections from, 565
Mole, 29
Molecular biology, history of, 19
Molecular defenses, 458–463
   acute phase response, 463
   complement system, 460–463
   interferon, 458–460
Molecular genetics, 198
Molecular mimicry, 361–362
Molecules, composition of, 27
Molla, A., 270
Molluscum contagiosum, 562
Monera, characteristics of, 231–232
Monilia fructicola, fruit contamination, 762
Monilias, 565
Monilia sitophila, bread mold, 761
Monoacylglycerols, 38
Monoclonal antibodies, 219, 495–497
   in deodorant, 496

production of, 496, 497
   uses of, 496
Monocular eyepiece, 57
Monocytes, 452, 453
Monod, Jacques, 178, 285
Monoglycerides. *See* Monoacylglycerols
Monolayers, 289
Monosaccharides, 35–36
Monosodium glutamate (MSG), 783
Monotrichous bacteria, 86
Montagu, Lady Ashley, 13
Montali, Richard, 370–371
Montezuma's revenge, 622
Moraxella, disease from from, 253
Moraxella catarrhalis, 513
Morbidity and Mortality Weekly Report (MMWR), 434
Morbidity rate, 417
Mordant, 65
Morphine, 15
Morphology, in bacterial taxonomy, 247
Mortality rate, 417–418
Mosaic laws of sanitation, 6
Mosquitoes
   bites of, 573
   diseases from, 14, 325–326, 669–670
Most probable number, 143, A3
Mother cell, 136
Motile proteins, 42
Mount St. Helens, 266
Mouth
   components of, 467
   normal microbiota of, 396
   See also Oral cavity diseases
Movement
   Brownian movement, 131
   flagella, 86–87, 95–96, 131
   methods of, 131
Moxalactam, 374
mRNA. *See* Messenger RNA
Mucin, 467
Mucociliary escalator, 465
Mucor, 565, 567, 762, 777
Mucous membranes, as defense against disease, 450
Mucus, 450
Mucus-secreting cells, 465
Multiple sclerosis (MS), 421–422
Multiple-tube fermentation method, 748
Mumps, 615
   complications of, 615
   vaccine for, 615
Murine typhus, 659
Murray, R. G. E., 248
Musca domestica, 325
Mushrooms, 313, 315
   spore prints from, 315
   toxins from, 632
Mustard gas, 599
Mutagens, 183–184,187–189
Mutants, auxotrophic, 182
Mutations, 180–189
   Ames test, 188–189
   chemical mutagens, 183–184
   DNA damage, repair of, 184–185
   fluctuation test, 186–187
   frameshift mutation, 181
   induced mutations, 183, 186
   mutagens, actions of, 183
   nature of, 165
   phenotypic variations, 181–182
   point mutations, 180–181
   problems in study of, 186
   and radiation, 184
   replica plating, 187
   spontaneous mutations, 182–183, 186
Mutualism, 392
Myasthenia gravis, characteristics of, 532
Mycelium, 308
Mycobacterium, characteristics of, 261
Mycobacterium avium–intracellulare, disease from, 261
Mycobacterium bovis, in milk, 764, 767
Mycobacterium fortuitum, 592
Mycobacterium kansasii, 592
Mycobacterium leprae
   armadillo in study of, 399
   disease from, 261, 684

Mycobacterium leprae (cont.)
    staining of, 67
Mycobacterium marinum, 592
Mycobacterium tuberculosis
    disease from, 261, 591
    fluorochrome for visualization, 59
    and host, 395
Mycology, 308
    study of, 5
Mycoplasma, characteristics of, 258
Mycoplasma fermentans, 258
Mycoplasma hominis, disease from, 714
Mycoplasmal infections, 714
    diagnosis of, 714
    organisms in, 714
    treatment of, 714
Mycoplasma pneumoniae, 150, 590
    disease from, 258, 582
Mycorrhizae, 310
Mycoses, 311
Mycostatin, 380
Myeloma cells, in monoclonal antibodies, 496
Myelomas, 219
Myiasis, 573
Myocardial infarction (heart attack), and bacterial
    enzymes, 397
Myocarditis, 646
Myocardium, 468
Myxococcus, motion of, 131

NAD. *See* Nicotinamide adenine dinucleotide
NADP. *See* Nicotinamide adenine dinucleotide
    phosphate
Naegleria fowleri, 684
Nafcillin, 369
Nail infections, 380, 381
Naked viruses, 272, 285, 287
Nalidixic acid, 377
Nanometer, 52
Narrow-spectrum drugs, 359
Nasal cavity, 465
Nasal diphtheria, 583
Nasal sinuses, 465
Nasopharyngeal carcinoma, 664
National Hansen's Disease Center, 685
National Vaccine Compensation Act, 588
Native Americans, herbal medicine of, 15
Naturally aquired immunity, 482
Natural killer (NK) cells, 455, 489
    immune function of, 489
Natural selection, 208
Necator americanus, 319, 635
Necrotic tissue, 457
Necrotizing fascitis, 557
Needham, John, 9
Negative stains, 68
Negative-sense RNA, 275
Negri, Adelchi, 681
Negri bodies, 681
Neisser, Albert, 706
Neisseria, transformation in, 199
Neisseria gonorrhoeae, 89
    culture of, 157
    disease from, 253, 706
    penicillin resistance, 208
Neisseria meningitidis
    disease from, 253, 679
    seasonal outbreaks, 443
Nelson, Douglas, 744–745
Nematodes, characteristics of, 316
Neomycin, 374
Neonatal chlamydial pneumonia, 713
Neonatal herpes, 717
Neoplasm, 293
Neoplastic disease, nature of, 400
Neoplastic transformation, 294
Nephrons, 472
Nerve diseases
    African sleeping sickness, 691–692
    botulism, 687–689
    Chagas' disease, 692–693
    Hansen's disease, 684–686
    poliomyelitis, 689–691
    tetanus, 686–687
Nerves, 471

Nervous system, 471–472
    components of, 471–472
    functions of, 471
Netilmicin, 374
Neural tissue, macrophages in, 453
Neuraminidase, 598, 601
Neurofibrillary tangles (plaques), 696
Neurons, 471
Neurospora, 19
    study of, 313
Neurosyphilis, 711
Neurotoxins, 404
Neutralization, in humoral immunity, 494
Neutralization reactions, 543
Neutral solution, 34
Neutrons, 27, 28
Neutrophiles, 145
Neutrophils, 145, 452–453
Newborn, hemolytic disease of the, 524–526
Niclosamide, 383
Nicotinamide, 377
Nicotinamide adenine dinucleotide (NAD), 113,
    116, 118, 119, 122, 123
Nicotinamide adenine dinucleotide phosphate
    (NADP), 118
Niel, C. B. van, 228
Nifuratel, 377
Nifurtimox, 383
Nigrosin, 68
Nisin, 770
Nitrates
    in foods, 770
    mutation and cancer from, 183
    from sewage treatment, 752
Nitrification, 733–734
Nitrifying bacteria, 733–734
Nitrites
    cancer and mutation from, 183
    in foods, 342, 770
Nitrobacter, 265
    nitrate production, 734
Nitrocellulose filters, 350
Nitrofurans, 377
    diseases treated with, 377
    types of, 377
Nitrofurantoin, 377
Nitrogen, and microbial growth, 149
Nitrogenase, 732
Nitrogen bacteria
    denitrifying bacteria, 734
    nitrifying bacteria, 733–734
    nitrogen-fixing bacteria, 731–733
Nitrogen cycle, 731–734
    denitrification, 734
    events in, 731
    nitrification, 733–734
    nitrogen fixation, 731–733
Nitrogen fixation, 731–733
Nitrogen-fixing bacteria, 731–733
Nitroimidazole, 691
Nitrosomonas, 264–265
    nitrate production, 733
NK cells, 489
Nobel Prize awards, for microbiology, 21
Nocardia, characteristics of, 261
Nocardia asteroides, disease from, 261, 598
Nocardia bradiliensis, disease from, 261
Nocardia lactamdurans, protoplast fusion, 214
Nocardioforms, 261
Nocardiosis, 598
Nocturia, 703
Nomarski microscopy, 60
Nomenclature. *See also* Taxonomy
    binomial, 227
Non-A, non-B (NANB) hepatitis.
    See Hepatitis C.
Noncommunicable disease, causes of, 400
Noncompetitive inhibitors, 114
Noncyclic photoreduction, 128
Nonfruiting bacteria, 263
Nongonococcal urethritis (NGU)
    chlamydial infections, 713–714
    mycoplasmal infections, 714
Nonindigenous organisms, nature of, 727

Noninfectious disease, meaning of, 399
Nonionizing radiation, 769
Nonpolar compounds, 30
Nonsense codons, 172
Nonspecific defenses, 448–449
Nonsynchronous growth, 139
Norfloxacin, 377
Normal flora. *See* Normal microbiota
Normal microbiota, 395–397
    resident microbiota, 396
    transient microbiota, 396–397
    types of, 396
Norwalk virus, 626
Nosocomial infection, 436–442
    and compromised hosts, 437
    control in burn unit, 440–441
    definition of, 436
    endogenous infections, 437
    equipment/procedures as causes of, 439, 442
    exogenous infections, 437
    prevention of, 439, 442
    sites of infection, 439
    transmission of, 437
    and Universal Precautions, 437–438
Notifiable disease, types of, 435, 436
Nuclear envelope, 74, 92
Nuclear pores, 92
Nuclear region, 84
Nucleases, 199
Nucleic acids, 43–44
    DNA, 166–170
    functions of, 42
    RNA, 43–44, 170–175
    structure of, 43
    of virus, 272
Nucleic acid synthesis, antibiotic effects, 361
Nucleocapsid, 272
Nucleoid, 84
Nucleoplasm, 92
Nucleoli, 92
Nucleoside analogues, 548
Nucleotides, 42–43
    ATP as, 42–43
    components of, 42
Nucleus
    of atom, 28
    cell, 91–92
Nuisance organisms, 749
Numerical aperture, microscopy, 53–54
Numerical taxonomy, 236
Nutrient broth, 154
Nutrients, adaption to limited, 151
Nutrition, and immune system, 501
Nutritional complexity, and microbial growth, 150
Nutritional deficiency disease, nature of, 399
Nutritional requirements of microorganisms,
    148–151
Nystatin, 380

Objective lens, of microscope, 57
Obligate aerobes, 147
    culture of, 156
Obligate anaerobes, 147
    culture of, 156
Obligate intracellular parasites, 270
Obligate organism, 145
Obligate parasites, 301
Obligate psychrophiles, 145
Obligate thermophiles, 145
Oceans, 742–743
    characteristics of ocean water, 742
    deep-sea vents, exploration of, 744–745
    organisms of, 742–743
Octet rule, 28
Ocular lens, of microscope, 57
Ocular micrometer, of microscope, 58
Oculinum A, 404
Oil spills, bioremediation for cleanup, 126,
    754–755
Oil spraying, in control of vector-transmitted dis-
    eases, 429
Okazaki fragments, 168
Oleic acid, 38
Olives, production of, 776–777
Onchocerca volvulus, 322, 325, 569

Preserved cultures, 158
Pressure
  hydrostatic, 147
  osmotic, 148
Presumptive test, water purity test, 748
Prevalence, meaning of, 417
Primaquine, 382
Primary atypical pneumonia, 590
Primary cell cultures, 289
Primary characteristic, 475
Primary immunodeficiencies, 517, 527
Primary infection, meaning of, 408
Primary response, 493
Primary sewage treatment, 750–751
Primary structure, of proteins, 40, 41
Primaxin, 374
Prions, 235, 292–293
  characteristics of, 293
Probes, 237
Processing of antigens, 489
Prochloron, 265
Prochlorophytes, characteristics of, 265
Prochlorothrix, 265
Proctitis, 258
Prodigiosin, 255
Prodromal phase, infectious disease, 409
Prodrome, 409
Producers, nature of, 727, 728
Productive infection, 405–406
Progeny, 164
Proglottids, 317
Progressive multifocal leukoencephalopathy, 683
Prokaryotae, 248–249
  divisions of, 249
Prokaryotes
  kingdom for, 231
  special taxonomic methods, 235
Prokaryotic cells, 74–90
  arrangements of groups of, 76–78
  axial filaments, 87
  cell membrane, 82–84
  cell wall, 78–82
  chromatophores, 84
  cytoplasm, 84
  endospores, 86
  endosymbiont theory, 102
  compared to eukaryotic cells, 76
  flagella, 86–87
  glycocalyx, 90
  inclusions, 84–86
  nuclear region, 84
  pili, 87, 89–90
  ribosomes, 84
  shapes of, 75–76
  size of, 75
  types of, 74
Promoter, 178
Prontosil, 18, 358
Propagated epidemic, meaning of, 419–420
Properdin, 461
Properdin pathway, 461
Prophage, 201, 284
Prophylactic treatment, 372, 381
Prophylactic use of antibiotics, 352, 442
Propionibacterium, 776
Propionibacterium acnes
  antibiotic treatment, 376, 557
  disease from, 260, 557
Propionibacterium freudenreichii, 260
Propionic acid, 337, 338
Propionic fermentation, 118
Prospective study, 422
Prostaglandins, 456
  in allergy, 518
Prostate gland, 473
Prostatitis, 700, 702
  routes of infection, 702
Protease, 548, 783
Protease inhibitors, 548
Protectin, 489
Protein profiles, production of, 237–238
Proteins, 40–42
  and amino acids, 40
  classification of, 42
  denaturation of, 42, 336

and enzymes, 42
enzymes for breakdown of, 151
metabolism of, 126–127
and peptide bonds, 40
structural proteins, 42
structure of, 40–42
Proteinaceous infectious particles, 235
Proteinases, 731
Protein synthesis, 170–175
  antibiotic effects, 361
  steps in, 175, 176–177
  transcription, 170–171
  translation, 174–175
  types of RNA in, 171–173
Proteolytic enzymes, 783
Proteus, diseases of, 254–255
Proteus mirabilis, 255
Proteus vulgaris, 154
Protista
  characteristics of, 232
  creation of kingdom, 228
Protists, 302–308
  animal-like protists, 305–307
  apicomplexans, 306–307
  ciliates, 307–308
  diatoms, 303
  dinoflagellates, 303–304
  euglenoids, 303
  funguslike protists, 304–305
  importance of, 302–303
  mastigophorans, 306
  parasitic diseases caused by, 303
  plantlike protists, 303–304
  sarcodines, 306
  slime molds, 304–305
  water molds, 304
Protoctista, 230
Proton motive force, 125
Protons, 27, 28
Proto-oncogene, 294
Protoplasmic cylinder, in spirochetes, 250
Protoplast fusion, 211, 214
Protoplasts
  nature of, 211
  survival of, 81
Prototheca, 406, 566
Prototrophs, 182
Protozoa, 305–308
  actions in disease, 406
  apicomplexans, 306–307
  characteristics of, 4, 305–306
  control of. *See* Antiprotozoan agents
  diseases caused by, 4, 306–308
  diseases of gastrointestinal tract, 628– 631
  immune response to, 508, 509
  mastigophornas, 306
  sarcodines, 306
  systemic diseases from, 667–672
Protozoology, study of, 5
Providencia, 254
Provirus, 278
Prowazek, Baron von, 658
Prusiner, Stanley, 293
Pseudocoelom, 316
Pseudocysts, 692
Pseudomembrane, 583
Pseudomembranous colitis, 624
  organism in, 624
Pseudomonads, 252–253
Pseudomonas, 252–253
  dentrification, 734
  diseases of, 253–254
  flagella of, 87
  fluorescence of, 58–59
  metabolic pathway of, 118
  in milk, 765
  as obligate aerobe, 147
  in ocean water, 742
  oil spill cleanup, 126
  plant infections from, 253
Pseudomonas aeruginosa
  in burn patients, 440–441, 557
  diseases of, 252, 405, 644, 702
  in whirlpools, 558
Pseudomonas cepacia, 340

Pseudomonas cocovenenans, disease from, 617
Pseudomonas denitrificans, 780
Pseudomonas fluorescens
  on fruits/vegetables, 761
  as pesticide, 218–219
Pseudomonas mephitica, meat contamination, 762
Pseudomonas putida, to degrade oil, 218
Pseudomonas syringae, 762
  frost-causing bacteria, 221
  genetic fusion, 211, 221
Pseudoplasmodium, 305
Pseudopodia, 96
Psittacosis, 596
Psychrophiles, 145
Public health organizations, 433–435
  Centers for Disease Control and Prevention (CDC), 434
  World Health Organization (WHO), 434–435
Puerperal fever, 12–13, 645–646
  symptoms of, 645
  treatment of, 646
Pulex irritans, 325
Pure culture
  anaerobic transfer chamber, 157
  aseptic techniques in, 158
  control of oxygen content of media, 156–157
  diagnostic media, examples of, 156
  differential medium for, 155
  discovery of, 11
  enrichment medium for, 155
  maintenance of cultures, 157–158
  meaning of, 153
  preserved cultures, 158, 228
  reference culture, 158
  selective medium for, 155
  stab cultures, 157
  stock culture, 157
  streak plate method, 153–154
Purine analogues, as antiviral agents, 381
Purines, 44, 381
Pus, 456
Pustules, 554
Pyelonephritis, 701, 702–703
  organisms in, 702–703
  treatment of, 703
Pygmies, 570
Pyoderma, 556
Pyridoxal, 377
Pyrimethamine, 383
Pyrimidine analogues, 381
Pyrimidines, 44, 130
Pyrogens, 410, 458
Pyrsonympha, 102
Pyruvate. *See* Pyruvic acid
Pyruvic acid, 116, 118

Q fever, 257, 325, 596–598
  cause of, 597
  symptoms of, 598
  treatment of, 598
Quarantine, disease control, 432
Quaternary ammonium compounds (quats), 336, 338–339
Quaternary structure, of proteins, 40
Quinacrine (Atabrine), 184, 629
Quinine, 382
Quinolones, 377
  diseases treated with, 377
  types of, 377
Quinones, 123

Rabies, 278, 429–430, 681–683
  progression of, 681–682
  transmission of, 681
  treatment/prevention of, 682
  vaccine, 682
Rabies virus, 278, 405, 681–682, 694
  detection of, 544
  virulence of, 395
Radial immunodiffusion, 541
Radiation, and mutations, 184
Radiation sterilization, 347–349
  ionizing radiation, 348–349
  microwave radiation, 349

PHOTO CREDITS

About the Author, p. iii Elizabeth Nightlinger

Chapter 1   Chapter Opening 1 James Gillroy/The Art Institute of Chicago 1.1 Jacquelyn S. Black 1.2a CNRI/Science Photo Library/Photo Researchers, Inc. 1.2b R. B. Taylor/Science Photo Library/Photo Researchers, Inc. 1.2c CBC/Phototake 1.2d Thomas Broker/CNRI/Phototake 1.2e M. Abbey/Visuals Unlimited 1.2f Cath Ellis/Science Photo Library/Photo Researchers, Inc. 1.3a U.S. Department of Agriculture 1.3b U.S. Department of Agriculture 1.3c David Parker/Science Photo Library/Photo Researchers, Inc. 1.3d U.S. Department of Agriculture 1.3e U.S. Department of Agriculture Box, p. 7, top Coulter Corporation Box, p. 7, bottom Marcel Miranda 1.4 Scala/Art Resource 1.5 Bettmann 1.7 Charles O'Rear/Westlight 1.8 The Granger Collection 1.9 Science Photo Library/Photo Researchers, Inc. 1.10a The Granger Collection 1.10b National Library of Medicine/Science Photo Library/Custom Medical Stock Photo 1.11 Edward Jenner/The Granger Collection 1.12 Stock Montage, Inc. 1.13a Omikron/Photo Researchers, Inc. 1.14 Bettmann p. 16 Frank LaBua/PH Archives p. 17 Frank LaBua/PH Archives 1.15 UPI/Bettmann 1.16 U.S. Department of Agriculture

Chapter 2   Chapter Opening 2 John Wilson/Science Photo Library/Photo Researchers, Inc. 2.6b Biophoto Associates/Photo Researchers, Inc. 2.18a, c Richard Weiss/Peter Arnold, Inc. 2.23 Richard Weiss/Peter Arnold, Inc. 2.24 Scala/Art Resource

Chapter 3   Chapter Opening 3 The Cleveland Museum of Art 3.1b John D. Cunningham/Visuals Unlimited 3.8 Runk/Schoenberger/Grant Heilman Photography Box, p. 56 Richard Megna/Fundamental Photographs 3.10c Runk/Schoenberger/Grant Heilman Photography 3.13a, b Jim Solliday/Biological Photo Service/Terraphotographics 3.14 Biophoto Associates/Photo Researchers, Inc. 3.15 Charles J. Wrobel, M.D. 3.16 Biological Photo Service/Terraphotographics 3.17a, b David M. Phillips/Visuals Unlimited 3.18b Jean Claude Revy/Phototake NYC 3.19a Eric V. Grave/Photo Researchers, Inc. 3.19b Biophoto Associates/Science Source/Photo Researchers, Inc. 3.20 John J. Cardamone, Jr., & B. A. Phillips/Biological Photo Service/Terraphotographics 3.22 Biological Photo Service/Terraphotographics 3.23a, b David M. Phillips/Visuals Unlimited 3.24a Visuals Unlimited 3.24b David Phillips/Visuals Unlimited 3.24c Dr. Anne Smith/Science Photo Library/Photo Researchers, Inc. 3.25b A. M. Siegelman/Visuals Unlimited 3.26a, b, c, d Raymond B. Otero/Visuals Unlimited 3.26e, f George A. Wistreich 3.27 John D. Cunningham/Visuals Unlimited 3.28 Jack Bostrack/Visuals Unlimited 3.29 George A. Wistreich 3.30 George A. Wistreich 3.31 Dr. Wolfgang M. Heckl

Chapter 4   Chapter Opening 4 CNRI/Science Photo Library/Photo Researchers, Inc. 4.2a, top George Musil/Visuals Unlimited 4.2a, bottom David M. Phillips/Visuals Unlimited 4.2b Cabisco/Visuals Unlimited 4.2c R. Kessel/G. Shih/Visuals Unlimited 4.2d Dr. Tony Brain/Science Photo Library/Photo Researchers, Inc. 4.2e George J. Wilder/Visuals Unlimited 4.6a Biological Photo Service/Terraphotographics 4.6b Dr. T. J. Beveridge/Biological Photo Service/Terraphotographics 4.6c Terrance J. Beveridge & T. Paul/Visuals Unlimited 4.9 Paul W. Johnson/Biological Photo Service/Terraphotographics Box, p. 85 R. Blakemore 4.10 Institut Pasteur/CNRI/Phototake 4.11a E. C. S. Chan/Visuals Unlimited 4.11b Jack M. Bostrack/Visuals Unlimited 4.11c E. C. S. Chan/Visuals Unlimited 4.11d John D. Cunningham/Visuals Unlimited 4.12b Julius Adler/Visuals Unlimited 4.14a CNRI/Science Photo Library/Photo Researchers, Inc. 4.14b Dr. Max Listgarten/University of Toronto 4.15 Charles C. Brinton, Jr., & Judith Carnahan 4.16 B. Berg/B. V. Hofstein/G. Petterson 4.18a, b Don W. Fawcett/Photo Researchers, Inc. 4.20b C. J. Flickinger 4.21b Dr. Kenneth R. Miller/Science Photo Library/Photo Researchers, Inc. 4.23a K. G. Murti/Visuals Unlimited 4.24b M. Abbey 4.31 Micrograph

by David Chase from *Early Life* by Lynn Margulis. Copyright 1984 by Jones and Bartlett Publishers Inc., Fig. 4–8, p. 90.

Chapter 5   Chapter Opening 5 Nancy M. Hamilton/Photo Researchers, Inc. 5.5a Tripos Associates, Inc. 5.15 George A. Wistreich p. 126 Rocky Kneten/TexStock Photo, Inc. 5.28 John D. Cunningham/Visuals Unlimited 5.30 Microbics Corporation

Chapter 6   Chapter Opening 6 R. Carentine/Visuals Unlimited 6.1b Visuals Unlimited 6.1c David M. Phillips/Visuals Unlimited 6.4 Wheaton Science Products, Inc. 6.7a SIM/Science/Visuals Unlimited 6.7b Biological Photo Service/Terraphotographics 6.10 Raymond B. Otero/Visuals Unlimited 6.11 Fisher Scientific 6.12 Stephen Trimble/DRK Photo 6.15b Michael Pasdzior/The Image Bank Box, p. 150, top U.S. Department of Agriculture Box, p. 150, bottom David M. Phillips/Visuals Unlimited 6.17a Soad Tabaqchali/Visuals Unlimited 6.17b Alfred Pasieka/Science Photo Library/Photo Researchers, Inc. 6.18b Christine Case/Visuals Unlimited p. 156, top George A. Wistreich p. 156, 2d from top Runk/Schoenberger/Grant Heilman Photography p. 156, 2d from bottom Cytographics, Inc./Visuals Unlimited p. 156, bottom George A. Wistreich Box, p. 157 Becton Dickinson 6.20 Jack M. Bostrack/Visuals Unlimited 6.21 SHELLAB 6.22a Becton Dickinson

Chapter 7   Chapter Opening 7 Dr. Gopal Murti/Science Photo Library/Photo Researchers, Inc. 7.5b NIH/Kakefuda/Science Source/Photo Researchers, Inc. 7.9b Tripos Associates 7.10a O. L. Miller, Jr., & B. R. Beatty, *J. Cell Physiol.*, Vol. 74 7.10b Dr. Barbara Hamkalo 7.21 Ken Greer/Visuals Unlimited Box, p. 186 D. Karentz/UCSF 7.24a Dr. Bruce N. Ames 7.25a Dr. Gerald Goldstein/Licking County Archeological and Landmarks Society 7.25b Licking County Archeological and Landmarks Society

Chapter 8   Chapter Opening 8 Hank Morgan/SS/Photo Researchers, Inc. 8.7 Dr. L. Caro/Science Photo Library/Photo Researchers, Inc. 8.11a K. G. Murti/Visuals Unlimited 8.14 Dr. Jeremy Burgess/Science Photo Library/Photo Researchers, Inc. Notebook, p. 212 Marty Katz/The Institute of Genomic Research Notebook, p. 213 Arthur M. Siegelman/Visuals Unlimited Box, p. 217 Lyons Photography 8.16 J. R. Adams/Science VU/Visuals Unlimited 8.17 Dr. Jeremy Burgess/Science Photo Library/Photo Researchers, Inc. 8.18 Dr. Trevor Suslow, DNAP

Chapter 9   Chapter Opening 9 Biophoto Associates/Photo Researchers, Inc. 9.1 Bettmann Box, p. 228 ATCC/Complete Phototographic Services 9.7 Dudley Foster/Woods Hole Oceanographic Institution 9.12a J. Alcock/Visuals Unlimited 9.12b Martin G. Miller/Visuals Unlimited 9.12c J. William Schopf 9.14 LI-COR 9.16a Cytographics, Inc./Visuals Unlimited p. 240 Biolog, Inc. 9.17 Dr. Edward J. Bottone/Mt. Sinai Hospital

Chapter 10   Chapter Opening 10 Manfred Kage/Peter Arnold, Inc. 10.1 Archives, Center of the History of Microbiology 10.3 CNRI/Science Photo Library/Photo Researchers, Inc. 10.4 Heather Davies/Science Photo Library/Photo Researchers, Inc. 10.5 David M. Phillips/Visuals Unlimited 10.6 C. P. Venae/Visuals Unlimited 10.8 Dr. Edward J. Bottone 10.9 Christine Case/Visuals Unlimited 10.11a H. Pol/CNRI/Science Photo Library/Photo Researchers, Inc. 10.11b Fred Hossler/Visuals Unlimited 10.13 Michael Gabridge/Visuals Unlimited 10.14 Dr. T. J. Beveridge/Biological Photo Service/Terraphotographics Box, p. 260 Ester R. Angert and Norman R. Pace 10.15 Eric V. Grave/Photo Researchers, Inc. 10.16 Science/Visuals Unlimited 10.17 F. Widdel/Visuals Unlimited 10.18 Cabisco/Visuals Unlimited 10.19 J. Poindexter/Science VU/Visuals Unlimited 10.20 D. Foster, WHOL/Science/Visuals Unlimited 10.21 Centers for Disease Control 10.22 David Weintraub/Photo Researchers, Inc.

Chapter 11   Chapter Opening 11 Science Photo Library/Custom Medical Stock Photo Box, p. 274, top A. Jones/Visuals Un-

limited Box, p. 274, bottom Science VU—Wayside/Visuals Unlimited Box, p. 275 Dr. Steve Patterson/Science Photo Library/Photo Researchers, Inc. 11.3a Omikron/Photo Researchers, Inc. 11.3b CNRI/Science Photo Library/Custom Medical Stock Photo 11.3c Tektoff—RM/CNIR/Science Photo Library/Photo Researchers, Inc. 11.3d David A. Wagner/Phototake 11.3e K. G. Murti/VU/Visuals Unlimited 11.4 Michael G. Rossmann 11.5a CDC/Photo Researchers, Inc. 11.5b CNRI/Science Photo Library/Photo Researchers, Inc. 11.5c EM Unit, CVL Weybridge/Science Photo Library/Photo Researchers, Inc. 11.6a Robley Williams 11.6b Biology Media/Photo Researchers, Inc. 11.9 Bruce Iverson 11.10 Dr. M. Wurtz/Photo Researchers, Inc. 11.13b Chris Bjornberg/Photo Researchers, Inc. 11.15b Centers for Disease Control and Prevention 11.16 N. G. Gabridge, Cyto Graphics, Inc./BPS 11.17 Centers for Disease Control 11.18a, b G. Steven Martin 11.19a Agricultural Research Service/U.S. Department of Agriculture 11.19b U.S. Department of Agriculture 11.20 Fred E. Cohen

Chapter 12   Chapter Opening 12 Andrew Syred/Science Photo Library/Photo Researchers, Inc. 12.1a Carolina Biological Supply Company 12.1b Andrew Syred/Science Photo Library/Photo Researchers, Inc. 12.1c David M. Phillips/Visuals Unlimited Box, p. 304 Dr. R. R. Colwell Box, p. 305 Daniel Gotshall/Visuals Unlimited 12.2a Dwight Kuhn Photography 12.2b Cabisco/Visuals Unlimited 12.3a Eric Grave/Science Source/Photo Researchers, Inc. 12.3b M. Abbey/Visuals Unlimited 12.3c Arthur M. Siegelman/Visuals Unlimited 12.3d M. Abbey/Visuals Unlimited 12.5 David Scharf/Peter Arnold, Inc. 12.6 J. Forsdyke/Science Photo Library/Photo Researchers, Inc. 12.8a Andrew Syred/Science Photo Library/Photo Researchers, Inc. 12.8b Bruce Iverson/Visuals Unlimited Box, p. 310 Dwight Kuhn Photography 12.9 Richard Thom/Visuals Unlimited 12.10a, b Dr. Michael Orlowski, Department of Microbiology, Louisiana State University 12.11 Bruce Iverson/Science Photo Library/Photo Researchers, Inc. 12.13a Grant Heilman/Grant Heilman Photography 12.13b S. Flegler/Visuals Unlimited Box, p. 314 Dr. Johann N. Bruhn Box, p. 315 Dwight Kuhn Photography 12.14a John D. Cunningham/Visuals Unlimited 12.14b G. Shih/R. Kessel/Visuals Unlimited 12.14c Fred Marsik/Science VU/Visuals Unlimited 12.14d George J. Wilder/Visuals Unlimited 12.19a L. West/Photo Researchers, Inc. 12.19b Cath Wadforth/Science Photo Library/Photo Researchers, Inc. 12.19c Runk/Schoenberger/Grant Heilman Photography 12.19d Richard Walters/Visuals Unlimited 12.19e A. M. Siegelman/Visuals Unlimited p. 320 The Carter Center p. 321 The Carter Center

Chapter 13   Chapter Opening 13 Philippe Plailly Eurelios/Science Photo Library/Photo Researchers, Inc. 13.2a, b Jack M. Bostrack/Visuals Unlimited 13.6 Centers for Disease Control and Prevention 13.7 AMSCO International, Inc. 13.8 AMSCO International, Inc. 13.9 Richard Hutchings/Photo Researchers, Inc. Box, p. 345 Barry L. Runk/Grant Heilman Photography 13.12 AMSCO International, Inc. 13.13 Link/Visuals Unlimited 13.14a, b FTS Systems 13.16 CEM Corporation 13.17a, b Millipore Corporation

Chapter 14   Chapter Opening 14 Dr. Jacquelyn G. Black 14.3a, b Antimicrobial Agents and Chemotherapy Box, p. 364 Paolo Koch/Photo Researchers, Inc. 14.9a Science VU—Miles/Visuals Unlimited 14.10a, b Vitek Systems Box, p. 370 Jessie Cohen/Office of Graphics and Exhibits/National Zoological Park/Smithsonian Institution Box, p. 371, top and bottom Jessie Cohen/Office of Graphics and Exhibits/National Zoological Park/Smithsonian Institution 14.12 Custom Medical Stock Photo Box, p. 376 New York University/Skin and Cancer Unit 14.14 Barts Medical Library/Phototake NYC 14.17 Heather Davies/Science Photo Library/Photo Researchers, Inc.

Chapter 15   Chapter Opening 15 The Granger Collection 15.1 George A. Wistreich 15.2 Daniel E. Snyder 15.3 David M. Phillips/Visuals Unlimited Box, p. 396 L. Migdale/Stock Boston

Box, p. 399 William J. Weber/Visuals Unlimited **15.7a** L. M. Pope and D. R. Grote, University of Texas, Austin/Biological Photo Service/Terraphotographics **15.7b** Michael Abbey/Science Source/Photo Researchers, Inc. Box, p. 404, left and right Dr. Albert W. Biglan **15.8a, b** Gail W. T. Wertz, University of Alabama/BPS

Chapter 16   Chapter Opening 16 UPI/Bettmann Box, p. 424 David Scharf Photography **16.12** Dr. Gary Settles/Photo Researchers, Inc. Box, p. 430 Dr. Jacquelyn G. Black **16.14** Centers for Disease Control **16.16** Centers for Disease Control and Prevention **16.17** Centers for Disease Control and Prevention **16.18a** D. Espinoza, Peru/World Health Organization **16.18b** Chang Hongen, China/World Health Organization Notebook, pp. 440, 441 Dr. Arthur Mason, U.S. Army Institute of Surgical Research

Chapter 17   Chapter Opening 17 Lennart Nilsson **17.4** NIBSC/Science Photo Library/Custom Medical Stock Photo **17.5a** Dr. D. F. Barnton **17.9b** From Bhakdi, Sucharit, et al., 1990, "Functions and relevance of the terminal complement sequence," *Blut* 60:309–16. **17.12** Jack M. Bostrack/Visuals Unlimited

Chapter 18   Chapter Opening 18 David M. Phillips/Science Source/Photo Researchers, Inc. **18.3b** R. Feldman/NIH/Science VU/Visuals Unlimited **18.17** Dr. A. Lipeins/Science Photo Library/Photo Researchers, Inc.

Chapter 19   Chapter Opening 19 National Library of Medicine **19.2a** SIU Biomed Comm/Custom Medical Stock Photo **19.2b** VU/SIU/Visuals Unlimited **19.3** Ralph C. Eagle/Photo Researchers, Inc. **19.4a** Scott Camazine/Photo Researchers, Inc. **19.4b** W. Ober **19.7** Larry Mulvehill/Science Source/Photo Researchers, Inc. **19.8c, d** From *Rh* by Edith Potter (1947) Chicago: Yearbook Medical Publishers **19.11** From Top, F. H., Sr.: *Communicable and Infectious Diseases*, ed. 6, St. Louis, 1968, The C. V. Mosby Co. **19.13a** Ed Reschke/Peter Arnold, Inc. **19.13b** Beckman/Custom Medical Stock Photo **19.14** National Medical Slide/Custom Medical Stock Photo **19.16** CNRI/Science Photo Library/Photo Researchers, Inc. **19.17** Ken Greer/Visuals Unlimited Box, p. 537 AP/Wide World Photos **19.21** St. Mary's Hospital Medical School/Science Photo/Photo Researchers, Inc. **19.26** George Whiteley/Photo Researchers, Inc. **19.27** SIU/Visuals Unlimited **19.29b** Leon J. Le Beau/Biological Photo Service **19.30d** George A. Wistreich

Chapter 20   Chapter Opening 20 George A. Wistreich **20.1** University of Virginia (Dermatology Department) **20.2** Ken Greer/Visuals Unlimited **20.3** Ken Greer/Visuals Unlimited **20.5** Wolfe/Yearbook **20.6a** Barts Medical Library/Phototake **20.6b** N. M. Hauprich/Photo Researchers, Inc. **20.7a** Kenneth E. Greer/Visuals Unlimited **20.7b** Ken Greer/Visuals Unlimited **20.8** Everett S. Beneke/Visuals Unlimited **20.9** NIH **20.10a** Ken Greer/Visuals Unlimited **20.10b** Lester V. Bergman & Associates, Inc. **20.11** Ed Rottinger/Custom Medical Stock Photo **20.12** National Medical Slide/Custom Medical Stock Photo **20.13** Science VU/Visuals Unlimited **20.14** Carl Purcell/Photo Researchers, Inc. **20.15** Science VU/AFIP/Visuals Unlimited **20.16** Science VU/Visuals Unlimited **20.17** William E. Ferguson **20.18** World Health Organization **20.19** World Health Organization Box, p. 578 George A. Wistreich

Chapter 21   Chapter Opening 21 UPI/Bettmann **21.1a** Biophoto Associates/Photo Researchers, Inc. **21.1b** SIU/Peter Arnold, Inc.

21.2 Fred Marsik/Visuals Unlimited **21.3** Science VU/Visuals Unlimited **21.6** Dr. S. Girod De Bentzmann **21.7a, b** David M. Phillips/Visuals Unlimited **21.8** CDC/Science Source/Photo Researchers, Inc. **21.10** Armed Forces Institute of Pathology Notebook, p. 595, top and bottom David Kaplan Box, p. 597 U.S. Department of Agriculture **21.11** Moredun Animal Health Ltd/Science Photo Library/Photo Researchers, Inc. **21.12** Frederick C. Skvara, M.D. **21.14b** CDC/Science Source/Photo Researchers, Inc. **21.15a** Winograd/Biological Photo Service/Terraphotographics **21.15b** Science VU/Visuals Unlimited **21.16** G. W. Willis/Biological Photo Service Terraphotographics **21.17** A. M. Siegelman/Visuals Unlimited **21.18** Iowa State University

Chapter 22   Chapter Opening 22 Bettmann **22.1a** Fred E. Hossler/Visuals Unlimited **22.1b** Dr. Ross P. Karlin **22.2d** Rosen, S., et al. "Dental carries in gnotobic rats," *J. Dent. Research,* 1968:47, p. 362 **22.3** Dr. Ross P. Karlin **22.4a** Dr. R. Gottsegen/Peter Arnold, Inc. **22.4b** Science VU/Max Listgarten/Visuals Unlimited **22.6** John D. Cunningham/Visuals Unlimited **22.7** Science Photo Library/Photo Researchers, Inc. **22.8** Veronika Burmeister/Visuals Unlimited **22.9** Veronika Burmeister/Visuals Unlimited **22.10** K. G. Murti/Visuals Unlimited **22.11** Alfred Pasieka/Peter Arnold, Inc. **22.12a** Jerome Paulin/Visuals Unlimited **22.13** Lauritz Jensen/Visuals Unlimited **22.14a** Larry Jensen/Visuals Unlimited **22.15a** John D. Cunningham/Visuals Unlimited **22.16a** W. Ormerod/Visuals Unlimited **22.16b** Forest W. Buchanan/Visuals Unlimited **22.17** Bruce Iverson/Science Photo Library/Photo Researchers, Inc. **22.18a** Stanley Flegler/Visuals Unlimited **22.18b** Bruce Iverson/Science Photo Library/Photo Researchers, Inc. **22.18c** Larry Jensen/Visuals Unlimited **22.19** James Solliday/Biological Photo Service/Terraphotographics **22.20** R. Calentine/Visuals Unlimited Box, p. 597 U.S. Department of Agriculture **22.21** Dr. Daniel H. Connor, Armed Forces Institute of Pathology **22.22** Arthur M. Siegelman/Visuals Unlimited **22.23a** Alejandro Balaguer/AP/Wide World Photos **22.23b** A. Balaguer/Sygma **22.23c** Reuters/Bettmann

Chapter 23   Chapter Opening 23 Bettmann **23.1** Ken Greer/Visuals Unlimited **23.3a** Goivaux Communication/Phototake NYC **23.3b** Sinclair Stammers/Science Photo Library/Photo Researchers, Inc. **23.3c** NIAID/NIH **23.4a** National Medical Slide/Custom Medical Stock Photo **23.4b** Fred Marsik/Visuals Unlimited Box, p. 650 R. Calentine/Visuals Unlimited **23.5** Science VU/Charles W. Stratton/Visuals Unlimited **23.7** The Granger Collection **23.9a** Eric Grave/Phototake **23.9b** National Medical Slide/Custom Medical Stock Photo **23.11b** John Radcliffe/Science Photo Library/Custom Medical Stock Photo **23.14a** R. Calentine/Visuals Unlimited **23.14b** Kenneth Greer/Visuals Unlimited **23.15** Armed Forces Institute of Pathology **23.16** G. Musil/Visuals Unlimited **23.17** Science VU/Visuals Unlimited Box, p. 665 Science Source/Photo Researchers, Inc. **23.18a** George A. Wistreich **23.18b** Science VU/Visuals Unlimited **23.20** Phototake NYC **23.21** Science Photo Library/Custom Medical Stock Photo **23.23** CDC/Science Source/Photo Researchers, Inc.

Chapter 24   Chapter Opening 24 J. F. Chretien/World Health Organization **24.1** Dr. Edward V. Bottone/Mt. Sinai Hospital **24.2** Science VU/Visuals Unlimited **24.4a** CNRI/Phototake NYC **24.4b** Ken Greer/Visuals Unlimited **24.5** University of Virginia (Dermatology Department) **24.6a** Dr. T. J. Beveridge/Biological Photo Service/Terraphotographics **24.6b** Biological Photo Service/Terraphotographics **24.7** By courtesy of The Royal College

of Surgeons of Edinburgh **24.9a** March of Dimes Birth Defects Foundation **24.9b** March of Dimes Birth Defects Foundation **24.11a** A. M. Siegelman/Visuals Unlimited **24.11b** R. Ashley/Visuals Unlimited **24.13a** Frederick C. Skvara, M.D. **24.13b** Frederick C. Skvara, M.D. **24.14** NIAID/NIH **24.15** U.S. Department of Agriculture

Chapter 25   Chapter Opening 25 Alfred Pasieka/Science Photo Library/Custom Medical Stock Photo **25.1** Jon Meyer/Custom Medical Stock Photo **25.3** CNRI/Science Photo Library/Photo Researchers, Inc. **25.4** C. A. Speigel, R. Amsel, and K. K. Holmes **25.5** Mike Rein **25.6a** A. M. Siegelman/Visuals Unlimited **25.6b** Centers for Disease Control **25.7a** Ken Greer/Visuals Unlimited **25.7b** Centers for Disease Control and Prevention **25.10a** Centers for Disease Control and Prevention **25.10b** University of Virginia (Dermatology Department) **25.11a** SIU/Visuals Unlimited **25.11b** Science VU/Visuals Unlimited **25.12a, b, c** University of Virginia (Dermatology Department) **25.14a** Science VU/Visuals Unlimited **25.14b** Centers for Disease Control **25.15** Centers for Disease Control **25.17** CNRI/Phototake **25.18** Centers for Disease Control and Prevention **25.19** Kenneth E. Greer/Visuals Unlimited **25.20** Centers for Disease Control and Prevention **25.21** Centers for Disease Control

Chapter 26   Chapter Opening 26 Tim Hauf/Visuals Unlimited **26.5a** David Hall/Science Photo Library/Photo Researchers, Inc. **26.5b** John D. Cunningham/Visuals Unlimited **26.5c** A. M. Siegelman/Visuals Unlimited **26.6** Paul W. Johnson/Biological Photo Service/Terraphotographics **26.8** F. Widdel/Visuals Unlimited **26.10a** Graseby Anderson **26.10b** Forma/Science VU/Visuals Unlimited **26.14** Dwight Kuhn Photography **26.15** Bruce F. Molnia/Biological Photo Service/Terraphotographics **26.16a** Manfred Kage/Peter Arnold, Inc. **26.16b** Paul W. Johnson/Biological Photo Service/Terraphotographics Notebook, p. 744 Rich Pebroncelli Notebook, p. 745, left and right D. C. Nelson **26.18a** Leon J. Le Beau/Biological Photo Service **26.18b** Raymond B. Otero/Visuals Unlimited **26.19** Slide Works **26.21** R. F. Ashley/Visuals Unlimited **26.22a** Judith F. M. Hoeniger, University of Toronto/Biological Photo Service/Terraphotographics **26.22b** Paul W. Johnson/Biological Photo Service/Terraphotographics **26.23** Larry Lefever/Grant Heilman Photography, Inc. **26.25** AP/Wide World Photos

Chapter 27   Chapter Opening 27 John Colwell/Grant Heilman Photography **27.1** R. Calentine/Visuals Unlimited **27.2** U.S. Department of Agriculture Box, p. 763 U.S. Department of Agriculture **27.3** Jane Latta/Photo Researchers, Inc. **27.4** Larry Lefever/Grant Heilman Photography Box, p. 770 Chris Keith/U.S. Department of Agriculture **27.5** Tim McCabe/U.S. Department of Agriculture **27.6** Biophoto Associates/Photo Researchers, Inc. **27.7a, b, c** John Colwell/Grant Heilman Photography **27.8** John Colwell/Grant Heilman Photography **27.9** David M. Phillips/Visuals Unlimited **27.10a** Bill Frantz **27.12a** Fred Lyon **27.12b** Sylvain Grandadam/Photo Researchers, Inc. **27.12c** Photo Researchers, Inc. **27.13a** Joseph Nettis/Photo Researchers, Inc. **27.13b** Science VU/Visuals Unlimited **27.15** Warren Gretz/National Renewable Energy Lab **27.16a** A. McClenaghan/Science Photo Library/Photo Researchers, Inc. **27.16b** Dan McCoy/Rainbow **27.17** Hank Morgan/Photo Researchers, Inc. **27.18a** Corale L. Brierley/Visuals Unlimited **27.18b** Kennecott **27.19** International Atomic Energy Association

VIRUSES

Virus	Group Family	Disease	Page	Virus	Group Family	Disease	Page
adenovirus	Adenoviridae	acute upper & lower respiratory tract distress, pharyngitis, pneumonia, follicular conjunctivitis, epidemic keratoconjunctivitis	276, 279, 569, 601–602	herpes simplex type 2	Herpesviridae	genital herpes, oral & whitlow	276, 279, 715
				herpesvirus	Herpesviridae	meningoencephalitis	717–718
arenavirus	Arenaviridae	Bolivian hemorrhagic fever	279, 666	human immunodeficiency (HIV)	Retroviridae	HIV disease, AIDS	272, 276, 278, 295, 501, 538–540
	Arenaviridae	Lassa fever	666				
bunyavirus	Bunyaviridae	encephalitis	279, 665–666	human papillomavirus	Papovaviridae	common warts (papillomas), genital warts (condylomas); associated with cervical cancer	276, 293, 562–563
canine parvovirus	Parvoviridae	severe vomiting & diarrhea	666				
Colorado tick fever	Reoviridae	encephalitis	325, 666	influenza	Orthomyxoviridae	influenza (flu)	278, 381, 504, 598–601, 606
coronavirus	Coronaviridae	colds, GI disturbances	275, 585				
coxsackie	Picornaviridae	common cold syndrome & pharyngitis; severe systemic illness of newborn; muscle pain & damage; diabetes; meningoencephalitis	585, 687	Marburg	Filoviridae	hemorrhagic fever	665
				measles	Paramyxoviridae	rubeola, sometimes subacute sclerosing panencephalitis (SSPE)	276, 278, 558–560
cytomegalovirus	Herpesviridae	mononucleosis, congenital cytomegalic inclusion disease, severe birth defects	291, 719–721	parainfluenza	Paramyxoviridae	rhinitis, pharyngitis, bronchitis, pneumonia, croup	278, 585–586
dengue	Flaviviridae	dengue fever (break-bone fever)	325, 326, 662–663	paramyxovirus (mumps)	Paramyxoviridae	mumps	278, 615
Eastern equine encephalitis	Togaviridae	encephalitis	276, 419, 683	poliovirus	Picornaviridae	poliomyelitis	276, 277, 689–690
Ebola	Filoviridae	hemorrhagic fever	276, 665	polyomavirus: BK	Papovaviridae	associated with renal transplant infection, immunosuppressed patients	684
enterovirus	Picornaviridae	acute hemorrhagic conjunctivitis	277, 569	polyomavirus: JC	Papovaviridae	mild respiratory illness	683–684
Epstein–Barr	Herpesviridae	Burkitt's lymphoma, infectious mononucleosis, nasopharyngeal carcinoma	293, 663–664	poxvirus group (unclassified)	?	molluscum contagiosum	280, 562
erythrovirus (B19)	Parvoviridae	aplastic crisis in sickle cell anemia fifth disease (erythema infectiosum)	276, 280, 666–667, 280, 666–667	rabies	Rhabdoviridae	rabies	276, 278, 429–430, 681–683
feline panleukopenia	Parvoviridae	decreased number of white blood cells with fever	666	respiratory syncytial	Paramyxoviridae	pneumonia in children under age 1, upper respiratory infection in older children & adults	601–602
Hantaan	Bunyaviridae	Korean hemorrhagic fever	279	rhinovirus	Picornaviridae	common cold	276, 277, 585
hantavirus	Bunyaviridae	hantavirus pulmonary syndrome	279, 295, 602, 666, 673	Rift Valley fever	Bunyaviridae	fever & hemorrhage	666
				rotavirus	Reoviridae	enteritis	276, 279, 626
hepatitis A	Picornaviridae	infectious hepatitis	276, 626, 627–628	rubella	Togaviridae	German measles, 3-day measles	558
hepatitis B	Hepadnaviridae	serum hepatitis	276, 626, 628	St. Louis encephalitis	Flaviviridae	encephalitis	683
hepatitis C	?	hepatitis C (non-A, non-B)	626–628	varicella-zoster	Herpesviridae	chickenpox, shingles	276, 279, 560
hepatitis D	?	hepatitis D (delta hepatitis)	626–628	Venezuelan equine encephalitis	Togaviridae	encephalitis	276, 683
hepatitis E	?	hepatitis E (enterically transmitted non-A, non-B non-C)	626–628	Western equine encephalitis	Togaviridae	encephalitis	276, 326, 419, 683
herpes simplex type 1	Herpesviridae	oral herpes, gingivostomatitis, herpes labialis (cold sores), keratoconjunctivitis, herpetic whitlow	276, 279, 715	yellow fever	Flaviviridae	yellow fever	276, 277, 295, 325, 663

FUNGI

Organism	Disease	Page	Organism	Disease	Page
Aspergillus sp.	aspergillosis, pneumonia in compromised patients, skin infections in burn patients, corneal & external ear infections	63, 313, 565, 604, 761	*Filobasidiella neoformans*	cryptococcosis	602
			Epidermophyton sp.	ringworm (tinea)	563–564
			Histoplasma capsulatum	histoplasmosis	313, 603–604
Blastomyces dermatitidis	blastomycosis	313, 564–565	*Microsporum* sp.	ringworm (tinea)	563–564
Candida albicans	candidiasis	565	*Mucor* sp.	zygomycosis	565
Claviceps purpurea	ergot poisoning	631–632, 761	*Pneumocystis carinii*	*Pneumocystis* pneumonia	604
			Rhizopus sp.	zygomycosis	565
Coccidioides immitis	coccidioidomycosis (valley fever)	602–603	*Sporothrix schenckii*	sporotrichosis	564
			Trichophyton sp.	ringworm (tinea)	563–564

PATHOGENS AND THE DISEASES THEY CAUSE (CONTINUED)

BACTERIA

Organism	Gram Stain*	Basic Morphology	Diseases	Page
Actinomadura sp.	+	rod, some filamentous forms	Madura foot (maduromycosis)	566
Actinomyces israelii	+	filamentous, diphtheroid, & coccal	actinomycosis, mouth & other lesions	261
Afipia felis	−	rod	cat scratch fever	572
Bacillus anthracis	+	rod, encapsulated	anthrax	130, 259, 405, 650, 740–741, 768
Bacillus cereus	+	rod, encapsulated	food poisoning	259, 405, 616, 617
Bacteroides sp.	−	small rod	mouth lesions, septicemia, abscesses, Vincent's angina	256, 644
Bartonella bacilliformis	−	curved or coccoid	Oroya fever (systemic form), verruga peruana (cutaneous form)	257, 661
Bartonella henselae	NA	coccobacillus	cat scratch fever	257, 572, 662
Bordetella pertussis	−	coccobacillus	whooping cough	253, 587
Borrelia burgdorferi	−	spiral	Lyme disease	252, 324, 325, 656
Borrelia recurrentis	−	large spiral	epidemic relapsing fever	252
Brucella sp.	−	coccobacillus	brucellosis (undulant fever or Malta fever)	253, 654–655
Calymmatobacterium granulomatis	−	rod, encapsulated	granuloma inguinale (donovanosis)	715
Campylobacter sp.	−	rod	gastroenteritis	147, 252, 622, 747
Chlamydia psittaci	NA	coccoid, very tiny	ornithosis (psittacosis)	258, 596
Chlamydia trachomatis	NA	coccoid, very tiny	conjunctivitis, trachoma, genital tract infection (nongonococcal urethritis), infant pneumonitis, lymphogranuloma venereum	258, 568, 713, 714
Clostridium botulinum	+	rod	food poisoning (botulism), wound infections, infant botulism	155, 259, 345, 346, 347, 404, 405, 616, 617, 687, 740, 764, 766
Clostridium difficile	+	rod	pseudomembranous colitis	387, 624
Clostridium perfringens	+	rod	gas gangrene, food poisoning	86, 259, 405, 616, 740, 762
Clostridium tetani	+	rod	tetanus	259, 405, 686–687, 740
Corynebacterium diphtheriae	+	rod, club-shaped, pleomorphic, forms palisades	diphtheria: pharyngeal, laryngeal & cutaneous	260, 405, 582–583
Coxiella burnetii	NA	coccobacillus	Q fever, pneumonia	767
Enterococcus faecalis	+	cocci	endocarditis	74
Escherichia coli	−	rod	urinary tract infections, "traveler's diarrhea," nosocomial infections	4, 21, 63, 79, 118, 130–131, 156, 166, 167, 178, 186, 201, 208, 254, 405, 622, 701–702, 747, 764
Francisella tularensis	−	small rod (coccobacillus)	tularemia	253, 324, 654
Gardnerella vaginalis	−	small rod	bacterial vaginitis (nonspecific), urethritis	704

Organism	Gram Stain*	Basic Morphology	Diseases	Page
Haemophilus aegyptius	−	coccobacillus	bacterial conjunctivitis	255
Haemophilus ducreyi	−	rod	chancroid	255, 712
Haemophilus influenzae	−	coccobacillus, some strains form capsules	meningitis in children under 5, epiglottitis, ear infections, pneumonia in elderly or compromised patients	208, 212, 255, 581, 584, 680
Helicobacter pylori	−	curved rod	chronic gastritis, peptic ulcer	252, 623–624
Klebsiella pneumoniae	−	rod, encapsulated	pneumonia, infant diarrhea, urinary tract infections	119, 120, 254, 732, 767
Legionella pneumophila	−	coccoid rod	Legionnaires' disease (pneumonia)	253, 590, 747
Leptospira interrogans	−	spiral	leptospirosis	89, 704
Listeria monocytogenes	+	rod	listeriosis, meningitis, abortion	260, 680, 767
Mycobacterium avium	A-F	rod	chronic pulmonary disease, opportunistic infections in immunosuppressed patients	261
Mycobacterium leprae	A-F	rod	Hansen's disease (leprosy)	67, 261, 399, 684
Mycobacterium tuberculosis	A-F	rod, branching forms	tuberculosis	59, 261, 395, 591
Mycoplasma pneumoniae	NA	too small to be visualized by light microscope	primary atypical bacterial pneumonia	150, 258, 582, 590
Neisseria gonorrhoeae	−	cocci in pairs	gonorrhea, ophthalmia neonatorum, meningitis, arthritis, keratitis	89, 157, 208, 253, 706
Neisseria meningitidis	−	cocci in pairs; capsules formed in young cells	meningitis, Waterhouse-Friderichson syndrome	253, 443, 679
Nocardia sp.	+	rod, some filamentous forms	nocardiosis, Madura foot (maduromycosis)	214, 261, 598
Porphyromonas gingivalis	−	rod	periodontal disease	614
Propionibacterium acnes	+	rod	acne	260, 376, 557
Providencia stuartii	−	rod	urinary tract infections, wound infections	254
Pseudomonas aeruginosa	−	rod	urinary tract infections, skin lesions, eye & ear infections, septicemia in immunocompromised patients	252, 405, 440–441, 557, 558, 644, 702
Rickettsia akari	NA	coccobacillus	rickettsialpox	661
Rickettsia prowazekii	NA	coccobacillus	epidemic typhus, Brill–Zinsser disease	658–659
Rickettsia rickettsii	NA	coccobacillus	Rocky Mountain spotted fever	660
Rickettsia tsutsugamushi	NA	coccobacillus	tsutsugamushi fever	257, 324, 659
Rickettsia typhi	NA	coccobacillus	endemic or murine typhus	659
Rochalimaea quintana	NA	coccobacillus	trench fever	257
Salmonella enteritidis, S. paratyphi, S. typhimurium	−	rod	salmonellosis (food poisoning)	617, 764
Salmonella typhi	−	rod	typhoid fever	155, 156, 618, 747
Serratia marcescens	−	rod	urinary tract infections, hospital epidemics, septicemia, peritonitis, arthritis, pneumonia	255, 644